Understanding Physics

Third Edition

D1216430

Understanding Physics

Third Edition

MICHAEL MANSFIELD AND COLM O'SULLIVAN

Physics Department
University College Cork
Ireland

WILEY

This edition first published 2020
© 2020 John Wiley & Sons Ltd

Edition History
John Wiley & Sons (1e, 2006)
John Wiley & Sons (2e, 2011)

All rights reserved. No part of this publication may be reproduced, stored in a retrieval system, or transmitted, in any form or by any means, electronic, mechanical, photocopying, recording or otherwise, except as permitted by law. Advice on how to obtain permission to reuse material from this title is available at http://www.wiley.com/go/permissions.

The right of Michael Mansfield and Colm O'Sullivan to be identified as the authors of this work has been asserted in accordance with law.

Registered Offices
John Wiley & Sons, Inc., 111 River Street, Hoboken, NJ 07030, USA
John Wiley & Sons Ltd, The Atrium, Southern Gate, Chichester, West Sussex, PO19 8SQ, UK

Editorial Office
John Wiley & Sons Ltd, The Atrium, Southern Gate, Chichester, West Sussex, PO19 8SQ, UK

For details of our global editorial offices, customer services, and more information about Wiley products visit us at www.wiley.com.

Wiley also publishes its books in a variety of electronic formats and by print-on-demand. Some content that appears in standard print versions of this book may not be available in other formats.

Limit of Liability/Disclaimer of Warranty

In view of ongoing research, equipment modifications, changes in governmental regulations, and the constant flow of information relating to the use of experimental reagents, equipment, and devices, the reader is urged to review and evaluate the information provided in the package insert or instructions for each chemical, piece of equipment, reagent, or device for, among other things, any changes in the instructions or indication of usage and for added warnings and precautions. While the publisher and authors have used their best efforts in preparing this work, they make no representations or warranties with respect to the accuracy or completeness of the contents of this work and specifically disclaim all warranties, including without limitation any implied warranties of merchantability or fitness for a particular purpose. No warranty may be created or extended by sales representatives, written sales materials or promotional statements for this work. The fact that an organization, website, or product is referred to in this work as a citation and/or potential source of further information does not mean that the publisher and authors endorse the information or services the organization, website, or product may provide or recommendations it may make. This work is sold with the understanding that the publisher is not engaged in rendering professional services. The advice and strategies contained herein may not be suitable for your situation. You should consult with a specialist where appropriate. Further, readers should be aware that websites listed in this work may have changed or disappeared between when this work was written and when it is read. Neither the publisher nor authors shall be liable for any loss of profit or any other commercial damages, including but not limited to special, incidental, consequential, or other damages.

Library of Congress Cataloging-in-Publication Data

Names: Mansfield, Michael, 1943- author. | O'Sullivan, Colm, author.
Title: Understanding physics / Professor, Michael Mansfield, University College Cork, Professor, Colm O'Sullivan, University College Cork, Department of Physics.
Description: Third edition. | Hoboken, NJ : Wiley, 2020. | Includes bibliographical references and index.
Identifiers: LCCN 2019051502 (print) | LCCN 2019051503 (ebook) | ISBN 9781119519508 (paperback) | ISBN 9781119519515 (adobe pdf) | ISBN 9781119519522 (epub)
Subjects: LCSH: Physics.
Classification: LCC QC23 .M287 2020 (print) | LCC QC23 (ebook) | DDC 530–dc23
LC record available at https://lccn.loc.gov/2019051502
LC ebook record available at https://lccn.loc.gov/2019051503

Cover Design: Wiley
Cover Images: Courtesy of Michael Mansfield and Colm O'Sullivan,
Background © MARK GARLICK/SCIENCE PHOTO LIBRARY/Getty Images

Set in 10/12pt, TimesLTStd by SPi Global, Chennai, India

Printed and bound by CPI Group (UK) Ltd, Croydon, CR0 4YY

10 9 8 7 6 5 4 3 2 1

Contents

[Red text indicates that these sections/subsections may be viewed on the internet and are downloadable in the form of .pdf files]

Preface to Third Edition

Goals and objectives

Understanding Physics is written primarily for students who are taking their first course in physics at university level. While it is anticipated that many readers will have some previous knowledge of physics or of general science, each topic is introduced from first principles so that the text is suitable for students without any prior background in physics. The book has been written to support most standard first year undergraduate university physics courses (and often beyond the first year) and can serve as an introductory text for both prospective physics majors and other students who will need to apply the principles and techniques of basic physics in subsequent courses. A principal aim of this book is to give the reader the foundation required to proceed smoothly to intermediate level courses in physics and engineering and to courses in the chemical, computer, materials, and earth sciences, all of which require a sound knowledge of basic physics.

Students with some previous knowledge of physics will find that they are already familiar with many of the topics covered in the early sections. These readers should note, however, that the treatment of these topics in Understanding Physics often differs from that given in school textbooks and is designed to lay the foundations for the treatment of new and more advanced topics. As authors, one of our aims is to integrate school physics more closely to that studied at university, encouraging students to appreciate the relevance of physics previously studied and to integrate it with the material encountered at university. For these reasons we hope that students with a previous knowledge of physics will take the opportunity to refresh and deepen their understanding of topics which they may regard as familiar.

Some knowledge of simple algebra, geometry, and trigonometry is assumed but differential and integral calculus, vector analysis, and other more advanced mathematical methods are introduced within the text as the need arises and are presented in the context of the physical problems which they are used to analyse. Historically, many mathematical techniques were developed specifically to address problems in physics and these can often be grasped more easily when applied to a relevant physical situation than when presented as an otherwise abstract mathematical concept. These mathematical asides are indicated throughout the text by a grey background and it is hoped that by studying these short sections, the reader will gain some insight into both the mathematical techniques involved and the physics to which the techniques are applied.

The mathematical asides, together with Appendix A (Mathematical Rules and Formulas), cannot, however, substitute for a formal course in mathematical methods, but rather they could be considered a mathematical 'survival kit' for the study of introductory physics. It is hoped that most readers will either have already taken or be studying an introductory mathematics course. In reality the total amount of mathematics required is neither large nor particularly demanding.

Approach

It is no longer credible to describe the discoveries and developments made during the early years of the twentieth century as 'modern physics'. This is not to deny the radical and revolutionary nature of these developments but rather is a recognition that they have long since become a part of mainstream physics. Quantum mechanics, relativity, and our picture of matter at the subatomic level will surely form part of the 'classical' tradition of twenty-first century physicists. On the other hand, the discoveries of the seventeenth, eighteenth, and nineteenth centuries have lost none of their importance. The majority of everyday experiences of the material world can be understood in a fully satisfactory manner in terms of classical physics. Indeed attempts to explain such phenomena in the language of twentieth century physics, while possible in principle, tend to be unnecessarily complicated and often confusing.

In *Understanding Physics*, 'modern' (twentieth century) topics are introduced at an earlier stage than is usually found in introductory textbooks and are integrated with the more 'classical' material from which they have evolved. Although many of the concepts which are basic to twentieth century physics are relatively easy to represent mathematically, they are not as intuitive as those of classical physics, particularly for students with an extensive previous acquaintance with 'classical' concepts. This book aims to encourage students to develop an intuition for relativistic and quantum concepts at as early a stage as is practicable. However, if instructors prefer to introduce relativity (Chapter 9) and quantum physics (Chapter 14) at a later stage, their introduction may be delayed until after Chapter 23.

Understanding Physics has been kept to a compact format in order to emphasise, in a fully rigorous manner, the essential unity of physics. At each stage new topics are carefully integrated with previous material. Throughout the text references are given to other sources where more detailed discussions of particular topics or applications may be found. In order to avoid breaking the flow and unity of the material within chapters, worked examples are placed at the end of each chapter. Indications are given throughout the text as to when a particular worked example might be studied.

The internationally agreed system of units (SI) is now adopted almost universally in science and engineering and is used uncompromisingly in this text. In addition, we have adhered rigorously to the recommendations of the International Union of Pure and Applied Physics (IUPAP) on symbols and nomenclature (Cohen and Giacomo, 1987). As noted below, this edition of *Understanding Physics* has been rewritten to conform with the revision of the SI which came into force in May 2019.

The text takes a reflective approach towards the scientific method at all stages – that is, while learning the fundamentals of physics the student should also become familiar with the scientific method. In keeping with the title of the text, emphasis is placed on understanding of and insight into the material presented. The book therefore seeks not merely to describe the discoveries and the models of physics but also, in the process, to familiarise readers with the skills and techniques which been have developed to analyse natural phenomena, skills and techniques which they can look forward to applying themselves. This book does not seek to reveal and explain all the mysteries of the physical universe but, instead, lays the foundations on which readers can build and (perhaps more importantly) encourages and equips readers to explore further.

Structure

Chapter 1 starts with a short overview of the way in which physics today describes the material universe, from the very smallest building blocks of matter up to large scale bulk materials. It is a remarkable fact that the same basic principles seem to apply over the full range of distance scales – from sub-nuclear to inter-galactic. The physical principles encountered in subsequent chapters are applied to systems on all of these scales, as the need arises. The basic ideas of calculus are introduced in Chapter 2 in the context of the description of motion in one dimension; readers with a good prior knowledge of this material may wish to skip this chapter, although such readers might find it profitable to use the chapter to refresh their memories.

Chapters 3 to 7 introduce the main themes of classical dynamics. This is followed by an introduction to relative motion (Chapter 8), which is an essential prerequisite to the study of the special theory of relativity (Chapter 9). Chapters 10 to 12, deal with the mechanical and thermal behaviour of matter. A sound knowledge of wave motion (Chapter 13), a very important part of physics in its own right, is essential for a proper understanding of quantum mechanics (Chapter 14). The seven subsequent chapters (15 to 21) cover the main aspects of classical electromagnetism and its application to wave and geometrical optics is covered in Chapters 22 and 23.

The final four chapters (24 to 27) – on atomic physics, on electrons in solids, on semiconductors and on nuclear and particle physics – are a little more specialised and detailed than the others. Depending on the subjects which the reader plans to pursue subsequently, significant amounts of all or some of these chapters might well be omitted.

Changes in the third edition

- This edition has been rewritten to conform with the revision of the SI which came into force in May 2019. In the revised system definitions are achieved by adopting fixed numerical values for certain fundamental constants of nature (see Appendix D for details).
- The electromagnetism chapters have been reorganised to emphasise the integration of the various topics into a view of physics as a unified whole. Particular emphasis is placed on the use of the concept of flux (and Gauss's law) as a basis for the analyses of gravitation, electricity and magnetism.
- More advanced sections, which were indicated by a blue background in previous editions may now be accessed through links to the *Understanding Physics* Website (described below). Problems are also accessed through the Website.
- Numerous detailed improvements have been made throughout the book following suggestions from instructors, students and from our own experience.

A message for students

You should not expect to achieve an instant understanding of all topics studied. The learning process starts through an understanding of concepts and then progresses.

New material may not be fully absorbed at first reading but only after more careful study. From our own personal experience, however, we can assure you that persistence will be rewarded and that initially challenging material will be revealed as being both simple and elegant.

We have deliberately not provided end-of-chapter summaries. We feel that it is an important part of the learning exercise that students create such summaries for themselves. To assist this process, however, we have adopted a range of specific highlighting styles throughout the book (indicating fundamental principles/laws, equations of state, definitions, important relationships, etc.). A key to the more important examples of the notations used is located inside the front cover.

Readers who are studying physics for the first time are starting on a great adventure; we hope that this book will help you to find the early stages of the journey both exciting and rewarding. We also hope that it will prove to be a source of continuing support for your subsequent studies.

Acknowledgements

Understanding Physics has benefitted greatly from the many contributions, comments, and criticism generously provided over many years by numerous individuals.

We wish to express our gratitude again to all those colleagues in University College Cork and elsewhere, to our students and to the staff of Praxis Publishing and of John Wiley and Sons whose help and advice was so important in the preparation of the first and second editions of *Understanding Physics*.

The third edition has benefitted particularly from the assistance of Tony Deeney, Stephen Fahy, Joe Lennon, and David Rea, and again from the support provided by the Physics Department of University College Cork under the leadership of John McInerney. The third edition has been brought to fruition through the professionalism of a number of people at John Wiley and Sons, in particular Jenny Cossham and Emma Strickland in Chichester, Shirly Samuel and Adalfin Jayasingh in India and Mary Malin of Transtype for copyediting.

Finally, and most importantly, we want to record our deep appreciation of the support we received from our wives, Madeleine and Denise, and our children Niamh, Eoin, Katie, Chris and Claire.

Colm O'Sullivan, Michael Mansfield
Cork, July 2019

The Understanding Physics Website

The Student Companion Website for this textbook may be found at http://up.ucc.ie

The site comprises a wide selection of problems for each Chapter and a number of additional sections and subsections covering somewhat more advanced material. These resources may be accessed either by specifying the relevant URL on a browser or via the QR codes in the book margins using a mobile phone, tablet or laptop with an appropriate QR reader. Documents may be downloaded in the form of .pdf files. The Website also contains a range of interactive software designed to enhance insight and understanding of various topics covered in the text. Any reported errata will be published on the Website.

Students are encouraged to enhance their understanding and insight by using the website in parallel with studying the text. The Website will continue to be developed. The authors wish to thank Lisa Faherty of the Physics Department and Peter Flynn and Noelette Hurley of IT Services UCC for their many important contributions to the Website.

Problems

More than 600 problems are available on the Website. For each problem a link is given to its answer. Each answer is then linked to a detailed solution. Students are encouraged to refrain from studying the detailed solution to a problem until they have made a number of serious attempts to find the solution for themselves.

1

Understanding the physical universe

AIMS

- to show how matter can be described in terms of a series of *models* (mental pictures of the structures and workings of systems) of increasing scale, starting with only a few basic building blocks
- to describe how, despite the great complexity of the material world, interactions between its building blocks can be reduced to no more than four distinct interactions
- to describe how natural phenomena can be studied methodically through observation, measurement, analysis, hypothesis, and testing (the *scientific method*)

1.1 The programme of physics

Humans have always been curious about the environment in which they found themselves and, in particular, have sought explanations for the way in which the world around them behaved. All civilisations have probably engaged in science in this sense but sadly not all have left records of their endeavours. It would seem, however, that sophisticated scientific activity was carried out in ancient Babylonian and Egyptian civilisations and, certainly, many oriental civilisations had expert astronomers – every appearance of Halley's comet over a time span of 1000 years was recorded by Chinese astronomers. Science as we know it today developed from the Renaissance in Europe which in turn owed much to the rediscovery of the work of the great Greek philosopher/scientists such as Aristotle, Pythagoras, and Archimedes, work that had been documented and further developed in the Islamic world between the seventh and sixteenth centuries particularly during the Golden Age of Islamic Science, circa 750 to 1250 CE.

Common to all scientific activity is the general observation that, in most respects, the physical world behaves in a regular and predictable manner. All other things being equal, an archer knows that if he fires successive arrows with the same strength and in the same direction they follow the same path to their target. Similar rules seem to govern the trajectories of stones, spears, discuses, and other projectiles. Regularities are also evident in phenomena involving light, heat, sound, electricity, and magnetism (a magnetic compass would not be much use if its orientation changed randomly!). The primary objective of physics is to discover whether or not basic 'rules' exist and, if they do, to identify as exactly as possible what these 'rules' are. As we shall see, it turns out that most of the everyday behaviour of the physical universe can be explained satisfactorily in terms of rather few simple 'rules'. These basic 'rules' have come to be called *laws of nature*, examples of which include the Galilean /Newtonian laws of motion (Sections 3.2, 3.3, 6.1), Newton's law of gravitation (Section 5.1) and the laws of electromagnetism associated with the names of Ampère (Section 18.5), Faraday (Section 20.1), Coulomb (Section 16.3) and Maxwell (Section 21.1). In addition to these basic laws there are also 'laws' of a somewhat less fundamental nature which are used to describe the general behaviour of specific systems. Examples of the latter include Hooke's law for helical springs (Section 3.5), Boyle's (or Mariotte's) law for the mechanical behaviour of gases (Section 10.10) and Ohm's law for the conductivity of metals (Section 15.4).

The objective in studying physics, therefore, is to investigate all aspects of the material world in an attempt to discover the fundamental laws of nature and hence to understand and explain the full range of phenomena observed in the physical universe. This programme must include a satisfactory explanation of the structure of matter in all its forms (for example solids, liquids, gases), which in turn requires an understanding of the interactions between the basic building blocks from which all matter is constituted. How these interactions are responsible for the mechanical, thermal, magnetic, and electrical properties of matter must also be explained. Such explanations, once discovered, can be applied to develop descriptions of phenomena ranging from the subatomic to the cosmic and to develop practical applications for the benefit of, and use by, society.

In the next three sections we will review the language and images currently used by physicists to describe the structure of matter and the fundamental interactions of nature.

Understanding Physics, Third Edition. Michael Mansfield and Colm O'Sullivan.
© 2020 John Wiley & Sons Ltd. Published 2020 by John Wiley & Sons Ltd.

1.2 The building blocks of matter

Fundamental particles

Our present view of the nature of matter is very different from that which prevailed even sixty years ago. All matter is currently viewed as comprising various combinations of two classes of elementary particles – the basic building blocks – called, respectively, **quarks** and **leptons**. We give below an introductory account of the terminology and models used in the quark/lepton description of matter. The quark/lepton model will be discussed in more detail in Sections 27.11 and 27.12

Quarks and leptons occur in three distinct **generations** but only those in the first generation are involved in ordinary stable everyday matter. The first generation comprises two quarks, the up quark (symbol u) and the down quark (d), and two leptons, the electron (e) and the electron neutrino (ν_e). Matter comprising particles of the second and third generations is invariably unstable and is normally only formed when particles collide at very high speeds, such as those prevailing at the beginning of the Universe or in experiments with particle accelerators.

Leptons can exist as free isolated particles. Quarks, on the other hand, do not exist in isolation and are only observed grouped together, usually in threes, to form the wide range of different **particles** which form ordinary matter or which are produced in high-speed collisions.

In this section we will describe how quarks and leptons, the basic building blocks of matter, combine to form larger building blocks which, in turn, combine to form even larger building blocks, etc. as summarised in Table 1.1. Let us consider each stage in more detail, starting with combinations of quarks.

Table 1.1 Building blocks of matter

Building block	Scale/m
Quarks	$<10^{-20}$
Particles	$\sim10^{-15}$
Nuclei	$\sim10^{-14}$
Atoms	$\sim10^{-10}$
Molecules	10^{-10} to 10^{-8}
Bulk matter	$>10^{-9}$

Nuclei

The simplest combinations of first-generation quarks which are observed are three-quark combinations called **nucleons**. As illustrated in Figure 1.1, two different types of nucleon are observed, namely the **proton** (p), which comprises two u quarks and one d quark, and the **neutron** (n), which comprises one u quark and two d quarks. The electric charge of the proton is $+e$ (e is called the fundamental electric charge), while that of the neutron is zero. While a proton is stable, a free neutron is not and decays radioactively to form a proton and two leptons. Further three quark combinations, involving quarks from other generations, will be considered when we come to discuss subnuclear particles in Section 27.11.

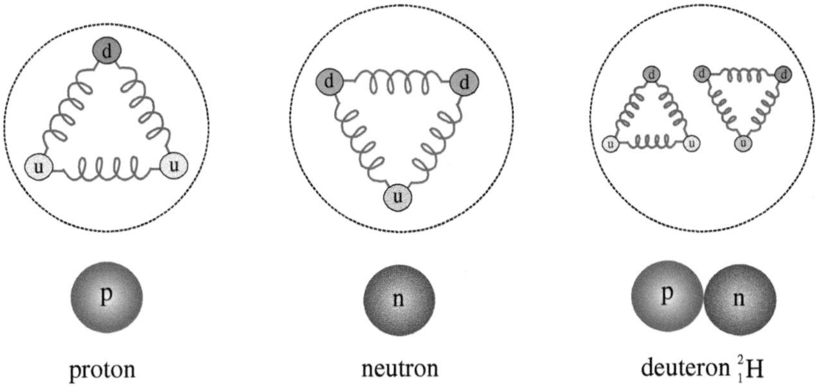

Figure 1.1. The quark and nucleon compositions of the proton ($^1_1$p), neutron ($^1_0$n) and deuteron ($^2_1$H).

The next simplest combination, also illustrated in Figure 1.1, comprises six quarks (uuuddd), equivalent to one p and one n. This combination occurs in the **nucleus** of the deuterium atom (discussed below) and is called the deuteron. The electric charge of the deuteron, like that of the proton, is $+e$. Two combinations of nine quarks, equivalent to pnn and ppn, are known; the first combination (pnn) is unstable (radioactive) and the second (ppn) stable. When we consider atoms below we will identify these combinations as nuclei of tritium and helium atoms, respectively. Hundreds of stable particles (nuclei), comprising various combinations of u and d quarks (or, equivalently, protons and neutrons), are the basis of ordinary matter and will be discussed in Chapter 27. A great many other combinations can be created artificially, for example in nuclear reactors, and, while these are unstable, their lifetimes are often sufficiently long for them to be studied in detail and put to practical use (Chapter 27).

Atoms and molecules

All nuclei have an electric charge of $+Ze$, where Z is an integer; Z can be thought of as the number of protons in the nucleus. We will discover later (Chapter 16) that positive and negative charges are attracted to one another by electrostatic attraction. Under normal conditions (by which is meant an environment which is not too hot and in which the matter density is not too low) the positively charged nuclei attract electrons to form electrically neutral systems called **atoms**. In atoms the electrons do not coalesce with the nuclei but, instead, may be thought of as moving around them in orbits with radii of the order of 10^{-10} m. This picture of an atom is something like that illustrated in Figure 1.2 – a very small nucleus of charge $+Ze$ surrounded by Z orbiting electrons, each of charge $-e$. Alternatively, as we will see in Chapter 24, the electrons may be considered as a cloud of negative charge surrounding the nucleus.

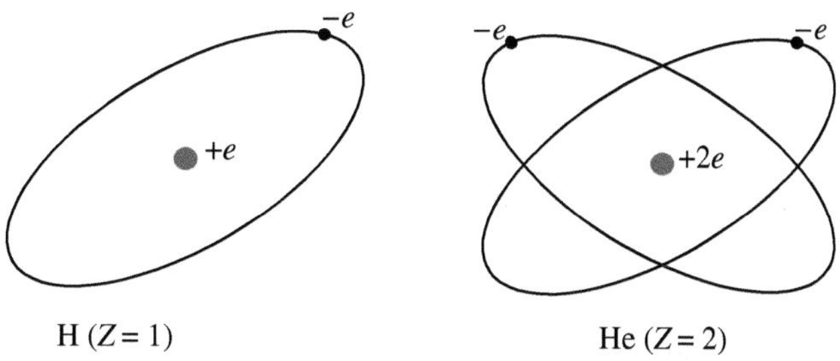

Figure 1.2. The electronic structure of the hydrogen and helium atoms.

The overall charge on the atom is thus zero; it is electrically neutral. The radius of an atom is 10 000 times greater than the radius of the nucleus (which is about 10^{-14} m). The electron is a very light particle, nearly 2000 times lighter than the proton, so nearly all the matter in an atom is concentrated in the nucleus.

As argued above, the electric neutrality of the atom requires that the nuclear charge $+Ze$ is balanced by the negative charge of Z electrons; Z therefore also gives the number of electrons in a neutral atom and is called the **atomic number**. The chemical properties of an atom are determined by the number of electrons it contains. An atom with $Z = 1$, that is with a single proton in its nucleus and hence containing a single electron, is known as a hydrogen atom (Figure 1.2). The hydrogen nucleus can also contain one or two neutrons. Such atoms are called deuterium or tritium atoms, respectively, and are known as **isotopes** of hydrogen because they are chemically identical. Helium atoms have $Z = 2$ (Figure 1.2); two different stable isotopes exist, ${}^{3}_{2}\mathrm{He}$ (two p and one n) and ${}^{4}_{2}\mathrm{He}$ (two p and two n). The chemical **elements,** listed in Appendix E, correspond to different values of Z ($Z = 3$ for lithium, $Z = 4$ for boron and so on). Note that the conventional notation used to specify an atomic nucleus (or **nuclide**) is ${}^{A}_{Z}\mathrm{X}$ where X is the chemical symbol for the particular element, Z is the atomic number (the number of protons in the nucleus) and A (the number of nucleons – that is protons plus neutrons – in the nucleus) is called the **mass number**. Isotopes of an element therefore have the same Z but different values of A.

If an atom loses or gains an electron it will end up with a net positive or negative electric charge and is called an **ion**. The number of electrons lost or gained is conventionally denoted by a suffix to the notation for the atomic nucleus, for example ${}^{A}_{Z}\mathrm{X}^{+}$(one electron lost), ${}^{A}_{Z}\mathrm{X}^{2+}$ (two electrons lost) or ${}^{A}_{Z}\mathrm{X}^{-}$(one electron gained).

When atoms come sufficiently close together that their electron systems begin to overlap, they may form stable groupings of two or more atoms which are called **molecules**. Representations of some common molecules are illustrated in Figure 1.3. Molecular sizes vary from atomic dimensions ($\sim 10^{-10}$ m) to dimensions which are many hundreds of times larger in the case of biological molecules such as proteins and nucleic acids.

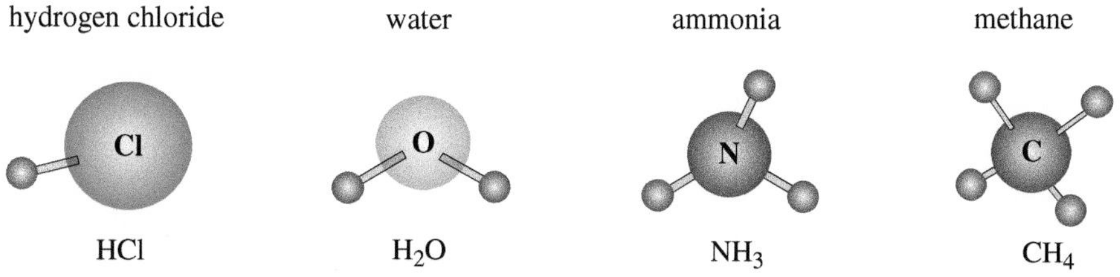

Figure 1.3. The atomic compositions of some common molecules – the smaller grey spheres represent hydrogen atoms.

The conventional notation for a molecule places the number of each type of atom in the molecule at the bottom right of the symbol for that atom. For example, a water molecule (a grouping of two atoms of hydrogen and one atom of oxygen) is denoted by the symbol H_2O (or ${}^{1}_{1}\mathrm{H}_2\,{}^{16}_{8}\mathrm{O}$, if the isotopic species of each atom is also to be shown). We will consider the various processes by which atoms can bind together to form molecules in Section 25.1.

The description of matter which we have outlined in this section is summarised in Figure 1.4.

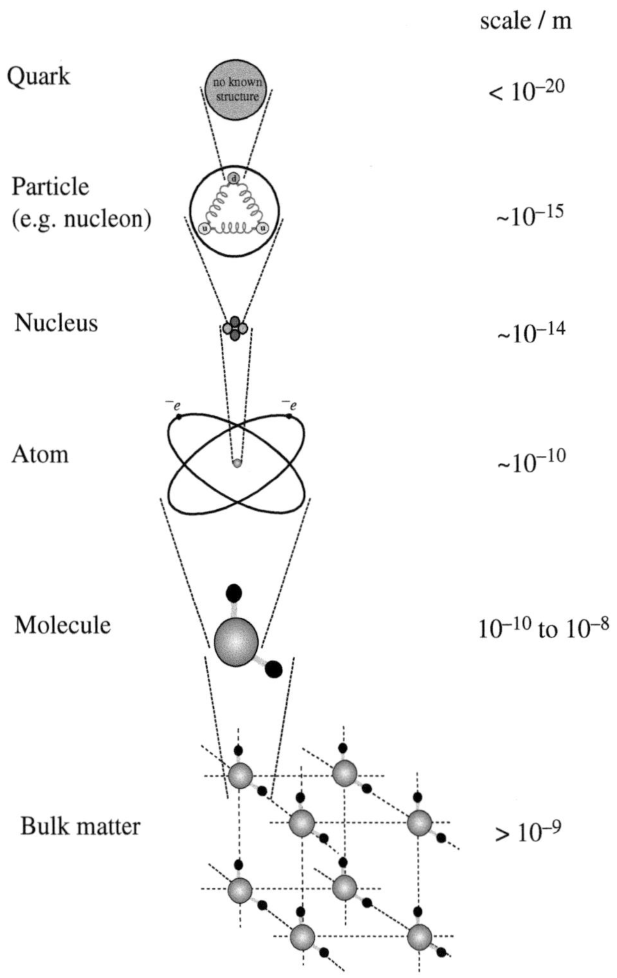

Figure 1.4. Models of the structure of matter — from the quark scale to the bulk matter scale.

1.3 Matter in bulk

When large numbers of atoms or molecules are bound closely together the atoms tend to arrange themselves in regular patterns, some examples of which are illustrated in Figure 1.5.

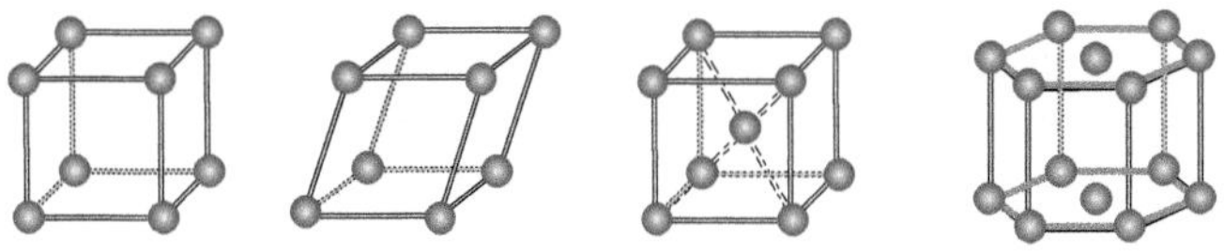

Figure 1.5. Some crystal lattice structures.

These patterns can extend over a very large number of atoms to form crystal lattices. Most **solids** are aggregates of crystals formed in this way and, if care is taken in their preparation, a solid may even be grown as one large single crystal.

Gases, on the other hand, comprise large numbers of molecules which are spaced so that the average distance between them is much greater than the molecular diameters. Molecules in gases move around rapidly and only interact with one another when they collide; otherwise they move in straight lines between collisions. The molecules in **liquids** are very close together but remain mobile and do not form crystal lattices. Thus liquids fall somewhere between gases and solids. Many materials, glass for example, do not fall into these simple categories and have properties which are somewhere between those of solids and liquids.

Our everyday experience of solids, liquids, and gases does not give any hint of their microscopic nature, that is of their molecular, atomic or sub-atomic composition. Indeed, matter in bulk appears continuous – most materials seeming to be uniform in their composition and properties at this level. Thus, if we are interested in answering questions such as 'where is a stone going to land if I throw it from the top of a cliff?' or 'how much will the air in a balloon compress if I squeeze it?', it hardly seems sensible to consider what happens to the atoms in the stone or to the quarks in the air! Questions like this are best addressed by employing **macroscopic models** (large scale pictures) of the systems being investigated rather than the **microscopic models** which we have outlined in Section 1.2. Clearly a range of different models is available to us and the choice as to which one is best to use depends on the question being asked. The criterion which we must use here is that of *simplicity* – in attempting to explain any phenomenon only those concepts necessary for the explanation should be included in the theory. This principle, which is central to all scientific endeavour, is known as *Occam's razor* after the medieval philosopher William of Occam (1285–1349), although the formulation in which it is normally stated (*entia non sunt multiplicanda praeter necessitatem* – entities are not to be multiplied unnecessarily) is attributed to John Ponce (1603–1661).

In this book we will adhere to this principle as far as possible. We will generally begin a discussion of a phenomenon from a macroscopic viewpoint. There will be many cases in which we are also able to discuss a phenomenon starting from a microscopic viewpoint (for example, kinetic theory in Section 10.11). An important test of the microscopic approach will be whether its predictions agree with those of the macroscopic treatment. We will find that when the two approaches agree we can be more confident that the microscopic approach is correct and, perhaps more importantly, we will gain some rewarding insights into the meaning of macroscopic concepts at a more basic level.

1.4 The fundamental interactions

We have seen that, despite the extraordinary complexity of the material world, all matter is made up from a relatively small number of basic building blocks. Equally remarkably, we find that the way in which these building blocks interact with one another can be reduced to no more than four distinct interactions, namely:

(a) ***The strong interaction:*** This is the force between quarks which keeps them bound together within a particle or an atomic nucleus. It is responsible for the force between nucleons in a nucleus, as described in Chapter 27. The range over which the strong interaction operates is very small – it has negligible effect if the distance between particles is much greater than 10^{-15} m.

(b) ***The electromagnetic interaction:*** This is the force which exists between all particles which have an electric charge, such as the force which keeps the electrons bound to the nucleus in an atom. The electromagnetic interaction is long range, extending in principle over infinite distances, but it is over 100 times weaker than the strong interaction within the range over which the strong interaction operates.

(c) ***The weak interaction:*** Leptons are not affected by the strong interaction but interact with one another and with other particles via a much weaker force called the weak interaction, whose strength is only 10^{-14} times that of the strong interaction. While all particles interact weakly, the effect is only noticeable in the absence of the strong and electromagnetic interactions. The weak interaction is very short range ($\sim 10^{-18}$ m) and plays a role only at the nuclear and sub-nuclear level.

(d) ***The gravitational interaction:*** By far the weakest of the fundamental interactions is the gravitational interaction, the interaction which, for example, gives a body weight at the surface of the Earth. Its strength is 10^{-38} times that of the strong interaction. All particles interact gravitationally and, like the electromagnetic interaction, the gravitational interaction operates over an infinite range.

Unification of the basic interactions

There is a long tradition in physics of attempting to unify theories which were originally distinct. For example, for a long time magnetism and electricity were considered to be quite different phenomena but during the nineteenth century the two areas were united in Maxwell's theory of electromagnetism (which will be described in Chapter 21). Over the past sixty years the theories covering the fundamental interactions have been undergoing a similar unification process. In the 1960s Weinberg, Salam, and Glashow showed that, when viewed at a more fundamental level, the electromagnetic and weak interactions can be seen to be manifestations of a single interaction (known as the **electroweak interaction**).

Since then considerable progress has been made towards the unification of the electroweak interaction with the strong interaction and this objective (known as **Grand Unified Theory**) is still being pursued. The relative strengths of the four basic interactions can be stated in terms of *coupling constants*. The values of the coupling constants of the electroweak and strong interactions vary with energy and tend to converge on the same value at very high energies, indicating that these interactions are indeed manifestations of a single interaction. A model known as the **standard model** has been developed to provide a theory of the electroweak and strong interactions and of the elementary particles that take part in these interactions. To date the results of high energy nuclear physics experiments are consistent with the standard model. In particular, in 2012, an important particle predicted by the standard model – the *Higgs boson* – which explains the existence of mass, was observed at CERN's Large Hadron Collider.

The final step in the unification of the fundamental interactions is to unify the gravitational interaction with the other fundamental interactions but, to date, even the possibility of such a single theory of all four fundamental interactions, a **Theory of Everything**, remains in the realm of speculation. Several possible lines of approach to this goal are being pursued.

1.5 Exploring the physical universe: the scientific method

Our aim in physics is to explore the physical universe, to observe, analyse and (hopefully) eventually understand the natural phenomena and processes which underlie the workings of the universe. In the process of achieving an understanding of natural phenomena we will often acquire an ability to predict their future course and hence an ability to apply our knowledge – to use it for practical purposes.

How then can we investigate natural phenomena? We outline below an approach known as the **scientific method**. It is a method which has proved its value over many centuries but it is important to note there is nothing particularly remarkable about it – it has not been handed to us on 'tablets of stone'. As we shall see it is merely a series of practical steps that anyone who wishes to study a natural phenomenon methodically might well devise on his or her own initiative. We outline these steps below.

Observation

The first step is simply to observe the phenomenon – to watch it unfold. Careful systematic observation leads us inevitably to take notes on what we see – to **record** our observations. With records we can later remind ourselves, or others, of what we have observed. The process of recording what we see in a thorough and rigorous manner leads us quickly to make measurements. For example, if we are observing the motion of a moving object we could describe its motion in words by stating that 'the object is first a long way from us, then not so far, then nearer and finally very near'. It is clear, however, that words alone soon become inadequate; they are not sufficiently precise and can be ambiguous. One person's idea of 'very near' may not be the same as that of the next person. Measurement is therefore the next step in the scientific method.

Measurement

In making measurements we must decide which (physical) quantities associated with the phenomenon that we are observing can be measured most conveniently and accurately. Note that the process is already becoming a little arbitrary. One person's idea of what can be measured conveniently may not be the same as that of the next person. As experience is built up, a consensus usually emerges on the best way to make a certain measurement. Sometimes, as we will see, technical developments can force a change in the consensus and hence even in the way in which physics is formulated. The development of physics has always been rooted strongly in empirical observation and hence in the process of measurement.

In making a measurement we inevitably have to choose a **unit** in which to make the measurement. In the case of a moving object we would naturally tend to measure its distance from us in metres because a unit of distance, the metre, has already been defined for us. Had it not been defined we would have had to invent some such unit. In choosing units for measurement it is also sensible to coordinate our choice with that of others, that is to choose agreed **measurement standards** and **systems of units**. This will enable us to communicate our observations to colleagues on the other side of the world in such a way that they will know precisely what we mean.

The internationally agreed system of units (SI), summarised in Appendix D, is now adopted almost universally in science and engineering and is used uncompromisingly in this book by following rigorously the recommendations of the General Conference on Weights and Measures. In particular we use the revised definitions of SI base units which came into force in May 2019. As we will see (for example in Section 3.4 where the definition of the metre is discussed) the revisions, which are based on the adoption of fixed numerical values for certain fundamental constants of nature, provide a good illustration of how technical developments can force a change in the way in which units are defined and physics is communicated.

Analysis and hypothesis

Having observed a phenomenon, and then having collected a set of measurements – our **experimental data** – the next step in the scientific method, in our attempt to understand the phenomenon, is to look for relationships between the quantities we have measured. For example in the case of a moving object we may have a set of measurements which gives the object's position at certain times. In comparing the measurements of position with those of time can we see any pattern? Can we put forward any **hypothesis** (inspired guess) which describes and accounts for the relationship between the quantities? Can we go further and put forward a **model** of the situation, an idealised picture of what is happening, usually based on situations we already understand – that is, on our experience?

At this stage the scientific method becomes arbitrary and personal. Different people from different backgrounds and with different experiences may see different patterns and may put forward different models. There is not necessarily any one correct interpretation. In time it may turn out that one approach is simpler and easier to follow than the others but it does not follow that this is the only correct approach. It is always wise to keep an open mind in studying natural phenomena – we are less likely to spot new patterns if we have already decided what we expect to see. We must always be on our guard against introducing prejudices when drawing on our experience.

A number of procedures may help us to identify patterns in our observations. As will be illustrated in Section 2.3 for the case of a moving object, we can assemble tables of data and can draw graphs of one measured quantity against another. We will see in Section 2.3 how analyses of tables and graphs often enable us to deduce relationships between observed quantities. Very general relationships that predict the behaviour of systems in nature are described as **laws of physics**. One of the things which makes physics such a rewarding subject to study is that not only are the fundamental laws few in number but they are also usually of relatively simple form. Because of the essential simplicity of the laws, the natural and most straightforward way to express them is through the language of mathematics.

When we are successful in identifying relationships between observed quantities we are usually able to express them as mathematical equations, which, as we will see in Section 2.3, are usually the most concise and unambiguous way of expressing relationships.

The description of relationships between quantities as 'laws' of physics is perhaps unfortunate because these laws should not be regarded as incontrovertible edicts. They are merely well-established principles based on the experimental evidence available. Sometimes, after further investigation, laws are found not to be as well established as was first believed. It is important, therefore, to **test** hypotheses and models regularly. This brings us to the final step in the scientific method.

Testing and prediction

It is now necessary to establish the range of applicability of any hypotheses and models which may have been proposed. We use these hypotheses and models, therefore, to *predict* results in situations in which measurements have not yet been made. We then make measurements in the new situations and see how well these measurements match predictions. Sometimes they do not match, although this does not necessarily mean that our previous hypotheses and models were wrong. It means that they are limited in their applicability and that we have to extend the hypotheses and models to cover the new situations.

As we shall see, developments in physics in the twentieth century have shown that many apparently universal laws of classical physics do not apply at velocities which approach the speed of light or to particles on the microscopic (atomic and nuclear) scale. It has been necessary to develop new more comprehensive theories, namely the special theory of relativity (Chapter 9) and quantum mechanics (Chapter 14), to interpret and understand these situations.

As is apparent from the account of the scientific method given above, there is nothing particularly remarkable about the method. It has been described quite simply as 'organised common sense', a method which a person without a scientific background might well adopt when faced with the task of trying to understand a physical process. In physics we have the advantage of a wealth of techniques for observation and analysis that have been developed by the scientific community over a long period of time. This gives us a head start in seeking to understand new phenomena, although we should always be aware of the possible limitations of established thinking.

In this book therefore we will not only describe the discoveries and the models which have been put forward by physicists, we will also, in the process, learn the skills and techniques which been have developed to analyse natural phenomena. We will then be able to apply these skills and techniques ourselves as we study the physical universe. The end product will be the ability to describe a whole range of apparently disconnected and complex phenomena in terms of an underlying simplicity of mathematically expressed structures. On many occasions we will see how advances in knowledge have led to new theories or models which replace a whole range of different models which were needed previously. This unifying process is one of the most satisfying aspects of physics. New understanding can actually simplify a situation, or a number of situations; we then feel instinctively that we are closer to the truth. The methods which we will uncover are powerful, intellectually satisfying, and useful. We will not be able to reveal all the mysteries of the physical universe in this book but we will take some steps along the way and, perhaps more importantly, we will emerge equipped to explore further ourselves.

1.6 The role of physics; its scope and applications

In Sections 1.2 to 1.4 we saw how physics describes the basic components of matter and their mutual interactions. We also saw how physics endeavours to describe the physical world on all its scales – from that of the quark to that of the universe. In this sense, physics provides the basic conceptual and theoretical framework on which other natural sciences are founded and may therefore be regarded as the most fundamental and comprehensive of the natural sciences.

The techniques which have been developed to analyse the physical world can be used in almost any area of pure and applied research. Physics provides an excellent testing ground for the scientific method. Moreover, in seeking to unify understanding of the natural world, physics can play an important simplifying role in science, reducing complex situations to more understandable forms. In doing so, physics can also counteract the fragmentation into separate disciplines which tends to accompany the ever expanding growth in scientific and technical knowledge.

Physics is at the basis of most present technology and is sure to be at the basis of much future technology, tackling problems as pressing and diverse as the development of new energy sources, of more powerful and less intrusive medical diagnostics and treatments and of more effective electronic devices. The growth of physics has spawned a multitude of technological advances which impact on almost all areas of science. Engineering practice must be revised regularly to take advantage of opportunities presented by the advance of physics.

In the previous section we noted that new and more comprehensive theories, namely the special theory of relativity and quantum mechanics, were developed in the last century to account for situations in which the laws of classical physics do not apply. The new theories have stimulated important new technologies, such as quantum engineering (the development of new microelectronic devices), laser technology, and nuclear technology, technologies which could hardly have been dreamt of at the beginning of the twentieth century.

A sound knowledge of physics is needed by scientists and technologists if they are to be able to understand and adjust to the rapidly changing world in which they find themselves. Moreover, this understanding should stimulate them to devise and initiate further advances.

2

Using mathematical tools in physics

AIMS

- to demonstrate the scientific method by applying it to the analysis of motion in a straight line
- to introduce the basic calculus methods used in this book and to demonstrate how they may be used in the analysis of physical phenomena
- to derive equations which describe some special cases of one-dimensional motion quantitatively and which can be used to predict their future courses

2.1 Applying the scientific method

In this chapter we shall illustrate the scientific method by using it to study certain types of motion. In doing so we shall introduce some important mathematical techniques which will enable us to analyze and represent physical processes in a concise and rigorous manner. At the same time we will introduce the physical quantities which are used to describe motion in a straight line and angular motion about a fixed axis.

While readers who are familiar with the analysis of linear motion and of angular motion, and who are also familiar with the use of elementary calculus in physics, may choose to proceed to Chapter 3, we recommend that they take the opportunity to refresh their understanding of these topics in this chapter.

2.2 The use of variables to represent displacement and time

We begin our investigation of motion by studying and characterising different types of motion. At this stage we are not concerned with the cause of motion, although the cause of motion is a topic which is of central interest in physics and will be investigated in detail in the next chapter. First we simply consider the behaviour of a moving object and decide which quantities associated with the motion we can measure. We will then see if there is any discernible pattern in a particular motion – whether we can establish any relationships between the measured quantities and whether we can establish any *model* for the motion.

A moving object is an object whose position changes with time. The obvious physical quantities to measure in recording the behaviour of a moving object are therefore its **position** and the **time** at which it is at that position. Let us first consider measurement of position.

We can specify the position of a point P by measuring its **displacement** with respect to some reference point O which we call the **origin**. We use the symbol r to represent the value of displacement, a variable quantity. Note however that in specifying the position of P relative to O it is not sufficient simply to state the distance from O to P. If, for example, we say that a point P is in the plane of this page and is at a distance r from O, P could be anywhere on a circle of radius r drawn around O (as illustrated in Figure 2.1). To avoid ambiguity in specifying the position of P we must also specify the direction of P relative to O. In this case this could be achieved by stating that P is directly to the right of O, as shown in Figure 2.1.

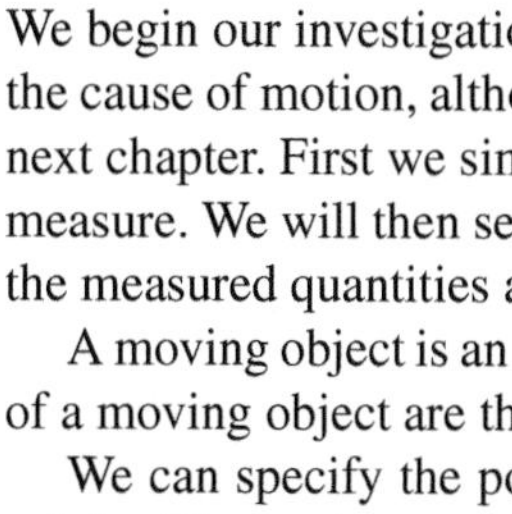

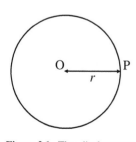

Figure 2.1. The displacement r of the point P from the origin.

To specify a displacement r unambiguously, therefore, we must specify both its magnitude (the distance from O to P) and its *direction* (the direction of the line OP). Later (Section 4.1) we will use the term *vector* to describe a quantity which has both magnitude and direction; we will also show that vectors must be handled using well defined methods. For our present purposes however, we can simplify the treatment of displacement by considering the special case of *linear (or one-dimensional) motion*, that is motion which is confined to a straight line. As illustrated in Figure 2.2, a linear displacement from the origin O along a straight line can be in one of only two directions so that a point which is a distance 2 cm from O can be at either of the two positions P or P′.

Understanding Physics, Third Edition. Michael Mansfield and Colm O'Sullivan.
© 2020 John Wiley & Sons Ltd. Published 2020 by John Wiley & Sons Ltd.

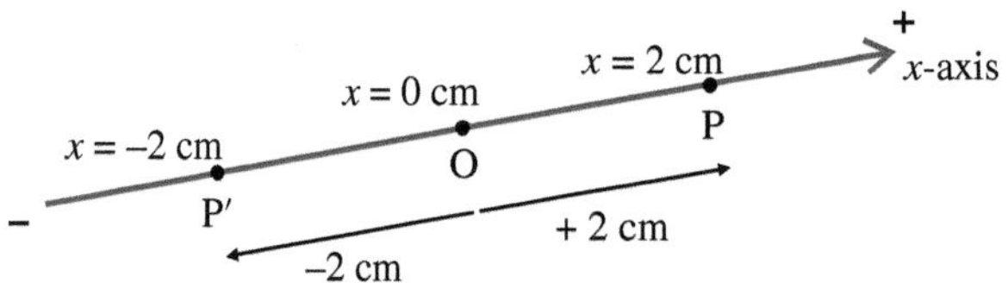

Figure 2.2. The x-coordinate axis, showing the displacements of P and P′ relative to O.

We distinguish between the two possible directions in linear motion by using a sign convention to specify the direction of the displacement. Displacement therefore can be represented by an **algebraic quantity**, namely a quantity which can be expressed in terms of its magnitude preceded by a plus or minus sign; thus the displacements of the points P and P′ are +2 cm and −2 cm from O, respectively.

The choice between the + and − labels for the two directions in Figure 2.2 is of course arbitrary. We could equally well have chosen the opposite sign labels. The important point is that, having adopted a **convention for signs**, we follow this convention consistently throughout our analysis.

In linear motion, displacements from the origin are usually represented by the variable quantity x. The straight line along which the motion occurs is then described as the ***x*-axis** and the algebraic value of the displacement, x, of a certain position from the origin O is the **coordinate** of this position. The position of a point on the straight line is specified unambiguously by stating the algebraic value of x provided a convention for positive x has been adopted. For example, based on the conventions adopted in Figure 2.2, the displacement of P is $x = +2$ cm and that of P′ is $x = -2$ cm.

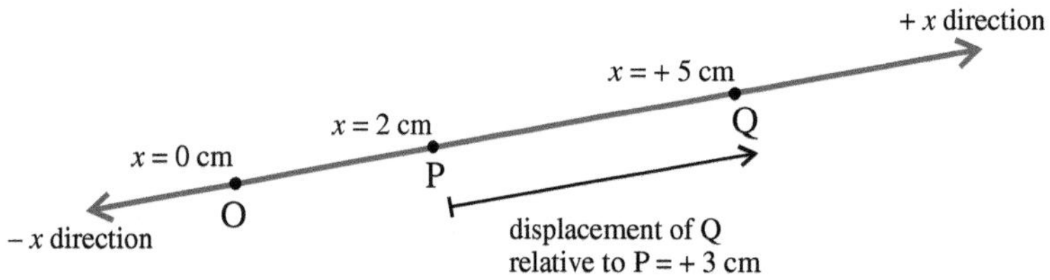

Figure 2.3. The displacements of P and Q relative to O, and of Q relative to P.

We can also define the displacement of a second point on the straight line, such as Q in Figure 2.3, *relative* to P.

If the displacement of P relative to O is +2 cm (that is, the x-coordinate of P is +2 cm) and the displacement of Q relative to O is +5 cm, we can easily deduce from an inspection of Figure 2.3 that the displacement of Q relative to P is $5 - 2 = +3$ cm, a positive displacement. Similarly, the displacement of P relative to Q, is $2 - 5 = -3$ cm, a negative displacement. Note how the signs of the algebraic quantities which represent relative displacements give the directions of the displacements.

The second quantity which we have decided to measure in our study of motion is **time**, denoted by the symbol t, which can also be represented by an algebraic quantity. Unlike displacement, t can only increase while we are making our observations – it can change in only one direction, which we define to be the positive direction. Like displacement, time is measured with reference to an origin, in this case the starting instant. Note that, although time can only change in the positive direction it is possible for t to be negative. For example, if we choose 10.00 a.m. as our starting instant the time 9.55 a.m. becomes −5 minutes.

2.3 Representation of data

Let us consider the case of an object which is only free to move along a straight line, the x-axis, as illustrated in Figure 2.4. As an example we will consider the motion of a train along a straight section of track. Suppose that we make a series of measurements of the train's position together with the corresponding times. We can display these measurements (our *data*) in a number of ways, the most obvious of which is the **tabular representation**, illustrated in Table 2.1 for a particular motion of the train which we call motion M.

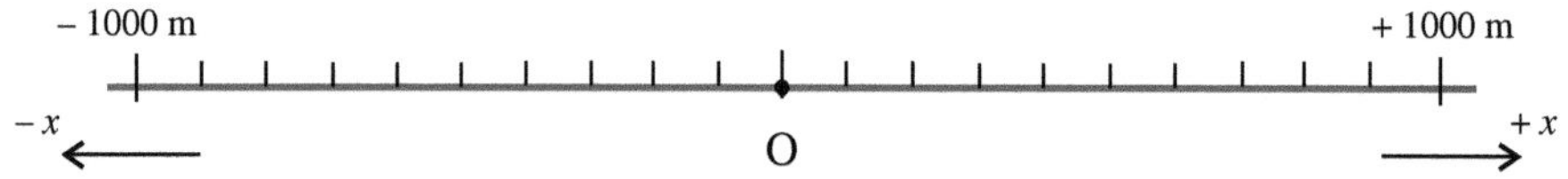

Figure 2.4. The x-axis for a moving train.

In the third column of Table 2.1, in order to make the relationship between displacement and time more obvious, times are also stated with reference to 10.00 a.m., the time at which we start observing the train's motion (our time origin). In this case a simple relationship between x and t can be deduced quite easily from an inspection of the numbers in the first and third columns of Table 2.1.

Table 2.1 Tabular representation of train motion M

displacement x/m (±200 m)	time (clock readings) (±0.2 min)	time (measured with reference to 10:00 a.m.) t/min (±0.2 min)
0	10:00.0 a.m.	0.0
+400	10:01.0	1.0
+800	10:02.0	2.0
+1200	10:03.0	3.0
+1600	10:04.0	4.0
+2000	10:05.0	5.0
+2400	10:06.0	6.0
+2800	10:07.0	7.0
...	...	...

A second method of representing the data for motion M, the **graphical representation**, is illustrated in Figure 2.5. A graph is drawn of x against t. The precision of any real measurement of displacement or time is limited by the precision of the equipment and the skill of the experimenter. This lack of precision – the *uncertainty* associated with the measurement – is indicated at the head of each column in Table 2.1, namely ±200 m for displacement measurements and ±0.2 min for time measurements. In Figure 2.5 these uncertainties are represented by the crosses which are drawn through each data point, the lengths of the vertical lines indicating the estimated range of uncertainty in the displacement measurements and the lengths of the horizontal lines indicating the estimated range of uncertainty in the time measurements.

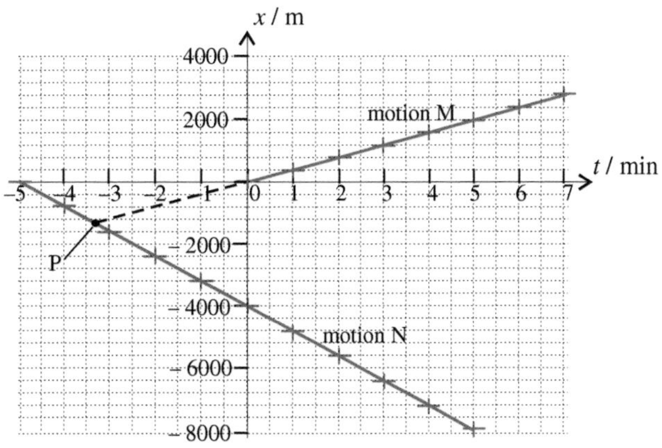

Figure 2.5. Graphs of displacement x against time t for train motions M (Table 2.1) and N (Table 2.2).

Within the stated uncertainties the plot of x against t for motion M may be represented by a straight line and we can use the plot to determine the position of the train at times when measurements are not available. When we move outside the range of the data points this process becomes more speculative and we need to test our hypothesis concerning the relationship between x and t by accumulating more data points.

A straight line can be represented mathematically by the general equation (Equation (2.1))

$$x = mt + c \tag{2.1}$$

(Appendix A.10.1 – with $y \rightarrow x$ and $x \rightarrow t$). A plot of x against t gives a straight line (Figure 2.6). $m = \frac{\text{CB}}{\text{AB}}$ is the **gradient** (or slope) of the line and c is its **intercept** on the x-axis. Note that the values of m and c remain unchanged for all values of t. A quantity like this, whose value does not change with time, is called a **constant**.

Comparing Figure 2.6 to the straight line graph through our data points for motion M of the train in Figure 2.5, that is *fitting* our experimental results to the mathematical function (2.1), it is evident that Equation (2.1) can represent our data points if $c = 0$ m and $m = 400$ m per minute, which is usually written 400 m min^{-1}. The equation which represents motion M is therefore

$$x = (400\,\text{m}\,\text{min}^{-1})t \tag{2.2}$$

where x is measured in m and t is measured in minutes. Equation (2.2) provides us with a third means of representing motion M. This is the **mathematical representation**, which in this case is an algebraic representation. The mathematical representation

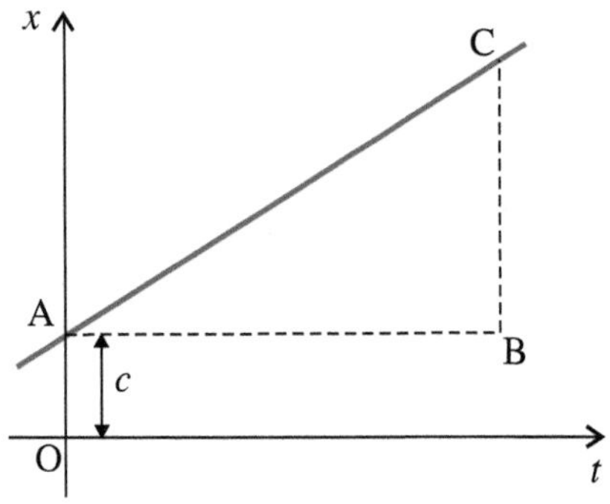

Figure 2.6. The representation of a straight line by the equation $x = mt + c$.

is clearly more concise than the tabular or graphical representations. Hence, in physics we will usually seek to represent relationships between quantities through mathematical equations. Mathematics provides us with a very valuable shorthand notation for expressing and processing our findings in physics. For a more complex set of data than that displayed in Table 2.1 and in Figure 2.5, the process of seeking a mathematical function to represent data is usually achieved through numerical curve fitting using a computer; the mathematical function is varied until the best fit of quantities such as m and c to the experimental data is obtained.

All three methods of representation can be valuable in identifying patterns – in identifying the relationships between measured quantities. Any of the three can be used determine values of x at times for which we have no data – they all have predictive power. A tabular representation of data – a *spreadsheet* – is probably best suited for input into a computer whereas the graphical and mathematical representations are generally more readily assimilated by the human mind.

At the end of the last section we noted that the algebraic quantities x and t can have negative values. Let us consider a motion of a second train along the straight line shown in Figure 2.4, a motion which we call motion N. In motion N, observation of the train's motion begins at 9.55 a.m. and the train moves in the negative x direction. Table 2.2 represents motion N in tabular form

Table 2.2 Tabular representation of train motion N.

displacement x/m (±200 m)	time (clock readings) (±0.2 min)	time (measured with reference to 10:00 a.m.) t/min
0	09:55.0 a.m.	−5.0
−800	09:56.0	−4.0
−1600	09:57.0	−3.0
−2400	09:58.0	−2.0
−3200	09:59.0	−1.0
−4000	10:00.0	0.0
−4800	10:01.0	+1.0
−5600	10:02.0	+2.0
−6400	10:03.0	+3.0
−7200	10:04.0	+4.0
...	...	...

Motion N is also represented graphically in Figure 2.5. A plot of x against t again produces a straight line although, unlike the line which represents motion M, the line which represents motion N has a negative slope. If we fit the straight line Equation (2.1) to the graph of motion N, we obtain $m = -800\ \text{m min}^{-1}$ and $c = -4000$ m so that we can write this equation as

$$x = -(800\ \text{m min}^{-1})t - 4000\ \text{m} \tag{2.3}$$

Equation (2.3) is therefore the mathematical representation of motion N. Note that, although the magnitudes and signs of the variable quantities x and t change during the motion, the equation which represents the motion mathematically remains the same.

We can use any of the representations described above to deduce useful information concerning the motions of the two trains, M and N. For example we may wish to know whether there is any danger of a collision. Put another way the question is "can the two trains be in the same place at the same time? – can they have the same value of x at the same time t?" In the graphical representation this is true if the two lines intersect and in the mathematical representation it is true if there are values of x and t which simultaneously satisfy Equations (2.2) and (2.3).

Solving the simultaneous equations $x = 400t$ and $x = -800t - 4000$, we obtain $x = -1333$ m and $t = -3.33$ min, corresponding to the point P in Figure 2.5. Our data for motion M, as presented in Table 2.1, does not however extend to negative values of x so that we do not know whether the above solution is valid physically. We need more data to test this conclusion. Specifically, we need to know the position of the train which is performing motion M at negative times (the dashed section of the line representing motion M in Figure 2.5).

The representation of experimental data is clearly a very important step in the scientific method. It helps us to identify patterns and to formulate these patterns mathematically. Representation of experimental data may also enable us to make predictions concerning the course of the phenomenon which we are observing, predictions which can then be tested by making new measurements.

2.4 The use of differentiation in analysis: velocity and acceleration in linear motion

Further properties of a moving object may be defined in terms of the quantities of displacement and time. Clearly some objects move more quickly than others – a greater change of displacement is achieved in a given time interval. From the data presented in Tables 2.1 and 2.2, it is clear that the train which is described in motion N is moving faster in the $-x$ direction than the train which is described in motion M is moving in the $+x$ direction. For either train this property of the motion can be described quantitatively by evaluating the ratio in Equation (2.4)

$$\frac{\text{change in displacement}}{\text{corresponding change in time}} \tag{2.4}$$

This gives the rate of change of displacement with respect to time, known as the **velocity** of the motion.

For motion M the ratio is $\dfrac{(2800-0)\text{m}}{(7-0)\text{min}} = 400\,\text{m min}^{-1}$

and for motion N it is $\dfrac{(-7200-0)\text{m}}{[4-(-5)]\text{min}} = -800\,\text{m min}^{-1}$

Note that the units of velocity follow directly from its definition. If displacement is measured in metres and time in seconds, velocity is given in m s^{-1}; if displacement is measured in metres and time in minutes, velocity is given in m min^{-1}.

From the data which is presented in Table 2.1, the train which is performing motion M is moving at a constant velocity of 400 m min^{-1} throughout its motion. In general however the velocity of a moving object is rarely constant. If it starts from rest, it usually takes some time to reach a steady velocity and will also take some time to reduce its velocity to zero when it comes to rest. In such a case the ratio defined in (2.4) only gives the *average velocity* of the object, that is

$$\langle v\rangle = \frac{\text{change in displacement}}{\text{corresponding change in time}}$$

Unless the object is moving with constant velocity throughout its motion, the average velocity does not tell us the velocity of the object at a particular time that we may be interested in. To specify the velocity of an object at a particular time we must define its **instantaneous velocity**, its average velocity over an infinitesimally small time interval, centred on the time of interest. Instantaneous velocity is defined in the following way.

Consider an object which is moving along a straight line, the x-axis illustrated in Figure 2.7. In travelling between two fixed points A and B on the line, the object is displaced by Δx. If the corresponding time interval is Δt, the average velocity between A and B is given by $\langle v\rangle = \dfrac{\Delta x}{\Delta t}$.

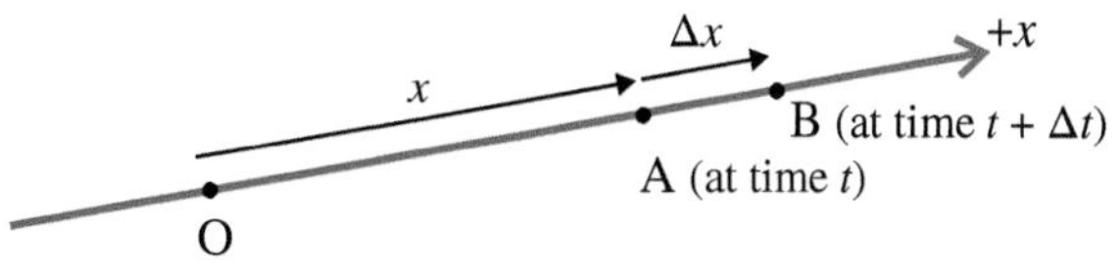

Figure 2.7. The displacements of a moving object at times t and $t + \Delta t$.

If the point B is so close to A that the two points almost coincide, Δt and Δx become infinitesimally small, although their ratio remains finite. The (instantaneous) **velocity** at A is then defined as the value of the ratio $\dfrac{\Delta x}{\Delta t}$ in the limit of $\Delta t \to 0$. Thus

$$\boxed{v := \underset{\substack{\Delta t\to 0\\ i.e.,\ B\to A}}{\text{limit}} \frac{\Delta x}{\Delta t}} \tag{2.5}$$

Equation (2.5) [1] is written in terms of limiting the value of the ratio of two quantities when they become infinitesimally small. In the notation of differential calculus this equation is written

$$\boxed{v := \frac{dx}{dt}} \tag{2.6}$$

where $\dfrac{dx}{dt}$ is known as the **derivative** of x with respect to t.

[1] Note: the symbol := is used here to denote a **defining equation** and will be used regularly in this book.

Differentiation

In general if f represents a quantity whose value depends on the value of another variable quantity u; for example, if $f = u^2 + 2u$, we describe f as a *function* of u and write $f(u) = u^2 + 2u$.

The derivative of f with respect to u is then defined as

$$\boxed{\frac{df}{du} := \lim_{\Delta u \to 0} \frac{\Delta f}{\Delta u}} \tag{2.7}$$

If f is plotted against u, the derivative of f at a particular value of u is the **slope** (or **gradient**) of the tangent to the curve at that point, $\frac{df}{du}$, as illustrated in Figure 2.8. As also indicated in this figure the slope, $\frac{df}{du}$, is equal to tan α.

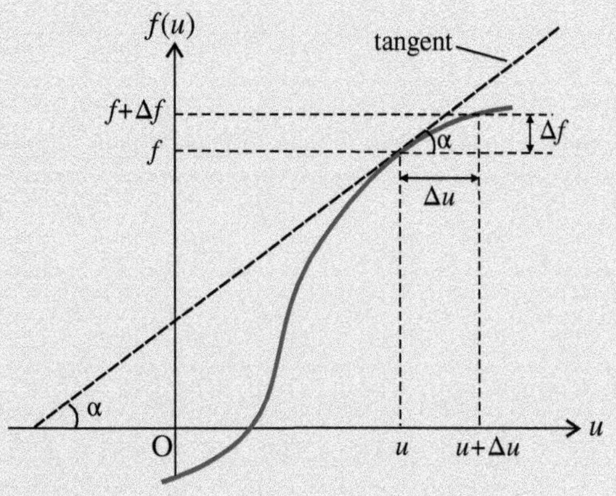

Figure 2.8. A plot of a general function $f(u)$ against u.

As long as the function f varies smoothly with u, that is, as long as there are no breaks in the plot, its derivative exists and can be determined using Equation (2.7), a process which is known as **differentiation**. Let us consider, as an example, differentiation of the function $f(u) = u^2$. As u increases from u to $u + \Delta u$, f increases to $f + \Delta f = (u + \Delta u)^2$

We can therefore write $\Delta f = (f + \Delta f) - f = (u + \Delta u)^2 - u^2$

$$= u^2 + 2u\Delta u + (\Delta u)^2 - u^2$$
$$= 2u\Delta u + (\Delta u)^2$$

and the derivative $\frac{df}{du} = \underset{\Delta u \to 0}{\text{limit}} \frac{\Delta f}{\Delta u}$ can be written

$$\underset{\Delta u \to 0}{\text{limit}} \frac{2u\Delta u + (\Delta u)^2}{\Delta u} = \underset{\Delta u \to 0}{\text{limit}} 2u + \underset{\Delta u \to 0}{\text{limit}} \Delta u$$

Δu is clearly zero in the limit $\Delta u \to 0$ and thus $\frac{df}{du} = \frac{d(u^2)}{du} = 2u$. We have obtained an equation which gives the value of $\frac{df}{du}$ at any value of u.

This is an example of a general rule for the differentiation of a function of the form u^n, namely

$$\frac{d(u^n)}{du} = nu^{n-1}$$

For u^2 this yields $\frac{d(u^2)}{du} = 2u^{2-1} = 2u$, in agreement with the result we have just obtained above.

Note that the differentiation of the derivative $\frac{df}{du}$ with respect to u, is written $\frac{d}{du}\left(\frac{df}{du}\right) = \frac{d^2f}{du^2}$ and is called the *second derivative* of f. Similarly, $\frac{d^3f}{du^3}$ is known as the third derivative, etc.

EXAMPLE: Consider a case in which the relationship between two quantities, represented by the variables x and y, is given by the equation

$$x(y) = ay^3 - by^2 + c$$

where a, b and c are constants.

If a third quantity z is defined through the equation $z = \frac{dx}{dy}$, we can use differentiation to derive an equation which gives z as a function of y. Thus

$$z(y) = \frac{dx}{dy} = 3ay^2 - 2by$$

Differentiation is a standard procedure in mathematics. The results of applying this procedure to functions of interest in this book are summarised in Table A.1 of Appendix A, together with some general rules for differentiating the sums and the products of functions (Appendix A.6.1).

Returning to the definition of velocity in Equation (2.5) note that, because Δt is always positive, v has the same sign as Δx. Hence v is in the same direction as Δx. Thus if the object is moving in the $+x$ direction v is positive, while if it is moving in the $-x$ direction v is negative.

If the mathematical function which describes the displacement of a point as a function of time for a particular motion is known, for example:

$$x = x(t) = \mathrm{A}t + \mathrm{B}t^2$$

where A and B are specified constants, v can be evaluated as a function of time simply by differentiating x with respect to t.

Thus, in this case
$$v = \frac{dx}{dt} = \mathrm{A} + 2\mathrm{B}t$$

For the motions of the two trains considered in the previous section, this process yields:

For motion M (Equation (2.2)):
$$x = (400\ \mathrm{m\ min^{-1}})t$$

Thus
$$v = \frac{dx}{dt} = +400\ \mathrm{m\ min^{-1}} \tag{2.8}$$

For motion N (Equation (2.3)):
$$x = -(800\ \mathrm{m\ min^{-1}})t - 4000\ \mathrm{m}$$

Thus
$$v = \frac{dx}{dt} = -800\ \mathrm{m\ min^{-1}} \tag{2.9}$$

Note that, as required in the algebraic representation, the sign of v gives the direction of the velocity with respect to the chosen positive direction,

Let us consider now a general case, in which v is not constant throughout the motion. In such a case a plot of x against t (an '*x–t graph*') does not produce a straight line as it did for the two cases which are represented in Figure 2.5. Instead we might obtain a graph of the type shown in Figure 2.9. In this case v is no longer constant – it varies with time. It is a function of time which we write as $v = v(t)$.

Consider two points on the graph A and B which are close enough together that the line AB may be considered to be a straight line. The average velocity between A and B,

$$\langle v \rangle = \frac{\Delta x}{\Delta t} = \frac{\mathrm{BC}}{\mathrm{AC}} \tag{2.10}$$

The value of $\frac{\mathrm{BC}}{\mathrm{AC}}$, the slope of AB, is given by $\tan \alpha$.

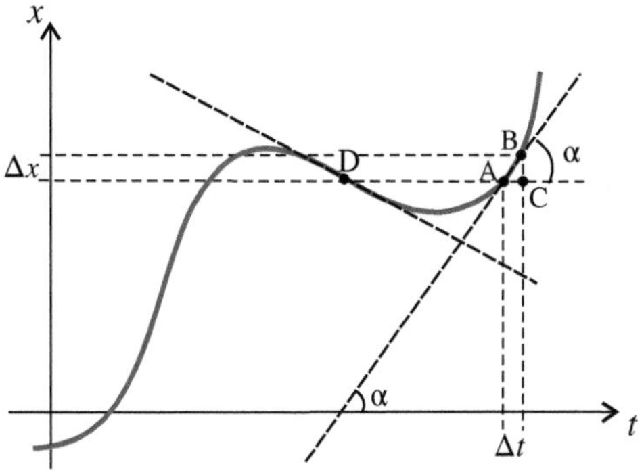

Figure 2.9. A plot of displacement x against time t (an 'x–t graph'), for a case in which the velocity v is not constant.

The value of v at any time t is given by the gradient of the tangent to the x–t graph at that time. We can therefore calculate the value of v at a particular time t_1 by determining the gradient of the x–t graph at $t = t_1$. We can achieve this either graphically (by drawing the tangent to the curve at $t = t_1$ and evaluating $\frac{\mathrm{BC}}{\mathrm{AC}}$ or $\tan \alpha$) or, if the functional dependence of x on t is known, using Equation (2.6), through differentiation of $x(t)$ with respect to t to give $v(t)$ followed by substitution of the value of t_1 into $v(t)$ to give $v(t_1)$, that is $v(t_1) = \left[\frac{dx}{dt}\right]_{t=t_1}$.

At a point such as D in Figure 2.9, where the gradient of the x–t graph is negative, v is negative.

Acceleration

A third characteristic of a moving object, in addition to its displacement and velocity, may be defined – its acceleration**,** the rate of change of velocity with respect to time. If, as illustrated in Figure 2.10, at two points A and B on the path of a moving object the displacements, times and velocities are (x, t, v) and $(x + \Delta x, t + \Delta t, v + \Delta v)$, respectively, we define the *average acceleration* of the object between A and B as

$$\langle a \rangle = \frac{\Delta v}{\Delta t}$$

and the (instantaneous) **acceleration**, the average acceleration over an infinitesimally small time interval, Δt, as

$$\boxed{a := \underset{\substack{\Delta t \to 0 \\ i.e.\ \mathrm{B} \to \mathrm{A}}}{\mathrm{limit}} \frac{\Delta v}{\Delta t} = \frac{dv}{dt}} \tag{2.11}$$

Note that
$$a = \frac{dv}{dt} = \frac{d}{dt}\left(\frac{dx}{dt}\right) = \frac{d^2x}{dt^2}$$

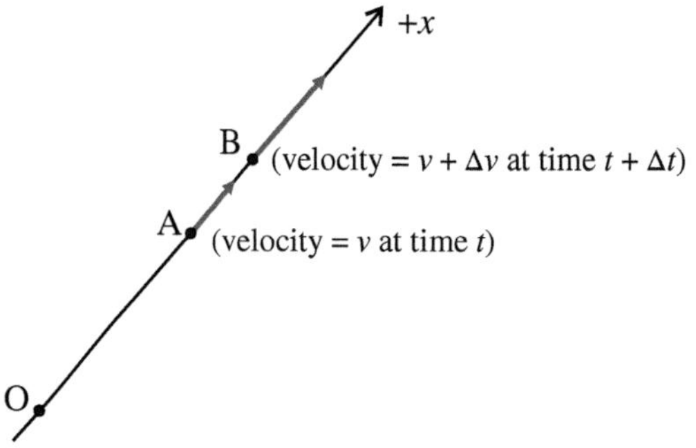

Figure 2.10. The displacements and velocities of a moving object at times t and $t + \Delta t$.

As indicated in definition (2.11), acceleration, like velocity, is defined in terms of a derivative. In linear motion acceleration, like displacement and velocity, is represented by an algebraic quantity. From definition (2.11) it follows that a has the same sign as Δv.

The units of acceleration follow directly from its definition (2.11). For example, if displacement is measured in metres and time in seconds (and velocity therefore in m s^{-1}), acceleration is given in m s^{-2}.

As illustrated in Figure 2.11, the slope of the tangent to the v–t graph, the value of $\frac{dv}{dt}$ at a particular value of t, gives the value of a at that time (just as the slope of the tangent to the x–t graph gives the value of v at a particular value of t), that is $a = \frac{dv}{dt}$, which is given by $\tan \alpha$.

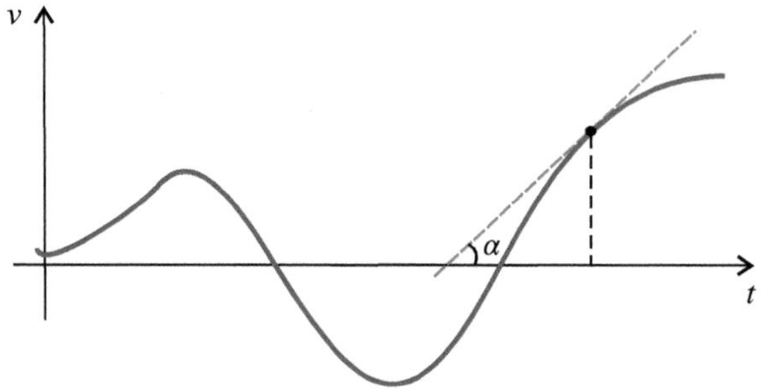

Figure 2.11. A plot of velocity v against time t (a 'v–t graph').

For the train motions M and N, which we examined in the previous section, the acceleration a can be determined by differentiating the velocity equations (2.8) or (2.9) with respect to time. In both cases v is constant and therefore

$$a = \frac{dv}{dt} = 0$$

In Worked Example 2.1 we show how differentiation may be used to evaluate velocity and acceleration in linear motion. Worked Example 2.2 gives a further example of the use of differentiation.

Study Worked Examples 2.1 and 2.2

For problems based on the material presented in this section visit up.ucc.ie/2/ and follow the link to the problems.

2.5 The use of integration in analysis

Consider a plot of v against t (a v–t graph), as illustrated in Figure 2.12.

From Equation (2.10) it follows that Δx, the change of displacement in the time interval Δt, is given by

$$\Delta x = \langle v \rangle \Delta t$$

As $\Delta t \to 0$, B $\to$ A and $\langle v \rangle \to v$, the average velocity between A and B becomes the (instantaneous) velocity at A.

In a v–t graph $\langle v \rangle \Delta t = \Delta x$ is the area of a rectangle. As indicated in Figure 2.12, as $\Delta t \to 0$ this rectangle approximates to the area (shaded) under the v–t graph between A and B; therefore this area equals Δx, the distance travelled in the time interval Δt.

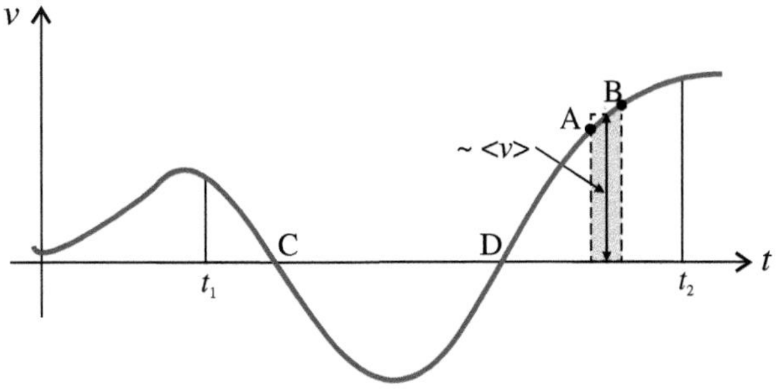

Figure 2.12. A v–t graph, showing $\langle v\rangle \Delta t$, the (shaded) area under the graph over a time interval Δt

We can break any time interval (such as t_1 to t_2 in Figure 2.12) into a very large number of very small time intervals Δt, as indicated in Figure 2.13. The total area under the v–t graph between t_1 and t_2 is then the sum of all the $v_i \Delta t_i = \Delta x_i$ elements.

$$v_1 \Delta t_1 + v_2 \Delta t_2 + v_3 \Delta t_3 + \dots + v_i \Delta t_i + \dots = \Delta x_1 + \Delta x_2 + \Delta x_3 + \dots + \Delta x_i + \dots \tag{2.12}$$

which is the total change of displacement between t_1 and t_2, namely $(x_2 - x_1)$. Note that when v is negative, as it is in the region CD in Figures 2.12 and 2.13, the $v_i \Delta t_i$ elements make negative contributions (corresponding to negative displacements Δx_i, etc.) to the net displacement.

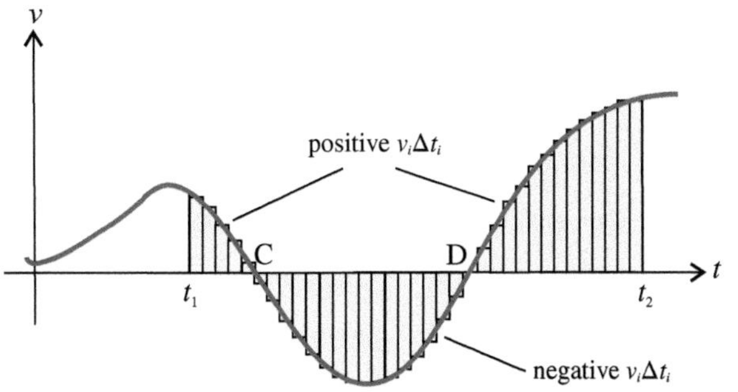

Figure 2.13. A v–t graph, indicating the total area under the graph between times t_1 and t_2

The process of summing an infinite number of infinitesimal elements is the mathematical procedure known as **integration**. We saw in the previous section how differentiation enables us to determine the instantaneous velocity of an object if we know its position as a function of time. Integration enables us to determine the displacement of an object if we know its instantaneous velocity as a function of time.

Integration

Like differentiation, integration is a standard procedure – it is in fact the reverse procedure to differentiation. If the differentiation of a function $f(u)$ produces the function $f'(u) = \dfrac{df}{du}$, integration enables us to reverse the process – to determine $f(u)$ from a knowledge of $f'(u)$. In general the integral of a function $f = f(u)$ between $u = u_a$ and $u = u_b$ (known as the *limits* of the integration) is written $\int_{u_a}^{u_b} f(u)du$ in the notation of calculus. The **integral** of f is defined as follows

$$\int_{u_a}^{u_b} f(u)du = F(u_b) - F(u_a) = [F(u)]_{u_a}^{u_b} = \lim_{\Delta u_i \to 0} [f(u_1)\Delta u_1 + f(u_2)\Delta u_2 + f(u_3)\Delta u_3 + \dots]$$

where $F(u_b)$ and $F(u_a)$ are the values of the integral at the limiting values of u. The integral function $F(u)$ is sometimes called the antiderivative of the function $f(u)$.

When written as above, with specified limits to the integration, the integral is known as a **definite integral**. The integral can also be written without specified limits in which case it is described as an **indefinite integral**

$$\int f(u)du = F(u) + C,$$

where C is a constant which can have any value because differentiation of $F(u) + C$ produces $f(u)$ whatever the value of C. Note that when limits are specified, C cancels out and the indefinite integral becomes definite.

$$\int_{u_a}^{u_b} f(u)du = F(u_b) + C - F(u_a) - C = F(u_b) - F(u_a)$$

EXAMPLE: If a quantity w is defined in terms of an equation such as $x(y) = \frac{dw}{dy}$ (a *differential equation*), we can use integration to write w as a function of y. In this case if $x(y) = ay^3 - by^2 + c$, we obtain

$$w(y) = \int xdy = \int (ay^3 - by^2 + c)\ dy = \frac{ay^4}{4} - \frac{by^3}{3} + cy + D,$$

where D is an arbitrary constant whose value may be fixed by initial conditions. We can use this result to evaluate the change in the value of w when the value of y changes, for example from y_1 to y_2.

$$w(y_2) - w(y_1) = \frac{a}{2}(y_2^4 - y_1^4) - \frac{b}{3}(y_2^3 - y_1^3) + c(y_2 - y_1) + D - D$$

This is usually denoted by $w(y_2) - w(y_1) = \left[\frac{ay^4}{4} - \frac{by^3}{3} + cy\right]_{y_1}^{y_2}$

It would have been difficult to determine such functions by other means.
The results of the integration of functions of interest in this book are summarised in Appendix A (Tables A.2 and A.3) together with some useful rules for the evaluation of integrals (Appendix A.7.1).

Thus we can write Equation (2.12) in terms of integrals between x_1 and x_2 and between t_1 and t_2

$$\int_{t1}^{t_2} v(t)dt = \int_{x_1}^{x_2} dx$$

$$\int_{t1}^{t_2} v(t)dt = x_2 - x_1 = (\text{the area under the } v\text{–}t \text{ graph}) \tag{2.13}$$

Worked Example 2.3 shows how integration may be used to evaluate displacement in linear motion.

Study Worked Example 2.3

We are now able to derive general equations for the position and velocity of a point which is performing linear motion in two important cases by applying the technique of integration to the definitions of velocity and acceleration. As we will see below, it is also possible, in these special cases, to derive the equations by evaluating the areas under curves directly – without using the technique of integration.

(a) ***Linear motion with constant velocity***

Using upper case V to denote the constant value of the velocity, that is

$$v = V = \text{a constant}$$

and from the definition of velocity (Equation (2.5)

$$V = \frac{dx}{dt}$$

so that

$$\Delta x = V(\Delta t)$$

We can write this in terms of indefinite integrals $\int dx = \int Vdt = V\int dt$

Integrating, we obtain $x = Vt + c_1$, where c_1 is an arbitrary constant.
If $x = x_0$ at $t = t_0$, then $x_0 = Vt_0 + c_1$, and, hence, $c_1 = x_0 - Vt_0$

Thus,

$$\boxed{x = x_0 + V(t - t_0)} \tag{2.14}$$

[Note: An equivalent procedure, which we will use frequently throughout the book, involves the use of definite integrals and primed dummy variables, x' and t', as follows.

If $x = x_0$ at $t = t_0$, $V = \frac{dx}{dt} \quad \rightarrow \int_{x_0}^{x} dx' = V\int_{t_0}^{t} dt' \quad \rightarrow \quad x = x_0 + V(t - t_0)$

The primed variables are used to avoid confusion with the variables x and t used to denote the upper limits of the definite integrals]

Motions M and N, which we examined in Section 2.3, are examples of linear motion with constant velocity. The Equations (2.2) and (2.3) which we used to describe these motions are special cases of Equation (2.14) with values of x_0 and t_0 which are set by the starting conditions. In Equation (2.2), $x_0 = t_0 = 0$ and $V = 400$ m min^{-1} and in Equation (2.3), $x_0 = 0$, $t_0 = -5$ min and $V = -800$ m min^{-1}.

Figures 2.14 and 2.15 and are x–t and v–t plots for the case of linear motion with constant velocity.

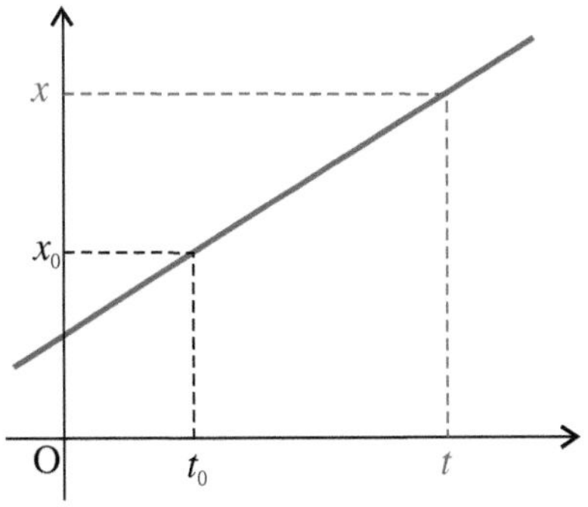

Figure 2.14. An x–t graph for linear motion with constant velocity.

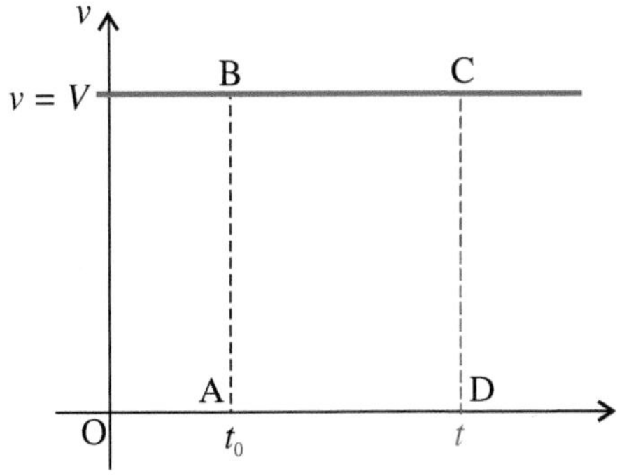

Figure 2.15. A v–t graph for linear motion with constant velocity.

Note that, because in this case (of constant velocity) the v–t graph (Figure 2.15) is a horizontal straight line, the displacement $x - x_0$ may be evaluated from the area under the graph using simple geometry. The area is that of the rectangle ABCD in Figure 2.15.

Thus, $\quad x - x_0 = (\text{AB} \times \text{BC}) = V(t - t_0) \quad$ (which is the same as Equation (2.14))

(b) ***Linear motion with constant acceleration***

The equations which describe this second special case are obtained by integrating Equation (2.11) twice with respect to time and by setting the starting conditions so that $x = x_0$ and $v = v_0$ at $t = t_0$. We use upper case A to denote the constant value of the acceleration, that is

$$a = A = \text{a constant}$$

Thus,
$$A = \frac{dv}{dt}$$

$$\Delta v = A(\Delta t)$$

If $v = v_0$ at $t = t_0$ (at $x = x_0$),
$$\int_{v_0}^{v} dv' = \int_{t_0}^{t} A dt' = A\int_{t_0}^{t} dt'$$

$$\boxed{v = v_0 + A(t - t_0)} \tag{2.15}$$

Note that this result may also be obtained by applying simple geometry to a plot of acceleration against time (Figure 2.16). From the definition of acceleration, Equation (2.11), $\Delta v = A(\Delta t)$, and the change in velocity $v - v_0$ is given by the area under the a–t graph, in the same way as change in displacement is given by the area under a v–t graph. In Figure 2.16 this area is that of the rectangle BCDE.

Thus, $\quad v - v_0 = (\text{BC} \times \text{BE}) = A(t - t_0), \quad$ (which is the same as Equation (2.15))

From Equations (2.6) and (2.15) $\quad v = \frac{dx}{dt} = v_0 + At - At_0$

Integrating this equation
$$\int_{x_0}^{x} dx' = \int_{t_0}^{t} v_0 dt' + \int_{t_0}^{t} At' dt' - \int_{t_0}^{t} At_0 dt'$$

$$\int_{x_0}^{x} dx' = v_0\int_{t_0}^{t} dt' + A\int_{t_0}^{t} t' dt - At_0\int_{t_0}^{t} dt'$$

$$x - x_0 = v_0(t - t_0) + \frac{1}{2}A(t^2 - t_0^2) - At_0(t - t_0)$$

$$= v_0(t - t_0) + \frac{1}{2}A(t - t_0)[(t + t_0) - 2t_0]$$

$$= v_0(t - t_0) + \frac{1}{2}A(t - t_0)^2$$

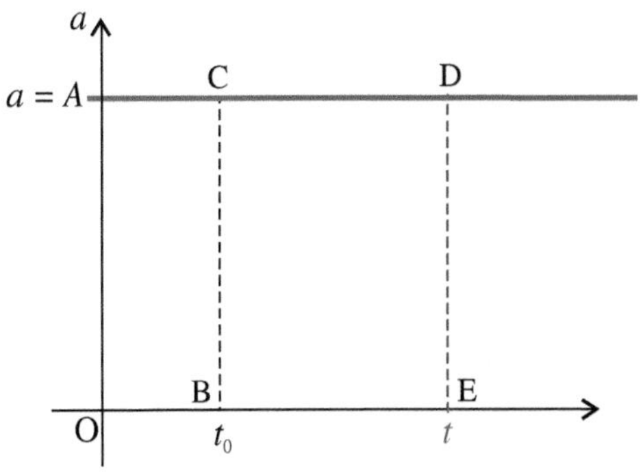

Figure 2.16. An a–t graph for linear motion with constant acceleration.

Thus,

$$\boxed{x = x_0 + v_0(t - t_0) + \frac{1}{2}A(t - t_0)^2} \tag{2.16}$$

By substituting $(t - t_0) = \dfrac{(v - v_0)}{A}$, from Equation (2.15), into Equation (2.16) we obtain a third equation which relates v to x,

$$(x - x_0) = v_0\frac{(v - v_0)}{A} + \frac{A}{2}\frac{(v - v_0)^2}{A^2}$$

Thus,

$$2A(x - x_0) = (2vv_0 - 2{v_0}^2) + (v^2 + {v_0}^2 - 2vv_0) = v^2 - {v_0}^2$$

which simplifies to

$$\boxed{v^2 = {v_0}^2 + 2A(x - x_0)} \tag{2.17}$$

For the case of an object which is performing linear motion with constant acceleration, Equation (2.16) enables us to predict the position of the object as a function of time. Equations (2.15) and (2.17) enable us to predict the velocity of the object as a function of time or as a function of position, respectively.

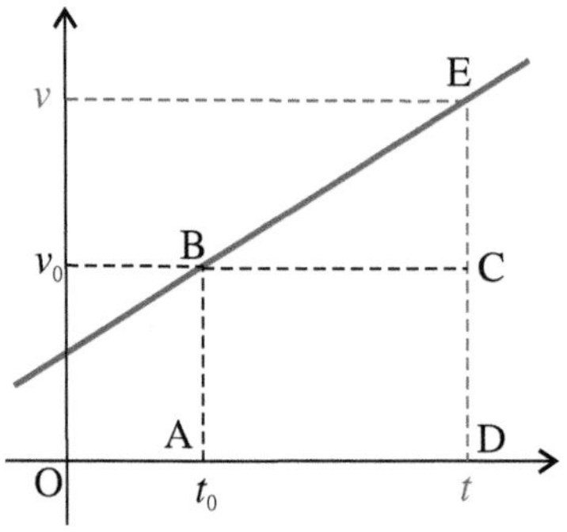

Figure 2.17. A v–t graph for linear motion with constant acceleration.

In Figure 2.17 a v–t graph is drawn for the case of a linear motion with constant acceleration. Note that, because in this case the v–t graph is a straight line, the displacement $x - x_0$ may be evaluated from the area under the graph from geometrical considerations. The area under the graph is that of the rectangle ABCD plus that of the triangle BCE in Figure 2.17.

Thus, $\quad x - x_0 = (\text{AB} \times \text{BC}) + \frac{1}{2}(\text{CE} \times \text{BC}) = v_0(t - t_0) + \frac{1}{2}(v - v_0)(t - t_0)$

Substituting $\quad v - v_0 = A(t - t_0)$ (from Equation (2.15))

this becomes $\quad x - x_0 = v_0(t - t_0) + \frac{1}{2}A(t - t_0)^2$ (which is the same as Equation (2.16))

In Figure 2.18(a) an x–t graph is drawn for the case of a linear motion with constant acceleration in which the object starts from rest (that is, $v_0 = 0$) when $x = x_0$ at the instant $t = t_0$. Note that, when $v_0 = 0$, Equation (2.16) tells us that the change in displacement $(x - x_0) = (\text{constant})(t - t_0)^2$. This is the equation of a parabola, as described in Appendix A.10.2 (iii).

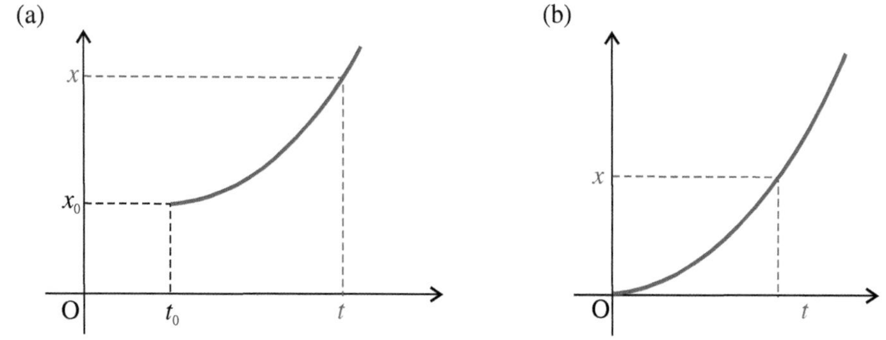

Figure 2.18. (a) An x–t graph for linear motion with constant acceleration starting from rest at $x = x_0$, $t = t_0$. (b) The same motion but with the origin moved to (0,0), that is $x_0 = 0$ and $t_0 = 0$.

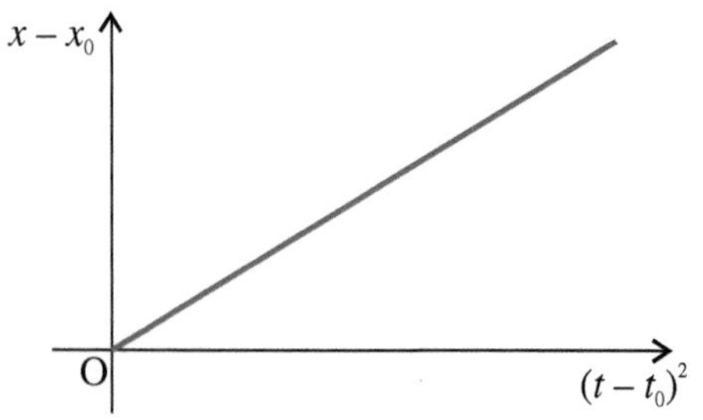

Figure 2.19. An graph of $(x - x_0)$ against $(t - t_0)^2$ for linear motion with constant acceleration, starting from rest at $x = x_0$, $t = t_0$.

Note also that a plot of $(x - x_0)$ against $(t - t_0)^2$ (Figure 2.19) gives a straight line; such a plot may be used to identify this type of motion.

It is often convenient to choose the origin and starting instant so that $x = 0$ when $t = 0$. In this case $x_0 = 0$ and $t_0 = 0$ (Figure 2.18(b)) and Equations (2.15), (2.16) and (2.17) simplify to become

$$v = v_0 + At \tag{2.18}$$

$$x = v_0t + \frac{1}{2}At^2 \tag{2.19}$$

and

$$v^2 = {v_0}^2 + 2Ax \tag{2.20}$$

Note that the equations which we have just derived to describe linear motion with constant velocity or with constant acceleration are special cases which follow directly from the definitions of velocity and acceleration; they are not based on experimental observations. In this sense their derivation is an exercise in applied mathematics rather than physics although, as will be seen in subsequent chapters, the equations which we have obtained are of great practical importance in physics.

The acceleration due to gravity

One of the most important cases of linear motion with constant acceleration is the case of an object which is falling freely under gravity near the Earth's surface. In this case the acceleration of the object, A in case (b) above, acts vertically downwards and is approximately constant. It is measured to be about 9.8 m s^{-2} and is given the special symbol g.

For problems based on the material presented in this section visit up.ucc.ie/2/ *and follow the link to the problems.*

2.6 Maximum and minimum values of physical variables: general linear motion

Let us consider a general case of linear motion, a case in which neither the velocity nor the acceleration of the moving object is constant. As an example we will examine the motion of an object whose displacement x as a function of time is described by the equation

$$x(t) = At^3 + Bt^2 + Ct + D \tag{2.21}$$

where A, B, C and D are constants.

We can easily verify that neither the velocity nor the acceleration is constant by applying the definitions of velocity (Equation (2.6)) and acceleration (Equation (2.11)) to Equation (2.21). From the definition of velocity

$$v = \frac{dx}{dt} = 3At^2 + 2Bt + C \tag{2.22}$$

and from the definition of acceleration

$$a = \frac{dv}{dt} = \frac{d^2x}{dt^2} = 6At + 2B \tag{2.23}$$

Both velocity and acceleration are functions of time in this case; neither are constant.

We can determine the values of velocity and acceleration at a particular time t_1 by substituting $t = t_1$ into Equations (2.22) and (2.23). We can also use differentiation to determine maximum and minimum values of the displacement x during the motion. As illustrated in Figure 2.20 (a x–t graph), the gradient $\frac{dx}{dt}$, that is the velocity, is zero at times when x is at a maximum or minimum value because the tangent is horizontal at these points. We can determine these times by equating Equation (2.22) to zero, that is

$$\frac{dx}{dt} = 3At^2 + 2Bt + C = 0$$

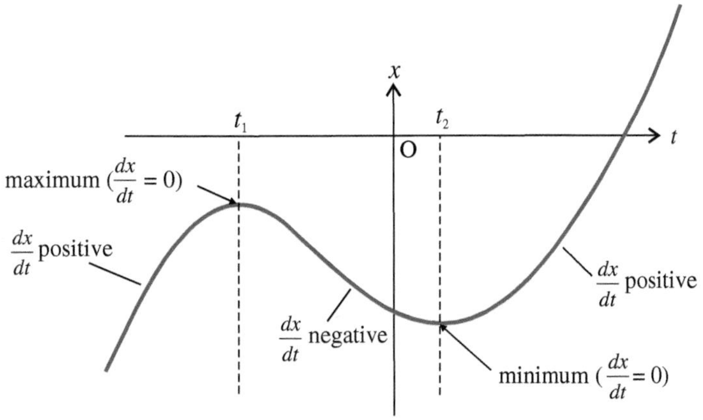

Figure 2.20. An x–t graph of a function of the type $x = At^3 + Bt^2 + Ct + D$.

and solving for t. In this case we have a quadratic equation with two solutions t_1 and t_2. Physically these times correspond to turning points in the motion, points at which the velocity is instantaneously zero when the object reverses its direction of motion.

Note that, as illustrated in Figure 2.20, as time advances towards a moment of maximum displacement, the gradient (the velocity) decreases, switching from positive to negative values at the maximum. In contrast at a minimum the gradient increases, switching from negative to positive values as t passes through the minimum. Thus we can use the values of the rate of change of velocity, the acceleration $\frac{dv}{dt}$ as given by Equation (2.23) at t_1 and t_2, to determine whether the displacement is at a maximum or minimum at these times. If $\frac{dv}{dt}$ is positive, that is $\frac{d^2x}{dt^2} > 0$, displacement is at a minimum, whereas if it is negative, that is $\frac{d^2x}{dt^2} < 0$, displacement is at a maximum.

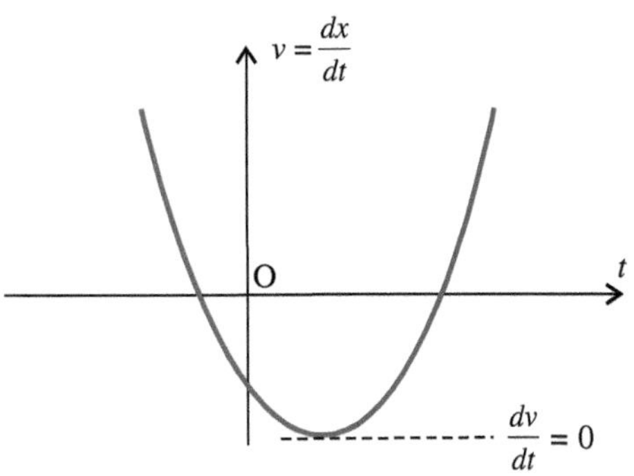

Figure 2.21. A v–t graph of a function of the type $v = 3At^2 + 2Bt + C$.

We can also use this technique to determine the maximum and minimum values of v in a v–t graph, as illustrated in Figure 2.21. Maximum and minimum values of v occur when the gradient $\frac{dv}{dt}$, the acceleration, is zero. We can determine these times by equating Equation (2.23) to zero,

$$\frac{dv}{dt} = 6At + 2B = 0$$

and solving for t. We can establish whether the solution corresponds to a maximum or minimum value of v by determining the value of $\frac{d^2v}{dt^2}$ at this time. If $\frac{d^2v}{dt^2}$ is positive the velocity is at a minimum; if it is negative it is at a maximum.

Worked Example 2.4 shows how calculus may be used to determine maximum and minimum values of displacement and velocity.

Note that, although we have confined our analysis of motion to the special case of motion along a straight line, the analysis can be extended easily to cover any motion along a *fixed* track or path, such as the motion of a train along curved railway lines. In this case the straight line displacement variable x is replaced by a variable s which represents displacement measured along the track, as illustrated in Figure 2.22. As in straight line motion there are only two options for the direction of the motion so that s, like x, is an algebraic quantity. The definitions and equations which we have derived for straight line motion may be applied to this more general case simply by replacing x by s throughout. In the next section we will see how the analysis of linear motion can also be applied to angular motion about a fixed axis.

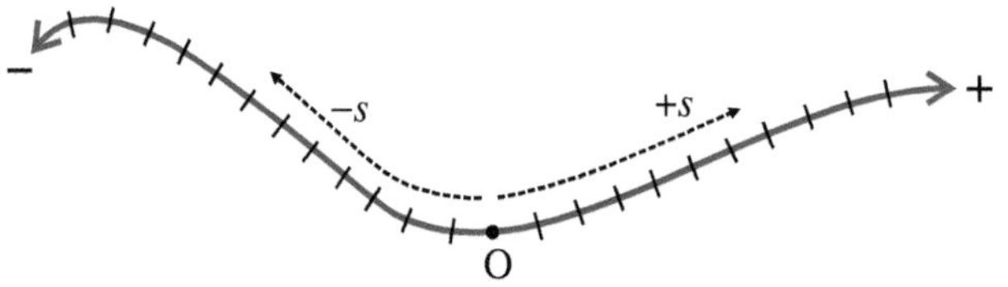

Figure 2.22. Displacement s along a fixed track.

Study Worked Example 2.4

For problems based on the material presented in this section visit up.ucc.ie/2/ *and follow the link to the problems.*

2.7 Angular motion: the radian

In the previous sections of this chapter we have developed a quantitative description of the motion of an object which is confined to move along a straight line. We were able to specify the position of the object in terms of a single quantity, its displacement, which we were able to represent by the algebraic quantity x. More specifically this type of displacement, in which two positions along a straight line are compared, is known as **translational** displacement.

Consider now the motion of the line CP which, as illustrated in Figure 2.23, is moving in such a way that one end of the line is fixed at C; that is, the line CP can rotate about C. In one revolution CP returns to its starting point so that there is no net translational displacement of the point P, although clearly there is motion. We can describe this type of motion by specifying the value of a single quantity – the **angular displacement**, θ, of CP from the line CO. The line CO has therefore been chosen as the origin for measurements of angular displacement, that is, CO is the line at which $\theta = 0$.

Also seen in Figure 2.23, two directions are possible for angular displacement from CO, namely clockwise and anticlockwise. While we are free, as in linear motion, to choose either direction as positive, the following convention is usually adopted for angular motion.

anticlockwise – defined as positive angular displacement, $\theta > 0$ in Figure 2.23
clockwise – defined as negative angular displacement, $\theta < 0$ in Figure 2.23

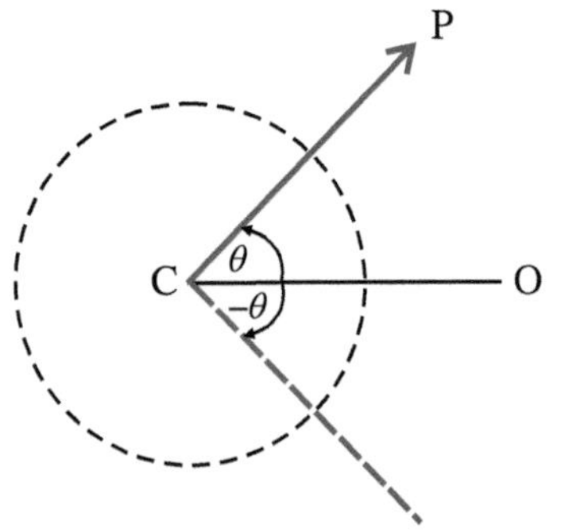

Figure 2.23. The angular displacement θ of CP relative to CO. By convention anticlockwise displacements are positive and clockwise angular displacements are negative.

We can also view the motion of CP around the point C as a rotational motion about a line which is drawn through C and is perpendicular to the plane of the page. This line is described as the **axis of rotation** of the motion. In this section we will only consider cases in which the direction of the axis of rotation is fixed during the motion, that is, cases in which CP does not move out of the plane of the page. The simplification achieved by considering only rotation about a fixed axis is equivalent to that achieved in the translational case by considering only motion along a fixed straight line. In each case only one variable, an algebraic quantity, x in the translational case or θ in the angular case, is needed to specify displacement so that both motions can be described as one dimensional.

The formalism which we now develop to describe angular motion will be used, in Chapter 4, to describe the path of a particle moving in a circle around a point and, in Chapter 7, to describe the rotation of an extended rigid object about a fixed axis.

Our approach in developing a quantitative description of angular motion will be to define angular equivalents of the quantities which we defined to describe translational motion. We will then derive a set of equations relating these angular quantities which will be entirely analogous to their translational equivalents.

The next step in this approach is to define the quantities **angular velocity** and **angular acceleration**. These are defined below and compared with the definitions of the equivalent translational quantities.

	rotational	**translational**
angular velocity	$\omega := \dfrac{d\theta}{dt}$	$v := \dfrac{dx}{dt}$
angular acceleration	$\alpha := \dfrac{d\omega}{dt} = \dfrac{d^2\theta}{dt^2}$	$a := \dfrac{dv}{dt} = \dfrac{d^2x}{dt^2}$

Because the definitions of θ, ω and α are entirely equivalent to those of x, v and a respectively, the equations which we will now derive to relate θ, ω and α in the special cases of rotational motion with constant ω, or with constant α are of the same form as the equations which were derived in Section 2.5 to relate x, v and a in translational motion for the special cases of constant v or of constant a.

As in the translational case the equations for rotation about a fixed axis with constant angular velocity and with constant angular acceleration follow directly from the definitions of angular velocity and angular acceleration. Not surprisingly the resulting equations are completely analogous to their translational equivalents.

Case 1: Motion with constant angular velocity

From the definition of angular velocity $\dfrac{d\theta}{dt}$ = constant, which we call Ω. Thus

$$d\theta = \Omega dt$$

If $\theta = \theta_0$ at $t = t_0$

$$\int_{\theta_0}^{\theta} d\theta' = \Omega \int_{t_0}^{t} dt'$$

$$\theta - \theta_0 = \Omega(t - t_0)$$

Thus,

$$\theta = \theta_0 + \Omega(t - t_0)$$

This result is completely analogous to the translational Equation (2.14), $x = x_0 + V(t - t_0)$.

Case 2: Motion with constant angular acceleration

From the definition of angular acceleration, $\dfrac{d\omega}{dt} = \dfrac{d^2\theta}{dt^2}$ = constant, which we call A'. The equations which describe this special case are obtained by integrating this equation twice with respect to time and by setting the starting conditions so that $\theta = \theta_0$ and $\omega = \omega_0$ at $t = t_0$. The procedure is analogous to that used to derive the equations for translation motion in one dimension with constant acceleration (Section 2.5) and will not be repeated here; θ replaces r, ω replaces v and A' replaces A, so that we obtain

rotational equations	***translational equivalents***
$\omega = \omega_0 + A'(t - t_0)$	$v = v_0 + A(t - t_0)$
$\theta = \theta_0 + \omega_0(t - t_0) + \frac{1}{2}A'(t - t_0)^2$	$x = x_0 + v_0(t - t_0) + \frac{1}{2}A(t - t_0)^2$
$\omega^2 = \omega_0^2 + 2A'(\theta - \theta_0)$	$v^2 = {v_0}^2 + 2A(x - x_0)$

The radian

We have not so far considered how angles are to be measured; we have not defined a unit for θ. We now define a unit of plane angle, called the **radian**, which will generally be used to measure angular displacement in this book**.** If, as illustrated in Figure 2.24, a point P travels along an arc length s while the line joining P to C sweeps out an angle θ, then the θ in radians is defined by

$$\theta := \frac{s}{R} \tag{2.24}$$

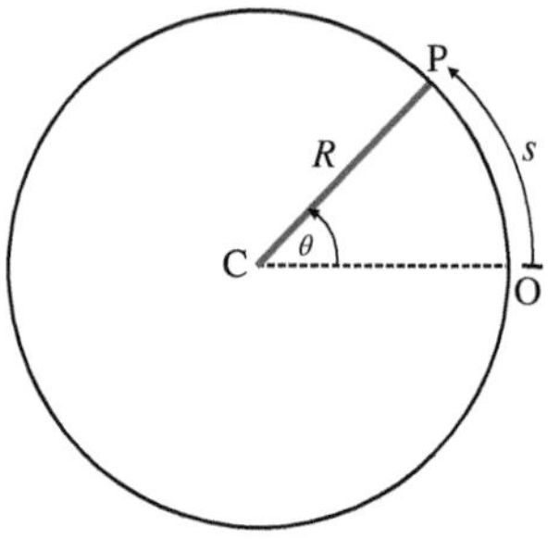

Figure 2.24. The radian, $\theta = \frac{s}{R}$, is used as a standard measure of angle.

Note that the unit of plane angle, as determined by definition (2.24), is $\frac{\text{length}}{\text{length}}$; it is dimensionless. However when an angle is used as a measure of angular displacement it is customary to treat the radian (abbreviation rad) as if it was a base unit. Thus ω and α are stated in terms of the units rad s^{-1} and rad s^{-2} respectively.

It is important to note that all measurements of angular displacement in this book are assumed to be made in radians rather than in other angular measures, such as degrees. The conversion factor from degrees to radians is easily deduced from the definition of the radian. In one complete revolution (360°), an arc length of $2\pi R$ is swept out. By definition (2.24), therefore, 360° corresponds to $\frac{2\pi R}{R} = 2\pi$ radius. The conversion factor between degrees and radians is thus $1° = \frac{\pi}{180}$ radians.

It will often be convenient to write functions in terms of trigonometric functions of plane angles, for example, $\sin\theta$, $\cos\theta$ or $\tan\theta$. Some standard relationships between trigonometric functions – trigonometric identities – are listed in Appendix A.3. The results of differentiation and integration of some simple trigonometric functions are listed in Tables A.1, A.2 and A.3 of Appendix A. Worked Example 2.5 illustrates the use of such procedures.

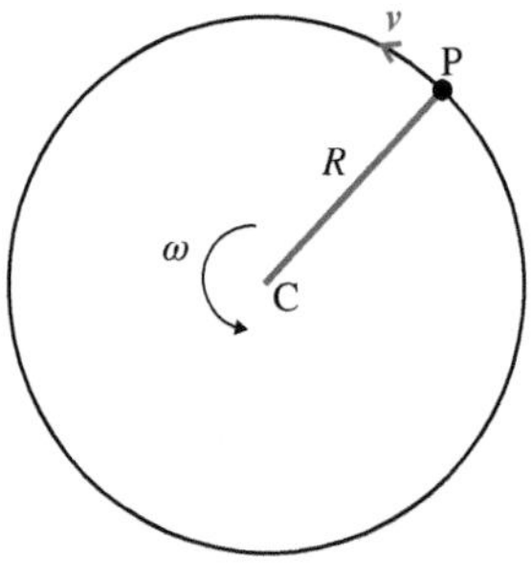

Figure 2.25. The translational and angular velocities (v and ω respectively) of a point as it moves along the arc of a circle.

We can use the definition of the radian to derive a direct relationship between the instantaneous angular velocity ω of a point P which is sweeping out an angle as it moves along the arc of a circle of radius R and the velocity of this point along the arc, $v = \frac{ds}{dt}$. Thus, since R is constant,

$$\omega = \frac{d\theta}{dt} = \frac{d}{dt}\left(\frac{s}{R}\right) = \frac{1}{R}\frac{ds}{dt} = \frac{v}{R}$$

and

$$\boxed{v = \omega R} \qquad (2.25)$$

Note that, as illustrated in Figure 2.25, v is always directed along a tangent to the circle and is in the direction of rotation of CP.

Study Worked Example 2.5

For problems based on the material presented in this section visit up.ucc.ie/2/ *and follow the link to the problems.*

2.8 The role of mathematics in physics

In this chapter we have shown that, if we know the relationship between two physical quantities such as displacement and time, calculus techniques may be used to derive relationships between quantities which are defined in terms of these two. We can also evaluate maximum and minimum values of any of these quantities. Although we have applied calculus techniques only to the study of motion, the techniques can be shown to be quite general and can be applied to any physical quantities whose relationship is known.

We will see many examples in this book of how calculus, and other mathematical methods which will be introduced as they are required, provide us with concise, accurate and efficient means of analysing physical phenomena. Effort spent in mastering these techniques will be well rewarded.

In the next and subsequent chapters we will begin to investigate the fundamental laws of physics, It turns out that these laws are not only few in number but are also usually of relatively simple form. Because of this, scientists have found that the easiest, most concise and least ambiguous way to express and describe phenomena in physics is by using the language of mathematics. The most powerful tool available to the physicist, therefore, is that of **mathematical modelling** often enhanced by computational tools.

The central role of mathematics in physics has been understood since the birth of modern science almost four centuries ago.

> *Philosophy is written in this grand book. I mean the universe which stands continually open to our gaze. But it cannot be understood unless one first learns to comprehend the language and interpret the characters in which it is written. It is written in the language of mathematics .... without which it is humanly impossible to understand a single word of it; without [this] one is wandering about in a dark labyrinth.*

Galileo Galilei, *The Assayer*, 1623.

Newton developed differential calculus primarily to provide the tools for the formulation of analytical dynamics. The extraordinary advances in man's understanding of the material universe since Newton's time have been strongly driven by mathematical modelling, a fact widely recognised by those who contributed to the developments.

The enormous usefulness of mathematics in the natural sciences is something bordering on the mysterious and there is no rational explanation for it...... The miracle of the appropriateness of the language of mathematics for the formulation of the laws of physics is a wonderful gift which we neither understand nor deserve.

Eugene P. Wigner, *Communications in Pure and Applied Mathematics*, 1960.

Every one of our laws is a purely mathematical statement. Why? I have not the slightest idea. ..... It is impossible to explain honestly the beauties of the laws of nature in a way that people can feel, without their having some deep understanding of mathematics. I am sorry, but this seems to be the case.

Richard P. Feynman, *The Character of Physical Law*, 1965.

Students starting out in the study of physics need to embrace the concept of mathematical modelling readily. Throughout most of this book, mathematical modelling means no more than using algebra, geometry, trigonometry, and calculus to represent physical situations. It is important to keep in mind that in physics, as distinct from pure mathematics, each graph and each mathematical expression 'tells a story' about the real world.

WORKED EXAMPLES

Worked Example 2.1: It is found that the displacement of a point which is moving along a straight line can be represented by the function $x = A + Bt^2 + Ct^3$, where A, B, C and D are constants with the values, $A = 6.0$ cm , $B = -3.2$ cm s^{-2} and $C = 4.7$ cm s^{-3}. Derive equations which give the velocity and acceleration of the point as functions of time and evaluate the displacement, velocity and acceleration of the particle at $t = -2$ s and at $t = +3$ s.

$$x = 6.0 - 3.2t^2 + 4.7t^3$$

Thus,

$$v = \frac{dx}{dt} = -6.4t + 14.1t^2$$

and

$$a = \frac{dv}{dt} = -6.4 + 28.2t$$

Substitution of $t = -2$ s gives $x = -44.4$ cm, $v = +69.2$ cm s^{-1} and $a = -62.8$ cm s^{-2}
and substitution of $t = +3$ s gives $x = +104.1$ cm, $v = +107.7$ cm s^{-1} and $a = +78.2$ cm s^{-2}

Worked Example 2.2: If the radius of a circle is increasing at a steady rate of 2 mm s^{-1} calculate the rate of increase of its area when its radius is 1.5 m.

The area of a circle is given by $A = \pi r^2$

To evaluate $\frac{dA}{dt}$, the rate of increase of A, we differentiate this equation to obtain

$$\frac{dA}{dt} = \frac{dA}{dr} \times \frac{dr}{dt} = 2\pi r\frac{dr}{dt}$$

We are given

$$\frac{dr}{dt} = 2 \text{ mm s}^{-1}$$

Thus,

$$\frac{dA}{dt} = 2 \times \pi \times 1.5 \times 0.002 = 0.019 \text{ m}^2 \text{ s}^{-1}$$

Worked Example 2.3: The velocity of a particle which is moving in a straight line is given by $v = -(6.0 \text{ m s}^{-1}) + (4.0 \text{ m s}^{-3})\, t^2$. Evaluate the change in displacement of the particle from $t = -2$ s to $t = +3$ s.

$$x = \int_{-2}^{3} v dt = \int_{-2}^{3} (-6.0 + 4.0t^2)dt$$

$$= \left[-6t + \frac{4}{3}t^3\right]_{-2}^{3} = 18 - 1.33 = 16.67 \text{ m}$$

Worked Example 2.4: It is found that the displacement of a particle which is moving along a straight line is described by the equation $x(t) = at^3 + bt^2 + ct + d$, where a, b, c and d are constants with the values, $a = 4.0$ m s^{-3}, $b = 6.0$ m s^{-2}, $c = -50$ m s^{-1}, $d = -38$ m. Determine maximum and minimum values of the displacement and the velocity of this particle.

First, we differentiate

$$x(t) = 4t^3 + 6t^2 - 50t - 38 \tag{2.26}$$

to obtain an equation for $\frac{dx}{dt} = v(t)$ which we equate to zero. Thus

$$\frac{dx}{dt} = 12t^2 + 12t - 50 = 0 \qquad (2.27)$$

Solving this quadratic equation using the method shown in Appendix A.4.2,

$$t = \frac{-12 \pm \sqrt{12^2 - (4 \times 12 \times (-50))}}{2 \times 12} \qquad \text{so that} \qquad t = +1.6 \text{ or } -2.6 \text{ s}$$

The values of x at these times are obtained by substitution into Equation (2.26). Thus at $t = 1.6$ s, $x = -86.3$ m and at $t = -2.6$ s, $x = 62.2$ m. To determine whether these times correspond to maxima or minima we differentiate Equation (2.27) and substitute our solutions for t. Thus,

$$\frac{d^2x}{dt^2} = \frac{dv}{dt} = 24t + 12 \qquad (2.28)$$

When $t = +1.6$ s, $\frac{d^2x}{dt^2}$ is positive, indicating a minimum, and, when $t = -2.6$ s, $\frac{d^2x}{dt^2}$ is negative, indicating a maximum. These results are illustrated in Figure 2.26, an x–t plot.

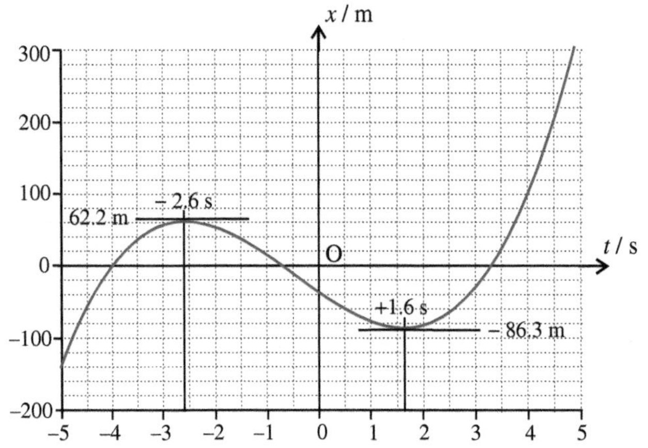

Figure 2.26. The x–t graph of $x = 4t^3 + 6t^2 - 50t - 38$ (Worked Example 2.4).

To determine where maximum and minimum values of $v(t)$ occur we equate $\frac{dv}{dt}$ (Equation (2.28)) to zero. Thus

$$\frac{dv}{dt} = 24t + 12 = 0 \qquad (2.29)$$

The solution of this equation is $t = -0.5$ s. The value of v at this time is obtained by substitution into Equation (2.27). This yields $v = -53$ m s^{-1}. To determine whether this time corresponds to a maximum or minimum value of v we differentiate Equation (2.29) and substitute our solution for t. Thus

$$\frac{d^2v}{dt^2} = 24 \text{ m s}^{-3}$$

Since $\frac{d^2v}{dt^2}$ is positive for all t, $t = -0.5$ s must correspond to a minimum. This result is illustrated in Figure 2.27, a v–t plot.

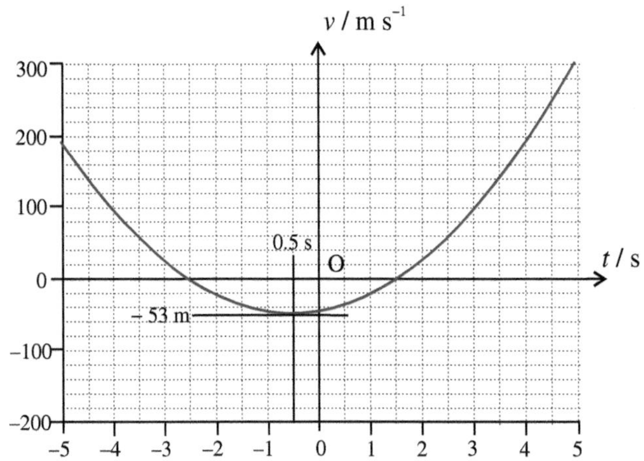

Figure 2.27. A v–t graph of $v = 12t^2 + 12t - 50$ (Worked Example 2.4).

Worked Example 2.5: The velocity of a particle, which is moving along a straight line, is found to be described by the equation $v = K\cos\omega t$ where K and ω are constants. If at $t = 0$ the displacement of the particle is zero, derive equations which give the acceleration and displacement of the particle as functions of time and show that, at any time, the acceleration of the particle, a, is related to its displacement, x, through the equation $a = -\omega^2 x$. If K and ω have the values 0.1 m s^{-1} and 2 rad s^{-1}, respectively, determine the displacement, velocity and acceleration of the particle at $t = 1$ s.

$$v = K\cos\omega t \tag{2.30}$$

Thus, using Table A.1 of Appendix A.6,

$$a = \frac{dv}{dt} = \frac{d}{dt}(K\cos\omega t) = -K\omega\sin\omega t \tag{2.31}$$

and, from Table A.2 of Appendix A.7,

$$x = \int v dt = \int (K\cos\omega t)dt = \frac{K}{\omega}\sin\omega t + C$$

$x = 0$ when $t = 0$. Thus $C = 0$ and we can write

$$x = \frac{K}{\omega}\sin\omega t \tag{2.32}$$

From Equations (2.31) and (2.32)

$$\sin\omega t = -\frac{a}{\omega K} = \frac{\omega x}{K}$$

Thus,

$$a = -\omega^2 x$$

Substituting $t = 1$ s, $K = 0.1$ m s^{-1} and $\omega = 2$ rad s^{-1} into Equations (2.30) to (2.32) we obtain $x = 0.045$ m, $v = -0.042$ m s^{-1} and $a = -0.182$ m s^{-2}.

PROBLEMS

For problems based on the material covered this chapter visit up.ucc.ie/2/ and follow the link to the problems.

3

The causes of motion: dynamics

AIMS

- to investigate the fundamental principles ('laws of motion') which describe how the motion of a body is related to the force producing the motion
- to show how these laws can be used to predict the motion of a body to which a known force is applied
- to introduce the system of units (SI) which will be used throughout this book
- to study two specific types of motion in one dimension which arise very often in nature, namely: (i) motion under constant forces, and (ii) a form of oscillatory motion called simple harmonic motion
- to introduce the quantities of mechanical work and energy and to show how they may be used to describe one-dimensional mechanical systems
- to study damped harmonic systems

3.1 The concept of force

Consider what happens if you push or pull an object. If you push a book along the surface of a table, for example, it takes a certain amount of an effort to get it moving, but, once it is in motion, the harder you push the faster you can get the book to move. The same sort of effect is observed if you attach a string to an object and pull. On the other hand, if you push on the wall of a room or pull on a rope which is rigidly attached to the ground nothing obvious may happen no matter how hard you push or pull. Pushing or pulling in this sense, whether motion is produced or not, we call 'exerting a force'. Here the word 'force' is used in a qualitative sense to describe processes such as pushing and pulling; a more quantitative definition will be required, however, if force is to prove to be a useful scientific concept.

When motion is produced by applying a force or forces to an object, it is clear that larger forces give rise to faster motion than smaller forces on the same object under the same conditions. Indeed, nature shows considerable regularity in this matter in that if the same force is applied to the same object a number of times under identical conditions, the effect observed will be essentially the same each time. Think of how pointless and impossible any ball game or field sport would be if this were not the case. There is clearly some underlying rule or 'law of nature' which determines how the motion produced by a force is related to the size of the force involved. The exact nature of this relationship engaged the minds of philosophers for thousands of years, but it was not until the seventeenth century that a workable scientific theory (called **dynamics**) began to be developed to explain such phenomenon.

Fortunately, it turns out that the basic rules ('laws') governing the motion of bodies produced by forces have rather simple forms. One reason why it took so long to discover these basic laws is the fact that most everyday phenomena involve quite complicated situations. Greek philosophers believed that all bodies naturally come to rest when no forces are applied to them and indeed this is also what our common sense seems to suggest. Doubt begins to creep in, however, when one considers situations involving very smooth surfaces, such as an ice rink or a road covered by oil. A skater knows that a small push on the side of the rink is enough to propel herself rapidly across the ice while the same push would get her nowhere across a rough surface. The complicating factor here is *friction* which, as we will see in Section 4.6, gives rise to subtle additional unseen forces.

Bodies in space, such as planets or satellites, do not experience friction. This is one reason why the study of astronomy by scientists such as Tycho Brahe, Kepler, Galileo and Newton was such an important stimulus to the development of the science of dynamics (as we shall see in Chapter 5).

Understanding Physics, Third Edition. Michael Mansfield and Colm O'Sullivan.
© 2020 John Wiley & Sons Ltd. Published 2020 by John Wiley & Sons Ltd.

3.2 The First law of Dynamics (Newton's first law)

As a first step towards attempting to discover the fundamental laws of dynamics we will try to imagine what would happen in the complete absence of frictional forces by considering the motion of bodies on very smooth surfaces. This approach is central to the scientific method, namely we are attempting in the first instance to study a very simplified, idealised situation. It is not possible, of course, to eliminate friction entirely from any bench top experiment but we can study situations in which friction is reduced or almost completely removed, such as on an ice surface or in outer space. Controlled laboratory experiments can be performed in the low-friction environments provided by air-tracks or air-tables, where bodies float on a cushion of air. Carefully controlled experiments of this kind enable us to infer what would happen in the complete absence of friction.

Let us return to the skater moving across the ice as a result of a small push against the side of the rink. She will, of course, gradually slow down but at a much slower rate than if the surface was less smooth. The smoother the surface the less slowing down will be observed and one is inevitably drawn to the conclusion that, if friction could be removed completely, there would be no slowing down at all. Thus, the Greek philosophers who believed that objects naturally came to rest in the absence of forces, came to the wrong conclusion: the 'natural' state of motion in the absence of forces would seem to be motion at constant velocity, in apparent contradiction to our everyday experience. This result, first proposed by Galileo Galilei (1564–1642) and formulated by Isaac Newton (1642–1727) as his first law of motion, can be stated as follows (**Newton's first law**):

> If no net force acts on a body it will move in a straight line at constant velocity, or will stay at rest if initially at rest

The reader may well point out that there is, in fact, a force acting on the skater even in the absence of friction, namely the force of gravity pulling her towards the centre of the Earth. This indeed is the case but the effect of gravity has been offset here by ensuring that the surface of the ice is horizontal; only if the ice surface sloped to any degree would the effect of the gravitational force be readily discernible from the fact that the skater would speed up (accelerate) down the slope.

It is not sufficient, however, to conclude that a law of nature is valid simply on the basis of general qualitative observations such as those outlined above. The scientific method requires that conclusions be confirmed by more careful quantitative experiments. One experimental arrangement in which the motion of a body can be studied in a low-friction environment is shown in Figure 3.1. The body under investigation in this case is a cart which is constrained to move along a very smooth straight horizontal track. An electronic sensor system, which is designed to detect the position of a specific point on the cart as it moves along the track, is connected via a data acquisition interface to a computer. The computer is programmed to sense the position of the cart on the air-track at regular intervals and hence to record its position as a function of time. We know from Chapter 2 how to interpret such distance–time plots for one dimensional motion, at least in some simple cases.

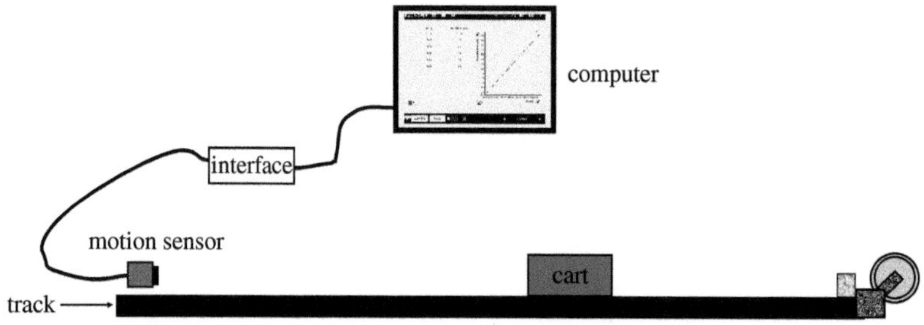

Figure 3.1. An example of an experiment to investigate Newton's first law. The cart moves along the smooth horizontal surface of a track. The position of the cart on the track is sensed electronically at regular time intervals and is recorded, together with the corresponding time, by the computer.

A typical output from a set of such experiments is shown in Figure 3.2, where the computer has plotted position–time data points for a cart, the cart being pushed off at a different velocity in each case. Clearly all plots are consistent with linear variation of position with respect to time, showing that the velocity was constant in each case (recall Figure 2.14). Hence Newton's first law is confirmed within the accuracy of the experiments performed. Note, however, that experiments like this do not 'prove' a law of nature but rather indicate that the law may be applied in a certain context and within a specific accuracy. More sensitive apparatus and more accurate measurements would enable the regime of validity of the law to be specified further.

While we will find in Chapter 8 that Newton's first law (and the laws of mechanics generally) may not be applied directly to the analysis of a body's motion when measurements are made relative to an accelerating origin, in this chapter and in Chapters 4 to 7, we will be considering only situations in which the laws may be applied directly.

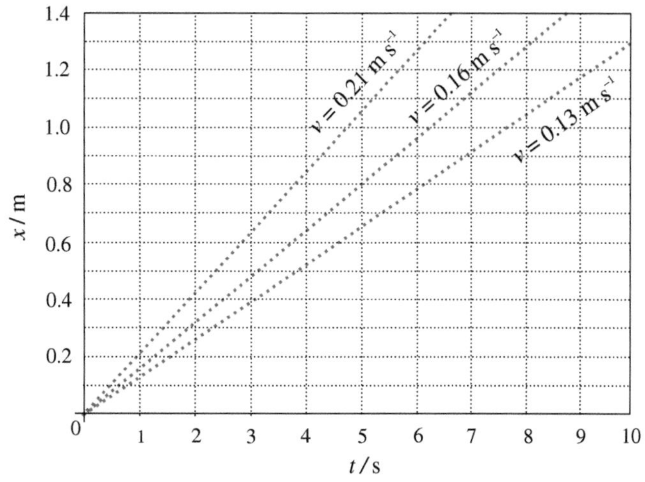

Figure 3.2. Position-time plots for a cart moving with constant velocity on a smooth horizontal track for three different velocities; the steeper the slope the larger the velocity.

3.3 The fundamental dynamical principle (Newton's second law)

An inference from Newton's first law is that, if a force does act on a body, its velocity and/or its direction of motion will change. If we are to be in a position to discover precisely how the state of motion of a body is related to the force applied to it, we need to have some way of determining how 'big' or how 'small' a particular force is. In other words, we need a way of measuring force, at least in principle.

There are many different ways in which this can be done and we will consider one such approach in the following experiment. Again, in order to avoid complications arising from friction and gravitational forces, a smooth (that is frictionless) horizontal track is used as shown in Figure 3.3. The body whose motion is under investigation, in this case, is a combination of the cart and the attached force sensor as shown. The force sensor is connected to the computer which monitors the force applied; that is, a number is returned by the computer which corresponds to the magnitude of the force. Different forces may be applied by attaching a piece of *light* string to the cart as shown, passing the string over a pulley and hanging an object from the end of the string so that the cart is pulled horizontally. The force sensor may be calibrated by holding the cart still and recording the reading returned by the computer when one, two (three, four, ...) *identical* small objects are hung together from the end of the string. It is clear from symmetry, in this case, that the force applied to the cart must be twice (three times, four times, ...) the force applied when one such object is attached.

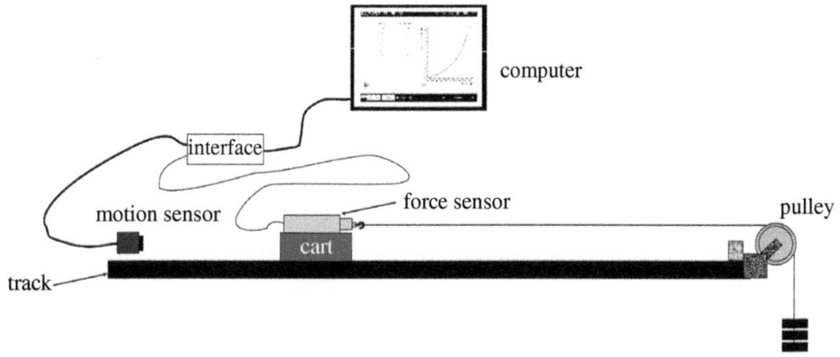

Figure 3.3. An experiment to investigate the relationship between the motion of a cart and the force applied to it. The cart slides on the smooth horizontal surface of the track under the influence of a force (horizontally to the right) provided by one (two, three, four, ...) identical objects hanging from the end of a string passing over a pulley. The computer records and plots the position of the cart as a function of time.

As before, the computer can record and plot the position of the cart as a function of time, in this case for each of the different numbers of identical objects applied to the cart and the corresponding force sensor reading in each case is logged.[1] Figure 3.4 shows an example of the data obtained when one, two, three, four and five units, respectively, are applied. None of the plots is a straight line so that clearly the velocity does not remain constant; in each case the velocity increases with time. The general shape of the data plots suggests, however, that it might be productive to ask the computer to find the best fit of each of the data sets to the function $x = v_0 t + \frac{1}{2}At^2$ (that is, using

[1] The force sensor reading may not be exactly the same as that recorded for the same arrangement when the cart was held fixed. We will return to this issue in Section 6.1 where we will see that the values will only be the same if the objects attached to the end of the string are much lighter than the combination of the cart and force sensor.

Equation (2.19) to test for constant acceleration in each case). The curves in Figure 3.4 are the 'best fit' to this equation generated by the computer and it can be seen from the figure that all the data is consistent with the assumption of constant acceleration within the accuracy of the experiment. The curve-fitting procedure also gives the best fit values of the parameters v_0 (the velocity at $t = 0$) and A (the acceleration).

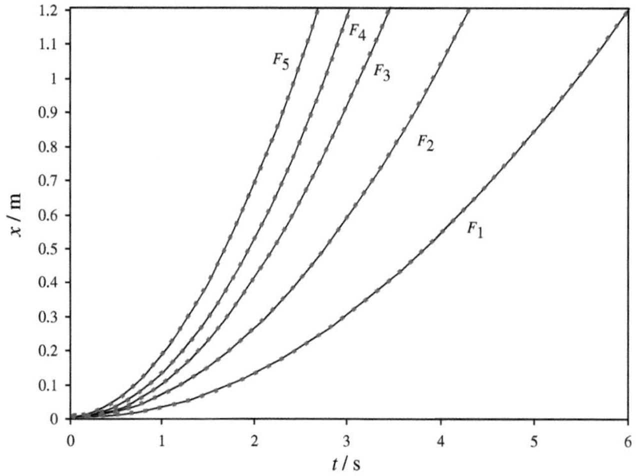

Figure 3.4. Position-time curves for a body to which different constant forces have been applied. Computer generated 'best fits' (black curves) indicate that the acceleration is constant in each case.

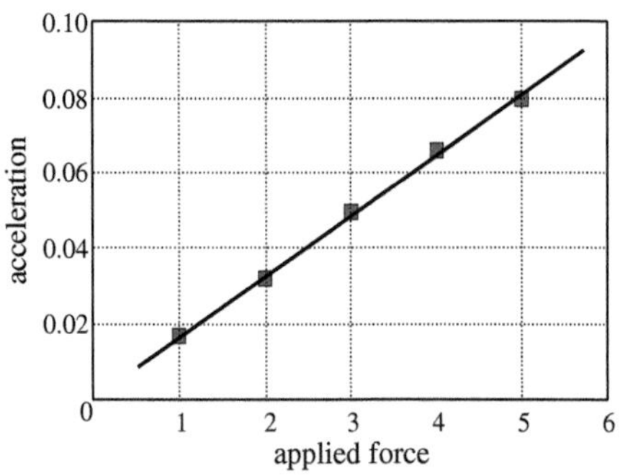

Figure 3.5. Experimental plot of acceleration as a function of applied force from the air-track experiment described in the text, indicating that the acceleration is directly proportional to the applied force.

Finally, when the measured value of the acceleration A in each case is plotted against the corresponding applied force (Figure 3.5) it is found that this relationship is linear. This experimental result implies that the acceleration is directly proportional to the applied force. Thus, the conclusions which can be drawn from the experiment are as follows:

1. A constant force applied to any body produces constant acceleration of that body ($F =$ constant $\Rightarrow$ $a =$ constant)
2. The acceleration produced by a force which is applied to a body is proportional to the magnitude of the force ($a \propto F$)

Note that if $F = 0$ then $a = 0 \Rightarrow v =$ constant; that is, if no force acts on a body it will remain at rest or continue to move at constant velocity, as expected from Newton's first law discussed in the previous section.

Mass

In the experiment which we have just described only the motion of a single body was considered. If two identical carts are joined together to make a body which contains twice the amount of material as in a single cart and the experiment repeated, it is found that the same force produces only half the acceleration. For three (four, five ...) identical carts joined together the acceleration turns out to be one-third (one-quarter, one-fifth ...) of the acceleration produced by the same force when applied to a single cart. We have used the term 'amount of material' here in a very loose sense. The precise characteristic property of a body which corresponds to the concept of 'amount of material' in the body is called the **mass** of the body (which we denote by m) but we are not quite in a position yet to define this term rigorously. Nevertheless, we can safely assert that bodies comprising one, two, three, four ... etc. identical carts can be considered as having one, two, three, four .... etc. mass units, respectively.

Thus, from experiments performed with different numbers of identical carts joined together as described above, one reaches the additional conclusion that

3. The acceleration produced by a given force is inversely proportioned to the mass of the body to which the force is applied ($a \propto \frac{1}{m}$, for fixed F).

Conclusions **2** and **3** can be combined in the single statement (**Newton's second law**)

$$a \propto \frac{F}{m} \tag{3.1}$$

since, if m is constant in Equation (3.1), $a \propto F$ and, if F is constant in Equation (3.1), $a \propto \frac{1}{m}$.

Conclusion **3** enables us to refine our concept of mass. If the same force is applied to two different masses m_1 and m_2 they will experience different accelerations a_1 and a_2, respectively. Comparing these accelerations allows us to compare the two masses since

$$\frac{m_1}{m_2} = \frac{a_2}{a_1}$$

Mass, therefore, is a measure of how a body responds to an applied force; it requires a larger force to give a large mass the same acceleration as a small mass. This property of matter is called *inertia*. Different masses may be compared experimentally by measuring the acceleration of each which is produced by the same applied force.

Multiple forces

Conclusions 1, 2 and 3 were arrived at by studying the effect of a *single force* which caused motion in one dimension. The ideas developed above need to be extended further if more than one force is acting on a body. Consider, for example, the case of two forces acting in opposite directions on a body of mass m, as shown in Figure 3.6. The force F_1 tends to accelerate the mass to the left while force F_2 tends to accelerate it to the right. An experimental investigation of this situation shows that the acceleration of the body in this case is the same as if a single force was acting on the body, the magnitude of which is the difference in the magnitudes of F_1 and F_2. Thus, for motion in one dimension, force can be treated as an *algebraic* quantity (recall Section 2.2) so that the force F above must now be understood as the *algebraic sum* of all the forces acting on the body. For the case indicated in Figure 3.6,

$$a \propto \frac{F_1 + F_2}{m}$$

where F_1 and F_2 are positive or negative quantities depending on whether they are directed in the $+x$- or $-x$-direction, respectively. If the $+x$-direction is chosen as indicated in Figure 3.6, then $F_2 > 0$ and $F_1 < 0$.

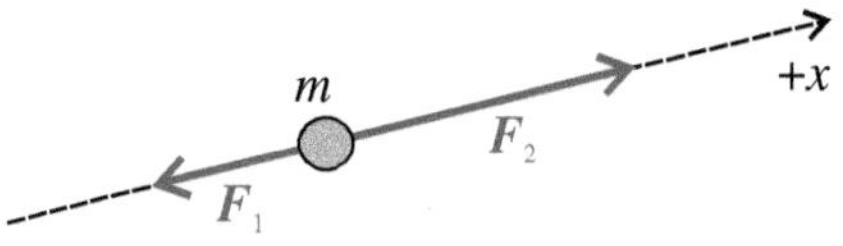

Figure 3.6. Two forces acting upon a mass m in opposite directions.

In dealing with one-dimensional dynamical systems, therefore, force must be treated as an algebraic quantity (having sign, magnitude, and unit) just like displacement, velocity, and acceleration. By definition, of course, mass is a positive quantity.

More generally, therefore, Newton's second law retains the form of (3.1) above, namely

$$a \propto \frac{F} {m} \tag{3.2}$$

but in this case F is the *algebraic sum of all the forces* acting on the mass m.

If $F = 0$, that is if the algebraic sum of all the forces on a body add up to zero, the acceleration will be zero – the body must either be at rest or move with constant velocity. In such circumstances the body is said to be **in equilibrium**.

Momentum

The product of the mass m of a body and its velocity v is defined as the momentum of the body, usually denoted by p, that is

$$\boxed{p := mv}$$

Since, for one dimensional motion, v is an algebraic quantity then, by this definition, p is also algebraic.

Newton originally formulated the second law in terms of momentum, stating that the rate of change of momentum was proportional to the applied force. It is easy to see that this form of the law is the same as Equation (3.1) since

$$\frac{dp}{dt} = \frac{d(mv)}{dt} = m\frac{dv}{dt} = ma$$

provided of course that m does not change with time. Hence Newton's second law can be stated equivalently as

$$\frac{dp}{dt} \propto F \tag{3.1a}$$

and in this form can be used to study situations in which the mass of the system varies (for example, the motion of rockets).

3.4 Systems of units: SI

While relationships (3.1) and (3.1a) are mathematical statements of Newton's second law, they are not very useful in this form because it does not enable us to predict, for example, what would happen if a specific force were to be applied to a particular object. For verifiable predictions to be made one needs an *equation* which relates measurable physical quantities to one another. Relationship (3.1) can be written in the form of an equation as follows

$$a = k\frac{F}{m}$$

but this is of no help since the constant k can take on any value whatsoever, at this stage, depending on how we choose to measure a, F and m.

As we saw in Chapter 2, to measure a velocity or acceleration we must first decide on a unit of length (for example, metre, foot, mile, etc.) and a unit of time (for example, second, hour, year, etc.). Units so chosen are usually called *base units*. Units of other quantities such as velocity (m s^{-1}, miles per hour, etc.) and acceleration (m s^{-2}, etc.) are *derived* from base units. Thus, the value of a in Equation (3.2) depends on the choice of base units adopted for length and time.

We have not, as yet, given any consideration to the units of force and mass which might be used. There are three possible ways forward (each strategy would give rise to a different 'system of units'):

(a) We could choose two further base units, for force and mass, in which case the value of k in Equation (3.2) would take on a fixed value in terms of the four base units chosen (for time, length, force, and mass) and would have to be determined experimentally.
(b) We could choose a base unit of force and assign an arbitrary numerical value to the constant k, in which case the unit of mass becomes a derived unit in terms of the base units chosen for time, length, and force
(c) We could choose a base unit for mass, assign an arbitrary numerical value to the constant k and, in this case, the unit of force is derived in terms of base units of time, length, and mass.

Each of the above strategies has been used historically and, as we will see in subsequent chapters, similar strategies are followed when developing systems of units in other areas of physics.

The system of units adopted by international agreement throughout most of the world today uses the third of these strategies. This system had its genesis in the formation of the Conférence Générale des Poids et Mesures (CGPM) in 1889 when the system using the metre, the kilogram, and the second as base units was formally adopted. This system has been refined, extended, and updated periodically in the intervening years and in 1960 the CGPM officially named it Système Internationale d'Unités (SI). SI is now used almost universally in science and engineering and will be adopted uncompromisingly in this book. As we shall see later there are seven base units in SI, but three base units will suffice for the study of dynamics; that is to say, the units of all physical quantities that we will encounter in the study of dynamics can be written in terms of the following three base units: **second, metre,** and **kilogram**. The current formal definitions of these units are given below; the reader should not be concerned if some of the technical terms used in these definitions are unfamiliar at this stage.

The SI unit of time, the second

Until the 1960s, the second was defined in terms of a specified fraction of the mean solar day (or sometimes of a year). The development of atomic clocks around that time, however, enabled more accurate definition to be made and, in 1967, the second was re-defined as the duration of 9 192 631 770 periods of the radiation corresponding to the transition between the two hyperfine levels of the ground state of the caesium-133 ($^{133}_{55}Cs$) atom. The current definition of the second is as follows:

The second, symbol s, is the SI unit of time. It is defined by taking the fixed numerical value of the caesium frequency $\Delta\nu_{Cs}$, the unperturbed ground-state hyperfine transition frequency of the caesium-133 atom, to be 9 192 631 770 when expressed in the unit Hz, which is equal to s^{-1}.

The details of caesium atomic clock standards need not concern us here; suffice it to say that governmental or intergovernmental standards agencies have developed devices containing caesium gas which can emit a string of very sharp pulses (at rates of, say, one thousand pulses per second or one pulse per second) accurately locked to the specified frequency of caesium atoms. These pulses can be used to calibrate very accurate timing devices (standard clocks) which can be used in turn to calibrate other timers.

GPS satellites carry highly accurate atomic clocks, so even a relatively inexpensive GPS receiver can be operated as a timing standard. Many such receivers provide a pulse per second output that can be used to calibrate a stopwatch or similar laboratory instrument.

The SI unit of length, the metre

The redefinition of the metre in 1983 provides a good example of how the definition of base units in SI is tied to practical considerations. Since 1960 the metre had been defined in terms of a wavelength of the radiation emitted by ^{86}Kr atoms. This wavelength is measurable to an accuracy of about 4 parts in 10^9 compared to an accuracy of 1 part in 10^{13} in the case of the second. This meant that, although it was possible to measure the time taken by a laser pulse to travel from the Earth to the Moon and back to an accuracy of 1 part in 10^{13}, it was not possible to express the Earth–Moon distance in metres with the same precision. The situation may be compared with attempting to measure the length of an object with millimetre accuracy using a ruler which has been graduated only in centimetres. The 1983 redefinition of the metre as the length of the path travelled by light in a vacuum during a time interval of $\frac{1}{299\,792\,458}$ of a second, removed this problem.

It is interesting to examine what exactly is going on here. The definition of the metre requires that the speed of light in vacuum (known to be a fundamental constant of physics, to be discussed further in Chapter 9) has the following *fixed* value:

$$c = 299\,792\,458 \text{ m s}^{-1}$$

It is clear from this that, since the second is already defined, *fixing c defines the metre*. Note that the fixed value chosen was based on the most accurate measured value of the speed of light in 1983 so that the re-definition ensured the continuity of the unit of length with the pre-1983 definition.

The formal SI definition of the metre is as follows:

The metre, symbol m, is the SI unit of length. It is defined by taking the fixed numerical value of the speed of light in vacuum c to be 299 792 458 when expressed in the unit m s^{-1}, where the second is defined in terms of the caesium frequency $\Delta\nu_{Cs}$.

Thus, in principle, any length (l) can be determined by measuring the time (t) taken for a light pulse to traverse that distance; that is, $l = ct$.

The concept of fixing the value of a fundamental constant of physics to define a base unit is invoked again in the definition of the kilogram and, as we will see later, in other SI base units.

The SI unit of mass, the kilogram

For well over a century the kilogram was defined as being equal to the mass of the international prototype of the kilogram. The international prototype kilogram was manufactured in the 1880s of an alloy of 90% platinum-10% iridium and was kept, with six official copies, in a vault at the Bureau International des Poids et Mesures near Paris. The kilogram was the last remaining base unit of the SI that was defined by a material artifact. Finally, after considerable effort over many years, this artifact definition has now been replaced by one based on a fundamental physical constant (with effect from 20 May 2019).

The particular fundamental constant chosen for the re-definition of the kilogram is called the Planck constant (h). The physical significance of this constant will be discussed at length in Chapter 14 but, for present purposes, it is only necessary to note that it has a fixed value and may be used to relate the kilogram to the metre and the second, the dimensions of h being those of the product of momentum and speed (that is, ML^2T^{-1}) which, in SI, are kg m^2 s^{-1}. The 2019 definition fixes the value of the Planck constant as

$$h = 6.626\,069\,57 \times 10^{-34}\ \text{kg m}^2\ \text{s}^{-1}$$

from which it follows that, since the second and the metre are already defined, *fixing h defines the kilogram*. Again, because of the accuracy to which the value of h was known, the re-definition ensures the continuity of the unit of mass with the previous definition.

From 2019, the formal definition of the kilogram is

The kilogram, symbol kg, is the SI unit of mass. It is defined by taking the fixed numerical value of the Planck constant h to be 6.626 070 15 × 10^{-34} when expressed in the unit J s, which is equal to kg m^2 s^{-1}, where the metre and the second are defined in terms of c and $\Delta\nu_{Cs}$.

Newton's Second law in SI Units

In SI, therefore, the unit of acceleration is m s^{-2} and the unit of mass is kg. Before a unit of force can be derived in terms of the base units, however, a numerical value for the (dimensionless) constant of proportionality k in Equation (3.2) must also be adopted as part and parcel of the system of units. The internationally agreed convention in almost all systems of units, including SI, is to set $k = 1$ in which case Newton's Second law, Equation (3.2), becomes

$$a = \frac{F}{m} \tag{3.3}$$

which, in terms of momentum, can also be stated as

$$\frac{dp}{dt} = F \tag{3.3a}$$

It is important to recognise that the form of relationships (3.3) and (3.3a) depends: (i) on experimental observations, and (ii) on the system of units adopted. Unlike equations derived in Chapter 2, it does not follow directly from the definitions of the quantities involved but rather is an empirical 'law of nature'.

The unit of force, when expressed in SI, is the product of the unit of mass and the unit of acceleration, namely kg m s^{-2}. This unit is given a special name, the newton, and the symbol for the newton is N; thus 1 newton = 1 N = 1 kg m s^{-2}.

Practical realisation of a unit from its definition

National Standards Laboratories in individual countries are responsible for creating and maintaining what are called 'primary standards' for fundamental physical units. The process of experimentally determining the unit from the definition and making a primary standard is called 'realising the unit'. No specific technique is prescribed for this but a number of recommended procedures (*mises en pratique*) are suggested. Primary standards never leave their Standards Laboratory but reference standards are calibrated from the primary standards for dissemination to industry, government and academia *via* physical artifacts ('secondary' or 'transfer' standards). It may be assumed that instruments used in physics laboratory classes have been calibrated, albeit very indirectly, with respect to such secondary standards.

Prefixes for decimal multiples and fractions

There is also international agreement on how to form prefixes for decimal multiples and decimal fractions of units (see Table 3.1). Different letters are prefixed to the symbols for units for each 10^3 of magnitude as shown in Table 3.1; for example, 1 μs = 10^{-6} s. Note, however, that there is one exception to this practice: in the case of the unit of mass, prefixes are attached to the symbol g (1 g = 10^{-3} kg) rather than to kg, that is 10^3 kg = 1 Mg (not 1 kkg).

Table 3.1 Decimal prefixes for units. Some older prefixes, such as d (for 10^{-1}) and c (for 10^{-2}), are sometimes used.

Prefix	Symbol	Multiple
yotta	Y	10^{24}
zetta	Z	10^{21}
exa	E	10^{18}
peta	P	10^{15}
tera	T	10^{12}
giga	G	10^{9}
mega	M	10^{6}
kilo	k	10^{3}
milli	m	10^{-3}
micro	μ	10^{-6}
nano	n	10^{-9}
pico	p	10^{-12}
femto	f	10^{-15}
atto	a	10^{-18}
zepto	z	10^{-21}
yocto	y	10^{-24}

Motion under a constant force

In many situations the force acting on a body is constant. An important case is the force due to gravity near the surface of the Earth. When a body is thrown vertically upward or dropped vertically from the top of a building, the force it experiences remains constant provided the distance it travels is small compared to the radius of the Earth. Newton's second law tells us that if the force on a body is constant it will move with constant acceleration. In fact, Galileo discovered that all bodies falling freely under gravity at a particular location on the surface of the Earth have the same acceleration, a point to which we shall return in later chapters. As noted at the end of Section 2.5, the magnitude of the acceleration common to all bodies near the surface of the Earth is called the acceleration due to gravity and is denoted by the symbol g. For most purposes it is sufficient to know the value of g to about 1% accuracy in which case it can be taken to be 9.8 m s^{-2}.

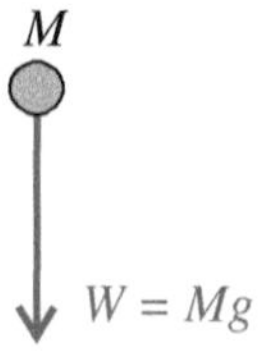

Figure 3.7. The weight of a body is the product of its mass and the acceleration due to gravity.

Weight: The force on a body due to gravity is called the weight of the body. Applying Newton's second law, we see that the weight of a body of mass M (Figure 3.7) is given by

$$\boxed{W = Mg}$$

For bodies moving vertically under gravity, therefore, as well as for other cases of constant acceleration in a straight line, the relationships derived in Chapter 2 for displacement and velocity under these circumstances, namely Equations (2.18), (2.19) and (2.20), may be applied.

Study Worked Example 3.1

For problems based on the material presented in this section visit up.ucc.ie/3/ and follow the link to the problems.

3.5 Time dependent forces: oscillatory motion

In all the cases which we have considered so far, the forces applied to a body have been constant. We must now ask the question of how the situation might change if a force applied to a body varies with time. One might expect that if the force F in Equation (3.3) is time dependent, the acceleration observed would exhibit the corresponding time dependence. This, as we shall see, is borne out by further observations; in other words, Newton's second law can be considered to hold *instantaneously*. An example of motion under an explicitly time-dependent force is discussed in Worked Example 3.2.

Another example of a non-constant force is one which depends on the velocity of an object. The drag on a body moving slowly through air ('air resistance') can be modelled quite well by a force which is directly proportional to the velocity of the body, as described in Worked Example 3.3.

Study Worked Examples 3.2 and 3.3

A commonly occurring type of motion in which the forces acting on a body are certainly not constant is *oscillatory motion*. Such oscillatory or vibrational motion occurs when an object, after being displaced from its equilibrium position, performs symmetrical oscillations about this position. There are many examples of oscillatory motion in nature. Three familiar examples are illustrated in Figure 3.8, namely a mass which is oscillating vertically at the end of a helical spring (Figure 3.8(a)), a swinging pendulum (Figure 3.8(b)), and a mass which is rolling to and fro along the inside of a spherical bowl (Figure 3.8(c)).

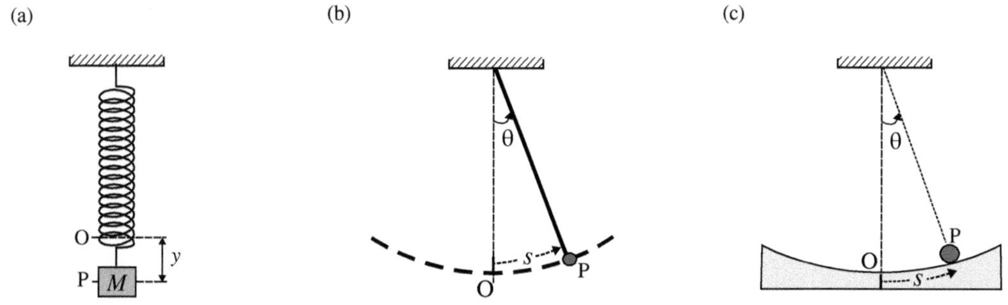

Figure 3.8. Examples of oscillatory motion: (a) a mass executing vertical oscillations at the end of a helical spring, (b) a mass at the end of a string executing oscillations in a vertical plane and (c) a sphere rolling on a concave spherical surface with the motion confined to a vertical plane.

In the first example, oscillatory motion can be initiated if the mass is displaced vertically from its equilibrium position and released. The mass then oscillates in such a way that its displacement from equilibrium is confined to the vertical direction. The motion is a linear translational motion in which displacement from the equilibrium position is characterised in Figure 3.8(a) by the variable y. In the other two examples, however, the displacement from equilibrium is not along a straight line but nevertheless is one-dimensional in that the displacement can be characterised by a single variable (either the angle θ or the distance s measured along the arc of the circle).

How then can we analyse oscillatory motion? Which quantities can we use to describe and characterise such a motion? An important characteristic of oscillatory motion, which is evident in each of the three cases illustrated in Figure 3.8(a)–(c), is that the motion is repeated at regular time intervals. A characteristic time interval of the motion can therefore be measured for a particular oscillatory motion, namely the time which elapses before the oscillating object again passes through a given position, such as P in Figure 3.8(a)–(c), travelling in the same direction and begins to repeat its motion. This time is known as the periodic time or **period** of the motion. Alternatively, the periodic nature of an oscillatory motion can be measured by counting the number of times that the motion repeats itself in one second, that is the number of cycles the motion goes through in one second. This quantity is known as the **frequency**. It follows from these definitions that the frequency of an oscillatory motion is given by

$$\boxed{f := \frac{1}{T}}$$

where T is the period. The SI unit for frequency is the hertz (symbol Hz), defined as one cycle per second.

A second characteristic quantity associated with a particular oscillatory motion is its **amplitude**, which is the maximum displacement of the oscillating object from its equilibrium position during the motion. The magnitude of the amplitude depends on how the body was set in motion in the first place.

We have now defined two measurable quantities, the period (or frequency) and the amplitude, which can be used to characterise a particular oscillatory motion. The next steps in the scientific method are representation of these measurements and a search for relationships between the measured quantities. As before, an electronic sensor system interfaced with a computer can be used to determine distance–time data experimentally (Figure 3.9) for such oscillatory motion. Our eventual aim is to explain the observed behaviour of the system. In other words, we would like to be able to predict the future course of the motion in terms of initial conditions as we were able to do in the special cases of linear motion which we considered in Section 2.5.

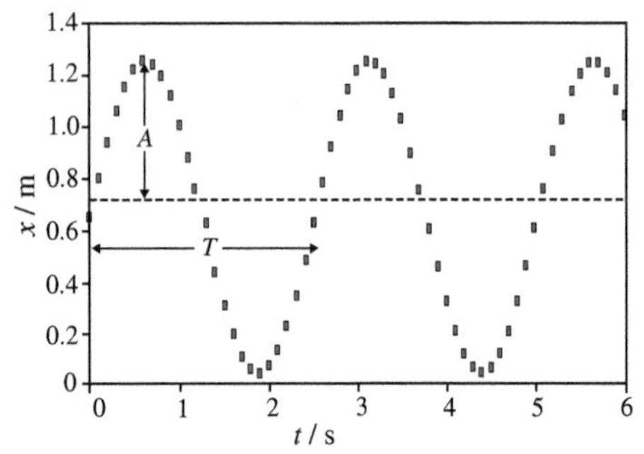

Figure 3.9. An example of a position—time plot for oscillatory motion in one dimension; the amplitude (A) and period (T) of the motion are indicated.

Let us consider again the situation, illustrated in Figure 3.8(a), in which a vertical helical spring is stretched by hanging a mass M from it. The spring produces a force F on the mass M which acts to return the mass to its original position – a *restoring force*. In equilibrium the restoring force balances the weight of the mass.

Experimental studies of helical springs show that, provided the extension is not too large, the restoring force exerted by an extended spring is directly proportional to the magnitude of the extension of the spring. Similarly, if a spring is compressed, the restoring force acting in the opposite direction is again proportional to the magnitude of the displacement from equilibrium. In either case, the restoring force exerted by a spring that has been extended or compressed by an amount x is given by

$$F = -kx$$

where x is the displacement from the equilibrium position ($x > 0$ for an extension and $x < 0$ for compression) and k is a positive constant (called the *spring constant*). This relationship, known as *'Hooke's law'* (after Robert Hooke (1635–1703)), represents a model which gives a reasonable description of the behaviour of helical springs under small extension or compression. The minus sign indicates that the force always acts in the opposite direction to the displacement from equilibrium. The value of k is characteristic of a particular spring and is a measure of how 'strong' or how 'weak' the spring is; the value of k depends on the dimensions of the particular spring and on the material from which it is made. The SI unit of k is N m^{-1}.

It follows that, in the case of the spring in Figure 3.8(a), if the mass M is displaced downward from the equilibrium position, the upward restoring force on the mass is directly proportional to the downward extension y of the spring. If the spring is compressed, that is given a negative (upward) displacement from equilibrium, the restoring force acting downwards to restore the mass to its equilibrium position is again proportional to y, that is, $F = -ky$.

Study Worked Example 3.4

Clearly in any case of oscillatory motion the force involved is not constant since it varies with position and hence with time. Thus if a body is displaced from equilibrium and released, its acceleration will not be a constant, so the relationships (2.18), (2.19) and (2.20) are **not applicable** to oscillatory motion.

Consider now the system illustrated in Figure 3.10 where a mass m on a smooth horizontal table is connected to a fixed wall via a helical spring. If the mass is displaced from equilibrium along the axis of the spring and then released, it will oscillate about the equilibrium position O. If we represent the displacement from equilibrium by the symbol x, the net force on the mass at that displacement is the restoring force ($-kx$) exerted by the spring. Applying Newton's second law in this case we get $F = -kx = ma$ and hence the motion is described by the following equation

$$a = -\frac{k}{m}x \tag{3.4}$$

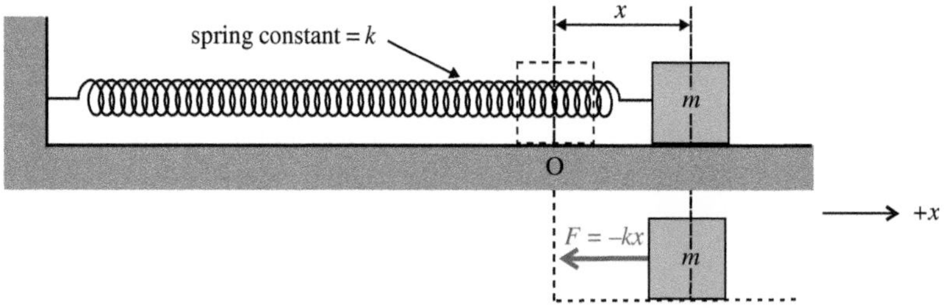

Figure 3.10. A block of mass m moving on a horizontal surface under the influence of a helical spring attached to a fixed support. The position of the mass when the spring is unextended (the 'equilibrium position', O) is chosen as the origin of displacement and displacements to the right are taken as positive.

Again, the negative sign indicates that the acceleration is always in the opposite direction to the displacement; that is, towards the equilibrium position.

Equation (3.4) describes a type of motion which arises quite commonly in nature. In the next section we will show how the course of such motion can be predicted in detail; for example, we will obtain relationships which give position, velocity and acceleration as functions of time.

For problems based on the material presented in this section visit up.ucc.ie/3/ and follow the link to the problems.

3.6 Simple harmonic motion

The case of a mass oscillating at the end of a helical spring, discussed above, is an example of a system in which the acceleration is always directed towards a fixed central point and increases linearly with displacement from that point (that is $a \propto -x$, for motion along the x-axis). Such systems, called *harmonic oscillators*, are very common in nature and involve a special type of oscillatory behaviour called **simple harmonic motion**. Thus simple harmonic motion is defined by the relationship

$$a = \frac{d^2x}{dt^2} = -\gamma x$$

or

$$\boxed{\frac{d^2x}{dt^2} + \gamma x = 0} \tag{3.5}$$

where γ is a positive constant characteristic of the particular system involved; for example, $\gamma = \frac{k}{m}$ in Equation (3.4).

Our principal task is to derive expressions for the displacement, velocity and acceleration as functions of time for a point which is moving with simple harmonic motion (analogous to relationships (2.18), (2.19), etc. for constant acceleration). There are formal mathematical methods for solving Equations such as (3.5), which are called 'differential equations', but careful inspection of the equation makes such formal treatment unnecessary in this case. Clearly what is needed is a functional dependence of x on t which is such that, on being differentiated twice with respect to t, the same function is returned multiplied by a negative constant. Examination of the table of derivatives in Appendix A (Table A.1) shows that there are a number of functions which behave in this way. The function $x = \sin t$, for example, satisfies this criterion, since

$$\frac{dx}{dt} = \cos t \text{ and hence } \frac{d^2x}{dt^2} = -\sin t = -x, \text{ that is } \gamma = 1$$

Other functions which would do equally well are $x = \cos t$ and $x = e^{\pm it}$ or indeed functions such as $x = \sin(\omega t)$, $x = \cos(\omega t)$, and $x = e^{\pm i\omega t}$, where ω is a constant in which case $\gamma = \omega^2$. Multiplying any of these functions by a constant, for example $x = A\sin(\omega t)$, would also work. A more general function which satisfies Equation (3.5) is

$$x = A\sin(\omega t + \phi) \tag{3.6}$$

where A, ω and ϕ are arbitrary constants (note that $x = A\cos(\omega t + \phi)$ or $x = Ae^{\pm i(\omega t+\phi)}$ would do equally well).

That $x = A\sin(\omega t + \phi)$ is indeed a solution of Equation (3.5) can be verified as follows. Differentiating Equation (3.6) successively with respect to time gives

$$\frac{dx}{dt} = \omega A\cos(\omega t + \phi)$$
$$\frac{d^2x}{dt^2} = -\omega^2 A\sin(\omega t + \phi) = -\omega^2 x$$

Thus Equation (3.6) gives the time dependence of the displacement of a point undergoing simple harmonic motion, described by Equation (3.5), if $\omega = \sqrt{\gamma}$. The required relationships for displacement, velocity and acceleration as functions of time follow directly, since $v = \frac{dx}{dt}$ and $a = \frac{dv}{dt} = \frac{d^2x}{dt^2}$.

$$x(t) = A\sin(\omega t + \phi) \tag{3.7a}$$
$$v(t) = \omega A\cos(\omega t + \phi) \tag{3.7b}$$
$$a(t) = -\omega^2 A\sin(\omega t + \phi) \tag{3.7c}$$

Note that, in these equations, ωt and ϕ are usually expressed in radians and hence ω is expressed in rad s^{-1}.

We can also obtain an expression for velocity as a function of displacement by using the trigonometric identity $\sin^2\theta + \cos^2\theta = 1$ (Appendix A.3.5(i)) as follows.

$$v^2 = \omega^2A^2\cos^2(\omega t + \phi) = \omega^2A^2(1 - \sin^2(\omega t + \phi)) = \omega^2A^2 - \omega^2A^2\sin^2(\omega t + \phi) = \omega^2A^2 - \omega^2x^2$$

and hence

$$\boxed{v(x) = \pm\omega\sqrt{A^2 - x^2}} \tag{3.7d}$$

Note that there are two velocity values for each x, corresponding to the two times at which the moving point passes through a given position during a cycle of the motion.

We know, from definition (3.5), that $a(x) = -\omega^2 x$, which expresses the acceleration as a function of displacement.

The constant A is the maximum value of x, that is the value of x when $\sin(\omega t + \phi) = 1$, and hence is the **amplitude** of the motion as defined in the previous section. Note that the maximum value of the velocity $v_{max} = \omega A$ occurs at $x = 0$ and that the maximum value of the acceleration $a_{max} = -\omega^2 A$ occurs at $x = \pm A$. Furthermore $a = 0$ at $x = 0$ and $v = 0$ at $x = \pm A$.

The value of the constant ϕ, called the **phase constant** (or 'phase angle') of the motion, is quite arbitrary, its value being determined by the choice of $t = 0$. For example, if we choose to measure time from an instant when $x = 0$ (that is, starting timing from the instant when the moving body is passing through the centre point of the motion) then

$$0 = A \sin(0 + \phi) = A \sin \phi \quad \Rightarrow \quad \phi = 0$$

with consequent simplification of Equations (3.7a)–(3.7c).

Alternatively, we might choose to measure time from the instant when the body was at an extremity of the motion, such as $x = A$ at $t = 0$. In this case $\phi = \frac{\pi}{2}$ and we get $x = A \cos \omega t = A \sin(\omega t + \frac{\pi}{2})$ (Appendix A.3.2).

Geometrical interpretation of simple harmonic motion

We can get some further insight into simple harmonic motion using the geometrical construction shown in Figure 3.11. Consider a point P moving with constant speed on a circle of radius A. The radius from the centre of the circle C to the point P sweeps around the circle with constant angular velocity ω. If we choose to measure time from the instant when the moving point is at F, the angle between CP and CF after a time t is ωt. Let Q be the projection of the moving point P on a line XX′ in the plane of the circle and let O be the projection of C on the same line. If x is the displacement of Q from O measured along XX′, we note that $x = OQ = PN = A \sin(\omega t + \phi)$. Thus, the motion of Q, the projection of the motion of P onto the line XX′, is simple harmonic motion. Note again that the value of ϕ depends on the choice of time origin and is given by the angle OP makes with OC at $t = 0$.

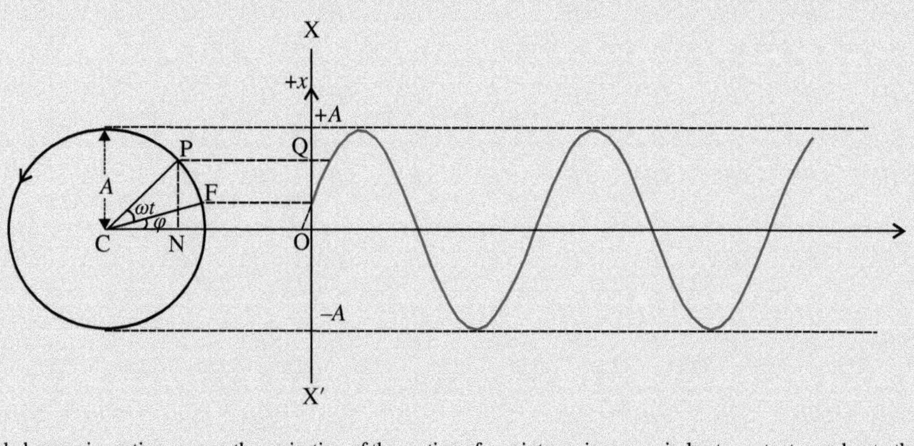

Figure 3.11. Simple harmonic motion seen as the projection of the motion of a point moving on a circle at constant speed onto the vertical line XX′.

We can draw theoretical plots (Figure 3.12) to show how the displacement, velocity and acceleration of a body undergoing simple harmonic motion vary with time. It is clear that the motion is periodic; that is, there is a characteristic periodic time after which the motion proceeds to repeat exactly. Stated more exactly, the period of the motion is such that x remains the same if $t \to t + T$, that is

$$A \sin[\omega(t + T) + \phi] = A \sin(\omega t + \phi)$$

or

$$\sin[\omega t + \phi + \omega T] = \sin(\omega t + \phi)$$

and thus $\omega T = 2\pi$ (since $4\pi, 6\pi \ldots$ corresponds to $2T, 3T \ldots$)

The period of the motion, therefore, is related to the constant ω as follows

$$\boxed{T = \frac{2\pi}{\omega}} \qquad (3.8)$$

and hence depends on the value of the constant of proportionality $\gamma = \omega^2$ in the defining equation for simple harmonic motion (3.5).

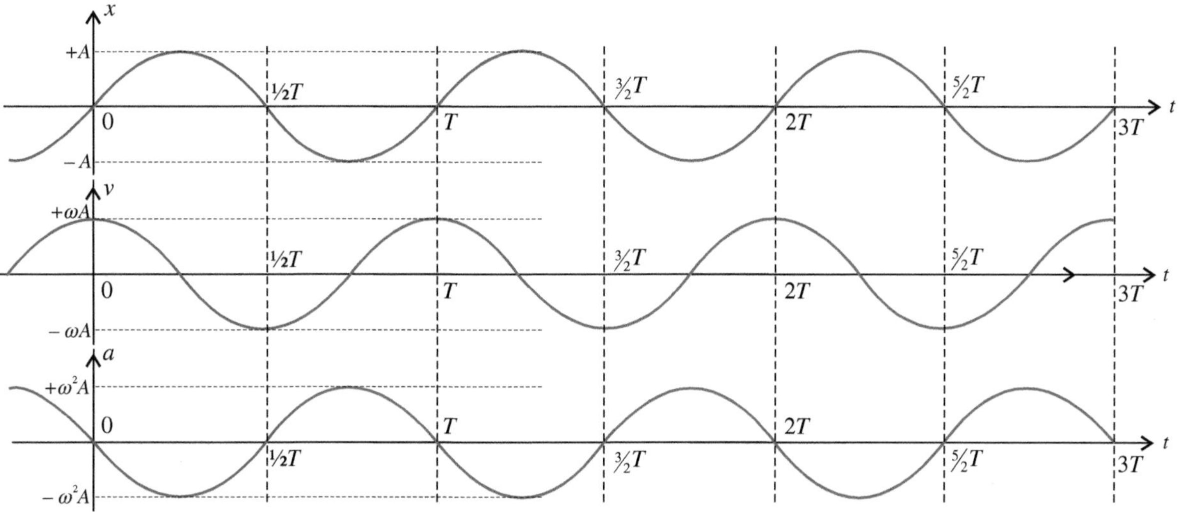

Figure 3.12. Plots of displacement (x), velocity (v) and acceleration (a) as functions of time for simple harmonic motion.

Note that, using Appendix A.3.2, $v = \omega A \cos \omega t$ can be written as $v = \omega A \sin(\omega t + \frac{\pi}{2})$. Comparing plots of v and of x as functions of ωt (Figure 3.13), we see that the phase angle of v is greater than that of x by $\frac{\pi}{2}$ rad. We therefore say that v *leads* x by $\frac{\pi}{2}$ rad, for example v is at maximum one quarter of a period before x is at maximum.

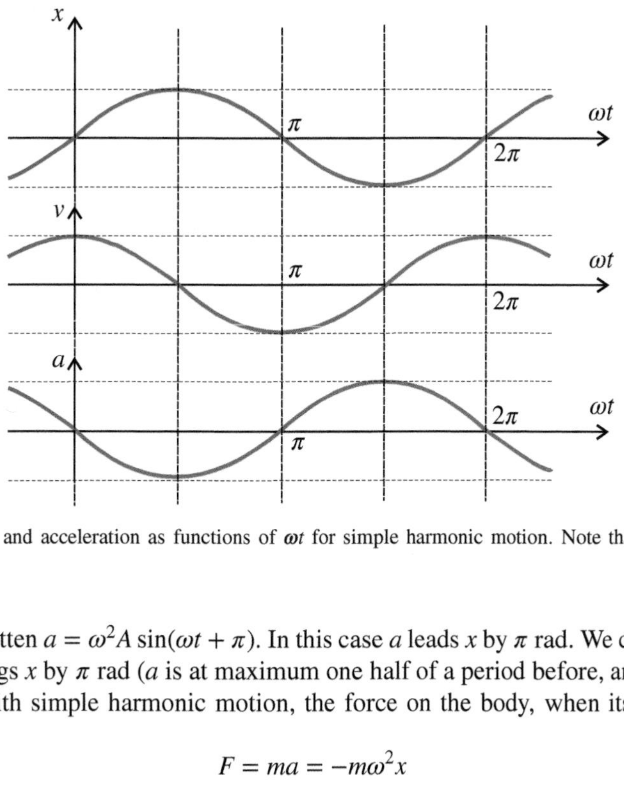

Figure 3.13. Plots of displacement, velocity, and acceleration as functions of ωt for simple harmonic motion. Note that velocity leads displacement by $\frac{\pi}{2}$ rad and acceleration leads displacement by π rad.

Similarly $a = -\omega^2 A \sin \omega t$ can be written $a = \omega^2 A \sin(\omega t + \pi)$. In this case a leads x by π rad. We can also write $a = A\omega^2 \sin(\omega t - \pi)$ so that we can equally well say that a lags x by π rad (a is at maximum one half of a period before, and after, x is maximum).

For a body of mass m moving with simple harmonic motion, the force on the body, when its displacement from the equilibrium position is x, is given by

$$F = ma = -m\omega^2 x \tag{3.9}$$

Mass oscillating at the end of helical spring

Returning, finally, to the case of the helical spring system illustrated in Figure 3.10 from which our discussion of simple harmonic motion started, we see that the complete behaviour of the system is determined by a knowledge of ω which, in turn, is determined from the application of Newton's Second law to the dynamical system involved. In the case of that system, $\gamma = \frac{k}{m}$ and hence $\omega = \sqrt{\frac{k}{m}}$ and the period of the oscillations, determined from Equation (3.8), is $T = 2\pi\sqrt{\frac{m}{k}}$. The corresponding frequency is given by

$$f = \frac{1}{2\pi}\sqrt{\frac{k}{m}} \tag{3.10}$$

As an illustration let us estimate what value of spring constant would be required if we want a mass of 100 g to vibrate at the end of a spring with a period of one second. In this case

$$k = \frac{4\pi^2 m}{T^2} = \frac{4\pi^2(0.1\text{kg})}{(1\text{ s})^2} = 3.9\text{ kg s}^{-2} = 3.9\text{ N m}^{-1}.$$

The time dependence of the displacement and the velocity, from Equations (3.7a) and (3.7b), are given by

$$x = A\sin\left(\sqrt{\frac{k}{m}}t + \phi\right)$$

$$v = \sqrt{\frac{k}{m}}A\cos\left(\sqrt{\frac{k}{m}}t + \phi\right)$$

where A is determined by the initial displacement given to the mass before release and ϕ is fixed by the (arbitrary) choice of $t = 0$.

Solving problems in simple harmonic motion

The general procedure for dealing with problems involving simple harmonic motion can be summarised as follows:

(i) Use Newton's second law to write down the equation of motion for the system.
(ii) Put the equation of motion in the form $a = -\gamma x = -\omega^2 x$.
(iii) From this obtain $\omega = \sqrt{\gamma}$ and hence, if required, determine the period $T = \dfrac{2\pi}{\omega}$.
(iv) Obtain A and ϕ from the initial conditions.
(v) Use Equations (3.7a–c) to determine displacement, velocity, *etc.* as functions of time.

Simple harmonic oscillations play a very important role in many different branches of physics, for example lattice vibrations in solids (Chapter 12), sources of waves (Chapter 13) and electromagnetic radiation (Chapter 21).

Study Worked Example 3.5

For problems based on the material presented in this section visit up.ucc.ie/3/ *and follow the link to the problems.*

3.7 Mechanical work and energy

The work-energy theorem

This section introduces two important physical quantities, *work* and *energy*. The terms work and energy, of course, are commonly used in everyday parlance but, in science, they are used in a precise and specific manner which is significantly different from their vernacular usage. When first encountered, these concepts may appear somewhat abstract. A better appreciation of their value will come with experience of their application, in particular to the study of motion in two and three dimensions in Chapter 5. For our present purposes, however, it is sufficient to consider work and energy in the more limited context of one-dimensional motion.

Consider the motion of the point mass m which is confined to the x-axis as shown in Figure 3.14. The particle experiences a force which may vary with position, that is

$$F = F(x)$$

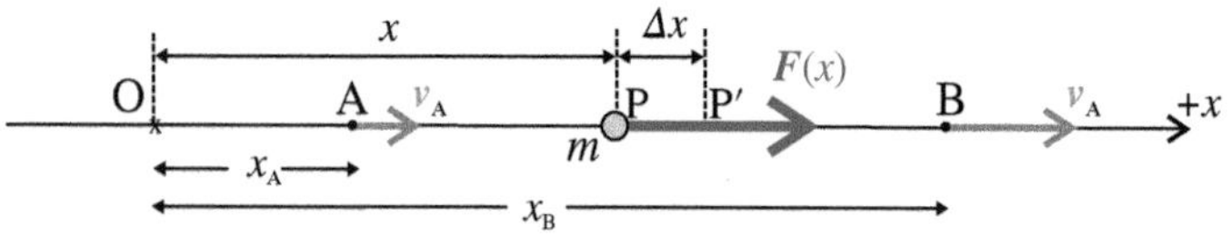

Figure 3.14. A mass m moves on the x-axis under the influence of a force that varies with x.

Let us first consider a small displacement of the mass from some arbitrary position P (displacement = x) to a position P′ (displacement = $x + \Delta x$) under the influence of the force $F(x)$. The element of **mechanical work done** *by the force F* in moving the mass through a displacement Δx is *defined* as

$$\boxed{(\Delta W) := F(\Delta x)}$$

The total mechanical work done in moving the mass from some point A ($x = x_A$) to point B ($x = x_B$) is the sum of all the (ΔW)s between A and B, that is (recalling the definition of an integral in Section 2.5)

$$W_{AB} := \int_{x_A}^{x_B} F(x)\,dx \tag{3.11}$$

From Newton's second law, $F(x) = ma = m\frac{dv}{dt}$, we see that

$$W_{AB} = \int_{x_A}^{x_B} F(x)dx = \int_{x_A}^{x_B} m\frac{dv}{dt}dx = m\int_{x_A}^{x_B} \frac{dv}{dt}\frac{dx}{dt}dt = m\int_{x_A}^{x_B} \frac{dv}{dt}v\,dt$$

Now $\frac{d}{dt}(v^2) = 2v\frac{dv}{dt} \quad \rightarrow \quad \frac{dv}{dt}v = \frac{1}{2}\frac{d(v^2)}{dt}$ and hence

$$W_{AB} = m\int_{x_A}^{x_B} \frac{dv}{dt}v\,dt = \frac{1}{2}m\int_{x_A}^{x_B} \frac{d(v^2)}{dt}dt = \frac{1}{2}m\int_{x_A}^{x_B} d(v^2) = \frac{1}{2}m(v_B^2 - v_A^2)$$

where v_A and v_B are the velocities of the mass at $x = x_A$ and $x = x_B$, respectively.

The last result, which states that the mechanical work done by $F(x)$ on the body in moving it from a point A to a point B in such a way that the velocity of the body changes from v_A to v_B, is given by

$$W_{AB} = \frac{1}{2}mv_B^2 - \frac{1}{2}mv_A^2 \tag{3.12}$$

If more than one force acts on the body, we know that the F in Newton's second law refers to the net force (the algebraic sum of all forces on m) and hence, in determining W_{AB}, the force used within the integral in Equation (3.11) must also be the net force. Nevertheless, when a number of forces act on a body simultaneously, it is sometimes convenient to also use Equation (3.11) to define the work done by any particular one of these forces.

Kinetic energy: conservation of mechanical energy

It is interesting to consider the conditions under which it might be possible to evaluate the integral in Equation (3.11) when detailed information on the force is known. If the dependence of F on x is a simple mathematical function, then the integral should be capable of evaluation as a definite integral (recall Section 2.5). As we will see shortly, this is not necessarily the case in many mechanical systems, but first we will consider situations in which the integral can be so evaluated. In such cases, a function $f(x)$ exists so that we can write

$$W_{AB} = \int_{x_A}^{x_B} F(x)dx = f(x_B) - f(x_A)$$

Invoking Equation (3.12), this gives

$$f(x_B) - f(x_A) = \frac{1}{2}mv_B^2 - \frac{1}{2}mv_A^2$$

We now define a function which is simply the negative of $f(x)$; this function is usually denoted by $U(x)$, that is $U(x) = -f(x)$ and $W_{AB} = \int_{x_A}^{x_B} F(x)dx = -U(x_B) + U(x_A)$.

Thus,

$$-U(x_B) + U(x_A) = \frac{1}{2}mv_B^2 - \frac{1}{2}mv_A^2$$

or

$$\frac{1}{2}mv_A^2 + U(x_A) = \frac{1}{2}mv_B^2 + U(x_B)$$

We have arrived at a very important general result which follows directly from Newton's second law and the condition that the integral in Equation (3.11) can be evaluated as a definite integral. Under these circumstances, the quantity $\frac{1}{2}mv^2 + U(x)$ remains a constant throughout the motion. This quantity is called the **total energy** of the system and is defined as

$$\boxed{E := \frac{1}{2}mv^2 + U(x)}$$

The second term in the expression is the function $U(x)$ which gives an energy associated with the *position* of the mass and is called the **potential energy function.** We have defined potential energy in terms of the work which must be done in moving a body to a certain position in the field. Potential energy can then be returned as work done; for example, a compressed spring or an object which is released near the Earth's surface can do work on any objects with which it subsequently interacts. Thus, potential energy is a measure of the ability of a body to do mechanical work by virtue of its *position*.

We also know that a moving object has the ability to do work on another object even if the moving object moves between two points of equal potential energy, as in a collision, for example. The first term in the defining expression for the total energy above represents an energy associated with the *motion* of the mass m and is called the **kinetic energy** defined as

$$K := \frac{1}{2}mv^2$$

The statement that, under the condition that the integral in Equation (3.11) can be evaluated as a definite integral, the total energy must be constant throughout the motion, that is

$$K + U = E = \text{constant} \tag{3.13}$$

is called the **principle of conservation of mechanical energy**. Equation (3.12) is sometimes called the **work-energy theorem.**

Forces for which the principle of conservation of mechanical energy is obeyed are called **conservative forces** and associated force fields are called **conservative fields.** In Section 3.11 we will encounter forces that are non-conservative. A more complete description of the difference between conservative and non-conservative systems requires a more general three dimensional treatment that we will meet in Chapter 5. For now, however, it is sufficient to say that drag forces such as air resistance or friction, forces that depend on the direction of motion or the speed of the body, are all non-conservative. The existence of non-conservative forces means that mechanical energy is lost; that is, it is dissipated out of the system and may appear as energy in other forms. The principle of conservation of mechanical energy states that, *in the absence of such dissipative forces*, the total energy of the system remains constant.

Note that, from their definitions, the units of mechanical work and energy must be the same. Since $(\Delta W) := F(\Delta x)$, the SI unit is N m which is also called the joule (J).

Determination of potential energy functions

Note that the potential energy function $U(x)$ is defined in terms of the *difference* in the value of the function between two points and hence an arbitrary constant can always be added. One is free, therefore, in any specific situation to choose arbitrarily a convenient point of zero potential energy. In the case of forces for which there are points where the force on the body is zero, it is conventional to choose such a point as a point of zero of potential energy; that is, $U(x_0) = 0$ if $F(x_0) = 0$. Once a zero has been defined, the value of $U(x)$ can be considered to be **the potential energy of the body** at the position x, that is

$$\int_{x_0}^{x} F(x')dx' = U(x_0) - U(x)$$

Thus,

$$U(x) = -\int_{x_0}^{x} F(x')dx' + U(x_0) = -\int_{x_0}^{x} F(x')dx' \tag{3.14}$$

Once again, we have used 'primed' variables inside the definite integrals, as we did in Section 2.5.

Note that the potential energy of a body is the work done *against* the applied force in moving a body from a point of zero potential energy to its present location, provided that the body is moved in a quasistatic manner, that is the force applied to the body is just sufficient to balance the applied force at all points.

We now proceed to derive expressions for the potential energy function in two important special cases.

Case 1: Potential energy function for a constant force

In the case in which a particle experiences a constant force ($F(x) = F_c$, say) there is no point at which the particle does not experience a force, so we must choose an arbitrary point ($x = x_0$) as the point of zero potential energy, that is $U(x_0) = 0$. Using Equation (3.14) we can determine the potential energy function in this case as follows

$$U(x) = -\int_{x_0}^{x} F(x')dx' = -\int_{x_0}^{x} F_c dx' = -F_c \int_{x_0}^{x} dx' = -F_c(x - x_0)$$

$$U(x) = -F_c(x - x_0)$$

It is often convenient to choose the point $x = 0$ as the point of zero potential energy (that is, $x_0 = 0$), in which case $U(x_0) = -F_c x$.

Case 2: Potential energy function for a helical spring

In this case (Figure 3.15) the force is zero at $x = 0$ so we can choose $U(0) = 0$. Thus the potential energy of a helical spring extended from equilibrium by a displacement x can be determined as follows

$$U(x) = -\int_{x_0}^{x} F(x')dx' = -\int_{x_0}^{x} (-kx')dx' = \frac{1}{2}kx^2$$

$$U(x) = \frac{1}{2}kx^2$$

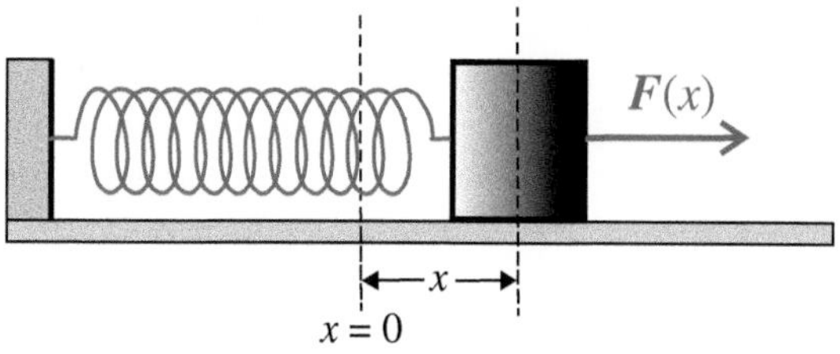

Figure 3.15. Determination of the potential energy function for a helical spring.

Derivation of the force from potential energy functions

If we know the appropriate potential energy function we can always determine the force on the body at any point. From the definition of the potential energy function, we can see that $\Delta U = -F(x)\Delta x$ and hence

$$F = -\lim_{\Delta x \to 0} \frac{\Delta U}{\Delta x} = -\frac{dU}{dx}$$

Thus the force can be derived from the potential energy; for example, for the two cases discussed above we get the expected results

Case 1: $$U(x) = -F_c(x - x_0) \quad \rightarrow \quad F(x) = -\frac{dU}{dx} = -\frac{d}{dx}\{-F_c(x - x_0)\} = +F_c\frac{d}{dx}(x - x_0) = F_c$$

Case 2: $$U(x) = \frac{1}{2}kx^2 \quad \rightarrow \quad F(x) = -\frac{dU}{dx} = -\frac{d}{dx}\left(\frac{1}{2}kx^2\right) = -\frac{1}{2}k(2x) = -kx$$

For problems based on the material presented in this section visit up.ucc.ie/3/ and follow the link to the problems.

3.8 Plots of potential energy functions

It can be very instructive to plot graphs of the variation of potential energy as a function of a position co-ordinate. Figure 3.16 gives a general example of such a plot in which the potential energy $U(x)$ varies as a function of x.

If the force is conservative, a plot of the total energy E of a body as a function x will produce a straight line which is parallel to the x-axis as shown in Figure 3.16. We also note that, when mechanical energy is conserved, the variation of K (the kinetic energy) as a function of x is represented by the difference between E and U at any value of x. Because kinetic energy, $K = \frac{1}{2}mv^2$, is always positive, for Equation (3.13) to be valid it follows that $E \geq U$ always. Referring to Figure 3.16, it is clear that this can only be true in the region between x_1 and x_2. The regions in which $x < x_1$ and $x > x_2$ are inaccessible, that is *forbidden* by the rules of classical mechanics.

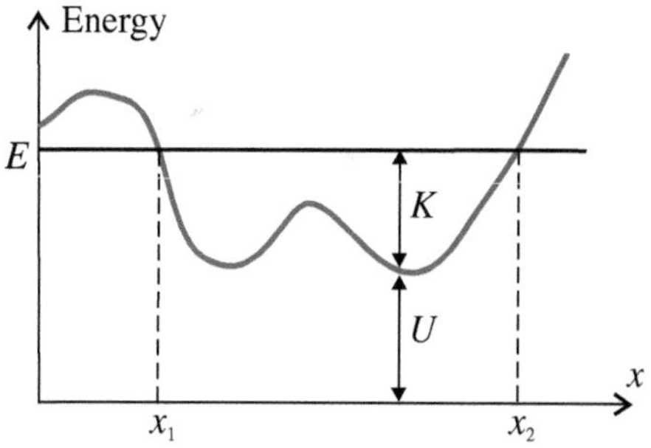

Figure 3.16. A plot of energy as a function of position (x). The potential energy function for the system is shown in blue. For a conservative system, the total mechanical energy (E) is constant and is represented by the horizontal line. At any point in the motion the total energy is divided between potential energy (U) and kinetic energy (K). In the case shown, the particle executes bounded motion between turning points $x = x_1$ and $x = x_2$.

It may be helpful to visualise Figure 3.16 as a plot of the behaviour of a small ball of mass which is free to roll without friction over hills and valleys. For a rolling ball, our practical experience enables us to understand this situation instinctively. If the ball has a fixed energy – provided for example by an initial push – it can climb only so far up a hill (to $x = x_2$ in Figure 3.16). The ball then rolls back until at $x = x_1$, it again encounters a point at which it has insufficient energy to proceed further and rolls back again, etc., etc. At the boundaries of the allowed region, $x = x_1$ and $x = x_2$, the kinetic energy of the ball falls to zero; it stops instantaneously as it changes direction. At the minimum of the potential plot, the kinetic energy is at a maximum – the ball travels at its greatest speed.

We showed above that force can be derived from a potential energy function using

$$F = -\frac{dU}{dx}$$

This equation tells us that the gradient of the $U(x)$ plot – the tangent to the curve at any point – gives the force on the body and thus, if the mass is known, the acceleration at that point.

We now illustrate the points made above by considering potential plots for some special cases.

A particle in a one-dimensional 'box'

In Figure 3.17 the potential energy of a particle which is confined to a limited region of space (between $x = 0$ and $x = a$, where the potential energy is defined to be zero) is plotted as a function of x.

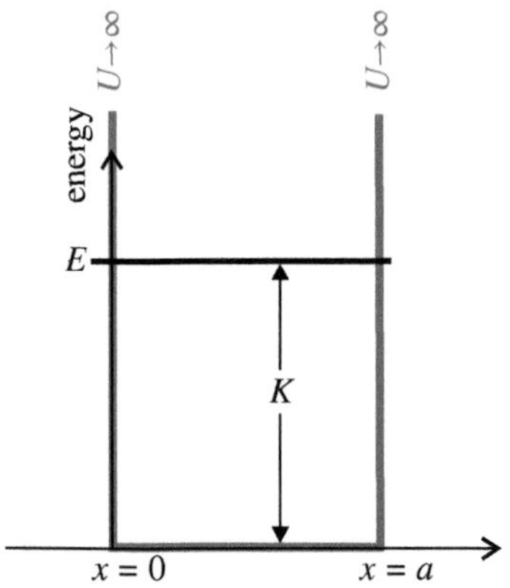

Figure 3.17. Energy plot for a particle confined in a one-dimensional 'box'.

Within the limited region of space (the 'box') the potential energy of the particle is always zero so that its energy E is entirely kinetic and constant. The particle therefore moves with a velocity of constant magnitude which changes direction each time the particle reaches a side of the box. The regions $x < 0$ and $x > a$ (the 'forbidden regions') are represented by taking the potential function to infinity at these values of x. The resulting potential (shown in blue in the figure) is described as an 'infinite square well'. The potential function, therefore, can be written

$$U(x) = \begin{cases} \infty & \text{for } x \leq 0 \text{ and } x \geq a \\ 0 & \text{for } 0 < x < a \end{cases}$$

Thus, within the box, the total energy is given by $E = K + U = K = \frac{1}{2}mv^2$ and hence the magnitude of the speed is $\sqrt{\frac{2K}{m}} = constant.$

A body moving under a constant force

The potential energy function has already been determined (*Case 1* of the previous section) to be

$$U(x) = -F_c(x - x_0)$$

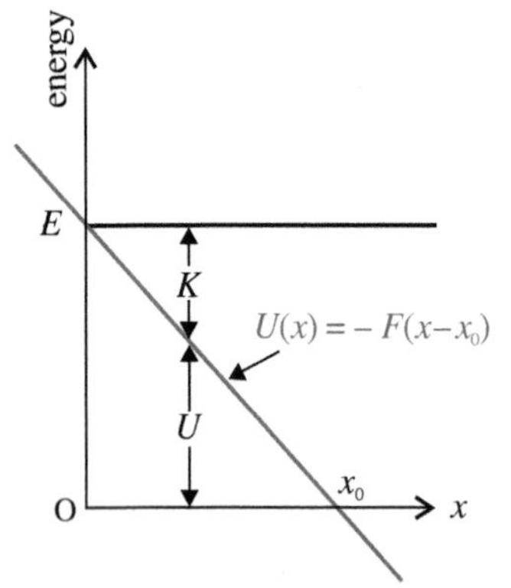

Figure 3.18. Energy plot for a mass moving under the influence of a constant force.

so that the plot is a straight line of slope $-F_c$ (recall Equation (2.1)) as shown in Figure 3.18.

The linear transfer of energy from the potential to the kinetic form with increasing x is clearly evident. The slope of the graph is negative and constant and since $F = -\frac{dU}{dx}$, the force on the mass, and hence its acceleration, is in the positive x direction and is constant – as expected for this case. In general, a straight line potential energy plot represents a uniform force in the region covered by the plot.

3.9 Power

The mechanical work done by a force in moving a body from one point to another does not depend on how quickly or slowly the process takes place. In many circumstances, however, it can be important to know the *rate* at which the work is being done. Let us consider again the situation illustrated in Figure 3.14 in which a body is moved from a point A to a point B under the influence of a force $F = F(x)$. Suppose that the body is at the point P at some instant t and at the point P′ at $t + \Delta t$ (Figure 3.19) and let the work done in moving the body from P to P′ (that is, in the time interval Δt) be ΔW. A physical quantity called **power**, which is a measure of the rate of doing work, is defined as

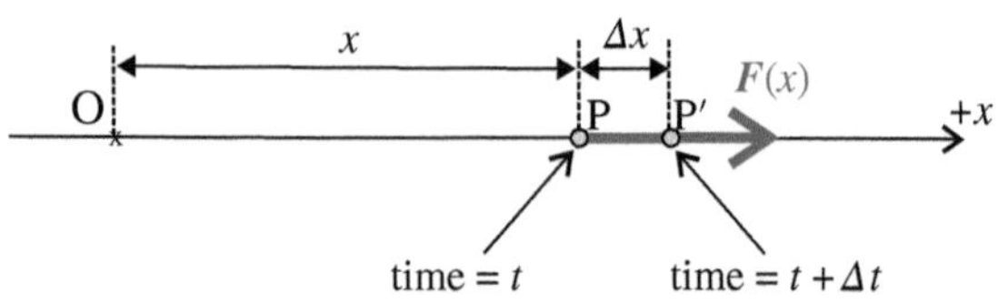

Figure 3.19. The rate at which work is done by a force is called the power generated by the force.

$$\boxed{P := \underset{\Delta t \to 0}{\text{limit}} \frac{\Delta W}{\Delta t} = \frac{dW}{dt}}$$

From the definition, the SI unit of power is J s^{-1}; 1 J s^{-1} = 1 watt = 1 W.

Suppose that it takes a time Δt for a force F to do the amount of work ΔW and that the corresponding change in displacement in this time interval is Δx. Recalling the definition of mechanical work, namely

$$\Delta W := F\Delta x$$

we see that the power generated by the force F is given by

$$P = \frac{dW}{dt} = \lim_{\Delta t \to 0} \frac{\Delta W}{\Delta t} = \lim_{\Delta t \to 0} \frac{F(\Delta x)}{\Delta t} = F \lim_{\Delta t \to 0} \frac{\Delta x}{\Delta t} = F\frac{dx}{dt} = Fv \qquad (3.15)$$

where v is the velocity of the body on which the force F is being exerted.

3.10 Energy in simple harmonic motion

As shown in Section 3.7 (*Case 2*), the potential energy function of a simple harmonic oscillator is given by $U(x) = \frac{1}{2}kx^2$. Note that, in general, k is not necessarily a spring constant; any simple harmonic oscillator can be described in this way. In general $k = m\omega^2$ (recall $\omega = \sqrt{\frac{k}{m}}$ from Section 3.6) and hence

$$U(x) = {}^1\!/\!_2 m\omega^2 x^2$$

The total mechanical energy, therefore, is

$$E = K + U = {}^1\!/\!_2\, mv^2 + {}^1\!/\!_2 kx^2$$

Equation (3.7d) relates displacement to velocity in simple harmonic motion, that is $v = \pm\omega\sqrt{A^2 - x^2}$, and hence we can write

$$E = {}^1\!/\!_2\, m\omega^2(A^2 - x^2) + {}^1\!/\!_2\, kx^2 = {}^1\!/\!_2\, kA^2 = {}^1\!/\!_2\, m\omega^2 A^2 \qquad (3.16)$$

Equation (3.16) tells us that if the total energy is constant – that is, if the system is conservative – the amplitude remains constant. We can thus interpret simple harmonic motion in energy terms as a motion in which there is a periodic exchange of energy between the kinetic and potential forms in such a way that their sum is constant.

When $x = 0$ (at the equilibrium position) the energy is entirely kinetic:

$$E = K + 0 = {}^1\!/\!_2\, kA^2 = {}^1\!/\!_2 m\omega^2 A^2$$

When $x = \pm A$ (at the end-points) the energy is entirely potential:

$$E = 0 + U = {}^1\!/\!_2\, kA^2 = {}^1\!/\!_2 m\omega^2 A^2$$

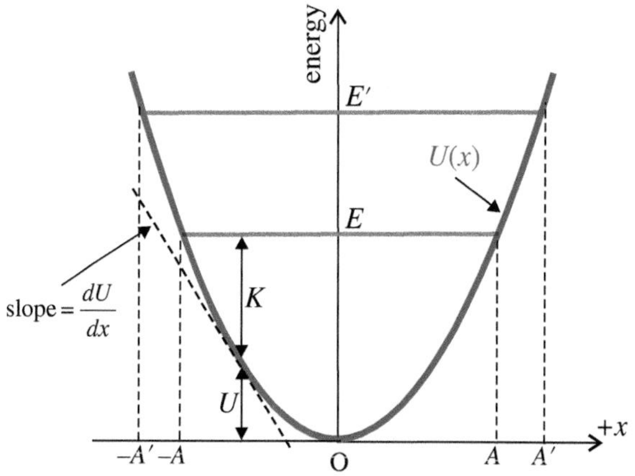

Figure 3.20. Energy plot of a simple harmonic oscillator.

In Figure 3.20 the potential energy of a simple harmonic oscillator is plotted as a function of displacement from equilibrium; since $U(x) \propto x^2$, this function is a parabola (Appendix A.10.2(iii)).

By plotting the total energy E on the graph – as before, for a conservative system, it has the same value for all x and is therefore a straight line parallel to the x-axis – we can see how, at any point in the motion, the kinetic and potential energies add to give E and also how they vary in accordance with the energy interpretation of simple harmonic motion given above. As indicated in Figure 3.20, increasing the total energy from E to E' increases the width of the region accessible to the motion; that is,the amplitude of the motion increases from A to A', in accordance with Equation (3.16).

Recalling that $F = -\frac{dU}{dx}$, we can also see how the slope of the graph of $U(x)$ at any point represents the variation of force with displacement which is characteristic of simple harmonic motion. Moving from left to right, the gradient decreases in magnitude as the equilibrium position O is approached and $-\frac{dU}{dx}$ changes direction from positive to negative as the object passes through the origin. This means that the magnitude of the force decreases as the origin is approached and changes from positive to negative as the displacement changes from negative to positive (Figure 3.21).

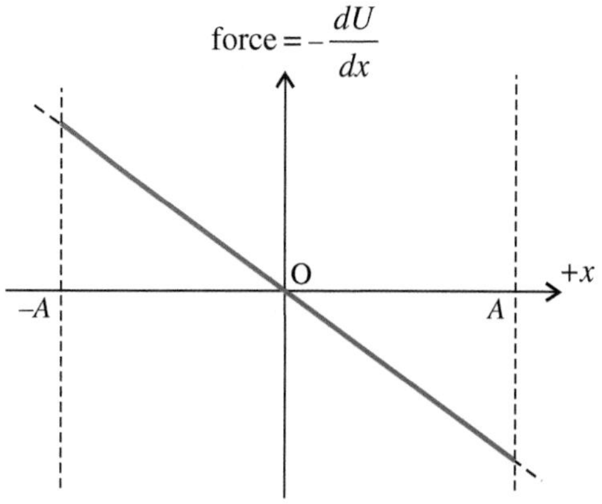

Figure 3.21. A plot of force as a function of displacement for a simple harmonic oscillator.

Worked Example 3.6 illustrates how quantities such as K, U and F may be determined in simple harmonic motion.

Study Worked Example 3.6

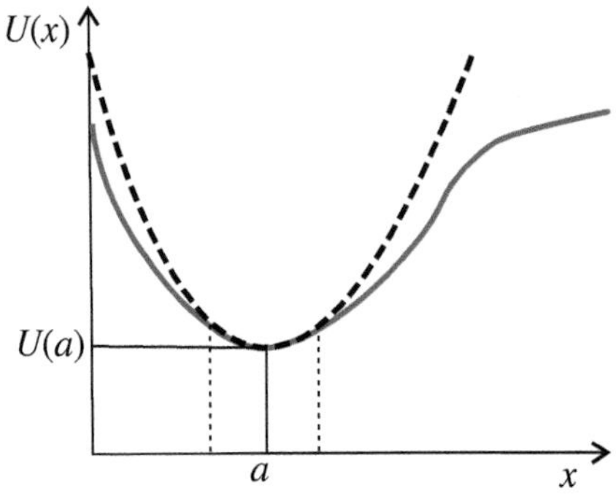

Figure 3.22. Any well behaved potential energy function can be approximated by a parabola (dashed curve) in the region close to the minimum of the function. Hence small amplitude oscillations about the minimum can be described approximately as simple harmonic motion.

One reason why simple harmonic motion plays such an important role in physics is that any well-behaved potential energy function can be approximated by a parabola near a minimum in the function. Consider, for example, the more complicated potential function represented by the blue curve in Figure 3.22. The region close to the minimum can be approximated to a parabola as long as the amplitudes of any oscillations are small.

This statement can be justified in mathematical terms using the Taylor series expansion (Appendix A.9.3) which states that any function, such as $U(x)$ can be written as a series of the following form:

$$U(x) = U(a) + \frac{(x-a)}{1!}\left[\frac{dU}{dx}\right]_{x=a} + \frac{(x-a)^2}{2!}\left[\frac{d^2U}{dx^2}\right]_{x=a} + \frac{(x-a)^3}{3!}\left[\frac{d^3U}{dx^3}\right]_{x=a} + \ldots$$

where $\left[\frac{dU}{dx}\right]_{x=a}$ represents the (constant) value of the first derivative of U at $x = a$, $\left[\frac{d^2U}{dx^2}\right]_{x=a}$ represents the value of the second derivative, at $x = a$ *etc*. If, as indicated in Figure 3.22, $U(a)$ corresponds to a minimum of the potential function, that is $\left[\frac{dU}{dx}\right]_{x=a} = 0$, the Taylor expansion becomes

$$U(x) = U(a) + \frac{(x-a)^2}{2!}\left[\frac{d^2U}{dx^2}\right]_{x=a} + \frac{(x-a)^3}{3!}\left[\frac{d^3U}{dx^3}\right]_{x=a} + \ldots$$

If we only consider the region in which $(x - a)$ is small (that is, small oscillations about $x = a$), the $(x - a)^3$ term and higher terms are negligible compared with the $(x - a)^2$ term and we can write

$$U(x) \approx U(a) + \frac{(x-a)^2}{2!}\left[\frac{d^2U}{dx^2}\right]_{x=a}$$

Thus $U(x) \approx U(a) + k(x - a)^2$, where $k = \frac{1}{2}\left[\frac{d^2U}{dx^2}\right]_{x=a} =$ *constant* and, as indicated by the dashed curve in Figure 3.22, the plot of the potential energy function is a parabola with its vertex at $(U = U(a), x = a)$. Thus, small oscillations of a particle about $x = a$ will be approximately simple harmonic.

We see, therefore, that simple harmonic motion solutions can often be applied to oscillations in complicated situations as long as the amplitudes of the oscillations are sufficiently small. The approximation of real potentials by harmonic potentials is used widely in all branches of physics.

For problems based on the material presented in this section visit up.ucc.ie/3/ and follow the link to the problems.

3.11 Dissipative forces: damped harmonic motion

In considering the energy aspects of simple harmonic motion in the previous section we assumed that the total energy of the oscillating system was conserved. In such circumstances the oscillations would continue indefinitely without any change of amplitude. Our everyday experience, however, is that real systems in the macroscopic world do not behave in this way; the amplitudes of real pendulums, for example, always decrease with time. This reduction of amplitude with time is called *damping* and arises from the existence of forces such as friction and air resistance, these so-called 'drag' forces being directed so as to oppose the motion. In such cases, the energy of the oscillator is not conserved but is instead transferred (dissipated) out of the system to the surroundings. We will see later in the book that this 'lost' mechanical energy can be transferred into other forms so that the total energy involved remains conserved.

Many dissipative forces of this type can be described by models which assume specific dependences of the magnitude of the drag force on the speed v of the body. To include such damping effects in the analysis of harmonic motion, therefore, we need to include an additional force $F_d = F_d(v)$ in the equation of motion. Each of the three models discussed below may be significant in different situations but it is important to be aware that damping of real oscillating systems may involve a combination of two or more of these effects.

In general, the energy lost during any cycle is equal to the work done in overcoming the friction force during that cycle. Thus, using Equations (3.16) and (3.11), we obtain for the energy lost over the nth cycle

$$\frac{1}{2}m\omega^2 A_n^2 - \frac{1}{2}m\omega^2 A_{n+1}^2 = \frac{1}{2}m\omega^2(A_n + A_{n+1})(A_n - A_{n+1}) = \int_{\text{cycle}} dW = -\int_{\text{cycle}} F_d dx$$

If the change in amplitude in a cycle is small compared to the amplitude itself, that is

$$A_n - A_{n+1} << A_n$$

we can assume, from the symmetry of the oscillations, that the energy loss is approximately the same over each of the four quarters of a cycle. Thus we can write

$$\frac{1}{2}m\omega^2(2A_n)(A_n - A_{n+1}) = -4\int_0^{A_n} F_d dx \tag{3.17}$$

MODEL A: *Constant drag ('dry friction')*. An example of this type of damping is illustrated in Figure 3.23. A block of mass M rests on a horizontal surface and is attached to a fixed support by a helical spring. If displaced from equilibrium and released the block would execute simple harmonic motion about the equilibrium position, provided there was no friction between the block and the surface. With friction, however, the block may be considered to experience an additional constant drag force (this particular model will be discussed in more detail in Section 4.6).

Because the direction of the frictional drag force reverses each time the block changes direction at an extremity of the motion, the resulting discontinuity means that a simple analytical solution for the variation of displacement with time cannot be found in this case. We can, however, calculate the loss of energy of the system in a single cycle. Since the friction force is constant, that is $F_d = -\alpha$ where α is a constant, we can use Equation (3.17) to determine the loss of energy per cycle.

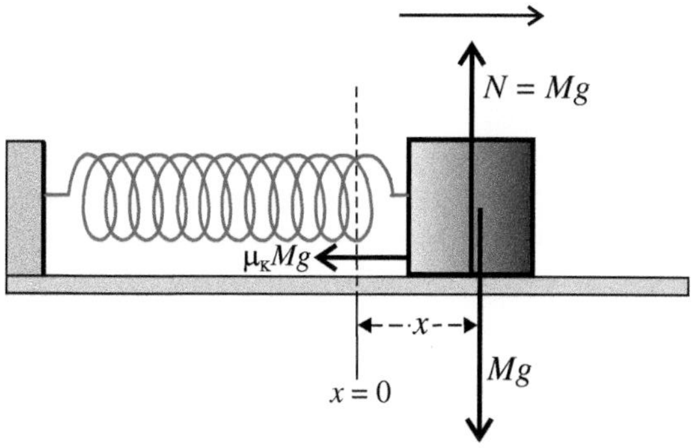

Figure 3.23. A block of mass M moving on a horizontal surface under the influence of a helical spring attached to a fixed support. A constant frictional force acts in the opposite direction to that of the motion of the block.

$$M\omega^2(A_n)(A_n - A_{n+1}) = -4\int_0^{A_n}(-\alpha)dx = 4\alpha A_n$$

$$\rightarrow \quad A_{n+1} = A_n - \frac{4\alpha}{M\omega^2} \tag{3.18A}$$

which shows that the amplitude decreases linearly in this case, assuming that no other dissipative forces are significant. Figure 3.24 shows experimental data for this kind of damped system.

MODEL B: *Linear aerodynamic drag*; that is, $F_d = -bv$, where v is the velocity of the body and b is a positive constant called the damping (or drag) coefficient. This type of drag arises, for example, in the case of bodies with streamline shapes moving through a fluid where the velocities involved are not too large. We will see below that, in this important case, the equation of motion can be solved reasonably easily when a damping term of this form is included.

To see how the amplitude decreases due to this kind of damping, we can follow the procedure used for model A above but with $F_d = -bv$ in this case. Thus, for a body of mass m, Equation (3.17) gives

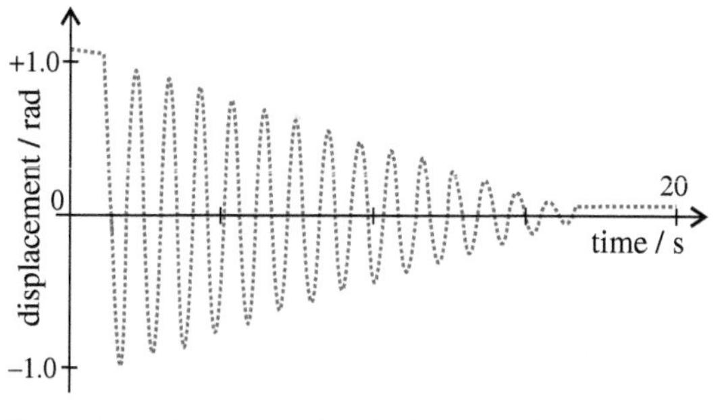

Figure 3.24. Experimental data for damped oscillation with dry friction.

$$m\omega^2 A_n(A_n - A_{n+1}) = -4\int_0^{A_n}(-bv)dx = 4b\int_0^{T/4} v\frac{dx}{dt}dt = 4b\int_0^{T/4} v^2 dt$$

If, as we have assumed, the change in amplitude per cycle is small, it is reasonable to suppose that the motion of the mass does not differ significantly from the undamped behaviour during a single cycle, that is $x = A_n \sin \omega t$ and hence $v = \omega A_n \cos \omega t$. In this case, therefore,

$$m\omega^2 A_n(A_n - A_{n+1}) = 4b\omega^2 A_n^2\int_0^{T/4}\cos^2(\omega t)dt = 4b\omega A_n^2\int_0^{T/4}\cos^2(\omega t)d(\omega t) = 4b\omega A_n^2\int_0^{\pi/2}\cos^2\vartheta d\vartheta$$

Evaluating the integral (Appendix A, Table A.2), we find

$$A_n - A_{n+1} = \frac{4b}{m\omega}A_n\left[\frac{1}{2}(\vartheta + \sin\vartheta\cos\vartheta)\right]_0^{\frac{\pi}{2}} = \frac{\pi b}{m\omega}A_n$$

$$\rightarrow \quad A_{n+1} = A_n - \frac{\pi b}{m\omega}A_n \tag{3.18B}$$

We will see below that this result implies that the amplitude falls off exponentially.

MODEL C: *Quadratic aerodynamic drag*. When a body moves through a fluid at larger speeds, turbulence may arise (see Section 10.7). In such circumstances, the system is better described by quadratic velocity dependence; that is, $F_d = -\beta v^2$, where β is a constant. We can use the same approach as above to determine the way in which the damping affects the amplitude. In this case

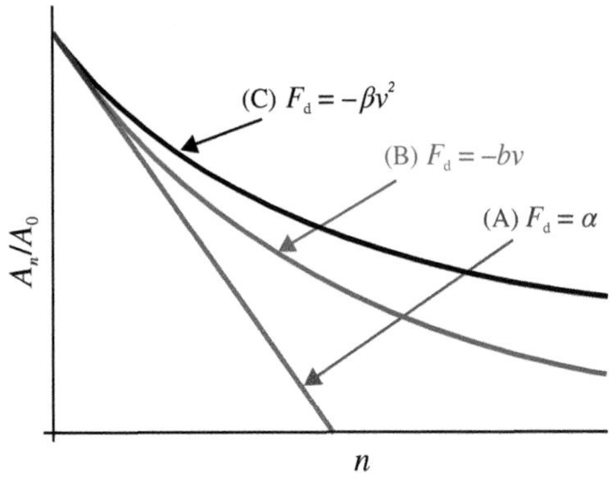

Figure 3.25. Decay of the amplitude for the three models of damped oscillations discussed in the text.

$$m\omega^2 A_n(A_n - A_{n+1}) = -4\int_0^{A_n}(-\beta v^2)dx = 4\beta\int_0^{T/4} v^3 dt = 4\beta\omega^3 A_n^3\int_0^{T/4}\cos^3(\omega t)d(\omega t)$$

$$= 4\beta\omega^2 A_n^3\int_0^{\pi/2}\cos^3\vartheta d\vartheta$$

Evaluating this integral (Appendix A, Table A.2), we get

$$A_n - A_{n+1} = \frac{4\beta}{m}A_n^2\left[\frac{1}{3}(3\sin\vartheta - \sin^3\vartheta)\right]_0^{\frac{\pi}{2}} = \frac{8\beta}{3m}A_n^2$$

$$\rightarrow \quad A_{n+1} = A_n - \frac{8\beta}{3m}A_n^2 \tag{3.18C}$$

Equations (3.18A) – (3.18C) can be used in a simple spreadsheet analysis to compare the three different types of damping discussed in this section. Figure 3.25 shows an example of the results of such a calculation.

Harmonic oscillator with linear drag

Of the three models discussed above, model B is particularly important as it has applications in other branches of physics, such as electromagnetism. Accordingly, we will discuss this model in some more detail before proceeding.

For the case of a mass m attached to a spring, of spring constant k, oscillating in one dimension but also experiencing a damping force which is directly proportional to the velocity ($F_d = -bv$, model B above), Newton's second law gives

$$F = -kx - bv = ma$$

or

$$m\frac{d^2x}{dt^2} + b\frac{dx}{dt} + kx = 0 \tag{3.19}$$

There are standard mathematical methods for solving differential equations such as (3.19), and the reader is encouraged to use such techniques if familiar with them. As shown in the *Understanding Physics* website at up.ucc.ie/3/11/1/, the equation can also be solved by guessing a trial solution, the method which was used to solve the equation of simple harmonic motion in Section 3.6.

The solution of Equation (3.19), giving the displacement at time t, is

$$x = Ae^{-\frac{b}{2m}t}\sin(\omega t + \varphi) \tag{3.20}$$

The reader can confirm the validity of this result by substitution into Equation (3.19); ω is the angular frequency of the oscillations, given by

$$\omega = \sqrt{\omega_0^2 - \frac{b^2}{4m^2}} \tag{3.21}$$

where $\omega_0 = \sqrt{\frac{k}{m}}$ is the angular frequency in the absence of damping. Note that $b = 0$ gives Equation (3.7a), as expected for no damping.

Thus the manner in which the displacement of a damped harmonic oscillator varies with time depends on the relative values of $\frac{k}{m}$ and $\frac{b^2}{4m^2}$. Three cases may be identified

Case 1: $\frac{b^2}{4m^2} < \frac{k}{m}$. The angular frequency $\omega = \sqrt{\omega_0^2 - \frac{b^2}{4m^2}}$ is real and the system oscillates. Figure 3.26(a) shows the displacement as a function of time of a damped system which has been displaced from equilibrium and released from rest at $t = 0$. The simple harmonic oscillations are modulated by an envelope — dashed curves in Figure 3.26(a) — which decays exponentially, that is according to the relationship $x \propto \exp\left(-\frac{b}{2m}t\right)$. In this case the system is said to be *underdamped*. It follows from Equation (3.21) that the period of the motion $\left(T = \frac{2\pi}{\omega}\right)$ is also affected slightly by the damping, being somewhat greater than if the motion was undamped.

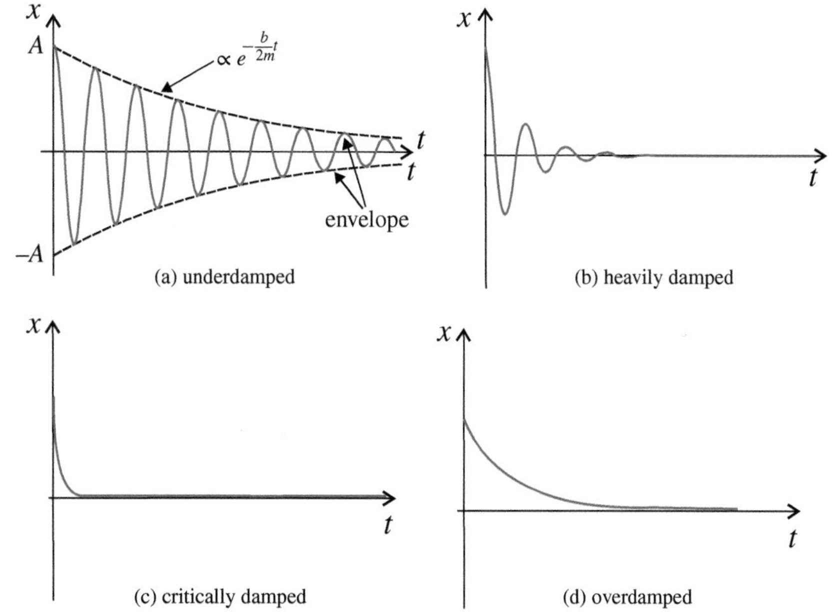

Figure 3.26. Displacement—time plots for the different cases of damped oscillations discussed in the text.

For larger values of the damping coefficient b the envelope decays more rapidly and the period increases. As the value of b approaches $\sqrt{4mk}$ the system becomes heavily damped as illustrated in Figure 3.26(b).

Case 2: $\frac{b^2}{4m^2} = \frac{k}{m}$, known as *critical damping*. In this case, from Equation (3.19), $\omega = 0$, the system does not oscillate and the displacement decreases exponentially as illustrated in Figure 3.26 (c).

Case 3: $\frac{b^2}{4m^2} > \frac{k}{m}$, known as *overdamping*. The value of $\omega = \sqrt{\omega_0^2 - \frac{b^2}{4m^2}}$ is the square root of a negative number and hence is imaginary (Appendix A.8)[2]. The system does not oscillate in this case; if displaced from equilibrium the body simply returns slowly to the equilibrium position (Figure 3.26(d)).

The underdamped case (Case 1) is interpreted in energy terms in Figure 3.27. As the total mechanical energy of the oscillator ($\propto |e^{-\alpha t}A|^2$, where $\alpha = \frac{b}{2m}$) decreases exponentially with time, the allowed region and thus the amplitude of the motion decreases accordingly.

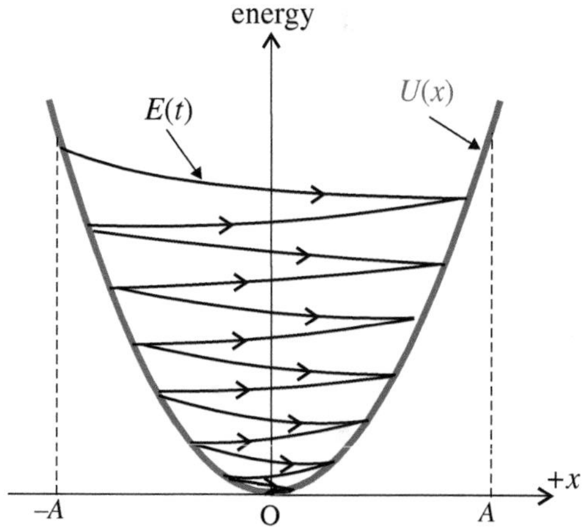

Figure 3.27. Energy plot for an underdamped harmonic oscillator: the total mechanical energy of the oscillator system, $E(t)$, is not conserved but decreases exponentially with time.

Study Worked Example 3.7

For problems based on the material presented in this section visit up.ucc.ie/3/ *and follow the link to the problems.*

3.12 Forced oscillations

In the case of the damped oscillator (discussed in the previous section) the system is disturbed initially but is allowed subsequently to evolve without further interference. In many physical systems, however, a time varying *external* force is applied continuously to the oscillator. In the *Understanding Physics* website at up.ucc.ie/3/12/ the important case of a one-dimensional damped oscillator, to which an external driving force of the form $F = F_0 \cos(\omega t)$ is also applied, is considered for the case of linear drag (model B of Section 3.11).

[2]The sine of an imaginary number is called the hyperbolic sine and is denoted by sinh; that is, $\sin(i\theta) = \sinh(\theta)$ (Appendix A.8.vi).

In general, the applied frequency ω is different from the natural undamped frequency of the oscillator $\left(\omega_0 = \sqrt{\frac{k}{m}}\right)$. The amplitude of the oscillation is found to increase, in some cases sharply, as the driving frequency approaches the natural frequency, a phenomenon known as **resonance**.

3.13 Non-linear dynamics: chaos

The equations which describe undamped and damped simple harmonic motion, respectively Equations (3.5) and (3.19) are *linear* differential equations in that they do not contain terms such as $x^2, x^3, \left(\frac{dx}{dt}\right)^2, \left(\frac{dx}{dt}\right)^3$, $\sin x, \cos x$, etc. Doubling the initial displacement, for example, will double the amplitude of the subsequent oscillations of such linear systems. In real physical oscillating systems, however, linear behaviour is normally observed only when the amplitude is kept small. For example, the equation of motion for an *anharmonic* oscillator, where the force depends on higher powers of x, for example $F(x) = -kx + \varepsilon_1 x^2 + \varepsilon_2 x^3$ where ε_1 and ε_2 are constants, is a **non-linear** differential equation.

In the *Understanding Physics* website at up.ucc.ie/3/13/, a system which exhibits the main features of non-linear behaviour is considered. It is noted that, in such a case, small differences in initial conditions can produce very different evolutionary paths, a phenomenon known as **dynamical chaos**.

3.14 Phase space representation of dynamical systems

A general technique for characterizing dynamical systems, and which is particularly useful in the case of non-linear systems, is called **phase space representation.** In the case of motion in one-dimension, the instantaneous state of a system is represented by point (x,p) on a p versus x diagram. The time evolution of the system develops as a curve on such a diagram (the 'phase space trajectory') starting from some initial conditions. In the *Understanding Physics* website at up.ucc.ie/3/14/ the technique is applied to a variety of oscillating systems including undamped, damped, forced and non-linear cases.

Further reading

Further introductory material on oscillating systems may be found in *Vibrations and Waves* by King (see BIBLIOGRAPHY for details).

WORKED EXAMPLES

Worked Example 3.1: A body of mass 20 kg is moving at a velocity of 12 m s^{-1} when it starts to experience a constant force of 60 N applied in the opposite direction to its direction of motion. How long after this force is first applied will it take before the body comes to rest?

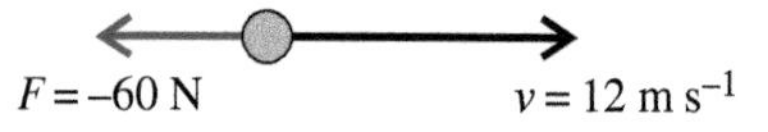

Figure 3.28. Worked Example 3.1.

Applying Newton's second law to the mass m in Figure 3.28: $a = \frac{F}{m} = \frac{-60\ \text{N}}{20\ \text{kg}} = -3.0\ \text{m s}^{-2}$.

Since acceleration is constant, we can determine the time taken for the body to come to rest by using Equation (2.18), namely

$$v = v_0 + At.$$

Letting $t = \tau$ when $v = 0$, we obtain $\quad 0 = 12\ \text{m s}^{-1} + (-3.0\ \text{m s}^{-2})\,\tau \quad \Rightarrow \quad \tau = 4.0\ \text{s}$

Worked Example 3.2: If the force applied to the body in Worked Example 3.1 is not constant but increases in magnitude from zero at a constant rate of 40 N s^{-1}, how long will it take for the body to come to rest in this case? Draw a sketch to show how the velocity of the body varies with time.

In this case $F = -\beta t$ where $\beta = 40\ \text{N s}^{-1}$

Newton's second law: $\quad a = \frac{F}{m} = \frac{-\beta t}{m} = -\frac{\beta}{m}t \quad \Rightarrow \quad \frac{dv}{dt} = -\frac{\beta}{m}t$

Integrating $\quad v = -\frac{\beta}{2m}t^2 + constant = -\frac{\beta}{2m}t^2 + v_0$ (where $v = v_0 = 12\ \text{m s}^{-1}$ at $t = 0$)

Thus

$$v = v_0 - \frac{\beta}{2m}t^2$$

Letting $t = \tau$ when $v = 0$, we obtain

$$0 = v_0 - \frac{\beta}{2m}\tau^2 \quad \rightarrow \quad \tau^2 = \frac{2mv_0}{\beta}$$

$$\rightarrow \quad \tau = \sqrt{\frac{2mv_0}{\beta}} = \sqrt{\frac{2(20\text{ kg})(12\text{ m s}^{-1})}{40\text{ N s}^{-1}}} = 2\sqrt{3}\text{ s} = 3.5\text{ s}$$

The required velocity–time plot is shown in Figure 3.29 and compared with the corresponding force–time plot.

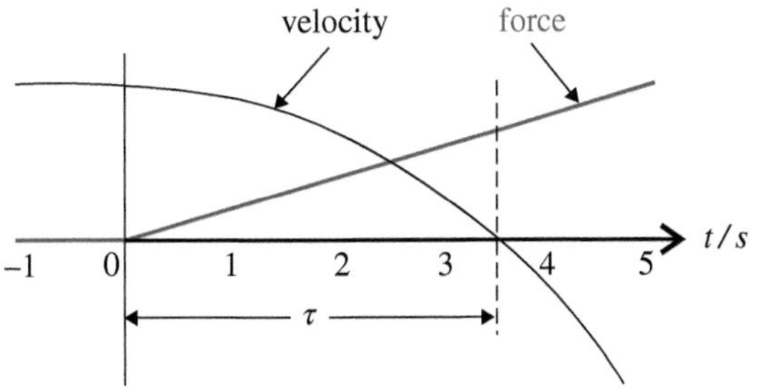

Figure 3.29. Worked Example 3.2.

Worked Example 3.3: An object of mass 150 g drops from a hovering helicopter and falls vertically downwards. As well as its weight the object experiences an upward force F_{res} due to air resistance which depends on the velocity v of the object in such a way that that $F_{res} = bv$, where $b = 0.024\text{ N (m s}^{-1})^{-1}$. What is the maximum velocity which the object can reach?

The net force downwards on the object is $W - F_{res} = Mg - bv = Ma$

As the object falls, v increases until the net force on the object falls to zero. The acceleration of the object is then zero so that its maximum velocity ($v = V_m$) has been reached.

At that point $a = 0 \rightarrow Mg - bV_m = 0 \rightarrow V_m = \frac{Mg}{b} = \frac{0.15 \times 9.8}{0.024} = 61\text{ m s}^{-1}$

Worked Example 3.4: (a) Two helical springs with spring constants k_1 and k_2, respectively, are connected end-to-end with their axes coinciding, as shown in Figure 3.30. What is the spring constant of the single spring which is equivalent to the combination of the two springs connected in this way ('in series')?

(b) If the two helical springs in (a) were to be connected 'in parallel', as indicated in Figure 3.31, what is the spring constant of the single spring which is equivalent to the combination?

(a) Springs in series

Consider the two joined springs as being attached to a fixed support as in the figure. Suppose that the combination is held extended by a fixed force F and let the corresponding extension produced be x. Thus, $F = -kx$, where k is the spring constant of the equivalent single spring.

Let the extending forces on the individual springs be F_1 and F_2, respectively. Applying Hooke's law (Section 3.5) we get

$$F_1 = -k_1x_1 \quad \text{and} \quad F_2 = -k_2x_2$$

Since each part of the system is in equilibrium, each spring must experience the same extending force F, that is

$$F = F_1 = F_2 \quad \text{or} \quad -kx = -k_1x_1 \quad \text{and} \quad -kx = k_2x_2 \quad \rightarrow \quad x_1 = \frac{k}{k_1}x \quad \text{and} \quad x_2 = \frac{k}{k_2}x$$

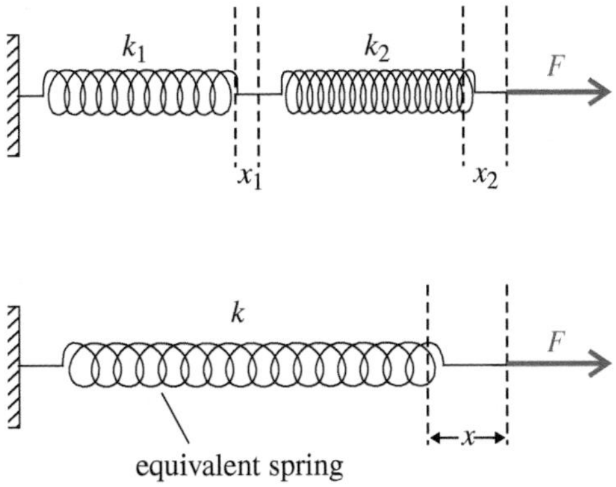

Figure 3.30. Worked Example 3.4.

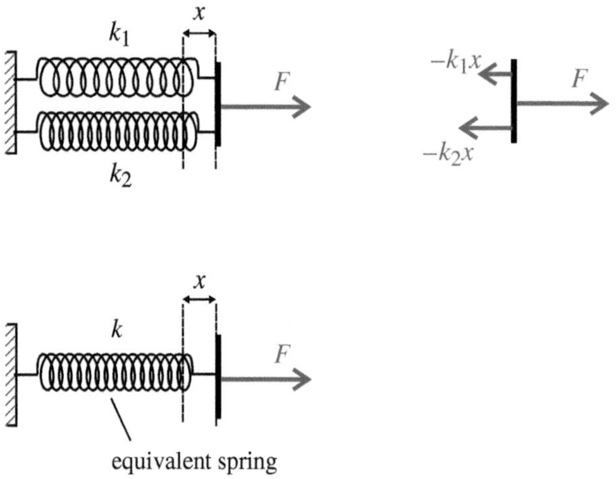

Figure 3.31. Worked Example 3.4.

The total extension x must be the sum of the individual extension of each spring, that is

$$x = x_1 + x_2 \quad \rightarrow \quad \frac{k}{k_1}x + \frac{k}{k_2}x = x \quad \rightarrow \quad \frac{1}{k_1} + \frac{1}{k_2} = \frac{1}{k}$$

The result may be extended easily to n helical springs connected in series.

$$\rightarrow \quad \frac{1}{k_1} + \frac{1}{k_2} + \frac{1}{k_3} \cdots + \frac{1}{k_n} = \frac{1}{k}$$

(b) Springs in parallel

In this case it is clear from the Figure 3.31 that the two restoring forces acting together balance the applied force F, that is

$$F = F_1 + F_2 = -k_1x - k_2x = -(k_1 + k_2)x$$

Treating the combination as a single equivalent spring we have $F = kx$ and hence, in this case,

$$k = k_1 + k_2$$

Again, the result may be extended easily to n helical springs connected in parallel.

$$\rightarrow \quad k = k_1 + k_2 + k_3 + \ldots \ldots + k_n$$

Worked Example 3.5: A point on the diaphragm of a loudspeaker oscillates with simple harmonic motion at a frequency of 512 Hz with an amplitude of 0.45 mm. What are (a) the maximum acceleration, (b) the maximum velocity, (c) the displacement and the velocity 1.2 ms after the point has passed through the equilibrium position and (d) the displacement and the velocity 3.0 ms after it has passed through the equilibrium position?

(a) $a_{max} = -\omega^2 A = 4\pi^2 f^2 A = 4\pi^2 (512\ \text{Hz})^2 (0.45 \times 10^{-3}\text{m}) = 4.66 \times 10^3\ \text{m s}^{-2}$

(b) $v_{max} = \omega A = 2\pi f A = 2\pi(512\ \text{Hz})(0.45 \times 10^{-3}\ \text{m}) = 1.45\ \text{m s}^{-1}$

(c) Taking $t = 0$ when $x = 0$ gives $\phi = 0$ and Equations (3.7a) and (3.7b) can be written

$$x = A\sin(\omega t) = A\sin(2\pi f t) = (0.45\ \text{mm})\sin[2\pi(512\ \text{Hz})(1.2 \times 10^{-3}\text{s})] = -0.30\ \text{mm}$$

$$v = \omega A\cos(\omega t) = 2\pi f A\cos(2\pi f t) = 2\pi(512\ \text{Hz})(0.45 \times 10^{-3}\text{m})\cos[2\pi(512\ \text{Hz})(1.2 \times 10^{-3}\ \text{s})] = -1.09\text{m s}^{-1}$$

(d) $x = A\sin(\omega t) = A\sin(2\pi f t) = (0.45\ \text{mm})\sin[2\pi(512\ \text{Hz})(3.0 \times 10^{-3}\ \text{s})] = -0.10\ \text{mm}$

$$v = \omega A\cos(\omega t) = 2\pi f A\cos(2\pi f t) = 2\pi(512\ \text{Hz})(0.45 \times 10^{-3}\ \text{m}))\cos[2\pi(512\ \text{Hz})(3.0 \times 10^{-3}\text{s})] = -1.41\ \text{m s}^{-1}$$

Worked Example 3.6: A mass of 1.5 kg is suspended from a vertical helical spring of spring constant 25 N m^{-1}. The mass is displaced by 150 mm from its equilibrium position and released. Derive expressions giving (a) the displacement from equilibrium and (b) the velocity of the mass as functions of time after release. Calculate (c) the maximum kinetic energy of the mass and (d) the kinetic and potential energies of the mass when it is 90 mm from the equilibrium position.

Draw a graph of the potential energy as a function of displacement from equilibrium and use the graph to confirm your results for the kinetic and potential energies in (d) above. Use the gradient of the potential energy at 90 mm to make a graphical estimate of the force on the mass at that point.

As noted in Section 3.6, if we measure time from the instant the mass is at the extremity of its motion, that is taking the release time as $t = 0$, $\phi = \frac{\pi}{2}$, and Equations (3.7a) and (3.7b) may be written

$$x = A\cos\omega t \quad \text{and} \quad v = \frac{dx}{dt} = -A\omega\sin\omega t.$$

(we need to know A and ω)

At $t = 0, x = A = 0.15$ m. $\omega = \sqrt{\frac{k}{m}} = \sqrt{\frac{25}{1.5}} = 4.08$ rad s^{-1}

Thus (a) $x = A\cos\omega t = (0.15 \text{ m})\cos(4.08t)$

(b) $v = \frac{dx}{dt} = -A\omega\sin\omega t = -(0.15)(4.08)\sin(4.08t) = -(0.61 \text{ m})\sin(4.1t)$

The maximum velocity and maximum kinetic energy occur at $t = 0 \rightarrow v_{max} = A\omega = -0.61$ m s^{-1}

(c) Maximum kinetic energy $= \frac{1}{2}mv_{max}^2 = \frac{1}{2}(1.5)(-0.61)^2 = 0.28 \text{ J} = E =$ total energy (3.22)

(d) At $x = 0.09$ m, $\cos\omega t = \frac{x}{A} \rightarrow \omega t = 0.93$ rad

$$v = -A\omega\sin\omega t = -(0.15)(4.08)\sin(0.93) = 0.49 \text{ m s}^{-1}$$

$\rightarrow$ kinetic energy $= \frac{1}{2}mv^2 = 0.18$ J and potential energy $= \frac{1}{2}kx^2 = 0.10$ J

[Note that, at $x = 0.09$ m, kinetic energy + potential energy $= 0.18 + 0.10 = 0.28$ J = total energy]

From the graph of $U(x) = \frac{1}{2}kx^2$ against x (Figure 3.32): At $x = 0.09$ m, $K \approx 0.18$ J and $U \approx 0.10$ J

$$\rightarrow \quad F = -\frac{dU}{dx} = -\frac{0.1}{0.090 - 0.045} = -2.2 \text{ N}$$

From theory: $F = -m\omega^2 x = -1.5 \times \frac{25}{1.5} \times 0.09 = -2.25$ N

(in reasonable agreement considering the accuracy of the graphical method)

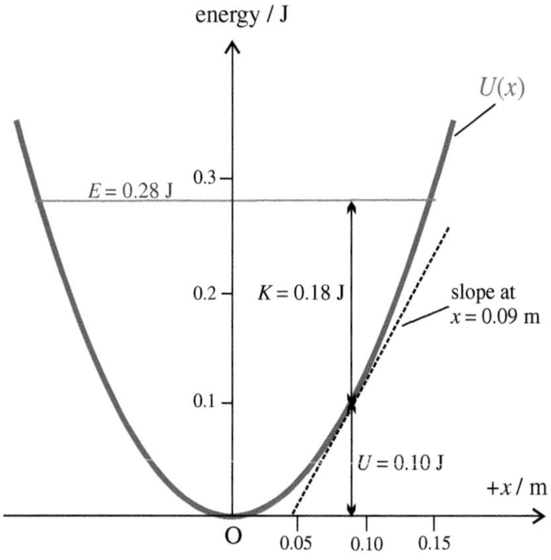

Figure 3.32. Worked Example 3.6.

Worked Example 3.7: A simple pendulum executes small oscillations in a fluid. The damping is such that the amplitude of the oscillations decreases by to 91% of the initial amplitude after one period. What percentage error would one have made by assuming that the period was the same as that of an undamped pendulum of the same length.

Equation (3.20), $x = A_0 e^{-\alpha t}\sin(\omega t + \phi)$ where $\omega = \sqrt{\omega_0^2 - \alpha^2} = \frac{2\pi}{T}$

At $t = 0$, the (initial) amplitude is A_0 and, at $t = T$, the amplitude is $A_0 e^{-\alpha T}$

so that the amplitude ratio, $\frac{A_T}{A_0} = f = e^{-\alpha T}(= 0.91) \rightarrow \alpha T = \ln f (= \ln(0.91) = -0.0943)$

$$\rightarrow \alpha^2 T^2 = (\ln f)^2 \rightarrow \alpha^2 \frac{4\pi^2}{\omega_0^2 - \alpha^2} = (\ln f)^2 \rightarrow \frac{\omega_0^2}{\alpha^2} = \frac{4\pi^2}{(\ln f)^2} + 1$$

But $\frac{4\pi^2}{(\ln f)^2} = 4.4 \times 10^3 \gg 1$ and hence $\frac{\omega_0}{\alpha} = \frac{2\pi}{|\ln f|} \rightarrow \alpha = \frac{|\ln f|}{2\pi}\omega_0 = 0.015\,\omega_0$

$$\rightarrow T = \frac{2\pi}{\sqrt{\omega_0^2 - \alpha^2}} = \frac{2\pi}{\omega_0}\frac{1}{\sqrt{1 - (0.015)^2}} = 1.00011 T_0$$

$$\rightarrow \frac{\Delta T}{T_0} = \frac{T - T_0}{T_0} = 0.00011 \sim 0.01\%.$$

PROBLEMS

For problems based on the material covered this chapter visit up.ucc.ie/3/ and follow the link to the problems.

4

Motion in two and three dimensions

AIMS

- to show how, in two and three dimensions, physical quantities can be represented by mathematical entities called vectors
- to rewrite the laws of dynamics in vector form
- to study how the laws of dynamics may be applied to bodies which are constrained to move on specific paths in two and three dimensions
- to describe how the effects of friction may be included in the analysis of dynamical problems
- to study the motion of bodies which are moving on circular paths

4.1 Vector physical quantities

The material universe is a three-dimensional world. In our investigation of the laws of motion in Chapter 3, however, we considered only one-dimensional motion, that is situations in which a body moves on a straight line and in which all forces applied to the body are directed along this line of motion. If a force is applied to a body in a direction other than the direction of motion the body will no longer continue to move along this line. In general, the body will travel on some path in three-dimensional space, the detail of the trajectory depending on the magnitude and direction of the applied force at every instant. Equation (3.3) as it stands is not sufficiently general to deal with such situations, for example the motion of a pendulum bob (Figure 4.1) or the motion of a planet around the Sun (Figure 4.2). Newton's second law needs to be generalised from the simple one-dimensional form discussed in Chapter 3.

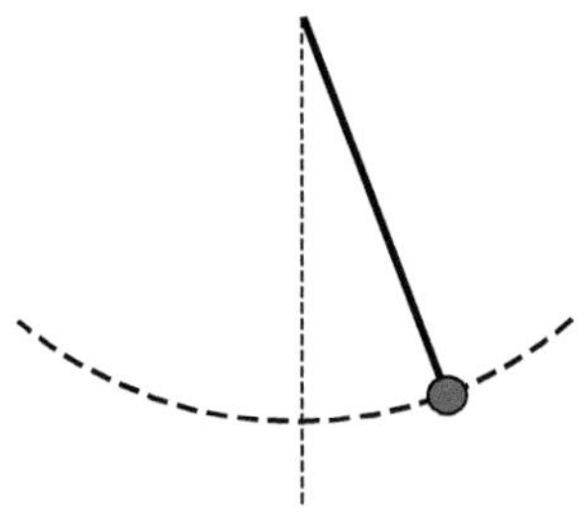

Figure 4.1. A pendulum comprising a mass attached to the end of a string; the mass can move on a path such that the distance from the fixed end of the string remains constant.

A similar problem arises if two or more forces act on a body simultaneously, for example when a number of tugs are manoeuvring a large ship (Figure 4.3). Even if the body is at rest initially, there is no way that we can deduce, from the techniques which we have acquired so far, the direction of motion let alone the magnitude of the acceleration resulting from a number of forces acting on a body in different directions. In other words, we do not yet have a rule which enables us to describe the effect of the superposition of forces which are not collinear (that is, not acting along the same straight line).

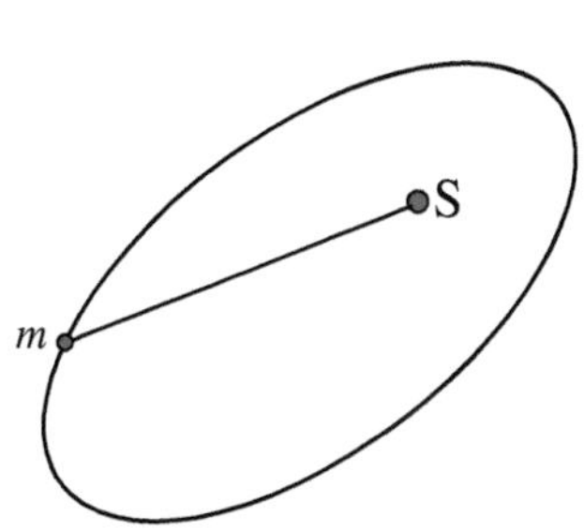

Figure 4.2. A planet of mass m in orbit around the Sun (this type of motion will be studied later in Sections 5.12 and 5.13)

To be able to analyse dynamical systems in two or three dimensions, therefore, we need a mathematical entity which can represent simultaneously both a magnitude and a direction. Mathematical elements called **vectors** turn out to be ideal for this purpose. Extensive use will be made, throughout the rest of this book, of the vector techniques introduced in this chapter.

Let us first consider an object which is free to move anywhere in a general plane such as the plane of this page. An example of such *two-dimensional* motion is illustrated in Figure 4.4. The motion is no longer confined to a straight track as it was in Figure 2.7 so that we can no longer use an algebraic representation to specify the direction of a displacement, an algebraic representation being limited to situations in which only two directions (positive or negative) are available. In two and three dimensions there is an infinite number of different possible displacement directions so that we must use other means, for example compass points, to specify direction. Thus the displacement OP in Figure 4.4 can be specified uniquely as 10 m at 40° west of south while the displacement OQ is 10 m due south. For two displacements to be equal they must have the same magnitude *and* be in the same direction; the two displacements of 10 m (OP and OQ) in Figure 4.4 take us to different places from O — they are clearly different displacements.

Understanding Physics, Third Edition. Michael Mansfield and Colm O'Sullivan.
© 2020 John Wiley & Sons Ltd. Published 2020 by John Wiley & Sons Ltd.

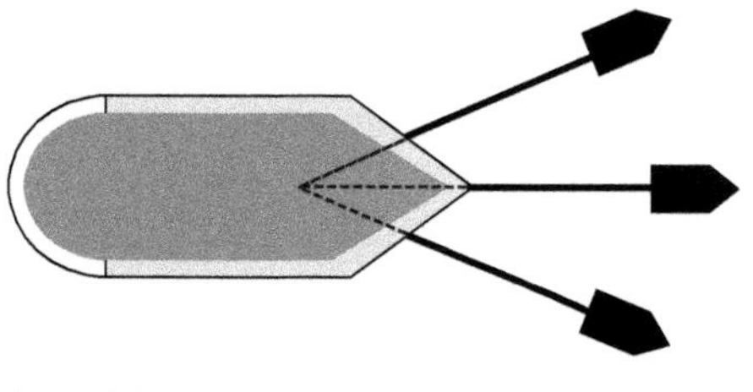

Figure 4.3. Three tugs manoeuvring a ship: each tug applies a different force in a different direction.

Since, in general, displacements cannot be simply added and subtracted arithmetically, different rules must be used for their addition and subtraction. As illustrated in Figure 4.5, a displacement of 3 m due east, followed by a displacement of 4 m due south does not take us 7 m from our starting point. The net effect of the two successive displacements in Figure 4.5 is in fact a displacement of 5 m at an angle of $\tan^{-1}\left(\frac{4}{3}\right) = 53°$ south of east. The 'net effect' of two successive displacements, in this sense, can be considered as the **sum** of the two displacements, thereby defining a rule for addition of displacements.

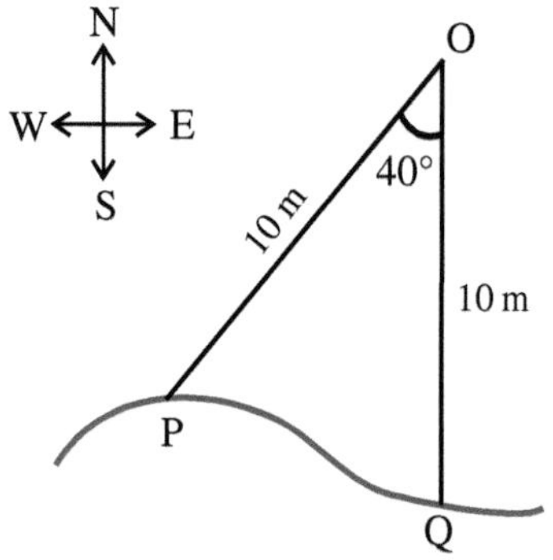

Figure 4.4. An example of general motion in two dimensions (along the blue path). The points P and Q are at the same distance from O but have different displacements relative to O.

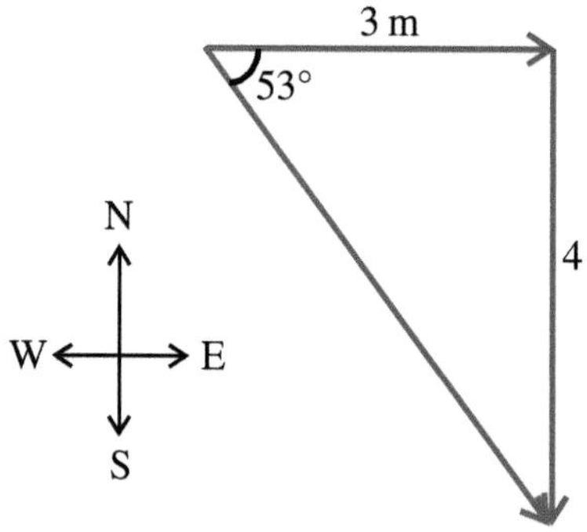

Figure 4.5. The combined effect of 3 m due east and 4 m due south is a displacement of 5 m at an angle of 53° ($\tan^{-1}\frac{4}{3}$) south of east.

Most of the physical quantities which will be encountered in this book come in two types, scalars and vectors. **Scalars** are quantities for which the ordinary rules of arithmetic addition and subtraction apply. Examples of scalar quantities are volume (for example, $2\,\text{mm}^3 + 7\,\text{mm}^3 = 9\,\text{mm}^3$) and mass (for example, 3 kg + 4 kg = 7 kg). **Vectors** are quantities for which the rule for addition is, *by definition*, the same as the rule for addition of displacements. Vector quantities, therefore, cannot usually be added arithmetically; different methods for adding vectors will be outlined below.

4.2 Vector algebra

Representation of vectors

In printed text **bold type** is usually used to denote vector quantities, for example $\boldsymbol{r}$ for displacement, $\boldsymbol{v}$ for velocity, etc. (in some texts an arrow drawn above the symbol is used to denote its vector nature, for example $\overrightarrow{OP}$). In handwriting, an arrow or a line may be drawn over the symbol or a line drawn under it. To represent the vector $\boldsymbol{r}$, the displacement of the point P with respect to O (Figure 4.6(a)), we draw a line which is $|\boldsymbol{r}|$ units long, where $|\boldsymbol{r}|$ denotes the **magnitude** (the length) of the $\boldsymbol{r}$ vector, with an arrowhead in the direction from O to P to indicate the direction. In representing a vector on a page we may scale its magnitude to our convenience, for example if $|\boldsymbol{r}| = 2$ m, we may choose to scale magnitudes in the ratio 100:1, as in Figure 4.6 (where the line representing $\boldsymbol{r}$ is 2 cm long), to allow the representation of $\boldsymbol{r}$ to fit conveniently on the page.

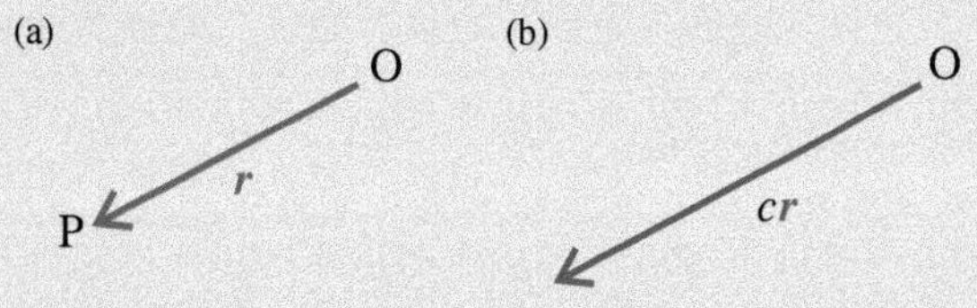

Figure 4.6. (a) The displacement of P relative to O is represented by the vector $\boldsymbol{r}$ which gives both the magnitude and direction of the displacement of P with respect to O. (b) The vector $c\boldsymbol{r}$ is a vector parallel to $\boldsymbol{r}$ which has magnitude c times the magnitude of $\boldsymbol{r}$.

The multiplication of a vector by a scalar produces a vector that is 'scaled' by the magnitude of the scalar. Thus, by definition, the vector $c\boldsymbol{r}$ is a vector which is parallel to the vector $\boldsymbol{r}$ and has magnitude c times the magnitude of $\boldsymbol{r}$, that is $|c\boldsymbol{r}| = c|\boldsymbol{r}|$ (for example, in Figure 4.6(b) where $c = 1.4$).

The addition of two vectors, $\boldsymbol{A}$ and $\boldsymbol{B}$, to form a *resultant* vector $\boldsymbol{C} = \boldsymbol{A} + \boldsymbol{B}$ whose magnitude and direction may be determined by either of two equivalent methods, the 'triangle method' or the 'parallelogram method'. In the *triangle method* the two vectors are drawn with the head of $\boldsymbol{A}$ joined to the tail of $\boldsymbol{B}$, as shown in Figure 4.7(a). The line drawn from the tail of $\boldsymbol{A}$ to the head of $\boldsymbol{B}$ then represents $\boldsymbol{C}$ in both magnitude and direction. This is the same approach as that used in Figure 4.5 above for displacement; in effect we treat displacement as the prototype vector. The *parallelogram method* is illustrated in Figure 4.7(b) — in this method $\boldsymbol{A}$ and $\boldsymbol{B}$ are drawn from the same point to form adjacent sides of a parallelogram. The diagonal of the parallelogram drawn from this point then represents $\boldsymbol{C}$.

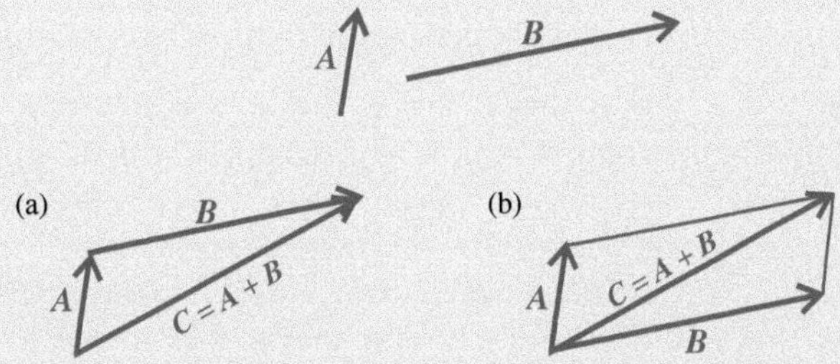

Figure 4.7. (a) The triangle method and (b) the parallelogram method for addition of vectors.

In either method the magnitude and direction of $\boldsymbol{C}$ may be determined graphically, using a ruler and protractor, or trigonometrically using the cosine rule (Appendix A.3.3) or the sine rule (Appendix A.3.4), as convenient.

Components of vectors

Consider the addition of two vectors $\boldsymbol{A}$ and $\boldsymbol{B}$ which are at right angles to each other (Figure 4.8(a)). By applying the parallelogram method to the addition of these two vectors (Figure 4.8(b)) to determine their sum, that is $\boldsymbol{C} = \boldsymbol{A} + \boldsymbol{B}$, we can easily deduce that $|\boldsymbol{C}|$, the magnitude of $\boldsymbol{C}$, is given by Pythagoras' theorem (Appendix A: Theorem A.2.1)

$$|\boldsymbol{C}|^2 = |\boldsymbol{A}|^2 + |\boldsymbol{B}|^2$$

Furthermore, we see that the angle θ between $\boldsymbol{C}$ and $\boldsymbol{A}$ is given by $\tan\theta = \dfrac{|\boldsymbol{B}|}{|\boldsymbol{A}|}$ and that the angle between $\boldsymbol{C}$ and $\boldsymbol{B}$ is given by $\phi = 90° - \theta$.

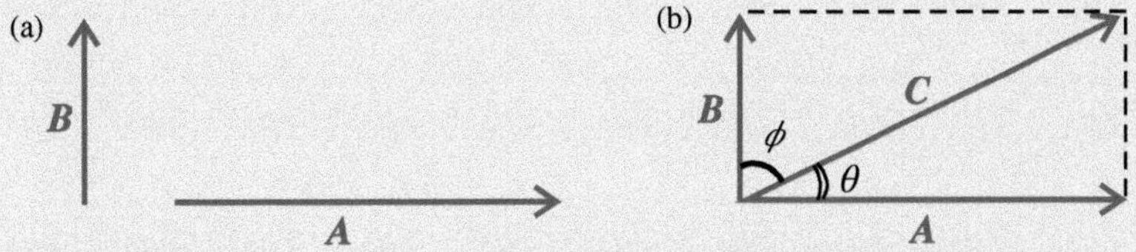

Figure 4.8. (a) Vectors $\boldsymbol{A}$ and $\boldsymbol{B}$ are at right angles to each other. (b) The vector $\boldsymbol{C}$, which is the sum of the vectors $\boldsymbol{A}$ and $\boldsymbol{B}$, is determined by the parallelogram rule.

Conversely we can argue that any vector $\boldsymbol{C}$ can be split into two vectors at right angles which add to give $\boldsymbol{C}$; the vector $\boldsymbol{C}$ is then said to have been **resolved** into two **component vectors**. For example, in Figure 4.9, two displacements along the x- and y-axes, $\boldsymbol{A}$ and $\boldsymbol{B}$ respectively, add to give $\boldsymbol{C}$. Note that there are an infinite number of ways in which this decomposition can be realised depending on how the axes are chosen.

The projection of a vector in any direction is called the **component** of the vector in that direction. Thus $\boldsymbol{C}$ can be resolved into two components as follows.

The ***x*-component** of $\boldsymbol{C}$ is defined as $C_x := |\boldsymbol{C}|\cos\theta$ and the ***y*-component** of $\boldsymbol{C}$ is defined as $C_y := |\boldsymbol{C}|\cos\phi = |\boldsymbol{C}|\sin\theta$.

A **unit vector** is a vector of unit magnitude which defines a direction. Thus, if we define $\hat{\boldsymbol{\imath}}$ and $\hat{\boldsymbol{\jmath}}$ to be unit vectors in the x- and y-directions respectively, as shown in Figure 4.9, we can write

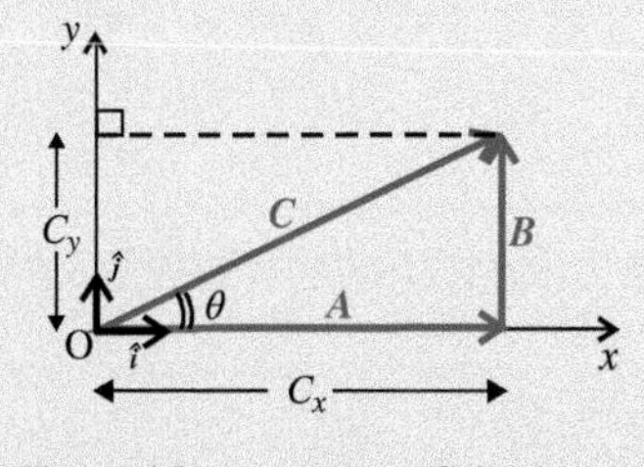

Figure 4.9. The vector $\boldsymbol{C}$ is decomposed into component vectors $\boldsymbol{A}$ and $\boldsymbol{B}$ in the x- and y-directions, respectively. The projections of $\boldsymbol{C}$ onto the x- and y-axes are called the x-component (C_x) and the y-component (C_y), respectively.

$$\boldsymbol{C} = C_x\hat{\boldsymbol{\imath}} + C_y\hat{\boldsymbol{\jmath}} = |\boldsymbol{C}|\cos\theta\,\hat{\boldsymbol{\imath}} + |\boldsymbol{C}|\sin\theta\,\hat{\boldsymbol{\jmath}}$$

where

$$\boxed{|\boldsymbol{C}|^2 = C_x^2 + C_y^2} \quad \text{and} \quad \boxed{\tan\theta = \frac{C_y}{C_x}}$$

In general, to determine the component vector of $\boldsymbol{C}$ in any direction, we need only evaluate

$$\boxed{C_i = |\boldsymbol{C}|\cos\theta}$$

where θ is the angle between the vector and the line along which the component is to be determined, as illustrated in Figure 4.10.

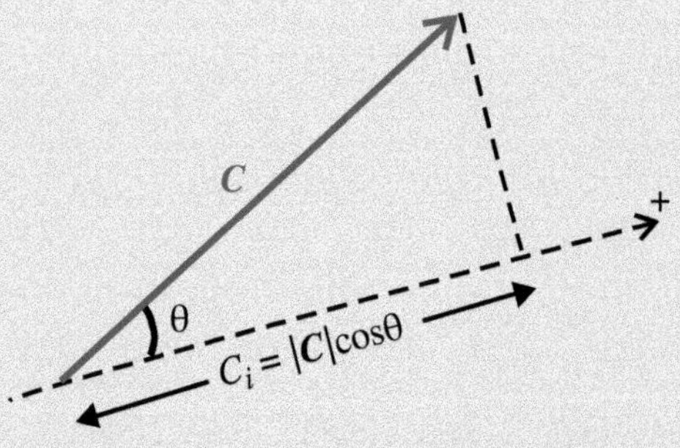

Figure 4.10. The component of a vector along any direction is the magnitude of the vector multiplied by the cosine of the angle between the direction of the vector and the positive direction along which the component is being taken.

Note that the component of a vector is an *algebraic* quantity (if $\theta < \frac{\pi}{2}$ then C_i is positive and if $\theta > \frac{\pi}{2}$ then C_i is negative) and is the projection of the vector onto the line along which its component is taken. The choice of which is the positive direction and which is the negative direction along this line is, of course, arbitrary. Once a positive direction has been adopted, however, the sign of the component of a vector can be determined by noting whether the projection of the vector onto the line is in a positive or negative sense.

Coordinate systems

The choice of axes (x, y) and of the origin O relative to which the components of the vector $\boldsymbol{C}$ are determined in Figure 4.9 is arbitrary, although in any given problem a well considered choice will often facilitate solution. A system of axes and the associated origin is an example of a **coordinate system**. The values of the components of a vector, therefore, depend on the coordinate system used. This can be seen in Figure 4.11, for example, where an alternative coordinate system (x', y') is shown which is rotated with respect to the (x, y) system. Note that the magnitude (the 'length') of the vector is the same in both coordinate systems; that is,

$$|\boldsymbol{C}| = \sqrt{C_x^2 + C_y^2} = \sqrt{C_x'^2 + C_y'^2}$$

Indeed, all scalar quantities have the same value in all coordinate systems: scalars are said to be *invariant* under a change of coordinate system.

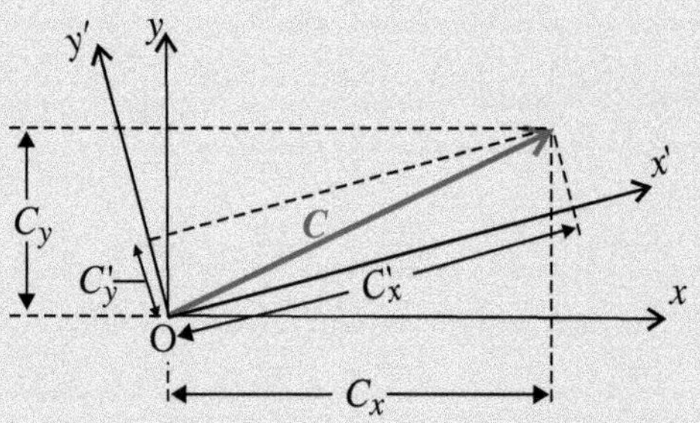

Figure 4.11. The components of a vector are different in two different coordinate systems, one rotated with respect to the other.

The scalar product

As we will see throughout this book, the product of the magnitude of two vectors multiplied by the cosine of the angle between them arises frequently in physics and mathematics. Such an entity, when it arises, can be expressed compactly in terms of a type of vector product called the **scalar product**, which, with reference to Figure 4.12, is defined as follows.

The **scalar product** (or 'dot product') of two vectors $\boldsymbol{A}$ and $\boldsymbol{B}$ is a scalar quantity, denoted by $\boldsymbol{A} \cdot \boldsymbol{B}$ ("A dot B"), and defined by

$$\boldsymbol{A} \cdot \boldsymbol{B} := |\boldsymbol{A}||\boldsymbol{B}| \cos\theta$$

where θ is the angle between the directions of $\boldsymbol{A}$ and $\boldsymbol{B}$.

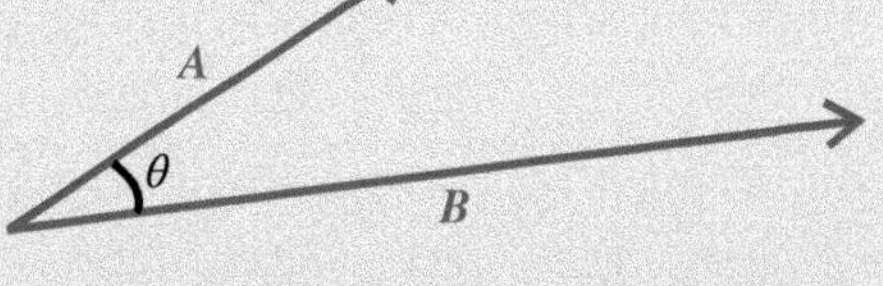

Figure 4.12. The scalar product of the vectors A and B is defined as $|\boldsymbol{A}||\boldsymbol{B}|\cos\theta$

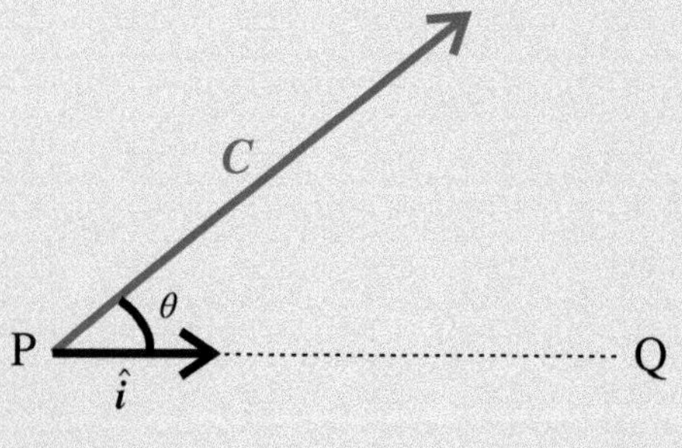

Figure 4.13. The component of the vector $\boldsymbol{C}$ along the direction given by the unit vector i is $\mathbf{C} \cdot \hat{\imath} = |\boldsymbol{C}| \cos\theta$.

The **component** of a vector along some arbitrary direction is the scalar product of the vector and a unit vector in that direction. Thus the component of the vector $\boldsymbol{C}$ in Figure 4.13 in the direction PQ is given by

$$\mathbf{C} \cdot \hat{\imath} = |\boldsymbol{C}||\hat{\imath}| \cos\theta = |\boldsymbol{C}| \cos\theta$$

Some useful properties of the dot product are given in Appendix A.11.6.

At the end of this chapter (Section 4.11) we will introduce a different type of product of two vectors (called the 'vector product') which also has many important applications in physics.

Solving physics problems in two and three dimensions

The technique of resolving a vector into components in *perpendicular* directions is a very valuable tool in the analysis and solution of problems involving vector quantities. In the case of a two-dimensional problem, for example, it enables us to resolve all vectors into their components in two perpendicular directions, which we can choose according to our convenience in solving a particular problem. This will reduce a problem of motion in a general plane (two-dimensional motion) to two one-dimensional problems which we can represent and solve algebraically. We can then reconstruct any two-dimensional vectors which we need to know simply by adding their components according to the rules of vector addition. This technique will be illustrated in Section 4.5 and encountered frequently throughout this book.

The arguments which we have applied to two-dimensional motion can be extended to general motion in space. In the three-dimensional case the vector $\boldsymbol{r}$ has components in three perpendicular directions, the x-, y- and z-axes as indicated in Figure 4.14 — an example of a three-dimensional coordinate system called a *Cartesian coordinate system.* If the vector $\overrightarrow{OP} = r = x\hat{\imath} + y\hat{\jmath} + z\hat{k}$, where $\hat{\imath}$, $\hat{\jmath}$ and $\hat{k}$ are unit vectors in the x-, y-, and z-directions, respectively, note that

$$|\boldsymbol{r}|^2 = \mathrm{OP}^2 = \mathrm{ON}^2 + \mathrm{NP}^2 = (x^2 + y^2) + z^2 = x^2 + y^2 + z^2$$

which is Pythagoras' theorem in three dimensions.

Note also, from the definition of the scalar product, that $\boldsymbol{i}\cdot\boldsymbol{j} = \boldsymbol{j}\cdot\boldsymbol{k} = \boldsymbol{k}\cdot\boldsymbol{i} = 0$ and $\boldsymbol{i}\cdot\boldsymbol{i} = \boldsymbol{j}\cdot\boldsymbol{j} = \boldsymbol{k}\cdot\boldsymbol{k} = 1$.

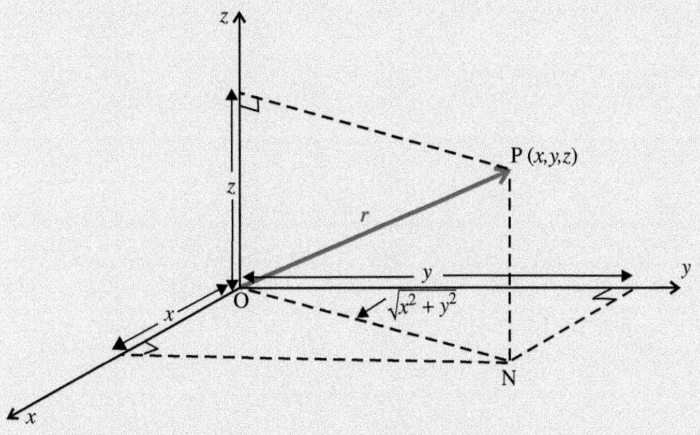

Figure 4.14. Example of components of a vector in three dimensions.

Thus, just as any vector in a plane can be broken into two component vectors in two perpendicular directions, any vector in space can be broken into three component vectors in three perpendicular directions. Vectors with two components are known as *2-vectors* and those with three components as *3-vectors* (we will even encounter *4-vectors* in Chapter 9). While the analysis of a three-dimensional problem is harder to visualise and represent graphically, and also generates more equations, the principles used to analyse and solve problems in three dimensions are exactly the same as those applied to the two-dimensional case.

The basic properties of vectors which we have discussed above are treated in more detail in Appendix A.11.1 – A.11.6. For an excellent survey of vector calculus and many good exercises and problems on the use of vectors see Halliday, Resnick and Walker, 10th edition (2013), Chapter 3. Some alternative coordinate systems are introduced in Appendix Section A.13.

Study Appendix A, Section A.11.1 - 6 and Worked Example 4.1

For problems based on the material presented in this section visit up.ucc.ie/4/ *and follow the link to the problems.*

4.3 Velocity and acceleration vectors

Velocity

Consider a point P which is moving in two dimensions (the plane of the page). As indicated in Figure 4.15, the displacements of P from O at two points A and B on the path of P and the corresponding times are $(\boldsymbol{r}, t)$ and $(\boldsymbol{r} + \Delta\boldsymbol{r}, t + \Delta t)$, respectively.

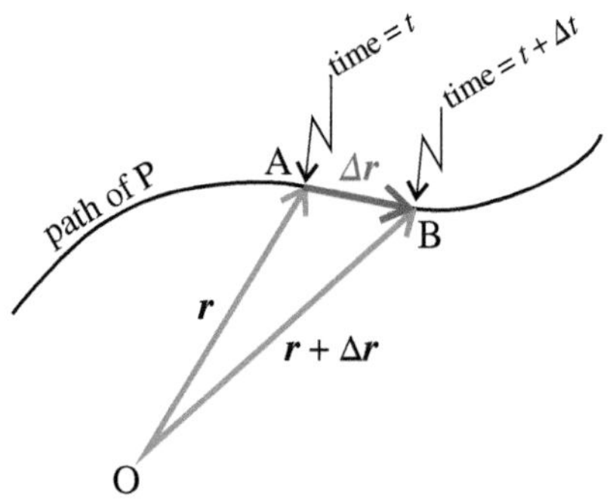

Figure 4.15. A point moves from A to B in a time Δt; its change of displacement during that time interval is $\Delta\boldsymbol{r}$.

We can now define **velocity** in vector form (recall Equation (2.5) for the one-dimensional version). The velocity of the point P at the instant it is at A is defined as follows

$$\boldsymbol{v} := \lim_{\substack{\Delta t \to 0 \\ \mathrm{B}\to\mathrm{A}}} \frac{\Delta\boldsymbol{r}}{\Delta t} = \frac{d\boldsymbol{r}}{dt}$$

The quantity $\frac{d\boldsymbol{r}}{dt}$ is a vector ($\Delta\boldsymbol{r}$) multiplied by a scalar $(\frac{1}{\Delta t})$ and hence $\boldsymbol{v}$ is a vector in the direction of $\Delta\boldsymbol{r}$ in the limit $\Delta\boldsymbol{r} \to 0$, that is in the direction of the tangent to the path at A. Thus the direction of the velocity vector is always tangential to the path of the moving point.

The magnitude of the velocity vector, denoted by $|\boldsymbol{v}|$, is called the *speed*. This is the only case in physics in which the magnitude of a vector is given a special name. Note that if an object is moving at constant speed but changes direction, for example from 30 km per hour due North to 30 km per hour due East, its velocity has changed although its speed has not.

Acceleration

In a similar way can define **acceleration** in vector form. As illustrated in Figure 4.16, if the velocities and corresponding times at two points A and B along a point's path of motion are $(\boldsymbol{v}, t)$ and $(\boldsymbol{v} + \Delta\boldsymbol{v}, t + \Delta t)$, respectively, the acceleration of the moving point at A at the instant t is defined as

$$\boldsymbol{a} := \lim_{\substack{\Delta t \to 0 \\ \mathrm{B}\to\mathrm{A}}} \frac{\Delta\boldsymbol{v}}{\Delta t} = \frac{d\boldsymbol{v}}{dt}$$

Note that the direction of $\boldsymbol{a}$ is in the direction defined by $\Delta\boldsymbol{v}$ in the limit $\Delta t \to 0$, which in general is not the same direction as that of $\boldsymbol{v}$. The velocity vector triangle, representing the addition $(\boldsymbol{v}+\Delta\boldsymbol{v}) = \boldsymbol{v} + \Delta\boldsymbol{v}$ (Figure 4.16(b)), demonstrates this point.

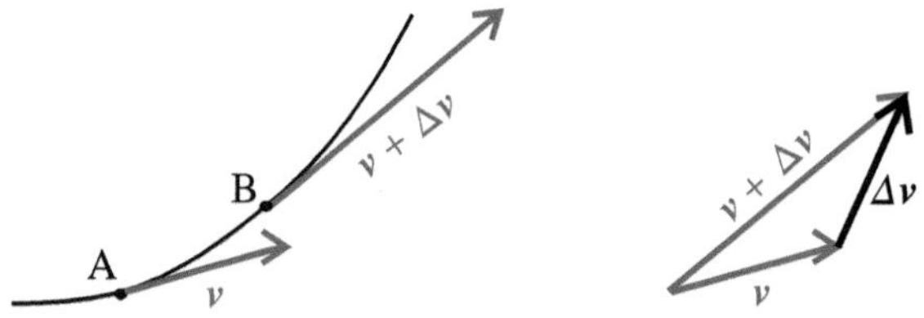

Figure 4.16. (a) The change in velocity of a point as it moves from A to B. (b) The direction of the change in velocity of a point as it moves from A to B in is determined by the triangle rule for vector addition. The direction of the acceleration is parallel to $\Delta\boldsymbol{v}$ in the limit $\Delta t \to 0$.

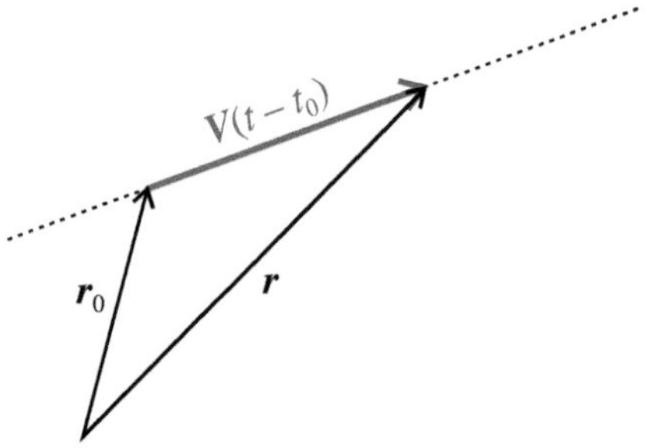

Figure 4.17. Motion with constant velocity. If the displacement of the moving point at $t = 0$ is $\boldsymbol{r}_0$, its displacement at any other time t is given by Equation (4.1).

Linear motion, in which displacement, velocity and acceleration are all directed along the same straight line, is a special case. Equations equivalent to those used in Section 2.5 to describe the special cases of constant velocity and constant acceleration in one dimension can be written in vector form by replacing the algebraic quantities x, v and a, respectively, by their vector equivalents $\boldsymbol{r}$, $\boldsymbol{v}$ and $\boldsymbol{a}$. Equation (2.14) for constant velocity becomes

$$\boldsymbol{r} = \boldsymbol{r}_0 + \boldsymbol{V}(t - t_0) \tag{4.1}$$

This equation describes motion in a straight line with constant (vector) velocity as illustrated in Figure 4.17.

For motion with constant acceleration, where the origins of coordinates and time are chosen such that $\boldsymbol{r}_0 = 0$ and $t_0 = 0$,

$$(2.18) \rightarrow \boldsymbol{v} = \boldsymbol{v}_0 + \boldsymbol{A}t \tag{4.2a}$$

$$(2.19) \rightarrow \boldsymbol{r} = \boldsymbol{v}_0 t + \tfrac{1}{2}\boldsymbol{A}t^2 \tag{4.2b}$$

Equation (4.2a) is represented as a vector addition triangle in Figure 4.18.

Thus,

$$\boldsymbol{v}\cdot\boldsymbol{v} = |\boldsymbol{v}|^2 = (\boldsymbol{v}_0 + \boldsymbol{A}t)\cdot(\boldsymbol{v}_0 + \boldsymbol{A}t) = |\boldsymbol{v}_0|^2 + 2\boldsymbol{v}_0\cdot\boldsymbol{A}t + (\boldsymbol{A}\cdot\boldsymbol{A})t^2$$
$$= |\boldsymbol{v}_0|^2 + 2\boldsymbol{A}\cdot\left(\boldsymbol{v}_0 t + \frac{1}{2}\boldsymbol{A}t^2\right)$$

Using Equation (4.2b), we can replace $\left(\boldsymbol{v}_0 t + \frac{1}{2}\boldsymbol{A}t^2\right)$ by $\boldsymbol{r}$ and hence

$$(2.20) \quad \rightarrow \quad |\boldsymbol{v}|^2 = |\boldsymbol{v}_0|^2 + 2\boldsymbol{A}\cdot\boldsymbol{r} \qquad (4.2c)$$

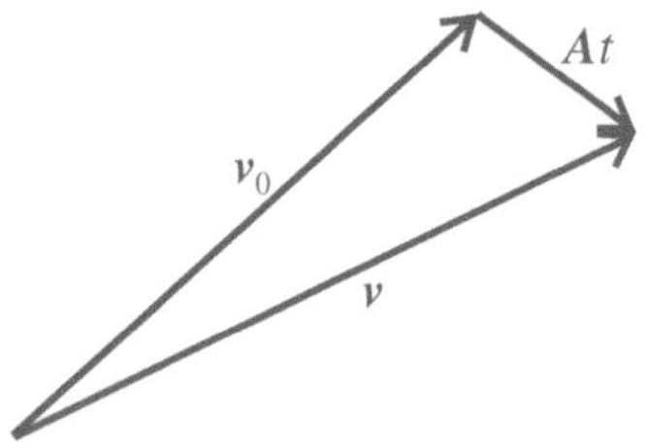

Figure 4.18. Motion with constant acceleration $\boldsymbol{A}$. The velocity $\boldsymbol{v}$ of the moving point at time t is given by Equation (4.2a), where $\boldsymbol{v}_0$ is its velocity at $t = 0$.

The use of the vector Equations (4.2a) – (4.2c) will be illustrated in the next section.

Study Worked Example 4.2

For problems based on the material presented in this section visit up.ucc.ie/4/ *and follow the link to the problems.*

4.4 Force as a vector quantity: vector form of the laws of dynamics

We have defined a vector as being any quantity for which the rule for addition is the same as the rule for addition of displacements (Section 4.2). Since, as we have seen, differentiating with respect to time involves multiplying by a scalar, both velocity and acceleration are vector quantities, by definition.

The situation is less obvious in the case of force, however. Consider the case of two forces acting in different directions on a body as illustrated in Figure 4.19. It would seem likely that if $\boldsymbol{F}_1$ and $\boldsymbol{F}_2$ were to be added together according to the rules of vector addition, the resulting vector sum $\boldsymbol{F}$ would be the equivalent of the two forces acting together, in the sense that a single force $\boldsymbol{F}$ would produce precisely the same acceleration as the two forces $\boldsymbol{F}_1$ and $\boldsymbol{F}_2$ acting simultaneously on the body. That such a 'principle of superposition' holds for forces in the real world is not self evident, however, and requires experimental confirmation.

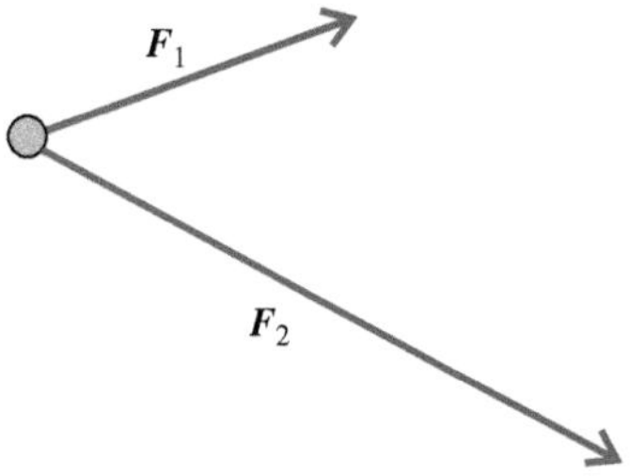

Figure 4.19. Two forces acting in different directions on a body.

One simple experiment that provides support for the idea that force is a vector is illustrated in Figure 4.20(a). Known masses are attached to the ends of strings which pass over pulleys as shown and the whole system is allowed to settle in equilibrium. Consider the forces on the point X where the three strings are tied together. As long as the pulleys operate reasonably smoothly, that is with negligible friction, it can be assumed that the force acting along any string is proportional to the mass attached to that string. Since there is no motion of the point X once equilibrium has been realised, the effect of any two of the forces must exactly balance the third. If, for example, the forces exerted by the two strings which pass over the pulleys are added together using the rule for addition of vectors (Figure 4.20(b)), it is found that their sum is equal and opposite to the force exerted by the third string vertically downward (within the accuracy of the experiment). Thus, only if the forces are added *as vectors* do we find that their net effect is zero, as required to maintain the point X in equilibrium.

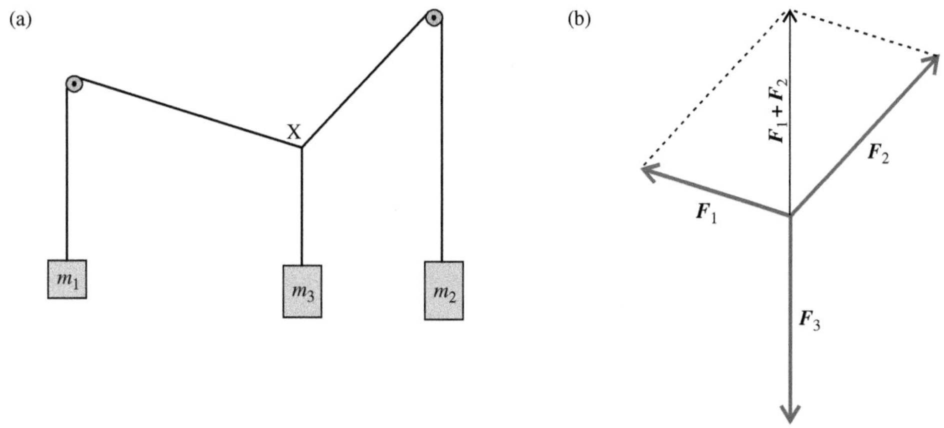

Figure 4.20. (a) An experiment which indicates that forces add according to the rule of vector addition. (b) The vector sum of the forces $\boldsymbol{F}_1$ and $\boldsymbol{F}_2$ is equal and opposite to the force $\boldsymbol{F}_3$.

Further experimental investigations confirm that it is valid to treat force as a vector and that if a number of forces act on a body, the body will behave in exactly the same manner as it would if a single force equal to the vector sum of the individual forces acted on it. Thus **Newton's Second law**, Equation (3.3), can be generalised to deal with situations involving many forces producing motion in three dimensions by writing it in the following vector form

$$\boldsymbol{a} = \frac{\boldsymbol{F}}{m} \tag{4.3}$$

where $\boldsymbol{F}$ is understood to be *the vector sum of all the forces* acting on the mass m.

We can also generalise the definition of **momentum**, given in Section 3.3, to take its vector nature into account as follows

$$\boxed{\boldsymbol{p} := m\boldsymbol{v}}$$

and thus, in terms of momentum, Newton's second law can also be written in vector form as

$$\frac{d\boldsymbol{p}}{dt} = \boldsymbol{F}$$

If the force acting on a body is a function of time, that is if its magnitude or direction or both vary with time ($\boldsymbol{F} = \boldsymbol{F}(t)$), the acceleration will change correspondingly with time ($\boldsymbol{a} = \boldsymbol{a}(t)$), in which case Equation (4.3) can be considered to hold instantaneously, that is

$$\boldsymbol{a}(t) = \frac{\boldsymbol{F}(t)}{m}$$

In many everyday situations the forces which act on bodies may be constant. One such example, as we have already observed, is the force due to gravity on a body near the surface of the Earth. As long as the mass does not move over too great a vertical distance, the pull of gravity vertically downwards on the body, which we called its **weight** ($\boldsymbol{W}$) in Section 3.4, does not change. A body of mass m falling freely under gravity, therefore, will experience acceleration given by

$$\boldsymbol{g} = \frac{\boldsymbol{W}}{m}$$

and thus the weight of any body is the product of its mass and the acceleration due to gravity, that is

$$W = mg$$

We will get a deeper understanding of the nature of weight in Chapter 5.

Whenever the resultant force on a body is constant, therefore, the body will move with constant acceleration and the resulting motion can be described by invoking Equations (4.2a) to (4.2c) or, if the motion is confined to a straight line, by invoking Equations (2.18) to (2.20).

Study Worked Example 4.3

For problems based on the material presented in this section visit up.ucc.ie/4/ *and follow the link to the problems.*

4.5 Constraint forces

Consider the forces acting on a book which is placed on a smooth horizontal table (Figure 4.21) or on an object which is hanging from a vertical string (Figure 4.23). If the only force acting on these bodies was the force due to gravity they would, of course, accelerate downwards. Since there is no acceleration, however, a second force must be acting on the body to cancel the effect of the weight in each case. These are examples of what are called **constraint forces**. In the case of the book on the table, the constraint force is a force exerted upwards *by* the table *on* the book as shown as $\boldsymbol{N}$ in Figure 4.21(a). Since this force always acts perpendicularly to the table surface it is called a *normal contact force* or *normal reaction*. It is important to note that it is the nature of constraint forces to adjust automatically in magnitude and/or direction to conform to the constraint; if the book had been twice as heavy, for example, the upward force of the table on the book would have been twice as large. Applying Newton's Second law to the forces on the book we get

$$\boldsymbol{W} + \boldsymbol{N} = M\boldsymbol{a} = 0 \text{ (since } \boldsymbol{a} \text{ is zero)}$$

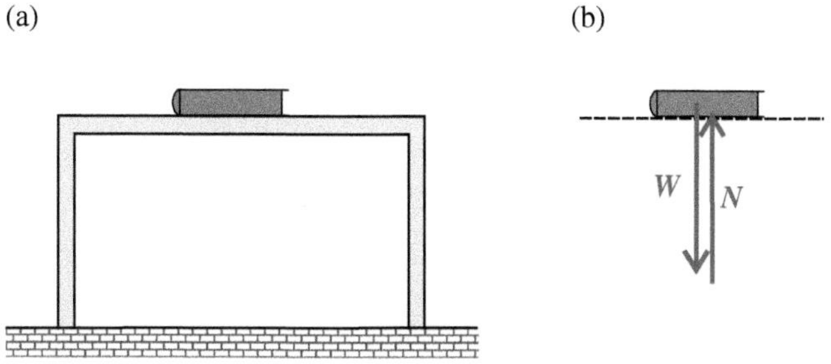

Figure 4.21. Example of a constraint force. (a) The book is constrained from moving downwards by the surface of the table. (b) The forces acting on the book in are (i) the weight $\boldsymbol{W}$ of the book and (ii) the normal reaction $\boldsymbol{N}$ exerted upwards by the table surface.

and hence

$$\boldsymbol{N} = -\boldsymbol{W} = -M\boldsymbol{g}$$

It should be noted that the magnitude of the normal contact force is not necessarily equal to the weight of the body, even on a horizontal surface. Consider, for example, the case of a block which is pulled along a horizontal surface by a force $\boldsymbol{P}$ which makes an angle α with the horizontal as shown in Figure 4.22(a). Taking vertical components of the forces (Figure 4.22(b)) in this case and applying Newton's second law, we find that

$$|\boldsymbol{N}| + |\boldsymbol{P}| \sin\alpha - Mg = 0$$

and hence the normal contact force in this case is given by

$$|\boldsymbol{N}| = Mg - |\boldsymbol{P}| \sin\alpha$$

Taking horizontal components we obtain $|\boldsymbol{P}|\cos\alpha = Ma \quad \Rightarrow \quad a = \dfrac{|\boldsymbol{P}|}{M}\cos\alpha$, showing that the block moves along the surface with constant acceleration.

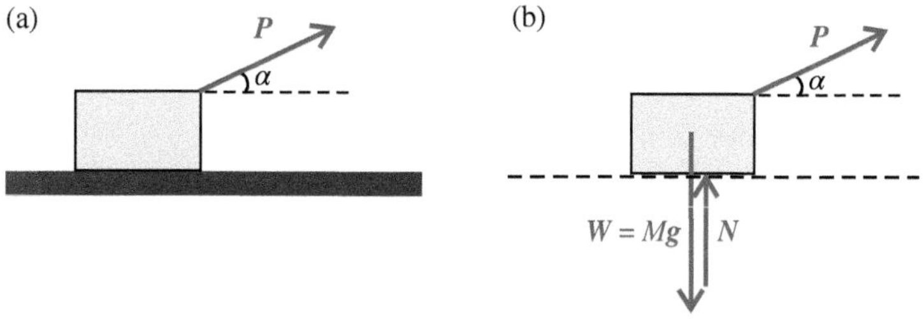

Figure 4.22. (a) A block being pulled along a horizontal surface by a force $\boldsymbol{P}$ directed at an angle α to the horizontal. The forces acting on the block are shown in (b).

Note that it is assumed in this chapter that bodies, such as the block considered here, are sufficiently small that they can be treated as 'particles'. The application of Newton's laws of motion to extended objects will be discussed in Chapter 7.

Turning now to the case of the mass which is constrained by a string (Figure 4.23), a diagram indicating the forces on the mass m (Figure 4.23(b)) shows that, in this case, the constraint force is provided by the tension force $\boldsymbol{T}$ in the string acting along the line of the string. Applying the laws of motion we get

$$\boldsymbol{T} = -\boldsymbol{W} = -m\boldsymbol{g}$$

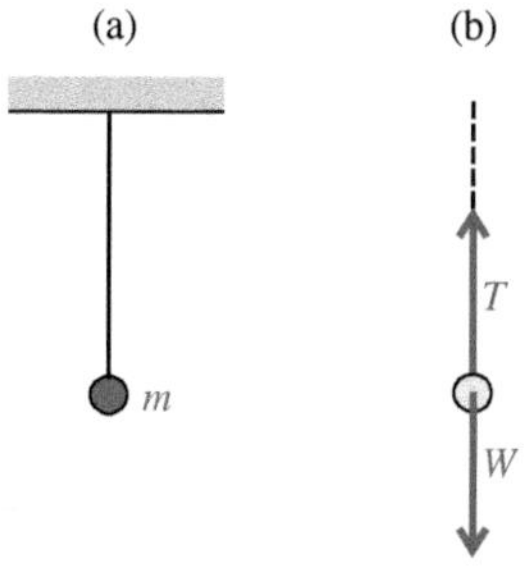

Figure 4.23. (a) An object of mass m hanging from the end of a string which is attached to a fixed support. (b) The mass m is held in equilibrium by its weight and the constraint force $\boldsymbol{T}$ exerted vertically upwards by the string.

Finally, let us consider what happens if the table shown in Figure 4.21 were to be tilted at some angle with respect to the horizontal, as illustrated in Figure 4.24. In this case, the table surface forms an inclined plane down which the book will slide, at least if we assume that the surface is very smooth. Looking at the forces on the book (Figure 4.24(b)) we see that in this case the two forces $\boldsymbol{W}\,(= M\boldsymbol{g})$ and $\boldsymbol{N}$ no longer add up to zero. The magnitude of $\boldsymbol{N}$ is determined by the criterion required to satisfy the constraint, which means that the acceleration, and hence the resultant (that is, the vector sum) of $\boldsymbol{W}$ and $\boldsymbol{N}$, must be directed along the plane (Figure 4.24(c)). The obvious strategy is to resolve the problem into components (i) along the plane and (ii) perpendicular to the plane. Since the component of the acceleration in the latter direction must be zero this gives rise to significant simplification of the resulting expression when Newton's second law is applied.

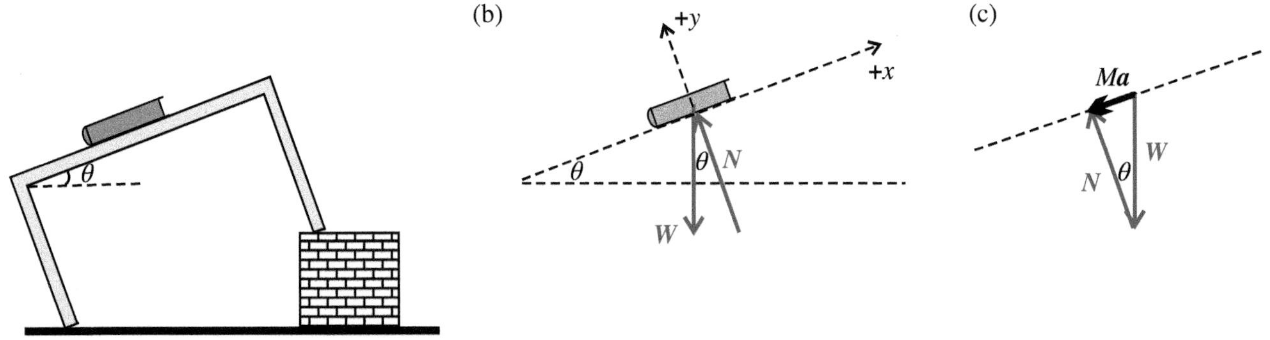

Figure 4.24. The table shown in Figure 4.21(a) is tilted so that its surface makes an angle of θ with the horizontal. (b) The forces acting on the book are its weight $\boldsymbol{W}$ and the normal reaction N. (c) The normal reaction must be such that the vector sum of $\boldsymbol{W}$ and $\boldsymbol{N}$ is directed along the surface of the table, giving rise to an acceleration in that direction determined by Newton's second law.

Taking the x- and y-directions as shown in Figure 4.24(b) and applying Newton's second law we get:

$$\begin{array}{llll}\text{(a) } x\text{-components:} & -|W|\sin\theta + 0 = Ma & \Rightarrow & a = -g\sin\theta \\ \text{(b) } y\text{-components:} & |N| - |W|\cos\theta = 0 & \Rightarrow & |N| = Mg\cos\theta\end{array}$$

The minus sign in the expression for the acceleration tells us that the acceleration is in the $-x$-direction, that is directed down the plane. Note that the acceleration is constant and, therefore, relationships (2.18) – (2.20) may be applied to describe the motion of the body down the plane.

Study Worked Example 4.4

For problems based on the material presented in this section visit up.ucc.ie/4/ *and follow the link to the problems.*

4.6 Friction

Situations such as those analysed in the previous section are rather idealised, as we noted in Section 3.2, in that it was assumed that the book slid smoothly along the surface of the table. The magnitude of the predicted acceleration ($g\sin\theta$) could be achieved only if frictional effects could be completely eliminated. If we are to be able to describe real situations we must develop models which allow us to include frictional forces in dynamical problems, as we did for one-dimensional simple harmonic oscillators in Section 3.11.

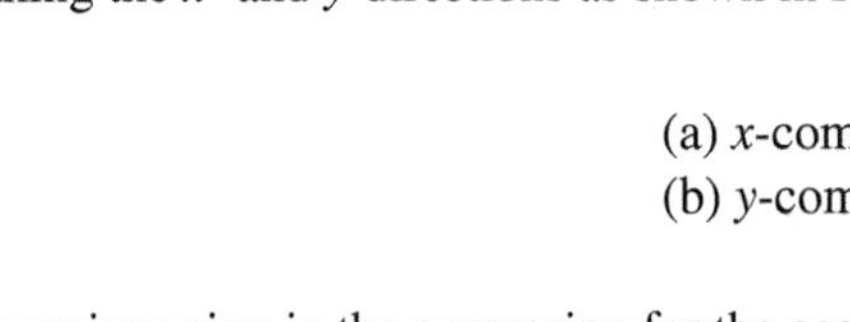

Figure 4.25. In real situations, when a force is applied to a body in an attempt to slide it along a surface, a force due to friction ($\boldsymbol{F}_\mathrm{R}$) acts on the body to oppose the motion.

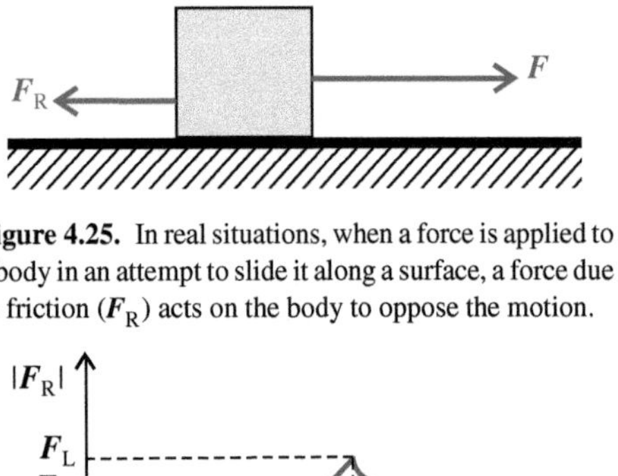

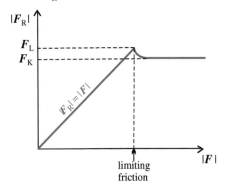

Figure 4.26. A simple model of dry friction; the frictional resistance plotted as a function of the force ($\boldsymbol{F}$) applied to the block.

Bodies sliding over rough surfaces can be described reasonably well by the 'dry friction' model encountered in Section 3.11 (model A) which we will develop further here. Consider the situation illustrated in Figure 4.25 in which a body such as a wooden block or a book lies on a horizontal surface. We can attempt to visualise what happens as a horizontal force $\boldsymbol{F}$ is applied to the body, starting as a very small force and gradually increased. When the force is small, no motion at all will be observed as the two surfaces remain stuck together. Since there is no motion there must be no *net* force and, therefore, there must be another force acting on the block, equal and opposite to $\boldsymbol{F}$, which opposes the tendency to move. As $\boldsymbol{F}$ is increased this frictional resistance ($\boldsymbol{F}_\mathrm{R}$) must also increase such that $\boldsymbol{F}_\mathrm{R} = -\boldsymbol{F}$. We know from experience, however, that the frictional resistance cannot continue to balance the force $\boldsymbol{F}$ as $\boldsymbol{F}$ is increased indefinitely. There comes a point where any further increase in $\boldsymbol{F}$ will cause the block to move; this is called *limiting friction,* at which point the frictional resistance has its maximum value, $\boldsymbol{F}_\mathrm{R} = \boldsymbol{F}_\mathrm{L}$

Once the block is in motion (*sliding friction*) the frictional resistance opposing the motion remains reasonably constant, independent of the speed of the block along the surface, that is $\boldsymbol{F}_\mathrm{R} = \boldsymbol{F}_\mathrm{K} =$ *constant*. For most surfaces the magnitude of the frictional force opposing the motion of a moving body is less than the magnitude of the frictional resistance at limiting friction, that is $|\boldsymbol{F}_\mathrm{K}| < |\boldsymbol{F}_\mathrm{L}|$.

This simple picture of the nature of frictional forces between two surfaces, summarised in Figure 4.26, proves perfectly adequate to describe many everyday situations involving friction. The model can be taken a step further if we realise that, while the magnitude of the frictional force is independent of the areas of the surfaces in contact, it depends on how strongly the two surfaces are being pushed together. The normal contact force of the table surface on the block is a measure of 'how strongly the surfaces are being pushed together'; for example, if the lower surface was sloped rather than horizontal, the normal contact force and hence the frictional force would be correspondingly reduced. It can be argued that the magnitude of the frictional force might be directly proportional to the normal contact force pushing the surfaces together but, in the last resort, this hypothesis needs to be tested experimentally.

Experiments show that this is indeed the case (at least approximately), for both limiting friction and sliding friction. Thus

at limiting friction:

$$|\boldsymbol{F}_{\mathrm{L}}| \propto |\boldsymbol{N}| \quad \Rightarrow \quad |\boldsymbol{F}_{\mathrm{L}}| = (\text{constant})|\boldsymbol{N}|$$

or

$$\boxed{|\boldsymbol{F}_{\mathrm{L}}| = \mu_{\mathrm{s}}|\boldsymbol{N}|}$$

for sliding friction:

$$|\boldsymbol{F}_{\mathrm{K}}| \propto |\boldsymbol{N}| \quad \Rightarrow \quad |\boldsymbol{F}_{\mathrm{K}}| = (\text{constant})|\boldsymbol{N}| = \mu_{\mathrm{K}}|\boldsymbol{N}|$$

or

$$\boxed{|\boldsymbol{F}_{\mathrm{K}}| = \mu_{\mathrm{K}}|\boldsymbol{N}|}$$

The values of the constants of proportionality are different in the two situations: μ_{S} is called the *coefficient of static friction* and μ_{K} is called the *coefficient of kinetic friction*. In our model these are constants which are characteristic of a particular pair of surfaces; for most surfaces $\mu_{\mathrm{K}} < \mu_{\mathrm{S}}$; that is, once motion has been initiated a smaller force is needed to maintain motion.

Friction has its origin in the microscopic properties of the surfaces in contact and, while some additional insight may be gained by considering the phenomenon at this level, no simple microscopic theory has been developed that can be used to make quantitative predictions.

Let us return to the problem of a block on an inclined plane which we considered in Section 4.5, but with frictional forces now included. We will consider this problem in some detail not only because it helps us to understand the nature of forces due to friction but also because it provides some general procedures for tackling problems in dynamics. The three possible situations that can arise have to be dealt with separately, as described below.

Case 1: Body at rest on an inclined plane

If the slope of the plane is small, the block will remain at rest with the frictional force adjusting to prevent motion. As the slope is gradually increased there will come a point ($\theta = \theta_{\mathrm{L}}$) where the block is just on the point of moving. The forces on the block at this point are shown in Figure 4.27, where M is the mass of the block and hence its weight is $\boldsymbol{W} = M\boldsymbol{g}$. Taking x- and y- directions parallel to and perpendicular to the plane, respectively, the conditions for equilibrium of the block are given by Newton's second law which we apply in component form, that is $F_x = Ma_x$ and $F_y = Ma_y$ with, since there is no motion, $a_x = a_y = 0$ in this case.

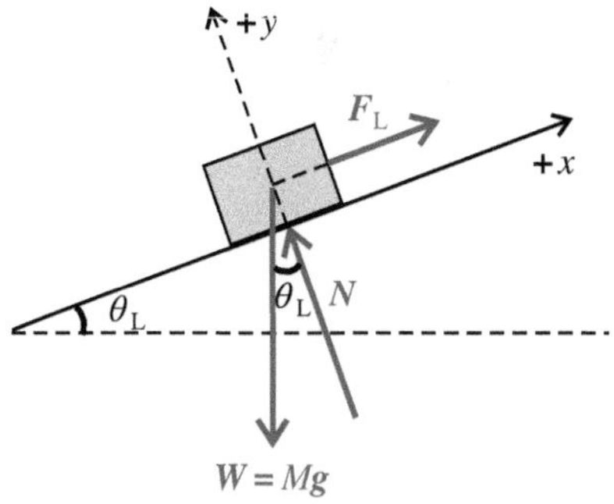

Figure 4.27. Forces on a block resting on an inclined plane where the slope of the plane is such that the block is just on the point of moving.

$$\text{(i) } x\text{-components:} \quad |\boldsymbol{F}_{\mathrm{L}}| - |\boldsymbol{W}|\sin\theta_{\mathrm{L}} = 0 \quad \Rightarrow \quad \mu_{\mathrm{S}}|\boldsymbol{N}| = Mg\sin\theta_{\mathrm{L}}$$

$$\text{(ii) } y\text{-components:} \quad |\boldsymbol{N}| - |\boldsymbol{W}|\cos\theta_{\mathrm{L}} = 0 \quad \Rightarrow \quad |\boldsymbol{N}| = Mg\cos\theta_{\mathrm{L}}$$

and hence

$$\theta_{\mathrm{L}} = \tan^{-1}\mu_{\mathrm{S}}$$

Thus the block will remain at rest on the plane for angles $0 < \theta < \tan^{-1}\mu_{\mathrm{S}}$. Note that this provides a convenient method for measuring μ_{S}, the coefficient of static friction between a pair of surfaces. The angle of the plane with respect to the horizontal is increased until the block is just on the point of moving, at which angle ($\theta = \theta_{\mathrm{L}}$) $\mu_{\mathrm{S}} = \tan\theta_{\mathrm{L}}$.

Case 2: Body sliding down an inclined plane

If the slope of the plane is increased so that $\theta > \theta_{\mathrm{L}}$, the block will no longer remain at rest. The frictional resistance $\boldsymbol{F}_{\mathrm{K}}$ is directed so as to oppose the motion of the block down the plane, as shown in Figure 4.28. In this case Newton's second law gives:

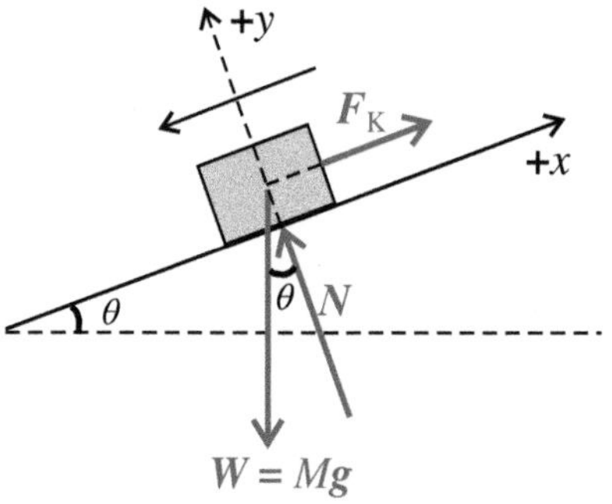

Figure 4.28. Forces on a block sliding down an inclined plane.

(a) x-components: $|\boldsymbol{F}_K| - |\boldsymbol{W}|\sin\theta = Ma \Rightarrow \mu_K|N| - Mg\sin\theta = Ma$
(b) y-components: $|\boldsymbol{N}| - |\boldsymbol{W}|\cos\theta = 0 \Rightarrow |N| = Mg\cos\theta$

Eliminating $|N|$, we obtain $\mu_K Mg\cos\theta - Mg\sin\theta = Ma$, from which we obtain the following expression for the acceleration of the body

$$a = -g(\sin\theta - \mu_K\cos\theta) \tag{4.4}$$

$\theta > \theta_K \rightarrow \tan\theta > \mu_S > \mu_K \rightarrow \sin\theta > \mu_K\cos\theta$ and hence the term inside the brackets is positive. Thus, the acceleration is directed in the $-x$-direction, that is down the plane. Note that the block moves with constant acceleration and that in the limit of no friction ($\mu_K \rightarrow 0$) the result becomes the same as that derived in Section 4.5.

Case 3: Body sliding up an inclined plane
The situation is significantly different when the block is projected up along the plane as a result, for example, of a push at the bottom of the slope. Since the friction force always opposes the motion, $\boldsymbol{F}_K$ must be directed down the plane (as shown in Figure 4.29) while the block is moving up the plane. Application of Newton's second law to this situation gives:

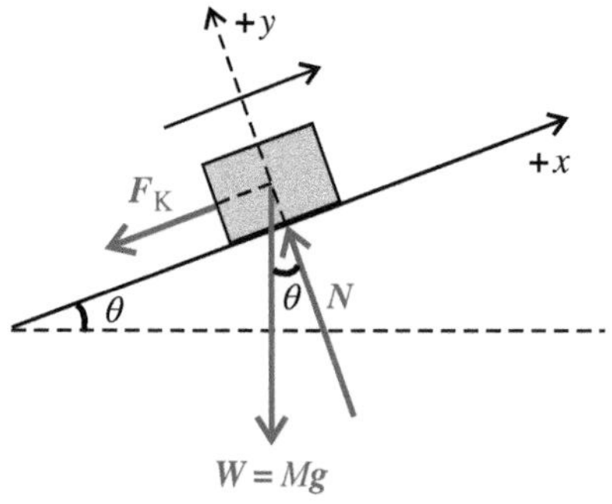

Figure 4.29. Forces on a block sliding up an inclined plane. The block, having been given a push at the bottom of the plane, decelerates as it moves up the plane.

(a) x-components: $-|\boldsymbol{F}_K| - |\boldsymbol{W}|\sin\theta = Ma \Rightarrow -\mu_K|N| - Mg\sin\theta = Ma$
(b) y-components: $|\boldsymbol{N}| - |\boldsymbol{W}|\cos\theta = 0 \Rightarrow |N| = Mg\cos\theta$

and hence,

$$a = -g(\sin\theta + \mu_K\cos\theta)$$

Again the acceleration is constant and is directed down the plane — the block slows down (decelerates) as it moves up the plane — but the magnitude of the deceleration is larger than that of the acceleration in Case 2 above (Equation (4.4)) for the same angle θ. After coming to rest, the block will begin to slide down the plane again, as described in Case 2, provided $\theta > \theta_L$.

Note that different dynamical situations yield different relationships and, as a result, it is always safer to start problems in dynamics from first principles, namely by (i) drawing an appropriate force diagram, (ii) resolving forces into appropriate perpendicular components (usually parallel and perpendicular, respectively, to the direction in which the body is constrained to move) and (iii) applying Newton's second law to the component motions. Learning derived relationships by rote is generally unproductive as these may not apply outside a very specific context.

Study Worked Example 4.5

For problems based on the material presented in this section visit up.ucc.ie/4/ *and follow the link to the problems.*

4.7 Motion in a circle: centripetal force

Consider the case of a block of mass M which is resting on a horizontal frictionless table and is connected to a fixed point C by a horizontal piece of string; Figure 4.30(a) shows the situation viewed from above while Figure 4.30(b) shows the same arrangement in elevation, that is viewed from the side. If the string is taut and the block is given a push at right angles to the string the block will move in a circle of radius R with the point C as centre. The weight of the block is balanced by the normal contact force exerted by the table so the net force on the block is the force $\boldsymbol{T}$ exerted by the string (Figure 4.31). Note that, as the block moves around in a circular path, the direction of the force $\boldsymbol{T}$ changes continuously so that it is always directed towards the centre C of the circle.

This is an example of a situation which arises in many contexts in nature in which a force on a moving body is directed towards a central point and is therefore called a **centripetal force** ('centre seeking force'). Invoking Newton's Second law in this case, the acceleration of the block is given by $\boldsymbol{a} = \dfrac{\boldsymbol{T}}{M}$ and hence the acceleration is always parallel to $\boldsymbol{T}$. Such an acceleration, directed towards the centre of a circle, is called a **centripetal acceleration**.

Many other examples of centripetal forces will be familiar to the reader; some of these will be discussed later in this chapter (conical pendulum, banked track and centrifuge in the next section and the simple pendulum in Section 4.10) and in subsequent chapters (planetary orbits in Sections 5.12 and 5.13, charged particles in magnetic fields in Section 19.6 and electrons in atoms in Chapter 24).

4.8 Motion in a circle at constant speed

One of the simplest and most important cases of motion in a circle, usually called 'uniform circular motion', is the case of a point which is moving on the circumference of a circle at constant speed. In this case the radius joining the centre of the circle to the point rotates with constant angular velocity about the centre of the circle. Even though the point is moving at constant speed it nevertheless has an acceleration because the velocity vector, which is tangential to the circle, is continually changing in direction as the point moves along the circumference of the circle.

To understand the nature of this acceleration, consider a point P which is moving on a circular path of radius R, as shown in Figure 4.32, with constant speed which, from Equation (2.25), is given by $V = |\boldsymbol{V}| = \omega R$. At some instant t the velocity is $\boldsymbol{V}$ and at some slightly later instant $t + \Delta t$ the velocity is $\boldsymbol{V}' = \boldsymbol{V} + \Delta\boldsymbol{V}$. Since the speed is constant, $|\boldsymbol{V}'| = |\boldsymbol{V}|$. Let the angle between $\boldsymbol{V}'$ and $\boldsymbol{V}$, which is also the angle between the corresponding radii of the circle, be $\Delta\theta$. To determine the change in velocity $\Delta\boldsymbol{V} = \boldsymbol{V}' - \boldsymbol{V}$, the triangle rule for addition of vectors may be used (Figure 4.33) from which it can be seen that, when $\Delta\theta$ is small, $|\Delta\boldsymbol{V}| = |\boldsymbol{V}|\Delta\theta$. Thus $|\Delta\boldsymbol{V}| = |\boldsymbol{V}|\omega\Delta t$ and hence, taking the limit as $\Delta t \to 0$ ($\Delta\theta \to 0$), the magnitude of the instantaneous acceleration of the point is given by

$$\boldsymbol{a} = \underset{\Delta t\to 0}{\text{limit}}\frac{\Delta|\boldsymbol{V}|}{\Delta t} = \underset{\Delta t\to 0}{\text{limit}}|\boldsymbol{V}|\frac{\Delta\theta}{\Delta t} = |\boldsymbol{V}|\underset{\Delta t\to 0}{\text{limit}}\frac{\Delta\theta}{\Delta t} = |\boldsymbol{V}|\omega = \frac{|\boldsymbol{V}|^2}{R} = \omega^2 R$$

Since $\Delta\theta \to 0$ as $\Delta t \to 0$, it can be seen from Figure 4.33 that $\Delta\boldsymbol{V}$ becomes perpendicular to $\boldsymbol{V}$ as $\Delta\theta \to 0$ and hence $\boldsymbol{a}$ is directed perpendicularly to $\boldsymbol{V}$ (that is radially inward) as $\Delta t \to 0$. Thus the acceleration is given by

$$\boxed{\boldsymbol{a}_{\mathrm{c}} = -\frac{|\boldsymbol{V}|^2}{R}\hat{\boldsymbol{r}} = -\omega^2 R\hat{\boldsymbol{r}}} \qquad (4.5)$$

where $\hat{\boldsymbol{r}}$ is a unit vector directed radially outwards from the centre of the circle (Figure 4.34). Thus $\boldsymbol{a}_{\mathrm{c}}$ is a centripetal acceleration as described in Section 4.7.

Different cases of uniform circular motion differ in the way that the centripetal force is provided, as illustrated in the following examples.

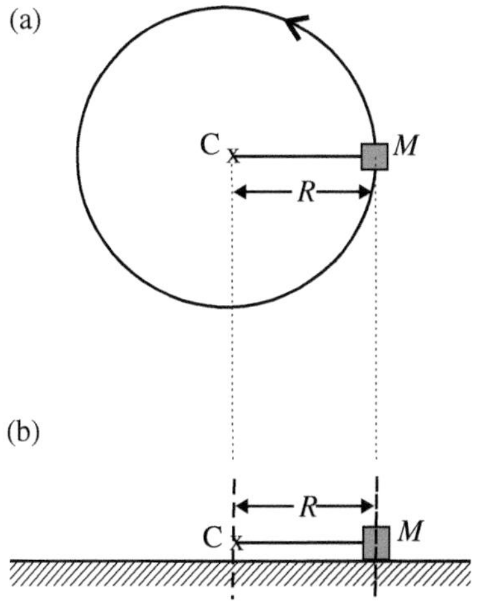

Figure 4.30. A block of mass M on a smooth horizontal surface is attached by a string to a fixed point C. The block is caused to move in a circular path of radius R by giving it a push horizontally perpendicular to the string. (a) The system viewed from above. (b) The same situation viewed from the side.

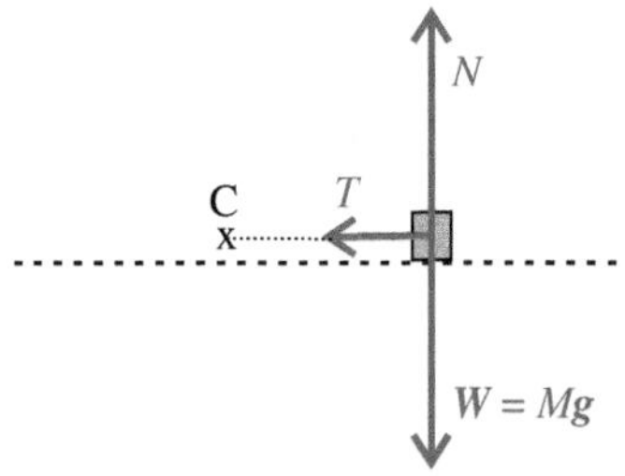

Figure 4.31. Forces acting on the block in Figure 4.30.

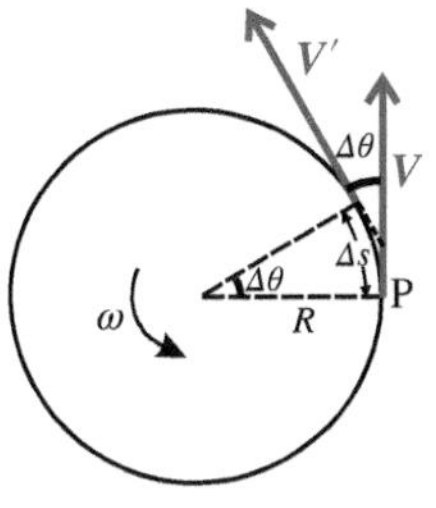

Figure 4.32. Motion in a circle at constant speed. The velocity of the moving point is $\boldsymbol{V}$ at time t and is $\boldsymbol{V}' = \boldsymbol{V} + \Delta\boldsymbol{V}$ at time $t + \Delta t$. The angle swept out by the radius in the time interval Δt is $\Delta\theta$.

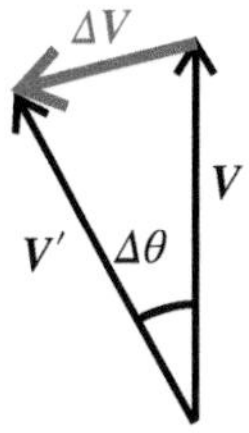

Figure 4.33. The change in velocity, $\Delta\boldsymbol{V}$, in the time Δt is determined by the triangle rule for addition of vectors.

Case 1: First let us return to the problem of the block which is moving in a circle on a horizontal frictionless surface at the end of a string. In this case the centripetal force (the tension in the string, indicated in Figure 4.31) is given by

$$\boldsymbol{T} = M\boldsymbol{a}_{\mathrm{c}} = -M\frac{|\boldsymbol{V}|^2}{R}$$

The tension in the string becomes greater as $|\boldsymbol{V}|$ increases as we would expect instinctively.

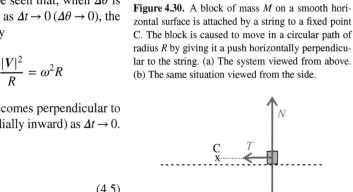

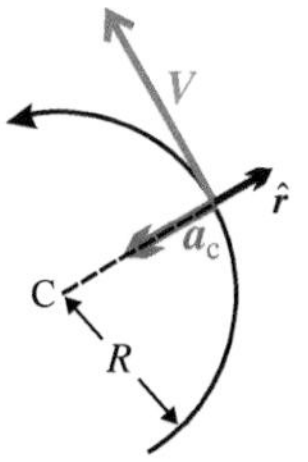

Figure 4.34. The direction of the acceleration of a point moving in a circle at constant speed is parallel to $\Delta\boldsymbol{V}$ in the limit $\Delta t \to 0$; that is, the direction of the acceleration $\boldsymbol{a}_{\mathrm{c}}$ is perpendicular to $\boldsymbol{V}$. Thus $\boldsymbol{a}_{\mathrm{c}}$ is directed towards the centre of the circle.

Case 2: In a *conical pendulum* (illustrated in Figure 4.35) the centripetal force is provided by the horizontal component of the tension in the string. Thus, applying Newton's second law to the mass m (force diagram Figure 4.35(b)), we obtain:

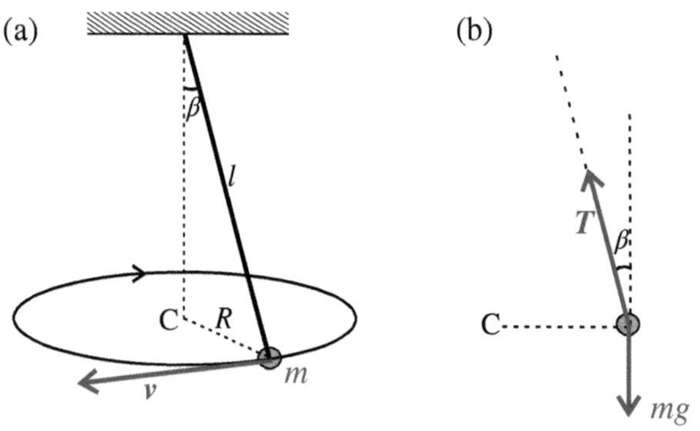

Figure 4.35. A conical pendulum. (a) A mass m at the end of a string of length l moves with speed $|v|$ in a horizontal circle of radius $R = l\sin\beta$ while the string sweeps out the surface of a cone. (b) the forces acting on the mass m.

$$\text{(a) centripetal components: } |\boldsymbol{T}|\sin\beta = m|\boldsymbol{a}_c| = m\frac{|\boldsymbol{v}|^2}{R} = m\frac{|\boldsymbol{v}|^2}{l\sin\beta}$$

$$\text{(b) vertical components: } |\boldsymbol{T}|\cos\beta - mg = 0$$

and hence, eliminating $|\boldsymbol{T}|$, we find that the speed which must be imparted to mass m to get it to move in a horizontal circle of radius R is given by

$$|v| = \sqrt{gl\sin\beta\tan\beta} = \sqrt{gR\tan\beta}$$

where $R = l\sin\beta$ is the radius of the circle.
The period of the pendulum is given by

$$T = \frac{\text{circumference}}{\text{speed}} = \frac{2\pi R}{v} = \frac{2\pi l\sin\beta}{\sqrt{gl\sin\beta\tan\beta}} = 2\pi\sqrt{\frac{l}{g}\cos\beta}$$

Case 3: A *centrifuge* is an important instrument in chemical, biological, medical and other sciences which is used to separate particles or large molecules of different masses. One such arrangement is illustrated in Figure 4.36 where the particles under investigation are suspended in a fluid in a test tube which can be rotated at high speeds around an axis perpendicular to the plane of the figure. The centripetal force required to keep a particle of mass m moving in a circular path is provided by the net force exerted on m by the surrounding fluid. A particle moving with speed v will remain on a path of radius r only if this force is equal to $\frac{mv^2}{r}$. In practice, at high rotational speeds, the fluid cannot fully support the particles; that is the force exerted by the surrounding fluid is a little less than $\frac{mv^2}{r}$ and the particles slowly drift ('sediment') towards the bottom of the tube. The force required to support the particles is proportional to m; thus heavier particles are separated from lighter ones.

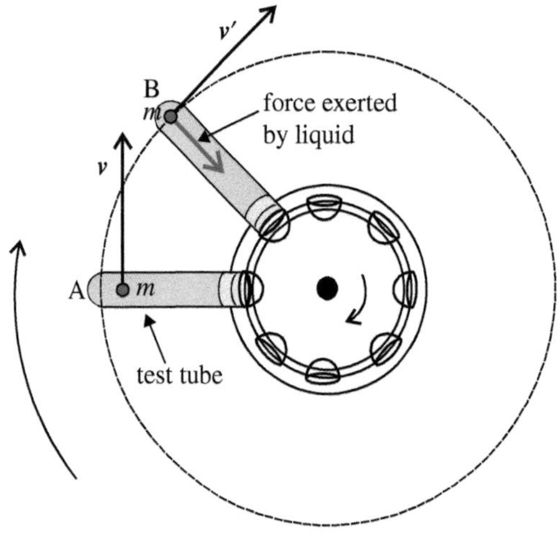

Figure 4.36. An example of a centrifuge. As it rotates, a particle in the body of the test tube (A) drifts towards the bottom of the tube (B).

Study Worked Example 4.6

For problems based on the material presented in this section visit up.ucc.ie/4/ *and follow the link to the problems.*

4.9 Tangential and radial components of acceleration

If the magnitude of the velocity of a point moving in a circle increases (or decreases) the angular velocity also increases (or decreases) as determined by Equation (2.25). The acceleration of a point moving in a circle can be broken into tangential and radial components. Consider the motion, illustrated in Figure 4.37, of a point along the circumference of a circle of radius R where the velocities of the moving point are $\boldsymbol{V}$ and $\boldsymbol{V}'$ at A and B, respectively, noting that $|\boldsymbol{V}| \neq |\boldsymbol{V}'|$ in this case. If the time taken by the point to travel between A and B is Δt, the instantaneous tangential acceleration is given by

$$a_{\mathrm{t}} = \underset{\Delta t \to 0}{\text{limit}} \frac{\Delta |\boldsymbol{V}|}{\Delta t} = \frac{d^2 s}{dt^2}$$

where $\Delta|\boldsymbol{V}| = |\boldsymbol{V}'| - |\boldsymbol{V}|$ is the change in *speed* in time Δt and s represents displacement along the arc of the circle. Since $|\boldsymbol{V}| = \omega R$

$$a_{\mathrm{t}} = R \underset{\Delta t \to 0}{\text{limit}} \frac{\Delta \omega}{\Delta t} = R\frac{d\omega}{dt} = R\alpha \tag{4.6}$$

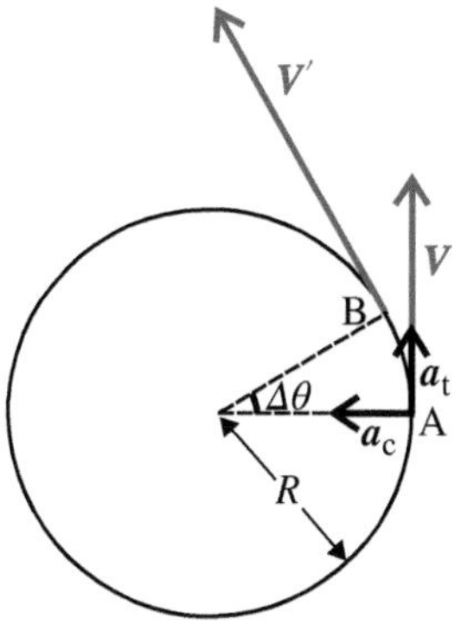

Figure 4.37. Motion of a point on a circular path in a situation where the speed is not constant. In this case, the acceleration can be decomposed into (i) a tangential component (a_{t}) and (ii) a centripetal component (a_{c}).

where α is the instantaneous angular acceleration (Section 2.7). Note that Equation (4.6) is a general result which may be applied to motion in a circle at any instant. For uniform circular motion $a_{\mathrm{t}} = 0$ and $\alpha = 0$.

A similar result applies even if the point is not moving on a circular path. The acceleration of a point which is moving with arbitrary velocity on a general curved trajectory can also be broken down into a centripetal component and a tangential component. This is illustrated in Figure 4.38 where the velocity at time t is $\boldsymbol{v}$ and the velocity at time $t + \Delta t$ is $\boldsymbol{v}' = \boldsymbol{v} + \Delta\boldsymbol{v}$. Now $\Delta\boldsymbol{v}$ can be resolved into perpendicular and tangential components as shown in the figure, that is $\Delta\boldsymbol{v} = \Delta\boldsymbol{v}_{\parallel} + \Delta\boldsymbol{v}_{\perp}$, where once again the + sign refers to vector addition. Hence the acceleration at any instant is given by

$$\boldsymbol{a} = \underset{\Delta t \to 0}{\text{limit}} \frac{\Delta \boldsymbol{v}}{\Delta t} = \underset{\Delta t \to 0}{\text{limit}} \frac{\Delta \boldsymbol{v}_{\parallel}}{\Delta t} + \underset{\Delta t \to 0}{\text{limit}} \frac{\Delta \boldsymbol{v}_{\perp}}{\Delta t} = \underset{\Delta t \to 0}{\text{limit}} \frac{\Delta |\boldsymbol{v}_{\parallel}|}{\Delta t}\hat{\boldsymbol{t}} + \underset{\Delta t \to 0}{\text{limit}} \frac{|\Delta \boldsymbol{v}_{\perp}|}{\Delta t}\hat{\boldsymbol{r}} = a_{\mathrm{t}}\hat{\boldsymbol{t}} + a_{\mathrm{c}}\hat{\boldsymbol{r}}$$

where $\hat{\boldsymbol{r}}$ is a unit vector directed outwards from the centre of the curvature and $\hat{\boldsymbol{t}}$ is a unit vector tangential to the trajectory and in the direction of motion of the point (as shown in Figure 4.38(c)). Thus a_{t} and a_{c} are, respectively, the tangential and centripetal components of the acceleration.

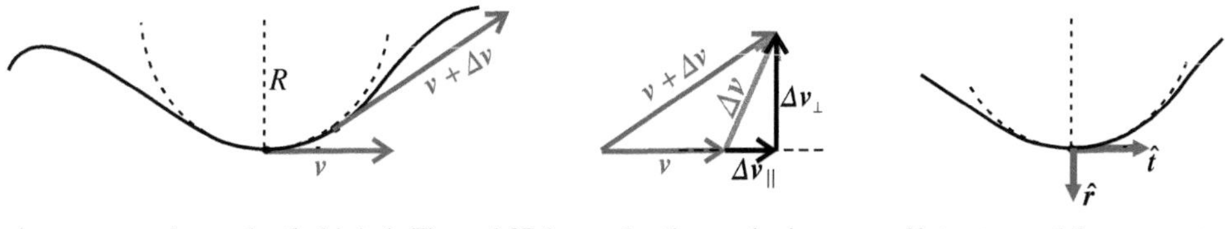

Figure 4.38. Motion of a point on a general curved path. (a) As in Figure 4.37 the acceleration can be decomposed into a tangential component and a centripetal component. (b) The corresponding vector diagram for the decomposition of $\Delta\boldsymbol{v}$. (c) The radial and tangential unit vectors.

A point which is moving on a curved path of arbitrary shape may be considered, at any instant, to be executing circular motion, where the radius of the circle is the radius of curvature of the path. In other words, Equations (4.5) and (4.6) can be applied to such general motion, provided that R is understood to be the radius of curvature of the path at the instantaneous position of the moving point. Thus

$$\boldsymbol{a} = a_{\mathrm{t}}\hat{\boldsymbol{t}} + a_{\mathrm{c}}\hat{\boldsymbol{r}} = \alpha R\hat{\boldsymbol{t}} - \omega^2 R\hat{\boldsymbol{r}} = \frac{d^2 s}{dt^2}\hat{\boldsymbol{t}} - \frac{|\boldsymbol{v}|^2}{R}\hat{\boldsymbol{r}} \tag{4.7}$$

For problems based on the material presented in this section visit up.ucc.ie/4/ and follow the link to the problems.

4.10 Hybrid motion: the simple pendulum

The simple pendulum (Figure 4.39) brings together nicely many of the ideas introduced in this chapter. The pendulum comprises a small mass m (called the 'bob') attached to a light inextensible string of length l, the other end of the string being connected to a fixed support. When the pendulum is at rest in equilibrium the string hangs vertically and, as we have seen, the tension in the string is mg. If the mass is drawn to one side, keeping the string taut, and then released the pendulum will execute a bounded oscillatory motion in a vertical plane with the mass constrained to move on a circular path.

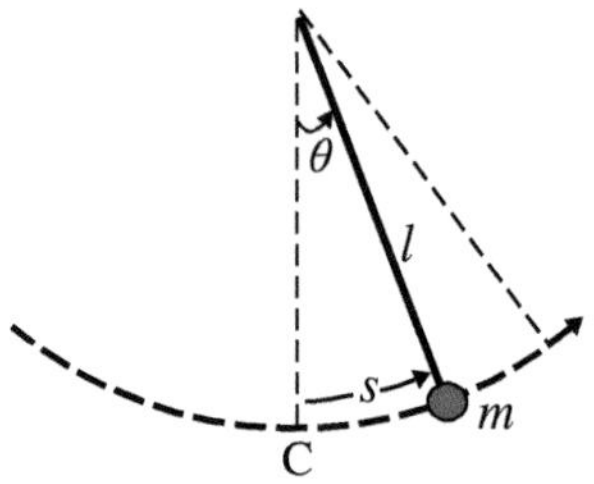

Figure 4.39. A simple pendulum. The motion can be characterised by either the variable θ (the angle the string makes with the vertical) or the variable s (the arc length measured along the circle from C).

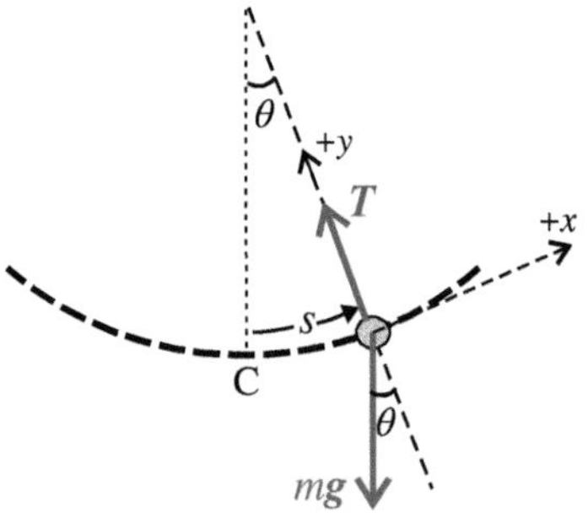

Figure 4.40. The forces acting on the pendulum bob at an arbitrary point in the motion are (i) the weight ***mg*** of the bob and (ii) the force ***T*** exerted by the string on the bob.

Figure 4.40 shows the forces acting on the mass m at some arbitrary point in the motion. Let the angle which the string makes with the vertical at this point be θ and the corresponding position of the mass measured along the arc of the circle be s. Taking components along the tangent to the circle (x-direction) and along the radius (y-direction), Newton's second law yields:

(a) x-direction: $$-mg\sin\theta = ma_x \tag{4.8}$$

(b) y-direction: $$|\boldsymbol{T}| - mg\cos\theta = ma_y \tag{4.9}$$

Note that, in this case, the acceleration has both tangential and radial (centripetal) components.

Equation (4.8) gives us an equation which describes the motion of the mass m on the arc of the circle

$$a_x = \frac{d^2s}{dt^2} = -g\sin\theta \tag{4.10}$$

and, while the motion is oscillatory, it is not simple harmonic since, in general, $\sin\theta$ is not proportional to s.

If the motion is confined to *small amplitude oscillations*, however, the equation takes on a more familiar appearance. If θ is small, $\sin\theta \approx \theta = \frac{s}{l}$, and Equation (4.10) reduces to

$$\frac{d^2s}{dt^2} = -\frac{g}{l}s$$

which is the differential equation for simple harmonic motion (recall Equation (3.5) with $\gamma = \omega^2 = \frac{g}{l}$). Hence the period of motion is given by

$$T = 2\pi\sqrt{\frac{l}{g}} \tag{4.11}$$

and the position (s) and velocity ($v = \frac{ds}{dt}$) of the bob as functions of time can be determined from Equations (3.7a) and (3.7b) using $\omega = \sqrt{\frac{g}{l}}$.

The motion of a simple pendulum is 'hybrid' in the sense that it exhibits features of both simple harmonic motion and circular motion. From Equation (4.9) and invoking Equation (4.5) one can determine an expression for the tension in the string at any point in the motion, that is

$$|\boldsymbol{T}| = ma_y + mg\cos\theta = mg\cos\theta + m\frac{v^2}{l}$$

The applicability or otherwise of the 'small angles' approximation to describe real pendulums depends on the accuracy required. For angles up to 10°, $\sin\theta$ is equal to θ in rad to 0.5% accuracy ($\sin 10° = 0.1736$ and $10° = 0.1745$ rad). An exact solution to Equation (4.10) for oscillations of arbitrary amplitude is beyond the scope of this book but, as described in the web section up.ucc.ie/4/10/1/, it is relatively easy to derive correction terms which can be added to the expression (4.11) for the period if greater accuracy is required.

For problems based on the material presented in this section visit up.ucc.ie/4/ and follow the link to the problems.

4.11 Angular quantities as vector: the cross product

For a particle which is moving on a circular path, where the motion is confined to a plane, it is sufficient to treat angular displacement, velocity and acceleration as algebraic quantities, as discussed in Section 2.7. In other situations, however, it proves useful to extend the definitions of these quantities so that they can be treated as vectors.

Before we can do this, an additional vector technique must be introduced, namely a second method of forming a product of two vectors, in addition to the scalar product introduced in Section 4.2.

The vector product

The product of two vectors, which is defined below, will prove to be a very powerful tool for dealing with angular motion and a wide range of other physical phenomena. Learning this definition carefully at this stage will avoid the need to learn a variety of different rules later on.

The **vector product** (or '**cross product**') of two vectors $\boldsymbol{A}$ and $\boldsymbol{B}$ is a *vector* $\boldsymbol{C}$ denoted by $\boldsymbol{C} = \boldsymbol{A} \times \boldsymbol{B}$ ("A cross B") such that (i) the magnitude of $\boldsymbol{C}$ is given by

$$|\boldsymbol{C}| = |\boldsymbol{A}||\boldsymbol{B}| \sin\theta$$

where θ is the smaller angle between the directions of $\boldsymbol{A}$ and $\boldsymbol{B}$, and (ii) the direction of $\boldsymbol{C}$ is defined as being perpendicular to the plane containing $\boldsymbol{A}$ and $\boldsymbol{B}$ and in the direction of advance of a right hand screw rotated from $\boldsymbol{A}$ to $\boldsymbol{B}$.

In Figure 4.41 this definition is applied to the vector products $\boldsymbol{C}_1 = \boldsymbol{A} \times \boldsymbol{B}$ and $\boldsymbol{C}_2 = \boldsymbol{B} \times \boldsymbol{A}$. Note that although $|\boldsymbol{C}_1| = |\boldsymbol{C}_2|$, $\boldsymbol{C}_1 = -\boldsymbol{C}_2$; the vectors $\boldsymbol{C}_1$ and $\boldsymbol{C}_2$ are in opposite directions and hence are distinctly different vectors. The order of multiplication is important in a vector product; in mathematical language, the vector product is *not commutative*. Note also that, if the direction of one vector in a vector product, such as $\boldsymbol{A}$ in the product $\boldsymbol{A} \times \boldsymbol{B}$, is reversed (indicated by $\boldsymbol{D} = -\boldsymbol{A}$ in Figure 4.42), the smaller angle between the vectors becomes $\theta' = 180^\circ - \theta$, as shown. While the magnitude of the cross product $|\boldsymbol{D}||\boldsymbol{B}| \sin\theta' = |\boldsymbol{A}||\boldsymbol{B}|\sin\theta' = |\boldsymbol{A}||\boldsymbol{B}|\sin(180-\theta) = |\boldsymbol{A}||\boldsymbol{B}|\sin\theta)$ is unchanged, its direction reverses because it is now given by an anticlockwise rotation instead of a clockwise rotation. When using the vector product, therefore, it is vitally important to make sure that the vectors are multiplied in the specified order.

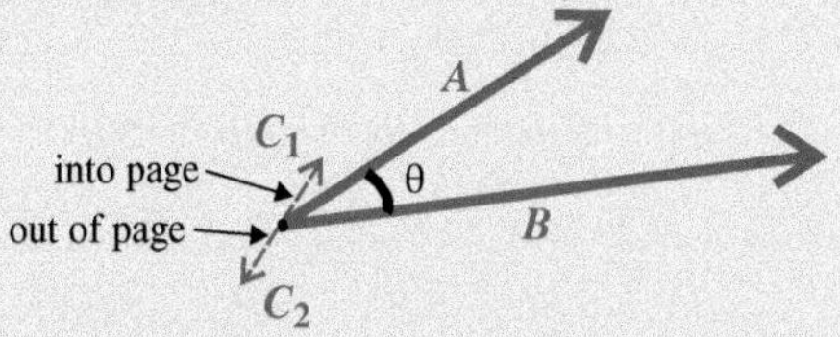

Figure 4.41. Direction of the vector product. If the vectors $\boldsymbol{A}$ and $\boldsymbol{B}$ are in the plane of the figure, the vector $\boldsymbol{A} \times \boldsymbol{B}$ is directed perpendicularly to the page inward while the vector $\boldsymbol{B} \times \boldsymbol{A}$ is directed perpendicularly to the page outward.

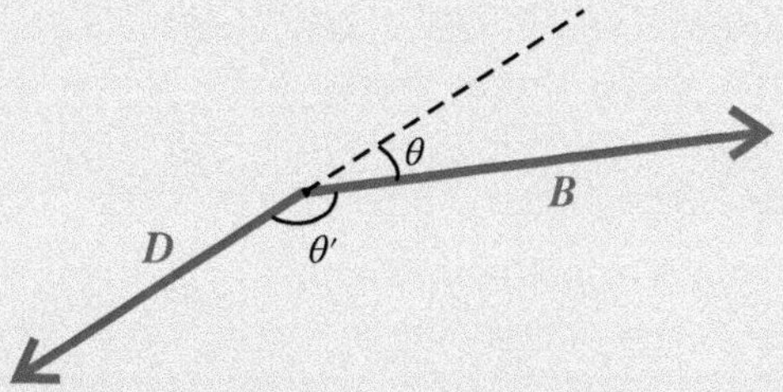

Figure 4.42. If the direction of one of the vectors in a vector product is reversed (for example $\boldsymbol{D} = -\boldsymbol{A}$), then the direction of the product vector is also reversed; $\boldsymbol{A} \times \boldsymbol{B}$ (into page in Figure 4.41) becomes $\boldsymbol{D} \times \boldsymbol{B}$ (anticlockwise rotation → out of page).

Area of a triangle

As an example of the use of the vector product, note that the area of a triangle can be expressed simply in terms of vectors representing two adjacent sides (Figure 4.43(a)) since

$$\text{area of triangle} = \frac{1}{2}|\boldsymbol{b}|h = \frac{1}{2}|\boldsymbol{b}||\boldsymbol{a}| \sin\phi = \frac{1}{2}|\boldsymbol{a} \times \boldsymbol{b}|$$

Similarly the area of a parallelogram in Figure 4.43(b) is $|\boldsymbol{a} \times \boldsymbol{b}|$.

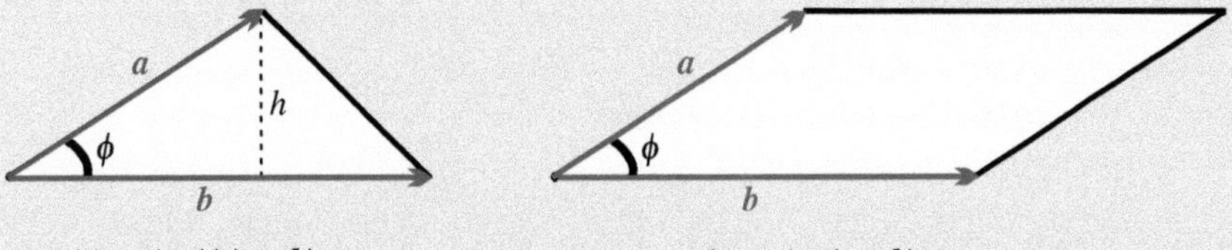

(a) $A = \frac{1}{2} |\boldsymbol{a} \times \boldsymbol{b}|$ (b) $A = |\boldsymbol{a} \times \boldsymbol{b}|$

Figure 4.43. The areas of a triangle and a parallelogram in terms of a vector product.

Area element as a vector

It follows from above that the magnitude of an element of area formed by the displacement elements $\Delta\boldsymbol{a}$ and $\Delta\boldsymbol{b}$ (Figure 4.44(a)) is $|\Delta\boldsymbol{a} \times \Delta\boldsymbol{b}|$. It is often useful to treat such an area element as a vector itself, defined by $\Delta\boldsymbol{A} := \Delta\boldsymbol{a} \times \Delta\boldsymbol{b}$, which is a vector perpendicular to the plane containing $\boldsymbol{\Delta a}$ and $\boldsymbol{\Delta b}$ directed along the direction defined by the cross-product rule. This definition can in turn be generalised to a plane area element of any shape as follows

$$\boxed{\boldsymbol{\Delta A} := |\boldsymbol{\Delta A}|\hat{\boldsymbol{n}}}$$

where, in this case, $\hat{\boldsymbol{n}}$ is a unit vector perpendicular to the area as indicated in Figure 4.44(b). In this case, the direction of $\hat{\boldsymbol{n}}$ is determined by the arbitrary choice of the sense in which the closed loop C which encloses the area ΔA is traced out (Figure 4.44(b)). To be consistent with the vector cross-product, therefore, the direction of $\hat{\boldsymbol{n}}$ must be perpendicular to plane of the element and in the direction of advance of a right hand screw rotated in the sense (or direction) chosen for the curve C. This selection of the way the arrow on the curve C points in Figure 4.44(c) is an arbitrary choice; technically speaking, one is adopting what is called an *orientation* for the curve

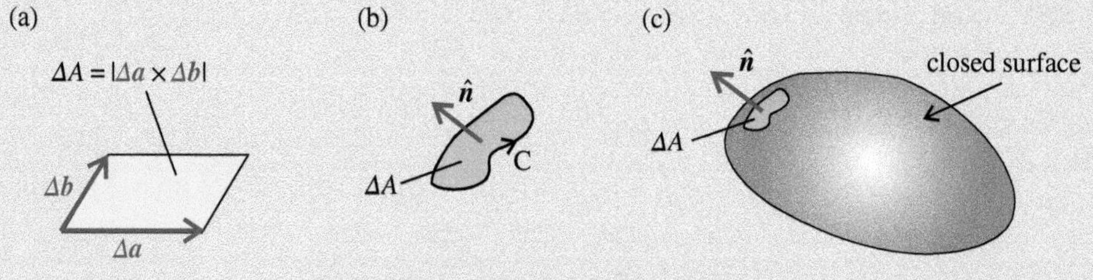

Figure 4.44. An area element as a vector quantity.

In the case of a closed surface it is conventional to choose the direction of $\hat{\boldsymbol{n}}$ as pointing outwards from the surface (Figure 4.44(c)).

Some useful properties of the cross product are given in Appendix A.11.7 and a good collection of problems may be found, for example, in Halliday, Resnick and Walker.

Vector element of plane angle

In Section 2.7 we defined the unit of plane angle (the radian — Equation (2.24)) such that an element of plane angle is given by $\Delta\theta = \dfrac{\Delta s}{R}$, where Δs is the element of arc of a circle of radius R swept out by the element of angle $\Delta\theta$ (Figure 4.45) and hence we hence we can write $\Delta s = R(\Delta\theta)$.We can generalise this result by defining a vector quantity, the element of angular displacement, $\boldsymbol{\Delta\theta}$, such that

$$\boldsymbol{\Delta s} := \boldsymbol{\Delta\theta} \times \boldsymbol{R}$$

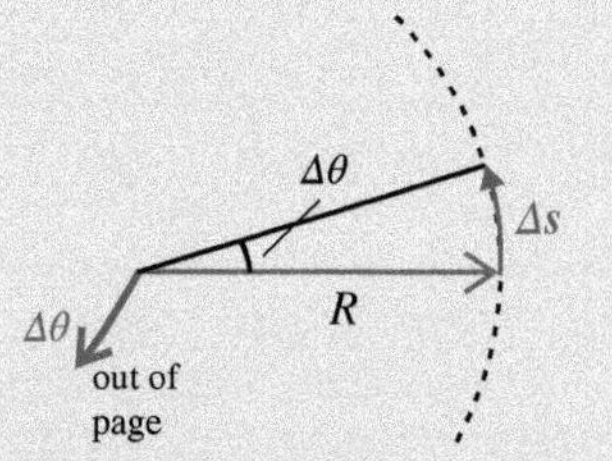

Figure 4.45. The element of plane angle $\boldsymbol{\Delta\theta}$ can be considered as a vector quantity defined through the vector product $\boldsymbol{\Delta s} = \boldsymbol{\Delta\theta} \times \boldsymbol{R}$.

where $\boldsymbol{\Delta s}$ is the (vector) element of arc length along the circumference of a circle of radius R and $\boldsymbol{R}$ is the displacement of a point on the circumference relative to the centre of the circle. From the definition of the cross product, it follows that the direction of the vector $\boldsymbol{\Delta\theta}$ in Figure 4.45 is perpendicular to the page outwards if the angle $\Delta\theta$ is described in an anticlockwise sense. On the other hand, if the angle $\Delta\theta$ had been swept out clockwise, the vector $\boldsymbol{\Delta\theta}$ would have been directed into the page.

Angular velocity as a vector

Consider a point which is moving on a circular path as shown in Figure 4.46. If $\boldsymbol{\Delta s}$ is the element of arc length swept out by the radius vector $\boldsymbol{R}$ in time Δt, the velocity of the moving point is given by

$$\boldsymbol{v} = \frac{d\boldsymbol{s}}{dt} = \underset{\Delta t \to 0}{\text{limit}} \frac{\boldsymbol{\Delta s}}{\Delta t} = \underset{\Delta t \to 0}{\text{limit}} \frac{\boldsymbol{\Delta\theta}}{\Delta t} \times \boldsymbol{R}$$

or

$$\boxed{\boldsymbol{v} = \boldsymbol{\omega} \times \boldsymbol{R}} \qquad (4.12)$$

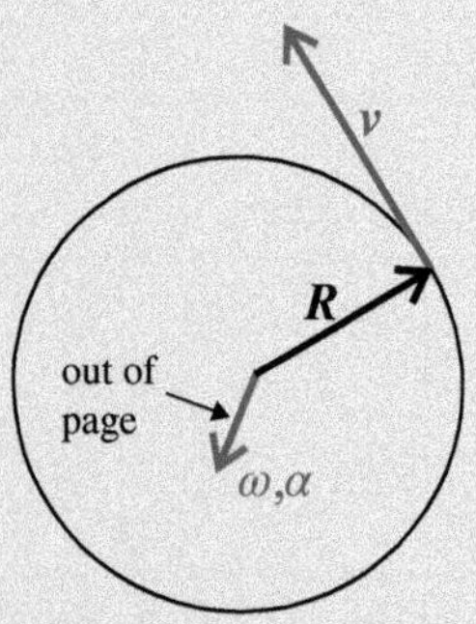

Figure 4.46. Angular velocity can be defined as a vector quantity using the vector product; that is, $\boldsymbol{\omega}$ is defined such that $\boldsymbol{v} = \boldsymbol{\omega} \times \boldsymbol{R}$.

which is the vector generalisation of Equation (2.25).

As illustrated in Figure 4.46, rotation of $\boldsymbol{\omega}$ into $\boldsymbol{R}$ according to the rule for the cross product gives the direction of $\boldsymbol{v}$. The advantage of the vector form of the equation is therefore that the direction of $\boldsymbol{v}$ is built into the equation unambiguously and does not have to be inferred through any additional rules.

Angular acceleration as a vector

In general, the velocity of a point which is moving on a circle can be changing in both magnitude and direction. The acceleration of the point is given by

$$\begin{aligned} \boldsymbol{a} = \frac{d\boldsymbol{v}}{dt} = \frac{d}{dt}(\boldsymbol{\omega}\times\boldsymbol{R}) = \frac{d\boldsymbol{\omega}}{dt}\times\boldsymbol{R} + \boldsymbol{\omega}\times\frac{d\boldsymbol{R}}{dt} &= \boldsymbol{\alpha}\times\boldsymbol{R} + \boldsymbol{\omega}\times\boldsymbol{v} \\ &= \boldsymbol{\alpha}\times\boldsymbol{R} + \boldsymbol{\omega}\times(\boldsymbol{\omega}\times\boldsymbol{R}) \end{aligned} \tag{4.13}$$

where $\boldsymbol{\alpha} = \dfrac{d\boldsymbol{\omega}}{dt}$ is the (vector) angular acceleration and we have used the result Appendix A.12.1 for the derivative of a cross product. Note that, in writing Equation (4.13) and elsewhere in this book, we follow the convention that multiplication is implemented before addition. Thus the vector products $\boldsymbol{\alpha}\times\boldsymbol{R}$ and $\boldsymbol{\omega}\times(\boldsymbol{\omega}\times\boldsymbol{R})$ are implemented before the two terms are added.

Note that in Figure 4.46 the directions of both $\boldsymbol{\omega}$ and $\boldsymbol{\alpha}$ are perpendicular to the plane of the circle and hence $\boldsymbol{\alpha}\times\boldsymbol{R}$ is tangential to the circle in the direction of $\boldsymbol{v}$. Furthermore $\boldsymbol{\omega}\times(\boldsymbol{\omega}\times\boldsymbol{R})$ is directed towards the centre of the circle and we can recognise Equation (4.13) as an alternative form of Equation (4.7) since

$$\boldsymbol{a} = \boldsymbol{\alpha}\times\boldsymbol{R} + \boldsymbol{\omega}\times(\boldsymbol{\omega}\times\boldsymbol{R}) = \alpha R\hat{\boldsymbol{t}} - \omega^2 R\hat{\boldsymbol{r}}$$

Unlike Equation (4.7), however, Equation (4.13) also applies when the motion is not confined to a plane. It, therefore, can accommodate situations in which the directions of $\boldsymbol{\Delta\theta}$, $\boldsymbol{\omega}$ and $\boldsymbol{\alpha}$ are not necessarily parallel to one another and can change direction in the course of the motion. In general in this book we will only need to consider cases in which $\boldsymbol{\Delta\theta}$, $\boldsymbol{\omega}$ and $\boldsymbol{\alpha}$ are all fixed in the same direction. In Section 7.11 (gyroscopic motion), however, we will consider a situation in which the direction of $\boldsymbol{\omega}$ changes; the advantages of the vector representation of angular variables will then be clear. In Section 8.5 (Coriolis force) the real advantage of a vector treatment of angular quantities will become evident.

Note on unit vector notation

Throughout this chapter we have used the standard notation for a unit vector, that is indicating a unit vector by placing a 'hat' over the corresponding vector (for example, $\hat{\boldsymbol{r}} := \dfrac{\boldsymbol{r}}{|\boldsymbol{r}|}$). We shall continue to use this notation for unit vectors throughout this book with four exceptions: provided that the context is unambiguous, it is conventional to drop the 'hat' for the three Cartesian coordinate unit vectors, ($\hat{\boldsymbol{i}}, \hat{\boldsymbol{j}}$ and $\hat{\boldsymbol{k}}$) and for the unit vector $\hat{\boldsymbol{n}}$, which conventionally represents a unit vector directed normal to a surface. Thus, we shall henceforth use $\boldsymbol{i}, \boldsymbol{j}, \boldsymbol{k}$ and $\boldsymbol{n}$ rather than $\hat{\boldsymbol{i}}, \hat{\boldsymbol{j}}, \hat{\boldsymbol{k}}$ and $\hat{\boldsymbol{n}}$ unless there is a risk of ambiguity.

For problems based on the material presented in this section visit up.ucc.ie/4/ *and follow the link to the problems.*

WORKED EXAMPLES

Worked Example 4.1: Use the scalar product of two vectors to derive the cosine rule (Appendix A.3.3), that is, referring to Figure 4.47, $c^2 = a^2 + b^2 - 2ab\cos\theta$

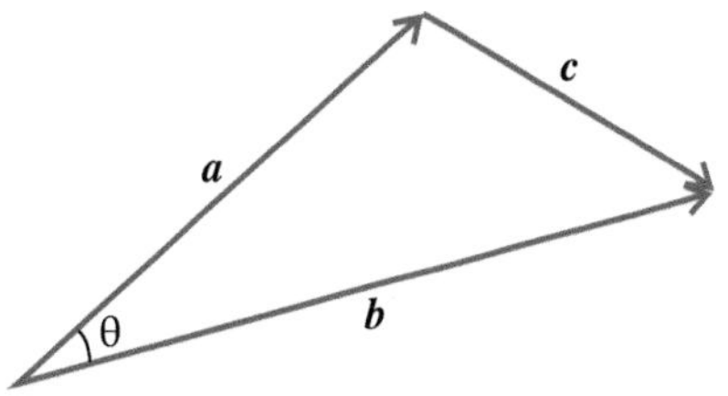

Figure 4.47. Worked Example 4.1.

Treating the sides of the triangle as vectors, as in Figure 4.47, $\boldsymbol{a} + \boldsymbol{c} = \boldsymbol{b}$ or $\boldsymbol{c} = \boldsymbol{b} - \boldsymbol{a}$. Thus

$$c^2 = |\boldsymbol{c}|^2 = \boldsymbol{c} \cdot \boldsymbol{c} = (\boldsymbol{b} - \boldsymbol{a}) \cdot (\boldsymbol{b} - \boldsymbol{a}) = |\boldsymbol{b}|^2 - 2\boldsymbol{b} \cdot \boldsymbol{a} + |\boldsymbol{a}|^2 = a^2 + b^2 - 2\boldsymbol{a} \cdot \boldsymbol{b} = a^2 + b^2 - 2ab\cos\theta$$

Worked Example 4.2: A 200 g mass, moving in a horizontal plane, continuously experiences a constant force of 2.0 N directed southward. If, at some instant $t = 0$, the mass is moving in a direction 30° north of east at a speed of 40 m s^{-1}, what will be the easterly and northerly components of (a) its position and (b) its velocity after 5.0 s?

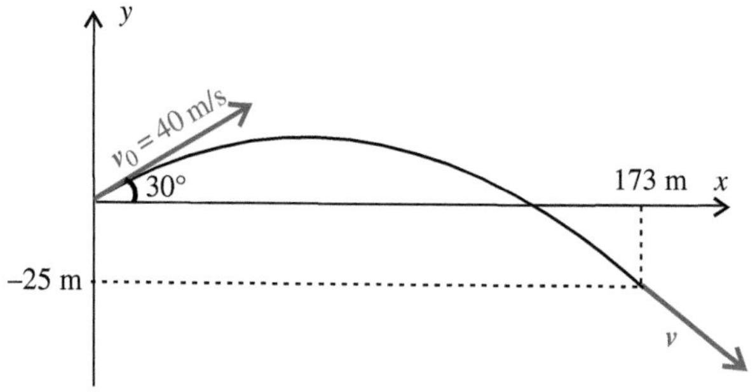

Figure 4.48. Worked Example 4.2.

Taking the + x-axis as east and the + y-axis as north as indicated in Figure 4.48:

$$(v_0)_x = (40\ \text{m s}^{-1})\cos 30° = 34.6\ \text{m s}^{-1}; \quad a_x = 0$$

$$(v_0)_y = (40\ \text{m s}^{-1})\cos 60° = 20.0\ \text{m s}^{-1} \quad a_y = -\frac{F_y}{m} = -\frac{2.0}{0.2} = -10\ \text{m s}^{-2}$$

Using Equations (4.2a) and (4.2b) and taking x- and y-components:

(a) $x = (v_0)_x t = (34.6\ \text{m s}^{-1})(5.0\ \text{s}) = 173\ \text{m}$

$y = (v_0)_y t + \frac{1}{2} a_y t^2 = (20.0\ \text{m s}^{-1})(5.0\ \text{s}) + ½(-10\ \text{m s}^{-2})(5.0\ \text{s})^2 = 100 - 125\ \text{m} = -25\ \text{m}$

(b) $v_x = (v_0)_x + a_x t = 34.6\ \text{m s}^{-1} + (0)(5.0\ \text{s}) = 34.6\ \text{m s}^{-1}$

$v_y = (v_0)_y + a_y t = (20.0\ \text{m s}^{-1}) + (-10\ \text{m s}^{-2})(5.0\ \text{s}) = 20 - 50 = -30.0\ \text{m s}^{-1}$

Note $|\boldsymbol{v}| = \sqrt{(34.6)^2 + (-30.0)^2} = 45.8\ \text{m s}^{-1}$

$\theta = \tan^{-1}\left(\frac{-30.0}{34.6}\right) = \tan^{-1}(-0.145) = -40.9°$ — that is, moving in a direction 40.9° S of E

Worked Example 4.3: A boat of mass 8500 kg is to be towed by two tugs as illustrated in Figure 4.49. Tug A pulls with a force of 3000 N in the NW direction while tug B pulls with a force of 1500 N in the direction 20° E of N. (a) What is the magnitude of the net force exerted on the boat by the tugs? (b) In which direction will the boat start to move? (c) If the acceleration of the boat is observed to be 0.13 m s^{-2}, what is the resistive force exerted by the water on the boat?

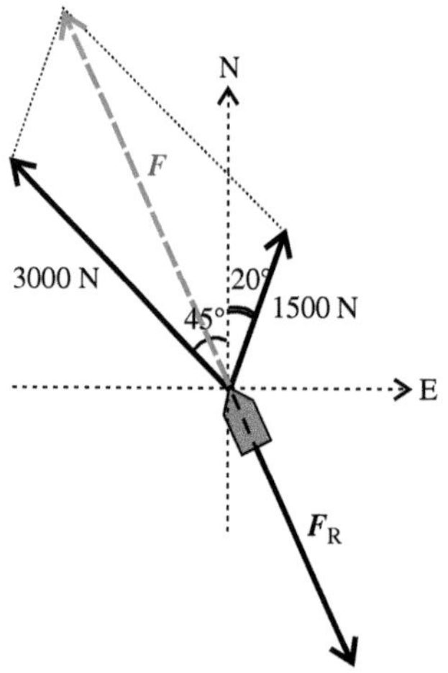

Figure 4.49. Worked Example 4.3.

Taking the $+x$-axis as east and the $+y$-axis as north as indicated in Figure 4.49:

$$F_x = -(3000\ \text{N})\cos 45^\circ + (1500\ \text{N})\cos 70^\circ = -1608\ \text{N}$$

$$F_y = (3000\ \text{N})\sin 45^\circ + (1500\ \text{N})\cos 20^\circ = 3531\ \text{N}$$

(a) $|\boldsymbol{F}| = \sqrt{(-1608)^2 + (3531)^2} = 3900\ \text{N}$

(b) $\tan^{-1}\left(-\frac{3531}{1608}\right) = -65^\circ$ or 115° — that is, $\boldsymbol{F}$ is directed at 65° N of W

(c) $3900\ \text{N} - F_R = Ma = (8500\ \text{kg})(0.13\ \text{m s}^{-2}) \rightarrow F_R = 3900 - 1100 = 2800\ \text{N}$

Worked Example 4.4: A 30 kg crate is pushed up a smooth plane, inclined at an angle of 12.5° with the horizontal, by means of a force of 200 N which is applied horizontally as indicated in Figure 4.50. What are (a) the acceleration of the crate and (b) the normal contact force exerted on the crate by the plane?

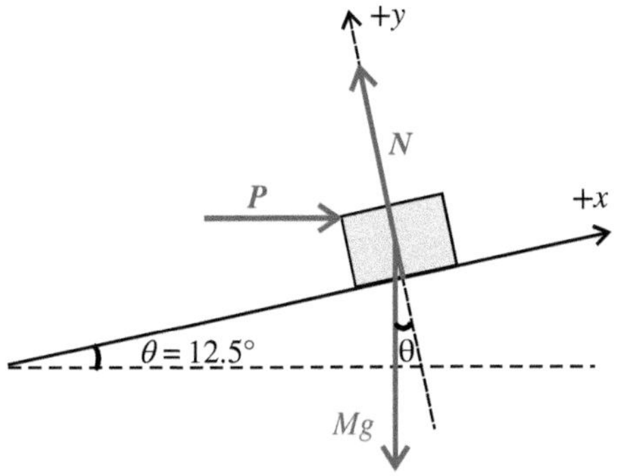

Figure 4.50. Worked Example 4.4.

Taking the $+x$- and $+y$-directions as indicated in the figure

(a) components along the plane (x-direction): $|\boldsymbol{P}|\cos\theta - Mg\sin\theta = Ma \quad \rightarrow \quad a = -g\sin\theta + \frac{|\boldsymbol{P}|}{M}\cos\theta \rightarrow a = 4.4\ \text{m s}^{-2}$.

(b) components perpendicular to plane (y-direction): $|\boldsymbol{N}| - Mg\cos\theta - |\boldsymbol{P}|\sin\theta = 0 \quad \rightarrow \quad |\boldsymbol{N}| = Mg\cos\theta + |\boldsymbol{P}|\sin\theta = 330\ \text{N}$

Worked Example 4.5: A block is projected up along an inclined plane with an initial speed of 5.5 m s^{-1}. The plane is inclined at an angle of 20° to the horizontal (Figure 4.51) and the coefficients of dynamic and static friction between the block and the plane are 0.20 and 0.25, respectively. (a) How far will the block travel up the plane? (b) Determine whether or not the block slides back down the plane and calculate the acceleration if it does. (c) What would be the answers to parts (a) and (b) if the slope of the plane had been 10°?

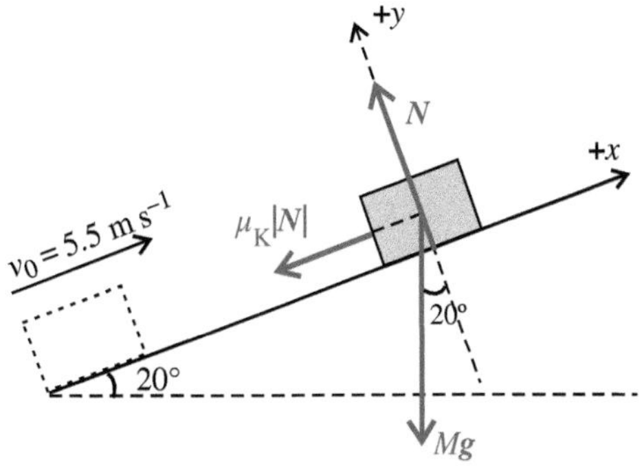

Figure 4.51. Worked Example 4.5.

(a) Components perpendicular to the plane (y-direction): $|\boldsymbol{N}| - Mg\cos 20^\circ = 0 \quad \rightarrow \quad |\boldsymbol{N}| = Mg\cos 20^\circ$

Components along the plane (x-direction): $-Mg\sin 20^\circ - (0.2)|\boldsymbol{N}| = Ma \rightarrow a = -g(\sin 20^\circ + (0.2)\cos 20^\circ) = -5.2\ \text{m s}^{-2}$

Using $v^2 = v_0^2 + 2Ax \quad \rightarrow \quad 0 = (5.5\ \text{m s}^{-1})^2 + 2(-5.2\ \text{m s}^{-2})x \quad \rightarrow \quad x = \frac{5.5^2}{2(5.2)} = 2.9\ \text{m}$

(b) $\tan^{-1}(\mu_s) = \tan^{-1}(0.25) = 14^\circ < 20^\circ$ so the block will slide back down the plane with acceleration given by $a = -g(\sin 20^\circ - (0.2)\cos 20^\circ) = -1.5\ \text{m s}^{-2}$

(c) Repeating (a) above with 10° instead of 20°

In this case $a = -g(\sin 10^\circ + (0.2)\cos 10^\circ) = -3.6\ \text{m s}^{-2}$

and (using $v^2 = v_0^2 + 2Ax$) $0 = (5.5\ \text{m s}^{-1})^2 + 2(-3.6\ \text{m s}^{-2})x \quad \rightarrow \quad x = \frac{5.5^2}{2(3.6)} = 4.2\ \text{m}$

Since $10^\circ < \tan^{-1}(0.25) = 14^\circ$, the body will not slide down the plane when it comes to rest.

Worked Example 4.6: A car is moving on a smooth banked track which is sloped at 30° to the horizontal. At what speed must the car travel to round a turn without slipping if the radius of curvature of the turn is 75 m?

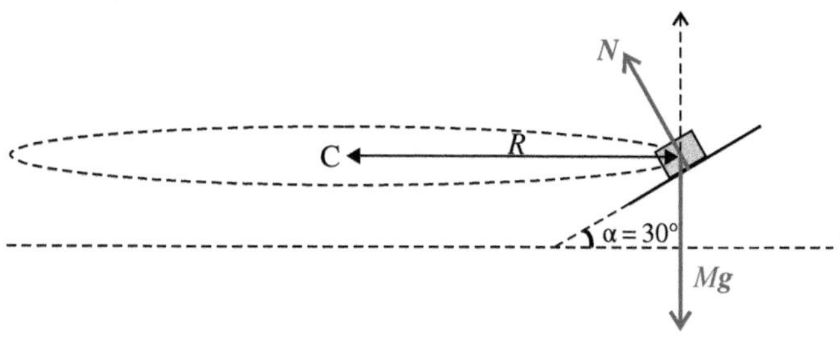

Figure 4.52. Worked Example 4.6.

The analysis of this problem is formally identical to that of the conical pendulum (Section 4.8). The body is moving in a horizontal circle of radius R and the centripetal force, in this case, is provided by the horizontal component of the normal contact force of the track on the moving body as indicated in Figure 4.52.

(i) horizontal (centripetal) components: $|N| \sin \alpha = M|\boldsymbol{a}_c| = M\dfrac{|\boldsymbol{v}|^2}{R}$

(ii) vertical components: $|N| \cos \alpha - Mg = 0 \;\rightarrow\; |N| = \dfrac{Mg}{\cos a} \;\rightarrow\; Mg \tan \alpha = M\dfrac{|\boldsymbol{v}|^2}{R} \;\rightarrow\; |\boldsymbol{v}| = \sqrt{gR \tan a} = 20.6 \text{ m s}^{-1}$

PROBLEMS

For problems based on the material covered this chapter visit up.ucc.ie/4/ and follow the link to the problems.

5

Force fields

AIMS

- to introduce the fundamental law which governs the attractive force between masses (*Newton's law of gravitation*)
- to describe how the concept of a force field may used to describe the behaviour of bodies under the influence of a variety of forces
- to extend the definition of mechanical work, energy, and power to two and three dimensions and to demonstrate the value of these quantities in the analysis of complicated mechanical systems.
- to introduce the concept of angular momentum and to see how it is applied in the description of the rotational movement of bodies about a fixed centre such as the orbits of planets about the Sun

5.1 Newton's law of universal gravitation

The development of mankind's understanding of the universe and, in particular, the role of gravity, from the early ideas of philosophers in ancient Greece to the proposal of **the law of gravitation** by Newton in the seventeenth century, is one of the most extraordinary stories in the history of human knowledge. No brief treatment here could do justice to the excitement and heroism of the tale; readers are encouraged to study any one of the many books on the subject (for example, Koestler, A. (1968) *The Sleepwalkers*; London: Hutchison).

What Newton proposed was that, between any two point masses, there is a force of attraction which acts along the line joining them and which is directly proportional to each of the masses involved and inversely proportional to the square of the distance between them. Thus, if a point mass m_2 has a displacement $\boldsymbol{r}$ from a point mass m_1 (Figure 5.1) the force on m_2 due to m_1 is given by

$$\boldsymbol{F} = -(constant)\frac{m_1 m_2}{r^2}\hat{\boldsymbol{r}} = -G\frac{m_1 m_2}{r^2}\hat{\boldsymbol{r}} \tag{5.1}$$

where $\hat{\boldsymbol{r}}$ is a unit vector (Section 4.2) in the $\boldsymbol{r}$-direction. In Equation (5.1) the magnitude of $\boldsymbol{F}$ depends on $\frac{1}{r^2}$ so that the law of gravitation is described as an *inverse square law*. The constant of proportionality G in Equation (5.1) is called the *gravitational constant* and its value must be determined experimentally. The minus sign indicates that the force is attractive (the force on m_2 is directed towards m_1). Because the gravitational force between laboratory masses is so small, very sensitive apparatus is required to obtain an accurate value of G. A laboratory technique developed in 1798 by Henry Cavendish (1731–1810) can be used for this purpose, the basic principles of which will be discussed in Chapter 7 (Section 7.5). The value of the gravitational constant in SI units (to three significant figures) is measured to be $6.67 \times 10^{-11}\ \mathrm{N\,m^2\,kg^{-2}}$.

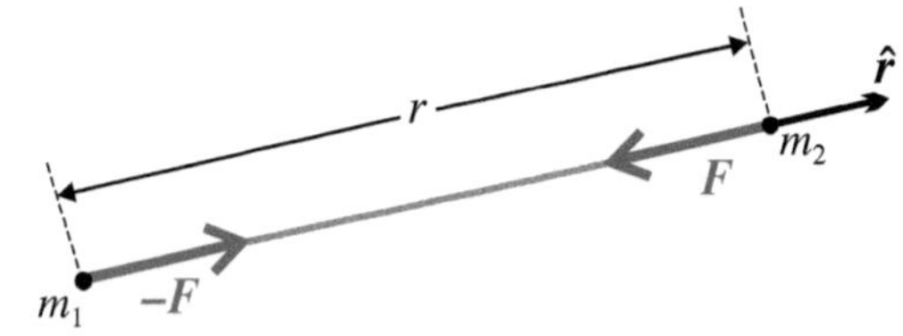

Figure 5.1. Gravitational interaction between two point masses.

As an illustration of the magnitude of the gravitational force let us estimate the gravitational attraction between the proton and the electron in a hydrogen atom. The masses of the electron and proton are, respectively, 9.1×10^{-31} kg and 1.67×10^{-27} kg and they are separated by 53 pm so the gravitational attraction between them is

$$|\boldsymbol{F}| = G\frac{m_e m_p}{r^2} = (6.67 \times 10^{-11})\frac{(9.1 \times 10^{-31})(1.67 \times 10^{-27})}{(53 \times 10^{-12})^2} \sim 10^{-47}\ \mathrm{N}$$

Understanding Physics, Third Edition. Michael Mansfield and Colm O'Sullivan.
© 2020 John Wiley & Sons Ltd. Published 2020 by John Wiley & Sons Ltd.

The strength of the gravitational force is over forty orders of magnitude less than the electrical force (see Sections 16.3 and 24.1) which keeps the electron bound to the proton.

For problems based on the material presented in this section visit up.ucc.ie/5/ and follow the link to the problems.

5.2 Force fields

Imagine that you are moving a mass to different positions in a room; wherever the mass is positioned it experiences the same force due to gravity, namely its weight, both in magnitude and in direction. We can think of this as a gravitational influence existing everywhere in the room. While the influence is present at all points in the room there is a force only at the location where the mass happens to be.

Consider now the region of gravitational influence around a point mass or a uniform sphere. Since the strength and direction of the gravitational attraction is different in different locations, the force experienced by a point mass in this case changes in both magnitude and direction as its location is changed. Again, the gravitational influence due to the source mass may be thought of as existing at every point in space, although this influence only becomes apparent when a mass is located at the point.

These 'regions of gravitational influence' are examples of **force fields**. The idea of a force field proves to be a very useful concept, particularly where '*action at a distance*' is involved; that is, when the influence is experienced at points distant from the source of the force even though there is no material medium (such as a string) through which the force can be transmitted.

Gravitational field lines

To get some idea of the 'shape' of a force field the concept of **field lines** proves useful. In the case of gravitational fields, field lines are defined as the lines along which a free point mass (a 'test particle') would move if released from rest. Some field lines around a single fixed-point mass M are indicated in Figure 5.2. Since the gravitational attraction between the test particle and the mass M is along the line joining them, the field lines are directed radially inwards as shown; the gravitational field around a single point mass or, as we will see later, around a uniform sphere are examples of a *radial field*. Field lines are, of course, simply geometrical constructions and do not exist as physical entities.

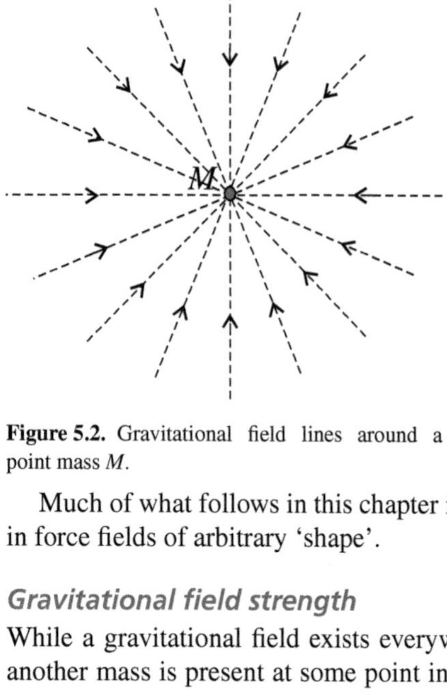

Figure 5.2. Gravitational field lines around a point mass M.

Force fields of more complicated 'shapes' are quite common in nature. Consider, for example, the 'shape' of the gravitational field due to the combined effects of the Sun and the planets at some point (say) between the orbits of Jupiter and Saturn. The gravitational field in regions like this is much more complicated than in the special case discussed above; the complex motion of comets and asteroids in this part of the solar system is evidence of this. Note that field lines never intersect one another.

Over reasonably small regions within any field, the field lines may be considered to be essentially parallel (Figure 5.3) and the field is said to be *uniform* in this case. The behaviour of bodies in uniform fields is important in many branches of physics; the motion of a mass in a uniform gravitational field will be discussed in Section 5.6.

Much of what follows in this chapter is directed towards developing and using a formalism which can deal with the motion of bodies in force fields of arbitrary 'shape'.

Gravitational field strength

While a gravitational field exists everywhere due to the presence of a mass or a configuration of masses, no force is exerted unless another mass is present at some point in the field. It would be useful, however, if a physical quantity could be defined which contains information on the 'strength' and the 'shape' of a field. To achieve this, the **gravitational field strength** at a point $\boldsymbol{r}$ in a gravitational field is defined as

$$\boldsymbol{g}(\boldsymbol{r}) := \frac{\boldsymbol{F}(\boldsymbol{r})}{m}$$

where $\boldsymbol{F}(\boldsymbol{r})$ is the force which would be exerted on a point mass m if placed at the point $\boldsymbol{r}$. It follows that the SI unit of gravitational field strength is N kg^{-1} or m s^{-2}. Knowing the value of $\boldsymbol{g}(\boldsymbol{r})$ – a vector function of position – everywhere gives a complete description of the field. Note that we use the notation $\boldsymbol{F}(\boldsymbol{r})$ to denote explicitly that $\boldsymbol{F}$ is a function of $\boldsymbol{r}$.

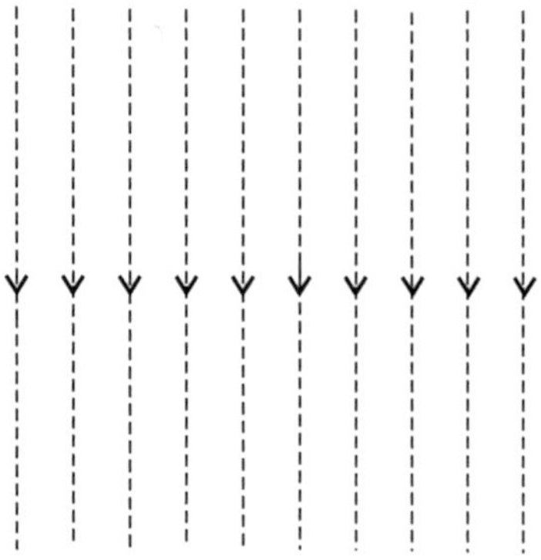
Figure 5.3. Field lines representing a uniform field.

The gravitational field strength functions for the special cases discussed so far are:

uniform field: $\boldsymbol{g}(\boldsymbol{r}) = \boldsymbol{g}_o = constant$ (vector)

radial field due to a point mass M: $\boldsymbol{g}(\boldsymbol{r}) = -G\dfrac{M}{r^2}\hat{\boldsymbol{r}}$

For problems based on the material presented in this section visit up.ucc.ie/5/ and follow the link to the problems.

5.3 The concept of flux

The gravitational force is just one example of a number of 'inverse square law' forces which arise in nature (we will meet further examples in Chapters 16 and 18). Whenever an inverse square law force is encountered, it turns out to be useful to introduce the concept of a **flux**. In such contexts, the concept is purely abstract but is adopted deliberately by analogy with examples of real fluxes that arise frequently in nature; for example, the flow of water from a garden hosepipe, the flow of light from an electric bulb, the flow of particles from a radioactive source, etc. It is conventional to *define* the total flux emanating from such a source as being equal to the strength of the source; for example, in the case of a light bulb the light flux can be measured in terms of the rate at which the bulb dissipates electrical energy, measured in watts. Other examples are given in Table 5.1

Table 5.1 Some examples of flux and flux density

Source	Total flux from source	Flux density at a point
water hose	litre per second	litre s^{-1} m^{-2}
light bulb	watt	watt m^{-2}
radioactive source	particles per second	particles s^{-1} m^{-2}
mass	kilogram (*)	kg m^{-2} (*)

(*) Note: In this case the flux is entirely abstract.

Gravitational flux density

The effect of a flux at a point some distance from its source can be described in terms of the flux per unit area in the neighbourhood of that point. As an example, consider a lawn being watered by a garden hose: the strength of the source, and hence the corresponding flux, may be considered as the volume of water per unit time emitted from the hosepipe. The effect at any point of the lawn (the amount of wetting) is determined by the volume per unit time *per unit area* landing at that point of the lawn. This latter measure is an example of a **flux density** (see the right hand column of Table 5.1 for other examples).

The treatment of gravitational flux is a little more complicated than the first three examples in Table 1, first because nothing actually flows. One can think, however, of a flow of 'gravitational influence' in the general direction indicated by the gravitational field lines. The second difference is that, since these field lines terminate on a mass, strictly speaking mass represents a *sink* of flux rather than a source. Gravitational flux, therefore, can be thought of as originating at infinity (source) and ending up on a mass (sink). The concept of an abstract flux may appear somewhat intangible when first encountered, but it repays careful study (particularly as it will reappear in the treatment of both electric and magnetic fields in Chapters 16 and 18).

Gravitational flux is a useful concept when a number of masses is involved (Figure 5.4). In this case the field lines and corresponding flux may have a more complicated shape. The total flux associated with all the masses in the figure is given by

$$\Phi_g = \Phi_1 + \Phi_2 + \Phi_3 = M_1 + M_2 + M_3$$

all of which passes through any surface completely enclosing the masses.

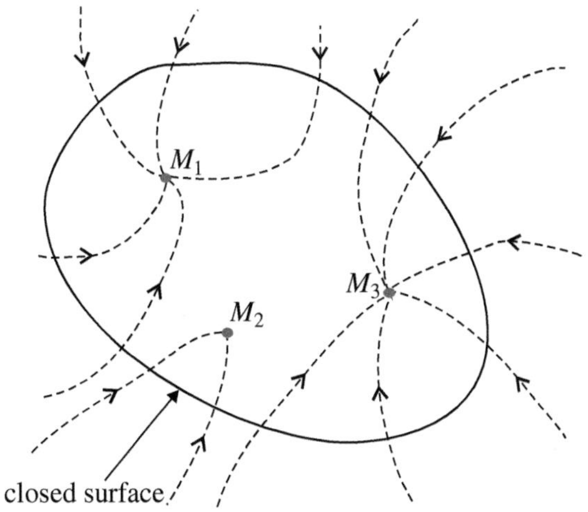

Figure 5.4. Field lines representing the gravitational flux due to three point masses; all the flux passes through any closed surface enclosing the masses.

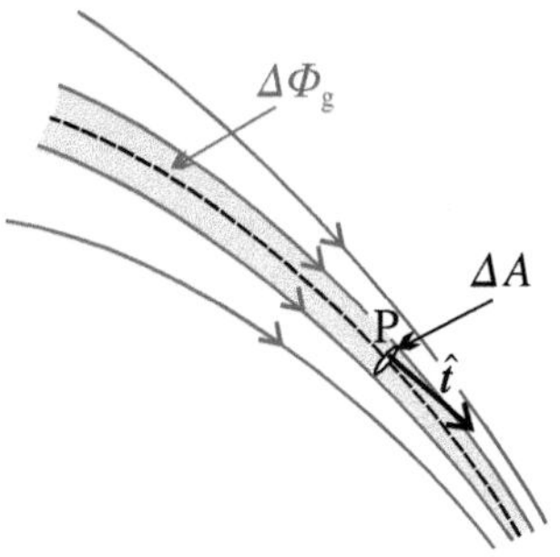

Figure 5.5. $\Delta\Phi_g$ (shaded) is the gravitational flux through the element of area ΔA around the point P.

We can now formally define the quantity **gravitational flux density** at a point P (Figure 5.5) in a gravitational field as follows

$$\boldsymbol{\Gamma}_g := \left[\underset{\Delta A \to 0}{\text{limit}} \frac{\Delta\Phi_g}{\Delta A}\right]\hat{\boldsymbol{\imath}}$$

where $\Delta\Phi_g$ is the amount of flux passing through an area element ΔA set up around P perpendicular to the flux and $\hat{\boldsymbol{\imath}}$ is a unit vector perpendicular to ΔA and directed along the direction of flow of flux. Note that $\hat{\boldsymbol{\imath}}$ is tangential to the local field line through P and, hence, the flux density vector is parallel to the field strength vector; that is, $\boldsymbol{\Gamma}$ is parallel to $\boldsymbol{g}$.

Flux density due to a point source

Consider a point source of flux of strength Φ as illustrated in Figure 5.6. One can invoke the symmetry of the situation to derive an expression for the flux density at an arbitrary point P, displaced $\boldsymbol{r}$ relative to the source. If one imagines a spherical surface of radius r centred on the source, it is clear that the flux per unit area must be the same at every point on the surface of the sphere (for example, a small light bulb placed at the centre of a hollow sphere will illuminate all points equally on the inside surface of the sphere). Thus, in this case, the flux density at the point P will be the total flux divided by the total surface area of the sphere, that is

$$\Gamma = \frac{\Phi}{4\pi r^2}$$

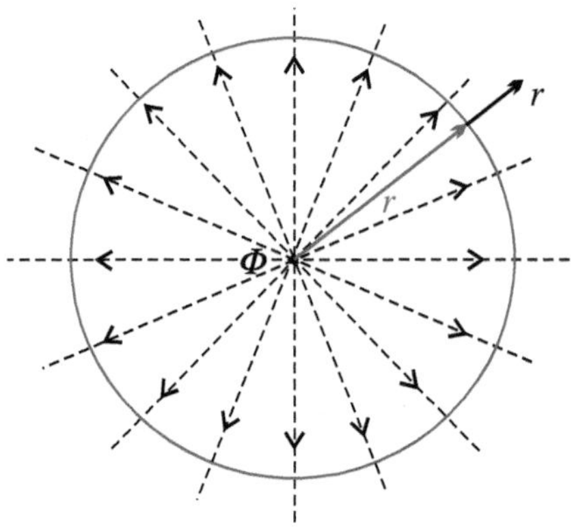

Figure 5.6. A point source of flux (e.g., a small light source) will give rise to the same flux density at all points on the surface of a sphere centred on the source.

In the case of the gravitational force, the flux is directed inwards, that is in the $-\boldsymbol{r}$ direction so the gravitational flux density at a point, displaced $\boldsymbol{r}$ from a single point mass M, is given by

$$\boldsymbol{\Gamma}_g(\boldsymbol{r}) = -\frac{M}{4\pi r^2}\hat{\boldsymbol{r}}$$

Recalling, from the end of Section 5.2, that the gravitational field strength is given by

$$\boldsymbol{g}(\boldsymbol{r}) = -G\frac{M}{r^2}\hat{\boldsymbol{r}}$$

we see that this is related to the gravitational flux density as follows

$$\boldsymbol{g}(\boldsymbol{r}) = 4\pi G\boldsymbol{\Gamma}_g(\boldsymbol{r}) \qquad (5.2)$$

5.4 Gauss's law for gravitation

As we have noted, the concept of flux proves to be a fruitful model only where inverse square law force fields are involved. As we will see, an entirely equivalent statement of Newton's law of gravitation, known as **Gauss's Law for gravitation**, may be expressed as follows:

The flux through any closed surface in a gravitational field is equal to the sum of the masses within the surface.

If one wishes to determine the flux through any plane area one needs to consider the angle that the direction of the flux makes with the plane. Consider, for example, the flow of a liquid (or light) through a rectangular area of side lengths l and h, as illustrated in Figure 5.7. When the plane of the area lh is perpendicular to the direction of the incident flux, the amount of flux passing through the area is at a maximum. If the area is rotated (or the direction of the flux changes) through an angle θ, the area perpendicular to the flow, that is the projection of lh onto the plane perpendicular to the direction of the flux, is given by $l\cos\theta \times h = lh\cos\theta$. Hence the amount of flux passing through the area is reduced by a factor $\cos\theta$. Thus, if the incident flux density is Γ, the flux through the blue rectangle in Figure 5.7 is $\Gamma lh\cos\theta$.

Similarly, in Figure 5.8, we see that the flux through the area element ΔA is given by

$$\Delta\Phi = |\boldsymbol{\Gamma}|(\Delta A)cos\theta = \boldsymbol{\Gamma}\cdot(\Delta A)\boldsymbol{n} = \boldsymbol{\Gamma}\cdot\Delta\boldsymbol{A}$$

where we have invoked the definition of the scalar product (Section 4.2) and the notation $\boldsymbol{\Delta A} := (\Delta A)\boldsymbol{n}$, where $\boldsymbol{n}$ is a unit vector perpendicular to the surface outward (recall Section 4.11). Since $\Delta\Phi$ is a scalar, the total flux through the whole surface S is the algebraic sum of the contributions from each such element, that is

$$\Phi = \sum_i (\Delta\Phi)_i = \lim_{\Delta A_i \to 0}[\boldsymbol{\Gamma}(\boldsymbol{r}_1)\cdot\boldsymbol{\Delta A}_1 + \boldsymbol{\Gamma}(\boldsymbol{r}_2)\cdot\boldsymbol{\Delta A}_2 + \boldsymbol{\Gamma}(\boldsymbol{r}_3)\cdot\boldsymbol{\Delta A}_3 + \ldots\ \boldsymbol{\Gamma}(\boldsymbol{r}_i)\cdot\boldsymbol{\Delta A}_i + \ldots]$$

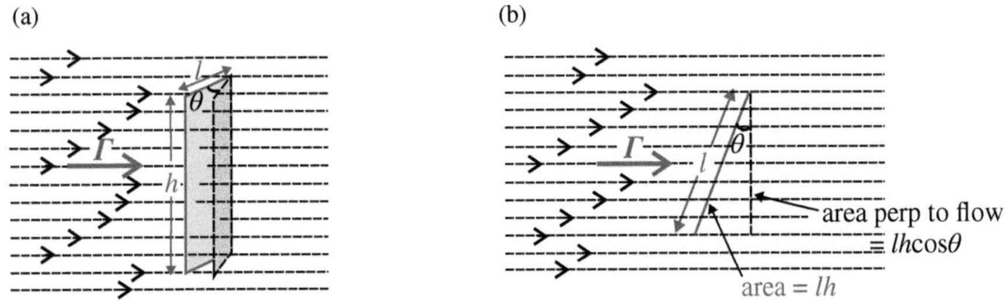

Figure 5.7. (a) A uniform flux passes through a rectangle of side lengths l and h; θ is the angle between the plane of the rectangle and a plane perpendicular to the direction of the incident flux. (b) Same as (a) but viewed from above (side of length h is perpendicular to the page in this case).

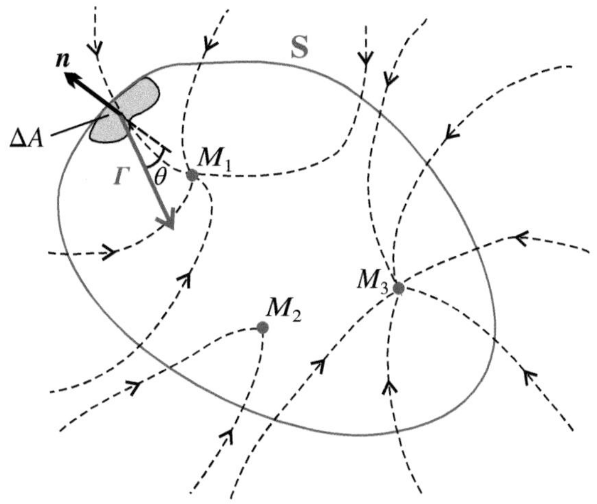

Figure 5.8. Gauss's law: the gravitational flux through the closed surface S is equal to the total mass within the surface.

Surface integrals

The expression on the right side of the equation above is the definition of what is known as a **surface integral** (see Appendix A.12.3), that is

$$\iint_{\text{surface S}} \boldsymbol{\Gamma}\cdot d\boldsymbol{A} := \lim_{\Delta A_i \to 0}[\boldsymbol{\Gamma}(\boldsymbol{r}_1)\cdot\boldsymbol{\Delta A}_1 + \boldsymbol{\Gamma}(\boldsymbol{r}_2)\cdot\boldsymbol{\Delta A}_2 + \boldsymbol{\Gamma}(\boldsymbol{r}_3)\cdot\boldsymbol{\Delta A}_3 + \ldots\ \boldsymbol{\Gamma}(\boldsymbol{r}_i)\cdot\boldsymbol{\Delta A}_i + \ldots]$$

This type of integral is a little different from the simple one-dimensional integrals usually encountered in that it involves vector functions and integration over two dimensions (hence the 'double integral' notation). The notation allows us to express the total flux through any surface S compactly as

$$\Phi = \iint_S \boldsymbol{\Gamma}\cdot d\boldsymbol{A}$$

Using the surface integral notation, Gauss's law can be stated mathematically as follows.

$$\oiint_{\substack{\text{closed}\\ \text{surface}}} \boldsymbol{\Gamma}\cdot d\boldsymbol{A} = \sum_{\substack{\text{inside}\\ \text{surface}}} M_i \qquad (5.3)$$

where the 'loop' indicates that the integration is over a closed surface.

It must be emphasised that Equations (5.1) and (5.3) are equivalent statements of the same basic law as can be seen from consideration of the two point masses shown in Figure 5.9.

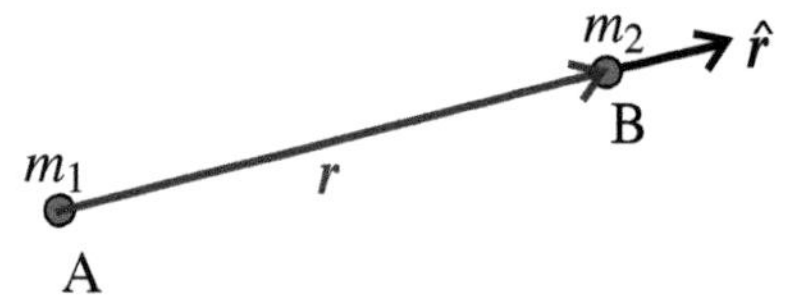

Figure 5.9. Derivation of Newton's law of gravitation from Gauss's law.

The gravitational flux density at the point B due to the mass m_1 is given by

$$\boldsymbol{\Gamma}_g(\boldsymbol{r}) = -\frac{m_1}{4\pi r^2}\hat{\boldsymbol{r}}$$

and hence, from (5.2), the gravitational field strength at B due to m_1 is

$$\boldsymbol{g}(\boldsymbol{r}) = 4\pi G\boldsymbol{\Gamma}_g(\boldsymbol{r}) = -4\pi G\frac{m_1}{4\pi r^2}\hat{\boldsymbol{r}} = -G\frac{m_1}{r^2}\hat{\boldsymbol{r}}$$

from which it follows that the force on m_2 due to m_1 is given by

$$\boldsymbol{F}(\boldsymbol{r}) = m_2\boldsymbol{g}(\boldsymbol{r}) = -G\frac{m_1 m_2}{r^2}\hat{\boldsymbol{r}}$$

which shows that Gauss's law and Newton's law of gravitation are equivalent.

For problems based on the material presented in this section visit up.ucc.ie/5/ *and follow the link to the problems.*

5.5 Applications of Gauss's law

Gauss's law proves to be a valuable tool in determining the gravitational field strength due to mass distributions which have a high degree of symmetry. As shown in the following two examples, Gauss's law may be used to derive expressions for the gravitational field at any point due to spherical distributions of mass.

Example 1: *Gravitational field due to a uniform spherical shell of matter*
Consider a mass M_S which is distributed uniformly in a thin spherical shell of radius R. Gauss's law may be applied to derive expressions for the gravitational field strength at any point due to this distribution of mass. We distinguish two cases: 1. a point located outside the shell ($r > R$) and 2. a point located inside the shell ($r < R$).

Case 1: To determine the gravitational field strength at a point P outside the shell, consider a spherical surface S of radius R centred on O as in Figure 5.10. The total mass inside the surface S is M_S. Let $|\boldsymbol{\Gamma}|$ be the magnitude of the gravitational flux density at P (and, by symmetry, at all points on the surface S, of area $4\pi r^2$). Applying Gauss's law

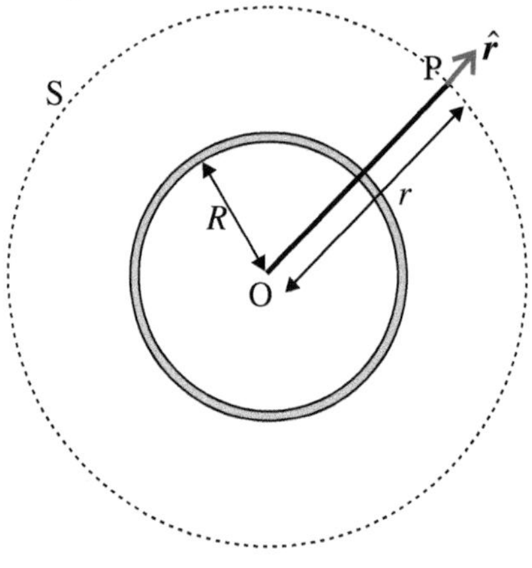

Figure 5.10. The gravitational field strength at any point *outside* a spherical shell is the same as if all the mass of the shell had been concentrated at its centre.

$$\oiint_S \boldsymbol{\Gamma}\cdot d\mathbf{A} = |\boldsymbol{\Gamma}|(4\pi r^2) = M_S \rightarrow |\boldsymbol{\Gamma}(\boldsymbol{r})| = -\frac{M_S}{4\pi r^2} \rightarrow \boldsymbol{\Gamma}(\boldsymbol{r}) = -\frac{M_S}{4\pi r^2}\hat{\boldsymbol{r}}$$

Thus, invoking Equation (5.2), the gravitational field strength at P is

$$\boldsymbol{g}(\boldsymbol{r}) = 4\pi G\boldsymbol{\Gamma}(\boldsymbol{r}) = -G\frac{M_S}{r^2}\hat{\boldsymbol{r}}$$

where the minus sign indicates that the field is directed *inwards* towards O.

Comparing this with the result obtained for the gravitational field strength due to a point mass at the end of Section 5.2, we see that the gravitational field strength at any point outside the shell is the same as if all the mass of the shell had been concentrated at O.

Case 2: Again, consider a spherical surface S of radius R centred on O but inside the shell as shown in Figure 5.11. The total mass inside the surface is zero. If $\boldsymbol{\Gamma}$ is the gravitational flux density at P, applying Gauss's law in this case we get

$$|\boldsymbol{\Gamma}|(4\pi r^2) = 0 \rightarrow \boldsymbol{\Gamma}(r) = 0 \rightarrow \boldsymbol{g}(r) = 0$$

showing that there is no gravitational field at any point inside a closed shell of matter.

Example 2: *Gravitational field due to a uniform sphere*
Consider a mass M distributed uniformly throughout the volume of a sphere of radius a. Again we consider the strength of the field at a point P in two distinct cases.

Case 1: *Point P outside sphere ($r > a$)*
Applying Gauss's law to a spherical surface of radius r concentric with the sphere (Figure 5.12), we find

$$\oint\!\!\oint_S \boldsymbol{\Gamma}\cdot d\mathbf{A} = |\boldsymbol{\Gamma}|(4\pi r^2) = M \;\rightarrow\; \boldsymbol{\Gamma}(r) = -\frac{M}{4\pi r^2}\hat{\boldsymbol{r}} \;\rightarrow\; \boldsymbol{g}(\boldsymbol{r}) = 4\pi G\boldsymbol{\Gamma}(\boldsymbol{r}) = -G\frac{M}{r^2}\hat{\boldsymbol{r}}$$

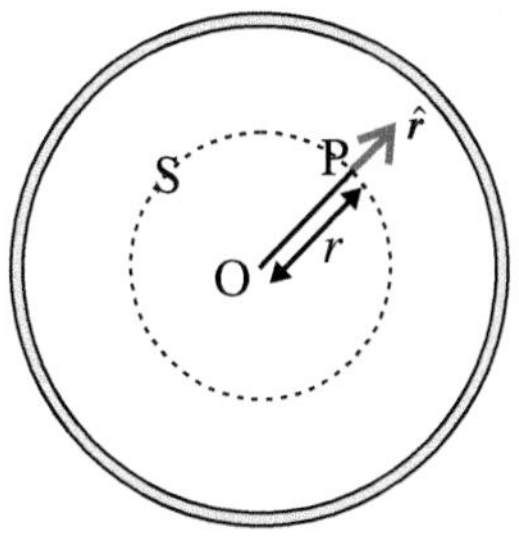

Figure 5.11. The gravitational field strength is zero at any point inside a spherical shell of matter.

which shows that, as in the case of a shell, the field outside the sphere is indistinguishable from that due to a point mass concentrated at the centre of the sphere.

Case 2: *Point P inside sphere ($r < a$)*
From Figure 5.13, $\oint\!\!\oint_S \boldsymbol{\Gamma}\cdot d\mathbf{A} = |\boldsymbol{\Gamma}|(4\pi r^2)$ = total mass inside the spherical surface S, that is $|\boldsymbol{\Gamma}|(4\pi r^2) = \left(\frac{4}{3}\pi r^3\right)\rho_0$, where ρ_0 is the (uniform) density of the material of the sphere. Thus the gravitational flux density at P is

$$|\boldsymbol{\Gamma}(r)| = \frac{\left(\frac{4}{3}\pi r^3\right)\rho_0}{4\pi r^2} = \frac{\left(\frac{4}{3}\pi a^3\right)\rho_0 r}{4\pi a^3} = \frac{Mr}{4\pi a^3}$$

and hence

$$\boldsymbol{g}(\boldsymbol{r}) = 4\pi G\boldsymbol{\Gamma}(\boldsymbol{r}) = -G\frac{Mr}{a^3}\hat{\boldsymbol{r}} = -G\frac{M}{a^3}\boldsymbol{r}$$

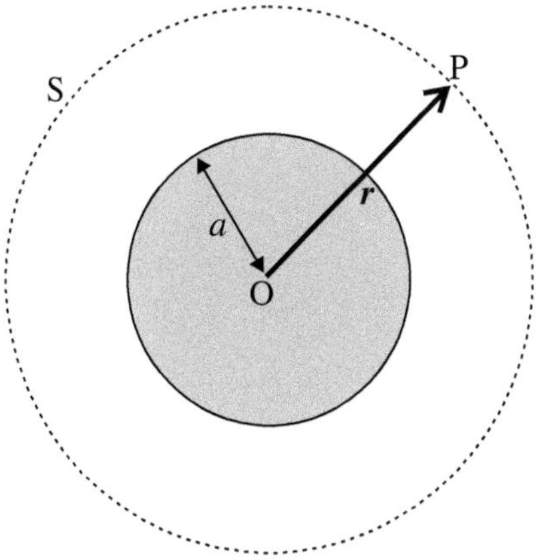

Figure 5.12. Uniform spherical distribution of matter: the gravitational field strength at a point P outside the sphere is the same as if all the mass had been concentrated at the centre of the sphere.

Note that both cases 1. and 2. give the same result for the gravitational field strength at the surface of the sphere, namely

$$\boldsymbol{g}(\boldsymbol{a}) = -G\frac{M}{a^2}\hat{\boldsymbol{r}}$$

In Figure 5.14 the magnitude of the gravitational field strength is plotted as a function of distance from the centre of the sphere.

In summary, the force experienced by a point mass m placed at a distance r from the centre of a uniform sphere of mass M and radius a (Figure 5.15), is given by

$$\boldsymbol{F}(\boldsymbol{r}) = \begin{cases} -G\dfrac{mM}{r^2}\hat{\boldsymbol{r}} & \text{when } r \geq a \\ -G\dfrac{mMr}{a^3}\hat{\boldsymbol{r}} = -G\dfrac{mM}{a^3}\boldsymbol{r} & \text{when } r \leq a \end{cases}$$

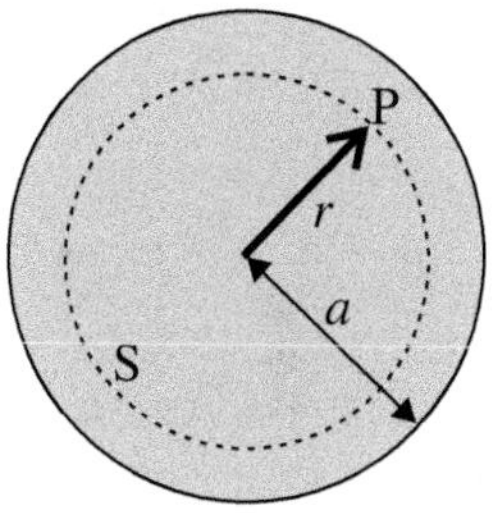

Figure 5.13. Uniform spherical distribution of matter: the gravitational field strength at a point P inside the sphere is the same as if all the mass inside the surface S had been concentrated at the centre of the sphere.

Weight

We are now in a position to account for the phenomenon of weight as simply the force due to gravity on a body (recall Section 3.4). In particular, for a mass m which is located near the surface of a spherical planet of radius R_p and mass M_p, the force experienced (Figure 5.16) is given by

$$\boldsymbol{F} = -G\frac{mM_p}{R_p^2}\hat{\boldsymbol{r}}$$

At the surface of the Earth (which is approximately spherical), this force is observed as the *weight* of the mass m which is given by

$$\boldsymbol{W} = m\boldsymbol{g} = -G\frac{mM_p}{R_p^2}\hat{\boldsymbol{r}}$$

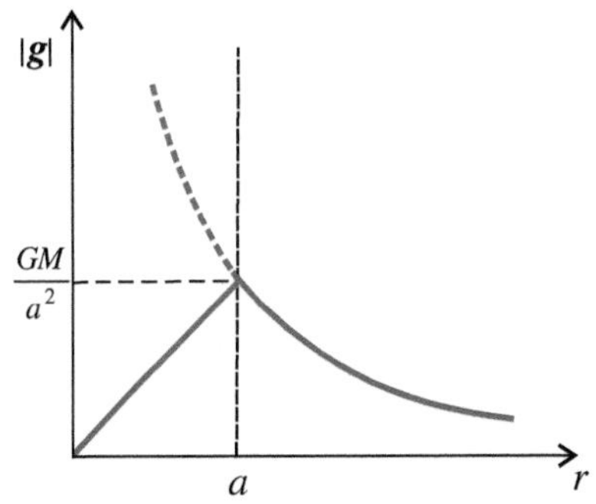

Figure 5.14. Plot (solid line) of the gravitational field strength as a function of distance from the centre of a uniform sphere of radius a. The dashed curve represents the corresponding field strength for the case in which all of the mass is concentrated at a point at the centre of the sphere.

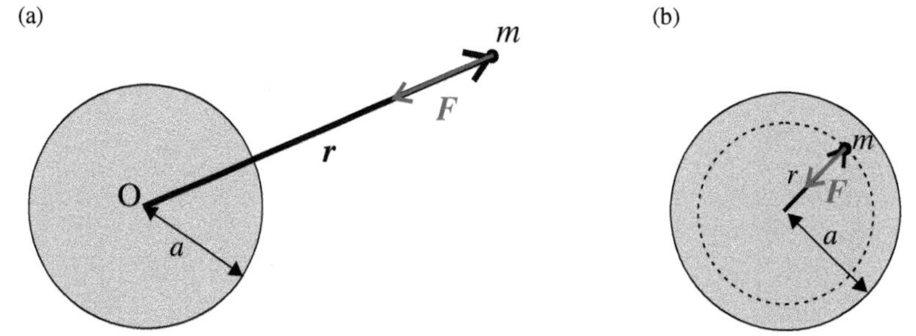

Figure 5.15. The gravitational attraction exerted on a point mass m by a uniform spherical distribution of matter. (a) When $r > a$, the force is the same as that due to a point mass M located at O, the centre of the sphere. (b) When $r < a$, the force is the same as that due to a point mass located as O where this mass is that of the part of the sphere within the dashed spherical surface shown.

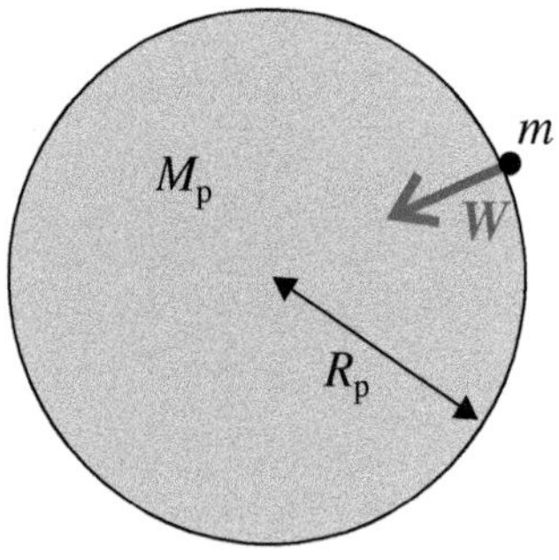

Figure 5.16. The weight of a body on a planet is the force of gravity exerted on the body by the planet.

Thus, assuming the Earth to be spherical,

$$g = \frac{GM_E}{R_E^2} \tag{5.4}$$

where M_E ($\approx 6.0 \times 10^{24}$ kg) and R_E ($\approx 6.4 \times 10^6$ m) are the mass and mean radius, respectively, of the Earth, giving $g \approx 9.8$ m s^{-2}. For the Moon $M_M \approx 7.4 \times 10^{22}$ kg and $R_M \approx 1.7 \times 10^6$ m so that the acceleration due to gravity at the surface of the Moon, $g_M \approx 1.7$ m s^{-2}. For the planet Jupiter $M_J \approx 1.9 \times 10^{27}$ kg and $R_J \approx 7.15 \times 10^7$ m so that the acceleration due to gravity at the top of the atmosphere of Jupiter, $g_J \approx 25$ m s^{-2}.

The measured value of the acceleration due to gravity at any specific point on the Earth's surface differs a little from the value calculated from Equation (5.4) due to (i) variation from spherical shape of the Earth, (ii) local inhomogeneities in the Earth's crust (regions of higher or lower than average density) and (iii) the rotation of the Earth. Two of these effects, (i) and (iii), will be examined quantitatively in Worked Example 8.3.

Because of the linear dependence of the gravitational force on the mass of the body which is experiencing the force, the acceleration of a body under gravity must be independent of the mass of the body. This follows from Newton's second law,

$$G\frac{mM}{r^2} = ma \quad \rightarrow \quad a = G\frac{M}{r^2}$$

Hence the acceleration a is independent of m. This aspect of the laws of dynamics, that the mass appearing in Newton's second law ('inertial mass') is the same as, or is at least directly proportional to, the mass appearing in the law of gravitation ('gravitational mass'), explains Galileo's discovery that at any given location all bodies have the same acceleration due to gravity.

5.6 Motion in a constant uniform field: projectiles

The gravitational field around the Earth is of a generally radial shape but over reasonably small distances, such as within a room or even over quite large distances on the surface, the field lines are essentially parallel (recall Figure 5.3); that is, the field can be considered to be *uniform* in this case.

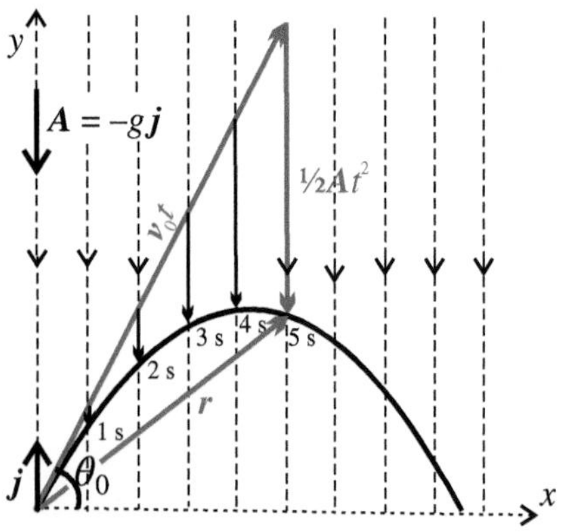

Figure 5.17. Motion of a projectile: if the velocity of a projectile at $t = 0$ is v_0, its displacement and velocity at any subsequent time are given by Equations (5.5a) and (5.5b).

We now consider the general motion of a body moving in such a field. A common example is the motion of an object such as a ball, stone or other projectile thrown at an angle with respect to the horizontal near the Earth's surface. If all other effects, such as air resistance, are neglected, the object will move under the influence of the force due to gravity only and the motion takes place entirely in a vertical plane.

Suppose that a stone is projected with an initial velocity $\boldsymbol{v}_0$ which makes an angle θ_0 with respect to the horizontal as shown in Figure 5.17. Since the acceleration $\boldsymbol{A}$ is constant we can invoke the (vector) Equations (4.2a) and (4.2b) to determine the displacement and velocity of the stone at a time t after it was projected. Thus

$$\boldsymbol{v} = \boldsymbol{v}_0 + \boldsymbol{A}t \tag{5.5a}$$

$$\boldsymbol{r} = \boldsymbol{v}_0 t + \tfrac{1}{2}\boldsymbol{A}t^2 \tag{5.5b}$$

where we have chosen the origin of coordinates to coincide with the point of projection of the stone (that is, $\boldsymbol{r}_0 = 0$ at $t = 0$)). The motion of the stone will be confined to the vertical plane defined by $\boldsymbol{v}_0$ and $\boldsymbol{A}$. Equations (5.5a) and (5.5b) may be used to determine the velocity and displacement at any time. Figure 5.17 indicates how Equation (5.5b) can be used to trace out the trajectory of the stone as time evolves by calculating the value of $\boldsymbol{r}$ (the vector sum of $\boldsymbol{v}_0 t$ and $\frac{1}{2}\boldsymbol{A}t^2$) at 1 s, 2 s, 3 s, etc. We shall see shortly that the trajectory turns out to have parabolic shape.

The acceleration vector is directed vertically downward everywhere, that is $\boldsymbol{A} = -g\boldsymbol{j}$ where $\boldsymbol{j}$ is a unit vector upwards as indicated in Figure 5.17. The displacement and velocity vectors change in both magnitude and direction in the course of the motion, so that in vector terms the motion appears to be complex. It may be analysed quite easily, however, if we resolve the vectors involved into components in the horizontal (x-direction) and in the vertical (y-direction). This will result in significant simplification in the case of x-components because there is no component of the acceleration in that direction. Thus Equations (5.5a) and (5.5b) reduce to

$$\text{(i) } v_x = |\boldsymbol{v}_0|\cos\theta_0 \qquad \text{(ii) } v_y = |\boldsymbol{v}_0|\sin\theta_0 - gt \tag{5.6a}$$

$$x = (|\boldsymbol{v}_0|\cos\theta_0)t \qquad y = (|\boldsymbol{v}_0|\sin\theta_0)t - \frac{1}{2}gt^2 \tag{5.6b}$$

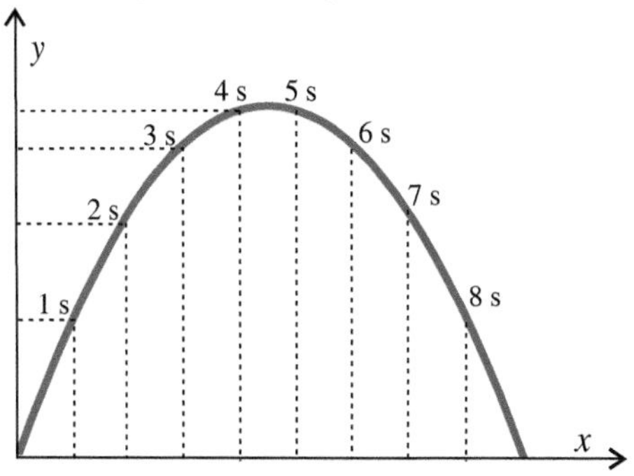

Figure 5.18. The trajectory of a projectile may be determined by evaluating Equations (5.6b) at regular time intervals. The projectile moves equal horizontal distances in equal times (showing that the horizontal component of velocity is constant) and the vertical component of the motion is governed by constant acceleration in the $-y$-direction.

and the original two-dimensional problem is reduced to two one-dimensional problems, namely (i) a constant velocity problem in the x-direction and (ii) a constant acceleration problem in the y-direction. The two one-dimensional problems are essentially independent, linked only by the common time (t). This point is illustrated in Figure 5.18, where the horizontal and vertical components of the displacement of a projectile are plotted at regular time intervals.

The **maximum height reached** in the motion (y_m in Figure 5.19) can be determined by noting that at the highest point on the trajectory the vertical component of the velocity (v_y in the second equation of (5.6a)) will be instantaneously zero, that is $0 = v_0\sin\theta_0 - gT$, where T is the time taken to reach the maximum height and $|\boldsymbol{v}_0|$ has been written as v_0 for simplicity. Thus the time taken to reach this height is given by

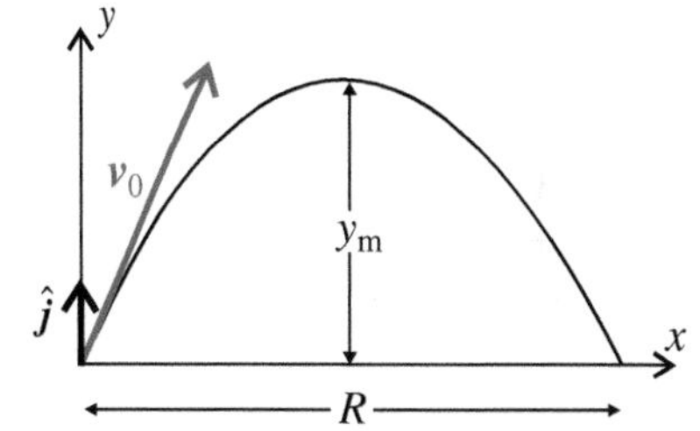

Figure 5.19. Projectile motion is confined to the vertical plane: y_m is the maximum height reached and R is the horizontal range.

$$T = \frac{v_0\sin\theta_0}{g}$$

and the maximum height reached can be determined by substituting $t = T$ and $y = y_m$ in the y-equation of (5.6b) and hence

$$y_m = v_0\sin\theta_0 T - \frac{1}{2}gT^2 = \frac{v_0^2\sin^2\theta_0}{g} - \frac{1}{2}g\frac{v_0^2\sin^2\theta_0}{g^2} = \frac{v_0^2\sin^2\theta_0}{2g} \tag{5.7}$$

From symmetry considerations, we conclude that, after a time $2T$, the stone will have returned to the same horizontal level from which it started. Using the x-equation of (5.6b), an expression for the **horizontal range** can be determined as follows

$$R = (v_0\cos\theta_0)(2T) = v_0\cos\theta_0\frac{2v_0\sin\theta_0}{g} = \frac{2v_0^2\sin\theta_0\cos\theta_0}{g} = \frac{v_0^2\sin 2\theta_0}{g} \tag{5.8}$$

The value of R will be maximum when $\sin(2\theta_0) = 1$. This occurs when $2\theta_0 = \frac{\pi}{2} \to \theta_0 = \frac{\pi}{4}$. As long as air resistance does not play a significant role, therefore, the maximum range of a projectile over horizontal ground is achieved by firing it at an angle of 45°. This result has important applications in many ball games and in field events.

To derive an expression for the **equation of the trajectory**, that is the relationship between y and x, we can use the first of the equations in Equation (5.6b) above to solve for t and substitute the result into the second equation. Thus

$$t = \frac{x}{v_0\cos\theta_0}$$

and hence

$$y = v_0\sin\theta_0\left(\frac{x}{v_0\cos\theta_0}\right) - \frac{1}{2}g\left(\frac{x}{v_0\cos\theta_0}\right)^2$$

$$\to \quad y = (\tan\theta_0)x - \frac{g}{2v_0^2\cos^2\theta_0}x^2$$

which can be rewritten as follows

$$y - y_m = -\frac{g}{2v_0^2\cos^2\theta_0}\left(x - \frac{v_0^2\sin 2\theta_0}{2g}\right)^2 \tag{5.9}$$

(the reader should check this result by multiplying out the right hand side of the above equation and substituting the value of y_m given by Equation (5.7)).

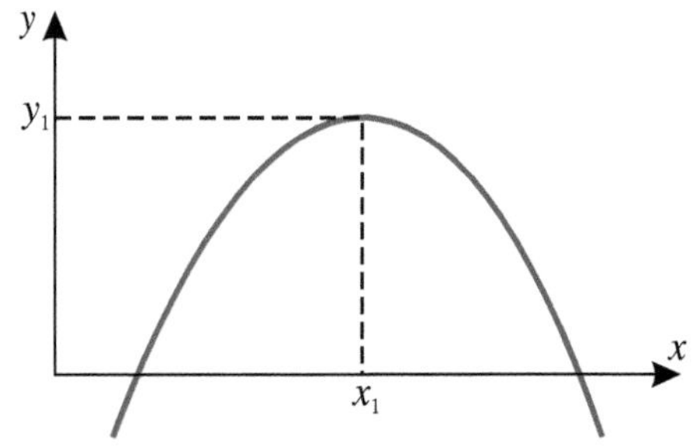

Figure 5.20. The general equation for a parabola is $y-y_1 = 4a(x-x_1)^2$; when a is negative, the parabola has the form shown with a maximum at the point (x_1, y_1).

Equation (5.9) can be recognised as the equation of a parabola by comparison with the general equation for a parabola (Appendix A.10.2 (iii)), namely

$$(y - y_1) = 4a(x - x_1)^2$$

Comparing this equation with Equation (5.9),

$$y_1 = y_m \qquad x_1 = \frac{v_0^2\sin 2\theta_0}{2g} \qquad \text{and} \qquad 4a = -\frac{g}{2v_0^2\cos^2\theta_0}$$

where the minus sign in the value for $4a$ indicates that the parabola is inverted with respect to the parabola of Figure A.15 (see Figure 5.20). Thus particles moving freely with constant acceleration execute parabolic trajectories.

Study Worked Example 5.1

For problems based on the material presented in this section visit up.ucc.ie/5/ *and follow the link to the problems.*

5.7 Mechanical work and energy

The concepts of mechanical work and energy that were developed in Chapter 3 for one-dimensional dynamical systems must be generalised so that they may be applied to three dimensions. In this section we will follow a similar approach to that used in Section 3.7 but with some important differences. In the process further insights into the energy concept will be encountered which were not apparent in the one-dimensional treatment. Furthermore, many algebraic quantities are replaced by their vector counterparts and, where appropriate, vector calculus must be applied.

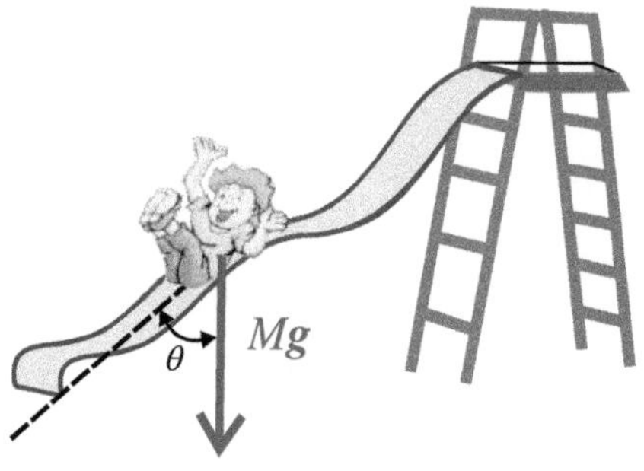

Figure 5.21. The acceleration of a child at any point on a frictionless slide depends on the component of the gravitational force tangential to the slide at that point ($Mg\cos\theta$).

Consider first a child who is playing on a slide (as illustrated in Figure 5.21). Starting from rest at the top she slides down speeding up as she goes. Her motion is brought about by the gravitational force field of the Earth but, unlike the situation studied in the previous section, the motion is restricted by the constraint provided by the slide. Her acceleration depends only on the component of the field force (the weight of the child) along the direction of motion. At any point, the tangential acceleration of the child of mass M along the slide is produced by a force $Mg\cos\theta$, where θ is the angle between the vertical and the direction of motion of the child at that point.

Let us consider, more generally, the situation illustrated in Figure 5.22 in which a body is moved from a point A to a point B under the influence of a general field force, the field lines of which are as shown in the figure. Let the infinitesimal change in displacement from P to P′ be $\boldsymbol{\Delta s}$. The element of **mechanical work done** *by the force field* in moving the body through a displacement $\boldsymbol{\Delta s}$ is *defined* as

$$\Delta W := |\boldsymbol{F}|\cos\theta|\boldsymbol{\Delta s}| \tag{5.10}$$

Using the definition of the scalar product of two vectors (Section 4.2), the element of mechanical work done by a force $\boldsymbol{F}$ in displacing a body by $\boldsymbol{\Delta s}$ (Equation (5.10)) may be written more compactly as

$$\boxed{\Delta W := \boldsymbol{F}\cdot\boldsymbol{\Delta s}} \tag{5.11}$$

To determine the total mechanical work done by a field force in moving a body from a point A to a point B, one can think of the path from A to B as being divided up into a very large number of very small elements $\boldsymbol{\Delta s}_1$, $\boldsymbol{\Delta s}_2$, $\boldsymbol{\Delta s}_3$, ... as indicated in Figure 5.23. Since ΔW is a scalar, the total mechanical work done is the algebraic sum of the contributions from all of the elements, namely

$$W_{AB} = \boldsymbol{F}_1\cdot\boldsymbol{\Delta s}_1 + \boldsymbol{F}_2\cdot\boldsymbol{\Delta s}_2 + \boldsymbol{F}_3\cdot\boldsymbol{\Delta s}_3 + \ldots\ldots\ldots\ldots\ldots = \sum_i \boldsymbol{F}_i\cdot\boldsymbol{\Delta s}_i$$

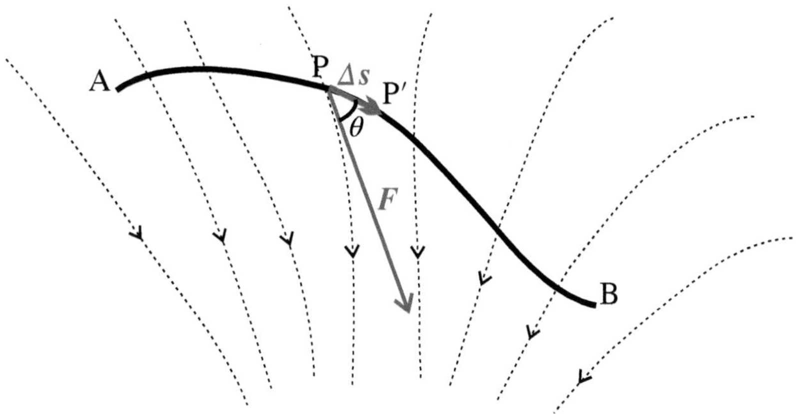

Figure 5.22. The mechanical work done by the field force in moving a body from P to P′ is the product of the component of the force along the path and the magnitude of the element Δs.

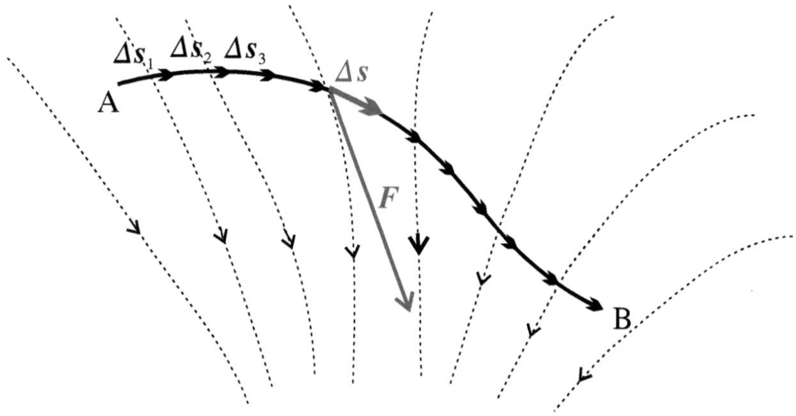

Figure 5.23. The mechanical work done by a field force in moving a body from A to B along the path indicated is the sum of the amounts of mechanical work done in moving the body through each element of the path.

This last equation, in the limit $\Delta s_i \to 0$, looks like the definition of an integral (Section 2.5 and Appendix A.7).

Line integrals

This type of integral, like the surface integral encountered in Section 5.4, involves vector functions but also depends on the path taken from A to B. Such integrals are called **line integrals** (see Appendix A.12.2) and are defined, for any vector function $\boldsymbol{F}(\boldsymbol{r})$ by

$$\int_{\mathrm{A}}^{\mathrm{B}} \boldsymbol{F}(\boldsymbol{r}) \cdot d\boldsymbol{l} = \underset{\Delta l_i \to 0}{\text{limit}} [\boldsymbol{F}(\boldsymbol{r}_1) \cdot \Delta \boldsymbol{l}_1 + \boldsymbol{F}(\boldsymbol{r}_2) \cdot \Delta \boldsymbol{l}_2 + \boldsymbol{F}(\boldsymbol{r}_3) \cdot \Delta \boldsymbol{l}_3 + \ \ldots \ \boldsymbol{F}(\boldsymbol{r}_i) \cdot \Delta \boldsymbol{l}_i + \ \ldots]$$

Thus the mechanical work done by a field force in moving a body between the points A and B in the field can be defined as

$$W_{\mathrm{AB}} := \int_{\mathrm{A}}^{\mathrm{B}} \boldsymbol{F} \cdot d\boldsymbol{s} \tag{5.12}$$

Potential energy: conservative fields

In contrast to motion in one dimension, the mechanical work done in moving a body from one point in a field to another is not defined uniquely, in that the value of the integral in Equation (5.12) depends on the particular path taken between the two points. However, in the important case of **conservative fields**, W_{AB} is *independent of the path* taken from A to B. As was the case in one-dimension, conservative fields are those in which there are no frictional or other dissipative forces (such as occur, for example, when the field force is velocity dependent as discussed in Section 3.11). In a conservative field, the work done depends *only on the coordinates of the end points of the path*, in which case we can express the work done as the difference in the values of a function at the end points of the path, that is

$$W_{AB} = \int_{A}^{B} \boldsymbol{F} \cdot d\boldsymbol{s} = U(\boldsymbol{r}_A) - U(\boldsymbol{r}_B) \tag{5.13}$$

The function, $U(\boldsymbol{r})$ defined by Equation (5.13), is called the **potential energy function** of the body in the field and is characteristic of the particular force field involved. The **potential energy difference** between two points in the field is defined as the work done in moving the body between the two points. If, in a conservative field, the motion between the points is reversed, the potential energy can be converted back into mechanical work done.

In the case of two points which are an infinitesimal distance $\Delta\boldsymbol{s}$ apart, the difference in potential energy is given by $\Delta U = -\boldsymbol{F} \cdot \Delta\boldsymbol{s}$.

Note that the potential energy function is defined in terms of the *difference* in the value of the function between two points and hence an arbitrary constant can always be added. One is free, as before, to choose arbitrarily a convenient point of zero potential energy in any specific situation. In the case of fields in which there are points where the force on the body is zero, however, it is conventional to choose such a point as the zero of potential energy. Once a zero has been defined, the value of $U(\boldsymbol{r})$ can be considered to be **the potential energy of the body** when at the position $\boldsymbol{r}$, that is

$$\int_{r_0}^{r} \boldsymbol{F}(\boldsymbol{r}') \cdot d\boldsymbol{s}' = U(\boldsymbol{r}_0) - U(\boldsymbol{r})$$

Thus

$$U(\boldsymbol{r}) = -\int_{r_0}^{r} \boldsymbol{F}(\boldsymbol{r}') \cdot d\boldsymbol{s}' + U(\boldsymbol{r}_0) = -\int_{r_0}^{r} \boldsymbol{F}(\boldsymbol{r}') \cdot d\boldsymbol{s}'$$

where $\boldsymbol{r}_0$ is the displacement of a point of zero potential ($U(\boldsymbol{r}_0) = 0$). Thus, the potential energy of a body is the work done *against* the field force in moving a body from a point of zero potential energy to its present location in the field, provided that the force applied to the body is just sufficient to balance the field force at all points along the path.

Examples of the determination of potential energy functions are given below for some important cases.

Case 1: Potential energy function of a body in a uniform field

For the case (Figure 5.24) of a uniform gravitational field $\boldsymbol{F}(\boldsymbol{r}) = m\boldsymbol{g} = -mg\boldsymbol{j}$ everywhere and thus

$$U(\boldsymbol{r}) = -\int_{r_0}^{r} \boldsymbol{F}(\boldsymbol{r}') \cdot d\boldsymbol{s}' = -\int_{r_0}^{r} -(mg\boldsymbol{j}) \cdot d\boldsymbol{s}' = mg\int_{r_0}^{r} \boldsymbol{j} \cdot (dx'\boldsymbol{i} + dy'\boldsymbol{j})$$

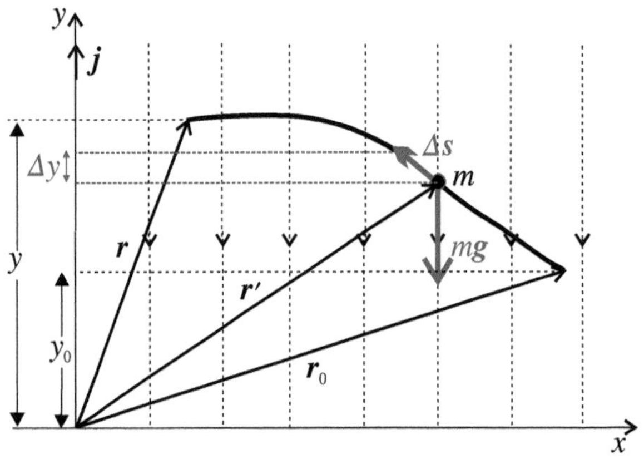

Figure 5.24. Determination of the potential energy function for a uniform field.

Noting that $\boldsymbol{i} \cdot \boldsymbol{j} = 0$ and $\boldsymbol{j} \cdot \boldsymbol{j} = 1$ (Appendix A.11.6), we see that

$$U(\boldsymbol{r}) = mg\int_{y_0}^{y} dy' = mg(y - y_0)$$

where we have chosen $U(y_0) = 0$

Thus,

$$U(\mathbf{r}) = U(x, y) = mg(y - y_0) \tag{5.14}$$

Thus if we take a point on the Earth's surface to be the zero of potential energy, the gravitational potential energy of a body near the surface is determined purely by $y - y_0$, its vertical height above the surface.

Case 2: Potential energy function of a body in an inverse square law field

In this case (Figure 5.25), the force is zero at $\boldsymbol{r} = \infty$, so we set $U(\infty) = 0$. Thus, for example, the potential energy of a point mass m displaced by $\boldsymbol{r}$ from a point mass M is given by

$$U(\boldsymbol{r}) = -\int_{r_0}^{r} \boldsymbol{F}(\boldsymbol{r}') \cdot d\boldsymbol{s}' = -\int_{\infty}^{r} -\left(\frac{GmM}{r'^2}\hat{\boldsymbol{r}}'\right) \cdot d\boldsymbol{s}'$$

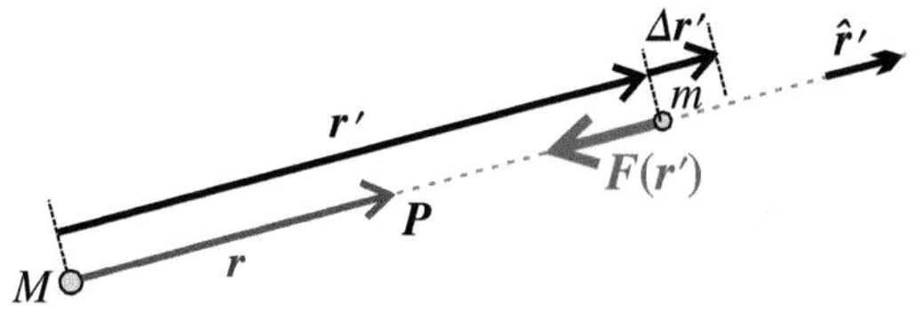

Figure 5.25. Determination of the potential energy function for an inverse square law field.

Since the field is conservative, the integration is independent of the path taken by m so that, for convenience, we can take a radial path from infinity to r and hence

$$U(\boldsymbol{r}) = \int_{\infty}^{r} \left(\frac{GmM}{r'^2}\hat{\boldsymbol{r}}'\right) \cdot d\boldsymbol{r}' = GmM \int_{\infty}^{r} \frac{dr'}{r'^2} = -\frac{GmM}{r} \tag{5.15}$$

In Worked Example 5.2 this result is used to determine the gravitational 'self energy' of a uniform spherical mass, which is equivalent to the energy required to overcome the gravitational attraction of the matter within the sphere and disperse it to infinity.

Study Worked Example 5.2

Derivation of force from potential energy functions

In three dimensions, in Cartesian coordinates,

$$U(\boldsymbol{r}) = U(x, y, z) \text{ and } \boldsymbol{F}(\boldsymbol{r}) = \boldsymbol{F}(x, y, z) \text{ and since } \Delta U = -\boldsymbol{F} \cdot \Delta\boldsymbol{s}$$

$$\Delta U = -(F_x\boldsymbol{i} + F_y\boldsymbol{j} + F_z\boldsymbol{k}) \cdot (\Delta x\boldsymbol{i} + \Delta y\boldsymbol{j} + \Delta z\boldsymbol{k})$$

Recalling that $\boldsymbol{i}\cdot\boldsymbol{j} = \boldsymbol{j}\cdot\boldsymbol{k} = \boldsymbol{k}\cdot\boldsymbol{i} = 0$ and $\boldsymbol{i}\cdot\boldsymbol{i} = \boldsymbol{j}\cdot\boldsymbol{j} = \boldsymbol{k}\cdot\boldsymbol{k} = 1$ (Appendix A.11.6), we obtain

$$\Delta U = -(F_x\Delta x + F_y\Delta y + F_z\Delta z)$$

If we consider a situation in which y and z are kept constant ($\Delta y = \Delta z = 0$) while varying x we see that F_x is the derivative of U with respect to x under these conditions. Thus, we can write

$$F_x = -\frac{\partial U}{\partial x}$$

where the notation represents the **partial derivative** of U with respect to x.

Partial derivatives

The **partial derivative** of a function $U = U(x,y,z)$ with respect to the variable x is defined as

$$\frac{\partial U}{\partial x} := \underset{\Delta x \to 0}{\text{limit}} \left\{ \frac{U(x + \Delta x, y, z) - U(x, y, z)}{\Delta x} \right\}$$

The partial derivative of a function of a number of variables, therefore, is the derivative of the function with respect to one variable while *keeping the other variables constant.* Bearing this latter maxim in mind, the ordinary rules for differentiation can be used to determine the partial derivative of a function. Partial differentiation proves to be a very useful tool in physics and can be applied quite easily by anyone who is familiar with ordinary differentiation. More details on partial derivatives, including some examples of their application, are provided in Appendix A.6.3.

Using the notation for partial derivatives we can also write expressions for the y- and z-components of force as follows

$$F_y = -\frac{\partial U}{\partial y} \quad \text{and} \quad F_z = -\frac{\partial U}{\partial z}$$

And hence,

$$\boldsymbol{F} = -\left(\frac{\partial U}{\partial x}\boldsymbol{i} + \frac{\partial U}{\partial y}\boldsymbol{j} + \frac{\partial U}{\partial z}\boldsymbol{k}\right) = -\nabla U \tag{5.16}$$

where ∇U (or grad$U$) denotes the ***gradient*** of a function $U = U(x,y,z)$ and is defined as

$$\nabla U := \frac{\partial U}{\partial x}\boldsymbol{i} + \frac{\partial U}{\partial y}\boldsymbol{j} + \frac{\partial U}{\partial z}\boldsymbol{k}$$

The gradient may be considered to be the three-dimensional generalisation of the quantity which we defined as the *slope* in one-dimension (recall Section 2.4). A short summary of the calculus of vector quantities is given in Appendix A.12.

Thus, in terms of the gradient, Equation (5.16) can be written in the more compact form

$$\boldsymbol{F} = -\nabla U$$

For example, if $U(\boldsymbol{r}) = mg(y - y_0)$ (*Case 1* above), as expected

$$\boldsymbol{F} = -\nabla U = 0\boldsymbol{i} + mg\boldsymbol{j} + 0\boldsymbol{k} = -mg\boldsymbol{j}$$

The advantages of using the concept of energy to analyse a problem are now becoming clearer. Energy is a scalar quantity and, in a conservative field, we need only calculate the value of a body's potential energy at the beginning and end of its motion to evaluate the motion in energy terms. We do not need to know details of the path taken between the two points. The analysis of motion in a field by other means, for example by applying Newton's laws in detail to motion along a path in space (which is a vector problem), would otherwise prove extremely difficult in a situation in which the force is varying in both magnitude and direction along the path. Moreover, with a knowledge of the potential energy function $U(\boldsymbol{r}) = U(x, y, z)$, we can determine the force (a vector quantity) at any point by using Equation (5.16).

Kinetic energy: conservation of mechanical energy

We can now extend the analysis in Section 3.7 to three dimensions. Starting with the definition (5.12) of mechanical work and applying Newton's second law (that is, substituting $m\boldsymbol{a}$ for $\boldsymbol{F}$) we obtain

$$W_{\text{AB}} = \int_{\text{A}}^{\text{B}} \boldsymbol{F}(\boldsymbol{r}) \cdot d\boldsymbol{s} = \int_{\text{A}}^{\text{B}} m\boldsymbol{a} \cdot d\boldsymbol{s} = m\int_{\text{A}}^{\text{B}} \frac{d\boldsymbol{v}}{dt} \cdot d\boldsymbol{s} = m\int_{\text{A}}^{\text{B}} \frac{d\boldsymbol{v}}{dt} \cdot \frac{d\boldsymbol{s}}{dt}dt = m\int_{\text{A}}^{\text{B}} \frac{d\boldsymbol{v}}{dt} \cdot \boldsymbol{v}dt$$

The integrand can be simplified since

$$\frac{d}{dt}(v^2) = \frac{d}{dt}(\boldsymbol{v} \cdot \boldsymbol{v}) = \frac{d\boldsymbol{v}}{dt} \cdot \boldsymbol{v} + \boldsymbol{v} \cdot \frac{d\boldsymbol{v}}{dt} = 2\frac{d\boldsymbol{v}}{dt} \cdot \boldsymbol{v}$$

and thus,

$$W_{\text{AB}} = m\int_{\text{A}}^{\text{B}} \frac{1}{2}\frac{d}{dt}(v^2)dt = \frac{1}{2}m\int_{\text{A}}^{\text{B}} d(v^2) = \frac{1}{2}m(v_{\text{B}}^2 - v_{\text{A}}^2)$$

or

$$W_{\text{AB}} = \frac{1}{2}mv_{\text{B}}^2 - \frac{1}{2}mv_{\text{A}}^2 \tag{5.17}$$

We recognise the quantity $\frac{1}{2}mv^2$ as being the **kinetic energy** (Section 3.7) of a particle of mass m which is moving with velocity $\boldsymbol{v}$. Equation (5.17) is the **work-energy theorem** already encountered in Section 3.7 (Equation (3.12)).

Comparing Equations (5.13) and (5.17) we see that, for a conservative field,

$$\frac{1}{2}mv_B^2 - \frac{1}{2}mv_A^2 = U(\boldsymbol{r}_A) - U(\boldsymbol{r}_B)$$

or

$$\boxed{\frac{1}{2}mv_A^2 + U(\boldsymbol{r}_A) = \frac{1}{2}mv_B^2 + U(\boldsymbol{r}_B)} \tag{5.18}$$

Since the points A and B can be any points in the field, the quantity $\frac{1}{2}mv^2 + U(\boldsymbol{r})$ must remain constant as the mass m moves in the field. This tells us that, while the potential energy of the particle $U(\boldsymbol{r})$ can change as the particle moves, its kinetic energy must change correspondingly to ensure that the sum of the two terms is constant, as was the case for motion in one dimension. The sum of the kinetic energy and the potential energy of a particle is called the **total mechanical energy** defined by

$$\boxed{E := K + U := ½\, mv^2 + U(\boldsymbol{r})}$$

and Equation (5.18) is a statement of the **principle of conservation of mechanical energy**, namely that **in a conservative field the total energy is constant.**

From the definition of the momentum of a particle $\boldsymbol{p} := m\boldsymbol{v}$, the kinetic energy can also be written as $\frac{p^2}{2m}$ and hence the total energy can be written

$$E = \frac{p^2}{2m} + U(\boldsymbol{r})$$

The expression for the total energy, written as a function of momentum and displacement, is called the **Hamiltonian function** or the Hamiltonian of the system (William Rowan Hamilton 1788–1856).

$$\boxed{H(\boldsymbol{p},\boldsymbol{r}) := \frac{p^2}{2m} + U(\boldsymbol{r})}$$

and the principle of conservation of mechanical energy can equally well be written

$$\boxed{H(\boldsymbol{p},\boldsymbol{r}) = E = constant}$$

The Hamiltonian function will play an important role in some later chapters.

Study Worked Example 5.3

For problems based on the material presented in this section visit up.ucc.ie/5/ *and follow the link to the problems.*

5.8 Power

Power, the *rate* at which mechanical work is being done on a body in a force field, can be defined as in Section 3.9. Let us consider again the situation, illustrated in Figure 5.22, in which a body is moved from a point A to a point B under the influence of a general field force. Suppose that the body is at the point P at some instant t and at the point P′ at $t + \Delta t$ (Figure 5.26) and let the work done in moving the body from P to P′ be ΔW. The physical quantity **power**, which is a measure of the rate of doing work, is defined as

$$\boxed{P := \underset{\Delta t \to 0}{\text{limit}} \frac{\Delta W}{\Delta t} = \frac{dW}{dt}}$$

If it takes a time Δt for a force $\boldsymbol{F}$ to do the amount of work ΔW and the corresponding change in displacement in this time interval is $\Delta \boldsymbol{s}$, recalling the definition (Equation (5.11)) of mechanical work, namely $\Delta W := \boldsymbol{F} \cdot \Delta \boldsymbol{s}$, we see that the power generated by the force $\boldsymbol{F}$ is given by

$$P := \frac{dW}{dt} = \underset{\Delta t \to 0}{\text{limit}} \frac{\Delta W}{\Delta t} = \underset{\Delta t \to 0}{\text{limit}}\, \boldsymbol{F} \cdot \frac{\Delta \boldsymbol{s}}{\Delta t} = \boldsymbol{F} \cdot \left(\underset{\Delta t \to 0}{\text{limit}} \frac{\Delta \boldsymbol{s}}{\Delta t}\right) = \boldsymbol{F} \cdot \frac{d\boldsymbol{s}}{dt} = \boldsymbol{F} \cdot \boldsymbol{v} \tag{5.19}$$

where $\boldsymbol{v}$ is the velocity of the body on which the force $\boldsymbol{F}$ is being exerted.

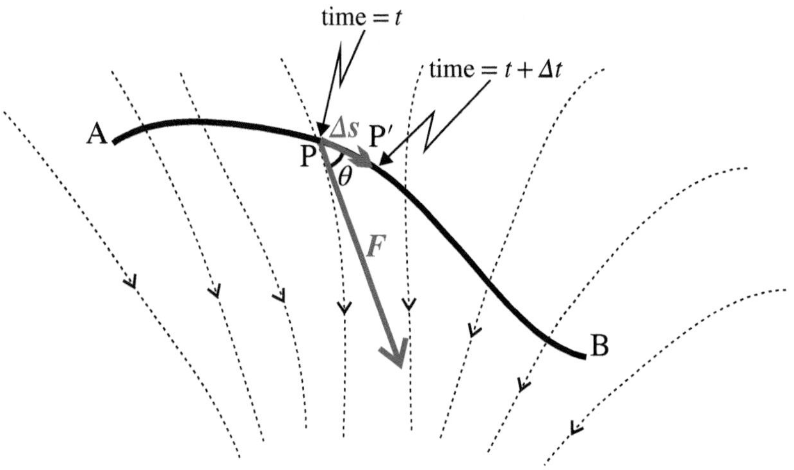

Figure 5.26. The rate at which work is done by a field force is called the power generated by the force.

5.9 Energy in a constant uniform field

We saw in Chapter 3 how plots of potential energy functions can be used to analyse dynamics problems in one-dimension. The same technique can also be effective in two and three dimensions in cases in which the potential energy function depends on only one variable. An example of such a case, potential energy in a constant uniform field, is discussed below. A further example, potential energy in inverse square law fields, will be treated in Section 5.10.

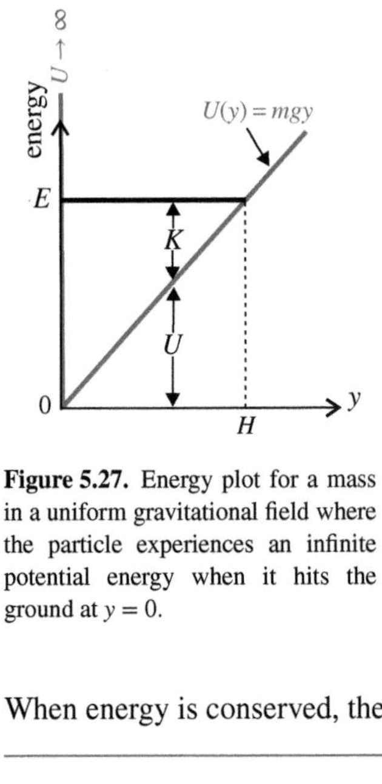

Figure 5.27. Energy plot for a mass in a uniform gravitational field where the particle experiences an infinite potential energy when it hits the ground at $y = 0$.

In Figure 5.27 the potential energy function of a mass which is moving in the gravitational field near the Earth's surface, as in projectile motion, is plotted as function of the vertical coordinate of its position. If potential energy is taken to be zero when the mass is on the surface (that is $U(0) = 0$), we see from Equation (5.14) that

$$U(y) = mgy$$

so that the graph is a straight line of (positive) slope mg.

The region where $y < 0$ is not accessible to the mass so, as in the case of the infinite square well (Figure 3.17), we represent such a situation by requiring that the potential energy function be infinite at that point.

If the mass were a ball that bounces perfectly elastically when it hits the ground, that is, rebounds without losing energy (in other words, if the system is conservative) so that $K + U = E =$ constant, the ball would continue bouncing along indefinitely between $y = 0$ and $y = H$, the height from which it is dropped. The allowed region for the ball is $0 < y < H$. The potential function is written

$$U(y) = \begin{cases} \infty & \text{for } y \le 0 \\ mgy & \text{for } y > 0 \end{cases}$$

When energy is conserved, the motion can be considered as a continuing exchange of energy between the kinetic and potential forms.

For problems based on the material presented in this section visit up.ucc.ie/5/ *and follow the link to the problems.*

5.10 Energy in an inverse square law field

Case 1: An attractive inverse square law force

Newton's Law of gravitation (Section 5.1) states that the attractive force between two masses varies inversely with the square of the distance between them. In Section 5.7 we showed that the potential energy of a mass m in the gravitational force field which surrounds another mass M varies with r, the radial co-ordinate of m with respect to M, as follows (Equation 5.15)

$$U(r) = -\frac{GMm}{r}$$

A plot of $U(r)$ as a function of r is shown Figure 5.28. Later in this book we will encounter other important examples of inverse square law forces whose potential energy functions have the same form as that shown in Figure 5.28.

If the mass has a positive total energy, such as E_1 in Figure 5.28, it is free to move anywhere in the region from $r = 0$ to $r = \infty$. If, however, the energy is negative with respect to the zero of potential energy (the value of $U(r)$ at $r = \infty$), such as E_2 in the figure, the mass is restricted to the region where $E \geq U$, that is between $r = 0$ and $r = r_1$ in Figure 5.28. The mass is said to be *bound* in the force field in this case. The energy E_2 is known as the *binding energy* of the mass; note that this energy is always negative so that, in a bound system, if the mass is to escape from the field, it must be given an injection of positive energy sufficient to cancel this negative energy.

The gradient of $U(r)$ is always positive and thus the force $F = -\dfrac{dU}{dr}$ is always in the negative r direction; that is, the force is always attractive. The magnitude of $\dfrac{dU}{dr}$, and thus of F, decreases with increasing r, as expected for an inverse square law force.

Note that in interpreting Figure 5.28 we have considered motion only along a radius – effectively motion in one dimension. To consider orbital motion around the origin of the force field, we will need to consider motion in two dimensions (Sections 5.12 and 5.13 below). The mass then has kinetic energy due to motion perpendicular to the $\mathbf{r}$ direction (that is perpendicular to the page) and we need to be more careful in our interpretation of one-dimensional potential plots such as Figure 5.28.

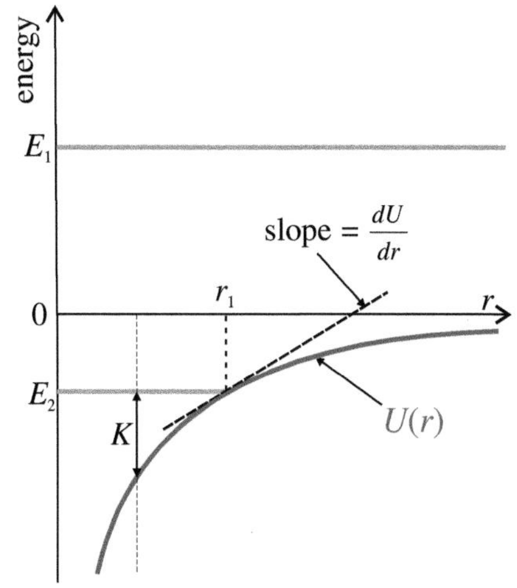

Figure 5.28. Energy plot for a particle in an attractive inverse square law field. Two cases are shown, namely $E = E_1 > 0$ and $E = E_2 < 0$. In the first case the particle is unbound but in the second case the particle is bound such that $r < r_1$. K is the kinetic energy of the particle at displacement r.

In Worked Example 5.4, the relationships in this section are used to determine the **escape speed** – the speed needed for a mass to escape the gravitational attraction of another mass. An example is the minimum speed required to launch a body from the surface of a planet so that it will escape the gravitational field of the planet.

Study Worked Example 5.4

Before we leave the case of an attractive inverse square law force, consider a small section of Figure 5.28, as illustrated in Figure 5.29. If the section Δr is small enough, the potential energy function may be considered to be a straight line over that interval which means that the field can be considered uniform within this small region.

The section Δr in Figure 5.29 is effectively the same as Figure 5.27 with r replacing y as the relevant co-ordinate. This is why we are able to treat the gravitational field near the Earth's surface as uniform. The treatment is valid as long as the spatial region involved is sufficiently small to allow the potential plot to be considered to be a straight line.

We can confirm this conclusion mathematically. The gravitational potential energy of a mass m near the Earth is given by Equation (5.15), namely

$$U = -\frac{GmM_E}{r}$$

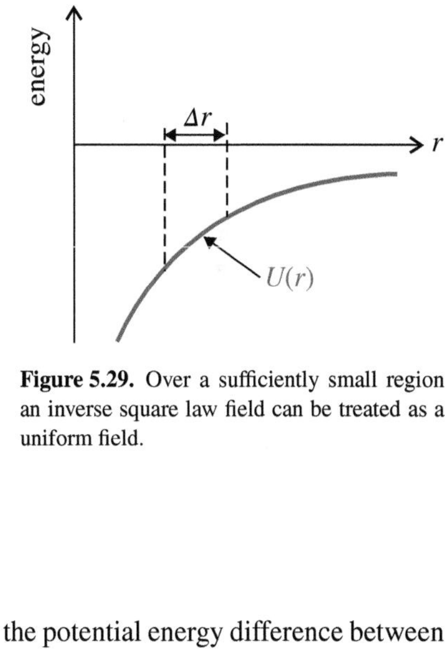

Figure 5.29. Over a sufficiently small region an inverse square law field can be treated as a uniform field.

where M_E is the mass of the Earth. Differentiating this equation we obtain

$$\frac{dU}{dr} = \frac{GmM_E}{r^2}$$

If two points are at distances r and $(r + \Delta r)$ respectively from the centre of the Earth we can write the potential energy difference between these points

$$\Delta U = \frac{dU}{dr}\Delta r = \frac{GmM_E(\Delta r)}{r^2}$$

In the case of a mass which is moved to a height y above the Earth's surface $\Delta r = y$ and $r = R_E$, the radius of the Earth, so that ΔU can be written

$$\Delta U = \frac{GmM_E y}{R_E^2} \quad (\text{for } y \ll R_E)$$

Recalling, from Equation (5.4), that $g = \dfrac{GM_E}{R_E^2}$, we can write the potential energy difference between the two points as

$$\Delta U = mgy$$

The expression for the potential energy of a mass m in the uniform gravitational field near the Earth's surface given in Equation (5.14) is, therefore, a special case of the general equation for potential energy in the Earth's gravitational field; it is valid if $y \ll R_E$.

Case 2: A repulsive inverse square law force

Although the gravitational inverse square law force can only be attractive, other inverse square law forces, such as the electrostatic force (Chapter 16), can be either repulsive or attractive. In the case of a repulsive inverse square law force, the potential energy varies with r according to the equation

$$U(r) = +\frac{C}{r}$$

where C is a constant. Such a function is plotted in Figure 5.30.

Note that in this case the restriction $E \geq U$ means that a particle which is subject to this force field is free to move anywhere between $r = r_1$ and $r = \infty$. As can be seen from Figure 5.30 the kinetic energy, and thus the velocity, of the particle increases as r increases. The gradient of the potential energy plot is always negative and decreases with r, thus confirming that the force is in the positive r direction (repulsion) and that it decreases with r.

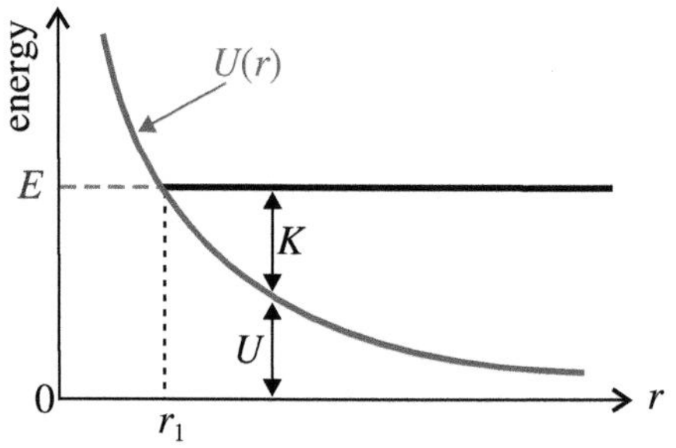

Figure 5.30. Energy plot for a particle in a repulsive inverse square law field.

It should be clear from the foregoing that potential energy plots can be very useful tools for representing the main features of force fields. They will be used frequently in this book, particularly when we come to consider and discuss such problems in quantum mechanics. The simplicity of the technique derives from the scalar nature of energy; potential energies due to independent sources of force fields may be added algebraically, as illustrated below.

Figure 5.31 represents a situation in which two force fields which are centred at different points (A and B) overlap. In the case illustrated in the figure, both fields are attractive inverse square force fields of equal strength. Potential energy is plotted as a function of position along an axis that passes symmetrically through A and B. The dashed blue lines in Figure 5.31 are potential energy plots for A and B individually. The solid blue line is the algebraic sum of the two and represents their net effect. The situation may be visualised as that encountered by a mass which is moving in the field of a binary star system – assuming that the mass can move through the star centres without encountering any additional forces (not a very realistic assumption).

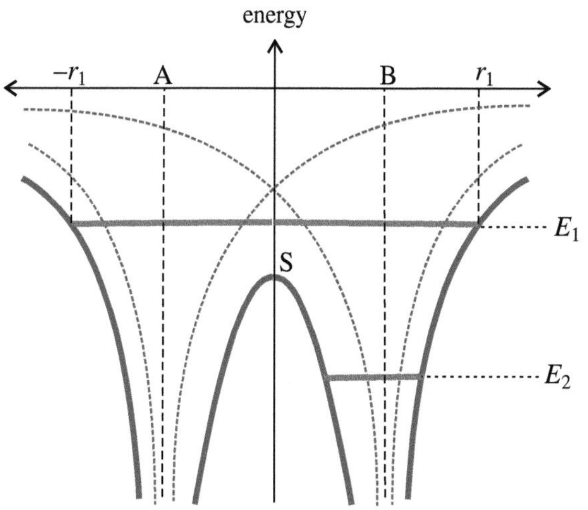

Figure 5.31. Energy plot for a particle in a combination of two attractive inverse square law fields. An example of such a system is the gravitational field due to the two stars in a binary system.

If the total energy of the particle in the force field is E_1, a negative energy which is less negative than the energy of the saddle-point (S in Figure 5.31) of the potential plot, the particle can move freely in the two fields but is confined to the region of space between $-r_1$ and $+r_1$. It is bound to the two field system. If, however, the energy of the particle is more negative than the energy of the saddle point (E_2, in the figure), the particle is confined to the region surrounding one of the centres of the force field. In the case illustrated, the particle is confined to the region around the point B.

For problems based on the material presented in this section visit up.ucc.ie/5/ and follow the link to the problems.

5.11 Moment of a force: angular momentum

In Section 4.4 we showed that Newton's second law for translational motion, Equation (4.3), can also be expressed in terms of momentum as

$$\frac{d\boldsymbol{p}}{dt} = \boldsymbol{F} \tag{5.20}$$

Our objective in this section is to derive a similar equation applicable to angular motion, namely a version of Newton's second law for angular motion.

Moment of a force

First we use the vector product to define a quantity called the **moment of a force**. The moment of a force $\boldsymbol{F}$ relative to some origin O (Figure 5.32) is defined as

$$\boxed{\boldsymbol{M} := \boldsymbol{r} \times \boldsymbol{F}}$$

where $\boldsymbol{r}$ is the displacement, relative to O, of *any* point on the line of action of $\boldsymbol{F}$.

If a perpendicular is dropped from O onto the line of action of the force in Figure 5.32, the length of the perpendicular b is $|\boldsymbol{r}|\sin\theta$. We can then write the magnitude of the moment of the force $\boldsymbol{F}$ as

$$|\boldsymbol{M}| = |\boldsymbol{r}||\boldsymbol{F}|\sin\theta = |\boldsymbol{F}|b,$$

which is independent of the point on the line of the force to which $\boldsymbol{r}$ is drawn. The magnitude of the moment, therefore, can be calculated as the product of the magnitude of the force and the perpendicular distance from the origin onto the line of action of the force; the direction of the moment is given by the definition of the vector product. In the case illustrated in Figure 5.32 the moment of the force is directed perpendicularly to and into the page. Note that if the point O lies on the line of action of the force $\boldsymbol{F}$, then $\boldsymbol{r} \parallel \boldsymbol{F}$ and the moment of $\boldsymbol{F}$ relative to O is zero. The SI unit of moment of force is N m.

By definition, the moment of a force is a vector quantity, being the vector (cross) product of $\boldsymbol{r}$ and $\boldsymbol{F}$. When a number of forces act on a particle, therefore, the net effect of these will be the same as if they were replaced by a single force the moment of which is the vector sum of the moments of the individual forces.

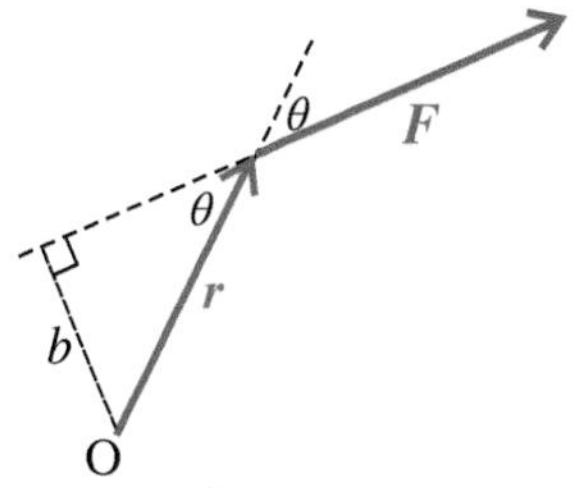

Figure 5.32. The moment of a force relative to an origin O is defined as $\boldsymbol{M} = \boldsymbol{r} \times \boldsymbol{F}$. The magnitude of the moment is the product of the magnitude of $\boldsymbol{F}$ and the perpendicular distance from O onto the line of action of F, that is $|\boldsymbol{M}| = |\boldsymbol{F}b$.

Angular momentum; Newton's second law for angular motion

The next step is to obtain an expression for the moment of the force $\boldsymbol{F}$ by taking the vector product of $\boldsymbol{r}$ with both sides of the Equation (5.20)

$$\boldsymbol{M} = \boldsymbol{r} \times \boldsymbol{F} = \boldsymbol{r} \times \frac{d\boldsymbol{p}}{dt} \tag{5.21}$$

Consider now the differentiation of the vector product $\boldsymbol{r} \times \boldsymbol{p}$. Using the result for the derivative of a cross product (Appendix A.12.1) we obtain

$$\frac{d}{dt}(\boldsymbol{r} \times \boldsymbol{p}) = \left(\boldsymbol{r} \times \frac{d\boldsymbol{p}}{dt}\right) + \left(\frac{d\boldsymbol{r}}{dt} \times \boldsymbol{p}\right)$$

We recognise $\frac{d\boldsymbol{r}}{dt}$ as the velocity $\boldsymbol{v}$ and, since $\boldsymbol{v}$ is always parallel to $\boldsymbol{p}$,

$$\frac{d\boldsymbol{r}}{dt} \times \boldsymbol{p} = \boldsymbol{v} \times \boldsymbol{p} = m\boldsymbol{v} \times \boldsymbol{v} = 0$$

and hence

$$\frac{d}{dt}(\boldsymbol{r} \times \boldsymbol{p}) = \left(\boldsymbol{r} \times \frac{d\boldsymbol{p}}{dt}\right)$$

so that we can write (5.21) as

$$\boldsymbol{M} = \frac{d}{dt}(\boldsymbol{r} \times \boldsymbol{p}) \tag{5.22}$$

Equation (5.22) takes a form similar to Newton's second law if we make the following definition.

The **angular momentum** of a mass m, at displacement $\boldsymbol{r}$ and moving with velocity $\boldsymbol{v}$, relative to an origin O (Figure 5.33), is defined as

$$\boxed{\boldsymbol{L} := \boldsymbol{r} \times m\boldsymbol{v} = \boldsymbol{r} \times \boldsymbol{p}}$$

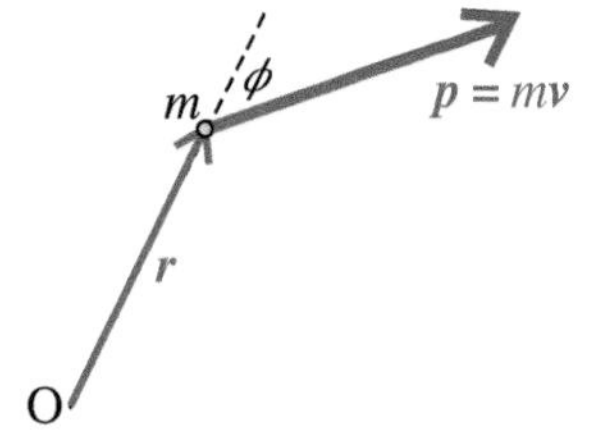

Figure 5.33. The angular momentum, relative to an origin O, of a particle of mass m is defined as $\boldsymbol{L} = \boldsymbol{r} \times \boldsymbol{p}$, where $\boldsymbol{p} = m\boldsymbol{v}$ is the momentum of the particle relative to O.

The magnitude and direction of $\boldsymbol{L}$ follow from the definition of the vector product. The magnitude of $\boldsymbol{L}$ is given by

$$|\boldsymbol{L}| = |\boldsymbol{r}||m\boldsymbol{v}|\sin\phi$$

The direction of $\boldsymbol{L}$ in the case illustrated in Figure 5.33 is perpendicular to the page pointing into the page. The SI unit for angular momentum is kg m^2 s^{-1}.

Using the definition of angular momentum, Equation (5.22) can be written as

$$\boxed{\boldsymbol{M} = \frac{d\boldsymbol{L}}{dt}} \tag{5.23}$$

which is similar to $\boldsymbol{F} = \frac{d\boldsymbol{p}}{dt}$ with $\boldsymbol{M}$ replacing $\boldsymbol{F}$ and $\boldsymbol{L}$ replacing $\boldsymbol{p}$. The full physical significance of the moment of a force and the angular momentum of a body will become clearer in Chapters 6 and 7 when the dynamics of many-body and rigid body systems are studied. In the next two sections we will apply these ideas to planetary orbits.

From Equation (5.23) it follows that if $\boldsymbol{M} = 0$, as for example when the force acts through the origin, then $\frac{d\boldsymbol{L}}{dt} = 0$ and $\boldsymbol{L}$ = constant. In other words, relative to an origin on the line of action of the force, the angular momentum remains constant.

Study Worked Example 5.5

For problems based on the material presented in this section visit up.ucc.ie/5/ and follow the link to the problems.

5.12 Planetary motion: circular orbits

We are now in a position to analyse the motion of a body in a familiar force field – namely a mass which is bound by gravitation in an orbit around a second, usually much larger, mass. Consider, for example, the motion of a planet of mass m in the gravitational field of the Sun (mass M).

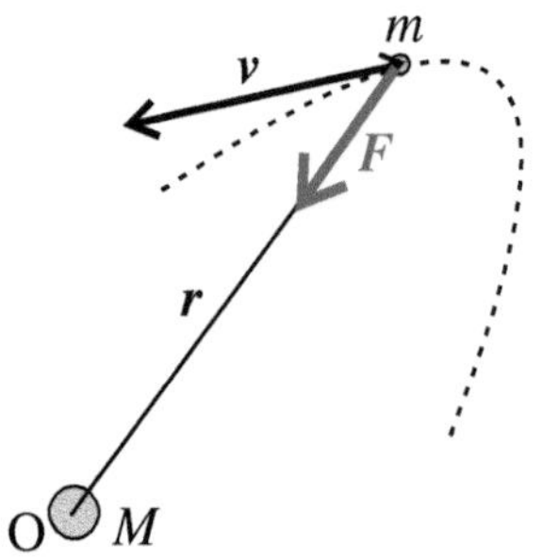

Figure 5.34. A planet (mass = m) is kept in orbit around the Sun (mass = M) by the gravitational force F which is directed along the line from the planet to the Sun.

We shall show in Section 6.7 that, because $m \ll M$, the Sun may be treated as fixed during the motion. Figure 5.34 shows the mass m moving under the influence of a gravitational force centred at O, the position of M. The force exerted on m is given by

$$\boldsymbol{F}(\boldsymbol{r}) = \frac{d\boldsymbol{p}}{dt} = -\frac{GMm}{r^2}\hat{\boldsymbol{r}}$$

Note that, relative to O, the moment of the force acting on the planet is zero. This is because the force $\boldsymbol{F}$ due to gravitation is always directed towards O and thus is anti-parallel to the radius vector $\boldsymbol{r}$ in Figure 5.34, in which case

$$\boldsymbol{M} = \boldsymbol{r} \times \boldsymbol{F} = 0$$

and hence, as pointed out at the end of the previous section, $\frac{d\boldsymbol{L}}{dt} = 0$ and thus $\boldsymbol{L} = constant$.

Circular orbits

In the next section we will analyse the general motion of a planet under the influence of this force, but we can immediately gain insight into the problem by considering the special case of a circular orbit (Figure 5.35); the orbits of most planets in the solar system are almost circular. When a mass is moving in a circular path ($r = R = constant$) the velocity is always perpendicular to the radius of the orbit and thus $|\boldsymbol{L}| = |\boldsymbol{r} \times \boldsymbol{p}| = |\boldsymbol{r}||m\boldsymbol{v}|\sin\frac{\pi}{2} = mvR$ or, invoking Equation (2.25), $|\boldsymbol{L}| = m\omega R^2$. Since for circular orbits R is constant, v and ω must also be constant and hence this is a case of uniform circular motion with the centripetal force supplied by the gravitational attraction. Applying Newton's second law

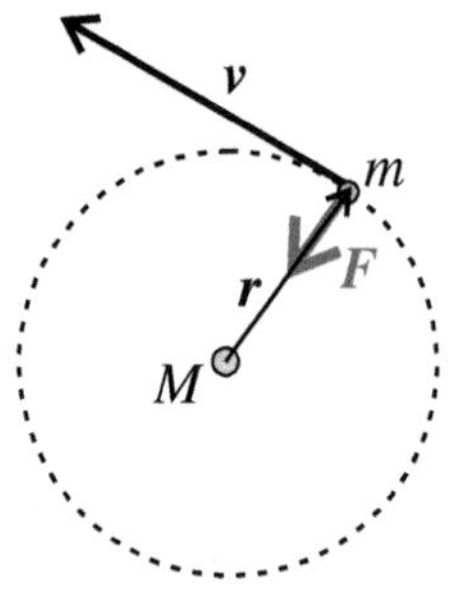

Figure 5.35. Planet in a circular orbit.

$$\frac{GMm}{R^2} = ma_c = m\frac{v^2}{R}$$

and hence the speed of the planet is given by

$$v = \sqrt{\frac{GM}{R}}$$

The kinetic energy of the planet is given by $K = \frac{1}{2}mv^2 = \frac{GMm}{2R}$ and its total energy can be determined, using Equation (5.15), as follows

$$E = K + U = \frac{GMm}{2R} - \frac{GMm}{R} = -\frac{GMm}{2R}$$

Note that the total energy is negative, as expected for a bound system (Section 5.10).

The period of the orbit is the time taken to complete one revolution, namely

$$T = \frac{2\pi R}{v} = 2\pi R\sqrt{\frac{R}{GM}}$$

and hence,

$$T^2 = \frac{4\pi^2 R^3}{GM}$$

Thus the square of the period of the planet's orbit around the Sun (one year in the case of the Earth) is proportional to the cube of the radius of its orbit. Johannes Kepler (1571–1630) noted that all planets obeyed this rule (called Kepler's 'law of periods' – the third of Kepler's laws which are listed in the next section). We have derived the law here for the special case of a circular orbit.

For problems based on the material presented in this section visit up.ucc.ie/5/ and follow the link to the problems.

5.13 Planetary motion: elliptical orbits and Kepler's laws

We will now analyse the more general situation of a planet of mass m which is moving in the gravitational field of a large mass M, but not necessarily in a circular orbit. There are three constants of the motion of the system (conserved quantities).

(i) *Conservation of energy:* Since the system is conservative, the total mechanical energy is conserved, that is the total energy (kinetic energy + potential energy) is given by

$$E = \frac{p^2}{2m} - \frac{GMm}{r} = constant$$

(ii) *Conservation of angular momentum:* As noted in the previous section, the force on m is always directed towards O, so that $\boldsymbol{M} = \boldsymbol{r} \times \boldsymbol{F} = 0$ always, and hence $\frac{d\boldsymbol{L}}{dt} = 0$. Thus, the angular momentum remains constant, that is

$$\boldsymbol{L} = m\boldsymbol{r} \times \boldsymbol{v} = constant.$$

Note that the vectors $\boldsymbol{r}$ and $\boldsymbol{v}$ define an instantaneous plane of the motion (for example, the plane of the page in Figure 5.34) and hence $\boldsymbol{L}$ is perpendicular to that plane. Since $\boldsymbol{L}$ is constant, the mass remains confined to that plane; that is, it moves in a planar orbit with $\boldsymbol{L}$ perpendicular to the plane of the orbit. Conservation of angular momentum is a result of the fact that a central force is involved; that is, $\boldsymbol{F}$ is antiparallel to $\boldsymbol{r}$ at all points of the orbit.

Consider the shaded area in Figure 5.36, which is the area swept out by the $\boldsymbol{r}$ vector in a time Δt. The magnitude of this area is given by $\Delta A = \frac{1}{2}|\boldsymbol{r} \times \Delta\boldsymbol{r}|$ (compare with Figure 4.43) and hence,

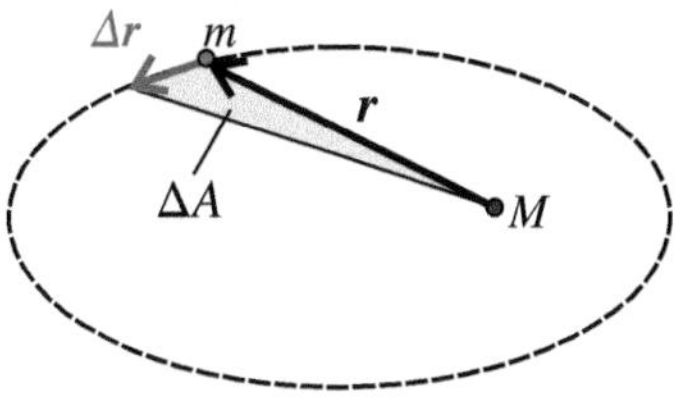

Figure 5.36. The line joining the two masses sweeps out the shaded area ΔA in a time Δt, that is while the orbiting mass moves from $\boldsymbol{r}$ to $\boldsymbol{r} + \Delta\boldsymbol{r}$.

$$\frac{dA}{dt} = \frac{1}{2}\left|\boldsymbol{r} \times \frac{d\boldsymbol{r}}{dt}\right| = \frac{1}{2}|\boldsymbol{r} \times \boldsymbol{v}| = \frac{|\boldsymbol{L}|}{2m} = constant \tag{5.24}$$

This result, that the line from the Sun to a planet sweeps out equal areas in equal times (the 'law of areas'), was discovered by Johannes Kepler based on an analysis of the astronomical observations of Tycho Brahe. Kepler's discovery, which predated Newton's laws of motion and the law of gravitation, was a direct observation of conservation of angular momentum – long before the concept of angular momentum had been developed.

(iii) *Conservation of the Runge-Lenz vector:* In the case of inverse square law forces a further conserved quantity appears, namely the vector $\boldsymbol{R} := (\boldsymbol{v} \times \boldsymbol{L}) - GMm\hat{\boldsymbol{r}}$, known as the Runge-Lenz vector. Note that, since $\boldsymbol{L}$ is perpendicular to the plane of the orbit and hence is perpendicular to $\boldsymbol{v}$, $\boldsymbol{v} \times \boldsymbol{L}$ is in the plane of the orbit and hence $\boldsymbol{R}$ is also in the plane of the orbit. Proof of the conservation of $\boldsymbol{R}$ is given in the *Understanding Physics* website at up.ucc.ie/5/13/1/ where it is also shown that the magnitude of $\boldsymbol{R}$ is given by

$$|\boldsymbol{R}| = \sqrt{G^2M^2m^2 - \frac{2L^2|E|}{m}}$$

which is constant since L and E are constant. It is important to note that the conservation of the Runge-Lenz vector arises from the special character of the inverse square law force.

Equation of the orbit

The equation of the orbit can be derived by taking the scalar product of $\boldsymbol{r}$ and $\boldsymbol{R}$.

$$\boldsymbol{r} \cdot \boldsymbol{R} = \boldsymbol{r} \cdot (\boldsymbol{v} \times \boldsymbol{L}) - GMm\hat{\boldsymbol{r}} \cdot \boldsymbol{r} = (\boldsymbol{r} \times \boldsymbol{v}) \cdot \boldsymbol{L} - GMmr = \frac{L^2}{m} - GMmr$$

and thus, $rR\cos\theta = \frac{L^2}{m} - GMmr$ where θ is the angle between $\boldsymbol{r}$ and the direction of $\boldsymbol{R}$. Solving for r, and after a little algebraic manipulation, we obtain the equation of the orbit in polar coordinates, namely

$$r = \frac{\alpha}{1 + \varepsilon\cos\theta}$$

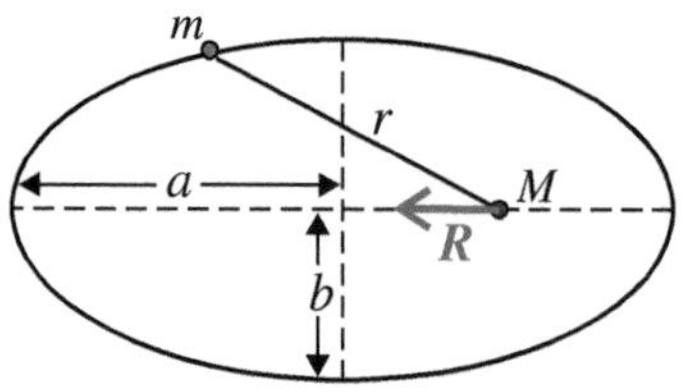

Figure 5.37. An elliptical orbit: $\boldsymbol{R}$, the Runge-Lenz vector, is directed along the major axis of the ellipse.

where $\alpha = \frac{L^2}{GMm^2}$ and $\varepsilon = \frac{R}{GMm} = \sqrt{1 + \frac{2L^2E}{G^2M^2m^3}}$ are constants.

This is the equation of an ellipse (Appendix A.10.3) whose semi-major and semi-minor axes, respectively, are given by $a = \frac{\alpha}{1-\varepsilon^2}$ and $b = \frac{\alpha}{\sqrt{1-\varepsilon^2}}$ where ε is the eccentricity.

This result, that planets move in elliptical orbits (Figure 5.37), was discovered empirically by Johannes Kepler (1571–1630).

Period of orbit

The period of the orbit can be obtained by integrating Equation (5.24) over a complete orbit as follows.

$$T = \oint dt = \frac{2m}{|\boldsymbol{L}|}\oiint dA = \frac{2m}{|\boldsymbol{L}|}(\text{area of ellipse}) \tag{5.25}$$

Since the area of an ellipse is πab (Appendix A.1 (iii)), we can write

$$T = \frac{2m}{|L|}(\pi ab) = \frac{2\pi m}{|L|}a\left(\frac{\alpha}{\sqrt{1-\varepsilon^2}}\right) = \frac{2\pi m}{|L|}a\left(\sqrt{\frac{\alpha}{1-\varepsilon^2}}\right)\alpha^{1/2} = \frac{2\pi m}{|L|}a^{3/2}\left(\frac{L^2}{GMm^2}\right)^{1/2} = \frac{2\pi}{\sqrt{GM}}a^{3/2}$$

We have therefore confirmed, for the general case of an elliptical orbit, that $T^2 \propto a^3$ (the 'law of periods'), as established by Kepler from astronomical observations. Note that this reduces to the result derived above for circular motion where $a = R$. The equation is of considerable practical importance in astronomy. With a knowledge of T and a for any planet and using the known value of G, we can calculate the mass of the Sun, an almost impossible measurement by any other means. The equation can also be applied to any situation in which two masses form a bound system, for example a satellite orbiting a planet. With knowledge of the period and of the semimajor axis of the orbit, the mass of the object at the centre of the orbit can be determined.

Kepler's laws

Historically, the laws of motion and of universal gravitation were deduced by Newton from a study of empirical laws which had been formulated by Kepler, the reverse of the treatment given in this section. Based on his analysis of the data of the Danish astronomer Tycho Brahe (1546–1601), Kepler had proposed the following rules (known as 'Kepler's laws') for planetary motion.

(a) Each planet moves in an elliptical orbit with the Sun at one focus of the ellipse.
(b) A line from the Sun to a planet sweeps out equal areas in equal times ('law of areas').
(c) The square of the period of each planet is proportional to the cube of the semimajor axis of its orbit ('law of periods')

Thus astronomical observations played a key part in stimulating Newton to postulate the laws of motion and gravitation. This is not surprising since laboratory observations tend to be dominated by the Earth's gravity, friction, and other forces, effects which can obscure simple direct observations of the basic laws of mechanics in operation. The motion of objects in space is largely free of such effects and has proved to be an excellent laboratory in which to study the laws of mechanics.

Other central forces

The vector $\boldsymbol{R}$ is directed along the semimajor axis of the ellipse and the fact that it is constant is the reason that we always find closed orbits in the case of inverse square law forces. If $n \neq 2$, however, the $\boldsymbol{R}$ vector rotates in the plane of the motion and the orbits are not closed. For example, the effect of the other planets, especially Jupiter and Saturn, on the motion of a planet around the Sun provides a small deviation from exact inverse square law behaviour which results in a small 'precession of the perihelion', as illustrated in Fig. 5.38.

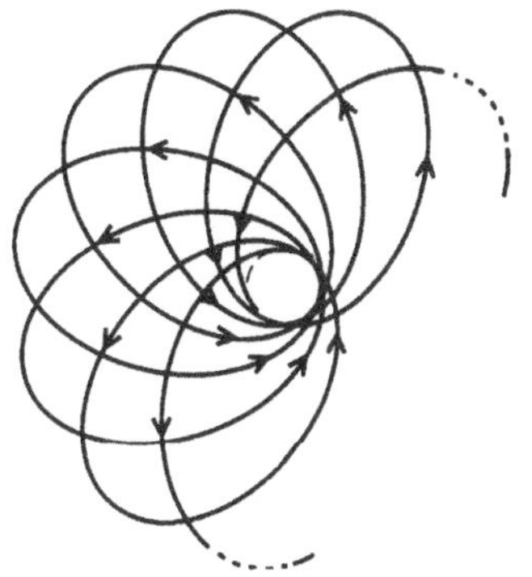

Figure 5.38. Precession of the perihelion is caused by variation from an exact inverse square law field.

For problems based on the material presented in this section visit up.ucc.ie/5/ and follow the link to the problems.

WORKED EXAMPLES

Worked Example 5.1: A boy throws a stone from the top of a 10 m high vertical cliff into the water below. The stone is projected at an angle of 25° above the horizontal at a speed of 15 m s^{-1}.

(a) How far from the foot of the cliff will the stone strike the water?
(b) How long after it is thrown will the stone strike the water?
(c) What is the velocity (in magnitude and direction) of the stone just before it strikes the water?

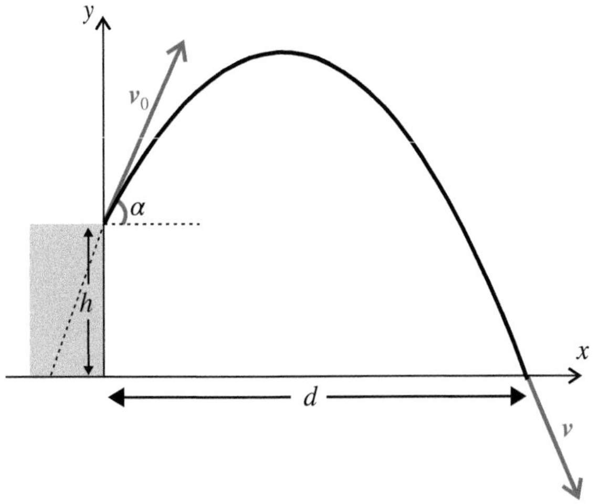

Figure 5.39. Worked Example 5.1 (not to scale).

We wish to know where the stone hits the water or, more specifically, the horizontal distance from the foot of the cliff to the point at which it hits the water. The velocity of the stone changes in both magnitude and direction during the motion but, by specifying the displacement of the stone along two perpendicular co-ordinate axes, the y-axis (vertical) and the x-axis (horizontal) as indicated in Figure 5.39, we are able to reduce the problem to the analysis of two special cases of linear motion which we have already analysed in Section 2.5.

Along the y-axis the stone falls downward with constant acceleration of 9.8 m s^{-2}, like any free object near the Earth's surface, while in the x-direction it moves with constant speed.

Horizontal motion:	$v_x = v_0 \cos \alpha = constant$	$x = v_x t = (v_0 \cos \alpha)t$
Vertical motion:	$v_y = v_0 \sin \alpha + At$	$y = v_{0y} t + \tfrac{1}{2}At^2 = (v_0 \sin \alpha)t + \tfrac{1}{2}At^2$

(a) When the stone hits the ground, at a time which we call T, $y = -h$ and $x = d$ where h and d are the height of the cliff and the horizontal distance travelled by the stone, respectively, as indicated in Figure 5.39 and the acceleration $A = -g = -9.8\ \mathrm{m\,s^{-2}}$. Thus,

$$d = (v_0 \cos\alpha)T \quad \rightarrow \quad T = \frac{d}{v_0 \cos\alpha} \quad \text{and}$$

$$-h = (v_0 \sin\alpha)T - \frac{g}{2}T^2 = (v_0 \sin\alpha)\frac{d}{v_0 \cos\alpha} - \frac{g}{2}\left(\frac{d}{v_0 \cos\alpha}\right)^2$$

Substituting $h = 10$ m, $v_0 = 15\ \mathrm{m\,s^{-1}}$, $\alpha = 25°$ and $g = -9.8\ \mathrm{m\,s^{-2}}$ we obtain $-10 = 0.466d - 0.0265d^2$ or $d^2 - 17.6d - 377 = 0$. The formula A.4.2 from Appendix A can be used to solve this quadratic equation as follows $d = \dfrac{+17.6 \pm \sqrt{(17.6)^2 - 4(-377))}}{2} = 30.1$ m (or $-$ 12.5 m). The second alternative corresponds to the point where the parabola intersects the x-axis for negative x (which does not represent a physically meaningful solution in this situation since it requires the stone to pass through the cliff). Hence, the answer to part (a) is 30.1 m.

(b) $T = \dfrac{d}{v_0 \cos\alpha} = \dfrac{30.1\ \mathrm{m}}{(15\ \mathrm{m\,s^{-1}})\cos 25°} = 2.21\ \mathrm{s}$

(c) At $t = T$: $v_x = v_0 \cos\alpha = (15\ \mathrm{m\,s^{-1}})\cos(25°) = 13.6\ \mathrm{m\,s^{-1}}$ and
$v_y = v_0 \sin\alpha - gT = (15\ \mathrm{m\,s^{-1}})\sin(25°) - (9.8\ \mathrm{m\,s^{-2}})(2.2\ \mathrm{s}) = -15.2\ \mathrm{m\,s_{-1}}$
$\Rightarrow v = \sqrt{(13.6)^2 + (-15.2)^2} = 20.4\ \mathrm{m\,s^{-1}}$ at an angle of $\tan^{-1}\left(\dfrac{15.2}{13.6}\right) = -48°$, that is 48° below the horizontal.

Worked Example 5.2: Derive an expression for the gravitational 'self energy' (as described in Case 2 of Section 5.7) of a uniform spherical distribution of matter of mass M and radius a. Use your result to estimate the gravitational potential energy of the Earth assuming it to be a sphere of uniform density.

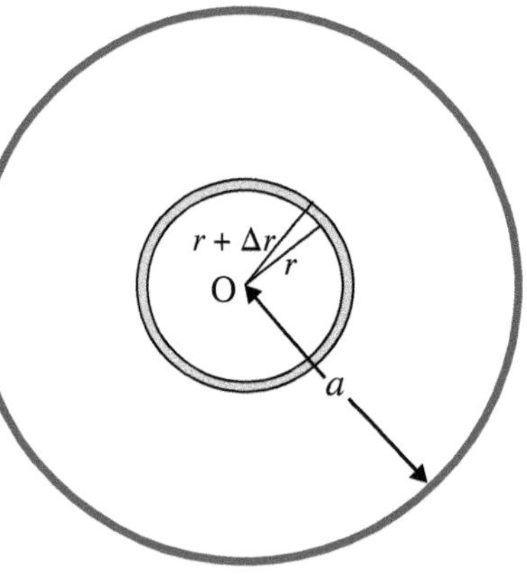

Figure 5.40. Worked Example 5.2.

Let ρ be the density of the matter in the sphere, that is $\rho = \dfrac{M}{\frac{4}{3}\pi a^3}$. Consider a shell of matter with radius between r and $r + \Delta r$ (Figure 5.40). The mass of this shell is given by $\Delta m = \rho(\Delta V) = \rho(4\pi r^2 \Delta r)$ and the gravitational potential energy of this shell is given by $\Delta U = -\dfrac{GM_{in}(\Delta m)}{r}$, where M_{in} is the mass of the material inside r (which the shell sees as a point mass concentrated at the centre – recall Section 5.5). Now, $M_{in} = \frac{4}{3}\pi r^3 \rho$ and hence,

$$\Delta U = -\frac{G\left(\frac{4}{3}\pi r^3 \rho\right)(4\pi r^2 \rho \Delta r)}{r} = -\frac{16\pi^2 G\rho^2}{3} r^4 \Delta r$$

Integrating over the sphere, we obtain an expression for the total potential energy of the system

$$U_{sphere} = \int_0^a dU = -\frac{16\pi^2 G\rho^2}{3}\int_0^a r^4 dr = -\frac{16}{3}\pi^2 G\rho^2 \frac{a^2}{5}$$

Since $M^2 = \frac{16}{9}\pi^2 \rho^2 a^6$, we can write $U_{sphere} = -\dfrac{3GM^2}{5a}$
For the Earth, $M = 6 \times 10^{24}$ kg and $a = 6.4 \times 10^6$ m and thus,

$$U_{earth} = \frac{3(6.67 \times 10^{-11})(6 \times 10^{24})^2}{5(6.4 \times 10^6)} \sim 10^{32}\ \mathrm{J}$$

Worked Example 5.3:

(a) A wooden block of mass M slides down a plane which is inclined at an angle of α to the horizontal. If the coefficient of kinetic friction between the block and the plane is μ_K, derive an expression for the mechanical energy lost by the block per unit length of path.

(b) A 20 kg block is projected up a wooden plank which is inclined at 25° to the horizontal. If the velocity imparted to the block at the bottom of the plank is 8.0 m s^{-1} and the coefficient of kinetic friction between the block and the plank is 0.20, how far up the plank will the block travel? How much mechanical energy is lost by the block when it is has travelled half of this distance?

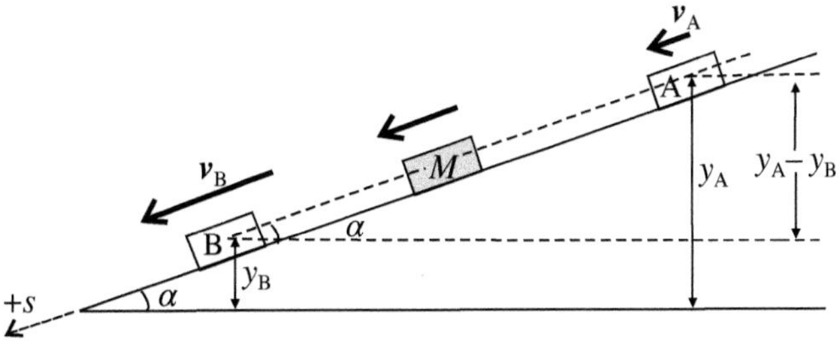

Figure 5.41. Worked Example 5.3(a).

(a) Taking down the plane as the positive direction in Figure 5.41, the acceleration is given by

$$a = g(\sin\alpha - \mu_K \cos\alpha) \qquad \text{(recall Section 4.6, Case 2)}$$

$$E_A - E_B = \left(Mgy_A + \frac{1}{2}Mv_A^2\right) - \left(Mgy_B + \frac{1}{2}Mv_B^2\right) = Mg(y_A - y_B) + \frac{1}{2}M(v_A^2 - v_B^2)$$

Now $v_B^2 = v_A^2 + 2as \rightarrow v_B^2 - v_A^2 = 2as$ and hence

$$E_A - E_B = Mg(y_A - y_B) + \frac{1}{2}M(-2as) = Mg(y_A - y_B) - Mg(\sin\alpha - \mu_K\cos\alpha)s$$

Since $y_A - y_B = s\sin\alpha \rightarrow E_A - E_B = (\mu_K Mg\cos\alpha)s$ (Note that this is the work done by the frictional resistance)

$$\rightarrow \quad \frac{dE}{ds} = \mu_K Mg\cos\alpha$$

(b) Taking up the plane as positive in Figure 5.42, the acceleration in this case is given by

$$a = -g(\sin\alpha + \mu_K\cos\alpha) \qquad \text{(recall Section 4.6, Case 3)}$$

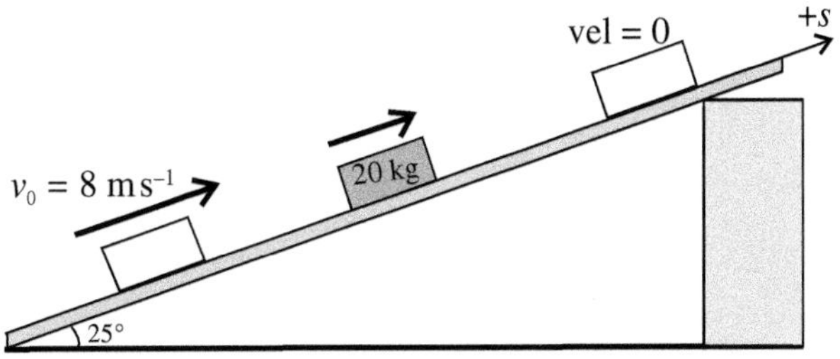

Figure 5.42. Worked Example 5.3(b).

Now $v^2 = v_o^2 + 2as = v_o^2 - 2g(\sin\alpha + \mu_K\cos\alpha)s \rightarrow (0)^2 = (8)^2 - 2(9.8)(\sin 25° + (0.2)\cos 25°)s$ and hence

$$\rightarrow \quad s = \frac{64}{2(9.8)(0.604)} = 5.41 \text{ m}$$

The loss in energy = the work done against friction, $\Delta E = (\mu_K mg\cos\alpha)s = (0.2)(20)(9.8)(\cos 25°)(5.41) = 192$ J

[Note that the loss in energy may also be calculated as $\frac{1}{2}mv_o^2 - mgy = \frac{1}{2}mv_o^2 - mgs\sin\alpha = \frac{1}{2}(20)(64) - (20)(9.8)(5.41)(\sin 25°) = 640 - 449 = 192$ J]

Worked Example 5.4: Estimate the minimum speed needed for a mass on the surface of the Earth to escape the Earth's gravity (this is called the **escape speed**).

The condition for a mass m to attain escape velocity is that its kinetic energy must exceed its potential (binding) energy in the Earth's gravitational field, that is $\frac{1}{2}mv^2 > \frac{GmM_E}{R_E}$. Thus the minimum velocity v_{esc} can be determined from $\frac{1}{2}mv_{esc}^2 = \frac{GmM_E}{R_E}$ and hence, using the values for M_E and R_E given in Appendix C.2,

$$v_{esc} = \sqrt{\frac{2GM_E}{R_E}} = \sqrt{\frac{2(6.67\times10^{-11})(6\times10^{24})}{6.4\times10^6}} = 1.12\times10^4\ \text{m s}^{-1}.$$

Note that the escape speed of a body is independent of its mass.

Worked Example 5.5: A stone of mass 0.22 kg is projected over horizontal ground with an initial velocity of 25 m s^{-1} at an angle of 30° to the horizontal. Determine the angular momentum of the stone relative to the point of projection (a) when it is at the highest point in its trajectory and (b) just before it hits the ground.

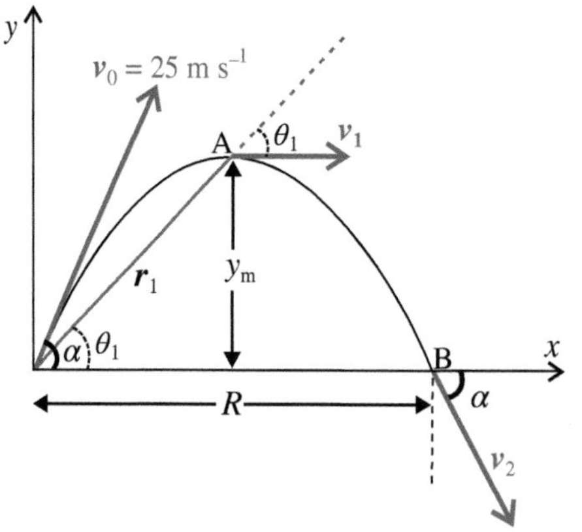

Figure 5.43. Worked Example 5.5.

(a) The angular momentum of the stone at the maximum height reached (point A in Figure 5.43) $= \boldsymbol{L}_A = m\boldsymbol{r}_1 \times \boldsymbol{v}_1$

$$\rightarrow \quad |\boldsymbol{L}_A| = mr_1v_1\sin\theta_1 = mv_1h = (mv_0\cos\alpha)h$$

Now, from Equation (5.7), the maximum height reached is given by $y_m = \frac{v_0^2\sin^2\alpha}{2g}$ and hence

$$|\boldsymbol{L}_A| = (mv_0\cos\alpha)\frac{v_0^2\sin^2\alpha}{2g} = 37\ \text{J s} \qquad \text{(perpendicular to and into the page)}$$

(b) The angular momentum of stone at point B $= \boldsymbol{L}_B = m\boldsymbol{R}\times\boldsymbol{v}_2 = m\boldsymbol{R}\times(v_0\cos\alpha\boldsymbol{i} - v_0\sin\alpha\boldsymbol{j}) = -mv_0\sin\alpha(\boldsymbol{R}\times\boldsymbol{j})$

From Equation (5.8), the horizontal range is given by $R = \frac{v_0^2\sin 2\alpha}{g}$ and hence

$$|\boldsymbol{L}_B| = mRv_0\sin\alpha = (mv_0\sin\alpha)\frac{v_0^2\sin 2\alpha}{g} = 150\ \text{J s} \qquad \text{(perpendicular to and into the page)}$$

PROBLEMS

For problems based on the material covered this chapter visit up.ucc.ie/5/ and follow the link to the problems.

6

Many-body interactions

AIMS

- to introduce Newton's third law of motion, the law which governs the interaction between two bodies
- to derive the general principles of conservation of momentum, energy and angular momentum as applied to many-body systems
- to develop techniques for analysing particle decays and collisions between particles
- to show how an appropriate choice of origin of a coordinate system makes some problems more amenable to easy solution, particularly when the CM coordinate system is used
- to introduce the principle of conservation of angular momentum of a system of particles

6.1 Newton's third law

Newton's second law provides us with a method of determining the state of motion of a single body, provided we know the magnitudes and directions of all forces acting on that body. The full analysis of dynamical systems which comprise two or more interacting bodies (such as two masses connected by a taut string, gravitationally bound bodies such as the solar system, electrically bound systems such as atoms or molecules, etc.) cannot be treated using Newton's second law alone. To deal with such systems requires knowledge of an additional law of nature which Newton called the third law of motion.

Consider the two bodies in Figure 6.1 which interact, that is a force is exerted on A by B ($\boldsymbol{F}_{\mathrm{B}\to\mathrm{A}}$) and a force is exerted on B due to A ($\boldsymbol{F}_{\mathrm{A}\to\mathrm{B}}$). **Newton's third law** states that the forces on each of these body due to the other are always equal in magnitude and opposite in direction, that is

$$\boldsymbol{F}_{\mathrm{A}\to\mathrm{B}} = -\boldsymbol{F}_{\mathrm{B}\to\mathrm{A}}$$

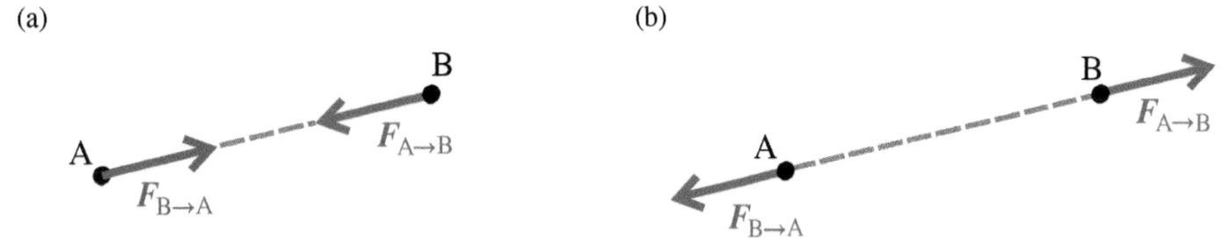

Figure 6.1. Newton's third law: the force exerted on B by A ($\boldsymbol{F}_{\mathrm{B}\to\mathrm{A}}$) is equal and opposite to the force exerted on A by B ($\boldsymbol{F}_{\mathrm{A}\to\mathrm{B}}$). The interaction in (a) is attractive while in (b) it is repulsive.

We now consider the applications of this law to a number of familiar situations.

Case 1: A body resting on horizontal table

As a simple example of the use of this law, recall (Section 4.5) the case of the book resting on a horizontal table (Figure 6.2). We showed in that section that the force exerted on the book by the table (the normal reaction, upwards — a *constraint force*) was given by

$$\boldsymbol{N} = -M\boldsymbol{g} = Mg\boldsymbol{k}$$

Understanding Physics, Third Edition. Michael Mansfield and Colm O'Sullivan.
© 2020 John Wiley & Sons Ltd. Published 2020 by John Wiley & Sons Ltd.

where $\boldsymbol{k}$ is a unit vector directed vertically upwards. If we are interested in the forces acting *on the table*, Newton's third law tells us that the force acting downwards on the table by the book is

$$-\boldsymbol{N} = +M\boldsymbol{g} = -Mg\boldsymbol{k}.$$

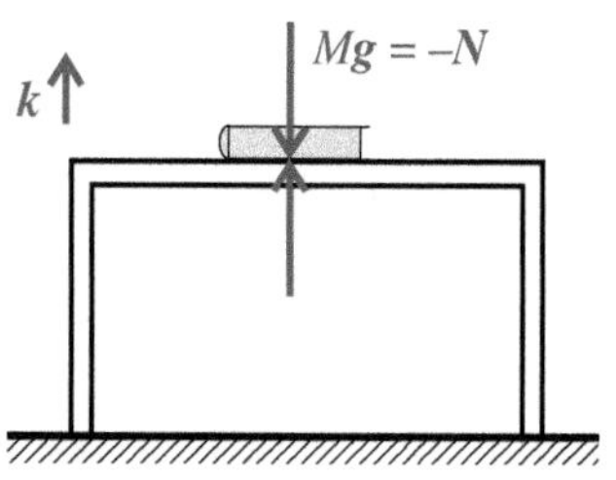

Figure 6.2. The force exerted by the book on the table is equal and opposite to the force exerted by the table on the book.

Case 2: Blocks in contact on horizontal table

Let us now consider the system illustrated in Figure 6.3(a) where a block of mass M_1 in contact with a block of mass M_2 is pushed along a horizontal table by a horizontal force $\boldsymbol{P}$. For simplicity, the effect of friction will be neglected (although frictional forces could be included in the usual way, if desired). Our objective is to determine the force which each block exerts on the other.

For this purpose a very useful analytical tool, called a **free-body diagram**, is introduced. A free-body diagram is a diagram which shows all the forces acting on *part* of the system under study. Figure 6.3(b) is a free-body diagram for the mass M_1 and the free-body diagram for the mass M_2 is shown in Figure 6.3(c). The value of the technique resides in the fact that Newton's second law can be applied to any part of a system as well as to the system as a whole. Applying Newton's second law to the mass M_1 (and taking the direction to the right as positive in Figure 6.3(b)) we get $|\boldsymbol{P}| - |\boldsymbol{R}_1| = M_1 a$, where $\boldsymbol{R}_1$ is the force on M_1 due to M_2. Similarly, from Figure 6.3(c), we get $|\boldsymbol{R}_2| = M_2 a$ where $\boldsymbol{R}_2$ is the force on M_2 due to M_1.

Newton's third law tells us that $\boldsymbol{R}_1 = -\boldsymbol{R}_2$ so that $|\boldsymbol{R}_1| = |\boldsymbol{R}_2|$ and therefore

$$|\boldsymbol{P}| - M_2 a = M_1 a$$

and the acceleration of the system is given by

$$a = \frac{P}{M_1 + M_2}$$

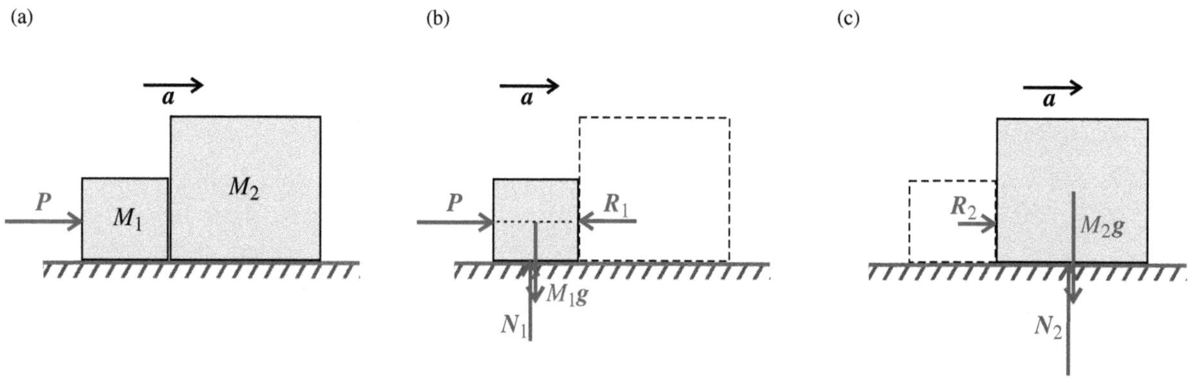

Figure 6.3. (a) Two blocks are accelerated along a smooth horizontal surface by a horizontal force $\boldsymbol{P}$ applied to the smaller block. (b) Free-body diagram showing the forces on the smaller block. (c) Free-body diagram showing the forces on the larger block.

Note that we could also have determined the last equation from an analysis of the system as a single body, but a free-body diagram is needed to determine $\boldsymbol{R}_1$ and $\boldsymbol{R}_2$.

Since $|\boldsymbol{R}_2| = M_2 a$ we can write $\boldsymbol{R}_2 = \dfrac{M_2}{M_1 + M_2}\boldsymbol{P}$ and $\boldsymbol{R}_1 = -\dfrac{M_2}{M_1 + M_2}\boldsymbol{P}$

Case 3: Bodies connected by a string

As a third example, we will look in some detail at the system illustrated in Fig. 6.4(a), which is similar to the experimental arrangement described in Section 3.3 except that the friction between the block of mass M and the horizontal table or track on which it slides will now be included in the analysis. We will assume that any friction at the 'pulley' can be neglected and that the mass m is large enough that static friction is overcome and, therefore, that the system is in motion.

Figure 6.4(b) and 6.4(c) are free-body diagrams for the masses M and m, respectively. The force $\boldsymbol{T}$ in Figure 6.4(b) is the force exerted on the block M by the end of the string that pulls the block horizontally to the right. The force $\boldsymbol{T'}$ in Figure 6.4(c) is the force exerted vertically on the mass m by the other end of the string. Note that we are not assuming that $|\boldsymbol{T'}|$ is equal to $|\boldsymbol{T}|$.

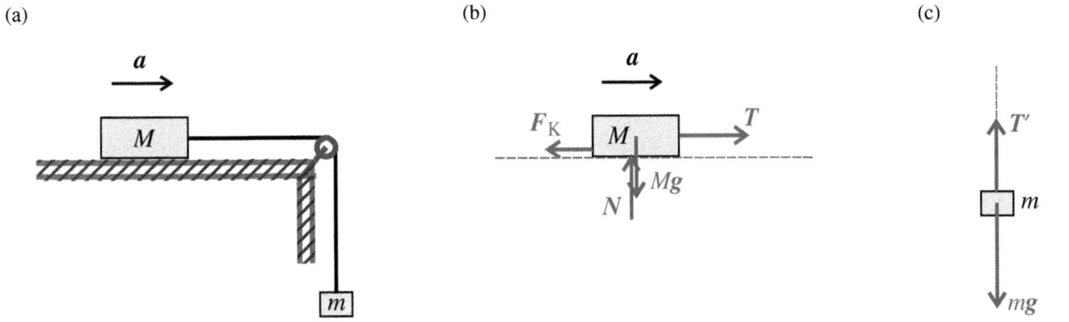

Figure 6.4. (a) A block of mass M on a horizontal table is connected to a mass m by a string which passes over a small frictionless pulley. (b) Forces acting on the mass M. (c) Forces acting on the mass m.

Applying Newton's second law to each of these masses in turn gives

$$|\boldsymbol{T}| - |\boldsymbol{F}_{\mathrm{K}}| = M|\boldsymbol{a}| \tag{6.1}$$

and

$$mg - |\boldsymbol{T}'| = m|\boldsymbol{a}'| \tag{6.2}$$

If the string is *inextensible* the distance between M and m measured along the string must remain constant. This implies that, at any instant, the speeds of each of the masses must be the same and hence that the magnitude of their accelerations must be the same also. That is, for an inextensible string

$$|\boldsymbol{a}'| = |\boldsymbol{a}| = a.$$

The dynamics of strings is a somewhat more complicated issue. Consider the forces acting on a small element Δm of the string, as shown in Figure 6.5. Ignoring the effect of gravity, we see that the element experiences a force $\boldsymbol{T}_2$ exerted by the part of the string to its right and a force $\boldsymbol{T}_1$ exerted by the part of the string to the left. Newton's second law applied to this element, therefore, gives

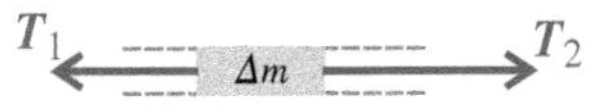

Figure 6.5. Forces acting on an element (mass = Δm) of the string.

$$|\boldsymbol{T}_2| - |\boldsymbol{T}_1| = (\Delta m)|\boldsymbol{a}|,$$

indicating that the tension in the string increases along its length to the right. In the case in which the mass of the string is negligible, however, $\Delta m \to 0$ and $\boldsymbol{T}_1 = -\boldsymbol{T}_2$. Thus, for a *massless* string the tension ($= |\boldsymbol{T}_1| = |\boldsymbol{T}_2|$) is the same everywhere along the string and, in particular, is the same at each end of the string, that is $|\boldsymbol{T}| = |\boldsymbol{T}'| = T$ (say).

Thus, in this case, Equations (6.1) and (6.2) can be written as

$$T - |\boldsymbol{F}_{\mathrm{K}}| = Ma$$

$$mg - T = ma \tag{6.3}$$

Eliminating T and writing $|\boldsymbol{F}_{\mathrm{K}}| = \mu_K Mg$ we obtain the following equation for the acceleration of the system

$$a = \frac{m - \mu_{\mathrm{K}} M}{M + m} g$$

Substituting this result into Equation (6.3) and solving for T, we obtain an expression for the tension in the string

$$T = \frac{mM}{M + m}(1 + \mu_{\mathrm{K}})g$$

In the absence of friction ($\mu_{\mathrm{K}} \to 0$) the tension becomes

$$T = \frac{mM}{M + m} g = \frac{mg}{1 + \frac{m}{M}}$$

Note the significance of this last result for the interpretation of the experiment from which Newton's Second law was deduced, as described in Section 3.3. In that experiment it was assumed that the force on the rider exerted by the string was proportional to the mass hanging from the free end of the string. We now see that this assumption is valid only if $m \ll M$. This condition had to be satisfied in that experiment in any event, since m had to be very small to keep the acceleration sufficiently small to be determined with any accuracy.

Study Worked Example 6.1

For problems based on the material presented in this section visit up.ucc.ie/6/ *and follow the link to the problems.*

6.2 The principle of conservation of momentum

Consider a general system of interacting particles as indicated in Figure 6.6. Each particle experiences forces due to all the other particles in the system together with a force due to sources outside the system. Thus the rate of change of the momentum of the first particle is given by

$$\frac{d\boldsymbol{p}_1}{dt} = (\boldsymbol{F}_1)_{ext} + \boldsymbol{f}_{2\to 1} + \boldsymbol{f}_{3\to 1} + \dots \tag{6.4a}$$

where $(\boldsymbol{F}_1)_{ext}$ is the force on particle 1 due to external sources and the $\boldsymbol{f}_{i\to 1}$ is the force on particle 1 due to the ith particle (where $i = 2, 3, 4, \dots$ — that is, the equation extends to include all the other particles in the system).

Thus,

$$\frac{d\boldsymbol{p}_2}{dt} = (\boldsymbol{F}_2)_{ext} + \boldsymbol{f}_{1\to 2} + \boldsymbol{f}_{3\to 2} + \dots \tag{6.4b}$$

$$\frac{d\boldsymbol{p}_3}{dt} = (\boldsymbol{F}_3)_{ext} + \boldsymbol{f}_{1\to 3} + \boldsymbol{f}_{2\to 3} + \dots \tag{6.4c}$$

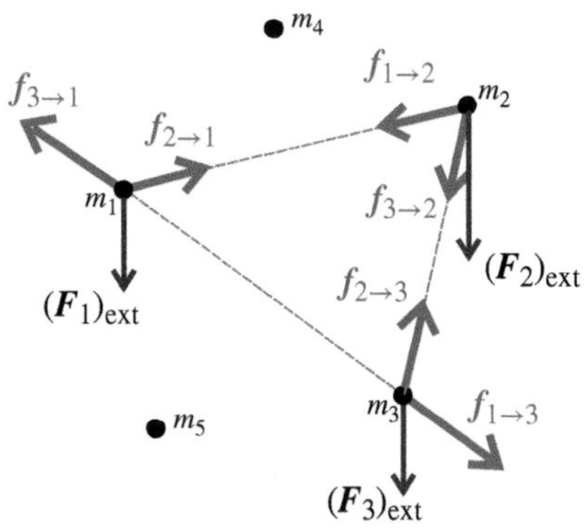

Figure 6.6. A system of interacting particles; each particle experiences a force due to an external field and forces exerted by each of the other particles in the system.

and so on, with a similar equation for each particle in the system.

We now *define* the **total momentum** of a system of particles as follows

$$\boldsymbol{P} := \boldsymbol{p}_1 + \boldsymbol{p}_2 + \boldsymbol{p}_3 + \boldsymbol{p}_4 + \boldsymbol{p}_5 \dots$$

The rate of change of the total momentum is then

$$\frac{d\boldsymbol{P}}{dt} = \frac{d\boldsymbol{p}_1}{dt} + \frac{d\boldsymbol{p}_2}{dt} + \frac{d\boldsymbol{p}_3}{dt} + \dots$$

which can be determined by adding up the right hand sides of Equations (6.4a), (6.4b), (6.4c) … etc. Because of Newton's third law, however, the contributions to this addition from the interactions between the particles will add up to zero, since $\boldsymbol{f}_{2\to 1} = -\boldsymbol{f}_{1\to 2}$, $\boldsymbol{f}_{1\to 3} = -\boldsymbol{f}_{3\to 1}$, and so forth. Thus

$$\frac{d\boldsymbol{P}}{dt} = (\boldsymbol{F}_1)_{ext} + (\boldsymbol{F}_2)_{ext} + (\boldsymbol{F}_3)_{ext} + \dots = \boldsymbol{F}_{EXT}$$

where $\boldsymbol{F}_{EXT}$ is the vector sum of all the *external* forces acting on the system.

Thus we have derived a result which describes the translational behaviour of a system of particles under the influence of an external force

$$\boxed{\frac{d\boldsymbol{P}}{dt} = \boldsymbol{F}_{EXT}} \tag{6.5}$$

which has exactly the same form as Newton's second law except that it relates the *total* momentum and the *total* external force. Note that Equation (6.5) follows directly from the laws of motion; no additional principles have been invoked.

An important consequence of Equation (6.5) is that, if the net external force on a system of particles is zero, then $\frac{d\boldsymbol{P}}{dt} = 0$ and hence the total momentum of the system must be constant. That is

$$\boldsymbol{F}_{EXT} = 0 \quad \Rightarrow \quad \boldsymbol{P} = \text{constant}.$$

This general rule, called the **principle of conservation of momentum**, asserts that

> *if no net external force acts on a system of particles, the total momentum of the system remains constant.*

This principle, which has followed directly from Newton's laws of motion, turns out to be extremely useful in situations in which the exact details of the interaction between bodies are not known. Some examples of its application are studied in Sections 6.4 to 6.7 below and in the worked examples at the end of this chapter.

As an illustration, let us consider the case of two vehicles which collide at an intersection as a result of which they remain entangled so that they behave as a single body after the collision (Figure 6.7). Suppose that one had a mass of 1000 kg and had been travelling northward at 45 km h^{-1} and that the other had a mass of 1600 kg and had been travelling eastward at 55 km h^{-1}. The total momentum before collision was $(1600 \times 55 \text{ kg km h}^{-1})\boldsymbol{i} + (1000 \times 45 \text{ kg km h}^{-1})\boldsymbol{j}$, where $\boldsymbol{i}$ and $\boldsymbol{j}$ are unit vectors in the eastward and northward directions, respectively. Since the total momentum is conserved this is also the momentum of the combined mass immediately after impact. Thus the two vehicles move as a single body with speed $\frac{\sqrt{(88000)^2 + (45000)^2}}{2600} = 38$ km h^{-1} in a direction which is $\tan^{-1}(\frac{45}{88}) = 27°$ north of east.

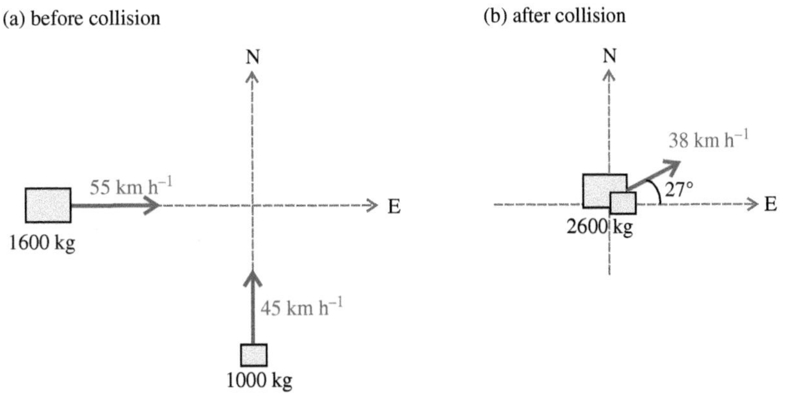

Figure 6.7. (a) Two vehicles, one moving eastward and the other moving northward, on a collision course. (b) The two vehicles moving as a single composite body after collision.

Note that if there is an external force on a system, the total momentum of the system is not conserved but it follows, from Equation (6.5), that the components of the total momentum perpendicular to the direction of that force (that is, perpendicular to the direction of $\boldsymbol{F}_{EXT}$) are conserved. Thus, if an object spontaneously breaks up into small fragments in outer space, the total momentum of all the fragments (the vector sum of their individual momenta) is the same as the momentum of the object before break up. On the other hand, if such an object were to break up near the surface of the Earth, where there is an external force (due to gravity) perpendicular to the surface, only the horizontal components of the total momentum would be conserved.

6.3 Mechanical energy of systems of particles

In discussing energy in Chapter 5 we considered only systems which comprise a single particle in a force field. For a system comprising a number of particles, the potential energy of the system may include a contribution arising from the forces between the particles as well as from the forces on the particles in the external field. Thus the total energy of a system of interacting masses $m_1,\ m_2,\ m_3 \ldots$ etc. in an external field would, in general, have the form

$$E = \frac{1}{2}m_1{v_1}^2 + \frac{1}{2}m_2{v_2}^2 + \frac{1}{2}m_3{v_3}^2 + \ \ldots\ + U(\boldsymbol{r}_1, \boldsymbol{r}_2, \boldsymbol{r}_3,\ \ldots)$$

and the Hamiltonian function (recall Section 5.7) would be given by

$$H(\boldsymbol{p}_1, \boldsymbol{p}_2, \boldsymbol{p}_3, \ \ldots \ \ \boldsymbol{r}_1, \boldsymbol{r}_2, \boldsymbol{r}_3, \ \ldots) = \frac{p_1^2}{2m_1} + \frac{p_2^2}{2m_2} + \frac{p_3^2}{2m_3} + \ \ldots \ \ U(\boldsymbol{r}_1, \boldsymbol{r}_2, \boldsymbol{r}_3, \ \ldots).$$

For the total mechanical energy of a system of particles to be conserved, therefore, it is necessary that the force fields experienced by each particle be conservative. If these conditions are fulfilled, that is in the absence of dissipative forces such as friction, the total mechanical energy of the system is conserved. Thus we can state the **principle of conservation of energy**, that is,

> *for a conservative system of particles, the total mechanical energy of the system remains constant, that is*
> $$H(\boldsymbol{p}_1, \boldsymbol{p}_2, \boldsymbol{p}_3, \ldots\, \boldsymbol{r}_1, \boldsymbol{r}_2, \boldsymbol{r}_3, \ldots) = E = \text{constant}$$

6.4 Particle decay

Consider what happens when a object breaks up (or 'decays') into a number of fragments. Examples of such processes are exploding shells, the decay of radioactive nuclei, dispersion of seeds from an exploding seed pod, etc. In these cases a certain amount of energy (which is of chemical, nuclear and mechanical origin, respectively, in these examples) is released in the form of kinetic energy of the fragments. We will see in Chapter 27 that the study of the decay products of a radioactive nucleus can give important information about the nature of the interactions inside the nucleus.

If no external force acts on the system, the total momentum is conserved. Thus, if the body was at rest before it broke up, that is if its initial momentum was zero, the sum of the momenta of the scattered fragments immediately afterwards must also be zero. For a body of mass M which breaks up into masses m_1, m_2, m_3, ..., etc. (Figure 6.8) we have

$$(M)(0) = 0 = m_1\boldsymbol{v}_1 + m_2\boldsymbol{v}_2 + m_3\boldsymbol{v}_3 + m_4\boldsymbol{v}_4 + \ldots$$

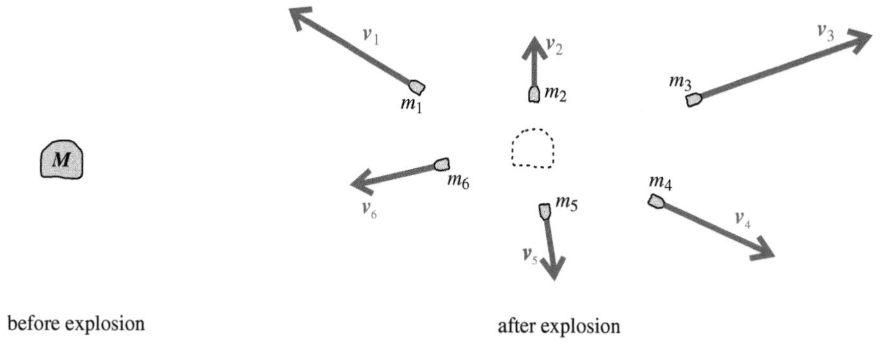

Figure 6.8. A body of mass M, initially at rest, explodes into a large number of fragments.

In the case in which a body at rest breaks into *two* parts, such as a bullet fired from a gun or a two-particle decay, the two parts must emerge in opposite directions as shown in Figure 6.9; otherwise the total momentum after break up could not be zero, since the sum of two vectors is zero only if the vectors are collinear.

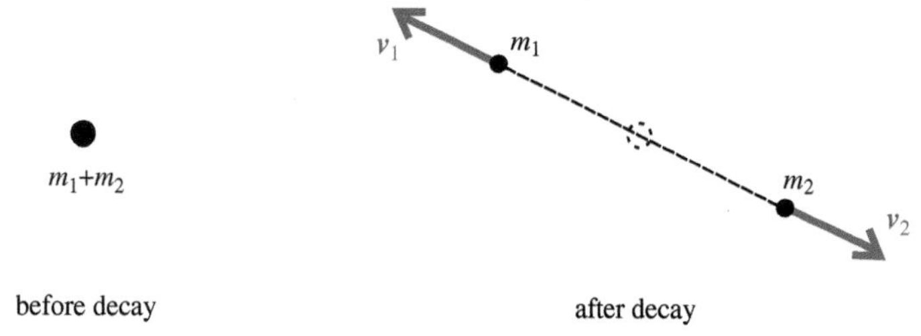

Figure 6.9. A body, initially at rest, breaks up into two parts (masses m_1 and m_2).

In this case, $m_1\boldsymbol{v}_1 + m_2\boldsymbol{v}_2 = 0$ and hence

$$\boldsymbol{v}_2 = -\frac{m_1}{m_2}\boldsymbol{v}_1 \tag{6.6}$$

The total mechanical energy released ΔE is distributed between the two fragments in such a way that

$$\Delta E = \frac{1}{2}m_1v_1^2 + \frac{1}{2}m_2v_2^2$$

Using Equation (6.6) and eliminating v_2 and v_1, in turn, we obtain the following expressions for the kinetic energies of the fragments

$$K_1 = \frac{1}{2}m_1v_1^2 = \frac{\Delta E}{1 + \frac{m_1}{m_2}} \quad \text{and} \quad K_2 = \frac{1}{2}m_2v_2^2 = \frac{\Delta E}{1 + \frac{m_2}{m_1}}$$

and hence

$$\frac{K_1}{K_2} = \frac{1 + \frac{m_2}{m_1}}{1 + \frac{m_1}{m_2}} = \frac{m_2}{m_1}$$

As an illustration of a two-body decay let us consider the α-decay of the americium-241 nucleus ${}^{241}_{95}\text{Am} \to {}^{237}_{93}\text{Np} + {}^{4}_{2}\text{He}$, using the notation for the Z and A of atomic nuclei introduced in Section 1.2. The energy released in this process is 9.02×10^{-13} J and this appears as kinetic energy of the α-particle (${}^{4}_{2}\text{He}$ nucleus) and the recoiling neptunium nucleus. Thus the energy carried away by the α-particle is $K_\alpha = \dfrac{\Delta E}{1+\dfrac{m_\alpha}{m_N}} = \dfrac{9.02 \times 10^{-13}}{1+\dfrac{4}{237}} = 8.87 \times 10^{-13}$ J and that carried away by the recoiling nucleus is $K_N = \dfrac{\Delta E}{1+\dfrac{m_N}{m_\alpha}} = \dfrac{9.02 \times 10^{-13}}{1+\dfrac{237}{4}} = 0.15 \times 10^{-13}$ J. Note that the lighter particle always carries more energy.
The corresponding speeds of the particles are

$$v_\alpha = \sqrt{\frac{2K_\alpha}{m_\alpha}} = \sqrt{\frac{2(8.87 \times 10^{-13})}{4(1.67 \times 10^{-27})}} = 1.6 \times 10^7 \text{ m s}^{-1}$$

and

$$v_N = \sqrt{\frac{2K_N}{m_N}} = \sqrt{\frac{2(0.15 \times 10^{-13})}{237(1.67 \times 10^{-27})}} = 2.8 \times 10^5 \text{ m s}^{-1}.$$

This situation, in which the kinetic energy of each particle after break up has a unique value, is specific to two-body decays. If three or more particles are produced, such as in a β-decay process (Section 27.5), it is not possible to predict their speeds and kinetic energies on the basis of conservation of momentum and conservation of energy alone. In such cases the energy of a decay product can have any value between zero and ΔE. We shall see in Section 27.5 how this effect led to the hypothesis that a previously unknown particle, the neutrino, is emitted in the β-decay of a nucleus.

Study Worked Example 6.2

For problems based on the material presented in this section visit up.ucc.ie/6/ *and follow the link to the problems.*

6.5 Particle collisions

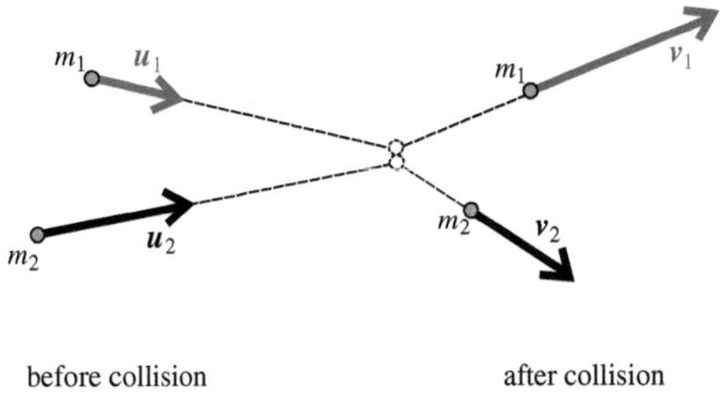

Figure 6.10. A two-particle collision.

In a 'decay' process, as we have seen, a single particle breaks up into two or more fragments. In a collision (illustrated in Figure 6.10) the motion of two particles approaching each other is modified as a result of the mutual interaction between them. The particles emerge from the collision with their individual momenta, and possibly their energies, changed from their values before collision. A familiar example is the collision of snooker balls, but note that a comet passing close to the Sun or an α-particle passing near another nucleus are also collisions in the same sense, even though in these cases the 'colliding' bodies never come into contact. Much of what was discovered in the 20th century about the sub-atomic world came from the study of collisions between fundamental particles such as electrons, protons, muons, neutrinos, etc. and their antiparticles. In high energy collisions between two particles, additional particles may be produced in the process, thus providing a rich source of information on the fundamental interactions at the sub-nuclear level. A proper treatment of particle collisions at very high energies, however, requires a knowledge of relativistic dynamics which must await discussion until Chapter 9. In this section we will consider only non-relativistic collisions, that is collisions between macroscopic bodies which do not move at speeds comparable to the speed of light.

Because the forces between particles (even between snooker balls) can be very complicated, it can prove difficult or impossible to describe the collision process completely and in detail. Collisions, however, must obey relevant general rules which follow from the laws of dynamics. In particular, if there are no external forces acting on the system, the principle of conservation of momentum must be obeyed. For the particles shown in Figure 6.10, therefore,

$$\boldsymbol{p}_1 + \boldsymbol{p}_2 = \boldsymbol{q}_1 + \boldsymbol{q}_2$$

where $\boldsymbol{p}_1$ and $\boldsymbol{p}_2$ are the momenta of the particles before the collision and $\boldsymbol{q}_1$ and $\boldsymbol{q}_2$ are their momenta after the collision.

Alternatively we can write the conservation of momentum in terms of velocities as

$$m_1\boldsymbol{u}_1 + m_2\boldsymbol{u}_2 = m_1\boldsymbol{v}_1 + m_2\boldsymbol{v}_2 \tag{6.7}$$

Note that while Equation (6.7) alone does not determine the state of system after the collision uniquely, it restricts the possible outcomes.

Energy, on the other hand, is not necessarily conserved in a collision process since the force of interaction between the particles may not be conservative. If the system is conservative, however, energy will be conserved, in which case the collision is said to be **elastic**. In elastic collisions, any energy temporarily stored in the system during the collision process (for example, in deformation of the colliding objects) is returned to the system after the collision entirely as kinetic energy of the bodies involved. Generally speaking, collisions between macroscopic bodies, such as between snooker balls, are **inelastic**, that is, not all of the stored potential energy reappears as kinetic energy.

If Figure 6.10 represents an *elastic* collision

$$\frac{1}{2}m_1u_1^2 + \frac{1}{2}m_2u_2^2 = \frac{1}{2}m_1v_1^2 + \frac{1}{2}m_2v_2^2 \tag{6.8}$$

the change of energy is zero, that is

$$\Delta E = \left(\frac{1}{2}m_1v_1^2 + \frac{1}{2}m_2v_2^2\right) - \left(\frac{1}{2}m_1u_1^2 + \frac{1}{2}m_2u_2^2\right) = 0$$

For *inelastic* collisions the change of energy is given by

$$\Delta E = \left(\frac{1}{2}m_1v_1^2 + \frac{1}{2}m_2v_2^2\right) - \left(\frac{1}{2}m_1u_1^2 + \frac{1}{2}m_2u_2^2\right) \tag{6.9}$$

where ΔE has a specific value, characteristic of the particular collision (depending, among other factors, on the elastic properties of the colliding bodies). If energy is lost in the collision, as it always is in macroscopic collisions, then ΔE is negative.

It is important to note that in the above equations the speeds involved are the values that the particles have when they are very far from each other, before (u_1 and u_2) and after (v_1 and v_2) the collision process, since during the collision the system will have some potential energy, albeit temporarily in the case of an elastic collision.

Even the combination of the information given by Equations (6.7) and (6.8) or (6.9) is still not sufficient to predict the state of the system after collision; that is to determine $\boldsymbol{v}_1$ and $\boldsymbol{v}_2$ in both magnitude and direction as there are more unknowns than there are equations. Further information, usually determined experimentally, is required to solve the problem.

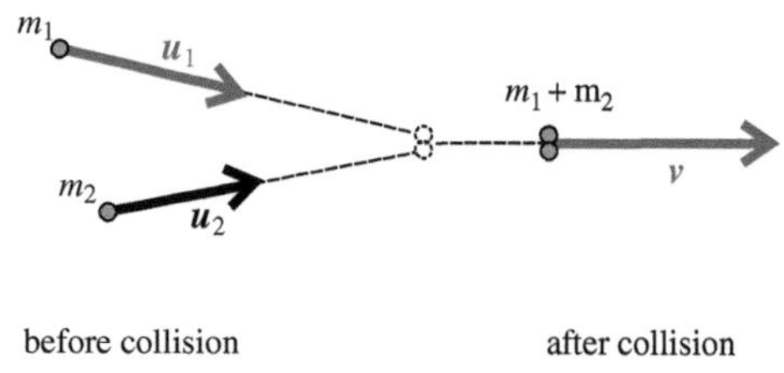

Figure 6.11. A totally inelastic collision.

There is one extreme situation, however, for which the problem is fully determined by momentum considerations alone. This is the case of a **totally inelastic collision** in which the two particles coalesce to form a single body after collision (Figure 6.11). Applying the principle of conservation of momentum in this case

$$m_1\boldsymbol{u}_1 + m_2\boldsymbol{u}_2 = (m_1 + m_2)\boldsymbol{v}$$

and hence the velocity of the composite body of mass m_1+m_2 after collision is given by

$$\boldsymbol{v} = \frac{m_1\boldsymbol{u}_1 + m_2\boldsymbol{u}_2}{m_1 + m_2} = \frac{\boldsymbol{p}_1 + \boldsymbol{p}_2}{m_1 + m_2} = \frac{\boldsymbol{P}}{m_1 + m_2}$$

where $\boldsymbol{P}$ is the total momentum of the two-body system.

LAB coordinates

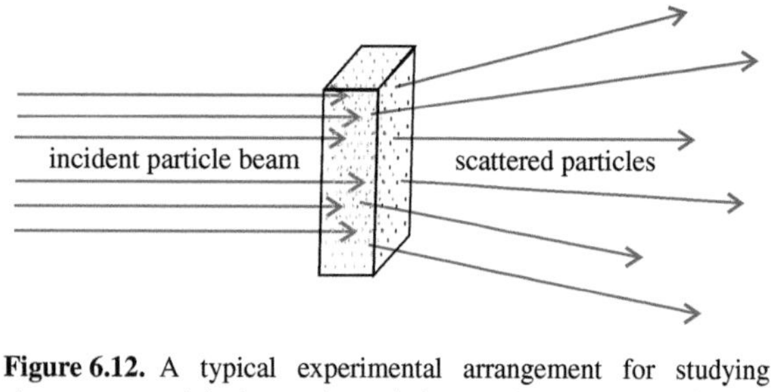

Figure 6.12. A typical experimental arrangement for studying elementary particle interactions. A beam of energetic particles is incident on a target which is at rest in the laboratory and some of the beam particles are scattered as a result of collisions with particles in the target.

In many experimental arrangements involving particle collisions, a beam of particles is incident on a target which is at rest in the laboratory (Figure 6.12). In this case it is natural to choose a point which is fixed with respect to the target as the origin of the coordinate system, relative to which displacements, velocities, momenta, *etc.* are measured. Such a system of coordinates is an example of what is usually called a laboratory or **LAB coordinate system** although, in general, none of the particles has to be at rest in the LAB system.

In this section we will use the symbol $\boldsymbol{p}_i$ to represent the momentum of the ith particle before collision and $\boldsymbol{q}_i$ to represent its momentum after collision. We consider below the cases of (1) an elastic collision and (2) a totally inelastic collision in the LAB coordinate system.

Case 1: Elastic collision

Consider a particle (mass $= m_1$) in a beam which is incident on a particle (mass $= m_2$) in a stationary target in the laboratory (Figure 6.13). The momentum of the mass m_1 before the collision is $\boldsymbol{p}_1$ and the mass m_2 is at rest ($\boldsymbol{p}_2 = 0$). Let the particle momenta after collision be $\boldsymbol{q}_1$ and $\boldsymbol{q}_2$, as indicated.

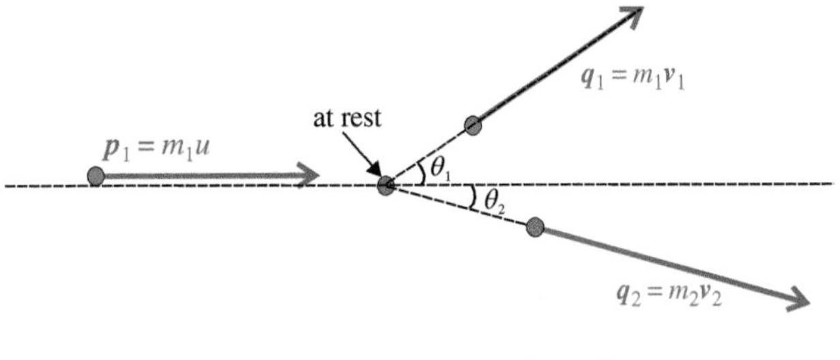

Figure 6.13. A moving particle (m_1) strikes a particle (m_2) which is at rest in the laboratory. After the collision both particles are scattered as shown; if the total mechanical energy of the system is unchanged, the collision is said to be *elastic*.

Applying conservation of momentum: $$\boldsymbol{p}_1 + 0 = \boldsymbol{q}_1 + \boldsymbol{q}_2$$

Applying conservation of energy: $$\frac{p_1^2}{2m_1} + 0 = \frac{q_1^2}{2m_1} + \frac{q_2^2}{2m_2}$$

As noted above, in general these equations (equivalent to three algebraic equations — two equations for momentum, when resolved in two perpendicular directions, and one equation for energy) are not sufficient to determine the outcome of the collision fully; that is to determine the four quantities q_1, q_2, θ_1 and θ_2. Further information, such as a knowledge of one of the angles, is required, as in the example below or in a one-dimensional collision (a 'head-on' collision in which $\theta_1 = \theta_2 = 0$).

Example:

Consider the case shown in Figure 6.14 in which a particle of mass m travelling with velocity $\boldsymbol{u}$ collides obliquely (that is not head-on) with a particle of equal mass at rest. After the collision, the incident particle is observed to have been deflected from its original direction through an angle θ_1 while the particle that was initially at rest moves at an angle θ_2 with respect to the direction of the incident particle.

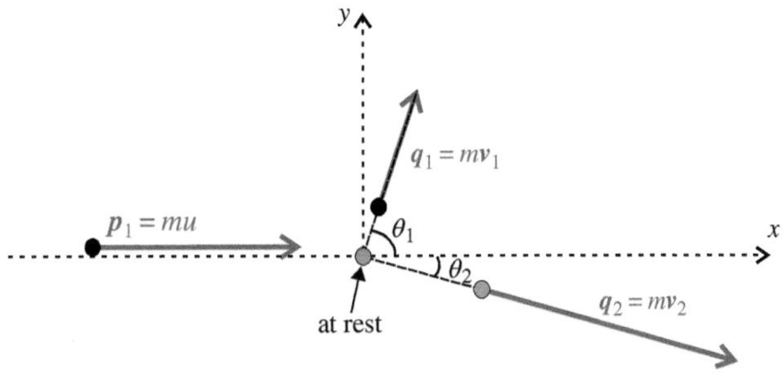

Figure 6.14. Example of an elastic collision between particles with the same mass.

As before, let the momenta of the particles after the collision be q_1 and q_2, respectively.

Conservation of momentum: $$\boldsymbol{p}_1 + 0 = \boldsymbol{q}_1 + \boldsymbol{q}_2 \tag{6.10}$$

x-components: $$|\boldsymbol{p}_1| = |\boldsymbol{q}_1|\cos\theta_1 + |\boldsymbol{q}_2|\cos\theta_2 \tag{6.11}$$

y-components: $$0 = |\boldsymbol{q}_1|\sin\theta_1 - |\boldsymbol{q}_2|\sin\theta_2 \tag{6.12}$$

Conservation of energy: $$\frac{p_1^2}{2m_1} + 0 = \frac{q_1^2}{2m_1} + \frac{q_2^2}{2m_2}$$

Since, in this case, $$m_1 = m_2 \quad \rightarrow \quad |\boldsymbol{p}_1|^2 = |\boldsymbol{q}_1|^2 + |\boldsymbol{q}_2|^2 \tag{6.13}$$

Because the particles have identical masses, an interesting special case arises. Squaring both sides of Equation (6.10)

$$|\boldsymbol{p}_1|^2 = |\boldsymbol{q}_1 + \boldsymbol{q}_2|^2 = |\boldsymbol{q}_1|^2 + 2\boldsymbol{q}_1 \cdot \boldsymbol{q}_2 + |\boldsymbol{q}_2|^2$$

comparison with Equation (6.13) shows that $\boldsymbol{q}_1 \cdot \boldsymbol{q}_2 = 0$ and hence $q_1 q_2 \cos\phi = 0$, where $\phi = \theta_1 + \theta_2$ is the angle between $\boldsymbol{q}_1$ and $\boldsymbol{q}_2$. Thus

$$\rightarrow \quad \cos\phi = 0 \quad \rightarrow \quad \phi = \frac{\pi}{2}$$

This result is a characteristic property of oblique elastic collisions between particles of the same mass when one of them is initially at rest.

In the case of a head-on elastic collision between identical particles, readers can check for themselves that the incident particle comes to rest after collision while the target particle moves forward at the same speed as that of the incident particle before collision. In this case, the kinetic energy of the incident particle is entirely transferred to the target particle.

Study Worked Examples 6.3 and 6.4

Case 2: *Totally inelastic collision*

In the case of a totally inelastic collision, already mentioned earlier in this section, the two particles coalesce to form a single particle. Momentum is conserved *but energy is not conserved* in this case. In the situation illustrated in Figure 6.15, the target particle (mass m_2) is at rest in the laboratory before the collision and is struck by an incident particle of mass m_1 with momentum $\boldsymbol{p}_1$. Let the momentum of the composite particle (of mass $m_1 + m_2$) be $\boldsymbol{q}$, as indicated in the figure.

Figure 6.15. A totally inelastic collision with the target particle at rest in the laboratory.

Conservation of momentum:

$$\boldsymbol{p}_1 = \boldsymbol{q}$$

$$\text{Change of energy} = \Delta E = \frac{q^2}{2(m_1+m_2)} - \frac{p_1^2}{2m_1} = \frac{p_1^2}{2(m_1+m_2)} - \frac{p_1^2}{2m_1}$$

$$= \frac{p_1^2}{2m_1}\left(\frac{m_1}{m_1+m_2} - 1\right) = -\left(\frac{m_2}{m_1+m_2}\right)\frac{p_1^2}{2m_1}$$

Thus

$$\Delta E = -\left(\frac{m_2}{m_1+m_2}\right)K_0$$

where $K_0 = \dfrac{p_1^2}{2m_1}$ is the kinetic energy before collision.

The change of energy (ΔE) is sometimes called the Q-value of the collision, that is $Q = -\left(\dfrac{m_2}{m_1+m_2}\right)K_0$ in this case. Note that in macroscopic collisions energy is lost in the process of achieving coalescence so that $Q < 0$. However, energy may actually be released in nuclear and sub-nuclear processes (discussed in Sections 9.12 and 27.7) with the result that Q may be positive.

Example:

As a specific example of a totally inelastic collision, consider the situation illustrated in Figure 6.16 in which two identical particles, each of mass m and travelling at right angles to each other with at equal speeds u, collide to form a single composite particle of mass $2m$ (neither particle is at rest in the laboratory in this case).

In this case $\quad |\boldsymbol{p}_1| = |\boldsymbol{p}_2| = p \quad$ and $\quad |\boldsymbol{u}_1| = |\boldsymbol{u}_2| = u$

Conservation of momentum (x-components): $\quad p_1\cos\theta_1 + p_2\cos\theta_2 = q,$

where $\theta_1 = \theta_2 = 45°$, so that $\frac{p}{\sqrt{2}} + \frac{p}{\sqrt{2}} = q$

$\rightarrow \quad q = \sqrt{2}p$ or, in terms of velocities, $v = \frac{u}{\sqrt{2}}$

$$\text{Change of energy} = \Delta E = \frac{q^2}{2(2m)} - \left(\frac{p_1^2}{2m} + \frac{p_2^2}{2m}\right) = \frac{(\sqrt{2}p)^2}{4m} - 2\frac{p^2}{2m}$$

$$\rightarrow \Delta E = -\frac{p^2}{2m} = -\frac{1}{2}mu^2$$

In this case half of the initial energy is lost in the collision.

We will return to the study of particle collisions in more detail in Section 8.3.

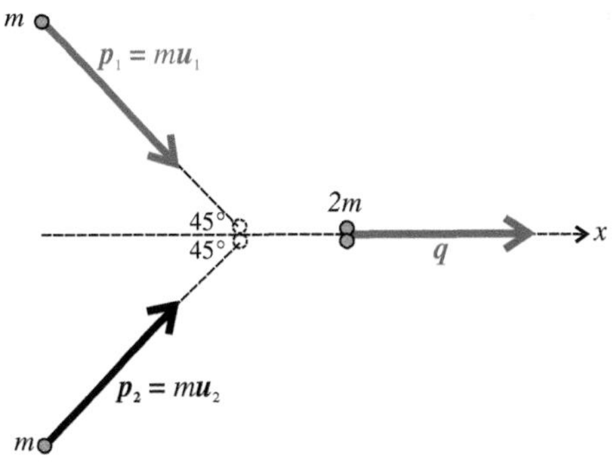

Figure 6.16. Example of a totally inelastic collision.

Study Worked Example 6.5

For problems based on the material presented in this section visit up.ucc.ie/6/ and follow the link to the problems.

6.6 The centre of mass of a system of particles

The centre of mass of a system of particles is a point in space that, in a loose sense, represents the average position of the particles in the system. More precisely, the **centre of mass** of a system comprising particles of masses m_1, m_2, m_3 ... etc. which are located, respectively, at positions $\boldsymbol{r}_1, \boldsymbol{r}_2, \boldsymbol{r}_3$, ... relative to an origin O, as illustrated in Figure 6.17, is defined as the point with displacement $\boldsymbol{R}_C$ relative to O such that

$$R_C := \frac{m_1\boldsymbol{r}_1 + m_2\boldsymbol{r}_2 + m_3\boldsymbol{r}_3 + \ldots}{m_1 + m_2 + m_3 + \ldots} \tag{6.14}$$

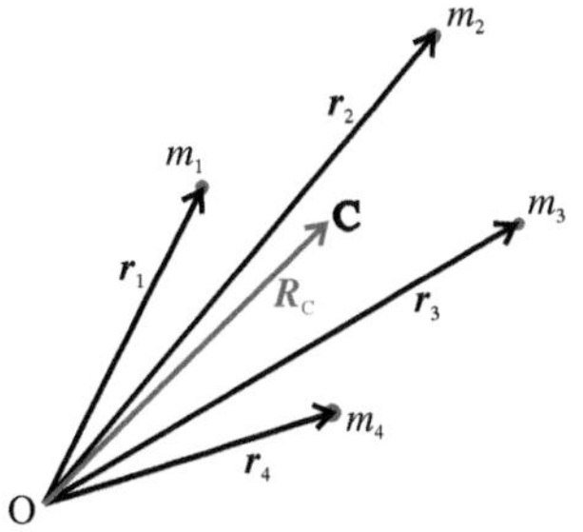

Figure 6.17. The centre of mass (C) of a system of particles.

In a rectangular (Cartesian) coordinate system,

$$\boldsymbol{R}_C = X_C\,\boldsymbol{i} + Y_C\boldsymbol{j} + Z_C\,\boldsymbol{k} \qquad \text{where} \qquad X_C := \frac{m_1x_1 + m_2x_2 + m_3x_3 + \ldots}{m_1 + m_2 + m_3 + \ldots},$$

with similar expressions for Y_C and Z_C.

Differentiating the defining Equation (6.14) with respect to time we obtain the following expression for the velocity of the centre of mass relative to O

$$V_C = \frac{m_1\boldsymbol{u}_1 + m_2\boldsymbol{u}_2 + m_3\boldsymbol{u}_3 + \ldots}{m_1 + m_2 + m_3 + \ldots} \tag{6.15}$$

where $\boldsymbol{u}_i = \frac{d\boldsymbol{r}_i}{dt}$ is the velocity of the ith particle ($i = 1, 2, 3$... etc.). Hence the total momentum of the system of particles, relative to O, is

$$\boldsymbol{P} = m_1\boldsymbol{u}_1 + m_2\boldsymbol{u}_2 + m_3\,\boldsymbol{u}_3 + \ldots = (m_1 + m_2 + m_3 + \ldots)\boldsymbol{V}_C = M\boldsymbol{V}_C \tag{6.16}$$

where M is the total mass of the system. Differentiating once more with respect to time we find that

$$\frac{d\boldsymbol{P}}{dt} = m_1\boldsymbol{a}_1 + m_2\boldsymbol{a}_2 + m_3\,\boldsymbol{a}_3 + \ldots = \frac{d\boldsymbol{p}_1}{dt} + \frac{d\boldsymbol{p}_2}{dt} + \frac{d\boldsymbol{p}_3}{dt} + \ldots = M\frac{d\boldsymbol{V}_C}{dt}$$

and hence, from Equation (6.5), the total external force on the system is given by

$$\boldsymbol{F}_{EXT} = (\boldsymbol{F}_1)_{ext} + (\boldsymbol{F}_2)_{ext} + (\boldsymbol{F}_3)_{ext} + \ldots = \frac{d\boldsymbol{P}}{dt} = M\frac{d\boldsymbol{V}_C}{dt} \tag{6.17}$$

Equations (6.16) and (6.17) have an obvious interpretation, namely that, in relation to external forces acting on a system of particles, the system can be treated as a single body whose total mass is located at a point at the centre of mass of the system. This justifies our previous treatment of extended objects as point masses.

CM coordinates

It turns out that it is often simpler or more convenient to solve certain many-body problems using a system of coordinates which is fixed relative to the centre-of-mass of the system. In these coordinates, the total momentum of the system must be zero. Such a system is called the **CM coordinate system**. We shall see later that the concept of centre of mass and the CM coordinate system plays an important role in the study of rigid bodies (Chapter 7) and of relative motion (Section 8.3). More immediately, we shall see in the next section how the concept is used in the analysis of a two-body system.

The summation notation

When dealing with many-body systems we often need to write summations over the total number of bodies in the system, such as in Equations (6.14) to (6.16) above. Such longwinded expressions can be time consuming to write and difficult to read. However, the mathematics can be kept in a more compact form by adopting the following notation for sums

$$\sum_{i=1}^{i=N} f_i = f_1 + f_2 + f_3 + f_4 + \ldots\ f_N$$

where N is the total number of particles in the system.

In what follows the explicit limits on the sum will often be dropped, that is we will write the above summation as $\sum_i f_i$, where the sum is understood to include all the particles in the system.

Using this notation, Equations (6.14) and (6.15) can be written, respectively, in the following more compact forms

$$\boldsymbol{R}_\mathrm{C} := \frac{\sum_i m_i \boldsymbol{r}_i}{\sum_i m_i} \quad \text{and} \quad \boldsymbol{V}_\mathrm{C} := \frac{\sum_i m_i \boldsymbol{u}_i}{\sum_i m_i}$$

Note that, if the origin O is chosen to coincide with the centre of mass (that is $\boldsymbol{R}_\mathrm{C} = 0$ or, equivalently, $X_\mathrm{C} = Y_\mathrm{C} = Z_\mathrm{C} = 0$), it follows that $\sum_i m_i \boldsymbol{r}_i = \sum_i m_i x_i = \sum_i m_i y_i = \sum_i m_i z_i = 0$.

For problems based on the material presented in this section visit up.ucc.ie/6/ and follow the link to the problems.

6.7 The two-body problem: reduced mass

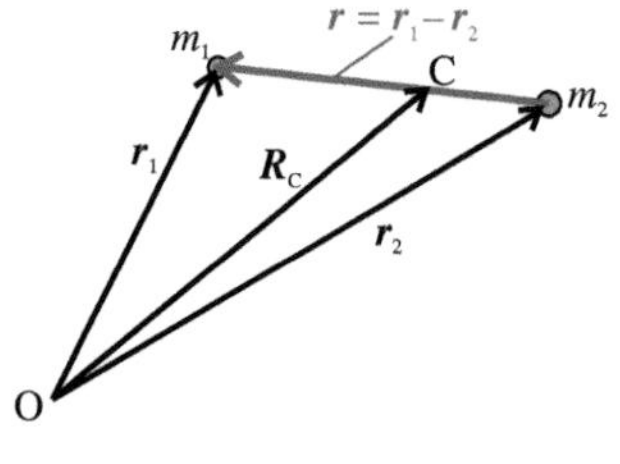

Figure 6.18. A system comprising two particles; $\boldsymbol{R}_\mathrm{C}$ is the displacement of the centre of mass relative to the origin and $\boldsymbol{r} = \boldsymbol{r}_1 - \boldsymbol{r}_2$ is the displacement of m_1 relative to m_2.

Consider a system, comprising two interacting bodies of masses m_1 and m_2, as shown in Figure 6.18. Newton's third law (Section 6.1) tells us that $\boldsymbol{F}_{2\to1} = -\boldsymbol{F}_{1\to2} = \boldsymbol{F}$ (say). In the absence of external forces, the accelerations of the particles are given by

$$\boldsymbol{a}_1 = \frac{d\boldsymbol{v}_1}{dt} = \frac{\boldsymbol{F}_{2\to1}}{m_1} = \frac{\boldsymbol{F}}{m_1} \quad \text{and} \quad \boldsymbol{a}_2 = \frac{d\boldsymbol{v}_2}{dt} = \frac{\boldsymbol{F}_{1\to2}}{m_2} = -\frac{\boldsymbol{F}}{m_2}$$

Thus,

$$\boldsymbol{a}_1 - \boldsymbol{a}_2 = \frac{\boldsymbol{F}}{m_1} - \left(-\frac{\boldsymbol{F}}{m_2}\right) = \left(\frac{1}{m_1} + \frac{1}{m_2}\right)\boldsymbol{F} \tag{6.18}$$

Now $\boldsymbol{a}_1 - \boldsymbol{a}_2 = \frac{d\boldsymbol{u}_1}{dt} - \frac{d\boldsymbol{u}_2}{dt} = \frac{d}{dt}(\boldsymbol{u}_1 - \boldsymbol{u}_2) = \frac{d^2}{dt^2}(\boldsymbol{r}_1 - \boldsymbol{r}_2) = \frac{d^2\boldsymbol{r}}{dt^2}$, where $\boldsymbol{r} = \boldsymbol{r}_1 - \boldsymbol{r}_2$ is the displacement of m_1 relative to m_2 (as indicated in Figure 6.18).

Thus we see that the velocity and acceleration, respectively, of m_1 relative to m_2, are given by

$$\boldsymbol{u}_1 - \boldsymbol{u}_2 = \frac{d\boldsymbol{r}}{dt} \text{ (which we shall denote as } \boldsymbol{u} \text{ in what follows) and}$$

$$\boldsymbol{a}_1 - \boldsymbol{a}_2 = \frac{d\boldsymbol{u}}{dt} = \frac{d^2\boldsymbol{r}}{dt^2} \text{ (which we shall denote as } \boldsymbol{a}\text{)}$$

An observer located at the particle m_2, therefore, will view the motion of m_1 under the influence of the force $\boldsymbol{F}$ as being governed by Equation (6.18); that is, the mass m_1 will be seen to accelerate relative to m_2 with an acceleration given by

$$\boldsymbol{a} = \left(\frac{1}{m_1} + \frac{1}{m_2}\right)\boldsymbol{F} = \left(\frac{m_1 + m_2}{m_1 m_2}\right)\boldsymbol{F}$$

This relationship can be written in the form of Newton's second law if we define a quantity, called the **reduced mass** of the system, as follows

$$\boxed{m_r := \frac{m_1 m_2}{m_1 + m_2}} \tag{6.19}$$

in which case the acceleration is given by

$$\boldsymbol{a} = \frac{\boldsymbol{F}}{m_\mathrm{r}}$$

Thus, to an observer located at the mass m_2, the motion of m_1 is equivalent to the motion of a particle of mass m_r under the influence of the force $\boldsymbol{F} = \boldsymbol{F}_{2\to 1}$. In effect, the two-particle problem in Figure 6.18 can be reduced to a problem of a single particle in a force field provided (i) that the reduced mass of the system is used for the mass of the particle and (ii) that the motion is described by the relative coordinates $\boldsymbol{r} = \boldsymbol{r}_1 - \boldsymbol{r}_2$, $\boldsymbol{u} = \boldsymbol{u}_1 - \boldsymbol{u}_2$, $\boldsymbol{a} = \boldsymbol{a}_1 - \boldsymbol{a}_2$, *etc.* Note that relative coordinates are independent of the choice of origin, provided that the origin chosen is at rest or is moving with constant velocity relative to the origin O. We shall return to this issue in Chapter 8. As an illustration, let us calculate the reduced mass of the electron-proton system which comprises the hydrogen atom, namely

$$m_\mathrm{r} = \frac{m_\mathrm{p} m_\mathrm{e}}{m_\mathrm{p} + m_\mathrm{e}} = \frac{m_\mathrm{e}}{1 + \frac{m_\mathrm{e}}{m_\mathrm{p}}}$$

Since $m_\mathrm{p} = 1.6726 \times 10^{-27}$ kg and $m_\mathrm{e} = 9.1094 \times 10^{-31}$ kg, $\dfrac{m_\mathrm{e}}{m_\mathrm{p}} = 5.44625 \times 10^{-4}$

and hence $m_\mathrm{r} = 0.999456 m_\mathrm{e} = 9.1044 \times 10^{-31}$ kg

We can apply Equation (6.14) to define the centre of mass of the two-particle system shown in Figure 6.18. The centre of mass of this system is a point whose displacement relative to O is given by

$$\boldsymbol{R}_\mathrm{C} := \frac{m_1 \boldsymbol{r}_1 + m_2 \boldsymbol{r}_2}{m_1 + m_2}$$

The velocity of the centre of mass relative to O is given by

$$\boldsymbol{V}_\mathrm{C} = \frac{m_1 \boldsymbol{u}_1 + m_2 \boldsymbol{u}_2}{m_1 + m_2} = \frac{\boldsymbol{P}}{m_1 + m_2} \tag{6.20}$$

where $\boldsymbol{P} = m_1 \boldsymbol{u}_1 + m_2 \boldsymbol{u}_2$ is the total momentum of the system.

Consider now

$$\boldsymbol{V}_C - \boldsymbol{u}_1 = \frac{m_1 \boldsymbol{u}_1 + m_2 \boldsymbol{u}_2}{m_1 + m_2} - \boldsymbol{u}_1 = \left(\frac{m_1}{m_1 + m_2} - 1\right)\boldsymbol{u}_1 + \frac{m_2 \boldsymbol{u}_2}{m_1 + m_2}$$
$$= -\frac{m_2}{m_1 + m_2}\boldsymbol{u}_1 + \frac{m_2}{m_1 + m_2}\boldsymbol{u}_2 = -\frac{m_2}{m_1 + m_2}(\boldsymbol{u}_1 - \boldsymbol{u}_2) = -\frac{m_2}{m_1 + m_2}\boldsymbol{u}$$

Hence,

$$\boldsymbol{u}_1 = \boldsymbol{V}_\mathrm{C} + \frac{m_2}{m_1 + m_2}\boldsymbol{u} \tag{6.21a}$$

and similarly,

$$\boldsymbol{u}_2 = \boldsymbol{V}_\mathrm{C} - \frac{m_1}{m_1 + m_2}\boldsymbol{u} \tag{6.21b}$$

Equations (6.21a) and (6.21b) express the velocities of the two particles in terms of (i) the velocity of the centre of mass of the system and (ii) the relative velocity ($\boldsymbol{u} = \boldsymbol{u}_1 - \boldsymbol{u}_2 = \dfrac{d\boldsymbol{r}}{dt}$, where $\boldsymbol{r} = \boldsymbol{r}_1 - \boldsymbol{r}_2$) of the two particles, specifically the velocity of m_1 relative to m_2 as before. Essentially, what we have done here is to express the displacements of the particles in terms of the displacement of their centre of mass relative to O ($\boldsymbol{R}_C$) and their relative displacement ($\boldsymbol{r}$) instead of $\boldsymbol{r}_1$ and $\boldsymbol{r}_2$ as in Figure 6.18. The strategy behind this change of variable will become clear from what follows.

Let us first determine the kinetic energy of the two-particle system in Figure 6.18 in terms of $\boldsymbol{V}_C$ and $\boldsymbol{u}$, that is

$$K = K_1 + K_2 = \frac{1}{2}m_1 u_1{}^2 + \frac{1}{2}m_2 u_2{}^2 = \frac{1}{2}m_1\left|\boldsymbol{V}_C + \frac{m_2}{m_1+m_2}\boldsymbol{u}\right|^2 + \frac{1}{2}m_2\left|\boldsymbol{V}_C - \frac{m_1}{m_1+m_2}\boldsymbol{u}\right|^2$$

$$= m_1 V_C{}^2 + \frac{m_1 m_2}{m_1+m_2}\boldsymbol{V}_C\cdot\boldsymbol{u} + \frac{1}{2}\frac{m_1 m_2^2}{(m_1+m_2)^2}u^2$$

$$+\frac{1}{2}m_2 V_C{}^2 - \frac{m_1 m_2}{m_1+m_2}\boldsymbol{V}_C\cdot\boldsymbol{u} + \frac{1}{2}\frac{m_1^2 m_2}{(m_1+m_2)^2}u^2$$

$$= \frac{1}{2}(m_1+m_2)V_C^2 + \frac{1}{2}\frac{m_1 m_2}{m_1+m_2}u^2$$

$$\rightarrow \quad K = \frac{1}{2}(m_1+m_2)V_C^2 + \frac{1}{2}m_r u^2 \tag{6.22}$$

where $m_r = \dfrac{m_1 m_2}{m_1+m_2}$ is again the reduced mass of the system as defined by Equation (6.19).

The total mechanical energy of the system is then given by

$$E_T = \frac{1}{2}(m_1+m_2)V_C^2 + \frac{1}{2}m_r u^2 + U(\boldsymbol{r}_1,\boldsymbol{r}_2)$$

where $U(\boldsymbol{r}_1,\boldsymbol{r}_2)$ is the potential energy function of the system. In most situations in physics, for example in the case of the gravitational interaction, the force between two particles acts along the line joining them and depends on the distance between them. In these circumstances, the potential energy depends on r only, that is

$$U(\boldsymbol{r}_1,\boldsymbol{r}_2) = U(|\boldsymbol{r}_1 - \boldsymbol{r}_2|) = U(|\boldsymbol{r}|) = U(r)$$

and hence

$$E_T = \frac{1}{2}(m_1+m_2)V_C^2 + \frac{1}{2}m_r u^2 + U(r)$$

Thus it can be seen that the total energy can be written as the sum of two expressions, one in terms of the coordinates of the centre of mass and the other in terms of the relative coordinates, that is

$$E_T = \left\{\frac{1}{2}(m_1+m_2)V_C^2\right\} + \left\{\frac{1}{2}m_r u^2 + U(r)\right\} = \text{constant}$$

Since this must hold for all possible positions of the masses, the two expressions in brackets must be constant independently, that is

$$\frac{1}{2}(m_1+m_2)V_C^2 = E_{CM} = \text{constant}$$

and

$$\frac{1}{2}m_r u^2 + U(r) = E = \text{constant} \tag{6.23}$$

where

$$E_{CM} + E = E_T = \text{constant}$$

The two-body problem, therefore, has been reduced to two one-body problems, namely

(i) that of a particle of mass m_1+m_2 located at the centre of mass which moves at constant velocity $\boldsymbol{V}_C$ (a 'free particle') and
(ii) that of a particle of mass m_r moving in a central force field described by the potential energy function $U(r)$.

If we choose the origin of the coordinate system to be at the centre of mass of the system, then $R_C = 0$ and $V_C = 0$ and only the problem in relative coordinates (Equation (6.23)) needs be solved. Such a system is called the centre-of-mass coordinate system or the **CM coordinate system**. We shall see in the next section and in Section 8.3 that, if collisions between particles are analysed in this coordinate system, the problem is often simplified considerably. Furthermore, in the CM coordinate system, $E_{CM} = 0$ and thus the total energy (E_T) is at a minimum in this coordinate system.

Clearly if $m_1 \ll m_2$ then $m_r \to m_1$ and the centre of mass is effectively co-located at the centre of the heavy particle. This was the assumption made in the treatment of planetary orbits in Sections 5.12 and 5.13. We can now see that the problem can be treated without this assumption if the mass m of the planet is replaced by the reduced mass of the system. In other words, the analyses presented in Sections 5.12 and 5.13 hold rigorously for any gravitationally bound two-body system provided m is replaced by $m_r = \dfrac{mM}{m+M}$ and M is replaced by $(m + M)$.

Study Worked Example 6.6

For problems based on the material presented in this section visit up.ucc.ie/6/ *and follow the link to the problems.*

6.8 Angular momentum of a system of particles

Consider again a general system of interacting particles, illustrated in Figure 6.19. As discussed in Section 6.2, each particle experiences forces due to all the other particles in the system together with forces due to sources outside the system. In considering the angular momentum of such a system we need to specify an origin relative to which angular momentum and the moments of forces are measured, such as the origin O in Figure 6.19. The rate of change of angular momentum of the first particle is equal to the sum of the moments of all the forces acting on this particle and, using Equation (5.23), is given by

$$\frac{d\boldsymbol{L}_1}{dt} = \boldsymbol{M}_1 = \boldsymbol{r}_1 \times (\boldsymbol{F}_1)_{\text{ext}} + \boldsymbol{r}_1 \times \boldsymbol{f}_{2\to1} + \boldsymbol{r}_1 \times \boldsymbol{f}_{3\to1} + \ldots \tag{6.24a}$$

where $\boldsymbol{f}_{i\to1}$ is the force on particle 1 due to the ith particle, $(\boldsymbol{F}_1)_{ext}$ is the force on particle 1 due to external sources and the sum is over all the particles in the system.

Similarly,

$$\frac{d\boldsymbol{L}_2}{dt} = \boldsymbol{M}_2 = \boldsymbol{r}_2 \times (\boldsymbol{F}_2)_{\text{ext}} + \boldsymbol{r}_2 \times \boldsymbol{f}_{1\to2} + \boldsymbol{r}_2 \times \boldsymbol{f}_{3\to2} + \ldots \tag{6.24b}$$

$$\frac{d\boldsymbol{L}_3}{dt} = \boldsymbol{M}_3 = \boldsymbol{r}_3 \times (\boldsymbol{F}_3)_{\text{ext}} + \boldsymbol{r}_3 \times \boldsymbol{f}_{1\to3} + \boldsymbol{r}_3 \times \boldsymbol{f}_{2\to3} + \ldots \tag{6.24c}$$

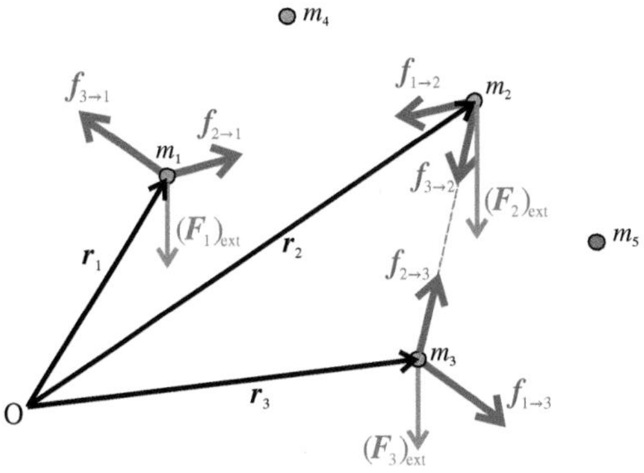

Figure 6.19. A system of interacting particles in an external field.

and so on, with a similar equation for each particle in the system.

We now define the **total angular momentum** of a system of particles as follows

$$\boldsymbol{L} := \boldsymbol{L}_1 + \boldsymbol{L}_2 + \boldsymbol{L}_3 + \boldsymbol{L}_4 + \boldsymbol{L}_5 + \ldots$$

The rate of change of the total angular momentum is

$$\frac{d\boldsymbol{L}}{dt} = \frac{d\boldsymbol{L}_1}{dt} + \frac{d\boldsymbol{L}_2}{dt} + \frac{d\boldsymbol{L}_3}{dt} + \ldots$$

which can be determined by adding up the right hand sides of Equations (6.24a), (6.24b), (6.24c) … etc. Using Newton's third law and the result that the moment of a force is equal to the product of the magnitude of the force and the perpendicular distance from the origin to the line of action of the force (Section 5.11), we find that $\boldsymbol{r}_1 \times \boldsymbol{f}_{2\to1} = -\boldsymbol{r}_2 \times \boldsymbol{f}_{1\to2}$ (see Figure 6.20), $\boldsymbol{r}_3 \times \boldsymbol{f}_{1\to3} = -\boldsymbol{r}_1 \times \boldsymbol{f}_{3\to1}$, and so forth. Thus the contributions from the interactions between the particles add up to zero and

$$\frac{d\boldsymbol{L}_1}{dt} = \boldsymbol{r}_1 \times (\boldsymbol{F}_1)_{\text{ext}} + \boldsymbol{r}_2 \times (\boldsymbol{F}_2)_{\text{ext}} + \boldsymbol{r}_3 \times (\boldsymbol{F}_3)_{\text{ext}} + \ldots = \boldsymbol{M}_{\text{EXT}}$$

where $\boldsymbol{M}_{\text{EXT}}$ is the total moment due to the external forces acting on the system.

Thus, as a direct consequence of the laws of dynamics, we have derived a result which describes the rotational behaviour of a system of particles under the influence of an external force

$$\boxed{\frac{d\boldsymbol{L}}{dt} = \boldsymbol{M}_{\text{EXT}}} \tag{6.25}$$

which is analogous to Newton's second law for translational motion with total angular momentum replacing (linear) momentum and $\boldsymbol{M}_{\text{EXT}}$ replacing force.

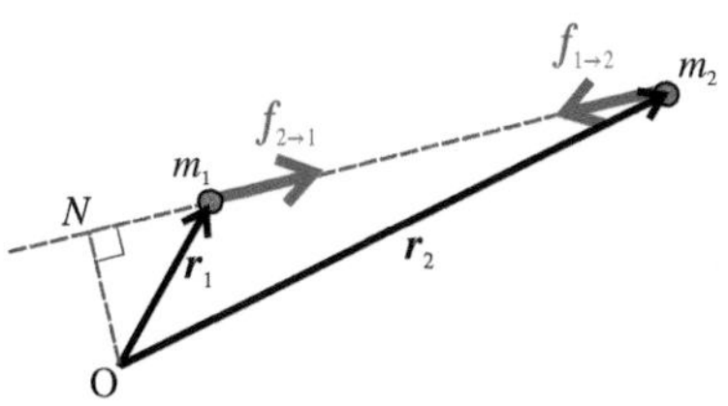

Figure 6.20. The moment of $\boldsymbol{f}_{2\to1}$ relative to O is equal and opposite to that of $\boldsymbol{f}_{1\to2}$.

Principle of conservation of angular momentum

From Equation (6.25) we see that if $\boldsymbol{M}_{\mathrm{EXT}} = 0$ then $\frac{d\boldsymbol{L}}{dt} = 0$ which requires that $\boldsymbol{L}$ is constant. This general rule, called the **principle of conservation of angular momentum**, asserts that

> *if the net moment due to external forces acting on a system of particles is zero, the total angular momentum of the system remains constant.*

This principle has wide applications in physics, ranging from the rotation of galaxies to that of atomic and molecular systems. It plays a particularly important role in the dynamics of rigid bodies, the subject of the next chapter.

For problems based on the material presented in this section visit up.ucc.ie/6/ and follow the link to the problems.

6.9 Conservation principles in physics

In this and in the previous chapter, three major conservation principles have been highlighted, namely the principles of conservation of energy, momentum and angular momentum. None contains any new physics in the sense that all three are derived from the laws of motion; they are, nevertheless, of profound practical and philosophical importance.

The practical importance of the conservation principles is that they enable us to make quantitative statements and predictions about complex dynamical systems even when we have little knowledge of the detailed workings of these systems. For example, the application of the principles of conservation of momentum and energy help to predict the outcome of collision processes despite a lack of knowledge of the details of the interactions between the particles involved. The fact that no net external force acts on a system is sufficient on its own to ensure that, whatever happens in the collision, momentum must be conserved. The requirement of conservation of energy is much less easily satisfied so that, while momentum is conserved in most collision problems, energy is conserved only in more restricted circumstances (namely in elastic collisions).

Conservation principles are a consequence of broad general features of the laws of motion. When the particular system under study has specific properties, such as the absence of a net external force, this restricts the possible evolution of the system. In what follows we shall see that those properties of dynamical systems which enable conservation laws to be invoked are intimately connected with **symmetries** in the system. Richard Feynman (1918–1988) has described these connections as being 'among the most beautiful and profound things in physics'. We outline some of these connections here without proof; for more details see Feynmann, Leighton and Sands, Volume I, pp. 52.1 – 52.12.

Geometrical symmetry

Conservation of energy and time reversal invariance: Consider what happens if the sign of time is reversed ($t \to -t$) in Newton's second law, $\boldsymbol{F} = M\frac{d^2\boldsymbol{r}}{dt^2}$. Since the derivative on the right hand side is of *second order* changing t into $-t$ will not change the relationship. Most of the examples of force which we have encountered so far are also unaffected by time reversal, in which case the equation of motion must be unchanged by $t \to -t$. An exception, however, is the case of damped harmonic motion when the damping force is velocity dependent (Section 3.11), in which case its sign is reversed by the transformation $t \to -t$ and a different equation of motion results. If a field force depends on velocity, the work done in moving from a point A to a point B in the field cannot be independent of the path taken and hence energy cannot be conserved.

This result can be stated more formally in terms of the Hamiltonian function of the system. Mechanical energy is conserved if *the Hamiltonian is unchanged as a result of time reversal*, that is $H(-t) = H(t)$. What this means is that if a video of a conservative mechanical system were to be made and then replayed backwards, the reversed motion would satisfy the same laws of mechanics. This would be true, for example, of an undamped simple pendulum but, on the other hand, a video of a damped pendulum played in reverse would show unfamiliar and unphysical behaviour; a spontaneous increase of amplitude with time is never observed.

Conservation of momentum and translational invariance: Total momentum is conserved in a mechanical system if *the Hamiltonian is unchanged as a result of translation,* that is $H(\boldsymbol{r} + \boldsymbol{a}) = H(\boldsymbol{r})$ where $\boldsymbol{a}$ is a constant displacement. This is equivalent to stating that the external force field experienced by the system is *homogeneous*; it is the same everywhere within the field. Note that if the Hamiltonian remains unchanged by a translation in a particular direction only, then only the component of the momentum in that direction is conserved.

Conservation of angular momentum and rotational invariance: Total angular momentum is conserved if *the Hamiltonian is unchanged as a result of rotation*; this is equivalent to stating that the external force field is *isotropic*. If the Hamiltonian is unchanged as a result of a rotation about a particular axis only, then only the component of angular momentum about that axis is conserved.

Dynamical symmetry

Two of the symmetries listed above (those giving rise to the conservation of energy and of angular momentum) were invoked in the analysis of planetary orbits in Section 5.13. The third conserved quantity (the Runge-Lenz vector) which arose in our treatment of that problem is an example of a somewhat different type of symmetry. We saw in Section 5.13 that this symmetry emerges from the particular character of inverse square law forces. Symmetries like this, which arises from the form of the force law involved in a particular context, are called *dynamical symmetries*.

WORKED EXAMPLES

Worked Example 6.1: The coefficients of static and kinetic friction between the 1.5 kg block and the plane in Figure 6.21 are 0.20 and 0.12, respectively. (i) For what range of values of m will the system remain at rest? (ii) If $m = 2.0$ kg determine the acceleration. You may assume that the string is massless and inextensible and that the pulley is massless and frictionless.

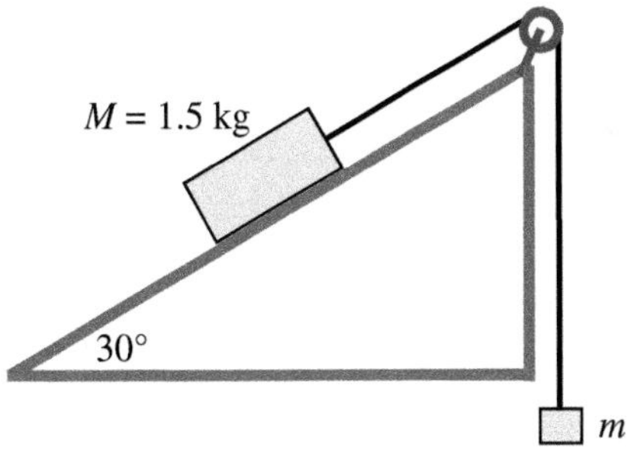

Figure 6.21. Worked Example 6.1.

(i) We consider two possible situations in which the block is at rest and about to move, namely (a) when the block is about to move up the plane (the positive x-direction) and (b) when it is about to move down the plane

(a) If the block is about to move up the plane, since it is at rest, the tension in the string $= |\boldsymbol{T}| = mg$

Forces on mass M (Figure 6.22a) x-components: $|\boldsymbol{T}| - |\boldsymbol{f}_R| - Mg\sin\alpha = 0$

y-components: $|\boldsymbol{N}| - Mg\cos\alpha = 0$

$\rightarrow\ mg - \mu_s Mg\cos\alpha - Mg\sin\alpha = 0$

$\rightarrow\ m = M(\sin\alpha + \mu_s\cos\alpha) = (1.5)(\sin 30° + (0.2)\cos 30°) = 1.0$ kg

(b) If the block is about to move down the plane the situation is as in (a) but with the frictional force reversed, that is $\boldsymbol{f}_R \rightarrow -\boldsymbol{f}_R$. The forces on the block are then given by

$mg + \mu_s Mg\cos\alpha - Mg\sin\alpha = 0$

$\rightarrow m = M(\sin\alpha - \mu_s\cos\alpha) = (1.5)(\sin 30° - (0.2)\cos 30°) = 0.49$ kg

Thus the block will move if $m \leq 1.0$ kg or if $m \geq 0.49$ kg and we conclude that the block will remain at rest if $1.0\,\text{kg} \geq m \geq 0.49$ kg

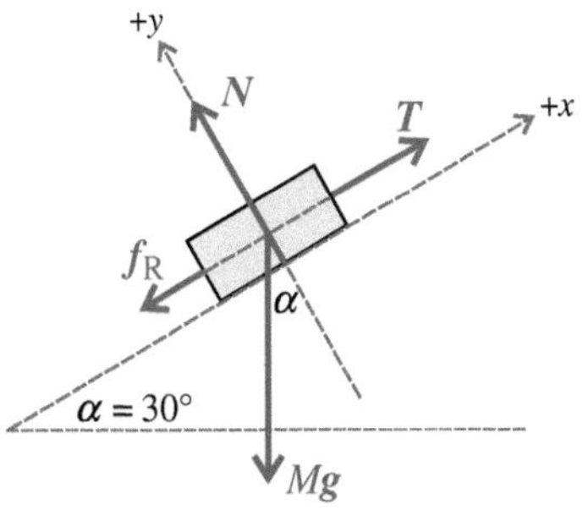

Figure 6.22 a. Worked Example 6.1.

(ii) When $m = 2.0$ kg the block accelerates up the plane

Forces on mass m (Figure 6.22a) x-components: $|\boldsymbol{T}| - |\boldsymbol{f}_R| - Mg\sin\alpha = M|\boldsymbol{a}|$

y-components: $|\boldsymbol{N}| - Mg\cos\alpha = 0$

$\rightarrow\ |\boldsymbol{T}| - \mu_K Mg\cos\alpha - Mg\sin\alpha = M|\boldsymbol{a}|$

Forces on mass m (Figure 6.22b) $mg - |\boldsymbol{T}'| = m|\boldsymbol{a}'|$

Since string is massless and inextensible $|\boldsymbol{T}| = |\boldsymbol{T}'| = T$ (say) and $|\boldsymbol{a}| = |\boldsymbol{a}'| = a$ (say). Hence, eliminating T, we obtain $(m - M\sin\alpha - \mu_K M\cos\alpha)g = (M + m)a$

$$\rightarrow \quad a = \frac{m - M(\sin\alpha + \mu_K\cos\alpha)}{M + m}g$$

When $m = 2.0$ kg, $M = 1.5$ kg, $\alpha = 30°$ and $\mu_K = 0.12$ we obtain $a = 3.1\ \text{m s}^{-2}$ (up the plane since $a > 0$)

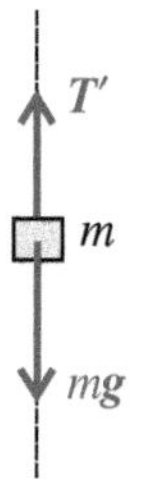

Figure 6.22 b. Worked Example 6.1.

Worked Example 6.2: A radioactive nucleus of mass 4.0×10^{-25} kg which is moving with speed $1.0 \times 10^{6}\ \text{m s}^{-1}$ breaks into two particles, one of mass 1.7×10^{-25} kg and the other of mass 2.3×10^{-25} kg. The lighter particle is observed to move along a path which makes an angle of 30° with the direction of motion of the original particle while the heavier one makes an angle of 60° with this direction (Figure 6.23). (a) Determine the speed of each of the particles. (b) What is the change in kinetic energy in the process?

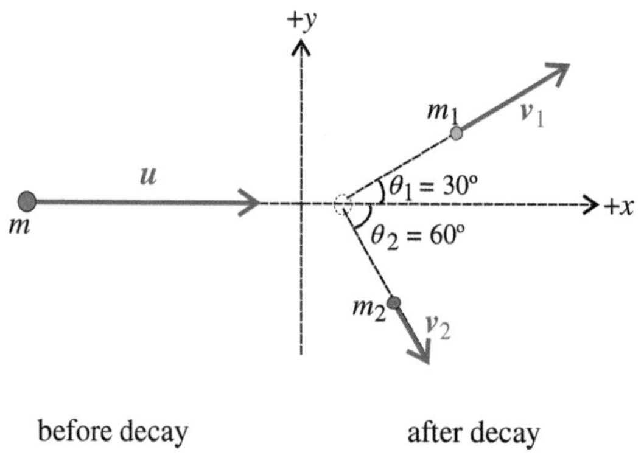

Figure 6.23. Worked Example 6.2.

(a) Momentum is conserved separately in both the x- and the y- directions

y-components: $0 = m_1 v_1 \sin\theta_1 - m_2 v_2 \sin\theta_2 \rightarrow v_1 = \dfrac{m_2 \sin\theta_2}{m_1 \sin\theta_1} v_2$

x-components: $mu = m_1 v_1 \cos\theta_1 + m_2 v_2 \cos\theta_2 \rightarrow mu = (m_1 \dfrac{m_2 \sin\theta_2}{m_1 \sin\theta_1} \cos\theta_1 + m_2 \cos\theta_2) v_2$

$\rightarrow v_2 = 0.87 \times 10^{6}\ \text{m s}^{-1}$ and hence $v_1 = 2.03 \times 10^{6}\ \text{m s}^{-1}$

(b) Change in energy $= (\frac{1}{2}m_1 v_1^2 + \frac{1}{2}m_2 v_2^2) - \frac{1}{2}mu^2 = 2.4 \times 10^{-13}$ J (an increase in energy, which, as noted in Section 6.5, is possible in a nuclear interaction).

Worked Example 6.3: A particle of mass m_1 travelling with velocity u_1 (momentum $= p_1 = m_1 u_1$) collides elastically head-on with a particle of mass m_2 at rest (Figure 6.24). Derive expressions for (i) the momentum of each particle after collision in terms of p_1, the momentum of the incident particle, (ii) the velocity of each particle after collision in terms of u_1, the velocity of the incident particle and (iii) the fraction of the kinetic energy of the incident particle that is transferred to the other particle by the collision.

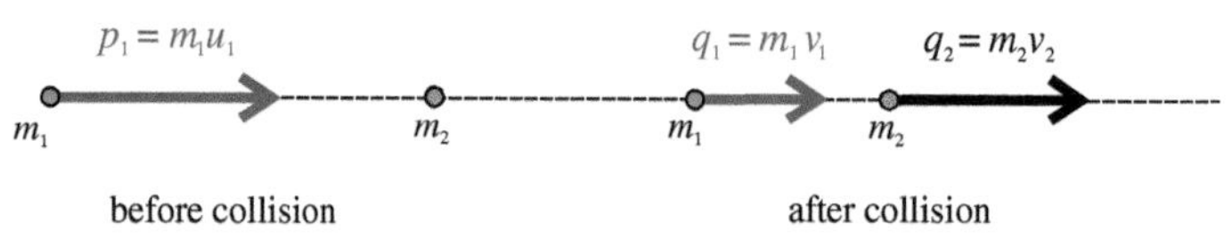

Figure 6.24. Worked Example 6.3.

Let the momenta of the particles after the collision be q_1 and q_2, respectively.

Conservation of momentum: $p_1 + 0 = q_1 + q_2 \rightarrow q_2 = p_1 - q_1$

Conservation of energy: $\dfrac{p_1^2}{2m_1} = \dfrac{q_1^2}{2m_1} + \dfrac{q_2^2}{2m_2} \rightarrow p_1^2 = q_1^2 + rq_2^2$, where $r = \dfrac{m_1}{m_2}$ is the ratio of the masses.

(i) $p_1^2 = q_1^2 + r(p_1 - q_1)^2 = q_1^2 + rp_1^2 - 2rp_1q_1 + rq_1^2$

$$\rightarrow (1+r)q_1^2 - 2rp_1q_1 - (1-r)p_1^2 = 0$$

$$\rightarrow q_1 = \frac{2rp_1 \pm \sqrt{4r^2p_1^2 - 4(1-r^2)p_1^2}}{2(1+r)} = \frac{2rp_1 \pm \sqrt{4p_1^2}}{2(1+r)} = \frac{r \pm 1}{r+1}p_1$$

$$\rightarrow q_1 = p_1 \text{ (and } q_2 = p_1 - q_1 = 0) \text{ or } q_1 = \frac{r-1}{r+1}p_1 \text{ (and } q_2 = p_1 - q_1 = \left(1 - \frac{r-1}{r+1}\right)p_1 = \frac{2}{r+1}p_1)$$

The first of the alternative solutions for q_1 and q_2 involves no change in either particle and hence is a trivial solution corresponding to no collision taking place. Hence the only physically meaningful solution is

$$\boxed{q_1 = \frac{r-1}{r+1}p_1} \text{ and } \boxed{q_2 = \frac{2}{r+1}p_1} \text{ (Note that } q_1 + q_2 = p_1 \text{ as expected)}$$

(ii) Since $q_1 = m_1v_1$ and $q_2 = m_2v_2$, $\boxed{v_1 = \dfrac{r-1}{r+1}u_1}$ and $\boxed{v_2 = \dfrac{2r}{r+1}u_1}$

(iii) The fractional transfer of energy is given by $\dfrac{\Delta E}{E_0} = \dfrac{\dfrac{q_2^2}{2m_2}}{\dfrac{p_1^2}{2m_1}} = \dfrac{m_1}{m_2}\dfrac{q_2^2}{p_1^2} \rightarrow \boxed{\dfrac{\Delta E}{E_0} = r\dfrac{4}{(r+1)^2}}$

It is interesting to examine these results in some special cases (see the table below). Note that the maximum (100%) energy transfer from the incident particle occurs when the masses are identical. As the masses become more unequal, for both the $m_1 \gg m_2$ and $m_1 \ll m_2$ cases, the energy transfer decreases.

	q_1	q_2	v_1	v_2	$\Delta E/E_0$
$m_2 \gg m_1$	$-p_1$	p_1	u_1	negligible	negligible
$m_2 = 3m_1$	$-\frac{1}{2}p_1$	$\frac{3}{2}p_1$	$-\frac{1}{2}u_1$	$\frac{1}{2}u_1$	$\frac{3}{4}$
$m_2 = 2m_1$	$-\frac{1}{3}p_1$	$\frac{4}{3}p_1$	$-\frac{1}{3}u_1$	$\frac{2}{3}u_1$	$\frac{8}{9}$
$m_1 = m_2$	0	p_1	0	u_1	1
$m_1 = 2m_2$	$\frac{1}{3}p_1$	$\frac{2}{3}p_1$	$\frac{1}{3}u_1$	$\frac{4}{3}u_1$	$\frac{8}{9}$
$m_1 = 3m_2$	$\frac{1}{2}p_1$	$\frac{1}{2}p_1$	$\frac{1}{2}u_1$	$\frac{3}{2}u_1$	$\frac{3}{4}$
$m_1 \gg m_2$	p_1	negligible	u_1	$2u_1$	negligible

Worked Example 6.4: A particle of mass 3.3×10^{-27} kg, moving at a speed of 6.0×10^7 m s^{-1}, strikes a stationary particle twice as massive as itself. After the collision the lighter particle is observed to be travelling in a direction at right angles to its original direction with a speed of 2.0×10^7 m s^{-1} (Figure 6.25). (a) In what direction and at what speed does the more massive particle move after the collision? (b) How much energy is lost in the collision?

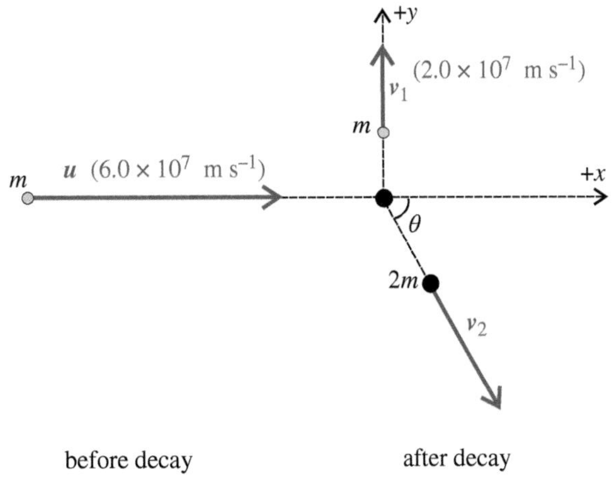

Figure 6.25. Worked Example 6.4.

x-components of total momentum: $mu = (2m)v_2 \cos\theta \rightarrow u = 2v_2 \cos\theta$

y-components of total momentum: $0 = mv_1 - (2m)v_2 \sin\theta \rightarrow v_1 = 2v_2 \sin\theta$

$\rightarrow \tan\theta = \frac{v_1}{u} = \frac{1}{3} \rightarrow \theta = 18.4° \text{ and } v_2 = \frac{u}{2\cos\theta} = \frac{6\times 10^7}{2(\cos 18.4°)} = 3.2\times 10^7 \text{m s}^{-1}$

$$\Delta E = \left(\frac{1}{2}mv_1{}^2 + \frac{1}{2}(2m)v_2{}^2\right) - \frac{1}{2}mu^2 = \frac{1}{2}m(v_1{}^2 + 2v_2{}^2 - u^2)$$

$$= \frac{1}{2}(3.3\times 10^{-27})((2.0\times 10^7)^2 + 2(3.2\times 10^7) - (6.0\times 10^7))^2$$

$$= -1.9\times 10^{-12}\text{ J}$$

Worked Example 6.5: A particle moving at a speed of 3.0×10^6 m s^{-1} collides with a particle which is three times as massive travelling in the opposite direction at a speed of 1.0×10^6 m s^{-1} (Figure 6.26). Each particle is scattered through 30° and 30% of the initial energy is lost in the collision process. Determine the velocity of each particle after collision.

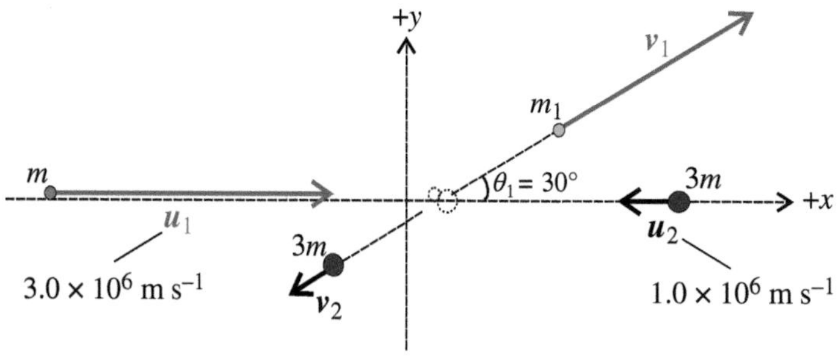

Figure 6.26. Worked Example 6.5.

Note that $m_1\boldsymbol{u}_1 = m(3.0\times 10^6 \text{ m s}^{-1})\boldsymbol{i} = -3m(1.0\times 10^6 \text{ m s}^{-1})\boldsymbol{i}$ which is equal to $-m_2\boldsymbol{u}_2 = \boldsymbol{p}$ (say) and thus the total momentum is zero; that is, the collision is described in the CM coordinate system. Thus, by conservation of momentum, $m_1\boldsymbol{v}_1 + m_2\boldsymbol{u}_2 = 0$ and thus $m_1\boldsymbol{v}_1 = -m_2\boldsymbol{u}_2 = \boldsymbol{q}$ (say)

The total energy after collision is given by $K_a = 0.7K_0$ and hence $\frac{q^2}{2m_1} + \frac{q^2}{2m_2} = 0.7\left(\frac{p^2}{2m_1} + \frac{p^2}{2m_2}\right)$

$$\rightarrow q = \sqrt{0.7}p = \sqrt{0.7}m_1u_1 = \sqrt{0.7}m(3\times 10^6 \text{ m s}^{-1})$$

$$\rightarrow v_1 = \frac{q}{m} = \sqrt{0.7}\times\ 3\times 10^6 \text{ m s}^{-1} = 2.5\times 10^6 \text{ m s}^{-1}$$

$$\rightarrow v_2 = \frac{q}{3m} = \sqrt{0.7}\times\ 10^6 \text{ m s}^{-1} = 0.84\times 10^6 \text{ m s}^{-1}$$

Worked Example 6.6: The vibrational frequency of a carbon monoxide ($^{12}C^{16}O$) molecule is 6.49×10^{13} Hz. Assuming that a vibrating CO molecule behaves as a simple harmonic oscillator as illustrated in Figure 6.27, estimate the equivalent 'spring constant' of the inter-atomic bond. For reasons that will be explained in Chapter 14, the minimum vibrational energy that a CO molecule can have is 2.1×10^{-20} J. Estimate the amplitude of the vibrations at this energy.

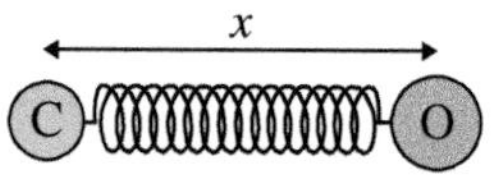

Figure 6.27. Worked Example 6.6.

In terms of relative coordinates (x in Figure 6.27), Equation (6.23) gives

$$\frac{1}{2}m_r v^2 + U(x) = \frac{1}{2}m_r v^2 + \frac{1}{2}kx^2 = E = constant$$

Differentiating with respect to time

$$\frac{1}{2}m_r(2v)\frac{dv}{dt} + \frac{1}{2}k(2x)\frac{dx}{dt} = 0$$

$$\rightarrow \quad m_r\frac{dv}{dt} = -kx \quad \rightarrow \quad \frac{d^2x}{dt^2} = -\frac{k}{m_r}x$$

Now $m_r = \dfrac{m_C m_O}{m_C + m_O} = \dfrac{12m_u \times 16m_u}{12m_u + 16m_u} = \dfrac{12 \times 16}{12 + 16} m_u = (6.86)(1.67 \times 10^{-27}\ \text{kg}) = 1.15 \times 10^{-26}\ \text{kg}$

Hence (recalling Section 3.6) $\omega^2 = \dfrac{k}{m_r}$

or $k = m_r\omega^2 = m_r(2\pi f)^2 = 4\pi^2 m_r f^2 = 4\pi^2(1.15 \times 10^{-26})(6.49 \times 10^{13})^2 = 1.9 \times 10^3\ \text{N m}^{-1}$

$\dfrac{1}{2}kA^2 = E$ for $E = 2.1 \times 10^{-20}$ J $\rightarrow$ $A = \sqrt{\dfrac{2E}{k}} = \sqrt{\dfrac{2(2.1 \times 10^{-20})}{1.9 \times 10^3}} = 4.7 \times 10^{-12}\ \text{m} = 4.7\ \text{pm}.$

(Note that the equilibrium separation of the atoms in a CO molecule is 114 pm, so the vibrational amplitude estimated above is only about 4% of this distance)

PROBLEMS

For problems based on the material covered this chapter visit up.ucc.ie/6/ *and follow the link to the problems.*

7

Rigid body dynamics

AIMS

- To show how the principles governing the motion of many-body systems, discussed in the previous chapter, can be applied to the motion of rigid bodies
- To develop a formalism for the treatment of the motion of a rigid body about a fixed axis
- To outline the formal equivalence between the principles which determine the motion of rigid bodies about a fixed axis and those which determine the motion of a point particle, as discussed in earlier chapters
- To consider the more complicated situation where the axis of rotation is not fixed, as in the case of *gyroscopic* motion

7.1 Rigid bodies

A rigid body is a many-body system in which the distance between each pair of particles remains fixed, that is, the system keeps its shape despite the action of any external forces. The motion of such a system under the influence of a net external force $\boldsymbol{F}_{\mathrm{EXT}}$ and of a net moment $\boldsymbol{M}_{\mathrm{EXT}}$ is determined by the Equations (6.5) and (6.25) which were derived in the previous chapter, namely

$$\frac{d\boldsymbol{P}}{dt} = \boldsymbol{F}_{\mathrm{EXT}} \tag{7.1}$$

and

$$\frac{d\boldsymbol{L}}{dt} = \boldsymbol{M}_{\mathrm{EXT}} \tag{7.2}$$

In Equation (7.1) $\boldsymbol{P}$ is the total momentum which is related to $\boldsymbol{V}_{\mathrm{C}}$, the velocity of the centre of mass of the rigid body, by Equation (6.16) and hence, from Equation (6.17),

$$\boldsymbol{F}_{\mathrm{EXT}} = \mathcal{M}\frac{d\boldsymbol{V}_{\mathrm{C}}}{dt} \tag{7.3}$$

where $\mathcal{M}$ is the mass of the body. Note that, in this and in some subsequent sections, the symbol $\mathcal{M}$ is used for the mass of a body to avoid any possible confusion with the symbol for moment ($\boldsymbol{M}$).

As an example of a rigid body, consider a body of arbitrary shape, as illustrated in Figure 7.1, which can be thought of as comprising a large number of particles of masses m_1, m_2, m_3, ... etc. If all of the body is near the Earth's surface so that $\boldsymbol{g}$ is effectively constant over its extent, then each component mass m_i of a body near the Earth's surface experiences a force $m_i\boldsymbol{g}$. In the absence of any other external forces, the net external force on the body is given by $\boldsymbol{F}_{\mathrm{EXT}} = \sum_i m_i\boldsymbol{g} = \mathcal{M}\boldsymbol{g}$, where $\mathcal{M} = \sum_i m_i$ is the mass of the body. The net external moment, relative to the origin O, is given by $\boldsymbol{M}_{\mathrm{EXT}} = \sum_i \boldsymbol{r}_i \times m_i\boldsymbol{g}$.

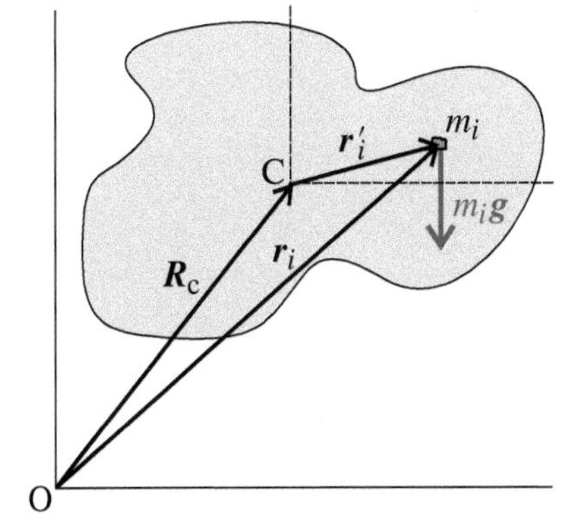

Figure 7.1. A rigid body can be considered to comprise a large number of particles, m_i being the mass of the ith particle. The displacement of the ith particle relative the origin O is $\boldsymbol{r}_i$ and its displacement relative to the centre of mass (C) is $\boldsymbol{r}'_i$. For a rigid body $|\boldsymbol{r}_i - \boldsymbol{r}_j|$ = constant for all i and j.

If C is the centre of mass of the rigid body, the displacement of C relative to O is defined as $\boldsymbol{R}_c = \sum_i m_i\boldsymbol{r}_i/\sum_i m_i$ (Section 6.6). As indicated in Figure 7.1, the displacement of the particle of mass m_i relative to O can be written as $\boldsymbol{r}_i = \boldsymbol{R}_{\mathrm{C}} + \boldsymbol{r}'_i$, where $\boldsymbol{r}'_i$ is the displacement of m_i

Understanding Physics, Third Edition. Michael Mansfield and Colm O'Sullivan.
© 2020 John Wiley & Sons Ltd. Published 2020 by John Wiley & Sons Ltd.

relative to C. Thus

$$M_{\text{EXT}} = \sum_i r_i \times m_i g = \sum_i (R_C + r'_i) \times m_i g = \sum_i R_C \times m_i g + \sum_i r'_i \times m_i g$$

$$= R_C \times \left(\sum_i m_i\right) g + \left(\sum_i m_i r'_i\right) \times g = R_C \times \mathcal{M} g + \left(\sum_i m_i r'_i\right) \times g$$

From the definition of centre of mass, $\sum_i m_i r'_i = 0$ and hence

$$M_{\text{EXT}} = R_C \times \mathcal{M} g$$

Thus we can treat the weight of an extended body as that of a point mass of mass $\mathcal{M} = \sum_i m_i$, located at the centre of mass of the body.

We shall consider the general motion of a rigid body later in this chapter but first we confine our attention to rigid bodies which are at rest.

7.2 Rigid bodies in equilibrium: statics

For a rigid body to be at rest (that is, $P = 0$ and $L = 0$) we see from Equations (7.1) and (7.2) that the following two conditions must be satisfied simultaneously.

(a) $F_{\text{EXT}} = 0$ (the vector sum of all external forces must be zero) and
(b) $M_{\text{EXT}} = 0$ (the vector sum of the moments of all the external forces, *relative to any origin*, must be zero).

A body for which these conditions are satisfied is said to be in *static equilibrium*. Note that conditions (a) and (b) are also satisfied if the centre of mass of the body is moving with constant velocity (P is constant) and/or if the body is rotating with constant angular velocity (L is constant).

The general procedures used to solve problems involving rigid bodies at rest are illustrated in the following two examples.

Example 1: Consider the case of a rigid beam which is suspended horizontally and which is able to rotate about a horizontal axis perpendicular to the page through the point O. A number of masses are hung from the beam at various points along its length in such positions that the beam remains horizontal, as shown in Figure 7.2.

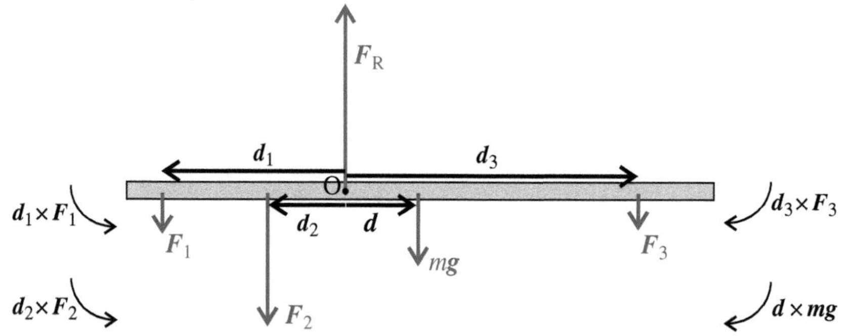

Figure 7.2. A rigid beam is suspended so that can rotate about a horizontal axis perpendicular to the page through the point O. The beam is in equilibrium due to the vertical forces shown. Since the reaction force F_R at the support acts through O, it has no moment about O. The direction of the moments of the other forces are determined by the rule for the vector product and are indicated in the insets.

Applying condition (a), we obtain

$$F_1 + F_2 + F_3 + mg + F_R = 0$$

where F_R is the reaction exerted upward on the beam by the support at O and mg is the weight of the beam which, as we have seen, can be treated as due to a point mass acting at the centre of mass of the beam. Applying condition (b), taking moments relative to O, gives

$$(d_1 \times F_1) + (d_2 \times F_2) + (d_3 \times F_3) + (d \times mg) + (F_R \times 0) = 0 \tag{7.4}$$

Note that we are free to take any point as the origin relative to which the moments of forces are determined. We can simplify the analysis, however, by taking as the origin a point through which one of the forces acts, particularly where the magnitude of this force is unknown and its value is not required. Thus the point O, through which F_R acts, is an obvious and convenient choice of origin in this case. We have seen (Section 5.11) that the moment of a force may be determined as the product of the magnitude of the force $\boldsymbol{F}$ and the perpendicular distance p from O to the line of action of each force, that is $|\boldsymbol{M}| = |\boldsymbol{r} \times \boldsymbol{F}| = |\boldsymbol{F}|p$. Adopting the convention that anticlockwise moments are positive and clockwise moments are negative (that is, choosing perpendicular to the page outwards as the positive direction), Equation (7.4) becomes

$$|\boldsymbol{F}_1|d_1 + |\boldsymbol{F}_2|d_2 - |\boldsymbol{F}_3|d_3 - m|\boldsymbol{g}|d = 0.$$

This result is sometimes called 'the principle of moments'; it states that for a rigid body to be in equilibrium, the sum of the anticlockwise moments about any point must balance the sum of the clockwise moments about the same point. The moments of forces $\boldsymbol{F}_1$ and $\boldsymbol{F}_2$ are both directed out of the page (recall the definition of the cross product in Section 4.11), as indicated in the diagrams on the left hand side of Figure 7.2; their sum cancels the effect of the moments of other two forces which are directed into the page, as indicated on the right hand side of the figure.

Example 2: Consider the equilibrium of a uniform ladder of length l and mass m which rests against a vertical wall, the ladder making an angle θ with the horizontal (Figure 7.3).

We assume, for simplicity, that the friction between the ladder and the vertical wall is negligible but that there is some friction between the ladder and the ground which stops the ladder from sliding. Since the ladder is not necessarily on the point of sliding at this angle, however, it is not possible to express the force due to friction (f_r) in terms of a coefficient of friction in this case (see Worked Example 7.1 for a situation in which a ladder is just on the point of slipping).

The forces on the ladder are shown in the figure; $\boldsymbol{R}$ and $\boldsymbol{N}$ are, respectively the normal reactions of the wall and ground. Assuming that the ladder is uniform (that is, mass is distributed uniformly along its length), the centre of mass can be taken to be at its mid point.

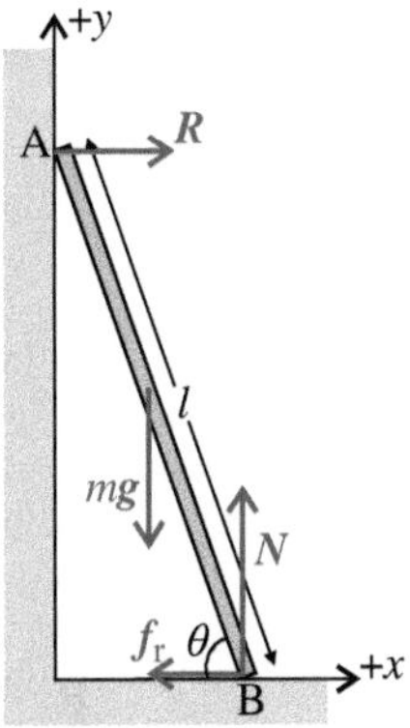

Figure 7.3. A uniform ladder leaning against a vertical wall is prevented from slipping by the frictional resistance (f_r) exerted on the ladder where it is in contact with the ground. For equilibrium, (i) the sum of the forces on the ladder must be zero and (ii) the sum of the moments of these forces about any point must be zero.

(a) Forces in x-direction (horizontal): $|\boldsymbol{R}| - |\boldsymbol{f}_r| = 0$

(b) Forces in y-direction (vertical): $|\boldsymbol{N}| - |m\boldsymbol{g}| = 0$

(c) Moments of forces, about B: $-|\boldsymbol{R}|l\sin\theta + |m\boldsymbol{g}|\left(\frac{1}{2}l\right)\cos\theta = 0$

Thus $|\boldsymbol{N}| = |m\boldsymbol{g}|$ and $|\boldsymbol{R}| = |\boldsymbol{f}_r| = \frac{1}{2}mg\cot\theta.$

Note that, in (c), the sum of the moments was taken relative to the point B for simplicity and convenience. Any other point could have been chosen equally well for this purpose but the choice of B gives the required result most directly, since the unknown forces $\boldsymbol{N}$ and $\boldsymbol{f}_r$ make zero contribution to the net moment in this case.

Study Worked Example 7.1

For problems based on the material presented in this section visit up.ucc.ie/7/ and follow the link to the problems.

7.3 Torque

Let us consider again a rigid body of arbitrary shape which can be thought of as comprising a large number of particles with masses $m_1, m_2, m_3,\ldots$ *etc.* which experience external forces $\boldsymbol{F}_1$, $\boldsymbol{F}_2, \boldsymbol{F}_3,\ldots$ etc., respectively, as illustrated in Figure 7.4.

The displacement of the ith particle relative to the origin O is $\boldsymbol{r}_i$ and its displacement relative to a second origin O′ is $\boldsymbol{r'}_i$. The total moment experienced by the rigid body due to all these forces is $\boldsymbol{M}_{\text{EXT}} = \sum_i \boldsymbol{r}_i \times \boldsymbol{F}_i$ relative to O and is $\boldsymbol{M'}_{\text{EXT}} = \sum_i \boldsymbol{r'}_i \times \boldsymbol{F}_i$ relative to O′. Now $\boldsymbol{r}_i = \boldsymbol{r'}_i + \boldsymbol{R}$, where $\boldsymbol{R}$ is the displacement of O′ relative to O, as indicated in the figure. Thus

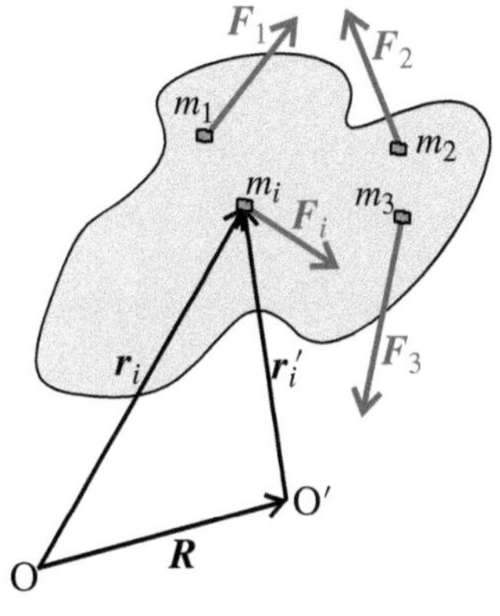

Figure 7.4. A rigid body in which each particle in the body experiences an external force. The external force on the ith particle is $\boldsymbol{F}_i$ and the displacement of the ith particle relative to O is $\boldsymbol{r}_i$ and its displacement relative to O′ is $\boldsymbol{r'}_i$.

$$\boldsymbol{M}_{\text{EXT}} = \sum_i \boldsymbol{r}_i \times \boldsymbol{F}_i = \sum_i (\boldsymbol{r'}_i + \boldsymbol{R}) \times \boldsymbol{F}_i = \left(\sum_i \boldsymbol{r'}_i \times \boldsymbol{F}_i\right) + \left(\sum_i \boldsymbol{R} \times \boldsymbol{F}_i\right)$$

$$= \left(\sum_i \boldsymbol{r'}_i \times \boldsymbol{F}_i\right) + \left(\boldsymbol{R} \times \sum_i \boldsymbol{F}_i\right) = \boldsymbol{M'}_{\text{EXT}} + \left(\boldsymbol{R} \times \sum_i \boldsymbol{F}_i\right)$$

that is

$$\boldsymbol{M}_{\text{EXT}} = \boldsymbol{M'}_{\text{EXT}} + (\boldsymbol{R} \times \boldsymbol{F}_{\text{EXT}}) \tag{7.5}$$

An important consequence follows from this result, namely

$$F_{EXT} = \sum_i F_i = 0 \;\Rightarrow\; M_{EXT} = M'_{EXT}$$

Thus, if no net force acts on a rigid body, the body experiences a net moment which is *independent of the origin of coordinates* chosen. In these circumstances the sum of the moments M_{EXT} is called the **torque** on the body[1]. We will usually denote torque by the symbol T.

The simplest example of a torque is that exerted by two equal and opposite forces, such as the forces $+F$ and $-F$ applied to the rigid body shown in Figure 7.5.

A combination of two equal and opposite forces applied in this way is called a **couple**. The net torque (the sum of the moments) exerted on the body in this case is given by

$$T = (r_1 \times F) + r_2 \times (-F)$$

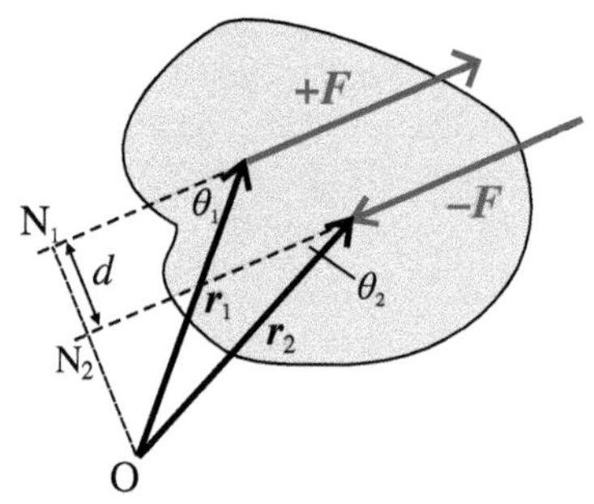

Figure 7.5. Two equal and opposite forces acting on a rigid body comprise a *couple*. The net moment due to the couple (that is, the torque exerted on the body) is independent of the choice of origin. The magnitude of this torque is equal to the product of the magnitude of F and the perpendicular distance between the two forces comprising the couple, that is $|T| = |F|d$.

that is $\quad |T| = |r_1||F|\sin\theta_1 - |r_2||F|\sin\theta_2 = |F|(\mathrm{ON}_1 - \mathrm{ON}_2) = |F|d$

where d is the perpendicular distance between the lines of action of the two forces. In this case we see clearly that the torque produced by a couple is independent of the choice of origin, as expected from the discussion above.

7.4 Dynamics of rigid bodies

Let us consider the situation, illustrated in Figure 7.6, in which the point C is the centre of mass of a rigid body. Applying Equation (7.5) to this case, we see that the sum of the moments of all the external forces on the body (relative to an arbitrary origin O) is given by

$$M_{EXT} = M'_{EXT} + (R_C \times F_{EXT})$$

where $M'_{EXT} = \sum_i r'_i \times F_i$ is the sum of the moments of the forces relative to C and F_{EXT} is the net external force on the body. Invoking Equations (7.1), (7.2) and (7.3) we see that

$$\frac{dP}{dt} = \mathcal{M}\frac{dV_C}{dt} = F_{EXT} \tag{7.6}$$

and

$$\frac{dL}{dt} = M_{EXT} = M'_{EXT} + (R_C \times F_{EXT}) = M'_{EXT} + \left(R_C \times \mathcal{M}\frac{dV_C}{dt}\right) \tag{7.7}$$

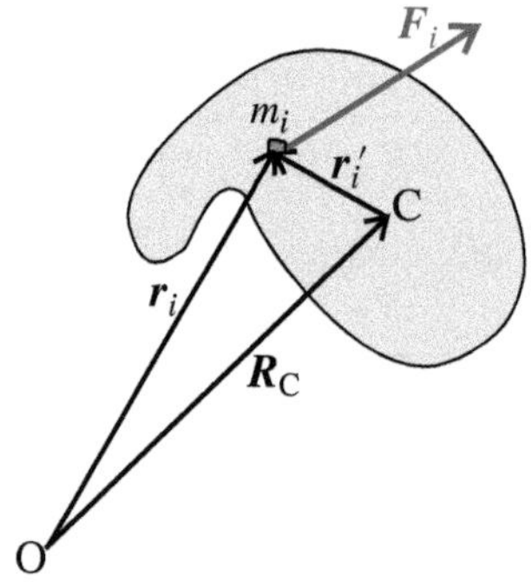

Figure 7.6. A rigid body in which the ith particle (mass = m_i) experiences a force F_i due to an external field. The displacement of the ith particle relative to O is r_i and its displacement relative to the centre of mass (C) is r'_i. The displacement of C relative to O is R_C.

Equations (7.6) and (7.7) can be interpreted as follows. The motion of a rigid body is seen to be a superposition of two independent motions (i) the motion of a point mass $\mathcal{M}$ located at the centre of mass of the body moving under the influence of a force F_{EXT} and (ii) the rotation of the body about the centre of mass due to the influence of a net moment M'_{EXT}. Figure 7.7 shows an example of such a motion.

As we have seen, when there is no net external force (that is $F_{EXT} = \sum_i F_i = 0$) a rigid body will experience an external torque only (that is, $M_{EXT} = M'_{EXT} = T_{EXT}$, independent of the origin of coordinates). In these circumstances the centre of mass of an unconstrained rigid body will remain at rest or move with constant velocity relative to an inertial reference system. The motion of the body relative to the centre of mass is then governed by Equation (7.7) which in this case becomes

$$\boxed{\frac{dL}{dt} = T_{EXT}} \tag{7.8}$$

The significance of torque in relation to the rotation of rigid bodies can be seen from the following example. Consider a horizontal wheel of radius a which is constrained to rotate about an axle through its centre C, as illustrated in Figure 7.8.

[1]In many texts the term 'torque' is used interchangeably with 'moment of a force'. In this book we follow the IUPAP recommendations and reserve the term torque for the net moment due to external forces in systems in which the vector sum of these forces is zero.

If we apply an external force to the rim of the wheel at any point there will be an equal and opposite reaction (a constraint force) applied to the wheel by the axle. How do we apply a force to the wheel to achieve maximum rotational effect, that is to produce the maximum rate of change of angular momentum? If a radial force is applied, that is a force $\boldsymbol{F}_r$ in Figure 7.8 such that the line of action of the force passes through the centre of the wheel, there will be no torque about the axis (that is, $|\boldsymbol{F}_r|d = 0$ since $d = 0$ in this case) and the force will be ineffective in providing a rotational effect. On the other hand, if a force of the same magnitude is applied at a tangent to the rim of the wheel, that is $\boldsymbol{F}_t$ in Figure 7.8, this will result in the maximum rate of change of angular momentum since, in this case, the torque $|\boldsymbol{F}_r|d = |\boldsymbol{F}_t|a$ is maximum.

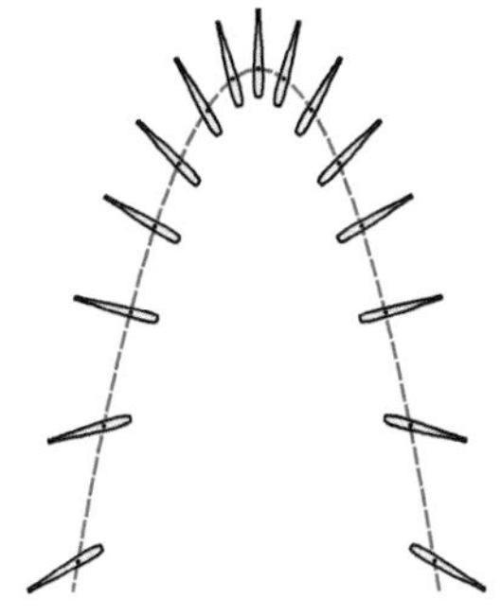

Figure 7.7. Motion of a rigid body projectile is seen as the superposition of two independent motions. The centre of mass of the body follows a parabolic path under the influence of gravity. The body rotates with constant angular velocity about the centre of mass (since there is no net moment relative to the centre of mass, in this case).

7.5 Measurement of torque: the torsion balance

Consider what happens when one end of a piece of wire or a metal rod is twisted about its axis while the other end is held fixed. This can be studied experimentally by fixing one end of a horizontal metal rod and attaching a solid disc to the other end as shown in Figure 7.9.

By winding a string around the circumference of the disc and hanging a mass m from the end of the string as shown, a known torque (magnitude mgR, where R is the radius of the disc) may be applied to twist the rod about its axis. It is observed that the rod provides a resistance to being twisted in this way; in other words, a twisted rod exerts a 'restoring torque' which tends to oppose the applied torque and tries to return the rod to its equilibrium (untwisted) position. This resistance increases with increasing angle of twist θ until an equilibrium is reached in which the restoring torque balances the applied torque. By varying the value of the mass hanging from the string, the dependence of the torque T_R exerted by the twisted rod on the angle of twist may be investigated.

It is found experimentally that, provided the angle θ is not too large, the restoring torque is directly proportional to the angle of twist and opposed to the angular displacement, that is

$$T_R \propto -\theta \quad \text{or} \quad \boxed{T_R = -c\theta}$$

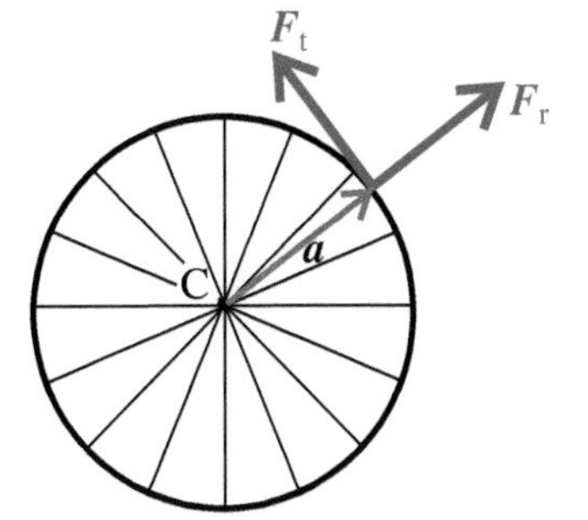

Figure 7.8. A wheel of radius a is free to rotate about a fixed axis through its centre C which is perpendicular to the plane of the wheel. A radial force ($\boldsymbol{F}_r$) will not contribute to the rotational motion of the wheel while a tangential force ($\boldsymbol{F}_t$) will have maximum rotational effect. In the latter case the rate of change of angular momentum is given by $\frac{d\boldsymbol{L}}{dt} = |\boldsymbol{F}_t|a$.

where c is a constant, called the *torsion constant*, characteristic of a particular rod. This result, which describes the behaviour of a wide variety of materials under torsion, may be considered as the rotational analogue of the behaviour described in Section 3.5 for helical springs, where the restoring force is directly proportional to the amount of translational extension or compression ($F = -kx$). The SI unit for torsion constant is N m rad^{-1}.

The torsion balance

The observed relationship between T_R and θ provides a convenient way to measure an unknown torque using an arrangement called a **torsion balance**. To illustrate the principle of the torsion balance we describe below a technique for measuring the value of the gravitational constant G, which is very similar to Cavendish's original method mentioned in passing in Section 5.1.

Two small lead spheres, each of mass m, are attached to the ends of a beam of negligible mass and of length a (Figure 7.10).

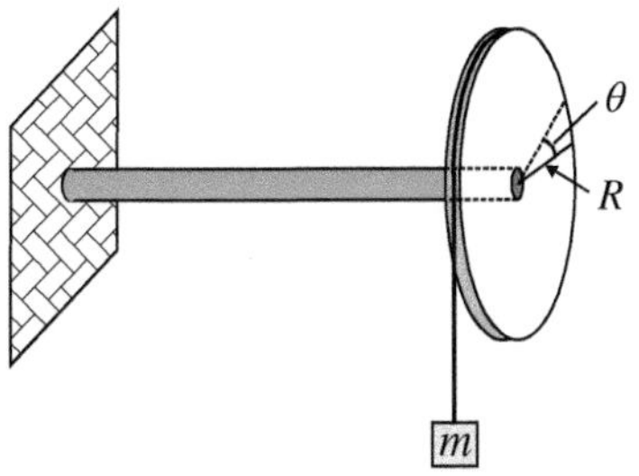

Figure 7.9. Experimental arrangement to study the torsion of a metal rod. The rod is held horizontally with one end fixed rigidly to a vertical wall. A torque is applied to the rod by hanging a mass from a string which passes over a disc attached to the other end of the rod as shown. The angle θ through which the disc is rotated is recorded for different applied torques (different masses attached to the string).

The beam is suspended by means of a thin fibre or wire which hangs vertically from a fixed support and is attached to the midpoint of the beam so that the beam lies in a horizontal plane. The finer the suspension wire the more sensitive is the torsion balance and hence the smaller the torque which can be measured. Because the gravitational interaction is so weak, a very sensitive torsion balance is required. Care must be taken to exclude draughts and other effects which might disturb the beam. In the absence of such disturbances, the beam will gradually settle down in an equilibrium position. A beam of light which is reflected onto a screen from a small mirror attached to the suspension wire enables any rotation of the beam to be measured.

Two larger lead spheres, each of mass M, are then brought close to the small spheres and held rigidly so that each large sphere is the same distance from the adjacent small sphere. Figure 7.11 shows the torsion balance viewed from above.

The gravitational force exerted on each small sphere by the nearby large sphere causes a couple to be exerted on the beam which gives rise to a rotation of the beam until it finally settles in an equilibrium position when the restoring torque exerted by the twisted suspension balances the torque of the couple due to the gravitational forces. The angle θ through which the beam has been rotated can be measured by observing the deflection of the light beam reflected from the mirror. The condition for equilibrium is

$$G\frac{mM}{d^2}a = c\theta \tag{7.9}$$

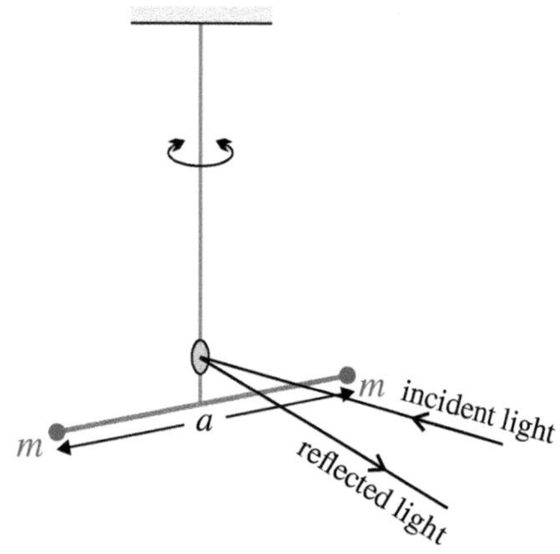

Figure 7.10. A torsion balance. Identical spheres, each of mass m, are attached to either end of a light beam of length a. The beam is suspended from a fixed point by means of an elastic suspension wire so that the beam can rotate in a horizontal plane about the axis of the wire. The angle through which the beam is twisted from its equilibrium position is measured by reflecting light from a small mirror attached to the suspension wire.

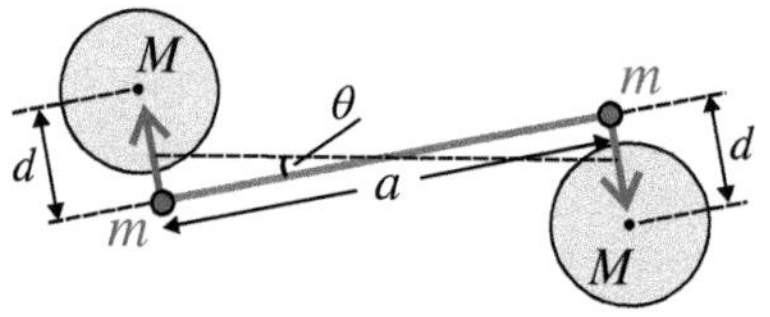

Figure 7.11. The torsion balance in Figure 7.10, viewed from above. The beam has rotated through an angle θ as a result of the gravitational attraction acting on the smaller spheres (m) by the two large spheres (mass = M) which have been brought close to the smaller spheres as shown.

where d is the centre-to-centre distance between a large and a small sphere. The torsion constant of the suspension c can be determined either from the dimensions and elastic properties of the suspension wire (which will be discussed in Chapter 10, Section 10.2) or may be measured experimentally (see Section 7.10 below). All other quantities in Equation (7.9) can be measured easily and hence the gravitational constant can be determined.

To estimate how sensitive a torsion balance must be to be used to measure the gravitational constant, let us estimate the value of c required to obtain a deflection of, say, 20° (0.35 rad) with typical values of the other quantities in Equation (7.9). For $m = 20$ g, $M = 1.0$ kg, $d = 50$ mm and $a = 400$ mm we find that

$$c = G\frac{mM}{d^2}\frac{a}{\theta} = (6.67\times 10^{-11})\frac{(0.02)(1.0)}{(0.05)^2}\frac{0.4}{0.35} \approx 6\times 10^{-10}\ \text{N m rad}^{-1}$$

7.6 Rotation of a rigid body about a fixed axis: moment of inertia

If a rigid body rotates about a fixed axis, such as an axis perpendicular to the page through A in Figure 7.12, each mass element will move in a circle with angular velocity ω about the axis of rotation, as indicated in the figure, where ω is the instantaneous angular velocity of the whole body. Let r_i be the radius of the circle described by the particle of mass m_i; that is, r_i is the perpendicular distance from m_i to the axis of rotation.

The instantaneous (translational) velocity of the particle is directed tangentially to the circle and is given by Equation (2.25), that is $v_i = \omega r_i$. The angular momentum of the ith particle is given by $L_i = m_i v_i r_i$ directed along the axis of rotation as required by the cross product. Thus

$$L_i = m_i r_i(\omega r_i) = \omega m_i {r_i}^2$$

The magnitude of the total angular momentum of the rotating body can be determined by adding algebraically the contributions from all the particles, since the individual angular momenta are all in the same direction (that is, along the axis of rotation). Thus the total angular momentum can be written as

$$L = \sum_i \omega m_i r_i^2 = \omega\left(\sum_i m_i r_i^2\right) \tag{7.10}$$

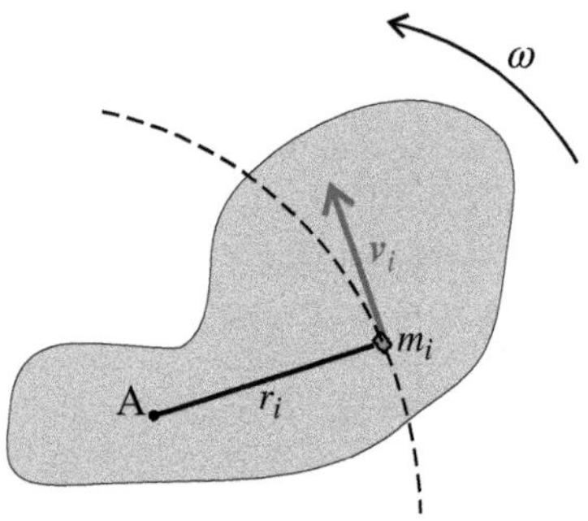

Figure 7.12. A rigid body rotating about an axis perpendicular to the page through A. A particle of mass m_i at a perpendicular distance r_i from the axis will have a speed $v_i = \omega r_i$, where ω is the angular velocity of the rigid body about the rotation axis.

Comparing this result with the definition of linear momentum ($p := mv$) we note that, in changing from linear to rotational motion, ω replaces v and $\sum_i m_i r_i^2$ replaces m; thus the quantity $\sum_i m_i r_i^2$ is the rotational equivalent of mass in the translational case. We define this quantity to be the **moment of inertia** of the body about the axis through A, that is

$$\boxed{I_A := \sum_i m_i r_i^2} \tag{7.11}$$

Equation (7.10), therefore, gives the angular momentum of a rotating body in terms of its angular velocity and its moment of inertia about a rotation axis through A, that is

$$L = I_A \omega$$

Note that the moment of inertia of a body is a characteristic of the mass distribution in the body. The value of the moment of inertia will differ with a different distribution of mass about the same axis, that is with a different set of values of m_i and r_i. In general, the moment of inertia of any given body will also be different if we take a different axis. For example if, as indicated in Figure 7.13, we take an axis of rotation through B at the edge of the body, the mass elements are, on average, further from the axis and consequently the moment of inertia is larger ($I_B > I_A$). It is clear therefore that, for a stated value of a moment of inertia to have any useful meaning, the axis about which it is determined must be specified.

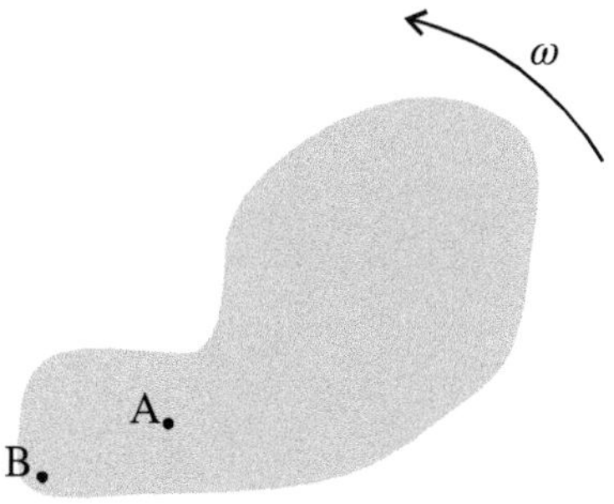

Figure 7.13. The moment of inertia of a rigid body about an axis perpendicular to the page through B is greater than its moment of inertia about a parallel axis through A.

Since we have defined both angular momentum and angular velocity as vector quantities and, in this case, both are directed along the axis of rotation, Equation (7.10) can be written as a vector equation as follows

$$\boldsymbol{L} = I_A \boldsymbol{\omega} \tag{7.12}$$

Comparing this with the definition of momentum as a vector ($\boldsymbol{P} := m\boldsymbol{v}$) we observe a significant analogy between moment of inertia in rotational motion and mass in translational motion. When the axis of rotation is fixed, moment of inertia may be treated as a scalar quantity, like mass in translational motion. In Chapter 3 we noted that mass represents a measure of how a body responds to an applied force, in the sense that it proves more difficult to give a large mass a translational acceleration than it is to give a small mass the same acceleration. In the same sense moment of inertia represents a measure of how a rigid body responds to an applied torque; for example, it requires a larger torque to produce the same angular acceleration in a wheel of large radius than in a wheel of the same mass but of smaller radius.

The SI unit for moment of inertia is kg m^2.

For problems based on the material presented in this section visit up.ucc.ie/7/ and follow the link to the problems.

7.7 Calculation of moments of inertia: the parallel axis theorem

So far we have treated a rigid body as an assemblage of discrete particles. Alternatively we may wish to consider a rigid body as a continuous distribution of matter, in which case the body may be thought of as divided into a very large number of infinitesimal mass elements (Δm in Figure 7.14 is one such element). The summations in Equations (7.10) and (7.11) are then replaced by integrals and the moment of inertia can be calculated by integrating over the whole body, that is $I = \iiint_V r^2 dm$, where r is the perpendicular distance from the element Δm to the axis. This is an example of a **volume integral** (Appendix A.12.4).

The moment of inertia about a symmetry axis of a body which has a relatively simple geometrical shape can be determined without undue difficulty as demonstrated by the following examples.

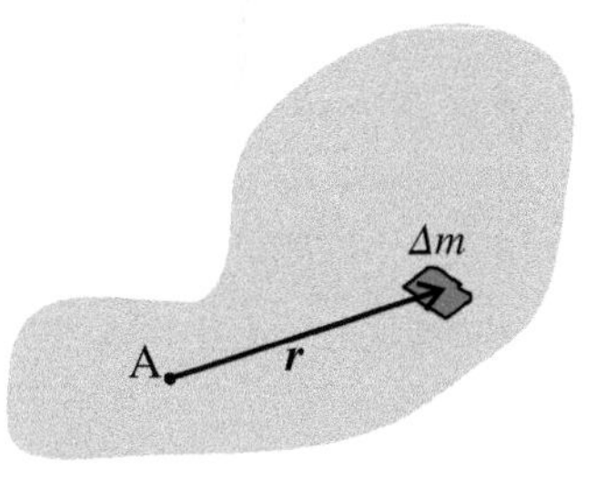

Figure 7.14. For a continuous distribution of matter in a rigid body, the body may be thought of as comprising a very large number of mass elements Δm (as distinct from a large number of discrete particles as considered heretofore). The method of analysis is the same, however, provided that sums over discrete particles are replaced by integrals.

Case 1: Moment of inertia of a solid cylinder about its axis

As a first example we will determine the moment of inertia about the symmetry axis of a uniform solid cylindrical body, which has mass M, radius R and length L. The body can be thought of as being divided up into a large number of cylindrical elements, for example the element of radius between r and $r + \Delta r$ which is shown shaded in Figure 7.15. The mass of this thin-walled cylindrical tube is $[(2\pi r L)\Delta r]\rho$, where ρ is the density of the solid material. The moment of inertia of the whole body about the axis XX′ is

$$I_{XX'} = \iiint_V r^2 dm = \int_0^R r^2 (2\pi r L \rho dr) = 2\pi L \rho \int_0^R r^3 dr = \frac{1}{2}\pi L \rho R^4 = \frac{1}{2}(\pi R^2 L \rho) R^2 = \frac{1}{2} M R^2$$

$$I_{XX'} = \frac{1}{2} M R^2 \tag{7.13}$$

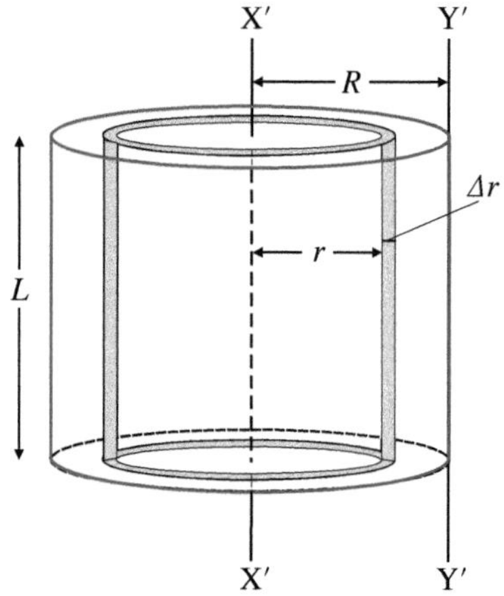

Figure 7.15. Determination of the moments of inertia of a solid cylinder about the axes XX′ and YY′.

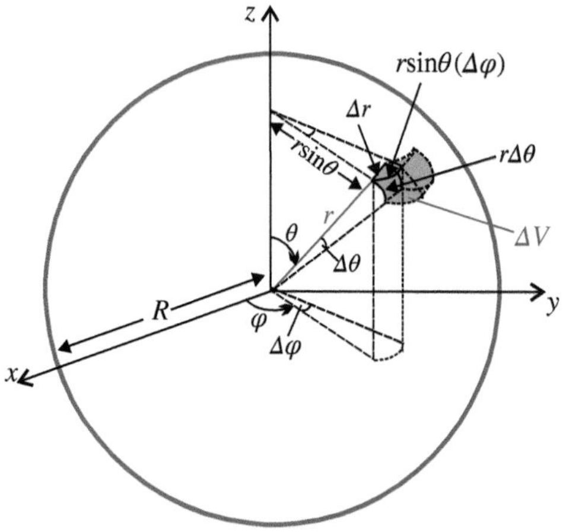

Figure 7.16. Determination of the moment of inertia of a solid sphere about an axis through its centre.

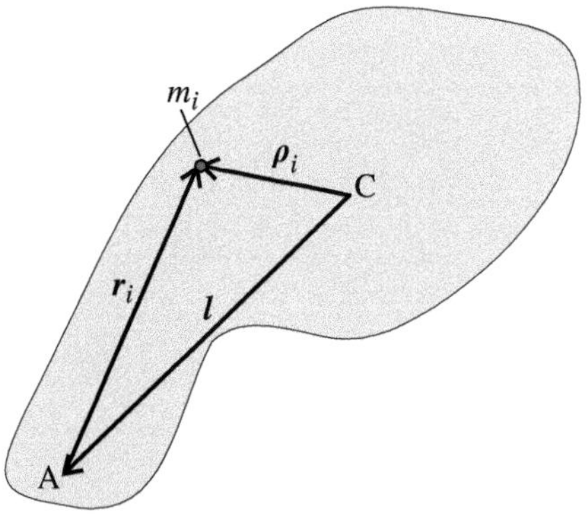

Figure 7.17. Proof of the parallel axis theorem: $\boldsymbol{l}$ is the perpendicular distance between the two parallel axes perpendicular to the page through the points C and A and $\boldsymbol{\rho}_i$ and $\boldsymbol{r}_i$ are the displacements of a particle of mass m_i from the points C and A, respectively.

Case 2: Moment of inertia of a uniform solid sphere about a diameter

Let us now consider the moment of inertia of a solid sphere about an axis through the centre of the sphere. Here we will use for the first time a coordinate system known as *spherical polar co-ordinates* (described in Appendix A.13.3), a system which is more suited to the symmetry of the sphere. Instead of (x, y, z), the variables in this case are (r, θ, φ) as indicated in Figure 7.16 and the volume element is the small cube (shaded in the figure) whose volume is $\Delta V = (\Delta r)(r\Delta\theta)(r\sin\theta\Delta\varphi) = r^2\sin\theta\Delta r\Delta\theta\Delta\varphi$. The mass of the corresponding element is $\Delta m = \rho\Delta V$ and thus the moment of inertia of the sphere about any axis through the centre (say, about the z-axis in Figure 7.16) is

$$I_Z = \iiint_V (r\sin\theta)^2 dm = \iiint (r\sin\theta)^2 \rho r^2 \sin\theta drd\theta d\varphi = \rho\int_0^R r^4 dr \int_0^\pi \sin^3\theta d\theta \int_0^{2\pi} d\varphi$$

Integrating over the sphere (radius $= R$) we see that the integral over r must be evaluated from $r = 0$ to $r = R$, the integral over θ from $\theta = 0$ to $\theta = \pi$ and the integral over φ from $\varphi = 0$ to $\varphi = 2\pi$. The r and φ integrals are simple to evaluate – the value of the r-integral is $\frac{1}{5}R^5$and that of the φ-integral is 2π.

To evaluate the θ-integral we change the variable to $\lambda = \cos\theta$ and write

$$\int_0^\pi \sin^3\theta\, d\theta = -\int_0^\pi \sin^2\theta\, d(\cos\theta) = -\int_0^\pi (1-\cos^2\theta)d(\cos\theta) = \int_{-1}^{+1}(1-\lambda^2)d\lambda$$

$$= 2\int_0^1 d\lambda - 2\int_0^1 \lambda^2 d\lambda = 2 - \frac{2}{3} = \frac{4}{3}$$

and hence $$I_Z = \rho\left(\frac{1}{5}R^5\right)\left(\frac{4}{3}\right)(2\pi) = \frac{4}{3}\pi R^3\rho\frac{2}{5}R^2 = \frac{2}{5}MR^2 \quad \rightarrow \quad I_Z = \frac{2}{5}MR^2$$

Parallel axis theorem

An important and useful relation between the moments of inertia of a rigid body about different but parallel axes can now be derived. The rigid body in Figure 7.17 has its centre of mass at the point C. Consider two axes perpendicular to the plane of the page, one through C and one through an arbitrary point A. Let the perpendicular distances between the two axes be $\boldsymbol{l} = \vec{\mathrm{CA}}$ and let the displacements of a particle of mass m_i be $\boldsymbol{\rho}_i$ and $\boldsymbol{r}_i$ relative to C and A, respectively, that is $\boldsymbol{\rho}_i = \boldsymbol{r}_i + \boldsymbol{l}$.

The moment of inertia of the rigid body about the axis through A is

$$I_A = \sum_i m_i r_i^2 = \sum_i m_i |\boldsymbol{\rho}_i - \boldsymbol{l}|^2 = \sum_i m_i(\rho_i^2 - 2\boldsymbol{\rho}_i\cdot\boldsymbol{l} + l^2)$$

$$= \sum_i m_i\rho_i^2 - 2\left(\sum m_i\boldsymbol{\rho}_i\right)\cdot\boldsymbol{l} + \left(\sum_i m_i\right)l^2$$

As we saw at the end of Section 6.6, the definition of the centre of mass requires that $\sum_i m_i\boldsymbol{\rho}_i = 0$, and hence we obtain

$$I_A = \sum_i m_i\rho_i^2 + \left(\sum_i m_i\right)l^2$$

$$\rightarrow \quad \boxed{I_A = I_C + Ml^2} \tag{7.14}$$

where $M = \sum_i m_i$ is the mass of the body. Equation (7.14) is known as the **parallel axis theorem**.

Note that, once an expression for the moment of inertia about an axis through the centre of mass has been determined, the parallel axis theorem can be used to obtain an expression for the moment of inertia of a rigid body about any axis which is parallel to that axis. For example, the moment of inertia of the cylinder in Figure 7.15 about the axis YY′ along its surface parallel to XX′ is given by

$$I_{YY'} = I_{XX'} + MR^2 = \frac{1}{2}MR^2 + MR^2 = \frac{3}{2}MR^2.$$

Study Worked Examples 7.2 and 7.3

For problems based on the material presented in this section visit up.ucc.ie/7/ and follow the link to the problems.

7.8 Conservation of angular momentum of rigid bodies

When a rigid body is constrained to rotate about a fixed axis, the application of a torque will cause it to rotate about that axis. In addition, as we have seen, the constraint force at the axis provides a reaction which balances any external forces on the body so that the net external force is zero. When an external torque directed along the axis of rotation is applied to a rigid body, the body will rotate about the same axis, either speeding up or slowing down depending on the direction of the torque. In this case, from Equation (7.8),

$$T_{EXT} = \frac{dL}{dt} = \frac{d(I_A\omega)}{dt} = I_A\frac{d\omega}{dt} = I_A\alpha \tag{7.15}$$

where α is the angular acceleration of the rigid body (as defined in Section 2.7).

In the absence of an external torque, we see that $\alpha = \frac{d\omega}{dt} = 0$ and hence ω is constant and the body rotates with constant angular velocity. This is, of course, a consequence of the principle of conservation of angular momentum, derived in Section 6.8, applied to the case of a rigid body. In other words

$$T_{EXT} = 0 \quad \Rightarrow \quad L = I_A\omega = \text{constant}$$

Let us consider a practical example of conservation of angular momentum. A skater who is spinning on ice and who wishes to increase the rate at which she is spinning – her angular velocity – uses the technique of drawing her arms and legs closer to her body, as illustrated in Figure 7.18. In this way she reduces her moment of inertia $I = \sum_i m_i r_i^2$ about her rotation axis and thus $I_2 < I_1$. Applying the principle of conservation of angular momentum,

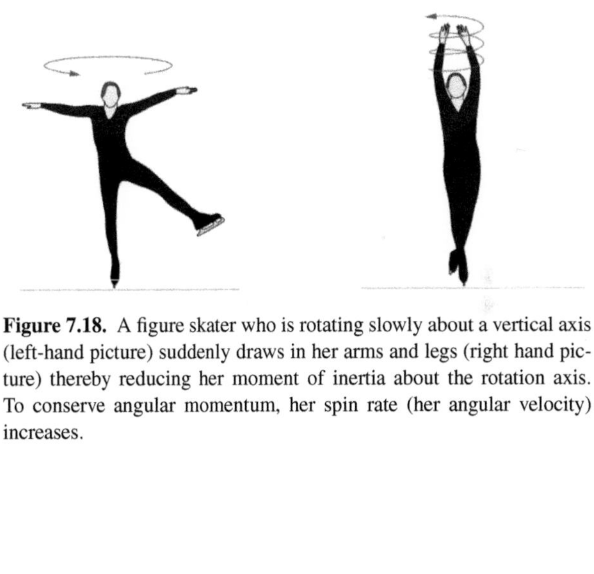

Figure 7.18. A figure skater who is rotating slowly about a vertical axis (left-hand picture) suddenly draws in her arms and legs (right hand picture) thereby reducing her moment of inertia about the rotation axis. To conserve angular momentum, her spin rate (her angular velocity) increases.

$$L = I_1\omega_1 = I_2\omega_2$$

so that if $I_2 < I_1$ then $\omega_2 > \omega_1$. The skater, therefore, has used conservation of angular momentum to increase her rate of spinning.

Finally we note that, since we have defined angular acceleration as a vector quantity (Section 4.11), we can write Equation (7.15) in vector form as follows

$$\boxed{\boldsymbol{T}_{EXT} = I_A\boldsymbol{\alpha}}$$

provided that the rotation is about a fixed axis.

Comparing this with Newton's second law, $\boldsymbol{F} = m\boldsymbol{a}$, we observe again the formal analogy between rotational motion and translational motion; torque and moment of inertia in rotational motion play the roles of force and mass, respectively, in translational motion.

For problems based on the material presented in this section visit up.ucc.ie/7/ and follow the link to the problems.

7.9 Conservation of mechanical energy in rigid body systems

We now derive an expression for the kinetic energy of a body due to its rotational motion. As in Section 7.3, we consider a rigid body as being divided into mass elements (m_i being the mass of the ith element). The total kinetic energy of the system is the sum of the translational kinetic energy of all the particles in the body. As before, when the body rotates the element of mass m_i within the body moves in a circle of radius r_i about the axis of rotation (perpendicular to the page through A in Figure 7.19).

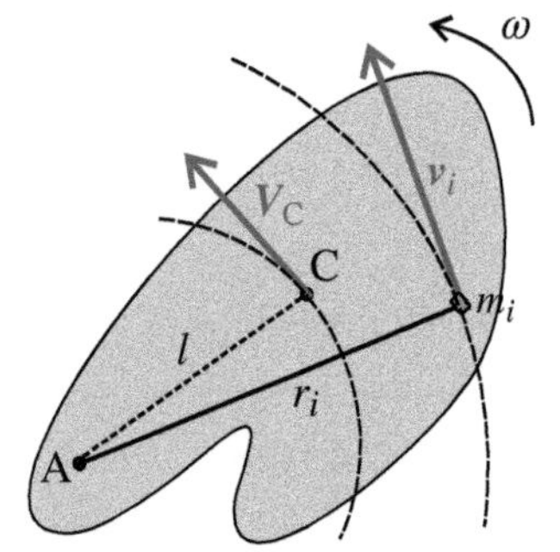

Figure 7.19. A rigid body rotating about an axis perpendicular to the page through A. A particle at a perpendicular distance r_i from the axis will have a speed $v_i = \omega r_i$ and the centre of mass will have speed $V_C = \omega l$, where ω is the (instantaneous) angular velocity of the rigid body about the rotation axis.

At any instant the magnitude of the velocity of the mass element (directed at a tangent to the circle) is given by $v_i = r_i\omega$ and therefore the mass m_i has kinetic energy given by

$$K_i = \frac{1}{2}m_i v_i^2 = \frac{1}{2}m_i r_i^2\omega^2.$$

Thus the total **rotational kinetic energy** of the body can be written as

$$K = \sum_i \frac{1}{2}m_i r_i^2\omega^2 = \frac{1}{2}\omega^2\sum_i m_i r_i^2 = \frac{1}{2}\left(\sum_i m_i r_i^2\right)\omega^2 = \frac{1}{2}I_A\omega^2 \quad (7.16)$$

or, in terms of the angular momentum ($L = I_A\omega$) of the rotating body, as $K = \dfrac{L^2}{2I_A}$.

Note once again the formal analogy between the expressions for rotational kinetic energy and the corresponding expressions for translational kinetic energy (Section 5.7), namely

$$K = \frac{1}{2}mv^2 \quad \text{and} \quad K = \frac{p^2}{2m}, \text{respectively.}$$

Applying the parallel axis theorem to Figure 7.19, we see that

$$K = \frac{1}{2}I_A\omega^2 = \frac{1}{2}(I_C + Ml^2)\omega^2 = \frac{1}{2}I_C\omega^2 + \frac{1}{2}M(l^2\omega^2)$$

where I_C is the moment of inertia about an axis perpendicular to the page through the centre of mass C and l is the perpendicular distance between the two axes. The instantaneous velocity of the centre of mass is given by $V_C = \omega l$ and thus

$$\boxed{K = \frac{1}{2}I_C\omega^2 + \frac{1}{2}MV_C^2}$$

In applying the principle of conservation of mechanical energy (Section 6.3) to rotating rigid bodies, we can consider the kinetic energy as comprising translational and rotational parts. Thus, in a conservative system, the sum of the translational and rotational kinetic energies and the potential energy is constant. For example, if a rigid body of mass M is rotating with angular velocity ω about a fixed axis through its centre of mass, the total energy is given by

$$E = \frac{1}{2}MV_C^2 + \frac{1}{2}I\omega^2 + U(\boldsymbol{R}_C) = \mathit{constant} \quad (7.17)$$

where $\boldsymbol{R}_C$ and $\boldsymbol{V}_C$ are the displacement and velocity, respectively, of the centre of mass of the body, I is the moment of inertia about the axis of rotation and $U(\boldsymbol{R}_C)$ is the potential energy of the rigid body when its centre of mass has displacement $\boldsymbol{R}_C$.

To write down the Hamiltonian function (Section 5.7) for a system with rotation, we must express the kinetic energy in terms of the momentum and the angular momentum, that is

$$H(\boldsymbol{p}, \boldsymbol{L}, \boldsymbol{r}) = \frac{p^2}{2M} + \frac{L^2}{2I} + U(\boldsymbol{r})$$

Example 1: *Wheel rolling on an inclined plane*

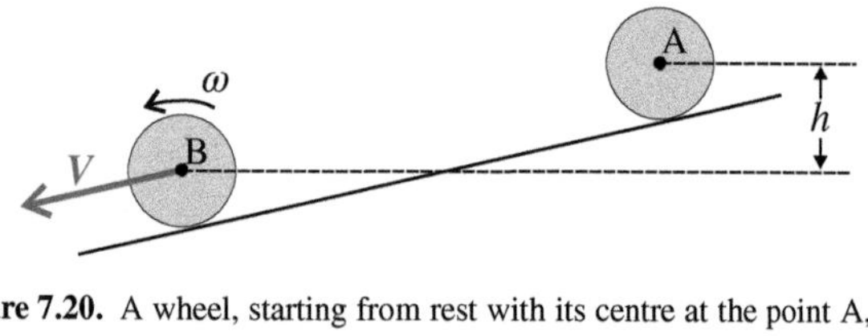

Figure 7.20. A wheel, starting from rest with its centre at the point A, rolls down an inclined plane. After moving through a vertical distance h, its angular velocity is ω and the translational velocity of the centre of the wheel is V.

Consider a wheel which is rolling down an inclined plane (Figure 7.20). Some friction is necessary to prevent slipping but negligible energy is dissipated by friction in this case.

If the wheel is released from rest at the point A (where $V_A = 0$, $\omega_A = 0$) and achieves translational and rotational velocities of V and ω, respectively, at the point B, application of the principle of conservation of energy, Equation (7.17), gives

$$0 + 0 + Mgh = +\frac{1}{2}MV^2 + \frac{1}{2}I\omega^2 + 0 \quad (7.18)$$

where h is the change in vertical displacement of the centre of mass of the wheel and I is the moment of inertia of the wheel about its symmetry axis. The loss in potential energy, therefore, is converted into both translational and rotational kinetic energy.

It can be seen from Figure 7.21 that the instantaneous speed of the centre of mass V and the angular velocity ω are related by Equation (2.25), namely $V = \omega R$, where R is the radius of the wheel. This is because, as the centre of mass moves forward a distance Δs, a point on the circumference of the wheel in contact with the plane will advance by an element Δs of arc, and hence

$$\Delta s = R(\Delta\theta) \quad \text{and} \quad V = \frac{ds}{dt} = R\frac{d\theta}{dt} = \omega R.$$

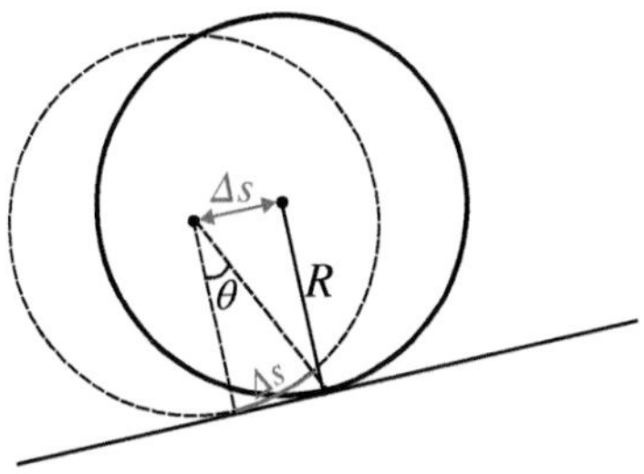

Figure 7.21. As the centre of the wheel advances by a small distance Δs, the point on the circumference of the wheel initially in contact with the plane will be an arc length Δs further along the edge of the wheel from the point which is now in contact with the plane.

Thus Equation (7.18) becomes

$$Mgh = \frac{1}{2}MV^2 + \frac{1}{2}I\frac{V^2}{R^2} = \frac{1}{2}\left(M + \frac{I}{R^2}\right)V^2$$

For the purpose of illustration, let us assume that the wheel is a solid cylinder of mass M and radius R. In this case, using Equation (7.13), we can write $I = \frac{1}{2}MR^2$ and hence

$$Mgh = \frac{1}{2}MV^2 + \frac{1}{4}MV^2 = \frac{3}{4}MV^2.$$

In the case of a solid cylinder rolling down an inclined plane, therefore, the potential energy is converted to translational and rotational kinetic energy in the ratio 2:1 and the translational velocity after descending a perpendicular height h is $\sqrt{\frac{4}{3}gh}$. Readers can check for themselves that, if there was no friction and the cylinder slid down the plane without rotation, the velocity would be $\sqrt{2gh}$.

Study Worked Example 7.4

Example 2: *The physical (or compound) pendulum*

If a rigid body is suspended so that it is free to rotate about a horizontal axis and is then displaced from equilibrium and released, it will oscillate back and forth under the influence of the gravitational field. In this case the rigid body behaves as a pendulum; we will now investigate which factors determine the period of oscillation of such a pendulum.

The rigid body shown in Figure 7.22 is free to rotate about a horizontal axis through the point A. If the perpendicular distance from its centre of mass C to the rotation axis is l, the point C will move in a vertical circle of radius l. At any arbitrary instant during the motion the total energy is the sum of the rotational kinetic energy about the axis through A and the gravitational potential energy. For the purpose of determining the gravitational potential energy, we may consider the mass of the rigid body as being concentrated at the centre of mass, that is $U = mgy$, where y is the vertical displacement of C above the lowest point in the motion (which we adopt as the point of zero potential energy). Thus, in this case, the principle of conservation of energy (Equation (7.17)) gives

$$\frac{1}{2}I_A\omega^2 + mgy = E = constant$$

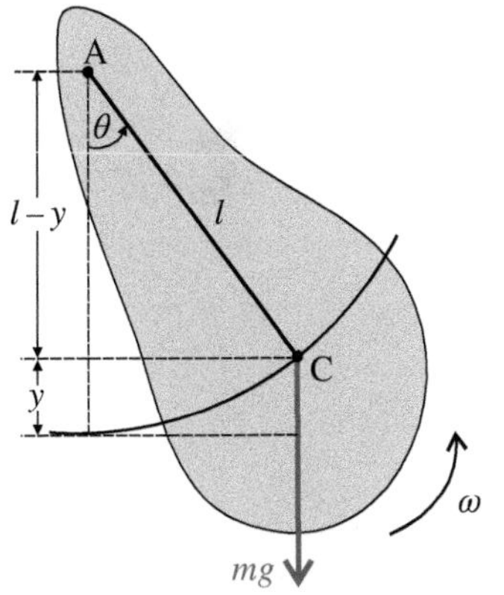

Figure 7.22. The physical pendulum. At some arbitrary instant in its motion, the perpendicular distance of the centre of mass (C) from the rotation axis (perpendicular to the page through A) makes an angle θ with the vertical.

At the instant when the line AC makes an angle θ with the vertical, we see from Figure 7.22 that $\cos\theta = \frac{l-y}{l}$ and hence $y = l(1 - \cos\theta)$. In terms of the angle θ, therefore, the statement that energy is conserved can be written as

$$\frac{1}{2}I_A\omega^2 + mgl(1 - \cos\theta) = E = constant$$

Differentiating with respect to time we get

$$\frac{1}{2}I_A\left(2\omega\frac{d\omega}{dt}\right) + mgl\sin\theta\frac{d\theta}{dt} = 0$$

Now $\omega = \frac{d\theta}{dt}$ and hence the equation governing the motion of the pendulum is

$$\frac{d^2\theta}{dt^2} + \frac{mgl}{I_A}\sin\theta = 0$$

This differential equation is similar to the non-linear Equation (4.10) which governs the behaviour of a simple pendulum with arbitrary amplitude. As we saw in Section 4.10, if the motion is confined to small angles ($\sin\theta \approx \theta$) the equation takes on a more familiar appearance, namely

$$\frac{d^2\theta}{dt^2} = -\frac{mgl}{I_A}\theta$$

which shows that, provided the amplitude of the oscillations is small, the pendulum oscillates with simple harmonic motion. The period of the simple harmonic oscillations is given by

$$T = 2\pi\sqrt{\frac{I_A}{mgl}} \tag{7.19}$$

Note that, for a simple pendulum of length l, $I_A = ml^2$ so that, as expected, Equation (7.19) reduces to Equation (4.11), that is

$$T = 2\pi\sqrt{\frac{ml^2}{mgl}} = 2\pi\sqrt{\frac{l}{g}}.$$

Study Worked Example 7.5

For problems based on the material presented in this section visit up.ucc.ie/7/ *and follow the link to the problems.*

7.10 Work done by a torque: torsional oscillations: rotational power

We will now determine the work done by a torque $\boldsymbol{T}$ in rotating a rigid body through a small angle $\Delta\theta$ about a fixed axis. Consider the torque as being due to a couple of equal and opposite forces $\boldsymbol{F}$ and $-\boldsymbol{F}$ as shown in Figure 7.23. Each point on the body will move in a circle about the axis which is perpendicular to the plane of the diagram through A. Let the forces make angles of ϕ_1 and ϕ_2, respectively, with the tangents to the circle at the points of application of each force.

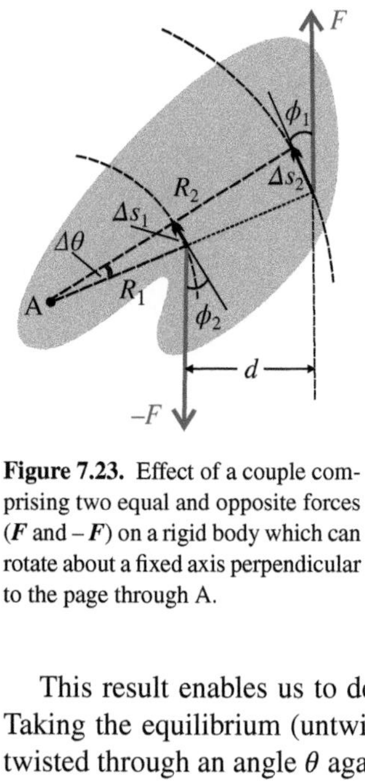

Figure 7.23. Effect of a couple comprising two equal and opposite forces ($\boldsymbol{F}$ and $-\boldsymbol{F}$) on a rigid body which can rotate about a fixed axis perpendicular to the page through A.

The mechanical work done by the force $\boldsymbol{F}$ in moving a point on the circumference of the circle of radius R_2 through a displacement $\Delta\boldsymbol{s}_2$ is $\boldsymbol{F}\cdot\Delta\boldsymbol{s}_2$. From the definition of the radian (Appendix A.3.1) we see that $|\Delta\boldsymbol{s}_2| = R_2\Delta\theta$, where $\Delta\theta$ is the angle indicated in the figure. Thus the mechanical work done by the couple in rotating the rigid body through $\Delta\theta$ is given by

$$\Delta W = \boldsymbol{F}\cdot\Delta\boldsymbol{s}_2 - \boldsymbol{F}\cdot\Delta\boldsymbol{s}_1 = |\boldsymbol{F}|\cos\phi_2 R_2\Delta\theta - |\boldsymbol{F}|\cos\phi_1 R_1\Delta\theta$$

For small $\Delta\theta$, $\phi_1 \approx \phi_2 = \phi$, so we can write

$$\Delta W = |\boldsymbol{F}|\cos\phi(R_2 - R_1)\Delta\theta = |\boldsymbol{F}|d\Delta\theta = |\boldsymbol{T}|\Delta\theta, \tag{7.20}$$

where d is the perpendicular distance between $\boldsymbol{F}$ and $-\boldsymbol{F}$ and $\boldsymbol{T}$ is the torque about A.

More generally, when $\boldsymbol{F}$ and $-\boldsymbol{F}$ are not in the plane perpendicular to the rotation axis, then $\boldsymbol{T}$ is not parallel to $\Delta\boldsymbol{\theta}$, in which case we write Equation (7.20) as a scalar product $\Delta W = \boldsymbol{T}\cdot\Delta\boldsymbol{\theta}$ where $\Delta\boldsymbol{\theta}$ is a vector quantity as defined in Section 4.11. The total work done in a finite rotation between angles θ_1 and θ_2 is given by

$$W = \int_{\theta_1}^{\theta_2} \boldsymbol{T}\cdot d\boldsymbol{\theta}$$

which is a rotational analogue of Equation (5.12).

This result enables us to derive an expression for the potential energy of a rod or wire under torsion as discussed in Section 7.5. Taking the equilibrium (untwisted) position to be the zero of potential energy, the potential energy in a rod or wire, which has been twisted through an angle θ against a restoring torque exerted by the rod or wire, is given by

$$U(\theta) = -\int_0^\theta T d\vartheta = -\int_0^\theta (-c\vartheta)d\vartheta = +c\int_0^\theta \vartheta d\vartheta = \frac{1}{2}c\theta^2$$

where c is the torsion constant (Section 7.5). This result is directly analogous to the expression $U(x) = \frac{1}{2}kx^2$ for a helical spring (Case 2, Section 3.7).

Torsional oscillations

Finally let us consider the oscillations which are observed when a rigid body is suspended from a vertical length of wire or rod (Figure 7.24), given a small twist in a horizontal plane and then released from rest.

When the angular displacement is θ, the restoring torque is $T = -c\theta$ and hence from Equation (7.12)

$$T = \frac{dL}{dt} = \frac{d}{dt}(I\omega) = I\frac{d\omega}{dt} = I\alpha = I\frac{d^2\theta}{dt^2} = -c\theta$$

where I is the moment of inertia of the rigid body about the axis of the wire or rod.

Thus

$$\frac{d^2\theta}{dt^2} = -\frac{c}{I}\theta$$

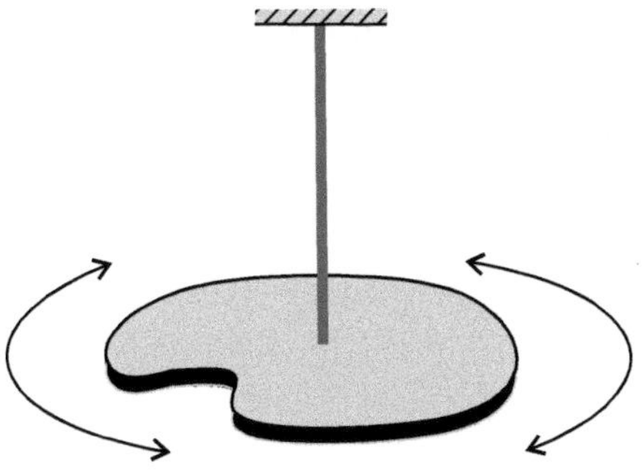

Figure 7.24. Torsional oscillations of a rigid body suspended by an elastic wire which is attached to a fixed point.

which we recognise as the equation of motion of a simple harmonic oscillator; that is, it is Equation (3.5) with $\gamma = \frac{c}{I}$. Thus, we see that the rigid body executes simple harmonic oscillations about the axis of the wire with a period given by $T = 2\pi\sqrt{\frac{I}{c}}$.

Note that this result could also have been obtained from energy considerations as follows. In the absence of damping forces, mechanical energy is conserved, that is

$$\frac{1}{2}I\omega^2 + \frac{1}{2}c\theta^2 = E = constant.$$

Differentiating with respect to time we get

$$\frac{1}{2}I\left(2\omega\frac{d\omega}{dt}\right) + \frac{1}{2}c\left(2\theta\frac{d\theta}{dt}\right) = 0 \quad \rightarrow \quad I\omega\frac{d^2\theta}{dt^2} + c\theta\omega = 0 \quad \rightarrow \quad I\frac{d^2\theta}{dt^2} = -c\theta$$

If the body has a regular shape so that its moment of inertia about the rotation axis can be determined analytically, an experimental measurement of the periodic time of torsional oscillations can be used to determine the torsion constant of a particular suspension, for example that used in the Cavendish experiment (Section 7.5) although the period of the oscillations can be very long in the latter case.

While the above discussion has referred specifically to torques arising from torsional forces, note that the analysis holds whatever the source of the restoring torque. Provided $|\boldsymbol{T}| \propto -\theta$, the expression for the period of oscillation is

$$2\pi\sqrt{\frac{\text{moment of inertia about the axis of rotation}}{\text{restoring torque per unit angular displacement}}}$$

For example, looking back at the discussion of the physical pendulum in the previous section, we see from Figure 7.22 that the restoring torque in that case is $Mg(l\sin\theta)$. For small angles this becomes $Mgl\theta$ and hence the restoring torque per unit angular displacement is Mgl, from which the expression for the period of the physical pendulum, Equation (7.19), follows immediately.

Power in rotational motion

Using Equation (7.20), we can also derive an equation for the power generated by a rotating body. Over a small time interval the torque $\boldsymbol{T}$ can be considered to be constant and so that we can write

$$P = \frac{dW}{dt} = |\boldsymbol{T}|\frac{d\theta}{dt} = |\boldsymbol{T}|\omega$$

or more generally, if $\boldsymbol{T}$ is not parallel to $\boldsymbol{\omega}$,

$$P = \boldsymbol{T}\cdot\boldsymbol{\omega}$$

which can be compared with the expression for the power generated in translational motion $P = \boldsymbol{F}\cdot\boldsymbol{v}$, derived in Section 5.8.

Study Worked Example 7.6

For problems based on the material presented in this section visit up.ucc.ie/7/ *and follow the link to the problems.*

7.11 Gyroscopic motion

In this book we will consider mainly cases of rotational motion in which the axis of rotation is fixed and, as we have seen, in these circumstances, quantities such as angular velocity, angular acceleration, angular momentum, torque etc. can be treated as algebraic quantities.

A full vector description is essential, however, whenever the direction of the axis of rotation changes during rotational motion. Following the procedures outlined in Sections 7.6 and 7.8, we can write vector forms of Equations (7.12) and (7.15) as follows.

$$\boldsymbol{L} = I_{\mathrm{A}}\boldsymbol{\omega} \quad \text{and} \quad \boldsymbol{T}_{\mathrm{EXT}} = I_{\mathrm{A}}\boldsymbol{\alpha}$$

We need to be careful in our use of these equations in this case, since I_{A} is not necessarily a scalar – it can be treated as a scalar, however, for rotations about a symmetry axis.

As an illustration of the value of the general vector description of rotational motion, we will consider an important practical example in which the direction of rotation changes, — *gyroscopic motion*. Uniform circular motion (Section 4.8) is an example of a situation in which there is an acceleration (centripetal acceleration) because the velocity of the point executing the motion is continually changing in direction while maintaining a constant magnitude. We shall now consider an analogous situation in rotational motion in which there is an angular acceleration due a change in the direction of angular velocity while the magnitude of the angular velocity remains constant.

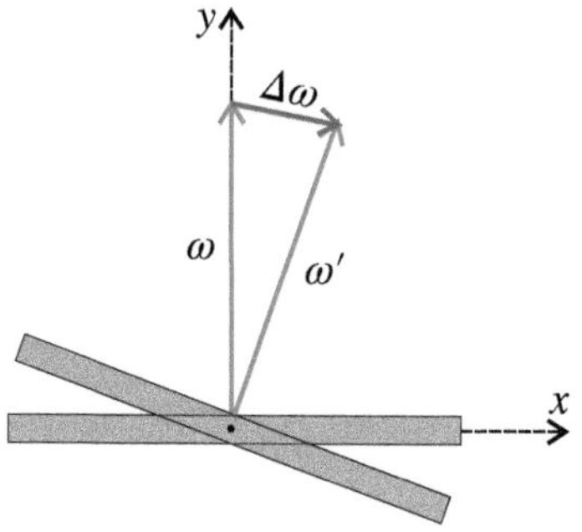

Figure 7.25. A disc, initially spinning about the y-axis, has a torque applied to it about an axis perpendicular to the page. As explained in the text, the net effect is a rotation about the x-axis.

Consider a disc which is spinning in the x-z plane with constant angular velocity $\boldsymbol{\omega}$ in the y-direction, as illustrated in Figure 7.25. A torque is applied to the disc about the z-axis (in the direction perpendicular to the page) causing $\boldsymbol{\omega}$ to change to $\boldsymbol{\omega}'$, a change in the direction of the angular velocity but not in its magnitude. Thus

$$|\boldsymbol{\omega}| = |\boldsymbol{\omega}'|$$

The change in $\boldsymbol{\omega}$ is represented by the vector $\boldsymbol{\Delta\omega}$ such that

$$\boldsymbol{\omega} + \boldsymbol{\Delta\omega} = \boldsymbol{\omega}'$$

Referring to Figure 7.25, we see that the resulting change in $\boldsymbol{\omega}$, namely $\boldsymbol{\Delta\omega}$, acts along the x-axis for small $\boldsymbol{\Delta\omega}$. Thus there is an angular acceleration along the x-axis so that the plane of the disc rotates about this axis. The end result of the application of a torque about the z-axis is therefore a rotation about the x-axis, an axis which is perpendicular to both the spinning axis and the axis about which the torque is applied. This effect can be demonstrated by spinning a wheel about a horizontal axis, the y-direction in Figure 7.26, and then attempting to turn the spinning wheel to the right (about the z-axis). The wheel in fact turns about the x-axis — it appears to have a 'will of its own', choosing to turn about an axis other than the axis about which the torque is being applied.

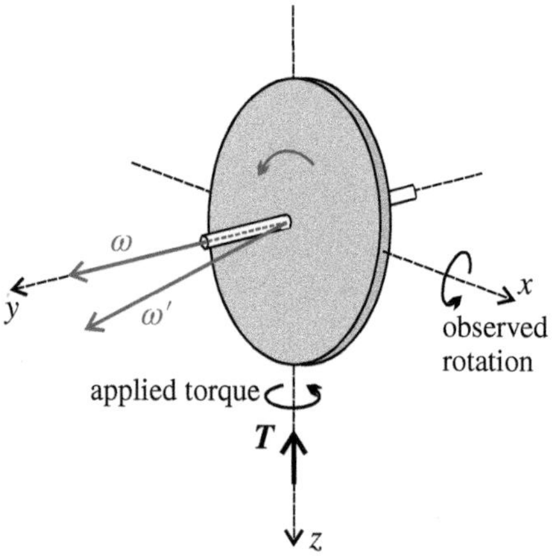

Figure 7.26. The effect of the applied torque on a wheel is to change its angular velocity from ω to ω', the net effect being to introduce a torque $\Delta\omega$ directed along the x-axis.

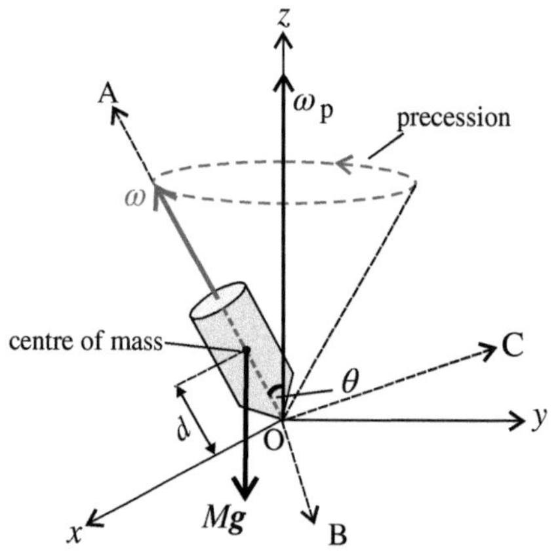

Figure 7.27. Precession of a spinning top whose axis makes an angle θ with the vertical. The constant torque due to its weight causes the top to precess about a vertical axis with constant angular velocity.

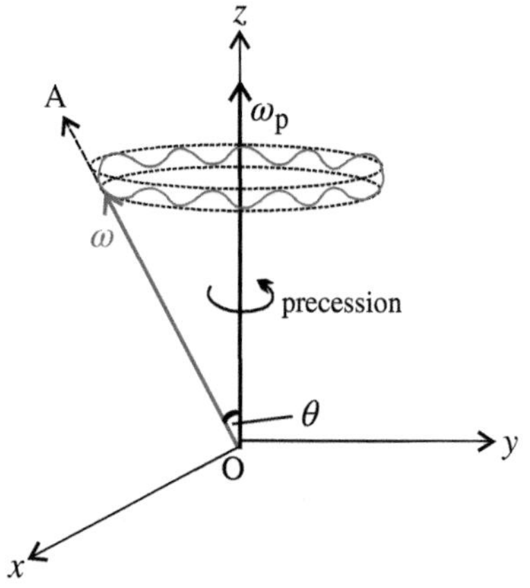

Figure 7.28. In addition to the precessional motion, the top undergoes an oscillatory motion called nutation.

When a steady torque is applied to a spinning object the result is that the spin axis rotates around a fixed axis, a motion known as **precession**. The motion of a toy spinning top, illustrated in Figure 7.27, provides an example of precessional motion. In this figure the z-axis is vertical and the top is spinning with angular velocity $\boldsymbol{\omega}$ about an axis OA, which makes an angle θ with the z-axis.

The top experiences a torque $|\boldsymbol{T}| = Mgd\sin\theta$, where Mg is its weight and d is the distance from its centre of mass to O. If the top were not spinning, this torque, which is directed along an axis OB in the x-y plane of Figure 7.27, would cause it to fall over. When the top is spinning as discussed above, however, the torque causes the spinning axis OA to rotate about the axis OC, which is perpendicular to both OA and OB, leading to a precession of OA about the z-axis, as illustrated in Figure 7.27. In the web subsection up.ucc.ie/7/11/1 the precessional angular velocity is shown to be given by $\omega_p = \frac{Mgd}{I\omega}$, where I is the moment of inertia of the top about the axis OA. Further, more detailed, analysis shows that the angle θ does not remain constant during precession, but oscillates between fixed maximum and minimum values, as illustrated in Figure 7.28. This oscillatory motion of the spin axis OA is known as **nutation**.

The *gyroscope* comprises a rotating disc which is supported in such a way that it is able to rotate freely about any direction in space – it has all three rotational degrees of freedom. This form of support means that any external torques which act on the support of the gyroscope cannot be transmitted to the disc. The disc therefore continues to rotate about the same axis in space; its axis of rotation always points in the same direction. The ability of a gyroscope to maintain a fixed axis in space is used in navigational instruments for aircraft, satellites, and ships (for example, the Hubble Space Telescope had six gyroscopes on board for navigation and sighting purposes).

7.12 Summary: connection between rotational and translational motions

Throughout this chapter attention has been drawn to the strong correspondence which exists between the formalisms of translational and rotational mechanics. Awareness of this connection can make the study of rotational systems much easier for anyone with a good understanding of particle dynamics. The formal connection is summarised below.

rotational mechanics		**translational mechanics**	
angular displacement:	θ	displacement:	$\boldsymbol{r}$
angular velocity:	$\boldsymbol{\omega} = \frac{d\boldsymbol{\theta}}{dt}$	velocity:	$\boldsymbol{v} = \frac{d\boldsymbol{r}}{dt}$
angular acceleration:	$\boldsymbol{\alpha} = \frac{d\boldsymbol{\omega}}{dt} = \frac{d^2\boldsymbol{\theta}}{dt^2}$	acceleration:	$\boldsymbol{a} = \frac{d\boldsymbol{v}}{dt} = \frac{d^2\boldsymbol{r}}{dt^2}$
moment of inertia:	$I\ (:= \Sigma_i m_i r_i^2)$	mass:	m
torque:	$\boldsymbol{T}\ (:= \boldsymbol{r} \times \boldsymbol{F})$	force:	$\boldsymbol{F}$
angular momentum:	$\boldsymbol{L} = I\boldsymbol{\omega}\ (:= \boldsymbol{r} \times \boldsymbol{p})$	momentum:	$\boldsymbol{p} = m\boldsymbol{v}$
equations of motion:	$\boldsymbol{T} = I\boldsymbol{\alpha} = \frac{d\boldsymbol{L}}{dt}$		$\boldsymbol{F} = m\boldsymbol{a} = \frac{d\boldsymbol{p}}{dt}$
conservation laws:	$\boldsymbol{T} = 0 \Rightarrow \boldsymbol{L} = const$		$\boldsymbol{F} = 0 \Rightarrow \boldsymbol{P} = const$
work done:	$W = \int \boldsymbol{T} \cdot d\boldsymbol{\theta}$		$W = \int \boldsymbol{F} \cdot d\boldsymbol{s}$
kinetic energy:	$K = \frac{1}{2}I\omega^2$		$K = \frac{1}{2}mv^2$
potential energy:	$U = \frac{1}{2}c\theta^2$		$U = \frac{1}{2}kx^2$

For problems based on the material presented in this section visit up.ucc.ie/7/ and follow the link to the problems.

WORKED EXAMPLES

Worked Example 7.1: A uniform ladder 5 m long and of mass 20 kg is placed against a smooth vertical wall so that it makes an angle of 60° with the horizontal (Figure 7.29). If the coefficient of static friction between the bottom of the ladder and the (horizontal) surface of the ground is 0.30, how far up the ladder can a 80 kg man climb before the ladder starts to slip?

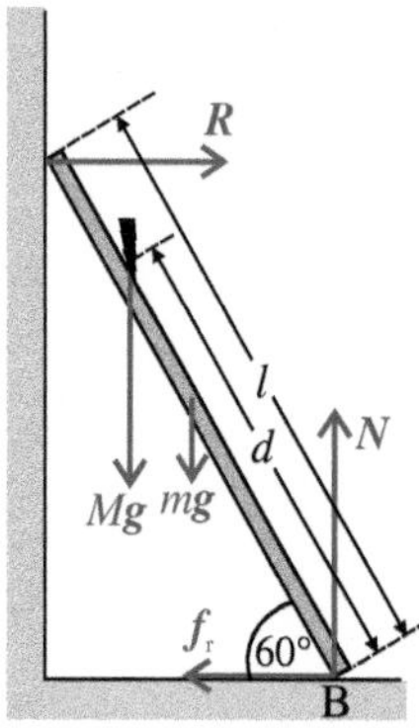

Figure 7.29. Worked Example 7.1 (similar to Figure 7.3).

Let the man (mass = M) have climbed a distance d along the ladder. The weight (mg) of the ladder acts through the mid point. $\boldsymbol{R}$ and $\boldsymbol{N}$ are, respectively, the normal reactions of the wall and ground. If the ladder is just on the point of slipping, the force due to friction is given by $|\boldsymbol{f}_r| = \mu_s|\boldsymbol{N}|$

Forces in x-direction (horizontal): $|\boldsymbol{R}| - |\boldsymbol{f}_r| = 0 \quad \rightarrow \quad |\boldsymbol{R}| = \mu_s|\boldsymbol{N}|$

Forces in y-direction (vertical): $|\boldsymbol{N}| - |\boldsymbol{Mg}| - |\boldsymbol{mg}| = 0 \quad \rightarrow \quad |\boldsymbol{R}| = \mu_s(M+m)g$

Torques about B: $-|\boldsymbol{R}|l\sin\theta + Mgd\cos\theta + \frac{1}{2}mgl\cos\theta = 0$

$$\text{Thus } \mu_s(M+m)gl\sin\theta = Mgd\cos\theta + \frac{1}{2}mgl\cos\theta$$

$$\rightarrow d = \mu_s\left(1+\frac{m}{M}\right)l\tan\theta - \frac{1}{2}\frac{m}{M}l$$

$$d = (0.3)\left(1+\frac{20}{80}\right)(5)\tan(60^\circ) - \frac{1}{2}\frac{20}{80}(5) = 2.6\text{ m}$$

Worked Example 7.2: Derive an expression for the moment of inertia of a long thin uniform rod of mass M and length l (a) about a perpendicular axis through its centre of mass and (b) about an axis through one end which is parallel to the axis in (a).

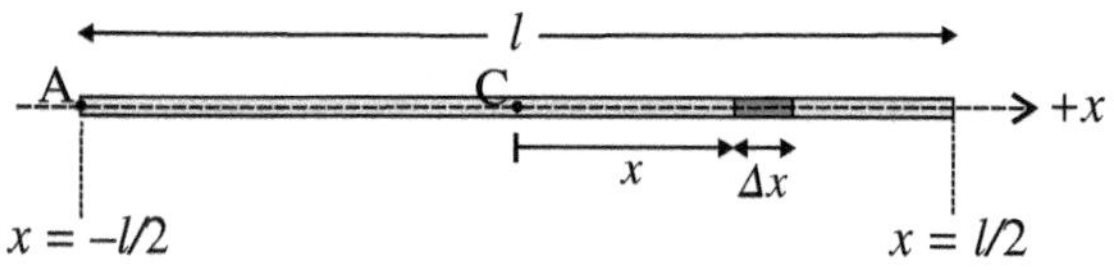

Figure 7.30. Worked Example 7.2.

If μ is mass per unit length of rod, the mass of an element $dm = \mu dx$.
Referring to Figure 7.30,

(a) $$I_C = \int x^2 dm = \int x^2 \mu dx = \mu\int x^2 dx$$

$$= \mu\int_{-\frac{l}{2}}^{\frac{l}{2}} x^2 dx = \mu\left[\frac{x^3}{3}\right]_{-\frac{l}{2}}^{\frac{l}{2}} = \mu\frac{l^3}{12} = \frac{1}{12}Ml^2$$

(b) Using the parallel axis theorem (Section 7.7), $I_A = I_C + M\left(\frac{l}{2}\right)^2 = \frac{1}{12}Ml^2 + \frac{1}{4}Ml^2 = \frac{1}{3}Ml^2$

Worked Example 7.3: Prove that the moment of inertia of the rectangular block of mass M shown in Figure 7.31 about the axis AA′ (a perpendicular axis through its centre of mass), is $\frac{1}{12}M(L^2+B^2)$

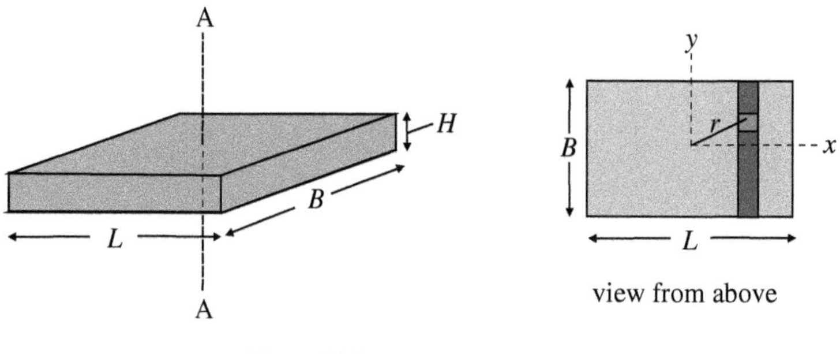

Figure 7.31. Worked Example 7.3.

Let H = dimension of block in direction of axis AA′ and let ρ = density of material in block

Moment of inertia about axis AA′ $= I = \int r^2 dm$, where $\Delta m = (\Delta x)(\Delta y)H\rho$

Thus

$$I = \rho H \iint (x^2 + y^2)dxdy$$

$$= \rho H \int_{-L/2}^{+L/2} dx \int_{-B/2}^{+B/2} (x^2 + y^2)dy = \rho H \int_{-L/2}^{+L/2} \left[Bx^2 + \frac{1}{12}B^3\right] dx$$

$$= \rho H \left[\frac{1}{12}BL^3 + \frac{1}{12}B^3L\right] = \frac{1}{12}\rho HBL[L^2 + B^2]$$

$$= \frac{1}{12}M[L^2 + B^2]$$

Worked Example 7.4: As illustrated in Figure 7.32, a uniform sphere of mass m and radius a rolls back and forth without slipping on a concave spherical surface of radius of curvature R so that it executes small amplitude oscillations in a vertical plane (Figure 7.32(a)). Show that these oscillations are approximately simple harmonic and determine an expression for the period of the motion.

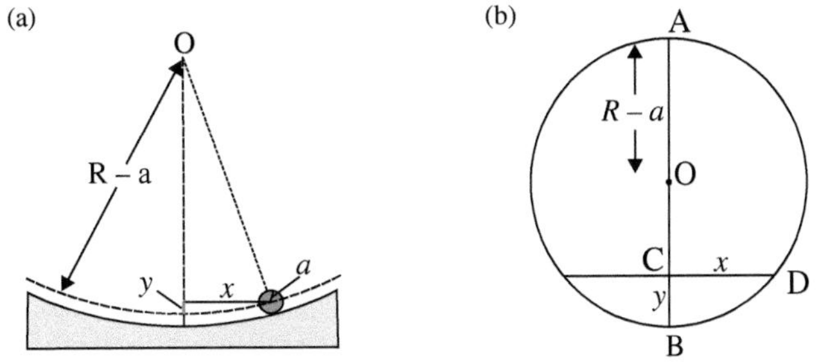

Figure 7.32. Worked Example 7.4.

The centre of mass of the sphere moves in a vertical circle of radius $(R - a)$. Applying the geometrical theorem, Appendix A.2.4, to Figure 7.32(b) we see that $(AC)(CB) = CD^2$ and hence

$$[2(R-a) - y][y] = x^2 \quad \rightarrow \quad 2(R-a)y - y^2 = x^2 \quad \Rightarrow \quad 2(R-a)y = x^2, \quad \text{for } y \ll R.$$

Applying conservation of energy, as discussed in Section 7.9

$$E = \frac{1}{2}mv^2 + \frac{1}{2}I\omega^2 + mgy = constant$$

where the velocity of the centre of mass of the sphere is given by $v = a\omega$.

Differentiating with respect to time we obtain

$$mv\frac{dv}{dt} + I\omega\frac{d\omega}{dt} + mg\frac{dy}{dt} = mv\frac{dv}{dt} + I\frac{v}{a^2}\frac{dv}{dt} + mg\frac{d}{dt}\left(\frac{x^2}{2(R-a)}\right) = mv\frac{dv}{dt} + I\frac{v}{a^2}\frac{dv}{dt} + mg\frac{xv}{R-a} = 0$$

$$\rightarrow \left(1 + \frac{I}{ma^2}\right)\frac{dv}{dt} = \left(1 + \frac{I}{ma^2}\right)\frac{d^2x}{dt^2} = -\frac{g}{R-a}x$$

The moment of inertia of the sphere about the axis of rotation (recall Section 7.7 - case 2) is $\frac{2}{5}ma^2$ and hence $1 + \frac{I}{ma^2} = \frac{7}{5}$ from which we obtain

$$\frac{d^2x}{dt^2} = -\frac{g}{\frac{7}{5}(R-a)}x$$

We recognise this as a differential equation for simple harmonic motion and hence the sphere oscillates with a period given by

$$T = 2\pi\sqrt{\frac{\frac{7}{5}(R-a)}{g}}$$

Worked Example 7.5: A pendulum comprises a uniform metal sphere of radius 50 mm at the end of a 950 mm long light string. Determine the fractional error in the period involved in treating this system as an ideal simple pendulum.

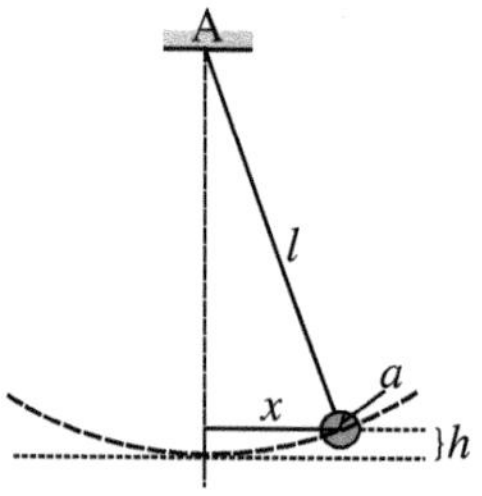

Figure 7.33. Worked Example 7.5.

The system can be considered as an extended body rotating about a horizontal axis through the point A in Figure 7.33 and hence can be treated as a physical pendulum. Let I_A be the moment of inertia of the system about this axis. As noted in Section 7.9, the period of oscillation is given by $T = 2\pi\sqrt{\dfrac{I_A}{mgl}}$

Using the parallel axis theorem and the expression for the moment of inertia of a sphere about a diameter (Section 7.7, case 2) we get

$I_A = I_C + ml^2 = \frac{2}{5}ma^2 + ml^2 = m\left(l^2 + \frac{2}{5}a^2\right)$

$$T = 2\pi\sqrt{\frac{m\left(l^2 + \frac{2}{5}a^2\right)}{mgl}} = 2\pi\sqrt{\frac{l}{g}\left(1 + \frac{2}{5}\frac{a^2}{l^2}\right)}$$

$$\rightarrow \quad \frac{T}{T_{ideal}} = \frac{2\pi\sqrt{\frac{l}{g}\left(l + \frac{2}{5}\frac{a^2}{l^2}\right)}}{2\pi\frac{l}{g}} = \sqrt{\left(l + \frac{2}{5}\frac{a^2}{l^2}\right)} = \sqrt{1.001} = 1.0005$$

$$\rightarrow \quad \frac{\Delta T}{T} = 0.0005 = 0.05\%$$

Worked Example 7.6: A cylindrical disc of mass 2.0 kg and radius 160 mm is suspended from a fixed support by means of a vertical wire connected to the centre of the disc as shown in Figure 7.34. The disc is rotated through an angle of 20° about the axis of the wire and then released. The disc is observed to execute simple harmonic oscillations about this axis in the horizontal plane. If the period of the oscillations is 1.5 s, determine (a) the torsion constant (the restoring torque per unit angle of twist) of the suspension wire, (b) the restoring torque exerted by the suspension wire at the extreme of a swing and (c) the angular momentum of the disc as it passes through the centre of the motion.

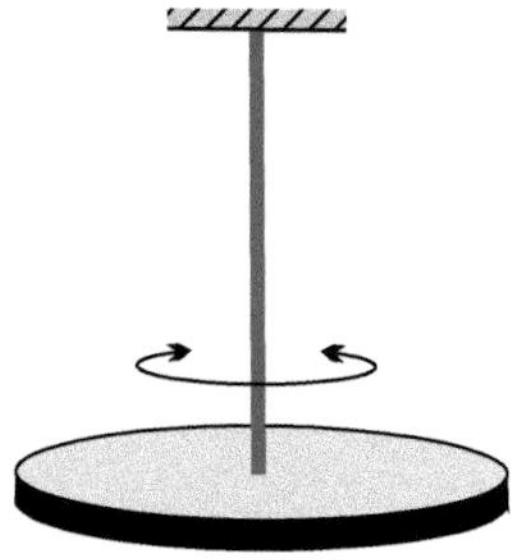

Figure 7.34. Worked Example 7.6.

(a) As noted in Section (7.10), Period $= T = 2\pi\sqrt{\dfrac{I}{c}} \rightarrow c = \dfrac{4\pi^2 I}{T^2}$.

Equation (7.13) $\rightarrow I = \dfrac{1}{2}MR^2 \rightarrow c = \dfrac{2\pi^2 MR^2}{T^2} = \dfrac{2\pi^2(2.0)(0.16)^2}{(1.5)^2} = 0.45\ \text{N m rad}^{-1}$

(b) Maximum restoring torque $= c\theta_{max} = (0.45\ \text{N m rad}^{-1})\left(20 \times \dfrac{\pi}{180}\right) = 0.16\ \text{N m}$

(c) Conservation of energy $\rightarrow$

total energy at the extreme of the motion = total energy at the centre of the motion

$$\rightarrow \quad \frac{1}{2}c\theta_{max}^2 + 0 = 0 + \frac{1}{2}I\omega_0^2 \quad \rightarrow \quad c\theta_{max}^2 = I\omega_0^2$$

Angular momentum $= L = I\omega_0 = \sqrt{I^2\omega_0^2} = (\sqrt{cI})\theta_{max} = \left(\sqrt{c\frac{1}{2}mR^2}\right)\theta_{max} = 0.04\ \text{kg m}^2\ \text{s}^{-1}$

PROBLEMS

For problems based on the material covered this chapter visit up.ucc.ie/7/ and follow the link to the problems.

8

Relative motion

AIMS

- to identify the conditions in which Newton's laws of motion may be applied and thereby to introduce the concept of *inertial reference frames*
- to derive procedures (known as the *Galilean transformation*) through which kinematic quantities, such as the coordinates of a point in one reference frame, can be converted to corresponding quantities in a second reference frame which is moving relative to the first
- to examine how motion can be analysed in situations in which Newton's laws of motion do not apply and hence to introduce the idea of the fictitious forces (*centrifugal force* and *Coriolis force*) which can be used to interpret such situations
- to consider how inertial frames may be defined for practical purposes

8.1 Applicability of Newton's laws of motion: inertial reference frames

Newton's first law of motion (Section 3.2) – which states that *if no net force acts on a body it will move at constant velocity or will stay at rest if initially at rest* – is a special case of the second law. Quite simply, according to the second law, if there is no force on a body there can be no change in its momentum, and hence no change in its velocity. The first law therefore seems unnecessary – it does not appear to tell us anything which cannot be deduced directly from the second law. Its purpose, as we demonstrate below, is more to 'set the scene' for the other laws – it describes the situations in which the second law may be applied.

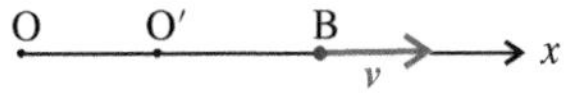

Figure 8.1. A moving object B viewed from two different origins O and O′.

In which circumstances then may Newton's laws of motion not be used? To answer this question we shall consider a body B which is moving freely along a straight line (the x-axis in Figure 8.1) at a constant velocity, v, relative to the origin O. The body does not experience any force so that this situation is in accord with Newton's first and second laws; there is no force and no acceleration.

Consider the motion of B as observed from O′, a point on the x-axis, in the following three situations:

Case 1 : O′ is at rest relative to O. When measured from O′, the velocity of B is still v – it is still constant. As viewed from O′, B is moving with constant velocity and is not experiencing any force so that this situation is in accord with Newton's second law.

Case 2 : O′ is moving along the x-axis with a constant velocity relative to O. When measured with respect to O′, the velocity of B is no longer v but it is still constant. As viewed from O′, B is moving with a constant velocity and is not experiencing any force so that this situation is also in accord with Newton's second law.

Case 3 : O′ accelerates (relative to O) along the x-axis. As viewed from O′, B is decelerating but is still not experiencing any force. We have acceleration without force. Newton's second law therefore is not obeyed if measurements are made from an origin O′ which is accelerating relative to O.

We conclude therefore that, if measurements are made with reference to an accelerating origin, the motion of the body does not obey Newton's second law. The key point is that, in a situation in which Newton's laws of motion are observed to hold, if we switch to making measurements relative to a point which is accelerating, we can no longer use Newton's laws directly.

Coordinate systems were introduced in Chapter 4 (recall Figure 4.14 – an example of a Cartesian coordinate system). Whenever a situation is observed from the point of view of two (or more) coordinate systems it is usual to refer to such systems as **reference frames.** Cases 1, 2 and 3 above are examples of different (one dimensional) reference frames

Figure 8.2 shows two such coordinate systems which we denote as the K and K′ frames. The K′ frame is moving with velocity $\boldsymbol{v}$ relative to the K frame. It is often convenient to choose the x'-, y'- and z'-axes so that they are parallel to the x-, y- and z-axes and to choose the origin of time so that the axes coincide at $t = 0$.

Understanding Physics, Third Edition. Michael Mansfield and Colm O'Sullivan.
© 2020 John Wiley & Sons Ltd. Published 2020 by John Wiley & Sons Ltd.

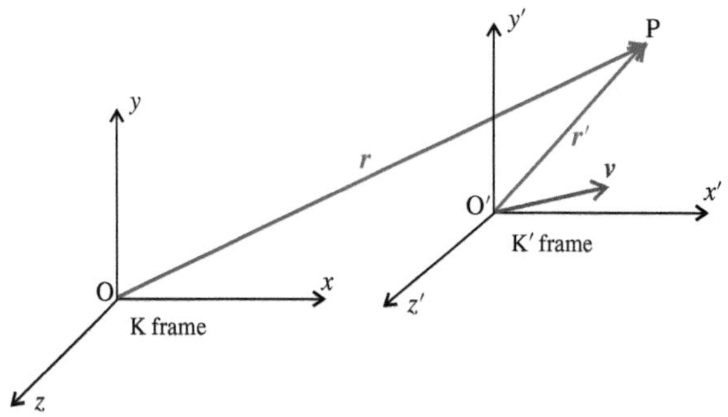

Figure 8.2. The coordinates of a point P in the Cartesian (x, y, z) system (K-frame) and in a second system (K′ frame) which is moving with velocity $\boldsymbol{v}$ relative to the first.

The position of a point at a specific time, described by four coordinates x, y, z, t is called an **event**. The procedure used to relate the coordinates of an event in one reference frame to the coordinates of the same event in another reference frame is called a **transformation**.

Inertial reference frames

An **inertial** reference frame is a frame in which Newton's first law of motion holds. In such a frame, any body which is in a state of rest or of uniform motion (that is, constant velocity, which may be translational, rotational or both) continues in that state without change of translational or angular velocity. Thus we see that Newton's first law provides a practical test of whether Newton's second law applies. We can therefore state that

> *Newton's second law of motion applies only in inertial reference frames.*

It follows that the general conservation principles (of momentum, angular momentum, and energy) also apply in inertial frames because they are derived directly from Newton's laws.

8.2 The Galilean transformation

A common practical requirement is the conversion of the coordinates of a point, as measured in one reference frame, to the coordinates of the same point, as measured in a second reference frame which is moving with constant velocity relative to the first. The equations which are used to achieve this conversion are known as the **Galilean transformation** (Galileo Galilei,1564–1642) and are derived below.

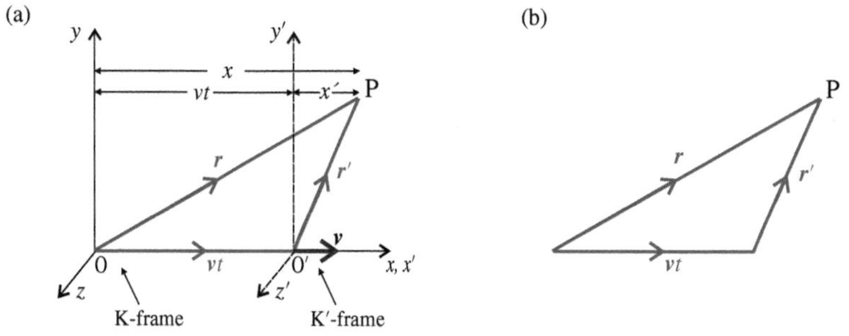

Figure 8.3. (a) The coordinates of a point P in the x—y plane as viewed from a fixed reference frame K (coordinate system (x, y)) and from a frame K′ (coordinate system (x', y')) which is moving with velocity $\boldsymbol{v}$ along the x-axis of the fixed system. (b) The Galilean displacement transformation of the displacement of P from the moving system to the fixed system, namely $\boldsymbol{r} = \boldsymbol{r}' + \boldsymbol{v}t$.

Consider the two reference frames, K and K′ (Cartesian coordinate systems), as illustrated in Figure 8.3. The second frame, which has axes x', y' and z', is moving so that its origin, O′, is moving at constant velocity, $\boldsymbol{v}$, along the x-axis of the first frame, which has axes x, y and z. The x'-axis of the second frame lies along the x-axis of the first frame. If the coordinates of an event are (x, y, z, t) and (x', y', z', t'), respectively, in the two frames and if the origins of the two frames, O and O′, coincide at $t = t' = 0$, we can, by inspecting Figure 8.3, write down equations which relate the coordinates of the point P as measured in each of the two frames.

$$x = x' + vt \tag{8.1}$$

$$y = y' \tag{8.2}$$

$$z = z' \tag{8.3}$$

and

$$t = t' \tag{8.4}$$

As illustrated in Figure 8.3 (b), we can also write down a vector equation which is equivalent to Equations (8.1) to (8.3)

$$\boldsymbol{r} = \boldsymbol{r}' + \boldsymbol{v}t \tag{8.5}$$

Equations (8.1) to (8.5), known as the **inverse Galilean transformation**, may be used to convert a set of coordinates in the (x', y', z', t') frame to coordinates in the (x, y, z, t) frame. The **Galilean transformation,** which converts coordinates in the (x, y, z, t) frame to their values in the (x', y', z', t') frame is obtained simply by rearranging Equations (8.1) to (8.5),

$$\boxed{\begin{aligned} x' &= x - vt \\ y' &= y \\ z' &= z \end{aligned}}$$

or

$$\boxed{\boldsymbol{r}' = \boldsymbol{r} - \boldsymbol{v}t}$$

Note that the Galilean transformation may be formally converted to the inverse Galilean transformation (and vice versa) by substituting $-\boldsymbol{v}$ for $\boldsymbol{v}$ and interchanging $t \leftrightarrow t'$ and $x \leftrightarrow x'$, etc.

Galilean velocity transformation

We can also derive transformation rules for the velocity of a point by differentiating Equations (8.1) to (8.4) with respect to t, remembering that v is constant. Thus:

$$\frac{dx}{dt} = \frac{dx'}{dt} + v \tag{8.6}$$

$$\frac{dy}{dt} = \frac{dy'}{dt} \tag{8.7}$$

$$\frac{dz}{dt} = \frac{dz'}{dt} \tag{8.8}$$

and

$$1 = \frac{dt'}{dt} \tag{8.9}$$

Using Equation (8.9), we can substitute dt' for dt in Equations (8.6) to (8.8) to obtain

$$\frac{dx}{dt} = \frac{dx'}{dt'} + v$$

$$\frac{dy}{dt} = \frac{dy'}{dt'}$$

and

$$\frac{dz}{dt} = \frac{dz'}{dt'}$$

which we write as

$$u_x = u'_x + v \tag{8.10}$$

$$u_y = u'_y \tag{8.11}$$

and

$$u_z = u'_z \tag{8.12}$$

where $u_x = \dfrac{dx}{dt}, u'_x = \dfrac{dx'}{dt'}$ etc.

The equivalent vector velocity transformation, illustrated in Figure 8.4, is obtained by differentiating Equation (8.5). Thus

$$\boldsymbol{u} = \boldsymbol{u}' + \boldsymbol{v} \tag{8.13}$$

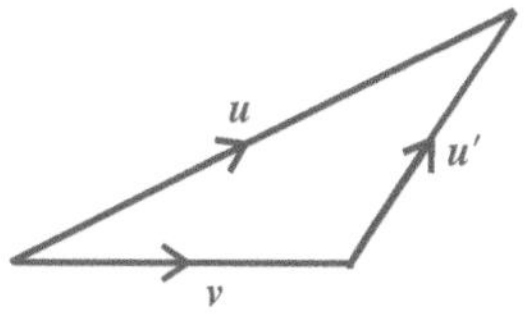

Figure 8.4. The Galilean velocity transformation $\boldsymbol{u} = \boldsymbol{u}' + \boldsymbol{v}$.

where $\boldsymbol{u} = \frac{d\boldsymbol{r}}{dt}$ and $\boldsymbol{u}' = \frac{d\boldsymbol{r}'}{dt'} = \frac{d\boldsymbol{r}'}{dt}$ are the velocities of P in the (x, y, z, t) and (x', y', z', t') frames respectively. Equations (8.10) to (8.13) are the **inverse Galilean velocity transformation.** The **Galilean velocity transformation** equations are obtained by rearranging Equations (8.10) to (8.13),

$$u'_x = u_x - v$$
$$u'_y = u_y$$
$$u'_z = u_z$$

or

$$\boldsymbol{u}' = \boldsymbol{u} - \boldsymbol{v} \tag{8.14}$$

We illustrate the use of the Galilean velocity transformation by considering the following two examples:

Example 1: A stone is thrown horizontally from a train, with velocity $\boldsymbol{u}'$ relative to the train as illustrated Figure 8.5(b). The origin, O, of the (x, y, z) coordinate frame (Figure 8.5(a)) is fixed to the track and the origin, O′, of the (x', y', z') frame (Figure 8.5(b)) is fixed to the train. The train is moving with constant velocity $\boldsymbol{v}$ along the x-axis of the track frame.

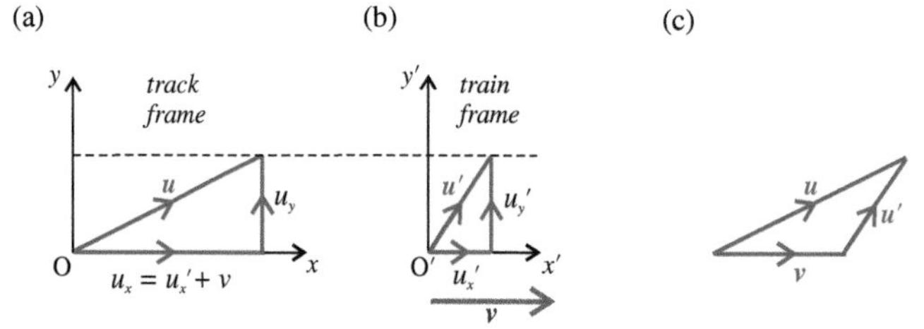

Figure 8.5. Example 1. (a) The components of the stone velocity $\boldsymbol{u}$ in the track frame, (b) the components of the stone velocity $\boldsymbol{u}'$ in the train frame and (c) the velocity transformation $\boldsymbol{u}' = \boldsymbol{u} - \boldsymbol{v}$.

We can resolve $\boldsymbol{u}'$, the velocity of the stone relative to the train, into two components u'_x and u'_y, as indicated in Figure 8.5(b). We then apply transformation Equations (8.10) and (8.11) to calculate u_x and u_y, the components in the horizontal plane of $\boldsymbol{u}$, the velocity of the stone relative to the ground and, using Figure 8.5(a), we can then determine the magnitude and direction of $\boldsymbol{u}$. The equivalent vector addition (8.13) is also shown in Figure 8.5(c). Worked Example 8.1 gives a numerical example of this type of calculation.

Example 2: In Section 13.13 we shall describe how a sound wave is transmitted through air. The velocity of the sound wave is a property of the medium (the air) which is transmitting the sound.

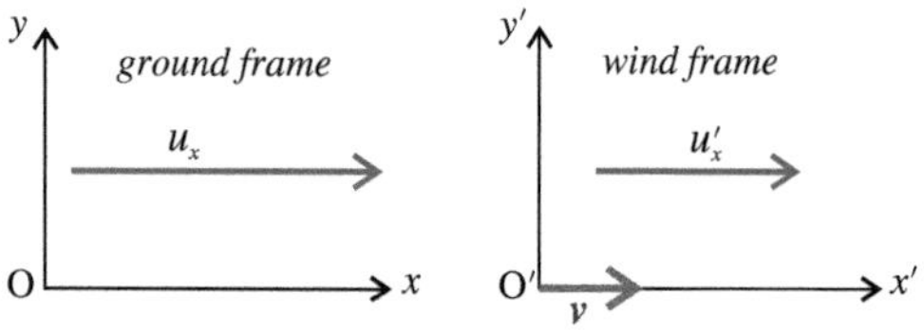

Figure 8.6. Example 2. The velocities of the sound wave in the ground and wind frames.

Let us consider a case in which sound is transmitted through a mass of air which is moving with a velocity v in the x-direction with respect to the ground; that is, v is the wind velocity.

If, as illustrated in Figure 8.6, the velocity of sound in still air is u'_x, the velocity of the sound wave with respect to the ground directly downwind from the source of sound, u_x, may be determined using Equation (8.10), as follows,

$$u_x = u'_x + v$$

Thus, the fact that the velocity of sound is different when it is measured in a reference frame which is moving relative to the frame in which the medium transmitting the sound is at rest follows inevitably from the Galilean transformation.

Galilean acceleration transformation

Returning to the general inverse Galilean velocity transformation Equations (8.10) to (8.12), we note that the acceleration transformation may be derived by differentiating these equations with respect to t. We thus obtain

$$\frac{d^2x}{dt^2} = \frac{d^2x'}{dt'^2}, \quad \frac{d^2y}{dt^2} = \frac{d^2y'}{dt'^2}, \quad \frac{d^2z}{dt^2} = \frac{d^2z'}{dt'^2},$$

that is,

$$a_x = a'_x, \quad a_y = a'_y, \quad a_x = a'_z$$

or, in vector terms,

$$\boldsymbol{a} = \boldsymbol{a}'$$

The acceleration of any object is therefore unchanged when it is transformed to another frame which is moving at constant velocity relative to the frame in which the acceleration is first measured. Thus, if Newton's laws hold in the first frame, they will hold in the other frame. This confirms in mathematical form our earlier deductions (Section 8.1) concerning the applicability of Newton's laws. It also confirms that, if we have an inertial frame, we can create any number of further inertial frames, each of which is moving with a constant translational velocity relative to the first. Conversely, a frame is inertial only if it has no acceleration relative to other inertial frames.

Note that, although the Galilean transformation tells us that the values of variable quantities (such as the x-coordinate of an object's position) may not be the same in all frames, any equations which describe the behaviour of physical systems such as Newton's second law of motion, $\boldsymbol{F} = m\boldsymbol{a}$ (Section 4.4), apply in the same form in all inertial frames. The values of variable quantities may change when the reference frame is changed but they must do so in such a way that the equation remains valid.

Study Worked Example 8.1

For problems based on the material presented in this section visit up.ucc.ie/8/ *and follow the link to the problems.*

8.3 The CM (centre-of-mass) reference frame

In Chapter 6 we saw that many-body problems could be studied from the points of view of different coordinate systems, specifically the LAB and CM coordinate systems. The conversion of displacement and velocity coordinates from LAB to CM coordinates, and vice versa, is an example of a Galilean transformation. If the LAB system is inertial then the CM system is also inertial. We shall henceforth refer to the LAB and CM coordinate systems, respectively, as the **LAB frame** and the **CM frame**.

The CM frame for a *single particle* is the frame which moves with the particle and, in this case, the CM frame is called the **rest frame** of the particle. The velocity of the particle is always zero in this frame and therefore the kinetic energy of the particle in its CM frame is also zero. The kinetic energy of a particle is greater than zero in any frame which is moving relative to its rest frame.

The treatment of various *many-body* problems proves to be simpler, and much insight can be gained, if the analysis is carried out in the CM frame of the system. In this case the origin of the coordinate system relative to which displacements, velocities, etc. are measured, is chosen to be a point O′ which is fixed with respect to the CM of the system. In what follows we will used primed variables to represent displacements, velocities and momenta of particles in the CM frame; that is $\boldsymbol{r}_i'$, $\boldsymbol{u}_i'$, $\boldsymbol{p}_i'$, $\boldsymbol{q}_i'$, etc. will refer to the ith particle with measurements made relative to O′.

To illustrate some of the advantages of solving problems in the CM frame, we revisit the two-body collisions analysed in Section 6.5. The velocity of the centre of mass relative to some arbitrary origin O is given (Equation (6.20)) by

$$\boldsymbol{V}_{\mathrm{C}} = \frac{m_1\boldsymbol{u}_1 + m_2\boldsymbol{u}_2}{m_1 + m_2} = \frac{\boldsymbol{P}}{m_1 + m_2}$$

where $\boldsymbol{P} = m_1\boldsymbol{u}_1 + m_2\boldsymbol{u}_2$ is the total momentum of the system relative to O. To convert the velocity of a particle from this frame to the CM frame, we use the Galilean velocity transformation (Equation (8.14))

$$\boldsymbol{u}_i' = \boldsymbol{u}_i - \boldsymbol{V}_{\mathrm{C}}$$

The velocity of the CM frame relative to O′ is zero, that is $\boldsymbol{V}'_{\mathrm{C}} = \dfrac{m_1\boldsymbol{u}'_1 + m_2\boldsymbol{u}'_2}{m_1 + m_2} = \dfrac{\boldsymbol{P}'}{m_1 + m_2} = 0$

Thus in the CM frame, the total momentum, $\boldsymbol{P}' = m_1\boldsymbol{u}_1' + m_2\boldsymbol{u}_2' = \boldsymbol{p}_1' + \boldsymbol{p}_2' = 0$ — the CM frame is a 'zero momentum frame' — in other words, for the two-body problem (Section 6.7),

$$\boldsymbol{p}_1' = -\boldsymbol{p}_2'\,(=\boldsymbol{p}', \text{say}) \qquad \text{and} \qquad \boldsymbol{q}_1' = -\,\boldsymbol{q}_2'\,(=-\boldsymbol{q}', \text{say})$$

This means that, in the CM frame, the particles have to be moving either towards one another or away from one another along the same straight line in order to conserve momentum.

We now re-analyse, in the CM frame, the cases and examples of elastic collisions (*Case 1*) and of totally inelastic collisions (*Case 2*) previously discussed in Section 6.5. The analysis is extended to include general elastic collisions (*Case 3*).

Case 1: Elastic collision

Figure 8.7 shows an elastic collision observed in the CM frame.

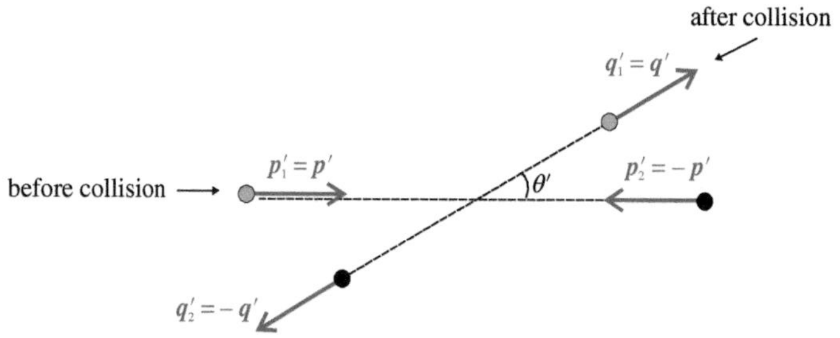

Figure 8.7. An elastic collision in the CM frame.

Conservation of momentum: $\boldsymbol{p}_1' + \boldsymbol{p}_2' = \boldsymbol{q}_1' + \boldsymbol{q}_2{}' = 0$

Conservation of energy: $\dfrac{p_1'^2}{2m_1} + \dfrac{p_2'^2}{2m_2} = \dfrac{q_1'^2}{2m_1} + \dfrac{q_2'^2}{2m_2}$

Let $\boldsymbol{p}' = \boldsymbol{p}_1' = -\boldsymbol{p}_2'$ and $\boldsymbol{q}' = \boldsymbol{q}_1' = -\boldsymbol{q}_2'$ as required by conservation of momentum. The statement of conservation of energy can then be written

$$\frac{p'^2}{2m_1} + \frac{p'^2}{2m_2} = \frac{q'^2}{2m_1} + \frac{q'^2}{2m_2} \rightarrow \frac{p'^2}{2}\left(\frac{1}{m_1} + \frac{1}{m_2}\right) = \frac{q'^2}{2}\left(\frac{1}{m_1} + \frac{1}{m_2}\right) \rightarrow \frac{p'^2}{2m_r} = \frac{q'^2}{2m_r}$$

where m_r is the reduced mass (Equation (6.19)).

Thus, we see that, for elastic collisions, $|\boldsymbol{p}'| = |\boldsymbol{q}'|$

Once again conservation of momentum and energy alone are not sufficient to determine completely the outcome of the collision but, in this frame, the magnitudes of the momenta after collision can be specified fully, being equal to the magnitudes of the momenta before collision. The indeterminacy in the directions of the momenta arises from our lack of specific knowledge of the angle θ' which depends on factors like the obliqueness of the contact, *etc*. A head-on collision is an example of a case in which θ' is known ($\theta' = 0$) and in which the problem is fully determined.

Example:

Consider an oblique (non head-on) elastic collision between two identical masses, one of which is at rest in the LAB frame, as illustrated in Figure 8.8(a), a reproduction of Figure 6.14. The velocity of the centre of mass in this case is in the $+x$-direction and is given by

$$V_C = \frac{m_1 u_1 + m_2 u_2}{m_1 + m_2} = \frac{mu + 0}{m + m} = \frac{1}{2}u$$

Using the Galilean velocity transformation (Equation 8.14), we get

$$u'_1 = u_1 - V_C = u - \frac{1}{2}u = \frac{1}{2}u$$

and

$$u'_2 = u_2 - V_C = 0 - \frac{1}{2}u = -\frac{1}{2}u$$

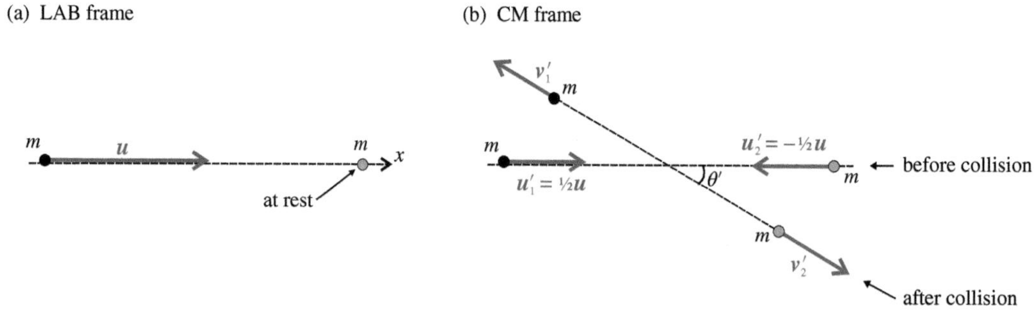

Figure 8.8. An elastic collision of identical masses: (a) before the collision in the LAB frame. (b) Before and after the collision in the CM frame. Since the collision is elastic, the magnitudes of the velocities after collision in the CM frame must be ½ $|\boldsymbol{u}|$.

Figure 8.8(b) shows this collision as viewed in the CM frame. In order to conserve momentum and energy the two particles must move with opposed velocities, each of magnitude $\frac{u}{2}$, after the collision. The CM velocities after the collision are at some angle θ' to the initial direction in that frame.

The velocities (magnitudes and directions) of the particles in the LAB frame after the collision, $\boldsymbol{v}_1$ and $\boldsymbol{v}_2$, can be determined using the inverse Galilean velocity transformation (Equation 8.13),

$$\boldsymbol{v}_1 = \boldsymbol{v}_1' + \boldsymbol{V}_C$$

and

$$\boldsymbol{v}_2 = \boldsymbol{v}_2' + \boldsymbol{V}_C$$

The vector additions are illustrated in Figure 8.9, where $\boldsymbol{V}_C$ is added to the particle velocities after the collision in the CM frame, $\boldsymbol{v}_1'$ and $\boldsymbol{v}_2'$, to give $\boldsymbol{v}_1$ and $\boldsymbol{v}_2$, respectively, (shown as black vectors in the figure).

Note the greater symmetry of the collisions when viewed from the CM frame. Since the total momentum of the system is zero, the momenta and energies of individual particles are, in general, smaller and are distributed more equally between the particles in the CM frame. As illustrated in Worked Example 8.2, this usually simplifies the conservation equations for momentum and energy which are used to analyse and solve collision problems.

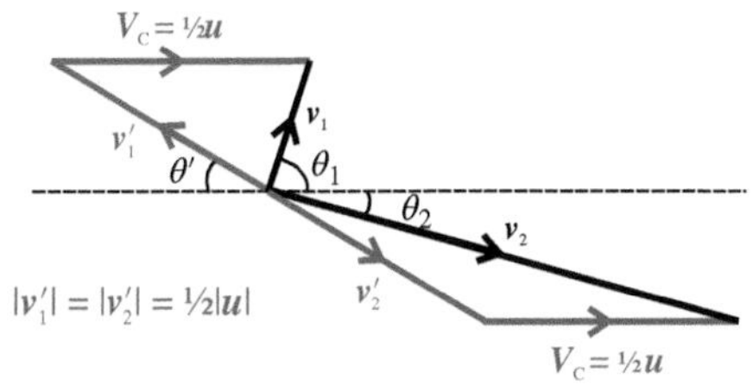

Figure 8.9. Transformation of velocities after the collision from the CM frame to the LAB frame.

Case 2: Totally inelastic collision

Figure 8.10 represents a totally inelastic collision as viewed in the CM coordinate system. The two particles coalesce on collision to form a single particle of mass $m_1 + m_2$ which must be at rest relative to the origin O′ (total momentum must be zero). Thus the total energy after collision must also be zero, so that the energy lost in a totally inelastic collision is the maximum that can be lost.

Conservation of momentum: $\boldsymbol{p}_1' + \boldsymbol{p}_2' = 0$ (so we can let $\boldsymbol{p}' = \boldsymbol{p}_1' = -\boldsymbol{p}_2'$, as before)

Change of energy $$\Delta E = 0 - \left(\frac{p_1'^2}{2m_1} + \frac{p_2'^2}{2m_2}\right) = -\left(\frac{p'^2}{2m_1} + \frac{p'^2}{2m_2}\right)$$

$$= -\frac{p'^2}{2}\left(\frac{1}{m_1} + \frac{1}{m_2}\right) = -\frac{p'^2}{2m_r}$$

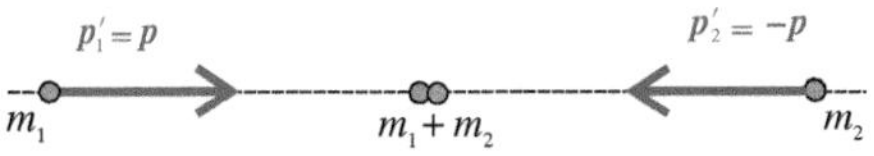

Figure 8.10. A totally inelastic collision in the CM frame. The composite mass $(m_1 + m_2)$ is at rest after collision.

where m_r is the reduced mass of the (two-body) system.

We can show that this energy loss is the same as that obtained when the problem is treated in LAB coordinates. From Equation (6.22), we know that the kinetic energy of the system before collision is given by

$$K = \frac{1}{2}(m_1 + m_2)V_C^2 + \frac{1}{2}m_r u^2$$

where $\boldsymbol{u}$ is the velocity of m_1 relative to m_2, that is $u = |\boldsymbol{u}_1 - \boldsymbol{u}_2|$. In the CM frame, $V_C = 0$ and $|\boldsymbol{u}_1' - \boldsymbol{u}_2'| = u' = |\boldsymbol{u}_1 - \boldsymbol{u}_2|$ and thus, in this frame, the kinetic energy before collision is

$$K'_0 = \frac{1}{2}m_r u'^2 = \frac{1}{2}m_r|\boldsymbol{u}_1 - \boldsymbol{u}_2|^2$$

Since, in the LAB frame, $\boldsymbol{u}_2 = 0$,

$$K'_0 = \frac{1}{2}m_r|\boldsymbol{u}_1|^2 = \frac{1}{2}m_r u_1^2 = \left(\frac{m_r}{m_1}\right)\frac{p_1^2}{2m_1} = \left(\frac{m_2}{m_1 + m_2}\right)K_0 = \frac{p'^2}{2m_r}$$

where K_0 is the kinetic energy before collision in the LAB frame.

The energy loss in the collision is $$\Delta E = 0 - K'_0 = -\left(\frac{m_r}{m_1}\right)\frac{p_1^2}{2m_1} = -\left(\frac{m_2}{m_1 + m_2}\right)K_0$$

Thus, as we anticipated, the energy loss is exactly the same as that calculated in the LAB system (Section 6.5, Case 2).

Example:

To illustrate the analysis of a totally inelastic collision in the CM frame, let us consider again the totally inelastic collision between two identical masses which we discussed in the Example in Case 2 in Section 6.5 (Figure 6.16).

The velocity of the centre of mass is given by $\boldsymbol{V}_C = \dfrac{\boldsymbol{p}_1 + \boldsymbol{p}_2}{m_1 + m_2} = \dfrac{\boldsymbol{q}}{2m}$ and thus, from the analysis carried out in Chapter 6, we get

$$|\boldsymbol{V}_C| = \frac{\sqrt{2}p}{2m} = \frac{p}{\sqrt{2}m} = \frac{u}{\sqrt{2}} \quad \rightarrow \quad \boldsymbol{V}_C = \frac{u}{\sqrt{2}}\boldsymbol{i}$$

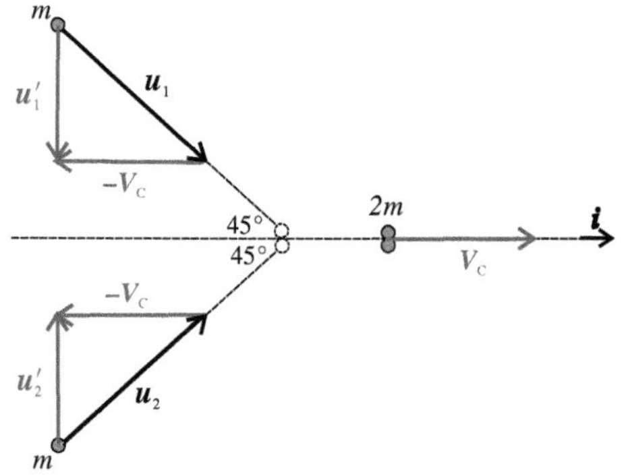

Figure 8.11. Transformation from LAB to CM frame for a totally inelastic collision (Example 2).

Transforming to the CM frame using the Galilean velocity transformation (Figure 8.11), this problem can be viewed in the CM frame. In this frame, the change of energy is given by

$$\Delta E = 0 - 2\frac{p'^2}{2m} = -\frac{p'^2}{m} = -mu'^2$$

Returning to the LAB frame,

$$\Delta E = -mu'^2 = -m|\boldsymbol{u} - \boldsymbol{V}_C|^2 = -m(u^2 - 2u \cdot \boldsymbol{V}_C + |\boldsymbol{V}_C|^2) = -\frac{1}{2}mu^2$$

we see that that the energy loss is the same as that calculated using LAB coordinates (Section 6.5) but that in the CM system this represents *all* of the kinetic energy of the system.

Case 3: General inelastic collision

Cases 1 and 2 represent two extreme situations, namely those in which no energy is lost (*Case 1*) or in which the maximum amount of energy is lost in the collision (*Case 2*). Once again the analysis of collisions from the standpoint of an observer fixed with respect to the CM provides the clearest perspective on this topic. In elastic collisions viewed in the CM frame (Figure 8.7), the magnitude of the momentum vectors after collision is the same as the magnitude of the momentum vectors before collision. At the other extreme, in totally inelastic collisions, the momenta after collision are zero (Figure 8.10). Most macroscopic collisions fall somewhere between these two extremes; if there is some loss of energy, the magnitude of the momentum vectors after collision must be less than that of the momentum vectors before collision. Such a situation (not considered in Section 6.5) is illustrated in Figure 8.12, viewed in the CM frame.

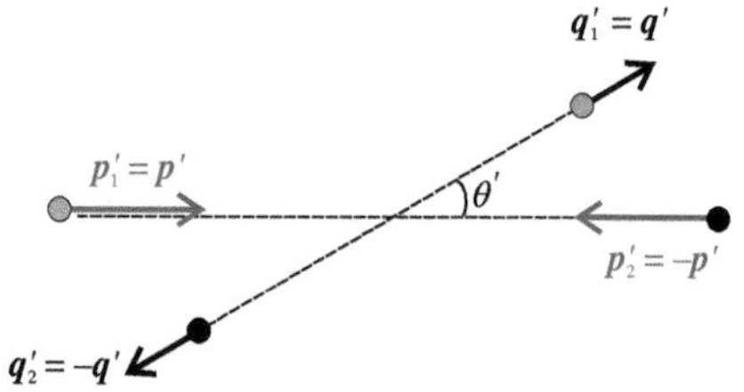

Figure 8.12. A general elastic collision viewed in the CM coordinate system.

The ratio of the magnitudes of the momentum vectors after collision to their magnitude before collision, therefore, gives a measure of how elastic or inelastic a particular collision is. We define this ratio as the **coefficient of restitution** of the collision as follows

$$\boxed{e := \frac{|\boldsymbol{q}'|}{|\boldsymbol{p}'|}} \tag{8.15}$$

Thus $|\boldsymbol{q}'| = e|\boldsymbol{p}'|$ where $1 \geq e \geq 0$; $e = 1$ for elastic collisions and $e = 0$ for totally inelastic collisions. We can see how the loss of energy is related to the coefficient of restitution as follows. The change of energy is given by

$$\begin{aligned}\Delta E &= \left(\frac{q_1'^2}{2m_1} + \frac{q_2'^2}{2m_2}\right) - \left(\frac{p_1'^2}{2m_1} + \frac{p_2'^2}{2m_2}\right) = \left(\frac{q'^2}{2m_1} + \frac{q'^2}{2m_2}\right) - \left(\frac{p'^2}{2m_1} + \frac{p'^2}{2m_2}\right) \\ &= \frac{q'^2}{2}\left(\frac{1}{m_1} + \frac{1}{m_2}\right) - \frac{p'^2}{2}\left(\frac{1}{m_1} + \frac{1}{m_2}\right) = \frac{q'^2}{2m_r} - \frac{p'^2}{2m_r} = \frac{(ep')^2}{2m_r} - \frac{p'^2}{2m_r} = (e^2 - 1)\frac{p'^2}{2m_r}\end{aligned}$$

Thus

$$\Delta E = -(1 - e^2)K_0'$$

where K_0' is the initial kinetic energy in the CM frame. This can be written in terms of the initial kinetic energy in the LAB system as follows

$$\Delta E = -(1 - e^2)\left(\frac{m_2}{m_1 + m_2}\right)K_0$$

since we have already shown, in Case 2 above, that $K_0' = \left(\dfrac{m_2}{m_1 + m_2}\right)K_0$

The fractional energy loss in a collision, therefore, is $(1 - e^2)$ in the CM frame and $\left(\dfrac{m_2}{m_1 + m_2}\right)(1 - e^2)$ in the LAB frame. Since $\Delta E = K'_a - K'_0$, where K'_a is the total kinetic energy after collision, $K'_a - K'_0 = -(1 - e^2)K'_0$ and hence $K'_a = e^2 K'_0$. Thus the ratio of the energy after collision to the energy before collision is given, in CM coordinates, by

$$\frac{K'_a}{K'_0} = e^2$$

The coefficient of restitution is characteristic of a particular collision event, in general a property of the materials of which the colliding bodies are made and of their velocities, and, as such, there is nothing particularly fundamental about it. A collision between the same two particles at a different energy will not necessarily have the same value of e.

Note: Many texts define the coefficient of restitution as the ratio of the relative velocity of the particles after collision to their relative velocity before collision, that is

$$\boxed{e := \frac{|\boldsymbol{v}_1 - \boldsymbol{v}_2|}{|\boldsymbol{u}_1 - \boldsymbol{u}_2|}}$$

To show that this is equivalent to the definition given above (Equation 8.15), recall that the relative velocity is the same in all frames and thus

$$\boldsymbol{u}_1 - \boldsymbol{u}_2 = \boldsymbol{u}'_1 - \boldsymbol{u}'_2 = \frac{\boldsymbol{p}'}{m_1} - \left(-\frac{\boldsymbol{p}'}{m_2}\right) = \frac{\boldsymbol{p}'}{m_r}$$

Similarly $$\boldsymbol{v}_1 - \boldsymbol{v}_2 = \boldsymbol{v}'_1 - \boldsymbol{v}'_2 = \frac{\boldsymbol{q}'}{m_r} \quad \text{and hence} \quad \frac{|\boldsymbol{v}_1 - \boldsymbol{v}_2|}{|\boldsymbol{u}_1 - \boldsymbol{u}_2|} = \frac{|\boldsymbol{q}'|}{|\boldsymbol{p}'|} = e$$

Study Worked Example 8.2

As we shall note in Chapter 27, the properties and structure of subnuclear particles can be investigated through experiments in which particles, which have been accelerated to very high energies, collide. The kinetic energy of the colliding particles – the *interaction energy* – can be used to create new particles, as will be discussed in Section 9.14.

If a collision takes place in the CM frame, all the initial kinetic energy is available as interaction energy. Consequently high energy particle experiments often employ collisions between particles, of identical masses and energies, which are moving in opposite directions (*colliding beam experiments*). The total momentum in the laboratory is then zero and the frame fixed in the laboratory is the same as the CM frame. Consequently, in such experiments, all the laboratory kinetic energy is available as interaction energy.

For problems based on the material presented in this section visit up.ucc.ie/8/ and follow the link to the problems.

8.4 Example of a non-inertial frame: centrifugal force

We shall now explore the applicability of Newton's laws further by considering a case in which they cannot be applied. In other words, we shall be studying non-inertial frames. In Section 8.1 we have already considered such a situation, namely Case 3, the observation of a moving particle relative to an accelerating origin.

The situation which we considered in Section 8.1, however, may seem a little contrived. It is quite easy to view the situation from outside – from the laboratory frame, an inertial frame – and to identify the problems caused by viewing the motion from an accelerating frame. It is more difficult, however, to view the situation from outside when we are attached to a rotating frame.

Figure 8.13 illustrates such a situation. The (x, y) frame is attached to the ground (the laboratory frame) and the (x', y') frame is rotating at an angular velocity ω with respect to the ground. Such a rotating frame can be visualised if we imagine ourselves to be on the edge of a fairground roundabout in the position of an observer O′ who is always facing outwards (the direction of the y'-axis) as the roundabout rotates.

Consider an object which is fixed at P at the edge of the roundabout. If the object were to be released it would move with the tangential velocity of P. The outside observer O, in the frame fixed to the ground (the LAB frame), would recognise the motion as motion in a straight line with constant velocity. There is no force on the object so such motion is in accordance with Newton's first law. The LAB frame is inertial.

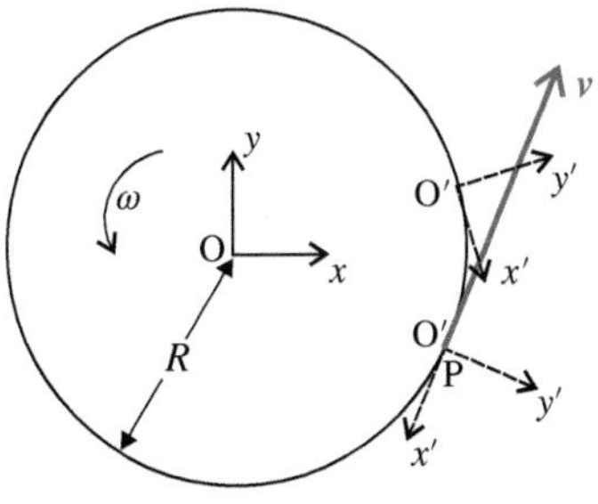

Figure 8.13. The motion of an object which is released from the edge of a rotating roundabout, as viewed in the ground (laboratory) frame (x, y). The figure also indicates the change in the object's position as the roundabout rotates, as viewed in the (x', y') frame, a frame which is fixed at the edge of the roundabout.

Consider the motion of the object as viewed by an observer O′ who is fixed at the edge of the roundabout, and whose reference frame rotates with the roundabout (as illustrated in Figure 8.13). On release from P, to this observer the object would suddenly start to move outwards, in the y'-direction,without any external force acting on it. Newton's first law would be violated in the rotating frame. This situation arises because, as shown in Section 4.8, in performing circular motion in the (inertial) LAB frame, O′ must always have a centripetal acceleration, $\omega^2 R$, directed towards O. The frame of O′ is accelerating relative to an inertial frame and cannot be inertial.

Centrifugal force

To the observer O′, fixed to the edge of the roundabout, any object released from the roundabout starts to move along the y'-axis – it appears to be subjected to a force outwards from the centre of the roundabout. We know however that this is not a real force; the acceleration appears only because we are viewing the situation from a non-inertial frame.

We can make the rotating frame appear to be inertial artificially by introducing a fictitious outward force of magnitude $m\omega^2 R$, along the y'-axis of the frame (where m is the mass of the object at P). With this force included, an object released from P obeys Newton's laws in the frame of O′. However it is important to note that this outward force, known as the **centrifugal force** (centre-fleeing force), is not a real force, it is merely a convenient way of making the O′ frame appear inertial, thus allowing us to apply Newton's laws in a rotating frame.

The Earth as a rotating frame: effective g

Situated, as we are, on the Earth's surface we are fixed to a rotating sphere. As illustrated in Figure 8.14, at a point P we are in effect attached to the edge of a very large rotating roundabout of radius r. At latitude λ the value of r is $R \cos\lambda$, where R is the radius of the Earth.

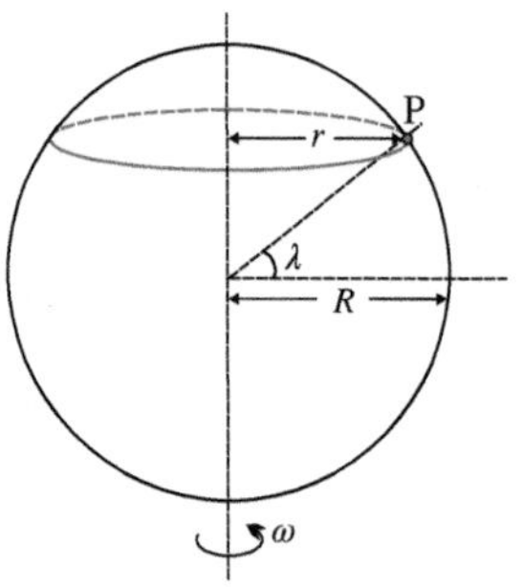

Figure 8.14. The path followed by a point P which is fixed to the Earth's surface at latitude λ, as the Earth rotates.

In this case it is very hard to view the situation from the outside. We naturally make measurements at the Earth's surface with reference to our surroundings, which are all attached to the Earth's surface and which are, therefore, performing uniform circular motion with us. We tend naturally to use an accelerating (non-inertial) frame.

The tendency of an object on the Earth's surface to travel in a straight line thus leads to an apparent outward centrifugal force $\boldsymbol{F}_c$ on any object when viewed by an observer who is fixed to the Earth's surface, as illustrated in Figure 8.15. The magnitude of the centrifugal force is given by

$$F_c = m\omega^2 r = m\omega^2 R \cos\lambda$$

where ω is the (constant) angular velocity of the Earth.

Thus, when $\lambda = 90°$ (at the poles), $F_c = 0$
and, when $\lambda = 0°$ (at the equator), $F_c = m\omega^2 R$ (directed vertically upwards from the Earth's surface)

In practice, centrifugal force is taken into account in the study of the motion of objects near the Earth's surface by allowing $\boldsymbol{g}$, the acceleration due to gravity at the Earth's surface, to vary with latitude. Thus we replace $\boldsymbol{g}$, by $\boldsymbol{g}^*$, the effective acceleration due to gravity near the Earth's surface. The magnitude of $\boldsymbol{g}^*$ has its maximum value, g, at the poles (where $F_c = 0$) but falls to $g - \omega^2 R$ at the equator (where $F_c = m\omega^2 R$).

As indicated in Figure 8.16, this also means that the direction of $\boldsymbol{g}^*$ is not perpendicular to the Earth's surface except at the equator (when $\lambda = 0°$), in which case $\boldsymbol{F}_c$ is antiparallel to $\boldsymbol{g}$, and at the poles (when $\lambda = 90°$), in which case $\boldsymbol{F}_c$ is zero.

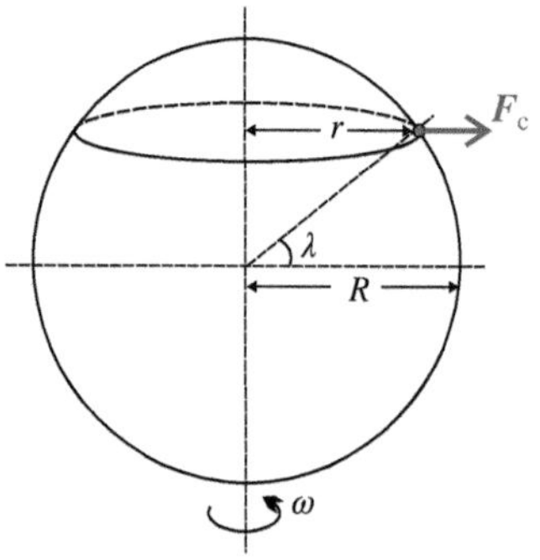

Figure 8.15. The centrifugal force $\boldsymbol{F}_c$ in the Earth's rotating frame on a mass which is fixed to the Earth's surface at latitude λ, as the Earth rotates.

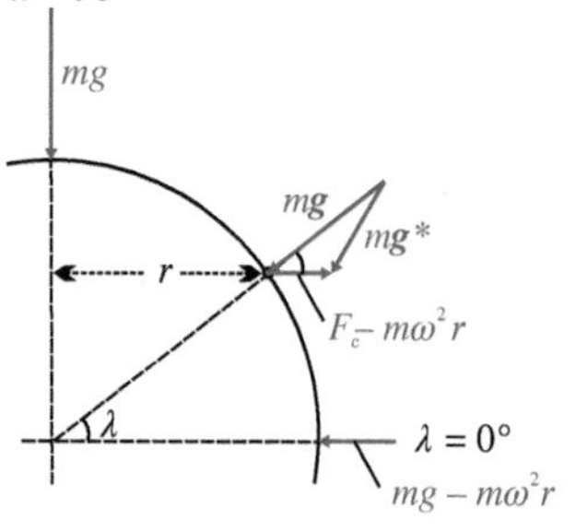

Figure 8.16. The addition of the centrifugal force $\boldsymbol{F}_c$ on a mass at latitude λ to its weight $m\boldsymbol{g}$ to give the effective weight of the mass in the Earth's rotating frame, $m\boldsymbol{g}^*$. Note that the magnitude of $\boldsymbol{F}_c$ relative to $m\boldsymbol{g}$ is exaggerated.

In effect we have created a pseudo-inertial reference frame at the Earth's surface by introducing a fictitious force. In practice, the magnitude of $\boldsymbol{g}^*$ changes very little between $\lambda = 0°$ and 90°,

$$\boldsymbol{g}^*\text{equator} - \boldsymbol{g}^*\text{pole} = \omega^2 R = 0.034 \text{ m s}^{-2},$$

so that the variation can often be ignored. However, if we wish to make an accurate calculation of an object's motion over a large distance near the Earth's surface, for example in launching a spacecraft, we need to take the effect into account.

As shown in Worked Example 8.3, the variation of $\boldsymbol{g}$ due to centrifugal force is comparable to the variation of $\boldsymbol{g}$ due to the non-sphericity of the Earth. As illustrated in Figure 8.17, the Earth bulges at the equator (as a result of the effect of the larger value of centrifugal force at the equator at the time when the Earth was still made up of molten material) so that R_e, the distance from the Earth's surface to its centre at the equator, is larger than R_p, the distance from the surface to the centre at the pole. The magnitude of $\boldsymbol{g}$ can be estimated approximately using Equation (5.4) for spherical planets. The value of g at the equator, $g_e = \dfrac{GM_E}{R_e^2}$, is therefore less than its value at the pole, $g_p = \dfrac{GM_E}{R_p^2}$.

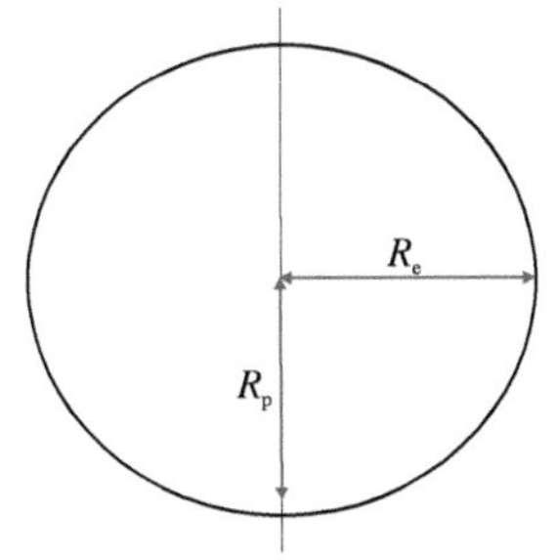

Figure 8.17. The non-sphericity of the Earth (exaggerated). The distance from the Earth's surface to its centre is larger at the equator than at a pole.

Study Worked Example 8.3

For problems based on the material presented in this section visit up.ucc.ie/8/ *and follow the link to the problems.*

8.5 Motion in a rotating frame: the Coriolis force

In Section 8.4 we considered the behaviour of an object which was released from rest at the edge of a rotating roundabout. To an observer who was fixed to the roundabout, the object experienced an apparent force, namely *centrifugal force*. In this section we broaden the discussion to include objects which are moving in the rotating frame. To understand the origin of the apparent force on a moving object consider the motion of an object which is moving with constant velocity in an inertial frame (the x, y frame) as it passes over a rotating roundabout, as illustrated in Figure 8.18(a). To an outside observer the object moves in a straight line but, when viewed by an observer who is fixed to the roundabout, looking outwards as it rotates (the x', y' frame), the particle follows a curved path, as illustrated in Figure 8.18(b). If the observer is unaware that the roundabout is rotating, he or she will ascribe the curvature to an apparent force perpendicular to the direction of motion. We now investigate this apparent force, called the *Coriolis force*.

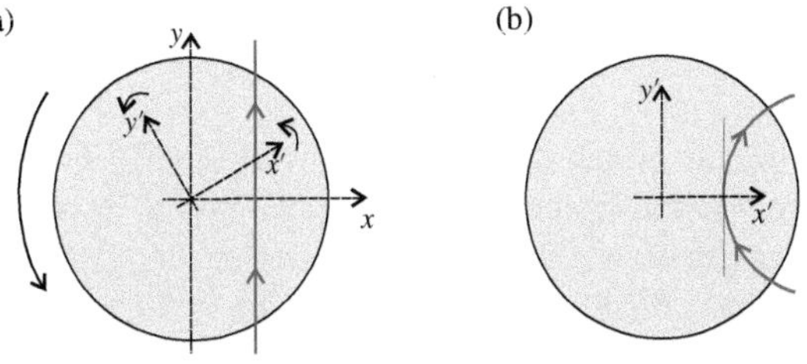

Figure 8.18. The path of a particle (blue line) moving with constant velocity in an inertial frame (a) as viewed in the inertial frame and (b) as viewed by an observer on a rotating roundabout frame.

In Figure 8.19 a coordinate frame (x', y', z') is rotating with constant angular velocity $\boldsymbol{\omega}$ about the z-axis of a fixed coordinate frame (x, y, z) with the z- and z'- axes and origins coinciding. The displacement of a body at a point P from the common origin (O,O′) is the same in the both frames so that $\boldsymbol{r} = \boldsymbol{r}'$.

While displacement from the origin of a fixed point is the same in both frames, a change of displacement is not the same when measured in the two frames. This is illustrated in Figure 8.20 which shows a displacement from P to Q, $\Delta\boldsymbol{r}$ in the fixed system. In the time Δt in which the body moves from P to Q, P moves to P′ so that the displacement of the body in the rotating frame, $\Delta'\boldsymbol{r}$, is not equal to $\Delta\boldsymbol{r}$.

The angular velocity of the rotating frame ω is constant so that PP′ $= (\boldsymbol{\omega} \times \boldsymbol{r})\ \Delta t$ and, as illustrated in Figure 8.20, we can write

$$\Delta\boldsymbol{r} = \Delta'\boldsymbol{r} + (\boldsymbol{\omega} \times \boldsymbol{r})\Delta t \qquad (8.16)$$

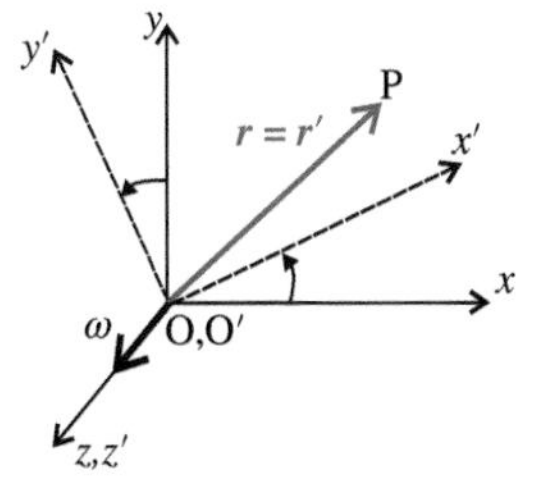

Figure 8.19. A coordinate frame (x', y', z') which is rotating with constant velocity $\boldsymbol{\omega}$ about the z-axis of a fixed coordinate frame (x, y, z) with the z- and z'- axes and origins coinciding. P is fixed in the rotating system.

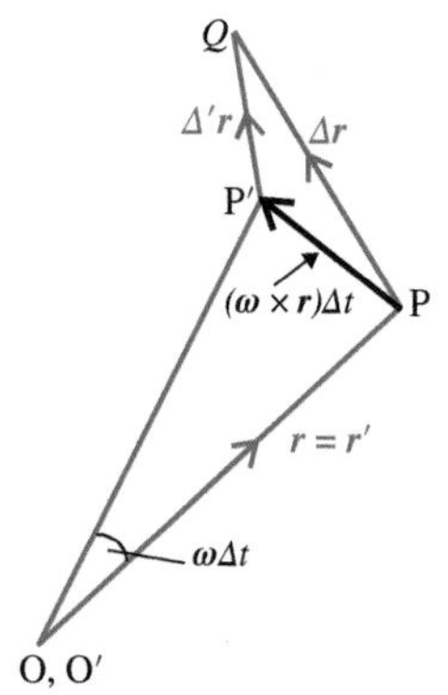

Figure 8.20. A change of displacement Δr, as viewed from a fixed frame, and $\Delta' \boldsymbol{r}$, as viewed from a rotating frame.

Thus, while $\boldsymbol{r} = \boldsymbol{r}'$, $\dfrac{\Delta \boldsymbol{r}}{\Delta t}$ is not equal to $\dfrac{\Delta' \boldsymbol{r}}{\Delta t}$ so that the time derivatives of displacement are different in the two frames. To distinguish between them, we use the notation $\dfrac{d\boldsymbol{r}}{dt}$ in the fixed frame and $\dfrac{d'\boldsymbol{r}}{dt}$ in the rotating frame. It follows from Equation (8.16) that

$$\boldsymbol{v} = \frac{d\boldsymbol{r}}{dt} = \frac{d'\boldsymbol{r}}{dt} + \boldsymbol{\omega} \times \boldsymbol{r} \tag{8.17}$$

where $\boldsymbol{v}$ is the velocity of the body in the fixed frame.

The procedure used to derive Equation (8.17) for the displacement vector may be applied to any vector A.

Thus
$$\frac{d\boldsymbol{A}}{dt} = \frac{d'\boldsymbol{A}}{dt} + \boldsymbol{\omega} \times \boldsymbol{r}$$

If $\boldsymbol{A}$ represents velocity $\boldsymbol{v}$, this becomes $\dfrac{d\boldsymbol{v}}{dt} = \dfrac{d'\boldsymbol{v}}{dt} + \boldsymbol{\omega} \times \boldsymbol{v}$

Substituting $\boldsymbol{v}$ from Equation (8.17), this equation becomes

$$\frac{d\boldsymbol{v}}{dt} = \frac{d'}{dt}\left(\frac{d'\boldsymbol{r}}{dt} + \boldsymbol{\omega} \times \boldsymbol{r}\right) + \boldsymbol{\omega} \times \left(\frac{d'\boldsymbol{r}}{dt} + \boldsymbol{\omega} \times \boldsymbol{r}\right)$$

$$= \frac{d'^2\boldsymbol{r}}{dt^2} + \left(\boldsymbol{\omega} \times \frac{d'\boldsymbol{r}}{dt}\right) + \left(\boldsymbol{\omega} \times \frac{d'\boldsymbol{r}}{dt}\right) + \boldsymbol{\omega} \times (\boldsymbol{\omega} \times \boldsymbol{r})$$

Thus
$$\frac{d\boldsymbol{v}}{dt} = \frac{d'^2\boldsymbol{r}}{dt^2} + 2\left(\boldsymbol{\omega} \times \frac{d'\boldsymbol{r}}{dt}\right) + \boldsymbol{\omega} \times (\boldsymbol{\omega} \times \boldsymbol{r}),$$

which we can write as

$$\boldsymbol{a} = \boldsymbol{a}' + 2(\boldsymbol{\omega} \times \boldsymbol{v}') + \boldsymbol{\omega} \times (\boldsymbol{\omega} \times \boldsymbol{r}) \tag{8.18}$$

where $\boldsymbol{a}$ and $\boldsymbol{a}'$ are, respectively, the accelerations of the body in the fixed and rotating frames and $\boldsymbol{v}'$ is the velocity of the body in the rotating frame.

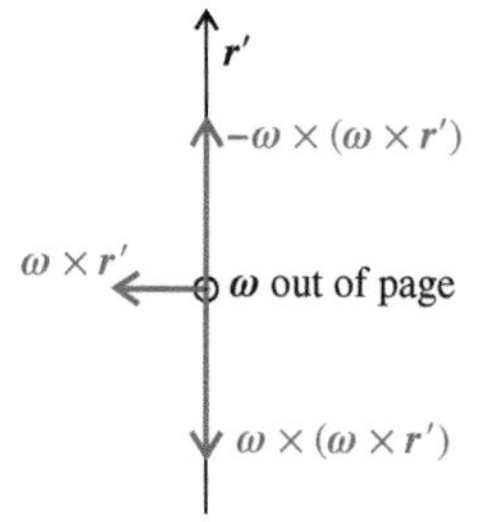

Figure 8.21. The vector multiplication of ω and $\omega \times r'$ to produce $\omega \times (\omega \times r')$.

Multiplying Equation (8.18) throughout by m, the mass of the body, we obtain an equation which relates $\boldsymbol{F}'$, the force experienced by the body in the rotating frame to $\boldsymbol{F}$, the force experienced in the fixed frame,

$$\boldsymbol{F} = \boldsymbol{F}' + 2m\,(\boldsymbol{\omega} \times \boldsymbol{v}') + m\boldsymbol{\omega} \times (\boldsymbol{\omega} \times \boldsymbol{r})$$

or
$$\boldsymbol{F}' = \boldsymbol{F} - 2m\,(\boldsymbol{\omega} \times \boldsymbol{v}') - m\boldsymbol{\omega} \times (\boldsymbol{\omega} \times \boldsymbol{r}) \tag{8.19}$$

The first point to note is that Equations (8.18) and (8.19) state clearly that, when we transfer from a fixed inertial frame to a frame which is rotating at a constant angular velocity $\boldsymbol{\omega}$ relative to the fixed frame, the mass appears to experience a change of acceleration, and hence an apparent change in the force acting on it. This confirms the statement made in the discussion of the roundabout in Section 8.4 to the effect that a frame which is performing uniform circular motion relative to an inertial frame cannot be inertial.

The double vector product $\boldsymbol{\omega} \times (\boldsymbol{\omega} \times \boldsymbol{r})$, which occurs in the third term on the right hand side of Equation (8.19), is evaluated by taking the vector cross product of $\boldsymbol{\omega}$ and the vector result of the product $(\boldsymbol{\omega} \times \boldsymbol{r})$. As noted in Section 4.11, cross products of vectors must be carried out in the prescribed order. Thus, as illustrated in Figure 8.21, the result of the vector multiplication $-m\boldsymbol{\omega} \times (\boldsymbol{\omega} \times \boldsymbol{r}')$ is a vector of magnitude $m\omega^2 r$ in the $+\boldsymbol{r}$ direction. The $-m\boldsymbol{\omega} \times (\boldsymbol{\omega} \times \boldsymbol{r}')$ term therefore represents centrifugal force and, as noted in Section 8.4, may be incorporated in the analysis of motion near the Earth's surface by adopting an effective value of $\boldsymbol{g}$, denoted by $\boldsymbol{g}^*$.

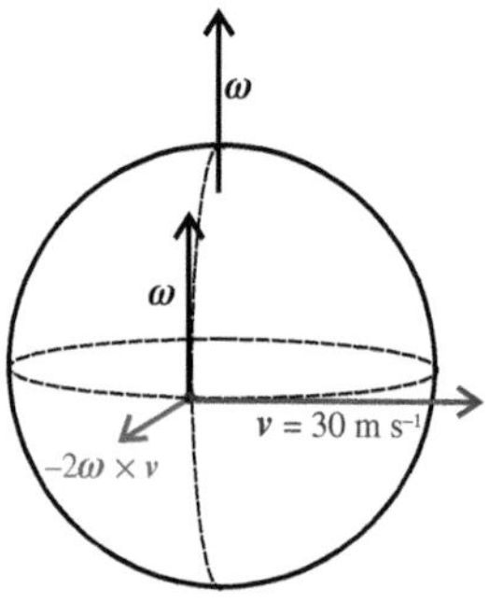

Figure 8.22. The direction of the Coriolis acceleration $-2\boldsymbol{\omega} \times \boldsymbol{v}$ of a mass which is moving eastwards along the Earth's equator with velocity $\boldsymbol{v}$.

The second term on the right-hand side of Equation (8.19), $-2m\,(\boldsymbol{\omega} \times \boldsymbol{v}')$, is new. It applies only when $\boldsymbol{v}' \neq 0$, that is, when the mass is moving in the rotating frame. It is known as the **Coriolis force**, after Gaspard de Coriolis (1792–1843), and, like centrifugal force, it is a fictitious force which is introduced to enable the motion of a mass in the rotating (non-inertial) frame to be described using Newton's second law of motion. An outside observer in an inertial frame sees no need for this term. To this observer the mass accelerates only if it is subjected to a force $\boldsymbol{F}$ in the inertial frame.

Figure 8.19 can be used to represent the rotation of the Earth if the z-axis is in the direction of the Earth's spin axis, that is, towards geographic north.

A Coriolis force applies to any moving mass when it is observed from the Earth's rotating frame. Consider, for example, the effect of the force on an object which is moving at 30 m s^{-1} (108 km hr^{-1}) eastwards along the equator, as illustrated in Figure 8.22. The Coriolis acceleration, $-2\boldsymbol{\omega} \times \boldsymbol{v}$, is directed away from the centre of the Earth and is of magnitude $-2\omega v$ where ω is the angular velocity of the Earth. The angular velocity of the Earth, ω, is given by $\frac{2\pi}{T}$, where T, the period of rotation, is one day. Thus $\omega = 7.3 \times 10^{-5}$ rad s^{-1}. The acceleration of this object due to the Coriolis force is therefore 0.004 m s^{-2}, which is small compared with g or centrifugal acceleration (estimated to be 0.034 m s^{-2} in Worked Example 8.3).

Coriolis force and the formation of weather systems

In Worked Example 8.4 the effect of the Coriolis force is estimated for a case in which a mass falls vertically through 100 m near a point on the Earth's equator. The net deflection is calculated to be 2.2 cm eastwards. While the effects of Coriolis forces are generally negligible for objects which travel short distances, they can be substantial for long range motion near the Earth's surface because the forces act over a long period of time. Coriolis force effects are particularly important in the formation and evolution of weather systems. As indicated in Figure 8.23(a), in the northern hemisphere the Coriolis force acts to deflect a body which is moving northwards to its right (eastwards).

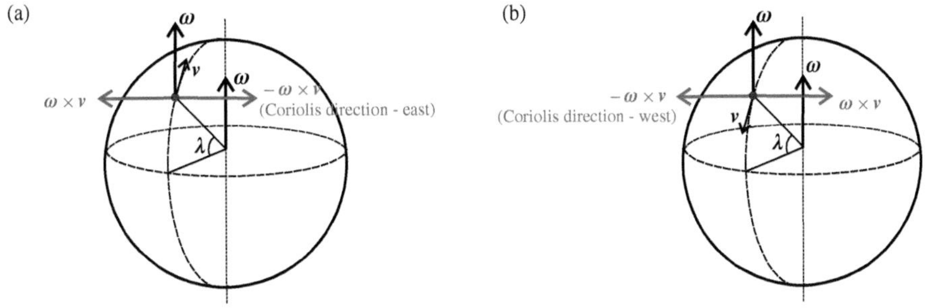

Figure 8.23. The direction of the Coriolis acceleration at latitude λ (a) for a body which is moving northwards and (b) for a body which is moving southwards.

For a southward moving body, the Coriolis force acts to deflect the body to its right (westwards), as shown in Figure 8.23(b).This motion in the Earth's reference frame occurs because the Earth's surface moves to the left (westwards) during the motion. In the southern hemisphere the Coriolis force acts to deflect both northward and southward moving bodies to their left.

Figure 8.24 describes the consequences of the Coriolis force for weather formation in the northern hemisphere. The winds which flow towards the centre of a low pressure region will be deflected to the right by the horizontal component of the Coriolis force so that they circulate in an anticlockwise direction to form a *cyclone*. On the other hand the deflection to the right of winds flowing outwards from a high pressure region causes them to circulate clockwise to form *anticyclones*. In the southern hemisphere the situation reverses so that winds circulate in a clockwise sense round low pressure (cyclonic) regions and in an anticlockwise sense around high pressure regions.

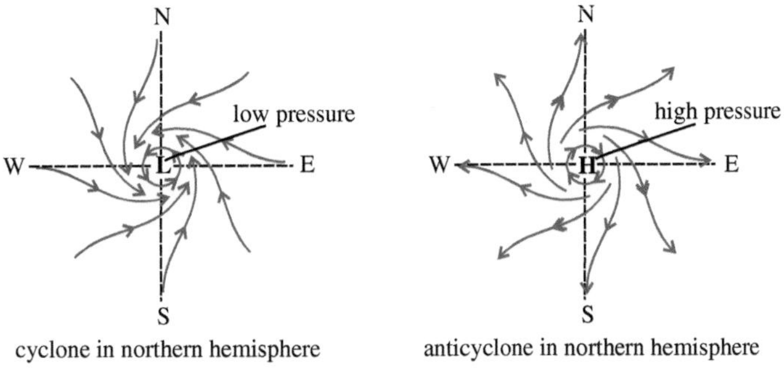

Figure 8.24. The formation of cyclones and anticyclones through the action of Coriolis forces in the northern hemisphere.

On a world-wide scale the action of the Coriolis force plays an important role in the circulation of the Earth's atmosphere and of ocean currents. For example, as illustrated in Figure 8.25, the air which moves into the equatorial regions to replace rising warm air is deflected towards the west in both hemispheres by Coriolis forces, producing the *trade winds*, prevailing winds from the northeast in the northern hemisphere and from the southeast in the southern hemisphere.

Coriolis forces have a particularly striking effect on the circulation of the atmosphere of the planet Jupiter. The rotation period of Jupiter is about 10 hours so that it rotates with an angular velocity which is more than twice that of the Earth. The atmosphere of Jupiter is observed to separate into a series of bands which, like the trade winds, circulate parallel to the planet's equator.

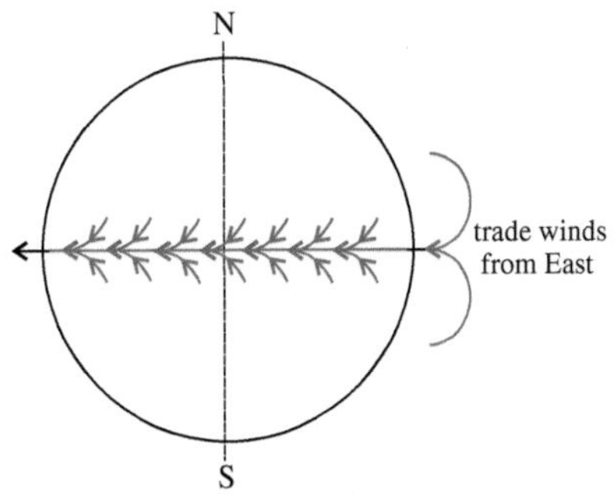

Figure 8.25. The formation of the trade winds in the equatorial region of the Earth.

Study Worked Example 8.4

For problems based on the material presented in this section visit up.ucc.ie/8/ *and follow the link to the problems.*

8.6 The Foucault pendulum

We now consider the effect of the rotation of the Earth on a situation which we have already analysed in Section 4.10, namely the simple pendulum. In performing simple harmonic motion in an inertial frame, pendulum oscillations are confined to a fixed vertical plane.

For small angle oscillations the motion of the bob may be approximated to horizontal motion in a straight line (Figure 8.26). By virtue of its velocity the bob will also experience a Coriolis force due to the rotation of the Earth.

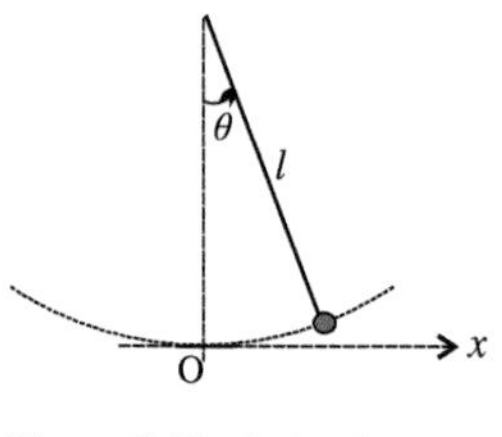

Figure 8.26. A simple pendulum.

Consider a pendulum which is released so that, initially, it oscillates in an east-west direction in the northern hemisphere, as illustrated in Figure 8.27. If the Earth were not rotating the pendulum would continue to oscillate between A and B. However, due to the Coriolis force its path is continuously deflected to the right, A to B′ to A′ to B″ to A″ etc., as shown in Figure 8.27. In consequence the plane of oscillation of the pendulum rotates clockwise in the northern hemisphere. In the southern hemisphere the rotation is anticlockwise.

This type of motion, in which the plane of the motion is rotating steadily, is known as a *precession*. In the web section up.ucc.ie/8/6/1/ it is shown that the period of the precession is a function of latitude λ, given by

$$T = \frac{2\pi}{\omega \sin \lambda}.$$

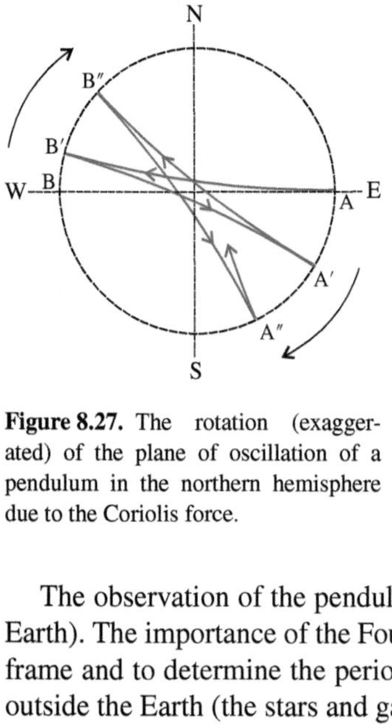

Figure 8.27. The rotation (exaggerated) of the plane of oscillation of a pendulum in the northern hemisphere due to the Coriolis force.

$\frac{2\pi}{\omega}$ is the period of rotation of the Earth which is one day so that the period of the precession, that is the period of rotation of the plane of oscillation of the pendulum, is $\frac{1}{\sin \lambda}$ days. At the poles ($\lambda = 90°$) the plane of oscillation precesses once in a day whereas when $\lambda = 45°$ the period of the precession is $\sqrt{2}$ days. The period increases to infinity at the equator ($\lambda = 0°$) so that there is no precession there. The effect at the poles is particularly simple to interpret; an outside observer will see the pendulum performing oscillations which are confined to the vertical plane in the fixed (inertial) frame while the Earth rotates beneath it once in a day. Thus, in the Earth's rotating frame, the plane of precession of the pendulum has rotated once in a day.

This effect can be demonstrated with a **Foucault pendulum** (named after Jean Bernard Leon Foucault (1819–1868)). This is a long simple pendulum which can oscillate in conditions which are stable enough to enable its motion to be observed over a long period of time (several days). Foucault carried out a series of experiments in early 1851, culminating in the use of a 67 m long pendulum suspended from the dome of the Pantheon building in Paris. The length of the Foucault pendulum ensures that the amplitude of oscillation is small and that the approximation to horizontal motion of the bob is valid. Foucault found that the plane of oscillation of the pendulum rotated at a rate of 11.25° per hour.

The observation of the pendulum's precession provides clear evidence that we are situated on a rotating reference frame (that of the Earth). The importance of the Foucault pendulum is that it would have enabled us to establish that we are situated in a rotating reference frame and to determine the period of rotation relative to an inertial frame, even if we had not been able to use observations of objects outside the Earth (the stars and galaxies, for example) as fixed reference points.

This last statement suggests an alternative approach to the characterisation of inertial reference frames. An acceptable working definition along the following lines could be adopted.

An inertial frame of reference is a coordinate system which is at rest or moving with constant velocity relative to the average spatial configuration of all the matter and energy of the universe.

From this perspective, Newton's first law becomes redundant and is simply an experimental observation that Newton's second law holds true in the particular set of frames defined in this way. It remains a largely open question as to why the laws of motion and, more generally, the laws of nature have their simplest forms in such frames

8.7 Practical criteria for inertial frames: the local view

In the previous sections we have seen that, in order to observe motion on the Earth's surface from an inertial reference frame, we must adopt a reference frame which is fixed in space outside the Earth. However we know that the Earth is moving around the Sun and that the Sun in turn moves in the galaxy, etc. so that, while the question of whether it is possible to establish an absolute inertial frame in the universe is of interest in cosmology, it is difficult to identify such a frame in practice.

We can, however, approach the problem from a practical viewpoint, using Newton's first law of motion. For most practical purposes a reference frame can be said to be effectively inertial throughout a certain region of space and time if a test particle which is released in that frame obeys Newton's first law within some specified accuracy. In other words, if any departures from Newton's first law due to the non-inertial nature of a frame are too small to detect, we can consider the frame to be inertial.

A frame can be considered to be inertial if all objects in the frame have the same acceleration with respect to an inertial frame. As an example, let us consider a lift which is descending with an acceleration $\boldsymbol{a}$ (Figure 8.28(a)), which is less than $\boldsymbol{g}$, the acceleration due to gravity. A test particle released from rest in the lift would not remain at rest but would accelerate downwards at $(\boldsymbol{g} - \boldsymbol{a})$ relative to the lift. It would not obey Newton's first law in relation to its surroundings so that a frame fixed to the lift would not be an inertial frame. A person in the lift would still be conscious of their (albeit reduced) weight because the floor of the lift is not moving away from them as quickly as they are falling under gravity.

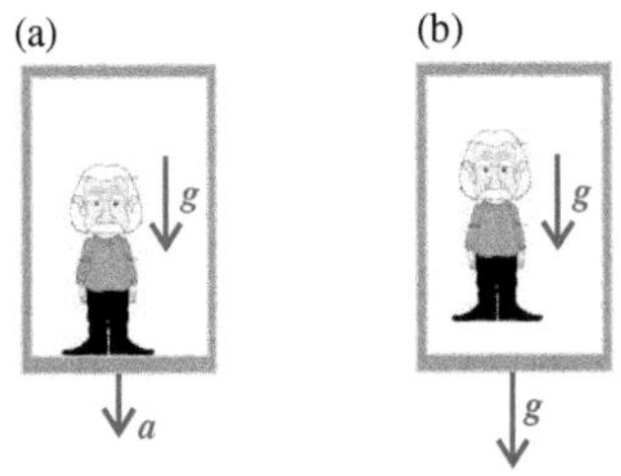

Figure 8.28. The acceleration of the occupant of a lift relative to the lift (a) when the lift is descending with an acceleration $\boldsymbol{a}$ which is less than $\boldsymbol{g}$ and (b) when the lift is in free fall, i.e. descending with an acceleration which is equal to $\boldsymbol{g}$.

If, however, the lift were allowed to fall freely, both the lift and its contents would have the same acceleration downwards, namely $\boldsymbol{g}$ (Figure 8.28(b)). A test particle released from rest would have zero acceleration relative to the lift and would obey Newton's first law in that frame. The lift frame could therefore be considered to be inertial, the acceleration due to gravity not being detectable relative to the lift. The occupant of the lift would be 'weightless' in this frame because the floor would move downwards with exactly the same acceleration as that with which he or she was falling. In the case of a freely falling lift this happy situation would quickly come to an abrupt end when the lift reached the ground. However, as illustrated in Figure 8.29, a similar, but more durable, situation applies in the case of a spaceship which is orbiting the Earth.

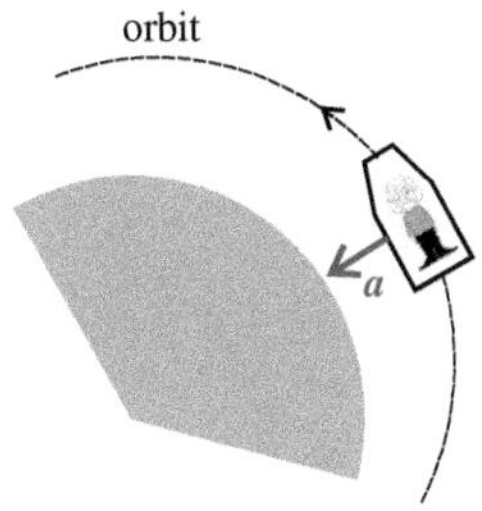

Figure 8.29. The acceleration a of a spaceship and its occupant when they are both in orbit around the Earth.

The spaceship and its contents have the same acceleration relative to the Earth, the centripetal acceleration provided by gravity which is needed to sustain their motion around the Earth. Hence a test particle, which is released from rest in the spaceship, obeys Newton's first law in the spaceship frame; the frame is inertial. An astronaut inside the spaceship experiences no acceleration relative to his or her immediate surroundings. The acceleration due to gravity cannot be detected in this frame. Consequently the occupants of the spaceship are weightless in the reference frame of the spaceship, the reference frame of their immediate surroundings.

In practice this view is valid if the region of space considered is not too large and is a long way from the centre of the force field which produces the acceleration (the centre of the Earth in this case). If, as illustrated in Figure 8.30, we consider a large region of space, the direction or magnitude of the acceleration due to gravity will vary detectably across the volume of the region of space and the existence of a varying external force will become apparent. The frame will then be non-inertial.

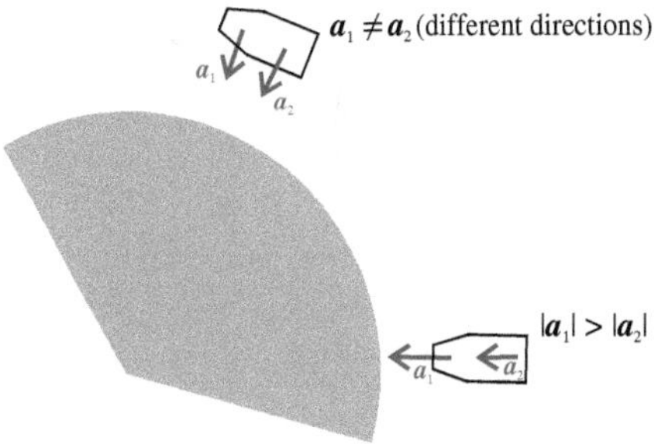

Figure 8.30. The variation of the acceleration due to gravity of particles which are released in a large region of space close to the Earth.

The question of whether a particular region can be considered to constitute an inertial reference frame is ultimately a practical matter. It depends on whether the methods of measurement available are accurate enough to detect departures from Newton's first law of motion. The more accurate the methods of measurement become or the closer to the centre of the force field the region moves, the smaller the inertial region becomes. This effect is examined quantitatively in Worked Example 8.5. The necessity of defining a practical inertial frame in order to analyse a situation, therefore, tends to limit the region of space considered – it leads us to take a **local view**.

Study Worked Example 8.5

For problems based on the material presented in this section visit up.ucc.ie/8/ and follow the link to the problems.

WORKED EXAMPLES

Worked Example 8.1: A train is travelling eastwards along a straight track at 100 km hr^{-1} when a stone is thrown from it horizontally with a velocity, relative to the train, of 30 km hr^{-1} in the direction 30° N of E relative to the train. Calculate the velocity (magnitude and direction) of the stone relative to the ground. If the stone is thrown from a height of 2 m above the ground how far does it travel before hitting the ground (i) in the frame of the track and (ii) in the frame of the train?

Let the (x, y) and (x', y') frames represent the track and train frames, respectively, with the x- and x'-axes pointing eastwards, as shown in Figure 8.31. We can apply the inverse Galilean velocity transformation Equations (8.10) and (8.11)

$$u_x = u_x' + v$$

$$u_y = u'_y$$

where v is the velocity of the train, u' is the velocity of the stone relative to the train and u is the velocity of the stone relative to the track (the ground). Thus $u_x' = 30\cos 30° = 26\ \text{km h}^{-1}$ and $u_y' = 30\sin 30° = 15\ \text{km h}^{-1}$.

The transformation equations therefore give

$$u_x = 1000 + 26 = 126\ \text{km h}^{-1}$$

and

$$u_y = 15\ \text{km h}^{-1}$$

As illustrated in Figure 8.31, the velocity of the stone relative to the ground (the track frame) is given by $u = \sqrt{u_x^2 + u_y^2} = 126.9\ \text{km h}^{-1}$ in the direction given by $\theta = \tan^{-1}\dfrac{u_y}{u_x} - \tan^{-1}\dfrac{15}{126} = 6.8°$ N of E.

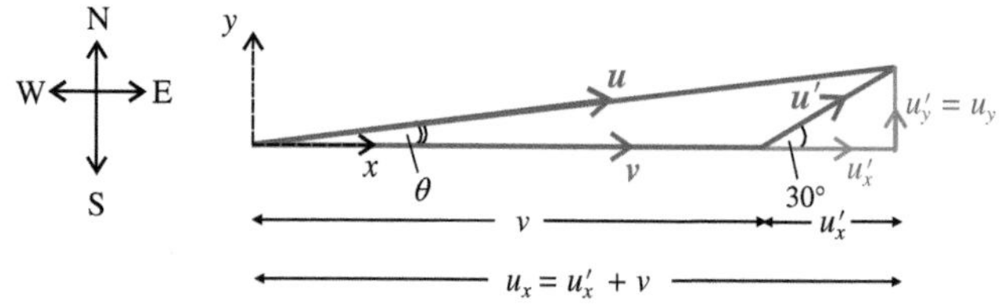

Figure 8.31. Worked Example 8.1. The components of $\boldsymbol{u}$, the velocity of the stone in the track frame, and of $\boldsymbol{u}'$, the velocity of the stone in the train frame.

The vertical component of the stone's motion, in either frame, is straight line motion with constant acceleration (g) so that the time taken for the stone to hit the ground is given by Equation (2.19). Taking downwards to be the positive direction,

$$x = v_0 t + \frac{1}{2}At^2 \quad \text{where } x = 2\ \text{m},\ v_0 = 0 \text{ and } A = g.$$

Thus

$$2 = \frac{1}{2} \times 9.8t^2 \quad \text{which gives } t = 0.64\ \text{s} = 1.78 \times 10^{-4}\ \text{h}$$

In this time the stone travels a distance ut in the track frame and $u't$ in the train frame. The distances travelled in two frames are therefore $126.9 \times 1000 \times 1.78 \times 10^{-4} = 22.6$ m and $30 \times 1000 \times 1.78 \times 10^{-4} = 5.3$ m, respectively.

Worked Example 8.2: A 1.0 kg mass travelling at 3.0 m s^{-1} collides elastically with a 2.0 kg mass which is travelling in the opposite direction at 2.0 m s^{-1}. Determine the velocities of the masses in the CM frame. If after the collision the 1.0 kg mass travels at an angle of 65° to its original direction in the CM frame, determine its velocity (magnitude and direction) in the laboratory frame.

From Equation (6.15) the velocity of the centre of mass of the system $V_C = \dfrac{m_1u_1 + m_2u_2}{m_1 + m_2} = \dfrac{(1.0 \times 3.0) + (-2.0 \times 2.0)}{3} = -0.33\ \text{m s}^{-1}$,

LAB frame

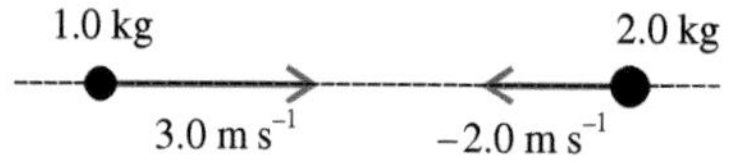

before collision

Figure 8.32. Worked Example 8.2. An elastic collision of 1.0 and 2.0 kg masses: velocities before the collision in the LAB frame.

The velocities of the masses in the CM frame are therefore given by the equations

$$u'_1 = u_1 - V_C = 3.00 - (-0.33) = 3.33\ \text{m s}^{-1}$$

and

$$u_2' = u_2 - V_C = -2.00 - (-0.33) = -1.67\ \text{m s}^{-1}, \text{ as illustrated in Figure 8.33.}$$

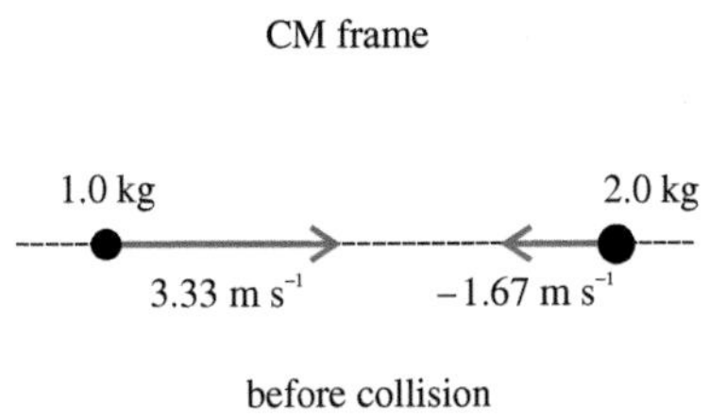

Figure 8.33. Worked Example 8.2. An elastic collision of 1.0 and 2.0 kg masses: velocities before the collision in the CM frame.

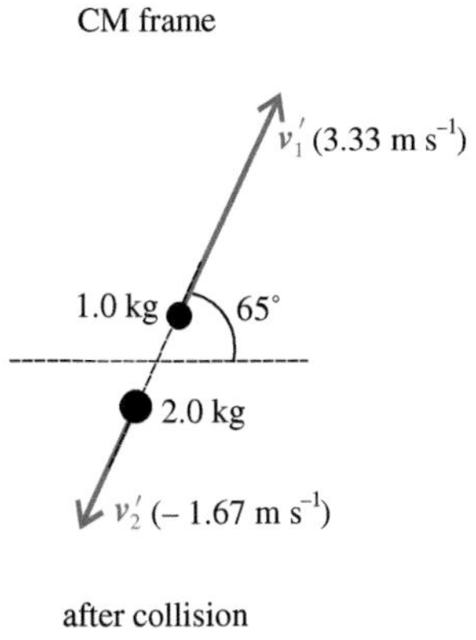

Figure 8.34. Worked Example 8.2. Velocities after the collision in the CM frame.

The velocities of the masses in the CM frame after the collision, $\mathbf{v}_1'$ and $\mathbf{v}_2'$, are then as indicated in Figure 8.34.

The velocity of the 1.0 kg mass after the collision in the LAB frame, $\mathbf{v}_1$, is obtained by adding $\mathbf{V}_C$ vectorially to its velocity in the CM frame, $\mathbf{v}_1'$ (3.33 m s^{-1}), $\mathbf{v}_1 = \mathbf{v}_1' + \mathbf{V}_C$, as shown in Figure 8.35.

From the cosine rule (Appendix A.3.3) $v_1^2 = (3.33)^2 + (0.33)^2 - (2 \times 3.33 \times 0.33)\cos 65°$, which gives $v_1 = 3.20$ m s^{-1} at an angle θ which is given by the sine rule (Appendix A.3.4) $\dfrac{\sin\theta}{3.33} = \dfrac{\sin 65°}{3.20}$.

Thus $\sin\theta = 0.943$ and $\theta = 70.6°$, as indicated in Figure 8.35.

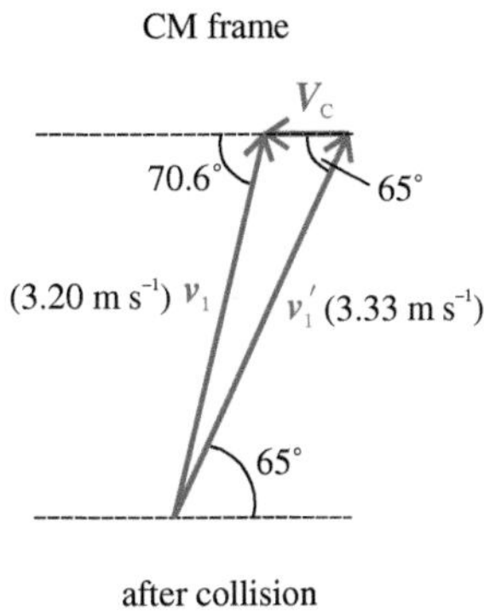

Figure 8.35. Worked Example 8.2. Transformation of the velocity of the 1.0 kg mass from the CM frame to the LAB frame, after the collision.

Worked Example 8.3: Estimate the variation in the value of g between the equator and the poles due to the non-sphericity of the Earth. The mass of the Earth is 5.98×10^{24} kg and the distance from the surface to the centre of the Earth is 6379 km at the equator and 6357 km at the poles.

Estimate also the variation in the value of the measured value of g between the equator and the poles of the Earth due to centrifugal force.

The value of g at the equator,
$$g_e = \frac{GM_E}{R_e^2} = \frac{6.67 \times 10^{-11} \times 5.98 \times 10^{24}}{(6379 \times 10^3)^2} = 9.80\ \text{m s}^{-2},$$

and its valueat the poles,
$$g_p = \frac{GM_E}{R_p^2} = \frac{6.67 \times 10^{-11} \times 5.98 \times 10^{24}}{(6357 \times 10^3)^2} = 9.87\ \text{m s}^{-2},$$

Thus the variation in g between the equator and the poles,due to the non-sphericity of the Earth, is 0.07 m s^{-2}.
The variation in g due to centrifugal effects is $\omega^2 R_e$ (Section 8.4) where ω is the angular velocity of the Earth, given by

$$\omega = \frac{2\pi}{T} = \frac{2\pi}{1\text{ day}} = 7.27 \times 10^{-5}\ \text{s}^{-1}.$$

Thus, the variation in g due to centrifugal effects is $(7.27 \times 10^{-5})^2 \times 6379 \times 10^3 = 0.034$ m s^{-2}.

Worked Example 8.4: A mass is released from rest 100 m directly above a point P on the Earth's equator. Neglecting any effects due to the variation of g with height and any effects of air resistance, determine how far from P the mass will hit the ground and the direction of the point of impact from P.

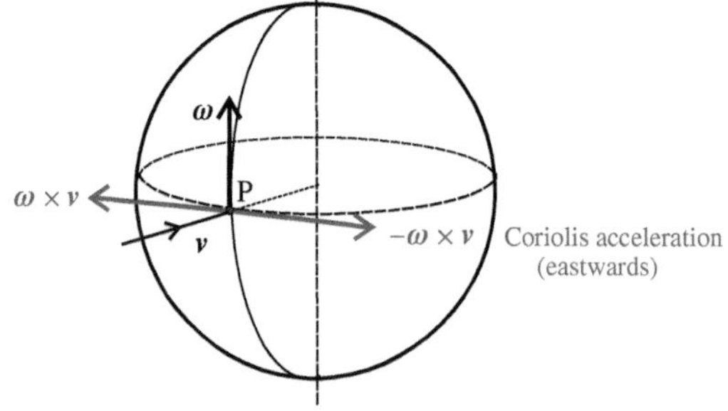

Figure 8.36. Worked Example 8.4. The direction of the Coriolis acceleration, $-\boldsymbol{\omega} \times \boldsymbol{v}$, of a mass which is released from rest directly above the Earth's equator.

The Coriolis acceleration, $\boldsymbol{a}_C = -2(\boldsymbol{\omega} \times \boldsymbol{v})$, is directed eastwards, as shown in Figure 8.36.

If the mass is released at $t = 0$ $\quad |\boldsymbol{a}_C| = 2\omega v$

$v = gt$ so that we can write $\quad |\boldsymbol{a}_C| = 2\omega g t$

We use integration to obtain an equation for the velocity due to the Coriolis force

$$|\boldsymbol{v}_c| = \int_0^t |\boldsymbol{a}_c|\, dt = \int_0^t (2\omega g t) dt = [\omega g t^2]_0^t = \omega g t^2$$

A further integration gives the displacement due to the Coriolis force

$$|\boldsymbol{r}_c| = \int_0^t |\boldsymbol{v}_c| dt = \int_0^t (\omega g t^2) dt = \left[\frac{1}{3}\omega g t^3\right]_0^t = \frac{1}{3}\omega g t^3$$

The time taken for the mass to reach the ground is given by $h = \frac{1}{2}gt^2$ where h is the height from which the mass is dropped. (Note that, in our approximation, $\boldsymbol{a}_c$ is perpendicular to $\boldsymbol{g}$ and does not affect the time taken to fall). Thus,

$$t = \sqrt{\frac{2h}{g}} = \sqrt{\frac{200}{9.80}} = 4.52\ \text{s}$$

and $r_c = \frac{1}{3}(7.27 \times 10^{-2}) \times 9.80 \times (4.52)^3 = 0.022$ m $= 2.2$ cm. The displacement is in the direction of the Coriolis force (eastwards), so that the point of impact is east of P.

Worked Example 8.5: (a) A cubic container of side length 10 m is released from rest when its bottom face is horizontal and at a height of 100 m from the Earth's surface. At the same time two particles are released from rest, one near the top and the other near the bottom so that the two particles are vertically aligned. How much further apart will they be when the container reaches the Earth?
(b) If, alternatively, the particles had been released from points on the sides of the container which are directly opposite each other horizontally, how much closer would the particles have been when the container reached the Earth? Use your results to discuss whether the container can be considered to be an inertial frame. [You may assume the Earth to be a uniform sphere of radius 6370 km]

(a) The change in the vertical separation of the two particles at the top and bottom of the container occurs because the acceleration due to gravity is smaller at the top of the container than it is at the bottom.

The time taken for the container to reach the Earth's surface is given by $h = \frac{1}{2}gt^2$, where h is the height through which the container falls. The value of g will change over the 100 m fall but the change is approximately the same for the two particles and is therefore neglected. Thus

$$g = \frac{GM_E}{R_E{}^2} = \frac{6.67 \times 10^{-11} \times 5.98 \times 10^{24}}{(6\,370\,000)^2} = 9.83\ \text{m s}^{-2} \quad \text{so that} \quad t = \sqrt{\frac{2h}{g}} = \sqrt{\frac{200}{9.83}} = 4.51\ \text{s}$$

In this time the particle at the top of the container falls through a height of $\frac{1}{2}g_t t^2$ whereas the particle at the top of the container falls through a height of $\frac{1}{2}g_b t^2$. The particles therefore separated by $\frac{1}{2}(g_b - g_t)t^2$.

For $h \ll R_E$, we can write

$$g_b - g_t = \frac{GM_E}{R_E^2} - \frac{GM_E}{(R_E+h)^2} \approx GM_E\left[\frac{(R_E+h)^2 - R_E^2}{R_E^4}\right] \approx GM_E\left[\frac{2R_E h}{R_E^4}\right] = 3\times10^{-5}\ \mathrm{m\,s^{-2}}$$

so that the separation is $\frac{1}{2}\times(3\times10^{-5})\times4.51^2 = 0.0003\ \mathrm{m} = 0.3\ \mathrm{mm}$

(b) For the particles which are directly opposite each other horizontally the movement towards one another is due to the change in the direction of g across the width of the container, $\Delta\theta$, as illustrated in Figure 8.37.

$$\Delta\theta = \frac{10}{6370100}\ \mathrm{rad}$$

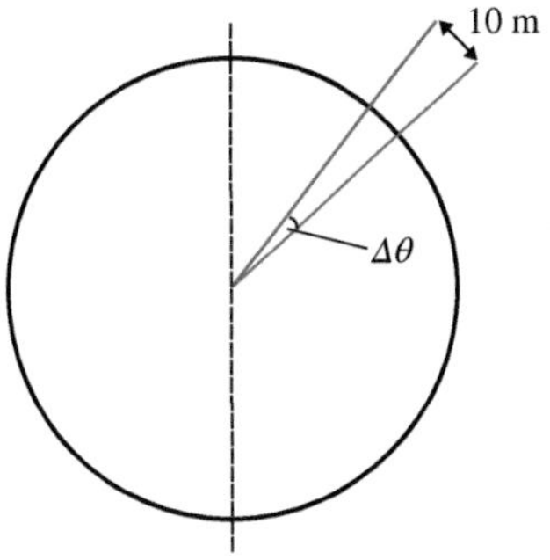

Figure 8.37. Worked Example 8.5. The angle $\Delta\theta$ subtended at the centre of the Earth by the width of the container.

When the container starts falling the two particles are separated by 10 m. When it hits the ground their separation is $R\Delta\theta = 6370000\times\left(\frac{10}{6370100}\right) = 9.99984$ m.

The particles are therefore 0.00016 m = 0.16 mm closer together when the container hits the ground. Thus, if equipment available cannot detect movements of less than 0.3 mm, the container can be considered to be an inertial frame.

PROBLEMS

For problems based on the material covered this chapter visit up.ucc.ie/8/ and follow the link to the problems.

9

Special relativity

AIMS

- to examine the consequences of the Principle of Relativity, in particular the invariance of the velocity of light in a vacuum with change of inertial reference frame
- to replace the Galilean transformation with a more general transformation (the *Lorentz transformation*) which is consistent with the direct connection between space and time coordinates which is introduced by the Principle of Relativity
- to use the Lorentz transformation to examine how the space and time separations of two events can be different in different inertial frames (the *time dilation* and *length contraction* phenomena)
- to redefine momentum and energy more generally so as to conform with the Principle of Relativity
- to examine the connection between momentum and energy in relativistic mechanics
- to examine the concept of relativistic energy and thus the equivalence of mass and energy, demonstrating that relativistic energy is a more basic and more comprehensive concept than (classical) Newtonian energy

9.1 The velocity of light

When we come to consider Maxwell's theory of electromagnetism in Chapter 21 we shall find that the velocity of light in a vacuum, or more generally the velocity of an electromagnetic wave in a vacuum, emerges from the theory with the same value ($c = 299\,792\,458\ \mathrm{m\,s^{-1}}$) regardless of the frame of reference in which the theory is formulated. The quantity c, therefore, is a fundamental constant of nature rather than simply a property of the medium (the vacuum) through which the light is passing. This implies that the value of c will not change when we change the inertial frame in which it is measured. The velocity of light is said to be **invariant** under such changes of frame.

This result is in direct conflict with the findings of the previous chapter, where the Galilean transformation showed us that the value of the velocity of an object *must* change when it is measured from a coordinate system which is moving. Specifically the Galilean velocity transformation, Equation (8.13), states that

$$\boldsymbol{u} = \boldsymbol{u}' + \boldsymbol{v},$$

where $\boldsymbol{u}'$ is velocity of the object in a frame which is moving with velocity $\boldsymbol{v}$ relative to the frame in which we wish to determine $\boldsymbol{u}$. In Example 2 of Section 8.2 it is shown that the equation applies correctly to the velocity of sound, which is a property of the medium which is transmitting the sound – the rest medium of the sound waves.

An important experiment was carried out by Michelson (Albert Michelson, 1852–1931) and Morley (Edward Morley, 1838–1923) in the early 1880's to investigate whether Equation (8.13) also applies to the motion of light (visit the web section up.ucc.ie/9/1/1 for details of this experiment). Michelson and Morley compared the velocity of light in the direction in which the Earth is orbiting the Sun with its value in a perpendicular direction. The experiment sought to reveal whether a rest medium (called the *ether*) existed for electromagnetic waves and, if it did exist, how the Earth was moving relative to this medium. The experiment can be compared with the Foucault pendulum experiment (Section 8.6) which established that the Earth is a rotating reference frame.

Michelson and Morley concluded that there was no difference between the values of c measured in reference frames moving parallel to and perpendicular to the Earth's motion and hence found no evidence for the existence of an ether. Later experiments with greater accuracy of measurement have confirmed this finding. Equation (8.13) does not apply when $\boldsymbol{v} = c$. The invariance of c is fundamentally at odds with the Galilean view of relativity.

Understanding Physics, Third Edition. Michael Mansfield and Colm O'Sullivan.
© 2020 John Wiley & Sons Ltd. Published 2020 by John Wiley & Sons Ltd.

The issue can be stated in more general terms. Two of the most successful theories of physics, namely classical mechanics and the theory of electromagnetism, are in conflict. Problems of this nature have recurred regularly throughout the history of physics. A valuable and well tested model fails when it is applied to a new situation. The model must be revised so that it can handle the new situation, while remaining valid in the contexts in which it is known to apply. As outlined in the next section, the conflict between classical mechanics and electromagnetism can be resolved by extending the formalism of relative motion so that it can accommodate the invariance of the velocity of light.

9.2 The principle of relativity

The starting point for a new approach to the treatment of relative motion is Einstein's **Principle of Relativity** (Albert Einstein 1879–1955). The Principle can be stated in the following terms:

The laws of Physics have the same form in all inertial reference frames

It follows from this principle that both the form of the laws and the numerical values of physical constants contained in the formulation of these laws are the same in all inertial frames. The invariance of c, a fundamental constant, will be built into the new theory from the outset, as a result of which there will be no difficulty in accommodating Maxwell's theory of electromagnetism within mechanics. Note that the Principle of Relativity restates a point which was made in the discussion of the Galilean Transformation (Section 8.2), namely that, while the values of variable quantities may change when the reference frame is changed, they must do so in such a way that any laws which relate such quantities remain valid. In particular, the values of fundamental constants must not change when the reference frame is changed.

The Principle of Relativity may seem an innocuous statement but, as we shall see in the next section, it leads us quickly to some remarkable but inescapable conclusions.

9.3 Consequences of the principle of relativity

We first illustrate some of the consequences of the Principle of Relativity by applying it to a particular situation. A more general treatment will be given in the next section.

The example which we now consider is the emission of a flash of light (which we call event A, specified by its position and time) and its detection at a point on the x-axis (event B) after reflection by a mirror which is at rest in the laboratory frame. In Figure 9.1 we consider these two events as they would be observed in two inertial reference frames, the laboratory frame and a moving frame (which we call the rocket frame).

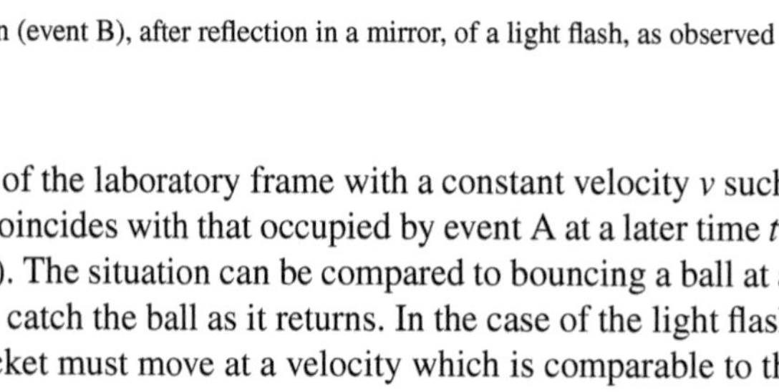

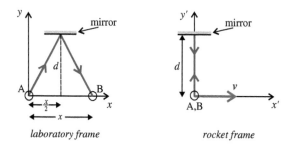

Figure 9.1. The emission (event A) and detection (event B), after reflection in a mirror, of a light flash, as observed in the laboratory (x, y) and rocket (x', y') frames.

The rocket frame moves along the x-axis of the laboratory frame with a constant velocity v such that, in the rocket frame, the position of event B (the detection of the light flash) coincides with that occupied by event A at a later time t'. In the rocket frame, therefore, events A and B occur at the same place (the origin). The situation can be compared to bouncing a ball at an angle off a wall while running parallel to the wall at exactly the speed needed to catch the ball as it returns. In the case of the light flash, in order to be in a position to catch the flash as it returns from the mirror, the rocket must move at a velocity which is comparable to the velocity of light ($\sim 3 \times 10^8$ m s^{-1}).

The velocity of the rocket frame relative to the laboratory frame,is given by

$$v = \frac{x}{t} \tag{9.1}$$

where x is the change in displacement of the origin of the rocket frame after the time t has elapsed in the laboratory frame.

Let us consider the values of some quantities as observed in the two frames.

	laboratory frame	rocket frame
Event coordinates		
Event A	$x = y = z = 0;\quad t = 0$	$x' = y' = z' = 0;\quad t' = 0$
Event B	$x = x,\ y = z = 0;\quad t = t$	$x' = y' = z' = 0;\quad t' = t'$
Distance travelled by the light flash (as illustrated in Figure 9.1)		
	$2\sqrt{d^2 + \left(\frac{x}{2}\right)^2}$ at velocity c	$2d$ at velocity c
Time taken $\left(\frac{\textit{distance}}{\textit{velocity}}\right)$	$t = \frac{2}{c}\sqrt{d^2 + \left(\frac{x}{2}\right)^2}$ (9.2)	$t' = \frac{2d}{c} = \frac{2}{c}\sqrt{d^2}$ (9.3)

Now $\sqrt{d^2 + \left(\frac{x}{2}\right)^2} > d$ so $t > t'$. The time elapsed is longer in the laboratory frame. This result can be understood quite simply. Because the light flash has travelled further in the laboratory frame and because *its velocity is the same in both frames*, it must take longer to get from A to B in the laboratory frame. We have shown that the time separation of the two events is not the same in the two frames; it is not an **invariant** quantity.

Although the time separation of two events is not invariant, a quantity can be defined which is invariant between the two frames. This quantity, called the **interval** between the two events, is defined as follows.

$$(\text{interval})^2 := (\text{time separation})^2 - \left(\frac{\text{space separation}}{c}\right)^2 = (t)^2 - \left(\frac{x}{c}\right)^2$$

In the laboratory frame:
$$(\textit{interval})^2 = \frac{4}{c^2}\left[d^2 + \left(\frac{x}{2}\right)^2\right] - \left(\frac{x}{c}\right)^2 = \frac{4d^2}{c^2} \tag{9.4}$$

In the rocket frame:
$$(\textit{interval})^2 = t'^2 - \left(\frac{x'}{c}\right)^2 = \frac{4d^2}{c^2} - 0 = \frac{4d^2}{c^2} \tag{9.5}$$

Thus, the interval is the same in either frame; it is invariant.

The definition of the interval can be generalised easily to three spatial dimensions as follows.

$$(\textit{interval})^2 = (t)^2 - \left(\frac{x}{c}\right)^2 - \left(\frac{y}{c}\right)^2 - \left(\frac{z}{c}\right)^2 = (t')^2 - \left(\frac{x'}{c}\right)^2 - \left(\frac{y'}{c}\right)^2 - \left(\frac{z'}{c}\right)^2 \tag{9.6}$$

We are now in a position to identify three notable consequences of the application of the Principle of Relativity to the case which we have been discussing, results which can be shown to be quite general.

First consequence: The time difference between two events changes when the reference frame is changed. We can state this result quantitatively using Equations (9.2) and (9.3).

$$\frac{t}{t'} = \frac{\frac{2}{c}\sqrt{d^2 + \left(\frac{x}{2}\right)^2}}{\frac{2d}{c}} = \sqrt{\frac{d^2}{d^2} + \left(\frac{x}{2d}\right)^2} = \sqrt{1 + \left(\frac{x}{2d}\right)^2}$$

Substituting $x = vt$ from Equation (9.1) and $2d = ct'$ from Equation (9.3),

we obtain $\left(\frac{t}{t'}\right)^2 = 1 + \left(\frac{vt}{ct'}\right)^2$ and thus $\left(\frac{t}{t'}\right)^2\left(1 - \frac{v^2}{c^2}\right) = 1,$

so that we can write
$$\frac{t}{t'} = \frac{1}{\sqrt{1 - \frac{v^2}{c^2}}} \tag{9.7}$$

Equation (9.7) tells us that when $v = 0$, $t = t'$, but, as v approaches c (the velocity of light in vacuum), $t \gg t'$. The time interval becomes much larger in the laboratory frame.

Second consequence: As we have demonstrated in Equations (9.4) and (9.5), the interval $\sqrt{t^2 - \frac{x^2}{c^2}}$ is invariant – it has the same value in any inertial reference frame.

The invariance of the interval can be compared to the invariance of s, the distance between two adjacent points in spatial coordinates. From Pythagoras' Theorem (Appendix A.2.1) we can deduce that, in two spatial dimensions,

$$s^2 = x^2 + y^2 = x'^2 + y'^2$$

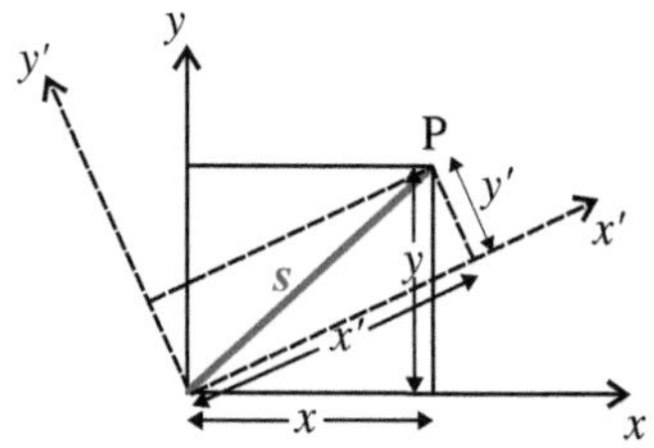

Figure 9.2. The invariance of distance s with change of reference frame.

As illustrated in Figure 9.2, s remains invariant when we change from the reference frame (x, y) to another frame whose axes (x', y') have been rotated through an angle with respect to the (x, y) axes. We will use the analogy between the invariance of distance in space and the invariance of the interval in space and time extensively in this chapter.

Three types of interval can be identified, namely (i) *spacelike* intervals in which $\frac{x}{t} > c$, (ii) *lightlike* intervals in which $\frac{x}{t} = c$ and (iii) *timelike* intervals in which $\frac{x}{t} < c$.

As we shall see in Section 9.9, all particles with mass travel at velocities which are smaller than c so that events along the path of a particle must be separated by timelike intervals. The timelike interval is given the name **proper time**, with the symbol τ. Thus

$$\boxed{\tau := \sqrt{t^2 - \frac{x^2}{c^2}}}$$

Note that, because for particles $\frac{x}{t} < c$, $\frac{x}{c} < t$ so that the proper time τ is always positive.

The proper time is the time separation of events A and B in the frame in which $x = 0$, that is, in the frame which moves with the particle whose motion we are studying; in other words, the frame in which the particle is at rest (the rocket frame in the case which we have just considered). As noted in Section 8.3, this frame is often called the **rest frame** of the particle. In all other reference frames the time separation between two events is longer because

$$t = \sqrt{\tau^2 + \frac{x^2}{c^2}} \quad \text{and therefore } t > \tau$$

Third consequence: the values of the space and time coordinates are directly connected through Equation (9.6)

$$(interval)^2 = (t)^2 - \left(\frac{x}{c}\right)^2 - \left(\frac{y}{c}\right)^2 - \left(\frac{z}{c}\right)^2$$

just as the three spatial coordinates of distance are connected through the equation

$$(distance)^2 = x^2 + y^2 + z^2$$

In either case, if one of the coordinates has a different value in a second frame, the other coordinates in the second frame must also change their values so as to maintain the invariance of the interval or distance. Thus, in relativity, it is usual to talk in terms of *four dimensions* (x, y, z, t), the **space–time coordinates**, instead of three dimensions (x, y, z).

Note that, in order to reach the three remarkable results which we have identified above, we have needed only the Principle of Relativity and some simple geometry, specifically Pythagoras' Theorem.

So far, we have illustrated the consequences of the Principle of Relativity by considering a special case, the emission and reception of a light flash. We now proceed to derive a general transformation which will enable us to convert coordinates between frames moving with constant relative velocity in a manner which is consistent with the Principle of Relativity.

For problems based on the material presented in this section visit up.ucc.ie/9/ and follow the link to the problems.

9.4 The Lorentz transformation

The Galilean Transformation (Section 8.2) gives the coordinates of a point in a moving frame (x', y', z', t') in terms of its coordinates in the laboratory frame (x, y, z, t) and v, the velocity of the moving frame along the x-axis of the laboratory frame. In deriving the Galilean Transformation, however, we assumed that time passed at the same rate in the two frames, specifically that $t = t'$ (Equation (8.4)).

We have found in the previous section (Equation (9.7)) that this assumption is not consistent with the Principle of Relativity. We must seek, therefore, a more general transformation which can replace the Galilean Transformation.

As in our treatment of the Galilean Transformation, we take the moving frame to be moving at a constant velocity v along the x-axis of the laboratory frame (Figure 9.3).

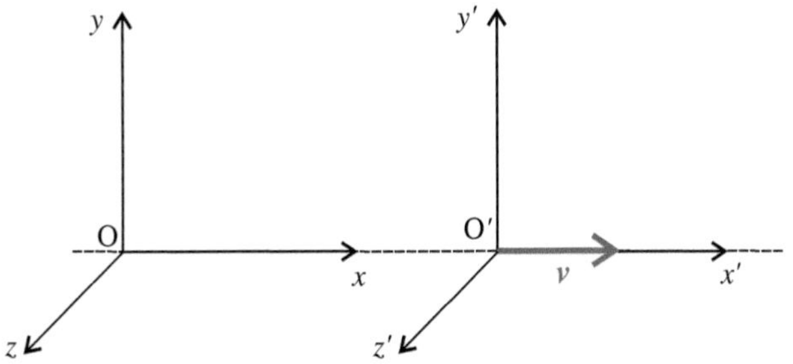

Figure 9.3. The Lorentz transformation. The frame (x', y') is moving with velocity $\boldsymbol{v}$ along the x-axis of the laboratory frame.

Since we can no longer set t equal to t' and since time and space coordinates are related, we now assume that the value of t depends on the values of both t' and x', that is that the transformation equations for t and x are of the general form,

$$t = Ax' + Bt' \tag{9.8}$$

and

$$x = Cx' + Dt' \tag{9.9}$$

where A, B, C and D are constants for a fixed value of v. We now have to determine how the values of these constants depend on v. Note that, in order to ensure that all solutions are single valued and physically meaningful, we have assumed that Equations (9.8) and (9.9) are linear – that is, that they do not involve higher powers of any of the variables (for example, x'^2).

We determine values for the coefficients A, B, C and D by substituting into Equations (9.8) and (9.9) the results for two special cases, (1) and (2) considered below, in which we already know what the relationships between x, t, x' and t' should be and by using the invariance of the interval in the two frames. For convenience we choose the origins of the two frames so that $x = t = 0$ coincides with $x' = t' = 0$.

Case 1: Consider the coordinates of a point which is at rest at the origin of the moving frame, that is a point for which $x' = 0$ at all times. The equation of motion of this point in the laboratory frame is therefore the equation of motion of the origin of the moving frame, that is

$$x = vt \tag{9.10}$$

By substituting $x' = 0$ in Equations (9.8) and (9.9) we obtain

$$t = Bt' \tag{9.11}$$

and

$$x = Dt' \tag{9.12}$$

The invariance of the interval gives

$$t^2 - \frac{x^2}{c^2} = t'^2 - 0 \tag{9.13}$$

Substituting $x = vt$ (Equation (9.10)) in Equation (9.13) we obtain

$$t^2 - \frac{v^2t^2}{c^2} = t'^2 \quad \rightarrow \quad t^2 = \frac{t'^2}{1 - \frac{v^2}{c^2}} \quad \rightarrow \quad t = \frac{t'}{\sqrt{1 - \frac{v^2}{c^2}}}$$

Because the quantity $\dfrac{1}{\sqrt{1 - \frac{v^2}{c^2}}}$, which depends only on the relative velocity of the frames, occurs regularly in relativistic relationships, we denote it by the symbol γ, that is

$$\boxed{\gamma := \frac{1}{\sqrt{1 - \frac{v^2}{c^2}}}}$$

and we can write
$$t = \gamma t' \tag{9.14}$$

and, from Equation (9.10),
$$x = vt = \gamma vt' \tag{9.15}$$

By comparing Equations (9.14) and (9.15) with (9.11) and (9.12), we can deduce that

$$B = \gamma \quad \text{and} \quad D = \gamma v$$

Hence Equations (9.8) and (9.9) can be written
$$t = Ax' + \gamma t' \tag{9.16}$$

and
$$x = Cx' + \gamma vt' \tag{9.17}$$

Case 2: In this more general case we consider a point which is not at the origin of the second frame, that is, $x' \neq 0$. The general formula for the invariance of the interval is
$$t^2 - \frac{x^2}{c^2} = t'^2 - \frac{x'^2}{c^2}$$

Substituting for t and x from Equations (9.16) and (9.17) we obtain

$$(Ax' + \gamma t')^2 - \frac{(Cx' + \gamma vt')^2}{c^2} = t'^2 - \frac{x'^2}{c^2}$$

Expanding the left-hand side of this equation

$$A^2x'^2 + \gamma^2t'^2 + 2A\gamma x't' - \frac{C^2x'^2}{c^2} - \frac{\gamma^2v^2t'^2}{c^2} - \frac{2C\gamma vx't'}{c^2} = t'^2 - \frac{x'^2}{c^2}$$

Thus
$$t'^2\left(\gamma^2 - \frac{\gamma^2v^2}{c^2}\right) + \frac{x'^2}{c^2}(A^2c^2 - C^2) + \left(2A\gamma - \frac{2C\gamma v}{c^2}\right)x't' = t'^2 - \frac{x'^2}{c^2}$$

Comparing the two sides of the equation and recalling, from the definition of γ, that $\gamma^2 - \dfrac{\gamma^2v^2}{c^2} = 1$, we find that the equation can only be satisfied if the coefficients of the terms in t'^2, x'^2 and $x't'$ are the same on both sides, that is if

$$A^2c^2 - C^2 = -1 \quad \text{which gives} \quad C^2 = A^2c^2 + 1 \tag{9.18}$$

and
$$2A\gamma - \frac{2vC\gamma}{c^2} = 0 \quad \text{which gives} \quad A = \frac{Cv}{c^2}$$

Substitution of $A = \dfrac{Cv}{c^2}$ into Equation (9.18) yields $C^2 = \dfrac{C^2v^2}{c^2} + 1$

and hence
$$C^2 = \frac{1}{1 - \dfrac{v^2}{c^2}} = \gamma^2$$

Thus
$$C = \gamma \quad \text{and} \quad A = \frac{\gamma v}{c^2}$$

We now have values for all four coefficients A, B, C and D and can write out Equations (9.8) and (9.9) in full:

$$x = \gamma x' + \gamma vt' = \gamma(x' + vt') \tag{9.19}$$
$$y = y' \tag{9.20}$$
$$z = z' \tag{9.21}$$
$$t = \frac{\gamma vx'}{c^2} + \gamma t' = \gamma\left(t' + \frac{vx'}{c^2}\right) \tag{9.22}$$

These equations are known as the **inverse Lorentz transformation** (after Hendrik Lorentz, 1853–1928). Note that if we apply Equation (9.22) to the emission and detection of a light flash, as considered in the previous section, we obtain, for $x' = 0$, $t = \gamma t'$, in agreement with Equation (9.7).

We can also write down the **Lorentz transformation**, which transforms coordinates from the laboratory frame to the moving frame. In the moving frame (x', y', z', t'), the laboratory frame is moving with a velocity $-v$ along the x'-axis. Hence, by substituting $-v$ for $+v$ and by interchanging $t \leftrightarrow t'$ and $x \leftrightarrow x'$ in the inverse Lorentz transformation (9.19) to (9.22) we obtain the Lorentz transformation,

$$x' = \gamma(x - vt) \tag{9.23}$$

$$y' = y \tag{9.24}$$

$$z' = z \tag{9.25}$$

$$t' = \gamma\left(t - \frac{vx}{c^2}\right) \tag{9.26}$$

Note that, when $v \to 0$, $\gamma \to 1$ and $\frac{v}{c^2} \to 0$, and consequently the Lorentz transformation reduces to the Galilean transformation of Section 8.2. The Galilean transformation, therefore, is adequate for most everyday applications. However, when $\frac{v}{c} \to 1$, $\gamma \to \infty$, the Lorentz transformation gives very different results from the Galilean transformation.

When $v \ll c$ we can use the binomial theorem (Appendix A.9.2), namely

$$(1 + x)^n = 1 + nx + n(n-1)\frac{x^2}{2!} + \ldots \text{ with } x = -\frac{v^2}{c^2} \quad \text{and} \quad n = -\frac{1}{2}$$

to write γ as $$\gamma = \frac{1}{\sqrt{1 - \frac{v^2}{c^2}}} = 1 + \frac{1}{2}\frac{v^2}{c^2} + \frac{3}{8}\frac{v^4}{c^4} + \text{higher terms.}$$

Thus, when $v \ll c$, the Lorentz transformation modifies the Galilean values of quantities such as x by adding a correction of order $\frac{1}{2}\frac{v^2}{c^2}$; for example, when $v = 0.1c$ the correction is about 0.5%. Whether or not a relativistic (Lorentz) treatment is required is ultimately a practical consideration. If the correction is greater than the accuracy of measurement in a particular situation, a relativistic treatment is required; if not it can be safely ignored.

We should note that we have not shown that the Galilean transformation is wrong; rather we have shown that it is limited in its range of applicability. In the next few sections we shall use the Lorentz transformation to show how the Principle of Relativity can change our view of the world radically when velocities approach the velocity of light.

For problems based on the material presented in this section visit up.ucc.ie/9/ and follow the link to the problems.

9.5 The Fitzgerald–Lorentz contraction

Consider a ruler which is strapped to a rocket which is moving with velocity v along the x-axis of the laboratory frame, as illustrated in Figure 9.4.

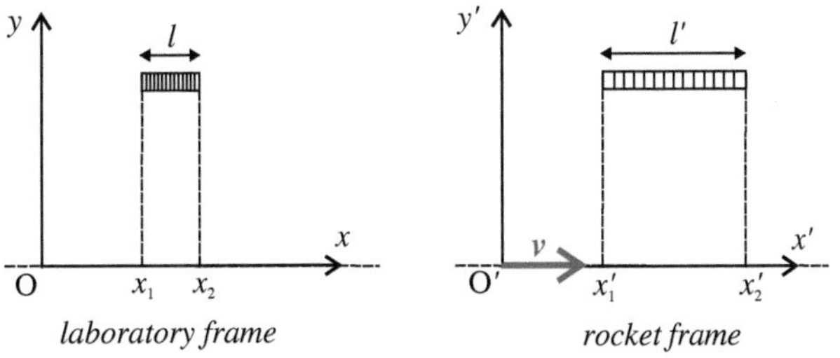

Figure 9.4. The Fitzgerald-Lorentz contraction of a ruler which is moving with velocity $\boldsymbol{v}$ along the x-axis of the laboratory frame. The length of the ruler is l' in its own frame and l in the laboratory frame.

The length of the ruler in the laboratory frame $\quad l = x_2 - x_1$

and its length in the rocket frame $\quad l' = x_2' - x_1'$

Substituting for x_1' and x_2', using the Lorentz transformation (9.23), we obtain

$$l' = \gamma(x_2 - vt_2) - \gamma(x_1 - vt_1)$$

Now, if both measurements x_2 and x_1 are made at the same instant in the laboratory frame, which requires the use of synchronised clocks (clocks which have been set so that they agree on time readings) at x_2 and x_1, then $t_2 = t_1 = t$ and we can write

$$l' = \gamma x_2 - \gamma vt - \gamma x_1 + \gamma vt = \gamma(x_2 - x_1) = \gamma l$$

Thus,

$$\boxed{l = \frac{l'}{\gamma} = l'\sqrt{1 - \frac{v^2}{c^2}}}$$

From its definition $\left(\gamma = \frac{1}{\sqrt{1 - \frac{v^2}{c^2}}}\right)$, $\gamma \geq 1$ and therefore $l \leq l'$ (with $\gamma = 1$ referring to the case when $v = 0$). The length of the ruler as observed in the frame in which the ruler is moving (the laboratory frame) is less than its length as measured in the frame in which it is at rest (the rocket frame). This effect is known as the **Fitzgerald–Lorentz (length) contraction**, after George Francis Fitzgerald (1851–1901).

Note that the Lorentz Transformation Equations (9.24) and (9.25) also tell us that $y = y'$ and $z = z'$; that is, there is no change of length in directions which are perpendicular to the direction in which the object is moving. The net effect, as illustrated in Figure 9.5 is that, when observed from a fixed frame, an object which is moving with a velocity which approaches that of light will appear to be compressed in its direction of motion. To make measurements of this effect one must also take account of other significant issues such as the different times taken by light from the two ends of the ruler to reach an observer.

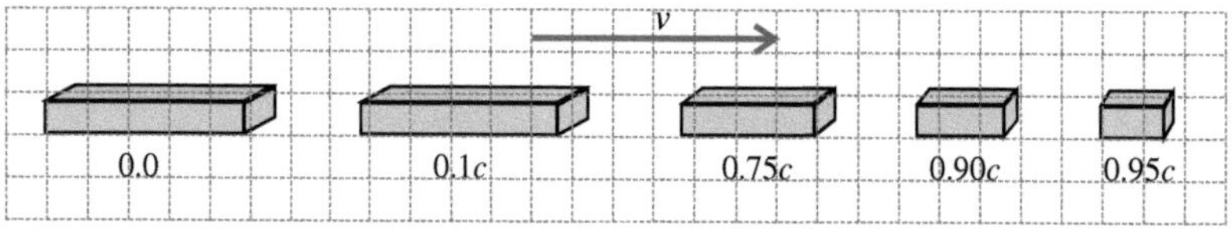

Figure 9.5. The compression of an object, as observed in the laboratory frame, as it moves with increasing velocity with respect to the laboratory frame.

For problems based on the material presented in this section visit up.ucc.ie/9/ *and follow the link to the problems.*

9.6 Time dilation

Consider a device which, while strapped to a rocket which is moving with velocity $\boldsymbol{v}$ along the x-axis of the laboratory frame, emits two signals (for example, successive light flashes), one at time t_1 and the other at time t_2. As illustrated in Figure 9.6, both signals are emitted at the origin ($x' = 0$) of the rocket frame.

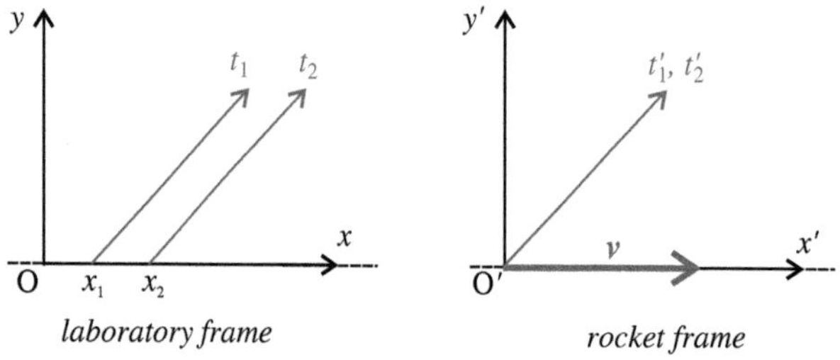

Figure 9.6. Time dilation. The emission of two signals which are emitted at the origin of the rocket frame, as observed in the laboratory and rocket frames.

In the laboratory frame the device changes its position between emitting each signal so that $x_1 \neq x_2$.

Using the inverse Lorentz transformation (9.22), we can determine the time interval in the laboratory frame $(t_2 - t_1)$ in terms of the time interval in the rocket frame $(t'_2 - t'_1)$. Thus

$$t_2 - t_1 = \gamma\left(t'_2 + \frac{vx'_2}{c^2}\right) - \gamma\left(t'_1 + \frac{vx'_1}{c^2}\right)$$

Now $x_1' = x_2' = 0$ at t_1' and t_2'.

Thus,

$$t_2 - t_1 = \gamma(t_2' - t_1') = \frac{t_2' - t_1'}{\sqrt{1 - \frac{v^2}{c^2}}} \tag{9.27}$$

so that, unless $v = 0$, $t_2 - t_1 > t_2' - t_1'$

The time interval in the laboratory frame is γ times longer than the time interval in the rocket frame, the rest frame of the device which is emitting the signals. Time is 'stretched' in the laboratory frame, the phenomenon known as **time dilation**. Because time intervals are longer in frames which are moving relative to the rest frame, such as the laboratory frame in the example we have just considered, clocks must run more quickly in such moving frames.

Time dilation for muons in the Earth's atmosphere

Experimental evidence of time dilation is provided by the observation of muons at the Earth's surface. Muons are fundamental particles (to be considered in Section 27.11) which are similar to electrons except that they are unstable and decay into other elementary particles. The half-life of the muon at rest (the time in which 50% of a group of muons decays) is 1.5 μs.

Muons are produced in large numbers by the interaction of cosmic rays with the top of the Earth's atmosphere. These interactions take place approximately 10 km above the Earth's surface. The velocity of a muon as it moves down through the atmosphere is very close to c (typically ~$0.9995c$), so that a calculation of the typical time taken by a muon to reach the Earth's surface, ignoring relativistic effects, yields $t = \frac{10^4}{3 \times 10^8} = 33 \times 10^{-6}$ s, which corresponds to $\frac{33}{1.5} = 22$ half-lives. The number of muons is reduced by half each time a half-life elapses so that we would expect only $\left(\frac{1}{2}\right)^{22} = 2 \times 10^{-7}$ of the muons which are produced at the top of the Earth's atmosphere to actually reach the Earth's surface. In practice, however, it is found that almost all muons reach the Earth's surface.

The discrepancy between our calculation and observation arises because the phenomenon requires a relativistic treatment, as we would expect for particles which travel with velocities which are so close to the velocity of light. The half-life of the muon, $t'_{1/2}$, which applies in the rest frame of the muon (the rocket frame), corresponds to a half-life, $t_{1/2}$, in the laboratory frame, the frame in which we make our observations. The muon half-life in the laboratory frame is given by the time dilation Equation (9.27) with $t_2 - t_1 = t_{1/2}$ and $t_2' - t_1' = t'_{1/2}$. Thus, $t_{1/2} = \gamma t'_{1/2}$

For a muon with velocity $v = 0.9995c$, $\gamma \approx 30$ and therefore $t_{1/2} = 30 \times 1.5 \times 10^{-6} = 45$ μs. This is the muon half-life in the laboratory frame. This time is significantly greater than the time which we estimated above was needed for a muon to reach the Earth's surface (33×10^{-6} s). Hence, a relativistic approach tells us that most muons will reach the Earth's surface, in accordance with observation.

For problems based on the material presented in this section visit up.ucc.ie/9/ and follow the link to the problems.

9.7 Paradoxes in special relativity

One of the questions raised by the phenomenon of time dilation is whether clocks, in particular biological clocks, run more slowly for observers who are moving at velocities which are close to that of light, that is, whether a person ages more slowly in such a time frame. The answer to this question is 'yes – a person does age more slowly in such a frame' but, as we shall see, if we are not careful in the way in which we treat such situations relativistically, we can encounter some apparent paradoxes. Consider, for example, the following tale of two twins, the so-called **twin (or clock) paradox**.

A pair of twins Terra and Astra are both 20 years old at the start of the tale. Astra makes a trip from the Earth (Terra's frame) at a velocity of $0.90c$. The trip takes five years in Astra's time frame, the frame of his biological clock. He then returns to the Earth at the same speed taking a further five years for the return trip.

Astra's biological age on his return to the Earth is therefore $20 + 5 + 5 = 30$ years

In Terra's frame (the Earth), which we take to be the fixed (inertial) frame, the time which has elapsed during Astra's outward journey is given by the inverse transformation (9.22) $t = \gamma\left(t' + \frac{vx'}{c^2}\right)$. Now $x' = 0$ always in Astra's rest frame, the rocket frame, and therefore $t = \gamma t'$. For $v = 0.90c$, $\gamma = 2.29$ and hence $t = 2.29 \times 5 = 11.45$ years

Similarly, a further 11.45 years elapse in Terra's frame, the frame of his biological clock, while Astra is making his return journey so that Terra's biological age on Astra's return is $20 + 11.45 + 11.45 = 42.9$ years. Thus, the twin who has been travelling at a velocity which is close to the velocity of light finds, on his return to the Earth, that he has aged more slowly than his twin who has stayed on Earth. Astra finds that he is in fact $42.9 - 30 = 12.9$ years younger than Terra on his return.

The following question arises, however. Could we not equally well calculate the time which has elapsed in Terra's frame during Astra's journey by using the Lorentz transformation (9.26), $t' = \gamma\left(t - \frac{vx}{c^2}\right)$? Now $x = 0$ in Terra's rest frame and therefore $t = \frac{t'}{\gamma} = \frac{5}{2.29} = 2.18$ years which would make Terra's age on Astra's return to be 20 + 2.2 + 2.2 = 24.4 years, in apparent contradiction to the answer of 42.9 years which we have just calculated above. This is the **twin paradox**. We seem to have obtained contradictory answers which depend on which frame we take to be the moving frame and which we take to be the fixed frame, an apparently arbitrary choice.

A *qualitative* explanation of the paradox is that, if we take Terra's frame to be the fixed (inertial) frame, Astra's (rocket) frame cannot also be inertial. When he reverses direction to return home he must accelerate, switching from a frame which is moving outwards from the Earth at $0.90c$ to another which is moving inwards at $0.90c$. Thus the second calculation which gave the answer of 24.4 years is not valid in Terra's frame because it assumes that Astra remains in the same inertial frame throughout his journey. Special relativity cannot be applied simply to a reference frame which is accelerating relative to an inertial frame because, as argued in Section 8.1, such a frame is not inertial.

To attempt to answer the paradox *quantitatively* we must consider the (lack of) time synchronisation between Astra's outward and inward frames. As explained in the web subsection up.ucc.ie/9/7/1/, there is a change in the standard of **simultaneity** (whether two events are simultaneous in Terra's and Astra's frames) when Astra switches frames for his return journey. Clocks in the two frames can no longer agree on the starting time, an effect which arises because the space and time coordinates are linked – they are components of the interval, an invariant quantity, so that a change in the value of the space component in a particular frame must lead to a change in the time component. In the web subsection up.ucc.ie/9/7/1/ it is shown that, when the change in the standard of simultaneity is taken into account, Terra's age, as calculated in Astra's frame, agrees with the value we have calculated in Terra's frame, that is 42.9 years.

In the cases which we consider in this book, such as the calculation of time dilation for the muon in the previous section, we do not have to switch inertial reference frames to complete a calculation so that the issue of simultaneity need not be brought into calculations explicitly. It is as well, however, to be aware of such considerations. There are many other examples of apparent paradoxes in special relativity, the origins of which lie in changes of standards of simultaneity.

9.8 Relativistic transformation of velocity

Consider an object which is moving with velocity u_x along the x-axis of the laboratory frame, as illustrated in Figure 9.7. What is its velocity, u'_x, in a rocket frame which is moving with velocity v along the x-axis of the laboratory frame? To answer this question we need to derive the Lorentz transformation for velocity.

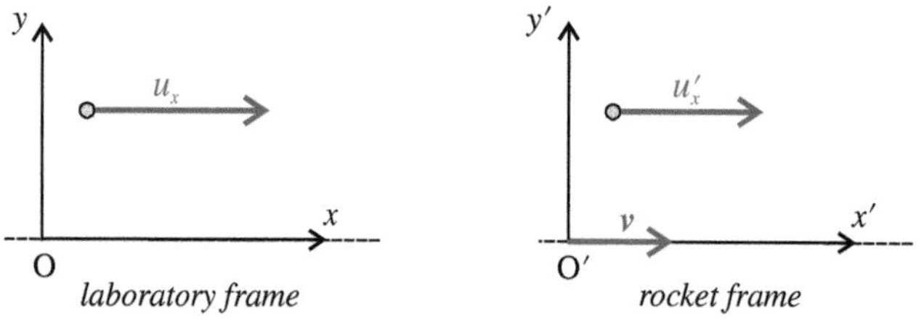

Figure 9.7. The velocity of an object which is moving along the x-axis of the laboratory frame, as viewed from the laboratory and rocket frames.

The Galilean transformation for velocity (Section 8.2) was obtained by differentiating the space transformation with respect to t. Thus the transformation $x' = x - vt$ yielded a velocity transformation

$$\frac{dx'}{dt} = \frac{dx}{dt} - v$$

so that we could write

$$u'_x = \frac{dx'}{dt'} = \frac{dx'}{dt} = u_x - v$$

However, this assumed that the time differential was the same in both frames, that is that $dt = dt'$, which we now know is not valid in the relativistic case. We must therefore use the Lorentz transformation to evaluate dt' in terms of dt

Differentiating the Lorentz transformation Equations (9.23) and (9.26) with respect to t, we obtain

$$\frac{dx'}{dt} = \frac{d}{dt}[\gamma(x - vt)] = \gamma\left(\frac{dx}{dt} - v\right)$$

and

$$\frac{dt'}{dt} = \frac{d}{dt}\left[\gamma\left(t - \frac{vx}{c^2}\right)\right] = \gamma\left(1 - \frac{v}{c^2}\frac{dx}{dt}\right) \tag{9.28}$$

Therefore,
$$u'_x = \frac{dx'}{dt'} = \frac{\dfrac{dx'}{dt}}{\dfrac{dt'}{dt}} = \frac{\dfrac{dx}{dt} - v}{1 - \dfrac{v}{c^2}\dfrac{dx}{dt}} = \frac{u_x - v}{1 - \dfrac{u_x v}{c^2}}$$

Similarly, because $dt \neq dt'$, we can no longer use the Galilean transformation equations for u_y and u_z, the velocity components perpendicular to v. We must derive transformation equations for u'_y and u'_z, as follows.

$$u'_y = \frac{dy'}{dt'} = \frac{\dfrac{dy'}{dt}}{\dfrac{dt'}{dt}} = \frac{u_y}{\gamma\left(1 - \dfrac{u_x v}{c^2}\right)} \quad \text{and similarly} \quad u'_z = \frac{u_z}{\gamma\left(1 - \dfrac{u_x v}{c^2}\right)}$$

The **Lorentz velocity transformation** is therefore

$$u'_x = \frac{u_x - v}{1 - \dfrac{u_x v}{c^2}} \tag{9.29}$$

$$u'_y = \frac{u_y}{\gamma\left(1 - \dfrac{u_x v}{c^2}\right)} \tag{9.30}$$

and

$$u'_z = \frac{u_z}{\gamma\left(1 - \dfrac{u_x v}{c^2}\right)} \tag{9.31}$$

Note that, as expected, when $v \ll c$ Equations (9.29) to (9.31) reduce to the Galilean transformation equations of Section 8.2.

The corresponding inverse transformation is obtained by reversing the sign of v and interchanging $t \leftrightarrow t'$ and $u_x \leftrightarrow u'_x$ in Equations (9.29) to (9.31). For example, for u_x the inverse transformation is

$$u_x = \frac{u'_x + v}{1 + \dfrac{u'_x v}{c^2}} \tag{9.32}$$

The Lorentz velocity transformation has some notable consequences which we illustrate below by considering two cases.

Case 1: *A particle which is moving with velocity $u'_x = 0.9c$ in a frame which, in turn, is moving with velocity $v = 0.9c$ along the x-axis of the laboratory frame*

The inverse Galilean velocity transformation (8.10) gives $u_x = 1.8c$, a velocity greater than the velocity of light but the inverse Lorentz velocity transformation (9.32) gives

$$u_x = \frac{0.9c + 0.9c}{1 + \dfrac{(0.9c)^2}{c^2}} = \frac{1.8c}{1.81} = 0.995c$$

Note that one cannot raise the velocity of a particle above that of light by viewing its motion from a moving reference frame. The velocity of a particle cannot exceed c in any frame.

Case 2: *An electromagnetic wave which is moving with velocity $u'_x = c$ in a frame which, in turn, is moving with velocity v along the x-axis of the laboratory frame*

The inverse Lorentz velocity transformation (9.32) gives

$$u_x = \frac{c + v}{1 + \dfrac{cv}{c^2}} = c$$

Thus, velocity of an electromagnetic wave has the same value, c, in any reference frame, an expected result because the Principle of Relativity was built into our derivation of the Lorentz transformation. Worked Example 9.1 illustrates further the use of the Lorentz velocity transformation.

Study Worked Example 9.1

For problems based on the material presented in this section visit up.ucc.ie/9/ and follow the link to the problems.

9.9 Momentum in relativistic mechanics

Consider a head-on elastic collision between a mass $2m$ which is moving with velocity u and a mass m which is at rest. As illustrated in Figure 9.8 the masses move with velocities v_1 and v_2, respectively, after the collision.

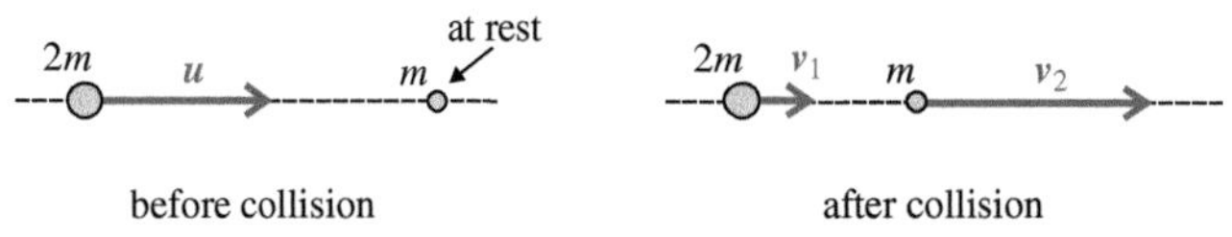

Figure 9.8. Before and after a head-on collision between masses $2m$ and m.

By applying the principles of conservation of momentum and energy, following the (non-relativistic) procedures used in Section 6.5, we obtain the equations

$$2mu = 2mv_1 + mv_2$$

and

$$\frac{1}{2}(2m)u^2 = \frac{1}{2}(2m){v_1}^2 + \frac{1}{2}m{v_2}^2$$

which we can solve to give $v_1 = \frac{1}{3}u$ and $v_2 = \frac{4}{3}u$. This tells us that the mass m moves with a velocity greater than u after the collision.

However, what if the $2m$ mass had been moving with a velocity close to of that of light, at $0.9c$ for example? We would then obtain $v_2 = 1.2c$; the mass m would be moving with a velocity which is greater than c after the collision, in conflict with the requirements of the Principle of Relativity. Thus the principle of conservation of momentum, as discussed in Section 6.2, is in direct conflict with the Principle of Relativity.

Since the principle of conservation of momentum arose as a consequence of Newton's laws of motion, the above discussion implies that the laws of motion themselves are in conflict with the Principle of Relativity. This is indeed the case, at least for Newton's second law as we have formulated it in earlier chapters. The Principle of Relativity requires that the law should have the same form in all inertial reference frames, that is if $\boldsymbol{F} = \frac{d\boldsymbol{p}}{dt}$ relative to an observer O then $\boldsymbol{F}' = \frac{d\boldsymbol{p}'}{dt'}$ relative to an observer O′, who is moving with constant velocity v relative to O. In particular, if no force is experienced by a particle, then the momentum of that particle must be constant in both inertial frames. It turns out to be impossible to satisfy these requirements and maintain the definition of momentum $\left(\boldsymbol{p} := m\frac{d\boldsymbol{r}}{dt}\right)$ which we have used hitherto (Sections 3.3 and 4.4). Newtonian mechanics is in fundamental conflict with the Principle of Relativity.

We have two choices: *either* we can abandon Newton's second law (and hence the principle of conservation of momentum as understood in Newtonian mechanics) *or* we can redefine the Newtonian definition of momentum so that Newton's second law and the principle of conservation of momentum are retained in a way which is consistent with the Principle of Relativity.

The choice is arbitrary. In principle, it would be possible to develop a workable formalism for relativistic mechanics based on the first option but we will follow convention and adopt the second approach which is to redefine momentum as $m\frac{d\boldsymbol{r}}{d\tau}$. Note that differentiation is now carried out with respect to *proper time* (as defined in Section 9.3).

We can state this definition in terms of the particle velocity $\boldsymbol{u} = \frac{d\boldsymbol{r}}{dt}$. Using the differential form of the inverse Lorentz transformation (9.22), namely $dt = \gamma\left(dt' + \frac{vdx'}{c^2}\right)$ to transform from the rest frame of the particle ($dx' = 0$, $dt' = d\tau$) to the frame in which the particle is moving with velocity $\boldsymbol{u}$ along the x-axis (that is, $v = u$ and $\gamma = \frac{1}{\sqrt{1-\frac{u^2}{c^2}}}$), we obtain

$$dt = \frac{1}{\sqrt{1-\frac{u^2}{c^2}}}d\tau \quad \rightarrow \quad \frac{dt}{d\tau} = \frac{1}{\sqrt{1-\frac{u^2}{c^2}}} \tag{9.33}$$

Thus $\frac{d\boldsymbol{r}}{d\tau} = \frac{d\boldsymbol{r}}{dt}\frac{dt}{d\tau} = \frac{\boldsymbol{u}}{\sqrt{1-\frac{u^2}{c^2}}}$ and we can write the definition of **relativistic momentum** as

$$\boxed{\boldsymbol{p} := m\frac{d\boldsymbol{r}}{d\tau} = \frac{m\boldsymbol{u}}{\sqrt{1-\frac{u^2}{c^2}}}} \tag{9.34}$$

Note that when $u \ll c, \mathbf{p} \rightarrow m\mathbf{u}$, so that this definition of momentum reduces to the Newtonian definition at low velocities. The revised definition does not invalidate the non-relativistic definition in a context in which the latter is applicable; rather it extends the definition into a regime in which u approaches the velocity of light.[1]

Note also that, if we rearrange Equation (9.34) as $p^2 = \dfrac{m^2u^2}{1-\dfrac{u^2}{c^2}}$, we can write $u^2 = \dfrac{p^2}{m^2}\left(1-\dfrac{u^2}{c^2}\right)$ and thus $u^2\left(1+\dfrac{p^2}{m^2c^2}\right) = \dfrac{p^2}{m^2}$ and hence the particle velocity u is given by $u^2 = \dfrac{\dfrac{p^2}{m^2}}{1+\dfrac{p^2}{m^2c^2}} = \dfrac{c^2}{1+\dfrac{m^2c^2}{p^2}}$. The denominator must be greater than or equal to unity, so that the relativistic definition of momentum (9.34) ensures that $u \leq c$ always. The velocity of the mass m, which we considered at the beginning of this section, cannot exceed c. Note also that as $m \rightarrow 0$, $u \rightarrow c$.

The revised definition of momentum allows us to retain the same form of Newton's second law which we have used in the previous chapters and, in particular, restores the principle of conservation of momentum, *provided* that it is understood that the momentum involved is that defined by Equation (9.34) and that the appropriate transformation equations for dynamical quantities are used when transforming between reference frames. These transformation equations are derived in Section 9.11 and in web section up.ucc.ie/9/11/1/.

9.10 Four-vectors: the energy–momentum 4-vector

We noted, in Section 4.1, that a vector in a plane (2 dimensions) can be resolved into two components and, therefore, can be described as a *2-vector* and, similarly a vector in space (3 dimensions) can be resolved into three components and is described as a *3-vector*.

A **4-vector** is a vector with four components. Although we cannot visualise a 4-vector as we can 2- or 3-vectors, we can manipulate a 4-vector and its components in much the same way as we do 2- or 3-vectors. We can use this procedure to extend special relativity to concepts such as momentum and energy.

We have already encountered one example of a 4-vector, the space–time interval in Section 9.3. An event in space–time (four dimensions – three spatial and one temporal) is specified by four coordinates (x, y, z, t) just as a position in space is specified by three coordinates (x, y, z). From Equation (9.6), a small space–time separation between two events, that is the interval between the events, is given by

$$(\textit{increment in interval})^2 = (\Delta\tau)^2 = (\Delta t)^2 - \left(\frac{\Delta x}{c}\right)^2 - \left(\frac{\Delta y}{c}\right)^2 - \left(\frac{\Delta z}{c}\right)^2 \tag{9.35}$$

where Δt, $\dfrac{\Delta x}{c}$, $\dfrac{\Delta y}{c}$ and $\dfrac{\Delta z}{c}$ are the four components of the space–time interval 4-vector. This can be compared to a change in the spatial separation between two points in space which is given by

$$(\textit{increment in distance})^2 = (\Delta s)^2 = (\Delta x)^2 + (\Delta y)^2 + (\Delta z)^2$$

where Δx, Δy and Δz are the three components of the spatial 3-vector between the points.

The interval of a 4-vector in space–time, like the change of distance for a 3-vector in space, is invariant; that is, it has the same value in all inertial reference frames.

As noted above, a vector in a plane, a 2-vector, is specified by two components. For example Δs, a change of displacement in the x-y plane (the plane of the page in Figure 9.9), is given by

$$(\Delta s)^2 = (\Delta x)^2 + (\Delta y)^2 \tag{9.36}$$

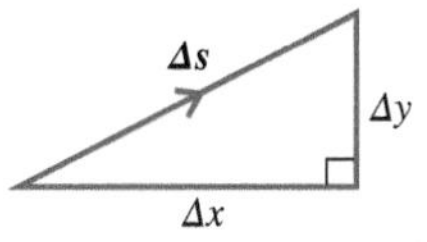

Figure 9.9. The components Δx and Δy of the displacement 2-vector $\boldsymbol{\Delta s}$.

We can multiply both sides of Equation (9.36) by the scalar quantity $\left(\dfrac{m}{\Delta t}\right)^2$ to obtain

$$\left(\frac{m\Delta s}{\Delta t}\right)^2 = \left(\frac{m\Delta x}{\Delta t}\right)^2 + \left(\frac{m\Delta y}{\Delta t}\right)^2$$

[1]In some treatments a *relativistic mass*, m', is defined such that $m' = \dfrac{m}{\sqrt{1-\dfrac{u^2}{c^2}}}$ and, in this case, m is called the **rest mass**. Momentum does not have to be redefined in these circumstances since $p = m'\dfrac{dr}{dt} = m'\mathbf{u} = \dfrac{m\mathbf{u}}{\sqrt{1-\dfrac{u^2}{c^2}}}$. Note that relativistic mass varies with velocity and the calculation of momentum through this equation involves differentiation with respect to time in the frame in which the momentum is calculated. In this book, however, we use only the definition of relativistic momentum given in (9.34) and do not require the additional concept of relativistic mass.

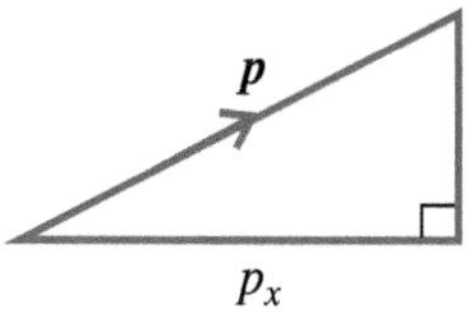

Figure 9.10. The components p_x and p_y of the momentum 2-vector $\boldsymbol{p}$.

As $\Delta t \to 0$, this becomes

$$\left(\frac{mds}{dt}\right)^2 = \left(\frac{mdx}{dt}\right)^2 + \left(\frac{mdy}{dt}\right)^2$$

which we can write as

$$(mu)^2 = (mu_x)^2 + (mu_y)^2 \quad \text{or} \quad p^2 = p_x{}^2 + p_y{}^2,$$

where p_x and p_y are the x- and y-components of the momentum 2-vector $\boldsymbol{p}$, as illustrated in Figure 9.10.

Thus, by multiplying Equation (9.36), which relates a displacement 2-vector to its components, by a scalar and taking the limit of small time intervals, we have obtained an equation which relates another 2-vector (a momentum vector) to its components.

We now apply a similar procedure to the space–time interval 4-vector. We multiply both sides of Equation (9.35) by the scalar quantity $\left(\frac{m}{\Delta\tau}\right)^2$. Note that we are dividing by $(\Delta\tau)^2$, the square of an increment in *proper time*. We then take the limit $\Delta\tau \to 0$ and obtain a new equation which relates a new 4-vector to its components.

$$\left(mc\frac{d\tau}{d\tau}\right)^2 = \left(mc\frac{dt}{d\tau}\right)^2 - \left(m\frac{dx}{d\tau}\right)^2 - \left(m\frac{dy}{d\tau}\right)^2 - \left(m\frac{dz}{d\tau}\right)^2 \tag{9.37}$$

We now identify each term in Equation (9.37). The term on the left-hand side of the equation is simply m^2c^2. Comparing the second term on the right-hand side of the equation, $\left(m\frac{dx}{d\tau}\right)^2$, to the definition of relativistic momentum given in Equation (9.34), namely $\boldsymbol{p} = m\frac{d\boldsymbol{r}}{d\tau}$, we can identify the second term as $-(p_x)^2$, where $(p_x)^2$ is the square of the x-component of the total relativistic momentum $\boldsymbol{p}$. Similarly, the third and fourth terms on the right-hand side are $-(p_y)^2$ and $-(p_z)^2$, respectively.

The first term on the right-hand side can be written $\frac{1}{c^2}\left(mc^2\frac{dt}{d\tau}\right)^2$. The quantity $\left(mc^2\frac{dt}{d\tau}\right)^2$, is new but has the dimensions of energy. We *define* this to be the relativistic energy, E.

Using Equation (9.33), we can write $\frac{dt}{d\tau}$ in terms of the particle velocity $\boldsymbol{u}$ so that the definition of **relativistic energy** can be written

$$\boxed{E := mc^2\left(\frac{dt}{d\tau}\right) = \frac{1}{\sqrt{1-\frac{u^2}{c^2}}}mc^2} \tag{9.38}$$

and, therefore, the first term on the right-hand side of Equation (9.37) can be written

$$\left(mc\frac{dt}{d\tau}\right)^2 = \left(\frac{E}{c}\right)^2$$

Relativistic energy has the dimensions of energy, but we will see in Section 9.12 that, unlike relativistic momentum, it does not reduce to the non-relativistic definition when $v \ll c$

Equation (9.37) can now be written

$$m^2c^2 = \frac{E^2}{c^2} - p_{x^2} - p_{y^2} - p_{z^2}$$

or

$$E^2 = c^2(p_x^2 + p_y^2 + p_z^2) + m^2c^4 \tag{9.39}$$

This equation relates the four components of a 4-vector, called the **energy–momentum 4-vector**, to one another, just as Equation (9.35) relates the four components of the space–time interval 4-vector.

Using $p^2 = p_x^2 + p_y^2 + p_z^2$, we can write Equation (9.39) as

$$\boxed{E^2 = c^2p^2 + m^2c^4} \tag{9.40}$$

Equation (9.40), the relativistic *energy–momentum relationship*, states the relationship between the energy and the momentum of a particle of mass m, as defined in relativistic mechanics. In relativistic mechanics, relativistic energy and momentum are connected directly just as space and time are. Mass, like the space–time interval, is invariant for transformation between reference frames. Note that in the limit $m \to 0$ a particle can still have energy and momentum. We shall encounter such a particle, the *photon* in Section 14.7. When $m = 0$ we can conclude from Equation (9.40) that energy and momentum are related by the equation

$$\boxed{E = cp}$$

and, as we have seen in the previous section, such a particle must always travel with velocity c.

Particle velocity

The velocity of a particle, $\boldsymbol{u}$, may be calculated directly from its relativistic energy and momentum since, from Equation (9.38), $E = \dfrac{mc^2}{\sqrt{1-\dfrac{u^2}{c^2}}}$ and, from Equation (9.34), $\boldsymbol{p} = \dfrac{m\boldsymbol{u}}{\sqrt{1-\dfrac{u^2}{c^2}}}$

Thus

$$\frac{\boldsymbol{p}}{E} = \frac{m\boldsymbol{u}}{mc^2} = \frac{\boldsymbol{u}}{c^2}$$

and we can write

$$\boxed{\boldsymbol{u} = \frac{c^2\boldsymbol{p}}{E}} \tag{9.41}$$

Note also that, by differentiating Equation (9.40), that is $E^2 = m^2c^4 + c^2p^2$, with respect to p,

we obtain

$$2E\frac{dE}{dp} = 2c^2p \quad \rightarrow \quad \frac{dE}{dp} = \frac{c^2p}{E}$$

which, from Equation (9.41), is the particle velocity. Thus we can also write

$$u = \frac{dE}{dp} \tag{9.42}$$

Study Worked Example 9.2

For problems based on the material presented in this section visit up.ucc.ie/9/ *and follow the link to the problems.*

9.11 Energy–momentum transformations: relativistic energy conservation

We can convert the Lorentz space–time interval transformation Equations (9.23) to (9.26) into energy–momentum transformation equations by multiplying them by m and differentiating with respect to τ. Thus $x' = \gamma(x - vt)$ becomes $m\dfrac{dx'}{d\tau} = \gamma\left(m\dfrac{dx}{d\tau} - mv\dfrac{dt}{d\tau}\right)$

Using the definitions of relativistic momentum (9.34) and energy (9.38), this can be written

$$p'_x = \gamma\left(p_x - v\frac{E}{c^2}\right). \tag{9.43}$$

Similarly,

$$p'_y = p_y$$

$$p'_z = p_z$$

and

$$E' = \gamma(E - vp_x)$$

This is the **Lorentz energy–momentum transformation**. The corresponding *inverse Lorentz energy–momentum transformation* may be derived from the space–time interval transformation Equations (9.19) to (9.22) through multiplication by m and differentiation with respect to τ. Alternatively the inverse Lorentz energy–momentum transformation may be obtained from Equations (9.43) by reversing the sign of v and interchanging $E \leftrightarrow E'$ and $p_x \leftrightarrow p'_x$, etc.

Consider a closed system of two particles which are not subject to any net external force. The principle of conservation of momentum applies but we would expect the principle of conservation of mechanical energy (Section 6.3) to apply only if any collisions between the particles are elastic.

Equation (9.43) may be used to relate the momenta of the particles in one frame, p'_{1x} and p'_{2x} to their momenta, p_{1x} and p_{2x}, and energies, E_1 and E_2, in a second frame. Thus

$$p'_{1x} = \gamma\left(p_{1x} - v\frac{E_1}{c^2}\right) \tag{9.44}$$

and

$$p'_{2x} = \gamma\left(p_{2x} - v\frac{E_2}{c^2}\right) \tag{9.45}$$

Adding both sides of (9.44) and (9.45) we obtain

$$(p'_{1x} + p'_{2x}) = \gamma\left[p_{1x} + p_{2x} - \frac{v}{c^2}(E_1 + E_2)\right]$$

which we can write as $$P'_x = \gamma\left[P_x - \frac{v}{c^2}E\right]$$

where $P'_x = (p'_{1x} + p'_{2x})$ is the x'-component of total (relativistic) momentum in the first frame, $P_x = (p_{1x} + p_{2x})$ is the x-component of total (relativistic) momentum in the second frame and $E = E_1 + E_2$, the total relativistic energy in the second frame.

Now both P_x and P'_x are conserved because, in Section 9.9, we have redefined momentum to ensure that it is conserved in relativistic mechanics. The quantities v and γ are both constants and hence E must also be conserved. The conservation laws for relativistic energy and (relativistic) momentum are intimately connected because they are components of the same 4-vector. Note that, unlike the principle of conservation of mechanical energy, the principle of conservation of relativistic energy may be applied to situations in which mechanical energy is converted into other forms of energy within the system under analysis.

We can generalise our findings to state that the total relativistic energy of a closed system is conserved, even in collisions which we previously regarded as inelastic. As we shall see in the following sections, relativistic energy is a more basic and more comprehensive concept than Newtonian mechanical energy.

Visit web section up.ucc.ie/9/11/1/ to see how a *force transformation* can be obtained by differentiating the momentum transformation equations, such as Equation (9.43), with respect to time.

For problems based on the material presented in this section visit up.ucc.ie/9/ and follow the link to the problems.

9.12 Relativistic energy: mass–energy equivalence

Equation (9.38) defines the relativistic energy of a particle to be $E = mc^2\dfrac{dt}{d\tau} = \dfrac{mc^2}{\sqrt{1-\dfrac{u^2}{c^2}}}$

We can expand $\dfrac{1}{\sqrt{1-\dfrac{u^2}{c^2}}}$ using the binomial theorem (Appendix A.9.2) to obtain

$$E = mc^2\left[1 + \frac{1}{2}\frac{u^2}{c^2} + \frac{3}{8}\frac{u^4}{c^4} + \ldots \text{ higher terms}\right] \tag{9.46}$$

Note that E is non-zero even when $u = 0$. The first term (mc^2) on the right-hand side of Equation (9.46) is called the rest energy of the particle.

The second term is $(mc^2)\dfrac{1}{2}\dfrac{u^2}{c^2} = \dfrac{1}{2}mu^2$, which is the Newtonian kinetic energy (Section 5.7).

The subsequent terms in Equation (9.46), namely $mc^2\left[\dfrac{3}{8}\dfrac{u^4}{c^4} + \ldots \text{ higher terms}\right]$, are relativistic in origin and become more important as the value of u approaches c.

If we now define the **relativistic kinetic energy**, $K := \left[\dfrac{1}{2}mu^2 + \dfrac{3}{8}\dfrac{mu^4}{c^4} + \ldots \text{ higher terms}\right]$, we can write Equation (9.46) as

$$\boxed{E = mc^2 + K} \tag{9.47}$$

Note that when $u \ll c$ Equation (9.47) does not reduce to the non-relativistic case in which the total mechanical energy, E, would have been equal to K. In relativistic mechanics rest energies must always be included in applying the principle of conservation of energy, which is now generalised to a **principle of conservation of mass–energy**.

Hence a major consequence of the theory of relativity is that energy and mass are essentially equivalent. For a particle in its rest frame (the frame in which $u = 0$), the total relativistic energy is called the **rest energy** of the particle and is given by

$$\boxed{E_0 = mc^2} \tag{9.48}$$

When $u \ll c$ the inclusion of rest energy, when applying the law of conservation of energy, generally produces the same result as the Newtonian case because the net result is the addition of the same rest energy term to both sides of the Newtonian conservation of energy equation.

We saw in Section 6.5 that, in a totally inelastic collision in which the colliding bodies stick together after the collision, Newtonian mechanical energy is not conserved – some of it is used to produce permanent deformation of the bodies, heat energy, etc. In relativistic mechanics, however, the total relativistic energy may be conserved in such a situation as long as no energy is transferred into or out of the system of colliding bodies. The loss of kinetic energy is accounted for by a change in the total rest energy of the system; indeed mass need not be conserved, as illustrated in the example described below.

Example 1: *A totally inelastic collision of two identical particles*
As illustrated in Figure 9.11, a particle of mass m with relativistic kinetic energy K collides with an identical particle which is at rest. After the collision the two particles stick together to form a composite mass M. The collision is inelastic. Non-relativistically we would expect that $M = 2m$ but we shall see that this is not true in relativistic mechanics.

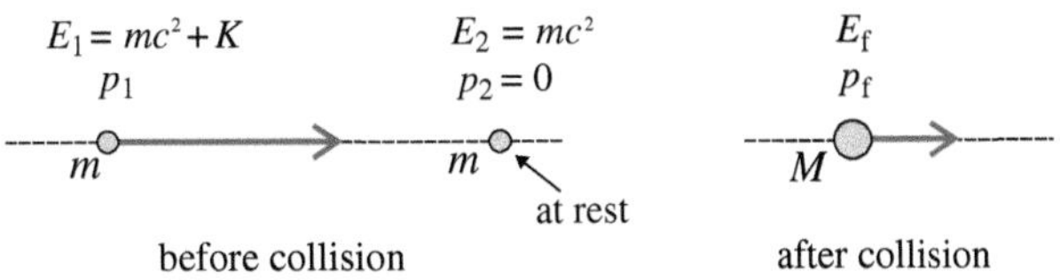

Figure 9.11. Before and after a totally inelastic collision between identical masses.

The mass after the collision, M, is given by equation (9.40). Thus

$$M^2c^4 = E_f^2 - p_f^2c^2 \tag{9.49}$$

where E_f is the final total relativistic energy and p_f is the final total (relativistic) momentum of the composite mass M. As described below, we can obtain equations for E_f and p_f, and hence for M, using the relativistic conservation laws for energy and momentum.

The total relativistic energy before the collision, $E_i = E_1 + E_2$, where the total relativistic energy of the incident mass, $E_1 = mc^2 + K$, and the rest energy of the target mass, $E_2 = mc^2$.

Relativistic energy conservation gives

$$E_f = E_i = mc^2 + K + mc^2 = K + 2mc^2 \tag{9.50}$$

Relativistic momentum conservation gives $p_f = p_1 + 0$, where p_1 is the momentum of the incident mass. Substituting for E_f and p_f in Equation (9.49), we obtain

$$M^2c^4 = (K + 2mc^2)^2 - {p_1}^2c^2 \tag{9.51}$$

The energy–momentum relation (9.40) for the incoming mass before the collision is

$${E_1}^2 = {p_1}^2c^2 + m^2c^4$$

Therefore,

$${p_1}^2c^2 = {E_1}^2 - m^2c^4 = (K + mc^2)^2 - m^2c^4 \tag{9.52}$$

Using Equation (9.52) to substitute for ${p_1}^2c^2$ in Equation (9.51), we obtain

$$\begin{aligned} M^2c^4 &= (K + 2mc^2)^2 - (K + mc^2)^2 + m^2c^4 \\ &= K^2 + 4mc^2K + 4m^2c^4 - K^2 - 2mc^2K - m^2c^4 + m^2c^4 \\ &= 4m^2c^4 + 2mc^2K \end{aligned}$$

Thus $M^2 = (2m)^2 + \dfrac{2mK}{c^2}$ so that, unless $K = 0$ (in which case there is no collision), $M > 2m$. The total mass after the collision is greater than the sum of the masses before.

The (relativistic) kinetic energy after the collision is given by

$$K_a = E_f - Mc^2$$

Substituting for E_f from (9.50)

$$K_a = K + 2mc^2 - Mc^2 \tag{9.53}$$

Thus, the change of kinetic energy in the collision is given by

$$K_a - K = 2mc^2 - Mc^2 = -(M - 2m)c^2 \tag{9.54}$$

where $(M - 2m)$ is the gain in mass in the collision. We have shown above that $(M - 2m)$ is positive so that $K_a - K$ must be negative; there is a loss of kinetic energy.

Therefore, we can write Equation (9.54) as *(loss of kinetic energy)* = *(gain of mass)*c^2
that is,

$$\Delta K = (\Delta m)c^2 \tag{9.55}$$

This equation tells us that the kinetic energy lost has been converted into mass; that is, any kinetic energy lost in the collision is accounted for by a net gain of mass. Conversely, a collision in which there is a net gain of kinetic energy will produce a net loss of mass.

For systems in which both kinetic and potential energy change, Equation (9.53) can be written, more generally, in terms of the change in the total energy of the system,

$$\boxed{\Delta E = (\Delta m)c^2} \tag{9.56}$$

In practice, because the c^2 factor in Equation (9.56) is such a large number (9×10^{16} m^2 s^{-2}), a change of mass can have an extremely small value when it is measured in kg. A change of energy of 1000 J for example corresponds to a mass change of only 10^{-14} kg.

While the above analysis is useful for collisions on a microscopic scale and, in principle can be extended to the collision of two macroscopic masses (masses on the scale of everyday objects), its application in the latter case is not usually realistic. Once we move up from the nuclear particle scale, in which velocities are usually relativistic, kinetic energies are too small to make the change of mass of the system a useful practical measure of changes of kinetic energy in the macroscopic system. Moreover, in the macroscopic case, it is more difficult to define the system exactly and therefore more difficult to take account of energy transfers into and out of the system. For these reasons, when we come to consider changes of energy in macroscopic systems in Chapter 11, we shall generally represent macroscopic energy through more classical concepts such as temperature. Note, however, that, when the energy of a macroscopic system changes, for example when a spring is compressed and thus its potential energy increases, its mass does increase in principle, even if the increase has no practical implications.

Note that invariance of mass in the energy–momentum 4-vector (Section 9.10) has not been violated by the change in the total mass of a system as measured before and after a collision. Invariance of mass applies to transformations between inertial frames. Mass is invariant between different frames, both before and after a collision, but its invariant value is changed by the collision. To illustrate this point consider, for example, the situation after the collision which we have just analysed, as viewed from two different frames, the rest frame of the composite mass M (the CM frame of the system as discussed in Sections 6.6 and 8.3) and the LAB frame (the rest frame of the target mass).

In the rest frame of the mass M

$$E' = Mc^2, \quad p' = 0$$

and thus, applying Equation (9.40) in this frame

$$(\text{mass})^2c^4 = E'^2 - p'^2c^2 = M^2c^4 - 0 = M^2c^4 \tag{9.57}$$

In the LAB frame, $E = \dfrac{Mc^2}{\sqrt{1 - \dfrac{u^2}{c^2}}}$ and $\boldsymbol{p} = \dfrac{M\boldsymbol{u}}{\sqrt{1 - \dfrac{u^2}{c^2}}}$, where $\boldsymbol{u}$ is the velocity of the mass.

In this frame,

$$(\text{mass})^2c^4 = E^2 - p^2c^2 = \frac{M^2c^4}{1 - \dfrac{u^2}{c^2}} - \frac{M^2u^2c^2}{1 - \dfrac{u^2}{c^2}} = \frac{Mc^2(c^2 - u^2)}{1 - \dfrac{u^2}{c^2}} = M^2c^4 \tag{9.58}$$

Comparison of Equations (9.57) and (9.58) confirms that the mass of the system after the collision has the same value in either of the two frames; it is invariant. Worked Example 9.3 gives a further example of an analysis of a relativistic collision.

In Table 9.1 we summarise and compare the definitions of dynamical quantities and transformation equations in Galilean and Special Relativity.

Table 9.1

GALILEAN RELATIVITY (Newton)	SPECIAL RELATIVITY (Einstein)
$x' = x + vt$	$x' = \gamma(x + vt), \quad \gamma = \dfrac{1}{\sqrt{1 - \dfrac{v^2}{c^2}}}$
$t' = t$	$t' = \gamma\left(t + \dfrac{v}{c^2}x\right)$
$l = l'$	$l = l'\sqrt{1 - \dfrac{v^2}{c^2}}$
$t = t'$	$t = \dfrac{t'}{\sqrt{1 - \dfrac{v^2}{c^2}}}$
$u_x = u'_x + v; \quad u_y = u'_y$	$u_x = \dfrac{u'_x + v}{1 + \dfrac{u'_x v}{c^2}}; \quad u_y = \dfrac{u'_y}{1 + \dfrac{u'_x v}{c^2}}$
$\boldsymbol{p} = m\boldsymbol{u}$	$\boldsymbol{p} = \dfrac{m\boldsymbol{u}}{\sqrt{1 - \dfrac{u^2}{c^2}}}$
$E = \dfrac{p^2}{2m} + mc^2$	$E = \dfrac{mc^2}{\sqrt{1 - \dfrac{u^2}{c^2}}} \quad \rightarrow \quad E^2 = c^2p^2 + m^2c^4$
$K = \dfrac{p^2}{2m}$	$K = E - mc^2$
$H(\boldsymbol{r}, \boldsymbol{p}) = \dfrac{p^2}{2m} + U(\boldsymbol{r})$	$H(\boldsymbol{r}, \boldsymbol{p}) = \sqrt{c^2p^2 + m^2c^4} + U(\boldsymbol{r})$

Study Worked Example 9.3

For problems based on the material presented in this section visit up.ucc.ie/9/ *and follow the link to the problems.*

9.13 Units in relativistic mechanics

In the study of atomic, nuclear and sub-nuclear systems, it is often convenient to use an alternative (non-SI) system of units. in which velocity, energy, and time are taken as base units. Velocities are measured in units of c, time is measured in seconds and energies are measured in electron volts (eV). The electron volt will be discussed further in Section 16.13, where it will be shown that 1 eV = 1.6×10^{-19} J. Thus the rest energy of the electron, in eV units, is $m_e c^2 = (9.1 \times 10^{-31}\ \text{kg})(3 \times 10^8\ \text{m s}^{-1})^2 = 8.2 \times 10^{-14}\ \text{J} = \dfrac{8.2 \times 10^{-14}}{1.6 \times 10^{-19}}\text{eV} = 0.51\ \text{MeV}$.

In this system, the unit of mass is $\dfrac{\text{energy equivalent}}{c^2}$, expressed in eV/$c^2$, and the unit of momentum is $\dfrac{\text{energy equivalent}}{c}$, expressed in eV/$c$.

This system of units is particularly convenient when using equations like (9.40), the relativistic equation which relates mass, momentum and energy

$$m^2c^4 = E^2 - p^2c^2$$

This can be written

$$p = \frac{1}{c}\sqrt{E^2 - m^2c^4} \tag{9.59}$$

If, for example, an electron has a relativistic kinetic energy of 0.30 MeV we can calculate its relativistic momentum, in eV/c, as follows

From Equation (9.47), $$E = m_e c^2 + K = 0.51 + 0.30 = 0.81\ \text{MeV}$$

Substituting for E (in MeV) and m (in MeV/c^2) in Equation (9.59)

we obtain $$p = \frac{1}{c}\sqrt{(0.81\ \text{MeV})^2 - (0.51\ \text{MeV})^2} = 0.63\ \text{MeV}/c$$

The velocity of a particle, $\boldsymbol{u}$, can be calculated directly from its relativistic energy and momentum using Equation (9.41). The velocity of the 0.30 MeV electron,

$$\boldsymbol{u} = \frac{\boldsymbol{p}c^2}{E} = \frac{0.63\ \text{MeV}/c}{0.81\ \text{MeV}}c^2 = 0.78c \text{ (in the same direction as } \boldsymbol{p}).$$

Further examples of the use of this system of units are given in Worked Example 9.4.

Study Worked Example 9.4

For problems based on the material presented in this section visit up.ucc.ie/9/ *and follow the link to the problems.*

9.14 Mass–energy equivalence in practice

Relativistic mechanics tells us that the mass of a closed system can change as a result of an interaction between its constituent particles. This of course runs contrary to our everyday experience; in fact in chemistry a law of conservation of mass is stated and is an important tool in the analysis of chemical reactions. In order to understand why no change of mass is detected in a chemical reaction let us estimate its magnitude.

Electrons are held in atoms and molecules by the attractive forces exerted by nuclei, a situation which we shall examine in detail in Chapters 24 and 25. The energy associated with this force, the **binding energy** of the atom or molecule (a form of potential energy), is of order 1 eV (1.6×10^{-19} J). In a chemical reaction, the binding energies of the electrons change and thus energies of order 1 eV are released or absorbed per reaction. When we calculate the change of mass which is equivalent to 1 eV, using Equation (9.56), we obtain a value of

$$1\ \text{eV} = 1.6 \times 10^{-19}\ \text{J} \quad \rightarrow \quad \Delta m = \frac{\Delta E}{c^2} = \frac{1.6 \times 10^{-19}}{9 \times 10^{16}} = 1.8 \times 10^{-36}\ \text{kg}$$

This mass is two millionths of the mass of the electron so that any changes in the mass of an atom or molecule as a result of a chemical reaction are undetectable, although they must occur. The law of conservation of mass should be regarded, therefore, as a useful practical tool rather than a fundamental law of nature.

We noted in Section 1.4 that the components of atomic nuclei are bound together by a force which we do not experience in the macroscopic world, the strong nuclear force. We do not experience this force because it is very short range in its effects — it drops to zero when the interacting particles are separated by more than about 2×10^{-15} m. However, as we shall see in Chapter 27, on the nuclear scale (less than 2×10^{-15} m) the strong nuclear force is extremely large and the associated binding energy is extremely high (of order 10^7 eV). Consequently, in a nuclear reaction or decay, when the components of new nuclei produced by the process rearrange with different binding energies, the total mass of the system can change significantly.

The mass change which is equivalent to a typical change of nuclear binding energy can be calculated using Equation (9.56)

$$10^7\ \text{eV} = 10^7 \times 1.6 \times 10^{-19}\ \text{J} \quad \rightarrow \quad \Delta m = \frac{\Delta E}{c^2} = \frac{1.6 \times 10^{-2}}{9 \times 10^{16}} = 1.8 \times 10^{-29}\ \text{kg}$$

Although this mass is very small by macroscopic standards, it is nonetheless equal to the mass of about 20 electrons and is easily measurable.

In a nuclear reaction or decay the difference in binding energy between the initial and final grouping of the nuclear particles is released as nuclear energy, usually as the kinetic energy of the product particles. If the energy released is large enough it can even be used to create additional particles; energy is converted into mass. Particles can also annihilate one another to produce energy; mass is thus converted into energy.

The analysis of nuclear and sub-nuclear reactions and decays through relativistic mechanics is the main testing ground of special relativity. A relativistic treatment is found to be essential if observed results are to be explained with any reasonable precision.

We noted in Section 1.2 that the proton is made up of three quarks (as illustrated in Figure 1.1). When we come to consider the structure of subnuclear particles in more detail in Section 27.12 we will find that the total mass of the quarks which make up the proton is substantially less than the measured mass of the proton; in fact quark masses account for only about 1% of the proton mass. The missing mass is accounted for by the binding energy associated with the interaction which holds the quarks together.

Special relativity was formulated because of a conflict between classical mechanics and electromagnetism. In resolving this conflict, a whole series of new insights has been obtained, culminating in a new interpretation of the concept of energy. Experiences of this sort are repeated frequently in physics. As concepts are unified our understanding becomes deeper and the techniques of analysis become more powerful.

9.15 General relativity

Special Relativity applies to inertial frames. In Section 8.7 we saw that, in practice, to consider a frame to be inertial we had to restrict ourselves to a limited region of space, a 'local view'. This view is appropriate when we are considering particles which, in principle, occupy points in space.

General Relativity relates motion in one inertial frame to motion in different but nearby frames. Thus, a universal view of the structure of space–time is built up.

The *Newtonian/Galilean* approach sets up a rectilinear (straight line) coordinate system throughout the universe. To explain the departures from straight line motion which every real free body exhibits in the presence of others, the concept of gravitation (Section 5.1) had to be introduced. However, it has many unsatisfactory aspects, for example it assumes that gravitational force is transmitted instantaneously, at a speed greater than the velocity of light.

In the *Einstein* approach (General Theory of Relativity, 1916), it is postulated that a free particle, moving in a region in which there are massive objects, does not experience any gravitational force and that such a particle departs from straight line motion because space–time is curved – warped by the presence of mass. Thus, a satellite orbiting the Earth is merely following the local structure of spacetime. Because the curvature is not apparent over a very short distance the local spacetime environment can be considered to be an inertial frame, as assumed in Special Relativity. This view does not mean that Newton's Law of Gravitation is wrong – it clearly works, it passes all tests. From this viewpoint, however, gravitational force may be considered to be a fictitious force, like centrifugal force (Section 8.4) or the Coriolis force (Section 8.5), which enables us to manage limitations in the coordinate system which we are using.

An important prediction of General Relativity is that light, which has energy but no mass, will follow the structure of space–time and will thus appear to be deflected in a gravitational field. This is not expected in Newtonian theory. For example, light from a star which passes the edge of the Sun, and which is observable during a solar eclipse, is deflected from its position as determined when it is not passing close to the Sun (Figure 9.12).

Figure 9.12. The deflection of light $\Delta\varphi$ (exaggerated) as it passes close to the Sun.

The observed displacement ($\Delta\varphi = 1.75$ seconds of arc) is exactly as predicted by Einstein's General Theory of Relativity. Other predictions of the theory have also been verified, as indicated in Worked Example 9.5.

Gravitational waves

As noted above, the general theory of relativity accounts for the phenomenon of gravity as a manifestation of the curvature of space–time caused by the presence of mass. As massive objects change their location in space–time they change this curvature, producing 'ripples in the fabric of space–time' which propagate as **gravitational waves**. When gravitational waves reach the Earth, they distort the local structure of space–time leading to the disturbance of masses they encounter. Because gravitational waves are so weak, however, by the time they reach the Earth, even when caused by some of the most violent and energetic events in the Universe, the disturbance is extremely difficult to detect.

The detection of gravitational waves, therefore, presented a formidable technical challenge but this was finally overcome in 2016 when the LIGO (Laser Interferometer Gravitational-Wave Observatory) detected gravitational waves which had been produced by the collision of two black holes. The observation required the LIGO interferometer to detect a disturbance which was thousands of times smaller than the size of an atomic nucleus.

The detection of gravitational waves has opened up a new method of studying the Universe — *Gravitational Wave Astronomy*. For more information visit the LIGO website, https://www.ligo.caltech.edu/page/learn-more.

Study Worked Example 9.5

WORKED EXAMPLES

Worked Example 9.1: A particle moves north at a speed of $0.85c$ relative to the Earth. What is the velocity of this particle as measured by an observer in a spaceship which is travelling east relative to the Earth at a speed of $0.9c$?

Let the (x, y, z, t) and (x', y', z', t') frames represent the Earth and spaceship respectively, with the x-and y-axes pointing due east and north respectively. The Lorentz velocity transformation equations for u'_x and u'_y are (9.29) and (9.30),

$$u'_x = \frac{u_x - v}{1 - \frac{u_x v}{c^2}} \quad \text{and} \quad u'_y = \frac{u_y}{\gamma\left(1 - \frac{u_x v}{c^2}\right)}$$

The particle velocity is directed northwards in the Earth's frame so that $u_y = 0.85c$ and $u_x = 0$. The spaceship is travelling eastwards so that $v = 0.9c$ and therefore $\gamma = \frac{1}{\sqrt{1-(0.9)^2}} = 2.29$

The Lorentz velocity transformation equations yield $u'_x = \frac{0 - 0.9c}{1} = -0.9c$ (westwards) and $u'_y = \frac{0.85c}{\gamma} = 0.37c$ (northwards). The resultant velocity of the particle in the frame of the spaceship is given by $u' = \sqrt{u'^2_x + u'^2_y} = \sqrt{(0.9c)^2 + (0.37c)^2} = 0.97c$, as illustrated in Figure 9.13.

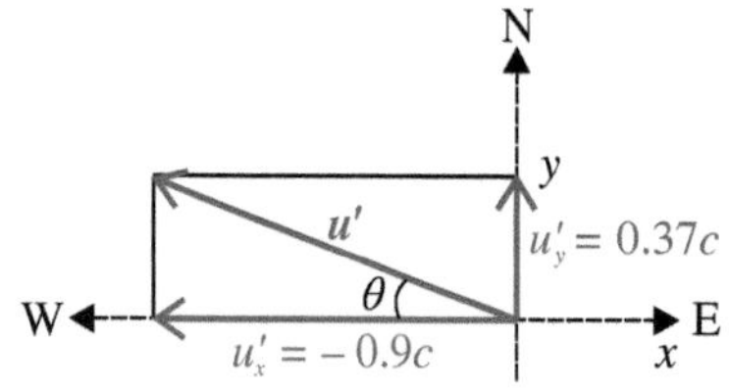

Figure 9.13. The components u'_x and u'_y of the particle velocity, as observed in the frame of the spaceship (Worked Example 9.1).

The direction of $\boldsymbol{u'}$ is given by the angle θ in Figure 9.13 where

$$\theta = \tan^{-1}\frac{u'_y}{u'_x} = \tan^{-1}\frac{0.37}{0.90} = 22.3^\circ$$

Thus,

$$u' = 0.97c \text{ at } 22.3^\circ \text{ N of W.}$$

Worked Example 9.2: An electron performs uniform circular motion of radius 20.0 m with a periodic time of 1.68 μs. Calculate the relativistic momentum and kinetic energy of the electron and estimate the percentage error made in using classical equations for these quantities.

The velocity of the electron, $u = \frac{2\pi R}{T} = \frac{2\pi \times 20}{1.68 \times 10^{-6}} = 7.48 \times 10^7 \text{m s}^{-1} = 0.249c$

Equation (9.34) → relativistic momentum $= \frac{mu}{\sqrt{1 - \frac{u^2}{c^2}}} = \frac{9.1 \times 10^{-31} \times 7.48 \times 10^7}{\sqrt{1 - 0.249^2}} = 7.03 \times 10^{-23} \text{ kg m s}^{-1}$

Equations (9.38) and (9.47) → relativistic kinetic energy $= \frac{mu^2}{\sqrt{1 - \frac{u^2}{c^2}}} - mc^2 = \left(\frac{1}{\sqrt{1 - \frac{u^2}{c^2}}}\right) mc^2$

$$= \left(\frac{1}{\sqrt{1 - 0.249^2}} - 1\right) 9.1 \times 10^{-31} \times (3 \times 10^8)^2 = 2.66 \times 10^{-15} \text{ J}$$

Classical momentum $= mu = 9.1 \times 10^{-31} \times 7.48 \times 10^7 = 6.81 \times 10^{-23}$ kg m s^{-1}, 3.1% below the relativistic value.
Classical kinetic energy $= \frac{1}{2}mu^2 = \frac{1}{2}m(7.48 \times 10^7)^2 = 2.54 \times 10^{-15}$J, 4.5% below the relativistic value.

Worked Example 9.3: A photon (a particle with zero mass as discussed at the end of Section 9.10) of energy 1.25 MeV makes a head-on collision with an electron which is at rest. Determine the energy of the photon after the collision. [The rest mass of the electron is 0.51 MeV/c^2]

The photon and electron energies before and after the collision are as indicated in Figure 9.14.

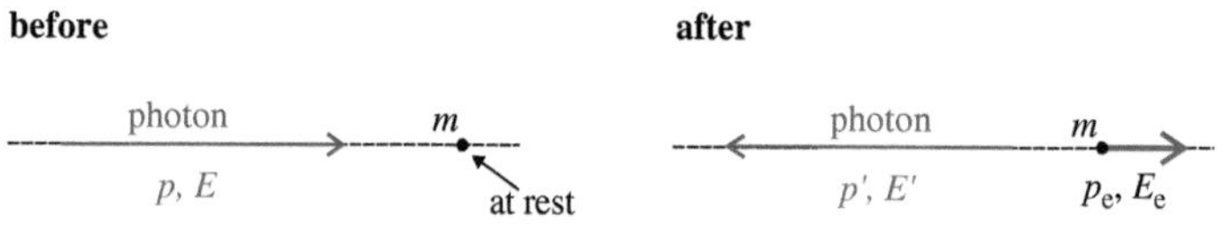

Figure 9.14. Before and after a head-on collision between a photon and an electron at rest (Worked Example 9.3).

Relativistic momentum conservation gives $$p = p_e - p' \tag{9.63}$$

and relativistic energy conservation gives $$E + mc^2 = E' + E_e \tag{9.64}$$

The energy-momentum relationships (Equation (9.40)) for the incoming and outgoing photon ($m = 0$) are, respectively, $E = pc$ and $E' = p'c$

and, for the outgoing electron, $$E_e^2 = m^2c^4 + p_e^2c^2 \tag{9.65}$$

We know m and E (and thus $p = \frac{E}{c}$). We want to know $E'(= p'c)$. To obtain an equation for p' we write

$$p_e = p + p' \quad \text{(from Equation (9.63))}$$

and $$E_e = E - E' + mc^2 = pc - p'c + mc^2 \quad \text{(from Equation (9.64))}$$

and substitute for p_e and E_e in Equation (9.65) to obtain

$$[(p - p')c + mc^2]^2 = m^2c^4 + (p + p')^2c^2$$

which can be written $$(p - p')^2c^2 + 2mc^3(p - p') + m^2c^4 = m^2c^4 + (p + p')^2c^2$$

Thus, $$p^2 + p'^2 - 2pp' + 2mcp - 2mcp' = p^2 + p'^2 + 2pp'$$

which reduces to $$mcp - mcp' = 2pp' \quad \rightarrow \quad p'(2p + mc) = mcp$$

The equation for p' is therefore $$p' = \frac{mcp}{2p + mc}$$

In the case considered here $m = 0.51\ \mathrm{MeV}/c^2$ and $p = \frac{E}{c} = 1.25\ \mathrm{MeV}/c$

so that $$p' = \frac{(0.51\ \mathrm{MeV}/c^2)c(1.25\ \mathrm{MeV}/c)}{2(1.25\ \mathrm{MeV}/c) + (0.51\ \mathrm{MeV}/c^2)c} = 0.21\ \mathrm{MeV}/c \text{ and thus } E' = p'c = 0.21\ \mathrm{MeV}$$

We will return to this problem – a photon-electron collision – when we consider the *Compton effect* in Section 14.7. We shall then analyze the collision more generally as a two-dimensional relativistic collision.

Worked Example 9.4: Given that the rest mass of the electron is 0.51 MeV/c^2 calculate (a) the relativistic kinetic energies, (b) the relativistic momenta (in MeV/c) and (c) the velocities of electrons of relativistic energies (i) 0.70 MeV (ii) 3.00 MeV and (iii) 250.00 MeV.

We use the methods described in Section 9.13.

(a) The relativistic kinetic energies are given by Equation (9.47) $K = E - mc^2$

Thus

$$\text{(i)}\ K = 0.70 - 0.51 = 0.19\ \mathrm{MeV}$$

$$\text{(ii)}\ K = 3.00 - 0.51 = 2.49\ \mathrm{MeV}$$

and

$$\text{(iii)}\ K = 250.00 - 0.51 = 249.49\ \mathrm{MeV}$$

(b) The relativistic momenta are given by Equation (9.59) $p = \frac{1}{c}\sqrt{E^2 - m^2c^4}$

Thus

$$\text{(i)}\ p = \frac{1}{c}\sqrt{0.70^2 - 0.51^2} = 0.48\ \mathrm{MeV}/c$$

$$\text{(ii)}\ p = \frac{1}{c}\sqrt{3.00^2 - 0.51^2} = 2.96\ \mathrm{MeV}/c$$

and

$$\text{(iii)}\ p = \frac{1}{c}\sqrt{250.00^2 - 0.51^2} = 249.9995\ \mathrm{MeV}/c$$

(c) The particle velocities are given by Equation (9.41) $u = \frac{pc^2}{E}$

Thus

$$\text{(i)}\ u = \frac{0.48\ \mathrm{MeV}/c}{0.70\ \mathrm{MeV}}c^2 = 0.69c$$

$$\text{(ii)}\ u = \frac{2.96\ \mathrm{MeV}/c}{3.00\ \mathrm{MeV}}c^2 = 0.987c$$

and

$$\text{(iii)}\ u = \frac{249.9995\ \mathrm{MeV}/c}{250.00\ \mathrm{MeV}}c^2 = 0.999998c$$

Note that when $E \gg mc^2$ (case (iii)), $pc \approx E$ and $u \approx c$, showing that a very energetic electron behaves in a similar manner to the (massless) photon.

Worked Example 9.5: Calculate the inertial mass equivalent (in kg) of a photon of energy 3.0 eV. Calculate the fractional change in the energy of such a photon when it 'falls' through 100 m near the surface of the Earth.

The photon energy $E = 3.0\text{ eV} = 4.8 \times 10^{-19}\text{ J}$

The inertial mass equivalent is $m = \frac{E}{c^2} = \frac{4.8 \times 10^{-19}}{9 \times 10^{16}} = 5.33 \times 10^{-36}\text{ kg}$

The energy change in falling through 100 m is $mgh = 5.33 \times 10^{-36} \times 9.8 \times 100 = 5.23 \times 10^{-33}\text{ J}$

The fractional change in energy is therefore $\frac{5.23 \times 10^{-33}}{4.8 \times 10^{-19}} = 1.09 \times 10^{-14}$

Such a change in photon energy is detectable. In 1960 Pound (Robert Pound 1919–2010) and Rebka (Glen Rebka 1931–2015), using photons of much higher energy, were able to detect such an energy change, thus providing evidence that the photon, although massless, is influenced by a gravitational field, as predicted by the general theory of relativity.

PROBLEMS

For problems based on the material covered this chapter visit up.ucc.ie/9/ and follow the link to the problems.

10

Continuum mechanics: mechanical properties of materials: microscopic models of matter

AIMS

- to describe mechanical properties of materials through models in which the materials involved are considered to be continuous throughout the bulk matter
- to explain the elastic nature of solids and the general properties of fluids
- to develop a simple model of gases at low pressure (called the *kinetic theory of gases*) which is based on their microscopic composition; to see how this model may be modified when the effect of the range of molecular forces and/or molecular size is taken into consideration
- to discuss microscopic models of liquids and solids which give further insight into their mechanical properties

10.1 Dynamics of continuous media

While we know from discussions in Chapter 1 that matter in bulk is formed from large collections of particles such as atoms or molecules, attempts to describe the macroscopic properties of materials on this basis can be difficult and non-productive. Our everyday experience of matter, in any event, is of continuous media; a block of metal appears continuous throughout, the air in the room shows no obvious evidence of being anything other than a continuous substance filling the room. Accordingly, we will begin our study of the properties of materials with a model in which matter is assumed to be continuous. Later in this chapter we shall see how many general properties, described at the macroscopic level, may be explained from a microscopic viewpoint; microscopic theories often give us insight into both the basic nature of these properties and the fundamental interatomic or intermolecular forces.

Materials seem to fall naturally into two broad categories – solids and fluids. Solids generally retain their shape under the influence of external forces, although they may show temporary distortions while such forces are applied. Fluids, as the name implies, are materials which can flow and are usually divided into two subcategories, liquids and gases. In the laboratory, a liquid settles to the bottom of the vessel into which it is poured – its volume does not change – while a gas, on the other hand, expands to fill the whole container to which it is confined. These broad categories of different states of matter are not quite as clear-cut as implied above; matter can also appear in other forms, such as vitreous (glassy) materials or plasmas (ionised gases), which do not fit into the simple categories of solids, liquids or gases.

One physical property that distinguishes one material from another is its **density** which is defined as the mass per unit volume. If the mass of a homogeneous material of volume V is m, then the density of the material is defined by:

$$\boxed{\rho := \frac{m}{V}}$$

The SI unit of density is kg m^{-3}. Note that density is a property *of the material* from which a body is made. The densities of some common substances are given in Table 10.1.

Understanding Physics, Third Edition. Michael Mansfield and Colm O'Sullivan.
© 2020 John Wiley & Sons Ltd. Published 2020 by John Wiley & Sons Ltd.

Table 10.1 Densities and specific volumes of some common substances

substance	density $\mathrm{kg\,m^{-3}}$	specific volume $\mathrm{m^3\,kg^{-1}}$
air (0 °C)	1.29	0.775
air (20 °C)	1.21	0.826
ice (0 °C)	9.17×10^2	1.09×10^{-3}
water (4 °C)	1.00×10^3	1.00×10^{-3}
sea water	$\sim 1.02 \times 10^3$	$\sim 9.76 \times 10^{-4}$
aluminium (20 °C)	2.70×10^3	3.70×10^{-4}
iron (20 °C)	7.87×10^3	1.27×10^{-4}
copper (20 °C)	8.96×10^3	1.12×10^{-4}
lead (20 °C)	1.14×10^4	8.77×10^{-5}
gold (20 °C)	1.93×10^4	5.18×10^{-5}

If a body is not homogeneous, the density will not be the same throughout the body, in which case the density will be a function of position in the body (Figure 10.1), that is

$$\rho = \rho(\boldsymbol{r}) := \frac{dm}{dV}$$

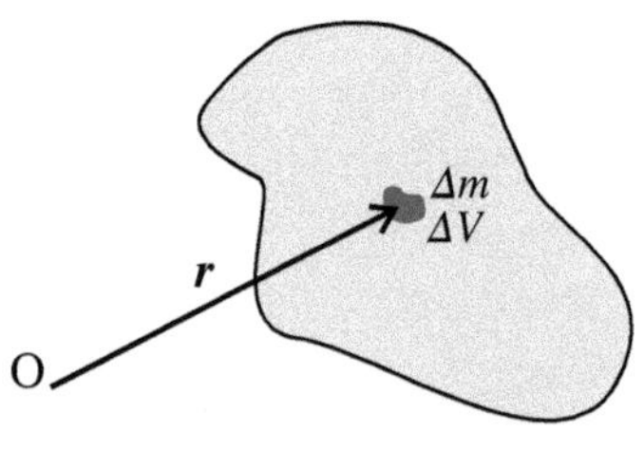

Figure 10.1. In a non-homogenous body the density may be different at different points. The density at the point r is defined as the limit of $\frac{\Delta m}{\Delta V}$ as $\Delta V \to 0$.

The density of the Earth's atmosphere, for example, varies with height above the surface of the Earth.

The reciprocal of density is called **specific volume**, defined as volume per unit mass

$$\boxed{v := \frac{1}{\rho} := \frac{dV}{dm}}$$

Specific volumes of common substances are also listed in Table 10.1

Note the use of the word 'specific' here; in modern conventional usage in physics 'specific' always means 'per unit mass'.

10.2 Elastic properties of solids

The shape and/or volume of a solid body can be changed by applying external forces to the body. Provided such forces are not too large, however, the body will return to its original shape and volume when the distorting force is removed. This phenomenon is called **elasticity**. The elastic behaviour of a solid body depends on the way in which the forces are applied and we will now describe three such ways, namely longitudinal, volumetric, and shearing deformations.

Longitudinal deformation

Consider first the case of a piece of wire or a metal bar of uniform cross-sectional area A under tension due to equal and opposite forces applied along its axis as indicated in Figure 10.2(a).

If the tensile force is changed from F to $F + \Delta F$, the axial length of the bar changes from L to $L + \Delta L$. Figure 10.2(b) shows the data from an experiment in which the dependence of the fractional change in length ('the longitudinal strain' $:= \frac{\Delta L}{L}$) on the force per unit area (called the 'longitudinal stress' $:= \frac{\Delta F}{A}$) producing the extension is studied. If the force is not too large, it is found that $\frac{\Delta L}{L}$ is directly proportional to $\frac{\Delta F}{A}$ (solid line in Figure 10.2(b)) and that when the force is removed the body retracts completely to its original length; such behaviour is called **elastic**. In the case of a solid rod a compressional force could be applied to reduce the length of the rod, in which case the same effect is observed, that is the (negative) strain is directly proportional to the (negative) stress.

Thus, in either case, stress is proportional to strain, that is

$$\frac{\Delta F}{A} = (constant)\frac{\Delta L}{L} = E\frac{\Delta L}{L},$$

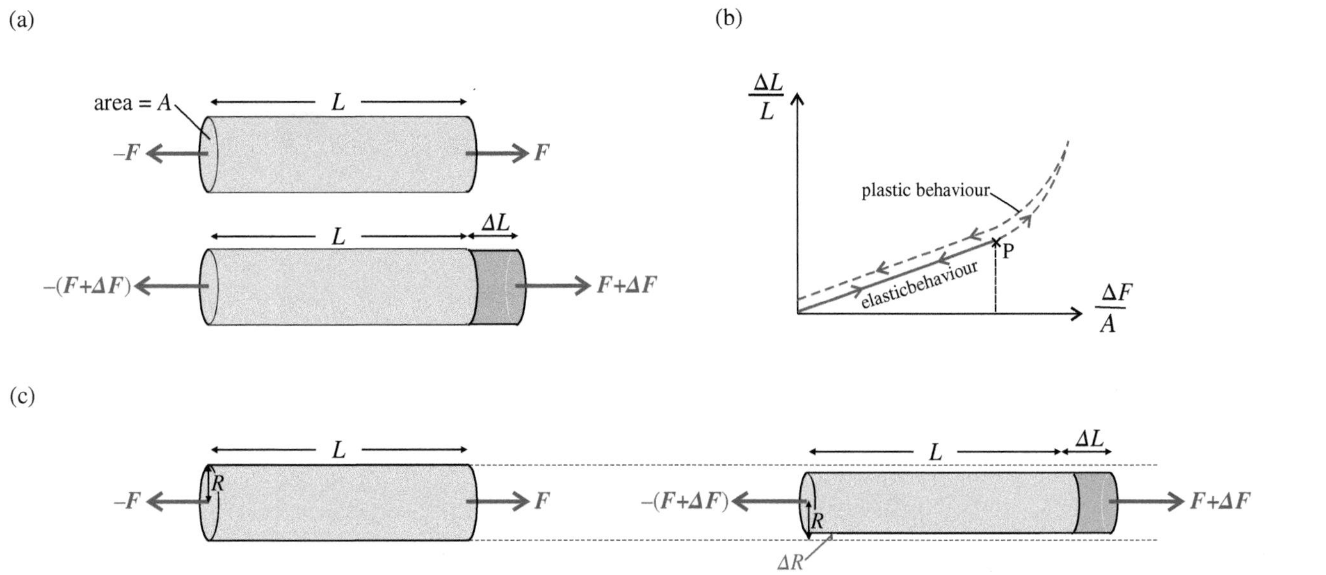

Figure 10.2. (a) A cylindrical specimen of elastic material is under tension due to the forces F and $-F$. When the forces are changed to $\boldsymbol{F + \Delta F}$ and $-\boldsymbol{(F + \Delta F)}$, the length of the specimen increases by ΔL. (b) A plot of longitudinal strain versus longitudinal stress for a metal specimen. For an applied stress below the elastic limit (point P), the strain is directly proportional to the stress and the body resumes its original length when the stress is removed (solid line). If the elastic limit is exceeded (dashed line – *plastic behaviour*) the strain depends on stress in a non-linear manner and the specimen remains permanently extended after the stress has been removed. (c) For most materials a longitudinal extension will be accompanied by a corresponding radial contraction.

where the constant of proportionality is characteristic of the material from which the rod or wire is made. This constant is called the **Young modulus** of the material (Thomas Young 1773–1829), defined by

$$\boxed{E := \frac{\text{longitudinal stress}}{\text{longitudinal strain}} = \frac{L}{A}\frac{dF}{dL}}$$

Values of the Young modulus of a number of common materials are given in Table 10.2.

If the forces applied to the wire or rod are sufficiently large, it may become permanently stretched ('plastic' behaviour – dashed line in Figure 10.2(b)). In this case the relationship between stress and strain is no longer linear and the process cannot be completely reversed by the removal of the force producing the distortion. The point at which the behaviour becomes non-linear and non-reversible is called the *elastic limit* (point P in Figure 10.2b).

Poisson ratio: For most elastic materials it is found that the cylinder in Figure 10.2(a) will contract (extend) radially when it is extended (compressed) axially (Figure 10.2(c)). Within the elastic limit, the ratio of transverse contraction strain to longitudinal extension strain is found to be constant. This ratio, which is characteristic of the material involved, is defined as the **Poisson ratio** of the material (Simeon Poisson 1781–1840) as follows.

$$\boxed{\upsilon := \frac{\text{transverse strain}}{\text{longitudinal strain}} = -\frac{\Delta R/R}{\Delta L/L} = -\frac{L}{R}\frac{dR}{dL}}$$

Note that an extended deformation is considered positive and a compressed deformation is considered negative so that the definition contains a minus sign to ensure that normal materials have a positive Poisson ratio. Values of the Poisson ratio of some common materials are given in Table 10.2.

Some rather exotic elastic substances known as *auxetic materials* display a negative Poisson's ratio. When subjected to positive strain in a longitudinal axis, the transverse strain in the material will actually be positive (that is, the cross-sectional area increases).

Volumetric deformation

Next, we consider the case in which the forces producing the extension of a solid body are applied to the body equally in all directions, as illustrated in Figure 10.3, to produce a change in volume. For simplicity we consider a solid cube in Figure 10.3 where the force ΔF, applied perpendicularly to all six faces as shown, produces a change in volume ΔV.

Table 10.2 Elastic constants (at 20 °C) of some common substances. ‡ denotes approximate values

substance	Young modulus N m^{-2}	shear modulus N m^{-2}	bulk modulus N m^{-2}	Poisson Ratio
aluminium	7.0×10^{10}	2.6×10^{10}	7.6×10^{10}	0.33
brass ‡	1×10^{11}	4×10^{10}	1×10^{11}	0.34
copper	1.3×10^{11}	4.8×10^{10}	1.4×10^{11}	0.33 – 0.36
gold	7.8×10^{10}	2.7×10^{10}	2.2×10^{11}	0.44
steel (mild)	2.1×10^{11}	8.2×10^{10}	1.7×10^{11}	0.30
water ‡	–	–	2×10^{9}	–
mercury	–	–	3×10^{10}	–

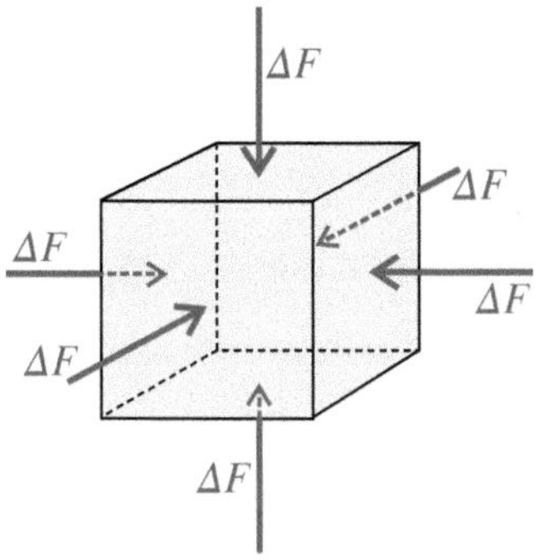

Figure 10.3. A volumetric stress applied to a solid cube.

Provided that the elastic limit is not exceeded, experiment shows that the volumetric strain $\frac{\Delta V}{V}$ is directly proportional to the volumetric stress $\frac{\Delta F}{A}$ producing that change in volume, where A is the area of an end face of the cube. Thus,

$$\frac{\Delta F}{A} = (constant)\frac{\Delta V}{V} = -K\frac{\Delta V}{V}$$

where K is a constant, called the bulk modulus, which is characteristic of the material under study. Thus, the **bulk modulus** is defined by

$$\boxed{K := \frac{\text{volumetric stress}}{\text{volumetric strain}} = -\frac{V}{A}\frac{dF}{dV}}$$

where the minus sign is included so that K is positive ($\frac{dF}{dV}$ is negative, since an increase in the applied force causes a decrease in volume). Values of the bulk modulus of a number of common materials are given in Table 10.2.

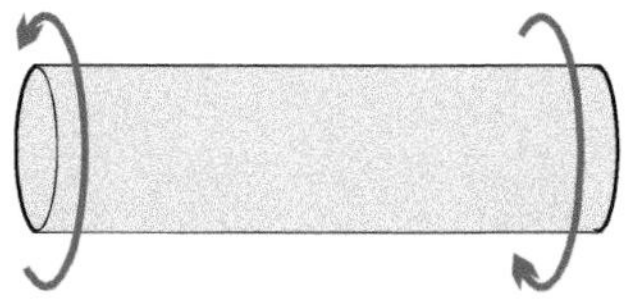

Figure 10.4. A shearing stress applied to a cylindrical rod by equal and opposite torques applied at the ends of the rod.

Shearing deformation

If, instead of equal and opposite longitudinal forces, as illustrated in Figure 10.2(a), equal and opposite torques are applied to a solid rod, as illustrated in Figure 10.4, the rod will show different elastic behaviour.

Provided the elastic limit is not exceeded, the rod returns to its original shape when the torques are removed. Indeed, this phenomenon has already been invoked in discussing the torsion balance in Section 7.5 and torsional oscillations in Section 7.10. This effect can be best understood by considering the distortion of a solid cube by equal and opposite forces which are applied tangentially to opposite faces as shown in Figure 10.5.

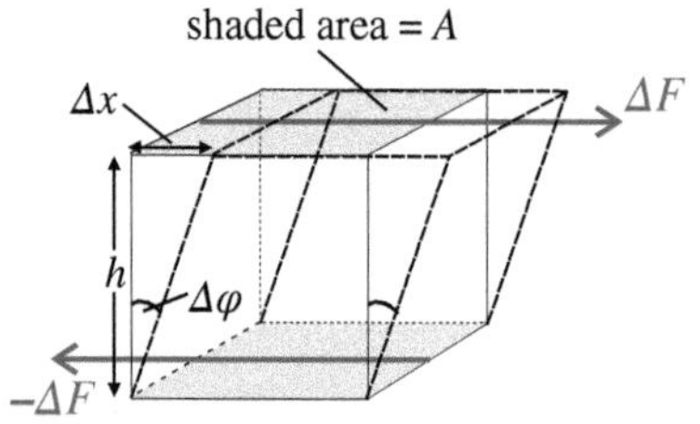

Figure 10.5. A shearing stress applied to a rectangular block by applying equal and opposite forces tangentially to the top and bottom faces.

The net effect, in this case, is that the square shape of the side faces of the cube becomes rhombic in shape, as shown in the figure. A force per unit area applied tangentially to this plane is called a *shearing stress* and the corresponding shearing strain is defined as $\frac{\Delta x}{h}$ or $\Delta\varphi$ for small Δx. Again, if the elastic limit is not exceeded, it is found that the stress is directly proportional to strain, that is

$$\frac{\Delta F}{A} = (\text{constant})\Delta\varphi$$

where the constant of proportionality, defined as

$$\boxed{G := \frac{\text{shearing stress}}{\text{shearing strain}} = \frac{1}{A}\frac{dF}{d\varphi}}$$

is called the **shear modulus** of the material of the body. Values of the shear modulus of some common materials are given in Table 10.2.

Since strain is dimensionless, the unit for all elastic moduli must be the same as that of stress; thus, the SI unit of any of the three elastic moduli defined above is N m^{-2}.

Application: Torsion of a cylindrical rod

We are now in a position to determine how the torsion constant (the restoring torque per unit angle of twist) of a rod or wire, discussed in Section 7.5, is related to the shear modulus of the material from which rod or wire is made. First we consider the twisting under equal and opposite torques of a cylindrical pipe of radius r, thickness Δr and length l as shown in Figure 10.6(a). The torque ΔT can be considered as being produced by a force ΔF applied tangentially to the shaded annular ring in the diagram, that is $\Delta T = (\Delta F)r$. The effect of the torque ΔT is to produce a shearing strain $\varphi = \frac{s}{l}$, as indicated in the figure, where $s = \varphi l = \theta r$ and θ is the corresponding angle of twist.

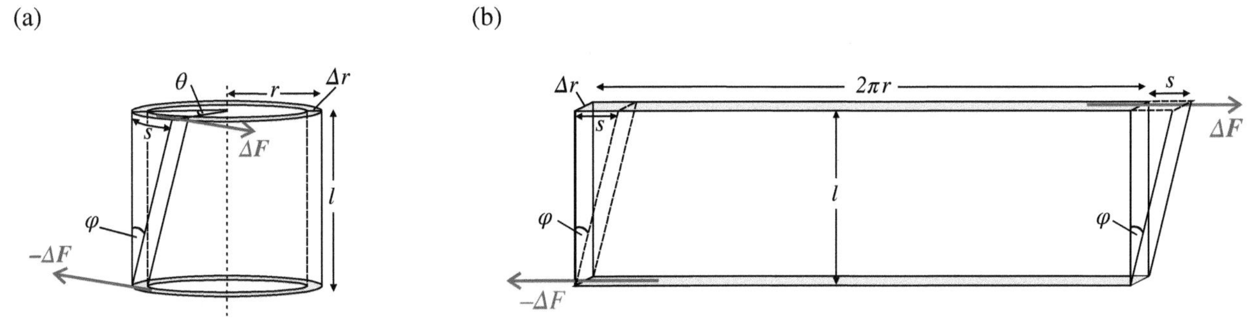

Figure 10.6. (a) A shearing stress applied to a cylindrical pipe. Equal and opposite torques of $(\Delta F)r$ and $-(\Delta F)r$ applied to the ends of the pipe cause the distortion shown. (b) If the cylindrical pipe in Figure 10.6(a) could be unwound it would form a rectangular block as in the figure. The distortion is similar to that shown in Figure 10.5.

If we now consider the cylindrical pipe as it would appear if 'unwound' to form a rectangular sheet of length $2\pi r$, height l and thickness Δr, as indicated in Figure 10.6(b), we obtain the following result (from the definition of the shear modulus).

$$\frac{\Delta F}{A} = \frac{\Delta F}{(2\pi r)(\Delta r)} = G\varphi = G\frac{s}{l} = G\frac{r\theta}{l}$$

$$\Rightarrow \quad \Delta F = G\frac{r}{l}(2\pi r\Delta r)\theta$$

$$\Rightarrow \quad \Delta T = r\Delta F = \frac{2\pi G}{l}\theta r^3 \Delta r$$

To determine the relationship between torque and angle of twist for a solid rod we simply add the torques on each of the concentric cylindrical pipes into which rod has been divided, that is we integrate the last result from $r = 0$ to $r = a$, where a is the radius of the rod. Hence,

$$T = \frac{2\pi G}{l}\theta \int_0^a r^3 dr = \frac{2\pi G}{l}\theta\frac{a^4}{4} = \frac{\pi G a^4}{2l}\theta = c\theta$$

Thus the torsion constant is given by $c = \frac{\pi G a^4}{2l}$ from which we see that for a sensitive torsion balance (such as that required for the Cavendish experiment, Section 7.5) a suspension wire with small shear modulus, small a and/or large l is required.

Study Worked Example 10.1

For problems based on the material presented in this section visit up.ucc.ie/10/ *and follow the link to the problems.*

10.3 Fluids at rest

A body immersed in a fluid experiences forces which are exerted by the surrounding fluid; for example, the buoyancy force which enables us to swim is a result of such forces. In a fluid at rest, any portion of the fluid itself is kept in equilibrium by forces exerted on it by the adjacent fluid around it. Consider the free body diagram of the arbitrary cube-shaped piece of fluid indicated by the shaded portion in Figure 10.7. This cube of fluid is kept in equilibrium as a result of forces exerted by the surrounding fluid; these forces act on all six faces as shown.

We shall see later that a fluid, just like a solid, is elastic when subjected to volumetric stress. Unlike a solid, however, a fluid cannot support longitudinal or shearing stresses – in fact this is what gives it its fluid properties. Thus the forces on any face in Figure 10.7 must be directed perpendicularly to that face, since any tangential component of an applied force would cause the fluid to flow – it would no longer be at rest.

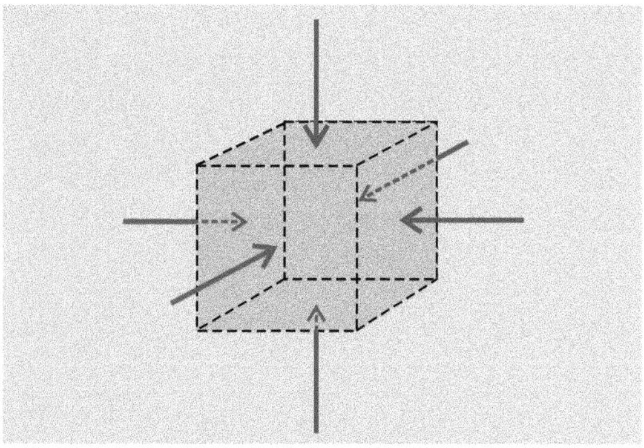

Figure 10.7. Any arbitrary cube of fluid at rest is kept in equilibrium by the forces exerted on it by the surrounding fluid.

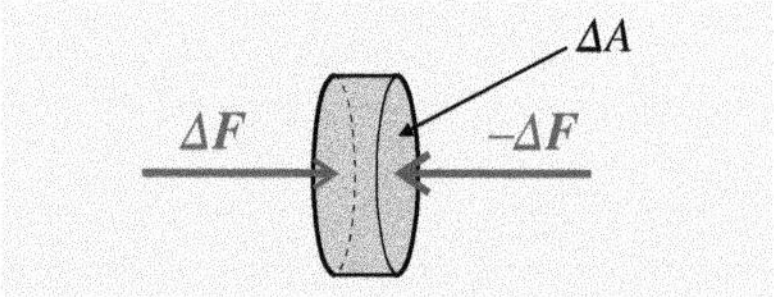

Figure 10.8. A thin 'pill-box' shaped element of fluid is kept in equilibrium by forces ΔF and $-\Delta F$ acting on opposite faces each of area ΔA.

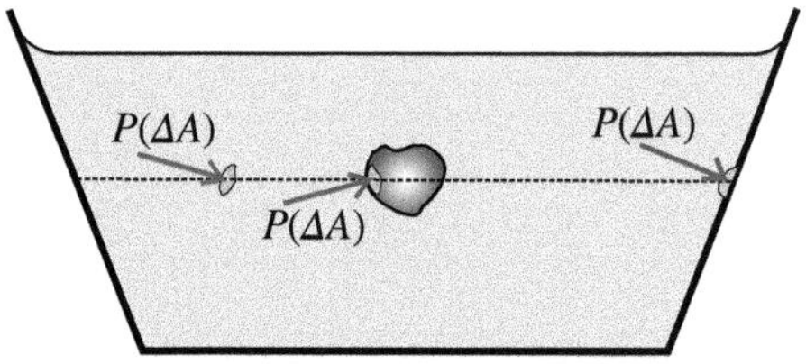

Figure 10.9. The average force exerted on an area ΔA is $P(\Delta A)$, where P is the pressure in the fluid at the area ΔA.

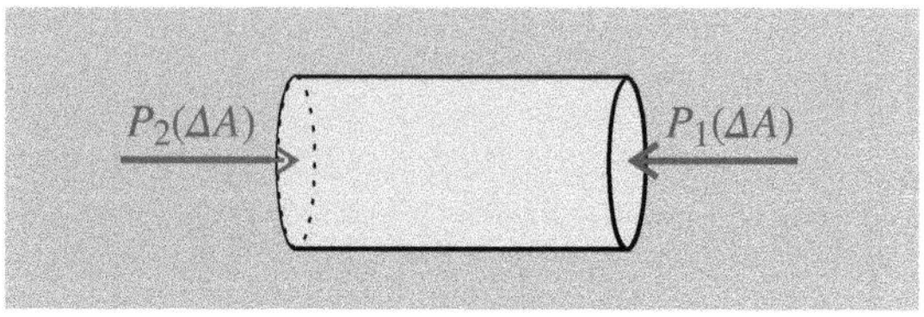

Figure 10.10. Equilibrium of a horizontal tube of fluid.

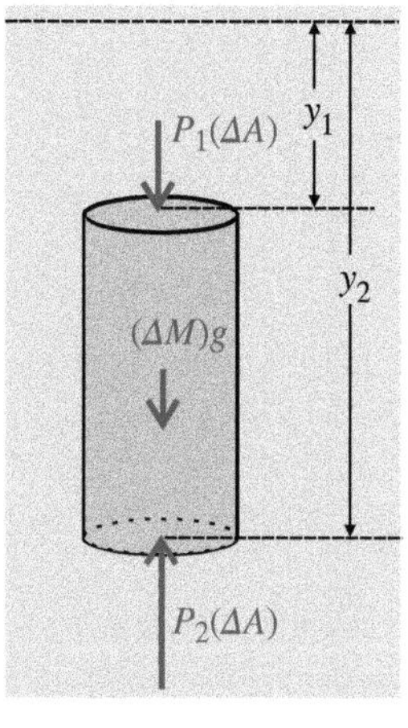

Figure 10.11. Equilibrium of a vertical tube of fluid in a uniform gravitational field.

Consider the forces acting on a very thin 'pill-box' shaped element of area ΔA in the fluid as shown in Figure 10.8. This element will be in equilibrium due to a force ΔF exerted by the surrounding fluid on one side and an equal and opposite force $-\Delta F$ exerted by the fluid on the other side. We define the **pressure** at a point in the fluid as follows

$$P := \underset{\Delta A \to 0}{\text{limit}} \frac{|\Delta F|}{\Delta A}$$

The SI unit of pressure is N m^{-2} which is called the pascal (Pa), that is 1 N m$^{-2} = 1$ Pa.

Note that pressure is a **scalar** quantity defined at any point in the fluid. Provided that the thickness of the element is very small, the definition is independent of the orientation of the pill-box element chosen. Thus the average force exerted by the fluid on a surface of area ΔA in the fluid, or on an area ΔA of a solid body immersed in the fluid or on an area ΔA of the wall of a container, as shown in Figure 10.9, is given by

$$|\Delta F| = P(\Delta A)$$

where P is the pressure in the fluid at the centre of at the element ΔA and the direction of ΔF is perpendicular to the area ΔA.

A number of conclusions can now be drawn concerning the pressure in a liquid at rest in a uniform gravitational field.

(a) Consider a horizontal tube of liquid of small cross-sectional area ΔA as shown in Figure 10.10. Let P_1 and P_2 be the pressures in the fluid at opposite ends of the tube. The net horizontal force on the tube of fluid $P_2(\Delta A) - P_1(\Delta A) = (P_2 - P_1)(\Delta A)$ must be zero because, otherwise, there would be motion of the fluid. Thus $P_1 = P_2$, that is *the pressure is the same at all points in a horizontal plane in a fluid at rest.*

(b) As a corollary of the above result, we conclude that *the surface of a fluid in a uniform gravitational field must be horizontal.*

(c) Let us now ask how the pressure in a fluid in a uniform gravitational field varies with depth. Consider a vertical cylindrical portion of fluid of cross-sectional area ΔA, as shown in Figure 10.11. Let P_1 be the pressure at a depth y_1 below some arbitrary horizontal level and let P_2 be the pressure at a depth y_2 below this level. For the fluid to be at rest there must be no net force on this cylinder of fluid, in particular the net vertical force must be zero. Thus the upward force on the bottom surface of the cylinder must be balanced by the net downward force, namely the force on the top surface of the cylinder plus the weight $(\Delta M)g$ of the cylinder of fluid. Thus,

$$P_2(\Delta A) - P_1(\Delta A) - (\Delta M)g = 0.$$

Now if the fluid is **homogeneous** (that is, ρ is independent of position) then

$$(\Delta M)g = \rho(\Delta V)g = \rho(\Delta A)(y_2 - y_1)g$$

and thus

$$P_2(\Delta A) - P_1(\Delta A) = \rho(\Delta A)(y_2 - y_1)g$$

$$\Rightarrow \quad \boxed{P_2 - P_1 = \rho g(y_2 - y_1)} \tag{10.1}$$

Thus, *the pressure in a homogeneous fluid in a uniform gravitational field increases linearly with depth.*

(d) Finally, let us consider a fluid which completely fills a closed container and is confined by two freely moving pistons as shown in Figure 10.12. A change in pressure exerted by one or both of the pistons can change the pressure at every other point in the fluid. We see from Equation (10.1), however, that pressure *differences* cannot change if the fluid is to remain static and thus *a change in the pressure anywhere within a fluid at rest in a closed container is transmitted undiminished to all points in the fluid.* This statement, first enunciated by Blaise Pascal (1623–1662), is known as **Pascal's Principle.**

If the force applied to one piston is changed, the force on the other piston must adjust appropriately to keep the fluid at rest. If the two pistons are kept at the same horizontal level, the pressure in the fluid just below the left-hand piston is equal to the pressure just below the right-hand piston, that is

$$\frac{F_1}{A_1} = \frac{F_2}{A_2} \quad \Rightarrow \quad \frac{F_1}{F_2} = \frac{A_1}{A_2}.$$

Hydraulic lifts and hydraulic brakes are based on this principle: if $A_1 \gg A_2$ then $F_1 \gg F_2$ and thus a small force applied to the smaller piston can be used to move a heavy object attached to the larger piston. The ratio $\frac{F_1}{F_2}$ is called the **mechanical advantage**.

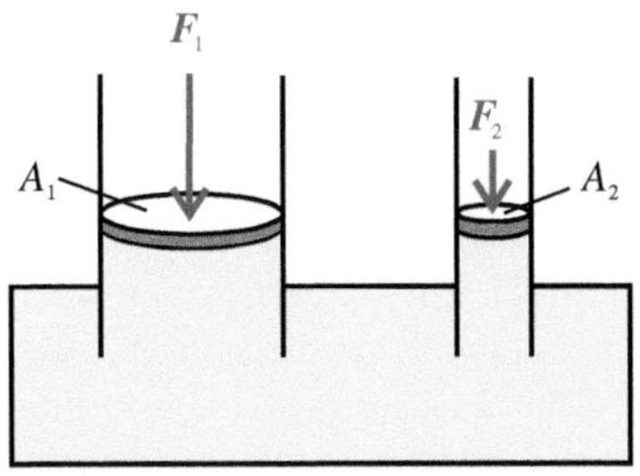

Figure 10.12. Pascal's Principle. A small force applied to the piston with small cross-section results in a large force being exerted by the piston of large cross-section. This is the principle used in hydraulic lifts, brakes, *etc*.

Study Worked Example 10.2

For problems based on the material presented in this section visit up.ucc.ie/10/ *and follow the link to the problems.*

10.4 Elastic properties of fluids

Fluids cannot withstand longitudinal or shearing stresses – as pointed out in the previous section, this property essentially defines a fluid. Thus the concepts of a Young modulus or a shear modulus are not applicable to fluids (in fact $E \to 0$ and $G \to 0$). The idea of a bulk modulus, however, can be extended simply to fluids by realising that a volumetric stress in a fluid is automatically provided by a change in pressure. Thus the bulk modulus of a liquid or a gas can be defined as

$$\boxed{K := -\frac{\Delta P}{\frac{\Delta V}{V}} = -V\frac{dP}{dV}}$$

As in the case of a solid, the bulk modulus of a liquid is reasonably constant over a wide range of pressures.

The bulk modulus of a gas, on the other hand, is much smaller than that of liquids and depends strongly on the pressure of the gas; in other words, gases are much more compressible than liquids or solids.

Mechanical work in expansion

Consider a fluid or a solid which initially has volume V and is then expanded to volume $V + \Delta V$ against a pressure P (Figure 10.13). We wish to obtain an expression for the mechanical work done in expanding the body by ΔV. First we divide ΔV into elements as shown in the figure. Consider the element, shown by blue shading, which has cross-sectional area δA and thickness δz. The force exerted on the area δA due to the pressure P is $P(\delta A)$ and hence the work done by this force in moving the surface of area δA through the distance δz is given by

$$\delta W = (\delta F)(\delta z) = (P\delta A)(\delta z) = P(\delta A)(\delta z) = P(\delta V)$$

where $\delta V = \delta A \delta z$.

Thus, the total work done in the expansion is given by

$$\boxed{\Delta W = P(\Delta V)} \qquad (10.2)$$

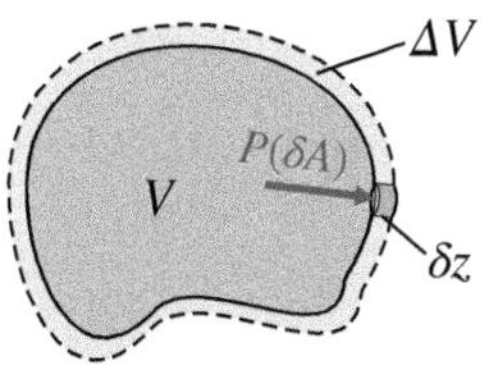

Figure 10.13. When a volume V of fluid is expanded by ΔV at a pressure P, the mechanical work done is $P(\Delta V)$.

For problems based on the material presented in this section visit up.ucc.ie/10/ *and follow the link to the problems.*

10.5 Pressure in gases

The Earth's atmosphere is a mixture of gases held in place by the gravitational field of the Earth. Because air is so compressible, its density is greatest at the Earth's surface and falls off with increasing altitude; that is, the atmosphere is a non-homogeneous fluid. The air in the atmosphere cannot be considered to be a static fluid; it is in continuous motion due to cyclones and anticyclones (Section 8.5) and its density, which varies with altitude, changes due to varying amounts of water vapour, etc. Thus the atmospheric pressure at any point depends on the state of local weather systems and its measurement is of vital importance in weather forecasting.

The density of the atmosphere varies with height above the earth's surface, that is $\rho = \rho(y)$. When a fluid is not homogeneous, Equation (10.1) is not applicable. Within a small horizontal slice between y and $y + \Delta y$, however, the density can be considered uniform and hence

$$\Delta P = \rho(y) g \Delta y$$

and the pressure at the surface of the earth, usually called the 'atmospheric pressure', is given by

$$B = \int dP = g \int_0^\infty \rho(y) dy$$

Since, in order to calculate B, $\rho(y)$ would need to be known at all heights, it is much simpler to determine atmospheric pressure through direct measurement, as described below.

Measurement of atmospheric pressure

A method of measuring atmospheric pressure is indicated in Figure 10.14. One end of a transparent tube is placed in a dish of high density liquid – mercury, being the liquid of greatest density, has traditionally been employed for this experiment (although no longer in general use because of the toxicity of mercury vapour). The other end of the tube is set up as indicated in the figure and connected to a vacuum pump. When the pump is switched on, the liquid rises in the tube until it settles in equilibrium with the pressure due to the column of liquid balancing the atmospheric pressure acting on the free surface of the liquid in the dish. Hence, invoking Equation (10.1) with $P_1 = 0$ and $P_2 = P_a$, the atmospheric pressure is given by

$$P_a = g \rho_L H$$

where ρ_L is the density of the liquid (13.6×10^3 kg m^{-3} in the case of the mercury) and H is the vertical height of the top of the column above the surface of the liquid in the open dish, as shown in Figure 10.14.

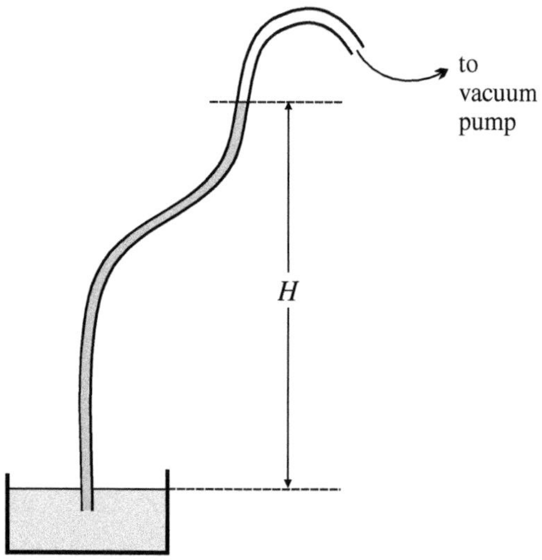

Figure 10.14. Experimental arrangement used to measure atmospheric pressure. The atmospheric pressure acting on the surface of the fluid in the open dish is balanced by a vertical height H of fluid in the tube.

Pressure units

In addition to the pascal, a number of other (non-SI) units are commonly used to measure pressures in gases. Sometimes atmospheric pressure is simply stated in terms of the height H of mercury in the mercury barometer described above; this unit is denoted by mmHg. Also used is a unit called bar which is defined as 10^5 Pa; most national weather services report atmospheric pressure in millibar (1 mbar = 10^2 Pa = 1 hectopascal). Pressure may also be stated in terms of a unit called the 'atmosphere' (which is defined as 1 atm = 101325 Pa) or in torr $\left(1 \text{ torr} = \frac{101325}{760} \text{ Pa}\right)$. Typically the atmospheric pressure at the Earth's surface is 10^5 Pa = 1000 mbar ≈ 1 atm ≈ 760 mmHg.

For problems based on the material presented in this section visit up.ucc.ie/10/ and follow the link to the problems.

10.6 Archimedes' principle

Every swimmer is familiar with the buoyancy force experienced when immersed in water, an effect which is noticeably greater in salt water than in fresh water. Clearly the forces exerted due to the pressure of the surrounding fluid act in such a way that they produce an upward force which 'buoys up' the swimmer in the water. This net upward buoyancy force is usually called an **upthrust**.

Consider a solid body of volume V immersed in a fluid as shown in Figure 10.15(a). The only forces acting on the body are (i) its weight acting vertically downward and (ii) the resultant force (the upthrust) on the body due to the surrounding fluid. The latter force is the vector sum of the forces on each area element ΔA of surface (directed perpendicularly to the element), that is the sum of all the $P(\Delta A)\mathbf{n}$ vectors where P is the pressure in the fluid adjacent to the element ΔA and $\mathbf{n}$ is a unit vector normal to the surface. Except possibly in the case of a body of very regular shape, the magnitude of this upthrust or buoyancy force would be difficult to determine in terms of the pressure in the fluid at all the points where the fluid is in contact with the body. The considerations which follow, however, show how the upthrust may be determined even in cases where the body has a very irregular shape.

(a) (b)

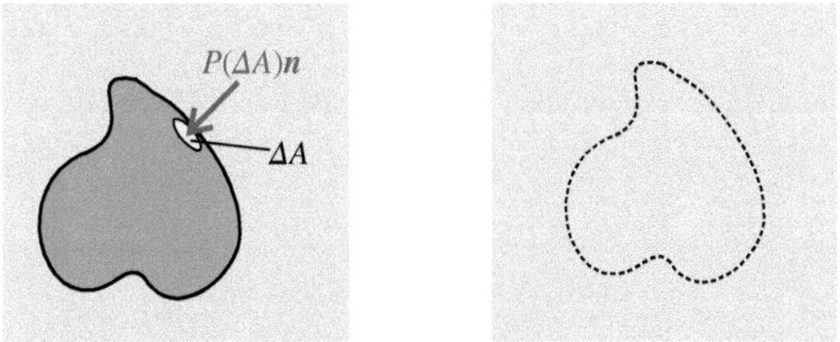

Figure 10.15. (a) A solid body immersed in a fluid. (b) The solid body in Figure 10.15(a) has been replaced by an equal volume of fluid. The forces exerted by the surrounding fluid on the fluid within the surface indicated by the dashed line must be the same as the forces acting on the solid body in Figure 10.15(a).

Suppose that the solid body has been removed from the fluid so that the space previously occupied by the body is filled instead with fluid. Consider the forces which act on the volume V of fluid which has replaced the body — indicated by the dashed line in Figure 10.15(b). Since the fluid is at rest, the fluid inside the surface indicated by the dashed line must be equilibrium. Thus, the forces exerted on the volume V of fluid by the fluid surrounding it must exactly balance the weight of the volume V of fluid. Clearly, however, the forces exerted on the volume V of fluid by the surrounding fluid are exactly the same as the forces which were previously acting on the solid body. Thus the upthrust experienced by the solid body of volume V must be equal to the magnitude of the weight of a volume V of fluid. This general principle, the **Principle of Archimedes** (Archimedes, 287–212 BC), may be stated as follows.

When a body is immersed in fluid, it experiences an upthrust which is equal in magnitude to the magnitude of the weight of fluid displaced by the body.

Note that this is a very general principle; we have not assumed that the fluid has uniform density, for example, and thus the principle is valid even though the density of the fluid may vary with depth. In most situations, in practice, the sizes of the objects involved are such that the variation of the density of the surrounding fluid over the dimensions of the body may be ignored safely. For very large bodies, however, the variation of fluid density with depth has to be taken into account.

In cases in which the fluid can be assumed to have uniform density (that is, density of fluid $= \rho_F =$ constant), the forces acting on a solid body of volume V made from material of density ρ_S are indicated in Figure 10.16. The net force is

$$U - W = \rho_F V g - \rho_S V g$$

where the upward direction has been taken as positive. Whether the net force is directed upwards or downwards depends on whether $\rho_F > \rho_S$ or $\rho_F < \rho_S$.

We have considered above only situations in which a body is fully submerged in the fluid but the same considerations apply, and hence Archimedes' Principle is applicable, even if the body is only partially immersed as shown in Figures 10.17(a) and 10.17(b). In this case the Archimedean upthrust is $\rho_F v g$ where v is the volume of the body below the surface. The net (upward) force on the body in this case is

$$U - W = \rho_F v g - \rho_S V g$$

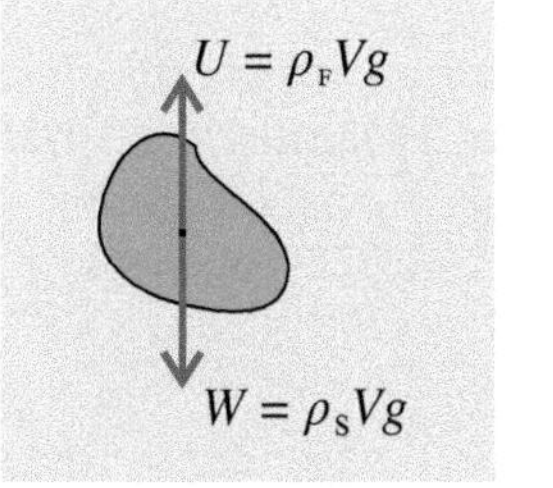

Figure 10.16. In addition to its own weight (W), a body immersed in a fluid experiences an upthrust (U) the magnitude of which is equal to the weight of the fluid displaced by the body.

Thus, the net force is upward if $\rho_F v > \rho_S V$ and downward if $\rho_F v < \rho_S V$. In cases where $\rho_S < \rho_F$ the body will sink in the fluid until it reaches a position where $\rho_F v = \rho_S V$ and the body will be in equilibrium, that is the body will float. The **condition for flotation**, therefore, is that the fraction of the body submerged is given by

$$\frac{v}{V} = \frac{\rho_S}{\rho_F} \leq 1$$

We see from the foregoing that the upthrust or buoyancy force on a body is proportional to the density of the fluid in which it is immersed. This explains, for example, why a swimmer finds it easier to float in salt water, the density of salt water being greater than that of fresh water.

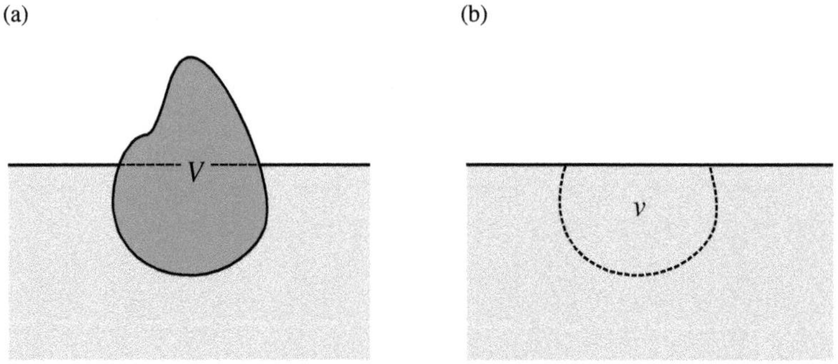

Figure 10.17. (a) A body partially immersed in a fluid. (b) The magnitude of the upthrust on the body in Figure 10.17(a) is equal to the weight of fluid within the volume v, as indicated by the dashed line.

Study Worked Example 10.3

For problems based on the material presented in this section visit up.ucc.ie/10/ *and follow the link to the problems.*

10.7 Fluid dynamics; the Bernoulli equation

The mechanics of general fluid flow is quite complicated and will not be discussed in this book. If the speed of flow is slow enough, however, a reasonably simple analysis is possible, as we shall see. We restrict our discussion to the special case, called **streamline flow**, illustrated in Figure 10.18.

In streamline flow, all particles of fluid which pass through any particular point in the fluid proceed along the same path or streamline. Particles passing through other points travel on different streamlines and distinct streamlines never intersect. If a tiny spot of dye is introduced at a point in a clear liquid, which is flowing slowly in a transparent pipe, a streamline can be made visible. If the rate of flow increases, however, vortices and eddies develop and the dye gets mixed through the liquid – the flow is no longer streamline and is said to have become **turbulent**.

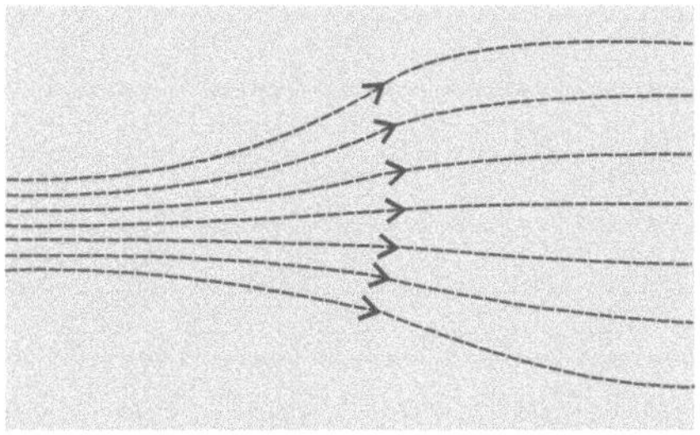

Figure 10.18. Streamlines in a slowly moving fluid.

Consider the streamline flow illustrated in Figure 10.19. The fluid flowing through some arbitrary area A_1 whose plane is perpendicular to the flow remains within a 'tube' delineated by the streamlines through the perimeter of A_1. Thus all the fluid entering this 'tube' through A_1 will later leave through A_2. The mass of the fluid passing through A_1 in any time interval must be equal to the mass of fluid passing through A_2 in the same time interval. Let Δx_1 be the distance travelled in a small time interval Δt by the fluid flowing through A_1 and let Δx_2 be the distance travelled in the same time interval by the fluid flowing through A_2. Clearly, $\Delta x_1 = v_1(\Delta t)$ and $\Delta x_2 = v_2(\Delta t)$ where v_1 and v_2 are the velocities of the fluid at A_1 and A_2, respectively. The volume of fluid flowing through A_1 in time (Δt), therefore, is the volume of a cylinder of cross-sectional area A_1 and length Δx_1. Hence the mass of fluid flowing through A_1 in time Δt is given by

$$\Delta m_1 = A_1 v_1 \rho_1 (\Delta t)$$

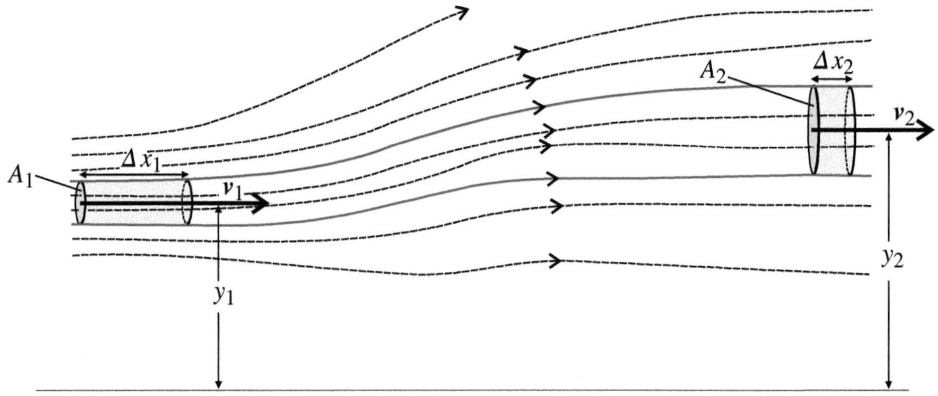

Figure 10.19. In a streamline flow, the fluid which flows through the area A_1 remains within the 'tube' indicated by the coloured lines.

where ρ_1 is the density of the fluid at A_1. Similarly, if ρ_2 is the density at A_2, the mass of fluid flowing through A_2 in time Δt is

$$\Delta m_2 = A_2 v_2 \rho_2 (\Delta t)$$

Now the mass of fluid in the 'tube' is conserved in streamline flow, that is $\Delta m_1 = \Delta m_2 = \Delta m$, and so we can write

$$\frac{dm}{dt} = A_1 v_1 \rho_1 = A_2 v_2 \rho_2$$

This last relationship is called *the continuity equation* which, equivalently, states that the mass of fluid flowing per unit area per unit time is $v\rho$.

In the case of an incompressible fluid $\rho_1 = \rho_2 = \rho$, in which case $\frac{dm}{dt} = A_1 v_1 \rho = A_2 v_2 \rho$ and the *volume* of fluid flowing per unit time is given by

$$\frac{dV}{dt} = \frac{d}{dt}\left(\frac{m}{\rho}\right) = A_1 v_1 = A_2 v_2 \tag{10.3}$$

The Bernoulli equation

We now consider the dynamics of the fluid flow described in Figure 10.19. The fluid in the 'tube' is being pushed as the result of the pressure exerted by the fluid to the left of A_1 and has to overcome a resistance to its motion due to the pressure exerted by the fluid to the right of A_2. Thus the net mechanical work done on the fluid in the 'tube' in a time Δt is given by Equation 10.2, namely

$$\Delta W = P_1 A_1 (\Delta x_1) - P_2 A_2 (\Delta x_2) = P_1 A_1 v_1 (\Delta t) - P_2 A_2 v_2 (\Delta t)$$

This work goes into changing the total mechanical energy of the system. Effectively, the fluid which entered the 'tube' through A_1 during the time interval Δt has been raised under gravity to a position near A_2 with a corresponding change in its velocity. Thus ΔW is equal to the change in mechanical energy of the mass of fluid Δm, namely

$$\left[\frac{1}{2}(\Delta m){v_2}^2 + (\Delta m) g y_2\right] - \left[\frac{1}{2}(\Delta m){v_1}^2 + (\Delta m) g y_1\right]$$

and thus

$$\frac{P_1 A_1 v_1 (\Delta t)}{\Delta m} - \frac{P_2 A_2 v_2 (\Delta t)}{\Delta m} = \frac{1}{2}({v_2}^2 - {v_1}^2) + g(y_2 - y_1)$$

or, using $\Delta m = A_1 v_1 \rho_1 (\Delta t) = A_2 v_2 \rho_2 (\Delta t)$,

$$\frac{P_1}{\rho_1} - \frac{P_2}{\rho_2} = \frac{1}{2}({v_2}^2 - {v_1}^2) + g(y_2 - y_1)$$

$$\Rightarrow \quad \frac{P_1}{\rho_1} + \frac{1}{2}{v_1}^2 + g y_1 = \frac{P_2}{\rho_2} + \frac{1}{2}{v_2}^2 + g y_2 \tag{10.4}$$

Thus, at any point in the fluid the pressure, density, velocity and vertical height are related in such a way that

$$\frac{P}{\rho} + \frac{1}{2}v^2 + gy = constant$$

If the fluid is incompressible ($\rho_1 = \rho_2 = \rho$), Equation (10.4) takes the following form

$$\boxed{P_1 + \frac{1}{2}\rho {v_1}^2 + g\rho y_1 = P_2 + \frac{1}{2}\rho {v_2}^2 + g\rho y_2} \tag{10.5}$$

Equation (10.5) is called the **Bernoulli equation** after Daniel Bernoulli (1700 – 1782).

Applications of the Bernoulli equation

Finally, in this section, we consider the consequences of the Bernoulli equation when applied to incompressible fluids in some specific situations

(a) *Fluid at rest*, that is $v_1 = v_2 = 0$

$$\text{Equation (10.4)} \quad \Rightarrow \quad P_1 + g\rho y_1 = P_2 + g\rho y_2$$
$$\Rightarrow \quad P_1 - P_2 = g\rho(y_2 - y_1)$$

which is, as we would expect, the result (10.1) already derived for fluids at rest in Section 10.3 (the change of sign arises because y is directed downwards in (10.1)).

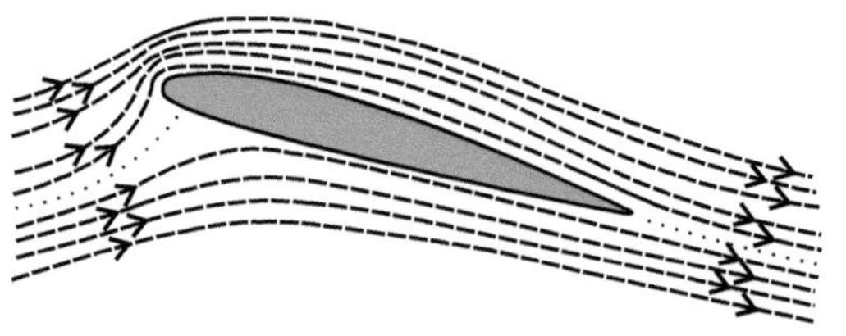

Figure 10.20. Dynamic pressure contributes to the lift provided by an aeroplane wing. The speed of the air flowing over the wing is greater than the speed of the air beneath the wing. Thus, from Equation (10.6), the upward pressure from below is greater than the downward pressure from above.

(b) *Horizontal flow*, that is $y_1 = y_2$

$$\text{Equation (10.5)} \quad \Rightarrow \quad P_1 + \frac{1}{2}\rho {v_1}^2 = P_2 + \frac{1}{2}\rho {v_2}^2 \qquad (10.6)$$

which shows that if $v_1 > v_2$, then $P_1 < P_2$; in other words, the faster the flow the lower the pressure in the fluid. The result has many important applications; for example, it contributes to the lift provided by the air flowing over aeroplane wings (Figure 10.20). The profile of a wing is such that the air has further to travel over the upper face and hence has to flow faster if streamline flow is to be maintained.

Readers may wish to observe for themselves the following very simple application of the Bernoulli equation. Hold two sheets of paper, one in each hand, so that they hang freely and vertically a few centimetres apart. Now blow horizontally between the two sheets and, perhaps contrary to your expectations, the sheets will tend to move *closer together*. The same effect is responsible for the tendency of two adjacent parallel boats to move closer together in fast flowing water.

Flowmeters

Bernoulli's equation is the basis of a technique that is used to measure the rate of flow of a fluid through a pipe. Suppose we want to measure the flow rate of an incompressible fluid in a horizontal pipe. A constriction may be introduced in the pipe as shown in Figure 10.21 and arrangements made (say by connecting a pressure gauge as indicated in the figure to measure the difference in pressure ($\Delta P = P_2 - P_1$) between the fluid in the main body of the pipe (cross-sectional area A_1) and that in the narrow section (area A_2). Combining Equations (10.6) and (10.3) we get

$$P_1 + \frac{1}{2}\rho {v_1}^2 = P_2 + \frac{1}{2}\rho {v_2}^2 = P_2 + \frac{1}{2}\rho\left(\frac{A_1}{A_2}v_1\right)^2 \qquad (10.7)$$

Note that, since $A_1 > A_2$, the pressure is less in the constriction where the fluid flow rate is greatest. The speed of the fluid in the main body of the pipe can be determined by solving for v_1 in Equation (10.7) and hence the volume and mass of fluid flowing in the pipe per

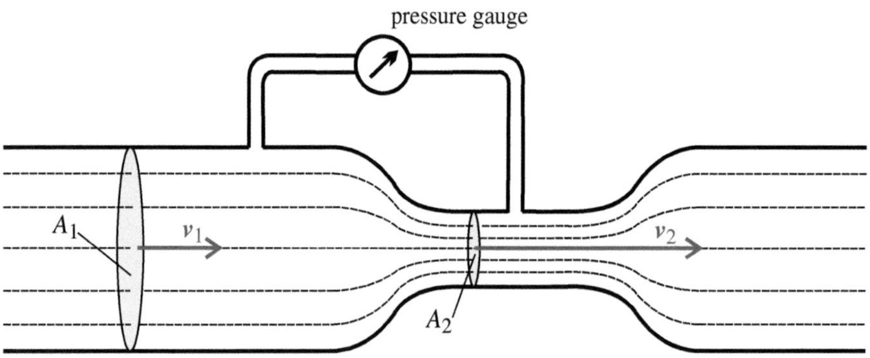

Figure 10.21. Example of a simple flowmeter.

unit time are given, respectively, by

$$\frac{dV}{dt} = A_1 v_1 \quad \text{and} \quad \frac{dM}{dt} = \rho A_1 v_1 \quad \text{where} \quad v_1 = \sqrt{\frac{2(P_1 - P_2)}{\rho\left[\left(\frac{A_1}{A_2}\right)^2 - 1\right]}}$$

We have assumed in the above analysis that there is no friction-like resistance to the flow of the fluid, an issue that will be addressed in the next section.

For problems based on the material presented in this section visit up.ucc.ie/10/ and follow the link to the problems.

10.8 Viscosity

As we have seen (Section 10.3), a fluid at rest is distinguished by the fact that, unlike a solid, it cannot withstand a shearing stress. In our considerations so far we have assumed that no force, other than that due to pressure, is required to keep a fluid in motion. In practice, however, the behaviour of real fluids deviates from this idealised model. For example, when a fluid flows in a pipe the layer of fluid adjacent to the wall of the pipe tends to adhere to the wall and hence moves at a lower speed than the fluid nearer to the centre of the pipe. Furthermore, a viscous fluid (like syrup) will flow much more slowly than a freely flowing liquid (such as water) under the same conditions.

Consider the situation, illustrated in Figure 10.22, in which a fluid is flowing over a large flat horizontal solid plate. As in the previous section, we deal only with fluids which are moving at sufficiently low velocities that turbulence has not set in. The fluid has a streamline type of motion but, in this case, the velocity of any layer of fluid increases with distance from the plate, that is there is a *velocity gradient* in the fluid. This type of flow is described as *laminar* – we think of the fluid as comprising a large number of thin layers or laminae, each layer of fluid moving with constant velocity in the steady state. As each layer flows over the adjacent (more slowly moving) layer it experiences a friction-like resistance to its motion. The shearing stress on any layer of fluid due to the layer immediately above must be equal and opposite to the shearing stress exerted by the layer immediately below; otherwise the net force would be non-zero and the fluid layer would not move with constant velocity.

Consider the layer of fluid between z and $z + \Delta z$, as shown in Figure 10.23 where z is measured perpendicular to the direction of flow, that is perpendicular to the surface of the flat plate. This layer of fluid experiences a shearing stress $\frac{dF}{dA}$ (Section 10.2) due to the fluid immediately above, where ΔF is the force exerted tangentially on the area ΔA and also experiences an equal and opposite shearing stress due to the layer of fluid immediately below it. It is found experimentally, provided the fluid velocities involved are not too large, that the velocity gradient $\frac{dv}{dz}$ is directly proportional to the shearing stress producing it, that is

$$\boxed{\frac{dF}{dA} = (constant)\,\frac{dv}{dz} = \eta\frac{dv}{dz}} \qquad (10.8)$$

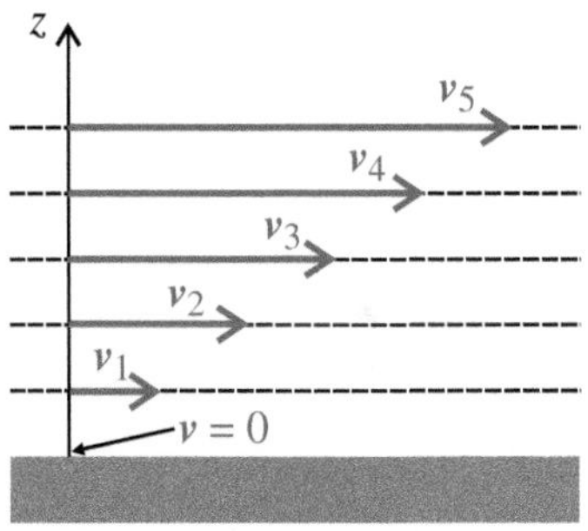

Figure 10.22. Streamline flow of a viscous fluid flowing over a solid surface. The layer of fluid in contact with the surface tends to adhere to the surface and the speed of the fluid increases with distance from the surface.

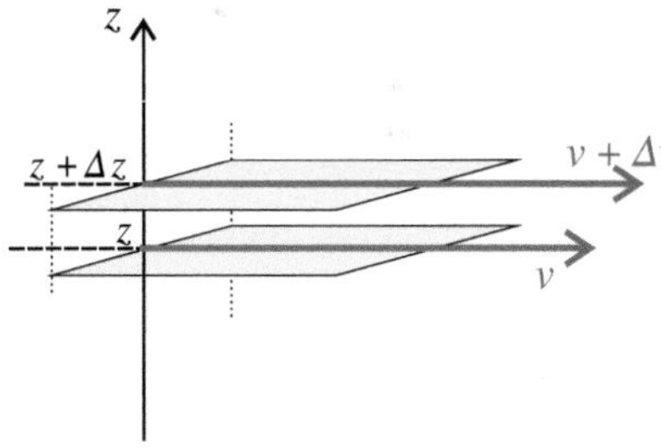

Figure 10.23. The speed of a layer of fluid is slightly less than that of the layer immediately above it and thus each layer experiences a shearing stress. Unlike solids, fluids cannot withstand a shearing stress and hence a velocity gradient $\frac{dv}{dz}$ is set up in the fluid.

where the constant of proportionality (η) is called the **viscosity** of the fluid. The value of the viscosity is characteristic of the particular fluid involved, being large for viscous fluids and small for freely flowing ones. The SI unit of viscosity, from the definition, is N s m^{-2} or kg m^{-1} s^{-1}. Fluids which obey the relationship (10.8) are often called 'Newtonian fluids'.

Application: flow of liquid in a cylindrical pipe

We now consider the effect of viscosity on the flow of liquid in a horizontal cylindrical pipe of radius a. We assume that the pressure difference between the ends of the pipe is constant and hence that a constant pressure gradient $\frac{dP}{dl}$ is set up throughout the length of the pipe.

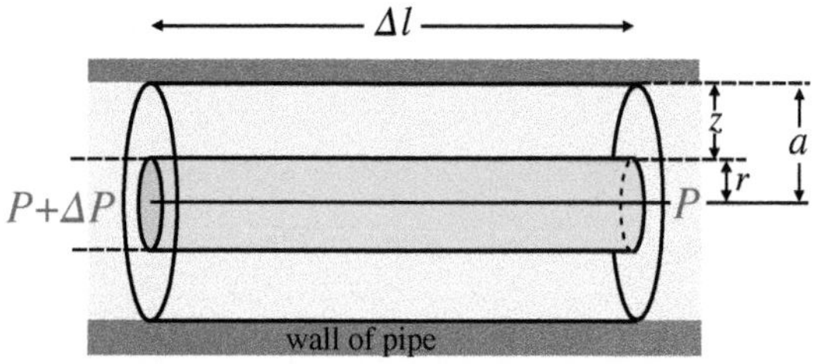

Figure 10.24. A small section of a cylindrical pipe in which a liquid is flowing. The shearing force which acts over the curved surface of the shaded portion of the fluid, that is at a distance r from the axis of the tube, is provided by the pressure difference across the section.

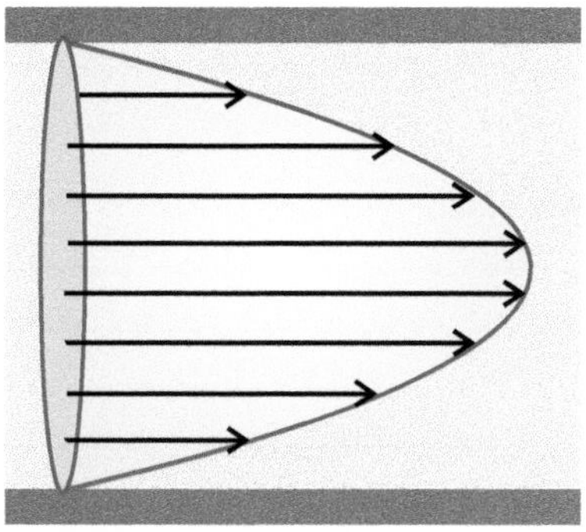

Figure 10.25. The speed of flow of liquid in a cylindrical pipe increases with distance from the wall of the pipe in a parabolic manner.

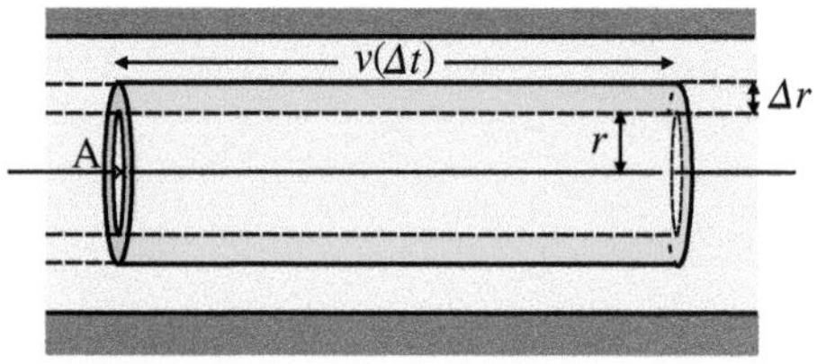

Figure 10.26. Determination of the rate of flow of liquid in a cylindrical pipe. The volume per second flowing through an annular ring between r and $r + \Delta r$ is determined first; integration from $r = 0$ to $r = a$ determines an expression for the rate of flow through the whole pipe.

Consider a small section of pipe of length Δl shown in Figure 10.24 and, in particular, consider how that part of the liquid (shaded blue) which is inside some arbitrary radius r moves relative to the remaining liquid in the pipe. The net force on the blue shaded cylinder of liquid is $(\Delta P)(\pi r^2)$ and this provides a shearing force acting over a surface of area $(2\pi r)(\Delta l)$. Thus the shearing stress on the liquid at a distance r from the axis is given by

$$\frac{dF}{dA} = \lim_{\Delta l \to 0} \frac{(\Delta P)(\pi r^2)}{2\pi r(\Delta l)} = \frac{r}{2}\frac{dP}{dl}$$

Thus $$\frac{dF}{dA} = \frac{r}{2}\frac{dP}{dl} = \eta\frac{dv}{dz} = -\eta\frac{dv}{dr}, \text{ (since } z = a - r \;\rightarrow\; dz = -dr),$$

and hence

$$\frac{dv}{dr} = -\frac{r}{2\eta}\frac{dP}{dl}$$

and $v = -\frac{1}{2\eta}\frac{dP}{dl}\int_a^r r dr$, from which we obtain an expression for the speed of the liquid at a distance r from the axis of the pipe

$$\boxed{v = \frac{1}{4\eta}\frac{dP}{dl}(a^2 - r^2)}$$

Thus, the shaded disc of liquid in Figure 10.25 will take the paraboloidal shape shown at some later instant.

We can now proceed to determine an expression for the rate of flow of liquid through the pipe. Consider first the liquid flowing between r and $r + \Delta r$ (Figure 10.26). The liquid in this annulus A at time t has moved forward a distance $v(\Delta t)$ at time $t + \Delta t$. Thus the volume ΔV of liquid passing through an annular ring between r and $r + \Delta r$ in time Δt is $[2\pi r(\Delta r)][v(\Delta t)]$ and hence the volume per unit time flowing through the pipe is given by

$$\frac{dV}{dt} = \int 2\pi r v dr = \frac{2\pi}{4\eta}\frac{dP}{dl}\int_0^a r\left(a^2 - r^2\right) dr = \frac{\pi}{2\eta}\frac{dP}{dl}\left[a^2\int_0^a r dr - \int_0^a r^3 dr\right]$$

$$\Rightarrow \quad \frac{dV}{dt} = \frac{\pi a^4}{8\eta}\frac{dP}{dl}$$

and the mass flowing per unit time is $\frac{\pi \rho a^4}{8\eta}\frac{dP}{dl}$. For a pipe of length L in which the pressure difference between its ends is $P_1 - P_2$, therefore, the mass flow per unit time is given by

$$\boxed{\frac{dM}{dt} = \frac{\pi \rho a^4}{8\eta}\frac{P_2 - P_1}{L}}$$

known as Poisseuille's formula (Jean Poisseuille (1797–1869).

For problems based on the material presented in this section visit up.ucc.ie/10/ *and follow the link to the problems.*

10.9 Surface properties of liquids

Water from a slowly dripping tap does not flow in a continuous stream but, instead, drops form at the tap outlet which increase in size before detaching and falling downwards under gravity. The liquid appears to be confined within the closed surface of the drop. This effect is most dramatic in the case of mercury which, if spilled in small amounts, can form into small, almost spherical, drops. In these circumstances, no container is needed to confine the liquid; clearly there are forces associated with the *surfaces* of liquids which play an important role in such situations.

These same surface effects are responsible for many other familiar phenomena. A light metal object such as a paper clip or sewing needle can 'float' on water if placed carefully on the surface even though its density is greater than that of water – many insects can move freely over a water surface using this effect. If a glass tube of small inside diameter is dipped in water, the water can rise up inside the tube by several centimetres before settling in equilibrium (Figure 10.27) – this effect, called 'capillarity', plays a role in enabling ground water near the roots of trees to be transported to the foliage at their top although, as shown Worked Example 10.4, this effect is not sufficient on its own to account for this phenomenon.

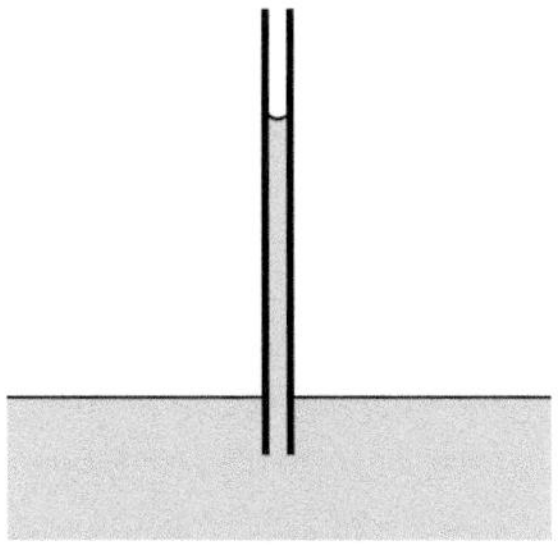

Figure 10.27. Capillary rise of a liquid in a narrow tube.

The shape of the surface of a liquid in a vessel, such as tea or coffee in a cup, deviates from horizontal near the side walls of the cup, forming what is called a 'meniscus'. Surface effects in liquids are most easily demonstrated by the use of soap solutions; a thin film is formed when a circular framework is dipped in soap solution and removed. Such a film can be considered as a body of liquid bounded by two adjacent parallel surfaces. Blowing gently on the film causes the formation of soap bubbles which generally take on a spherical shape.

How can surface phenomena such as those outlined above be explained in a simple way? The most convenient model that we can adopt is to consider the surface of a liquid as a sort of elastic membrane under tension, somewhat like a rubber sheet. For example, in this model a drop of liquid forming at the outlet of a tap is thought of as being confined within a stretched membrane which is somehow attached to the metal of the tap, as illustrated in Figure 10.28 rather as air is confined within a rubber balloon. Tension in an elastic membrane is similar to tension in an elastic string except that it is a two-dimensional effect. Thus any arbitrary line element of length Δl along the surface of the membrane is in equilibrium due to equal and opposite forces $\boldsymbol{\Delta T}$ and $-\boldsymbol{\Delta T}$ exerted perpendicular to Δl by the parts of the membrane on either side (Figure 10.29).

Figure 10.28. Liquid drop formation at the outlet of a tap. The surface tension model visualises the surface of the liquid as behaving somewhat like an elastic membrane.

The magnitude of this tensile force *per unit length* is called the **surface tension** and defined as

$$\boxed{\gamma := \underset{\Delta l \to 0}{\text{limit}} \frac{|\boldsymbol{\Delta T}|}{\Delta l}}$$

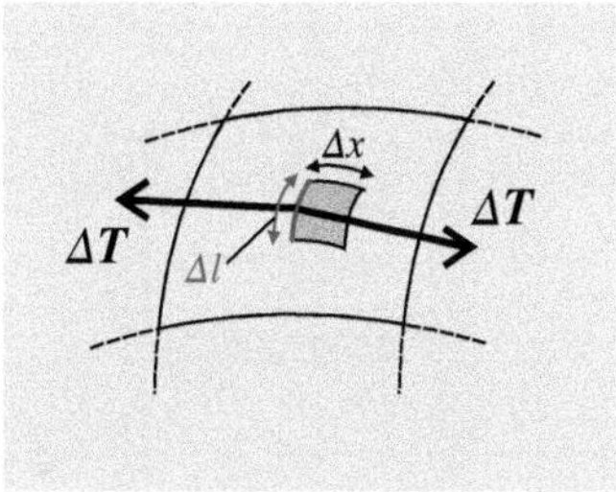

Figure 10.29. Surface tension: any line element on the surface of the 'membrane' is in equilibrium due to equal and opposite forces exerted by the parts of the 'membrane' on either side.

The value of the surface tension for any particular liquid is characteristic of that liquid. The SI unit of surface tension is N m^{-1}.

This simple model of liquid surfaces, developed by analogy with elastic membranes, proves very convenient in explaining phenomena such as those mentioned above, but the analogy cannot be taken too far. One major difference is that when a real elastic membrane is stretched further the tensile force is correspondingly increased, as in the one-dimensional case of an elastic string or a spring. The surface tension in a liquid, on the other hand, does not change as the liquid surface is 'stretched'. Furthermore, the model is applicable only to liquid surfaces for which the liquid is at rest; the model is of little help in explaining the dynamics of how the liquid reached this equilibrium state in the first place.

Suppose that the tension force $\boldsymbol{\Delta T}$ on a line element Δl on the surface of a liquid were to move the line element through a small distance Δx along the surface perpendicular to Δl as shown in Fig.10.29. The mechanical work done in this case is given by

$$\Delta W = (\Delta T)(\Delta x) = \gamma(\Delta l)(\Delta x) = \gamma(\Delta A)$$

where ΔA is the shaded area in the figure. Thus $\gamma = \dfrac{dW}{dA}$, that is the surface tension can also be understood as the energy per unit area of surface. Hence, γ can be called the *surface energy* as well as the surface tension (note that 1 N m^{-1} = 1 J m^{-2})

The size and shape of a liquid surface in any particular situation is determined by the requirement that the surface energy be at a minimum. Thus a liquid surface in equilibrium will tend to have minimum area consistent with whatever external forces and solid surfaces may be present. Thus, for example, the liquid drop attached to the tap illustrated in Figure 10.28 deviates from spherical shape due to the effect of gravity and the adhesive force between the liquid surface and the mouth of the tap. A soap bubble will tend to have a spherical shape because the effect of gravity on the soap film will be negligible but to avoid collapse of the bubble there must be an excess pressure inside the bubble above the ambient pressure outside (see Worked Example 10.5). While a liquid can be confined to a container in a gravitational field, in a freely falling (weightless) environment, such as that in an orbiting satellite, only the liquid's surface tension can keep a liquid confined as a cohesive unit.

Study Worked Examples 10.4 and 10.5

For problems based on the material presented in this section visit up.ucc.ie/10/ and follow the link to the problems.

10.10 Boyle's law (or Mariotte's law)

A liquid in a gravitational field can be confined in an open vessel and, as we have seen in the previous section, even in the absence of a gravitational field it will be confined by its own surface tension. Gases, on the other hand, usually have to be kept in *closed* containers because a gas tends to expand if free to do so. Of course a gas can be confined by a very strong gravitational field; for example, as we saw in Section 10.5, the Earth's atmosphere is kept in place by the gravitational pull of the Earth and the hot gas/plasma which comprises the Sun is confined by its own gravity.

Another important difference between gases and liquids is the fact that, since gases are very light, the variation of pressure with vertical height in a gas can usually be ignored over moderate differences in height. There are, of course, situations in which this assumption cannot be made such as the obvious cases of the terrestrial and solar atmospheres cited above. In most situations in which a gas is confined to a container, the pressure of the gas will differ little from one point to another. In these circumstances we can talk of the *pressure of the gas* which can thus be considered to be a property of the system.

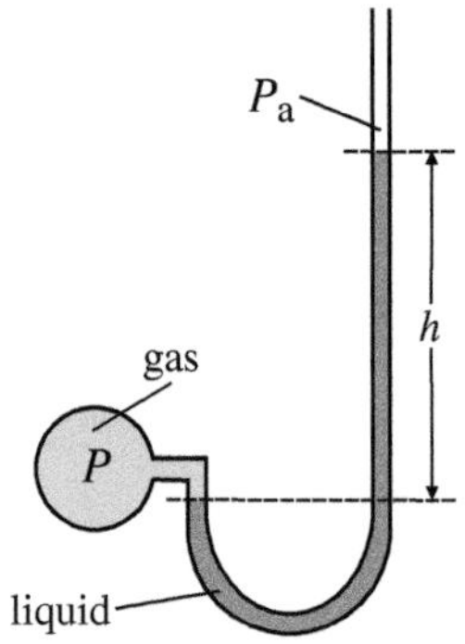

Figure 10.30. A manometer is used to measure the pressure of a gas.

A manometer, illustrated in Figure 10.30, provides a convenient way of measuring the pressure of a gas in a container. The apparatus simply comprises a U-tube connected to the gas as indicated in the figure and filled with a liquid whose density, ρ_L, is known. If the pressure P of the gas in the container is greater than the atmospheric pressure P_a, the liquid will be pushed down in the left-hand arm and up in the right-hand arm of the manometer and vice versa if $P < P_a$. In equilibrium, the pressure of the gas is exerted down on the liquid in the left-hand arm and, therefore, is equal to the pressure at the same horizontal level in the right-hand arm. Thus

$$P = P_a + g\rho_L h$$

where h is the difference in the height of the liquid in the two arms. The difference between the pressure of the gas and atmospheric pressure, $P - P_a = g\rho_L h$, is often called the *gauge pressure*.

A range of convenient instruments which measure gauge pressure directly and which do not involve the use of liquids are designed for specific purposes (Aneroid barometers, Bourdon pressure gauges, etc.). These usually employ the elastic properties of thin walled metal containers which flex due to pressure changes, thus causing a pointer to move along a scale. Such instruments, of course, have to be calibrated in terms of some instrument which measures pressure in an absolute manner, such as the barometer described in Section 10.5 for measuring atmospheric pressure.

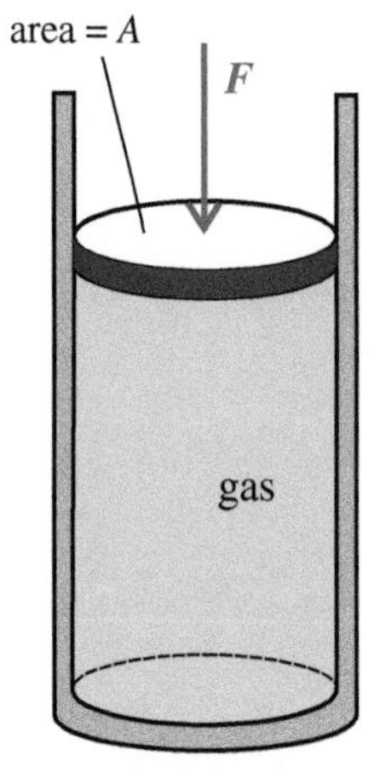

Figure 10.31. Idealised experiment to investigate how the pressure of a gas varies with its volume.

Boyle's law / Mariotte's law

Consider the situation illustrated in Figure 10.31 in which a fixed mass of gas is confined to a cylindrical container by means of a tightly fitting but freely moving piston of negligible weight.

Unless a force is maintained on it, the piston is pushed outwards by the pressure exerted by the gas. Suppose a force F is applied to the piston as shown in the figure and that the area of the piston, and hence the cross-sectional area of the cylinder, is A. The system will settle down in equilibrium when the force exerted upwards by the gas on the piston is equal and opposite to F, that is $PA = F$ and hence the pressure of the gas is given by

$$P = \frac{F}{A}$$

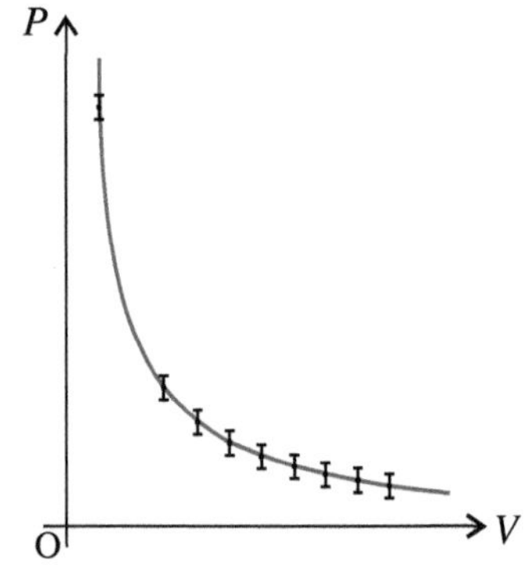

Figure 10.32. A plot of pressure against volume for a fixed mass of air at room temperature.

The pressure of the gas can be changed by varying the force applied to the piston. If the force is increased, the volume (V) of the gas in the new equilibrium situation decreases and its pressure increases. Similarly a decrease in force will result in a decrease in pressure and an increase in volume. Exactly how the volume of the gas varies with pressure is an issue which must be studied experimentally; Figure 10.32 shows data from such an experiment.

Figure 10.33 is based on the same data but, in this case the reciprocal of the volume is plotted against the corresponding pressure.

The linear nature of this plot indicates that, at least within the accuracy of the experiment, the pressure of a gas is inversely proportional to its volume, that is

$$P \propto \frac{1}{V} \quad \text{or} \quad P = (constant)\,\frac{1}{V} = \frac{C}{V}$$

$$\Rightarrow \quad \boxed{PV = C} \qquad (10.9)$$

where C is a constant which depends (among other factors) on the mass of gas trapped in the container. This result, which states that the pressure and volume of a given mass of gas must vary in such a way that their product PV remains constant, is known as **Boyle's law** (Robert Boyle 1627–1691) or **Mariotte's law** (Edme Mariotte 1620–1684). It should be noted that the temperature of the gas must be held constant for Equation (10.9) to be valid.

The importance of Boyle's law lies in the fact that **all gases** obey Equation (10.9) in the limit of low pressures – at higher pressures different gases show a variety of deviations from the simple behaviour described by Boyle's law, deviations which will be considered in Section 10.13. A gas which obeys Boyle's law over the full range of pressure and volume is called an **ideal gas**. The universality of Equation (10.9) indicates that all gases have fundamental features in common – an issue which we shall return to in the next section.

We can use Equation (10.9) to derive an expression for the bulk modulus of an ideal gas. Since $P = \frac{C}{V}$, we can write $\frac{dP}{dV} = -\frac{C}{V^2}$ and hence, from the definition of bulk modulus (Section 10.4), we obtain

$$K = -V\frac{dP}{dV} = -V\left(-\frac{C}{V^2}\right) = \frac{C}{V} = P$$

Thus, the bulk modulus of an ideal gas depends on pressure and at atmospheric pressure is approximately 10^5 N m^{-2}.

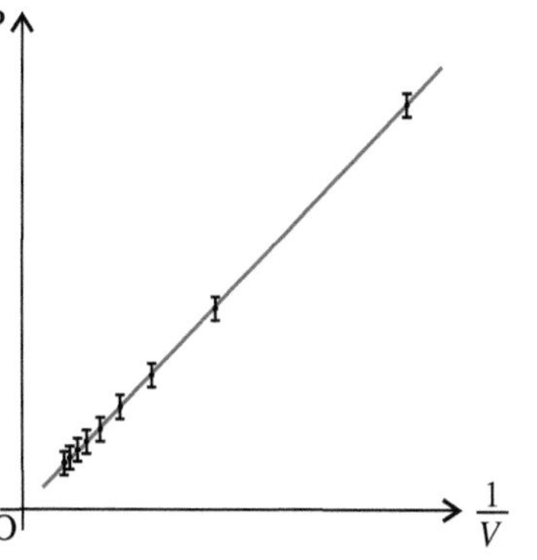

Figure 10.33. A plot of pressure against the reciprocal of volume using the same data as in Figure 10.32.

For problems based on the material presented in this section visit up.ucc.ie/10/ and follow the link to the problems.

10.11 A microscopic theory of gases

While different gases can have radically different chemical properties, the physical properties of all gases are very similar and, as we have seen, all gases show identical behaviour at very low pressures. This suggests strongly that all gases have some underlying attributes in common which one must suspect arises from their microscopic nature, mentioned in Section 1.3. In this section we will develop in some detail a microscopic model of a gas – called **the kinetic theory of gases**. The success of the theory lies in its ability to predict the low pressure behaviour of gases in general and, in particular, to explain Boyle's law in terms of microscopic characteristics.

Qualitative evidence of the microscopic nature of gases is shown by an effect called 'Brownian motion', so named after the English botanist Robert Brown (1773–1858) who first observed the phenomenon in the 1820s. The effect is most easily seen when smoke particles suspended in air are viewed using a microscope. Such particles can be seen to exhibit a continuous random zigzag motion which can be explained as being the result of the bombardment of the smoke particles by air molecules which are too small to be visible themselves.

The basic hypotheses of the kinetic theory of gases are as follows:

(a) Any volume of a gas contains a very large number of small entities (molecules).
(b) The size of a molecule is very much less than the average distance between molecules.
(c) Collisions between molecules and collisions of molecules with the walls of the container are perfectly elastic (that is, there is no loss of energy).
(d) All molecules in a given gas have the same mass.
(e) Forces between molecules have very short range ⇒ molecules move in straight lines between collisions.
(f) Molecules are rigid spheres.

As we shall see, these postulates are sufficient to explain the observed macroscopic properties of an ideal gas. In Section 10.13 we shall relax some of these hypotheses in an attempt to extend the regime of applicability of the model.

We now discuss the dynamical behaviour of the molecules in a gas in a container and, in particular, we consider the forces exerted on the walls of the container due to the continuous bombardment of the walls by the molecules. Our primary objective is to link such microscopic behaviour to macroscopically observed properties, such as pressure.

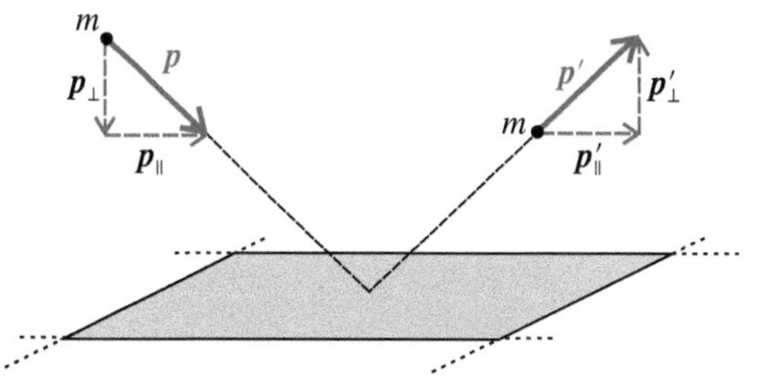

Figure 10.34. An elastic collision of a molecule with the wall of a container. The component of the momentum of the molecule perpendicular to the wall is reversed in the collision while the component parallel to the wall is unchanged.

Consider the collision of a molecule of mass m with the wall of a container. Let $\boldsymbol{p}$ be the momentum of the molecule before impact and $\boldsymbol{p}'$ its momentum after impact. We can resolve both of these momenta into components perpendicular and tangential to the wall, as indicated in Figure 10.34, that is $\boldsymbol{p} = \boldsymbol{p}_\parallel + \boldsymbol{p}_\perp$ and $\boldsymbol{p}' = \boldsymbol{p}'_\parallel + \boldsymbol{p}'_\perp$.

Since the collision is elastic there are no friction-like forces and hence the force on the molecule at impact is perpendicular to the wall; that is, a normal reaction. Thus the components parallel to the wall remain unchanged, $\boldsymbol{p}'_\parallel = \boldsymbol{p}_\parallel$. For elastic collisions the kinetic energy of the molecule must be the same before and after collision, that is $\frac{p'^2}{2m} = \frac{p^2}{2m}$, and hence $|p'| = |p|$. Thus we can see that the effect of the collision must be to reverse the direction of the perpendicular component of the momentum, that is $\boldsymbol{p}'_\perp = -\boldsymbol{p}_\perp$. The change in momentum resulting from the collision is therefore given by

$$\Delta \boldsymbol{p} = \boldsymbol{p} - \boldsymbol{p}' = \boldsymbol{p}_\perp - \boldsymbol{p}'_\perp = 2\boldsymbol{p}_\perp$$

For simplicity we consider the gas to be confined to a rectangular container, as illustrated in Figure 10.35.

We will consider first the effect of collisions with the bottom face of the container which has an area A and is in the x–y plane. The z-axis is chosen perpendicular to this plane and the height of the container in this direction is h. The change in momentum due to a collision of a single molecule of mass m with this face is $2p_z$. If we ignore, for the moment, collisions between the molecules themselves (it will be shown below that this does not affect any conclusions drawn), we can determine the frequency with which this molecule collides with the bottom face (collisions with the side walls do not change p_z), The time between successive collisions with the bottom wall, therefore, is $\frac{2h}{v_z}$, where v_z is the z-component of the velocity, and hence there will be $\frac{v_z}{2h}$ such collisions per unit time. Thus the rate of change of momentum arising as a result of the collisions between this molecule and the bottom wall is given by

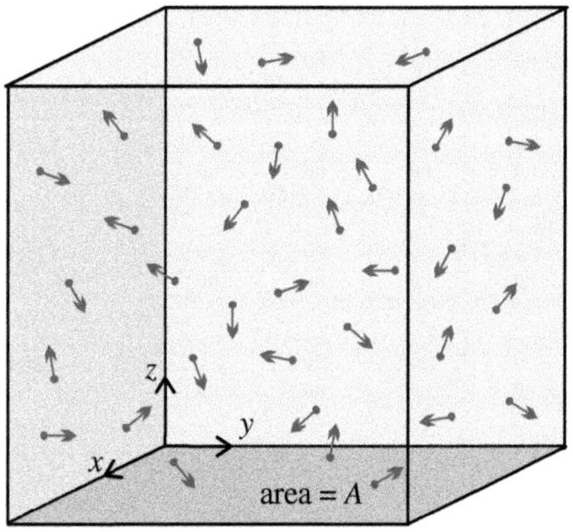

Figure 10.35. Gas molecules in a rectangular container; the molecules continually collide with one another and with the walls of the container.

$$\frac{dp}{dt} = 2p_z \frac{v_z}{2h} = \frac{p_z v_z}{h}$$

The continuous bombardment of this wall by all the molecules in the container gives rise to an average rate of change of momentum which appears as a force exerted on the bottom wall. The magnitude of this force is equal to the total rate of change of momentum with the bottom wall due to all the molecules in the container and is given by

$$F = \frac{(p_1)_z (v_1)_z}{h} + \frac{(p_2)_z (v_2)_z}{h} + \frac{(p_3)_z (v_3)_z}{h} + \frac{(p_4)_z (v_4)_z}{h} + \dots \quad \text{etc,}$$

where $(p_i)_z$ and $(v_i)_z$ are, respectively, the z-components of the momentum and velocity of the ith molecule.

Multiplying the right-hand side of this equation above and below by N, the number of molecules in the container, we get

$$F = \frac{N}{h}\left[\frac{(p_1)_z(v_1)_z + (p_2)_z(v_2)_z + (p_3)_z(v_3)_z + (p_4)_z(v_4)_z \cdots}{N}\right]$$

Now, by definition, the expression within the square brackets is the average value of $p_z v_z$, which we will denote by $\langle p_z v_z \rangle$. Thus

$$F = \frac{N}{h}\langle p_z v_z \rangle$$

Now for any particle the scalar product $\boldsymbol{p} \cdot \boldsymbol{v} = p_x v_x + p_y v_y + p_z v_z$ and hence the average value of $\boldsymbol{p} \cdot \boldsymbol{v}$ is given by

$$\langle \boldsymbol{p} \cdot \boldsymbol{v} \rangle = \langle p_x v_x \rangle + \langle p_y v_y \rangle + \langle p_z v_z \rangle$$

From symmetry (assuming there are no preferred directions in the system such as might be produced by an external field)

$$\langle p_x v_x \rangle = \langle p_y v_y \rangle = \langle p_z v_z \rangle = \frac{1}{3}\langle \boldsymbol{p} \cdot \boldsymbol{v} \rangle$$

and thus

$$F = \frac{N}{3h}\langle \boldsymbol{p} \cdot \boldsymbol{v} \rangle \tag{10.10}$$

Equation (10.10) was derived using the assumption that the molecules collide with the walls only and not with one another; in practice this will only be the case at very low densities. Since the number of molecules in the container is very large, however, for every molecule whose z-component of momentum is changed as a result of a collision there will be another molecule which experiences a collision elsewhere in the container such that its z-component of momentum is changed to precisely the value formerly associated with the first particle. Thus, the average force on a face of the container is unchanged by collisions between the molecules within the body of the gas. Since it is the average value of $\boldsymbol{p} \cdot \boldsymbol{v}$ which appears in Equation (10.10), the result is quite general. Furthermore, the result is not confined to rectangular containers but may be applied to containers of any shape.

The force per unit area on a wall of a container is the macroscopic quantity P, the pressure of the gas (Section 10.3), which we can now write as

$$P = \frac{F}{A} = \frac{N}{3hA}\langle \boldsymbol{p} \cdot \boldsymbol{v} \rangle = \frac{1}{3}\frac{N}{V}\langle \boldsymbol{p} \cdot \boldsymbol{v} \rangle \tag{10.11}$$

where $V = hA$ is the volume of the container. Since N, the total number of molecules in the container, is constant and, if energy is conserved, $\langle \boldsymbol{p} \cdot \boldsymbol{v} \rangle$ is also constant, we see that $P \propto \frac{1}{V}$. Thus, Boyle's law follows as a direct consequence of the hypotheses of the kinetic theory of gases.

Now for a gas of non-relativistic molecules, $\boldsymbol{p} = m\boldsymbol{v}$ and thus relationship (10.11) can be rewritten as follows

$$P = \frac{1}{3}\frac{N}{V}\langle mv^2 \rangle = \frac{2}{3}\frac{N}{V}\left\langle \frac{1}{2}mv^2 \right\rangle \tag{10.12}$$

Since $\left\langle \frac{1}{2}mv^2 \right\rangle$ is the average kinetic energy of a molecule, multiplying by N gives the total mechanical energy, which we call the **internal energy** of the system, defined as

$$\boxed{U := N\left\langle \frac{1}{2}mv^2 \right\rangle} \tag{10.13}$$

Thus, Boyle's law can be written in terms of the internal energy as follows

(for non-relativistic gases)

$$\boxed{PV = \frac{2}{3}U} \tag{10.14}$$

where U is constant as long as no energy enters or leaves the system. We can estimate, from Equation (10.14), a value for the internal energy of an ideal gas; for example, for ten litres of a gas at atmospheric pressure,

$$U = \frac{3}{2}PV = 1.5(10^5\ \text{N m}^{-2})(10^{-2}\ \text{m}^3) = 1.5 \times 10^3\ \text{N m} = 1.5\ \text{kJ}.$$

For a gas of molecules moving with extremely relativistic velocities ($v \to c$; $cp \gg mc^2$; $E^2 = c^2p^2 + m^2c^4 \approx c^2p^2$) the momentum of a particle is given by $p \to \frac{E}{c}$ in which case $\langle \boldsymbol{p} \cdot \boldsymbol{v} \rangle = \langle pv \rangle \to \left\langle \frac{E}{c}c \right\rangle = \langle E \rangle$. In this case Equation (10.11) becomes $P = \frac{1}{3}\frac{N}{V}\langle E \rangle = \frac{1}{3}\frac{U}{V}$ or

(for extremely relativistic gases)

$$\boxed{PV = \frac{1}{3}U} \tag{10.15}$$

Before completing this section perhaps it is worth reflecting on what has been achieved. We have shown that a macroscopic 'law', in this case an experimentally derived rule relating macroscopic quantities, can be deduced from a microscopic model. The laws of mechanics were applied successfully to the microscopic system to develop the theory. The net result is that our confidence in our microscopic model has been reinforced and we have gained a significant new insight into the macroscopic picture.

Study Worked Example 10.6

For problems based on the material presented in this section visit up.ucc.ie/10/ *and follow the link to the problems.*

10.12 The SI unit of amount of substance; the mole

We have seen that a gas is a system containing a very large number of molecules. Other microscopic models involve large numbers of other entities such as atoms, ions, electrons or nuclei. In these cases, it would be particularly convenient if we could measure the amount of substance involved by counting the number of entities in the system in groups of a specified large number. The strategy here is analogous to counting entities, like eggs or loaves of bread, in dozens but poses the question as to how we can adopt a suitable number which can be applied easily to matter in bulk.

We saw in Chapter 1 (Section 1.2) that the mass of an atom or a nucleus depends essentially on the number of nucleons in its nucleus, that is the mass of a nucleus is approximately equal to Am_N, where m_N is the mean mass of a nucleon and A is the (integer) mass number of the nucleus involved. Thus we see that the number of nucleons in 1 g of matter $\approx$ the number of atoms in 1 g of ^{1}H $\approx$ the number of atoms in 4 g of ^{4}He $\approx$ the number of atoms in 12 g of ^{12}C $\approx$ the number of atoms in 16 g of ^{16}O, etc. This number, which is of the order of

10^{23}, is not exactly the same for every atomic species but a number of this magnitude would be suitable for counting elementary entities like atoms, ions, etc. For the purpose of defining an SI base unit for the amount of substance, the exact number $6.022\,140\,76 \times 10^{23}$ is chosen and the amount of substance containing that number of elementary entities is defined as one mole (see below). This effectively fixes the value of a fundamental physical constant, called the Avogadro constant, to be

$$N_A = 6.022\,140\,76 \times 10^{23}\ \text{mol}^{-1}$$

The formal definition of **the mole** is as follows:

The mole, symbol mol, is the SI unit of amount of substance. One mole contains exactly $6.022\,140\,76 \times 10^{23}$ elementary entities. This number is the fixed numerical value of the Avogadro constant, N_A, when expressed in the unit mol^{-1} and is called the Avogadro number. The amount of substance, symbol n, of a system is a measure of the number of specified elementary entities. An elementary entity may be an atom, a molecule, an ion, an electron, any other particle or specified group of particles.

The effect of this definition is that the mole is the amount of substance of a system that contains $6.022\,140\,76 \times 10^{23}$ specified elementary entities.

The amount of substance measured in this way is given by the number of moles in the system, usually denoted by $n = \dfrac{N}{N_A}$; that is, the number of entities in the system is given by $N = nN_A$.

The amount of substance in a system may be specified in terms of the number of moles or in terms of its mass in kg. These are linked via the mean mass of a nucleon which turns out to be a very convenient unit of mass when dealing with microscopic entities. This (non-SI) unit is called the (unified) atomic mass unit and is denoted by u, where

$$1\ \text{u} = m_u = \frac{1}{10^3 N_A}\text{kg mol}^{-1} \approx 1.66 \times 10^{-27}\ \text{kg}.$$

Analogous to the 'specific' values (values per unit mass, recall Section 10.1) of physical quantities, we can now define *molar* quantities as the value per mole; molar quantities are denoted by lower case symbols with a subscript m. Examples of molar quantities are *molar volume* $:= v_m := \dfrac{V}{n}$ (SI unit: $\text{m}^3\ \text{mol}^{-1}$) and *molar internal energy* $:= u_m := \dfrac{U}{n}$ (SI unit: J mol^{-1}). Using these definitions, equation (10.13) can be written as $u_m = N_A\langle \frac{1}{2}mv^2\rangle$ and Equations (10.14) and (10.15), respectively, become

(for non-relativistic gases)

$$Pv_m = \frac{2}{3}u_m \tag{10.16}$$

and (for extremely relativistic gases)

$$Pv_m = \frac{1}{3}u_m \tag{10.17}$$

For problems based on the material presented in this section visit up.ucc.ie/10/ and follow the link to the problems.

10.13 Interatomic forces: modifications to the kinetic theory of gases

In the kinetic theory of gases the interaction between two molecules was considered to be similar to that between two rigid spheres, that is the force between the two is zero unless the molecules are in contact in which case a very large repulsive force comes into play. The potential energy plot for such an interaction is sketched in Figure 10.36 as a function of the distance r between molecules.

This potential energy function may be described as follows.

$$U(r) = \begin{cases} \infty & \text{for} \quad r \le R_0 \\ 0 & \text{for} \quad r > R_0 \end{cases}$$

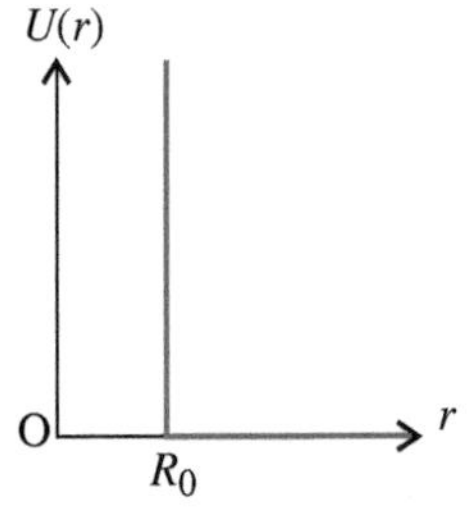

Figure 10.36. Plot of potential energy as a function of displacement between two rigid spheres (R_0 is the sum of the radii of the spheres).

where R_0 is the sum of the radii of the two molecules. As we have seen, this model works reasonably well for gases at low pressure but needs to be modified to deal with higher density systems such as gases at high pressure, liquids or solids.

In general, interatomic and intermolecular forces can be quite complicated and a proper explanation of such phenomena requires a knowledge of quantum mechanics (Chapter 14) and/or electrostatics (Chapter 16). A qualitative understanding of intermolecular interactions can be obtained, however, by assuming that the potential energy function is similar to that shown by the dashed curve in Figure 10.37.

A typical intermolecular interaction is a *short range* force which is attractive for $R_0 < r < R$ but has a strong repulsive core for $r < R_0$, where r is the distance between the centres of the molecules. For many purposes an intermolecular interaction can be approximately represented by a *'square-well' potential* (solid curve in Figure 10.37) such that

$$U(r) = \begin{cases} \infty & \text{for} \quad r < R_0 \\ -E & \text{for} \quad R_0 < r < R \\ 0 & \text{for} \quad r > R \end{cases}$$

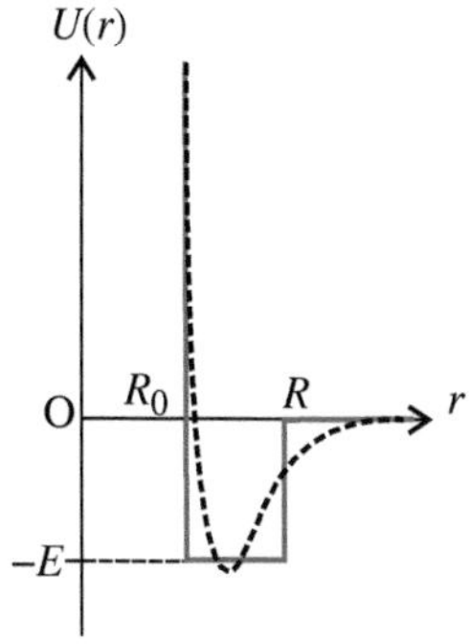

Figure 10.37. The potential energy function for a real intermolecular interaction (dashed curve) can be approximated by a 'square well' potential energy function (solid blue curve).

Range of intermolecular forces

In order take the effect of intermolecular forces into account in the kinetic theory of gases we must modify Equations (10.14) and (10.16). Consider an intermolecular force whose potential energy function is as indicated by the dashed curve in Figure 10.37. A molecule within the body of the gas will experience attractive forces due to its nearest neighbour molecules (Figure 10.38) but, since these are distributed equally in all directions, there is no net force.

A molecule approaching a wall, on the other hand, will feel a small net force directed back towards the body of the gas which will tend to slow it down just before it collides with the wall. Essentially the attractive part of the force gives rise to an effective decrease in the pressure of the gas on the walls of the container, that is $P \to \dfrac{\frac{2}{3}u_m}{v_m} - P_i$, where P_i is called the 'internal pressure'.

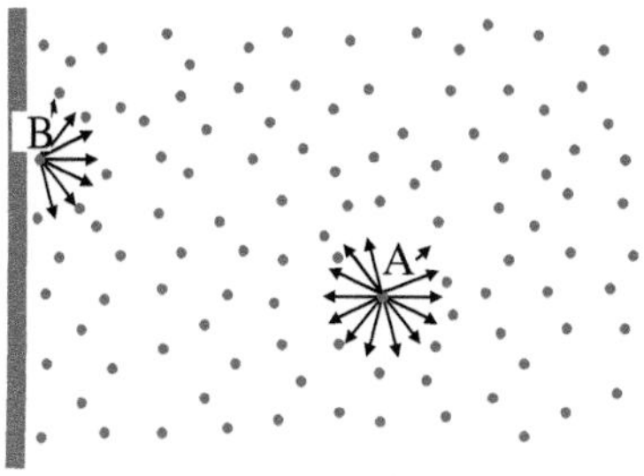

Figure 10.38. The net force on a molecule (such as A) within the body of the gas exerted by the surrounding molecules is zero, by symmetry. For a molecule (such as B) near the wall of a container, however, interactions with the other molecules provide a net force directed away from the wall.

In an attempt to take internal pressure into consideration, we assume that the intermolecular interaction can be approximated by a square-well potential as discussed above. Each molecule has around it an 'interaction volume' (shaded region in Figure 10.39) such that if the centre of another molecule is within this volume the two molecules interact.

The volume of the interaction region is $\frac{4}{3}\pi(R^3 - R_0^3)$ and hence the probability of interaction between any two molecules in a gas of N molecules in a container of volume V is proportional to $\dfrac{N(R^3 - R_0^3)}{V} = \dfrac{N}{V}(R^3 - R_0^3)$. The probability that a molecule approaching the wall of the container will experience a force due to any of the other $N-1$ molecules, therefore, is proportional to $(N-1)\dfrac{N}{V}(R^3 - {R_0}^3) \propto \dfrac{N^2}{V}$ since $N \gg 1$. Thus Equation (10.10) can be modified as follows to take into account the effect of the range of intermolecular forces

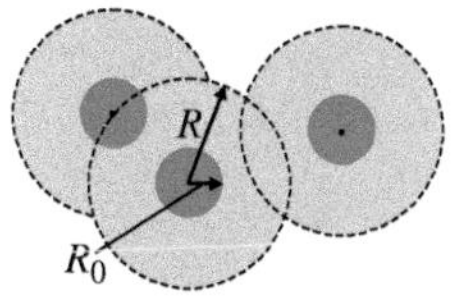

Figure 10.39. Each molecule has an 'interaction volume' (shaded) surrounding it such that two molecules will interact only if the centre of one is within the interaction volume of the other.

$$F = \frac{N}{3h}\langle \boldsymbol{p} \cdot \boldsymbol{v} \rangle - \alpha_1 \frac{N^2}{V}$$

where α_1 is a constant such that $\alpha_1 \to 0$ (that is, $R \to R_0$) for an ideal gas. Thus

$$P = \frac{F}{A} = \frac{N}{3hA}\langle \boldsymbol{p} \cdot \boldsymbol{v} \rangle - \alpha_1 \frac{N^2}{VA} = \frac{N}{3hA}\langle \boldsymbol{p} \cdot \boldsymbol{v} \rangle - \alpha_1 \frac{N^2 h}{VAh} = \frac{1}{3}\frac{N}{V}\langle \boldsymbol{p} \cdot \boldsymbol{v} \rangle - \alpha\left(\frac{N}{V}\right)^2$$

where $\alpha = \alpha_1 h$ is a constant which, like α_1, is zero for an ideal gas. Hence

$$P + \frac{\alpha N^2}{V^2} = \frac{1}{3}\frac{N}{V}\langle \boldsymbol{p} \cdot \boldsymbol{v} \rangle \quad \text{or} \quad P + \frac{\alpha N_A^2}{v_m^2} = \frac{1}{3}\frac{N_A}{v_m}\langle \boldsymbol{p} \cdot \boldsymbol{v} \rangle$$

and thus, by comparison with Equation (10.11), Equation (10.16) becomes

$$\left(P + \frac{a}{v_m^2}\right) v_m = \frac{2}{3}u_m, \quad \text{where } a = \alpha N_A^2.$$

The effect of molecular size

The volume occupied by the molecules was considered to be negligible in the kinetic theory of gases. This is a reasonable assumption at low densities. However, as the density is increased and the molecules are squeezed closer together, the proportion of the container volume occupied by the molecules can become significant. This means that the effective volume of the container is less than V by an amount equal to the volume occupied by the N molecules, that is, $V \to V - N\beta$, where β is the volume of a molecule. The molar volume, therefore, must be modified so that $v_{\mathrm{m}} \to v_{\mathrm{m}} - b$, where $b = N_{\mathrm{A}}\beta$ is a constant. In other words Equation (10.16) should be modified to give the following equation (called the *Clausius equation of state*)

$$P(v_{\mathrm{m}} - b) = \frac{2}{3}u_{\mathrm{m}}$$

Taking the two effects which we have discussed above into account together we get

$$\left(P + \frac{a}{v_{\mathrm{m}}^2}\right)(v_{\mathrm{m}} - b) = \frac{2}{3}u_{\mathrm{m}} \qquad (10.18)$$

where the constants a and b are characteristic of the particular gas involved and whose values have can be determined experimentally. Equation (10.18), known as the *van der Waals equation of state,* is a significant improvement on Equation (10.16) for describing real gases, particularly at higher pressures.

For problems based on the material presented in this section visit up.ucc.ie/10/ and follow the link to the problems.

10.14 Microscopic models of condensed matter systems

As the density in a gas is increased the average separation of the molecules becomes smaller and smaller. When the mean distance between the molecules is reduced to a value near the range of the attractive force ($r \approx R_{\mathrm{e}}$ in Figure 10.37), groups of molecules can become bonded together to form *bound aggregates*. Bound molecules will arrange themselves so that each one oscillates about the equilibrium position, that is $r = R_{\mathrm{e}}$, the minimum energy position in the potential energy plot in Figure 10.40.

When a number of atoms are bonded in this way they tend to form ordered arrays so that different states of what is called **condensed matter** can be formed. If the ordering is short-range (a few atoms in extent) a liquid or amorphous solid is formed – the dynamics of such systems is quite complex and will be discussed only qualitatively below. At even higher pressures the atoms become arranged in highly ordered three-dimensional lattice structures, giving rise to crystalline systems (recall Section 1.3). It should be noted, however, that many important materials cannot be labelled uniquely as either solids or liquids – such materials can have a mixture of solid and liquid characteristics.

Long-range order

The nature and structure of crystalline solids can be very varied depending on the type of the interatomic bonds and energy considerations; we shall discuss such issues in Section 25.1. We can construct an approximate microscopic model of solids, however, by thinking of adjacent atoms as being joined by small helical springs each with an 'interatomic spring constant' k, as shown in Figure 10.41.

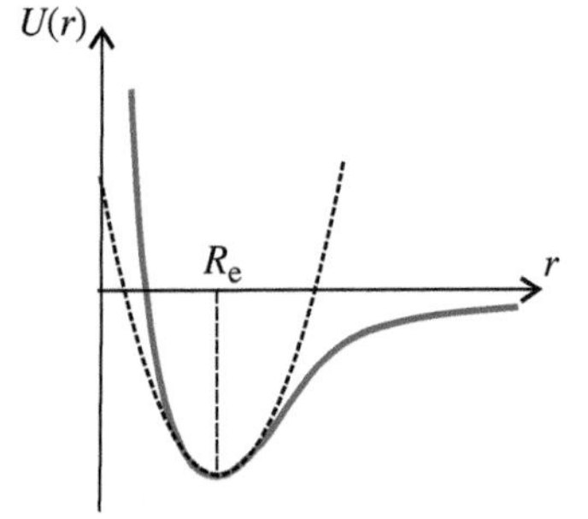

Figure 10.40. Potential energy plot (solid curve) for the intermolecular force. Molecules bound together will oscillate about the equilibrium position ($r = R_{\mathrm{e}}$, the minimum of the potential energy curve). Near the minimum the curve may be approximated by a parabola (dashed line) and hence, for small amplitude, bound particles execute simple harmonic motion.

This is not as wild an assumption as it might seem because, if the displacement of an atom from its equilibrium position is small then the part of the potential energy plot in which it moves can be approximated by a parabola (dashed curve in Figure 10.40). In fact, as noted in Section 3.10, in the limit of small displacements from equilibrium, the motion of a particle governed by any smooth attractive interaction can always be approximated by that of a simple harmonic oscillator.

The elastic properties of solids can be understood in terms of the crystal lattice model. If, for example, a stretching force is applied along one of the lattice directions in Figure 10.41 the interatomic spacing in that direction will change from a to $a + \Delta a$. Consider now the expansion of a unit cell along that direction (Figure 10.42).

Let ΔF be the force applied to the face of the cell of area bc which produces an extension Δa. There are, of course, four springs involved but each of these is shared between four adjacent

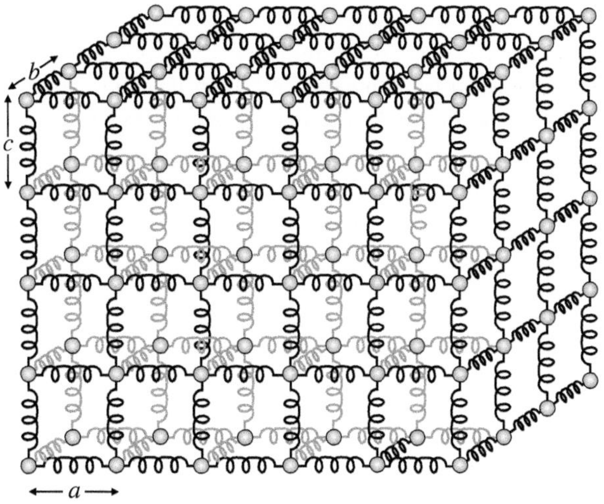

Figure 10.41. A simple macroscopic model of a solid. The atoms take up long-range order in the form of a crystal in which the forces between adjacent atoms can be represented by macroscopic 'helical springs' as shown.

cells so that the effective spring constant is k. Thus $\Delta F = k(\Delta a)$ and hence $\frac{\Delta F}{bc} = \frac{k(\Delta a)}{bc} = \frac{ka}{bc}\frac{\Delta a}{a}$ and, since $\frac{\Delta l}{l} = \frac{\Delta a}{a}$, the Young modulus is given by $E = \frac{a}{A}\frac{dF}{da}$ so that $E = \frac{ka}{bc}$. The other elastic constants (Section 10.2) can be understood in a similar manner.

In general, it is found that macroscopic properties of solids are determined by microscopic quantities which describe the interaction between the atoms or molecules from which the solid is constituted. Study of the macroscopic properties of materials, therefore, can provide information about their structure at a microscopic level. As an example, let us make an estimate of the magnitude of the interatomic spring constant in a metal. Typically $a \sim b \sim c \sim 10\,\text{nm}$ and $E \sim 10^{11}\ \text{N m}^{-2}$ and hence $k = \frac{bc}{a}E \sim (10^{-8})(10^{11}) \sim 1000\ \text{N m}^{-1}$ (equivalent to a very strong macroscopic spring).

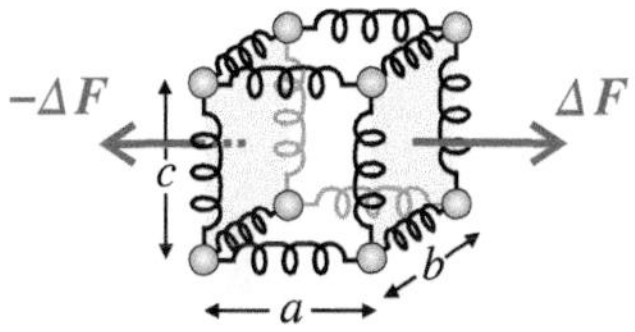

Figure 10.42. The elastic properties of solids can be understood in terms of the microscopic model of Figure 10.41. A longitudinal stress produces a small extension of each microscopic 'spring'.

Short-range order

The average distance between molecules in a liquid is usually somewhat larger than that pertaining in the solid phase (although there are exceptions such as water/ice). Nevertheless, the average molecular separation is still close to the range of the intermolecular forces but only a small number of these molecules are actually bound together at any time. Thus in many respects a liquid behaves like a gas but can also show some properties more usually associated with solids. Individual molecules and small bound aggregates of molecules move around with random velocities colliding with one another in a similar manner to molecules in a gas. Indeed the van der Waals equation of state can be applied, reasonably successfully, to many liquids. As a result of collisions with other molecules, however, the bound aggregates of liquid molecules are continually breaking up and reforming.

The (attractive) intermolecular forces are responsible for the *cohesion* of the liquid. The phenomenon of surface tension, discussed from a macroscopic viewpoint in Section 10.9, can be understood qualitatively as arising from these cohesive forces. Whereas a molecule in the bulk of a liquid will experience no net force, the net force on a molecule or group of bound molecules near the surface of the liquid will be directed in towards the body of the liquid. This will tend to keep all surface molecules attached to the liquid thereby sustaining the surface. When a liquid comes in contact with a solid, such as the walls of a container, the additional *adhesive* forces which arise from interactions between the liquid molecules and the molecules of the solid may be greater or smaller than the cohesive forces and may produce a deformation of the liquid surface near the walls, an effect which we described as a meniscus in Section 10.9.

Viscosity in a fluid (liquid or gas), which was discussed from a macroscopic viewpoint in Section 10.8, can also be understood qualitatively from a microscopic viewpoint. The adhesive forces on those molecules very close to a solid surface will tend to hold these molecules at rest at the surface. The lack of long-range order explains why liquids, like gases, cannot support a shearing stress. As a fluid is forced past the solid surface that part of the fluid immediately adjacent to the surface will be slowed down and a velocity gradient will be set up such that the bulk speed of the fluid increases with distance from the surface.

Many of the effects discussed above depend quite strongly on the temperature of the system involved so further discussion of the microscopic behaviour of condensed matter must await a microscopic theory of heat, a topic that will be addressed in the next chapter.

Further reading

Further material on some of the topics covered in this chapter may be found in *Properties of Matter* by Flowers and Mendoza (See BIBLIOGRAPHY for book details).

WORKED EXAMPLES

Worked Example 10.1: A piece of steel wire which is 3.0 m long and of 1.0 mm diameter hangs vertically from the ceiling of a laboratory (Figure 10.43). What is the extension of the wire when a 5.0 kg mass is attached to the free end?

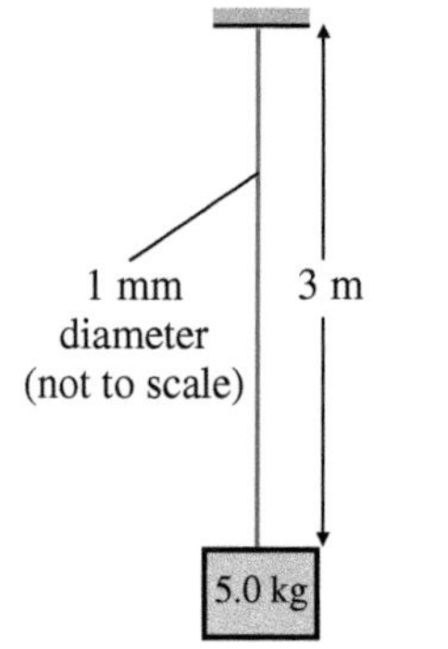

Figure 10.43. Worked Example 10.1.

From the definition of the Young modulus, $E := \frac{L}{A}\frac{\Delta F}{\Delta L} \rightarrow \Delta L = \frac{L}{EA}\Delta F = \frac{L}{E(\pi r^2)}Mg$

Using the value of E for steel given in Table 10.2, $\rightarrow \Delta L = \frac{3\text{ m}}{(2.1\times 10^{11}\text{ N m}^{-2})\pi(0.5\times 10^{-3}\text{ m})^2}(5.0\text{ kg})(9.8\text{ m s}^{-2}) = 0.9\text{ mm}$

Worked Example 10.2: The depth of water at the face of a dam (illustrated in Figure 10.44 is h and the horizontal width of the dam is w. Derive expressions for (a) the net horizontal force on the dam and (b) the moment about the line XX′ along the base of the dam due to the pressure exerted by the water. (c) At what depth below the surface should a single force, equivalent to that calculated in (a), be applied to get the same effect (that is, to provide the same moment about XX′)?

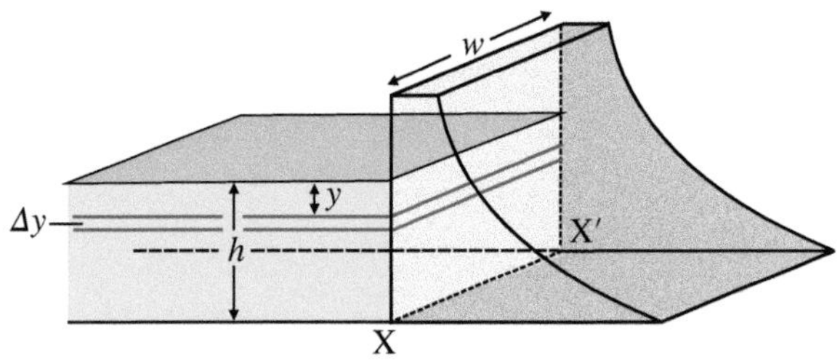

Figure 10.44. Worked Example 10.2.

(a) Consider the layer of water of thickness Δy at depth y below the surface.

Pressure at this depth $= P(y) = g\rho y \rightarrow$ Force on dam due to this layer $= \Delta F = g\rho yw(\Delta y)$

Total force on dam $= F = \int_0^h g\rho yw dy = g\rho w\int_0^h y dy = \frac{1}{2}g\rho wh^2$

(b) Moment of ΔF about XX′ $= \Delta M = (\Delta F)(h-y) = (h-y)g\rho yw(\Delta y)$

Total moment about XX′ $= g\rho wh\int_0^h y dy - g\rho w\int_0^h y^2 dy = \frac{1}{2}g\rho wh^3 - \frac{1}{3}g\rho wh^3 = \frac{1}{6}g\rho wh^3$

(c) Distance below surface at which force F must be applied to provide the same moment $= \frac{1}{3}h$

Worked Example 10.3: The 5.0 kg mass in Worked Example 10.1 is made from iron (density = 7.9×10^3 kg m^{-3}). By how much is the extension determined in Worked Example 10.1 reduced if the mass is totally immersed in water?

The Archimedean upthrust on the mass shown in Figure 10.45 $= \Delta F = \rho_W Vg$, where V is the volume of the 5.0 kg mass, that is $V = \frac{M}{\rho_I}$ (ρ_w = density of water; ρ_I = density of iron)

$\rightarrow$ change in force on wire, $\Delta F = Mg - \rho_W\frac{M}{\rho_I}g = \left(1-\frac{\rho_W}{\rho_I}\right)Mg$

$\rightarrow$ change in extension, $\Delta L = \frac{L}{EA}\Delta F = \frac{L}{E(\pi r^2)}\left(1-\frac{\rho_W}{\rho_I}\right)Mg = \left(1-\frac{\rho_W}{\rho_I}\right)\frac{L}{E(\pi r^2)}Mg$

$= \left(1-\frac{\rho_W}{\rho_I}\right)(0.9\text{ mm})$

$\rightarrow \Delta L = \left(1-\frac{1.0\times 10^3}{7.9\times 10^3}\right)(0.9\text{ mm}) = 0.79\text{ mm} \Rightarrow$ extension reduced by 0.11 mm

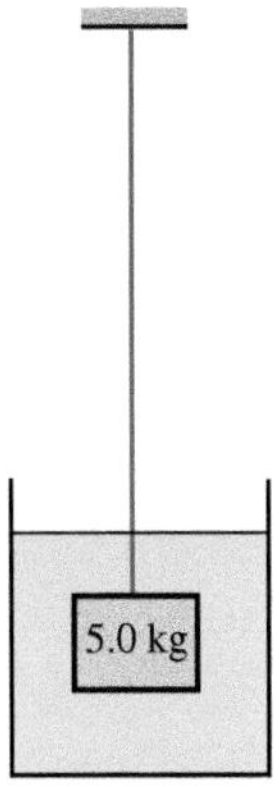

Figure 10.45. Worked Example 10.3.

Worked Example 10.4: What would the average diameter of the capillaries in a tree have to be for surface tension to account for the ability of water to climb to the top of a 40 m tree? Take the surface tension of water in this context to be 0.07 N m^{-1}.

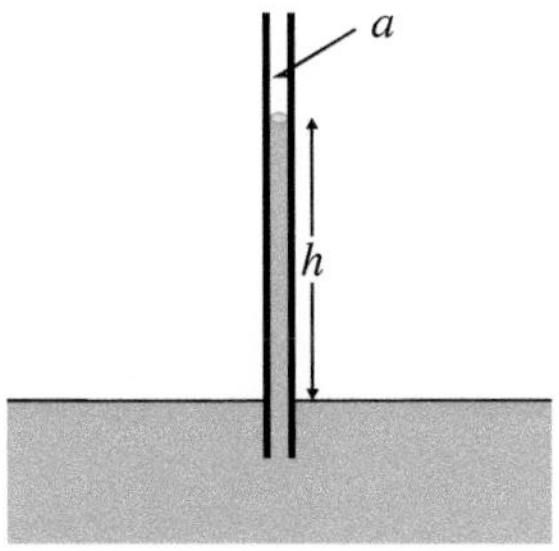

Figure 10.46. Worked Example 10.4.

For equilibrium, the weight of the liquid in the column of height h (Figure 10.46) must be balanced by the upward force due to surface tension. Thus $mg = \gamma(2\pi a)$, where a = radius of capillary and γ = surface tension.

$$\rightarrow \quad \pi a^2 h\rho g = 2\pi a\gamma \rightarrow \quad a = \frac{2\gamma}{\rho h g} = \frac{2(0.07)}{(10^3)(40)(9.8)} = 0.36\ \mu\text{m}$$

and thus the capillary diameter $\approx 0.7\ \mu$m.

Note that capillary diameters in trees are not much less than 1 mm, so this simple approach fails to explain the phenomenon.

Worked Example 10.5: What is the excess pressure inside a soap bubble of diameter 5.0 cm if the surface tension of the soap solution is 0.025 N m^{-1}?

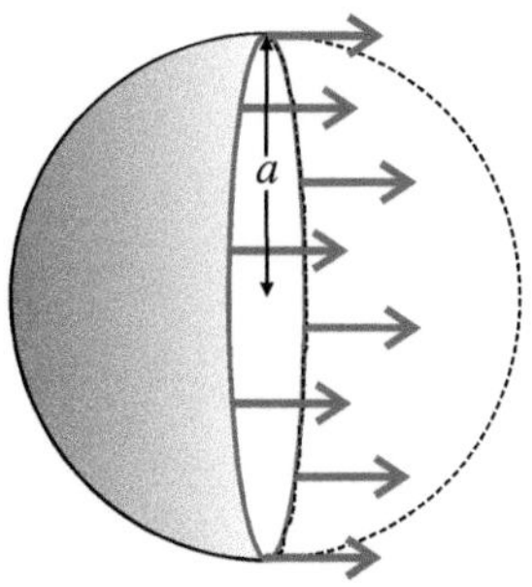

Figure 10.47. Worked example 10.5.

Consider the equilibrium of a hemisphere of the bubble (radius a) as shown in Figure 10.47.

The net force to the right due to surface tension acting around the circumference $2\pi a$ is $2\gamma(2\pi a)$ (the factor of two arises because there are two surfaces – the inner and outer surfaces – forming the soap film). The force due to surface tension is balanced by the horizontal component of the force exerted by the excess pressure P

$$\rightarrow \quad 2\gamma(2\pi a) = P(\pi a^2) \quad \rightarrow \quad P = \frac{4\gamma}{a} = \frac{4(0.025))}{2.5 \times 10^{-2}} = 4.0\ \text{N m}^{-2}$$

Worked Example 10.6: The square root of the average value of v^2 is called the *root mean square* (rms) speed, that is $v_{rms} = \sqrt{\langle v^2 \rangle}$. Estimate the rms speed of the molecules in helium gas at a pressure of 1 atm when the density of helium is 0.18 kg m^{-3}.

Equations (10.13) and (10.14) $\rightarrow\ PV = \frac{2}{3}U$ (where $U = N\langle\frac{1}{2}mv^2\rangle$)

$$\rightarrow\ P = \frac{1}{3}\frac{Nm}{V}\langle v^2\rangle = \frac{1}{3}\rho\langle v^2\rangle \text{ (where } \rho = \frac{Nm}{V} = \text{density of the gas)}$$

$$\rightarrow\ \langle v^2\rangle = \frac{3P}{\rho}$$

$$\rightarrow\ v_{rms} = \sqrt{\frac{3(10^5\,\text{Pa})}{(0.18\ \text{kg m}^{-3})}} = 1290\ \text{m s}^{-1}$$

PROBLEMS

For problems based on the material covered this chapter visit up.ucc.ie/10/ and follow the link to the problems.

11

Thermal physics

AIMS

- to introduce the *kelvin*, the SI unit of thermodynamic temperature
- to investigate the connection between the amount of mechanical energy generated in a body by friction and the corresponding change in its temperature
- to study the phenomena of thermal conduction, convection, radiation and expansion
- to develop a formalism (called *thermodynamics*) to describe the macroscopic behaviour of systems when their temperature changes and, simultaneously, mechanical work is done by the system on the surroundings or vice versa
- to study transformations (fusion and vaporisation) between different phases of matter

11.1 Friction and heating

We saw in Chapter 5 that mechanical energy is conserved only in the absence of friction. In Worked Example 5.3, for example, it was found that the loss of mechanical energy by a body sliding down a rough inclined plane increases linearly with distance travelled along the plane. Investigation of systems like these, in which friction-like forces are involved, reveals a further general observation, namely that loss of mechanical energy due to friction is always accompanied by the production of heat in the system. Furthermore, it is observed that the greater the loss of mechanical energy the hotter the system becomes.

We have used terms such as 'production of heat' and 'the hotter the system becomes' in a rather loose and non-scientific way. These somewhat colloquial terms correspond loosely to particular physical quantities; one of our first tasks in this chapter will be to define such physical quantities so that we can develop a formalism which will enable the quantitative study of thermal systems to be undertaken. Two distinct concepts will be introduced, namely: (1) the idea of a body getting 'hotter' (or 'colder') will be described as an increase (or decrease) of temperature, and (2) the taking in or release of 'heat' will be associated with a change of the energy content of the body.

The phenomenon of the conversion of mechanical energy into heat was studied in detail by James Prescott Joule (1818–1889) in a series of important experiments in the 1840s from which he established clearly that a given loss of mechanical energy always gives rise to the same 'quantity of heat'. One version of the apparatus used by Joule in his experiment is shown in Figure 11.1. In these experiments, friction was produced in water contained in a copper vessel by the rotation of a brass paddle wheel inside the vessel. The paddle wheel was rotated by allowing the lead weights to descend, thus unwinding the strings around the drum which in turn produced the rotation. The mechanical work done during the process was essentially equal to the loss of (gravitational) potential energy of the weights; the motion was sufficiently slow that kinetic energy could be neglected. The increase in the temperature of the water in the vessel was measured using accurate thermometers. While the increase in temperature which Joule observed was very small, he succeeded in establishing that the increase in temperature was directly proportional to the mechanical work done. A somewhat more modern version of Joule's experiment will be discussed in Section 11.2.

The importance of Joule's results was that they enabled the concept of energy to be extended to include heat as a form of energy. The mechanical work done by frictional forces which produce heat could now be considered to result in an increase in the heat energy or **internal energy** of a system, such an increase being exhibited as an increase in temperature. In describing the observed behaviour in this way, we are in essence proposing that mechanical work has somehow been absorbed into the system as energy. In this model, a rise (or fall) in the temperature of a body is seen as being a result of an increase (or decrease) in the internal energy of the body, irrespective of what caused the change in temperature. In other words, if a body is heated by mechanical (frictional), chemical, electrical, or other processes, the same increase in temperature is interpreted as corresponding to the same increase in internal energy. We will see in the next chapter, when we come to interpret these thermal quantities from a microscopic viewpoint, that internal energy is understood as

Understanding Physics, Third Edition. Michael Mansfield and Colm O'Sullivan.
© 2020 John Wiley & Sons Ltd. Published 2020 by John Wiley & Sons Ltd.

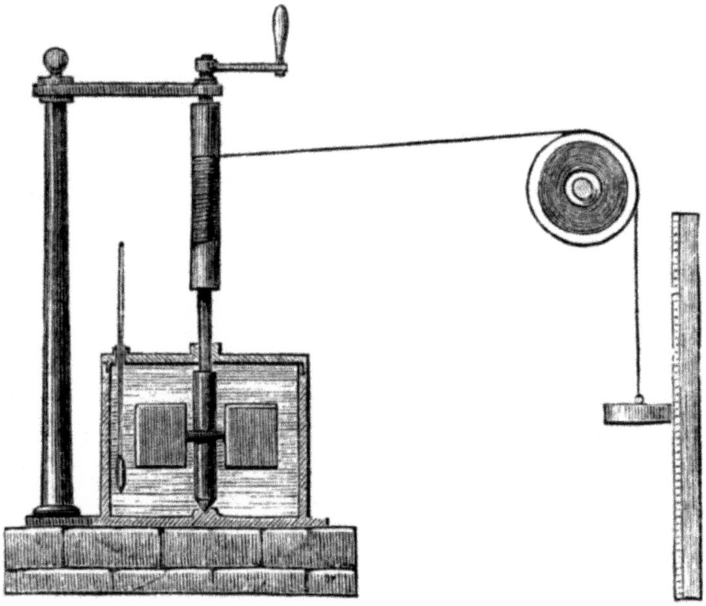

Figure 11.1. Engraving of Joule's apparatus for measuring the mechanical equivalent of heat, in which gravitational potential energy from the weight on the right is converted into heat by stirring the water in the container on the left (from *Harper's New Monthly Magazine*, August 1869).

mechanical energy at the microscopic level. The macroscopic concept of temperature is linked to aspects of this microscopic mechanical energy by a fundamental physical constant known as the Boltzmann constant (k).

Before we can proceed to develop a formalism which can be used to describe thermal systems, we must first develop our ideas of temperature in a more precise and exact way; qualitative descriptions in terms of adjectives like 'hot' and 'cold' are inadequate. The first task, therefore, is to adopt a quantitative scale of temperature.

11.2 The SI unit of thermodynamic temperature, the kelvin

The fundamental constant chosen to define the SI unit of temperature (the kelvin – symbol K) in the 2019 redefinition (see Appendix D) is the Boltzmann constant (k). In any system of units, k is measured in energy per unit temperature; thus, in SI, it has dimensions of joule per kelvin. The 2019 definition fixes the value of the Boltzmann constant at exactly

$$k = 1.380\,649 \times 10^{-23}\ \mathrm{J\,K^{-1}}.$$

Now $\mathrm{J} = \mathrm{kg\,m^2\,s^{-2}}$ and since the kilogram, metre and second are already defined, *fixing k defines the kelvin.* Because of the accuracy to which the value of k was known, the value of k chosen ensures the continuity of the unit with the previous SI definition.

The formal definition of the kelvin is as follows.

The kelvin, symbol K, is the SI unit of thermodynamic temperature. It is defined by taking the fixed numerical value of the Boltzmann constant k to be $1.380\,649 \times 10^{-23}$ when expressed in the unit $\mathrm{J\,K^{-1}}$, which is equal to $\mathrm{kg\,m^2\,s^{-2}\,K^{-1}}$, where the kilogram, metre and second are defined in terms of h, c and $\Delta\nu_{Cs}$.

The effect of this definition is that one kelvin is equal to the change of thermodynamic temperature that results in a change of thermal energy (kT) of exactly $1.380\,649 \times 10^{-23}$ J.

The Celsius scale

The unit of temperature in the Celsius scale is the degree Celsius, symbol °C. The Celsius temperature scale is defined in terms of the kelvin such that

$$\text{temperature}/^\circ\mathrm{C} = T/\mathrm{K} - 273.15$$

Thus, the unit of temperature difference on the Celsius scale is the same as the kelvin; that is, a change of 1 °C is equal to a change of 1 K.

11.3 Heat capacities of thermal systems

Figure 11.2 shows a more modern variation of Joule's apparatus discussed in Section 11.1. Two truncated cones of brass are machined so that one, C_1, which is solid, fits closely inside the other, C_2, as shown. A thermometer is inserted into a narrow hole drilled in the solid cone so that the temperature of the two-cone system can be measured. A circular drum of radius R is attached to the solid cone and a length of string is wound around the drum and then passed over a smooth (that is, frictionless) pulley P. A mass m is attached to the end of the string so that it hangs vertically, as shown in the figure. If the outer brass cone were to be held fixed, release of the mass m would cause the string to unwind and the drum, and hence the inner brass cone, to rotate. The mechanical work done as the mass m fell through a height h would then be $mgh - \frac{1}{2}mv^2$, where v is the velocity achieved by the mass m after falling through h.

In the version of the experiment shown in the figure the lower cone can be rotated by turning a wheel and the number of rotations made by the cone is recorded on a counter. If the surfaces of the cones in contact with one another are lubricated by means of a thin film of suitable oil or petroleum jelly, it is possible to rotate the wheel smoothly at such a rate as to keep the mass m fixed in space. In this case, the forces due to the friction between the cones, transferred via the string, exactly balance the force of gravity on the mass m. In this way a large number of rotations (N, say) can be completed which is equivalent to the mass m falling through a very large distance $h = (2\pi R)N$. Note that in this arrangement the mass m is stationary and hence does not possess kinetic energy at any stage. The corresponding amount of mechanical work done, therefore, is $\Delta W = mgh = 2\pi RmgN$ and the rate at which mechanical work is being done on the system is given by

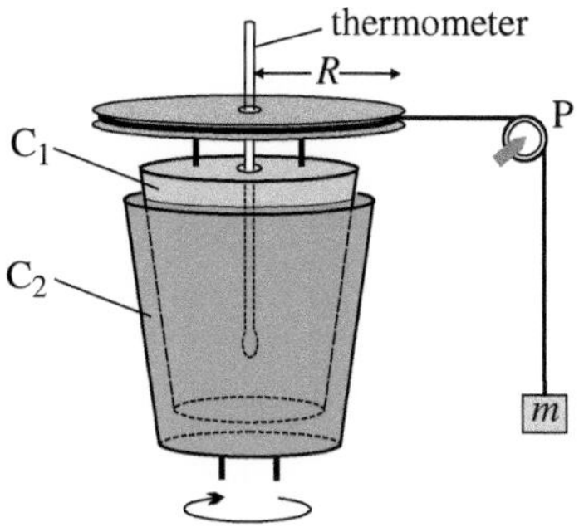

Figure 11.2. Experimental measurement of heat capacity. The outer metal cone (C_2) is rotated so that the mass m remains in the same position. Heat energy is generated by friction due to the rotation of the outer cone relative to the inner cone (C_1).

$$\frac{dW}{dt} = 2\pi Rmg\frac{dN}{dt}$$

Where $\frac{dN}{dt}$ is the number of rotations per unit time of the moving cone relative to the fixed cone which is required to keep the mass m stationary.

If the cones are thermally insulated so that the energy lost due to cooling during the experiment is negligible, it can be assumed that all of the mechanical work done goes into increasing the internal energy U of the system (recall Section 10.11). Thus the rate of increase in the internal energy of the cones is given by

$$\frac{dU}{dt} = \frac{dW}{dt} = 2\pi Rmg\frac{dN}{dt} \tag{11.4}$$

It is observed that the change in internal energy is accompanied by a corresponding increase in temperature.

Figure 11.3 shows the results of such an experiment in which the temperature of the cones is monitored throughout, and the temperature is plotted against time. In this case the outer cone was rotated at a constant rate between times t_1 and t_2; that is, energy was added to the system at a constant rate during this interval. It is clear from this that the corresponding rate of change of temperature is also constant, that is

$$\frac{dU}{dt} = \frac{dW}{dt} \propto \frac{dT}{dt}$$

This experiment shows that the change in internal energy is directly proportional to the corresponding change in temperature, at least over the change in temperature involved in the experiment, that is $\Delta U \propto \Delta T$. This suggests the definition of a physical quantity, which we call the **heat capacity** of the system, defined as

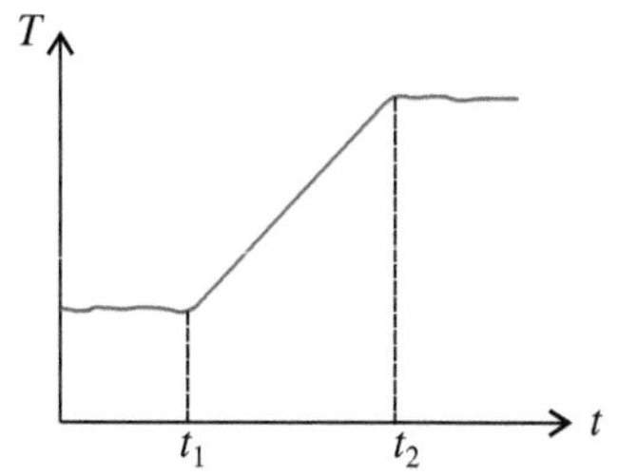

Figure 11.3. A temperature-time plot for the experiment shown in Figure 11.2. The outer cone is rotated at a constant rate between times t_1 and t_2.

$$\boxed{C := \lim_{\Delta T \to 0} \left(\frac{\Delta U}{\Delta T}\right)}$$

The SI unit of heat capacity is, from the definition, J K^{-1}.

The heat capacity of a thermal system relates the energy added to (or taken from) the system to the corresponding increase (or decrease) in temperature as follows,

$$\Delta U = C(\Delta T) \tag{11.5}$$

or

$$\frac{dU}{dt} = C\frac{dT}{dt}$$

Combining this with Equation (11.4) we obtain

$$2\pi Rmg\frac{dN}{dt} = C\frac{dT}{dt}$$

and thus by measuring $\frac{dT}{dt}$, the slope of the linear portion of the temperature–time graph in Figure 11.3, the value of C can be determined since all the other terms in the equation are easily measured. The experiment described, therefore, can be seen to be an absolute method of determining the heat capacity of the thermal system comprising the two brass cones.

While the value of C determined in this experiment appears to be constant within the accuracy of the measurements made and over the small range of temperature involved, in general the heat capacity of a system depends on temperature, that is $C = C(T)$. In many practical situations, however, the variation of heat capacity with temperature is sufficiently small that C may be considered to be constant.

When a thermal system comprises a single material only, as in the case of the experiment with brass cones just described, the heat capacity is proportional to the mass M of the material involved. The heat capacity per unit mass is a characteristic property of the material called the **specific heat capacity**. It is defined as

$$\boxed{c := \frac{C}{M}} = \frac{1}{M}\frac{dU}{dT}$$

It follows from the definition that the SI unit of specific heat capacity is J kg^{-1} K^{-1}.

The specific heat capacity of brass could be determined in the experiment described above simply by dividing the measured value of the heat capacity by the mass of brass in the two cones.

When a system comprises a number of different materials, the heat capacity can be determined from the masses and specific heat capacities of the individual components as follows

$$C = m_1c_1 + m_2c_2 + m_3c_3 + \ldots \tag{11.6}$$

where m_1 and c_1, are respectively, the mass and specific heat capacity of the first component, etc.

For problems based on the material presented in this section visit up.ucc.ie/11/ *and follow the link to the problems.*

11.4 Comparison of specific heat capacities: calorimetry

As explained in the previous section, it is possible to determine the specific heat capacity of a specimen of a material by converting a known amount of mechanical work into heat energy of the specimen and observing the corresponding increase in temperature. Such 'absolute' measurements are difficult to perform, however, and are often intrinsically inaccurate. It is not necessary to carry out this procedure for all materials, because, once the specific heat capacity of one material has been determined it is relatively easy to measure the specific heat capacities of other materials by 'comparative' methods. This technique, known as **calorimetry**, is based on the general principle that, if a system is thermally insulated, the total heat energy of the system must be conserved. Thus heat energy lost by one component part of the system must be balanced by a gain in the remainder of the system. The technique can best be illustrated by describing a simple calorimetric experiment, for example the measurement of the ratio of the specific heat capacity of a liquid (say water) to that of a solid (say brass).

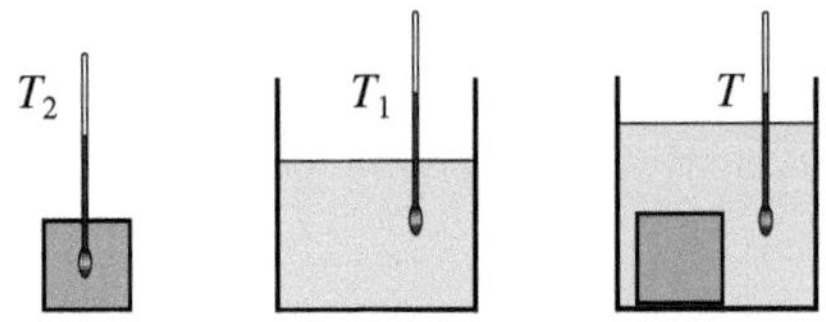

Figure 11.4. A simple calorimetric experiment. A hot metal object (temperature = T_2) is placed in a calorimeter containing cold water (temperature = T_1).

The basic apparatus used is illustrated in Figure 11.4. Water is placed in a well insulated brass container, called a *calorimeter* in this context, and the temperature (T_1) of the system is measured. The heat capacity of this system, as given by Equation (11.6), is $M_Cc_B + M_Wc_W$ where M_C and M_W are the masses of the metal calorimeter and of the water, respectively, and c_B and c_W are the specific heat capacities of brass and water, respectively. A small amount of brass, of mass m, is heated to a temperature T_2, which is significantly greater than T_1, and is then placed in the water in the calorimeter. Heat transfers from the hot brass to the water and calorimeter: that is, the internal energy of the hot brass decreases and the internal energy of the water and the calorimeter increases.

The system will finally settle down at a temperature T which is intermediate between T_1 and T_2. Assuming that the system is thermally insulated so that no heat energy is lost to the surroundings, the decrease in internal energy of the mass m of brass must be equal to the increase in the internal energy of the calorimeter and the water. Thus, using Equations (11.5) and (11.6),

$$\Delta U = C(\Delta T) = (M_Cc_B + M_Wc_W)(T - T_1) = mc_B(T_2 - T)$$

and hence,

$$m(T_2 - T) = \left(M_C + \frac{c_W}{c_B}M_W\right)(T - T_1)$$

Since the masses and temperature differences in this equation can all be measured, the value of $\frac{c_W}{c_B}$ can be determined. Using accurate thermometers and sensitive balances, calorimetric experiments such as this can be very precise. If the specific heat capacity of one of the materials is known from other experiments (for example, c_B might have been determined as described in the previous section) the specific heat capacity of the other material can be calculated. The specific heat capacities of a number of common substances are listed in Table 11.1.

Table 11.1 Specific heat capacities and thermal conductivities of some common substances

substance	specific heat capacity (at ~ 25 °C) J kg^{-1} K^{-1}	thermal conductivity (at ~ 20 °C) W m^{-1} K^{-1}
lead	130	35
silver	240	420
copper	390	400
glass	840	~ 1.0
aluminium	900	240
water	4180	0.6
air (0 °C, sea level)	~ 10^3	0.026
ice (–10 °C)	2000	2.3

In calorimetric methods, such as that described above, it is assumed that the system is insulated sufficiently well thermally that no energy is transferred to the surroundings. When the system comprises metals and liquids, the heat is transferred quite rapidly through the metallic components (as discussed in the next section) and can be transferred efficiently throughout the liquids by stirring. In these circumstances, it is sufficient to provide thermal insulation by enclosing the system in a simple lagging jacket. If the transfer of the heat energy within the system is slow, however, thermal insulation may be insufficient and corrections must be made for heat losses to the surroundings in this case.

Study Worked Example 11.1

For problems based on the material presented in this section visit up.ucc.ie/11/ *and follow the link to the problems.*

11.5 Thermal conductivity

If one end of a solid bar is connected to a cold body whose temperature is kept fixed at temperature T_1 and the other end is connected to a hot body which is kept at temperature T_2 (as illustrated in Figure 11.5), the rate at which energy flows from the hot end to the cold end depends on the material from which the bar is made as well as on the shape and size of the bar and the values of T_1 and T_2.

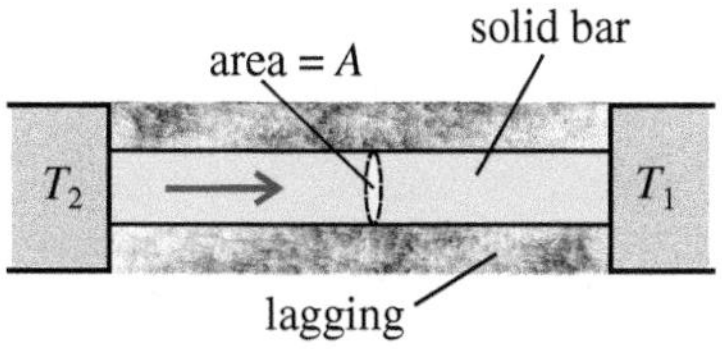

Figure 11.5. Heat conduction through a thermally insulated solid bar.

Consider first the situation deep within the body of a solid of arbitrary shape through which heat is flowing (Figure 11.6); non-intersecting surfaces of constant temperature exist within the solid, such surfaces are indicated by dashed lines in the figure. The direction of heat flow is perpendicular to these surfaces of constant temperature and in the direction from high temperature to low. The rate of heat energy flow through the area ΔA, indicated in the figure, will be proportional to ΔA and to the temperature gradient $\frac{\partial T}{\partial x}$, that is $\Delta P \propto \Delta A \frac{\partial T}{\partial x}$. Thus, the power transferred per unit area is given by

$$\boxed{\frac{\partial P}{\partial A} = \lambda \frac{\partial T}{\partial x}} \qquad (11.7)$$

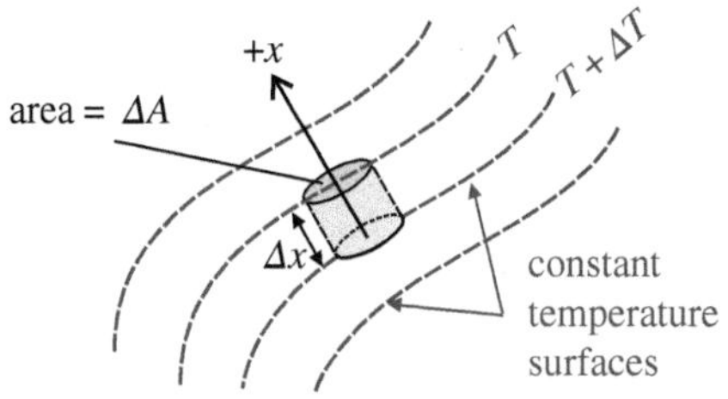

Figure 11.6. Heat conduction in a solid. The coloured dashed lines represent surfaces of constant temperature and the flow of heat energy is perpendicular to these surfaces. The rate of flow of energy is proportional to the temperature gradient $\frac{\partial T}{\partial x}$.

where the constant λ is characteristic of the material and is called the **thermal conductivity** of the material. From this definition, the SI unit of thermal conductivity is seen to be W m^{-1} K^{-1}.

Returning now to the experiment illustrated in Figure 11.5, if the cross-sectional area of the bar in the figure is A, then $\frac{\partial P}{\partial A} = \frac{1}{A}\frac{\partial Q}{\partial t}$ and the rate of heat energy flow through the bar (assuming that no heat flows out through the sides of the bar) is given by

$$\frac{\partial Q}{\partial t} = \lambda A \frac{\partial T}{\partial x}.$$

The heat flow rate $\frac{\partial Q}{\partial t}$ could be measured, for example, by measuring the rise in temperature of water which is made to flow across the end of the bar to keep the temperature of the cold end fixed. After the experiment is first set up it will take a short period of time for the system to reach an equilibrium state but, once such steady state conditions have been reached, $\frac{\partial T}{\partial x}$ is constant and is given by $\frac{\partial T}{\partial x} = \frac{T_2 - T_1}{L}$ so that

$$\frac{\partial Q}{\partial t} = \lambda A \frac{T_2 - T_1}{L}$$

Values of thermal conductivity of a range of substances can be determined using experimental arrangements which are variations of that shown in Figure 11.5. These experiments differ from one another only in the techniques used to measure $\frac{\partial Q}{\partial t}$. The thermal conductivities of some common substances are listed in Table 11.1.

For problems based on the material presented in this section visit up.ucc.ie/11/ and follow the link to the problems.

11.6 Convection

The density of most fluids decreases as the fluid is heated, a well-known consequence being that hot air rises. A hot solid object surrounded by a cooler fluid will heat the fluid immediately adjacent to it causing the heated fluid to rise and to be replaced by cooler fluid. The result of this process is that currents, called **convection currents**, are set up in the fluid. These have the effect of transporting heat energy away from the body, thus cooling the body. Cooling in this manner is called **natural convection** and is usually the dominant mechanism in the cooling of dry solid bodies in air. More rapid cooling can be achieved by blowing air, for example from a fan, directly at the body; this process is called **forced convection**. In either case, the fluid very close to the body will tend to adhere to the surface (as mentioned in the discussion of viscosity in Section 10.8) and will not participate in the convection currents. Thus, the heat energy must pass through this still layer of fluid by *conduction* before being carried away by the convection currents. The thermal conductivity of fluids can be quite small and hence the rate of heat loss from a solid surrounded by fluid is determined primarily by the conduction through the fluid which adheres to the surface of the body.

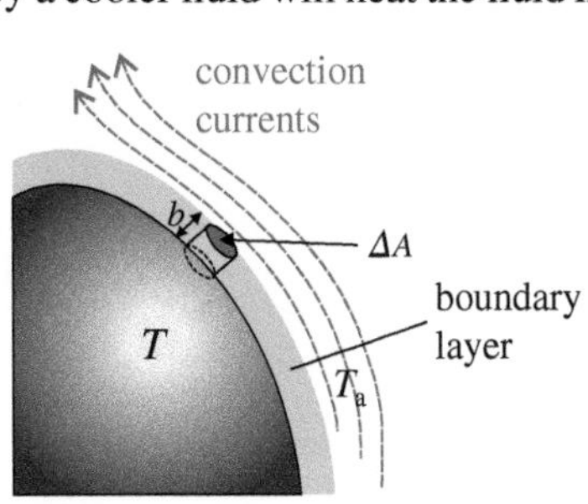

Figure 11.7. Simple model of cooling by convection. Heat energy from the solid surface is conducted through a layer of still fluid (called the 'boundary layer') before being carried away by convection currents outside the boundary layer.

We can consider, therefore, a heated body in a fluid to be surrounded by a thin film of motionless fluid, called the **boundary layer**, through which heat has to be conducted before being carried away by the convection currents outside the boundary layer. The boundary layer is not sharply delineated nor will its thickness be the same at all points around body. Nevertheless, the cooling of a hot body by convection can be represented quite well by the model illustrated in Figure 11.7. Here it is assumed that the boundary layer has a well defined thickness b, inside which the fluid is perfectly still and outside which heat energy is carried away with 100% efficiency by the convection currents. The rate at which heat energy leaves an area ΔA of surface is determined by the rate of conduction through the boundary layer. Thus, from Equation (11.7), the power leaving unit area of surface is given by

$$\frac{\partial P}{\partial A} = \lambda_f \frac{T - T_a}{b}$$

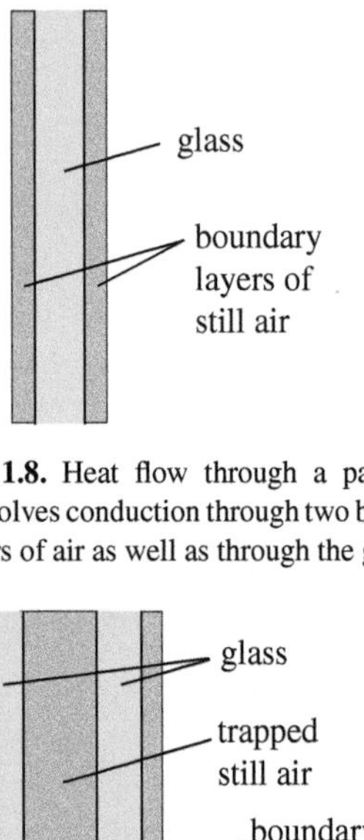

Figure 11.8. Heat flow through a pane of glass involves conduction through two boundary layers of air as well as through the glass.

where T is the temperature of the body, T_a is the temperature of the fluid outside the boundary layer and λ_f is the thermal conductivity of the fluid. Thus the rate of heat energy loss from a body of surface area A is given by

$$\frac{\partial Q}{\partial t} = \frac{\partial P}{\partial A} A = \frac{\lambda_f}{b} A(T - T_a) = hA(T - T_a) \tag{11.8}$$

where h is a constant usually called the **heat transfer coefficient**. This result, which states that the rate of heat energy loss per unit area is directly proportional to the temperature difference between the body and the surrounding fluid, turns out to be a reasonably good description of both natural and forced cooling. Cooling which obeys Equation (11.8) is called **Newtonian cooling**. The average thickness of the boundary layer, and hence the value of h, depends on the cooling conditions and on the shape of the body, b being smaller for larger convection currents.

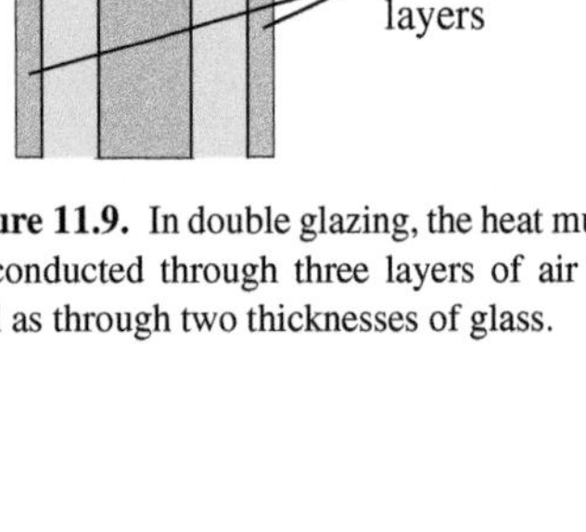

Figure 11.9. In double glazing, the heat must be conducted through three layers of air as well as through two thicknesses of glass.

The thermal conductivity of still air at room temperature is approximately 0.025 W m^{-1} K^{-1} and, for natural convection, the thickness of a boundary layer is usually of the order of 1 mm. Thus the cooling of a room through a single pane of glass is determined more by conduction through the two boundary layers on either side of the glass (Figure 11.8) than by conduction through the glass itself, as shown in Worked Example 11.2. If a strong wind is blowing on the outside of the glass, however, the thickness of the boundary layer on that side can be very much reduced and the rate of cooling increased.

In double glazing (Figure 11.9) a layer of still air or other gas a few millimetres thick is trapped between the two glass panes and this reduces heat loss significantly compared to that through a single pane. The analysis of this situation is similar to that of a glass pane sandwiched between two layers of air, as considered in Worked Example 11.2.

Study Worked Example 11.2

For problems based on the material presented in this section visit up.ucc.ie/11/ *and follow the link to the problems.*

11.7 Thermal radiation

Even in the absence of a surrounding fluid or of any solid connections through which heat could be conducted, a body can still lose heat energy. An isolated body in a vacuum cannot lose heat by conduction or convection but may lose heat by **radiation**. This phenomenon is electromagnetic in origin; electromagnetic radiation will be discussed in detail in Chapter 21. The energy which we receive directly from the Sun and most of the heat which we feel from an open fire are examples of the transfer of heat energy by radiation.

The rate of radiant heat loss from a body increases very rapidly with increasing temperature of the body. The phenomenon was studied experimentally by Josef Stefan (1835–1893) who, in a paper published in 1879 based on earlier observations by John Tyndall (1820–1893), suggested that the rate of heat loss per unit area by radiation from a body is proportional to the fourth power of its temperature (in kelvin). The same conclusion was reached in 1884 by Ludwig Boltzmann (1844–1906) from theoretical considerations. Thus the power transferred per unit area of surface of a body of surface temperature T by this process is given by

$$\frac{dP}{dA} = (constant)T^4$$

or

$$\frac{dP}{dA} = \varepsilon\sigma T^4$$

Where ε is a dimensionless number ($0 < \varepsilon < 1$) called the **emissivity**, which depends on the nature of the radiating surface, and σ is a fundamental constant of nature called the **Stefan-Boltzmann constant**. The most efficient radiators, therefore, are surfaces for which $\varepsilon \to 1$ while highly polished surfaces are inefficient radiators and have low emissivities ($\varepsilon \to 0$). The value of the Stefan-Boltzmann constant is found from experiment to be (to three significant figures)

$$\sigma = 5.67 \times 10^{-8}\ \mathrm{W\,m^{-2}\,K^{-4}}$$

Blackbody radiation

In 1859 Gustav Kirchhoff (1824–1887) had shown that the absorption of thermal radiation is found to be the exact reverse of the emission process. Highly polished surfaces ($\varepsilon \to 0$), for example, reflect the radiation rather than absorb it. A surface which is a very efficient emitter ($\varepsilon \to 1$), on the other hand, will absorb radiation with equal efficiency. In the case of a surface for which $\varepsilon = 1$ all radiation which falls on it will be absorbed. Since such a surface absorbs light with the same efficiency as it does thermal radiation, it appears black, at least provided it is not too hot, and hence is called a **blackbody** surface.

Observation of the range of frequencies – the *spectrum,* to be discussed in Section 21.4 – emitted by solids show that the radiation extends over a continuous range, described as a *continuum.* While the detailed form of the continuum emitted by a body depends to some extent on the composition of the body, Kirchhoff showed that for blackbodies the thermal emission spectrum depends only on temperature; that is, the emitted intensity is a function of temperature (T) and frequency (f) only. Kirchhoff challenged experimental and theoretical physicists to discover the precise functional dependence on temperature and frequency. While the Stefan-Boltzmann law provided a partial answer, the search for an explanation of the frequency dependence at a given temperature engaged physicists for the subsequent half century with, as we shall see in Section 14.2, revolutionary consequences.

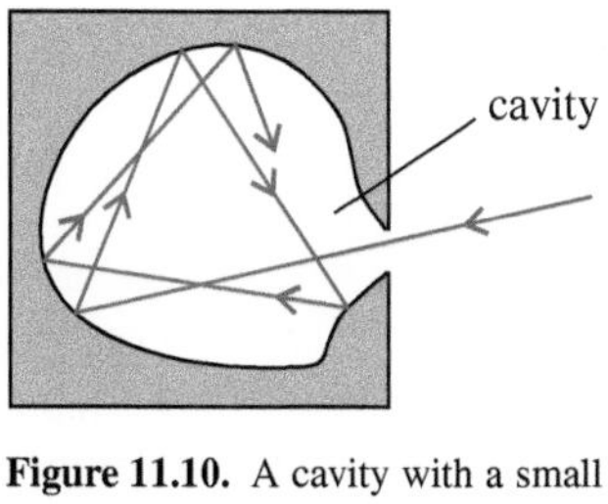

Figure 11.10. A cavity with a small aperture as an example of a blackbody. Electromagnetic radiation entering the cavity is absorbed after successive reflections so that, in effect, the cavity is a perfect absorber of electromagnetic radiation.

In practice, emitting solids are never perfect absorbers and, therefore, never perfect blackbodies. However, in principle, a metallic cavity with a small aperture, as illustrated in Figure 11.10, should approach the absorption properties required of a blackbody because, after successive reflections within the cavity, any radiation entering the cavity is likely to be absorbed. The emission spectrum (a plot of intensity against frequency) of a cavity should therefore resemble closely that of an ideal blackbody.

The observed temperature dependence of blackbody radiation is illustrated in Figure 11.11. As indicated in the figure the maximum intensity of emitted blackbody radiation occurs at a frequency, f_{max}, which increases with temperature T. In 1863 Wilhelm Wien (1864–1928) showed, on theoretical grounds, that $f_{max} \propto T$ (the *Wien displacement law*), a result confirmed by experiment. This law accounts for the changes in the colour of a solid with temperature. At low temperatures, f_{max} lies in the (non-visible) infrared region of the electromagnetic spectrum (Section 21.4) so that the object radiates heat energy without a change of colour. As the temperature of the object increases, f_{max} moves into the red region of the spectrum — the solid glows red. With further increases in temperature f_{max} moves towards the blue end of the visible part of the electromagnetic spectrum so that the colour of the object changes to orange and finally to white (that is, all visible frequencies).

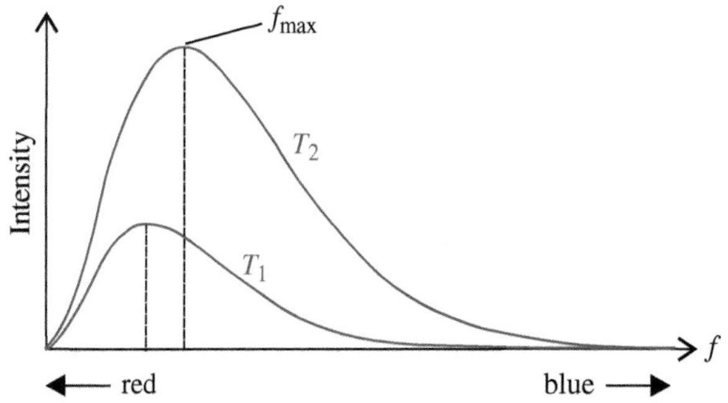

Figure 11.11. Blackbody spectra. Plots of the intensity of the electromagnetic radiation emitted by a blackbody as a function of frequency f for five different temperatures of the body.

In Chapter 13 (at the end of Section 13.8) we will describe another attempt to explain the shape of the blackbody continuum. This, and indeed all classical models of cavity radiation, failed to successfully predict the observed fall-off in intensity at high frequencies (on the blue side of Figure 11.11). Attempts, in the early years of the twentieth century, to explain the observed spectrum led to the development of a radical new theory (quantum mechanics) as we will see in Section 14.2.

Exchange of radiant energy with surroundings

Usually a radiating body will also absorb radiant energy from surrounding bodies which are themselves radiating. Thus the net power lost by the body will be equal to the total power emitted less the total power absorbed and, hence, the net rate at which heat energy is lost from a body of area A at a temperature T is given by

$$\frac{dQ}{dt} = \varepsilon\sigma A(T^4 - T_s^4)$$

where T_s is the average temperature of the surroundings.

The surface of the Earth approximates to a blackbody surface and cools at night by radiation. On a cloudy night the clouds, which are at a temperature similar to that of the ground, radiate back to the Earth at a rate comparable to the rate of emission ($T \approx T_s$). On a cloudless night, on the other hand, $T_s \sim 250$ K and the rate of heat loss from the ground can be very high. This explains why, for the most part, frost appears only on clear calm nights; in this sense clouds act as an insulating blanket which keeps the ground warm at night. On a clear night, the rate of heat loss per unit area by radiation from a region where the surface temperature is 27 °C (300 K), assuming the surface radiates like a blackbody, is approximately $(5.67 \times 10^{-8}\ \text{W m}^{-2}\ \text{K}^{-4})(300^4 - 250^4)\text{K}^4 \approx 200\ \text{W m}^{-2}$. To get an idea of the scale of this effect note that the average flux of radiation from the Sun at sea level is approximately 240 W m^{-2}.

For problems based on the material presented in this section visit up.ucc.ie/11/ and follow the link to the problems.

11.8 Thermal expansion

To a greater or lesser extent most substances expand on being heated; for example, the expansion of the liquid in a glass thermometer. While the amount of expansion of a solid or a liquid can be relatively small, gases can expand considerably even for small rises in temperature. When the rise in the temperature of a body is not too large, it is found experimentally that the fractional increase in the volume is directly proportional to the rise in temperature, that is

$$\frac{\Delta V}{V} \propto \Delta T$$

or

$$\frac{\Delta V}{V} = (constant)\Delta T \qquad (11.9)$$

the constant of proportionality being characteristic of the material involved. Since changes in volume can also be produced by changing the pressure on the body (elastic deformation, as discussed in Section 10.2), it is clear that the experimental result quoted above can only be valid if ΔV is caused by the change in temperature only; that is if the pressure is held constant throughout the experiment. The constant of proportionality in Equation (11.9) is called the **cubic expansion coefficient** of the material and is defined as

$$\alpha_V := \frac{1}{V}\left(\frac{\partial V}{\partial T}\right)_{P=const}$$

Note the use of the partial derivative notation (Appendix A.6.3) to indicate the change in volume with respect to a change in one variable (temperature in this case) while another (pressure) is held constant. From its definition, the SI unit of α_V can be seen to be K^{-1}.

The experimental determination of α_V for solids and liquids is relatively straightforward and, since the measurement is usually performed at atmospheric pressure, the pressure is automatically kept constant. In the cases of gases, however, provision must be made for maintaining a constant pressure, such as a movable piston or other suitable device.

In general, the value of α_V depends on the temperature of the material, that is $\alpha_V = \alpha_V(T)$, but in the case of solids and liquids the variation with temperature is small enough that cubic expansion coefficient can be considered to be constant provided the temperature changes involved are not too large.

In the case of solids, particularly when in the form of long rods, it is often convenient to work with a quantity called the **linear expansion coefficient** of a solid material defined as

$$\alpha_L := \frac{1}{L}\left(\frac{\partial L}{\partial T}\right)_{P=const}$$

where L is the length of the rod.

The linear and cubic expansion coefficients of some common substances are given in Table 11.2.

Table 11.2 Thermal expansion coefficients of some common substances at ~ 20 °C

substance	linear expansion coefficients K^{-1}	cubic expansion coefficients K^{-1}
iron	1.1×10^{-5}	3.3×10^{-5}
steel	$1.1 - 1.3 \times 10^{-5}$	$3.3 - 3.9 \times 10^{-5}$
copper	1.7×10^{-5}	5.1×10^{-5}
aluminium	2.3×10^{-5}	6.9×10^{-5}
lead	2.9×10^{-5}	8.7×10^{-5}
mercury	-	1.8×10^{-4}
water	-	2.1×10^{-4}
ice (at ~ 0 °C)	5.1×10^{-5}	1.5×10^{-4}

The relationship between α_L and α_V can be established by considering how a rectangular block (Figure 11.12) expands on being heated from T to $T + \Delta T$. Now $V = LBH$ and $V + \Delta V = (L + \Delta L)(B + \Delta B)(H + \Delta H) = LBH + LB(\Delta H) + LH(\Delta B) + BH(\Delta L) + L(\Delta B)(\Delta H) + B(\Delta L)(\Delta H) + H(\Delta L)(\Delta B) + (\Delta H)(\Delta L)(\Delta B)$. For small changes terms such as $L(\Delta B)(\Delta H)$ and $(\Delta H)(\Delta L)(\Delta B)$ are negligible compared to the first four terms and hence we can write $\Delta V = LB(\Delta H) + LH(\Delta B) + BH(\Delta L)$

Thus, dividing by LBH,

$$\frac{\Delta V}{V} = \frac{\Delta L}{L} + \frac{\Delta B}{B} + \frac{\Delta H}{H}$$

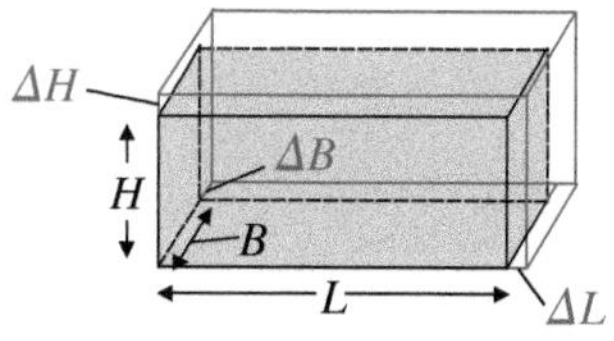

Figure 11.12. Thermal expansion of a rectangular specimen of solid material.

and finally, dividing by ΔT and taking the limit as $\Delta T \to 0$, we obtain

$$\alpha_V := \frac{1}{V}\frac{\partial V}{\partial T} = \frac{1}{L}\frac{\partial L}{\partial T} + \frac{1}{B}\frac{\partial B}{\partial T} + \frac{1}{H}\frac{\partial H}{\partial T} = \alpha_1 + \alpha_2 + \alpha_3.$$

For an *isotropic* solid, namely one whose properties are the same in all directions, $\alpha_1 = \alpha_2 = \alpha_3 = \frac{1}{3}\alpha_V$ and therefore

$$\alpha_V = 3\alpha_L.$$

Thus, in the case of isotropic solids, a measurement of α_L enables a value of α_V to be determined.

For problems based on the material presented in this section visit up.ucc.ie/11/ *and follow the link to the problems.*

11.9 The first law of thermodynamics

In defining heat capacity in Section 11.3, we assumed that the volume of a body did not change as it was heated and, therefore, that all the energy put into a system went into increasing its internal energy. In general, as we know, bodies expand on being heated, at least unless care is taken to ensure that this does not happen. It is clear that if energy is added to a system, this energy can contribute simultaneously (a) to increasing the internal energy and (b) to doing mechanical work as the body expands against an external force or pressure. Thus, if an amount of heat energy ΔQ is added to (or taken from) a thermal system we have

$$\Delta Q = \Delta U + \Delta W \tag{11.10}$$

where ΔU is the corresponding change in internal energy and ΔW is the mechanical work done on or by the surroundings. Equation (11.10) is simply a statement of the principle of conservation of energy applied to systems where the energy can appear in both mechanical and thermal forms – this statement is known as the **first law of thermodynamics**.

We saw in Section 10.4 that, when a body expands against a pressure P, the mechanical work done is given by $\Delta W = P\Delta V$, where ΔV is the change in volume. In this case the first law of thermodynamics can be written as:

$$\Delta Q = \Delta U + P\Delta V \tag{11.11}$$

The concept of the heat capacity of a thermodynamic system must be extended from the simple view taken in Section 11.3 to include situations in which the system expands. The general definition for the heat capacity of a system is

$$C = \underset{\Delta T \to 0}{\text{limit}} \frac{\Delta Q}{\Delta T} \tag{11.12}$$

which, in principle, can give rise to an infinite number of differently defined heat capacities which depend on the details of the response of the system to the change in energy ΔQ. The quantity defined as 'the heat capacity' in Section 11.3 was a rather special case in that it was assumed that the volume remained constant. We must be more careful, therefore, and describe the quantity defined in Section 11.3 more specifically as the **heat capacity at constant volume** which we henceforth denote by

$$C_V := \left(\underset{\Delta T \to 0}{\text{limit}} \frac{\Delta Q}{\Delta T}\right)_{V=const} = \left(\frac{\partial U}{\partial T}\right)_{V=const}$$

the latter equality following from Equation (11.11) since $\Delta V = 0$ in this case. Note that we have used the partial derivative notation again here; the state of a thermodynamic system depends on the values of three variables, namely P, V, and T and it is important to specify which variables change and which are held constant in any process.

Two heat capacities are of particular importance, namely (i) the heat capacity at constant volume, C_V defined above, and (ii) the **heat capacity at constant pressure** which is defined as

$$C_P := \left(\underset{\Delta T \to 0}{\text{limit}} \frac{\Delta Q}{\Delta T}\right)_{P=const}$$

Invoking Equation (11.11) we can write

$$\left(\frac{\Delta Q}{\Delta T}\right)_{P=const} = \left(\frac{\Delta U}{\Delta T}\right)_{P=const} + P\left(\frac{\Delta V}{\Delta T}\right)_{P=const}$$

and hence, taking the limit $\Delta T \to 0$, we see that

$$C_P = \left(\frac{\partial U}{\partial T}\right)_{P=const} + P\left(\frac{\partial V}{\partial T}\right)_{P=const}$$

If the internal energy *depends on the temperature only*, that is, if

$$\left(\frac{\partial U}{\partial T}\right)_{P=const} = \frac{dU}{dT} = \left(\frac{\partial U}{\partial T}\right)_{V=const} = C_V$$

then

$$C_P = C_V + P\left(\frac{\partial V}{\partial T}\right)_{P=const}.$$

Furthermore, invoking the definition of the cubic expansion coefficient $\alpha_V = \frac{1}{V}\left(\frac{\partial V}{\partial T}\right)_{P=const}$, we finally obtain

$$\boxed{C_P = C_V + P\alpha_V V} \tag{11.13}$$

Specific heat capacities

The **specific heat capacity at constant volume** and the **specific heat capacity at constant pressure** of a material are defined, respectively, by

$$\boxed{c_V := \frac{C_V}{M}} \quad \text{and} \quad \boxed{c_P := \frac{C_P}{M}}$$

where M is the mass of the material. Hence, from Equation (11.13),

$$c_P = c_V + P\alpha_V v = c_V + \frac{P\alpha_V}{\rho}$$

where $v = \frac{V}{m} = \frac{1}{\rho}$ is the specific volume (Section 10.1) of the material.

Typically, for a solid or a liquid at atmospheric pressure,

$$c_p \approx 10^3\ \mathrm{J\,kg^{-1}\,K^{-1}},\ P \approx 10^5\ \mathrm{N\,m^{-2}},\ \alpha_V \approx 10^{-5}\ \mathrm{K^{-1}} \quad \text{and} \quad \rho \approx 10^3\ \mathrm{kg\,m^{-3}} \quad \Rightarrow \quad \frac{P\alpha_V}{\rho} \approx 10^{-3}\ \mathrm{J\,kg^{-1}\,K^{-1}}$$

and hence $c_P \approx c_V$. Accordingly, for solids and liquids, we can talk simply, as we did in Sections 11.3 and 11.4, about a single specific heat capacity, without making any distinction as to whether the volume or the pressure is kept constant. On the other hand, as we shall see, the distinction between c_P and c_V is very important in the case of gases.

Enthalpy

When dealing with processes in which the pressure is kept constant, it is useful to work with a thermodynamic quantity called **enthalpy** which is defined as follows

$$\boxed{H := U + PV}$$

Recalling the first law of thermodynamics in the form given by Equation (11.11), namely

$$\Delta Q = \Delta U + P\Delta V$$

we see that if P is constant this can be written as

$$\Delta Q = \Delta(U + PV) = \Delta H$$

and hence

$$C_P = \left(\frac{\partial H}{\partial T}\right)_{P=const}$$

Note that the unit of enthalpy is the same as that of energy.

Study Worked Example 11.3

For problems based on the material presented in this section visit up.ucc.ie/11/ and follow the link to the problems.

11.10 Change of phase: latent heat

As we have seen, when a solid is heated its internal energy increases. Since the heat capacity of a solid is temperature dependent, the temperature does not increase linearly, but nevertheless it increases steadily with increasing internal energy. This process does not continue indefinitely, however, as can be seen from Figure 11.13, which shows how the temperature of a typical substance varies as energy is added steadily. At some temperature, the melting point T_M of the material, the solid will begin to melt. While the material is

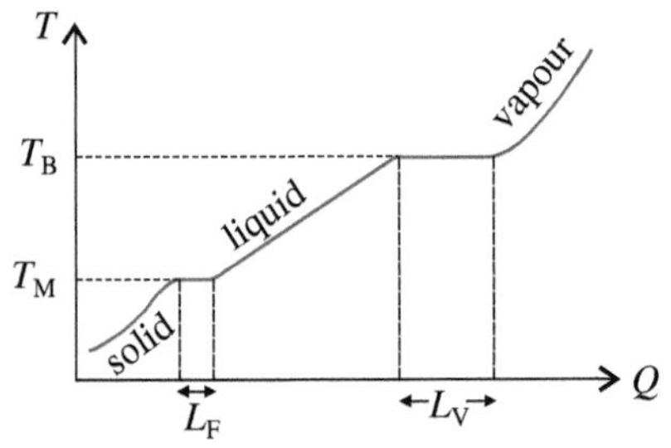

Figure 11.13. A plot of the variation of temperature as a function of heat energy added for a fixed mass of substance. While the body is melting or boiling, the heat energy absorbed goes into changing the phase and the temperature remains constant until all the material is melted/vaporised.

melting, heat is continually added to the system but no corresponding increase in temperature is observed. The melting process is an example of a **change of phase**; in this case the material is changing from the solid phase to the liquid phase. The heat energy required to melt the body is called the **latent heat of fusion** (L_F in Figure 11.13). Once the body has fully melted and heat energy continues to be added, the temperature will begin to rise once more, although not generally at the same rate because the value of the specific heat capacity of the body changes when it becomes liquid. Again the temperature will cease rising when the liquid begins to boil, that is as it changes phase from liquid to vapour at its boiling point T_B. The heat energy required to change the body from the liquid phase to the vapour phase is called the **latent heat of vaporisation** (L_V in Figure 11.13). The term **enthalpy of transformation** can be used instead of the term latent heat.

The values of L_F and L_V are proportional to the mass M of the body. The quantities **specific latent heat of fusion** and **specific latent heat of vaporisation** are defined, respectively, as follows

$$\boxed{l_F := \frac{L_F}{M}} \quad \text{and} \quad \boxed{l_V := \frac{L_V}{M}}$$

The quantities l_F and l_V are characteristic of the material undergoing the change of phase. The SI unit of specific latent heat is J kg^{-1}. In the case of water, for example,

$$l_F = 3.3 \times 10^5 \text{ J kg}^{-1} \quad \text{and} \quad l_V = 2.3 \times 10^6 \text{ J kg}^{-1}$$

For problems based on the material presented in this section visit up.ucc.ie/11/ and follow the link to the problems.

11.11 The equation of state of an ideal gas

We saw in Section 10.10 that all gases obey Boyle's law in the limit of low pressure; that is, for a fixed mass of gas having a volume V at pressure P, the product PV is constant (provided that the temperature T is held constant). Furthermore, it can also be observed that, in the low pressure limit, $\frac{P}{T}$ = constant (provided the volume V is kept constant); this a consequence of the temperature scale that we have adopted. Thus, the observed behaviour of gases in the limit of low pressure yields the general result that for a fixed mass of an ideal gas $\frac{PV}{T} = K$ or

$$\boxed{PV = KT} \tag{11.14}$$

where K is a constant

This equation, which describes the behaviour of all gases in the limit $P \to 0$, is called the **equation of state of an ideal gas**. The state of any system is determined by the values of the three variables P, V and T, each of which can vary but only within restrictions which are expressed as a functional relationship between P, V and T. In the case of an ideal gas the restriction states that the quantity $\frac{PV}{T}$ does not change. Real gases can vary somewhat in their behaviour from that given by Equation (11.14) but alternative equations of state can be found which give a better description of the behaviour of particular gases (see Section 12.1).

Once established, the equation of state can be used to evaluate certain physical properties of a system. For example, the cubic expansion coefficient of an ideal gas at some temperature T can be determined from Equation (11.14) as follows

$$\alpha_V := \frac{1}{V}\left(\frac{\partial V}{\partial T}\right)_{P=const} = \frac{1}{V}\left(\frac{\partial}{\partial T}\left(\frac{KT}{P}\right)\right)_{P=const} = \frac{1}{V}\frac{K}{P} = \frac{K}{PV} = \frac{1}{T}$$

Thus, in particular, the cubic expansion coefficient of an ideal gas at 0 °C is

$$\frac{1}{273 \text{ K}} \approx 0.0037 \text{ K}^{-1}.$$

Replacing α_V by $\frac{1}{T}$ in Equation (11.13) we can write $C_P - C_V = PV\left(\frac{1}{T}\right) = K$ and hence the equation of state of an ideal gas can also be written as

$$\boxed{PV = (C_P - C_V)T} \tag{11.15}$$

For problems based on the material presented in this section visit up.ucc.ie/11/ and follow the link to the problems.

11.12 Isothermal, isobaric and adiabatic processes: free expansion

As mentioned in the previous section, the equation of state of a thermodynamic system relates the three variables P, V and T; that is, an equation of state is a functional relationship $F(P, V, T) = constant$. We could represent the state of a system, therefore, as a point in a three-dimensional graph where the pressure, volume and temperature are plotted on three orthogonal axes. The equation of state then generates a surface on this diagram (as illustrated in Figure 11.14), each point on the surface representing a possible state of the system. Since the three variables (P, V, T) which describe a system are linked by an equation of state, they are not independent of one another. The equation of state, however, may allow any one variable to be written as a function of the other two, at least in principle. For example, in the case of an ideal gas, the temperature can be written as a function of pressure and volume, that is

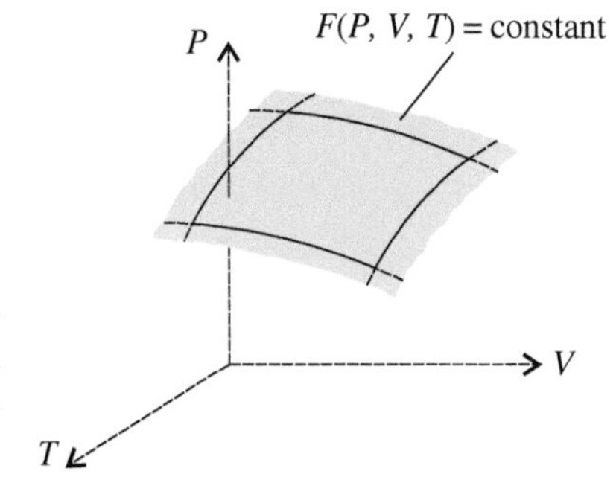

Figure 11.14. An equation of state may be represented by a surface on a P-V-T diagram.

$$T = f(P, V) = \frac{PV}{K}$$

in which case P and V can be treated as independent variables in the manner outlined below.

Isothermal processes

An alternative approach is to describe the system using a P–V diagram. In this case the system can be represented by an infinite family of P–V curves, one for each value of T. All possible states of the gas are represented by a family of such curves as indicated in Figure 11.15. Such constant temperature curves on P–V diagrams are called **isotherms** and a process which takes place at constant temperature, and is thus confined to a single isotherm, is called an *isothermal process*. For an ideal gas at a fixed temperature T_0, for example, we see from Equation (11.14) that $P = \frac{KT_0}{V}$.

Suppose a gas is expanded isothermally at a temperature T_0 from some volume V_1 to a volume V_2. This process can be indicated graphically by the path from A to B on the T_0 isotherm in Figure 11.15. To determine the mechanical work done in expanding the gas in this manner we can invoke Equation (10.2) and thus the work done is given by

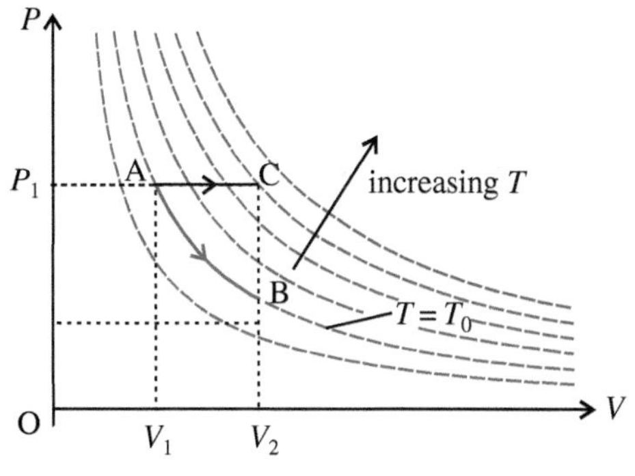

Figure 11.15. P–V curves for a fixed mass of an ideal gas. Each curve (isotherm) corresponds to a different temperature. The change in the system represented by the solid curve from A to B is called an isothermal expansion while that represented by the line from A to C is called an isobaric (constant pressure) expansion.

$$W_{A\to B} = \int_A^B PdV = \int_{V_1}^{V_2} \frac{KT_0}{V} dV = KT_0 \int_{V_1}^{V_2} \frac{dV}{V} = KT_0 \ln\left(\frac{V_2}{V_1}\right), \qquad (11.16)$$

where we have used the equation of state (11.14) to replace P in the first integral by $\frac{KT_0}{V}$.

The fact that materials expand on being heated means that we have to be more careful in the way we define elastic constants (Sections 10.2 and 10.4). Just as in the case of heat capacities, when defining bulk modulus, for example, we need to state which quantity is kept constant. In the case of an isothermal process, the relevant quantity is the **isothermal bulk modulus** which is defined as

$$K_T := -V\left(\frac{\partial P}{\partial V}\right)_{T=const}$$

For an ideal gas at constant temperature $PV = KT_0 = constant = C$ (say) and hence

$$K_T = -V\left(\frac{\partial P}{\partial V}\right)_{T=const} = -V\frac{\partial}{\partial V}\left(\frac{C}{V}\right) = -V\left(-\frac{C}{V^2}\right) = \frac{C}{V} = P$$

which is the same as the result derived in Section 10.10, where it was implicitly assumed that the temperature was kept constant.

Isobaric processes

Consider the process represented by the path A to C in Figure 11.15. In this case the gas is expanded from V_1 to V_2 keeping the pressure constant (such a constant pressure process is called an *isobaric process*). After the expansion the system ends up on a different (higher) temperature isotherm – clearly heat has to be supplied by the environment during the expansion. The mechanical work done in this case is given by

$$W_{A\to C} = \int_{V_1}^{V_2} PdV = P_1 \int_{V_1}^{V_2} dV = P_1(V_2 - V_1)$$

Adiabatic processes

Let us look again at the first law of thermodynamics in the form given by Equation (11.11), namely

$$\Delta Q = \Delta U + P\Delta V$$

We saw in Section 11.9 that, when the internal energy depends on the temperature only such as in the case of an ideal gas, $C_V = \frac{dU}{dT}$ and hence $\Delta U = C_V \Delta T$ so that the first law of thermodynamics can be written as

$$\Delta Q = C_V \Delta T + P\Delta V$$

We now consider changes in thermodynamic systems in which no energy is exchanged with the surroundings, that is processes in which $\Delta Q = 0$. Such processes are called *adiabatic*; examples are changes in thermally isolated systems or changes which take place so rapidly that there is no time for the system to exchange energy with its environment. Thus, for an adiabatic process

$$0 = C_V \Delta T + P\Delta V \quad \text{or} \quad \Delta T = -\frac{P}{C_V}(\Delta V)$$

Differentiating the equation of state for an ideal gas, $PV = (C_P - C_V)T$, with respect to T (using rule (iii) from Appendix A.6.1) we get

$$P\frac{\partial V}{\partial T} + \frac{\partial P}{\partial T}V = (C_P - C_V)$$

and thus,

$$P\Delta V + V\Delta P = (C_P - C_V)\Delta T = -(C_P - C_V)\frac{P}{C_V}(\Delta V).$$

$$\Rightarrow \quad P\Delta V + V\Delta P = -\frac{C_P}{C_V}P(\Delta V) + P(\Delta V)$$

Hence, for an adiabatic process,

$$\frac{C_P}{C_V}P\Delta V + V\Delta P = 0 \quad \text{or} \quad \frac{\Delta P}{P} + \frac{C_P}{C_V}\frac{\Delta V}{V} = 0.$$

Integrating this result we obtain

$$\int \frac{dP}{P} + \frac{C_P}{C_V}\int \frac{dV}{V} = 0$$

and hence

$$\ln P + \frac{C_P}{C_V}\ln V = \ln P + \gamma \ln V = \textit{constant},$$

where the quantity $\gamma := \frac{C_P}{C_V}$ is the **ratio of the principal heat capacities.**

Thus $\ln(PV^\gamma) = \textit{constant}$ and hence, for an adiabatic process in an ideal gas,

$$\boxed{PV^\gamma = K_1} \qquad (11.17)$$

where K_1 is a constant.

Adiabatic curves (solid lines) for an ideal gas are shown on a P–V diagram in Figure 11.16 where they have been superimposed on isothermal curves (dashed lines). It can be seen readily that if an ideal gas is expanded adiabatically, for example from A to D in the figure, the system will end up on a lower temperature isotherm; that is, the gas cools as a result of an adiabatic expansion and heats up as a result of an adiabatic compression. Note that because PV^γ is constant and $\frac{PV}{T}$ is constant, then $TV^{\gamma-1}$ is also constant ($=K_2$, say) in adiabatic processes, that is

$$\boxed{TV^{\gamma-1} = K_2} \qquad (11.18)$$

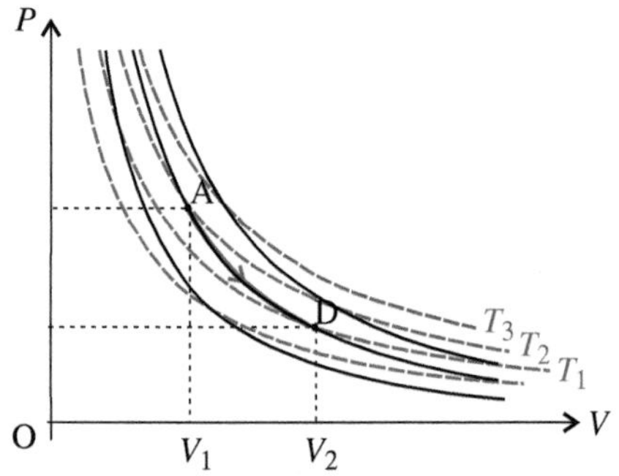

Figure 11.16. Adiabatic curves (black) are shown superimposed on isotherms (blue). An adiabatic expansion, for example A → D, causes a drop in temperature of the gas.

We can derive an expression for the work done in the adiabatic expansion of an ideal gas from volume V_1 to V_2 (represented by path A → D in Figure 11.16) as follows.

$$W_{A\to D} = \int_A^D PdV = \int_{V_1}^{V_2} \frac{K_1}{V^\gamma}dV = K_1\left(\frac{V_2^{\,1-\gamma}}{1-\gamma} - \frac{V_1^{\,1-\gamma}}{1-\gamma}\right) = \frac{1}{1-\gamma}(K_1 V_2^{1-\gamma} - K_1 V_1^{1-\gamma})$$

Since, from Equation (11.17), $P_1V_1^{\gamma} = P_2V_2^{\gamma} = K_1$ we obtain

$$W_{A\to D} = \frac{P_2V_2 - P_1V_1}{1-\gamma}$$

The **adiabatic bulk modulus** of a material is defined as

$$K_S := -V\left(\frac{\partial P}{\partial V}\right)_{\Delta Q=0}$$

Using the result derived above, that $PV^{\gamma} = K_1$ for an ideal gas in an adiabatic process, we see that the adiabatic bulk modulus of an ideal gas is

$$K_S = -V\frac{\partial}{\partial V}\left(\frac{K_1}{V^{\gamma}}\right) = -K_1V(-\gamma V^{-\gamma-1}) = \gamma\frac{K_1}{V^{\gamma}} = \gamma P$$

Free expansion

A significantly different thermodynamic process is shown in Figure 11.17. Here a gas is initially confined to a volume V_1 on one side of a thermally insulated container (of total volume V_2) by means of a partition that can be removed. When the partition is withdrawn (Figure 11.17(b)), the gas expands freely into the vacuum and ultimately settles back into equilibrium filling the whole container (Figure 11.17(c)). A state of thermal equilibrium does not exist except at the end points of the process (points A and B on the P—V diagram in Figure 11.17(d)), so a path cannot be drawn between A and B to represent this *irreversible and non-equilibrium process*. In this case, unlike the isothermal, isobaric and adiabatic processes discussed above, no mechanical work is done ($\Delta W = 0$)

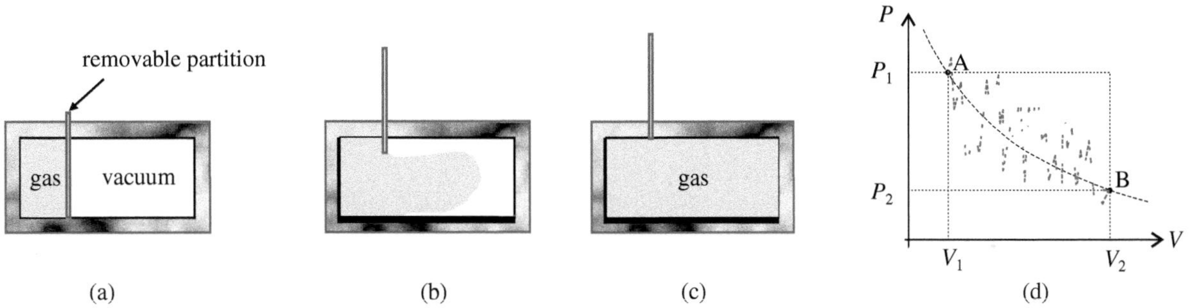

Figure 11.17. (a) A gas is initially confined to the left hand side of a container. (b) The gas expands to fill the container when the partition is removed. (c) The gas fills the whole container (d) Only the initial and final states on a P—V diagram are well defined; the blue points are intended to indicate that a path describing the process cannot be drawn from A to B.

Applying the first law of thermodynamics in the form of Equation (11.10), that is

$$\Delta Q = \Delta U + \Delta W$$

we see that, since the system is insulated, $\Delta Q = 0$ and, since no work is done, for a free expansion

$$\Delta W = 0 \quad \text{and} \quad \Delta U = 0$$

In 1845, Joule performed a series of experiments in which a gas at low density (that is, close to an ideal gas) underwent free expansion in a calorimeter. From his experimental observations, Joule concluded that the temperature of an ideal gas undergoing free expansion remains constant (thus points A and B in Figure 11.15(d) lie on the same isotherm). Since $\Delta U = 0$, the volume — and, by extension, the pressure — of the gas changed in the process, it follows that *the internal energy of an ideal gas must depend only on the temperature,* that is $U = U(T)$.

Study Worked Example 11.4

For problems based on the material presented in this section visit up.ucc.ie/11/ *and follow the link to the problems.*

11.13 The Carnot cycle

As can be seen from Figure 11.18(a), the mechanical *work* done in any thermodynamic process P → Q, $W_{P\to Q} = \int_P^Q PdV$, is equal to the area under the *P–V* curve which represents the process. Note that $W_{Q\to P} = -W_{P\to Q}$. In the case of a cyclic process, such as the closed path P → Q → P in Figure 11.18(b), the net mechanical work done in the cycle is given by

$$W_{\text{cycle}} = \oint_{\substack{\text{closed}\\ \text{cycle}}} PdV = \underset{\text{path 1}}{\int_P^Q PdV} + \underset{\text{path 2}}{\int_Q^P PdV} = \underset{\text{path 1}}{\int_P^Q PdV} - \underset{\text{path 2}}{\int_P^Q PdV}$$

$$= \text{area enclosed within the cycle on } P\text{–}V \text{ diagram.}$$

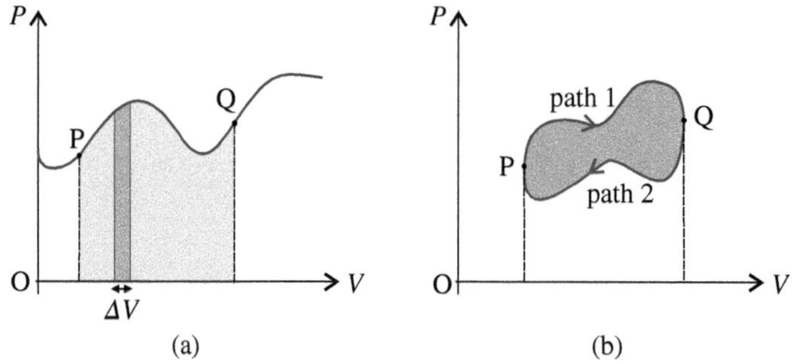

Figure 11.18. (a) Mechanical work done in a thermodynamic process is equal to the area under the *P–V* curve for the process. (b) In a cyclic process, the mechanical work done is equal to the area enclosed by the cycle on the *P–V* diagram representing the process.

An important cyclic process, known as the Carnot cycle (after Nicolas Leonard Sadi Carnot, 1796–1832), is illustrated schematically in Figure 11.19. An ideal gas, initially in state A (P_A, V_A, T_2), goes through four successive reversible stages as follows.

1. A → B: isothermal expansion to volume V_B at constant temperature T_2
2. B → C: adiabatic expansion to volume V_C; temperature drops to T_1
3. C → D: isothermal compression to volume V_D at constant temperature T_1
4. D → A: adiabatic compression to the original state of the system.

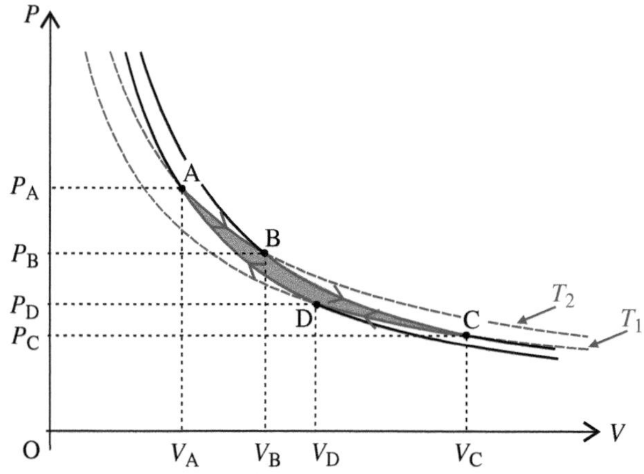

Figure 11.19. A Carnot cycle represented on a *P–V* diagram.

As a result of the process a net amount of mechanical work is done which is equal to the area enclosed by the cycle ABCD on the *P–V* diagram. This mechanical work is achieved by the absorption of an amount of heat energy ΔQ_2 from the surroundings at temperature T_2 during the isothermal expansion A → B which exceeds the amount of energy ΔQ_1 deposited in the environment at temperature T_1 during the isothermal compression C → D. Since the system ends up in exactly the same state as it was in at the start of the process, there can be no change in the internal energy of the system and hence (from the first law of thermodynamics)

$$\Delta Q_2 - \Delta Q_1 = \Delta W$$

By continuously repeating the cycle it is possible to do mechanical work by continuously taking energy from a heat reservoir at a higher temperature T_2 and depositing a smaller amount of energy in a reservoir at a lower temperature T_1. Such a device is called a **heat engine**. If the process is reversed, that is by cycling the system A → D → C → B → A in Figure 11.19, mechanical work is put into the system with the result that heat energy is taken from the low temperature reservoir and deposited in the higher temperature reservoir; in this case the device is a **refrigerator**. The heat energy ΔQ_2 absorbed by the gas from the higher temperature reservoir is converted to the mechanical work done in the isothermal expansion at temperature T_2.

The **efficiency** of a heat engine is defined as the ratio of the net mechanical work done to the heat energy withdrawn from the higher temperature reservoir, that is

$$\boxed{\eta := \frac{\Delta W}{\Delta Q_2}} = \frac{\Delta Q_2 - \Delta Q_1}{\Delta Q_2} = 1 - \frac{\Delta Q_1}{\Delta Q_2}$$

In the case of a Carnot cycle, an ideal gas is involved so we can use Equation (11.16) to determine $|\Delta Q_2|$ (the energy taken from the higher temperature reservoir) and $|\Delta Q_1|$ (the energy deposited at the lower temperature reservoir) during the isothermal parts of the cycle, thus

$$|\Delta Q_2| = KT_2 \ln\left(\frac{V_B}{V_A}\right) \quad \text{and} \quad |\Delta Q_1| = KT_1 \ln\left(\frac{V_C}{V_D}\right).$$

From Equation (11.18) we find that

for the process B → C $\quad T_2 V_B^{\gamma-1} = T_1 V_C^{\gamma-1}$

for the process D → A $\quad T_1 V_D^{\gamma-1} = T_2 V_A^{\gamma-1}$

and hence
$$\left(\frac{V_D}{V_A}\right)^{\gamma-1} = \frac{T_2}{T_1} = \left(\frac{V_C}{V_B}\right)^{\gamma-1} \quad \Rightarrow \quad \frac{V_D}{V_A} = \frac{V_C}{V_B} \quad \Rightarrow \quad \frac{V_B}{V_A} = \frac{V_C}{V_D}$$

Thus
$$\frac{|\Delta Q_1|}{|\Delta Q_2|} = \frac{KT_1 \ln\left(\frac{V_C}{V_D}\right)}{KT_2 \ln\left(\frac{V_B}{V_A}\right)} = \frac{T_1}{T_2}$$

and hence we can write
$$\boxed{\frac{|\Delta Q_1|}{T_1} = \frac{|\Delta Q_2|}{T_2}} \tag{11.19}$$

Thus, the efficiency of a Carnot cycle is $1 - \dfrac{T_1}{T_2}$.

11.14 Entropy and the second law of thermodynamics

The Carnot cycle, as described in the previous section, is an example of a **reversible** engine (or a reversible refrigerator if the cycle is taken in the anti-clockwise direction on the *P–V* diagram). For a process to be reversible, it must take place without friction and sufficiently slowly that the process can be made to proceed in the reverse direction at any stage by making an infinitesimal change in the environment of the system (for example, an infinitesimal change in pressure). The result expressed in Equation (11.19) is characteristic of all reversible cycles since a reversible cycle can be thought of as comprising a very large number of very 'thin' Carnot cycles, as illustrated in Figure 11.20.

For any of the small cycles $\dfrac{|\Delta Q_1|}{T_1} = \dfrac{|\Delta Q_2|}{T_2}$ and thus for the whole reversible cycle we can write $\sum_i \dfrac{\Delta Q_i}{T_i} = 0$. Note that in the last statement we have treated ΔQ as an algebraic quantity with the convention that, if heat is added to a system ΔQ is positive and if heat is taken from a system, ΔQ is negative. In the limit of $\Delta Q \to 0$, all the Carnot cycles becoming infinitesimally small, we can write $\oint \dfrac{d'Q}{T} = 0$ for reversible processes.

Note that here we have written $d'Q$ rather than dQ since, strictly speaking, $d'Q$ is an example of what is called an 'imperfect differential' which simply means that $\int_A^B d'Q$ is not independent of the path taken from A to B, where A and B are any two points on the *P–V* diagram. In other words, Q (and W) cannot be treated as thermodynamic variables in the sense that P, V, T, *etc.* are used.

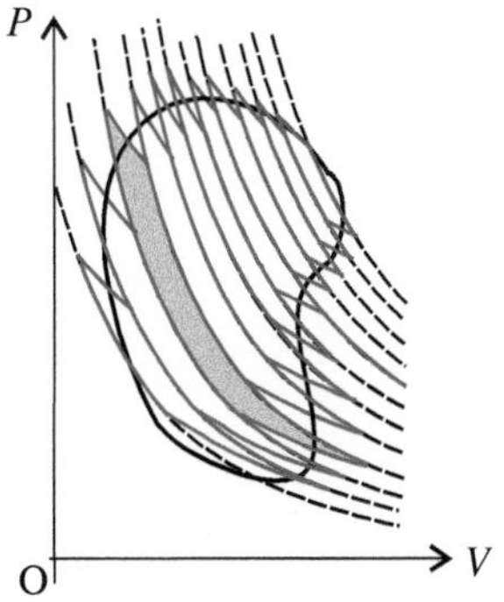

Figure 11.20. A reversible cycle can be considered to be equivalent to a large number of 'thin' Carnot cycles.

On the other hand, at least for a reversible process, $\int_A^B \frac{d'Q}{T}$ *is* independent of the path and this suggests that we might profitably define a new thermodynamic variable S, called the **entropy** of the system, defined as follows

$$\boxed{\Delta S := \frac{\Delta Q}{T}}$$

The SI unit of entropy is J K^{-1}.

It should be noted that only *differences* in the entropy of a system are defined (just as only differences were defined in the case of potential energy in Section 5.7). The difference between the entropy of a system when it is in state B and when it is in state A is given by

$$S_B - S_A = \int_A^B \frac{d'Q}{T} = \int_A^B dS$$

Note that the change of entropy in an adiabatic process ($\Delta Q = 0$) is zero; indeed an adiabatic process may now be redefined as one which takes place at *constant entropy* (hence, for example, the subscript S in the notation for adiabatic bulk modulus, namely K_s).

The definition of entropy allows us to rewrite the first law of thermodynamics, as given by Equation (11.11), in the following useful form.

$$T\Delta S = \Delta U + P\Delta V \tag{11.20}$$

Entropy has been introduced in this section as a somewhat abstract concept, which indeed it is. It is not possible to provide a concrete physical meaning for entropy as a macroscopic concept but a feeling for the role of the concept in thermodynamic processes can be developed by studying situations which involve entropy. We will confine ourselves here to two simple examples.

Example 1: *Change of entropy in an expansion of an ideal gas*

Using Equation (11.15) we can determine the change of entropy in the isothermal expansion of an ideal gas from volume V_1 to V_2 as follows.

$$S_B - S_A = \int_A^B \frac{d'Q}{T} = \int_{V_1}^{V_2} \frac{PdV}{T} = (C_P - C_V)\int_{V_1}^{V_2} \frac{dV}{V} = (C_P - C_V)\ \ln\left(\frac{V_2}{V_1}\right)$$

More generally, the change of entropy in the expansion of an ideal gas is given by

$$S_B - S_A = \int_A^B \frac{d'Q}{T} = \int_{V_1}^{V_2} \frac{dU}{T} + \int_{V_1}^{V_2} \frac{PdV}{T}$$

$$= C_V \int_{T_1}^{T_2} \frac{dT}{T} + (C_P - C_V)\int_{V_1}^{V_2} \frac{dV}{V} = C_V\ \ln\left(\frac{T_2}{T_1}\right) + (C_P - C_V)\ \ln\left(\frac{V_2}{V_1}\right)$$

By definition, there is no change of entropy in an *adiabatic expansion.*

In the case of *free expansion* of a thermally insulated system there is no well defined path over which the integral $\int_A^B \frac{d'Q}{T}$ can be determined even though the initial and final states A and B are specified. Since, however, the change in entropy is independent of the path from A to B, any path may be used. Furthermore, since there is no change of temperature, it is convenient to use the isothermal path from A to B for this purpose and so the change of entropy is the same as that determined above for an isothermal process.

Example 2: *Change of entropy in a phase transition*

The change of entropy in a phase transition, $\Delta S = \int \frac{d'Q}{T} = \frac{L}{T_t} = \frac{ml}{T_t}$

where L, l and T_t are the latent heat, specific latent heat and transition temperature, respectively. Thus $\Delta s = \frac{l}{T_t}$ where $s = \frac{S}{m}$ is the **specific entropy** (entropy per unit mass).

Thus, the change in specific entropy when ice melts is $\frac{3.35 \times 10^5\ \text{J kg}^{-1}}{273\ \text{K}} = 1230\ \text{J kg}^{-1}\ \text{K}^{-1}$

The second law of thermodynamics

In practice, reversible processes are idealised situations; in the real macroscopic world, friction can never be completely eliminated and infinitesimal changes in the environment are not usually achievable. Real thermodynamic processes are irreversible.

As an example of an irreversible process, let us consider the case of two identical blocks each of mass m made from a metal of specific heat capacity c. Initially one block is at temperature T_1 and the other at temperature T_2 and the whole system is thermally insulated from the surroundings and the blocks are insulated from one another (Figure 11.21(a)). If the two blocks are then brought into contact (Figure 11.21(b)), heat will flow from the hotter to the colder and the system will finally settle down at a temperature $T = \frac{1}{2}(T_1 + T_2)$. The change in entropy in this process is

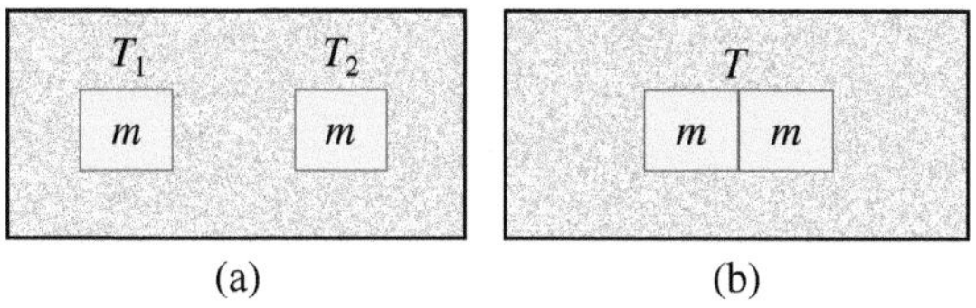

Figure 11.21. Example of an irreversible process. (a) Two identical conducting bodies at different temperatures are thermally insulated from one another and their surroundings. (b) If the bodies in Figure 11.21(a) are brought in contact, their temperature settles down at the average of their original temperatures.

$$\Delta S = \int_{T_1}^{T} \frac{d'Q}{T} + \int_{T_2}^{T} \frac{d'Q}{T} = \int_{T_1}^{T} mc\frac{dT}{T} + \int_{T_2}^{T} mc\frac{dT}{T} = mc\ln\left(\frac{T}{T_1}\right) + mc\ln\left(\frac{T}{T_2}\right)$$

$$= mc\ln\left(\frac{T^2}{T_1T_2}\right) = 2mc\ln\left(\frac{T}{\sqrt{T_1T_2}}\right) = 2mc\ln\left(\frac{\frac{1}{2}(T_1+T_2)}{\sqrt{T_1T_2}}\right)$$

We can show that the change in entropy ΔS is positive as follows.

Since $(T_1 + T_2)^2 - (T_1 - T_2)^2 = 4T_1T_2$, $(T_1 + T_2)^2 > 4T_1T_2 \quad \Rightarrow \quad \frac{1}{2}(T_1 + T_2) > \sqrt{T_1T_2}$ and thus

$$\frac{\frac{1}{2}(T_1+T_2)}{\sqrt{T_1T_2}} > 1 \;\Rightarrow\; \ln\left(\frac{\frac{1}{2}(T_1+T_2)}{\sqrt{T_1T_2}}\right) > 0 \quad \text{and hence} \quad \Delta S > 0.$$

Investigation of other similar processes in isolated systems shows that in all such cases $\Delta S > 0$. The reverse processes, for which $\Delta S < 0$, simply do not occur spontaneously in isolated systems; for example, two metal blocks in thermal contact at the same temperature will not change spontaneously to different temperatures.

This result, that the entropy of an isolated macroscopic system can never decrease, is of fundamental importance. This experimentally observed principle is known as the **second law of thermodynamics** which states that

for an isolated thermodynamic system $\Delta S \geq 0$

where the equality sign refers to reversible processes.

In general, for a system which is connected to the environment, the entropy can increase, decrease or remain the same depending on the details of the process involved. In such cases, however, the second law of thermodynamics requires that the entropy of the {system + environment} cannot decrease.

The deeper significance of the concept of entropy and of the second law of thermodynamics lies in the underlying microscopic nature of matter, an issue which we shall return to in Section 12.1.

For problems based on the material presented in this section visit up.ucc.ie/11/ *and follow the link to the problems.*

11.15 The Helmholtz and Gibbs functions

As we saw in Section 11.12, the three variables P, V and T of a thermodynamic system are inter-related by the equation of state. Since P, V and T are linked in this way, they are not independent of one another and the equation of state allows any one variable to be written as a function of the other two. In general, any thermodynamic function can be considered as a function of any two of the variables. For example, for some function $f = f(V, P)$ we can determine how a small change in f results from small simultaneous changes in V and P by using the expression for the differential of a function of a number of independent variables in Appendix A.6.3, namely

$$\Delta f = \frac{\partial f}{\partial P}\Delta P + \frac{\partial f}{\partial V}\Delta V$$

In particular, note that if $f = PV$ then $\Delta f = \Delta(PV) = V\Delta P + P\Delta V$. Similar statements can be made for the cases where $f = f(V, T)$ and $f = f(P, T)$.

Two thermodynamic functions which prove particularly useful will now be defined. The first of these, called the **Helmholtz function** (or the **free energy**) is defined as follows.

$$F := U - TS$$

Thus, a small change in F is given by $\Delta F = \Delta U - \Delta(TS) = \Delta U - T\Delta S - S\Delta T$

Now, invoking the first law of thermodynamics, as stated in Equation (11.20), we get

$$\Delta F = \Delta U - (\Delta U + P\Delta V) - S\Delta T$$

or

$$\Delta F = -P\Delta V - S\Delta T$$

which is an equivalent statement of the first law in a form which may be conveniently applied when the relevant variables are volume and temperature.

The second function is the **Gibbs function** (or **free enthalpy**) which is defined as

$$\boxed{G := H - TS}$$

where $H := U + PV$ is the enthalpy, defined in Section 11.9 above. A small change in G is given by

$$\Delta G = \Delta H - \Delta(TS) = \Delta U + \Delta(PV) - T\Delta S - S\Delta T = \Delta U + P\Delta V + V\Delta P - T\Delta S - S\Delta T$$

Again, invoking Equation (11.20), we get

$$\Delta G = V\Delta P - S\Delta T$$

which is an equivalent statement of the first law in a form which may be conveniently applied when the relevant variables are pressure and temperature. The Gibbs function plays a particularly important role in chemical reactions and in situations involving two phases (for example, solid and liquid) in equilibrium.

Further reading

Further material on some of the topics covered in this chapter may be found in *Statistical Physics* by Mandl (See BIBLIOGRAPHY for book details).

WORKED EXAMPLES

Worked Example 11.1: A copper cup of mass 200 g contains 90.1 g of water. The temperature of the cup and water is 15.5 °C. Aluminium filings of mass 20.0 g at a temperature of 100 °C are dropped into the cup. What is the temperature of the system when thermal equilibrium has been established if no heat is transferred to the surroundings? (See Table 11.1 for the heat capacities of aluminium, copper and water)

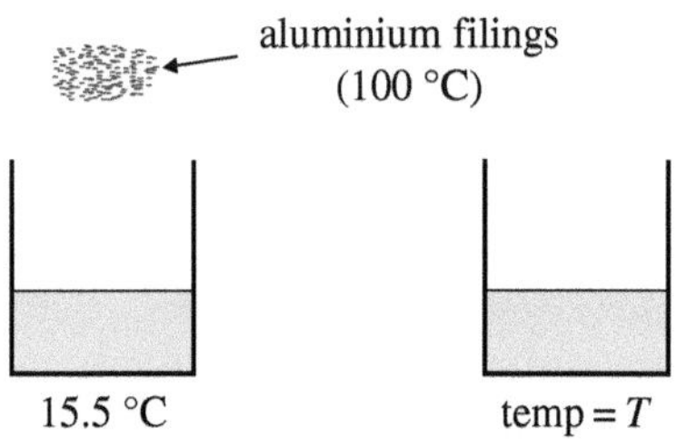

Figure 11.22. Worked Example 11.1.

Applying the procedure outlined in Section 11.4 to the process indicated in Figure 11.22,

$$(0.02)(900)(100 - T) = [(0.2)(390) + (0.091)(4180)](T - 15.5)$$

$$\rightarrow \quad 18(100 - T) = 458(T - 15.5)$$

$$\rightarrow \quad 8899 = 476T$$

$$\rightarrow \quad T = 18.7^\circ\mathrm{C}$$

Worked Example 11.2: The windows of a certain room have a single pane of glass 4.0 mm thick. When the temperature inside the room is 25 °C and that on the outside is 0 °C there is a 1.5 mm boundary layer inside the glass and 0.80 mm boundary layer outside. Estimate the rate of heat loss per unit area of window once a steady state of heat flow has been achieved, assuming that the thermal conductivity of glass is 1.1 W m^{-1} K^{-1} and that of still air is 0.025 W m^{-1} K^{-1}.

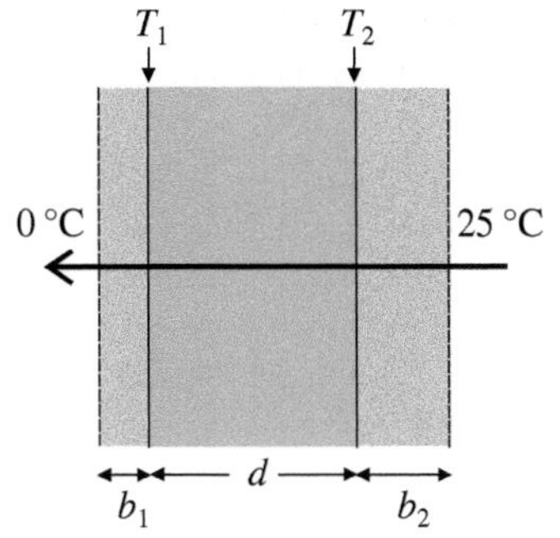

Figure 11.23. Worked Example 11.2.

Applying the final relationship derived in Section 11.5 in turn to each of the layers in Figure 11.23, where $b_1 = 0.8 \times 10^{-3}$ m, $b_2 = 1.5 \times 10^{-3}$ m and $d = 4.0 \times 10^{-3}$ m

$$\frac{1}{A}\frac{\partial Q}{\partial t} = \lambda_a \frac{T_1 - 0}{b_1} = \lambda_g \frac{T_2 - T_1}{d} = \lambda_a \frac{25 - T_2}{b_2}$$

$$\rightarrow \quad \left(\frac{\lambda_a}{b_1} + \frac{\lambda_g}{d}\right) T_1 - \frac{\lambda_g}{d} T_2 = 0 \quad \text{and} \quad -\frac{\lambda_g}{d} T_1 + \left(\frac{\lambda_g}{d} + \frac{\lambda_a}{b_2}\right) T_2 = \frac{25\lambda_a}{b_2}$$

$$\rightarrow \quad 306T_1 - 275T_2 = 0 \text{ W m}^{-2} \quad \text{and} \quad 275T_1 + 292T_2 = 417 \text{ W m}^{-2}$$

$$\rightarrow \quad (292 \times 306)T_1 - (292 \times 275)T_2 = 0$$

$$\rightarrow \quad -(275 \times 275)T_1 + (292 \times 275)T_2 = 417 \times 275$$

$$\rightarrow \quad T_1 = 8.35^\circ\text{C} \quad \text{and} \quad T_2 = 9.29^\circ\text{C}$$

$$\rightarrow \quad \frac{1}{A}\frac{\partial Q}{\partial t} = \frac{\lambda_a T_1}{b_1} = \frac{0.025 \times 8.35}{0.8 \times 10^{-3}} = 260 \text{ W m}^{-2}$$

Worked Example 11.3: A thermal system, initially with volume V_1 at pressure P_1, is expanded to volume V_2 at pressure P_2 in two different ways indicated by the paths $A \rightarrow B \rightarrow C$ (process 1) and $A \rightarrow D \rightarrow C$ (process 2) in Figure 11.24. The heat energy flowing into the system during process 1 is 9.5 kJ and the work done by the expanding system is 4.4 kJ. If the work done by the system in process 2 is 1.3 kJ, how much heat flows into the system in this case?
The system is compressed back to its original pressure and volume by a process $(C \rightarrow A)$ in which the work done on the system is 2.5 kJ. How much heat is absorbed or released in this process?

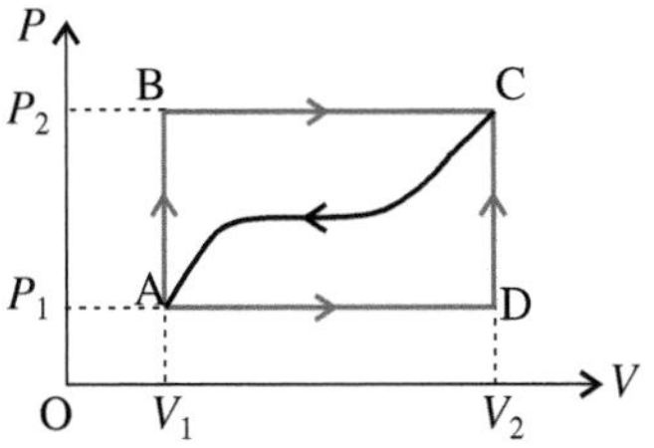

Figure 11.24. Worked Example 11.3.

From the first law of thermodynamics

$$\Delta Q = \Delta U + \Delta W$$

$$A \rightarrow B \rightarrow C: \quad 9.5 \text{ kJ} = \Delta U + 4.4 \text{ kJ}$$

$$\Delta U = U_C - U_A = 5.1 \text{ kJ}$$

$$A \rightarrow D \rightarrow C: \quad \Delta Q = 5.1 \text{ kJ} + 1.3 \text{ kJ} = 6.4 \text{ kJ}$$

$$C \rightarrow A: \quad \Delta Q = -5.1 \text{ kJ} - 2.5 \text{ kJ} = -7.6 \text{ kJ}$$

(the minus sign indicates that energy is released)

Worked Example 11.4: In an experiment (originally devised by Clément and Desormes) to measure the ratio of the principal heat capacities of a gas, the gas is confined to a vessel of volume V_0 (Figure 11.25(a)) at room temperature at a pressure P_0 somewhat above atmospheric pressure (P_a). A tap is opened for a short period during which the gas expands adiabatically and some escapes from the vessel until the gas remaining inside reaches atmospheric pressure, at which point the tap is closed again (Figure 11.25(b)). The temperature of the gas drops during the adiabatic expansion so the gas in the vessel takes in heat from the surroundings until it reaches room temperature again. In this last process the pressure rises to a value P_1. Show that

$$\gamma = \ln\left(\frac{P_0}{P_a}\right)\Big/\ln\left(\frac{P_0}{P_1}\right)$$

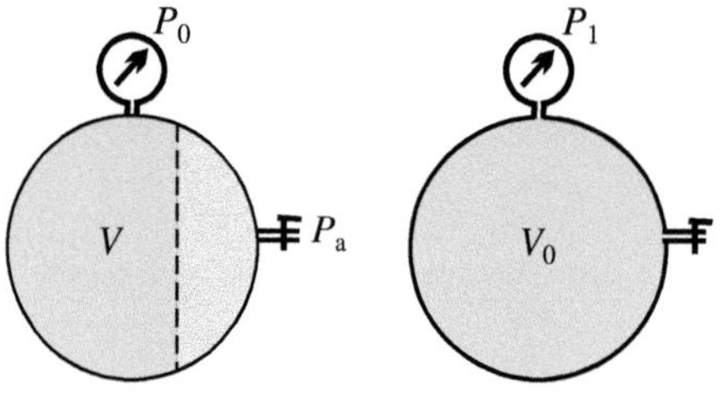

Figure 11.25. Worked Example 11.4.

Consider the mass of gas that remains in the vessel after expansion. This expands first adiabatically from volume V (the volume which it occupied before the expansion, that is, the volume to the left of the dashed line in Figure 11.25) to V_0 (the volume of the container), the process A → B in Figure 11.26. The gas then absorbs heat from the room at constant volume, the process B → C in the figure. The two processes are equivalent to a single isothermal expansion at room temperature (A → C).

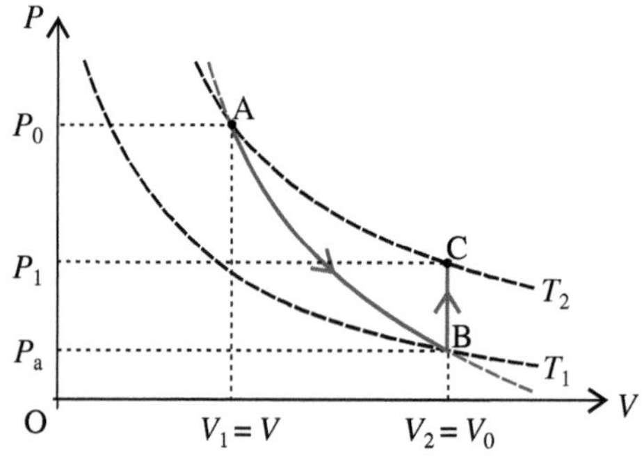

Figure 11.26. Worked Example 11.4.

For the adiabatic expansion (Section 11.12):

$$P_0 V^\gamma = P_a V_0^\gamma \quad \rightarrow \quad \left(\frac{P_0}{P_a}\right) = \left(\frac{V_0}{V}\right)^\gamma$$

Since A and C are on the same isotherm:

$$P_0 V = P_1 V_0 \quad \rightarrow \quad \left(\frac{V_0}{V}\right) = \left(\frac{P_0}{P_1}\right)$$

$$\rightarrow \quad \left(\frac{P_0}{P_a}\right) = \left(\frac{P_0}{P_1}\right)^\gamma \quad \rightarrow \quad \ln\left(\frac{P_0}{P_a}\right) = \gamma \ln\left(\frac{P_0}{P_1}\right)^\gamma \quad \rightarrow \quad \gamma = \ln\left(\frac{P_0}{P_a}\right)\Big/\ln\left(\frac{P_0}{P_1}\right)$$

PROBLEMS

For problems based on the material covered this chapter visit up.ucc.ie/11/ and follow the link to the problems.

The values of thermal and mechanical properties of the various materials involved in the problems may be found in Tables 10.1, 10.2, 11.1 and 11.2. The specific latent heats of water are given in Section 11.10. See Appendix C for astrophysical and geophysical data.

12

Microscopic models of thermal systems: kinetic theory of matter

AIMS

- to interpret thermodynamic behaviour in terms of microscopic models
- to explain how the temperature, the internal energy and the heat capacity of a gas may be understood from a microscopic perspective
- to describe how statistical distributions are used in the study of thermal systems (an introduction to *statistical mechanics*)
- to present a simple microscopic interpretation of the heat capacity of solids

12.1 Microscopic interpretation of temperature

In Chapter 10 we saw that the mechanical behaviour of gases at low pressure, essentially Boyle's law, could be explained by a microscopic model, called the kinetic theory of an ideal gas. In particular, we found (Equation (10.12)), that the product PV could be written in terms of microscopic quantities as follows

$$PV = \frac{2}{3}N\left\langle\frac{1}{2}mv^2\right\rangle$$

where N is the number of molecules, each having a mass m, in the system and $\left\langle\frac{1}{2}mv^2\right\rangle$ is the average kinetic energy per molecule. Comparing this with the equation of state of an ideal gas as given by Equation (11.14),

$$PV = KT$$

we see that the theoretical treatment of the microscopic model predicts a relationship which has the same general form as the experimentally based equation of state which relates macroscopic quantities. Furthermore, a comparison of the two equations leads us to the conclusion that the macroscopic concept of temperature can be interpreted as a direct measure of the average kinetic energy of the molecules (a microscopic quantity). In other words, we now conclude that

$$\left\langle\frac{1}{2}mv^2\right\rangle \propto T$$

$$\Rightarrow \left\langle\frac{1}{2}mv^2\right\rangle = \frac{3}{2}kT \qquad (12.1)$$

where k is a constant called the **Boltzmann constant** (the factor of $\frac{3}{2}$ arises because the Boltzmann constant was originally introduced in a somewhat different context). Historically the value of the Boltzmann constant had to be measured experimentally (for example, see Worked Example 12.2); however, the revised definition of the kelvin (Section 11.2) now fixes its value at exactly $k = 1.380\,649 \times 10^{-23}\ \mathrm{J\,K^{-1}}$. The Boltzmann constant links the macroscopic and microscopic worlds; it enables us to relate the temperature of a system in kelvin to the corresponding energy per molecule in joule.

The important insight represented by Equation (12.1) allows us to extend the application of the kinetic theory of gases to thermal systems with, as we shall see, important consequences for our understanding of such systems.

Understanding Physics, Third Edition. Michael Mansfield and Colm O'Sullivan.
© 2020 John Wiley & Sons Ltd. Published 2020 by John Wiley & Sons Ltd.

We can now write the equation of state of an ideal gas as

$$\boxed{PV = NkT} \tag{12.2}$$

Here the pressure is understood in terms of the microscopic considerations discussed in Chapter 10, namely the average force per unit area exerted by the molecules on the walls of the container while temperature is interpreted as discussed above. Note, in this context, the significance of the choice of temperature scale adopted. If a different scale of temperature had been adopted, the relationship between average molecular kinetic energy and temperature would not necessarily have been exactly linear with obvious complications for the theory.

From Equation (12.1) we see that $\langle v^2 \rangle = \dfrac{3kT}{m}$ and the square root of this quantity gives a measure of the average speed of the molecules. Such a quantity, the square root of the average of the square of a variable, is called the **root mean square** value (or 'rms' value). The root mean square velocity of the molecules in an ideal gas at temperature T, therefore, is given by

$$\boxed{v_{\text{rms}} = \sqrt{\langle v^2 \rangle} = \sqrt{\frac{3kT}{m}}} \tag{12.3}$$

Using (12.3) we see that the rms velocity of the molecules in, for example, helium gas at 0°C is $\sqrt{\dfrac{3(1.38 \times 10^{-23})(273)}{4 \times 1.67 \times 10^{-27}}} \approx 1300 \text{ m s}^{-1}$.

Internal energy

The concept of internal energy now has an obvious microscopic interpretation, namely it is understood to be the total mechanical energy of the system of N molecules, that is

$$U = N\left\langle \frac{1}{2}mv^2 \right\rangle = N\left(\frac{3}{2}kT\right) = \frac{3}{2}NkT \tag{12.4}$$

Thus we can see that the concept of heat energy, which in macroscopic systems seemed to be distinctly different from mechanical energy, can now be seen to be mechanical energy after all, albeit at a microscopic level.

Heat capacity

The heat capacity at constant volume of an ideal gas is given by

$$C_V = \frac{dU}{dT} = \frac{3}{2}Nk$$

Comparing Equations (11.15) and (12.2) we see that $(C_P - C_V) = Nk$ and hence

$$C_P = \frac{5}{2}Nk$$

Molar internal energy and molar heat capacities of an ideal gas

The equation of state of an ideal gas (12.2) can be written in terms of the number of moles, n (Section 10.12), as $PV = nN_A kT = nRT$ where $R = N_A k = 8.31 \text{ J K}^{-1} \text{ mol}^{-1}$ is called the **molar gas constant**. Thus, the equation of state expressed in terms of the molar volume, $v_m := \dfrac{V}{n}$ (Section 10.12), is

$$\boxed{Pv_m = RT}$$

and the **molar heat capacities** are defined as

$$\boxed{c_{V_m} := \frac{C_V}{n} = \frac{3}{2}R} \quad \text{and} \quad \boxed{c_{P_m} := \frac{C_P}{n} = \frac{5}{2}R}$$

Equation of state for real gases

Finally, we note that considerations of finite molecular sizes and of the finite range of inter-molecular forces, as discussed in Section 10.13, give rise to modifications to the equation of state of an ideal gas which help to describe real gases. In particular, taking the finite size of molecules into account, the molar equation of state in terms of molar volume, the *Clausius equation of state* becomes

$$P(v_m - b) = RT$$

and when allowance is also made for the effect of intermolecular forces, the *van der Waals equation of state* (Equation 10.18), becomes

$$\left(P + \frac{a}{v_{\mathrm{m}}^2}\right)(v_{\mathrm{m}} - b) = RT$$

The 'arrow of time'

Many everyday experiences are governed by the second law of thermodynamics. Heat is conducted from a hot region to a cold region without the intervention of an outside agent but the reverse process is never observed to happen (for example, heat flow from the cold to the hot end of the bar in Figure 11.5 cannot occur spontaneously). Mechanical energy can be converted directly into heat energy by friction, for example when a block slides down a rough inclined surface (Worked Example 5.4) or when the amplitude of an oscillating pendulum decreases due to friction. On the other hand, the reverse of these processes is never observed to occur in nature without some intervention from the outside. Until now we had no reason in principle to believe that a pendulum could not oscillate with ever increasing amplitude; such a phenomenon would be fully consistent with the laws of mechanics. We now know that such behaviour is forbidden by the second law of thermodynamics.

We can try to picture what kind of a world we would observe around us if the second law of thermodynamics was not applicable. All processes would be reversible; a pendulum with ever increasing amplitude would be as common as a pendulum with decreasing amplitude. In such a world it would be quite impossible to tell which way time was evolving; there would be no ageing process. This effect is often summarised by saying that the second law of thermodynamics gives us the *'arrow of time'*.

Microscopic interpretation of entropy

Consider the system indicated in Figure 12.1 in which a gas is confined to one section of a container by means of a partition. If the partition is removed the gas expands (irreversibly) to fill the container. Looking at this process from a microscopic viewpoint, we see that all the molecules are confined to a small region in the container initially, the rest of the container being empty of molecules. In the final state the molecules have distributed themselves uniformly throughout the container. The system, therefore, may be thought of as having evolved from a somewhat ordered configuration to a more disordered one. Clearly, the reverse process cannot happen spontaneously — there is negligible probability of molecules starting out with a uniform distribution and ending up gathered together at one side of the container.

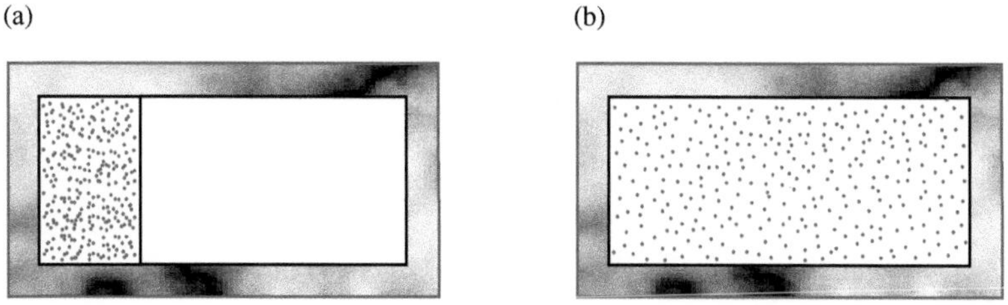

Figure 12.1. (a) All the molecules of a gas are confined to one end of the container by means of a removable partition. (b) When the partition is removed, the molecules spontaneously distribute themselves uniformly throughout the container. The reverse process does not occur spontaneously.

One can think of entropy (Section 11.14), therefore, as a quantitative measure of the **degree of disorder** in a system. A system can evolve spontaneously from a relatively ordered state to a more disordered state but not *vice versa*. Increasing disorder corresponds to an increase in entropy. In any real macroscopic process (for example, heating by friction) the conversion of mechanical energy to heat energy entails a net conversion of energy in an ordered form to energy in a disordered form, with a consequential increase in entropy.

A decrease in entropy (increased order) cannot occur without some outside intervention and in such a case the entropy of the {system + environment} must either remain unchanged or increase. In simple microscopic systems involving few particles, entropy is not a relevant concept and the second law of thermodynamics does not apply.

Study Worked Example 12.1

For problems based on the material presented in this section visit up.ucc.ie/12/ *and follow the link to the problems.*

12.2 Polyatomic molecules: principle of equipartition of energy

The kinetic theory of gases, when extended to thermal systems as described in the preceding section, proves very successful in explaining much of the thermodynamics which has been discussed in the previous chapter. In particular, it predicts the equation of state of an ideal gas which had been deduced from experimental studies of gases at low pressure. On closer comparison between the predictions of the theory and experimental observations on gases, however, a difficulty appears which requires modification to the theory developed so far. The discussion in the previous section predicts a very definite value for the ratio of the principal heat capacities of an ideal gas, namely

$\gamma = \frac{C_P}{C_V} = \frac{5}{3} = 1.67$. Now the value of γ for a gas can be determined easily and accurately by experiment (see, for example, Worked Example 11.4). The result of one such experiment in which the values of γ for molecular hydrogen and helium were measured over a range of temperatures is summarised in Figure 12.2 and it is clear that at higher temperatures the results for hydrogen are in significant conflict with the theory while those for helium are consistent with the theory we have developed so far.

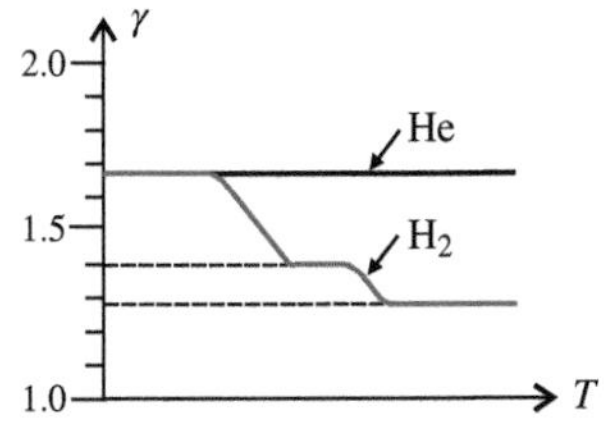

Figure 12.2. The value of the ratio of the principal heat capacities plotted against temperature for hydrogen and helium gases.

The explanation for the discrepancy between theory and experiment lies in the fact that so far we have assumed that the molecules behave like rigid spheres (hypothesis (f) in Section 10.11), a reasonable hypothesis for a monatomic gas such as helium. Hydrogen molecules, however, are diatomic and accordingly have a 'dumb-bell' shape as shown in Figure 12.3. Such molecules have rotational as well as translational kinetic energy and thus energy can be transferred from the rotational form to the translational form and *vice versa* as a result of collisions. The basic hypothesis adopted in our microscopic interpretation of temperature in the previous section was that temperature is proportional to the *translational* kinetic energy. Since this predicted the correct equation of state it must be essentially correct. In the case of a diatomic molecule, therefore, the basic hypothesis that $\left\langle \frac{1}{2}mv^2 \right\rangle \propto T$ is still required but, in this case, v is the velocity *of the centre of mass* of the molecule; it is the rms value of this velocity which determines the temperature.

Rotational kinetic energy, while having no effect on temperature, will nevertheless contribute to the total internal energy of the system. Thus, when the internal energy of a gas is increased, some of the added energy will go into increasing the translational kinetic energy (and hence will contribute to raising the temperature) while more will go towards increasing the average rotational kinetic energy. At higher temperatures still, collisions between the molecules will result in vibrations being set up along the line joining the atoms, again increasing the internal energy in a way that does not appear as a rise in temperature.

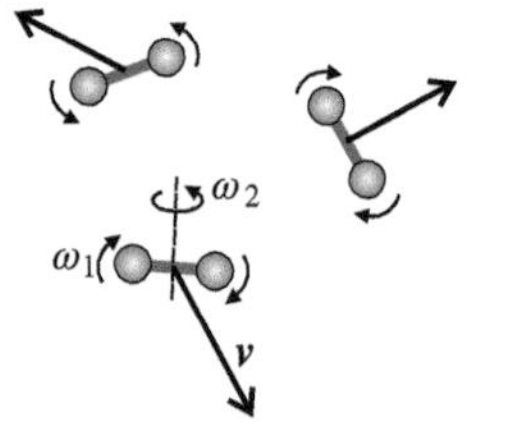

Figure 12.3. Diatomic molecules have rotational kinetic energy in addition to their translational kinetic energy.

Any variable which can have an associated energy is called a **degree of freedom**. In the case of monatomic (spherical) molecules there are three degrees of freedom v_x, v_y, v_z and the average energy of a molecule

$$E = \left\langle \frac{1}{2}mv^2 \right\rangle = \left\langle \frac{1}{2}mv_x^2 \right\rangle + \left\langle \frac{1}{2}mv_y^2 \right\rangle + \left\langle \frac{1}{2}mv_z^2 \right\rangle$$

was found to be $\frac{3}{2}kT$, that is $\frac{1}{2}kT$ per degree of freedom per molecule. In the case of diatomic (dumb-bell shaped) molecules, there are two further degrees of freedom associated with rotation about two axes (because the collisions are frictionless, rotation about the axis defined by a line joining the two atoms of the dumbbell cannot arise as a result of collisions). The total energy of a rotating molecule (Figure 12.3), recalling Section 7.9, is given by

$$E = \frac{1}{2}mv_x^2 + \frac{1}{2}mv_y^2 + \frac{1}{2}mv_z^2 + \frac{1}{2}I_1\omega_1^2 + \frac{1}{2}I_2\omega_2^2$$

where ω_1 and ω_2 are the components of the angular velocity about two mutually perpendicular axes in the plane at right angles to the axis of the molecule. At higher temperatures, where both rotation and vibration take place, two additional terms must be added to take into consideration, namely the kinetic and potential energy associated with the simple harmonic oscillator (Section 3.10). In this case the total energy is given by

$$E = \frac{1}{2}mv_x^2 + \frac{1}{2}mv_y^2 + \frac{1}{2}mv_z^2 + \frac{1}{2}I_1\omega_1^2 + \frac{1}{2}I_2\omega_2^2 + \frac{1}{2}m_r\left(\frac{d\xi}{dt}\right)^2 + \frac{1}{2}k\xi^2$$

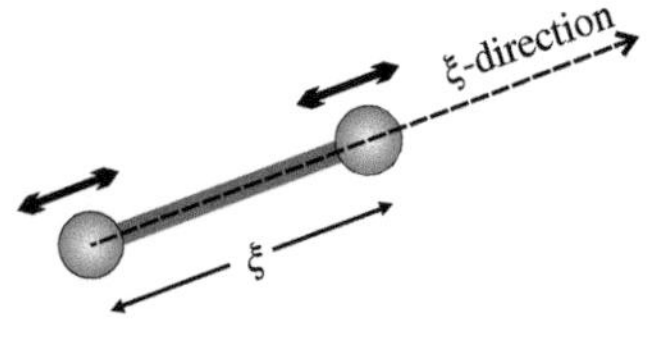

Figure 12.4. Vibration of a diatomic molecule.

where ξ is the separation between the atoms and m_r is the reduced mass of the two-body system (Figure 12.4).

The experimental results illustrated in Figure 12.2 can be explained if it is assumed that the energy of a gas is distributed equally between the different degrees of freedom associated with the molecules. On the basis of the success of the kinetic theory in explaining the behaviour of a monatomic ideal gas we can further assume that the average energy per degree of freedom per molecule is $\frac{1}{2}kT$. Thus for a system of diatomic molecules which have rotational and translational degrees of freedom (five degrees of freedom in all)

$$\left\langle \frac{1}{2}mv_x^2 \right\rangle = \left\langle \frac{1}{2}mv_y^2 \right\rangle = \left\langle \frac{1}{2}mv_z^2 \right\rangle = \left\langle \frac{1}{2}I_1\omega_1^2 \right\rangle = \left\langle \frac{1}{2}I_2\omega_2^2 \right\rangle = \frac{1}{2}kT.$$

and hence the internal energy must be given by $U = \frac{5}{2}NkT$, $C_V = \frac{5}{2}Nk$, $C_P = \frac{7}{2}Nk$ and $\gamma = \frac{7}{5} = 1.40$, in agreement with the experimental evidence (Figure 12.2) at intermediate temperatures. If vibrational degrees of freedom are also excited and the further two degrees of freedom become available, the internal energy is $U = \frac{7}{5}NkT$, $C_V = \frac{7}{2}Nk$, $C_P = \frac{9}{2}Nk$ and $\gamma = \frac{9}{7} = 1.29$ in agreement with the experimental results at high temperatures.

The general assumption invoked here, namely that the energy is distributed equally so that there is $\frac{1}{2}kT$ per degree of freedom per molecule, is called **the principle of equipartition of energy**. The principle clearly removes the discrepancies between theory and experiment mentioned at the start of this section, but it cannot be applied universally; one limitation to its application will be discussed in Section 12.7.

In general, however, the principle can be applied to a gas of polyatomic molecules with an arbitrary number of degrees of freedom f, in which case $U = \frac{f}{2}NkT$ and

$$c_{Vm} = \frac{f}{2}R \qquad c_{Pm} = \frac{f+2}{2}R \qquad \text{and} \qquad \gamma = \frac{f+2}{f}$$

The reader may well ask why rotational and vibrational degrees of freedom do not contribute to the internal energy of a gas at lower temperatures as seen from Figure 12.2. Indeed, all that we have learned of dynamics up to this would suggest that they should. In fact, experimental results such as those represented in Figure 12.2 cannot be explained satisfactorily by standard Newtonian mechanics. A proper treatment of the dynamics of microscopic systems requires a revised formulation of dynamics (namely quantum mechanics) which we shall encounter in Chapter 14. This specific issue will be treated in Section 14.3.

12.3 Ideal gas in a gravitational field: the 'law of atmospheres'

As mentioned in Section 10.10, in most practical applications it can be assumed that the effect of gravity on a gas can be neglected, but this is not the case when the gas occupies a large region in space. The distribution of molecules in the Earth's atmosphere, for example, varies considerably with height above the surface due to the effect of the Earth's gravity. In situations like this we can no longer assume that the gas is isotropic; clearly there is a preferred direction in the system, namely the vertical axis along which gravity acts.

We consider now the case of an ideal gas in a uniform gravitational field (Figure 12.5) and we assume that the temperature is the same everywhere in the gas. The pressure in the gas as a function of height z can be determined from the equation of state $PV = NkT$ and is given by

$$P(z) = \frac{N(z)}{V}kT = n(z)kT = \frac{mn(z)}{m}kT = \frac{\rho(z)}{m}kT$$

where m is the mass of a molecule, $n(z)$ is the number of molecules per unit volume and $\rho(z) = mn(z)$ is the density of the gas at the height z.

The difference in pressure between z and $z + \Delta z$ can be deduced from Equation (10.1) and is given by

$$\Delta P = -g\rho(z)\Delta z = -gmn(z)\Delta z = -mgn(z)\Delta z$$

where we have considered the density to be uniform over the small region Δz. Hence $n(z)$ can be regarded as constant over this region.

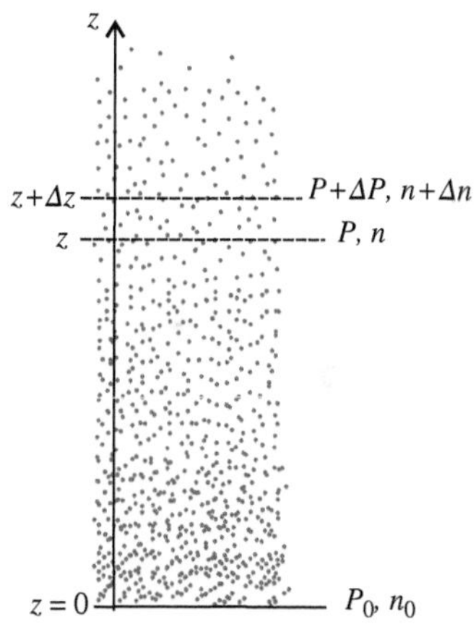

Figure 12.5. Distribution of molecules of an ideal gas in a uniform gravitational field. The density and the pressure decrease exponentially with height.

Thus $$\frac{\Delta P}{P} = \frac{\Delta n}{n} = -\frac{mgn(z)\Delta z}{n(z)kT} = -\frac{mg}{kT}\Delta z$$

and integrating we obtain $$\int \frac{dn}{n} = -\frac{mg}{kT}\int dz$$

$$\rightarrow \quad \ln n = -\frac{mg}{kT}z + \ln(const)$$

Let $n = n_0$ at $z = 0$ $$\rightarrow \quad \ln n = -\frac{mg}{kT}z + \ln n_0$$

$$\rightarrow \quad \ln\frac{n}{n_0} = -\frac{mg}{kT}z$$

and thus $$n(z) = n_0 e^{-\frac{mgz}{kT}} \qquad (12.5)$$

Hence the variation of pressure with height in the gas is given by

$$P(z) = P_0 e^{-\frac{mgz}{kT}}$$

where P_0 is the pressure at $z = 0$. This result is known as the 'law of atmospheres', although it cannot be applied to systems in which temperature does not remain constant with altitude, such as the Earth's atmosphere.

The quantity mgz is of course the gravitational potential energy of a molecule at a height z, that is $U(z) = mgz$, so that we can also write

$$n(z) = n_0 e^{-\frac{U(z)}{kT}}$$

Note that this result shows that it is the ratio of the mechanical to the thermal energy that determines the variation of pressure with height.

In passing we might note that, in 1908, the French physical chemist Jean Perrin (1870–1942) invoked a similar result for the distribution of colloidal particles suspended in liquids in one of the first experimental measurements of the Boltzmann constant (see Worked Example 12.2).

Study Worked Example 12.2

For problems based on the material presented in this section visit up.ucc.ie/12/ *and follow the link to the problems.*

12.4 Ensemble averages and distribution functions

We saw in Section 12.1 how the average of the square of the velocity of the molecules in an ideal gas is related to the temperature of the gas. In many situations, however, more detailed information is required. It is clear, for example, that for a gas in which only a few molecules have very large speeds and most molecules have low speeds, the rms velocity per molecule could be the same as that for a gas in which most of the molecules have intermediate speeds. To calculate the average value of quantities which are functions of the molecular velocity, one needs to know how the different velocity values are distributed among the molecules. Similarly, if we wished to determine how quickly a certain gas escapes from the Earth's atmosphere at a given temperature we would need to know what proportion of the molecules have speeds in excess of the escape velocity needed to overcome the Earth's gravitation (recall Worked Example 5.4). What we need to know in such cases is *the distribution of molecular velocities* in the gas but, before proceeding to develop this idea, we must first look again at how averages are determined in general.

Distribution functions

Suppose that we wish to calculate the average of the following twenty numbers

$$1, 1, 2, 2, 2, 2, 2, 4, 5, 5, 5, 5, 6, 7, 8, 8, 8, 9, 9, 11.$$

The direct method, the one we used when determining the average value of $\boldsymbol{p} \cdot \boldsymbol{v}$ in Section 10.11, is simply to add up all the numbers and divide by 20, that is

$$\frac{1+1+2+2+2+2+2+4+5+5+5+5+6+7+8+8+8+9+9+10}{20} = 5.05.$$

If we note that 1 appears twice, 2 appears five times, 4 once, 5 four times, 6 once, 7 once, 8 three times, 9 twice and 10 once, we can equally well calculate the average thus

$$\frac{(2\times 1)+(5\times 2)+(0\times 3)+(1\times 4)+(4\times 5)+(1\times 6)+(1\times 7)+(3\times 8)+(2\times 9)+10}{20} = 5.05.$$

This second approach can be considerably more efficient if the data set is very large. We can go one step further if we consider that the number 1 is represented in one tenth of the data, 2 in one quarter of the data, 3 is not represented at all, 4 in one twentieth of the data, 5 in one fifth, *etc.* and thus determine the average as follows

$$\left(\frac{1}{10}\times 1\right)+\left(\frac{1}{4}\times 2\right)+(0\times 3)+\left(\frac{1}{20}\times 4\right)+\left(\frac{1}{5}\times 5\right)+\left(\frac{1}{20}\times 6\right)+\left(\frac{1}{20}\times 7\right)+\left(\frac{3}{20}\times 8\right)+\left(\frac{1}{10}\times 9\right)+\left(\frac{1}{20}\times 10\right) = 5.05.$$

We can develop this approach in a more formal manner by defining the average of N values of a quantity as

$$\langle x \rangle := \frac{\sum_i N_i x_i}{N} = \sum_i \frac{N_i}{N} x_i = \sum_i f_i x_i$$

where N_i is the number of times the value x_i appears, $f_i = \frac{N_i}{N}$ is the fraction of times the value x_i appears and the summation notation $\sum_i$ has been used to indicate a sum over all N values.

A system (such as a gas) which contains a large number of entities (such as molecules) is called an ensemble. For an ensemble of N entities, in which N_i of these entities have a value x_i for some quantity x, the **ensemble average** of the quantity x is defined as

$$\boxed{\langle x \rangle := \sum_i \frac{N_i}{N} x_i = \sum_i f_i x_i} \tag{12.6}$$

where $f_i = \frac{N_i}{N}$ is the fraction of entities which have $x = x_i$. If an experiment were to be performed to measure the value of x for a single entity (for example for a single molecule) in the ensemble taken at random, then f_i is the **probability** of the measurement yielding the value x_i.

In an ensemble which is in thermal equilibrium all possible values of any quantity x will be represented, since the number of entities is very large. In this situation the discrete distribution discussed above can be approximated to a continuous distribution in the following way; if ΔN_i is the number of entities which have values of x between x_i and $x_i + \Delta x_i$, then $\Delta f_i = \frac{\Delta N_i}{N}$ is the fraction of entities with values of x between x_i and $x_i + \Delta x_i$. If Δx_i is small compared to x_i, but still contains a large number of entities, the summation in the definition (12.6) of the ensemble average can be approximated by an integral, that is

$$\langle x \rangle := \sum_i \frac{N_i}{N} x_i \quad \Rightarrow \quad \int x \frac{dN}{N} \quad \Rightarrow \quad \int x df$$

Since Δx_i is small, the fraction of entities with values of x between x_i and $x_i + \Delta x_i$ is proportional to Δx_i, that is

$$\Delta f_i = \frac{\Delta N_i}{N} \propto \Delta x_i \quad \text{or} \quad \Delta f_i = P(x_i)\Delta x_i$$

where the function $P(x)$ is defined as the **distribution function** or the **probability density**. $P(x)$ is the probability that a quantity has a value between x and $x + dx$. The ensemble average of the quantity x is given in terms of the distribution function by

$$\boxed{\langle x \rangle := \int x P(x) dx} \tag{12.7}$$

Note that since $\Delta f = P(x)\Delta x$ is dimensionless (being simply a fraction), the probability density $P(x)$ has dimensions which are the inverse of the dimensions of x.

Study Worked Example 12.3

For problems based on the material presented in this section visit up.ucc.ie/12/ *and follow the link to the problems.*

12.5 The distribution of molecular velocities in an ideal gas

It was pointed out by Boltzmann that, for an ideal gas, the decrease in molecular density with height in a uniform gravitational field, discussed in Section 12.3, can be understood in terms of the distribution of the velocities of the molecules at lower levels in the gas. As we saw in the treatment of the kinetic theory of gases in Section 10.11, we can ignore, without loss of generality, the possibility of collisions between the molecules. In these circumstances, one can see (Figure 12.6) that molecules leaving some level $z = 0$ with a z-component of velocity less than v_z such that $\frac{1}{2}mv_z^2 = mgz$ will fail to reach the height z. Similarly, only those molecules with z-component of velocity greater than $v_z + \Delta v_z$ will manage to get up higher than $z + \Delta z$. Thus the number of molecules per unit volume which have a z-component of velocity between v_z and $v_z + \Delta v_z$ should be the same as the difference Δn between the number per unit volume at height z and the number per unit volume at $z + \Delta z$. Differentiating Equation (12.5) with respect to z we obtain

$$\frac{dn}{dz} = n_0 \left(-\frac{mg}{kT} \right) e^{-\frac{mgz}{kT}}$$

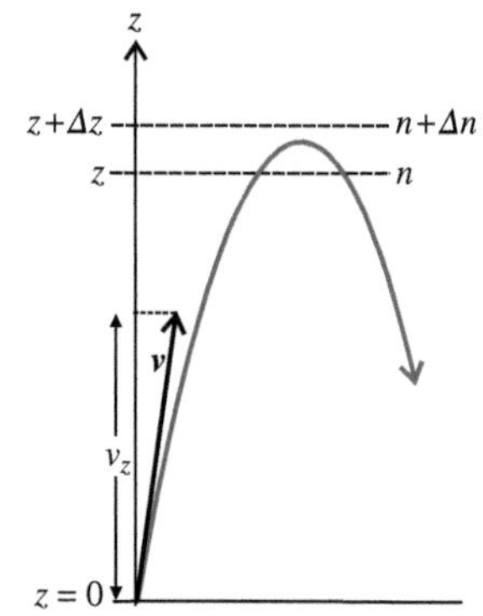

Figure 12.6. Molecules projected from $z = 0$ with z-components between v_z and $v_z + \Delta v_z$ reach a maximum height between z and $z + \Delta z$. Thus the number of molecules per unit volume with z-components of velocity between v_z and $v_z + \Delta v_z$ is the same as the difference in particle density (Δn) between z and $z + \Delta z$.

or

$$\Delta n = -\frac{mg}{kT} n_0 e^{-\frac{mgz}{kT}} \Delta z$$

Since, from conservation of energy, $mgz = \frac{1}{2}mv_z^2$, the number of molecules per unit volume at temperature T with z-component of velocity between v_z and $v_z + \Delta v_z$ is proportional to $e^{-\frac{\frac{1}{2}mv_z^2}{kT}}$. Boltzmann reasoned that this distribution should be the same whether or not the gas is in a gravitational field and therefore that the fraction of molecules in a gas at temperature T with z-component of velocity between v_z and $v_z + \Delta v_z$ should be given by

$$P_z(v_z)\Delta v_z = Ae^{-\frac{\frac{1}{2}mv_z^2}{kT}} \Delta v_z$$

where A is a constant. Since all of such fractions must add up to one then $\int_{-\infty}^{+\infty} P_z(v_z)\Delta v_z = 1$ and hence A can be determined as follows

$$A\int_{-\infty}^{+\infty} e^{-\frac{mv^2}{2kT}} dv_z = 1 \quad \rightarrow \quad A\sqrt{\frac{2kT}{m}} \int_{-\infty}^{+\infty} e^{-\xi^2} d\xi = 1 \quad \rightarrow \quad A = \sqrt{\frac{m}{2\pi kT}}$$

where we have made the substitution $\xi = \sqrt{\frac{mv_z^2}{2kT}}$ and evaluated the integral as follows $\int_{-\infty}^{+\infty} e^{-\xi^2} d\xi = 2\int_0^{+\infty} e^{-\xi^2} d\xi = \sqrt{\pi}$ (Appendix A, Table A.3). Hence the fraction of molecules with z-component of velocity between v_z and $v_z + \Delta v_z$ is given by

$$P_z(v_z)\Delta v_z = \sqrt{\frac{m}{2\pi kT}} e^{-\frac{\frac{1}{2}mv_z^2}{kT}} \Delta v_z$$

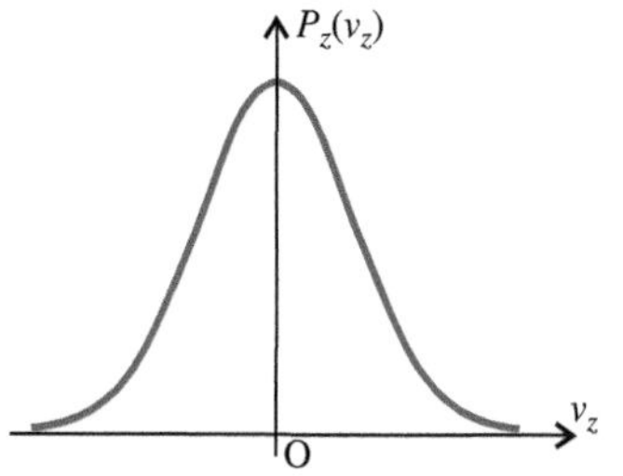

Figure 12.7. Distribution of z-components of molecular velocities in an ideal gas.

The distribution function $P_z(v_z)$ is plotted in Figure 12.7. This is in fact a well known statistical function, called the *gaussian*, which arises whenever random distributions are studied. The corresponding distribution functions for v_x and v_y are

$$P_x(v_x)\Delta v_x = \sqrt{\frac{m}{2\pi kT}} e^{-\frac{\frac{1}{2}mv_x^2}{kT}} \Delta v_x \quad \text{and} \quad P_y(v_y)\Delta v_y = \sqrt{\frac{m}{2\pi kT}} e^{-\frac{\frac{1}{2}mv_y^2}{kT}} \Delta v_y$$

We can now write down an expression for the fraction of molecules in a gas at temperature T with x-component of velocity lying between v_x and $v_x + \Delta v_x$ and with y-component of velocity between v_y and $v_y + \Delta v_y$ *and* with z-component of velocity between v_z and $v_z + \Delta v_z$, namely the product of the three component distribution functions, as follows.

$$P(v_x, v_y, v_z)\,\Delta v_x\,\Delta v_y\,\Delta v_z = \left(\frac{m}{2\pi kT}\right)^{\frac{3}{2}} e^{-\frac{\frac{1}{2}mv^2}{kT}} \Delta v_x \Delta v_y \Delta v_z \tag{12.8}$$

The function

$$P(v_x, v_y, v_z) = \left(\frac{m}{2\pi kT}\right)^{\frac{3}{2}} e^{-\frac{\frac{1}{2}mv^2}{kT}}$$

is known as **the Maxwell–Boltzmann velocity distribution function**.

12.6 Distribution of molecular speeds

The velocity of a single molecule in a gas at temperature T may be represented by a point having coordinates (v_x, v_y, v_z) on a velocity diagram. In this way the velocity states of the whole system of N molecules is represented by N points in velocity space, as in Figure 12.8. Thus the *number* of molecules with an x-component of velocity lying between v_x and $v_x + \Delta v_x$ *and* with a y-component of velocity between v_y and $v_y + \Delta v_y$ *and* with a z-component of velocity between v_z and $v_z + \Delta v_z$, using Equation (12.8), is

$$N\left(\frac{m}{2\pi kT}\right)^{\frac{3}{2}} e^{-\frac{\frac{1}{2}mv^2}{kT}} \Delta v_x \Delta v_y \Delta v_z$$

and is the number of points in the rectangular element of 'volume' $\Delta v_x\,\Delta v_y\,\Delta v_z$ shown in the figure. The number of molecules with speeds between v and $v+\Delta v$, therefore, is the number of points in the spherical shell between radius v and radius $v+\Delta v$, that is

$$\left[N\left(\frac{m}{2\pi kT}\right)^{\frac{3}{2}}e^{-\frac{\frac{1}{2}mv^2}{kT}}\right](4\pi v^2\Delta v)=4\pi v^2 N\left(\frac{m}{2\pi kT}\right)^{\frac{3}{2}}e^{-\frac{\frac{1}{2}mv^2}{kT}}\Delta v$$

Dividing by N we see that the fraction of molecules in a gas at temperature T with speed between v and $v+\Delta v$ is given by

$$P(v)\Delta v=4\pi\left(\frac{m}{2\pi kT}\right)^{\frac{3}{2}}v^2e^{-\frac{\frac{1}{2}mv^2}{kT}}\Delta v \tag{12.9}$$

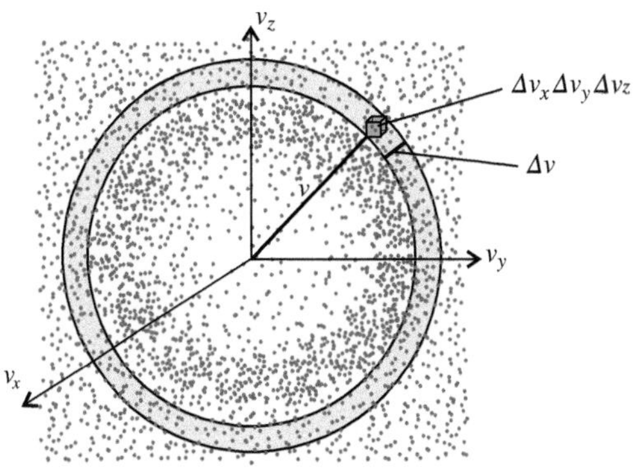

Figure 12.8. Plots of molecular velocities as points in velocity space.

The function $P(v)$ gives the **Maxwell–Boltzmann distribution of molecular speeds** and is sketched in Figure 12.9 for three different temperatures. The use of these functions to determine average values of microscopic quantities is demonstrated in derivations (a) and (b) which follow and also in Worked Example 12.4.

(a) *average speed*: The average speed of the molecules is defined by Equation (12.7) namely $\langle v\rangle=\int_0^\infty vP(v)dv$ and hence, using the distribution (12.9), we get

$$\langle v\rangle=4\pi\left(\frac{m}{2\pi kT}\right)^{\frac{3}{2}}\int_0^\infty v^3e^{-\frac{mv^2}{2kT}}dv=\sqrt{\frac{8kT}{\pi m}}\approx 1.6\sqrt{\frac{kT}{m}}$$

where, again, the value of the integral has been taken from Table A.3 in Appendix A.

(b) *rms speed*: The average value of the square of the speed can be determined as follows.

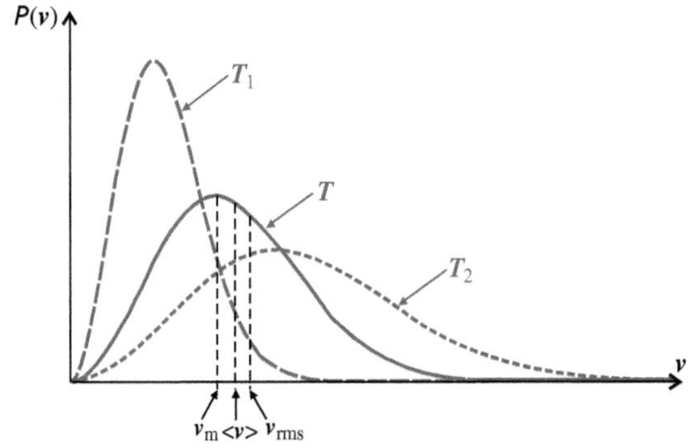

Figure 12.9. Distribution of molecular speeds in an ideal gas for three different temperatures where $T_2 > T > T_1$.

$$\langle v^2\rangle=\int_0^\infty v^2P(v)\,dv=4\pi\left(\frac{m}{2\pi kT}\right)^{\frac{3}{2}}\int_0^\infty v^4e^{-\frac{mv^2}{2kT}}dv=\frac{3kT}{m}$$

and hence the rms speed is given by

$$v_{rms}=\sqrt{\langle v^2\rangle}=\sqrt{\frac{3kT}{m}}\approx 1.7\sqrt{\frac{kT}{m}}$$

which is the same as Equation (12.3), obtained in Section 12.1 above.

(c) *most probable speed*: The most probable speed v_m of a molecule corresponds to the maximum of the speed distribution curve; that is to say, $v=v_m$ when $v^2e^{-\frac{mv^2}{2kT}}$ is maximum. Using the technique outlined in Appendix A.6.2, we see that v is maximum/minimum when

$$\frac{d}{dv}\left(v^2e^{-\frac{mv^2}{2kT}}\right)=0 \text{ or } 2ve^{-\frac{mv^2}{2kT}}-v^2\left(\frac{mv}{kT}\right)e^{-\frac{mv^2}{2kT}}=0 \text{ or } ve^{-\frac{mv^2}{2kT}}\left(2-\frac{mv^2}{kT}\right)=0$$

The two solutions $v=0$ and $v=\infty$ obviously correspond to minima in Figure 12.9 and hence the third solution, given by $2-\frac{mv^2}{kT}=0$, must correspond to the maximum value of $P(v)$, that is

$$v_m=\sqrt{\frac{2kT}{m}}\approx 1.4\sqrt{\frac{kT}{m}}$$

The relative magnitudes of $\langle v\rangle$, v_{rms} and v_m at the temperature T are indicated on Figure 12.9.

Study Worked Examples 12.4 and 12.5

For problems based on the material presented in this section visit up.ucc.ie/12/ and follow the link to the problems.

12.7 Distribution of molecular energies; Maxwell–Boltzmann statistics

It is also instructive to write the Maxwell–Boltzmann speed distribution in terms of energy. The kinetic energy of a molecule is given by $K = \frac{1}{2}mv^2$ and hence $v^2 = \frac{2K}{m}$ and $v = \sqrt{\frac{2K}{m}}$. From this we can derive

$$\frac{dK}{dv} = \frac{1}{2}m(2v) = mv \quad \text{or} \quad dv = \frac{dK}{mv} = \frac{dK}{\sqrt{2mK}}$$

Substituting these results into Equation (12.9) we can replace the speed distribution by an **energy distribution** as follows. We will write $p(K)\Delta K$ to represent the fraction of molecules with kinetic energies between K and $K + \Delta K$ and, since the same number of molecules must have speeds between v and $v + \Delta v$, that is $P(v)\Delta v = p(K)\Delta K$, we obtain

$$p(K)\Delta K = 4\pi\left(\frac{m}{2\pi kT}\right)^{\frac{3}{2}}\frac{2K}{m}e^{-\frac{K}{kT}}\frac{\Delta K}{\sqrt{2mK}} = \frac{2}{\sqrt{\pi}}(kT)^{-\frac{3}{2}}K^{\frac{1}{2}}e^{-\frac{K}{kT}}\Delta K$$

From this we obtain the average kinetic energy per molecule in a monatomic ideal gas at temperature T, namely

$$\langle K\rangle = \int_0^\infty Kp(K)dK = \frac{2}{\sqrt{\pi}}(kT)^{-\frac{3}{2}}\int_0^\infty K^{\frac{3}{2}}e^{-\frac{K}{kT}}dK = \frac{3}{2}kT$$

This, as we would expect, is equivalent to the result which we obtained in Section 12.1 and which led to Equation (12.4), the expression for the internal energy $U = \frac{3}{2}NkT$.

For systems with additional degrees of freedom, the above results can be generalised. For each additional degree of freedom there will be an additional distribution function; for example, for rotation of a molecule about an axis

$$P(\omega)\Delta\omega = \sqrt{\frac{I}{2\pi kT}}\,e^{-\frac{\frac{1}{2}I\omega^2}{kT}}\Delta\omega$$

and once again the average kinetic energy associated with this degree of freedom turns out to be $\frac{1}{2}kT$.

We find that the average energy per degree of freedom is always $\frac{1}{2}kT$ provided that the expression for the energy is *quadratic* in the corresponding variable ($\frac{1}{2}mv_x^2, \frac{1}{2}mv_y^2, \frac{1}{2}mv_z^2, \frac{1}{2}I_1\omega_1^2, \frac{1}{2}I_2\omega_2^2$, *etc.*). This gives us the conditions which determine whether or not the principle of equipartition of energy may be applied; for example, in the case of an ideal gas in a gravitational field, where the potential energy is linear in z, the principle does not hold.

Maxwell–Boltzmann statistics

The results derived in this section are special cases of a type of statistical distribution called Maxwell–Boltzmann statistics. Essentially, Maxwell–Boltzmann statistics assert that, in a system containing a large number of entities at temperature T, the probability of an entity having a particular energy E is proportional to $e^{-\frac{E}{kT}}$. Thus, we define the **Maxwell–Boltzmann distribution function** as follows:

$$F_{\text{M-B}}(E) = Ae^{-\frac{E}{kT}} \tag{12.10}$$

where A is a constant. This means that, for systems comprising a total of N entities obeying Maxwell–Boltzmann statistics, the fraction of these entities with energy E_i is given by

$$\frac{N_i}{N} = Ae^{-\frac{E_i}{kT}}$$

Since $\sum_i N_i = N$, a simple calculation shows that $A = \frac{1}{Z}$ where $Z = \sum_i e^{-\frac{E_i}{kT}}$ is called the *partition function*; in German, *Zustandssumme* (sum over states function).

Thus,

$$\frac{N_i}{N} = \frac{1}{Z}e^{-\frac{E_i}{kT}} = \frac{e^{-\frac{E_i}{kT}}}{\sum_i e^{-\frac{E_i}{kT}}} \tag{12.11}$$

Maxwell–Boltzmann statistics play an essential role in the understanding of classical systems (that is systems governed by Newtonian mechanics) which comprise large numbers of entities such as atoms, molecules, ions, *etc.* in thermal equilibrium. In particular, Maxwell–Boltzmann statistics enables us to predict, at least in principle, the thermodynamic behaviour of any macroscopic system for which we have a microscopic model.

For problems based on the material presented in this section visit up.ucc.ie/12/ and follow the link to the problems.

12.8 Microscopic interpretation of temperature and heat capacity in solids

In the microscopic model of solids, proposed in Section 10.14, each individual atom is considered as being bound to its nearest neighbours by short range forces which we approximate by helical springs. The general motion of an atom, therefore, can be thought of as the resultant of independent vibrational motion in each of the three directions indicated by the x-, y- and z-directions in Figure 12.10. As long as the amplitude of vibration is not too large, the motion in each of these directions is that of a simple harmonic oscillator. As we saw in Section 3.10, the energy of a simple harmonic oscillator is divided between kinetic energy and potential energy and is given by

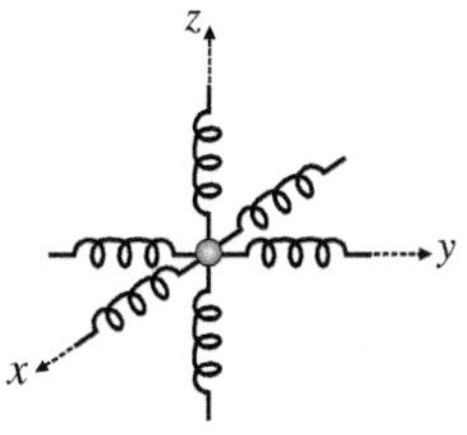

Figure 12.10. An atom in a solid is bound to its six nearest neighbours by spring-like forces. The motion of the atom is seen as a combination of three independent modes of vibration. Each mode of vibration has two degrees of freedom (corresponding to kinetic and potential energy), giving rise to a total of six degrees of freedom.

$$E = \frac{1}{2}mv^2 + \frac{1}{2}kx^2 = constant \tag{12.12}$$

Applying the principle of equipartition of energy to an ensemble of such we get

$$\left\langle \frac{1}{2}mv^2 \right\rangle = \left\langle \frac{1}{2}kx^2 \right\rangle = \frac{1}{2}E = \left\langle \frac{1}{4}kA^2 \right\rangle$$

where A is the amplitude of the oscillations. If we now apply the hypothesis that temperature is the macroscopic manifestation of average translational kinetic energy at the microscopic level, we obtain

$$T \propto \left\langle \frac{1}{2}mv^2 \right\rangle \propto \left\langle \frac{1}{4}kA^2 \right\rangle \propto \left\langle A^2 \right\rangle$$

In other words, the temperature of a solid is proportional to the average of the square of the amplitude of the atomic vibrations. The distribution in the energies of these vibrations is given by the Maxwell–Boltzmann distribution function, Equation (12.10).

We see from Equation (12.12) that there are two degrees of freedom (corresponding to kinetic energy and potential energy) for each of the three orthogonal directions, that is six degrees of freedom in all. Applying the principle of equipartition of energy to this case we find that the internal energy is given by $U = \frac{f}{2}NkT = \frac{6}{2}NkT = 3NkT$ and hence the heat capacity of a solid is $C_V = 3Nk$. Thus, the microscopic model we have adopted predicts that the molar heat capacity at constant volume of solids is given by

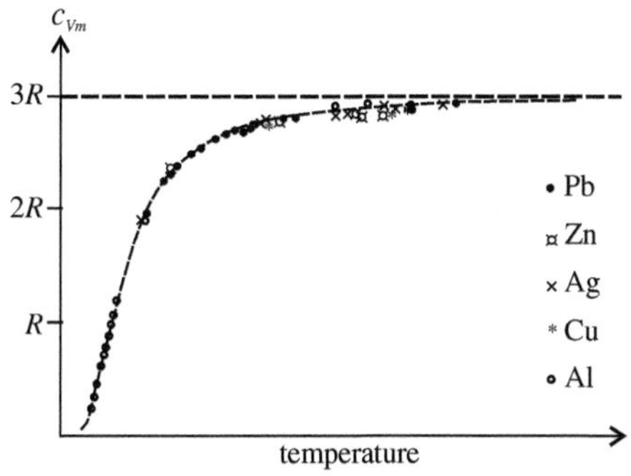

Figure 12.11. Experimental data showing how the molar heat capacities of different solids vary with temperature.

$$c_{V\mathrm{m}} = 3N_{\mathrm{A}}k = 3R$$

This last result, sometimes called the Dulong and Petit law, is obeyed quite well by solids at high temperatures. On the other hand, the measured values of molar heat capacities of solids at lower temperatures (< 1000 K for most solids) show significant variations from the Dulong–Petit prediction (as shown in Figure 12.11) and indeed it is found that for all solids $c_{Vm} \to 0$ as $T \to 0$ K. This problem arises not from the limitations of our simple microscopic model, as we might have suspected, but from limitations of the dynamical theory (Newtonian mechanics) that we have been using. In Section 14.4 we will see that the explanation of the observed specific heat capacities of solids requires a quantum mechanical treatment.

Further reading

Further material on some of the topics covered in this chapter may be found in *Statistical Physics* by Mandl (See BIBLIOGRAPHY for book details).

WORKED EXAMPLES

Worked Example 12.1: A helium bubble containing 4.00 mol of helium is submerged in water. When the temperature of the water and helium is increased by 15.0 °C at constant pressure the bubble expands. Treating the helium as an ideal gas, determine

(a) how much heat energy is added to the helium during the temperature increase,
(b) the change in the internal energy of the helium during the temperature increase and
(c) how much work is done by the helium as it expands against the pressure of the water during the temperature increase.

(a) In Section 12.1 the molar heat capacity at constant pressure is written in terms of the molar gas constant

$$c_{Pm} = \frac{5}{2}R$$

so that the heat energy added to the helium may be calculated from

$$\Delta Q = nc_{Pm}\Delta T = n\left(\frac{5}{2}R\right)\Delta T = 4.00 \times 2.5 \times 8.3 \times 15.0 = 1{,}245\ \text{J}$$

(b) The change in the internal energy may be calculated using Equation (12.4)

$$\Delta U = \frac{3}{2}Nk\Delta T = \frac{3}{2}nR\Delta T = 1.5 \times 4.00 \times 8.3 \times 15.0 = 747\ \text{J}$$

(c) The work done can then be calculated from the first law of thermodynamics (Equation 11.10),

$$\Delta W = \Delta Q - \Delta U = 1{,}245 - 747 = 498\ \text{J}.$$

Worked Example 12.2: In an experiment designed to measure the Boltzmann constant, Perrin studied the distribution of small particles of gum resin colloidally suspended in a liquid. The colloidal particles move in the liquid due to collisions with the liquid molecules ('Brownian motion') and may be considered to behave in a manner similar to that of an ideal gas in a gravitational field. Show that the number of particles per unit volume varies with height z in the liquid according to $n(z) = n_0 \exp\left(-\frac{(\rho_P - \rho_L)vghz}{kT}\right)$, where v is the volume of a particle and ρ_P and ρ_L are the densities of the resin and the liquid, respectively.

From the 'law of atmospheres' (Section 12.3), the number of particles per unit volume at a height z is given by

$$n(z) = n_0 \exp\left(-\frac{U(z)}{kT}\right)$$

where $U(z)$ is the potential energy at height z. In vacuum the potential energy would be mgz, but, in this case, the Archimedean upthrust must be taken into consideration in which case the effective weight is $mg - \rho_L vg$ and $U(z) = (\rho_P vg - \rho_L vg)z = (\rho_P - \rho_L)vgz$.

Hence

$$n(z) = n_0 \exp\left(-\frac{(\rho_P - \rho_L)vgz}{kT}\right)$$

Worked Example 12.3: In an exam the distribution of marks (that is the number of candidates with a mark x) is found to be described approximately by the following continuous function

$$N(x) = 33.6x \quad \text{for } 0 < x < 50$$

$$N(x) = 3780 - 42x \quad \text{for } 50 < x < 90$$

$$N(x) = 0 \quad \text{for } 90 < x < 100.$$

Calculate (i) the total number of candidates who took the exam and (ii) the average mark achieved.

(i) The total number of candidates taking the exam,

$$N_T = \int_0^{100} N(x)dx = \int_0^{50} 33.6x dx + \int_{50}^{90} (3780 - 42x)dx + \int_{90}^{100} 0 dx$$

$$\rightarrow \quad N_T = \left[16.8x^2\right]_0^{50} + \left[3780x - 21x^2\right]_{50}^{90} = 42{,}000 + [340{,}200 - 170{,}100] - [189{,}000 - 52{,}500] = 75{,}600$$

(ii) The average mark achieved,

$$\langle x\rangle = \frac{\int_0^{100} xN(x)dx}{N_T} = \frac{\int_0^{50} 33.6x^2 dx + \int_{50}^{90}(3780x - 42x^2)dx + \int_{90}^{100} 0 dx}{N_T}$$

$$= \frac{1}{75{,}600}\left([11.2x^3]_0^{50} + [1890x^2 - 14x^3]_{50}^{90}\right)$$

$$= \frac{1}{75{,}600}(1{,}400{,}000 + [15{,}309{,}000 - 10{,}206{,}000] - [4{,}725{,}000 - 1{,}750{,}000]) = 46.7$$

Worked Example 12.4: Using the Maxwell speed distribution, derive an expression (in the form of a definite integral) for the probability that a molecule will escape from the gravitational field of the planet. Show that this probability depends only on the ratio of the escape speed to the rms speed of the molecules.

Estimate $\frac{v_{esc}}{v_{rms}}$ for (a) hydrogen, (b) oxygen and (c) carbon dioxide molecules near the surface of the Earth at a temperature of 300 K.

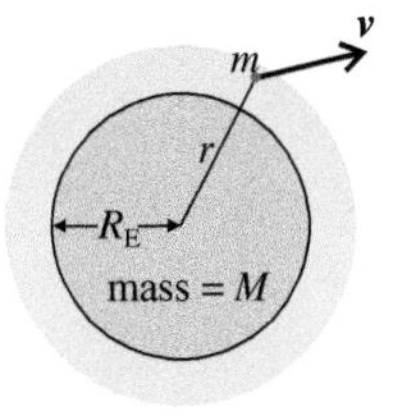

Figure 12.12. Worked Example 12.4.

The Maxwell speed distribution function (Equation (12.9)) is

$$P(v)\Delta v = 4\pi\left(\frac{m}{2\pi kT}\right)^{\frac{3}{2}} v^2 e^{-\frac{\frac{1}{2}mv^2}{kT}}\Delta v$$

The probability of a particle having sufficient speed to escape from the gravitational field (that is $v > v_{esc}$) is given by

$$P_{>v_{esc}} = 4\pi\left(\frac{m}{2\pi kT}\right)^{\frac{3}{2}}\int_{v_{esc}}^{\infty} v^2 e^{-\frac{\frac{1}{2}mv^2}{kT}}dv$$

Changing the variable in the integral as follows $\frac{mv^2}{2kT} = u^2;\quad v^2 = \frac{2kT}{m}u^2;\quad dv = \sqrt{\frac{2kT}{m}}du$, we get

$$P_{>v_{esc}} = 4\pi\left(\frac{m}{2\pi kT}\right)^{\frac{3}{2}}\left(\frac{2kT}{m}\right)^{\frac{3}{2}}\int_{u_0}^{\infty} u^2 e^{-u^2}du = \frac{4}{\sqrt{\pi}}\int_{u_0}^{\infty} u^2 e^{-u^2}du,\quad \text{where } u_0 = \sqrt{\frac{m}{2kT}}v_{esc}$$

Now

$$v_{rms} = \sqrt{\frac{3kT}{m}} \quad\rightarrow\quad \sqrt{\frac{m}{2kT}} = \frac{\sqrt{\frac{3}{2}}}{v_{rms}} \quad\rightarrow\quad u_0 = \sqrt{\frac{3}{2}}\frac{v_{esc}}{v_{rms}}$$

$$\rightarrow\quad P_{>v_{esc}} = \frac{4}{\sqrt{\pi}}\int_{u_0}^{\infty} u^2 e^{-u^2}du$$

Since the value of the integral depends on the limits only, the probability of escape from the gravitational field depends on the value of v_{esc}/v_{rms} only. Worked Example 5.4 shows that the escape speed of a molecule from the surface of a spherical planet of mass M and radius R_p is given by $v_{esc} = \sqrt{\frac{2GM}{R_p}}$ and, hence, the value of the ratio for a particle of mass m at a distance r from the centre of a spherical planet, such as the Earth (Figure 12.12), is given by

$$\frac{v_{esc}}{v_{rms}} = \frac{\sqrt{\frac{2GM}{r}}}{\sqrt{\frac{3kT}{m}}} \xrightarrow[M=M_E]{r=R_E} \sqrt{\frac{2GM_E}{3kTR_E}}\sqrt{m} = \sqrt{\frac{2(6.67\times10^{-11})(5.98\times10^{24})}{3(1.38\times10^{-23})(300)(6.37\times10^{6})}}\sqrt{m} = \begin{cases} 5.8 & \text{for } H_2 \\ 23.2 & \text{for } O_2 \\ 27.2 & \text{for } CO_2 \end{cases}$$

Worked Example 12.5: Using the Maxwell speed distribution, derive an expression for the corresponding distribution of momenta.

We can write down a Maxwell–Boltzmann distribution of momenta by replacing v in Equation (12.9) by $\frac{p}{m}$. Thus the fraction of molecules with momenta between p and $p + \Delta p$ is given by

$$\mathscr{P}(p)\Delta p = \frac{4\pi}{(2m\pi kT)^{\frac{3}{2}}}p^2 e^{-\frac{p^2}{2mkT}}\Delta p$$

PROBLEMS

For problems based on the material covered this chapter visit up.ucc.ie/12/ and follow the link to the problems.

13

Wave motion

AIMS

- to define quantities which can be used to characterise any one of the diverse range of wave motions which are observed in natural phenomena
- to derive an equation which describes the displacement produced by a simple wave – a *simple harmonic travelling wave* in one direction – at any position and time
- to show how waves from different sources can be superposed to produce cancellation or reinforcement of wave motion and thus to analyse *interference* of waves
- to examine how the wave frequency changes when there is relative motion between the source and/or the observer of the waves and the medium through which they are propagating – the *Doppler* effect
- to examine how the velocity of a wave motion depends on the properties of the medium through which it is propagating
- to describe how any wave form – continuous or pulsed – may be represented by a superposition of simple waveforms of different amplitudes and frequencies – *Fourier analysis*

13.1 Characteristics of wave motion

There are very many examples of wave motion in nature. One of the first to spring to mind is probably the wave motion which is observed when a water surface is disturbed. In this case waves move outwards across the water surface from the point of disturbance. Another example is wave motion along a string. If the end of a taut string is displaced rapidly and returned to its original position, as illustrated in Figure 13.1, the disturbance travels along the string, away from its source, as a single waveform – which we call a wave *pulse*. In both cases a disturbance is transmitted; this is an essential feature of any wave motion.

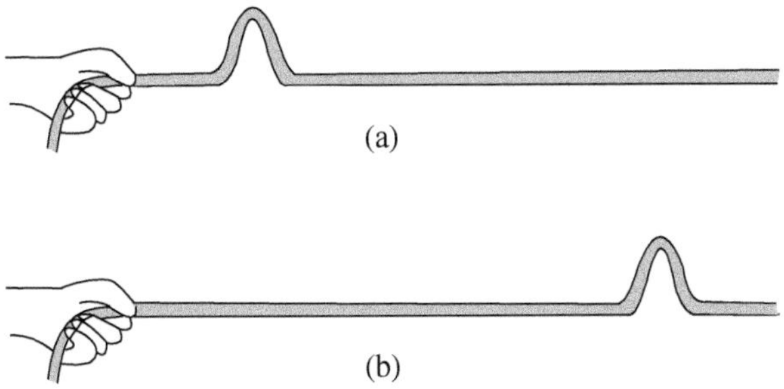

Figure 13.1. A wave pulse along a string.

For the disturbance to be transmitted in the two examples of wave motion which we have just considered, each particle in the medium (water or string) in which the wave motion is taking place must 'pass the message on' to its neighbour. For this to happen there have to be interactions between the particles which make up the medium. The velocity with which waves are propagated depends largely on the strength of these interactions.

In an elastic medium, a particle which has been disturbed by a wave pulse returns to its equilibrium position after the wave has passed. There is, therefore, no net motion of particles for a wave. This can be demonstrated in the case of a water wave by placing a cork on a

Understanding Physics, Third Edition. Michael Mansfield and Colm O'Sullivan.
© 2020 John Wiley & Sons Ltd. Published 2020 by John Wiley & Sons Ltd.

still water surface. As a wave passes the cork, it bobs up and down – oscillates – but it does not travel in the direction of the wave. In the case of the wave pulse which moves along the string, there is no net motion of string particles along the string. Although a wave does not transmit material, it clearly transmits energy. As the disturbance moves through the medium, the energy associated with the oscillation of particles in the medium also moves.

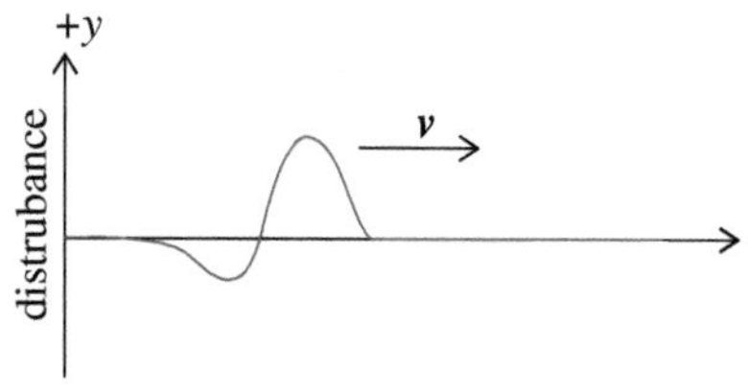

Figure 13.2. A transverse wave. The disturbance is perpendicular to the direction of the wave velocity.

Two categories of waves may be distinguished – *transverse and longitudinal*. In a **transverse** wave (Figure 13.2) the disturbance from equilibrium, y, is perpendicular to the direction in which the wave is propagating, that is, it is perpendicular to the direction of the **wave velocity**, $\mathbf{v}$. Waves across a water surface or waves along a string are both examples of transverse waves and Figure 13.2 can represent either phenomenon. As indicated in the figure, the disturbance from equilibrium can be positive or negative.

In a **longitudinal** wave the disturbance from equilibrium is parallel to the direction of propagation of the waves, the direction of their velocity. Sound waves in air, compression waves as illustrated in Figure 13.3, are longitudinal waves. The compression is caused by the source of the sound (for example, a vibrating membrane such as a vocal chord or a loudspeaker) and is carried as a compression wave; it can be detected when it disturbs another membrane (for example, an ear drum or a microphone). The disturbance and the direction of propagation of compression waves are therefore along the same axis but note that, as with transverse waves, there is no net motion of particles in the transmitting medium. Air molecules return to their equilibrium positions after a sound wave has passed. As might be expected, the transmission of sound waves depends on the elasticity of the air when it is compressed, that is, on its bulk modulus (Section 10.4). We shall investigate this dependence in Section 13.13.

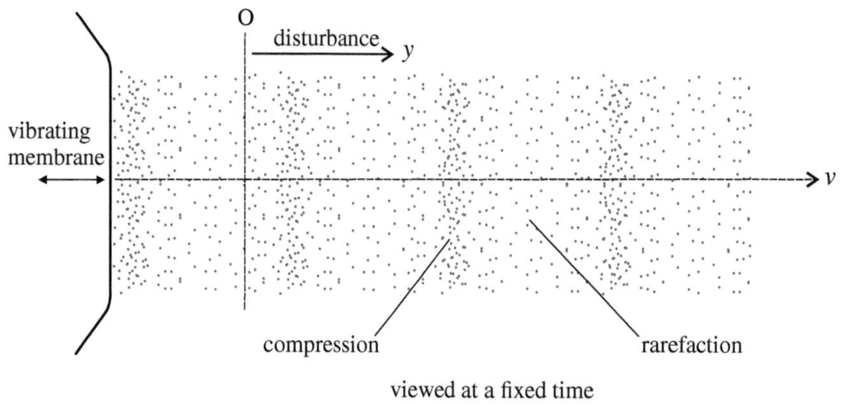

Figure 13.3. A longitudinal wave. As a sound wave travels through air, the disturbance of the air molecules is parallel to the direction of the wave velocity.

Although waves can be transverse or longitudinal, the formalism which we shall develop to describe wave motion may be applied equally well to either type of wave.

Sinusoidal waves

Initially, we shall find it easier to analyse wave motion by considering a continuing disturbance rather than a single wave pulse. Eventually, in Section 13.15, we will be able to describe a wave pulse in terms of a combination of continuing disturbances of a particular type, namely the steady sinusoidal wave motion which is produced by a source which is oscillating with simple harmonic motion. We begin by considering this type of motion.

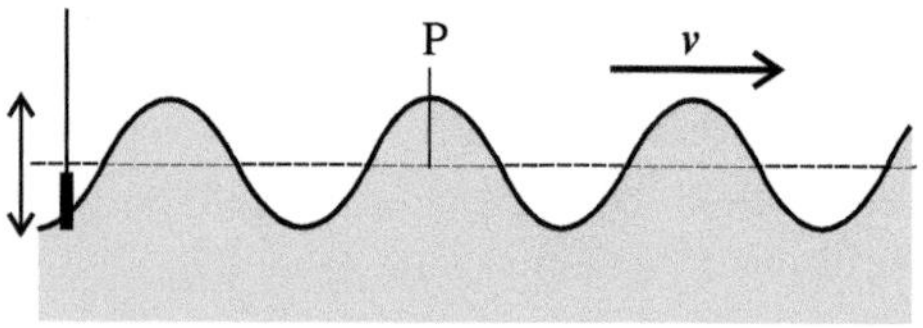

Figure 13.4. A sinusoidal wave across a water surface, viewed at a fixed time. The waves are produced by a plunger which is oscillating vertically with simple harmonic motion.

In the case of water waves, one source of sinusoidal waves could be a plunger which is oscillating vertically in the water surface as illustrated in Figure 13.4. We must first decide which features of the wave can be measured to characterise this type of wave motion.

Figure 13.5 is a plot of the disturbance of the water surface as a function of time, as observed at a fixed position, P in Figure 13.4, for example by observing the motion of a cork which is placed at P. Wave crests pass P at regular time intervals which, as in simple harmonic motion (Section 3.6), can be measured as the **period** of the wave, T, or as its **frequency**, f. As in any oscillatory motion (Section 3.5) it follows from their definitions that T and f are related through the equation

$$f = \frac{1}{T}$$

At the source the oscillatory motion is simply that of the source itself; at other points this motion is repeated, but at later times. At all points, therefore, the frequency of the waves is equal to the frequency of the source.

We can also plot the disturbance of the water surface as a function of position, as it is observed at a fixed time. Figure 13.6 is such a plot, a 'snapshot' of the profile of the water surface. In this case the characteristic quantity which can be defined to describe the observed periodicity of the wave motion is its **wavelength**, λ. The wavelength is the distance between successive points after which the waveform is repeated, for example between successive crests or successive troughs.

One complete wavelength, λ, passes the point P in Figure 13.4 in one period, T. We can therefore express the velocity of the waves in terms of these two quantities as follows:

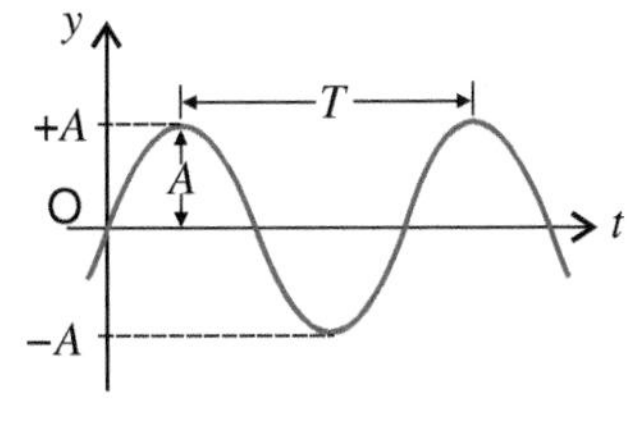

Figure 13.5. A sinusoidal wave across a water surface plotted as function of time. The wave is observed at a fixed position.

$$\boxed{v = \frac{\lambda}{T} = \lambda f} \tag{13.1}$$

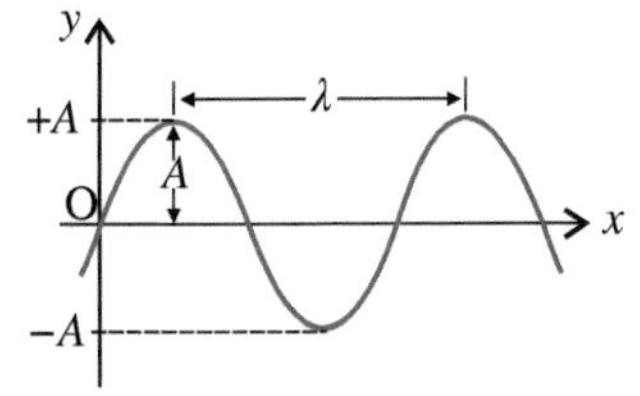

Figure 13.6. A sinusoidal wave across a water surface plotted as a function of position. The wave is observed at a fixed time.

As noted earlier, the wave velocity depends on interactions between the particles which make up the medium through which the waves are passing – it is therefore a characteristic of the medium and is fixed for a given medium in given conditions. The frequency is determined by the oscillation frequency of the source of the disturbance, with the wavelength adjusting to satisfy Equation (13.1).

A further characteristic of a wave motion is its **amplitude**. As in Section 3.5, the amplitude is defined as the maximum disturbance from equilibrium, indicated by A in Figures 13.5 and 13.6.

We have defined measurable quantities which can be used to characterise a wave motion. Our next task is to derive an equation which relates these quantities for the type of wave motion that is produced by a source which is oscillating with simple harmonic motion. We will then be able to predict the disturbance produced by the wave at any position and time.

Although we shall illustrate wave phenomena by considering wave motions with a wide variety of origins in nature, the description which will emerge may be applied to any type of wave motion. Light waves are an example of transverse electromagnetic waves; electromagnetic waves in general will be discussed in Chapters 21 and light in particular in Chapter 22. In this chapter electromagnetic wave phenomena will be used occasionally as examples of wave phenomena.

13.2 Representation of a wave which is travelling in one dimension

As waves pass through a medium along the x-axis, the disturbance y, which can be either transverse or longitudinal, varies with both position, x, and time, t; it can therefore be written as a function of these two quantities, that is $y = y(x, t)$.

In order to determine the form of the function $y(x, t)$ let us consider a wave in a taut string which is generated by a source which is oscillating with simple harmonic motion. The source could, for example, be a mass oscillating at the end of a spring, as considered in Section 3.6. Sinusoidal waves will be propagated along the string, as illustrated in Figure 13.7.

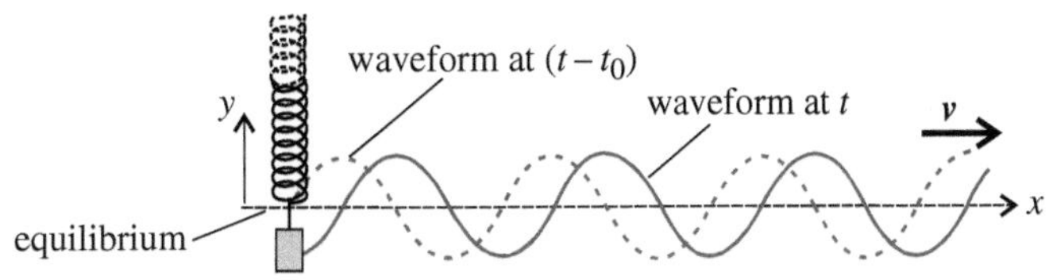

Figure 13.7. A sinusoidal wave along a taut string, as observed at time t (solid line) and at an earlier time ($t - t_0$, dashed line).

We already have an equation which gives the disturbance of the string at $x = 0$. This is the equation of simple harmonic motion (Equation (3.7a)). At $x = 0$, $y(x, t)$ becomes

$$y(0, t) = A\sin(\omega t + \phi)$$

where $\omega = \dfrac{2\pi}{T}$ (Equation (3.8)) which can be written in terms of frequency as $\omega = 2\pi f$. The value of the **phase constant** (or **phase angle**) ϕ (recall Section 3.6) is fixed by the choice of starting instant, that is by the value of y when $t = 0$.

The disturbance takes a time $t_0 = \dfrac{x}{v}$ to travel the distance x along the string. At any time t, therefore, the disturbance at a point x is the same as it was at $x = 0$ at the earlier time $(t - t_0)$. In mathematical terms we can write this condition as

$$y(x, t) = y(0, t - t_0)$$

Now

$$y(0, t - t_0) = A\sin[\omega(t - t_0) + \phi]$$

and thus

$$y(x,t) = A\sin[\omega(t - t_0) + \phi] = A\sin\left[\omega t - \omega\left(\frac{x}{v}\right) + \phi\right] = A\sin\left[\omega t - \left(\frac{\omega}{v}\right)x + \phi\right] \quad (13.2)$$

The quantity $\frac{\omega}{v}$ is termed the wave number or the propagation constant of the wave, that is, the **wave number** is defined as

$$\boxed{k := \frac{\omega}{v} = \frac{2\pi f}{v} = \frac{2\pi}{\lambda}} \quad (13.3)$$

and Equation (13.2) can be written as

$$\boxed{y(x,t) = A\sin(\omega t - kx + \phi)} \quad (13.4)$$

This is the equation we have been seeking. It is the **equation of a travelling (or progressive) simple harmonic (sinusoidal) wave in one dimension** – the $+x$ direction in this case. Equation (13.4) is very general and can be applied to any one-dimensional wave of any type which is produced by a source executing simple harmonic motion. If we know the values of A, ω, k and ϕ for a particular case, we can calculate the disturbance y at any position and time (see Worked Example 13.1).

Note that, using the definition of k which is given above, we can write Equation (13.1) for the wave velocity in terms of ω and k

$$v = f\lambda = \frac{\omega}{2\pi}\frac{2\pi}{k} = \frac{\omega}{k} \quad (13.5)$$

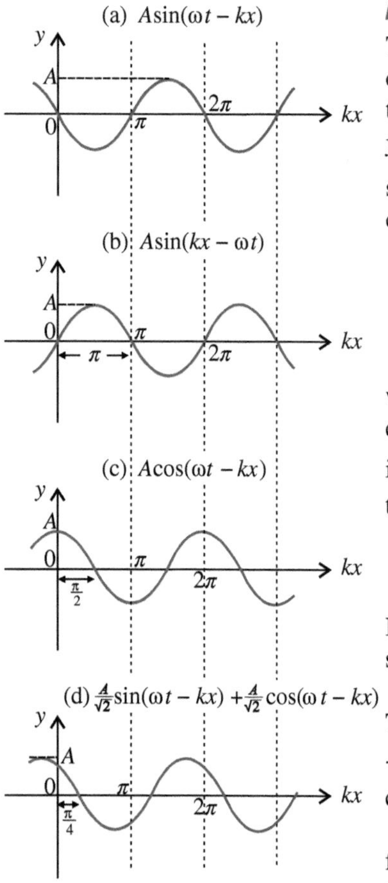

Figure 13.8. The waveform of a simple harmonic travelling wave in one dimension at $t = 0$ as represented by the following functions, (a) $A\sin(\omega t - kx)$, (b) $A\sin(kx - \omega t)$, (c) $A\cos(\omega t - kx)$ and (d) $A'\sin(\omega t - kx) + B'\cos(\omega t - kx)$, for the case in which $A'^2 = B'^2 = \frac{1}{2}A^2$.

Phase of a wave

The value of the phase constant ϕ in Equation (13.4) is determined by the initial conditions, that is by the value of y at $x = 0$ and $t = 0$. With a different choice of initial conditions, Equation (13.4) can equally well be represented by a cosine function, that is $y = A\cos(\omega t - kx + \phi')$, where $\phi' = \phi + \frac{\pi}{2}$. Furthermore, using the trigonometric identity, $\sin(\alpha + \beta) = \sin\alpha\cos\beta + \cos\alpha\sin\beta$ (Appendix A.3.5(viii)), where $\alpha = \omega t - kx$ and $\beta = \phi$, we can write Equation (13.4) as

$$y = (x,t) = A\sin(\omega t - kx + \phi) = A\sin(\omega t - kx)\cos\phi + A\cos(\omega t - kx)\sin\phi$$
$$= A'\sin(\omega t - kx) + B'\cos(\omega t - kx)$$

where $A' = A\cos\phi$ and $B' = A\sin\phi$. Thus we can also write the travelling wave equation as a linear combination of sine and cosine functions without using the constant ϕ, incorporating the phase information in the ratio of the coefficients of the sine and cosine functions, $\frac{A'}{B'} = \tan\phi$. In that case the amplitude is given by

$$A^2 = A^2\cos^2\phi + A^2\sin^2\phi = A'^2 + B'^2$$

In Figure 13.8 the equation of a travelling wave in one dimension is plotted at $t = 0$ as it is represented by four different functions.

The basic waveform is the same in each case and differs only in the starting point – the phase. The phase constant ϕ, as given in the general Equation (13.4), is 0 in case (a), $+\pi$ rad in case (b), $+\frac{\pi}{2}$ rad in case (c) and $+\frac{\pi}{4}$ rad in case (d). As is evident from Figure 13.8, any one of the waveforms can be made to coincide with any of the other three waveforms by adding an appropriate value of ϕ.

More generally, a travelling wave may be represented by any periodic function; that is, by any function which obeys the equation $f(kx \pm \omega t) = f(kx \pm \omega t \pm a)$ where a is the interval after which the waveform is repeated.

In deriving Equation (13.4), namely $y = (x, t) = A\sin(\omega t - kx + \phi)$, we assumed that the waves were travelling in the $+x$ direction, as indicated in Figure 13.7. We can confirm that this is the case by considering the motion of a point of constant disturbance, for example, a wave crest. For a constant value of the disturbance, $y(x, t)$ is constant and hence, from Equation (13.4), $(\omega t - kx + \phi)$ must be constant.

Thus $(\omega t - kx + \phi) = \text{constant}$ and, differentiating with respect to time we get

$$\omega - k\frac{dx}{dt} = 0 \Rightarrow v = \frac{dx}{dt} = +\frac{\omega}{k}, \text{ a velocity in the } +x \text{ direction.}$$

The same argument may be applied to any other form of the travelling wave equation. The general rule is that to represent waves in the positive x-direction, the kx and ωt terms in the oscillatory function must have opposite signs, as is the case for the functions (a) to (d) considered in Figure 13.8.

Similar arguments show that, in cases such as

$$y(x,t) = A\sin(\omega t + kx + \phi), \tag{13.6}$$

in which the kx and ωt terms in the oscillatory function have the same sign, that is $\omega t + kx + \phi = \text{constant} \Rightarrow v = \frac{dx}{dt} = -\frac{\omega}{k}$, the wave travels in the $-x$ direction.

Note that at a *fixed time* $t = t_1$, Equation (13.4) becomes

$$y = (x, t_1) = A\sin(\omega t_1 - kx + \phi) = A\sin(\theta_1 - kx + \phi)$$

where $\theta_1 = \omega t_1$, a constant. Thus at $t = t_1$ the waveform is shifted by a fixed amount θ_1 from the waveform at $t = 0$, namely $y = \sin(-kx + \phi)$, as illustrated in Figure 13.9; θ_1 is the phase shift of the wave at $t = t_1$, relative to $t = 0$.

Similarly at some *fixed position* $x = x_2$, Equation (13.4) becomes

$$y(x_2, t) = A\sin(\omega t - kx_2 + \phi) = A\sin(\omega t + \theta_2 + \phi)$$

where $\theta_2 = kx_2$, a constant. At $x = x_2$, therefore the wave exhibits simple harmonic motion but with a phase which is shifted by θ_2 from the simple harmonic motion performed at $x = 0$, namely $y = A\sin(\omega t + \phi)$. As waves travel along the string, particles in the string perform simple harmonic motion with the same amplitude and frequency but with different phase shifts whose values are determined by the value of x.

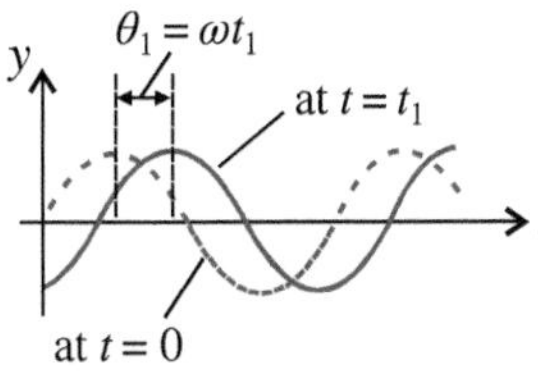

Figure 13.9. The waveforms of a simple harmonic travelling wave in one dimension, as observed at $t = 0$ (dashed line) and at a later time $t = t_1$ (solid line). θ_1 is the phase shift between the two waveforms.

Study Worked Example 13.1

For problems based on the material in this section visit up.ucc.ie/13/ *and follow the link to the problems.*

13.3 Energy and power in wave motion

As waves travel along a string or across a water surface, they cause particles in the medium to oscillate; they carry energy to such particles.

Masses which are travelling with a linear velocity can also carry energy (as kinetic energy). There are, however, notable differences between the ways in which waves and particles carry and thus transfer energy.

1. When waves transfer energy there is no transfer of matter. When particles with kinetic energy transfer energy, mass is also transferred.
2. When particles transfer energy, the energy is localised at the point occupied by each particle. When waves transfer energy, the energy is distributed over the space occupied by the wave disturbance.

We now derive an expression which gives the rate at which energy is transferred by *sinusoidal* waves – the *power* of the wave motion.

The power in the wave is the energy which passes a given point in the medium per unit time. It can be written as

$$P = (\text{wave energy in one wavelength}) \times (\text{number of wavelengths which pass per unit time})$$
$$= (\text{energy per wavelength}) \times (\text{frequency})$$

or

$$P = E_\lambda f \tag{13.7}$$

In Section 3.8 we showed (Equation (3.16)) that when a particle of mass m is performing simple harmonic motion with angular frequency ω and amplitude A, its total energy is constant and is given by $E = \frac{1}{2}m\omega^2A^2$. Hence the total energy of a particle of mass m, which is oscillating as a component part of a wave motion, is also $\frac{1}{2}m\omega^2A^2$. For waves along a string, for example, if μ is the mass per unit length of the string, the mass in one wavelength is $\mu\lambda$ and the total energy in one wavelength is $E_\lambda = \frac{1}{2}\mu\lambda\omega^2A^2$. Thus, from Equation (13.7), the power carried by such waves is given by $P = \frac{1}{2}\mu\lambda\omega^2A^2f$

Now $v = \lambda f$ (Equation (13.1)) so that

$$P = \frac{1}{2}\mu\omega^2 A^2 v$$

The power carried by the waves is therefore proportional to the square of the wave amplitude and proportional to the wave velocity, that is,

$$P \propto vA^2 \tag{13.8}$$

This is a general result which is applicable to any simple harmonic wave motion.

A similar result applies in the case of a uniform stream of mono-energetic particles. If there are n particles per unit length and each particle has velocity v and energy E_p, the energy, ΔE, passing any point in a time Δt is the energy within an element of length $v(\Delta t)$, that is $\Delta E = nE_p v(\Delta t)$ or

$$P = \frac{dE}{dt} = vnE_p$$

so, in this case,

$$P \propto vn$$

This suggests that, in the case of a stream of particles, the particle density plays a similar role to the (amplitude)^2 in the case of a sinusoidal wave.

13.4 Plane and spherical waves

In three dimensions, a travelling wave which propagates in one direction only is known as a **plane wave** because, as illustrated in Figure 13.10, the **wavefronts** (surfaces of constant phase, wave crests for example) are parallel planes perpendicular to the direction of propagation of the wave, the direction in which energy is carried. For plane waves energy may be considered to be carried in straight lines, as in the case of particles. For light waves this model of wave propagation leads to the idea of a light ray, a concept which is used extensively in geometrical optics (Chapter 23)

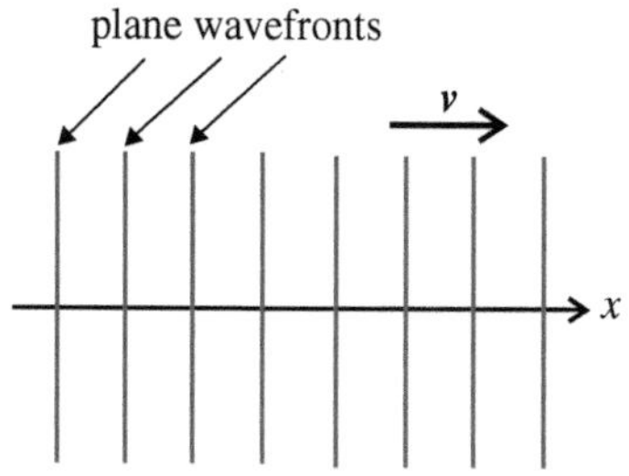

Figure 13.10. The plane wavefronts of a (one dimensional) travelling wave. The wavefront planes are perpendicular to the page and to $\boldsymbol{v}$.

In other situations, the waves spread out uniformly in three dimensions from a localised source, producing concentric spherical wavefronts, as illustrated in Figure 13.11. The flow of wave energy is an example of a flux, as discussed in Section 5.3.

In Section 5.3 total flux was defined in terms of the strength of the source so that the total flux from a source of waves can be equated to the power of the source, P, measured in watts. As illustrated in Figure 13.11, the **wave intensity** at a point X (that is, the flux density at X) is the proportion of P which passes through a unit area perpendicular to the direction of propagation of the waves – the direction of the wave velocity – at that point.

Thus, for spherical waves, the intensity at a distance r from the point source is given by

$$I(r) = \frac{P}{4\pi r^2}$$

The wave intensity of spherical waves, therefore, falls off with the inverse square of distance from the source. The SI unit of wave intensity is W m^{-2}.

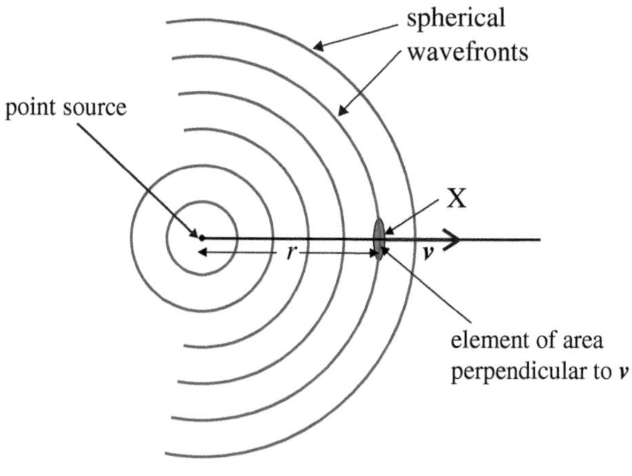

Figure 13.11. Concentric spherical wavefronts from a point source, viewed at a fixed time as they propagate in three dimensions.

The intensity of sound waves

In the case of sound waves a unit of relative intensity is defined to accommodate the fact that the human ear responds to relative increases in sound intensity rather than to absolute increases; the apparent loudness of a sound depends on the level of background sound. If the intensity of sound waves increases from I_1 to I_2, the relative intensity is defined in **bel** (B) as $\log_{10}\frac{I_2}{I_1}$. A logarithmic scale is used because the human ear can respond to sound over a large range of intensities, from the threshold of hearing (10^{-12} W m^{-2}) to 1 W m^{-2}. In practice, sound intensity is stated with reference to the threshold of hearing and in decibel units (dB), where 1 B = 10 dB. Thus relative sound intensity in dB is given by $10\log_{10}\frac{I}{I_0}$, where $I_0 = 10^{-12}$ W m^{-2}. The threshold of hearing sound is therefore stated as 0 dB, and the sound intensity of normal conversation, typically 10^{-6} W m^{-2}, corresponds to a relative intensity of about 60 dB.

For problems based on the material in this section visit up.ucc.ie/13/ and follow the link to the problems.

13.5 Huygens' principle: the laws of reflection and refraction

The progress of a wave motion can be analyzed using **Huygens' principle** (Christiaan Huygens, 1629–1695) which states that:

each point on a wavefront can be considered as a point source which is responsible for the subsequent progress of the wave.

The principle is illustrated in Figure 13.12, where a wavefront is produced at the envelope of (the common tangent to) all the wavefronts which are produced by point sources on the previous wavefront.

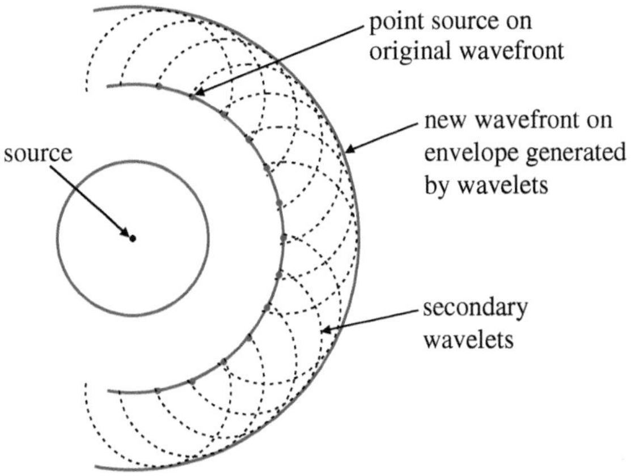

Figure 13.12. The progress of waves from a point source as described by Huygens' principle. A wavefront is produced along the envelope (solid blue line) which is generated by wavelets from all points on the previous wavefront.

We now proceed to use Huygens' principle to derive the *laws of reflection and refraction*, the laws which describe how waves behave when they encounter a barrier or an interface between two media.

The law of reflection

Consider plane waves which are incident on a plane surface, as illustrated in Figure 13.13. The *angle of incidence* θ_i is the angle between the direction of propagation of the waves and the normal to the surface. The wave front AB is perpendicular to the direction of propagation of the waves so that $\theta_i + \angle NAB = 90°$; $\angle NAB + \angle BAC = 90°$ and therefore $\angle BAC = \theta_i$.

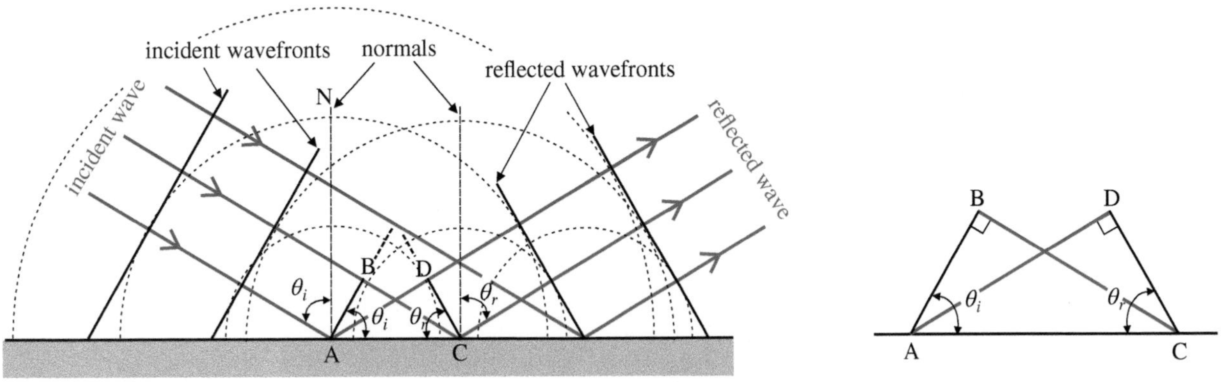

Figure 13.13. The reflection of a plane wave at a plane surface.

According to Huygens' principle, as each point on the wavefront BA reaches the plane surface, secondary wavelets are produced which intersect to form an envelope CD, as indicated in Figure 13.13; CD is the wavefront of the reflected waves. The *angle of reflection* θ_r is the angle between the direction of propagation of the reflected waves and the normal to the surface. Using arguments similar to those used above to show that $\angle BAC = \theta_i$, we can show that $\angle DCA = \theta_r$, the angle of reflection.

Because the incident and reflected waves are travelling in the same medium, and therefore with the same velocity, waves from B will reach C at the same time as waves from A reach D. Hence BC = AD and the triangles BAC and DCA are congruent. It follows that

$$\boxed{\theta_i = \theta_r} \tag{13.9}$$

We can now state the **law of reflection**:

> *The angles of incidence and reflection of waves which are incident on a plane surface are equal, with the reflected waves propagating in the plane which is defined by the incident wave and the normal to the reflection surface at the point of incidence (that is, in the plane of the page in Figure 13.13).*

The law of refraction

Next, we consider plane waves which are incident at an angle θ_1 on a plane interface between two media 1 and 2 and pass into the second medium, as illustrated in Figure 13.14. The wavefront AB is perpendicular to the direction of propagation of the waves so that $\angle BAC$ is equal to θ_1.

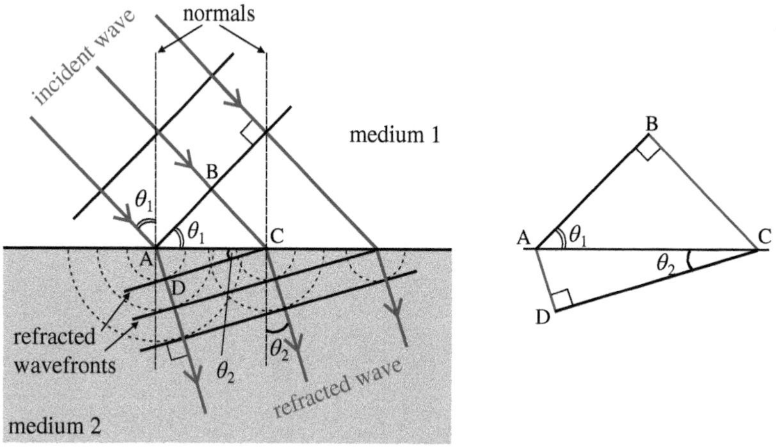

Figure 13.14. The refraction of a plane wave at a plane interface between two transparent media.

According to Huygens' principle, as each point in turn on the wavefront AB reaches the interface, secondary wavelets are produced which intersect to form an envelope at CD as indicated in Figure 13.14; CD, therefore, is a wavefront of the refracted waves (the waves transmitted into the second medium). The angle of refraction θ_2 is the angle between the direction of propagation of the refracted waves and the normal to the interface; therefore θ_2 is equal to $\angle ACD$.

In this case the velocities of the waves in the two media, v_1 and v_2 respectively, are not the same so that BC $\neq$ AD. However, if t is the time taken for wavelets from B to reach C, this is also the time taken for wavelets from A to reach D, so that we can write

$$\frac{BC}{AD} = \frac{v_1 t}{v_2 t} = \frac{BC}{AC}\Big/\frac{AD}{AC} = \frac{\sin\theta_1}{\sin\theta_2} \quad \rightarrow \quad \frac{\sin\theta_1}{v_1} = \frac{\sin\theta_2}{v_2}$$

We can therefore state **the law of refraction** as follows:

> *The direction of the refracted wave lies in the plane defined by the incident wave and the normal to the interface at the point of incidence and the angle of incidence, θ_1, and the angle of refraction, θ_2, are related through the equation*
>
> $$\frac{\sin\theta_1}{v_1} = \frac{\sin\theta_2}{v_2} \tag{13.10}$$

This law was first discovered by the scientist and engineer Ibn Sahl (c. 940–1000). It is also known as Snell's law (Willebrand Snellius 1580–1626) or Descartes' law (René Descartes (1596–1650).

We will return to the laws of reflection and refraction when we consider the propagation of light waves in Chapter 23.

13.6 Interference between waves

We now consider a situation in which a medium is subjected to two wave motions simultaneously. This can be analysed using the **principle of superposition** which states that:

> *if disturbances are produced by two wave motions, the resultant (total) disturbance produced by the two acting together at a point is the algebraic sum of the disturbances acting on their own.*

Thus, when two wave motions, $y_1(x, t)$ and $y_2(x, t)$, are superposed, the total disturbance is given by

$$y(x,t) = y_1(x,t) + y_2(x,t) \tag{13.11}$$

When we come to consider the *wave equation* in Section 13.10 we will see that the principle of superposition follows from the form of that equation.

Two sources vibrating in phase

In Figure 13.15 and Figure 13.16 we illustrate two cases in which a fixed point is subjected, simultaneously, to two sinusoidal waves of the same amplitude and frequency.

In the first case (Figure 13.15) the two waves are in phase at that point so that addition of the disturbances produces a resultant disturbance of the same frequency as the component waves but with twice the amplitude. This effect is known as **constructive interference.**

In the second case (Figure 13.16) the two waves are out of phase by half a period. The disturbances cancel, producing no net disturbance at the fixed point. This effect is known as **destructive interference**.

Destructive interference can also occur if the sources of the waves oscillate in phase (that is, they have the same value of ϕ at $t = 0$) but are positioned so that their distances from the fixed point differ by half a wavelength. Waves from the two sources are then out of phase by half a period at the fixed point. As illustrated in Figure 13.17, the type of interference observed at a point P depends on $(x_2 - x_1)$, the difference between the lengths of the paths travelled by the waves from the two sources S_1 and S_2.

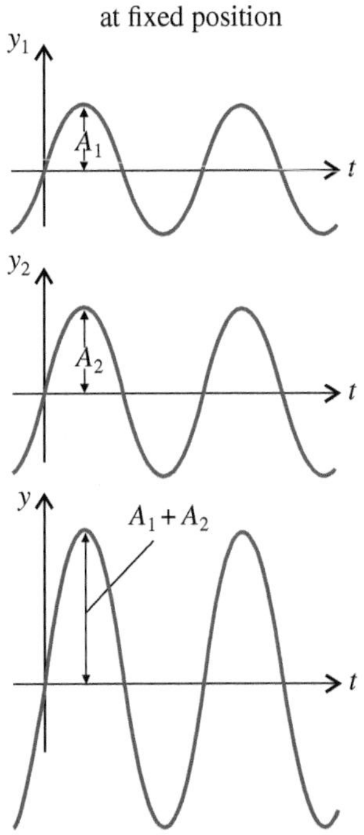

Figure 13.15. Constructive interference between two waves which are in phase at a fixed point. The net disturbance $y = y_1 + y_2$.

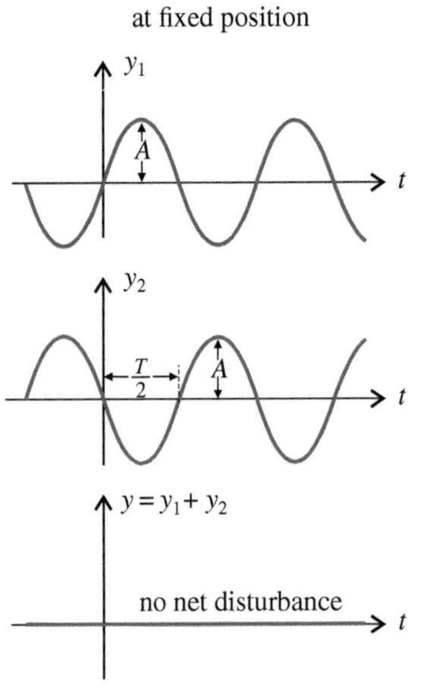

Figure 13.16. Destructive interference between two waves which are out of phase by half a period at a fixed point. The net disturbance $y = y_1 + y_2 = 0$.

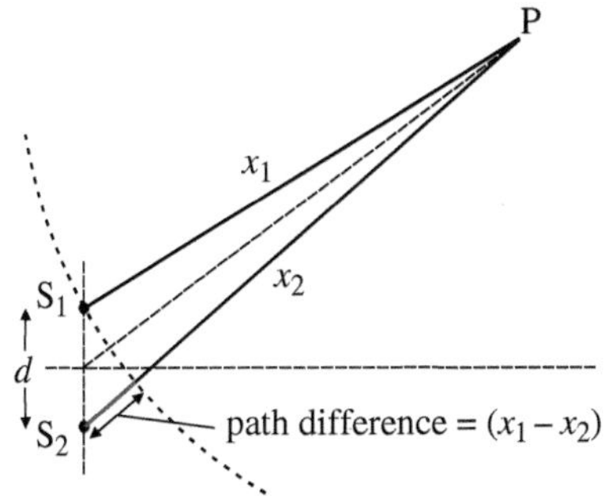

Figure 13.17. The difference in the lengths of the paths travelled to a point P by waves from two sources S_1 and S_2.

Thus, a path difference of one wavelength corresponds to a phase difference of 2π while a path difference of half a wavelength corresponds to a phase difference of π.

Note that colliding (superposed) particles cannot cancel one another to produce nothing so that interference is exclusively a wave property. When a phenomenon exhibits interference effects this is clear evidence of its wave nature.

Interference pattern produced by two synchronous sources

Let us now consider the interference produced by two point sinusoidal sources, S_1 and S_2 in Figure 13.17, of the same amplitude, frequency and phase – known as *synchronous* sources. The disturbances produced at the point P by waves from S_1 and S_2 may be represented by travelling wave equations (Section 13.2). Applying Equation (13.4) these are, respectively,

$$y_1 = A\sin(kx_1 - \omega t + \phi)$$

and

$$y_2 = A\sin(kx_2 - \omega t + \phi)$$

Applying the superposition principle, the total disturbance at P is given by

$$y = y_1 + y_2 = A[\sin(kx_1 - \omega t + \phi) + \sin(kx_2 - \omega t + \phi)]$$

Using the trigonometric identity Appendix A.3.5(xi), namely $\sin\alpha + \sin\beta = 2\sin\left(\frac{\alpha+\beta}{2}\right)\cos\left(\frac{\alpha-\beta}{2}\right)$,
where $\alpha = (kx_2 - \omega t + \phi)$ and $\beta = (kx_1 - \omega t + \phi)$
we can write

$$y = 2A\sin\left[\frac{1}{2}k(x_2 + x_1) - \omega t + \phi\right]\cos\left[\frac{1}{2}k(x_2 - x_1)\right] \tag{13.12}$$

or, using Equation (13.3),

$$\begin{aligned} y &= 2A\sin\left[\frac{\pi}{\lambda}(x_2 + x_1) - \omega t + \phi\right]\cos\left[\frac{\pi}{\lambda}(x_2 - x_1)\right] \\ &= \left\{2A\cos\left[\frac{\pi}{\lambda}(x_2 - x_1)\right]\right\}\sin\left[\frac{x}{\lambda}(x_2 + x_1) - \omega t + \phi\right] \end{aligned}$$

which represents a travelling wave of amplitude

$$A' = 2A\cos\left[\frac{\pi}{\lambda}(x_2 - x_1)\right] = 2A\cos\left[\frac{1}{2}k(x_2 - x_1)\right] \tag{13.13}$$

Since, from Equation (13.8), the intensity of a wave is proportional to the square of its amplitude, the intensity varies as a (cosine)^2 function of $(x_2 - x_1)$, namely

$$I \propto 4A^2\cos^2\left(\frac{1}{2}k(x_2 - x_1)\right)$$

Consider the form of Equation (13.13) in the following two situations:

Situation (a) When the path difference is an integral number of wavelengths, that is,

$$(x_2 - x_1) = n\lambda = \frac{2\pi n}{k} \text{ where } n = 0, 1, 2\ldots \text{ (an integer)}$$

In this case $A' = 2A\cos\left[\frac{\pi}{\lambda}n\lambda\right] = 2A\cos n\pi = \pm 2A$

The magnitude of the amplitude of the travelling wave is twice that of either of the original waves. For situation (*a*), there is constructive interference and the intensity at a point P is given by

$$I_P \propto [A']^2 = 4A^2$$

Situation (b) When the path difference is a half integral number of wavelengths,

$$(x_2 - x_1) = (2n+1)\frac{\lambda}{2} = (2n+1)\frac{\pi}{k} \text{ where } n = 0, 1, 2\ldots \text{ (an integer)}$$

In this case $A' = 2A\cos\left[\frac{\pi}{\lambda}(2n+1)\frac{\lambda}{2}\right] = 2A\cos\left[(2n+1)\frac{\pi}{2}\right] = 0$

Thus $y = 0$ always. Situation (*b*), therefore, results in destructive interference.

In Figure 13.18, the positions of points of constructive and destructive interference, as obtained in situations (*a*) and (*b*) for the path difference $(x_2 - x_1)$, are plotted for two sources, S_1 and S_2, of the same amplitude, frequency and phase. On the right of the figure the wave intensity is plotted in the *z*-direction, a direction perpendicular to the line OO′, which is equidistant from the two sources. Alternating maxima and minima of intensity are seen – an **interference pattern** – a characteristic feature of interference between waves.

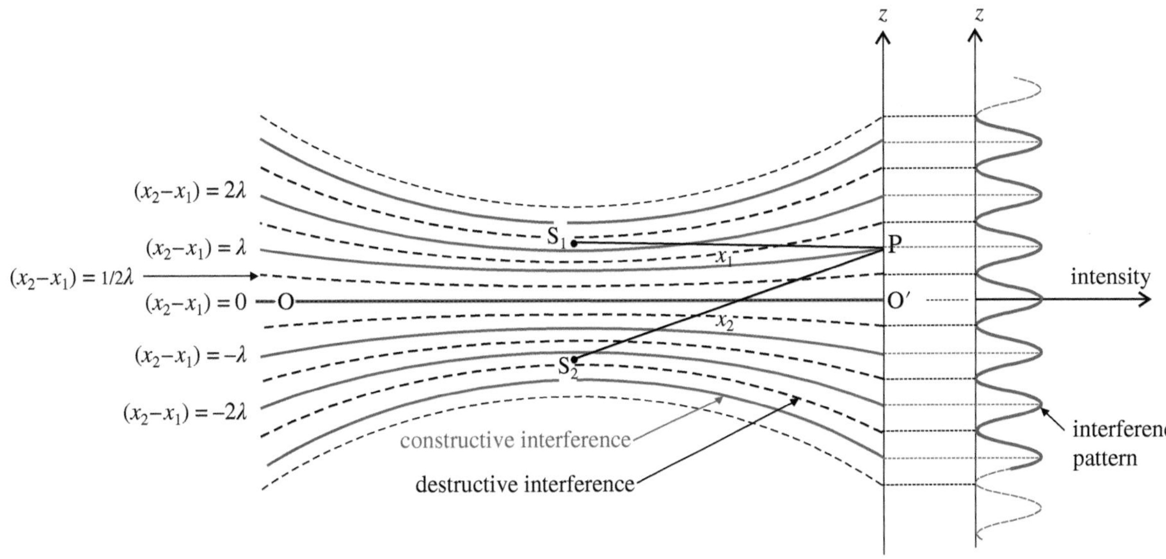

Figure 13.18. The interference pattern produced by two synchronous sources. The solid coloured lines show where constructive interference occurs and the dashed lines show where destructive interference occurs. On the right of the figure wave intensity is plotted as a function of z, the distance along a line which is perpendicular to OO′, the line which is equidistant from each source.

Synchronous sources viewed at a distance

We now derive an expression for the condition that constructive interference occurs at a point at a large distance away from the pair of point synchronous sources S_1 and S_2. Consider the point P in Figure 13.19(a) which is at a distance r from the mid-point C between the sources. The line from C to P makes an angle θ with respect to the normal to the line joining the sources, as shown in the figure.

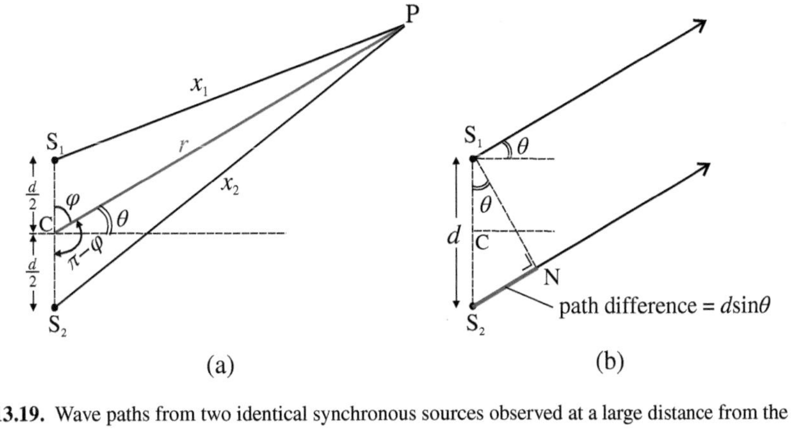

Figure 13.19. Wave paths from two identical synchronous sources observed at a large distance from the sources.

Applying the cosine rule (Appendix A.3.3) to the triangle PS_1C, we obtain

$$x_1^2 = r^2 + \left(\frac{d}{2}\right)^2 - 2r\left(\frac{d}{2}\right)\cos\varphi = r^2\left(1 - 2\left(\frac{\frac{d}{2}}{r}\right)\cos\varphi + \frac{\left(\frac{d}{2}\right)^2}{r^2}\right)$$

where $\varphi = \frac{\pi}{2} - \theta$, as indicated in the figure. Since we are interested in the case in which $d \ll r$, we can neglect the term $\frac{d^2}{r^2}$, which is small compared to $\frac{d}{r}$. Thus

$$x_1^2 \approx r^2\left(1 - 2\left(\frac{\frac{d}{2}}{r}\right)\cos\varphi\right) \quad \Rightarrow \quad x_1 \approx r\sqrt{1 - \frac{d}{r}\cos\varphi}$$

Expanding the square root term as a binomial series (Appendix A.9.2) and dropping terms of order $\frac{d^2}{r^2}, \frac{d^3}{r^3}$, etc. we get

$$\left(1-\frac{d}{r}\cos\varphi\right)^{\frac{1}{2}} = 1+\left(\frac{1}{2}\right)\left(-\frac{d}{r}\cos\varphi\right)+\frac{1}{2!}\left(\frac{1}{2}\right)\left(-\frac{1}{2}\right)\left(-\frac{d}{r}\cos\varphi\right)^2 \ldots \approx 1-\frac{1}{2}\frac{d}{r}\cos\varphi$$

and hence

$$x_1 = r\left(1-\frac{1}{2}\frac{d}{r}\cos\varphi\right)$$

Similarly, using the triangle PS_2C,

$$x_2 = r\left(1-\frac{1}{2}\frac{d}{r}\cos(\pi-\varphi)\right) = r\left(1+\frac{1}{2}\frac{d}{r}\cos\varphi\right)$$

Thus, the path difference is given by

$$x_2 - x_1 = r\left(1+\frac{1}{2}\frac{d}{r}\cos\varphi\right) - r\left(1-\frac{1}{2}\frac{d}{r}\cos\varphi\right) = d\cos\varphi = d\sin\theta$$

and the condition for constructive interference at the point P can be written as

$$\boxed{d\sin\theta = n\lambda} \qquad (n = 0, \pm1, \pm2, \pm3, \ldots) \tag{13.14}$$

Note that this result may be obtained more directly when the interference pattern is viewed from an infinite distance from the sources. In this case, the paths taken by the waves from each source may be considered to be effectively parallel, as illustrated in Figure 13.19(b). Since $\sin\theta = \frac{NS_2}{S_1S_2}$, the difference in the paths travelled to the observer by the two waves is $d\sin\theta$, as obtained above.

As noted earlier in this section, the intensity of the interference pattern varies with the square of the amplitude A', given by Equation (13.13), so that we can write

$$\text{intensity} \propto A^2\cos^2\left[\frac{1}{2}k(x_2-x_1)\right] = A^2\cos^2\left(\frac{1}{2}kd\sin\theta\right)$$

Linear array of synchronous sources

Let us now consider the situation in which, instead of the two synchronous sources considered above, we have a linear array of such sources, equally spaced as illustrated in Figure 13.20.

If the inter-source spacing is d, we can see from the figure that, if we view the waves emitted at an angle θ to the normal to the line of the array at a large distance from the source, the path difference between the waves from any two adjacent sources will be $d\sin\theta$. Thus the waves from all the sources interfere constructively when observed at the angle θ if $d\sin\theta = n\lambda$ (n is an integer), which is the same condition (Equation (13.14)) as that derived above for two sources with the same separation d.

There is, however, a significant difference in the interference pattern observed for a linear array comprising many point sources compared to that observed for just two sources. In the case of two sources, if the intensity is observed at an angle slightly different from that for maximum intensity, the observed intensity is not much less than that at the maximum; in other words, the interference between the two waves is still almost constructive and thus the intensity falls off relatively slowly. We can understand the different interference pattern that is observed when there are many sources by considering what happens if the angle θ in Figure 13.20 is changed very slightly. While the waves emanating from adjacent sources will still be almost in phase, waves from more distant sources along the array will be increasingly out of phase. The net effect, when the number of sources is large, is that the waves from almost all the sources will interfere destructively with one another everywhere except at angles θ *exactly* satisfying the condition for maximum intensity, namely $d\sin\theta = n\lambda$.

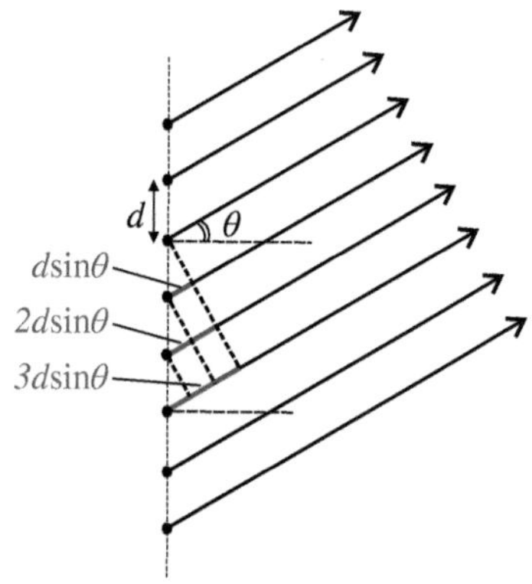

Figure 13.20. Linear array of equidistant sources vibrating in phase.

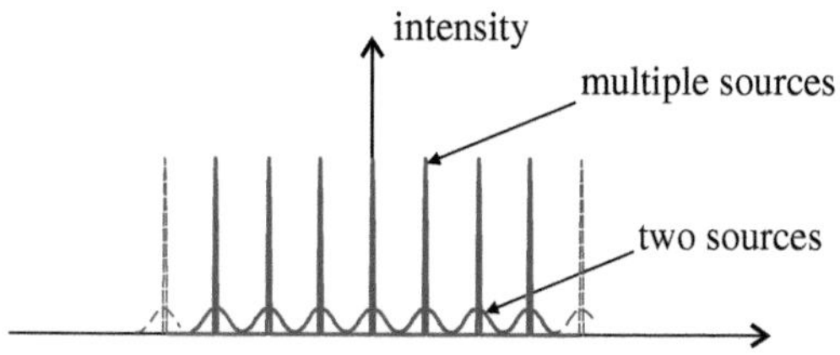

Figure 13.21. Comparison of the intensity pattern resulting from interference between waves from two identical synchronous point sources with that from a linear array of such sources with the same inter-source separation.

Thus, in the case of an array of sources, the observed intensity maxima are very sharp and narrow (Figure 13.21) compared with that observed for just two sources. Between these sharp maxima a very weak interference pattern occurs but it is only observable when the number of sources is small.

The geometrical derivation given above was confined to the plane defined by the points P, S_1 and S_2 (Figure 13.19). Point sources, however, radiate in all directions so that constructive interference will also occur at points not confined to that particular plane, provided that the path difference remains an integral number of wavelengths. As shown in Figure 13.22, because of symmetry about the line of the array, interference maxima corresponding to different values of n lie on circles of fixed latitude on a spherical surface.

For an array of point sources to vibrate in phase, they must be driven into oscillation by an appropriate source of energy applied in a synchronous manner; for example, an array of small loudspeakers driven from the same audio source behaves in this way. An entirely equivalent situation arises if an array of non-energised oscillators are set in vibration through exposure to an incident plane wave. An example of interference produced in this way will be encountered when we examine interference between electron waves in Section 14.8 and study X-ray diffraction in Section 22.6. Note that interference will occur in the backward as well as the forward direction relative to the incident plane wave beam.

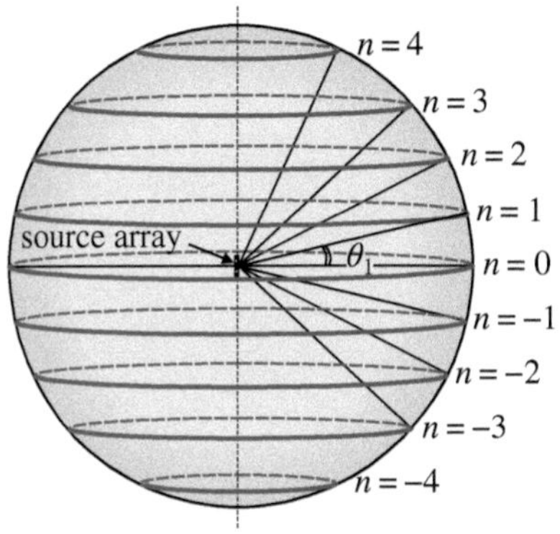

Figure 13.22. Location of the interference maxima, for an array of equally spaced point sources, on a sphere whose radius is very much greater than the distance between adjacent sources.

Study Worked Example 13.2

For problems based on the material in this section visit up.ucc.ie/13/ and follow the link to the problems.

13.7 Interference of waves passing through openings: diffraction

Two-slit interference

In some cases (in particular for light waves) it is not always possible to construct two independent sources of waves which operate in phase. In such cases it is possible, however, to produce two sources which oscillate with the same phase starting with a single source. Consider the progress of a wave motion emitted from a single primary source when it encounters two very narrow openings in a plane barrier which are equidistant from the source, as illustrated in Figure 13.23.

According to Huygens' principle (Section 13.5), the wavefronts which reach the openings act as secondary sources. Because the secondary sources follow any phase changes in the primary source faithfully, they are always in phase and can produce interference effects.

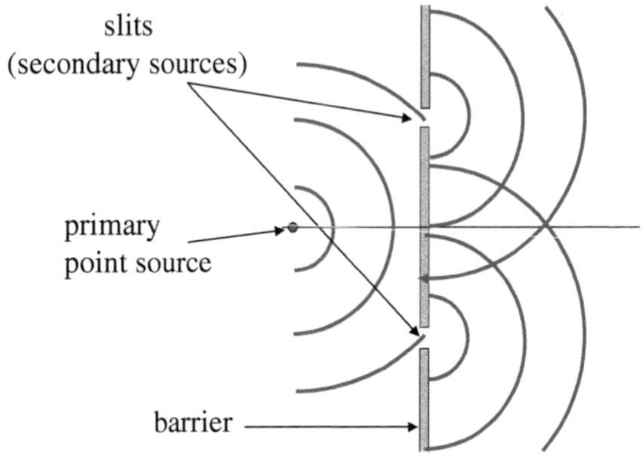

Figure 13.23. A double-slit arrangement: the slits act as two sources oscillating in phase.

At any position to the right of the barrier in Figure 13.23, the phase difference between waves is determined by the difference in the path lengths from the two secondary sources. The phase difference at any position, therefore, is fixed and the secondary sources produce an interference pattern which is similar to that produced by two independent sources of coherent waves (Figure 13.18). Such interference effects may be observed, for example, when water waves approach a barrier in which there are two openings; waves striking the barrier will be reflected while those passing through the openings will produce an interference pattern on the other side of the barrier.

For simplicity, we consider only cases in which the incident waves are plane waves (that is, the primary source is a long way from the slits) with wavefronts parallel to the barrier, as illustrated in Figure 13.24, and in which the interference effects are observed at a large distance from the source. The situation is equivalent to that shown in Figure 13.19 and, if the distance between the slits is d and the angle at which the transmitted waves are observed is θ, the condition for constructive interference is again given by Equation (13.14).

$$d\sin\theta = n\lambda \quad (n = 0, \pm 1, \pm 2, \pm 3, \ldots)$$

As before, the wave intensity varies as a (cosine)^2 function of path difference and alternating maxima and minima of intensity are observed. Note that, if the wavelength $\lambda \ll d$, the angle θ is very small for all n so that it can be difficult to resolve adjacent maxima in the interference pattern. Interference effects in waves are observed most easily when wavelengths are of the order of, but not greater than, the slit separation. Note that $\lambda > d$ gives $\sin\theta > 1$ for all n so that, in this case, interference maxima are not observed.

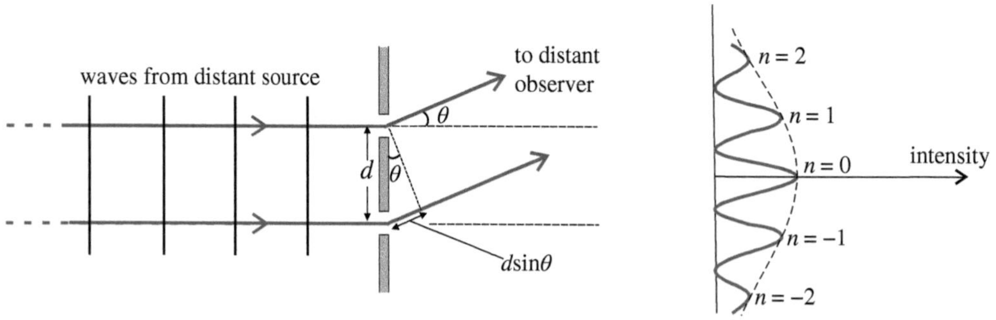

Figure 13.24. A double-slit arrangement and the interference pattern observed at a large distance from the slits.

In general, when waves encounter apertures or obstacles of dimensions which are much smaller than their wavelength interference effects are negligible. The wave motion continues to propagate as though it had not 'seen' the obstacle

Diffraction

We have seen that, according to Huygens' principle, each point on a wavefront acts as a point source.We now consider what happens when waves are incident on an opening (an aperture) of finite size in a barrier. Account must be taken of interference effects between waves emanating from the *different points on a wavefront* as it passes through a slit, as illustrated for plane waves in Figure 13.25.

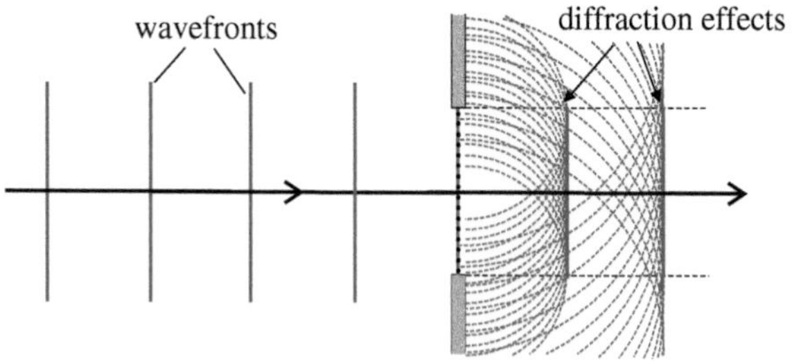

Figure 13.25. Diffraction at a slit. The dashed lines indicate the region accessible without diffraction.

This effect is known as *diffraction* because it causes waves to spread out after passing through the slit. Diffraction effects can be substantial when waves encounter apertures of dimensions which are comparable to their wavelengths so that, as illustrated in Figure 13.25, waves can propagate into regions which would not otherwise be accessible (that is, without diffraction). The wavelengths of audible sound waves range from centimetres to metres so that, when sound waves enter a room, for example through a doorway, sound is heard throughout the room.

Diffraction will be included in the detailed analysis of two slit interference of light waves to be given in Section 22.4. The analysis will show that the two-source (cosine)^2 interference pattern of Figure 13.18 is modulated in the manner indicated by a dashed line on the right of Figure 13.24. As noted above, diffraction becomes more important as the width of the slit approaches the wavelength of the waves, so that for light waves, whose wavelengths range between 400 and 700 nm, diffraction is important when they encounter slit widths of order 100 nm.

Multiple–slit interference

A multiple slit arrangement, comprising a large number of slits, equally spaced with separation *d*, is illustrated in Figure 13.26.

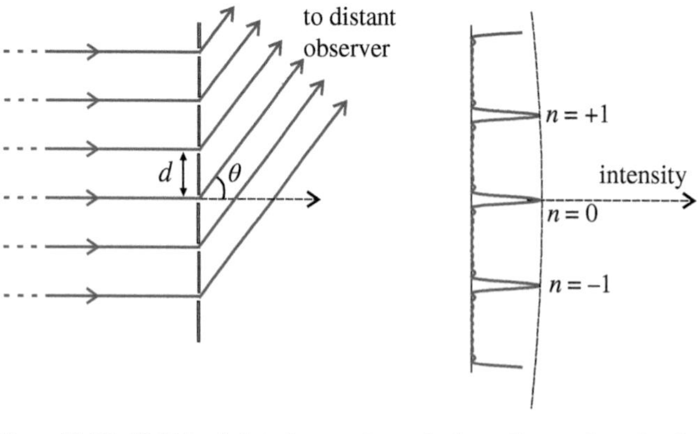

Figure 13.26. Multiple slit interference observed a large distance from the slits.

A typical multiple slit diffraction pattern is shown on the right of Figure 13.26. As in the two-slit case, the pattern is modulated by a diffraction envelope, as indicated by a dashed line on the right in Figure 13.26. As in the case of a linear array of synchronous sources, discussed in the previous section, intensity maxima become sharper and narrower as the number of sources increases and principal intensity maxima again follow the two slit condition

$$d\sin\theta = n\lambda \quad (n = 0, \pm1, \pm2, \pm3, \dots)$$

A full treatment of multiple slit diffraction effects will be given when we come to consider the *diffraction grating* in Section 22.5.

13.8 Standing waves

If a wave propagating in a medium encounters a point where the particles of the medium are held fixed so that they are unable to vibrate, the waves reflect at this point, as illustrated in Figure 13.27 for the case of waves along a string.

We can represent the incident wave (in the $-x$ direction) in Figure 13.27 by the travelling wave Equation (13.6),

$$y_1 = A\sin(kx + \omega t + \phi)$$

The reflected wave, which has the same amplitude and frequency but travels in the $+x$ direction, may be represented by the equation

$$y_2 = A\sin(kx - \omega t + \phi)$$

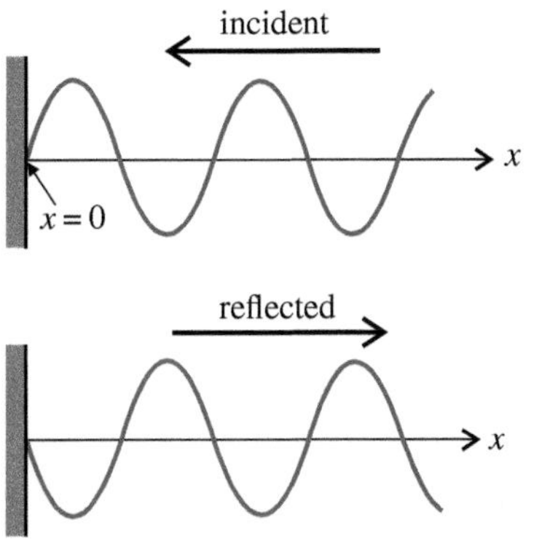

Figure 13.27. Travelling waves from the right are reflected at a fixed point ($x = 0$) so that the reflected waves travel *to* the right.

In practice the phase of a wave usually changes on reflection. The addition of such a phase to the equation for y_2 does not however change the essential features of the results derived below.

We now apply the principle of superposition (Section 13.6) to determine the resultant disturbance due to the combined effect of the incident and reflected waves at any point on the string.

$$y = y_1 + y_2 = A\sin(kx + \omega t + \phi) + A\sin(kx - \omega t + \phi)$$

Using the trigonometric identity (Appendix A.3.5(xi)), namely $\sin\alpha + \sin\beta = 2\sin\left(\frac{\alpha+\beta}{2}\right)\cos\left(\frac{\alpha-\beta}{2}\right)$, we can write

$$y = 2A\sin(kx + \phi)\cos\omega t$$

To simplify the analysis we place the origin of the x-axis, $x = 0$, at the point at which the string is fixed so that, at that point, $y = 0$ at all times. Thus $\phi = 0$ and the equation becomes

$$y = (2A\sin kx)\cos\omega t \tag{13.15}$$

Note that the sine and cosine terms in x and t, respectively, are now independent of one another. The wave motion does not progress through the medium – there is no travelling wave. The waves which are generated in this situation are described therefore as **standing (or stationary) waves**. Equation (13.15) can be regarded as a simple harmonic motion equation

$$y = A'\cos\omega t$$

in which the amplitude, $A' = 2A\sin kx$, varies with position along the string. At a fixed position the disturbance oscillates with time with a fixed amplitude A'. The resulting wave pattern is shown in Figure 13.28.

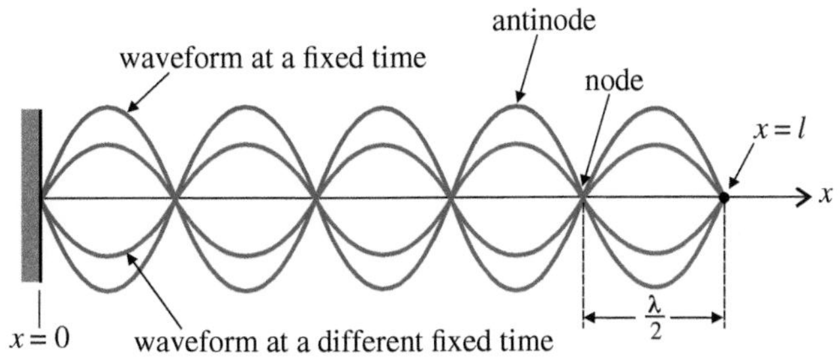

Figure 13.28. The standing wave pattern formed by the superposition of the incident and reflected waves shown in Figure 13.27.

Note that at certain positions, those for which $kx = n\pi$ (where $n = 0, 1, 2\dots$), the amplitude, A', is always zero. These positions are called **nodes**. The condition for a node can be written as $\frac{2\pi x}{\lambda} = n\pi$ so that nodes occur when $x = \frac{n\lambda}{2}$ – that is, at each half wavelength along the string, as indicated in Figure 13.28.

At certain other positions, those for which $kx = \frac{2n+1}{2}\pi$, (where $n = 0, 1, 2, \ldots$), and thus $x = \frac{2n+1}{2}\lambda$, the amplitude is $2A$. These positions, corresponding to maximum amplitude, are called **antinodes** and occur halfway between the nodes.

If the string is fixed at both ends, the ends of the string must serve as nodes in any standing wave pattern. We have chosen the origin of the x-axis so that one node is at $x = 0$. If the length of the string is l, as indicated in Figure 13.28, we can apply the node condition $kx = n\pi$ to the other end of the string where $x = l$.

Thus $$kx = kl = n\pi$$

and $$\frac{2\pi}{\lambda}l = n\pi$$

We can therefore write $$\lambda = \frac{2l}{n} \tag{13.16}$$

For a fixed value of l, therefore, standing waves are obtained only for certain wavelengths, and hence only for certain frequencies, given by

$$f = \frac{v}{\lambda} = \frac{nv}{2l} \tag{13.17}$$

where n is a *non-zero* integer. It follows from Equation (13.17) that these frequencies – known as **resonant** frequencies – are determined by the length, l, of the string and the velocity, v, at which a travelling wave travels along the string. In Section 13.11 we shall show that the velocity of a wave along a string, and hence the resonant frequency when the string is clamped at each end, depends on the tension in the string and its mass per unit length (effectively its thickness if the material used for each string has the same density). This is the basic principle used in stringed musical instruments. To change the musical note – the resonant frequency – produced by a string, the tension and/or the length of the string must be changed.

Wind instruments use standing sound waves in air columns to produce musical notes. The value of the resonant frequency then depends on the length of the air column in the instrument. As illustrated in Figure 13.29 for a simple pipe, the column of air can be closed at both ends, in which case there are nodes at both ends, or open at one end and closed at the other, in which case there is a node at one end and an antinode at the other.

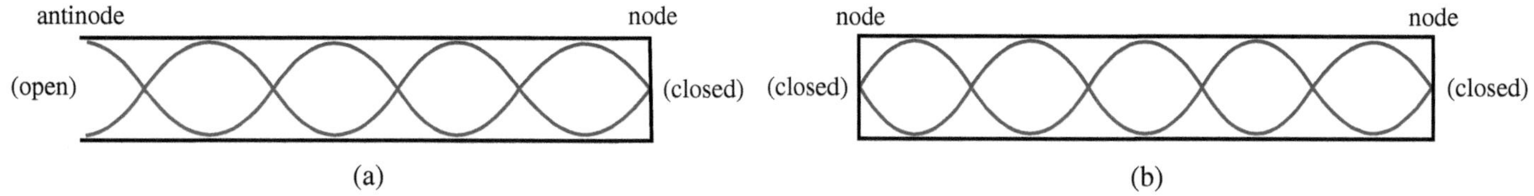

Figure 13.29. Standing waves (a) in a pipe which is closed at one end and (b) in a pipe which is closed at both ends.

The waveforms of the notes produced by musical instruments are rarely simple sine waves. In addition to a wave at the **fundamental** frequency, for which $n = 1$ in Equation (13.17), other waves (**harmonics**), corresponding to larger values of n, are superposed. Different musical instruments add different mixes of harmonics to the fundamental frequency to produce their characteristic sound quality. We will discuss the superposition of harmonics further in Section 13.15.

Counting standing waves in a cavity

We shall now derive expressions which give the number of standing waves, as a function of frequency, (a) in a one-dimensional cavity and (b) in a three-dimensional cavity. The three-dimensional case was considered by Rayleigh (John Strutt, Lord Rayleigh, 1842–1919) who determined how the frequency density of standing sound waves in a cubical room varied as a function of frequency. As will be seen in Section 14.2, the equation derived also played a part in the development of quantum theory.

(*a*) ***Standing waves in a one-dimensional cavity***

The allowed frequencies of standing waves in a cavity of length l are given by Equation (13.17) namely $f = \frac{nv}{2l}$ where v is the wave velocity and $n = 1, 2, 3, \ldots$ (an integer).

We represent allowed values of n by points distributed uniformly along an n-axis, as illustrated in Figure 13.30. The distance from any point to the origin, in units of n, is $\frac{2l}{v}f$.

The number of allowed values of f in the range f to $f + df$ is then given by the number of points on the n-axis which fall between these frequencies, that is

$$N(f)df = \frac{2l}{v}df$$

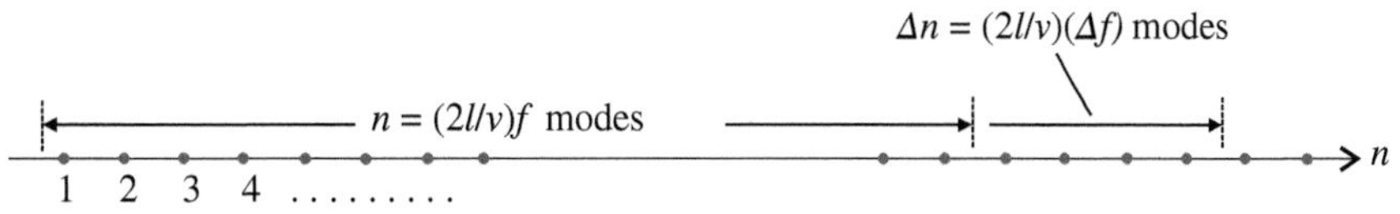

Figure 13.30. Allowed values of n for standing waves in a one dimensional cavity of length l.

***(b)* Standing waves in a three-dimensional cubical cavity**

Consider a three-dimensional standing wave of wavelength λ and frequency f which propagates in a cubical cavity of side length l. Its three components, in the x-, y- and z-directions, obey standing conditions given by Equation (13.16), namely

$$\frac{2x}{\lambda_x} = n_x, \text{ where } n_x = 1, 2, 3, \ldots; \quad \frac{2y}{\lambda_y} = n_y, \text{ where } n_y = 1, 2, \ldots; \quad \text{and} \quad \frac{2z}{\lambda_z} = n_z, \text{ where } n_z = 1, 2, \ldots$$

The standing wave condition for a three dimensional standing wave can be shown (up.ucc.ie/13/8/1/) to be

$$\frac{2l}{\lambda} = \sqrt{n_x^2 + n_y^2 + n_z^2} \tag{13.18}$$

We can now count the number of standing waves in a frequency interval, extending the method employed in (*a*). In the three dimensional case, the one dimensional plot of Figure 13.30 is replaced by a three-dimensional array of points which are distributed uniformly along each of three orthogonal n-axes, the n_x, n_y and n_z axes ('*n-space*'), as illustrated in Figure 13.31.

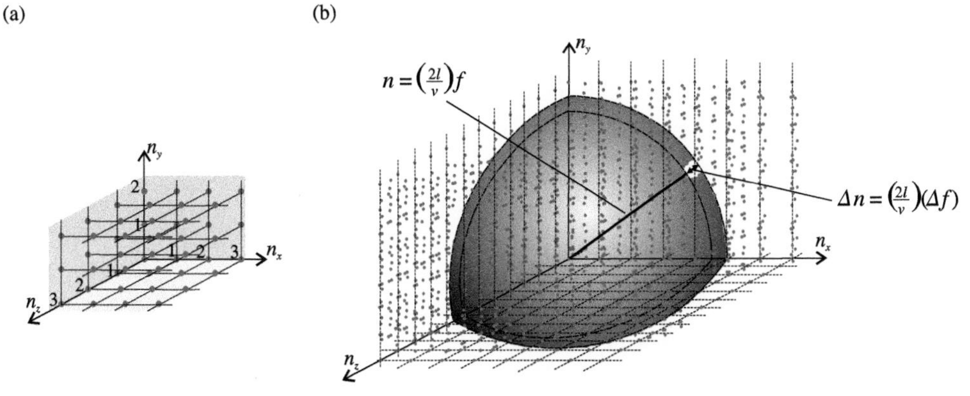

Figure 13.31. The allowed frequencies of standing waves in a cubical cavity.

Each point in the array corresponds to an allowed frequency ('mode') of a three-dimensional standing wave. In Figure 13.31 the distance from a point in n-space to the origin is given by $n^2 = n_x^2 + n_y^2 + n_z^2$ so that we can write the standing wave condition (Equation (13.18)) in terms of n, as $\frac{2l}{\lambda} = n$, which, in turn, can be written in terms of frequency as

$$f = \frac{v}{\lambda} = \frac{nv}{2l} \tag{13.19}$$

The number of allowed frequencies in the frequency interval f to $f + df$ is equal to the number of points contained in that part of the volume between two spheres, of radii n and $n + dn$, which lies in the octant corresponding to positive n in Figure 13.31.

Thus,

$$N(n)dn = \frac{1}{8}4\pi n^2 dn$$

which, using Equation (13.19), can be expressed as a function of frequency

$$N(f)df = \frac{\pi}{2}\left(\frac{2l}{v}\right)^3 f^2 df = \frac{4\pi V}{v^3} f^2 df, \tag{13.20}$$

where $V = l^3$ is the volume of the cavity.

Study Worked Example 13.3

For problems based on the material in this section visit up.ucc.ie/13/ and follow the link to the problems.

The blackbody radiation problem revisited

Following Kirchhoff's challenge to the physics community in 1859 (recall Section 11.7), techniques to study the frequency distribution of the radiation emitted from a hot cavity had gradually improved to the extent that it was possible to begin to distinguish different theoretical models. In particular, experimental results from a number of investigators in Berlin provoked certain theoretical developments.

At the turn of the twentieth century Rayleigh attempted to explain the observed blackbody spectrum by considering the behaviour of electromagnetic waves within a cavity. It was known from electromagnetic theory that, because metallic surfaces are not able to support electromagnetic fields, the waves must have nodes at the walls of the cavity — they must be standing waves. In 1900, Rayleigh attempted to apply his earlier treatment of sound waves in a room (part (*b*) of the previous subsection) to standing electromagnetic waves in a cavity.

In the case of electromagnetic waves, Equation (13.20) needs to be modified to take into consideration the fact that there are two possible polarisation states (to be discussed further in Section 21.5) and that, in vacuum, $v = c$. Thus Rayleigh, with additional input from James Jeans (1877–1946), proposed that the distribution of frequencies in cavity radiation would be given by

$$N(f)df = \frac{8\pi V}{c^3} f^2 df$$

Rayleigh then assumed that the radiation in the cavity was in thermal equilibrium with the oscillators in the walls that were the sources of the radiation. To determine the total energy of the radiation within the cavity, $N(f)df$, must be multiplied by the average energy per standing wave frequency mode ($\langle E\rangle$) which depends on the temperature T of the cavity walls. The frequency distribution of the radiation energy density (energy per unit volume) is thus obtained by dividing $\langle E\rangle N(f)df$ by the volume V. Hence, the energy density of waves with frequency between f and $f + df$ is given by

$$w(f,T)df = \frac{8\pi}{c^3} \langle E\rangle f^2 df \tag{13.21}$$

How to determine $\langle E\rangle$ and, in particular how it depends on T, was a matter of some debate. Despite his own and others' reservations about its applicability in this situation, Rayleigh applied the principle of equipartition of energy (Sections 12.2 and 12.8) to the standing wave oscillations assuming two degrees of freedom, each contributing $\frac{1}{2}kT$. Thus, $\langle E\rangle = kT$ and the frequency distribution of energy density in the cavity at temperature T is given by

$$w(f,T)df = \frac{8\pi kT}{c^3} f^2 df \tag{13.21a}$$

which is now known as the *Rayleigh-Jeans formula* for cavity radiation.

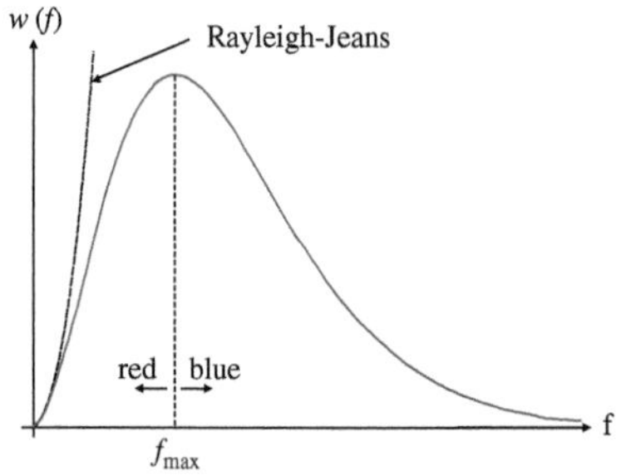

Figure 13.32. Comparison of the observed cavity radiation spectrum at a given temperature (blue) with the Rayleigh-Jeans prediction (13.21b) (dashed line).

The energy distribution of cavity radiation predicted by the Rayleigh-Jeans formula (13.21a) is compared with observation in Figure 13.32. While the equation accounts for the observed blackbody spectrum at low frequencies, it fails completely to predict the observed fall-off at high frequencies.

As noted, Section 11.7, radiation from a cavity is of particular interest because the radiative properties of a cavity are expected to approach those of a blackbody and thus the failure of the Rayleigh-Jeans analysis is also expected to apply to blackbody radiation. The implications are considerable because the Rayleigh-Jeans equation predicts that, at any temperature, the radiated energy tends to infinity at high frequencies. If this were the case, all objects would quickly cool to 0 K through emission of high frequency electromagnetic radiation and the universe as we know it would not be viable. This fundamental failure of classical theory is described as the *ultraviolet catastrophe*. We will see in Chapter 14 that attempts to resolve this issue contributed to momentous developments in physics (quantum mechanics) in the early years of the twentieth century.

13.9 The Doppler effect

When the source and/or the receiver (observer) of periodic waves are in relative motion with respect to the medium through which the waves are propagating, the wave frequency, as measured by the observer, differs from the source frequency. This change of frequency is known as the **Doppler effect** (Christian Doppler, 1803–1853).

The effect is illustrated in Figure 13.33 where the wavefronts produced by a source of sinusoidal waves are shown (a) when the source is stationary and (b) when the source is moving to the right. In case (a), the wavefronts are concentric circles but, if we consider the wavefronts emitted at successive positions in case (b), namely A, B, C, D, ..., it is clear that they are no longer concentric. Wave crests are more closely spaced on the side towards which the source is travelling. Thus, an observer on the x-axis of Figure 13.33(b) to the right of the source detects a shorter wavelength (and hence a higher frequency) than in case (a). If the observer is moving to the left as the source approaches, the observed wavelength will be even shorter.

We shall derive equations which give the change of frequency in two situations (i) a moving source and a stationary observer, and (ii) a stationary source and a moving observer. We shall then combine the results to write the general Doppler effect equation which applies when both source and observer are moving simultaneously.

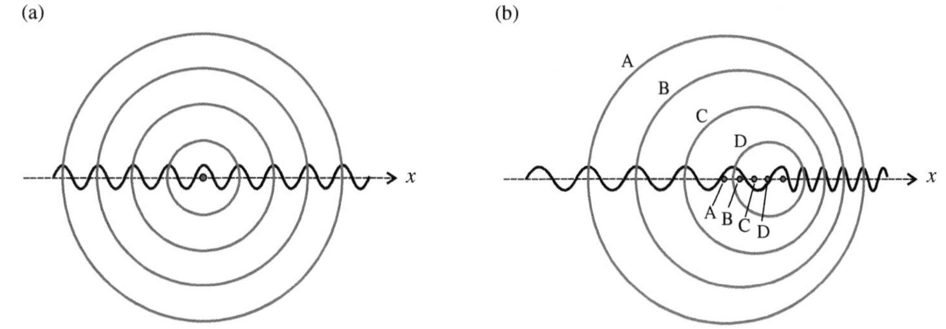

Figure 13.33. The Doppler effect. (a) The wavefronts produced by a stationary source and (b) the wavefronts produced by a source which is moving to the right. In case (b) the wavefronts A, B, C, and D are emitted at the successive positions of the source A, B, C, and D, as indicated on the x-axis.

(a) Moving source and stationary observer

In what follows f is the frequency of the source and v is the wave speed relative to the medium transmitting the waves when this medium is stationary; for example, for sound waves v would be the speed of sound waves in still air.

We consider a case in which the source is approaching the observer with speed v_s. $T\left(=\frac{1}{f}\right)$ is the time between the emission of successive wavefronts, shown as W_1 and W_2 in Figure 13.34.

During the time T, W_1 moves a distance vT and the source moves v_sT. The distance between W_1 and W_2, that is the wavelength λ' detected by the observer O, is then $T(v - v_s)$. The observer detects a frequency

$$f' = \frac{v}{\lambda'} = \frac{v}{T(v - v_s)} = \frac{v}{(v - v_s)}f \tag{13.22}$$

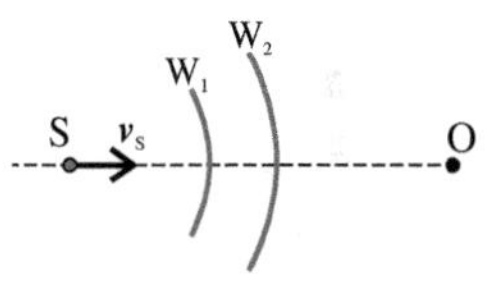

Figure 13.34. Successive wavefronts W_1 and W_2, emitted by a moving source S.

Thus $f' > f$. The frequency is higher when the source is approaching while, on the other hand, if the source is moving away from the observer v_s is negative and $f' < f$. This explains why the frequency (the *pitch* of the note) of an approaching train whistle drops as the train passes a stationary observer.

(b) Stationary source and moving observer

In this case the observer is moving towards the source with speed v_o and encounters more wavefronts than a stationary observer. A stationary observer receives ft wave fronts in a time t. In the same time the moving observer travels v_ot which corresponds to $\frac{v_ot}{\lambda}$ wavefronts. The total number of wavefronts received by the observer is therefore $ft + \frac{v_ot}{\lambda}$. The frequency detected by the observer is the number of wavefronts divided by the time over which they are received and is thus given by

$$f' = \frac{ft + \frac{v_ot}{\lambda}}{t} = f + \frac{v_o}{\lambda} = \left(1 + \frac{v_o}{f\lambda}\right)f = \frac{v + v_o}{v}f \tag{13.23}$$

We can now write an equation for the case in which both source and observer are moving. Replacing f in Equation (13.23) with f' from Equation (13.22) (that is, the frequency associated with the motion of the observer) gives the general Doppler effect equation

$$\boxed{f' = \frac{(v + v_o)}{(v - v_s)}f} \tag{13.24}$$

This equation was derived for the case in which the observer was moving towards the source and in which the source was moving towards the observer; v_o and v_s are algebraic quantities so that when the observer is moving away from the source v_o is negative and, similarly, when the source is moving away from the observer v_s is negative.

We will return to the Doppler effect in Section 22.1 where we will consider the Doppler effect for light waves, including the case in which the source is moving with a velocity approaching that of light (the relativistic Doppler effect).

Study Worked Example 13.4

For problems based on the material in this section visit up.ucc.ie/13/ and follow the link to the problems.

13.10 The wave equation

We have shown, in Section 13.2 (Equation (13.4)), that a travelling plane wave along the x-axis may be represented by the equation

$$y(x,t) = A\sin(\omega t - kx + \phi)$$

Let us consider the second derivatives of this equation with respect to x and t, respectively. Since it is an equation in two variables that we must use partial derivatives (Appendix A.6.3) to differentiate this equation. We obtain

$$\frac{\partial y}{\partial x} = -kA\cos(\omega t - kx + \phi)$$

$$\frac{\partial^2 y}{\partial x^2} = -k^2 A\sin(\omega t - kx + \phi) \tag{13.25}$$

$$\frac{\partial y}{\partial t} = \omega A\cos(\omega t - kx + \phi)$$

$$\frac{\partial^2 y}{\partial t^2} = -\omega^2 A\sin(\omega t - kx + \phi) \tag{13.26}$$

Dividing Equation (13.25) by (13.26) we obtain

$$\frac{\partial^2 y}{\partial x^2}\bigg/\frac{\partial^2 y}{\partial t^2} = \frac{k^2}{\omega^2} = \frac{1}{v^2} \quad \text{(from Equation (13.5))}$$

It can be shown that this result applies to any periodic function, that is a function for which $f(kx \pm \omega t) = f(kx \pm \omega t \pm a)$, where a is a constant. The functions used to represent a travelling plane wave in Figure 13.8 are examples of periodic functions. Thus, the following relationship, known as the **wave equation**, can be stated for any travelling plane wave in one dimension.

$$\boxed{\frac{\partial^2 y}{\partial x^2} = \frac{1}{v^2}\frac{\partial^2 y}{\partial t^2}} \tag{13.27}$$

The wave equation is an example of a *linear* equation in $y(x, t)$ as discussed in Section 3.13. For a linear equation, if $y_1(x, t)$ and $y_2(x, t)$ are two different solutions then their sum, $y_1(x, t) + y_2(x, t)$, also satisfies the equation. Thus the superposition principle (Section 13.6) follows directly from the form of the wave equation.

The wave Equation (13.27) can be generalised to spherical waves (waves in three dimensions with spherical wavefronts) as

$$\frac{\partial^2 \xi}{\partial x^2} + \frac{\partial^2 \xi}{\partial y^2} + \frac{\partial^2 \xi}{\partial z^2} = \frac{1}{v^2}\frac{\partial^2 \xi}{\partial t^2} \quad \text{or} \quad \nabla^2 \xi = \frac{1}{v^2}\frac{\partial^2 \xi}{\partial t^2}$$

where the symbol ∇^2 denotes the Laplacian (Appendix A.13.1).

Note also that, in this case, to avoid confusion with the y coordinate, we have used the symbol ξ (instead of y) to represent disturbance.

Equation (13.27) tells us that if, for any medium, we can express the second space derivative of disturbance as a constant C multiplied by the second time derivative, that constant must equal $\frac{1}{v^2}$; that is, if $\frac{\partial^2 y}{\partial x^2} = C\frac{\partial^2 y}{\partial t^2}$ then $v = \frac{1}{\sqrt{C}}$.

In the next three sections we derive expressions which relate $\frac{\partial^2 y}{\partial x^2}$ to $\frac{\partial^2 y}{\partial t^2}$ for the special cases of (i) waves along a string, (ii) waves in a solid rod and (iii) sound waves in a gas and thus identify which properties of the medium carrying the wave (the string, the rod and air, respectively) determine the wave velocities in each of these media.

13.11 Waves along a string

We now analyse the factors which govern the propagation of transverse waves along a string which is under a constant tension T, the situation which we used to derive the equation of a travelling wave in Section 13.2. Consider an element of the string, Δs, as illustrated in Figure 13.35.

The forces acting on the element Δs are shown as arrows in Figure 13.36.

At the point A the y-(transverse) component of the force on the string is $T\sin\theta \approx T\tan\theta = T\frac{\partial y}{\partial x}$ for small displacements.

We can use the Taylor series (Appendix A.9.3)

$$f(x+\Delta x) = f(x) + \frac{\partial f}{\partial x}\Delta x + \frac{1}{2!}\frac{\partial^2 f}{\partial x^2}(\Delta x)^2 + \ldots$$

to write the transverse force at B as $T\frac{\partial y}{\partial x} + \frac{\partial}{\partial x}\left(T\frac{\partial y}{\partial x}\right)\Delta x$ + negligible terms.

Therefore the net force on the element of string Δs in the y-direction is

$$\left\{T\frac{\partial y}{\partial x} + \frac{\partial}{\partial x}\left(T\frac{\partial y}{\partial x}\right)\Delta x\right\} - T\frac{\partial y}{\partial x} = \frac{\partial}{\partial x}\left(T\frac{\partial y}{\partial x}\right)\Delta x = T\left(\frac{\partial^2 y}{\partial x^2}\right)\Delta x$$

Applying Newton's second law, we can also write the net force on Δs as the product of its mass and its acceleration in the transverse direction. If μ is the mass per unit length of the string, the net transverse force is $\Delta m\frac{\partial^2 y}{\partial t^2} = \mu\Delta s\frac{\partial^2 y}{\partial t^2}$.

Equating the two expressions for the net force we obtain $T\frac{\partial^2 y}{\partial x^2}\Delta x = \mu\Delta s\frac{\partial^2 y}{\partial t^2}$

For small amplitude oscillations $\Delta s \approx \Delta x$ so that

$$T\frac{\partial^2 y}{\partial x^2} = \mu\frac{\partial^2 y}{\partial t^2}$$

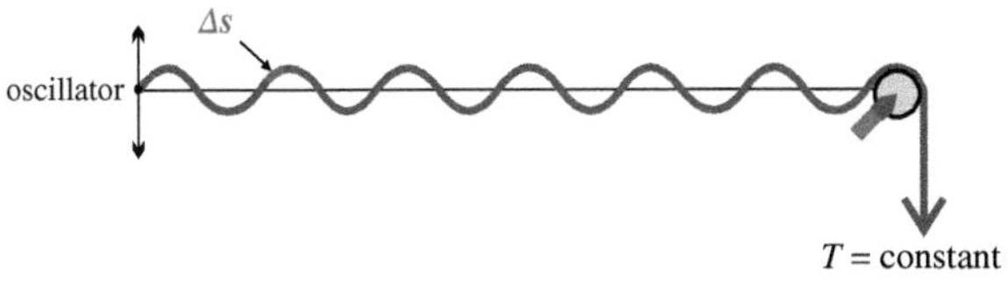

Figure 13.35. The propagation, along a taut string, of waves which are produced by oscillation of the string.

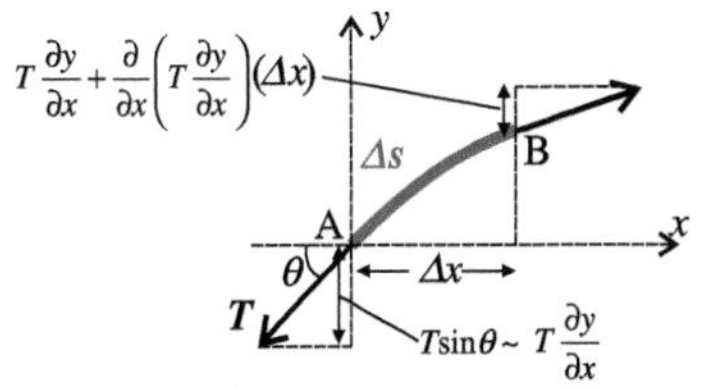

Figure 13.36. The forces acting on the element of string Δs, shown in Figure 13.35.

Comparing this result with the wave Equation (13.27) we see that we have derived a wave equation from which we can identify the wave velocity as being given by

$$\boxed{v = \sqrt{\frac{T}{\mu}}}$$

Thus, the velocity of a transverse wave along a string is determined by the tension in the string and its mass per unit length.

When we considered standing waves on a string in Section 13.8, we showed that standing waves could be established if the length of the string was $\frac{n\lambda}{2}$, where n is an integer, and that the frequencies of such waves are given by Equation (13.17), namely

$$f_n = \frac{v}{\lambda} = \frac{nv}{2l}$$

which we can now write as

$$f_n = n\left(\frac{1}{2l}\right)\sqrt{\frac{T}{\mu}}.$$

The lowest (fundamental) frequency of the string is obtained when $n = 1$ and, therefore, is given by $f_1 = \left(\frac{1}{2l}\right)\sqrt{\frac{T}{\mu}}$.

To increase the frequency of the note produced by a string in a musical instrument we can either (a) decrease its length or (b) increase its tension or (c) replace the string with another which has a smaller value of μ (in practice this usually means a thinner string).

For problems based on the material in this section visit up.ucc.ie/13/ and follow the link to the problems.

13.12 Waves in elastic media: longitudinal waves in a solid rod

Consider an element, length Δx, of a solid rod which has cross-sectional area A as illustrated in Figure 13.37.

As a longitudinal wave passes along the rod, material in the rod is displaced along the axis of the rod. As illustrated in Figure 13.38, the point P moves to P′ and the point Q moves to Q′; the displacement of the medium varies with x. If we write the displacement PP′ as y, we can again use the Taylor series (Appendix A.9.3) to write a good approximation to the displacement QQ′ as $y + \frac{\partial y}{\partial x}\Delta x$, where $\frac{\partial y}{\partial x}$ is the rate of change of displacement with respect to x (the displacement gradient).

The increase in length, the extension of Δx, as the wave passes is thus

$$y + \frac{\partial y}{\partial x}\Delta x - y = \frac{\partial y}{\partial x}\Delta x$$

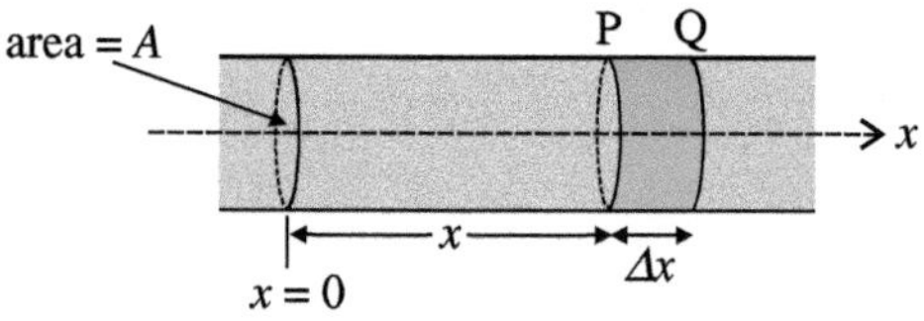

Figure 13.37. An element Δx of a solid rod.

The ratio of the extension to the original length Δx is therefore $\frac{\partial y}{\partial x}$. In Section 10.2 we defined this ratio to be the longitudinal strain.

In Section 10.2 we also defined the Young modulus E to be $\frac{\text{longitudinal stress}}{\text{longitudinal strain}}$ so that in the present case the longitudinal stress, the force per unit area on the element, is $E\frac{\partial y}{\partial x}$.

Therefore, the force on Δx from the left, due to the rest of the rod on the left, is given by

$$F(x) = \text{stress} \times A = EA\frac{\partial y}{\partial x}$$

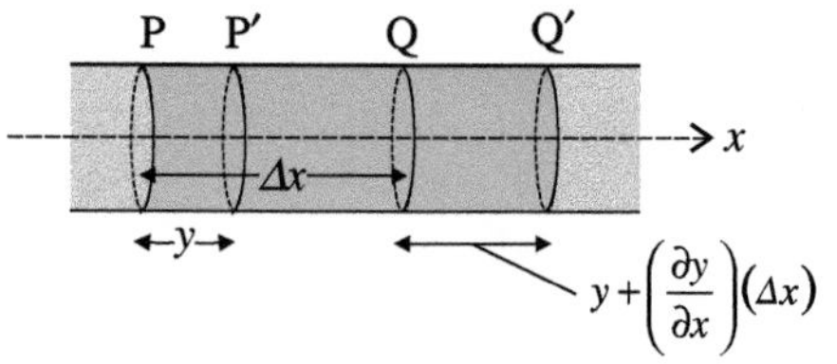

Figure 13.38. The displacement of the element Δx (as shown in Figure 13.37) which is produced by the passage of a longitudinal wave along the rod.

Differentiating with respect to x, we obtain $\frac{\partial F(x)}{\partial x} = EA\frac{\partial^2 y}{\partial x^2}$

so that the force on Δx from the right, due to the rest of the rod on the right, can be written

$$F(x + \Delta x) = F(x) + \frac{\partial F(x)}{\partial x}\Delta x = F(x) + EA\frac{\partial^2 y}{\partial x^2}\Delta x$$

As illustrated in Figure 13.39, the net force on Δx is $F(x + \Delta x) - F(x) = EA\frac{\partial^2 y}{\partial x^2}\Delta x$

We can equate this net force to the product of mass and acceleration for the element Δx, in other words, $\Delta m\frac{\partial^2 y}{\partial t^2} = \rho A \Delta x\frac{\partial^2 y}{\partial t^2}$, where ρ is the density of the material of the rod. Thus

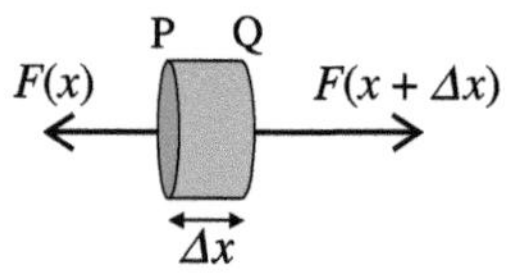

Figure 13.39. The forces on the element Δx.

$$EA\frac{\partial^2 y}{\partial x^2}\Delta x = \rho A \Delta x\frac{\partial^2 y}{\partial t^2}$$

which can be written

$$\frac{\partial^2 y}{\partial x^2} = \frac{\rho}{E}\frac{\partial^2 y}{\partial t^2}$$

Comparing this result with Equation (13.27) we see that we have derived a wave equation of the same form with, in this case,

$$v = \sqrt{\frac{E}{\rho}}.$$

Thus, the velocity of a longitudinal wave along a solid rod is determined by the Young modulus and the density of the material of the rod.

For problems based on the material in this section visit up.ucc.ie/13/ and follow the link to the problems.

13.13 Waves in elastic media: sound waves in gases

Consider a volume element of gas through which a plane sound wave is travelling in the x-direction. The length of the volume element is Δx and its cross-sectional area is A, as illustrated in Figure 13.40.

As the sound wave travels through this volume element, longitudinal displacements are produced. The point P moves to P′ and the point Q moves to Q′. The longitudinal displacement of the medium, y, varies with x. The analysis that follows is similar to that of the previous section, with the bulk modulus of the gas replacing the Young modulus as the appropriate elastic modulus.

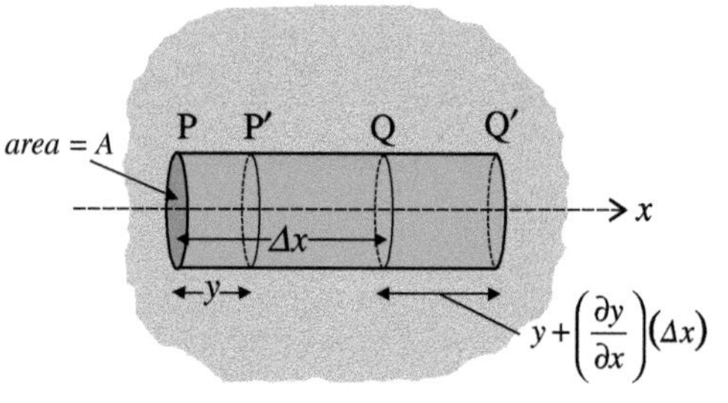

Figure 13.40. A volume element (of length Δx) of a gas.

If we write the displacement, PP′ as y we can write the displacement QQ′ as $y + \frac{\partial y}{\partial x}\Delta x$ where $\frac{\partial y}{\partial x}$ is again the rate of change of the displacement with respect to x.

The original volume of the element $= V = A\Delta x$ and the change of volume $= \Delta V = A\frac{\partial y}{\partial x}\Delta x$

Thus we can write the volumetric strain (Section 10.2) as $\frac{\Delta V}{V} = \frac{\Delta V}{A\Delta x} = \frac{\partial y}{\partial x}$.

The disturbance of the gas, which is caused by the passage of a sound wave, occurs so quickly that the process is adiabatic. The adiabatic bulk modulus of the gas is defined in Section 11.12 as $K_s := -V\left(\frac{\partial P}{\partial V}\right)_{\Delta Q=0}$. Therefore, we can write the pressure difference across the volume element as

$$\Delta P = -K_s\frac{\Delta V}{V} = -K_s\frac{\partial y}{\partial x} \tag{13.28}$$

As the pressure changes, the gas is compressed and expanded, that is concentrated and rarefied. If ΔP is the net excess pressure on this volume element at P, as indicated in Figure 13.41, its value at Q is given by $\Delta P + \frac{\partial P}{\partial x}\Delta x$, where $\frac{\partial P}{\partial x}$ is the excess pressure gradient.

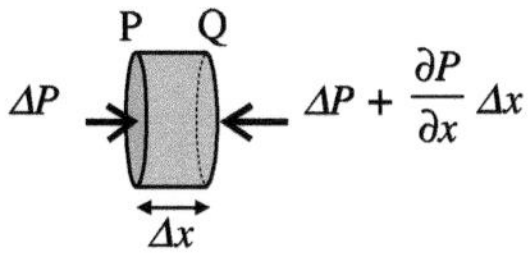

Figure 13.41. The pressure difference across the volume element shown in Figure 13.40 which is produced by the passage of a sound wave.

Using Equation (13.28) we can write $\frac{\partial P}{\partial x}\Delta x$ as $-K_s\frac{\partial^2 y}{\partial x^2}\Delta x$ so that the net excess pressure is

$$\Delta P = -K_s\frac{\partial^2 y}{\partial x^2}\Delta x$$

and, therefore, the net force on this volume element is $K_s A\frac{\partial^2 y}{\partial x^2}\Delta x$. This force can also be expressed as the product of mass and acceleration; that is, $\Delta m\frac{\partial^2 y}{\partial t^2} = \rho A\Delta x\frac{\partial^2 y}{\partial t^2}$ where ρ is the density of the gas.

Equating the two expressions for force $K_s A\frac{\partial^2 y}{\partial x^2}\Delta x = \rho A\Delta x\frac{\partial^2 y}{\partial t^2}$

and thus $$\frac{\partial^2 y}{\partial x^2} = \frac{\rho}{K_s}\frac{\partial^2 y}{\partial t^2}$$

Comparing this result with the Equation (13.27), we see that this is a wave equation for which

$$\boxed{v = \sqrt{\frac{K_s}{\rho}}}$$

Thus, the velocity of a sound wave in a gas depends on the adiabatic bulk modulus and density of the gas.

Using the relationship between the adiabatic bulk modulus and the pressure which we derived in Section 11.12, namely $K_s = \gamma P$, the equation for the velocity of a sound wave in a gas can be written in terms of the pressure and density of the gas as follows,

$$v = \sqrt{\frac{\gamma P}{\rho}}$$

where γ is the ratio of the principal heat capacities.

In the case of an ideal gas $PV = NkT$ (Equation 12.2). Since $\rho = \frac{M}{V}$ where M is the mass of gas involved, the expression for the speed of the waves in the gas reduces to

$$v = \sqrt{\frac{\gamma kT}{\langle m\rangle}}$$

where $\langle m\rangle$ is the average molecular mass.

Realising the kelvin (acoustic gas thermometry)

An absolute experimental measurement based directly on the definition of a base unit is called 'realising the unit'. The above equation for the speed of acoustic waves in a gas can be used to show, *in principle*, how the kelvin may be so realised. For simplicity here, we consider the case of an ideal gas confined in a cubical metal cavity such as that described in Section 13.8. The frequency of each standing wave mode is given by Equation (13.19), that is

$$f_n = \frac{v}{\lambda} = n\frac{v}{2l}$$

Using the expression for the speed of sound in an ideal gas derived above we find that, in this case, the temperature of the gas is given by

$$T = \frac{4\langle m\rangle l^2}{\gamma k}\frac{f_n^2}{n^2}$$

Now the Boltzmann constant k is fixed by the SI definition of the kelvin (Section 11.2), γ and $\langle m \rangle$ are known for a particular ideal gas and l is the side-length of the cavity, so the identification and measurement of the acoustic frequencies in a resonant cavity enables the temperature of the gas to be determined. In practice, of course, things are a bit more complicated. For example, corrections must be made for the facts that the gas is not ideal and that in reality the cavity is usually not cubic (often quasi-spherical) so a somewhat different theoretical treatment is required.

For problems based on the material in this section visit up.ucc.ie/13/ and follow the link to the problems.

13.14 Superposition of two waves of slightly different frequencies: wave and group velocities

Consider the superposition of two travelling plane waves of the same amplitude A and starting phase ϕ, that is, $y_1 = y_2 = A\sin\phi$ at $x = 0$ and $t = 0$, but with slightly different frequencies and thus slightly different values of ω and k,

$$y_1 = A\sin[kx - \omega t + \phi]$$

and

$$y_2 = A\sin[(k + \Delta k)x - (\omega + \Delta\omega)t + \phi]$$

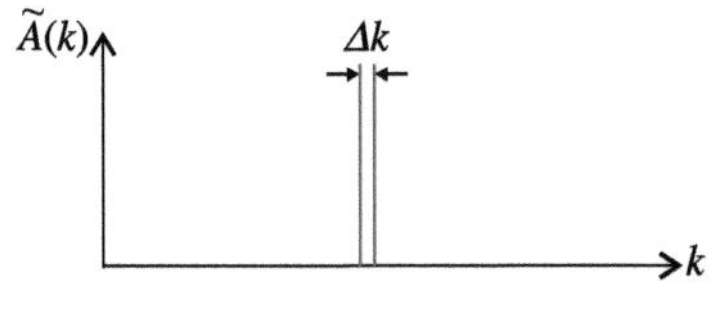

Figure 13.42. The wave number spectrum of two travelling waves of equal amplitude but with slightly different wave numbers.

In Figure 13.42 the wave number spectrum, which is a plot of wave amplitude $\tilde{A}(k)$ against k, is displayed for the two waves.

The net disturbance produced by the two waves is their algebraic sum, that is:

$$y = y_1 + y_2 = A\sin[kx - \omega t + \phi] + A\sin[(k + \Delta k)x - (\omega + \Delta\omega)t + \phi]$$

Using the trigonometric identity $\sin\alpha + \sin\beta = 2\cos\left(\frac{\alpha - \beta}{2}\right)\sin\left(\frac{\alpha + \beta}{2}\right)$ (Appendix A.3.5(xi)), we can write

$$y = 2A\cos\left(\frac{\Delta\omega}{2}t - \frac{\Delta k}{2}x\right)\sin\left[\frac{(2k + \Delta k)}{2}x - \frac{(2\omega + \Delta\omega)}{2}t + \phi\right]$$

When $\Delta k \ll k$ and $\Delta\omega \ll \omega$, as in the case we are currently considering, this can be written

$$y = \left\{2A\cos\left(\frac{\Delta\omega}{2}t - \frac{\Delta k}{2}x\right)\right\}\sin(kx - \omega t + \phi) \tag{13.29}$$

Equation (13.29) is the equation of a travelling wave in the $+x$ direction, $y = A'\sin(kx - \omega t + \phi)$, whose amplitude $A' = 2A\cos\left[\frac{\Delta\omega}{2}t - \frac{\Delta k}{2}x\right]$ varies periodically in both position and time. The resulting wave pattern at a fixed time is shown in Figure 13.43. The amplitude of the basic travelling wave of wave number k and angular frequency ω is modulated within the envelope shown in the figure. The regions in which the amplitude A' is large – and within which energy is carried – are known as **beats**. The beat phenomenon will be familiar to anyone who has tuned a musical instrument by comparing two sound waves of slightly different frequencies. The intensity of the basic note of wavenumber k is modulated at a low frequency and becomes lower as the frequencies of the two waves approach one another (that is, as $\Delta k \to 0$).

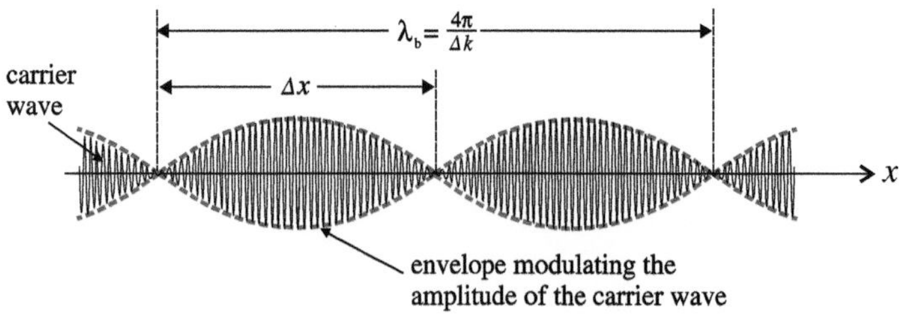

Figure 13.43. The waveform, viewed at a fixed time, which is produced by the superposition of two travelling waves of equal amplitude but with slightly different wavenumbers (as indicated in the wavenumber spectrum shown in Figure 13.42). The amplitude of the carrier wave is modulated by the wave packet envelope.

A comparison of the modulated amplitude $2A\cos\left(\frac{\Delta\omega}{2}t - \frac{\Delta k}{2}x\right)$ with the equation of a sinusoidal travelling wave in one dimension, which can be written $y = A\sin[\omega t - kx + \phi]$, shows that the modulated amplitude is itself a travelling wave with a wavenumber $k_b = \frac{\Delta k}{2}$

and an angular frequency, $\omega_b = \frac{\Delta\omega}{2}$. The beat wavelength, λ_b, which is indicated in Figure 13.43, is given by

$$\lambda_b = \frac{2\pi}{k_b} = 2\pi \Big/ \frac{\Delta k}{2} = \frac{4\pi}{\Delta k}$$

The modulated amplitude, $A' = 2A\cos\left(\frac{\Delta\omega}{2}t - \frac{\Delta k}{2}x\right)$, travels in the $+x$ direction in Figure 13.43.

Wave and group velocities

Two characteristic wave velocities can be defined for the waveform shown in Figure 13.43. The **phase velocity** or **wave velocity**

$$v = f\lambda = \frac{\omega}{k}$$

is the velocity of the basic wave, 'the carrier wave'.

The **group velocity** is the velocity of the beats which is

$$v_g = \frac{\omega_b}{k_b} = \frac{\Delta\omega}{\Delta k} \tag{13.30}$$

This is the velocity of the envelope modulating the amplitude, the velocity with which energy is carried by the wave.

Note that the energy of the wave is no longer distributed uniformly throughout the space occupied by the unmodulated wave motion; it is localised within the beats. The width of a beat Δx is $\frac{\lambda_b}{2}$. Thus

$$\Delta x = \frac{\lambda_b}{2} = \frac{2\pi}{\Delta k}$$

and

$$(\Delta k)(\Delta x) = 2\pi \tag{13.31}$$

As described in the next section, it is possible to increase this localisation of the wave energy by superposing further waves with slightly different frequencies, thus producing narrower **wave packets**.

For problems based on the material in this section visit up.ucc.ie/13/ *and follow the link to the problems.*

13.15 Other wave forms: Fourier analysis

Periodic waveforms

So far we have confined our discussion of periodic waves almost exclusively to simple harmonic (sinusoidal) waveforms such as,

$$y = A\sin(kx - \omega t + \phi)$$

Such a wave can be characterised by a single frequency $f = \frac{\omega}{2\pi}$ and, thus, a single wavenumber k. Many other examples of periodic waveforms are frequently encountered, some examples of which are illustrated in Figure 13.44.

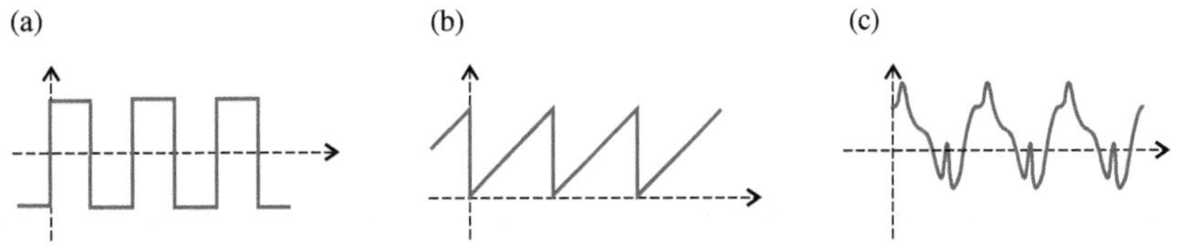

Figure 13.44. Some examples of periodic waveforms, viewed at a fixed time: (a) a square wave, (b) a 'sawtooth' wave and (c) a general periodic waveform.

Fourier analysis

We are able to describe any of these waveforms quantitatively using the principle of superposition (Section 13.6) and a mathematical technique known as **Fourier analysis**, named after Joseph Fourier (1768–1830). Fourier analysis uses the fact that any periodic function, $f(w) = f(w + W)$, where w may represent position or time and W is the appropriate periodicity (wavelength λ for position or period T for time), may be represented by a series of periodic functions – known as a *Fourier Series*.

For a periodic function of position x, the series is

$$f(x) = \frac{1}{2}a_0 + a_1\cos kx + a_2\cos 2kx \ldots + a_n\cos nkx + b_1\sin kx + b_2\sin 2kx \ldots + b_n\sin nkx + \ldots$$

that is

$$f(x) = \sum_n a_n\cos nkx + \sum_n b_n\sin nkx,$$

where $k = \dfrac{2\pi}{\lambda}$ is the wavenumber of the function.

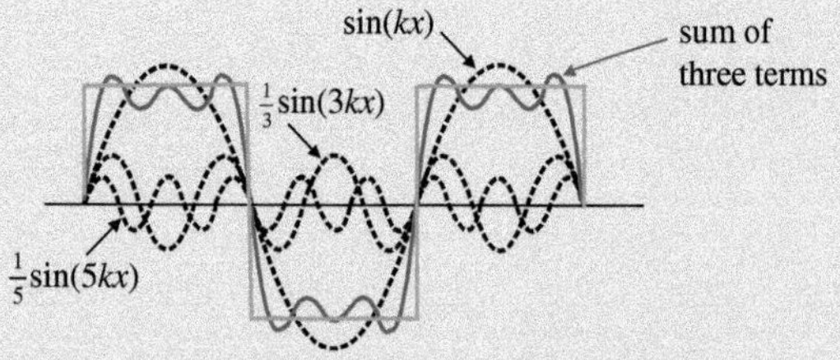

Figure 13.45. Fourier analysis. The generation of an approximation to a square waveform through the superposition of the following three sinusoidal waveforms, $\sin kx + \frac{1}{3}\sin 3kx + \frac{1}{5}\sin 5kx$.

The series comprises a constant term, $\frac{1}{2}a_0$, plus sine and cosine terms of different amplitudes, a_1, a_2, b_1, b_2, etc. The sine and cosine terms represent the addition of terms whose wavenumbers are multiples of the wavenumber k of the fundamental term; they are *harmonics*, as discussed in Section 13.8. To represent a particular periodic waveform, values of the coefficients, a_1, a_2, b_1, b_2, etc., are determined by integration of $f(x)$ over one period, 2π, as follows:

$$a_n = \frac{1}{\pi}\int_{-\pi}^{\pi} f(x)\cos nkx dx \quad n \geq 0$$

$$b_n = \frac{1}{\pi}\int_{-\pi}^{\pi} f(x)\sin nkx dx \quad n \geq 1$$

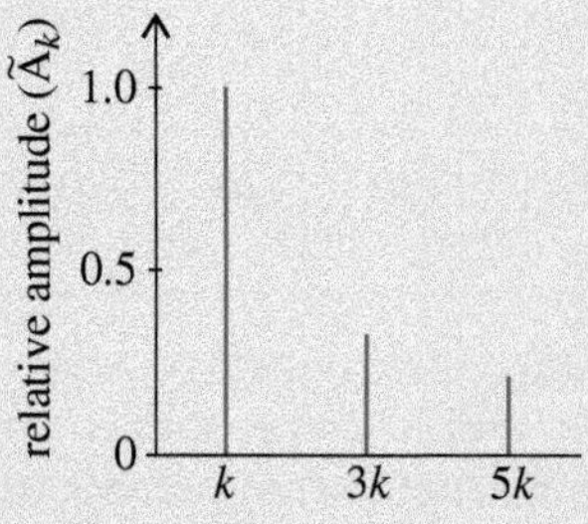

Figure 13.46. Wave number spectrum of the waves used to generate the waveform shown in Figure 13.45.

The first term in the Fourier series, $\frac{1}{2}a_0$, therefore, is $\dfrac{1}{2\pi}\displaystyle\int_{-\pi}^{\pi} f(x)dx$. This is simply the average value of $f(x)$ over one period.

We illustrate the Fourier technique by considering the following three terms of a Fourier series

$$f(x) = \sin kx + \frac{1}{3}\sin 3kx + \frac{1}{5}\sin 5kx$$

The $\sin kx$, $\sin 3kx$ and $\sin 5kx$ terms are each plotted in Figure 13.45, together with their sum (dark blue line).

Examination of this figure shows that the sum is beginning to approximate to a square wave (light blue line). The wavenumbers used to represent the square wave are shown as a wave number spectrum – a plot of amplitude against wave number – in Figure 13.46.

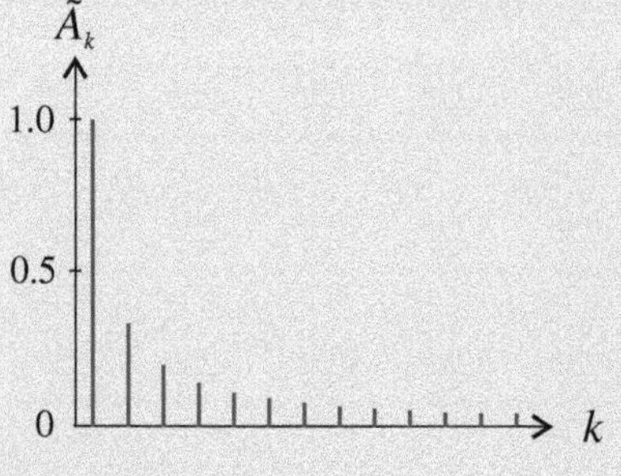

Figure 13.47. Wave number spectrum of the waves used to generate a square waveform.

The addition of further terms, $\frac{1}{7}\sin 7kx + \frac{1}{9}\sin 9kx + \ldots$, to the series, as indicated in the wavenumber spectrum shown in Figure 13.47, builds up the 'squareness' of the wave – the sides become sharper – until a very good representation of a square wave is obtained.

The Fourier series for a periodic function of time t is $y(t) = \sum_n a_n\cos n\omega t + \sum_n b_n\sin n\omega t$, where $\omega = \dfrac{2\pi}{T}$ is the angular frequency of the function.

Using Fourier analysis, the description of wave motion which we have built up for sinusoidal waves in this chapter may be extended to other periodic waveforms.

In discussing standing waves in Section 13.8, we noted that the waveforms of the notes produced by musical instruments are rarely simple sine waves. They include those harmonics of the fundamental frequency which satisfy the condition for standing waves, with different instruments showing different mixes of harmonics. Wave number or frequency spectra, such as those shown in Figures 13.46 and 13.47, may be used to describe this characteristic feature of a musical instrument.

Wave packets

Fourier analysis can also be used to represent wave pulses such as those we used as our first examples of wave motion in Section 13.1. To produce a wave pulse we must superpose sinusoidal waves with a *continuous* range of wave numbers. Such bundles of waves are called ***wave packets***.

We can begin to understand how a wave packet forms by considering as an example the superposition of seven sinusoidal waves of the general form $y = \tilde{A}_k\cos(kx - \omega t)$ at $t = 0$, where the wave numbers k have the values 27, 30, 33, 36, 39, 42 and 45, and the corresponding amplitudes $\tilde{A}_k$ are 0.25, 0.33, 0.5, 1.0, 0.5, 0.33 and 0.25, respectively. $\tilde{A}_k$ is plotted against k in Figure 13.48, a wave number spectrum.

The waveforms of the component waves and the resultant wave at $t = 0$ are shown in Figure 13.49.

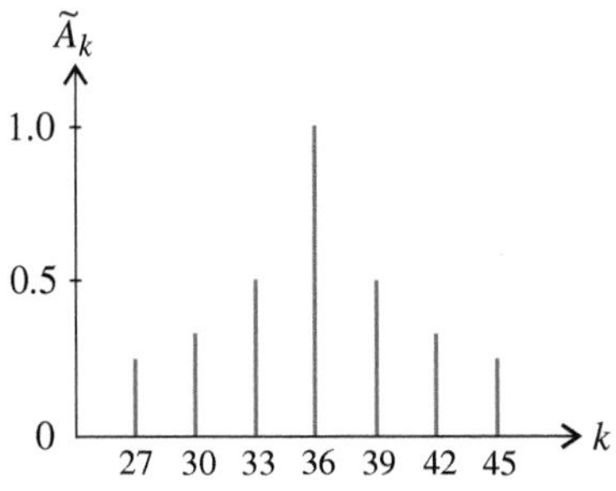

Figure 13.48. The wave number spectrum used to generate the wave packet shown in Figure 13.49.

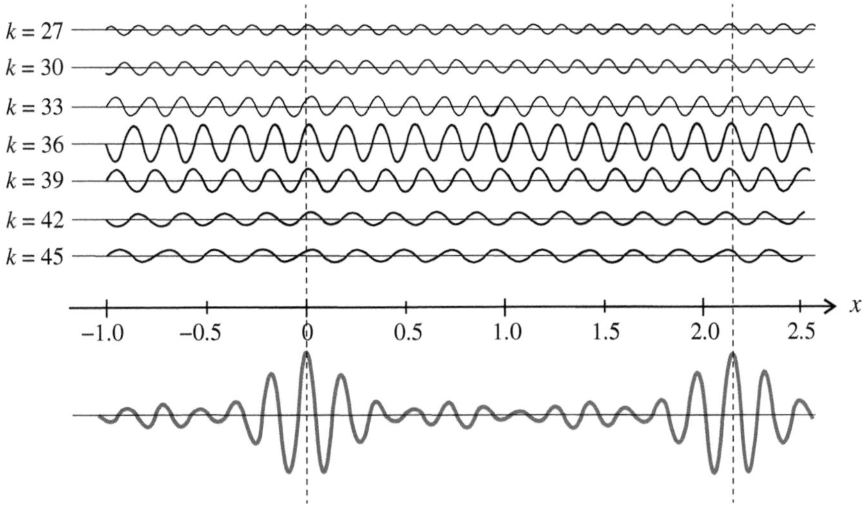

Figure 13.49. The wave groups (lower trace) generated by the superposition of the seven waveforms shown at the top of the figure.

The waves are in phase at $x = 0$ but become out of phase as x increases from this value in either the positive or negative directions. The superposition produces a wave group of finite extent in space. A further wave group occurs to the right in Figure 13.49 because we have used only a finite number of component waves, namely seven. When an infinitely large number of component waves with wave numbers which cover a continuum of k-values is used, only a single wave packet is produced, as illustrated in Figure 13.50. This is because there is no other finite length of the x-axis within which constructive interference occurs; the waves never get back in phase. Note that, in the case of a continuous distribution, the 'size' of a packet is denoted by its full width at half maximum (FWHM) – Δk and Δx in the figure.

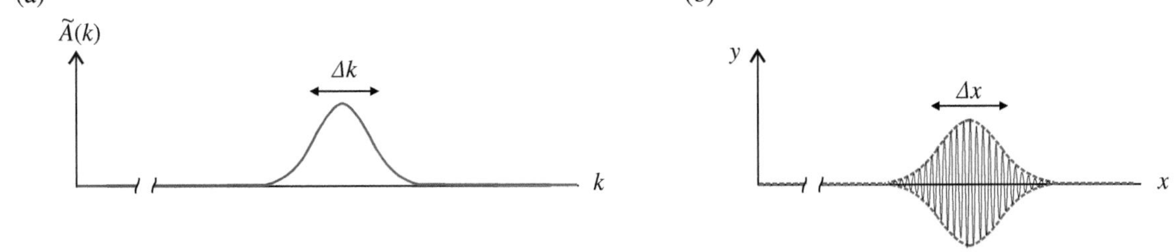

Figure 13.50. (a) The wavenumber spectrum of a group of waves which has a continuous range of wavenumbers of spread Δk. (b) The wave packet produced by the superposition of these waves.

In Figure 13.51 a similar plot is shown for a broader spread of wavenumbers. Comparing Figure 13.51 with Figure 13.50, we see that, as the continuous wave number spectrum becomes broader, that is Δk becomes larger, the width of the wave packet Δx becomes smaller; it becomes more localised.

Note that it follows that relation (13.8), which relates the power carried by a plane sinusoidal wave to the square of its amplitude may also be applied to the component waves of a wave packet and thus to the (varying) amplitude within a wave packet.

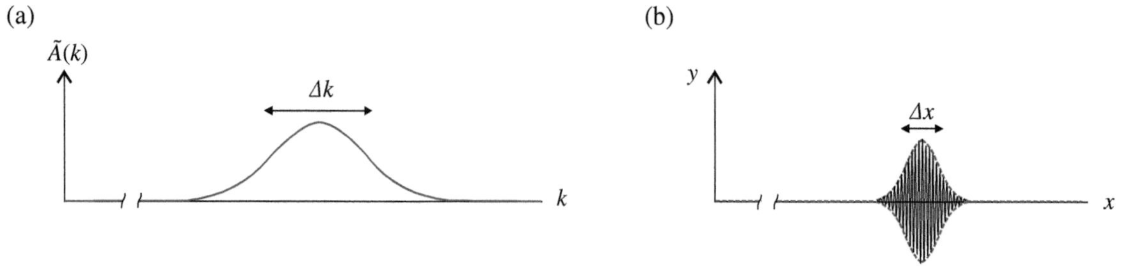

Figure 13.51. (a) The wavenumber spectrum and (b) the wave packet produced when Δk, the wavenumber spread shown in Figure 13.50, becomes broader.

Note also that, when the wavenumber of the component waves form a continuum over a finite interval, the definition of group velocity (Equation 13.30) can be generalised to

$$v_g := \lim_{\Delta k \to 0} \frac{\Delta\omega}{\Delta k} = \frac{d\omega}{dk}$$

If the range of k is not too large, one can consider the carrier wave to have a phase velocity given by

$$v_{ph} = \frac{\omega}{k}$$

where k is now the central frequency in the wave packet.

The superposition of an infinite number of waves is represented by a Fourier integral instead of a Fourier series. Thus at a fixed time a wave group is given by

$$y(x) = \int_0^\infty \tilde{A}(k) \cos kx dk \tag{13.32}$$

where $\tilde{A}(k)$, the amplitude, varies continuously with k. The Fourier integral [1] can be used to show that Δk, the width of the wavenumber spectrum, and the width of the wave packet, Δx, are related through the inequality

$$(\Delta x)(\Delta k) \geq \frac{1}{2} \tag{13.33}$$

Note that the relationship between Δx and Δk as given by the inequality (13.33) is the same, except for a factor 4π, as that obtained for two waves of slightly different frequencies (Equation (13.31)).

When applied to a wave pulse (a wave packet in time), an equivalent Fourier analysis in time can be used to derive an inequality which relates $\Delta\omega$, the width of the frequency spectrum of the wave pulse to Δt, its pulse duration. The inequality is

$$(\Delta t)(\Delta\omega) \geq \frac{1}{2} \tag{13.33a}$$

Dispersion

The velocity of a sinusoidal (harmonic) wave is the ratio of the angular frequency to the wave number, that is, $v = v_{ph} = \frac{\omega}{k}$. In general, it is found that ω depends on k, that is $\omega = \omega(k)$, and that the specific functional relationship depends on the medium in which the wave is travelling. Thus, the speed of a harmonic wave in such a medium will be different for different frequencies, that is $v = v(k)$. This phenomenon is known as **dispersion** and will be encountered again in Chapters 21 and 23. Only in the case where $\omega \propto k$ (v = constant) is there no dispersion; that is, in a non-dispersive medium, $v_g = v_{ph}$.

In the case of a wave packet or pulse, the effect of dispersion is that the packet broadens ('disperses') as it progresses. In the case of a non-dispersive medium, however, each Fourier component in the pulse travels at the same speed so such broadening does not occur.

Worked Example 13.5 gives two examples in which dispersion does occur and the relationship between group velocity and phase velocity is different in each case.

[1] A more detailed description of Fourier techniques and their applications may be found in Chapter 10 of Pain (2005).

Further reading

Further introductory material on wave motion may be found in *Vibrations and Waves* by King.
A more advanced treatment of vibrations and waves may be found in *Physics of Vibrations and Waves* by Pain.
See BIBLIOGRAPHY for book details.

Study Worked Example 13.5

For problems based on the material in this section visit up.ucc.ie/13/ *and follow the link to the problems.*

WORKED EXAMPLES

Worked Example 13.1: A transverse sinusoidal wave of frequency 25 Hz travels along a string at a speed of 30 m s^{-1}. At a given instant a certain point in the string is displaced by 15 mm from its equilibrium position and has a velocity of 1.5 m s^{-1} perpendicular (transverse) to the string. Taking the given instant to be $t = 0$ and the point to be at $x = 0$, derive an expression which gives the displacement of any point in the string as a function of position and time. Calculate the transverse displacement of a point on the string, which is + 0.6 m from the point described above, 20 ms after the starting instant.

The general equation of a travelling wave in one dimension can be written (Section 13.2),

$$y = A\sin(kx - \omega t + \phi) \tag{13.34}$$

where, in this case, $\lambda = \dfrac{v}{f} = \dfrac{30}{25} = 1.2\text{ m.}$ Thus $k = \dfrac{2\pi}{\lambda} = 5.2\text{ m}^{-1}$ and $\omega = 2\pi f = 160\text{ s}^{-1}$

The velocity of the displaced point is obtained by differentiating Equation (13.34),

$$\frac{dy}{dt} = -A\omega\cos(kx - \omega t + \phi) \tag{13.35}$$

Substituting conditions at $x = 0$, $t = 0$ into equations (13.34) and (13.35) we obtain

$$y = 0.015 = A\sin\phi \quad \text{and} \quad \frac{dy}{dt} = 1.5 = -A\omega\cos\phi$$

Solving these two simultaneous equations, we obtain $\phi = -\ 1.01$ rad and $A = -\ 0.018$ m
Thus, the equation of the travelling wave is $y = -\ 0.018\sin(5.2x - 160t - 1.0)$
Note that all angles must be expressed in radians.
When $x = +\ 0.6$ m and $t = 20 \times 10^{-3}$s

$$y = -0.018\ \sin[(5.2 \times 0.6) - (160 \times 20 \times 10^{-3}) - 1.01] = -0.018\sin(-1.09) = 0.016\text{ m.}$$

Worked Example 13.2: Two sinusoidal waves, identical except for a phase difference φ, travel in the same direction along a taut string. The two waves produce a resultant wave with an amplitude which is 25% greater than that of either of the original two waves. Determine the value of the phase difference.

The two waves may be represented by the travelling wave equations

$$y_1 = A\sin(kx - \omega t) \quad \text{and} \quad y_2 = A\sin(kx - \omega t + \varphi),$$

From the principle of superposition, the resultant wave is given by

$$y = y_1 + y_2 = A\sin(kx - \omega t) + A\sin(kx - \omega t + \varphi),$$

which, using the trigonometric identity $\sin\alpha + \sin\beta = 2\sin\frac{1}{2}(\alpha+\beta)\cos\frac{1}{2}(\alpha-\beta)$ (Appendix A.3.5 (xi)), can be written $y = 2A\cos\frac{\varphi}{2}\sin\left(kx - \omega t + \frac{\varphi}{2}\right)$.
The amplitude of this wave, $2A\cos\frac{\varphi}{2}$, is 1.25 times that of either of the original two waves so

$$2A\cos\frac{\varphi}{2} = 1.25A \quad \Rightarrow \quad \cos\frac{\varphi}{2} = 0.625 \quad \Rightarrow \quad \frac{\varphi}{2} = 0.896\text{ rad} \quad \Rightarrow \quad \varphi = 1.79\text{ rad.}$$

Worked Example 13.3: A 512 Hz tuning fork is used to set up standing waves on a taut string which is fixed at both ends. The standing waves are observed to have four antinodes of amplitude 2.5 mm. If the wave speed along the string is 450 m s^{-1}, determine the length of the string.
Write down an equation which gives the displacement of the string as a function of distance and time.

As indicated in Figure 13.28, the observation of four antinodes in the standing wave pattern means that the string length, L, corresponds to two wavelengths.

$$\text{Now } \lambda = \frac{L}{2} = \frac{v}{f} \Rightarrow L = \frac{2v}{f} = \frac{2 \times 450}{512} = 1.76 \text{ m.}$$

The displacement of the string as a function of position and time is given by Equation (13.15)

$$y(x,t) = (2A\sin kx)\cos\omega t,$$

where $A = 2.5$ mm, $k = \frac{2\pi}{\lambda} = \frac{2\pi f}{v} = \frac{2\pi \times 512}{450} = 7.15 \text{ m}^{-1}$ and $\omega = 2\pi f = 3220 \text{ rad s}^{-1}$, so the equation is $y(x,t) = (5.00 \text{ mm})\sin(7.15 \text{ m}^{-1})x\cos(3220 \text{ rad s}^{-1})t$.

Worked Example 13.4: An ambulance, emitting a sound of frequency 1600 Hz, passes a cyclist who is travelling at 20.0 km h^{-1}. If the frequency heard by the cyclist, as the ambulance approaches, is 1640 Hz determine (i) the speed of the ambulance and (ii) the frequency heard by the cyclist after the ambulance has passed. Assume that the ambulance and cyclist are both travelling at constant velocities and take the speed of sound in air to be 1230 km h^{-1}.

(i) The frequency of the sound as the ambulance (the source) approaches the cyclist (the observer) is given by Equation (13.24) $f' = \frac{(v + v_O)}{(v - v_S)}f$ where $v_o = -20.0 \text{ km h}^{-1}$ (negative because the observer is moving away from the source).

Thus $$1640 = \frac{1230 - 20}{1230 - v_S} \times 1600 \Rightarrow 1230 - v_S = \frac{1210 \times 1600}{1640} \Rightarrow v_S = 50 \text{ km h}^{-1}$$

(ii) The frequency after the ambulance has passed is again given by Equation (13.24) although v_o is now positive because the observer is moving towards the source and v_S is negative because the source is moving away from the observer.

Thus $$f' = \frac{1230 + 20}{1230 + 50} \times 1600 \Rightarrow f' = 1560 \text{ Hz}$$

Worked Example 13.5: The dispersion of surface waves in water depends on the surface tension (T) and density (ρ) of the liquid and on the acceleration due to gravity (g) as follows

(i) for $\omega = \sqrt{\frac{Tk^3}{\rho}}$ for small wavelengths and (ii) for $\omega = \sqrt{gk}$ for long wavelengths.

Derive expressions for the phase velocity and the group velocity in each of these cases.

$$\text{(i)}\ \omega = \sqrt{\frac{T}{\rho}}k^{\frac{3}{2}} \rightarrow v_{ph} = \frac{\omega}{k} = \sqrt{\frac{T}{\rho}}k^{\frac{1}{2}} \text{ and } v_g = \frac{d\omega}{dk} = \frac{3}{2}\sqrt{\frac{T}{\rho}}k^{\frac{1}{2}} \rightarrow v_g = \frac{3}{2}v_{ph}$$

$$\text{(ii)}\ \omega = \sqrt{g}k^{\frac{1}{2}} \rightarrow v_{ph} = \frac{\omega}{k} = \sqrt{g}k^{-\frac{1}{2}} \text{ and } v_g = \frac{d\omega}{dk} = \frac{1}{2}\sqrt{g}k^{-\frac{1}{2}} \rightarrow v_g = \frac{1}{2}v_{ph}$$

PROBLEMS

For problems based on the material covered this chapter visit up.ucc.ie/13/ and follow the link to the problems.

14

Introduction to quantum mechanics

AIMS

- to describe how, at the beginning of the twentieth century, a series of phenomena had been observed which could not be explained in terms of classical physics
- to explain how experimental evidence that, in certain circumstances, waves can behave like particles and particles can behave like waves — *wave-particle* duality — led to an understanding of these unexplained phenomena
- to show how the wave and particle models can be reconciled by considering the properties of wave packets and to show how this leads to a fundamental uncertainty in specifying the position and momentum of a particle simultaneously - the *Heisenberg uncertainty principle*
- to develop a formalism - *quantum mechanics* - which enables physical situations to be analysed in a manner which is consistent with the uncertainty principle
- to use quantum mechanics to analyse some simple one-dimensional problems and to show how quantum mechanical analyses frequently produce results which are very different from those obtained from classical treatments of the same situations
- to show how the main features of quantum mechanical solutions of more complex situations can be anticipated from quantum mechanical solutions of simpler problems

14.1 Physics at the beginning of the twentieth century

At the end of the nineteenth century, physics was at its most confident. Classical physics, as formulated in Newton's laws of mechanics and in Maxwell's theory of electromagnetism (which we shall study in Chapter 21) had proved very successful in solving almost every problem to which it had been applied. These theories appeared to be *universal* in their range of application; for example, we have seen in Chapter 12 how Newton's laws, which were developed to account for the motion of the planets, can be used equally well to interpret the motion of molecules on the microscopic scale.

A further feature of the laws of physics, as viewed at that time, was that, if all the initial conditions were known exactly for a physical system in a particular situation, the subsequent evolution of the system could be predicted with complete certainty. While it was clear that it was usually impractical to expect to know all initial conditions, such as the position and velocity of each of the millions of molecules in a gas, with certainty, the important point was that, in principle, these conditions were knowable and that, therefore, there could be only one possible outcome of any precisely defined initial situation.

At the end of the nineteenth century many believed that there was no question concerning the material universe for which physics could not provide an answer, at least in principle. There remained, however, a number of simple phenomena which classical physics had persistently failed to explain, some of which we have already encountered in Sections 11.7, 12.2, 12.8 and 13.8. While many expected that these problems would ultimately yield to classical solutions given the time and patience needed to unravel the complexity of real physical situations, such optimism proved unfounded.

The formulation of classical physics which we have developed in the preceding chapters has been based closely on our everyday experiences in the macroscopic world. The instincts which we have developed, therefore, are those of classical physics. Even when studying classical physics, however, we have had to set aside our intuition in advancing our understanding. In Section 3.1, for example, we noted that, while our experience of the world leads us to believe that bodies come to rest when no forces are applied to them, we had to abandon this belief in order to develop the laws of motion. Consequently, it should be no surprise if we find it difficult, initially, to accept the concepts on which quantum physics is based. In many ways these also seem to run counter to our intuition and to common sense. In this respect, our position is similar to that of physicists at the end of the nineteenth century. The concepts which quantum physics introduced seemed so strange and arbitrary to physicists of the time that they took many years to gain acceptance; in fact, many of the

Understanding Physics, Third Edition. Michael Mansfield and Colm O'Sullivan.
© 2020 John Wiley & Sons Ltd. Published 2020 by John Wiley & Sons Ltd.

physicists who played major roles in promoting this revolution in thinking were reluctant to accept it fully because it seemed so foreign to the traditions of classical physics. Even today the implications of quantum physics are probably not yet fully appreciated.

Thus, like the early twentieth century physicists, we too must build up confidence in quantum physics slowly by examining the evidence for the limitations of the classical approach and by appreciating how quantum physics accounts for situations in which classical physics fails. The next six sections give an historical outline of some of the phenomena which classical physics failed to explain and which, when examined carefully, provide evidence for quantum effects. This will lead us to reassess the distinctions between waves and particles which we have noted in Section 13.3 and will also introduce a fundamental uncertainty into our description of microscopic phenomena. We then develop **quantum mechanics**, a method of analysing physical situations which can accommodate this uncertainty. Finally, we shall apply quantum mechanics to some important physical situations, drawing attention to the remarkable differences which can arise between the results of classical and quantum dynamical analyses.

14.2 The blackbody radiation problem: Planck's quantum hypothesis

We saw in Sections 11.7 and 13.8 that all attempts by Wien, Rayleigh, Jeans and others to explain the spectrum of the radiation emitted from a cavity, and by implication by blackbodies, had ended in failure. Quantum physics is often considered to have been born with the presentation of a paper on the emission spectrum of cavity radiation by Max Planck (1858–1947) in late 1900. As we shall see below, in order to explain the failure of classical physics to account for the observed blackbody spectrum, Planck found it necessary to introduce a very new hypothesis governing the behaviour of harmonic oscillators – the *quantum hypothesis* – which was completely at variance with classical physics.

In 1900 Planck was professor of physics at the Friedrich Wilhelm University, Berlin and was aware of the results of experiments being performed in laboratories in the city. In seeking an explanation of the observed frequency distribution of cavity radiation he treated the problem from a very different viewpoint to that of Rayleigh. He concentrated more on the oscillators in the cavity walls which he addressed through his own version of the probabilistic dynamical approach to the study of heat which had been developed by Boltzmann in the 1870s. We will not describe Planck's approach here but confine ourselves to pointing out that underlying his treatment of the problem was the hypothesis that the oscillators, such as atoms in the walls of the cavity, can take only certain discrete energy values. These energies are given by **Planck's hypothesis** which states that *a system which is performing simple harmonic motion of frequency f can only have values of energy which satisfy the equation*

$$E_n = nhf \tag{14.1}$$

where $n = 0, 1, 2, 3, \ldots$ (integers) and h is a constant (known as the **Planck constant**). This is the constant whose value, $h = 6.626\,070\,15 \times 10^{-34}$ kg m^2 s^{-1} (or J s), was chosen to be fixed for the purpose of defining the kilogram in SI, as discussed in Section 3.4.

Rayleigh had shown (Equation (13.21a)) that the energy density of the waves in a cavity with frequencies between f and $f + df$ is given by

$$w(f,T)df = \frac{8\pi}{c^3}\langle E\rangle f^2 df \tag{14.2}$$

and had assumed, applying the principle of equipartition of energy, that average energy was given by $\langle E\rangle = kT$. However, if the Planck hypothesis (Equation (14.1)) is applied to the oscillations in the cavity it gives rise to a different expression for the average energy. Assuming that the Maxwell-Boltzmann statistics is applicable, $\langle E\rangle$ can be determined using Equation (12.11) as follows.

$$\langle E\rangle = \frac{\sum\limits_{n=0}^{\infty} E_n e^{-\frac{E_n}{kT}}}{\sum\limits_{n=0}^{\infty} e^{-\frac{E_n}{kT}}} = \frac{\sum\limits_{i=0}^{\infty} (nhf) e^{-\frac{nhf}{kT}}}{\sum\limits_{n=0}^{\infty} e^{-\frac{nhf}{kT}}}$$

Recognising that the denominator is a sum to infinity of a geometric progression whose sum is $\dfrac{1}{1 - e^{-\frac{hf}{kT}}}$ (Appendix A.9.5) and that the numerator is the negative of the derivative of the denominator with respect to $\dfrac{1}{kT}$, we find, after a few steps of algebra, that

$$\langle E\rangle = \frac{hf}{e^{\frac{hf}{kT}} - 1} \tag{14.3}$$

Replacing $\langle E \rangle$ in Equation (14.2) by this expression we get

$$w(f,T)df = \frac{8\pi h f^3}{c^3\left(e^{\frac{hf}{kT}}-1\right)}df \qquad (14.4)$$

When this expression is plotted as a function of frequency at different temperatures (Figure 14.1) it is found to be a good fit to the experimental results for cavity radiation, as described in Section 11.7.

Note that at frequencies which are sufficiently low that $hf \ll kT$, we can write $e^{\frac{hf}{kT}} \approx 1 + \frac{hf}{kT}$ and Equation (14.4) reduces to the classical Equation (14.2). Moreover, both the Wien displacement law (recall Section 11.7) and the Stefan-Boltzmann law (also Section 11.7) can be shown to follow from Equation (14.4), results which are derived in Worked Examples 14.1 and 14.2, respectively

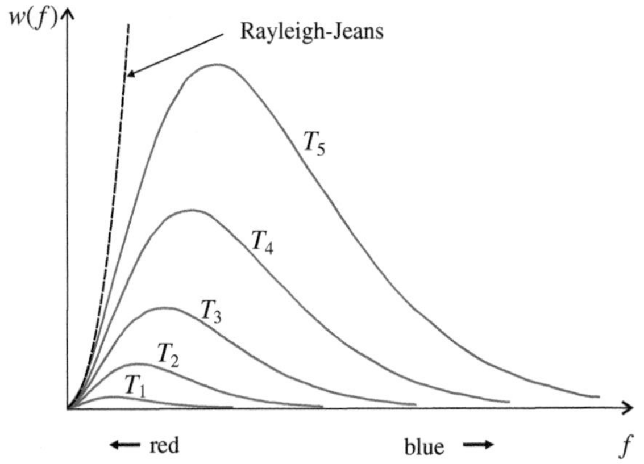

Figure 14.1. Energy density spectra of cavity radiation predicted by Planck's quantum hypothesis for five different temperatures ($T_5 > T_4 > T_3$, etc.). The Rayleigh-Jeans plot corresponding to the highest of these temperatures (T_5) is indicated.

Study Worked Examples 14.1 and 14.2

The Planck hypothesis clearly works — it predicts the observed spectrum of cavity radiation. However, it was introduced in an *ad hoc* manner and runs contrary to our previous understanding of the energy of a harmonic oscillator. Classically, we know that the total energy of such an oscillator is proportional to the square of the amplitude of the motion (Equation (3.16)). In that treatment, the energy of the oscillator, like its amplitude, can vary continuously from zero to infinity. Planck's hypothesis arbitrarily dismisses this aspect of the classical analysis and states that the energy of the oscillator can take only certain values, $E_n = nhf$, as indicated on the left in Figure 14.2. A consequence of the Planck hypothesis is that any losses of energy must occur in amounts which are multiples of hf. These very small quantities of energy are known as *quanta* (singular *quantum*). Planck himself was unhappy with the arbitrary nature of his hypothesis and it took a number of years before he and others accepted its radical nature and its consequences for physics.

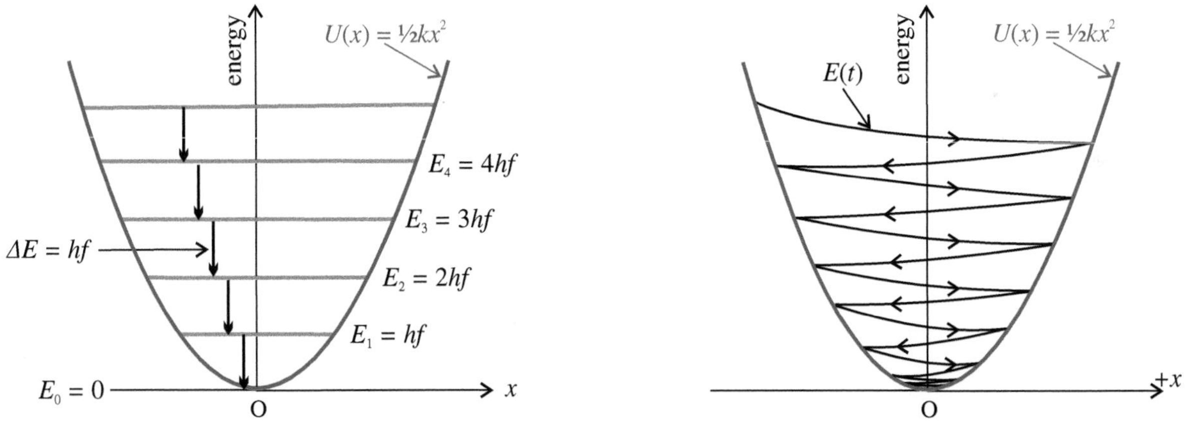

Figure 14.2. The quantised energies $E_n = nhf$ of a simple harmonic oscillator as proposed in the Planck (left) compared to the energy lost by a classical damped harmonic oscillator, as shown in Figure 3.27.

While, at this stage, we cannot avoid the arbitrary nature of Planck's quantum hypothesis, we can at least reconcile the hypothesis with our experience of the macroscopic world. The following question arises, for example. If an oscillator can lose energy (and thus amplitude) only in quantum jumps of hf, as predicted by Planck's hypothesis, why is this not evident in the behaviour of a macroscopic oscillator, such as the mass oscillating at the end of a helical spring (studied in Chapter 3)? We can answer this question by estimating the size of such jumps in energy in a typical case of (say) a 1 kg mass which is performing simple harmonic oscillations of amplitude 0.1 m at the end of a spring of spring constant $k = 100\ \mathrm{N\,m^{-1}}$.

The total energy of the motion is given by Equation (3.16) $E = \frac{1}{2}kA^2 = 0.5$ J and the frequency of the oscillation is given by Equation (3.10), $f = \frac{1}{2\pi}\sqrt{\frac{k}{m}} = 1.6$ Hz, so that Planck's hypothesis gives the separation of adjacent energy levels, ΔE in Figure 14.2, to be $hf \approx 10^{-33}$ J.

From Planck's hypothesis, therefore, we can conclude that the oscillator loses energy in quantum jumps of 10^{-33} J, an energy which is 10^{33} times smaller than its total energy. Such changes are far too small to observe and Planck's hypothesis has no observable effects on the motion of a macroscopic oscillator, although the consequences of the hypothesis may be noticed in macroscopic effects which depend ultimately on microscopic behaviour, such as the black-body spectrum.

Quantum effects are not observable in the motion of macroscopic objects and this is one reason why we find it difficult to visualise such effects — they are outside our everyday experience. Because the Planck constant is so small, quantum effects are noticeable only in microscopic systems. If, however, h had been thirty orders of magnitude larger, quantum effects would be an important part of our everyday experience.

For problems based on the material presented in this section visit up.ucc.ie/14/ *and follow the link to the problems.*

14.3 The specific heat capacity of gases

It was pointed out in Section 12.2 that the observed dependence of the molar heat capacity of a diatomic gas on temperature, such as that shown in Figure 14.3, cannot be explained satisfactorily by standard Newtonian mechanics. All that we have learned of classical dynamics would suggest that rotational and vibrational degrees of freedom should contribute to the internal energy of a gas. While this is indeed the case at high temperatures, the observed behaviour at lower temperatures indicates otherwise.

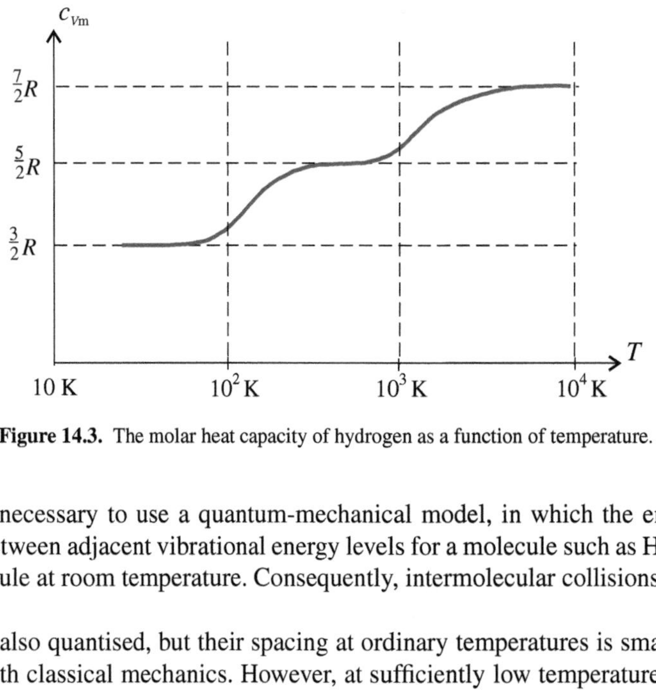

Figure 14.3. The molar heat capacity of hydrogen as a function of temperature.

For a satisfactory description it is necessary to use a quantum-mechanical model, in which the energy of an individual molecule is quantised. The energy separation between adjacent vibrational energy levels for a molecule such as H_2 is about ten times greater than the average kinetic energy of the molecule at room temperature. Consequently, intermolecular collisions do not provide sufficient energy to excite vibrational states.

The rotational energy levels are also quantised, but their spacing at ordinary temperatures is small compared with kT so the system normally behaves in accordance with classical mechanics. However, at sufficiently low temperatures (typically less than 50 K), where kT is small compared with the spacing between rotational levels, intermolecular collisions may not be sufficiently energetic to alter the rotational states. Collisions between molecules at low temperatures, therefore, do not provide enough energy to change the vibrational state of the molecule so vibrational energy does not contribute to the molar specific heats of molecules at these temperatures.

14.4 The specific heat capacity of solids

One scientist who seems to have grasped the significance of Planck's ideas at an early stage was Albert Einstein. In 1907 he applied the Planck hypothesis to explain the departures of measured molar heat capacities of solids from the predictions of the classical microscopic model of a solid.

We saw, in Section 12.8, that the classical model predicted that the specific heat capacity of a solid did not vary with temperature (Dulong and Petit law — horizontal line in Figure 12.11). This arose from the application of the principle of equipartition of energy to the oscillators in the model; that is, an energy of $\frac{1}{2}kT$ is associated with each of the six degrees of freedom in this case. Thus, the model yields an internal energy of

$$U(T) = 6N\left(\tfrac{1}{2}kT\right) = 3NkT \tag{14.5}$$

and molar heat capacity at constant volume is given by

$$C_{Vm} = 3N_A k = 3R \tag{14.6}$$

While the model works well at high temperatures, experimental measurements show significant variation from the Dulong and Petit behaviour at lower temperatures (as shown in Figure 12.11).

Einstein applied the Planck hypothesis to the oscillators in the model arguing that, if the energies of the oscillators are quantised, the right hand side of Equation (14.5) should be replaced by $3N\langle E\rangle$, where $\langle E\rangle$ is the average energy of a one-dimensional oscillator given in Equation (14.4). Thus Equation (14.5) should be replaced by

$$U(T) = 3N\langle E\rangle = 3N\frac{hf}{e^{\frac{hf}{kT}} - 1} = 3NkT\frac{\frac{hf}{kT}}{e^{\frac{hf}{kT}} - 1}$$

The heat capacity of the system is obtained by differentiating this expression with respect to T which, after rearranging terms and dividing by the number of moles, yields a molar heat capacity of

$$c_{Vm} = \frac{1}{n}\frac{dU}{dT} = 3N_A k\frac{\left(\frac{hf}{kT}\right)^2 e^{\frac{hf}{kT}}}{e^{\frac{hf}{kT}} - 1} \tag{14.7}$$

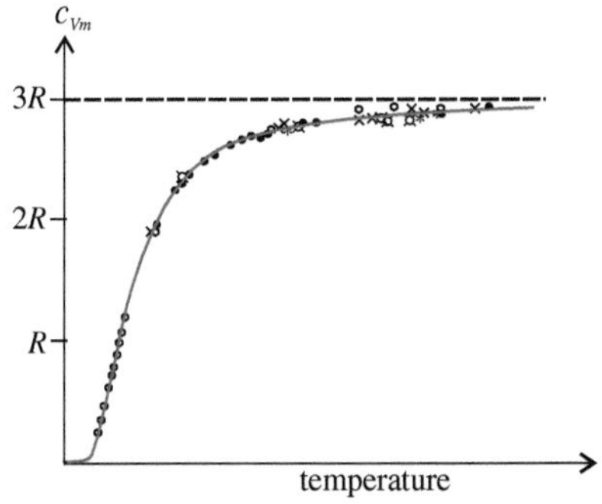

Figure 14.4. Einstein's 1907 theory of the specific heat capacities of solids (blue line) compared with the experimental data of Figure 12,11.

This expression is plotted as a function of temperature in Figure 14.4 and, as can be seen, gives a reasonably good fit to the experimental data at all temperatures.

Not all of Einstein's assumptions that led to this result turned out to be exactly correct (we will see, for example, in Section 14.18 that the quantised energy levels shown in Figure 14.2 are not quite correct). A subsequent modification by the Dutch physicist Peter J.W. Debye (1884–1966) in 1912 gave an even better fit to the data. Nevertheless, the publication of Einstein's theory, explaining the low temperature behaviour of the specific heat of solids, played an important role in gaining wider acceptance for the Planck hypothesis.

14.5 The photoelectric effect

The photoelectric effect was discovered in 1887 by the German physicist Heinrich Hertz (1857–1894). In connection with work on radio waves, Hertz observed that, when ultraviolet light shines on two metal electrodes with a voltage applied across them, the light changes the voltage at which sparking takes place. This relationship between light and electricity (hence *photoelectric*) was clarified in 1902 by Philipp Lenard (1862–1947). He demonstrated that electrically charged particles were liberated from an illuminated metal surface and identified these particles as electrons. When such a light beam (an electromagnetic wave) strikes a metal surface, as illustrated in Figure 14.5, the energy delivered by the light must be sufficient to overcome the attractive fields of atoms in the surface of the metal. This energy is known as the *work function* of the metal and is a property of the particular metal used. The work function will be discussed further in Section 26.4.

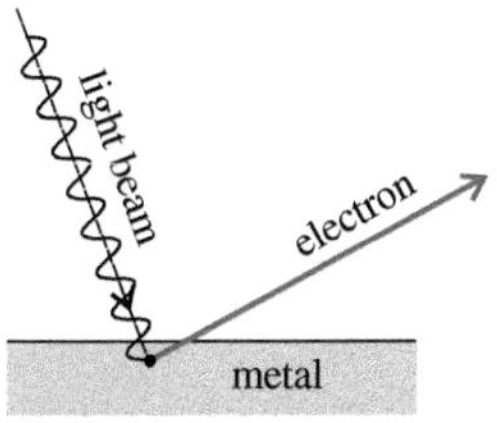

Figure 14.5. The photoelectric effect. A light beam releases an electron from a metal surface.

In experiments to study this effect (see up.ucc.ie/14/5/1 for details) the maximum kinetic energy of the electrons emitted from a metal surface is measured for different frequencies of the incident radiation. Typical results from such an experiment are shown in Figure 14.6 where maximum kinetic energy (K_m) is plotted as a function of frequency.

The following results emerge from such an experiment:

(a) There is a limiting frequency f_c (which is material dependent) below which no electrons are emitted, however high the intensity of the light.
(b) For any fixed value of the frequency greater than f_c, K_m is independent of the intensity of the light.
(c) K_m is found to be related linearly to f.
(d) Electrons are found to be emitted instantaneously when the light falls on the metal surface, even when the intensity of the incident light is very low.

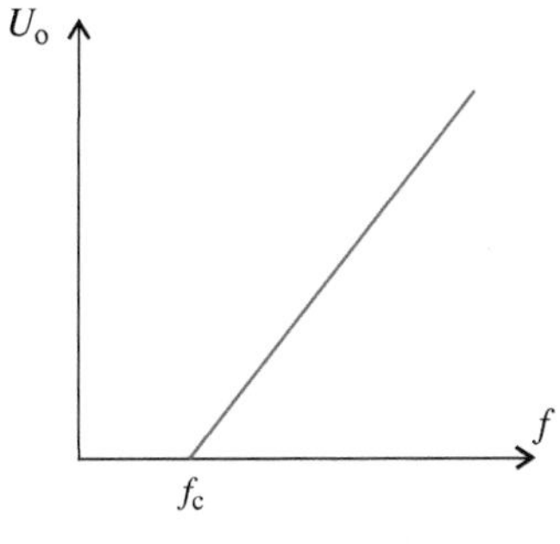

Figure 14.6. The photoelectric effect. The observed dependence of K_m (the maximum kinetic energy of an emitted electron) on frequency.

Let us consider these experimental results within the context of our (classical) understanding of waves, as developed in the previous chapter.

We noted, in Section 13.3, that wave energy is distributed over the space occupied by a wave and that the intensity of a wave is proportional to the square of its amplitude. From classical considerations, therefore, we expect the mechanism of the photoelectric effect to be as follows. The electromagnetic wave causes electrons in the metal to oscillate. If the energy of the oscillation is greater than ϕ, the work function of the cathode metal, electrons will be emitted with a maximum kinetic energy $K_m = E_w - \phi$, where E_w is the energy delivered by the wave. Thus, we would expect the photoelectric effect to exhibit the following features:

- Whatever the frequency of the light wave, f, it will eventually deliver enough energy to release an electron, if the light intensity is high enough. There should not be therefore a limiting frequency, f_c, as indicated by experimental result (a) above.
- Because the energy carried by a wave is delivered continuously and incrementally, there should be a finite time lapse before the energy, ϕ, needed to free an electron can be accumulated at the point occupied by the electron. Photoemission should not occur instantaneously when light falls on the cathode. Experimental result (d) above is therefore in direct disagreement with our expectations from classical wave theory.

The observed results were explained in 1905 by Einstein, who postulated that the energy of an electromagnetic wave is delivered by a stream of localised small *energy bundles* (which we will call **photons** in Section 14.7) and that the energy of each bundle, E, is determined by the wave frequency f through the equation

$$E = hf \tag{14.8}$$

where h is again the Planck constant. This postulate bears obvious similarities to the Planck hypothesis (Equation (14.1)) but note that Einstein goes a step further than Planck; quantisation is now presented as a property of the electromagnetic wave itself, rather than of the oscillating charges which produce the wave.

In the Einstein model of the photoelectric effect, the incident light beam interacts with the electrons in the metal as a beam of localised energy bundles, each of energy hf, rather than as waves. Increasing the intensity of the light beam, therefore, increases the number of energy bundles but not the energy of each bundle. When one energy bundle interacts with an electron in the metal its energy hf is used to provide the energy needed to release the electron (ϕ) and any excess energy is used to give kinetic energy to the electron (K_m). Thus

$$K_m = hf - \phi \tag{14.9}$$

This equation accounts for the experimental results (a) to (d) noted above:

(a) When $hf < \phi$, the equation cannot be satisfied. The energy bundles cannot deliver sufficient energy to release an electron, however great the intensity of the light — that is, however many energy bundles there are in the light beam. The equation $f_c = \dfrac{\phi}{h}$ therefore defines a limiting frequency, corresponding to $K_m = 0$, beneath which no electrons can be emitted.

(b) If f is fixed, K_m is fixed by the equation and is independent of the intensity of the light.

(c) The equation predicts that K_m is linearly related to f. A plot of K_m against f (Figure 14.7) produces a straight line graph in accordance with observation (Figure 14.6). As indicated in Figure 14.7, and as illustrated in Worked Example 14.3, such a graph can be used to determine h (from the slope of the graph) and ϕ (from f_c, the intercept on the f axis).

(d) The process is instantaneous because electrons do not have to build up energy from a wave whose energy is distributed in space.

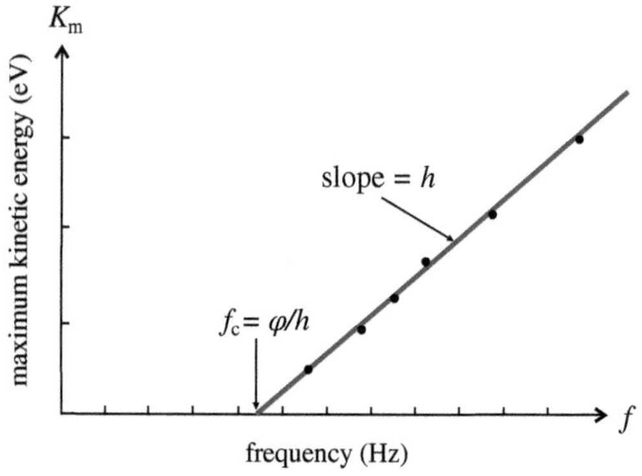

Figure 14.7. The photoelectric effect. The dependence of K_m (the maximum kinetic energy of an electron emitted at the cathode) on frequency, as predicted by the photoelectric effect equation (14.9).

The essence of the Einstein explanation is that the photoelectric effect is treated as a particle-particle collision, the particles being an energy bundle and an electron in the metal surface. A single energy bundle with energy $hf > \phi$ can release an electron immediately.

An apparently contradictory picture of the nature of electromagnetic waves is therefore emerging. In certain circumstances an electromagnetic wave can behave like a particle — that is, in an apparently un-wavelike manner. As indicated at the beginning of this chapter, quantum physics will often defy the understanding of the physical world which we have developed previously. It is interesting to note that Planck, in many ways the instigator of quantum physics, was not comfortable with the implications of the revolution in thinking which he had set in motion. In recommending Einstein for membership of the Prussian Academy of Sciences in 1914 he said:

> "Summing up, we may say that there is hardly one of the great problems in which modern physics is so rich to which Einstein has not made an important contribution. That he may sometimes have missed the target in his speculations as, for example, in his hypothesis of light quanta, cannot really be held against him, for it is not possible to introduce fundamentally new ideas, even in the most exact sciences, without taking a risk".

It is hardly surprising, therefore, if we also find the underlying concepts difficult to accept. We shall eventually reconcile the apparently contradictory behaviour of electromagnetic waves which is emerging from our analysis, but first let us consider further evidence of quantum behaviour.

Study Worked Example 14.3

For problems based on the material presented in this section visit up.ucc.ie/14/ *and follow the link to the problems*

14.6 The X-ray continuum

When electrons which are travelling at very high velocities collide with a metal target, their rapid deceleration produces high frequency electromagnetic radiation known as **X-rays**. This process is described by the German word *bremsstrahlung*, meaning 'braking radiation'.

A typical spectrum of the radiation emitted by an X-ray tube is shown in Figure 14.8. The X-ray spectrum displays two main features:

- Discrete emission lines with frequencies which are characteristic of the target material. These lines, which result from the excitation of electrons in target atoms are also of quantum origin and will be discussed further in Section 24.13.
- An emission continuum whose profile is found to be independent of the target material. The maximum frequency f_{max} of the observed X-ray continuum is found (the Duane-Hunt law of 1915) to be directly proportional the kinetic energy K of the electrons as they strike the metal target, that is, $f_{max} \propto K$.

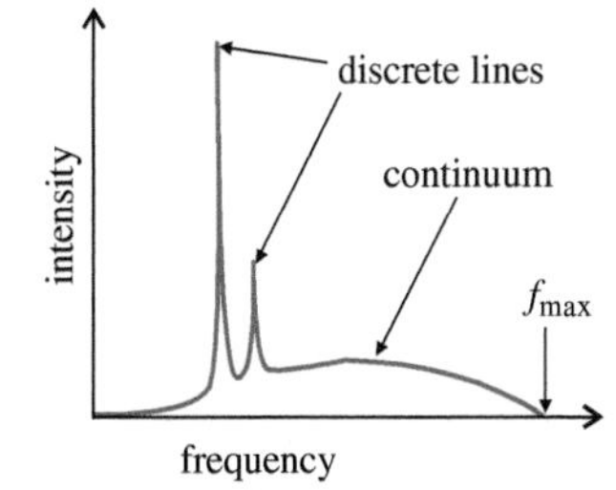

Figure 14.8. A typical spectrum of an X-ray source. Discrete emission lines are superposed on a background of continuous emission. Note that no X-rays are emitted with frequencies greater than a critical value f_{max}.

The occurrence of a maximum frequency requires a quantum explanation. The emission of electromagnetic radiation of frequency f_{max} corresponds to the case in which an electron loses all of its energy in a single collision, with this energy being released as an energy bundle of electromagnetic waves. In the production of X-rays, the kinetic energy of an electron is converted into a bundle of electromagnetic energy so that the process may be regarded as the opposite process to the photoelectric effect.

The energy of the energy bundle, given by the Einstein equation (14.8), is $E = K = hf_{max}$ and thus

$$f_{max} = \frac{K}{h} \tag{14.10}$$

Quantum theory, therefore, predicts that $f_{max} \propto K$, as observed, thereby providing further evidence of the validity of Einstein's postulate. The X-ray continuum at frequencies less than f_{max} corresponds to cases in which an electron loses its energy in a number of collisions in the target. Such collisions produce X-ray energy bundles whose energies correspond to frequencies which are less than f_{max}.

For problems based on the material presented in this section visit up.ucc.ie/14/ *and follow the link to the problems.*

14.7 The Compton effect: the photon model

As a final example of an experiment that established evidence of the quantum nature of electromagnetic radiation we consider the *Compton effect*, observed by the US physicist Arthur Holly Compton (1892–1962) in 1923. We shall see that, to analyse this effect, we must apply the principles of conservation of energy and momentum to the energy bundle. We have to treat the energy bundle quite unambiguously as a particle with a well specified energy and momentum. Such a particle is called a **photon**.

The Compton effect may be demonstrated using the experimental arrangement illustrated in Figure 14.9. Monochromatic X-rays of wavelength λ fall on a graphite target. The wavelengths of the X-rays which are scattered from the target are measured by an X-ray spectrometer as a function of the angle θ through which they are scattered. The observed X-ray spectrum shows two distinct peaks, one at λ, the wavelength of the incident X-rays, and a second at a longer wavelength λ' (Figure 14.10). It is found experimentally that the difference between these two wavelengths, $\Delta\lambda = \lambda' - \lambda$, varies with the angle θ according to the relation:

$$\Delta\lambda \propto (1 - \cos\theta)$$

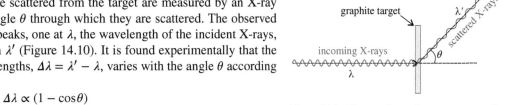

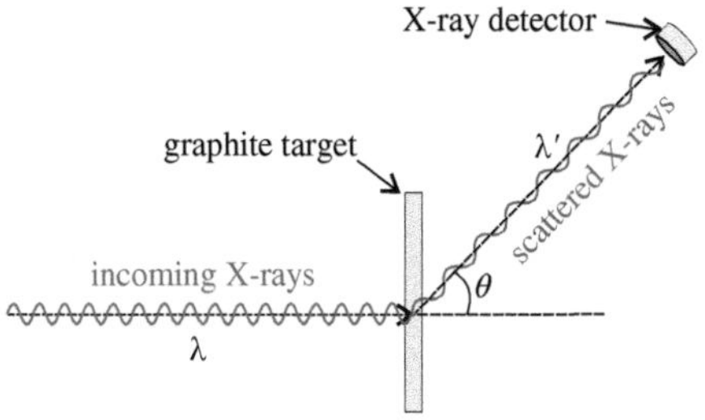

Figure 14.9. The experimental arrangement used to investigate the Compton effect.

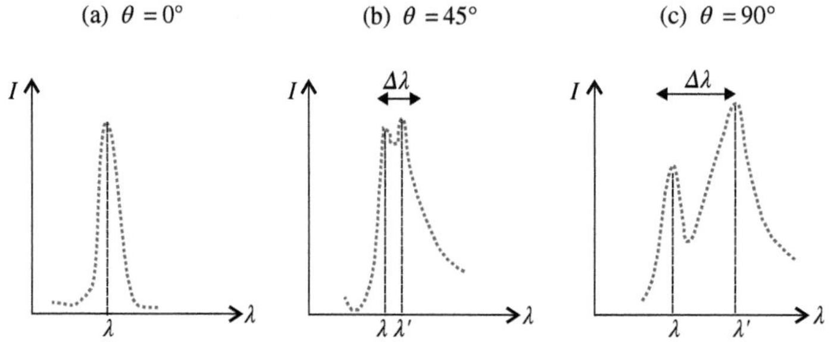

Figure 14.10. The intensities of the scattered X-rays in a Compton effect experiment, plotted as a function of wavelength (a) for $\theta = 0°$, (b) for $\theta = 45°$ and (c) for $\theta = 90°$.

This wavelength shift is the Compton effect. Classically, we expect electrons in the target to oscillate in the electromagnetic field of the incident X-rays and that their oscillation will produce radiation of the same frequency (and wavelength) as the incident radiation (*scattered* X-rays). The angle at which the scattered X-rays emerge may be expected to vary but their wavelength should not change. To explain the wavelength change in the Compton effect, we must treat the X-rays as photons which undergo relativistic particle-particle collisions with electrons in the target.

The photon model

Before we treat the Compton effect as a particle-particle collision we need to develop the photon model of electromagnetic radiation further, in particular we need to ascribe a momentum to the photon. Einstein's postulate (Equation (14.8)) allows us to ascribe an energy to the photon in terms of its frequency $E = hf$. The photon has no mass but, as already indicated in Section 9.10, we can use the relativistic relationship between mass, energy and momentum (Equation (9.40)),

$$E^2 = m^2c^4 + c^2p^2$$

to determine the momentum of the photon. To achieve this, we substitute $m = 0$ and $E = hf$ into this equation to obtain

$$E^2 = h^2f^2 = c^2p^2$$

and, hence,

$$E = hf = cp \tag{14.11}$$

Thus

$$\boxed{p = \frac{hf}{c} = \frac{h}{\lambda}} \tag{14.12}$$

Furthermore, we can use Equation (9.41), $v = \dfrac{c^2p}{E}$, to determine the velocity of the photon.

Substituting $p = \dfrac{h}{\lambda}$ and $E = hf$, we obtain $$v = \frac{hc^2}{fh\lambda} = \frac{c^2}{f\lambda} = c.$$

As expected for a massless particle, the photon travels at the speed of light. In summary, a photon is a massless particle of energy hf and momentum $\dfrac{h}{\lambda}$ which always travels with velocity c in a vacuum, whatever its energy.

Analysis of the Compton effect

We are now in a position to analyse the Compton effect as a relativistic collision between two particles, a photon (X-ray) and an electron. The collision is illustrated in the laboratory frame in Figure 14.11. The incoming photon, with momentum $p = \dfrac{h}{\lambda}$ and energy $E = hf$, strikes an electron which is at rest. After the collision the photon moves with momentum $p' = \dfrac{h}{\lambda'}$ and energy $E' = hf'$ at an angle θ to its incoming direction and the electron moves with momentum p_e and energy E_e at an angle ϕ to the direction of the incoming photon.

Application of the principle of relativistic energy conservation (Section 9.11) yields the equation

$$hf + mc^2 = hf' + E_e \tag{14.13}$$

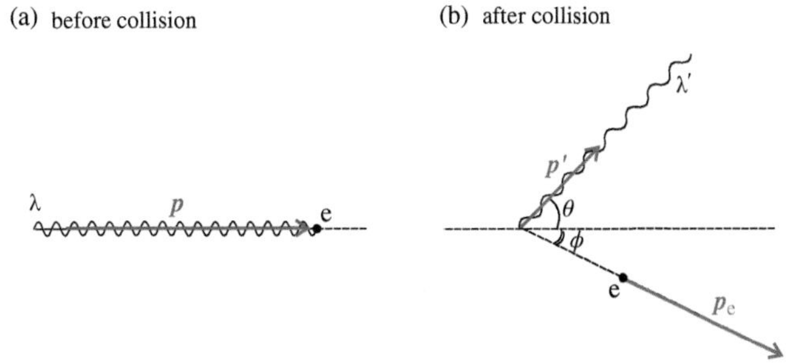

Figure 14.11. The interacting photon and electron in a Compton effect experiment (a) before they collide and (b) after they have collided.

where mc^2 is the rest energy of the electron, and the electron energy is related to its momentum through the Equation (9.40)

$$E_e{}^2 = m^2c^4 + p_e{}^2c^2 \tag{14.14}$$

The principle of relativistic momentum conservation (Section 9.9) yields the vector equation

$$\boldsymbol{p} = \boldsymbol{p}_e + \boldsymbol{p}'$$

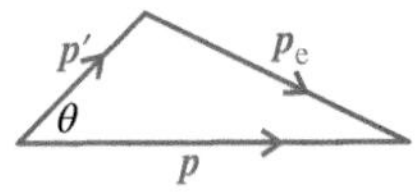

Figure 14.12. Conservation of momentum in a Compton effect experiment $\boldsymbol{p} = \boldsymbol{p}_e + \boldsymbol{p}'$.

which is illustrated in Figure 14.12. Thus $\boldsymbol{p}_e = \boldsymbol{p} - \boldsymbol{p}'$ which we can square (that is, the scalar product of $\boldsymbol{p}_e$ with itself) to give

$$\boldsymbol{p}_e \cdot \boldsymbol{p}_e = (\boldsymbol{p} - \boldsymbol{p}') \cdot (\boldsymbol{p} - \boldsymbol{p}') = \boldsymbol{p} \cdot \boldsymbol{p} + \boldsymbol{p}' \cdot \boldsymbol{p}' - 2\boldsymbol{p} \cdot \boldsymbol{p}'$$

which can be written

$$p_e^2 = p^2 + p'^2 - 2pp' \cos\theta \tag{14.15}$$

Equation (14.15) may also be obtained by applying the principle of conservation of momentum to the components of momentum in the direction of the incoming photon and in the perpendicular direction, and by solving the two equations to eliminate the angle ϕ. Note, however, the greater directness of the vector multiplication method.

The conservation equations have given us two equations (14.13) and (14.15). To account for the Compton effect we need an equation which relates $p \left(= \frac{h}{\lambda}\right)$ to $p' \left(= \frac{h}{\lambda'}\right)$. We therefore proceed to eliminate p_e. First, we write Equation (14.13) as

$$E_e = h(f - f') + mc^2$$

Substituting $hf = cp$ and $hf' = cp'$ (from Equation (14.11)), this becomes

$$E_e = c(p - p') + mc^2$$

We now square this equation and equate the result to E_e^2 as given by Equation (14.14). Thus

$$E_e^2 = c^2(p - p')^2 + m^2c^4 + 2mc^3(p - p') = m^2c^4 + p_e^2c^2$$

which simplifies to give

$$p_e^2 = (p - p')^2 + 2mc(p - p') \tag{14.16}$$

By equating p_e^2 in equations (14.15) and (14.16), we finally eliminate p_e

$$p^2 + p'^2 - 2pp' \cos\theta = p^2 + p'^2 - 2pp' + 2mc(p - p')$$

which simplifies to

$$mc(p - p') = pp'(1 - \cos\theta)$$

or

$$\frac{1}{p'} - \frac{1}{p} = \frac{1}{mc}(1 - \cos\theta) \tag{14.17}$$

Using $p = \frac{h}{\lambda}$ and $p' = \frac{h}{\lambda'}$, this equation can be written in terms of wavelength

Thus,

$$\lambda' - \lambda = \Delta\lambda = \frac{h}{mc}(1 - \cos\theta) \tag{14.18}$$

This is the Compton effect equation which predicts the observed wavelength shift and the dependence of the shift on θ. We have been able to obtain this result only by treating electromagnetic waves as particles in their interaction with other particles. Although this finding is at variance with our earlier (classical) understanding of the distinction between the ways in which waves and particles transport energy (Section 13.3), the evidence for the photon model is now compelling.

The unshifted peak at wavelength λ in the observed Compton spectrum (Figure 14.10) may be accounted for by considering collisions between photons and carbon atoms in the graphite target. The mass of a carbon atom is approximately 22,000 times greater than the electron mass. When applying the Compton effect Equation (14.18) to carbon atoms we should therefore use this larger value of m, leading to a $\frac{1}{22{,}000}$ reduction in the wavelength shift. The reduction in the shift is even smaller when we consider that all the carbon atoms in the graphite lattice are rigidly connected in a single mass. Such a small shift is not observable so that, for practical purposes, the wavelengths of photons which are scattered from the target are unshifted.

For problems based on the material presented in this section visit up.ucc.ie/14/ and follow the link to the problems.

14.8 The de Broglie hypothesis: wave-particle duality

We have shown, in the photon model, that electromagnetic waves can behave as particles but we are still treating particles only as particles. There is a lack of symmetry in our treatment and symmetry arguments can often be instructive in physics. To restore symmetry to the treatment of particles and waves, Louis de Broglie (1892–1987) advanced the following hypothesis in his 1924 PhD thesis at the University of Paris (Sorbonne).

> *The motion of a particle is governed by an associated 'matter wave', the wavelength and frequency of which are given by* $\lambda = \frac{h}{p}$ *and* $f = \frac{E}{h}$, *where p and E are, respectively, the momentum and energy of the particle.*

We can also write the de Broglie relationships in terms of the wave number, $k = \frac{2\pi}{\lambda}$ and angular frequency, $\omega = 2\pi f$, which we used to characterise waves in Chapter 13. Thus

$$p = \frac{h}{\lambda} = \hbar k \tag{14.19}$$

and the Einstein postulate (Equation (14.8) may be written

$$E = hf = \hbar\omega \tag{14.20}$$

where

$$\hbar = \frac{h}{2\pi}$$

According to the de Broglie hypothesis, therefore, particles can display wavelike properties so that both particles and waves possess a dual nature, a phenomenon known as **wave particle duality**. We will discuss and interpret wave particle duality in the next section but first let us examine the practical implications of the de Broglie hypothesis.

If particles can display wave properties why then do we not observe the wave nature of particles in our experience of the macroscopic world? To answer this question let us apply the de Broglie hypothesis to two cases, the first on the macroscopic scale and the second on the microscopic scale.

(a) *Macroscopic case*: a 10 kg mass moving with velocity 10 m s^{-1}

The momentum of the mass, $p = 100$ kg m s^{-1} and therefore its de Broglie wavelength, $\lambda = 6.63 \times \frac{10^{-34}}{100} = 6.6 \times 10^{-36}$ m. When we discussed interference and diffraction effects in Section 13.7, we noted that such effects become noticeable only when waves encounter objects of dimensions (for example, slit separations or slit widths) which are comparable to the wavelength of the wave.

The length 10^{-36} m is considerably less than any atomic, or even sub-atomic, particle dimension so that we would not expect the wave nature of macroscopic objects to produce any observable interference or diffraction effects.

(b) *Microscopic case*: an electron (mass 9.1×10^{-31} kg) moving with velocity 4.4×10^6 m s^{-1} (kinetic energy 8.8×10^{-18} J)

The momentum of the electron, $p = 9.1 \times 10^{-31} \times 4.4 \times 10^6 = 4.0 \times 10^{-24}$ kg m s^{-1} and its de Broglie wavelength, $\lambda = \dfrac{6.63 \times 10^{-34}}{4.0 \times 10^{-24}} \approx 1.65 \times 10^{-10}$ m. Although this length is small, it is comparable to the spacing of atoms or molecules in a crystal so that, according to the de Broglie hypothesis, we might expect interference or diffraction effects to be observable when electrons interact with an array of atoms in a crystal.

The Davisson-Germer experiment: electron waves

Experiments performed in 1925-27 by Clinton Davisson (1881–1958) and Lester Germer (1896–1971) at Bell Labs, then located in the West Village, Lower Manhattan, demonstrated such effects and thus provided direct confirmation of the validity of the de Broglie hypothesis. Also in 1927, George Paget (G.P.) Thomson (1892–1975) obtained similar results in experiments carried out at the University of Aberdeen.

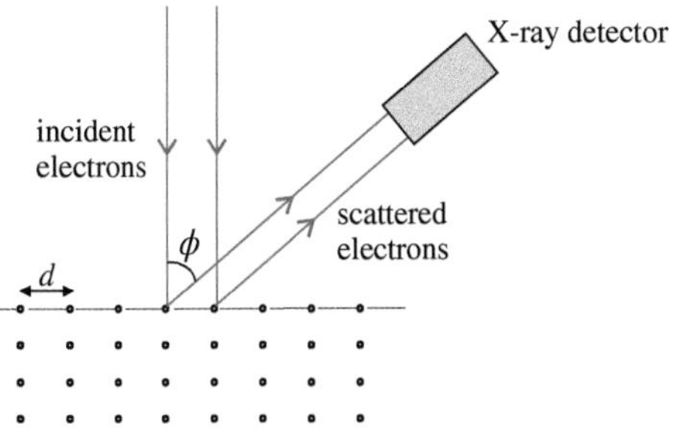

Figure 14.13. The Davisson and Germer experiment (demonstrating electron interference).

An experimental arrangement used by Davisson and Germer is illustrated in Figure 14.13. A beam of electrons strikes a nickel crystal at normal incidence and the intensity of the scattered electrons in the plane of the crystal atoms (the plane of the page) is measured as a function of the scattering angle ϕ.

Results of the Davisson and Germer experiment for a beam of incident electrons of energy 8.8×10^{-18}J are illustrated in Figure 14.14, in which the variation of distance from the origin O, plotted as a function of ϕ, represents the variation of intensity with ϕ (known as a *polar plot*).

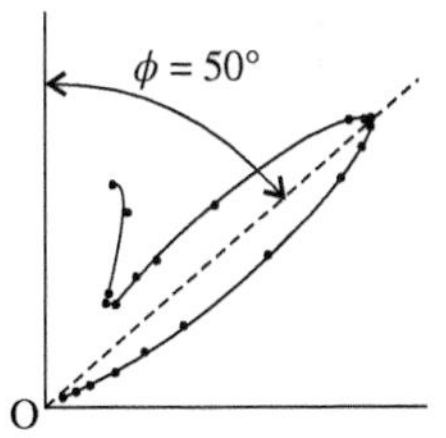

Figure 14.14. The intensity of the scattered electrons in the Davisson and Germer experiment (represented by distance from O), plotted as a function of the angle ϕ for an incident electron energy of 8.8×10^{-18} J.

Figure 14.14 shows clearly that, a maximum of intensity is observed when $\phi = 50°$. As noted in Section 13.6, the occurrence of such a maximum is qualitative evidence of interference — a wave effect. We showed above (case (b)) that the de Broglie wavelength of an electron of energy 8.8×10^{-18} J is 1.65×10^{-11} m. We now investigate whether the observed maximum corresponds quantitatively to interference at this wavelength.

The electron energy used, 8.8×10^{-18} J, is small so that little penetration of the Ni crystal is expected beyond its surface. This allows us to analyse the Davisson and Germer experiment by considering only scattering of electrons by the row of atoms at the surface of the crystal. If electrons behave like waves we expect to see the interference similar to that pattern produced by a linear array of synchronous sources, as discussed in Section 13.6. The interference condition for such an array is given by Equation (13.14), with $\theta = \phi$ in this case; that is $n\lambda = d\sin\phi$.

In this equation d is given by the spacing of atoms on the crystal surface, which was 0.215 nm in the Davisson and Germer experiment. For first order ($n = 1$), Equation (13.14) tells us that the observed maximum at $\phi = 50°$ corresponds to interference involving waves of wavelength $2.15 \times 10^{-10} \times \sin 50° = 1.65 \times 10^{-10}$ m. As noted above, this is the de Broglie wavelength of electrons of energy 8.8×10^{-18} J.

The electrons therefore exhibit the interference pattern expected of electron waves of their de Broglie wavelength. A more complete analysis which takes account of scattering of electrons by the array of atoms in the bulk of the Ni crystal (*Bragg scattering* – Section 22.6), confirms this finding.

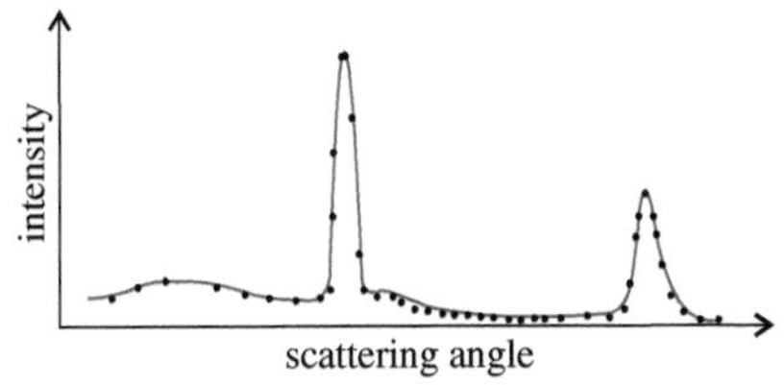

Figure 14.15. The diffraction pattern produced when monoenergetic neutrons are incident on a MnO crystal.

The wave nature of other particles may be demonstrated through scattering experiments. Figure 14.15 shows the pattern produced when neutrons are incident on a crystal. Intensity maxima are observed at particular scattering angles, evidence of interference and of the wave nature of neutrons. Scattering of electrons or neutrons (*electron or neutron diffraction*) is used as a standard technique in the analysis of crystalline structure. The techniques are equivalent to the X-ray diffraction method (to be discussed in Section 22.6).

The wave nature of electrons is also exploited in the *electron microscope*. As we will note in Section 23.7, the replacement of visible light in the optical microscope by an electron beam of much shorter (de Broglie) wavelength in the electron microscope allows a much better angular resolution to be achieved.

For problems based on the material presented in this section visit up.ucc.ie/14/ and follow the link to the problems.

14.9 Interpretation of wave particle duality

The principle of wave-particle duality asserts that particles can exhibit properties which are normally associated with waves and *vice versa*. While this might not appear to be consistent with the discussion (Section 13.3) of the different ways in which waves and particles transmit energy, we can begin to reconcile the pictures if we compare the relation obtained for the power transmitted by a sinusoidal wave with that for the power carried by a uniform stream of monoenergetic particles.

For the wave: $P \propto vA^2$ (relation (13.8)), where v is the wave velocity and A is the wave amplitude,
For the particles: $P \propto vn$, where n is the particle density
The comparison suggests that the particle density in a wave plays a similar role to that of $(\text{amplitude})^2$ in the case of a sinusoidal wave.

The concept of *wave packets* (Section 13.15) gives us a mechanism which enables a wave to deliver energy in discrete bundles. For wave packets, a superposition of sinusoidal waves with a continuous distribution of wavenumbers (k), illustrated in Figure 14.16, produces a localised wave-packet. As noted in Section 13.15, relation (13.8) may also be applied to the component waves of the wave packet so that it follows that the (varying) amplitude within a wave packet is related to particle density in the particle picture.

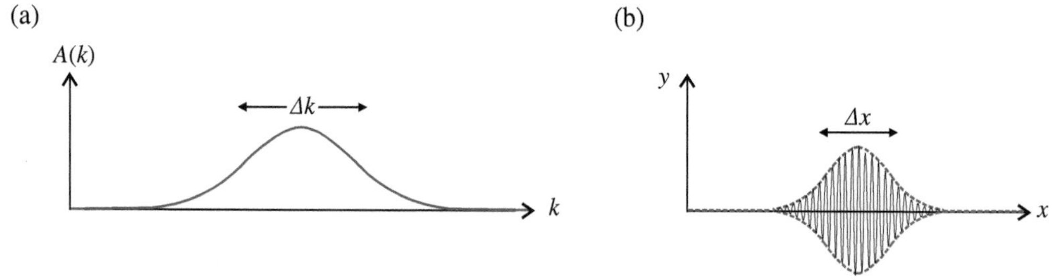

Figure 14.16. (a reproduction of Figure 13.51). A superposition of sinusoidal waves with a continuous distribution of wavenumbers (a) gives rise to a localised wave-packet (b).

This thinking leads us to no longer consider a particle (such as an electron) to be a localised point mass as in Newtonian mechanics, but instead to envisage the particle as a **matter wave** spread out over some small but finite region in space. Similarly, a photon can be represented by a packet of electromagnetic waves.

The group velocity of a matter wave

In Section 13.15 we defined the *group velocity* of a wave packet to be

$$v_g = \frac{d\omega}{dk}$$

Applying the de Broglie relationships $E = \hbar\omega$ and $p = \hbar k$ (Equations (14.19) and (14.20)) to a matter wave packet, we can write

$$v_g = \frac{d\omega}{dk} = \frac{d(\hbar\omega)}{d(\hbar k)} = \frac{dE}{dp} \tag{14.22}$$

In Section 9.10 we showed that the velocity of a relativistic particle is given by Equation (9.42), that is

$$u = \frac{dE}{dp}$$

Comparing Equations (14.22) and (9.42), we can interpret ***the group velocity of a matter wave packet as the relativistic velocity of the associated particle***, that is,

$$u = v_g$$

The progress of the particle is therefore the same as that of the equivalent wave packet of matter waves.

The phase velocity of a matter wave

Again using Broglie equations, the *phase velocity* (the velocity of the 'carrier' wave), can also be written in terms of E and p as follows

$$v_{ph} = f\lambda = \frac{\omega}{k} = \frac{E}{p} = \frac{E}{cp}c = \sqrt{1 + \frac{m^2c^2}{p^2}}\,c \tag{14.23}$$

Since $v_{ph} > c$, the phase velocity cannot correspond to any physical property of the particle.

Note that

$$(v_g)(v_{ph}) = \left(\frac{pc^2}{E}\right)\left(\frac{E}{p}\right) = c^2 \tag{14.24}$$

For a **massless particle**, such as a photon, $m \to 0$, in which case $v_{ph} = v_g = c$, confirming that a massless particle always travels at the velocity of light.

For problems based on the material presented in this section visit up.ucc.ie/14/ *and follow the link to the problems.*

14.10 The Heisenberg uncertainty principle

The discussion of the previous section suggests that, for the special case of a monoenergetic beam of particles, the square of the amplitude of the equivalent matter wave can be regarded as a measure of the probability of finding a particle. Extension of this hypothesis to a single particle, represented by a matter wave packet, suggests that the square of the amplitude of the matter wave at a certain position is a measure of the probability of finding the particle at that position.

The matter wave description of a particle, therefore, inevitably introduces uncertainty in the position of the particle, an uncertainty which is determined by the extent of the wave packet; it is conventional to express the extent of a wave packet by its full width at half maximum (FWHM), as indicated by Δx and Δk in Figure 14.16.

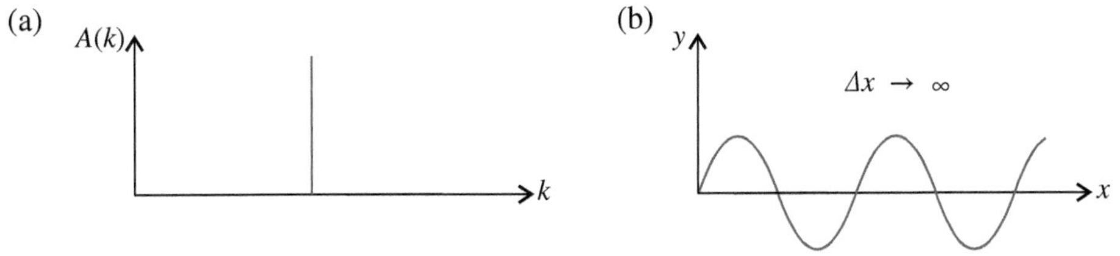

Figure 14.17. When the momentum of a particle is known precisely, as indicated in (a), its position is completely unlocalised, as indicated in (b) where the wave amplitude is unchanged between $-\infty$ and $+\infty$.

It follows from the de Broglie postulate (14.19), $p = \hbar k$, that, for matter waves, a spread of wavenumbers Δk represents a spread — an uncertainty — in momentum which is given by $\Delta p = \hbar \Delta k$. We saw in Figures 13.50 and 13.51 (reproduced in Figure 14.18) how, as Δk becomes larger, the width Δx of a wave packet becomes smaller.

In the case of a sinusoidal wave (Figure 14.17, recall Section 13.2) in which k (and thus p) is known precisely so that $\Delta p = 0$, the particle has the same probability at all points along the x-axis. Thus Δx is effectively infinite — the wave is *non-localised.* This is the special case of a monoenergetic beam considered at the beginning of the previous section.

On the other hand, when the position of the wave packet is known with high precision there is a large uncertainty in the value of k and hence in its momentum (as illustrated in Figure 14.18).

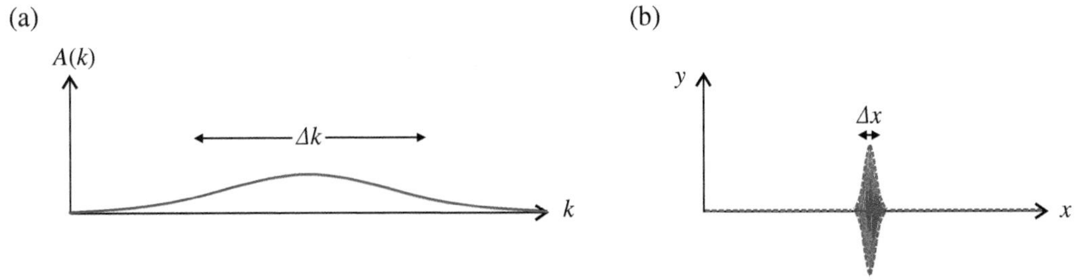

Figure 14.18. When the spread in the momentum of a particle is large, as indicated in (a), its position is localised as indicated in (b).

These results tell us that, as the wave number (and hence the momentum) of a particle becomes more uncertain, the position of the particle becomes more certain and *vice versa.* In Section 13.15 we noted (Equation (13.33)) that, for a wave packet with a continuous distribution in wave numbers

$$(\Delta k)(\Delta x) \geq \frac{1}{2}$$

For matter waves this equation can be written

$$(\Delta p)(\Delta x) \geq \frac{\hbar}{2} \qquad (14.25)$$

that is,

$$(\text{uncertainty in momentum}) \times (\text{uncertainty in position}) \geq \frac{\hbar}{2}$$

Equation (14.25) is known as the **Heisenberg uncertainty principle** (Werner Heisenberg, 1901–1976), a statement which is of fundamental importance in quantum physics. It tells us that we cannot know both the position and the momentum of a particle simultaneously with complete certainty. Note that it is not a statement of experimental limitations; this is the result which would apply if measurements were to be made by perfect experimenters using perfect equipment. Any uncertainties which represent experimental errors must be added.

We have shown that the Heisenberg uncertainty principle follows inevitably from the wave nature of particles. Despite its importance in physics, because of the particular value of the Planck constant, the macroscopic consequences of the principle are too small to be observed. This point is illustrated in the following two examples.

Macroscopic case: a person of mass 60 kg who is moving with a velocity of 1.5 m s^{-1}
We can use the uncertainty principle (Equation (14.25)) to estimate the minimum uncertainty in the person's position Δx_{min} if we take the maximum uncertainty in his or her momentum Δp_{min} to be of the same order as his or her momentum, that is, $\Delta p \sim p$. Thus

$$\Delta x_{min} \geq \frac{\hbar}{2(\Delta p_{min})} \sim \frac{\hbar}{p} = \frac{6.6 \times 10^{-34}}{2\pi \times 60 \times 1.5} \approx 10^{-36} \text{ m}$$

A minimum uncertainty of 10^{-36} m in position is clearly negligible in the macroscopic world. For practical purposes, the position of the person can be known with a precision which is limited only by uncertainties associated with the technique used to measure his or her position.

Microscopic case: an electron which is moving with a velocity of 2.2×10^6 m s^{-1}
Again, taking the maximum uncertainty in the momentum to be p,

$$\Delta x_{min} \sim \frac{\hbar}{p} = \frac{6.6 \times 10^{-34}}{2\pi \times 9.1 \times 10^{-31} \times 2.2 \times 10^6} \approx 5 \times 10^{-11} \text{ m}$$

This distance is comparable to the size of the atom and, as we will see in Section 24.4, 2.2×10^6 m s^{-1} represents a typical electron velocity in an atom. Thus, a minimum uncertainty of the order of 10^{-9} m in the position of the electron will have major implications for the manner in which we describe the behaviour of electrons in atoms (Section 24.5).

An uncertainty principle relating energy and time may be derived by applying the Einstein postulate, that is $\omega = \frac{E}{\hbar}$ to Equation (13.33a), the equation that relates $\Delta\omega$, the width of the frequency spectrum of a wave pulse, to Δt, the duration (time width) of the pulse.

Thus,

$$(\Delta\omega)(\Delta t) \geq \frac{1}{2} \quad \text{gives} \quad (\Delta E)(\Delta t) \geq \frac{\hbar}{2} \qquad (14.26)$$

This equation may be used to relate the *lifetime* of an excited atomic or nuclear state which decays through photon emission to the uncertainty in the energy of the photon, ΔE. The lifetime, represented by Δt, is the average time between the time of excitation and the time at which the state decays. ΔE is the measured spread of the observed energies of the photon – its energy *width*. Equation (14.26) tells us that a broad energy width denotes a short lifetime whereas a narrow energy width denotes a long lifetime. Worked Example 14.4 illustrates the use of this equation.

In summary, consideration of wave packets has shown that waves can deliver energy in the same (localised) way as a particle. It has reconciled the particle and wave models but at the same time it has introduced basic uncertainties in the determination of the position and the momentum of a particle. We can no longer expect physics to predict the outcome of a situation with complete precision (as discussed in Section 14.1). Our description of microscopic phenomena will have to be stated in terms of probabilities and average values. The exact prediction of the evolution of all physical quantities – an implicit objective of classical physics — is no longer feasible, not because the laws of physics have become imprecise but because we cannot know the initial conditions with complete precision. This imprecision, summarised in the Heisenberg uncertainty principle, will be built into the quantum mechanical formalism which is developed in subsequent sections.

Study Worked Example 14.4

For problems based on the material presented in this section visit up.ucc.ie/14/ and follow the link to the problems.

14.11 The Schrödinger (wave mechanical) method

Despite over two decades of experimental evidence that understanding nature at a microscopic level demanded a quantum explanation, it took de Broglie's wave-particle duality insight in 1924 to trigger the development of any theoretical formulation of quantum mechanics. In 1925 the first logically consistent formulation of quantum mechanics (based on matrix algebra) was developed by Heisenberg and his colleagues at the University of Göttingen, Max Born (1882–1970) and Pascual Jordan (1902–1980). An alternative formulation was proposed by Erwin Schrödinger (1887–1961), then at the University of Zurich, in a series of papers published throughout 1926. Schrödinger's approach, which we follow in this book, was based on formulation of the basic laws of quantum mechanics in the form of differential equations. It turned out later that the two approaches ('matrix mechanics' and 'wave mechanics') are formally identical.

As we have seen, the de Broglie hypothesis and the Einstein postulate show how a particle with momentum $p = \hbar k$ and energy $E = \hbar\omega$ can be described by a travelling wave, and such a free localised particle can then be represented by a wave packet, a superposition of many such waves. Schrödinger addressed how this could be extended to the treatment of a particle in a force field and postulated that the state of any dynamical system is mathematically described by a **wave function**, subject to initial and/or boundary conditions. Such a wave function is dependent on position and time, that is,

$$\Psi = \Psi(\boldsymbol{r}, t)$$

A knowledge of the wavefunction is necessary to be able to predict the behaviour of the particle, within limitations set by the uncertainty principle. What is also needed, of course, is the equivalent of an 'equation of motion', which in the Schrödinger approach is in the form of a differential equation — the **Schrödinger equation** — which, in general, involves the derivatives of Ψ with respect to x, y, z and t. It turns out that the Schrödinger equation contains imaginary terms which means that, in general, the wave function Ψ is complex, a point that will be illustrated in Section 14.13 where the wavefunction of a free particle is derived.

We present below a prescription for determining the Schrödinger equation for a non-relativistic particle in a general force field described by a potential energy function that depends on coordinates and time. As before, in introducing quantum concepts, we describe the method in an apparently arbitrary manner, without justification, but demonstrate its validity *post hoc* by showing that it gives results consistent with experimental evidence.

In this chapter we need consider only one-dimensional single-particle systems so the Schrödinger equation will involve only derivatives with respect to x and t. The extension to three-dimensional systems will be considered in Chapter 24.

Step (a) We write down an expression for the total energy of the system in terms of the non-relativistic Hamiltonian function $H(x,p)$, described in Section 5.7, that is

$$E = H(x, p, t) = \frac{p^2}{2m} + U(x, t) \tag{14.27}$$

Step (b) We replace E and $H(p, x, t)$ by their **operator equivalents** as defined below and apply these operators to the wave function, following the form of Equation (14.27), to obtain

$$H_{\text{op}}\,\Psi(x,t) = E_{\text{op}}\Psi(x,t)$$

that is

$$\frac{p_{\text{op}}^2}{2m}\Psi(x,t) + U_{\text{op}}\Psi(x,t) = E_{\text{op}}\Psi(x,t) \tag{14.28}$$

where

$$\boxed{p_{\text{op}} := -i\hbar\frac{\partial}{\partial x}} \tag{14.29}$$

$$\boxed{U_{\text{op}} := U(x,t)}$$

and

$$\boxed{E_{\text{op}} := i\hbar\frac{\partial}{\partial t}} \tag{14.30}$$

Note that p_{op} and E_{op} are *differential operators* acting on the wave function so that, when operator equivalents are substituted in equations such as (14.28), they must be placed on the left of the wave function. For example,

$$p_{\text{op}}(e^{-ix}) = -i\hbar\left[\frac{\partial}{\partial x}(e^{-ix})\right] = -i\hbar[-ie^{-ix}] = -\hbar e^{-ix}$$

In the next section we will show how, for the case of a free particle, the definitions of the operator equivalents of p (14.29) and E (14.30) are consistent with the de Broglie hypothesis.

Using these operator equivalents, Equation (14.28) becomes

$$\frac{1}{2m}\left(-i\hbar\frac{\partial}{\partial x}\right)\left(-i\hbar\frac{\partial}{\partial x}\right)\Psi(x,t)+U(x,t)\Psi(x,t)=i\hbar\frac{\partial}{\partial t}\Psi(x,t)$$

which can be written

$$-\frac{\hbar^2}{2m}\frac{\partial^2\Psi(x,t)}{\partial x^2}+U(x,t)\Psi(x,t)=i\hbar\frac{\partial\Psi(x,t)}{\partial t} \tag{14.31}$$

This is the **time dependent Schrödinger equation** which lies at the heart of the wave mechanical method used in non-relativistic dynamics. To apply it to a particular dynamical situation we substitute the relevant form of the classical potential energy function $U(x,t)$ into the time dependent Schrödinger equation and solve the resulting equation to obtain $\Psi(x, t)$.

Relativistic quantum mechanics

The Schrödinger equation (14.31) is non-relativistic. The operator equivalents, however, can also be applied to relativistic energy-momentum relationships. For example, for the case of a free particle, $E^2 = c^2p^2 + m^2c^4$, in terms of operators

$$E_{\text{op}}^2\Psi = c^2p_{\text{op}}^2\Psi + m^2c^4\Psi$$

or

$$\left(i\hbar\frac{\partial}{\partial t}\right)\left(i\hbar\frac{\partial\Psi}{\partial t}\right)=c^2\left(-i\hbar\frac{\partial}{\partial x}\right)\left(-i\hbar\frac{\partial\Psi}{\partial x}\right)+m^2c^4\Psi$$

whence

$$-\hbar^2\frac{\partial^2\Psi}{\partial t^2}=-\hbar^2c^2\frac{\partial^2\Psi}{\partial x^2}+m^2c^4\Psi$$

which is the **Klein-Gordon equation** for a free particle, after the Swedish physicist Oskar Benjamin Klein (b. 1894–1977) and the German Walter Gordon (1893–1939).

Application to the photon

For an extremely relativistic particle ($cp \gg mc^2$) and, in particular, if $m = 0$ (photon),

Equation (9.40) becomes

$$E^2 = c^2p^2$$

and

$$-\hbar^2\frac{\partial^2\Psi}{\partial t^2}=-\hbar^2c^2\frac{\partial^2\Psi}{\partial x^2}$$

or

$$\frac{\partial^2\Psi}{\partial x^2}=\frac{1}{c^2}\frac{\partial^2\Psi}{\partial t^2}$$

which is the wave equation (13.27). It is interesting to note this indicates that, in some sense, the *classical* equations governing photons (electromagnetic radiation) are already consistent with quantum mechanics. The result also gives us confidence as to the validity of the operator equivalent equations (14.29) and (14.30).

14.12 Probability density; expectation values

As we have seen, the wave nature of particles gives rise to the uncertainty principle which prevents us from determining simultaneously the precise values of quantities such as x and p. Thus, the best that we can achieve is to determine the average values of a physical quantity. Since $\Psi(x, t)$ is a complex function, it cannot represent a physical quantity. This issue was not addressed in Schrödinger's original paper. How to deal with dynamical variables that can be measured (observables) took another important contribution by Max Born. In a paper published in the middle of 1926, he pointed out (in a footnote added at the last minute) that the square modulus of a wave function is always real. The square modulus of $\Psi(x, t)$ is defined as

$$|\Psi(x,t)|^2 = \Psi^*(x,t)\Psi(x,t)$$

where $\Psi^*(x, t)$ is the complex conjugate of $\Psi(x, t)$, as defined in Appendix A.8. The complex conjugate of $Ae^{i(kx-\omega t)}$, for example, is $A^*e^{-i(kx-\omega t)}$.

In the footnote Born suggested that the quantity $|\Psi(x,t)|^2 = \Psi^*(x,t)\Psi(x,t)$ can be seen to be proportional to the *probability* of a particle being located at x at the instant t. Note that this accords with our earlier (Section 14.9) interpretation of the square of the amplitude of a matter wave. Now the probability of finding the particle anywhere in space (that is, between $x = -\infty$ and $x = +\infty$) must be unity. We therefore adjust the amplitude of $\Psi(x, t)$ to satisfy the condition

$$\boxed{\int_{-\infty}^{+\infty} |\Psi(x,t)|^2 dx = 1}$$

a procedure known as **normalisation**. Examples of normalisation will be given in Section 14.15 and in Worked Example 14.5.

Working with *normalised* wave functions allows us to define a **probability density** function

$$P(x,t) := \Psi^*(x,t)\Psi(x,t) \tag{14.32}$$

such that the probability of finding a particle within an interval between x and $x + \Delta x$ at time t is given by

$$P(x,t)\Delta x = \Psi^*(x,t)\Psi(x,t)\Delta x$$

The probability density plays a similar role to a distribution function, discussed in Section 12.4. As in the case of a distribution function, the average value of a quantity can be calculated by integrating the product of the quantity and the probability density function over all space. Thus the average value (called the **expectation value**) of a particle's position, for example, can be calculated as follows:

$$\langle x \rangle = \int_{-\infty}^{+\infty} xP(x,t)dx$$

where

$$\int_{-\infty}^{+\infty} P(x,t)dx = 1$$

Using definition (14.32), we can write $\langle x \rangle$ in terms of a normalised wave function

$$\langle x \rangle = \int_{-\infty}^{+\infty} x|\Psi(x,t)|^2 dx$$

For reasons that will be become clearer below, it is more convenient to write this equation as

$$\langle x \rangle = \int_{-\infty}^{+\infty} \Psi^*(x,t)x\Psi(x,t)dx$$

The **expectation value** of any quantity, $Q(x, t)$, which is a function of x and t can be evaluated through a similar calculation

$$\boxed{\langle Q \rangle := \int_{-\infty}^{+\infty} \Psi^*(x,t)Q(x,t)\Psi(x,t)dx} \tag{14.33}$$

Expectation value of momentum

Determining the expectation value of momentum or a quantity that depends on momentum is a little more complicated. Because we cannot write momentum as a function of x — this would require a precise knowledge of p and x at the same time, in violation of the uncertainty principle — we cannot evaluate the expectation value of momentum simply by substituting p for $Q(x,t)$ in the above equation. We can, however, use the operator equivalent of p to evaluate.

Thus,

$$\langle p \rangle = \int_{-\infty}^{+\infty} \Psi^*(x,t)p_{\text{op}}\Psi(x,t)dx = \int_{-\infty}^{+\infty} \Psi^*(x,t)\left(-i\hbar\frac{\partial}{\partial x}\right)\Psi(x,t)dx$$

Note that, because the operator equivalent p_{op} contains a differential operator, it is important that it be placed between the wave function and its complex conjugate, as $Q(x,t)$ is in Equation (14.33) above. Any physical observable that is a function of momentum, such as kinetic energy $T = \frac{p^2}{2m}$, may be treated in a similar manner. Further justification for the use of $\left(-i\hbar\frac{\partial}{\partial x}\right)$ as the operator equivalent for momentum is given on the *Understanding Physics* website at up.ucc.ie/14/12/1/.

Determination of the wavefunction provides the key to the analysis of any situation to which quantum considerations are to be applied. With a knowledge of the wave function, measurable quantities such as the probability densities and expectation values can be calculated. Worked Example 14.5 gives an example of how probabilities and expectation values may be calculated from wavefunctions.

In subsequent sections of this chapter we apply the wave mechanical method, which we have described above, to some simple, but important, physical situations.

Study Worked Example 14.5

For problems based on the material presented in this section visit up.ucc.ie/14/ *and follow the link to the problems.*

14.13 The free particle

A free particle is a particle that does not experience any force so that $F = -\frac{\partial U}{\partial x} = 0$; $U(x, t)$ is constant in position as well as time. The choice of a zero point for potential energy is arbitrary so that for convenience we set $U = 0$. The Schrödinger Equation (14.31) for the free particle case then becomes

$$\frac{-\hbar^2}{2m}\frac{\partial^2\Psi(x,t)}{\partial x^2} = i\hbar\frac{\partial\Psi(x,t)}{\partial t} \tag{14.34}$$

Our task is to solve this equation. Let us first try a travelling wave solution, as discussed in Section 13.2.

$$\Psi(x,t) = A\sin(kx - \omega t)$$

Substitution into Equation (14.34) gives

$$\frac{-\hbar^2}{2m}[-k^2A\sin(kx-\omega t)] = -i\hbar\omega A\cos(kx-\omega t)$$

which simplifies to

$$\tan(kx - \omega t) = -\frac{2im\omega}{\hbar k^2}$$

This equation can only be true for a specific value of $(kx - \omega t)$. It is not a general solution. Similarly substitution of $\Psi(x, t) = A\cos(kx - \omega t)$ into Equation (14.34) shows that this too is not a general solution. We can, however, construct a linear combination of the sine and cosine solutions of the type $\Psi(x, t) = A'\sin(kx - \omega t) + B'\cos(kx - \omega t)$, as discussed in Section 13.2. If we let $B' = A$ and $A' = iA$ this becomes

$$\Psi(x,t) = A\cos(kx-\omega t) + Ai\sin(kx-\omega t)$$

which, using Appendix A.8(i), can be written, $\Psi(x, t) = = Ae^{i(kx - \omega t)}$

When this function is substituted into Equation (14.34), the following result is obtained

$$-\frac{\hbar^2}{2m}[-k^2Ae^{i(kx-\omega t)}] = \hbar\omega Ae^{i(kx-\omega t)}$$

which simplifies to

$$\frac{\hbar^2k^2}{2m} = \hbar\omega \tag{14.35}$$

Therefore, the solution $\Psi(x, t) = Ae^{i(kx - \omega t)} + Be^{-i(kx - \omega t)}$ is a general solution for a free particle if Equation (14.35) is satisfied. We can also obtain Equation (14.35) by substituting equations (14.19) and (14.20) for E and p into the kinetic energy equation,

Thus, $E = \frac{p^2}{2m}$ becomes $\hbar\omega = \frac{\hbar^2k^2}{2m}$

Therefore Equation (14.35) is essentially a re-statement of the de Broglie hypothesis.

Note that k, and hence $E = \frac{\hbar^2k^2}{2m}$, can take any value for a free (unbound) particle; a *continuum* of energy states is available to a free particle.

The wave function $\Psi(x, t) = Ae^{i(kx - \omega t)}$ is a travelling wave moving in the $+x$ direction with an amplitude A which is independent of x. The particle can be found, therefore, with equal probability at any position on the x-axis — it is completely unlocalised. To describe a localised free particle we need a wave with an amplitude which is non zero for only a small range of x, that is a wave packet as described, for example, by Equation (13.32)

$$y(x) = \int A(k)\cos kx dx$$

A wave packet is formed by adding travelling waves with different values of A and k (recall the Fourier analysis technique, Section 13.15). We have noted above that k can take any value for a free particle so that the wave packet will also be a solution of the free particle Schrödinger Equation (14.34). For brevity, we shall discuss free particles in terms of the unlocalised solution,

$$\Psi(x,t) = Ae^{i(kx-\omega t)} \tag{14.36}$$

where A is a constant (possibly complex). Note, however, that a localised solution can always be constructed by adding further travelling wave solutions, that is by writing

$$\Psi(x,t) = \int A(k)e^{i(kx-\omega t)}dx$$

While $\Psi(x,t)$ has an imaginary component, observable quantities are always real because they involve the square modulus (Appendix A.8) of the wavefunction.

Thus, $|\Psi|^2 = \Psi^*(x,t)\Psi(x,t) = A^*e^{-i(kx-\omega t)}A^*e^{+i(kx-\omega t)} = A^*A = |A|^2$, which is always real.

We now apply the method described in previous sections to determine the expectation values of the momentum and energy of a free particle whose wavefunction is given by Equation (14.36).

First we note that
$$-i\hbar\frac{\partial}{\partial x}\Psi(x,t) = -i\hbar\frac{\partial}{\partial x}[A\, e^{i(kx-\omega t)}] = \hbar k[A\, e^{i(kx-\omega t)}] = \hbar k\Psi(x,t) = p\Psi(x,t)$$

which confirms that, in the case of the free particle, the $-i\hbar\frac{\partial}{\partial x}$ operator has exactly the same effect as multiplying by p.

The expectation value of p is given by

$$\langle p\rangle = \int_{-\infty}^{+\infty}\Psi^*(x,t)\left(-i\hbar\frac{\partial}{\partial x}\right)\Psi(x,t)dx = \int_{-\infty}^{+\infty}A^*e^{-i(kx-\omega t)}\hbar kAe^{+i(kx-\omega t)}dx = \hbar k\int_{-\infty}^{+\infty}A^*Adx$$

Now $\int_{-\infty}^{+\infty}\Psi^*(x,t)\Psi(x,t)dx = \int_{-\infty}^{+\infty}A^*e^{-i(kx-\omega t)}Ae^{+i(kx-\omega t)}dx = \int_{-\infty}^{+\infty}A^*Adx$ which is the probability of finding the particle somewhere on the x-axis. Note, however, that in the case of an unlocalised particle, whose wavefunction amplitude A is constant everywhere (that is, from $x=-\infty$ to $x=+\infty$), it is not possible to normalise $\Psi(x,t)$ by evaluating A from $\int_{-\infty}^{+\infty}A^*Adx = 1$. The reason for this is that, in such a case, the integral $\int_{-\infty}^{+\infty}A^*Adx = |A|^2\int_{-\infty}^{+\infty}dx$ is infinite so that A cannot be determined. There are three possible ways of overcoming this difficulty, as described below:

(a) While the wavefunction cannot be normalised between infinite limits it can be normalised by making the limits of integration very large but finite. We can confine the particle within a 'box' (for example, within an infinite square well, as will be considered in Section 14.15) and can then apply the solutions to a situation in which the box becomes infinitely large.
(b) We can work with wave packets (localised solutions). The problem does not arise for a localised particle because the amplitude of its wave function (a wave packet) is non zero for only a finite range of x, allowing the normalisation integral $\int_{-\infty}^{+\infty}\Psi^*(x,t)\Psi(x,t)dx = 1$ to be evaluated. This description is actually closer to most real physical situations than the unlocalised description since, in producing an energetic particle, it is not possible to create an exact momentum state (and wave number).
(c) Equation (14.36) can be interpreted as representing a uniform beam of particles with a uniform density along the x-axis, and hence an equal probability of finding a particle at any position on the x-axis. If we do not need to know the probability of finding a particle in any finite region, as will be the case when we interpret the flow of probability density as being proportional to the beam flux in Section 14.16, no problems will arise.

Assuming that, in the context of the above discussion, $\int_{-\infty}^{+\infty}A^*A\,dx = 1$ can be evaluated, for a free particle we can write $\langle p\rangle = \hbar k$.

This result shows that the expectation value of the momentum of a free particle is consistent with the de Broglie hypothesis (Equation (14.19)) although, because the expectation value is only a statement of the average value of p, we should note that we have not proved that p can have only this precise value. We can, however demonstrate that p can have only this value by evaluating $\langle p^2\rangle$, the mean (expectation) value of p^2. If $\langle p^2\rangle = \langle p\rangle^2$, p cannot fluctuate, because averaging a higher power of p, such as p^2, gives more weight to cases in which p is greater than $\langle p\rangle$, and less weight to the equally numerous cases in which p is less than $\langle p\rangle$.

We therefore evaluate

$$\begin{aligned}\langle p^2\rangle &= \int_{-\infty}^{+\infty} \Psi^*(x,t)p_{\text{op}}^2\Psi(x,t)dx \\ &= \int_{-\infty}^{+\infty} A^* e^{-i(kx-\omega t)}\left(-i\hbar\frac{\partial}{\partial x}\right)\left[\left(-i\hbar\frac{\partial}{\partial x}\right)A\, e^{i(kx-\omega t)}\right]dx \\ &= \int_{-\infty}^{+\infty} A^* e^{-i(kx-\omega t)}\left(-i\hbar\frac{\partial}{\partial x}\right)[\hbar k A\, e^{i(kx-\omega t)}]dx \\ &= \hbar^2 k^2\int_{-\infty}^{+\infty} A^*A\, dx = \hbar^2k^2\end{aligned}$$

We have shown that $\langle p^2\rangle = \langle p\rangle^2$. Thus the momentum of a particle described by the wave function (14.36) cannot fluctuate and can have only the precise value $\hbar k$, as stated in the de Broglie hypothesis (Equation (14.19)). This result is expected since, as we know, such a particle is completely unlocalised (that is $\Delta p \to 0$ corresponds to $\Delta x \to \infty$).

Similarly, we can show that when the energy operator $i\hbar\frac{\partial}{\partial t}$ is applied to the free particle wavefunction $\Psi(x, t) = A\, e^{i(kx-\omega t)}$, it has the same effect as multiplying by $\hbar\omega = E$

$$i\hbar\frac{\partial}{\partial t}\Psi(x,t) = \hbar\omega\Psi(x,t) = E\Psi(x,t) \tag{14.37}$$

It follows directly that the expectation value of the energy of a free particle

$$\langle E\rangle = \hbar\omega,$$

a result which is consistent with the Einstein postulate, as stated in Equation (14.20). By showing that $\langle E^2\rangle = \langle E\rangle^2$, we can further conclude that E can only have the precise value $\hbar\omega$.

For problems based on the material presented in this section visit up.ucc.ie/14/ and follow the link to the problems.

14.14 The time-independent Schrödinger equation: eigenfunctions and eigenvalues

As noted in Section 14.11, the time-dependent Schrödinger equation for a particle which is free to move in one dimension is a partial differential equation in the two variables x and t. A standard technique for solving certain types of such equations is *separation of the variables*. The solution, a function of two variables, is written as a product of two functions, each of which is a function of a single variable. In the present case, therefore, we seek a solution of the form

$$\Psi(x,t) = \psi(x)f(t)$$

The technique of separating the variables works if the potential function is a function of position only, that is if it is independent of time and can be written as $U(x)$ in the time-dependent Schrödinger equation (14.31). This is true for many potential energy fields, namely conservative systems such as those considered in Chapters 3 and 5 and in this chapter. The Schrödinger equation then becomes

$$\frac{-\hbar^2}{2m}\frac{\partial^2\Psi(x,t)}{\partial x^2} + U(x)\Psi(x,t) = i\hbar\frac{\partial\Psi(x,t)}{\partial t} \tag{14.38}$$

substituting $\Psi(x, t) = \psi(x)f(t)$ we obtain

$$\frac{-\hbar}{2m}f(t)\frac{d^2\psi(x)}{dx^2} + U(x)\psi(x)f(t) = i\hbar\psi(x)\frac{df(t)}{dt}$$

Note that, by separating the variables in the wavefunction, we have been able to dispense with the partial derivative notation.

If we divide the above equation throughout by $\psi(x)f(t)$, the variables can be taken to opposite sides of the equation (separated) to give

$$\frac{-\hbar^2}{2m}\left[\frac{1}{\psi(x)}\frac{d^2\psi(x)}{dx^2}\right] + U(x) = i\hbar\frac{1}{f(t)}\frac{df(t)}{dt} \tag{14.39}$$

This equation can be satisfied only if each side is equal to the same constant, known as the *separation constant, G* (with the dimensions of energy). Equation (14.39) can then be written as two ordinary differential equations (equations in only one variable) namely

$$\frac{-\hbar^2}{2m}\frac{1}{\psi(x)}\frac{d^2\psi(x)}{dx^2} + U(x) = G \tag{14.40}$$

and

$$i\hbar\frac{1}{f(t)}\frac{df(t)}{dt} = G \tag{14.41}$$

Equation (14.41) can be written

$$\frac{df(t)}{dt} = -\frac{iG}{\hbar}f(t)$$

with the solution

$$f(t) = e^{-iGt/\hbar} \tag{14.42}$$

a result which is verified easily by substitution. Using the identity $e^{-ix} = \cos x - i\sin x$ (Appendix A.8) we can can write

$$f(t) = \cos\left(2\pi\frac{G}{h}t\right) - i\sin\left(2\pi\frac{G}{h}t\right)$$

thus identifying *f*(*t*) as an oscillatory function of frequency $f = \frac{G}{h}$. Now, according to the Einstein postulate (14.8) the frequency must also be given by $f = \frac{E}{h}$, where *E* is the total energy of the particle associated with the wave function corresponding to *f*(*t*). Comparing the two expressions for frequency, we can identify the separation constant *G* as the total energy of the particle *E* and Equation (14.40) can be written:

$$\frac{-\hbar^2}{2m}\frac{d^2\psi(x)}{dx^2} + U(x)\psi(x) = E\psi(x) \tag{14.43}$$

and Equation (14.42) becomes $f(t) = e^{-iEt/\hbar}$ so that the complete wavefunction is

$$\Psi(x,t) = \psi(x)e^{-iEt/\hbar}$$

where $\psi(x)$ is the solution to Equation (14.43), the **time-independent Schrödinger equation**.

Equation (14.43) can be written as

$$H_{op}\psi(x) = E\psi(x)$$

where H_{op} is the time independent hamiltonan operator for the system being studied. Equations in this form are called 'eigenvalue equations' and their solutions *are known as* ***eigenfunctions***.

Newton's second law and quantum mechanics (Ehrenfest theorem)

We have seen that the expectation value of the momentum of a particle is

$$\langle p\rangle = \int \Psi^*(x)\left(-i\hbar\frac{\partial}{\partial x}\right)\Psi(x)dx$$

If the particle experiences a field force, one would expect the momentum to be varying with the rate of change of momentum given by

$$\frac{d\langle p\rangle}{dt} = \int\left\{-i\hbar\frac{\partial\Psi^*}{\partial t}\frac{\partial\Psi}{\partial x} - i\hbar\Psi^*\frac{\partial\Psi}{\partial t\partial x}\right\}dx = \int\left\{\left(-i\hbar\frac{\partial\Psi}{\partial t}\right)^*\frac{\partial\Psi}{\partial x} + \Psi^*\frac{\partial}{\partial x}\left(-i\hbar\frac{\partial\Psi}{\partial t}\right)\right\}dx$$

Now the time-dependent Schrödinger equation in the form in (14.38) can be used to substitute $-\frac{\hbar^2}{2m}\frac{\partial^2\Psi}{\partial x^2} + V(x)\Psi$ for $i\hbar\frac{\partial\Psi}{\partial t}$ in the above, where *V*(*x*) is the potential energy function of the field involved. Some further steps of algebra (provided on the *Understanding Physics* website at up.ucc.ie/14/14/1/) yield

$$\frac{d\langle p\rangle}{dt} = -\int\frac{dV}{dx}|\Psi|^2dx = -\int\Psi^*\frac{dV}{dx}\Psi dx = -\left\langle\frac{dV}{dx}\right\rangle$$

or, in three-dimensions,

$$\frac{d\langle \boldsymbol{p}\rangle}{dt} = -\langle \nabla V\rangle = \langle \boldsymbol{F}\rangle$$

This represents the quantum counterpart of Newton's second law; the one-dimensional version was first published in 1927 by the Austrian Paul Ehrenfest (1880–1927), a theoretical physicist at Leiden University. It can be shown that, if the potential energy function is sufficiently slowly varying that it does not change on scales of the size of the particle's wave packet, the Ehrenfest theorem reduces to

$$\frac{d\langle p\rangle}{dt} = -\frac{dV(\langle x\rangle)}{d\langle x\rangle}$$

which is the direct equivalent of Newton's second law with the expectation value $\langle x\rangle$ playing the role of the particle displacement. Furthermore, we may infer that, provided that the potential energy function is slowly varying[1], classical mechanics (derived from Newton's second law) may be applied to the expectation values of dynamical quantities by replacing variables in dynamical equations by their corresponding expectation values.

Acceptable wavefunctions

In a particular situation, in which $U(x)$ has a known form, we can solve Equation (14.43) by requiring (as in the case of solutions to differential equations in classical physics) that the functions $\psi(x)$ are *well-behaved* (as defined below). This may mean that only certain functions $\psi(x)$ are physically acceptable. With each $\psi(x)$ we find that there is a corresponding value of the total energy E known as an **eigenvalue**. We shall see that, for bound systems, such as the infinite square well considered in the next section, the requirement that $\psi(x)$ be well-behaved leads directly to quantisation of energy. Energy quantisation is no longer an arbitrary postulate, as it was in the Planck hypothesis (Section 14.2) or the Einstein postulate (Section 14.5), but is a direct consequence of the Schrödinger equation and the physical requirement that solutions be well-behaved.

For eigenfunctions to be well-behaved they must possess the following properties.

For all values of x, both $\psi(x)$ and $\dfrac{d\psi(x)}{dx}$ should be

(a) *finite*, that is, not as illustrated in Figure 14.19(a)
(b) *single-valued*, that is, not as illustrated in Figure 14.19(b), and
(c) *continuous*, that is, not as illustrated in Figure 14.19(c).

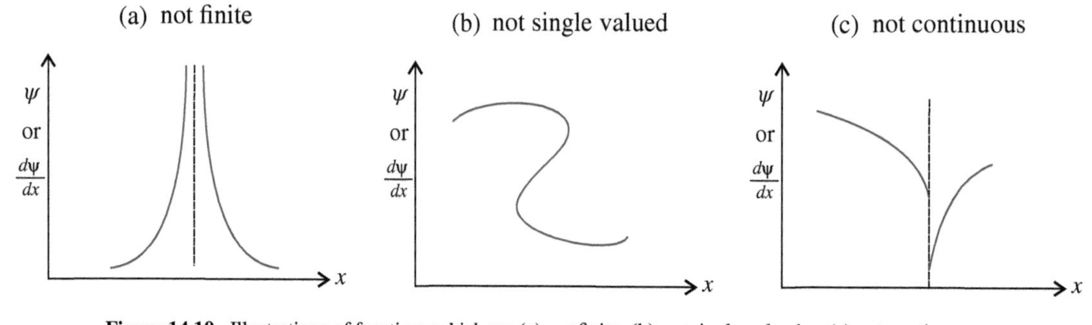

Figure 14.19. Illustrations of functions which are (a) *not* finite, (b) *not* single-valued or (c) *not* continuous.

These properties are required to ensure that eigenfunctions can represent real physical situations, that is, so that measurable quantities determined from eigenfunctions, for example expectation values such as $\langle x\rangle = \int_{-\infty}^{+\infty} \psi^*(x)x\psi(x)dx$ or $\langle p\rangle = \int_{-\infty}^{+\infty} \psi^*(x)\left(-i\hbar\frac{\partial}{\partial x}\right)\psi(x)dx$, are physically meaningful.

The time-independent Schrödinger Equation (14.43) can be written

$$\frac{-\hbar^2}{2m}\frac{d^2\psi(x)}{dx^2} = [E - U(x)]\psi(x)$$

If $\dfrac{d\psi(x)}{dx}$ is not finite and continuous everywhere $\dfrac{d^2\psi(x)}{dx^2}$ would be infinite which would mean that the right hand side of the equation would also become infinite at discontinuities in $\dfrac{d\psi(x)}{dx}$ which, in turn, would mean that either $U(x)$ or E would have to be infinite; this is not physically possible.

[1]Specifically, the width of the position probability distribution must be small compared with the distance over which V varies appreciably.

Discontinuous boundaries

In the examples which we consider in the next two sections we use the requirements (a) to (c) to establish **boundary conditions**, continuity equations for $\psi(x)$ and $\dfrac{d\psi(x)}{dx}$ at those values of x at which there is a discontinuity in the potential energy function $U(x)$. These conditions will enable us to determine eigenfunctions — solutions of the time-independent Schrödinger equation for the cases under consideration.

14.15 The infinite square potential well

Consider a particle of mass m which is held within a potential well which has infinitely high vertical sides; that is, it experiences a discontinuous potential energy given by

$$U = 0 \quad \text{when} \quad 0 < x < a$$

and

$$U \to \infty \quad \text{when} \quad x \leq 0 \text{ and } x \geq a$$

as illustrated in Figure 14.20.

This is called a 'square well' potential and is clearly a bound system. Whatever its energy, the particle can only be in the region between $x = 0$ and $x = a$. The regions in which $x \leq 0$ and where $x \geq a$ are inaccessible to the particle so that its eigenfunction $\psi(x)$ is zero in these regions.

Classically, the particle could have any energy in such a system and a continuum of energy levels is possible but, as shall see, in quantum mechanics the energy can only take certain values — it is *quantised*.

Inside the potential well $U = 0$ so that the time independent Schrödinger equation in this region is

$$\frac{-\hbar^2}{2m}\frac{d^2\psi(x)}{dx^2} = E\psi(x)$$

which can be written as

$$\frac{d^2\psi(x)}{dx^2} + \frac{2mE}{\hbar^2}\psi(x) = 0$$

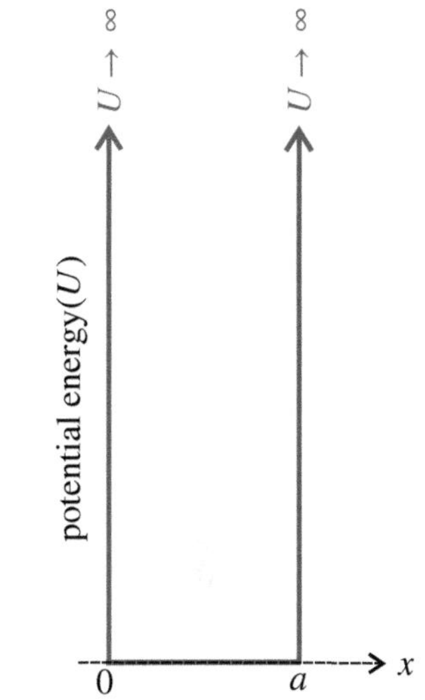

Figure 14.20. The potential function for a particle in an infinite square well.

Letting $k^2 = \dfrac{2mE}{\hbar^2}$, this becomes

$$\frac{d^2\psi(x)}{dx^2} + k^2\psi(x) = 0$$

This is the equation which we must solve to obtain the eigenfunctions $\psi(x)$ for the infinite square well. Let us first try the function $\psi(x) = e^{\alpha x}$ as a solution. Substitution gives $\alpha^2 + k^2 = 0$. Thus $\alpha = \pm\, ik$ and solutions are $\psi(x) = e^{ikx}$ and $\psi(x) = e^{-ikx}$. Note that these are the x dependent parts of the free-particle solutions, as discussed in Section 14.13.

The general solution, a linear combination of these two solutions, is then

$$\psi(x) = ae^{ikx} + be^{-ikx} \quad \text{where } a \text{ and } b \text{ are constants.}$$

Using Appendix A.8(i), we can write this as $\psi(x) = (a + b)\cos kx + i(a - b)\sin kx$ or

$$\psi(x) = A\cos kx + B\sin kx \quad \text{where } A \text{ and } B \text{ are constants.}$$

We now apply the conditions of finiteness and single valuedness, outlined in the previous section, at the boundaries $x = 0$ and $x = a$ so that we can deduce which values of the constants A, B and k produce acceptable functions $\psi(x)$.

At $x = 0$, $\psi(x) = 0$. This can only be true if $A = 0$ and thus

$$\psi(x) = B\sin kx \tag{14.44}$$

At $x = a$, $\psi(x) = 0$. This can only be true if $0 = B\sin ka$ which is possible either if $B = 0$ or if $\sin ka = 0$. The first option, $B = 0$, is not a useful solution because it would mean that $\psi(x)$ is zero everywhere — there would be no particle. The other possibility, $\sin ka = 0$ requires that $k = \dfrac{n\pi}{a}$ where $n = 1, 2, 3\ldots$ (an integer). The $n = 0$ possibility has not been included because $\psi(x) = B\sin\dfrac{n\pi x}{a}$ would yield $\psi(x) = B\sin 0 = 0$, meaning that there is no particle. We will also find below that the $n = 0$ possibility would not be consistent with the uncertainty principle.

The solutions (that is, the eigenfunctions) of the infinite potential well are therefore

$$\boxed{\psi_n(x) = B\sin\frac{n\pi x}{a}} \quad \text{where } n = 1, 2, 3, \ldots \tag{14.45}$$

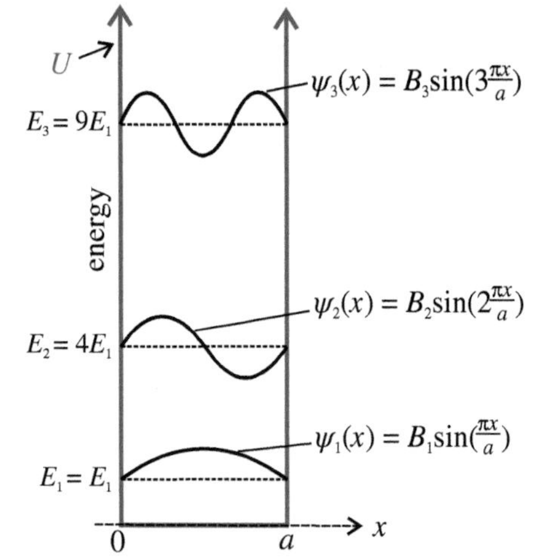

Figure 14.21. The first three energy levels (eigenvalues) of a particle in an infinite square well. The corresponding eigenfunctions have been superimposed.

For each n there is a corresponding energy given by $E_n = \frac{p^2}{2m} = \frac{\hbar^2 k^2}{2m} = \frac{n^2\hbar^2\pi^2}{2ma^2}$. These are the eigenvalues of the system, corresponding to the eigenfunctions ψ_n. As expected for a bound system, the energy of the particle is quantised — it can only take certain values which are given by

$$\boxed{E_n = n^2\left(\frac{\hbar^2\pi^2}{2ma^2}\right)} \quad \text{where } n = 1, 2, 3 \ldots . \tag{14.46}$$

Thus, the lowest energy state of the system has energy given by $E_1 = \frac{\hbar^2\pi^2}{2ma^2}$. The first few energy *levels* (quantised energy states) and corresponding eigenfunctions, at a fixed time, of the infinite square well are illustrated in Figure 14.21.

Note the resemblance of the eigenfunctions to standing waves (Section 13.8) in this case.

The condition $k = \frac{n\pi}{a}$ can be written $\qquad k = \frac{2\pi}{\lambda} = \frac{n\pi}{a}$

which becomes $n\frac{\lambda}{2} = a$, the condition for standing waves with nodes at $x = 0$ and $x = a$ (Equation (13.16)).

The boundary conditions in a bound system with infinite sides require there to be nodes in the eigenfunctions at the edge of the well. This imposes quantisation; energy quantisation in bound systems is a very general phenomenon of quantum mechanics and contrasts with the free particle case (Section 14.13) in which a continuum of energy states is accessible.

The ground state of the infinite square potential well

Note that the energy, E_1, of the lowest state, known as the **ground state** which corresponds to $n = 1$, is not zero as would be expected classically. This is a consequence of the uncertainty principle because, if the lowest energy were to be zero, p would also be zero and we would know p exactly, that is, $\Delta p = 0$, which would only be allowable if $\Delta x = \infty$. However, Δx, the maximum uncertainty in the position of the particle in the infinite well, is set by a, the width of the well, so that the energy solution $E = 0$ is unacceptable. This reasoning applies to any bound system, so that a non-zero ground state energy is a completely general quantum mechanical result for a bound system.

The constant B in Equation (14.45) is determined through normalisation.

For the $n = 1$ state $\qquad \Psi_1(x,t) = \psi_1(x)f(t) = B\sin\left(\frac{\pi x}{a}\right)e^{-iE_1 t/\hbar}$

Normalisation requires that $\qquad \int_0^a \Psi_1^*(x,t)\Psi_1(x,t)dx = \int_0^a B^2\sin^2\left(\frac{\pi x}{a}\right)(e^{iE_1 t/\hbar})(e^{-iE_1 t/\hbar})dx = 1$

Thus, $\qquad \left(\frac{a}{\pi}\right)\int_0^a B^2\sin^2\left(\frac{\pi x}{a}\right)d\left(\frac{\pi x}{a}\right) = 1$

Writing $\xi = \frac{\pi x}{a}$, this becomes $\frac{a}{\pi}\int_0^\pi B^2\sin^2\xi d\xi = 1$. The value of this integral is $\frac{\pi}{2}$ (Table A.2 of Appendix A) and thus

$$\left(\frac{a}{\pi}\right)B^2\left(\frac{\pi}{2}\right) = 1 \quad \text{and therefore} \quad B = \sqrt{\frac{2}{a}}$$

and the normalised ground state eigenfunction is $\psi_1(x) = \sqrt{\frac{2}{a}}\sin\frac{\pi x}{a}$

Knowing $\psi_1(x)$ we now determine the expectation values of the position, momentum, and energy of a particle in the ground state of an infinite square potential well.

(a) *average position*: $\langle x\rangle = \int \psi_1^*(x) x \psi_1(x) dx$

$$= \frac{2}{a}\int_0^a \sin^2\left(\frac{\pi x}{a}\right) x dx = \frac{2}{a}\frac{a^2}{\pi^2}\int_0^a \left(\frac{\pi x}{a}\right)\sin^2\left(\frac{\pi x}{a}\right) d\left(\frac{\pi x}{a}\right)$$

which we can write as $\langle x\rangle = \frac{2}{a}\frac{a^2}{\pi^2}\int_0^\pi \xi \sin^2\xi \, d\xi$, where $\xi = \frac{\pi x}{a}$.

This integral can be evaluated, using Table A.3 of Appendix A, to obtain $\langle x\rangle = \frac{a}{2}$.

This result is expected from the symmetry of $\psi_1(x)$, as plotted in Figure 14.22. For the ground state the most likely position of the particle is at the centre of the well.

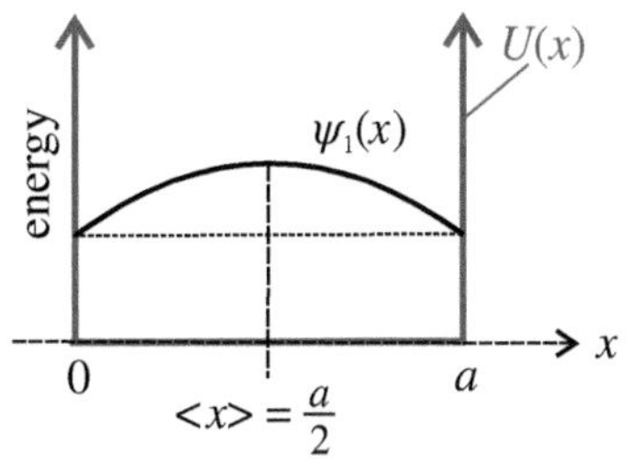

Figure 14.22. The ground state energy of a particle in an infinite square well showing $\langle x\rangle$, the expectation value of the particle's position in this state; the ground state eigenfunction is also shown.

(b) *average momentum*:

$$\langle p\rangle = \int_0^a \psi_1^*(x)\left(-i\hbar\frac{\partial}{\partial x}\right)\psi_1(x)dx$$

$$= \frac{2}{a}\int_0^a \sin\left(\frac{\pi x}{a}\right)\left(-i\hbar\frac{\partial}{\partial x}\right)\sin\left(\frac{\pi x}{a}\right)dx$$

$$= \frac{2}{a}\int_0^a \sin\left(\frac{\pi x}{a}\right)\left[\left(-i\hbar\frac{\pi}{a}\right)\cos\left(\frac{\pi x}{a}\right)\right]dx$$

$$= \frac{2}{a}(-i\hbar)\left[\frac{1}{2}\sin^2\left(\frac{\pi x}{a}\right)\right]_0^a = 0$$

and therefore $\langle p\rangle = 0$

This result is expected because the particle is as likely to be moving to the left as it is to the right.

(c) *average energy*:

The energy operator involves differentiation with respect to time so that we must use the full wave function, including time dependence $\Psi_1(x,t) = \psi_1(x)e^{-iE_1t/\hbar}$

$$\langle E\rangle = \int_0^a \Psi_1^*(x,t) i\hbar\frac{\partial}{\partial t}\Psi_1(x,t)dx$$

$$= \int_0^a \psi_1^*(x)e^{+iE_1t/\hbar} i\hbar\frac{\partial}{\partial t}\psi_1(x)e^{-iE_1t/\hbar}dx$$

$$= \int_0^a \psi_1^*(x)e^{+iE_1t/\hbar}\, i\hbar\,\psi_1(x)\left(\frac{-iE_1}{\hbar}\right)e^{-iE_1t/\hbar}dx$$

Thus,

$$\langle E\rangle = E_1\int_0^a \psi_1^*(x)\psi_1(x)dx = E_1,$$

a result which is consistent with Equation (14.46), which states that the $n = 1$ state can only have this energy.

For problems based on the material presented in this section visit up.ucc.ie/14/ and follow the link to the problems.

14.16 Potential steps

Consider a particle, of energy E, which is moving along the x-axis from left to right and which, at $x = 0$, encounters a sharp rise in potential energy (a *potential step*).

As illustrated in Figure 14.23, $U = 0$ when $x \le 0$

and $U = U_0$ when $x > 0$

We analyse the problem by distinguishing the two cases, (i) $E < U_0$, (Figure 14.23) and (ii) $E > U_0$, (Figure 14.25).

First, we consider how a particle would be expected to behave in these two cases according to classical physics, that is how such a mass would behave in the everyday macroscopic world. We then re-examine the same two cases, using quantum mechanical methods.

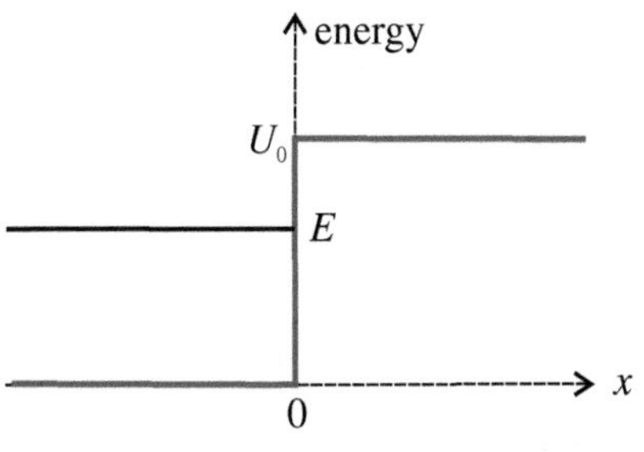

Figure 14.23. A potential step of height U_0.

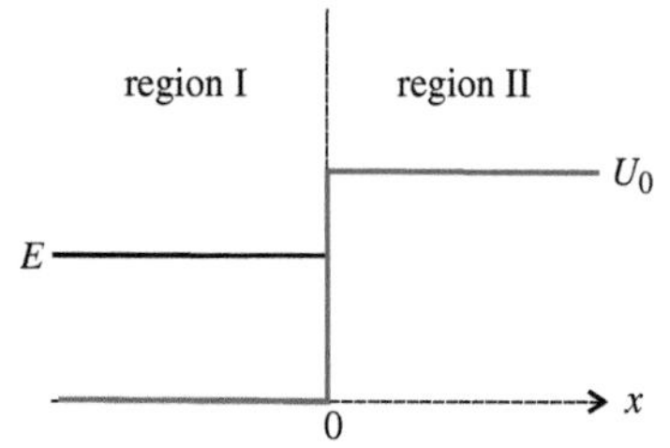

Figure 14.24. The potential step in *Case 1*, in which the particle energy E is less than the step height U_0.

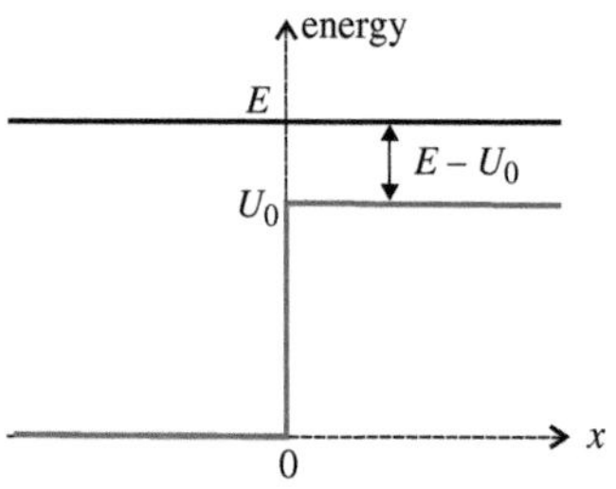

Figure 14.25. The potential step in *Case 2*, in which the particle energy E is greater than the step height U_0.

Classical treatment

Case 1: $E < U_0$

As illustrated in Figure 14.24, classically the particle can only be in the region $x \leq 0$, the region in which its (kinetic) energy is given by

$$E = \frac{p^2}{2m}$$

The momentum of the particle is thus given by $p = \pm\sqrt{2mE}$. The positive solution represents the incoming particle as it approaches the step from the left and the negative solution represents the particle as it travels to the left after reflection at the step. Classically, there is a 100% probability that the particle will be *reflected* at the step, and therefore a 0% probability of *transmission* of the particle past the step.

Case 2: $E > U_0$

As illustrated in Figure 14.25, when $x \leq 0$ the kinetic energy of the particle is E and its momentum, $p = \sqrt{2mE}$. When $x > 0$ the kinetic energy of the particle reduces to $(E - U_0)$ and its momentum reduces to $p = \sqrt{2m(E - U_0)}$. Classically, therefore, the probability of reflection at the barrier is 0% and the probability of transmission is 100%.

We now analyse these two cases quantum mechanically. We shall find that the results differ significantly from the classical results.

Quantum mechanical treatment

Case 1: $E < U_0$

When $x \leq 0$, the (time independent) Schrödinger equation is

$$-\frac{\hbar^2}{2m}\frac{d^2\psi}{dx^2} = E\psi \qquad (14.47)$$

This is the free particle case which we solved in Section 14.13.

When $x > 0$, the Schrödinger equation becomes

$$-\frac{\hbar^2}{2m}\frac{d^2\psi}{dx^2} = (E - U_0)\psi \qquad (14.48)$$

Note that $(E - U_0)$ is negative.

We solve the Schrödinger equation for each region and then ensure that the eigenfunction is well-behaved at $x = 0$, the boundary between the two regions. For the region $x \leq 0$, which we call region I in Figure 14.24, Equation (14.47) can be written

$$-\frac{d^2\psi_I}{dx^2} = k^2\psi_I \quad \text{where} \quad k^2 = \frac{2mE}{\hbar^2} \quad \text{and} \quad k = \frac{p}{\hbar}$$

The general solution of this equation (the free particle eigenfunction) is

$$\psi_I(x) = Ae^{ikx} + Be^{-ikx} \qquad (14.49)$$

The corresponding wave function, including time dependence, is

$$\Psi_I(x,t) = \psi_I(x)e^{-iEt/\hbar} = Ae^{i(kx-Et/\hbar)} + Be^{i(-kx-Et/\hbar)}$$

We have included time dependence so that we can determine in which directions the waves are travelling. In Section 13.2 we noted that, when the kx and ωt terms in the oscillatory function which represents a travelling wave have opposite signs, the wave is travelling in the $+x$ direction and, when they have the same sign, it is travelling in the $-x$ direction. Thus, $\Psi_+(x, t) = Ae^{i(kx - Et/\hbar)}$ represents a wave travelling in the $+x$ direction with momentum $p = \hbar k$ and $\Psi_-(x, t) = Be^{i(-kx - Et/\hbar)}$ represents a wave travelling in the $-x$ direction with momentum $p = -\hbar k$

Equation (14.48) (for the region $x > 0$, which we call region II in Figure 14.24), can be written

$$\frac{d^2\psi_{II}}{dx^2} = K^2\psi_{II} \quad \text{where} \quad K^2 = \frac{2m(U_0 - E)}{\hbar^2}$$

The solution of this equation is

$$\psi_{II}(x) = Ce^{Kx} + De^{-Kx} \tag{14.50}$$

Note that, because $U_0 > E$, K is real and positive. The Ce^{Kx} solution, in which the eigenfunction and hence the probability of finding the particle increases to infinity at large values of x is not physically realistic — the eigenfunction is not well-behaved because it is not finite when $x \to \infty$. We must set $C = 0$ to ensure that the eigenfunction is well-behaved. The wavefunction for $x > 0$ (region II) then becomes

$$\Psi_{II}(x,t) = \psi_{II}(x)\, e^{-iEt/\hbar} = De^{-Kx}\, e^{-iEt/\hbar}$$

We now apply boundary conditions of continuity to ψ and $\frac{d\psi}{dx}$ at $x = 0$, that is

$$\psi_I(0) = \psi_{II}(0) \quad \text{and} \quad \left[\frac{d\psi_I}{dx}\right]_{x=0} = \left[\frac{d\psi_{II}}{dx}\right]_{x=0}$$

At $x = 0$, Equation (14.49) becomes $\psi_I(0) = A + B$ and Equation (14.50) becomes $\psi_{II}(0) = D$. Continuity therefore requires that

$$A + B = D \tag{14.51}$$

We evaluate $\left[\frac{d\psi_I}{dx}\right]$ at $x = 0$ by differentiating Equation (14.49) and substituting $x = 0$. Thus $\left[\frac{d\psi_I}{dx}\right]_{x=0} = ikA - ikB$. Similarly, differentiation of Equation (14.50) and substitution of $x = 0$ gives $\left[\frac{d\psi_{II}}{dx}\right]_{x=0} = -KD$. Continuity of $\frac{d\psi}{dx}$ at $x = 0$ therefore requires that

$$ikA - ikB = -KD \tag{14.52}$$

Solving the simultaneous equations (14.51) and (14.52) to express A and B in terms of D, we obtain

$$A = \frac{(k + iK)D}{2k} \tag{14.53}$$

and

$$B = \frac{(k - iK)D}{2k} \tag{14.54}$$

and hence the solution (the eigenfunction) to *Case 1* is

for $x \leq 0$,

$$\psi_I(x) = \frac{D}{2k}[(k + iK)e^{ikx} + (k - iK)e^{-ikx}] \tag{14.55}$$

and for $x > 0$,

$$\psi_{II}(x) = De^{-Kx} \tag{14.56}$$

To interpret these solutions let us consider what it tells us about probability densities $P(x,t) = \psi^*(x)e^{iEt/\hbar}\,\psi(x)e^{-iEt/\hbar} = \psi^*(x)\psi(x)$ in the two regions. First when $x \leq 0$, for the incident particle

$$P(x,t) = A^*e^{-ikx}Ae^{ikx} = A^*A = \frac{(k + iK)D}{2k}\frac{(k - iK)D}{2k} = \frac{(k^2 + K^2)D^2}{4k^2}$$

The incident particle wave is therefore a travelling wave with constant probability density throughout the $x \leq 0$ region. The same result is obtained for the reflected particle wave.

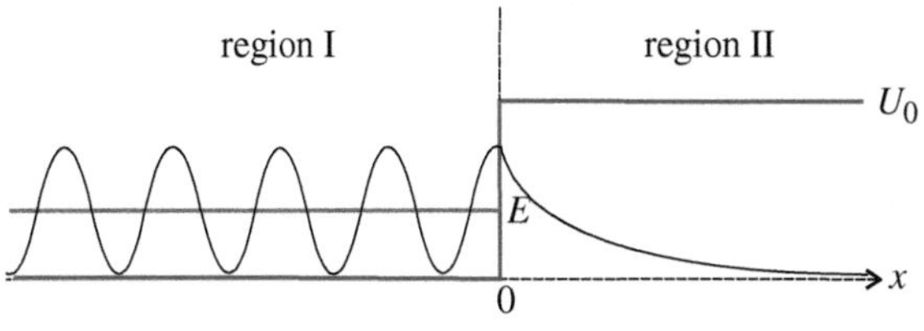

Figure 14.26. The real part of the wavefunction in *Case 1*, in which the particle energy E is less than the step height U_0.

When $x > 0$ $\quad P_{II}(x,t) = (De^{-Kx})^2 = D^2e^{-2Kx}$.

The qualitative features of the wavefunction in the two regions are indicated in Figure 14.26, a plot of the real part of the wavefunction in the two regions.

The most remarkable feature of Figure 14.26 is that there is a significant probability that the particle will penetrate into the region where $x > 0$, a region which is forbidden classically. This result may be understood in terms of the uncertainty principle. If we take the uncertainty in the momentum of the particle, Δp, to be of the same order as its momentum, $p = \hbar k$, we can estimate the uncertainty in the particle's position, Δx, from the uncertainty principle

$$(\Delta p)(\Delta x) \sim \frac{\hbar}{2}$$

$$\Delta x \sim \frac{\hbar}{2p} = \frac{\hbar}{2\hbar K} = \frac{1}{2K}$$

If we define the **penetration depth** x_p, to be the distance in which $P_{II}(x, t)$ falls to $\frac{1}{e}$ of its value at $x = 0$, then since the value of $P_{II}(x, t)$ at $x = 0$ is D^2, we can write

$$\frac{D^2}{e} = D^2e^{-1} = D^2e^{-2Kx_p} \quad \text{so that} \quad \boxed{x_p = \frac{1}{2K}}$$

Note that this distance is the same as the uncertainty in the particle's position, Δx, which we estimated above.

Note that, when $U_0 \gg E$, $K = \frac{\sqrt{2m(U_0 - E)}}{\hbar}$, is very large, and hence Δx is very small but if U_0 is only just greater than E, K is small and Δx is large — there is a deep penetration of the step.

For a beam of particles which encounters the step, the probability of reflection at the step is determined by the *reflection co-efficient*, the ratio of the number of particles which are reflected per second to the number which are incident per second. The number of particles per second per unit area is measured using the *probability flux*, the probability per second per unit area of a particle passing a point. In Section 12.3 we showed that the power, the energy flux, carried by a wave is proportional to the product of its velocity and the square of its amplitude. Similarly, the probability flux of a matter wave is proportional to the product of the particle velocity, v, and the square of the amplitude of its matter wave, $\Psi^*\Psi$; it is therefore given by $\Psi^*\Psi v$.

The reflection coefficient is given by $R = \frac{\Psi_-^*\Psi_- v_B}{\Psi_+^*\Psi_+ v_A} = \frac{B^*Bv_B}{A^*Av_A}$, where v_A and v_B are the velocities of the incident and reflected particles, respectively. Particle velocity is given by $v = \frac{p}{m} = \frac{\hbar k}{m}$. Both the incident and reflected particles propagate in the same region, region I in Figure 14.26, and therefore have the same wave number. Thus $v_A = v_B$ and, using equations (14.53) and (14.54), we can write

$$R = \frac{B^*B}{A^*A} = \frac{(k + iK)(k - iK)}{(k - iK)(k + iK)} = 1$$

This is the same as the classical result. Note that this result is not inconsistent with penetration of the step because, even when the particle penetrates it has to return, since the probability of the particle reaching $x = \infty$ is zero.

Case 2: $E > U_0$

In this case, illustrated in Figure 14.27, when $x \le 0$ (region I) the kinetic energy of the particle is E and the time-independent Schrödinger equation is exactly the same as in *Case 1* for $x \le 0$, namely

$$-\frac{\hbar^2}{2m}\frac{d^2\psi_I}{dx^2} = E\psi_I$$

This is again a free particle case. Defining $k_1^2 = \frac{2mE}{\hbar^2}$, the equation becomes

$$-\frac{d^2\psi_I}{dx^2} = k_1^2\psi_I$$

with the solution $$\psi_I(x) = Ae^{ik_1x} + Be^{-ik_1x} \qquad (14.57)$$

The corresponding wavefunction is $$\Psi_I(x, t) = \psi_I(x)e^{-iEt/\hbar} = A\,e^{i(k_1x - Et/\hbar)} + B\,e^{i(-k_1x - Et/\hbar)}$$

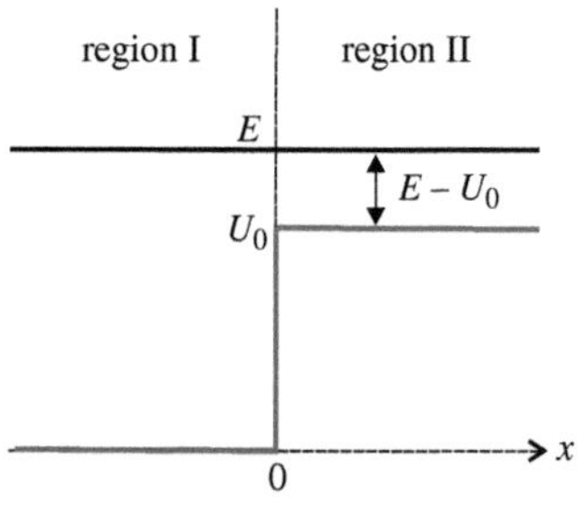

Figure 14.27. The potential step in *Case* 2, in which the particle energy E is greater than the step height U_0.

As before, the first term represents a particle travelling to the right with momentum $\hbar k_1$ (the incident wave) and the second represents a particle travelling to the left with momentum $-\hbar k_1$ (the reflected wave).

For $x > 0$ (region II), the kinetic energy is $(E - U_0)$, which is positive. This is still a free particle case, but with a smaller kinetic energy.

Defining $k_2^2 = \frac{2m(E - U_0)}{\hbar^2}$, the time-independent Schrödinger equation becomes

$$-\frac{d^2\psi_{II}}{dx^2} = k_2^2\psi_{II}$$

with the solution (eigenfunction) $\psi_{II}(x) = Ce^{ik_2x} + De^{ik_2x}$ (14.58)

The corresponding wavefunction is

$$\Psi_{II}(x,t) = \psi_{II}(x)e^{-iEt/\hbar} = Ce^{i(k_2x - Et/\hbar)} + De^{i(-k_2x - Et/\hbar)}$$

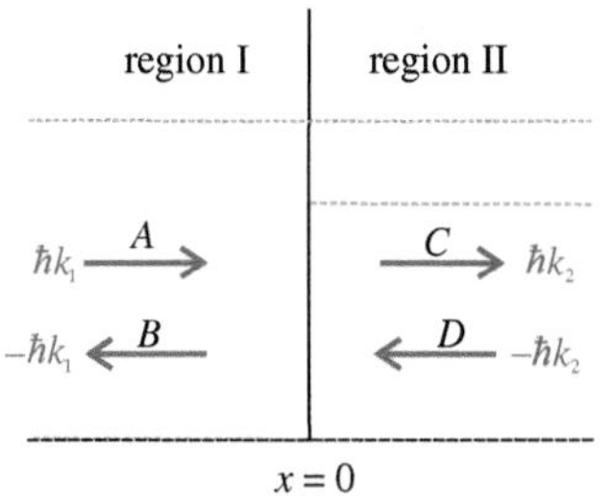

Figure 14.28. The directions of the components of the wave solutions given in equations (14.57) and (14.58).

The first term represents a particle travelling to the right with momentum $\hbar k_2$ and the second a particle travelling to the left with momentum $-\hbar k_2$.
The directions of the various components of the solutions given in equations (14.57) and (14.58) are summarised in Figure 14.28.
In this problem we are interested only in particles which enter from the left so that we set $D = 0$ in Equation (14.58).

As before, we apply boundary conditions to $\psi(x)$ and to $\frac{d\psi(x)}{dx}$ at $x = 0$. At $x = 0$, Equation (14.57) becomes $\psi_I(0) = A + B$ and Equation (14.58) becomes $\psi_{II}(0) = C$. Continuity therefore requires that

$$A + B = C \qquad (14.59)$$

By differentiating Equations (14.57) and (14.58) and substituting $x = 0$, we obtain

$$\left[\frac{d\psi_I}{dx}\right]_{x=0} = ik_1A - ik_1B \text{ and } \left[\frac{d\psi_{II}}{dx}\right]_{x=0} = ik_2C$$

Continuity of $\frac{d\psi}{dx}$ at $x = 0$ therefore requires that

$$k_1A - k_1B = k_2C \qquad (14.60)$$

Solving equations (14.59) and (14.60) to express A and B in terms of C, we obtain

$$A = \frac{(k_1 + k_2)C}{2k_1} \qquad (14.61)$$

and

$$B = \frac{(k_1 - k_2)C}{2k_1} \qquad (14.62)$$

The reflection coefficient is given by $R = \frac{B^*Bv_B}{A^*Av_A}$. As in *Case 1*, which we considered above, $v_A = v_B$ so that we can write

$$\boxed{R = \frac{B^*B}{A^*A} = \frac{(k_1 - k_2)^2}{(k_1 + k_2)^2}}$$

Note the departure from the classical result (that is, $R = 0$). Unless $k_1 = k_2$ (the case in which there is no potential step), $0 < R < 1$.

In calculating the probability of transmission past the step, T, we must take account of the fact that particles travel more slowly on the right hand side of the step, where their kinetic energy is $E - U_0$, than on the left hand side of the step, where their kinetic energy is E. The transmission probability, T, is given by

$$T = \frac{\text{probability flux after the step}}{\text{probability flux before the step}} = \frac{C^*Cv_2}{A^*Av_1}$$

where v_1 and v_2 are the particle velocities before and after the step, respectively.

Now

$$\frac{v_2}{v_1} = \frac{p_2}{p_1} = \frac{k_2}{k_1}$$

and, from Equation (14.61),

$$\frac{C}{A} = \frac{2k_1}{k_1 + k_2}$$

Thus,

$$T = \left[\frac{2k_1}{(k_1 + k_2)}\right]^2 \frac{k_2}{k_1} = \frac{4k_1 k_2}{(k_1 + k_2)^2}$$

Note that

$$R + T = \frac{(k_1 - k_2)^2}{(k_1 + k_2)^2} + \frac{4k_1 k_2}{(k_1 + k_2)^2} = 1$$

As expected, the probability of the particle being either reflected or transmitted is 100%; particle flux is conserved at the step but note that probability density is not, that is, $CC^* + BB^* \neq AA^*$, as may be demonstrated using Equations (14.61) and (14.62).

The similarities and differences between the quantum mechanical and classical results are highlighted by consideration of the following two cases.

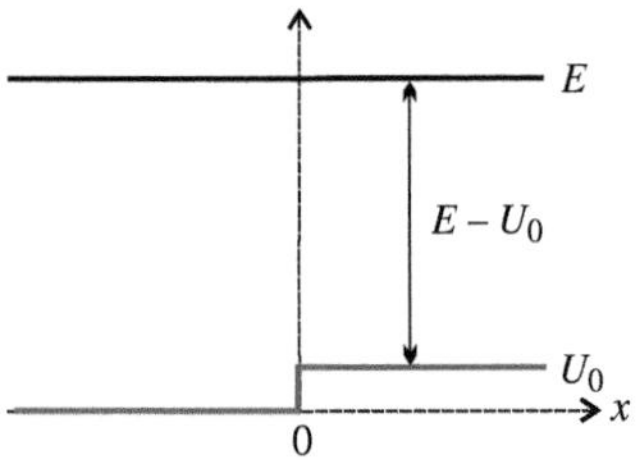

Figure 14.29. The potential step for a case in which the particle energy E is much greater than the step height U_0.

Case (a): $E \gg U_0$, a very small step, such that $E \approx E - U_0$, as illustrated in Figure 14.29.

In this case

$$k_1 \approx k_2$$

so that

$$R = \frac{(k_1 - k_2)^2}{(k_1 + k_2)^2} \approx 0 \quad \text{and} \quad T = \frac{4k_1 k_2}{(k_1 + k_2)^2} \approx 1$$

This result is very similar to the classical case. There is very little probability of reflection.

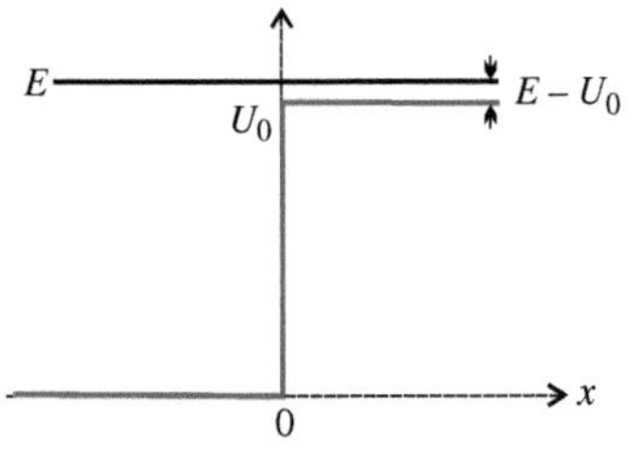

Figure 14.30. The potential step in a case in which the particle energy E is only just greater than the step height U_0.

Case (b): $E \gg E - U_0$, the energy of the particle is only just greater than the step height, U_0, as illustrated in Figure 14.30.

In this case

$$k_1 \gg k_2$$

so that

$$R = \frac{(k_1 - k_2)^2}{(k_1 + k_2)^2} \approx 1 \quad \text{and} \quad T = \frac{4k_1 k_2}{(k_1 + k_2)} \approx 0$$

This tells us that the particle is very likely to be reflected even though its energy is slightly greater than the step height, a result which is completely contrary to our classical experience of particle behaviour.

The quantum mechanical results for the potential step are summarised in Figure 14.31 where the values of the reflection and transmission coefficients are plotted as functions of the ratio $\frac{E}{U_0}$. The departures from the classical picture are most striking in the region in which $\frac{E}{U_0}$ is just greater than 1.

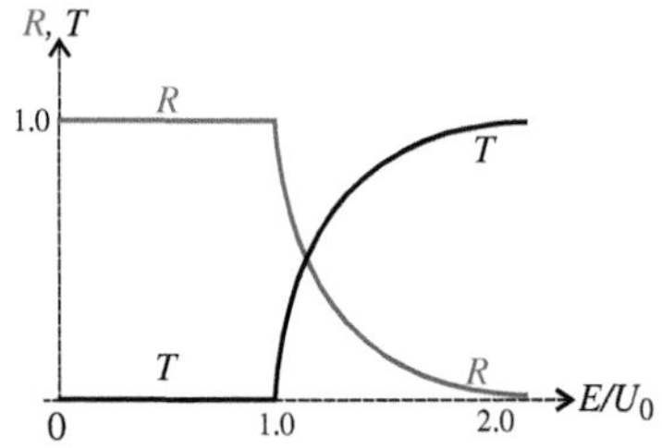

Figure 14.31. A summary of the quantum mechanical results for the potential step. Plots of the reflection (blue line) and transmission (black line) coefficients, against the ratio $\frac{E}{U_0}$ for a particle incident on the step.

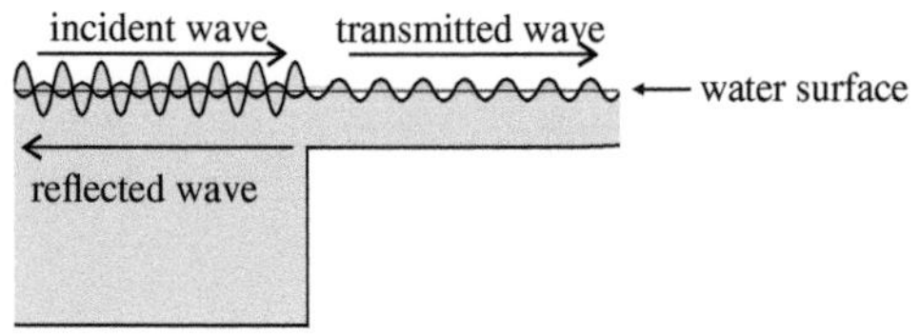

Figure 14.32. The reflection of water waves in a ripple tank on encountering a sharp decrease in water depth.

Although reflection of a particle at a discontinuity is a not a classical particle phenomenon, it is a well-known wave phenomenon. It may be observed for example in a ripple tank in which surface waves propagate across shallow water. The waves may be seen to reflect when they encounter a sharp decrease in the depth of the water, as illustrated in Figure 14.32. A *sharp* change of depth is defined as one in which the depth changes significantly over a wavelength of the water wave.

The potential drop

The analysis of the potential drop, illustrated in Figure 14.33 is identical to that given above for the potential step except that $k_1 > k_2$ instead of $k_2 > k_1$. Thus the equations for the reflection and transmission coefficients:

$$R = \frac{(k_1 - k_2)^2}{(k_1 + k_2)^2} \quad \text{and} \quad T = \frac{4k_1 k_2}{(k_1 + k_2)^2},$$

are the same as for the potential step. This leads to the remarkable result that if the energy of the particle is only just greater than the potential drop, as shown in Figure 14.33, that is, if $(E - U_0) \ll E$, there is a very high probability of reflection at the point at which the potential drops. This is because, if $(E - U_0) \ll E$, $k_2 \gg k_1$ and $R \approx 1$.

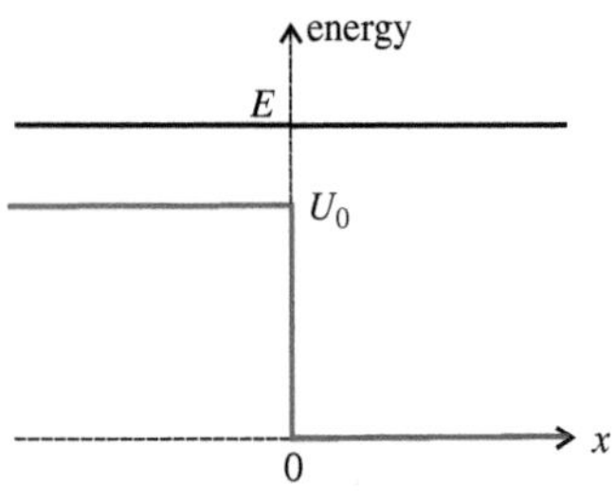

Figure 14.33. A potential drop.

For the quantum mechanical step (or drop), partial reflection occurs for $E > U_0$ only if the potential energy function varies significantly over a distance of the order of the de Broglie wavelength of the particle. The probability of reflection is small if the potential energy function varies slowly over the de Broglie wavelength.

This result runs contrary to our classical experience of particle behaviour but, as already noted, reflection of a wave at a discontinuity is a well-known wave phenomenon. As illustrated in Figure 14.34, when surface waves which are travelling across a water surface in a ripple tank encounter a region in which the water depth increases sharply, reflection may be observed.

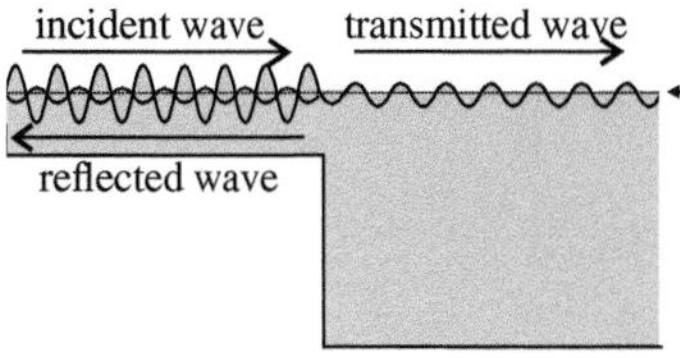

Figure 14.34. The reflection of surface water waves at a sharp increase in water depth in a ripple tank.

For problems based on the material presented in this section visit up.ucc.ie/14/ and follow the link to the problems.

14.17 Other potential wells and barriers

We now review the solutions of further one-dimensional potential wells and barriers, namely, (a) the potential barrier and (b) the finite square well. In each case we shall state the quantum mechanical solutions and interpret them using the results for the free particle, the infinite potential well and the potential step, which we have derived in Sections 14.13, 14.15 and 14.16.

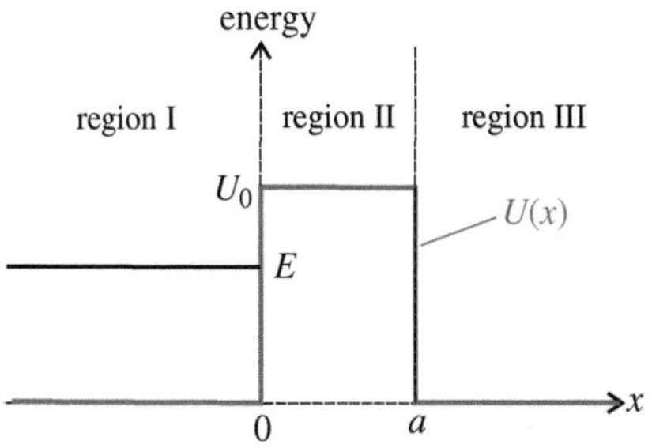

Figure 14.35. The potential barrier in a case in which the particle energy E is less than the barrier height U_0.

The potential barrier: tunnelling

The potential barrier, as illustrated in Figure 14.35, consists of a potential step followed by a potential drop.

The full quantum mechanical analysis of this problem involves the solution of the time-independent Schrödinger equation in the three regions I, II and III in Figure 14.35, the application of two sets of boundary conditions and, while straightforward, involves a substantial amount of algebra. We can discuss the results qualitatively, however, using our knowledge of the behaviour of a particle at potential steps and drops.

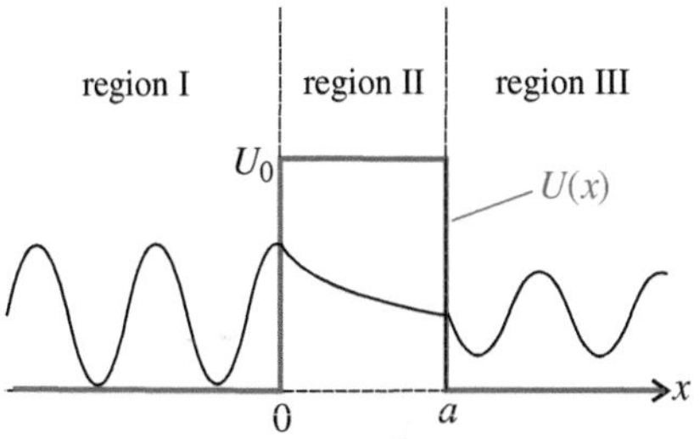

Figure 14.36. Tunnelling, illustrated by a plot of the real part of the wavefunction in a case in which the particle energy is less than the barrier height.

When $E < U_0$, a classical treatment leads us to expect a 100% probability of reflection at the barrier. As we have seen, however, quantum mechanically we know that there is some probability that the particle will penetrate a certain distance into the region beyond the potential step at $x = 0$. In region II the probability falls exponentially with increasing penetration, as illustrated by the plot of the real part of the wavefunction in Figure 14.36 but, if the barrier is sufficiently narrow, there may be a significant probability that the particle will reach the other side of the barrier, where it can propagate as a free particle. Quantum mechanically, therefore, there is a probability, which depends on the width of the barrier a and the energy of the particle relative to the barrier height $\frac{E}{U_0}$, that the particle will re-appear on the other side of the barrier, a phenomenon known as **tunnelling**.

Tunnelling is a wave phenomenon. A comparable phenomenon, known as *frustrated total internal reflection,* may be observed for light waves. As will be described in Section 23.4, if a light wave travels from a glass block into air in such a way that its angle of incidence at the glass surface is greater than a certain angle, it is totally reflected back into the glass (100% probability of reflection). If, however, a second glass block is brought close enough to the glass surface that the gap between the two blocks is comparable to the wavelength of the light, the light waves pass across the gap into the second block. Total internal reflection is frustrated by 'tunnelling' by the light waves.

There are many important examples of tunnelling by particles. One such is *α−particle* decay, the emission of $^4_2$He nuclei from atomic nuclei, a nuclear phenomenon which we shall consider in more detail in Section 27.5. Inside the nucleus the α−particle is bound by the strong nuclear force (Section 1.4), an attractive force which is exerted by the other nuclear particles. As noted in Section 1.4, the strong nuclear force is very short range in its effect so that, outside the nucleus ($r > r_1$), the longer range Coulomb repulsion, between the positively charged α−particle and the positively charged nucleus which it has left behind, dominates. The net result is a potential barrier as shown in Figure 14.37, a plot of the potential energy of the α−particle as a function of distance, r, from the centre of the nucleus.

Classically, the α−particle, whose energy, E, is less than the barrier height, would be expected to be bound in the nucleus forever – the system would be stable. Quantum mechanics, however, tells us that there is a probability that the α−particle will tunnel through the

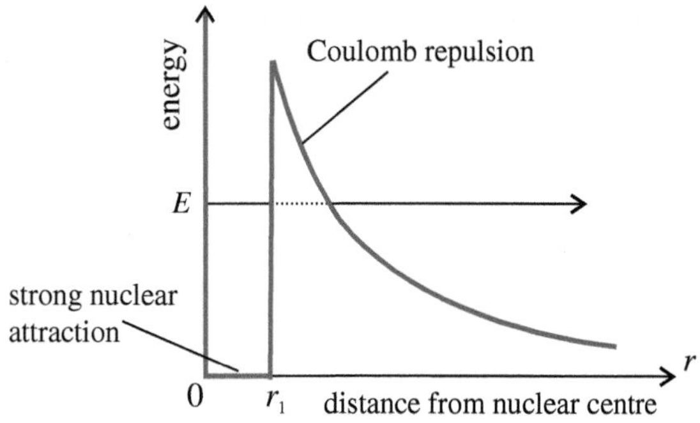

Figure 14.37. α−particle decay. The tunnelling of an α−particle through the potential barrier at the surface boundary of a nucleus of radius r_1.

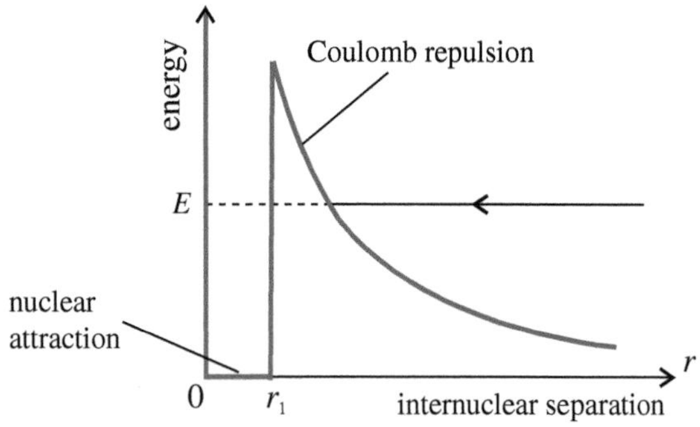

Figure 14.38. Fusion. The tunnelling of two deuterium nuclei through the repulsive Coulomb barrier between them.

barrier and escape. The escape of α−particles (α−decay) is therefore a random process which is governed by the probability of barrier penetration. The rate of α-decay from a nucleus can be analysed successfully using quantum mechanics.

A further example of tunnelling is nuclear fusion (the merging of two nuclei), to be considered in more detail in Section 27.8. Nuclear fusion takes place, for example, when two deuterium ($^2_1$H) nuclei, which are positively charged, come close enough to each other for the short range strong nuclear force to overcome Coulomb repulsion and bind three of the nucleons together in a single $^3_2$He or $^3_1$H (tritium) nucleus.

that is
$$^2_1\text{H} + ^2_1\text{H} \rightarrow ^3_2\text{He} + ^1_0\text{n}$$

or
$$^2_1\text{H} + ^2_1\text{H} \rightarrow ^3_1\text{H} + ^1_1\text{H}$$

The protons and neutrons in the $^3_2$He or $^3_1$H nuclei are more strongly bound after the fusion reaction – the total energy of the system is lower. The excess energy, which is carried by the product particles as kinetic energy, is considerable, as noted in the discussion of the energy released in nuclear processes in Section 9.14.

The situation is represented in potential energy terms in Figure 14.38. This figure is very similar to Figure 14.37, with the strong nuclear force dominating at small internuclear distances and the Coulomb repulsion barrier taking over at large distances. In this case, however, the process starts outside the barrier. Classically, the barrier would prevent fusion of the deuterium nuclei completely at room temperature but, quantum mechanically, there is the possibility of tunnelling.

The two deuterium nuclei in a D_2 molecule are separated by 7.4 × 10^{-11} m and the probability of fusion by tunnelling through the coulomb barrier between them is about 10^{-70} s^{-1} at room temperature. Although greater than the classical probability, which is zero, for practical purposes the quantum result is little different. It means that, if all $^2_1$H atoms in the sea (deuterium comprises one part in 6000 of the hydrogen in sea water) were to be held in D_2 molecules, fusion would be most unlikely to have occurred once in the history of the Earth. If, however, the $^2_1$H nuclei could be packed more closely together so that the internuclear distance is halved, the probability of tunnelling at room temperature increases to 10^{-23} s^{-1}, at which point fusion would take place with sufficient frequency to be observed. This was the idea behind cold fusion, a topic which received much attention in the late 1980's. However, the viability of such a process, to be discussed further in Section 27.10, has yet to be demonstrated.

A third example of barrier tunnelling — the tunnel diode, an important device in electronics — will be described in Section 26.9.

The finite potential square well

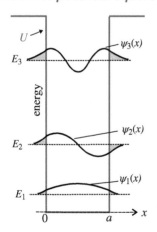

Figure 14.39. The energy levels (eigenvalues) and corresponding eigenfunctions of a particle in a finite potential square well.

The finite potential square well, illustrated in Figure 14.39, is similar to the infinite potential well (Section 14.15) but, in this case, the walls of the well are of finite height; they are no longer infinitely greater than the particle energy.

The finite potential well is a bound region of space with sides which are defined by two potential steps. As in the case of the potential barrier, the problem is solved by constructing wavefunctions in each of the three regions and by applying boundary conditions. The energy levels are quantised and the eigenfunctions are similar to those of the infinite potential well (standing waves) but with two main differences:

- Because the well is not infinitely deep it can support only a finite number of quantised energy levels (eigenvalues), E_1 etc., as illustrated in Figure 14.39. The number of eigenvalues depends on the depth and breadth of the well.
- Because the potential steps at the sides of the well are not infinitely larger than the particle energy, the eigenfunctions ($\psi_1(x)$ etc., also shown in Figure 14.39), although still standing waves, are not confined sharply to the well. There is some probability of penetration beyond the confines of the well. Note that, as indicated by the shaded portions of the wavefunctions in Figure 14.39, the penetration depth increases with the energy of the bound state, that is, as the energy of the particle approaches the step height.

Although the finite potential well is a more realistic representation of most physical systems than the infinite well, the simpler solutions of the infinite well (Section 14.15) can often be used as good approximations for finite wells.

For problems based on the material presented in this section visit up.ucc.ie/14/ *and follow the link to the problems.*

14.18 The simple harmonic oscillator

The application of the Schrödinger equation to the simple harmonic oscillator, the system to which quantum ideas were first applied when Planck interpreted the observed cavity radiation spectrum (Section 14.2), is an important problem but is beyond the scope of this textbook. The derivation of the ground state wavefunction and energy is discussed on the *Understanding Physics* website at up.ucc.ie/14/18/1/.

The eigenvalues and eigenfunctions of the simple harmonic oscillator are illustrated in Figure 14.40.

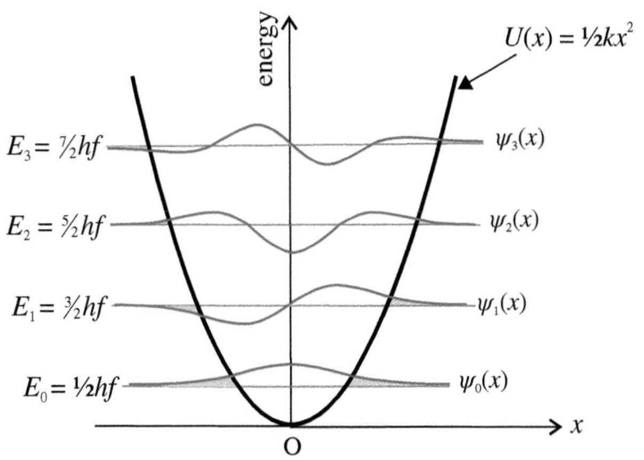

Figure 14.40. The energy levels (eigenvalues) and corresponding eigenfunctions of the four lowest energy states of a simple harmonic oscillator.

The energy levels (eigenvalues) are quantised, as expected for a bound potential; their values are given by the equation

$$E_n = \left(n + \frac{1}{2}\right) hf = \left(n + \frac{1}{2}\right) \hbar\omega \tag{14.63}$$

where $n = 0, 1, 2$ etc.…

The separation between adjacent energy levels is thus hf, as given by the Planck equation (14.1), namely $E_n = nhf$, but note that the energy of the lowest state is now $E_0 = \frac{1}{2}hf$ rather than zero, as suggested by Planck. As noted earlier (Section 14.15), the non-zero minimum energy of bound systems is a very general consequence of the uncertainty principle. This means that, even at 0 K, electrons and atoms in solids (recall the kinetic theory of solids, Section 12.8) still have kinetic energy.

The eigenfunctions (shown in Figure 14.40) are similar to those obtained for the infinite and finite square wells — they are again standing waves — and, as with the finite square well, because potential energy is not infinite at the positions at which the quantised energies intersect the $U(x)$ curve, some penetration of the walls occurs.

14.19 Further implications of quantum mechanics

We have seen that the uncertainty principle places a limitation on what we are able to know concerning the position and momentum of a particle. The following question arises however, "might there not still be an underlying reality – an underlying 'clockwork' – in which the particle has definite values of position and momentum even if they cannot be revealed?"

This problem concerned Einstein, in particular, and in 1935 together with Podolsky and Rosen he devised the following experiment – a 'thought experiment' – to highlight the implications of this aspect of quantum theory.

If two particles A and B interact with each other and then fly apart, the uncertainty principle would allow an experimenter to measure precisely the combined momentum of the pair at the time of collision and later the momentum of particle A and the position of B. The momentum of B could then be determined through the principle of conservation of momentum so that both the position and momentum of B would be known. The only way to preserve quantum uncertainty in this situation would be if particle A were to be disturbed by the measurement of particle B, not allowing the momentum of B to be determined precisely. The question then is how can this disturbance be communicated from particle B to particle A? There are two possibilities:

(a) Particle B could act on particle A instantaneously *at a distance*, requiring communication at a speed which is greater than that of light.
(b) The necessary information could be carried in the underlying clockwork which reveals this information as circumstances require.

Einstein preferred the second mechanism. The original Einstein, Podolsky and Rosen 'thought' experiment was devised to highlight the implications of quantum theory and was not a practical experimental proposition but, more recently, experiments have been devised by John Bell (1928–1990) to discover through which of the two possible mechanisms quantum mechanics operates. Experimental evidence to date favours mechanism (a), action at a distance.

The implication of this result is that, through action at a distance, particles are in some sense part of a single quantum system. The scope of quantum mechanics therefore seems greater than was at first appreciated and is not yet understood. It is important to note, however, that the somewhat philosophical question which we have just considered does not in any way bring into question the results which we have obtained in the preceding sections of this chapter.

In Chapters 24 to 27, we shall apply many of the results of this chapter to the behaviour of electrons in atoms and solids and to the behaviour of nucleons in nuclei. We shall find that the behaviour of such systems can only be interpreted satisfactorily in quantum mechanical terms. In these cases, a three-dimensional generalisation of the Schrodinger equation (14.31) and of the probability density definition (14.32) will be required, namely

$$-\frac{\hbar^2}{2m}\nabla^2\Psi(\boldsymbol{r},t) + U(\boldsymbol{r},t)\Psi(\boldsymbol{r},t) = i\hbar\frac{\partial\Psi(\boldsymbol{r},t)}{\partial t}$$

and

$$P(\boldsymbol{r},t) := \Psi^*(\boldsymbol{r},t)\Psi(\boldsymbol{r},t) \qquad \text{(probability per unit volume)}$$

Note that, among other applications, the latter definition allows quantities such as the charge density of a particle to be determined from the wave function, for example; the charge density of an electron can be written as

$$\rho(r) = -e\Psi^*(\boldsymbol{r},t)\Psi(\boldsymbol{r},t)$$

Further reading

Further material on some of the topics covered in this chapter may be found in *Introduction to Quantum Mechanics* by Phillips and in *Quantum Mechanics* by Mandl (See BIBLIOGRAPHY for book details).

WORKED EXAMPLES

Worked Example 14.1:

(a) Use the Planck cavity radiation Equation (14.4) to derive the Wien displacement law $\lambda_{max}T = 0.2014\frac{hc}{k}$, where λ_{max} is the wavelength at which $w(\lambda, T)$ is at a maximum at temperature T.
(b) Given that the energy density of the radiation emitted by the Sun has its maximum value at 510 nm, estimate the temperature of the Sun. [Note: The non-zero solution of the equation $e^{-x} + \frac{x}{5} = 1$ is $x = 4.965$]

We must first write Equation (14.4), $w(f,T)df = \dfrac{8\pi hf^3}{c^3\left(e^{\frac{hf}{kT}} - 1\right)}df$, in terms of wavelength.

For a given temperature T, we write $w(f, T)$ as $w_T(f)$. Note that $w_T(\lambda)d\lambda = -w_T(f)df$, the minus sign indicating that, although $w_T(\lambda)$and $w_T(f)$ are both positive, $d\lambda$ and df have opposite signs (an increase in frequency corresponds to a decrease in wavelength).

Differentiating $f = \frac{c}{\lambda}$, we obtain $df = -\frac{c}{\lambda^2}d\lambda$

so that substituting for f and df, we obtain

$$w_T(\lambda)d\lambda = -w_T(f)df = w_T(f)\frac{c}{\lambda^2} = \frac{8\pi hc}{\lambda^5}\left(\frac{d\lambda}{e^{\frac{hc}{\lambda kT}} - 1}\right)$$

$w_T(f)$ is at a maximum when $\frac{dw}{d\lambda} = 0$. We therefore differentiate $w_T(f)$ with respect to λ. Setting $\frac{hc}{\lambda kT} = x$, the function becomes $w(x) = \frac{Cx^5}{e^x - 1}$ where C is a constant. We differentiate $w(x)$ by parts using rule (iii) of Appendix A.6.1, namely

$$\frac{d(fg)}{dx} = \left[f\frac{dg}{dx} + g\frac{df}{dx}\right] \qquad (14.64)$$

where $f = x^5$ and $g = \frac{1}{e^x - 1}$. Now $\frac{df}{dx} = 5x^4$ and $\frac{dg}{dx}$ is obtained as follows:

Let $h = e^x - 1$ so that $g = \frac{1}{h}, \frac{dg}{dh} = -\frac{1}{h^2}$ and $\frac{dh}{dx} = e^x$

Thus $\frac{dg}{dx} = \frac{dg}{dh}\frac{dh}{dx} = -\frac{1}{h^2}e^x = -\frac{e^x}{(e^x - 1)^2}$

Substituting for f, g, $\frac{df}{dx}$ and $\frac{dg}{dx}$ in Equation (14.64) $\quad \frac{dw}{dx} = C\left[x^5\left(\frac{-e^{-x}}{(e^x - 1)^2}\right) + \left(\frac{1}{(e^x - 1)}\right)5x^4\right]$

This is zero when $\frac{-xe^x}{e^x - 1} + 5 = 0$, which can be written $e^{-x} + \frac{x}{5} = 1$.

We can check the solution to this equation, which is given above ($x = 4.965$), by substitution.

$$e^{-x} + \frac{x}{5} = e^{-4.965} + \frac{4.965}{5} = 0.00698 + 0.993 \approx 1$$

Thus $x = \frac{hc}{\lambda kT} = 4.965$ when $\lambda = \lambda_{max}$, the value of λ which corresponds to w_{max}.

$$\Rightarrow \quad \lambda_{max}T = \frac{0.2014hc}{k}, \quad \text{which is the Wien displacement law}$$

(b) For the Sun, $\lambda_{max} = 510 \times 10^{-9}$nm Thus $T \sim 5700$ K

Worked Example 14.2: Derive Stefan's law (Section 11.7) from Planck's Equation (14.4) for the distribution of frequencies of electromagnetic waves in a cavity at temperature T and hence determine an expression for the Stefan-Boltzmann constant in terms of fundamental physical constants. [Note that the radiated power per unit area emitted through a hole in the cavity is given by $\frac{dP}{dA} = \frac{1}{4}cw_{total}$, where w_{total} is the total radiation density in the cavity and c is the speed of electromagnetic waves in vacuum.]

$$\text{Equation (14.4)} \quad \rightarrow \quad w(f,T)df = \frac{8\pi hf^3}{c^3\left(e^{\frac{hf}{kT}} - 1\right)}df$$

To determine the total radiation density we integrate over all frequencies,

$$w_{total} = \int_0^{\infty} w(f,T)df = \frac{8\pi h}{c^3}\int_0^{\infty}\frac{f^3 df}{e^{\frac{hf}{kT}} - 1}$$

Changing the variable to $\xi = \frac{hf}{kT}$ this becomes

$$w_{total} = \frac{8\pi h}{c^3}\left(\frac{kT}{h}\right)^4\int_0^{\infty}\frac{\xi^3 d\xi}{e^{\xi} - 1} = \frac{8\pi k^4T^4}{c^3h^3}\left(\frac{\pi^4}{15}\right)$$

where the value of the integral has been obtained from Table A.3 in Appendix A.

Thus the radiated power per unit area is $\frac{dP}{dA} = \frac{1}{4}cw_{total} = \frac{2\pi^5k^4}{15c^2h^3}T^4$

This is Stefan's law, $\frac{dP}{dA} = \sigma T^4$, where the Stefan-Boltzmann constant has the value

$$\sigma = \frac{2\pi^5k^4}{15c^2h^3} = \frac{\pi^2k^4}{60c^2\hbar^3} = 5.64 \times 10^{-8}\ \text{W m}^{-2}\ \text{K}^{-4}$$

in good agreement with the experimental value (Section 11.7).

Worked Example 14.3: The maximum kinetic energy of the electrons released when a metal surface is illuminated by light of wavelength 492 nm is measured to be 9.9×10^{-20} J. This energy is found to drop to 3.8×10^{-20} J when the wavelength of the light is changed to 579 nm. Use this data to estimate (i) the value of the Planck constant and (ii) the work function of the metal.

The frequencies of light of wavelengths 492 and 579 nm are $\frac{3 \times 10^8}{492 \times 10^{-9}} = 6.10 \times 10^{14}$Hz and $\frac{3 \times 10^8}{579 \times 10^{-9}} = 5.18 \times 10^{14}$ Hz, respectively.

Applying Equation (14.9), $hf = \phi + K$, first to the 6.10×10^{14} Hz light, and secondly to the 5.18×10^{14} Hz light, yields

$$h(6.10 \times 10^{14}) = \phi + 9.9 \times 10^{-20}$$

and

$$h(5.18 \times 10^{14}) = \phi + 3.8 \times 10^{-20}\ \text{J}$$

Solving these simultaneous equations, we obtain $h = 6.63 \times 10^{-34}$ J s and $\phi = 3.05 \times 10^{-19}$ J = 1.9 eV.

Worked Example 14.4: An neon atom in an excited state releases its excess energy by emitting a photon of wavelength 632.8 nm. If the average time between excitation of the atom and the time at which it emits a photon is 7×10^{-9} s, estimate the uncertainty in the frequency of the photon, Δf.

From Equation (14.26)

$$\Delta E \geq \frac{\hbar}{2\Delta t}$$

Thus

$$\Delta E \geq 8 \times 10^{-27}\ \mathrm{J}$$

which corresponds to an uncertainty in frequency which is given by differentiating Equation (14.20),

$$\Delta E = h\Delta f$$

Thus

$$\Delta f = \frac{\Delta E}{h} \geq 12 \times 10^{6}\ \mathrm{Hz} = 12\ \mathrm{MHz}$$

The frequency of the 632.8 nm photon is $\dfrac{3 \times 10^{8}}{632.8 \times 10^{-9}} \approx 5 \times 10^{14}$ Hz. The fractional uncertainty in the frequency of the photon due to the uncertainty principle, $\dfrac{\Delta f}{f}$, is therefore about 10^{-8} which is generally much smaller than the spread of frequencies which is produced by other effects such as collisions and the Doppler effect, as will be discussed for light in Section 22.1.

Worked Example 14.5: A particle is confined to move along the x-axis between $x = 0$ and $x = 1$. Its wavefunction is given by $\psi = Cx$ within this region and is zero outside. Determine the value of C needed to normalise ψ. Calculate (a) the probability that the particle be found between $x = 0.0$ and 0.2, (b) the probability that the particle be found between $x = 0.4$ and 0.6 and (c) the expectation value of the particle's position.

The wavefunction is normalised if

$$\int_0^1 \psi^* \psi\, dx = 1 \text{ which gives}$$

$$\int_0^1 C^2 x^2 dx = C^2 \left[\frac{x^3}{3}\right]_0^1 = 1.$$

Thus $C^2 = 3$ and the normalised wavefunction is

$$\psi = \sqrt{3}x$$

(a) The probability that the particle be found between $x = 0.0$ and 0.2 is given by

$$\int_{0.0}^{0.2} \psi^* \psi\, dx = 3\int_{0.0}^{0.2} x^2 dx = 3\left[\frac{x^3}{3}\right]_{0.0}^{0.2} = 3\,[0.0027 - 0.0] = 0.008$$

(b) The probability that the particle be found between $x = 0.4$ and 0.6 is given by

$$\int_{0.4}^{0.6} \psi^* \psi\, dx = 3\int_{0.4}^{0.6} x^2 dx = 3\left[\frac{x^3}{3}\right]_{0.4}^{0.6} = 3\,[0.072 - 0.021] = 0.153$$

(c) The expectation value of x is given by

$$\langle x \rangle = \int_0^1 \psi^* x \psi\, dx = 3\int_0^1 x^3 dx = 3\left[\frac{x^4}{4}\right]_0^1 = \frac{3}{4}$$

PROBLEMS

For problems based on the material covered this chapter visit up.ucc.ie/14/ and follow the link to the problems.

15

Electric currents

AIMS

- to investigate some phenomena associated with electric currents and to define a unit (the *ampere*) which is used as a base unit in SI for the strength of an electric current
- to introduce concepts (*emf, resistance* and *potential difference*) which can be used to explain the observed behaviour of steady currents in electric circuits
- to introduce rules which predict the response of a particular class of components (*ohmic* components) in electric networks
- to introduce procedures (using *Kirchoff's rules*) which facilitate the analysis of circuits containing more than one current loop
- to study the characteristics (*resistivity* and *conductivity*) which determine the ability of different materials to conduct electricity

15.1 Electric currents

Electric cells and batteries are familiar everyday objects used, for example, in flashlights, watches, and calculators. The first such electric 'cell' was developed at the end of the eighteenth century by the Italian scientist Count Alessandro Volta (1745-1827) who placed a piece of moist cloth between a copper plate and a zinc plate. Strictly, a battery is a combination of cells, but a single cell is sometimes also referred to as a battery. Figure 15.1 shows an example of a typical modern alkaline cell. When the terminals of a cell or battery are joined together using a metal wire, energy is released. The energy (which lights the bulb, drives the watch, etc.) is chemical in origin in this case. Equivalent sources of electrical energy (power supplies) may be derived from the mains electricity supply.

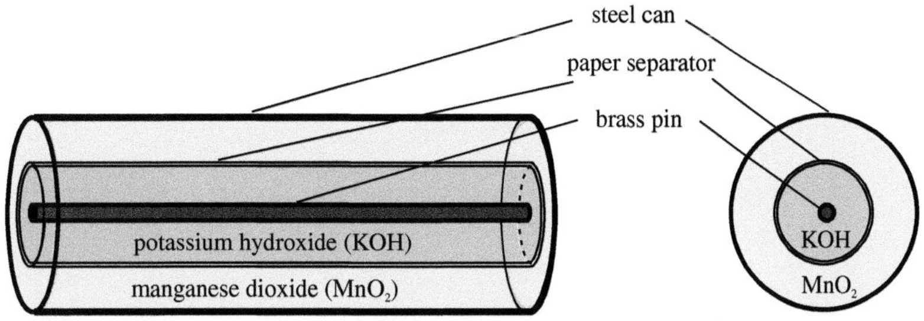

Figure 15.1. Arrangement in a typical alkaline cell: side-on view (left) and end-on view (right).

Whatever type of battery or power supply is used, the following three notable effects are observed when the terminals are joined by a piece of metal wire.

1. ***Heating effect:*** The wire joining the terminals is observed to get warm and this heat production continues for as long as the connection is made. The 'stronger' the power supply the greater is the heating effect observed. The effect will be more dramatic if part of the wire is very thin, in which case the thin part of the wire can glow red or white hot – this is the principle of the incandescent light bulb. Some heating of the power supply itself may also be observed.
2. ***Force effect:*** If one part of a flexible wire joining the terminals is brought close to another part, as shown in Figure 15.2, a force of interaction between these parts of the wire can be observed. This will be a repulsive force when the two parts of the wire are arranged as shown in Figure 15.2. This effect will be examined quantitatively in Section 19.2.
3. ***Magnetic effect:*** If a magnetic compass is brought close to any part of the wire joining the battery terminals its needle can be observed to deflect. Thus, there is clearly a region of magnetic influence (a 'magnetic field'), at least in the immediate vicinity of the wire.

Understanding Physics, Third Edition. Michael Mansfield and Colm O'Sullivan.
© 2020 John Wiley & Sons Ltd. Published 2020 by John Wiley & Sons Ltd.

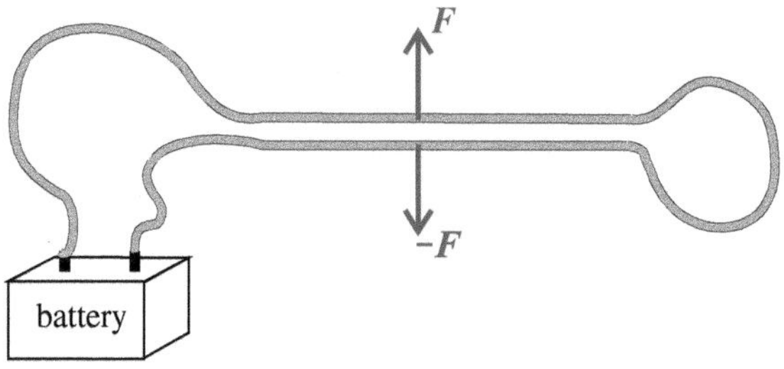

Figure 15.2. When the terminals of a power supply or battery are connected by a metal wire, different parts of the wire are observed to exert forces on one another.

The magnetic effect can be observed in more detail using an experimental arrangement such as that illustrated in Figure 15.3. Part of the wire is held vertically and passed through a hole in the centre of a horizontal platform. If a number of small magnetic compasses are placed on the platform, it is observed that the compass needles take up specific orientations. If the wire passing through the hole in the platform is long and straight, the compass needles are observed to align tangentially to circles concentric with the wire, as indicated in the figure. If the connections to the battery in the experiment are reversed, it is observed that the direction in which each compass needle points also reverses.

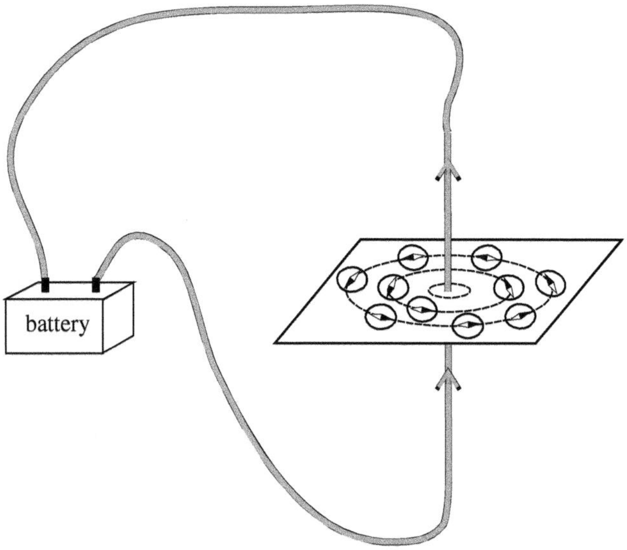

Figure 15.3. Magnetic effects may be observed in the region around a wire by placing small compasses nearby.

Magnetic effects will be discussed in detail in Chapter 18. For now, we confine ourselves to introducing the following terminology: we call the direction in which a magnetic compass needle points the 'direction of the magnetic field' at the location of the compass.

15.2 The electric current model; electric charge

Electric current

It is important to note that all three of the effects described above disappear if the connection between the terminals is broken. Thus, the effects can be switched on or off by making or breaking the connection; that is, the effects exist only while there is a complete *circuit*. This gives rise to the *model* which is used to explain these phenomena, namely that of an **electric current**. In this model, the wire joining the terminals is thought of as a type of pipe through which the current can flow. At this stage it is useful to think of an electric current as being like a 'fluid' flowing in the 'pipe' which is driven by a 'pump' (the battery or power supply). The existence or nature of this 'fluid' is not an issue at this point; a very satisfactory explanation of most electric current phenomena can be given based on this simple 'hydraulic' model. Explanations of other phenomena require refinement of the model, at which point the question "what exactly is it that flows?" needs to be addressed.

The metal wire, or any other piece of material through which a current can flow, is called a **conductor**. For simple closed loop circuits as shown in Figures 15.2 and 15.3, the model requires that the current be the same at all points in the circuit; that is, there is no 'leakage' of current at any point in the loop. Thus the nature of conductors is such that all the 'fluid' is confined within the 'pipe'.

Convention for direction of current flow

There is nothing in the current model to suggest in which direction the current should be considered to be flowing. The choice is arbitrary so, by international agreement, the direction of the magnetic field (as indicated, for example, by the compass needles in Figure 15.3) is used to adopt a convention for this purpose. The agreed convention is as follows (right hand rule for direction of current flow – Figure 15.4):

If the wire is grasped in the right hand in such a way that the fingers are pointing in the direction of the magnetic field, then the thumb will point along the direction of flow of the current.

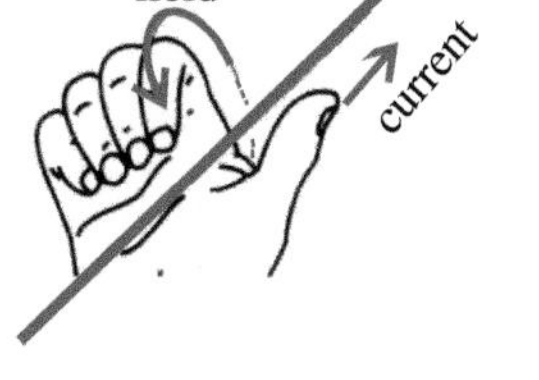

Figure 15.4. The right hand rule for direction of current flow.

Figure 15.5 shows how the rule is applied — the wire carrying the current is perpendicular to the plane of the diagram and passes through the centre of concentric circles on which compasses are placed. The arrows in Figure 15.3 also show the direction of current flow according to this convention.

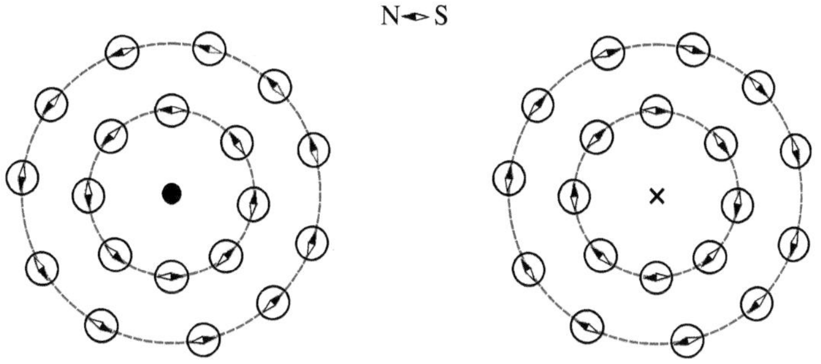

Figure 15.5. Magnetic compass needles indicate an anticlockwise magnetic field when the current is directed out of the page and a clockwise magnetic field when the current is directed into the page.

The rule adopted for the direction of flow of electric current also determines a corresponding convention for the *polarity of power supplies*. The terminal out of which the current flows is considered to be positive (denoted by +) while that into which the current flows is the negative terminal (denoted by –).

Electric charge

The 'fluid' involved in the current model is called **electric charge** which, as we will see in Chapter 16, can be either positive or negative. Current carrying conductors, such as metal wires, are usually electrically neutral so that when there is a current in such a conductor there must be either (a) positive charge which is moving with respect to an equal quantity of fixed negative charge, (b) negative charge which is moving with respect to an equal quantity of fixed positive charge or (c) a combination of mobile positive and negative charges which are moving in opposite directions in a fixed background of charge so that the overall charge on the conductor is zero. Positive charge moving to the right, say, has the same effect as an equal amount of negative charge moving to the left at the same rate and so, in terms of the current model, (a), (b) and (c) are completely equivalent.

If a current I flows in a conductor, we define the 'quantity of electricity' or the electric charge which passes any cross-section of the conductor in a time Δt as

$$\Delta Q := I(\Delta t)$$

or

$$\boxed{I := \frac{dQ}{dt}} \tag{15.1}$$

Note that Q is an *algebraic quantity* with a sign which is specified by the convention that $Q > 0$ for positive charge and $Q < 0$ for negative charge.

Suppose that the conductor shown in Figure 15.6 has uniform cross-sectional area A. The amount of mobile charge per unit volume of the conductor is called the mobile **charge density** and is denoted by the symbol ρ. Suppose that the mobile charge is positive and moves to the right at a constant velocity v, known as the *drift velocity*. To derive an expression for the current flowing in the conductor in terms of the mobile charge within it, we first determine the amount of charge in a volume of cross-sectional area A and length $v(\Delta t)$, namely

$$\Delta Q = A[v(\Delta t)]\rho = \rho v A(\Delta t)$$

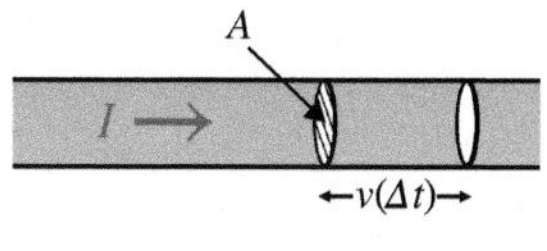

Figure 15.6. Electric current in a conductor of uniform cross-section.

and the electric current in the conductor is given by

$$I = \frac{dQ}{dt} = \rho v A \tag{15.2}$$

Current density

The **current density** at a point in a conductor is defined as

$$\boxed{\boldsymbol{J} := \underset{\Delta A \to 0}{\text{limit}} \frac{\Delta I}{\Delta A}\boldsymbol{n}}$$

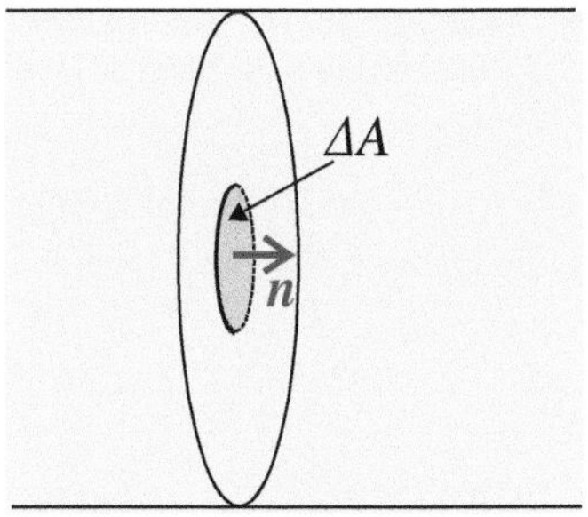

Figure 15.7. The current density at a point in a current carrying conductor is the current per unit area in the direction of current flow at that point.

where ΔI is the current flowing through the area element ΔA and $\boldsymbol{n}$ is a unit vector in the direction of current flow, that is normal to the area ΔA (see Figure 15.7). The current flowing through a conductor of cross-sectional area A, therefore, is given by

$$I = \iint_{\text{area } A} \boldsymbol{J} \cdot d\boldsymbol{A}$$

and the corresponding current density is

$$J = \frac{I}{A} = \rho v \tag{15.3}$$

Current density can be a useful concept as it does not depend on the geometry of the conductor and can be applied to situations where the current is not uniform; we shall use the concept extensively in later chapters.

If both types of charge are mobile in the conductor, with positive charge of charge density ρ_+ moving to the right with speed v_+ and negative charge of charge density ρ_- moving to the left with speed v_-, the current is given by

$$I = \rho_+ v_+ \text{A} + \rho_- v_- A \tag{15.2a}$$

and the current density by

$$J = \rho_+ v_+ + \rho_- v_- \tag{15.3a}$$

Note that both terms in Equations (15.2a) and (15.3a) have the same sign since, taking 'to the right' as the positive direction, ρ and v are both negative while ρ_+ and v_+ are both positive.

Note that Equation (15.3) can be written in vector form as

$$\boxed{\boldsymbol{J} = \rho \boldsymbol{v}}$$

and the corresponding current through any element of area ΔA is given by

$$I = \rho(\Delta A)\boldsymbol{v} \,.\, \boldsymbol{n}$$

where $\boldsymbol{n}$ is a unit vector directed perpendicularly to ΔA.

All the relationships in this section hold true no matter what system of units is involved. We now proceed to define the SI units used for electricity.

15.3 The SI unit of electric current; the ampere

In Section 1.2 we came across a fundamental physical constant associated with electric charge, namely the elementary charge e (the magnitude of the electric charge of fundamental particles like the electron, proton, etc). If the value of this constant is fixed this effectively defines the SI units of current, the **ampere** (symbol A), and charge (the ampere second which is called the **coulomb**, symbol C). The value chosen for the elementary charge to maintain continuity with older definitions of the ampere is exactly

$$e = 1.602\,176\,634 \times 10^{-19}\,\text{C}.$$

The formal definition of the ampere is as follows:

The ampere, symbol A, is the SI unit of electric current. It is defined by taking the fixed numerical value of the elementary charge *e* to be 1.602 176 634 × 10^{-19} when expressed in the unit C, which is equal to A s, where the second is defined in terms of $\Delta\nu_{Cs}$.

The effect of this definition is that the ampere is the electric current corresponding to the flow of 1/(1.602 176 634 10^{-19}) elementary charges per second.

To estimate how fast charge moves in a typical current carrying wire, let us consider the case of 1 A flowing in a copper wire of diameter 1 mm. The mobile (negative) charge density in copper is known from experiment to be approximately 10^{10} C m^{-3} and hence the drift velocity at which this charge moves is given by

$$v = \frac{I}{\rho\pi a^2} \approx \frac{1\text{A}}{\pi(10^{10}\,\text{C}\,\text{m}^{-3})(5\times10^{-4}\,\text{m})^2} \approx 10^{-4}\,\text{m}\,\text{s}^{-1} \approx 0.1\,\text{mm}\,\text{s}^{-1}$$

and the corresponding current density is $(10^{10}\,\text{C m}^{-3})(10^{-4}\,\text{m s}^{-1}) = 10^6\,\text{A m}^{-2}$.

It is instructive to consider the mechanism by which an electric light is switched on in view of the very small value obtained for the drift velocity. Taking the length of wire between a switch and a light bulb to be about 10 m, we find that charge would take $[10/(10^{-4})]\sim10^5$ s (over a day) to travel this distance. Clearly this is not the mechanism through which electric current is transmitted. Rather, the light switches on almost instantaneously because relevant information is propagated down the wire at a velocity close to the velocity of light causing charge to flow in the bulb without any noticeable delay. This is an example of transmission of information by an electromagnetic wave; electromagnetic radiation will be discussed in Chapter 21.

For problems based on the material presented in this section visit up.ucc.ie/15/ and follow the link to the problems.

15.4 Heating effect revisited; electrical resistance

As we have seen, when an electric current flows in a piece of metal wire, the wire is observed to get warm. This effect can be studied by means of the experiment illustrated in Figure 15.8. A length of coiled wire is immersed in water in a thermally insulated calorimeter (Section 11.4) and an electric current is passed through it. The wire is coiled to increase the length which can be fitted into the calorimeter. The strength of the current flowing in the coil is measured with a current measuring instrument (A in Figure 15.8) such as an ammeter or galvanometer. The heat energy generated in the coil causes the calorimeter and its contents to heat up. The temperature of the system (the calorimeter and its contents) can be monitored as a function of time by a thermometer which is inserted in the calorimeter.

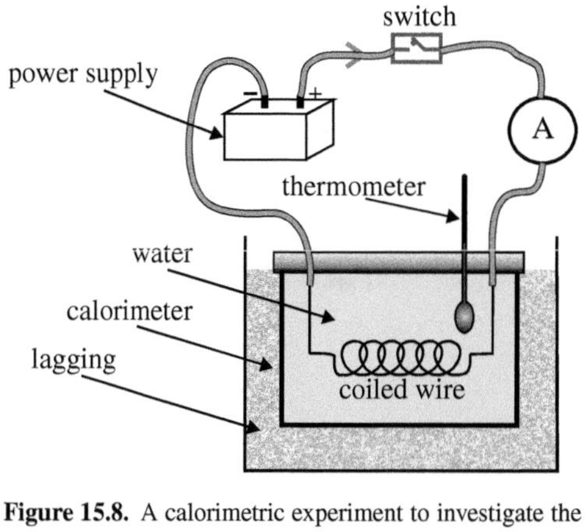

Figure 15.8. A calorimetric experiment to investigate the heating effect of an electric current. The water is heated by the electric current flowing in all of the wire inside the calorimeter.

A typical graph of temperature against time obtained in such an experiment is shown in Figure 15.9.

The first thing to notice is that the rate of increase in temperature is constant while the current is flowing, provided that the current remains steady and that the system is thermally insulated (no heat transfer in or out of the system). This means that the thermal power generated in the coil, given by $P = C\frac{dT}{dt}$ where C is the heat capacity of the system (recall Section 11.3), is constant within the accuracy of the experiment. Knowing the heat capacity, the power generated can be determined by measuring the slope $\left(\frac{dT}{dt}\right)$ of the temperature–time curve in the region of the graph corresponding to the time during which the current was flowing in the coil.

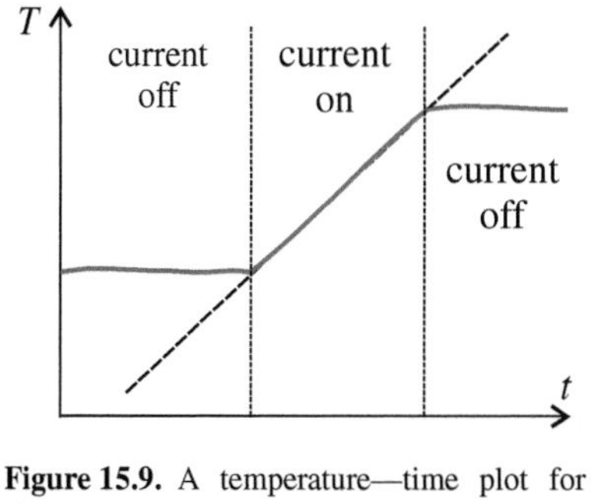

Figure 15.9. A temperature—time plot for the experiment illustrated in Figure 15.8. The rate of increase in temperature is constant while a steady current flows in the coil.

This experiment can be repeated for different currents by using different power supplies and/or by suitably modifying the circuit outside the calorimeter. Thus, the power generated can be recorded for different values of the current flowing in the coil. When the power is plotted against the square of the current, a linear relationship is obtained (Figure 15.10).

Thus, for any particular coil, $P \propto I^2$, that is

$$\frac{P}{I^2} = constant \tag{15.4}$$

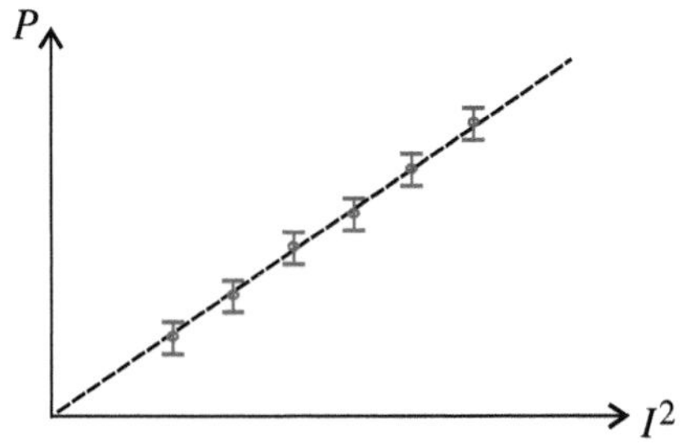

Figure 15.10. Experimental data for power generated in the coil in Figure 15.8 plotted against the square of the current flowing.

where the value of the constant is characteristic of the coil used. In other words, the value of the constant can be considered to be a property of the piece of wire making up the coil and determines the rate at which heat is generated in the wire when a given current is flowing in it. It must be pointed out, however, that the ratio of P to I^2 is constant (that is independent of I) only for a restricted but very important class of conductors, namely when the coils used in the experiment are made of *metal* wire. Furthermore, the rise in temperature during the experiment should not be too large, otherwise deviation from the behaviour described by Equation (15.4) would be observed (we will return to this latter issue in Section 15.15 below).

Equation (15.4), therefore, is an 'equation of state' which describes the behaviour of a certain category of electrical conductors. A conductor which obeys Equation (15.4), that is a conductor whose resistance is independent of the current flowing in it, is said to be **ohmic**. As stated above, only metallic conductors held at constant temperature are truly ohmic materials in practice.

The experimental results summarised by Equation (15.4) suggest that it would be useful to define a new physical quantity to describe the ability of a conductor to generate power when an electric current flows through it. Thus, we *define* the **resistance** of a conductor in an electric circuit as *the power generated per unit current squared* flowing in it. Resistance is usually denoted by the symbol R and, by definition,

$$R := \frac{P}{I^2} \tag{15.4A}$$

and hence the power generated by a current I flowing in a conductor of resistance R is given by

$$P = RI^2 \tag{15.5}$$

The SI unit of resistance is the watt per ampere squared (W A^{-2} or J s^{-1} A^{-2}) and one W A^{-2} is called one **ohm** (= 1 Ω). A conductor which has a fixed value of resistance is usually called a **resistor**.

In the special case where R is constant, that is independent of I, Equation (15.5) describes the behaviour of ohmic materials. In these circumstances, Equation (15.5) is often referred to as **Ohm's law** (or sometimes as **Joule's law**). Ohm's law can be stated equivalently as follows:

for metals at constant temperature, R = constant

where R is defined by Equation (15.4A).

Non-ohmic materials

Not all the conducting materials which are used as components in electrical circuits are ohmic. As we shall see in Section 15.7 and, in more detail, when we come to study semiconductors in Chapter 26, the behaviour of many commonly used components is distinctly non-ohmic. The concept of resistance, which was defined above for the restricted class of conductors which satisfy Equation (15.4), can be extended to cover non-ohmic components. As before resistance is defined as

$$R := \frac{P}{I^2}$$

but, in the case of a non-ohmic component, R is not constant but rather depends on the current flowing in it, that is $R = R(I)$.

Internal resistance

When an electric current is drawn from a power supply, the power supply can also heat up showing that the power supply itself has resistance. In many cases this effect is small compared to the heating of the other conductors, in which case it usually may be ignored, but in some situations the thermal power generated internally within the power supply itself may make a significant contribution to the total heat energy generated. This resistance of a power supply is described as its **internal resistance**. We shall discuss the effects of internal resistance of power supplies on the currents in electrical circuits in Section 15.8.

Electrical symbols

Most electrical or electronic instruments and appliances are made up of complicated networks of conductors having fixed resistance (resistors) and other components which may or may not have negligible resistance (power supplies, ammeters, etc.). In these circumstances, a symbol is needed for each component in order to simplify the representation and interpretation of such networks. Drawings of electrical networks using such symbols are variously called circuit diagrams, schematic diagrams or *schematics*. Figure 15.11 shows some of the internationally agreed schematic symbols; others will be introduced later as the need arises.

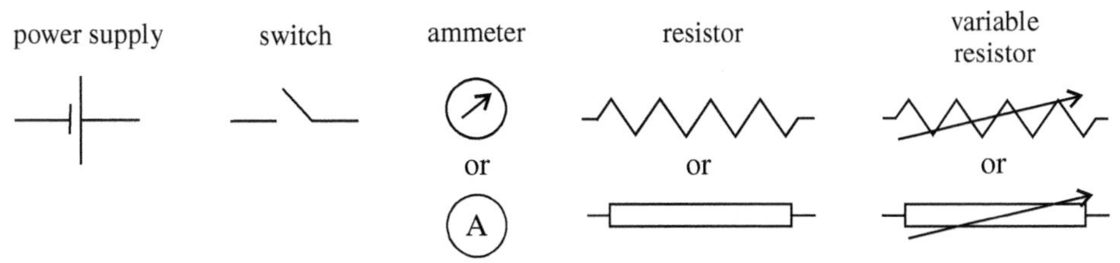

Figure 15.11. Some conventional electrical symbols. Note that in the case of the symbol for a power supply, the longer vertical line corresponds to the positive terminal. Conductors of negligible resistance are represented by solid lines.

Note that, in the discussion so far, it has always been assumed that any power supply used is such that the current drawn from it remains constant, provided that the resistances in the circuit do not vary. Such a current is called a *direct current* (d.c.) and a power supply which produces such a current is called a d.c. power supply. Time varying power supplies and currents will be discussed in Chapter 20.

Resistors connected in series

The three resistors shown in Figure 15.12(a) are said to be connected *in series*. The following question now arises: "What is the resistance of the single equivalent resistor, that is of the single resistor which can replace the three series resistors and has exactly the same effect?" The same effect, in this case, means that the single equivalent resistor (R in Figure 15.12(b)) draws the same current from the power supply and produces the same heating effect as the three series resistors together.

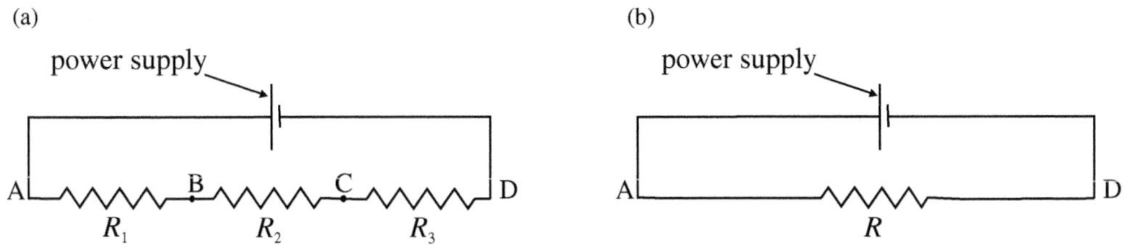

Figure 15.12. The single resistor in (b) is equivalent to the combination of the three resistors connected in series in (a).

The total power generated between the points A and D in Figure 15.12(a) is

$$P_{AD} = P_{AB} + P_{BC} + P_{CD} = R_1 I^2 + R_2 I^2 + R_3 I^2 = (R_1 + R_2 + R_3) I^2$$

and, since

$$R = \frac{P_{AD}}{I^2},$$

$$\boxed{R = R_1 + R_2 + R_3} \qquad (15.6)$$

This result can be extended to any number of resistors connected in series.

15.5 Strength of a power supply; emf

The heat energy generated in an electrical circuit will appear everywhere throughout the circuit, not only in the resistors but also within power supplies, connecting wires, current measuring instruments, and other components. To measure the total power generated, therefore, the whole circuit, including the power supply, should be immersed in the liquid in a calorimeter, at least in principle. It is often the case, however, that most of the heat will be generated in one or two components so, in that case, we need only determine experimentally the power generated in these components (for example, by immersing them in the liquid of a calorimeter) if we wish to measure the total power generated throughout the circuit. As in the previous section, this is the product of the heat capacity of the system and the rate of

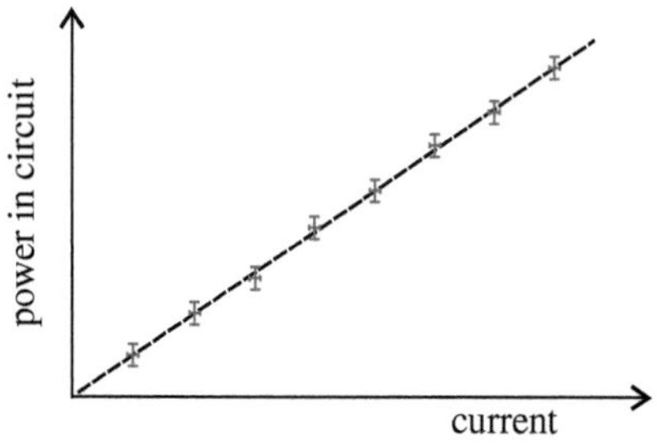

Figure 15.13. Experimental data for the total power generated in an electric circuit plotted against the current drawn from the power supply.

increase of temperature, assuming that the system is thermally insulated. That is to say, the total power generated in the whole circuit is given by

$$P_{\text{circ}} = C\frac{dT}{dt}$$

By changing the length of resistance wire immersed in the calorimeter, different currents can be made to flow in the circuit. If this experiment is carried out, it is usually found that the power generated throughout the circuit by any given power supply is directly proportional to the current drawn from the power supply, as indicated by the data plotted in Figure 15.13. Thus, for most types of power supply

$$P_{\text{circ}} = (constant)I$$

The value of the constant of proportionality is different for different power supplies, the 'stronger' the power supply the larger is the value of the constant. This quantity (which is the slope of P_{circ}—I graph) can be used as a convenient measure of the strength of the supply. Thus, we *define* the **emf**[1] of a power supply as the *total power generated in a circuit, including in the power supply, per unit electric current* drawn from the power supply.

The usual symbol for emf is E and, by definition,

$$\boxed{E := \frac{P_{\text{circ}}}{I}}$$

The SI unit of emf is the watt per ampere, called the **volt** (symbol V), named after Alessandro Volta (1745–1827), that is, $1\ \text{V} = 1\ \text{W A}^{-1}$.

15.6 Resistance of a circuit

From the definition of resistance (Section 15.4), the resistance of a circuit is given by

$$R_{\text{circ}} = \frac{P_{\text{circ}}}{I^2}$$

where I is the current flowing from the power supply. In general, R_{circ} is not constant, that is the value of R_{circ} may depend on I but in many situations, in particular when all the components in the circuit are metallic, R_{circ} will indeed be constant irrespective of the current flowing, provided the temperature is kept constant.

From the definition of emf in the previous section, the power generated in the circuit is

$$P_{\text{circ}} = EI = R_{\text{circ}}I^2$$

and hence

$$\boxed{I = \frac{E}{R_{\text{circ}}}}$$

This is an important result in that it predicts the current that will flow in any given situation in which R_{circ} is independent of current. It was first proposed by Georg Simon Ohm (1787–1854) and was known as **Ohm's law**. As we have already seen, however, the term 'Ohm's law' has been extended to cover also parts of a circuit in which resistance is constant, although historically it was proposed in the above more limited context.

15.7 Potential difference

If a liquid being pumped through a pipe encounters a section of pipe which provides a frictional resistance to the flow, a pressure gradient across that section is required to overcome the resistance (recall, for example, the discussion of flowmeters in Section 10.7).

[1]The term arises from an older expression 'electromotive force' which will not be used in this book.

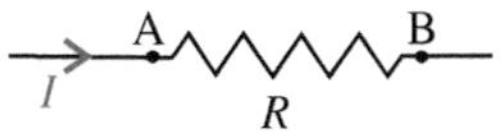

Figure 15.14. When an electric current passes through a resistor, a potential difference is developed across the resistor.

A quantity which is somewhat analogous to such pressure differences in liquids, and which proves to be very useful in the study of current electricity, can be introduced in the case of electrical circuits. This quantity, called the **potential difference** between two points A and B in a circuit (Figure 15.14), is defined as

$$\boxed{V := \frac{P}{I}}$$

where P is the power generated between the points A and B.

By comparing this definition with the definition of emf, it is evident that the potential difference can be thought of as the emf of an 'equivalent power supply'. That is to say, if the rest of the circuit was replaced by a power supply of emf $=V$ with negligible internal resistance and this power supply was connected across the resistor of resistance R, then this power supply would drive exactly the same current through R.

From the definition of resistance in Section 15.4, we can write the resistance between the points A and B as $R := \frac{P}{I^2} = \frac{VI}{I^2} = \frac{V}{I}$ and thus $R := \frac{V}{I}$ can be considered an equivalent definition of resistance. The relationship

$$\boxed{V = RI} \tag{15.7}$$

which is equivalent to Equation (15.5) when R is kept constant, determines the value of the potential difference which is developed across a resistor when a current flows through it. Equation (15.7) is also referred to as Ohm's law if R is constant.

Voltage-current behaviour of non-ohmic components

As was pointed out in Section 15.4, not all components in electrical circuits are ohmic. A commonly used component whose behaviour is distinctly non-ohmic is a *diode*. A diode (to be described in Section 26.6) has the property that when a current flows through it in one direction it has relatively small resistance, which depends on the current, but if the direction of the current is reversed the resistance of the diode is very large. Other components behave in even more complicated ways as illustrated in Figure 15.15. In this figure the variation of the potential difference developed across a component with the current flowing through it is plotted (a) for an ohmic component and (b), (c) and (d) for various non-ohmic components. The exact way in which the potential difference depends on current can be quite complicated in non-ohmic devices and it is often impossible to describe this behaviour by an algebraic relationship (an analytical equation of state). In these circumstances, it is more convenient to provide a description of the voltage-current behaviour of such devices in graphical or tabular form.

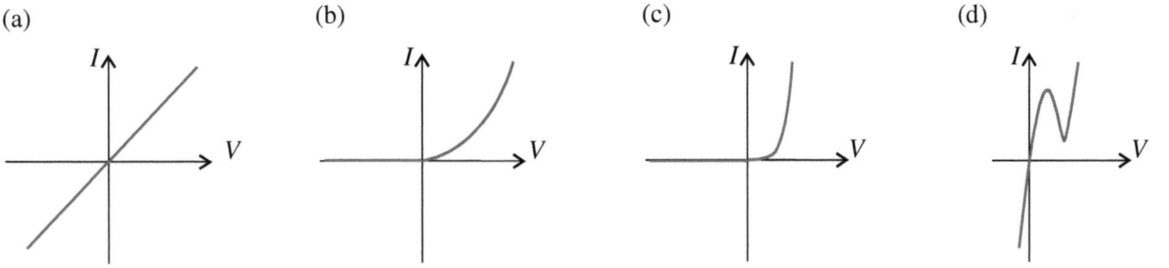

Figure 15.15. Current—voltage plots for different types of electrical components: (a) metal wire (ohmic); (b) thermionic diode; (c) junction diode; (d) tunnel diode. Diodes will be discussed in Section 26.6.

Summary

We can now summarise the most important results which are needed for the analysis of single loop circuits (such as that illustrated in Figure 15.16) with ohmic resistors.

- The current in the circuit is determined by $I = \frac{E}{R_{\text{circ}}}$
- The potential difference across a resistance R is given by $V = RI$.
- The total power generated in the circuit $= P_{\text{circ}} = EI = R_{\text{circ}}I^2 = \frac{E^2}{R_{\text{circ}}}$
- The power generated in a resistor (of resistance R) $= P = VI = RI^2 = \frac{V^2}{R}$

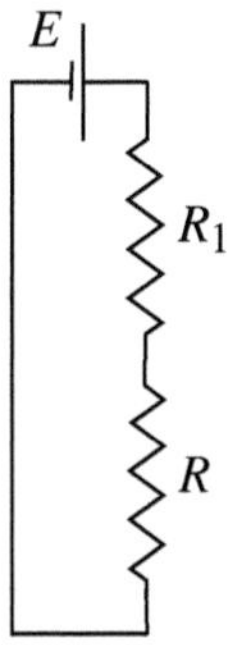

Figure 15.16. The potential divider circuit.

The potential divider

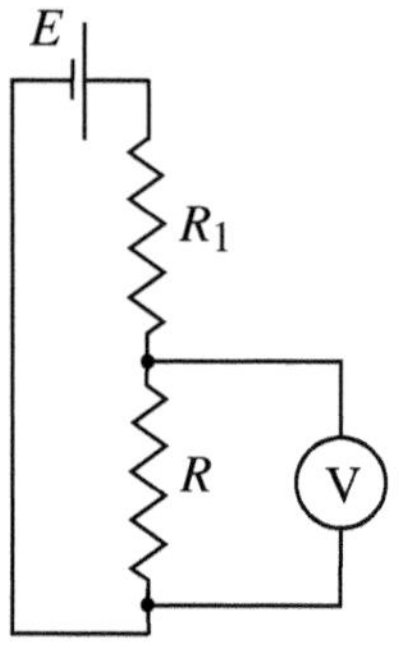

Figure 15.17. The potential difference across the resistor R is measured by a voltmeter (V).

Assuming that the internal resistance of the power supply is negligible, the potential difference across R in Figure 15.16 is $V = RI = R\frac{E}{R+R_1} = \frac{R}{R+R_1}E$ and, similarly, the potential difference across R_1 is $\frac{R_1}{R+R_1}E$. Thus, the potential difference across each resistor is a set fraction of the emf E. Such an arrangement is called a *potential divider*. By varying the ratio $\frac{R}{R+R_1}$ one can produce, across R, any potential difference between 0 and E which one might require. Note, however, that this source of voltage has an internal resistance which is dependent on the values of R_1 and R. The potential divider has many important applications as a source of variable potential difference.

To measure the potential difference across a component through which a current is flowing we must connect an instrument, such as a voltmeter, across the component. The symbol for a voltmeter is indicated in Figure 15.17, where the voltmeter, indicated by V, is shown connected across the resistor R.

Study Worked Example 15.1

For problems based on the material presented in this section visit up.ucc.ie/15/ and follow the link to the problems.

15.8 Effect of internal resistance

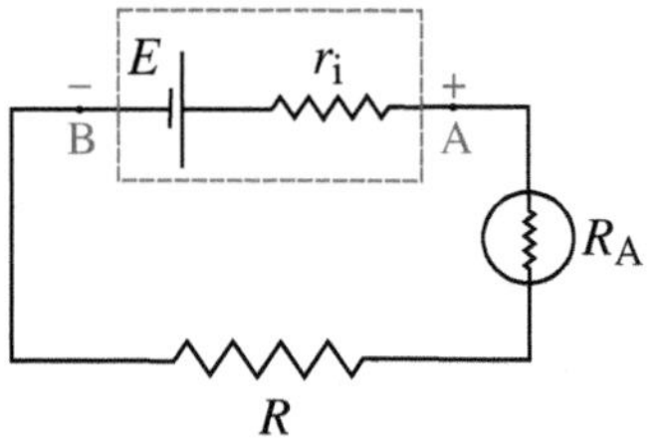

Figure 15.18. A simple circuit comprising a power supply (indicated by the dashed lines) with terminals A and B, a resistor (resistance = R) and an ammeter (resistance = R_A). The internal resistance (r_i) of the power supply can be treated as a resistor in series with an 'ideal' power supply.

As we have seen, when considering the total power generated in a circuit, the heat generated within the power supply must also be taken into account. In Section 15.4 we saw that any component, in which heat is generated when a current flows through it, has a resistance and that in the case of a power supply this resistance is called its *internal resistance*.

The simplest (but not unique) way of treating the internal resistance of a power supply is to consider it as a resistance in series with the supply (resistance r_i in Figure 15.18). Thus, the current flowing in the circuit in Figure 15.18 is given by

$$I = \frac{E}{R_{circ}} = \frac{E}{R + R_A + r_i} \tag{15.8}$$

where R_A is the resistance of the ammeter.

The potential difference between the points A and B (that is, across the terminals of the supply) is given by

$$V_{AB} = (R + R_A)I = (R + R_A)\frac{E}{R + R_A + r_i} \tag{15.9}$$

Thus, the potential difference across the terminals of a power supply will be less than the emf of the supply while a current is being drawn from it.

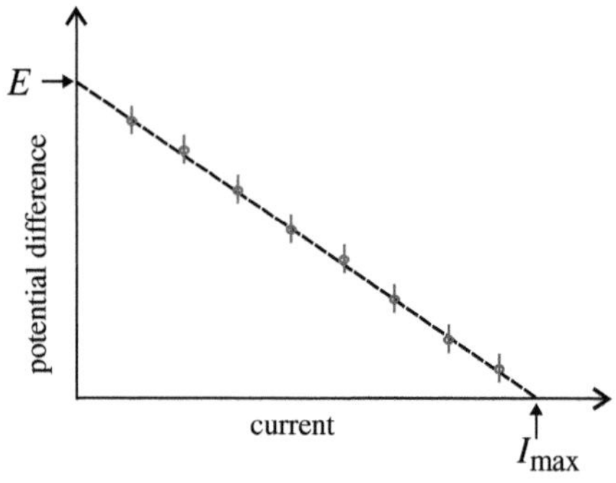

Figure 15.19. The potential difference across the terminals of a power supply plotted against the current drawn from the supply. The magnitude of the slope of the line is equal to the internal resistance of the power supply.

Now, from Equations (15.8) and (15.9)

$$E = (R + E_A + r_i)I = (R + R_A)I + r_iI = V_{AB} + r_iI$$

and hence

$$V_{AB} = E - r_iI \tag{15.10}$$

Equation (15.10) tells us that the potential difference across the terminals of the power supply decreases linearly with the current which is drawn from the supply, as illustrated in Figure 15.19, a plot of V_{AB} against I, where the slope of the straight line is $-r_i$. Note that the maximum current which can be drawn from the supply corresponds to the condition $V_{AB} = 0$, that is

$$I_{max} = \frac{E}{r_i} \tag{15.10a}$$

This is an example of *current limiting*. The existence of the resistance r_i, in series, limits the current that the power supply can provide; I_{max} is called the 'short circuit' current of the power supply. This is the current produced if the terminals of a power supply are connected by a wire of negligible resistance. Note that if r_i is small then I_{max} can be very large indeed. When using low internal resistance sources of current, therefore, great care must be taken to ensure that there is always sufficient resistance in the circuit. Because of the large current involved, a short circuit will often cause permanent damage to a power supply or battery.

Equation (15.9) can be written as $V_{AB} = \dfrac{E}{1 + \dfrac{r_i}{R + R_A}}$

so that $V_{AB} \to E$ as $R \to \infty$ (that is, $I \to 0$). This condition, in which the external resistance between the terminals of a power supply is effectively infinite, for example where there is a break in the circuit, is described as 'open circuit'.

Study Worked Examples 15.2 and 15.3

For problems based on the material presented in this section visit up.ucc.ie/15/ *and follow the link to the problems.*

Power supplies connected in series

Consider the situation in which a number of power supplies are connected in series, as shown in Figure 15.20(a). The three power supplies of emf E_1, E_2, and E_3 and internal resistance r_1, r_2 and r_3, respectively, are connected with the positive terminal of one joined to the negative terminal of the next, and so on. If the current flowing in the circuit due to all three working together as shown is I, the question "what are the emf and internal resistance of a single power supply which could replace all three existing power supplies so as to provide the same current I?" arises.

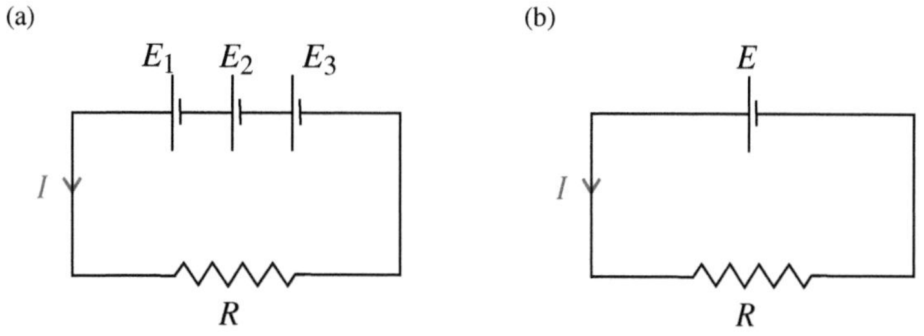

Figure 15.20. The single power supply in (b) is equivalent to the combination of three power supplies connected in series in (a).

First suppose that only the power supply with emf E_1 were to be connected into the circuit with the other two power supplies replaced by resistances of r_2 and r_3. In this case the current flowing would be $I_1 = \dfrac{E_1}{R_{circ}}$, where $R_{circ} = R + r_1 + r_2 + r_3$

Similarly, $I_2 = \dfrac{E_2}{R_{circ}}$ (E_2 alone with resistances r_1 and r_3 in series) and

$I_3 = \dfrac{E_3}{R_{circ}}$ (E_3 alone with resistances r_1 and r_2 in series).

Thus, when all three supplies are connected as shown, the total current flowing will be given by

$$I = I_1 + I_2 + I_3 = \frac{E_1}{R_{circ}} + \frac{E_2}{R_{circ}} + \frac{E_3}{R_{circ}} = \frac{E_1 + E_2 + E_3}{R + r_1 + r_2 + r_3}$$

Comparing this with the expression for the current drawn in the equivalent circuit (Figure 15.20(b)), that is

$$I = \frac{E}{R + r}$$

we conclude that the emf of the equivalent power supply is given by

$$E = E + E_2 + E_3 \qquad (15.11)$$

and its internal resistance is given by $r = r_1 + r_2 + r_3$.

Note that if one of the power supplies had been connected the other way around (+ and – terminals reversed), the corresponding current would be reduced and might flow in the opposite direction, depending on the values of E_1, E_2 and E_3. This situation can be incorporated without changing Equation (15.11) if emf is considered to be an *algebraic* quantity. That is to say, if we adopt a convention of signs for positive emf (say, power supplies which tend to drive a current anticlockwise are considered positive) then E_1, E_2 and E_3 can be considered as representing positive or negative values depending on the polarity of the terminals in the circuit. Internal resistance is unaffected by reversal of the terminals of a power supply.

15.9 Comparison of emfs; the potentiometer

The absolute determination of the emf of a power supply would require that a calorimetric experiment be carried out in which the power generated in all parts of the circuit would have to be measured. Clearly, it would be both difficult and expensive to carry out an accurate version of such an experiment. In practice, the emf of a power supply is more conveniently determined by *comparing* it to a *standard cell*. Standard cells whose emfs are very stable and known to considerable accuracy are commercially available.

To measure the emf of a power supply, we need to measure the potential difference between its terminals while no current is being drawn (recall that $V_{AB} \to E$ as $I \to 0$). A simple experimental arrangement for measuring the ratio of the emfs of two cells is shown in Figure 15.21(a). AB is a long straight length of uniform wire, that is, it has constant cross-sectional area and hence, from symmetry, the resistance of any length is proportional to that length. When connected across the power supply E_0, as shown, the potential difference between A and C, V_{AC} is proportional to AC.

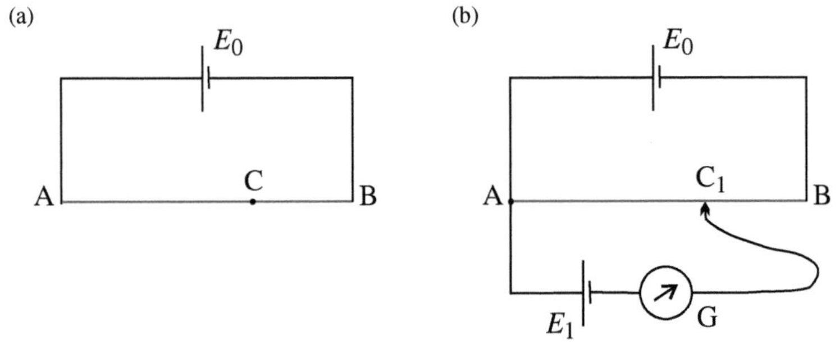

Figure 15.21. (a) A power supply connected across a length of uniform wire (AB). Since the wire is uniform, the potential difference between A and any point C is proportional to the length AC. (b) A simple potentiometer experiment: the movable contact is caused to slide along the wire AB until a point (C_1) is found such that no current flows in the galvanometer G.

One of the cells E_1 is connected as shown in Figure 15.21(b) and a movable contact C caused to slide along the wire until the point C_1 is found, such that no current is flowing in the galvanometer G (an accurate current measuring device). Thus the emf of that cell is given by $E_1 = V_{AC_1} \propto AC_1$. If the cell of emf E_1 is replaced by a second cell of emf E_2 and the new balance point C_2 is found (again no current in G), then $E_2 = V_{AC_2} \propto AC_2$ and thus

$$\frac{E_2}{E_1} = \frac{AC_2}{AC_1}$$

Note that, for both balance points to be found, E_0 must be chosen so that E_0 is greater than both E_1 and E_2. The ratio of the emfs is determined by the ratio of two lengths of wire. If one of the cells used is a standard cell of known emf, then the value of the emf of the other cell can be determined.

Very accurate potentiometers, based on the general principle described above, are commercially available. Note also that, when balanced, no current is drawn from either cell, an important consideration if the accuracy and shelf life of the standard cell is to be maintained.

A potentiometer may also be used to measure the potential difference between any two points in a circuit by comparison with the emf of a standard cell, that is the potentiometer can be used as a voltmeter. In certain applications, this technique has a great advantage over other voltmeters in that, since no current is drawn from the source of potential difference, the potentiometer acts as a voltmeter of infinite resistance and hence does not in any way affect the circuit to which it is connected.

For problems based on the material presented in this section visit up.ucc.ie/15/ *and follow the link to the problems.*

15.10 Multiloop circuits

All the circuits studied so far have had one single loop, that is, just one path along which the current must flow from the + to the – terminal of the power supply. In these cases the electric current model which we have adopted requires that the current is the same in every part of the circuit and is the full current drawn from the supply; in other words there can be no loss of 'fluid' out of the 'pipe' through which it flows.

The situation is more complicated if there are junctions involved, such as the points J and K in Figure 15.22. Note that the convention used to indicate a junction is to place a dot (sometimes referred to as a 'node' or a 'meatball') at the junction — if two lines are shown to cross over without such a dot this indicates that there is no connection. When the current flowing through R reaches the junction, it divides into two components, I_1 flowing through R_1 and I_2 through R_2. The current model requirement that there be no loss of 'fluid' means that *current is conserved* at a junction, that is

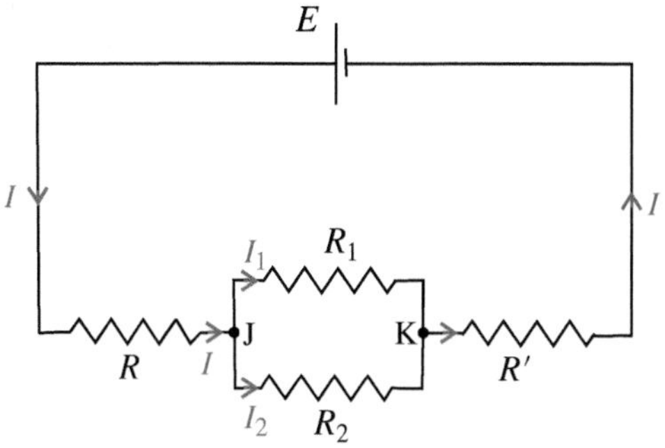

Figure 15.22. Example of a two-loop network. The current I flowing into the junction at J divides such that I_1 flows in the upper path and I_2 flows in the lower path so that a current $I = I_1 + I_2$ flows out at the junction K.

$$I = I_1 + I_2$$

Note that the currents I_1 and I_2 rejoin at the junction K and the current flowing through R' is again I.

Resistors connected in parallel

The resistors R_1 and R_2 shown in Figure 15.22 are said to be connected *in parallel*. As in the previous case of resistors connected in series (Section 15.4), here we wish to know how to determine the resistance of part of a circuit made up of resistors in parallel. In other words we need to answer the question "what is the value of the single resistor which is equivalent to the three (say) resistors connected in parallel in Figure 15.23(a)?"

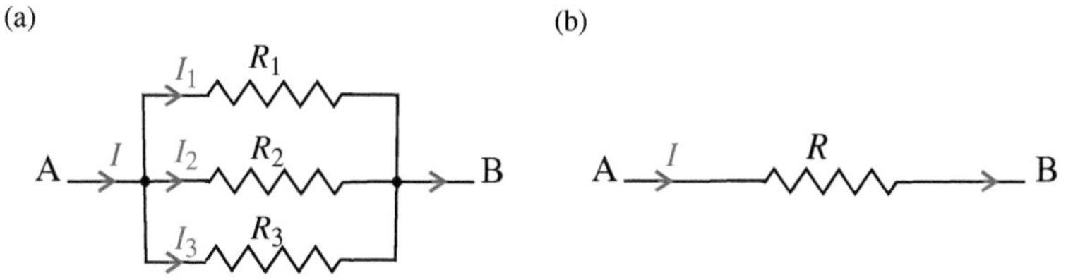

Figure 15.23. The single resistor in (b) is equivalent to the combination of three resistors connected in parallel in (a).

Since the current has three paths between A and B, the potential difference between A and B is given by $V_{AB} = R_1 I_1 = R_2 I_2 = R_3 I_3$

and so
$$I_1 = \frac{V_{AB}}{R_1}; \quad I_2 = \frac{V_{AB}}{R_2}; \quad I_3 = \frac{V_{AB}}{R_3}.$$

Therefore
$$I = I_1 + I_2 + I_3 = \frac{V_{AB}}{R_1} + \frac{V_{AB}}{R_2} + \frac{V_{AB}}{R_3} = V_{AB}\left\{\frac{1}{R_1} + \frac{1}{R_2} + \frac{1}{R_3}\right\}$$

But, in the equivalent circuit (Figure 15.23b),
$$I = \frac{V_{AB}}{R} \text{ so that}$$

$$\boxed{\frac{1}{R} = \frac{1}{R_1} + \frac{1}{R_2} + \frac{1}{R_3}} \tag{15.12}$$

This result, which states that the reciprocal of the equivalent resistance is equal to the sum of the reciprocals of the individual resistances connected in parallel, can be extended to any number of resistors connected in this manner.

Applying Equation (15.12) to the pair of resistors in parallel in Figure 15.22 we see that, if the equivalent resistance of the parallel combination is R_p, then $\frac{1}{R_p} = \frac{1}{R_1} + \frac{1}{R_2}$ and hence

$$R_p = \frac{R_1 R_2}{R_1 + R_2}$$

Applying the relationship for resistors connected in series, Equation (15.6), to the complete network in Figure 15.22, we can deduce that the resistance of the circuit in this case is

$$R + R_p + R' = R + \frac{R_1 R_2}{R_1 + R_2} + R'$$

Study Worked Example 15.4 and 15.5

For problems based on the material presented in this section visit up.ucc.ie/15/ and follow the link to the problems.

15.11 Kirchhoff's rules

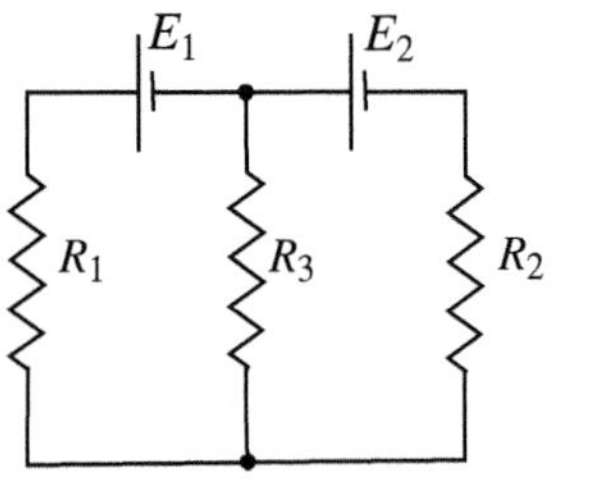

Figure 15.24. A multiloop network. Kirchhoff's rules provide a systematic method of analysing such networks.

We now have all the information needed to analyse even the most complicated networks of d.c. power supplies and resistors. Nevertheless, looking at a network such as that shown in Figure 15.24, it is not easy to see how to proceed. To deal with these more complicated situations in a systematic manner, a set procedure would be desirable. Such an approach is provided by Kirchhoff's rules (Gustav Robert Kirchhoff 1824–1887) given below. No new physics is involved here; Kirchhoff's rules simply incorporate what has already been developed in a more formal way for systematic application.

> **Kirchhoff's junction rule:** The sum of the currents flowing into a junction is equal to the sum of the currents leaving the junction.

Thus, if the currents are as shown in Figure 15.25, $I_1 + I_2 = I_3 + I_4$.

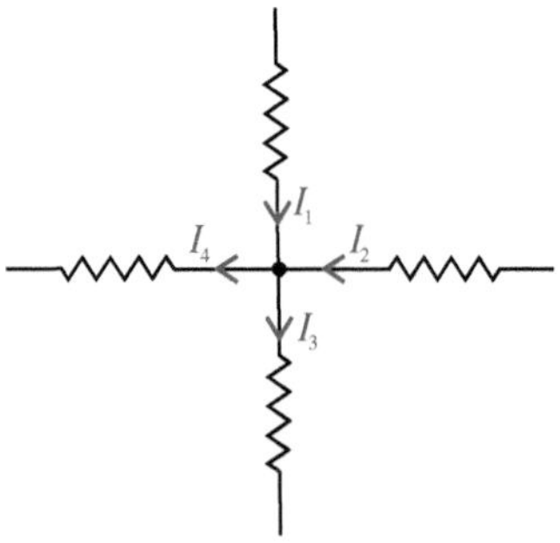

Figure 15.25. Application of Kirchhoff's junction rule.

> **Kirchhoff's loop rule:** The algebraic sum of the emfs in any loop is equal to the algebraic sum of all the *IR* drops around the loop.

Thus, if the currents are flowing as indicated in Figure 15.26 and taking the anticlockwise sense around the loop as positive then Kirchhoff's loop rule can be applied as follows

$$E_1 + E_2 = R_1 I_1 + R_2 I_2 + R_3 I_3 + R_4 I_4 + R_5 I_5$$

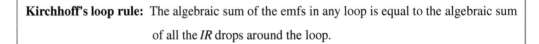

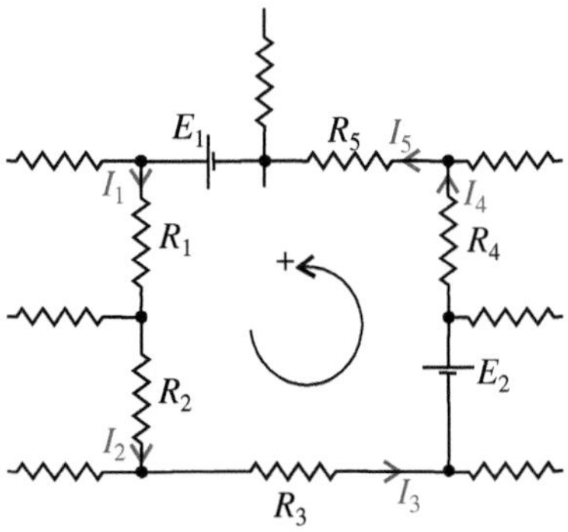

Figure 15.26. Application of Kirchhoff's loop rule.

Note that the rule implies a convention of signs for both emfs and potential differences. As when dealing with power supplies in series (recall Equation (15.11)), we treat emf as an algebraic quantity; that is, if any one of the power supplies in Figure 15.26 were to be reversed (opposing the current flowing), the corresponding emf will be negative. To be consistent, we must also choose positive *IR* values to be those for which the current is flowing in the positive sense around the loop.

As an illustration of the application of Kirchhoff's rules let us analyse the network in Figure 15.24. If we try to apply the rules to this case, it becomes immediately clear that we do not know *ab initio* in which sense the currents are flowing. In particular, in this case, depending on the values of the emfs and the resistances, the current in R_3 could be flowing in either sense. We do not need this information, however, since we can choose arbitrarily (in other words guess) a positive sense for the purpose of making the calculation – if we guess incorrectly, the value determined for the current after solving all the simultaneous equations will turn out to be negative.

Suppose that we have chosen the anticlockwise sense as positive in the case of both loops and that we have assumed initially that the currents flow as indicated in Figure 15.27.

Applying Kirchhoff's rules:

- junction rule applied to junction A (or B): $I_3 + I_2 = I_1$ (i)
- loop rule applied to loop on left: $E_1 = R_1 I_1 + R_3 I_3$ (ii)
- loop rule applied to loop on right: $E_2 = R_2 I_2 - R_3 I_3$ (iii)

These are three independent simultaneous linear equations which can be solved for the three unknowns, the currents I_1, I_2 and I_3. Note that although we could use Kirchhoff's rules to generate further equations, for example by taking the large loop through E_1, R_1, R_3 and E_2, the additional equations will not give us any additional information.

Equations (i) and (ii) give $E_1 = R_1(I_3 + I_2) + R_3 I_3$ and eliminating I_2 between this and (iii), we obtain (after some manipulation)

$$I_3 = \frac{R_2E_1 - R_1E_2}{R_1R_2 + R_2R_3 + R_3R_1}$$

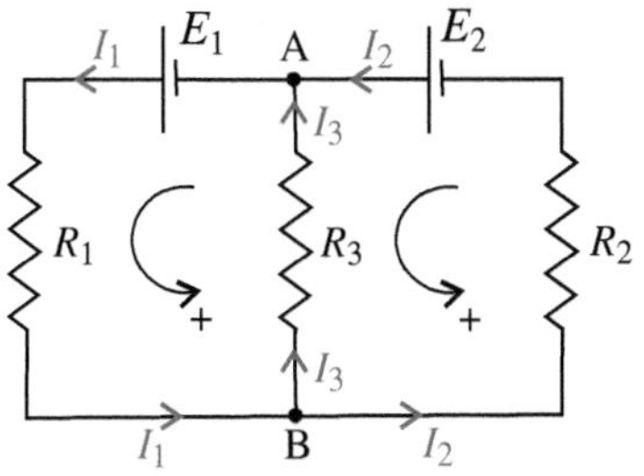

Figure 15.27. Application of Kirchhoff's rules to the network in Figure 15.24. Arbitrary choices of positive senses around the loops have been adopted and an initial guess at the sense of the current flow in each section have been made.

Note that if $R_2E_1 > R_1E_2$, the current I_3 flows as indicated in the figure but, on the other hand, if $R_2E_1 < R_1E_2$, it flows in the opposite sense.

The currents I_1 and I_2 can be determined in a similar manner.

Complex networks of current loops can generate a large number of simultaneous linear equations, the solution of which can be time consuming and tedious using simple algebra so that solutions are usually obtained using computer programs for solving simultaneous linear equations. For an illustration of this approach see Worked Example 15.6 where it is applied to the bridge network described in the next section.

For problems based on the material presented in this section visit up.ucc.ie/15/ and follow the link to the problems.

15.12 Comparison of resistances; the Wheatstone bridge

Since calorimetric experiments can be difficult to perform with sufficient accuracy, resistance measurements need to be carried out in practice by more simple, fast, and convenient methods. An example of such a technique, which involves comparison of resistances, will now be described.

Consider the electrical network (called a bridge) shown in Figure 15.28. If the values of the resistances are adjusted until no current is flowing between C and D, the bridge is said to be 'balanced' and the galvanometer G does not register any current. In this situation the current from the power supply divides at the junction A into a current I flowing in the upper path and a current I' flowing in the lower one. At balance $V_{CD} = 0$ and so, applying Kirchhoff's loop rule to the loops ACDA and BDCB, respectively,

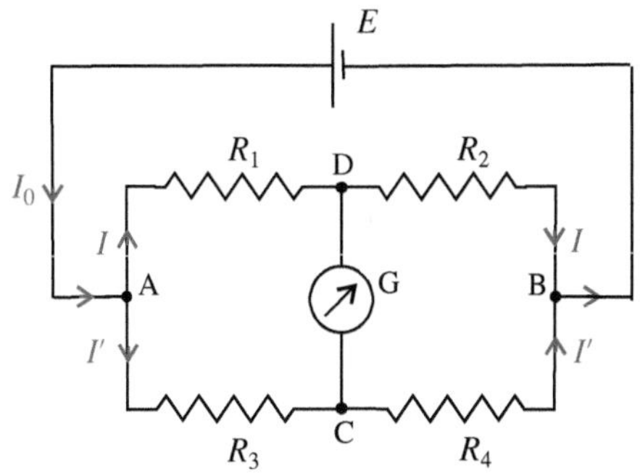

Figure 15.28. A Wheatstone bridge network.

$$-R_1I + R_3I' = 0 \quad \text{and} \quad -R_2I + R_4I' = 0$$

$$\Rightarrow \quad R_1I = R_3I' \quad \text{and} \quad R_2I = R_4I'$$

and hence,

$$\frac{R_1}{R_3} = \frac{I'}{I} = \frac{R_2}{R_4}$$

or

$$\boxed{\frac{R_1}{R_2} = \frac{R_3}{R_4}}$$

This provides a method (due to Charles Wheatstone 1802–1875) for measuring unknown resistances in terms of known (or standard) resistances. For example, if the value of R_4 is known and also the ratio R_1/R_2, then R_3 can be determined. Today modern digital ohmmeters, which are generally more convenient and are reasonably accurate, are usually used for resistance measurements. The Wheatstone bridge, however, is still used in a variety of important industrial applications, including in the monitoring of equipment and in the design of sensors.

Study Worked Example 15.6

15.13 Power supplies connected in parallel

When two power supplies are connected in parallel the situation is a little more complicated than in the case of the series connection discussed in Section 15.8. For one thing, as we shall see, the equivalent emf of the combination depends on the internal resistance of each power supply.

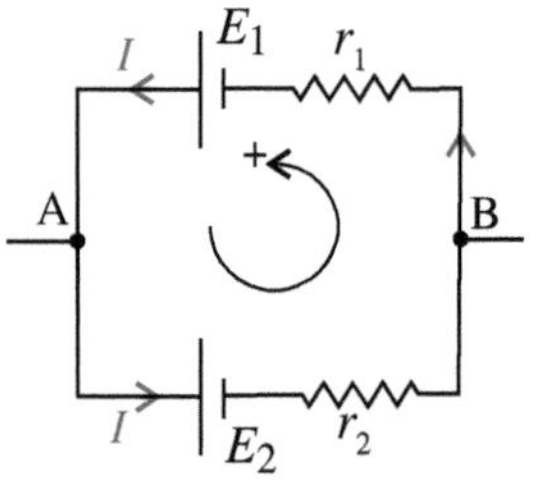

Figure 15.29. Two power supplies connected in parallel (r_1 and r_2 are internal resistances).

Consider the situation, illustrated in Figure 15.29, of two power supplies, one with emf E_1 and internal resistance r_1 and the other with emf E_2 and internal resistance r_2, which are connected in parallel. Here the symbols E_1 and E_2 represent the magnitude of each emf so an appropriate plus or minus sign must be included when applying Kirchhoff's loop rule. The emf E of the 'power supply' which is equivalent to the parallel combination is the potential difference between A and B (the 'terminals' of the equivalent power supply) when no current flows *externally* between A and B. If the current flowing internally around the loop in the sense (anticlockwise) shown in the figure is I, then applying Kirchhoff's loop rule

$$E_1 - E_2 = +r_1 I + r_2 I$$

and hence

$$I = \frac{E_1 - E_2}{r_1 + r_2}$$

Now,

$$E = V_{AB} = E_1 - r_1 I = E_1 - \frac{r_1}{r_1 + r_2}(E_1 - E_2) = \frac{r_2}{r_1 + r_2}E_1 + \frac{r_1}{r_1 + r_2}E_2$$

Note that if $E_2 = E_1$, then $E = E_2 = E_1$, independent of the values of r_1 and r_2.

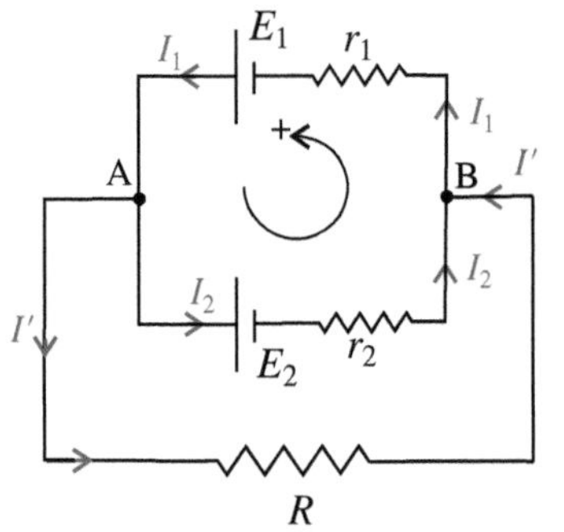

Figure 15.30. A current I' is drawn from the parallel combination of power supplies shown in Figure 15.29 by connecting a resistance R across the 'terminals' (A and B) of the combination.

To determine the internal resistance of the equivalent power supply, consider what happens when a current I' flows externally through a resistance R connected across the 'terminals' of the equivalent power supply (Figure 15.30). The currents flowing through each power supply (I_1 and I_2 in Figure 15.30) will be different in this case and the potential difference between A and B will be given by

$$V_{AB} = E_1 - r_1 I_1 = E_2 + r_2 I_2$$

Since $I' = I_1 - I_2$, we can write $E_1 - E_2 + r_2 I' = (r_1 + r_2)I_1$

and hence

$$I_1 = \frac{E_1 - E_2}{r_1 + r_2} + \frac{r_2}{r_1 + r_2}I'$$

Thus

$$V_{AB} = E_1 - r_1 I_1 = E_1 - \frac{r_1}{r_1 + r_2}(E_1 - E_2) - \frac{r_1 r_2}{r_1 + r_2}I'$$

Since when $I' \to 0$, $V_{AB} \to E$,

$$E = E_1 - \frac{r_1}{r_1 + r_2}(E_1 - E_2)$$

and

$$V_{AB} = E - \frac{r_1 r_2}{r_1 + r_2}I'$$

By comparing this result with Equation (15.10) for the potential difference across the terminals of a power supply, we see that the effective internal resistance of the equivalent power supply is given by

$$r_i = \frac{r_1 r_2}{r_1 + r_2} \quad \text{or} \quad \frac{1}{r_i} = \frac{1}{r_1} + \frac{1}{r_2}$$

as might have been anticipated from the result derived for resistors in parallel in Section 15.10.

It is not a good practice, in general, to connect power supplies in parallel if they have unequal emfs, because energy is lost due to the continuous current I around the loop containing the two emfs. Since, as we have already noted, if $E_2 = E_1$ then $E = E_2 = E_1$ independent of the values of r_1 and r_2, batteries are often constructed by connecting *identical* power supplies in parallel; this produces a lowering of internal resistance and, from Equation (15.10a), an increase in the maximum current available.

For problems based on the material presented in this section visit up.ucc.ie/15/ and follow the link to the problems.

15.14 Resistivity and conductivity

We now take a closer look at the model of electric current which we have used throughout this chapter. Figure 15.31 shows a length l of wire made from uniform material of uniform cross-sectional area A through which a uniform current I flows as a result of the application of a potential difference V across the length l.

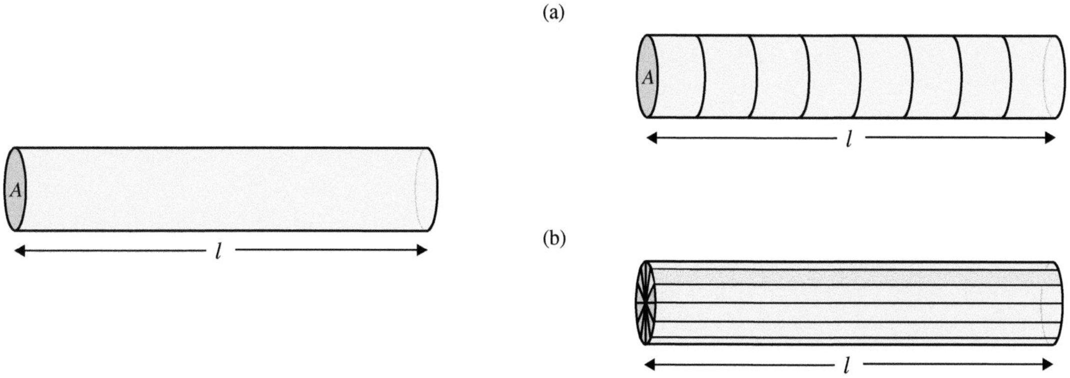

Figure 15.31. A piece of uniform conducting material of length l and cross-sectional area A (a) divided into sections of identical length and (b) divided into sections of identical cross-sectional area.

Consider first the wire as being divided into n parts of equal length each having a resistance R_1 (Figure 15.31(a)). The total resistance of the length l is the sum of n resistances R_1 connected in series and, from Equation (15.6), is given by $R = R_1 + R_1 + R_1 + . \dots = nR_1$. Hence

$$R \propto l$$

If, on the other hand, we think of the wire being divided into n sections all having equal cross-sectional area (Figure 15.31(b)), this can be considered as n resistances connected in parallel. In this case the resistance R of the piece of wire is related to the resistance of each section (R_2, say) by Equation (15.12), that is

$$\frac{1}{R} = \frac{1}{R_2} + \frac{1}{R_2} + \frac{1}{R_2} + . \dots \dots = \frac{n}{R_2}$$

In this case, $R = \frac{1}{n}R_2$ and hence

$$R \propto \frac{1}{A}$$

Thus the resistance R of the length l depends on the geometry of the wire as follows

$$R \propto \frac{l}{A} \quad \Rightarrow \quad R = (constant)\frac{l}{A}$$

and we can write

$$\boxed{R = \rho\frac{l}{A}}$$

where the constant of proportionality ρ is characteristic of the material from which the resistance is made and is defined as the **resistivity** of the material.[2] The unit of ρ in SI is ohm metre (Ω m). As in the case of resistance, the value of ρ depends on temperature; on the other hand, for metals at constant temperature ρ = constant (Ohm's law).

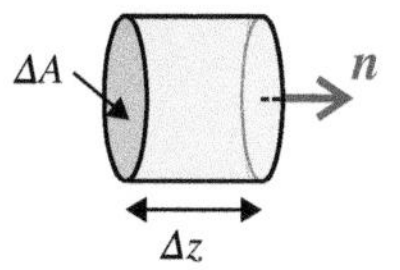

Figure 15.32. An element of conducting material through which a current is flowing. The unit vector $\boldsymbol{n}$ gives the direction of flow of electric current (the $+z$-direction).

The reciprocal of resistivity is called the **conductivity**, usually denoted by $\sigma \left(= \frac{1}{\rho}\right)$ and the corresponding SI unit is Ω^{-1} m^{-1}.

We now consider the element of length Δz and cross-sectional area ΔA within a conductor shown in Figure 15.32.

The potential difference across the element which carries a current ΔI is given by

$$\Delta V = (\Delta R)(\Delta I) = \rho\frac{\Delta z}{\Delta A}(\Delta I) = \rho(\Delta z)J$$

where J is the current density (Section 15.2).

[2]Note that the symbol (ρ) used to represent resistivity here is the same as that used for charge density in Section 15.2. In the unlikely event that both physical quantities are required in the same analysis, care should be taken to avoid confusion.

If the $+z$-direction is taken as the direction of current flow, then $\frac{dV}{dz}$ is negative and hence

$$\frac{dV}{dz} = -\rho J$$

or, in terms of conductivity,

$$\boxed{\boldsymbol{J} = -\sigma \frac{dV}{dz}\boldsymbol{n}} \tag{15.13}$$

The value of the resistivity (or conductivity) of a particular material depends on the microscopic properties of the substance. The conductivity of materials will be discussed from a quantum viewpoint in Chapter 25.

Values of resistivity at around 20° C are given in Table 15.1 for a variety of materials. Of particular importance are those materials whose conductivities are much less than those of metals but much larger than those of insulators. These materials are called **semiconductors** and are of considerable practical and industrial importance, being the basis of microelectronic devices (to be discussed in Chapter 26) and being widely used in telecommunications and computer applications.

Table 15.1 Typical values of resistivity (ρ) and temperature coefficient of resistance (α) at around 20 °C.

substance	$\frac{\rho}{\Omega \mathrm{m}}$	$\frac{\alpha}{\mathrm{K}^{-1}}$
silver	1.6×10^{-8}	3.8×10^{-3}
copper	1.7×10^{-8}	4.0×10^{-3}
gold	2.4×10^{-8}	3.4×10^{-3}
aluminium	2.7×10^{-8}	4.0×10^{-3}
tungsten	5.6×10^{-8}	4.5×10^{-3}
iron	9.7×10^{-7}	5×10^{-3}
constantan (60% Cu, 40% Ni)	4.9×10^{-8}	1×10^{-5}
nichrome (60% Ni, 24% Fe, 18% Cr)	1.1×10^{-6}	4×10^{-4}
carbon	$\sim3\times10^{-5}$	-5×10^{-4}
germanium	~0.5	
silicon	$\sim10^{3}$	
glass	$10^{10} - 10^{15}$	
sulphur	10^{15}	

Study Worked Example 15.7

For problems based on the material presented in this section visit up.ucc.ie/15/ and follow the link to the problems.

15.15 Variation of resistance with temperature

The resistivity of a sample of any material is observed to vary with temperature, the effect being much more pronounced in some materials than others. The resistivity of a metal, for example, usually increases with increasing temperature; the effect is quite small, however, and would not be noticeable in an experiment of the type described in Section 15.4. The size of this effect for any material is given by the **temperature coefficient of resistance** (α) for that material and is defined as the fractional increase in resistance per unit rise in temperature, that is

$$\boxed{\alpha := \frac{1}{R}\frac{dR}{dT}}$$

or, in terms of resistivity,

$$\alpha := \frac{1}{\rho}\frac{d\rho}{dT}$$

Thus, the SI unit of α is K^{-1}.

In general, α depends on temperature but, if the range of temperature involved is not too large, α is found to be reasonably constant for many materials. The temperature coefficient of resistance is characteristic of the particular material from which the resistor is made and, as such, carries information on the microscopic nature of the material. Values of α (at around 20° C) for a variety of common materials are given in Table 15.1.

Resistance thermometers

The dependence of resistance on temperature can be used to make devices which, once calibrated, enable temperature to be determined by measuring resistance. 'Resistance thermometers' can prove very useful in situations in which other types of thermometer cannot be used. Such thermometers can be quite sensitive if constructed from a material with a large temperature coefficient of resistance. **Thermistors**, which are made from semiconductor materials, are exceptionally useful for this purpose. Pure semiconductor materials usually have negative temperature coefficients of resistance but thermistors with positive coefficients can also be fabricated. Thermistors can be designed to have a large variation in α, and hence to be very sensitive, over a particular temperature range. Semiconductors will be discussed in Chapter 26.

For problems based on the material presented in this section visit up.ucc.ie/15/ and follow the link to the problems.

Further reading

Further material on some of the topics covered in this chapter may be found in *Electromagnetism* by Grant and Phillips (See BIBLIOGRAPHY for book details).

WORKED EXAMPLES

Worked Example 15.1: Three resistors of resistance 15 Ω, 30 Ω and 35 Ω are connected in series across a 2.0 V power supply (Figure 15.33). Assuming that the power supply has negligible internal resistance, determine (a) the current in the circuit, (b) the potential difference across each resistor and (c) the power generated in each resistor.

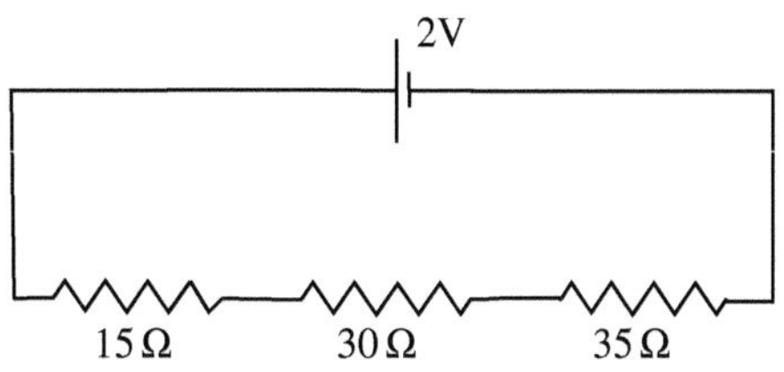

Figure 15.33. Worked Example 15.1.

(a) $$I = \frac{E}{R_{circ}} = \frac{E}{R_1 + R_2 + R_3} = \frac{2}{15 + 30 + 35} = \frac{2}{80}\,\text{A} = 25\ \text{mA}$$

(b) $$V_1 = R_1 I = (15\ \Omega)(0.025\ \text{A}) = 0.375\ \text{V};\quad V_2 = (30)(0.025) = 0.75\ \text{V};\quad V_3 = (35)(0.025\ \text{A}) = 0.875\ \text{V}$$

(c) $$P_1 = R_1 I^2 = (15\ \Omega)(0.025\ \text{A})^2\ \text{W} = 9.38\ \text{mW};$$

$$P_2 = (30)(0.025)^2\ \text{W} = 18.8\ \text{mW};$$

$$P_3 = (35)(0.025)^2\ \text{W} = 21.9\ \text{mW}$$

Check: $P_1 + P_2 + P_3 = 50\ \text{mW};\ R_{circ} I^2 = (80\ \Omega)(0.025\ \text{A})^2\ \text{W} = 50\ \text{mW}$

Worked Example 15.2: A variable resistor is connected directly across the terminals of a power supply. Prove that the power generated in this resistor is maximum when its resistance is equal to the internal resistance of the power supply.

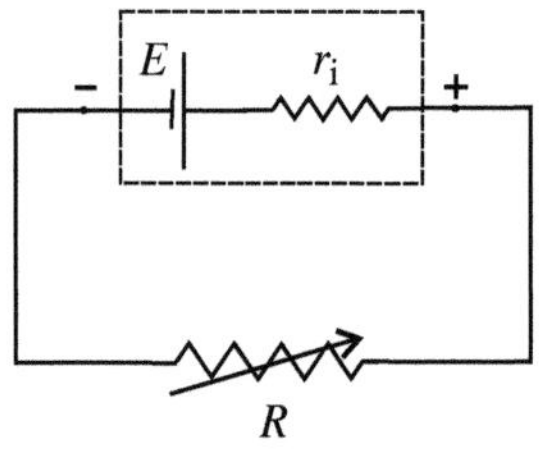

Figure 15.34. Worked Example 15.2.

The current in the circuit in Figure 15.34 is given by $I = \frac{E}{R_{circ}} = \frac{E}{R + r_i}$

The power in the resistor in the figure is given by $P_R = RI^2 = R\left(\frac{E}{R + r_i}\right)^2 = \frac{RE^2}{(R + r_i)^2}$

The condition for maximum power is $\frac{dP_R}{dR} = 0$ (Appendix A.6.2)

But

$$\frac{dP_R}{dR} = \frac{E^2}{(R + r_i)^2} - 2\frac{RE^2}{(R + r_i)^3} = \frac{E^2}{(R + r_i)^2}\left(1 - \frac{2R}{R + r_i}\right) = \frac{E^2}{(R + r_i)^2}\frac{r_i - R}{R + r_i} = \frac{(r_i - R)E^2}{(R + r_i)^3}$$

Thus

$$\frac{dP_R}{dR} = 0 \quad \Rightarrow \quad R = r_i$$

(the other solution ($R = \infty$) corresponds to minimum power)

Note that $R = r_i$ corresponds to a maximum, since $\frac{d^2P_R}{dR^2} = -\frac{E^2}{(R + r_i)^3} - 3\frac{(r_i - R)E^2}{(R + r_i)^4}$

and hence $\left.\frac{d^2P_R}{dR^2}\right|_{\text{at } R = r_i} = -\frac{E^2}{(2r_i)^3} < 0$

Worked Example 15.3: A 6 V power supply has an internal resistance of 0.2 Ω. What current flows when the terminals of the supply are connected by a wire of resistance of 4 Ω? What is the potential difference across the terminals? How much thermal power is being generated in the power supply?

With reference to Figure 15.35:

$$I = \frac{E}{R_{circ}} = \frac{6}{4 + 0.2} = \frac{6}{4.2} = 1.43 \text{ A}$$

$$V = RI = (4\,\Omega)(1.43 \text{ A}) = 5.75 \text{ V}$$

$$P_{supply} = r_i I^2 = (0.2\,\Omega)(1.43 \text{ A})^2 = 0.41 \text{ W}$$

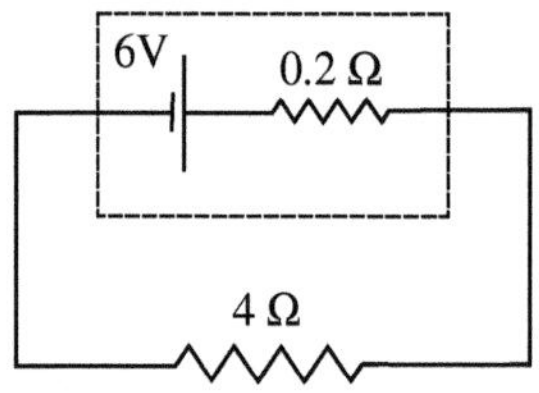

Figure 15.35. Worked Example 15.3.

Worked Example 15.4: Three identical resistors can be connected in three different combinations as shown in Figure 15.36. What are the values of the three resistances which can be obtained by combining three 6 Ω resistors in these three ways.

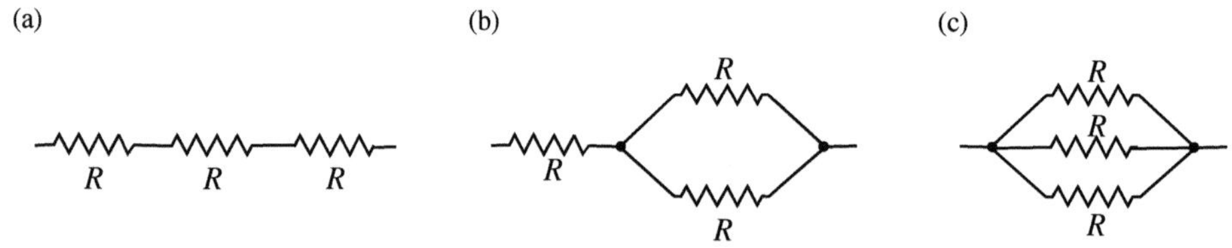

Figure 15.36. Worked Example 15.4.

(i)

$$R_1 = R + R + R = 3R = 18\ \Omega$$

(ii)

$$R_2 = R + R_p = R + \frac{(R)(R)}{R+R} = R + \frac{1}{2}R = \frac{3}{2}R = 9\ \Omega$$

(iii)

$$\frac{1}{R_3} = \frac{1}{R} + \frac{1}{R} + \frac{1}{R} = \frac{3}{R} \quad \rightarrow \quad R_3 = \frac{1}{3}R = 2\ \Omega$$

Worked Example 15.5: Show that the thermal power generated in a combination of two resistors connected in parallel (Figure 15.37) distributes itself so that the total power generated is at a minimum.

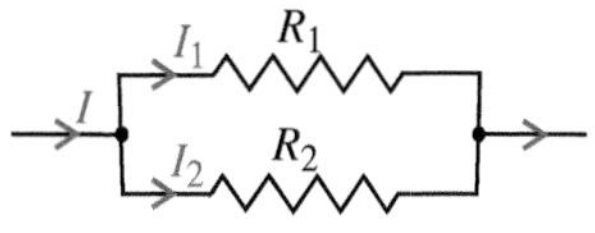

Figure 15.37. Worked Example 15.5.

Power generated in parallel combination $= P_c = R_1 I_1{}^2 + R_2 I_2{}^2 = R_1 I_1{}^2 + R_2(I - I_1)^2$

For minimum total power, $\dfrac{dP_c}{dI_1} = 0$.

Now $\dfrac{dP_c}{dI_1} = 2R_1 I_1 - 2R_2(I - I_1)$ and

$$\frac{dP_c}{dI_1} = 0 \quad \rightarrow \quad R_1 I_1 - R_2 I + R_2 I_1 = 0 \quad \Rightarrow \quad (R_1 + R_2)I_1 = R_2 I \quad \Rightarrow \quad I_1 = \frac{R_2}{R_1 + R_2} I$$

The potential difference across the combination $= V = R_1 I_1 = \dfrac{R_1 R_2}{R_1 + R_2} I$

Treating the combination as a single equivalent resistor of resistance R, $V = RI$

$$\rightarrow \quad R = \frac{R_1 R_2}{R_1 + R_2}$$

This relationship is identical to that derived at the end of Section 15.10 for the resistance of two resistors connected in parallel. Thus the general relationship for resistors connected in parallel, Equation (15.12), is equivalent to the statement that the power generated in a parallel combination of resistors is minimum. This result has significance in the context of the generalisation of the definition of electric potential which will be discussed in Section 16.10.

Worked Example 15.6: The galvanometer in the Wheatsone bridge arrangement shown in Figure 15.38 has a resistance of 2.5 Ω. What current flows through it when the resistance R is 10 Ω less than its value when the bridge is balanced?

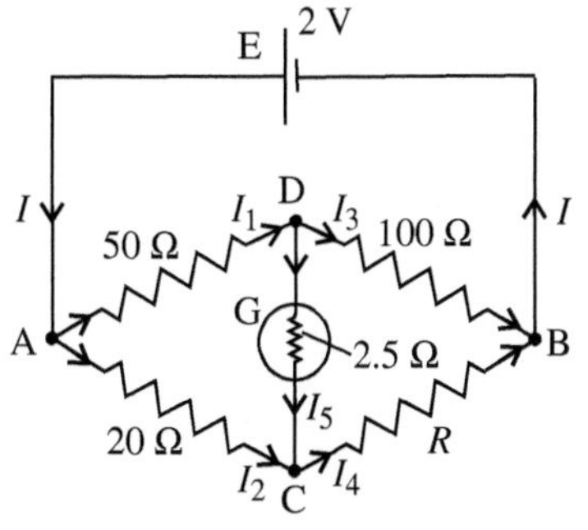

Figure 15.38. Worked Example 15.6.

When the bridge is balanced, $\dfrac{50}{100} = \dfrac{20}{R} \quad \rightarrow \quad R = 40\ \Omega$ and $(R - 10\ \Omega) = 30\ \Omega$

Junction rule At D: $I_1 = I_3 + I_5$ .... (i); At C: $I_2 + I_5 = I_4$ .... (ii)

Loop rule EADBE: $2 = 50I_1 + 100I_3$ .... (iii); EACBE: $2 = 20I_2 + 30I_4$ .... (iv)

ADCA: $50I_1 + 2.5I_5 - 20I_2 = 0$ .... (v)

(iii) $\rightarrow$ $I_3 = \frac{2}{100} - \frac{I_1}{2}$ (iv) $\rightarrow$ $I_4 = \frac{2}{30} - \frac{2}{3}I_2$ (v) $\rightarrow$ $20I_1 + I_5 - 8I_2 = 0$ .... (vi)

(i) $\rightarrow$ $I_1 = \frac{2}{100} - \frac{I_1}{2} + I_5$ $\rightarrow$ $I_1 = \frac{4}{300} + \frac{2}{3}I_5$ (ii) $\rightarrow$ $I_2 + I_5 = \frac{2}{30} - \frac{2}{3}I_2$ $\rightarrow$ $I_2 = \frac{6}{150} - \frac{3}{5}I_5$

(vi) $\rightarrow$ $20\left(\frac{4}{300} + \frac{2}{3}I_5\right) + I_5 - 8\left(\frac{6}{150} - \frac{3}{5}I_5\right) = 0$ $\rightarrow$ $\frac{287}{15}I_5 = \frac{16}{300}$ $\rightarrow I_5 = 2.8 \times 10^{-3}\text{ A} = 2.8\text{ mA}$

The five simultaneous equations (i) to (v), in five unknowns, I_1, I_2, I_3, I_4 and I_5, may also be solved using a standard computer program for solving simultaneous linear equations, such as the *Intemodino System* linear equations solver https://intemodino.com/en/equation-solver/solve-systems-of-equations/

Application of this program to equations (i) to (v) yields the solutions

$I_1 = 15.2$ mA, $I_2 = 38.3$ mA, $I_3 = 12.4$ mA, $I_4 = 41.1$ mA and $I_5 = 2.8$ mA, in agreement with the solution obtained above for I_5.

Worked Example 15.7: What is the resistance of a 1.5 m length of uniform copper wire of cross-sectional area 2.8 mm²? If this piece of wire was drawn out to make a uniform wire which is three times as long (the volume remaining constant), what would be its resistance?

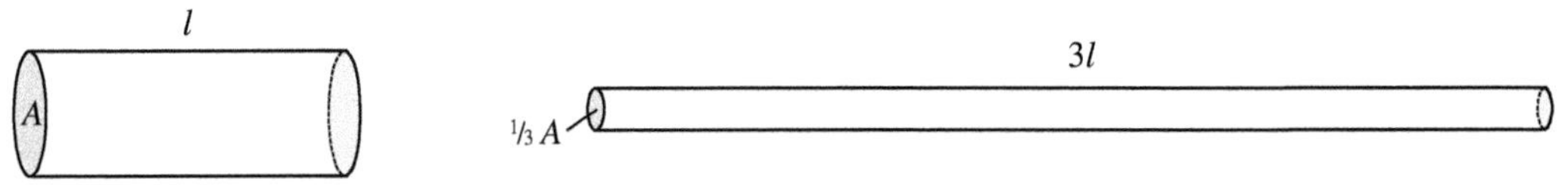

Figure 15.39. Worked Example 15.7.

Resistivity of copper (from Table 15.1) is $1.7 \times 10^{-8}\ \Omega\text{ m}$

For a wire of length l and cross-sectional area A (Figure 15.39), the resistance is given by

$$R = \rho\frac{l}{A} = (1.7 \times 10^{-8}\ \Omega\text{ m})\frac{1.5\text{ m}}{2.8 \times 10^{-6}\text{ m}^2} = 9.1 \times 10^{-3}\ \Omega = 9.1\text{ m}\Omega$$

For a wire of length $3l$ and cross-sectional area $A/3$, the resistance is given by

$$R' = \rho\frac{3l}{A/3} = 9R = 82\text{ m}\Omega$$

PROBLEMS

For problems based on the material covered this chapter visit up.ucc.ie/15/ and follow the link to the problems.

16

Electric fields

AIMS

- to show how force fields around electric charges may be used to describe the forces which are observed to exist between charges
- to introduce the concept of electric flux and to show how the consideration of the flux from electric charges may be used determine electric field strength
- to show how the force field of a single point charge may be used to analyse the behaviour of more complicated systems of charges by treating these as the superposition of the contributions from point charges
- to introduce Gauss's law and to show how it may be used to determine the electric field strength in situations with a high degree of symmetry
- to extend the idea of potential difference to develop an alternative formalism to describe electrostatic force fields

16.1 Electric charges at rest

The word 'electric' comes from the Greek word for amber, ηλεκτρον. The ancient Greeks knew that when a piece of amber had been rubbed with a cloth, it attracted light objects such as small pieces of dust or straw. Many familiar everyday phenomena provide further examples of this same effect — a nylon comb can pick up small pieces of paper after it has been used to comb dry hair; you may feel a small spark as you touch a metal object after walking across certain types of carpet.

These effects can be studied in a series of experiments with plastic rods, as illustrated in Figure 16.1(a)–(d). A rod of perspex (plexiglass) is suspended by a thread so that it is free to rotate in a horizontal plane. A second perspex rod is held so that one of its ends is near an end of the suspended rod. No effect is observed even if the ends of the two rods are brought very close together. If, however, one end of each rod is rubbed with a silk cloth and the ends which were rubbed are moved nearer together, a clear force of repulsion is observed (Figure 16.1(a)). A very similar effect is observed if the experiment is repeated using two polythene rods (Figure 16.1(b)). If, on the other hand, one of the rods is polythene and the other is perspex, the force between the rubbed ends is observed to be attractive (Figure 16.1(c)).

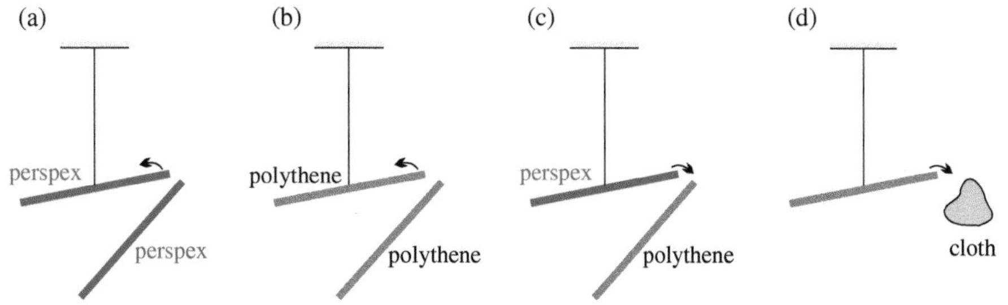

Figure 16.1. Experimental investigation of electrification by friction of plastic rods. In each case the arrows indicate the direction in which the suspended rod tends to rotate.

To explain the effects just described, the existence of a physical quantity, called **electric charge,** was postulated; this charge appears on the rods when they are rubbed with the cloth. Furthermore, the results of the experiments described in Figure 16.1(a)–(c) require that two distinct types of electric charge exist and that *like charges repel and unlike charges attract one another.* As we know from Section 15.2, such charge is identified as that which flows in an electric current.

Understanding Physics, Third Edition. Michael Mansfield and Colm O'Sullivan.
© 2020 John Wiley & Sons Ltd. Published 2020 by John Wiley & Sons Ltd.

It is also observed that the cloth used to rub the rod acquires an electric charge. The force between the rubbed end of the rod and the cloth is found to be attractive (as illustrated in Figure 16.1(d)). This suggests that when an uncharged rod is rubbed with an uncharged cloth the net result is to transfer charge from one to the other leaving a different type of electric charge on each.

A similar effect can be observed in an experiment of the type illustrated in Figure 16.2(a) and (b). Here a metal sphere is connected by a wire to the positive terminal of a power supply of reasonably high emf (> 1000 V) and the sphere is brought close to the charged end of a suspended perspex rod. A repulsive force is observed between the metal sphere and the charged end of the rod. If a metal sphere is connected to the negative terminal of the power supply and brought close to the rubbed end of the perspex rod the force observed is attractive.

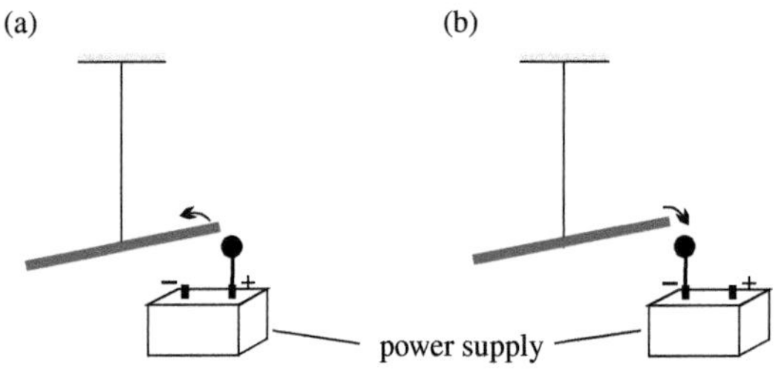

Figure 16.2. Forces between charged plastic rods and terminals of high voltage power supplies.

The type of charge on the metal sphere attached to the positive terminal is clearly of the same type as that left on a perspex rod after it has been rubbed with a cloth while that on the sphere which is attached to the negative terminal of the supply is of the opposite type. The two types of charge, therefore, can be classified as **positive** and **negative** — positive charge being that on the sphere attached to the positive terminal of the power supply and negative charge being the type of charge found on a sphere connected to the negative terminal. Thus we can identify the net charge left on perspex and polythene rods when they are rubbed with a cloth as positive and negative, respectively. Note that the choice of which type of charge is positive and which is negative is set by the (arbitrary) choice of convention for direction of current flow (Section 15.2).

Implicit in the thinking that has given rise to the idea of electric charge is the idea that uncharged bodies, such as the perspex rod before it had been rubbed, contain equal quantities of positive and negative charge. In the 'charging' process, whether it involves rubbing with a cloth or attaching to a power supply, electric charge is transferred between the two bodies leaving an excess of one type of charge on the first body and an excess charge of the other type on the second body. In the case of insulators, such as the perspex or polythene rods, the excess charges are fixed and therefore remain where they were deposited on the body.

In the case of bodies made from conducting materials, such as the metal sphere and piece of wire attached to a terminal of a power supply in Figure 16.2, the electric charges are free to move within the conductors. Indeed, when an electric charge moves at constant speed we observe exactly those effects which we observed to be associated with electric currents in Chapter 15, confirming that electric charge as appears in the experiments described here is the same as that which flows in an electric current.

When a body made from a conducting material is given a net electric charge (either positive or negative) the individual charges, being free to move, repel one another and migrate to the surface of the conductor. Exactly how the charge is distributed over the surface of the conductor depends on the shape of the conductor (as indicated in Figure 16.3(a)) and the presence or otherwise of nearby charged bodies (illustrated in Figure 16.3(b)). If a conductor with no net charge is brought near a charged body (as shown in Figure 16.3(c)) the distribution of charge on the conductor is changed, although its overall charge remains zero. In this case we say that a charge is *induced* on the surface of the uncharged conductor by the presence of the nearby charged body.

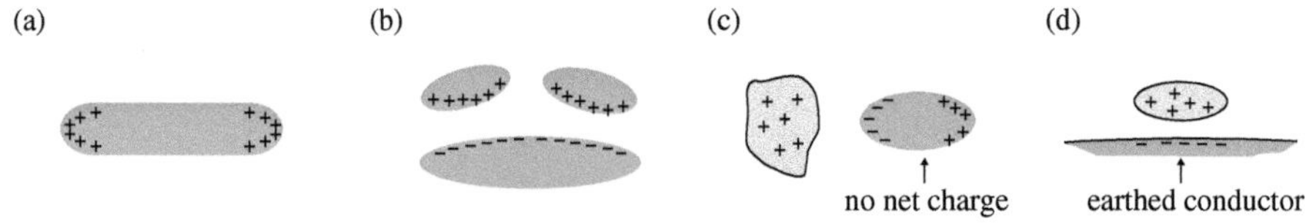

Figure 16.3. Distribution of electric charge on conductors (blue) in various situations.

Note that the distribution of charges on the surface of a conductor is such that there is no component of the net force on any charge in a direction tangential to the surface at the location of that charge. This is because, were a charge to experience a net tangential component, it would be moved along the surface until it found a position in which the tangential component of the force is zero. Thus the force on any charge on a conductor due to all other charges must be directed perpendicularly to the surface of the conductor and is balanced by a 'normal reaction' which is exerted by the force that confines the charge within the conductor.

The Earth may be considered to be a very large conducting body which is effectively neutral electrically. When a charged body is brought near an earthed conductor, an equal and opposite charge is induced on the part of the earthed conductor nearest to the charged body, as indicated in Figure 16.3(d). The positive charge on the Earth which balances the induced negative charge can be considered to have redistributed itself over the entire Earth, where it has negligible effect. If a charged conductor is connected directly to the Earth, by a piece of wire, the charge on the body flows to the Earth where it is neutralised by the induced charge on the Earth. Such an electrically neutral body is said to be 'earthed' or connected to 'ground'.

The electrical symbol for a ground connection is shown in Figure 16.4.

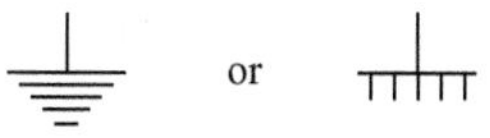

Figure 16.4. The conventional electrical symbols for ground (earth) connection.

16.2 Electric fields: electric field strength

The electrostatic force — the force that exists between charges at rest — is clearly an example of 'action at a distance'. The charged perspex or polythene rods which we discussed in Section 16.1 are not in contact with each other. Nevertheless, there is a force of attraction or repulsion between them even if the space in between is evacuated. The concept of a force field which proved to be a useful tool in mechanics, particularly in the study of gravitation (Section 5.2), can be invoked again in this case. The region around and between the charged conductors in Figure 16.3(b), for example, can be considered as a force field since a charge placed anywhere in this region will experience a force. In general the field is strongest near charged bodies and becomes weaker as one moves away.

Electric field lines

As in the case of gravitation (Section 5.1), we can introduce the idea of field lines to describe the 'shape' of the field. By convention, an **electric field line** is defined as the line along which a free *positive* point charge would move if released from rest. Some examples of the field line patterns produced by different electrostatic sources (electric charges at rest) are shown in Figure 16.5(a)–(c). Field lines are simply geometrical constructions and cannot be observed as physical entities. They comprise an infinite family of lines; that is, the space in between can always be filled with further field lines. Note that, while force fields due to multiple charges can be complicated, field lines never intersect one another.

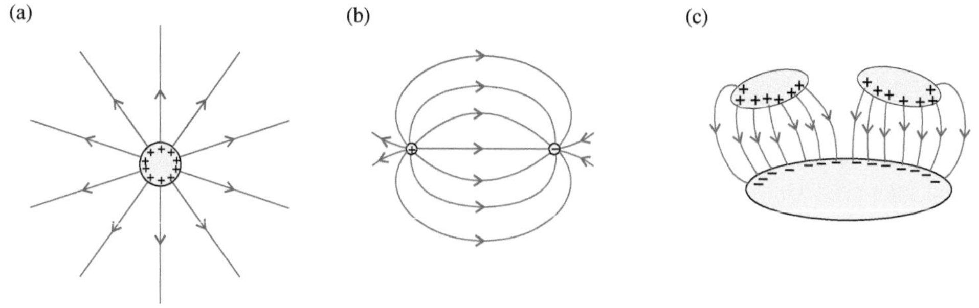

Figure 16.5. Electric field line maps: (a) outside a uniformly charged sphere, (b) due to a pair of small charges of opposite sign and (c) in the region around three charged conductors, two positively charged and one negatively charged.

The overall pattern of the field lines gives an indication of the shape of the field. Field lines emanate from positive charges and terminate on negative charges. In the case of an isolated charge, for example Figure 16.5(a), the field lines emanate from or terminate at infinity. Note that the field lines originating from or terminating on a conductor are perpendicular to the surface in the immediate neighbourhood of the conductor because, as noted in Section 16.1, the electrostatic force on a charge on the surface of a conductor must be normal to the surface.

The field lines around a single fixed point positive charge q are illustrated in Figure 16.6. Since the repulsion between a test positive charge and the charge q is along the line joining them, the field lines are directed radially outwards as shown (a *radial field*). When field are parallel over a particular region of space (as illustrated in Figure 16.7), the field is said to be *uniform* in that region. For any field, the field lines may be considered to be essentially parallel if the region of space considered is sufficiently small.

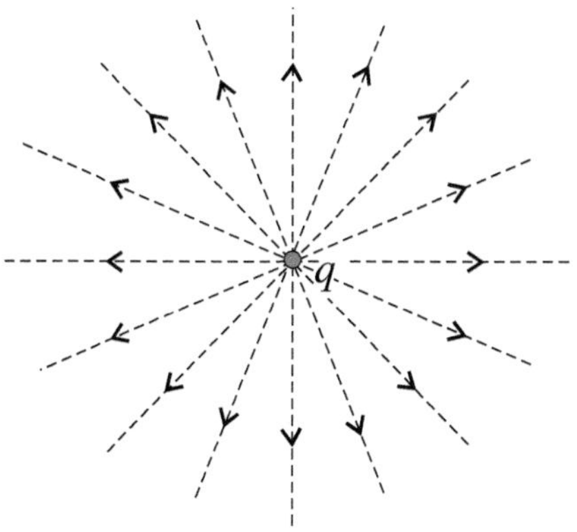

Figure 16.6. Field lines around a point positive charge.

Electric field strength

While a force field exists everywhere due to the presence of a charge or a configuration of charges, no force is exerted unless another charge is present at some point in the field. It would be useful, however, if a physical quantity could be defined which contains information on the 'strength' and the 'shape' of a field. To achieve this, the **electric field strength** at a point $\boldsymbol{r}$ in an electric field is defined as

$$\boldsymbol{E} := \frac{\boldsymbol{F}}{q},$$

where $\boldsymbol{F}$ is the force that would be experienced by a point positive charge q if placed at the point $\boldsymbol{r}$. Note that since $\boldsymbol{F}$ is a vector then, from its definition and since q is a scalar, $\boldsymbol{E}$ is also a vector quantity in the same direction as $\boldsymbol{F}$. Thus, the direction of $\boldsymbol{E}$ at any point is tangential to the electric field line through that point. The force on a charge q located at a point in a field at which the electric field strength is $\boldsymbol{E}$ is given by

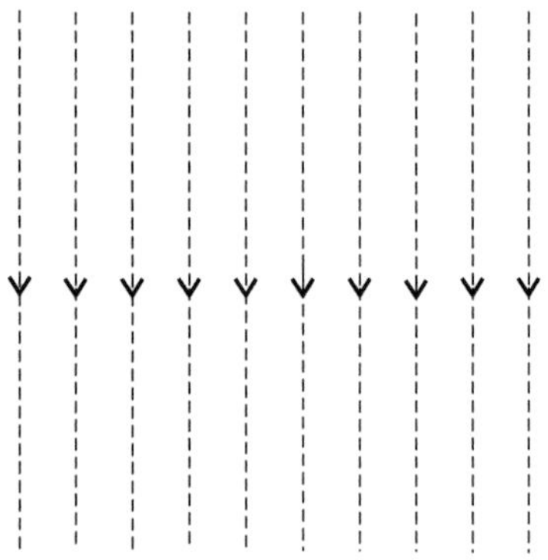

Figure 16.7. Field lines representing a uniform field.

$$\boldsymbol{F} := q\boldsymbol{E}$$

From the definition of $\boldsymbol{E}$ it follows that the SI unit of electric field strength is N C^{-1}. Recalling that 1 V = 1 W A^{-1}, 1 J = 1 N m and 1 C = 1 A s, we can write

$$1\ \mathrm{N\,C^{-1}} = 1\ \mathrm{J\,m^{-1}\,C^{-1}} = 1\ \mathrm{J\,m^{-1}\,A^{-1}\,s^{-1}} = 1\ \mathrm{W\,m^{-1}\,A^{-1}} = 1\ \mathrm{V\,m^{-1}},$$

so that electric field strengths are commonly stated in volt per metre.

For problems based on the material presented in this section visit up.ucc.ie/16/ and follow the link to the problems.

16.3 Forces between point charges: Coulomb's law

The nature of the force between electric charges exercised the minds of scientists throughout the second half of the eighteenth century. By then, Newton's law of gravitation was already well established and many suspected that the force between point charges might also obey an inverse square law. The French scientist Charles-Augustin de Coulomb (1736–1806), who is credited with inventing the torsion balance (Section 7.5), used such a device to determine how the force between charges depends on the distance between them.

The principles of Coulomb's torsion balance are shown in Figure 16.8. The 'pith balls' (pith is a light non-conducting material obtained from certain vascular plants) were electrified by momentary contact with a charged rod. One of the pith balls was fixed while the other was attached to the end of a very light horizontal needle suspended from a very thin vertical filament of silver. The paper disc was used to provide a counterweight to the pith ball on the needle with a secondary purpose of damping oscillations. The resulting torque due to the mutual repulsion of the charged pith balls caused the horizontal needle to rotate until balanced by the restoring torque exerted by the silver filament. The torsion micrometer was turned to restore the needle to its original position and, since the torsion system had been previously calibrated, the torque could be measured.

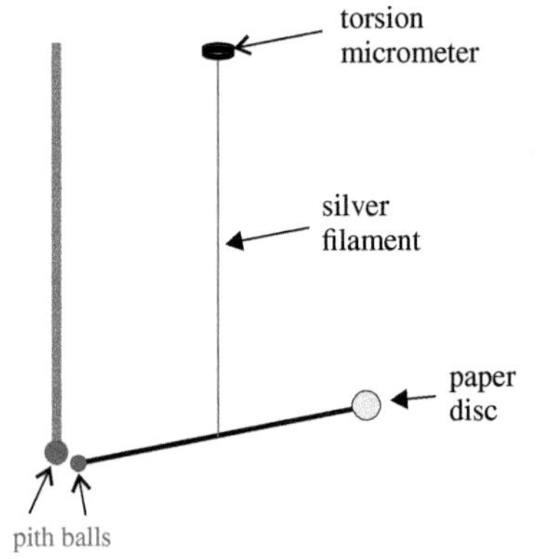

Figure 16.8. Experimental arrangement for investigating Coulomb's law.

Coulomb presented the results of his investigations to the Paris Academy of Sciences in June 1785 demonstrating that electrostatic repulsion does indeed vary inversely with the square of the distances between the charges. Today, when applied to point charges, this result is known as **Coulomb's law** which states that the force $\boldsymbol{F}$ between two point charges q_1 and q_2 which are separated by a displacement $\boldsymbol{r}$ (Figure 16.9) is given by

$$\boldsymbol{F} = (constant)\frac{q_1 q_2}{r^2}\hat{\boldsymbol{r}}$$

where the constant of proportionality depends on the system of units used.

Note that, if q_1 and q_2 have the same sign, $\boldsymbol{F}$ (the force on q_2 in the figure) is in the $+\boldsymbol{r}$ direction, directed away from q_1. Newton's third law applied to this situation tells us that the force on q_1 due to q_2 is $-\boldsymbol{F}$, directed away from q_1 (that is, 'like charges repel'). If, however, the charges have opposite signs, the product $q_1 q_2$ is negative and the forces reverse in direction ('unlike charges attract').

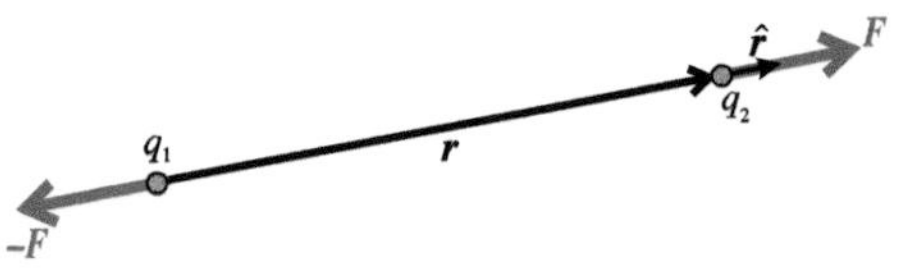

Figure 16.9. Force between point charges.

16.4 Electric flux and electric flux density

Electric field lines start from positive charges and end up either on negative charges or at an infinite distance from the positive charges. Since, as we have seen, the force between point electric charges is of the 'inverse square law' type, it will again prove useful, as in the case of gravitation (Section 5.3), to introduce the idea of a flux. In this case, we think of a positive charge as being a *source* of **electric flux** and a negative charge as a *sink* of electric flux. The flux concept can be carried over to the case of electric fields simply by defining the total electric flux (Φ_E) from a charge Q as

$$\Phi_E := Q$$

Hence, in SI, the unit of electric flux is the same as the unit of electric charge, namely the coulomb.

It is important to note that electric flux, defined in this way, is independent of the medium in which the sources (charges) are embedded.

Electric flux density

We define the intensity of any flux at a point some distance from the source as the flux through unit area perpendicular to the flux at that point. Thus, we define the **electric flux density** at a point in an electric field as

$$\boxed{\boldsymbol{D} := \left(\underset{\Delta A \to 0}{\text{limit}} \frac{\Delta \Phi_E}{\Delta A}\right) \hat{\boldsymbol{\imath}}} \qquad (16.1)$$

where $\Delta\Phi_E$ is the electric flux through an area ΔA which is perpendicular to the direction of the flux and $\hat{\boldsymbol{\imath}}$ is a unit vector along the direction of the flux (that is, perpendicular to the element ΔA and tangential to the local field line as illustrated in Figure 16.10). The electric flux density, which is also called the **displacement**, is a vector quantity; the SI unit of electric flux density is C m^{-2}.

The total electric flux[1] through any area is given by the integral of $\boldsymbol{D}$ over the area or, more strictly, the integral of the perpendicular component of $\boldsymbol{D}$ over the area, that is

$$\Phi_E = \iint_{\text{area}} \boldsymbol{D} \cdot d\boldsymbol{A}$$

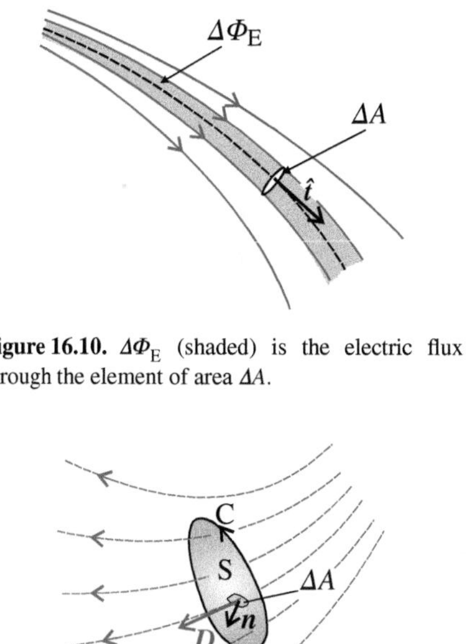

Figure 16.10. $\Delta\Phi_E$ (shaded) is the electric flux through the element of area ΔA.

where $d\boldsymbol{A} = (dA)\boldsymbol{n}$ and $\boldsymbol{n}$ is a unit vector perpendicular to the element dA (Figure 16.11). Note that, defined in this way, electric flux is an *algebraic quantity* in that Φ_E represents a sign as well as a magnitude and unit. The relevant convention of signs is determined by the choice of positive direction for the vector area element $d\boldsymbol{A} = (dA)\boldsymbol{n}$ and this, in turn, depends on the choice of positive sense of travel around the curve (C in Figure 16.11) that delineates the surface S.

The electric flux within any 'tube' delineated by field lines is a fixed quantity. As Maxwell (James Clerk Maxwell (1831–1879)) pointed out, this is exactly analogous to streamline flow in a fluid where there is continuity of flow across any area in a 'tube' delineated by streamlines (Section 10.7). Indeed, the definition of any flux quantity and, by extension, that of the corresponding flux density are chosen so that the total flux in such a tube is unchanged when passing from one medium to another.

It will probably have occurred to the reader that the electric flux density $\boldsymbol{D}$ provides a measure of a quantity which is very similar to the electric field strength $\boldsymbol{E}$. Indeed, as we have introduced it, $\boldsymbol{D}$ is parallel to $\boldsymbol{E}$, so that we can write $\boldsymbol{D} = (constant)\boldsymbol{E}$, that is

$$\boxed{\boldsymbol{D} = \varepsilon \boldsymbol{E}} \qquad (16.2)$$

where the constant of proportionality ε is characteristic of the medium in which the charges are immersed. The quantity ε is called the (electric) **permittivity** of the medium — the measurement of permittivity will be discussed in Section 17.13. The permittivity of vacuum, denoted as ε_0, is measured (to three significant figures) to be 8.85×10^{-12} C V^{-1} m^{-1}.

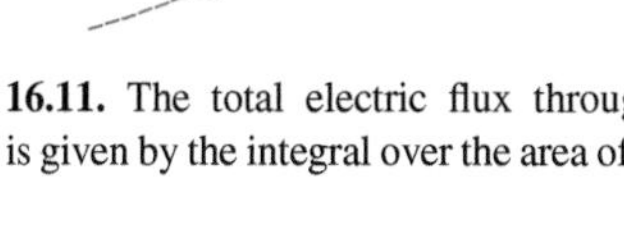

Figure 16.11. The total electric flux through any surface is given by the integral over the area of ($\boldsymbol{D}\cdot\boldsymbol{n}$).

[1]Note that electric flux may be defined differently in other texts; for example, in many physics textbooks the electric flux Φ_E is defined as $\int \boldsymbol{E} \cdot d\boldsymbol{A}$ rather than $\int \boldsymbol{D} \cdot d\boldsymbol{A}$ as adopted here. In this book we follow the BIPM SI brochure and the recommendations of the International Union of Pure and Applied Physics (IUPAP) and the International Union of Pure and Applied Chemistry (IUPAC). This is also more consistent with the definition used in electrical engineering texts and with the use of the flux model in magnetism (Chapter 18).

Strictly, ε is constant only in a *linear, isotropic and homogeneous medium*. In an inhomogeneous medium $\varepsilon = \varepsilon(\boldsymbol{r})$ while in a non-linear medium $\varepsilon = \varepsilon(\boldsymbol{E})$. In a non-isotropic medium, such as a crystalline material, $\boldsymbol{D}$ may not be parallel to $\boldsymbol{E}$. In such a case the relationship between $\boldsymbol{D}$ and $\boldsymbol{E}$ is somewhat more complicated than stated in Equation (16.2) and may be described by the relationship

$$D_i = \sum_{j=1}^{j=3} \varepsilon_{ij} E_j \tag{16.3}$$

where the subscripts 1, 2, and 3 refer to three orthogonal directions in the medium and $\boldsymbol{D}$ and $\boldsymbol{E}$ are written in terms of their components in these directions, that is

$$E = E_1\boldsymbol{i} + E_2\boldsymbol{j} + E_3\boldsymbol{k} \quad \text{and} \quad D = D_1\boldsymbol{i} + D_2\boldsymbol{j} + D_3\boldsymbol{k}$$

In this case the electric permittivity is said to be a **tensor** quantity; since only isotropic media are considered in this book, however, it will not be necessary to consider tensor permittivities again.

In the remaining sections of this chapter it will be assumed that all charges are either in a vacuum or embedded in an infinite linear isotropic homogeneous medium; the issue of the effect of interfaces between media of different permittivities will be considered in the next chapter.

16.5 Electric fields due to systems of charges

Single point charge

Let us now consider the case of a single isolated point charge Q as a source of flux, illustrated in Figure 16.12. In the absence of any other charges, the electric field lines radiate from Q as shown. The field lines in the figure have been drawn assuming that the charge is positive; if Q is negative, the arrows on the field lines would be reversed (that is, the charge would be a sink of flux). The field is clearly radial, the flux flowing outward from a positive charge or inward towards a negative charge. From the symmetry of the situation, the magnitude of the electric flux density must be the same at all points at equal distances from the charge Q. Thus, the magnitude of the electric flux density at any point at a distance r from a point source (Figure 16.13) must be equal to the total flux from the charge divided by the total surface area of a sphere of radius r. Therefore the electric flux density at the point P in a non-conducting medium is given by

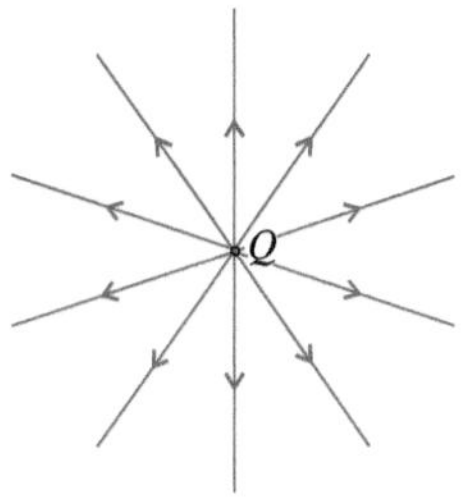

Figure 16.12. Electric field lines from a point positive charge; if the charge were negative the direction of the field lines would be reversed.

$$\boldsymbol{D} = \boldsymbol{D}(\boldsymbol{r}) = \frac{\Phi_E}{4\pi r^2}\hat{\boldsymbol{r}} \tag{16.4}$$

This derivation is entirely analogous to that used for the gravitational flux density due to a point mass in Section 5.3, except that in the case of a positive charge the point charge is a *source* of flux ($\boldsymbol{D}$ radially outward). A negative charge is a sink of flux ($\boldsymbol{D}$ radially inward, similar to the case of a point mass).

Using Equation (16.2), the electric field strength at a distance $\boldsymbol{r}$ from a point charge Q may be written

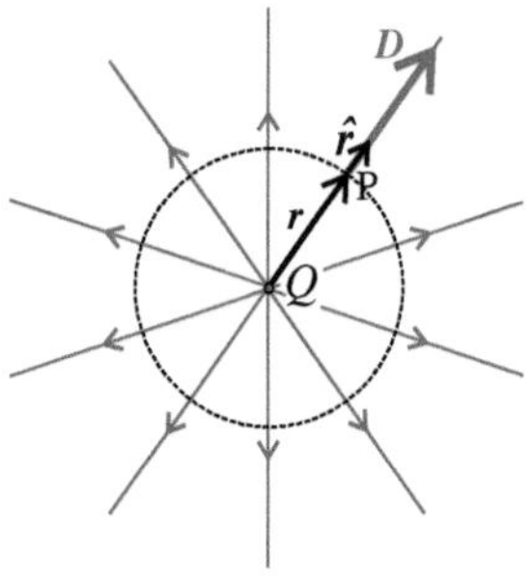

$$\boldsymbol{E} = \boldsymbol{E}(\boldsymbol{r}) = \frac{\boldsymbol{D}}{\varepsilon}$$

or, using Equation (16.4),

$$\boxed{\boldsymbol{E} = \frac{Q}{4\pi\varepsilon r^2}\hat{\boldsymbol{r}}} \tag{16.5}$$

Figure 16.13. The electric flux density due to a point charge is the same at all points on the surface of a sphere of radius r. The flux density vector D is directed radially outwards from the source if the charge is positive and radially inward if the charge is negative.

where ε is the permittivity of the medium in which the charge is immersed. Note that if Q is positive then $\boldsymbol{E}$ is in the $+\boldsymbol{r}$-direction and that if Q is negative then $\boldsymbol{E}$ is in the $-\boldsymbol{r}$-direction, in keeping with our convention for the direction of an electric field line. Note also that, while $\boldsymbol{D}$ at any point is independent of the medium, $\boldsymbol{E}$ depends on the permittivity of the medium.

Coulomb's law

Consider two point charges q_1 and q_2 which are separated by a displacement $\boldsymbol{r}$ as shown in Figure 16.14. From Equation (16.5), the electric field strength due to q_1 at the point occupied by q_2 is given by

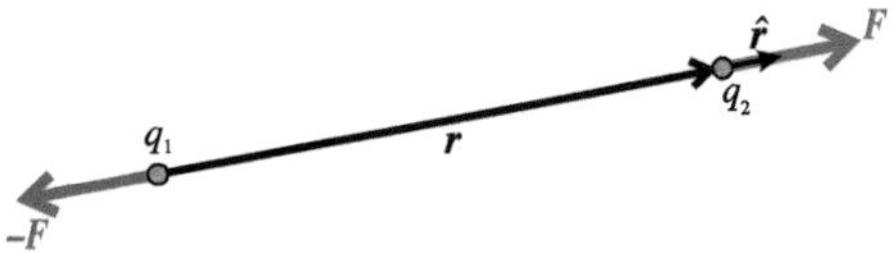

Figure 16.14. Force between point charges. Note that the directions of the forces are as shown (that is, the interaction is repulsive) if both charges have the same sign but that the directions are reversed if one charge is positive and the other is negative.

$$\boldsymbol{E} = \frac{q_1}{4\pi\varepsilon r^2}\hat{\boldsymbol{r}}$$

and hence the force on q_2 due to q_1 is given by $\boldsymbol{F} = q_2\boldsymbol{E}$, or

$$\boldsymbol{F} = \frac{q_1 q_2}{4\pi\varepsilon r^2}\hat{\boldsymbol{r}} \qquad (16.6)$$

which gives Coulomb's law (in SI units) for the force between point charges in an infinite linear homogeneous isotropic medium, in accordance with the experimentally deduced statement of the law given in Section 16.3. In the case of charges in a vacuum

$$\boldsymbol{F} = \frac{q_1 q_2}{4\pi\varepsilon_0 r^2}\hat{\boldsymbol{r}}$$

Coulomb's law (Equation (16.6)) was derived above from Equation (16.4) which, in turn, arose directly from the concept of electric flux. It must be remembered, however, that all flux models (gravitational flux, electric flux, magnetic flux) automatically lead to inverse square law force fields from simple geometric considerations and, conversely, the concept of flux is useful only in the context of inverse square law fields.

General charge distribution

The importance of equations (16.4) and (16.5), which refer to single isolated point charges, lies in their application to systems comprising many charges such as that illustrated in Figure 16.15. The net electric flux density or electric field strength at any point is the *vector sum* of the contributions from each point charge (a so-called Principle of Superposition), that is

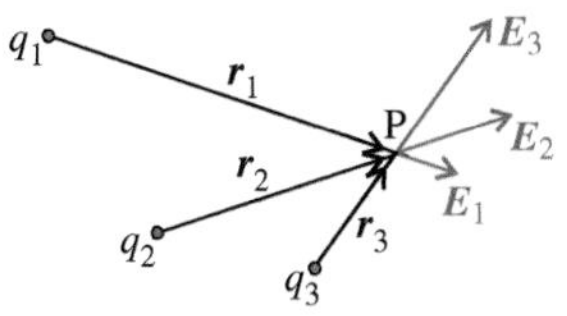

Figure 16.15. The net electric field strength at P due to a number of point charges is the *vector* sum of the individual contributions from each point charge.

$$\boldsymbol{D} = \boldsymbol{D}_1 + \boldsymbol{D}_2 + \boldsymbol{D}_3 + \ldots \text{ where } \boldsymbol{D}_i = \frac{q_i}{4\pi r_i^2}\hat{\boldsymbol{r}}_i \; (i = 1, 2, 3, \ldots)$$

and

$$\boldsymbol{E} = \boldsymbol{E}_1 + \boldsymbol{E}_2 + \boldsymbol{E}_3 + \ldots \text{ where } \boldsymbol{E}_i = \frac{q_i}{4\pi\varepsilon r_i^2}\hat{\boldsymbol{r}}_i$$

that is,

$$\boldsymbol{D} = \sum_i \frac{q_i}{4\pi r_i^2}\hat{\boldsymbol{r}}_i \quad \text{and} \quad \boldsymbol{E} = \sum_i \frac{q_i}{4\pi\varepsilon r_i^2}\hat{\boldsymbol{r}}_i \qquad (16.7)$$

Readers may have noticed that, although we have derived equations (16.4)–(16.6) and subsequent relationships above from first principles, we did not strictly have to do so. The treatment involved is formally identical to the analysis of similar situations in the case of gravitation (Chapter 5). All relationships derived for point masses, with no loss of rigour, can be converted to the corresponding relationships for point charges by using the following transformations

$$m_i \rightarrow -q_i; \quad G \rightarrow \frac{1}{4\pi\varepsilon} \quad \text{and} \quad \boldsymbol{g}(\boldsymbol{r}) \rightarrow \boldsymbol{E}(\boldsymbol{r})$$

Note, however, that charge can take both positive and negative values while mass can only be positive.

Continuous charge distribution

The same idea can be used for a continuous distribution of charge where the charge can be broken up into a very large number of infinitesimal elements, such as the element Δq of volume ΔV illustrated in Figure 16.16(a). The electric **charge density** at a point $\boldsymbol{r}$ is defined as

$$\rho := \lim_{\Delta V \to 0} \frac{\Delta q}{\Delta V} := \frac{dq}{dV}$$

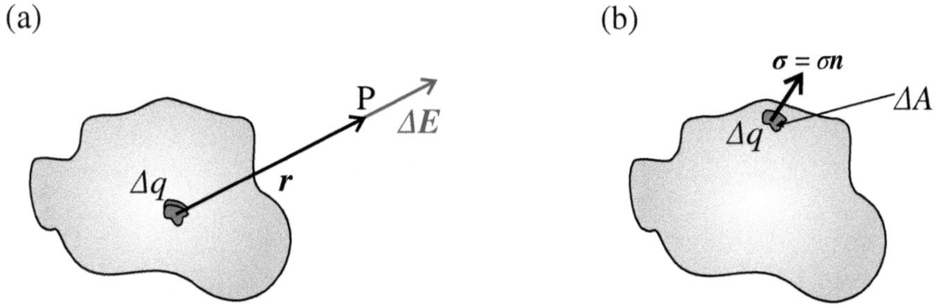

Figure 16.16. (a) The net electric field strength at P due to a continuous distribution of charge is the integral of the individual contributions from each volume element of charge. (b) The definition of surface charge density.

In the limit, each charge element can be considered to be a point charge and hence the contribution to the flux density at a point P from the element Δq is given, from Equation (16.4), by $\Delta \boldsymbol{D}(\boldsymbol{r}) = \frac{\Delta q}{4\pi r^2}\hat{\boldsymbol{r}}$ and the total electric flux density at P is then given by

$$\boldsymbol{D}(\boldsymbol{r}) = \int \frac{dq}{4\pi r^2}\hat{\boldsymbol{r}} = \frac{1}{4\pi}\iiint \frac{\rho(\boldsymbol{r})}{r^2}\hat{\boldsymbol{r}}dV \tag{16.8}$$

and

$$\boldsymbol{E}(\boldsymbol{r}) = \int \frac{dq}{4\pi\varepsilon r^2}\hat{\boldsymbol{r}} = \frac{1}{4\pi\varepsilon}\iiint \frac{\rho(\boldsymbol{r})}{r^2}\hat{\boldsymbol{r}}dV \tag{16.8a}$$

where the integral is over the whole volume containing charge (note that this is the limit of a *vector* summation).

Note also that $\rho(\boldsymbol{r})$ here is equivalent to the mobile charge density defined in Section 15.2 although, in this case, it refers to the distribution of net static charge.

In the context of continuous distributions of charge it is also useful to define an electric *surface charge density* at a point on the surface of a body. This quantity is defined as the *charge per unit area* at that point, that is

$$\sigma := \frac{dq}{dA}\boldsymbol{n}$$

where $\boldsymbol{n}$ is the unit vector normal to the surface at the point (Figure 16.16(b)).[2]

Study Worked Example 16.1

For problems based on the material presented in this section visit up.ucc.ie/16/ *and follow the link to the problems.*

16.6 The electric dipole

An electric dipole consists of two point charges of equal and opposite sign which are separated by some fixed distance (for example, $2l$ in Figure 16.17). The electric flux density at some arbitrary point P in an infinite non-conducting medium is given by

$$\boldsymbol{D}(\boldsymbol{r}) = \frac{(+q)}{4\pi r_+^2}\hat{\boldsymbol{r}}_+ + \frac{(-q)}{4\pi r_-^2}\hat{\boldsymbol{r}}_- = \boldsymbol{D}_+ + \boldsymbol{D}_- \tag{16.9}$$

where, as shown in Figure 16.17, $\hat{\boldsymbol{r}}_+$ and $\hat{\boldsymbol{r}}_-$ are the unit displacement vectors from the $+q$ and $-q$ charges, respectively, to P. The electric field strength at P can then be determined by dividing $\boldsymbol{D}$ by ε of the medium.

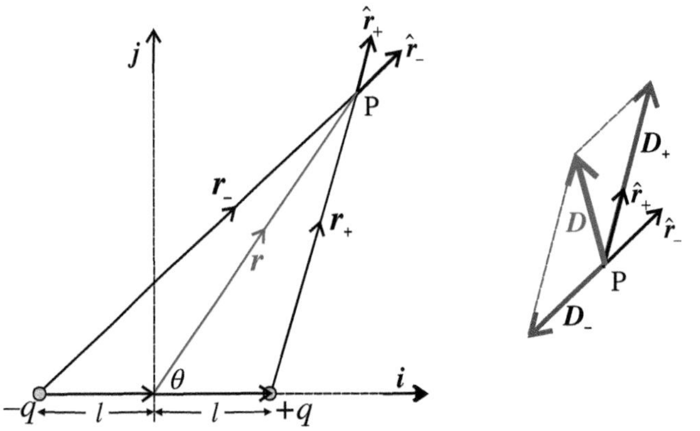

Figure 16.17. Determination of an expression for the electric field strength at P due to an electric dipole. The unit vector $\boldsymbol{i}$ is taken along the dipole axis.

[2]Note that the symbol σ used for surface charge density here is the same as the symbol used for conductivity in Section 15.14 although confusion between the two is unlikely. Care needs to be taken, however, not to confuse surface charge density with (volume) charge density ρ defined above.

We now derive an expressions for the electric flux density and field strength at a point at a distance r from the centre of a dipole of length $2l$ for the special case in which $r \gg l$. Using the cosine rule (Appendix A.3.3) we obtain

$$r_+{}^2 = r^2 + l^2 - 2rl\cos\theta = r^2\left(1 - 2\frac{l}{r}\cos\theta + \frac{l^2}{r^2}\right) \approx r^2\left(1 - 2\frac{l}{r}\cos\theta\right)$$

$$\rightarrow \quad r_+ \approx r\sqrt{1 - 2\frac{l}{r}\cos\theta}$$

and

$$r_-{}^2 = r^2 + l^2 + 2rl\cos\theta = r^2\left(1 + 2\frac{l}{r}\cos\theta + \frac{l^2}{r^2}\right) \approx r^2\left(1 + 2\frac{l}{r}\cos\theta\right)$$

$$\rightarrow \quad r_- \approx r\sqrt{1 + 2\frac{l}{r}\cos\theta}$$

From the rule for vector addition we can write $\boldsymbol{r}_+ = \boldsymbol{r} - l\boldsymbol{i}$ and $\boldsymbol{r}_- = \boldsymbol{r} + l\boldsymbol{i}$ from which we can determine expressions for $\hat{\boldsymbol{r}}_+ = \frac{\boldsymbol{r}_+}{|\boldsymbol{r}_+|}$ and $\hat{\boldsymbol{r}}_- = \frac{\boldsymbol{r}_-}{|\boldsymbol{r}_-|}$ as follows.

$$\hat{\boldsymbol{r}}_+ = \frac{\boldsymbol{r} - l\boldsymbol{i}}{r\sqrt{1 - 2\frac{l}{r}\cos\theta}} = \frac{\hat{\boldsymbol{r}} - \frac{l}{r}\boldsymbol{i}}{\sqrt{1 - 2\frac{l}{r}\cos\theta}} \quad \text{and} \quad \hat{\boldsymbol{r}}_- = \frac{\hat{\boldsymbol{r}} + \frac{l}{r}\boldsymbol{i}}{\sqrt{1 + 2\frac{l}{r}\cos\theta}}$$

Substituting these results in Equation (16.9) we obtain

$$\boldsymbol{D}(\boldsymbol{r}) = \frac{q}{4\pi r^2}\frac{\hat{\boldsymbol{r}} - \frac{l}{r}\boldsymbol{i}}{\left(1 - 2\frac{l}{r}\cos\theta\right)^{\frac{3}{2}}} - \frac{q}{4\pi r^2}\frac{\hat{\boldsymbol{r}} + \frac{l}{r}\boldsymbol{i}}{\left(1 + 2\frac{l}{r}\cos\theta\right)^{\frac{3}{2}}}$$

Expanding each of the terms in brackets in the denominators in a binomial series (Appendix A.9.2) and neglecting terms of order $\frac{l^2}{r^2}, \frac{l^3}{r^3}$, etc. which are small compared to $\frac{l}{r}$ we get

$$\left(1 - 2\frac{l}{r}\cos\theta\right)^{-\frac{3}{2}} = 1 + \left(-\frac{3}{2}\right)\left(-2\frac{l}{r}\cos\theta\right) + \frac{1}{2!}\left(-\frac{3}{2}\right)\left(-\frac{5}{2}\right)\left(-2\frac{l}{r}\cos\theta\right)^2 \approx 1 + \frac{3l}{r}\cos\theta$$

and

$$\left(1 + 2\frac{l}{r}\cos\theta\right)^{-\frac{3}{2}} \approx 1 - \frac{3l}{r}\cos\theta$$

$$\begin{aligned}
\rightarrow \quad \boldsymbol{D}(\boldsymbol{r}) &= \frac{q}{4\pi r^2}\left[\left(1 + \frac{3l}{r}\cos\theta\right)\left(\hat{\boldsymbol{r}} - \frac{l}{r}\boldsymbol{i}\right) - \left(1 - \frac{3l}{r}\cos\theta\right)\left(\hat{\boldsymbol{r}} + \frac{l}{r}\boldsymbol{i}\right)\right] \\
&= \frac{q}{4\pi r^2}\left[\left(\frac{3l}{r}\cos\theta\hat{\boldsymbol{r}} - \frac{l}{r}\boldsymbol{i}\right) - \left(-\frac{3l}{r}\cos\theta\hat{\boldsymbol{r}} + \frac{l}{r}\boldsymbol{i}\right)\right] \\
&= \frac{q}{4\pi r^3}(6l\cos\theta\hat{\boldsymbol{r}} - 2l\boldsymbol{i}) = \frac{2ql}{4\pi r^3}(3\cos\theta\hat{\boldsymbol{r}} - \boldsymbol{i})
\end{aligned}$$

Finally, since $\hat{\boldsymbol{r}} = \cos\theta\boldsymbol{i} + \sin\theta\boldsymbol{j}$, we get $\quad (3\cos\theta\hat{\boldsymbol{r}} - \boldsymbol{i}) = (3\cos^2\theta - 1)\boldsymbol{i} + 3\sin\theta\cos\theta\boldsymbol{j}$

and hence

$$\boldsymbol{D}(\boldsymbol{r}) = \frac{2ql}{4\pi r^3}((3\cos^2\theta - 1)\boldsymbol{i} + 3\sin\theta\cos\theta\boldsymbol{j}) \tag{16.10}$$

$$\rightarrow \quad \boldsymbol{E}(\boldsymbol{r}) = \frac{2ql}{4\pi\varepsilon r^3}((3\cos^2\theta - 1)\boldsymbol{i} + 3\sin\theta\cos\theta\boldsymbol{j})$$

or

$$\boldsymbol{E}(\boldsymbol{r}) = \frac{|\boldsymbol{p}|}{4\pi\varepsilon r^3}((3\cos^2\theta - 1)\boldsymbol{i} + 3\sin\theta\cos\theta\boldsymbol{j}) \tag{16.11}$$

where the quantity $\boldsymbol{p}$ is the **electric dipole moment** of the dipole which is defined as

$$\boxed{\boldsymbol{p} := 2ql\boldsymbol{n}}$$

where $\boldsymbol{n}$ is a unit vector directed along the axis of the dipole from negative to positive (for example, the $\boldsymbol{i}$ direction in Figure 16.17).

Torque on an electric dipole in an electric field

Consider an electric dipole in a uniform electric field of strength $\boldsymbol{E}$, as shown in Figure 16.18. The dipole experiences the effect of equal and opposite forces of magnitude qE acting on each pole; that is, a torque (Section 7.3) of $(qE)x = qE(2l\sin\theta)$, where θ is the angle between the magnetic axis and the direction of the field. This can be written in vector form as

$$\boldsymbol{T} = 2ql\boldsymbol{n} \times \boldsymbol{E} = \boldsymbol{p} \times \boldsymbol{E} \tag{16.12}$$

where $\boldsymbol{p}$ is the electric dipole moment as defined above.

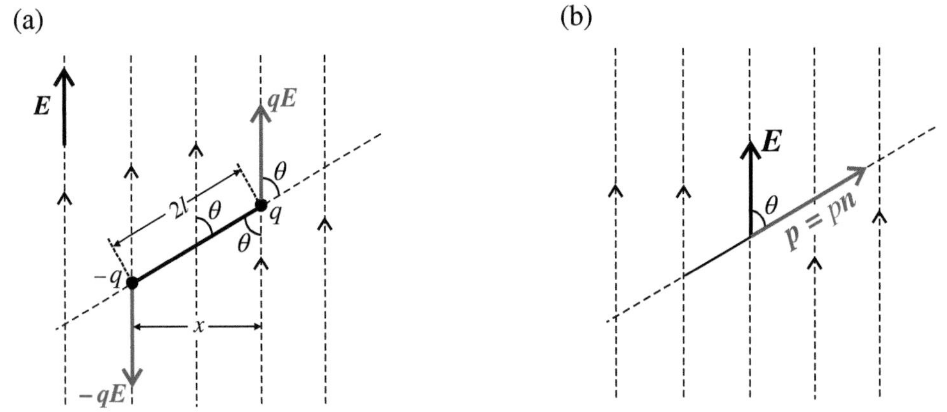

Figure 16.18. Torque on an electric dipole in a uniform electric field.

Potential energy of an electric dipole in an electric field

Finally in this section, we derive an expression for the potential energy of an electric dipole in an electric field which we assume is uniform over the dimensions of the dipole. The mechanical work done by a torque $\boldsymbol{T}$ in rotating a body through an angle θ is the scalar product $\Delta W = \boldsymbol{T} \cdot \Delta\boldsymbol{\theta}$ (Section 7.10), where $\Delta\boldsymbol{\theta}$ is a vector quantity as defined in Section 4.11. Thus, the total work done in a finite rotation between angles θ_1 and θ_2 is given by

$$W = \int_{\theta_1}^{\theta_2} \boldsymbol{T} \cdot d\boldsymbol{\vartheta}$$

where ϑ represents the angle between the direction of the electric dipole moment $\boldsymbol{p}$ and the electric field strength $\boldsymbol{E}$ at some point in the rotation,

The potential energy of the dipole when it has been turned through an angle θ from the position when it is aligned along the field direction ($\vartheta = 0$), against a restoring torque exerted by the field, is given by

$$U(\theta) = \int_0^{\theta} \boldsymbol{T} \cdot d\boldsymbol{\vartheta} = \int_0^{\theta} (\boldsymbol{p} \times \boldsymbol{E}) \cdot d\boldsymbol{\vartheta} = \int_0^{\theta} pE\sin\vartheta d\vartheta = -pE[\cos\vartheta]_0^{\theta} = pE(1 - \cos\theta)$$

Note that, when the angle θ is small, $\cos\theta \approx 1 - \dfrac{\theta^2}{2}$ and $U(\theta) = \dfrac{1}{2}pE\theta^2$.

As noted in Section 3.7, an arbitrary constant may always be added (or subtracted) to define the zero of potential energy. In this case the zero of potential energy may be fixed at $\vartheta = \dfrac{\pi}{2}$ (when the dipole moment is aligned perpendicularly to the field) by subtracting pE from the equation for the potential energy which becomes

$$U(\theta) = pE(1 - \cos\theta) - pE = -pE\cos\theta$$

so that the potential energy may be written

$$\boxed{U(\theta) = -\boldsymbol{p} \cdot \boldsymbol{E}} \tag{16.13}$$

Study Worked Examples 16.2

For problems based on the material presented in this section visit up.ucc.ie/16/ *and follow the link to the problems.*

16.7 Gauss's law for electrostatics

We saw in Section 5.4 that Newton's law of gravitation could be stated equivalently in the form of Gauss's law. The same argument shows, that in the case of electrostatics, an entirely equivalent statement of Coulomb's law is the following:

> The electric flux through any closed surface in an electric field is equal to the algebraic sum of the electric charges within the surface

This statement is known as **Gauss's Law** (for electrostatics) and, as in the corresponding case of gravitation (Section 5.4), the law maybe stated using surface integral notation (with reference to Figure 16.19) as follows:

$$\oiint_{\substack{\text{closed}\\\text{surface S}}} \boldsymbol{D}\cdot d\boldsymbol{A} = \sum_{\text{inside S}} q_i \tag{16.14}$$

where we have used the notation $\oiint_S$ to indicate that the integration is over a *closed surface* S. Note that we have denoted area here as a vector quantity, that is $\Delta\boldsymbol{A} := (\Delta A)\boldsymbol{n}$ where ΔA and $\boldsymbol{n}$ are as shown in Figure 16.19. Note also that q_i can represent either a positive or a negative quantity.

In the case of a continuous distribution of charge, Gauss's law can be written as

$$\oiint_S \boldsymbol{D}\cdot d\boldsymbol{A} = \iiint_V \rho(\boldsymbol{r})dV \tag{16.15}$$

where $\rho(\boldsymbol{r})$ is the charge density at a point $\boldsymbol{r}$ and the integral on the right hand side is over the volume enclosed by the surface S.

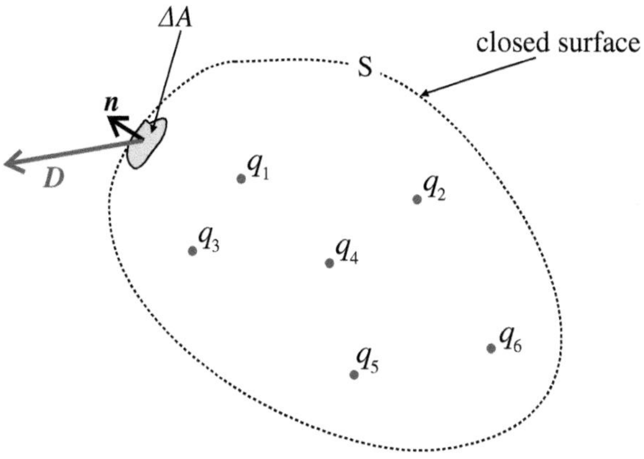

Figure 16.19. Gauss's law: the electric flux $\boldsymbol{D}$ through the closed surface is equal to the total charge enclosed within the surface.

Relation to Coulomb's law

For the case of a single point charge Q and taking the surface S to be a sphere of radius r, Equation (16.15) becomes

$$\oiint_{\text{sphere}} \boldsymbol{D}\cdot d\boldsymbol{A} = |\boldsymbol{D}|(4\pi r^2) = Q$$

which leads to Equation (16.4), $\boldsymbol{D}(\boldsymbol{r}) = \dfrac{\Phi_E}{4\pi r^2}\hat{\boldsymbol{r}}$, from which Coulomb's law was derived. Thus, Coulomb's law (Equation (16.6)) and Gauss's law (Equation (16.14)) are seen to be equivalent statements of the same basic law.

16.8 Applications of Gauss's law

Again, as was the case in our study of gravitation in Section 5.4, Gauss's Law proves to be a valuable tool in determining electric field strengths due to charge distributions which have a high degree of symmetry. For example, Gauss's law may be applied to derive expressions for the electric field strength at any point due to spherical distributions of electric charge. The derivations are formally identical to those utilised for the corresponding situations in the case of gravitation in Section 5.5.

(a) ***Uniformly charged spherical shell***

Consider a charge Q_s distributed uniformly in a thin spherical shell of radius R in a medium of permittivity ε (Figure 16.20). We distinguish two cases (a) a point located outside the shell ($r > R$) and (b) a point located inside the shell ($r < R$).

From the result derived for a spherical mass shell in Section 5.5 and using the appropriate transformations indicated in Section 16.5, $\left[\text{namely } (m_i \to -q_i; G \to \dfrac{1}{4\pi\varepsilon}; \boldsymbol{g}(\boldsymbol{r}) \to \boldsymbol{E}(\boldsymbol{r})\right]$, where ε is the permittivity of the (infinite) medium in which the sphere is embedded we can write down directly the expressions for the electric field strength in the two cases,

At a point outside the shell: $$\boldsymbol{E}(\boldsymbol{r}) = \frac{Q_s}{4\pi\varepsilon r^2}\hat{\boldsymbol{r}}$$

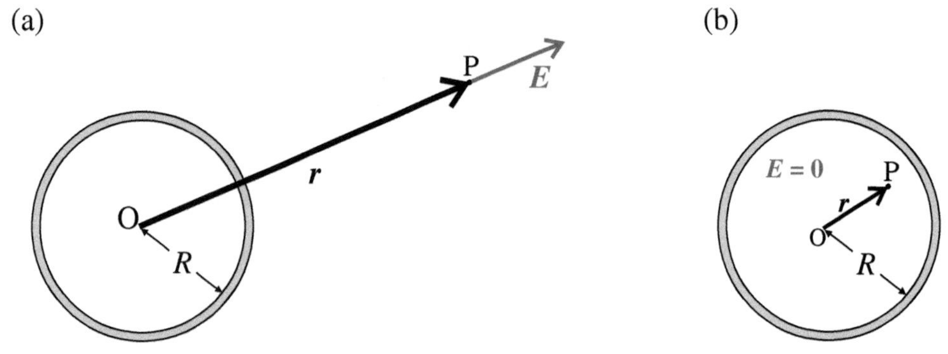

Figure 16.20. A uniformly charged spherical shell. (a) point outside shell (b) point inside shell.

which is the same as if all the charge was concentrated at the centre of the shell, O.

At a point inside the shell: $$\boldsymbol{E}(\boldsymbol{r}) = 0$$

showing that there is no electric field at any point inside a closed charged shell.

(b) *Uniform spherical charge distribution*

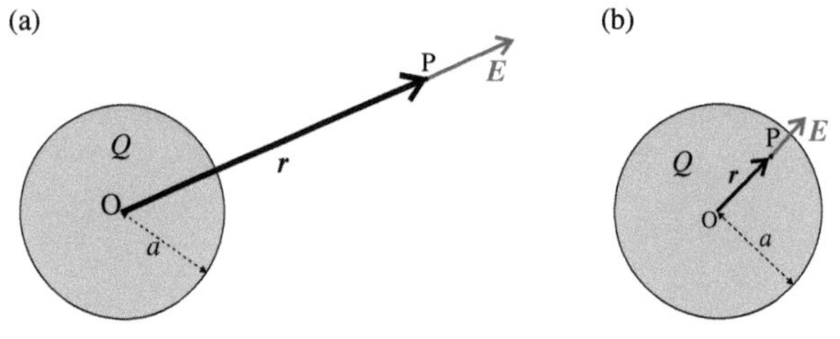

Figure 16.21. A uniformly spherical distribution of charge. (a) point outside shell (b) point inside shell.

Let us consider next how the electric field strength due to a charged sphere, in which the charge is uniformly distributed throughout its volume (Figure 16.21), varies with distance from the centre of the sphere. Let Q be the total charge which is distributed uniformly throughout the sphere of radius a; that is, the charge density is $\dfrac{Q}{\frac{4}{3}\pi a^3}$ everywhere inside the sphere. Again, by invoking the results already determined for gravitation in Section 5.5, we can write the expressions for the electric field strength in the two different cases, namely,

$$\boldsymbol{E}(\boldsymbol{r}) = \begin{cases} \dfrac{Qr}{4\pi\varepsilon a^3}\hat{\boldsymbol{r}} = \dfrac{Q}{4\pi\varepsilon a^3}\boldsymbol{r} & \text{when} \quad 0 < r < a \\ \dfrac{Q}{4\pi\varepsilon r^2}\hat{\boldsymbol{r}} & \text{when} \quad r > a \end{cases} \tag{16.16}$$

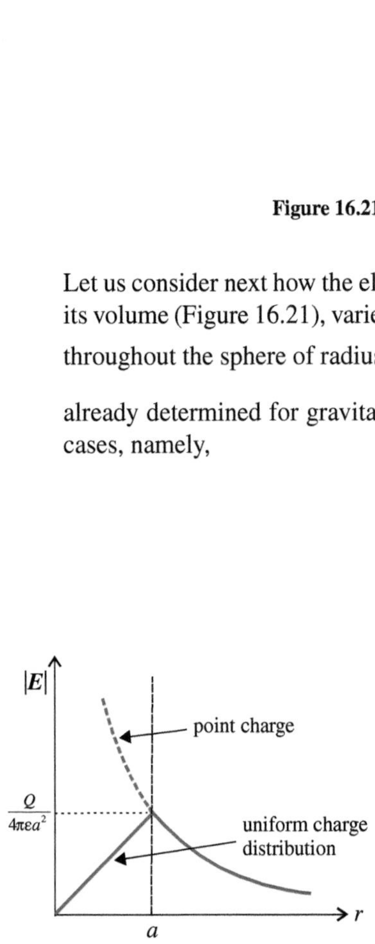

Figure 16.22. The electric field strength due to a uniform spherical distribution of charge plotted against the distance from the centre of the sphere. Note that, for $r \geq a$, the electric field strength is the same as that for a point charge Q at the centre of the sphere.

where ε is the permittivity of the (infinite) medium in which the sphere is embedded. Note that both cases yield the same value of $\boldsymbol{E}$ on the surface of the sphere ($r = a$), that is $\boldsymbol{E}$ is continuous at the surface of the sphere — Figure 16.22, analogous to Figure 5.14 for the gravitational case, shows how $|E|$ varies with r.

The result derived for the case $r > a$ shows that the electric field strength *outside* the sphere is the same as if all the charge had been located at a point at the centre of the sphere. Note that the derivation does not require that the charge density be uniform but only that it be spherically symmetric.

For heavy nuclei the nuclear charge $Q = Ze$ can be considered to be spherically (in fact almost uniformly) distributed throughout the nucleus and thus the result derived above has important applications in nuclear physics. To get an idea of the order of magnitude of the electric fields involved in this case let us estimate the magnitude of the electric field strength at the surface of a gold nucleus (for which the atomic number is 79 and the nuclear radius is 6.2 fm). In this case

$$|\boldsymbol{E}| = \frac{Ze}{4\pi\varepsilon_0 a^2} = \frac{79 \times (1.6 \times 10^{-19}\,\text{C})}{4\pi(8.85 \times 10^{-12}\,\text{C V}^{-1}\text{m}^{-1})(6.2 \times 10^{-15}\,\text{m})^2} \approx 3 \times 10^{21}\,\text{V m}^{-1}$$

Note that this is more than fifteen orders of magnitude greater than the largest electric field that can be created in the laboratory. We can also compare this field with that experienced by an electron in the ground state of the hydrogen atom, that is, at a distance of 53 pm (the 'Bohr radius', Section 24.4) from the nucleus, in which case

$$|\boldsymbol{E}| = \frac{e}{4\pi\varepsilon_0 a_0^2} = \frac{1.6\times10^{-19}\,\mathrm{C}}{4\pi(8.85\times10^{-12}\,\mathrm{C\,V^{-1}\,m^{-1}})(53\times10^{-12}\,\mathrm{m})^2} \approx 5\times10^{11}\,\mathrm{V\,m^{-1}}$$

(c) ***Electric field strength between two oppositely charged infinite planes***

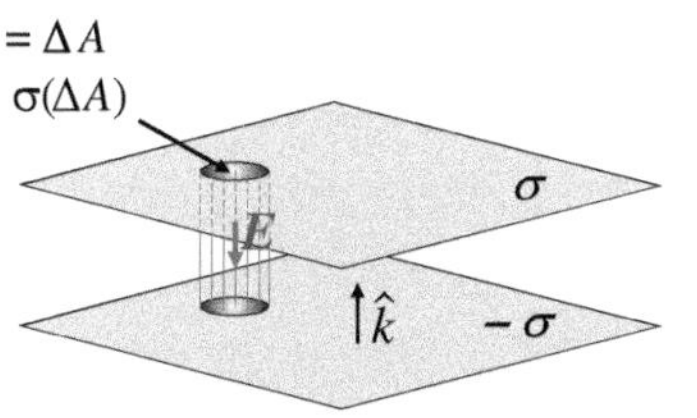

Figure 16.23. Determining the electric field strength between two oppositely charged infinite planes.

Consider the electric field in the space between the two charged planes shown in Figure 16.23. The upper plane is charged uniformly with a positive charge per unit area (surface charge density σ) while the lower plane has an equal but opposite negative charge. The space in between is filled with a medium of permittivity ε. From symmetry, if the extent of the planes is infinite, the field between the plates must be uniform.

The electric flux emanating from any area ΔA on the upper plane is $\sigma(\Delta A)$ and thus the magnitude of the electric flux density at any point between the planes is $|\boldsymbol{D}| = \sigma$ and, hence, the electric field strength at such a point is given by

$$\boldsymbol{E} = -\frac{\sigma}{\varepsilon}\hat{\boldsymbol{k}} \tag{16.17}$$

where $\hat{\boldsymbol{k}}$ is a unit vector as indicated in the figure.

(d) ***Electric field on the axis of a charged ring***

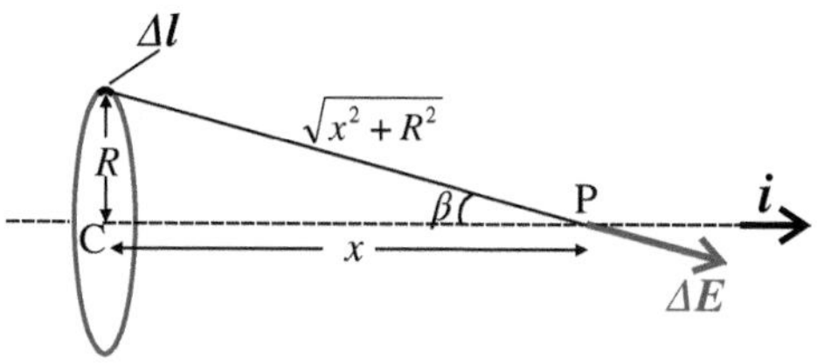

Figure 16.24. Determining the electric field strength at a point on the axis of a uniformly charged ring.

As another example, we consider the case of a charge Q which is distributed uniformly over a circular ring of radius R. We wish to determine an expression for the electric field strength at a point a distance x from the centre C of the ring, measured along the axis through C which is perpendicular to the plane of the ring.

Consider an element Δl of the ring as indicated in Figure 16.24.

The electric charge on Δl is $\dfrac{\Delta l}{2\pi R}Q = \dfrac{Q}{2\pi R}\Delta l$, and the contribution to the electric field strength at the point P is given by

$$\Delta E = \frac{\dfrac{Q}{2\pi R}\Delta l}{4\pi\varepsilon(x^2+R^2)}$$

where ε is the permittivity of the medium in which the ring is embedded.

The component of ΔE in the $+\boldsymbol{i}$-direction is

$$(\Delta E)\cos\beta = \frac{Q}{8\pi^2\varepsilon R}\frac{\Delta l}{(x^2+R^2)}\cos\beta = \frac{Q}{8\pi^2\varepsilon R}\frac{\Delta l}{(x^2+R^2)}\frac{x}{\sqrt{x^2+R^2}}.$$

On integrating over the ring, the components of ΔE perpendicular to the axis cancel and we obtain

$$E(x) = \frac{Q}{8\pi^2\varepsilon R}\frac{1}{(x^2+R^2)}\frac{x}{\sqrt{x^2+R^2}}\oint dl = \frac{Q}{8\pi^2\varepsilon R}\frac{x}{(x^2+R^2)^{3/2}}(2\pi R)$$

or

$$\boldsymbol{E}(x) = \frac{Q}{4\pi\varepsilon}\frac{x}{(x^2+R^2)^{3/2}}\boldsymbol{i} \tag{16.18}$$

(e) ***Electric field on the axis of a uniformly charged disc***

Figure 16.25 shows a circular disc of radius R which has an electric charge uniformly distributed over the disc with a charge per unit area σ. As illustrated in the figure, the charge distribution on the disc may be broken up into a series of concentric rings of charge. Consider first the contribution to the electric field strength at a point P on the axis from an annular ring of radius r and thickness Δr. The electric field strength at x due to a uniformly charged ring is given by Equation (16.18) above where, in this case, $Q \to \Delta Q = \sigma(2\pi r\Delta r)$. The components of the electric field parallel to the disc will cancel, so only the component perpendicular to the disc need be considered. Since all points on the ring are the same distance from the point P, we can use the result from the previous example to write the contribution to the electric field at P from the ring as

$$\Delta E(x) = \frac{2\pi\sigma\, r(\Delta r)}{4\pi\varepsilon}\frac{x}{(x^2 + r^2)^{3/2}}$$

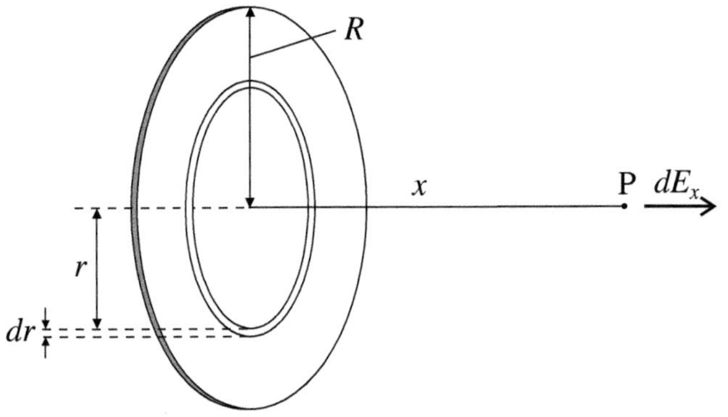

Figure 16.25. Determining the electric field strength at a point on the axis of a uniformly charged disc.

The total field at P due to all rings is obtained by integrating the latter expression from $r = 0$ to $r = R$, remembering that x is constant during the integration, Thus

$$E(x) = \int_0^R \frac{2\pi\sigma r dr}{4\pi\varepsilon}\frac{x}{(x^2 + r^2)^{3/2}} = \frac{\sigma x}{2\varepsilon}\int_0^R \frac{r dr}{(x^2 + r^2)^{3/2}}$$

The integral may be evaluated by substituting $\alpha = x^2 + r^2$ so that $d\alpha = 2rdr$. Thus

$$E(x) = \frac{\sigma x}{4\varepsilon}\int_{x^2}^{x^2+R^2} \frac{d\alpha}{\alpha^{3/2}} = \frac{\sigma x}{4\varepsilon}\left[-\frac{2}{\alpha^{1/2}}\right]_{x^2}^{x^2+R^2}$$

$$\rightarrow \quad E(x) = \frac{\sigma x}{2\varepsilon}\left[\frac{1}{x} - \frac{1}{\sqrt{x^2 + R^2}}\right] = \frac{\sigma}{2\varepsilon}\left[1 - \frac{x}{\sqrt{x^2 + R^2}}\right] \qquad (16.19)$$

or

$$E(x) = \frac{\sigma}{2\varepsilon}\left[1 - \frac{1}{\sqrt{1 + \dfrac{R^2}{x^2}}}\right]$$

Study Worked Eaxmple 16.3

For problems based on the material presented in this section visit up.ucc.ie/16/ *and follow the link to the problems.*

16.9 Potential difference in electric fields

Consider a small electric charge δq which is moved from a point A to the point B in an electric field along the path indicated in Figure 16.26. If the charge is moved in a quasistatic way (so that its acceleration is negligibly small), the force which is applied at any point P must balance the field force $(\delta q)\boldsymbol{E}$ at that point. The mechanical work done in moving the charge δq between these two points is given by

$$\Delta W_{AB} = \int_A^B \boldsymbol{F}\cdot d\boldsymbol{s} = \int_A^B (\delta q)\boldsymbol{E}\cdot d\boldsymbol{s} = -\int_B^A (\delta q)\boldsymbol{E}\cdot d\boldsymbol{s} \qquad (16.20)$$

If the time taken to move the charge in this manner is δt, the average power generated,

$$\langle P\rangle = \frac{\Delta W}{\delta t} = -\int_B^A \frac{\delta q}{\delta t}\boldsymbol{E}\cdot d\boldsymbol{s}$$

If the charge is moved at a constant speed throughout the path, $\frac{\delta q}{\delta t} = I = constant$, so we can write

$$\langle P \rangle = -I \int_{B}^{A} \boldsymbol{E} \cdot d\boldsymbol{s}$$

or

$$\frac{\langle P \rangle}{I} = -\int_{B}^{A} \boldsymbol{E} \cdot d\boldsymbol{s} \tag{16.21}$$

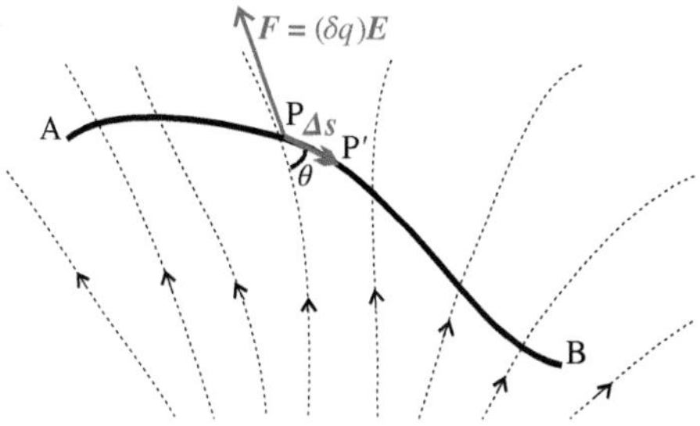

Figure 16.26. A small electric charge δq is moved from A to B in an electric field.

In Section 15.7, the power generated per unit current between two points in an electric circuit was defined as the potential difference between the two points, that is

$$V := \frac{P}{I}$$

and hence we can identify the right-hand side of Equation (16.21) as the potential difference between the points A and B in an electric field $\boldsymbol{E}$. By *defining* the **potential difference** between the points in this way, that is

$$\boxed{V := -\int_{B}^{A} \boldsymbol{E} \cdot d\boldsymbol{s}} \tag{16.22}$$

we have *generalised* the definition given in Section 15.7 to include cases in which the moving charge is not constrained to move along a particular fixed path between the two points (as it is, for example, along a wire in the case of electric circuits).

If a charge q is moved from A to B we find, from Equation (16.20), that the corresponding work done is given by

$$W_{AB} = -q \int_{B}^{A} \boldsymbol{E} \cdot d\boldsymbol{s}$$

from which we see that an alternative definition of the potential difference between the points B and A is

$$V := \frac{W_{AB}}{q} \tag{16.23}$$

In other words, the potential difference between any two points in an electric field is the *work done per unit charge* in moving a charge between the two points. Note that W_{AB}, and hence V, is an algebraic quantity which can be either positive or negative depending on whether the work done is positive or negative.

The above generalisation also implies that an alternative definition of the emf of a circuit (Section 15.5), expressed in terms of the electric field $\boldsymbol{E}$ at each point in the circuit, is

$$\boxed{\mathcal{E} := \oint_{\text{circuit}} \boldsymbol{E} \cdot d\boldsymbol{l}}$$

Study Worked Examples 16.4

For problems based on the material presented in this section visit up.ucc.ie/16/ and follow the link to the problems.

16.10 Electric potential

Because of the absence of dissipative forces (that is, in a conservative field), the electric field is conservative and hence the mechanical work done W_{AB} in moving a charge from the point A to the point B in Figure 16.27 is independent of the path taken from A to B. The potential difference between the points A and B in an electric field depends only on the position co-ordinates of A and B. Thus, in this case, the potential difference between B and A is also independent of the path over which the integral in definition (16.22) is taken and depends only on the end points. That is

$$V := -\int_{B}^{A} \boldsymbol{E} \cdot d\boldsymbol{s} = V(\boldsymbol{r}_B) - V(\boldsymbol{r}_A)$$

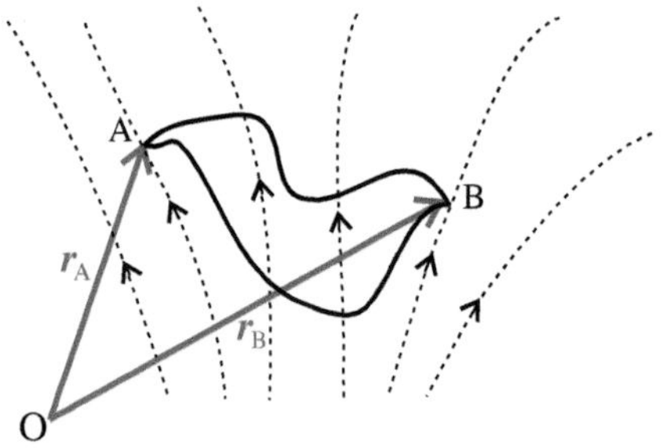

Figure 16.27. The potential difference between the points A and B in an electric field depends only on the co-ordinates of A and B.

where $V(\boldsymbol{r}_B)$ and $V(\boldsymbol{r}_A)$ are the values of the integral evaluated at $\boldsymbol{r} = \boldsymbol{r}_B$ and $\boldsymbol{r} = \boldsymbol{r}_A$, respectively (see Figure 16.27). Indeed, if the fields involved were not conservative the concept of potential difference would not be very useful. Note that, for a closed path in a conservative field,

$$\oint_C \boldsymbol{E} \cdot d\boldsymbol{s} = 0 \tag{16.24}$$

where, as before, the notation $\oint_C$ indicates a line integral over a **closed path** C.

In dealing with electric currents, the question of the independence of the path does not arise in the case of single loop circuits because only one path is possible. In the case of multiloop circuits, the condition that the current divides so as to give minimum power (Worked Example 15.5) is equivalent to the condition in electrostatics that the field be conservative.

From the definition of potential difference, it follows that the difference in potential energy of a charge q located at two different points in an electric field is the product of the charge and the potential difference, that is

$$U(\boldsymbol{r}_B) - U(\boldsymbol{r}_A) = qV(\boldsymbol{r}_B) - qV(\boldsymbol{r}_A)$$

Potential difference, like potential energy, is defined in terms of the difference in the value of a function at two points and hence an arbitrary constant can always be added to the function without changing the potential difference. Thus, one is free in any situation to specify an arbitrary point as being the zero of potential. In other words, we can define the **electric potential** at a point in a field as follows[3]

$$\boxed{\varphi(\boldsymbol{r}) := -\int_{\boldsymbol{r}_0}^{\boldsymbol{r}} \boldsymbol{E} \cdot d\boldsymbol{s}}$$

where $\boldsymbol{r}_0$ is the displacement of the point which has been specified such that $\varphi(\boldsymbol{r}_0) = 0$. It is conventional, where possible, to choose a point where the electric field strength is known to be zero as being the point of zero potential, that is $\varphi = 0$ when $\boldsymbol{E} = 0$. It is often convenient to choose a point electrically connected to the Earth as the point of zero potential; such a point is said to be at *ground potential*.

Example 1: *Electric potential in a uniform field*

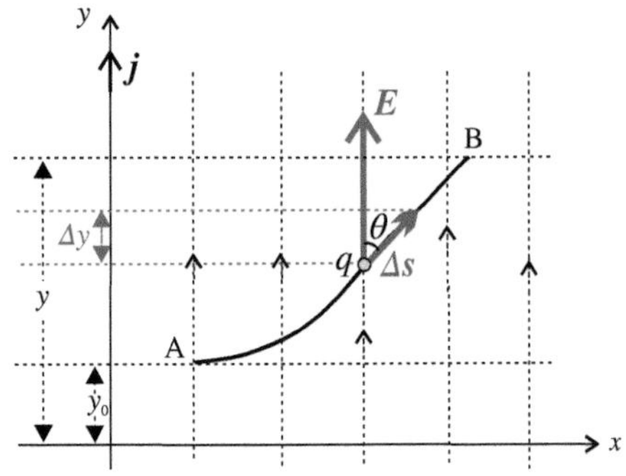

Figure 16.28. Determination of an expression for the potential difference between two points in a uniform electric field.

Consider the case of a uniform electric field, that is $\boldsymbol{E} = constant = \boldsymbol{E}_0$ (say) $= E_0\boldsymbol{j}$, as indicated in Figure 16.28. In this case

$$\varphi(\boldsymbol{r}) = -\int_A^B \boldsymbol{E} \cdot d\boldsymbol{s} = -E_0\int_A^B \boldsymbol{j} \cdot d\boldsymbol{s} = -E_0\int_A^B |\boldsymbol{j}| \cos\theta |d\boldsymbol{s}| = -E_0\int_{y_0}^{y} dy = -E_0(y - y_0)$$

and the potential energy of a charge q at position y in the uniform field is given by

$$U(\boldsymbol{r}) = U(y) = -qE_0(y - y_0)$$

This is a further example of the general potential energy function in a uniform field as discussed in Section 5.7 for the gravitational field. Comparison with Equation (5.14) shows, once more, the formal analogy between uniform gravitational and electrostatic fields $(mg \to -qE_0)$.

Example 2: *Electric potential due to a single point charge*

We now determine an expression for the electric potential at a distance r from a single point charge Q; that is, we wish to know the potential at an arbitrary point P, indicated in Figure 16.29, where the field lines have been drawn for the case when Q is a positive charge. The electric field strength is given by Equation (16.5), that is

$$\boldsymbol{E}(\boldsymbol{r}) = \frac{Q}{4\pi\varepsilon r^2}\hat{\boldsymbol{r}}$$

Now $\boldsymbol{E} = 0$ when $r = \infty$, so it is conventional to take points at $r = \infty$ as points of zero potential ($\varphi(\infty) = 0$).

[3]Note that the function φ is used to represent electric potential, reserving V for potential difference.

Thus $\varphi(\boldsymbol{r}) = -\int_{\infty}^{r} \boldsymbol{E}\cdot d\boldsymbol{s} = -\int_{\infty}^{r} \frac{Q}{4\pi\varepsilon\, r'^2}\hat{\boldsymbol{r}}'\cdot d\boldsymbol{s}$. Since the integral is independent of the path, we may choose any path to carry out the integration. Choosing the radial path indicated by the $\hat{\boldsymbol{r}}'$-direction in the figure so that $d\boldsymbol{s}$ becomes $d\boldsymbol{r}'$, we obtain

$$\varphi(\boldsymbol{r}) = -\int_{\infty}^{r} \frac{Q}{4\pi\varepsilon\, r'^2}\hat{\boldsymbol{r}}'\cdot d\boldsymbol{r}' = -\int_{\infty}^{r} \frac{Q}{4\pi\varepsilon\, r'^2}|\hat{\boldsymbol{r}}'||d\boldsymbol{r}'| = -\int_{\infty}^{r} \frac{Q}{4\pi\varepsilon}\frac{dr'}{r'^2} = -\frac{Q}{4\pi\varepsilon}\left[-\frac{1}{r'}\right]_{\infty}^{r}$$

and thus

$$\varphi(r) = \frac{Q}{4\pi\varepsilon\, r} \tag{16.25}$$

Once again, we note that this result could have been obtained directly from Equation (5.15) by replacing M by $-Q$ and G by $\frac{1}{4\pi\varepsilon}$.

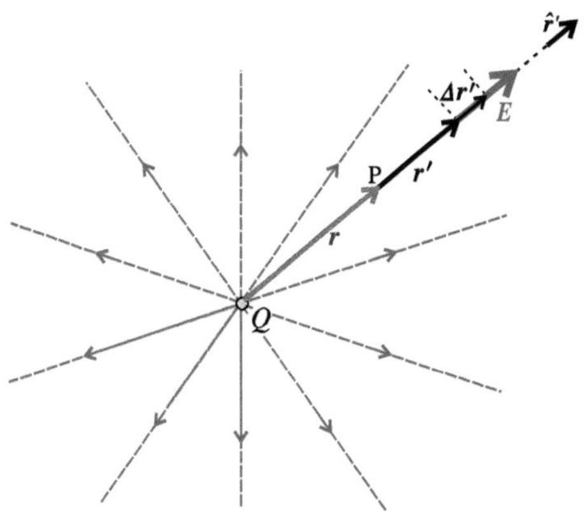

Figure 16.29. Determination of an expression for the electric potential at a point which is located at a displacement r from a point charge. Note that the direction of the field lines are indicated for the case where Q is positive; the arrows would be reversed if Q were negative.

Electric potential due to a system of point charges

Note that $\varphi(r)$ is a scalar quantity but is algebraic in the sense that $\varphi < 0$ if $Q < 0$. If we wish to determine the electric potential at a point resulting from a system of point charges (Figure 16.30) we simply determine the algebraic sum of the contributions from all point charges, that is

$$\varphi = \frac{q_1}{4\pi\varepsilon\, r_1} + \frac{q_2}{4\pi\varepsilon\, r_2} + \frac{q_3}{4\pi\varepsilon\, r_3} + \ldots$$

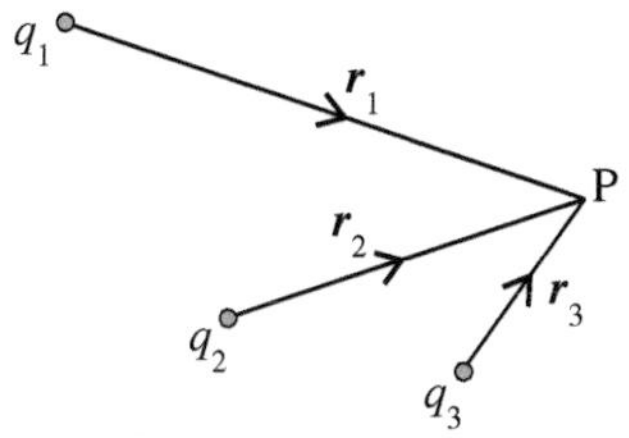

Figure 16.30. The electric potential at P due to a number of point charges is the *algebraic* sum of the individual contributions from each point charge.

Example 3: *Electric potential due to a uniform spherical charge distribution*

In Section 16.8 we determined the electric field strength at a distance r from the centre of a sphere in which the charge is distributed uniformly throughout the volume (uniform charge density). Outside the sphere ($r > a$) the system behaves as if all of its charge is concentrated at the centre; thus for $r > a$ the potential is the same as that due to a point charge, as determined in Example 2 above, given by Equation (16.25).

Inside the sphere ($0 < r < a$) the electric potential is given by

$$\varphi(\boldsymbol{r}) = -\int_{\infty}^{r} \boldsymbol{E}\cdot d\boldsymbol{s} = -\int_{\infty}^{a} \boldsymbol{E}\cdot d\boldsymbol{s} - \int_{a}^{r} \boldsymbol{E}\cdot d\boldsymbol{s}$$

Using the result (16.16) for the electric field strengths outside and inside the sphere and, for convenience, integrating along the x-axis (Figure 16.31) we obtain for $r < a$

$$\varphi(\boldsymbol{r}) = -\int_{\infty}^{a} \frac{Q}{4\pi\varepsilon x^2}dx - \int_{a}^{r} \frac{Q}{4\pi\varepsilon a^3}x dx = \frac{Q}{4\pi\varepsilon}\frac{1}{a} - \frac{Q}{4\pi\varepsilon a^3}\frac{r^2 - a^2}{2}$$

$$\rightarrow \quad \varphi(r) = \frac{Q}{4\pi\varepsilon}\left(\frac{3}{2a} - \frac{r^2}{2a^3}\right) \tag{16.26}$$

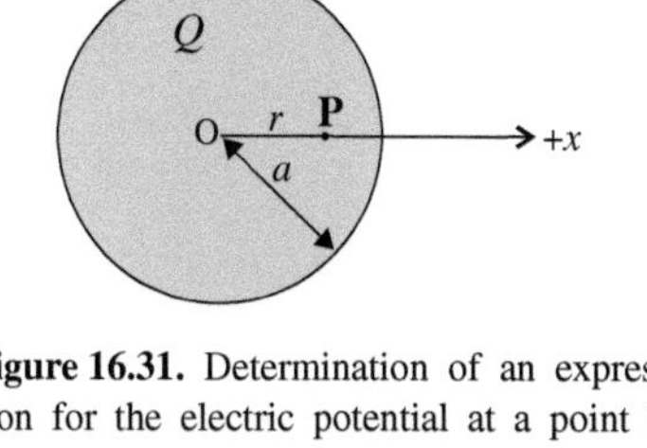

Figure 16.31. Determination of an expression for the electric potential at a point P located inside a uniform spherical distribution of charge.

Note that at $r = a$, $\varphi(a) = \frac{Q}{4\pi\varepsilon}\left(\frac{3}{2a} - \frac{1}{2a}\right) = \frac{Q}{4\pi\varepsilon a}$, as expected.

For problems based on the material presented in this section visit up.ucc.ie/16/ *and follow the link to the problems.*

16.11 Equipotential surfaces

An equipotential surface is defined as the locus of all points which have the same potential (φ = constant); that is, the potential difference between any two points on an equipotential surface is zero. There will be an infinite family of such non-intersecting surfaces for any static distribution of charge, each corresponding to a different fixed value of the electric potential. From the definition of potential difference, Equation (16.22), the potential difference between two points which are an infinitesimal displacement $\boldsymbol{\Delta s}$ apart is given by

$$\Delta V = \Delta\varphi = -\boldsymbol{E}\cdot\boldsymbol{\Delta s} \tag{16.27}$$

If $\boldsymbol{\Delta s}$ is an increment of path along an equipotential surface, then $\Delta\varphi = 0$ and hence $\boldsymbol{E}\cdot\boldsymbol{\Delta s} = 0$, that is $\boldsymbol{E}$ is perpendicular to $\boldsymbol{\Delta s}$ (as shown in Figure 16.32). Thus, equipotential surfaces are everywhere perpendicular to the electric field lines. Some examples of equipotential surfaces for simple systems are shown in Figure 16.33 where the intersections of the equipotential surfaces with the plane of the figure appear as equipotential curves.

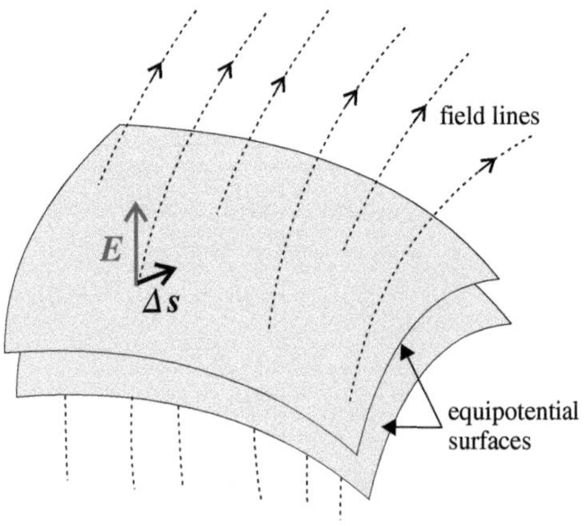

Figure 16.32. Two adjacent equipotential surfaces. Electric field lines are always perpendicular to equipotential surfaces.

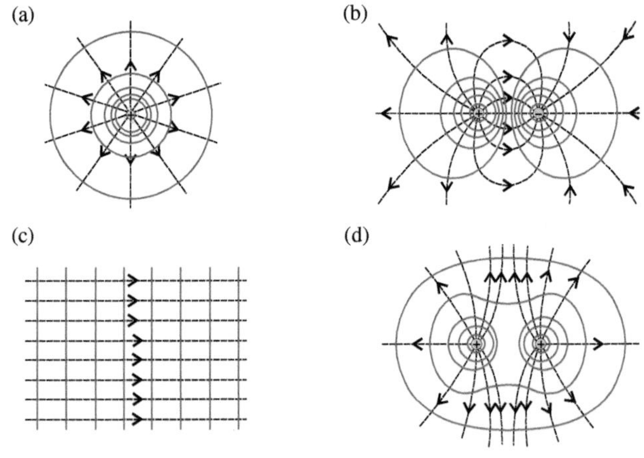

Figure 16.33. Electric field lines (black) and equipotential surfaces (blue) for a number of simple cases: (a) point charge, (b) two point charges of opposite signs, (c) uniform field and (d) two point positive charges.

Note that the surfaces of conductors must be equipotential surfaces since, as observed in Section 16.2, electric field lines are perpendicular to the surface at conductor surfaces. Indeed, since no work is done when a test charge is moved anywhere in a perfect conductor, all points in such a conductor must be at the same potential.

Electric field in a metal cavity

It follows that the electric field strength must be zero everywhere in an empty cavity within a conductor. Figure 16.34 shows such a cavity, for example a region totally enclosed in a metal surround. We know from Equation (16.24) that the integral of $\boldsymbol{E}\cdot\boldsymbol{ds}$ around any closed path must be zero, that is

$$\oint_C \boldsymbol{E}\cdot\boldsymbol{ds} = 0$$

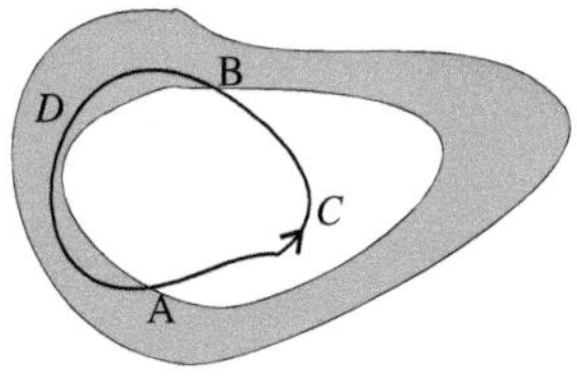

Figure 16.34. The electric field strength everywhere inside a cavity within a conductor is zero.

where C is the closed curve shown. Since the integral over that part of the path BDA passing through the conductor must be zero, then $\int_A^B \boldsymbol{E}\cdot\boldsymbol{ds} = 0$, where the integral is over the remainder of the path (AB), that is, through the cavity. This must hold for all possible paths AB from which it follows that $\boldsymbol{E} = 0$ everywhere in the cavity. Equipment that may be sensitive to external electric fields is often *screened* by surrounding it in a metal enclosure (known as a *Faraday cage*).

16.12 Determination of electric field strength from electric potential

From Equation (16.27) we can write

$$\Delta\varphi = -\boldsymbol{E}\cdot\boldsymbol{\Delta s} = -(E_x\boldsymbol{i} + E_y\boldsymbol{j} + E_z\boldsymbol{k})\cdot(\Delta x\boldsymbol{i} + \Delta y\boldsymbol{j} + \Delta z\boldsymbol{k}) = -\{E_x(\Delta x) + E_y(\Delta y) + E_z(\Delta z)\}$$

where $\boldsymbol{E}$ and $\boldsymbol{\Delta s}$ have been written in terms of their components in an arbitrary Cartesian x-, y-, and z-coordinate system.

Using partial derivative notation, as described in Appendix A.6.3,

that is $$E_x = -\frac{\partial\varphi}{\partial x},\ E_y = -\frac{\partial\varphi}{\partial y} \text{ and } E_z = -\frac{\partial\varphi}{\partial z},$$

we can write
$$E = -\frac{\partial\varphi}{\partial x}\boldsymbol{i} - \frac{\partial\varphi}{\partial y}\boldsymbol{j} - \frac{\partial\varphi}{\partial z}\boldsymbol{k} = -\nabla\varphi \tag{16.28}$$

where $\nabla\varphi$ (known as grad φ — Appendix A.12,1(i)) is the 3-dimensional gradient of the scalar function φ. $\nabla\varphi$ is given expressed in Cartesian co-ordinates in Appendix A.13.1.

It is often simpler to analyse problems in terms of potentials (which are scalar quantities) than in terms of field strengths (which are vector quantities). Electric potential in a conservative field may be evaluated at any point by evaluating $-\int \boldsymbol{E}\cdot d\boldsymbol{s}$ along any arbitrary path from infinity (point at zero potential) to that point. Electric potentials arising from different sources of field can be added algebraically. Differentiation using Equation (16.28) may then be used to recover the expression for electric field strength from the corresponding expression for the potential.

Note that if the potential is a function of only one variable, for example $\varphi = \varphi(x)$, Equation (16.28) simplifies to $E_x = -\frac{d\varphi}{dx}$, a one–dimensional gradient. When E_x is constant the field is uniform, producing the parallel field lines of Figure 16.33(c).

As an example of recovering an electric field vector from an electric potential function, we apply Equation (16.28) to the expression for the potential inside a uniform spherical charge distribution (Equation (16.26)). This was derived as Example 3 of Section 16.10, namely

$$\varphi(r) = \frac{Q}{4\pi\varepsilon}\left(\frac{3}{2a} - \frac{r^2}{2a^3}\right)$$

The spherical symmetry of the situation indicates that analysis will be simpler in spherical polar co-ordinates. Equation (16.28) can be written in spherical polar form using the equation for ∇ in spherical polar co-ordinates (Appendix A.13.3). The electric field strength is then
$$\boldsymbol{E} = -\nabla\varphi = -\frac{\partial\varphi}{\partial r}\hat{\boldsymbol{r}} - \frac{1}{r}\frac{\partial\varphi}{\partial\theta}\hat{\boldsymbol{\theta}} - \frac{1}{r\sin\theta}\frac{\partial\varphi}{\partial\phi}\hat{\boldsymbol{\phi}} \tag{16.29}$$

The potential function that we are now considering, $\varphi(r)$, is independent of θ and φ so that the application of Equation (16.29) to this function simplifies to
$$\boldsymbol{E}(\boldsymbol{r}) = -\frac{d\varphi(r)}{dr}\hat{\boldsymbol{r}} = -\frac{d}{dr}\left[\frac{Q}{4\pi\varepsilon}\left(\frac{3}{2a} - \frac{r^2}{2a^3}\right)\right]\hat{\boldsymbol{r}} = -\frac{Q}{4\pi\varepsilon}\left(\frac{-2r}{2a^3}\right)\hat{\boldsymbol{r}} = \frac{Q}{4\pi\varepsilon a^3}\boldsymbol{r},$$

in agreement with our earlier derivation of $\boldsymbol{E}(\boldsymbol{r})$, namely Equation (16.16) for $0 < r < a$. Note that, since $\varphi(r)$ depends only on r, there are no components of $\boldsymbol{E}$ perpendicular to $\boldsymbol{r}$.

Electric field strength and current density

In dealing with electric currents it should be noted that, unlike electrostatic situations, the electric field strength inside the conducting medium is not zero. Indeed an electric field is required to maintain the current.

In Section 15.14 we showed that the current density at a point in a conducting material is related to the gradient of the potential difference and hence potential by Equation (15.13), thus

$$\boldsymbol{J} = -\sigma\frac{\partial V}{\partial z}\boldsymbol{n} = -\sigma\frac{\partial\varphi}{\partial z}\boldsymbol{n}$$

where the z-direction is along $\boldsymbol{n}$, the direction of the flow of current, and σ is the conductivity of the material. We can now recognise $-\frac{\partial\varphi}{\partial z}$ as the electric field strength which gives rise to the current and hence can write

$$\boxed{\boldsymbol{J} = \sigma\boldsymbol{E}} \tag{16.30}$$

Equation (16.30) is equivalent to the relation $I = \frac{V}{R}$ and may be considered to be yet another form of 'Ohm's law'. Like Ohm's law, Equation (16.30) can be considered to be an equation of state for ohmic materials when σ is constant.

For problems based on the material presented in this section visit up.ucc.ie/16/ and follow the link to the problems.

16.13 Acceleration of charged particles

We know that at the microscopic level many particles (electrons, protons, ions, etc.) have electric charge. Suppose that a particle of mass m and of charge q is released from rest at some point X in an electric field, say very close to the positively charged conductor in Figure 16.35. Let the speed of the particle be v when it reaches the point Y near the grounded conductor and let V be the potential difference between the conductors.

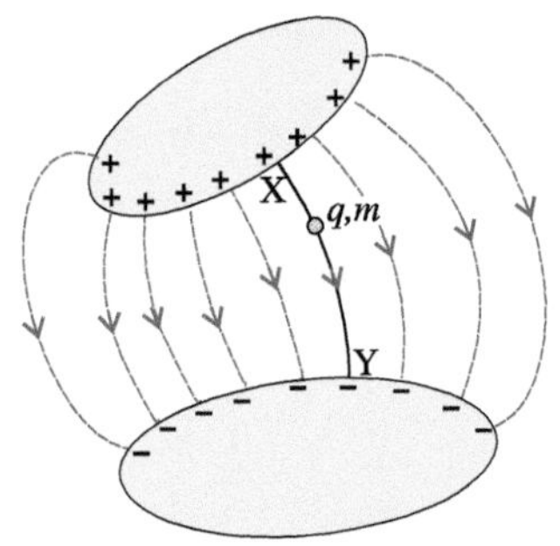

Figure 16.35. A positively charged particle (of charge q and mass m), starting from rest at X, accelerates as it moves along the field line from X to Y.

From Equation (16.23), the work done by the field force in accelerating the particle from X to Y is qV. Assuming that no other forces are involved, apart from the (conservative) electrostatic force, the gain in kinetic energy of the particle is equal to the work done by the field, that is

$$\frac{1}{2}mv^2 = qV \tag{16.31}$$

Hence the speed of the particle at Y is

$$v = \sqrt{\frac{2qV}{m}}$$

Let us estimate the speed of a proton which has been accelerated from rest across a potential difference of 10 kV. In this case

$$v = \sqrt{\frac{2(1.6 \times 10^{19}\ \text{C})(10^4\ \text{V})}{1.67 \times 10^{27}\ \text{kg}}} \approx 1.4 \times 10^6\ \text{m s}^{-1}$$

The electron volt

Equation (16.31) indicates that the product of charge and potential difference has the unit of energy. A (non-SI) unit, called the electron volt (eV), a convenient measure of energy on atomic and nuclear scales, is based on this result. The work done in moving one elementary charge e through a potential difference V is $(e)(V)$ and when $V = 1$ volt the work done is $(1.6 \times 10^{-19}\ \text{C})(1\ \text{V}) = 1.6 \times 10^{-19}$ J. Thus, we can use the following equation to convert energies from eV to joules.

$$1\ \text{eV} = 1.6 \times 10^{-19}\ \text{J}$$

Particle accelerators

Acceleration of charged particles across a potential difference is the basis of the so-called 'electron gun' which provides the narrow beam of rapidly moving electrons used in devices such as older types of television sets and visual display units. Similar acceleration of charged particles across very large potential differences is the basic principle underlying certain types of **particle accelerators** which have been used in the study of the fundamental interactions that operate at nuclear and sub-nuclear levels. The technique was used as a nuclear probe by John Cockroft (1897–1967) and Ernest Walton (1903–1995) in 1932. In their experiment, protons were accelerated across about half a million volts and were directed at a lithium target. Modern equivalents of the Cockroft-Walton accelerator use the same principle in a variety of ways; for more details see Martin and Shaw (2019)

For problems based on the material presented in this section visit up.ucc.ie/16/ *and follow the link to the problems.*

16.14 The laws of electrostatics in differential form

We have seen that the fundamental law of electrostatics is described by Coulomb's law or, equivalently, by Gauss's law. In Section 16.14 on the *Understanding Physics* website (up.ucc.ie/16/14/) it is shown that it follows that

$$\nabla \cdot \boldsymbol{D} = \rho$$

This is essentially yet another version of Gauss's law for electrostatics, but one which relates the derivative of the electric flux density $\boldsymbol{D}$ *at a point in an electric field* to the charge density ρ *at the same point.*

The fact that the electrostatic field is conservative gives rise to an empirical statement concerning the electric field strength *at a point* that can be expressed (up.ucc.ie/16/14/) in differential form, namely,

$$\nabla \times \boldsymbol{E} = 0$$

Both of these expressions relate the strength of the field at every point in an electrostatic field to the sources of that field indicating the value of writing these (and other laws of nature) in differential form.

Further reading

Further material on some of the topics covered in this chapter may be found in *Electromagnetism* by Grant and Phillips (See BIBLIOGRAPHY for book details)

WORKED EXAMPLES

Worked Example 16.1: Derive expressions for (i) the electric flux density and (ii) the electric field strength at the point P in Figure 16.36, giving magnitude and direction in each case. Calculate the magnitude of the electric field strength at P if $e = 1.6 \times 10^{-19}$ C and $a = 5.6$ nm and the charges are in vacuum.

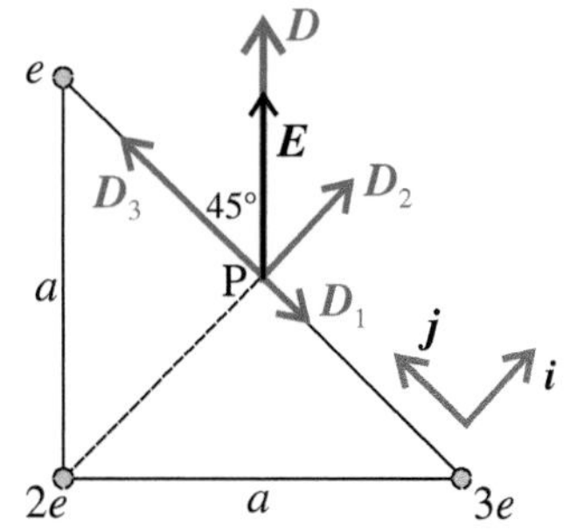

Figure 16.36. Worked Example 16.1.

The net electric flux density at P is the vector sum of the contributions from each charge, that is

$$\boldsymbol{D} = \boldsymbol{D}_1 + \boldsymbol{D}_2 + \boldsymbol{D}_3,$$

where $$|\boldsymbol{D}_1| = \frac{e}{4\pi(a/\sqrt{2})^2} = \frac{e}{2\pi a^2};\ |\boldsymbol{D}_2| = \frac{2e}{4\pi(a/\sqrt{2})^2} = \frac{e}{\pi a^2};\ |\boldsymbol{D}_3| = \frac{3e}{4\pi(a/\sqrt{2})^2} = \frac{3e}{2\pi a^2}$$

Thus $\boldsymbol{D} = -\frac{e}{2\pi a^2}\boldsymbol{j} + \frac{e}{\pi a^2}\boldsymbol{i} + \frac{3e}{2\pi a^2}\boldsymbol{j} = \frac{e}{\pi a^2}\boldsymbol{i} + \frac{e}{\pi a^2}\boldsymbol{j}$, where $\boldsymbol{i}$ and $\boldsymbol{j}$ are unit vectors directed as indicated in the diagram. Since the $\boldsymbol{i}$ and $\boldsymbol{j}$ components are equal, $\boldsymbol{D}$ must be directed vertically upwards. Hence $|\boldsymbol{D}| = \sqrt{2}\frac{e}{\pi a^2}$ and, since $\boldsymbol{D} = \varepsilon \boldsymbol{E}$, $|\boldsymbol{E}| = \frac{\sqrt{2}e}{\pi\varepsilon a^2}$.

Thus $$E_{\mathrm{P}} = \frac{\sqrt{2}(1.6 \times 10^{-19})}{\pi(8.85 \times 10^{-12})(5.6 \times 10^{-9})^2} = 2.6 \times 10^8\ \mathrm{V\ m^{-1}}$$

Worked Example 16.2: (a) Derive exact expressions for the electric flux density at the points A $(x, 0)$ and B $(0, y)$ due to the dipole illustrated in Figure 16.37. (b) Write down the corresponding expressions in the limiting cases $x \gg l$ and $y \gg l$. (c) Calculate the values of the electric field strengths at A and B for the case in which the dipole electric charge (q) is 8.0×10^{-19} C, the length of the dipole ($2l$) is 20 nm and $x = y = 30$ μm.

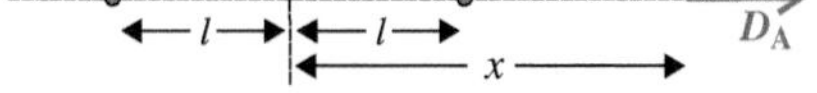

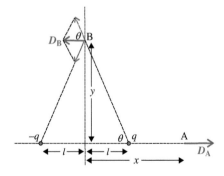

Figure 16.37. Worked Example 16.2.

(a) At the point A the electric flux density is in the $+x$-direction and is given by

$$D_{\mathrm{A}} = \frac{q}{4\pi(x-l)^2} - \frac{q}{4\pi(x+l)^2} = \frac{4qlx}{4\pi(x-l)^2(x+l)^2} = \frac{4qlx}{4\pi(x^2-l^2)^2}$$

At the point B, the components of the field in the y-direction are equal and opposite, so the net contribution to $\boldsymbol{D}_B$ is the sum of the x-components only and is in the $-x$-direction

$$D_B = 2\frac{q}{4\pi(y^2 + l^2)}\cos\theta = \frac{2q}{4\pi(y^2 + l^2)}\frac{l}{\sqrt{y^2 + l^2}} = \frac{2ql}{4\pi(y^2 + l^2)^{\frac{3}{2}}}$$

Thus, we can write
$$\boldsymbol{D}_A = \frac{4qlx}{4\pi(x^2 - l^2)^2}\boldsymbol{i} \quad \text{and} \quad \boldsymbol{D}_B = -\frac{2ql}{4\pi(y^2 + l^2)^{\frac{3}{2}}}\boldsymbol{i}$$

(b) For very short dipoles or at very great distances from the dipole, x, $y \gg l$, in which case

$$\boldsymbol{D}_A \underset{x\gg l}{\longrightarrow} \frac{4ql}{4\pi x^3}\boldsymbol{i} \quad \text{and} \quad \boldsymbol{D}_B \underset{y\gg l}{\longrightarrow} -\frac{2ql}{4\pi y^3}\boldsymbol{i}$$

The corresponding electric field strengths are

$$\boldsymbol{E}_A \underset{x\gg l}{\longrightarrow} \frac{4ql}{4\pi\varepsilon x^3}\boldsymbol{i} \quad \text{and} \quad \boldsymbol{E}_B \underset{y\gg l}{\longrightarrow} -\frac{2ql}{4\pi\varepsilon y^3}\boldsymbol{i}$$

(c)
$$E_A = \frac{4(8\times10^{-19})(10^{-8})}{4\pi(8.85\times10^{-12})(3\times10^{-5})^3} = 10.7\,\text{mV m}^{-1}; \quad E_B = 2\frac{(8\times10^{-19})(10^{-8})}{4\pi(8.85\times10^{-12})(3\times10^{-5})^3} = 5.3\,\text{mV m}^{-1}$$

Worked Example 16.3: Two identical small spheres, each of mass m and each carrying charge q, are suspended from a fixed point by two light non-conducting threads, each of length l. The charged spheres settle down in equilibrium, separated by a distance x as shown in Figure 16.38. Derive an expression for the angle θ that the strings make with the vertical and show that $x = \sqrt[3]{\frac{q^2 y}{2\pi\varepsilon_0 mg}}$, where y is the vertical distance of the spheres below the point of support. What is the magnitude of q if $x = 4$ cm, $l = 15$ cm and $m = 20$ g?

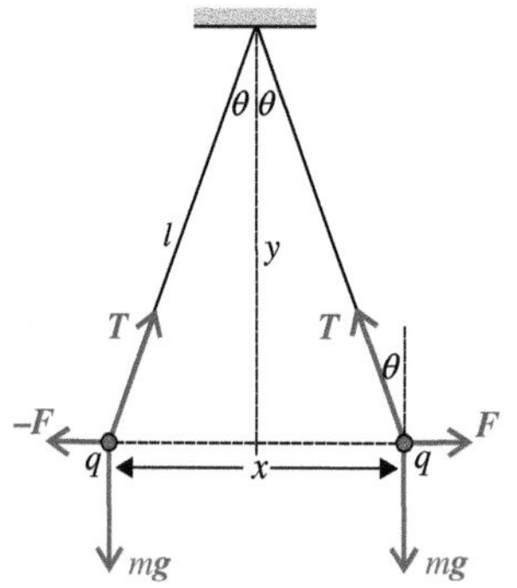

Figure 16.38. Worked Example 16.3.

Consider the forces acting on one of the spheres as indicated in Figure 16.38. The electrostatic repulsive force is given by Coulomb's law, that is $|\boldsymbol{F}| = \frac{q^2}{4\pi\varepsilon_o x^2}$ and, if $|\boldsymbol{T}|$ is the magnitude of the tension in each string,

$$|\boldsymbol{T}|\sin\theta = \frac{q^2}{4\pi\varepsilon_0 x^2} \quad \text{and} \quad |\boldsymbol{T}|\cos\theta = mg \quad \Rightarrow \quad \tan\theta = \frac{q^2}{4\pi\varepsilon_0 mgx^2} \quad \Rightarrow \quad \theta = \tan^{-1}\left(\frac{q^2}{4\pi\varepsilon_0 mgx^2}\right)$$

$$\tan\theta = \frac{x/2}{y} = \frac{x}{2y} \quad \Rightarrow \quad \frac{x}{2y} = \frac{q^2}{4\pi\varepsilon_0 mgx^2} \quad \Rightarrow \quad x^3 = \frac{q^2 y}{2\pi\varepsilon_0 mg} \quad \Rightarrow \quad x = \sqrt[3]{\frac{q^2 y}{2\pi\varepsilon_0 mg}}.$$

Now
$$q^2 = 4\pi\varepsilon_0 mgx^2\tan\theta = 4\pi(8.85\times10^{-12})(2\times10^{-2})(9.8)(0.04)^2\frac{0.02}{\sqrt{0.15^2 - 0.02^2}} = 4.7\times10^{-15}\,\text{C}^2 \Rightarrow q \approx 6.85\times10^{-8}\,\text{C} = 6.85\,\text{nC}.$$

Worked Example 16.4: Two concentric spherical conducting shells of radii a and b, where $b > a$, have a common centre. The permittivity of the medium between the shells is given by

$\varepsilon = \varepsilon_1$ when $a < r < R$

$\varepsilon = \varepsilon_2$ when $R < r < b$, where r is the distance from the common centre.

If the inner shell is given a charge Q and the outer shell is earthed, derive an equation for the electric field strength between the shells as a function of r. Derive also an equation for the potential of the inner shell.

Equation (16.16) gives the electric field strength outside a charged sphere. In the region between the shells,

$$\boldsymbol{E}_1 = \frac{Q}{4\pi\varepsilon_1 r^2}\hat{\boldsymbol{r}} \quad \text{when} \quad a < r < R \quad \text{and} \quad \boldsymbol{E}_2 = \frac{Q}{4\pi\varepsilon_2 r^2}\hat{\boldsymbol{r}} \quad \text{when} \quad R < r < b$$

With the outer shell earthed, the potential of the inner shell is the potential difference between the two shells, given by Equation (16.22)

$$V = -\int_A^B \boldsymbol{E}\cdot\boldsymbol{dr} = -\int_a^R \boldsymbol{E}_1\cdot\boldsymbol{dr} + \left(-\int_R^b \boldsymbol{E}_2\cdot\boldsymbol{dr}\right) = -\int_a^R \frac{Q}{4\pi\varepsilon_1 r^2}dr - \int_R^b \frac{Q}{4\pi\varepsilon_2 r^2}dr = \frac{Q}{4\pi\varepsilon_1}\left[\frac{1}{r}\right]_a^R + \frac{Q}{4\pi\varepsilon_2}\left[\frac{1}{r}\right]_R^b = \frac{Q}{4\pi\varepsilon_1}\left(\frac{1}{R} - \frac{1}{a}\right) + \frac{Q}{4\pi\varepsilon_2}\left(\frac{1}{b} - \frac{1}{R}\right)$$

so that

$$V = \frac{Q}{4\pi}\left[\left(\frac{a-R}{\varepsilon_1 aR}\right) + \left(\frac{b-R}{\varepsilon_2 bR}\right)\right]$$

PROBLEMS

For problems based on the material covered this chapter visit up.ucc.ie/16/ *and follow the link to the problems.*

17

Electric fields in materials; the capacitor

AIMS

- to investigate the behaviour of electric charges within and in the neighborhood of conducting materials
- to understand the response of non-conducting (dielectric) materials to electric fields
- to explore some properties of permanently polarised materials
- to introduce the *capacitor,* an important component of many electrical circuits
- to show how energy is stored in a capacitor
- to understand what happens when a capacitor is charged or discharged through a resistor

17.1 Conductors in electric fields

In electrostatics the electric field strength within the body of a conductor must always be zero since, if such a field existed, it would have the effect of moving the mobile charges to positions where the electric field, and hence the force on them, would be zero. Thus, there can be no net mobile charge anywhere within the body of a conductor since this would give rise to an electric field; hence all excess charge must reside on the surface of a conductor. The distribution of these charges on the surface of a conductor, therefore, must be such that the net effect is to produce zero electric field everywhere inside the conductor.

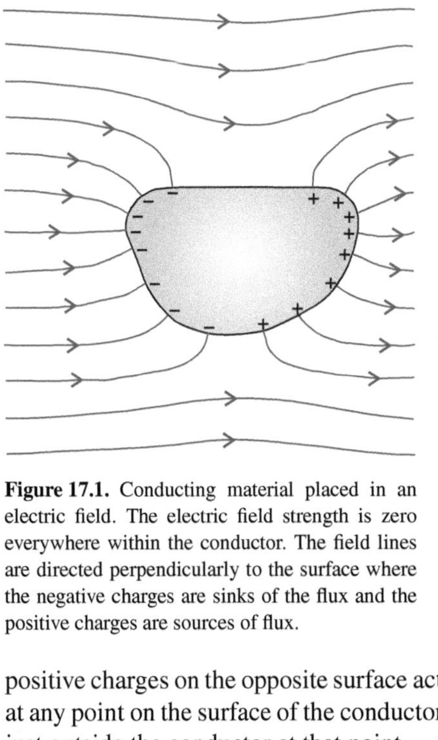

Figure 17.1. Conducting material placed in an electric field. The electric field strength is zero everywhere within the conductor. The field lines are directed perpendicularly to the surface where the negative charges are sinks of the flux and the positive charges are sources of flux.

Consider the case of a body made from a good conductor (such as a metal) placed in an electric field as illustrated in Figure 17.1. Once the external field is switched on, the mobile charges are driven to the surface and are distributed so that the electric field within the conductor is maintained at zero as required. The electric field lines on the left of the body terminate on the negative charges induced on the surface while the positive charges on the other side of the slab act as sources of field lines on that side.

The net electric field strength $\boldsymbol{E}$ at any point within the body of the conductor in Figure 17.1 can be considered to be the result of contributions from two independent sources, namely (i) an electric field strength $\boldsymbol{E}_0$ due to the external field and (ii) the electric field strength $\boldsymbol{E}_{\text{surface}}$ arising from the contribution from the excess charges located at each surface, that is

$$\boldsymbol{E} = \boldsymbol{E}_0 + \boldsymbol{E}_{\text{surface}}$$

Since within the conductor $\boldsymbol{E} = 0$, the electric field due to the induced charges at the surface must balance the external field at every point, that is

$$\boldsymbol{E}_{\text{surface}} = -\boldsymbol{E}_0$$

As noted in Section 16.2, the field lines must be perpendicular to the surface at the point of contact.

Finally, we note there is no electric flux within the conductor since the induced negative charges on the surface act as a sink for all of the flux coming from the left while the induced positive charges on the opposite surface act as a source of an equal amount of flux to the right. The *surface charge density* (σ, Section16.5) at any point on the surface of the conductor must sink/source the incident flux; in other words, $\sigma = |\boldsymbol{D}|$ where $\boldsymbol{D}$ is the electric flux density just outside the conductor at that point.

Understanding Physics, Third Edition. Michael Mansfield and Colm O'Sullivan.
© 2020 John Wiley & Sons Ltd. Published 2020 by John Wiley & Sons Ltd.

Electric field of a point charge in the neighbourhood of a grounded conducting plane

Figure 17.2(a) shows a point charge placed above a grounded (earthed) plane conductor of infinite extent. In this case, negative mobile charges in the conductor will be attracted to the surface and will distribute themselves in such a way as to keep the conductor at zero electrical potential. The requirement that every point in the surface plane of the conductor be at zero potential is an example of a 'boundary condition'.

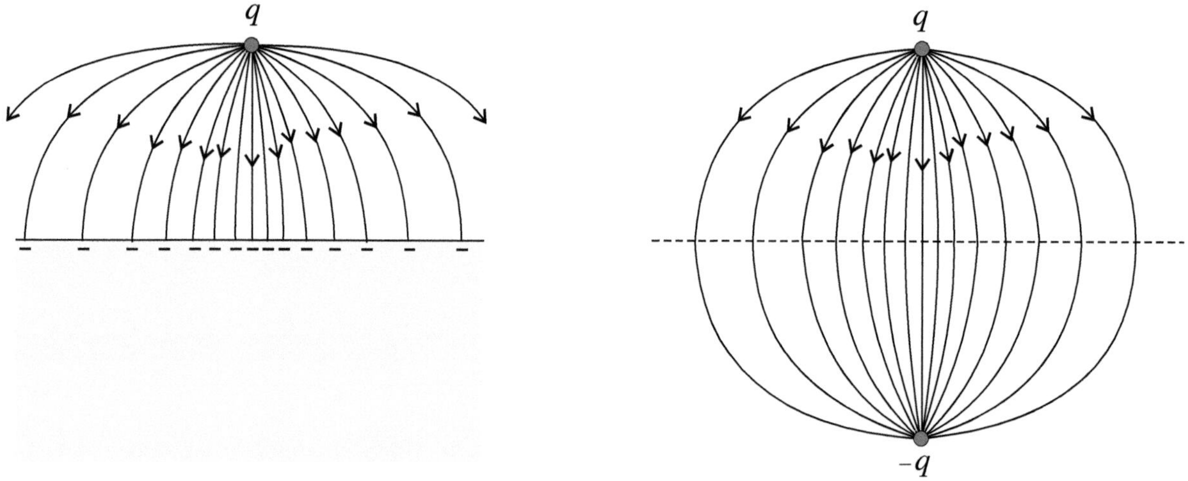

Figure 17.2. (a) A point charge directly above an earthed conducting plane. (b) The field of the corresponding dipole.

In the case of Figure 17.2(a), it is clear that the electric flux density and hence the surface charge density on the conductor is not uniform, that is, $\sigma = \sigma(x, y)$, where x and y are coordinates on the surface of the conductor. If we wish to determine the electric field strength at any point above the conducting plane, we can use a technique known as '*the method of images*'. The objective is to find a configuration of point charges which would satisfy the same boundary conditions as in the original problem. The simplest example that fulfils the requirement in this case is that of the dipole system shown in Figure 17.2(b). From symmetry the electric potential must be zero at all points in the x-y plane (represented by the dashed line in the figure). Thus, the electric field in the upper half-plane in Figure 17.2(b) is the same as that above the conducting plane in Figure 17.2(a) and the dipole field described by Equation (16.9) in Section 16.6 is a valid solution in this situation.

The analysis described here is called the 'method of images' because the charge $-q$ in Figure 17.2(b) is considered to be an 'image' of q formed by the conducting plane. Ensuring that the solution found by this method is *unique* is a more subtle issue but it does indeed turn out to be the case; proof of this 'uniqueness theorem' is beyond the scope of this book but readers with the appropriate mathematical background will find useful treatments in many more advanced texts, for example Griffiths (3rd ed., Section 3.1.5) or Jackson (3rd ed., Section 1.9).

Study Worked Example 17.1

For problems based on the material presented in this section visit up.ucc.ie/17/ *and follow the link to the problems.*

17.2 Insulators in electric fields; polarization

Figure 17.3 shows the electric field due to a number of charges arbitrarily embedded in an insulating medium. The electric flux density at any point P depends only on the way the charges are distributed and is given by Equation (16.7), that is

$$\boldsymbol{D} = \sum_i \frac{Q_i}{4\pi r_i^2}\hat{\boldsymbol{r}}_i$$

or, in the case of a continuous distribution of charge Equation (16.8),

$$\boldsymbol{D} = \iiint \frac{\rho(\boldsymbol{r})}{4\pi r^2}\hat{\boldsymbol{r}}dV$$

where $\rho(\boldsymbol{r})$ is the charge density at the point $\boldsymbol{r}$. It is clear that the electric flux density is independent of the nature of the material in the insulator and is the same as would be observed if the charges were in vacuum. On the other hand, we know that the electric field strength $\boldsymbol{E} = \frac{\boldsymbol{D}}{\varepsilon}$ does depend on the medium and, since $\varepsilon > \varepsilon_0$, it is always less than the corresponding field strength in vacuum $\frac{\boldsymbol{D}}{\varepsilon_0}$.

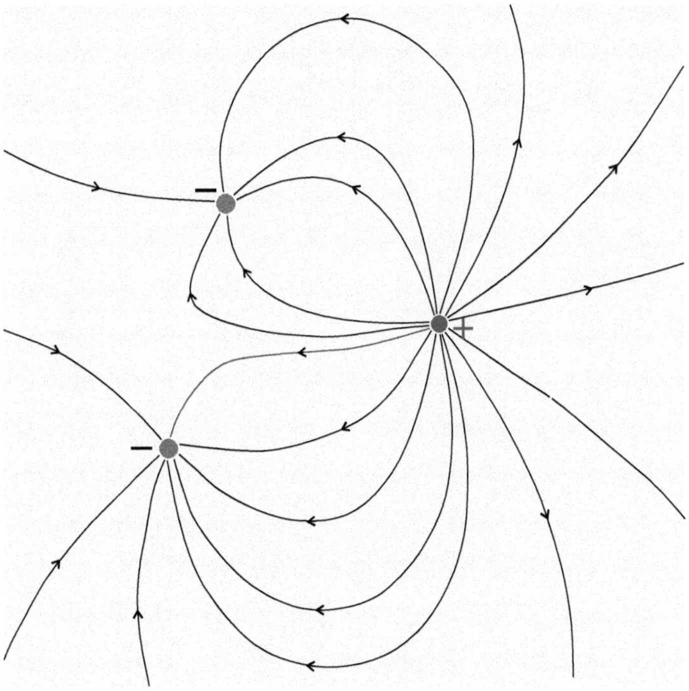

Figure 17.3. Electric field lines in an insulator due to an arbitrary configuration of electric charges gives rise to an electric flux density (displacement) at any point which is independent of the medium.

Electric polarisation

When an insulating material experiences an electric field, the material remains electrically neutral and there are no free mobile charges as there are in the case of a conductor, as discussed in the previous section. Nevertheless, within any element of volume (Figure 17.4), there will be a small displacement of positive charge in the direction of the field and of negative charge in the opposite direction; that is, the element has an electric dipole moment. The material is said to be **polarised** and a substance that can be polarised by an electric field in this way is called a **dielectric material**.

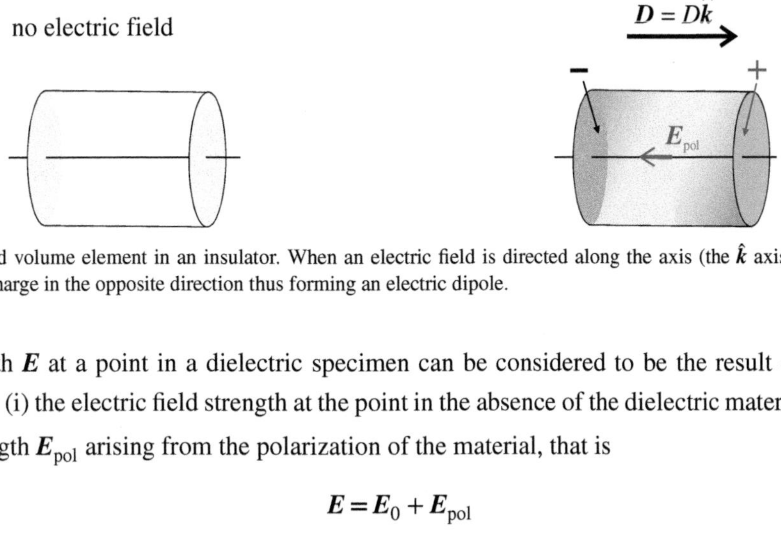

Figure 17.4. A cylindrically shaped volume element in an insulator. When an electric field is directed along the axis (the $\hat{k}$ axis), positive charge is displaced in the direction of the field and negative charge in the opposite direction thus forming an electric dipole.

The net electric field strength $\boldsymbol{E}$ at a point in a dielectric specimen can be considered to be the result of the contributions from two independent sources, namely (i) the electric field strength at the point in the absence of the dielectric material, that is $\frac{\boldsymbol{D}}{\varepsilon_0} = \boldsymbol{E}_0$, and (ii) an additional electric field strength $\boldsymbol{E}_{pol}$ arising from the polarization of the material, that is

$$\boldsymbol{E} = \boldsymbol{E}_0 + \boldsymbol{E}_{pol}$$

In dielectric materials $\boldsymbol{E}_{pol}$ opposes $\boldsymbol{E}_0$ and thus the magnitude of the electric field strength at any point within the specimen is less than that which would be observed in the absence of the specimen that is

$$|\boldsymbol{E}| = |\boldsymbol{E}_0| - |\boldsymbol{E}_{pol}|$$

with $|\boldsymbol{E}_{pol}| < |\boldsymbol{E}_0|$.

If the cylindrical volume element in Figure 17.4 is sufficiently small that the electric field inside the element may be considered uniform, the distribution of charge within the element may be considered to arise from two uniformly charged discs with equal and opposite charge at each end of the cylinder (Figure 17.5). From a macroscopic perspective such charges are 'fictitious'[1] in that they do not contribute to electric flux (they are cancelled by the opposite charge on the end face of the adjacent element). Nevertheless, for the purpose of determining the contribution to the electric field due to polarization, such charges can be treated as the source of $\boldsymbol{E}_{\text{pol}}$.

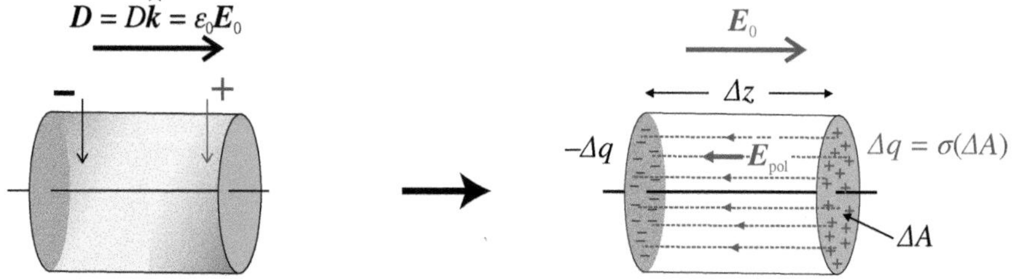

Figure 17.5. The cylindrically shaped volume element in Figure 17.4. In the limit of the cylinder being small, the field within the cylinder is uniform and the dipole can be considered to be equivalent to two uniformly charged discs of opposite charge.

The volume element has an electric dipole moment $\boldsymbol{p} = 2ql\hat{\boldsymbol{n}}$ (recall Section 16.6). The dipole length $2l = \Delta z$ so we can write

$$\Delta\boldsymbol{p} = (\Delta q)(\Delta z)\hat{\boldsymbol{k}} = \frac{\Delta q}{\Delta A}(\Delta z)(\Delta A)\hat{\boldsymbol{k}} = \sigma(\Delta V)\hat{\boldsymbol{k}}$$

or

$$\lim_{\Delta V\to 0}\frac{\Delta\boldsymbol{p}}{\Delta V} = \frac{d\boldsymbol{p}}{dV} = \sigma\hat{\boldsymbol{k}} \tag{17.2}$$

where $\sigma = \dfrac{dq}{dA}$ is the electric surface charge density. The quantity $\dfrac{d\boldsymbol{p}}{dV}$ in Equation (17.2), *the electric dipole moment per unit volume*, is called the **electric polarization**; that is, the electric polarization at a point in a medium is defined as

$$\boxed{\boldsymbol{P} := \frac{d\boldsymbol{p}}{dV}}$$

We now consider the polarization of the dielectric material (shaded) in Figure 17.6. The source of the electric field which causes the polarization of the material need not concern us here; it could be due to electric charges embedded elsewhere in the material as in Figure 17.3 or to an external electric field. As we have seen, the electric field produces a small redistribution of charge within the material such that any volume element will have a net electric dipole moment but, nevertheless, any such volume remains electrically neutral.

Consider the cylindrical element in Figure 17.6 where the axis of the cylinder is oriented parallel to the direction of the electric field, that is along the local flux density vector $\boldsymbol{D}$ indicated in the figure. As discussed above, the contribution to the electric field strength ($\boldsymbol{E}_{\text{pol}}$) arising from polarization at any point in a dielectric may be modelled by that due to two equally and oppositely charged discs in vacuum.

Since the field is uniform the polarization surface charge density of the discs is uniform (σ = constant) and the polarization electric field strength can be determined in a similar manner to that used in Example (c) of Section 16.8; that is, $\boldsymbol{E}_{\text{pol}} = -\dfrac{\sigma}{\varepsilon_0}\boldsymbol{k}$

and, using Equation (17.2),

$$\boldsymbol{E}_{\text{pol}} = -\frac{1}{\varepsilon_0}\frac{d\boldsymbol{p}}{dV}\boldsymbol{k} = -\frac{\boldsymbol{P}}{\varepsilon_0}$$

Thus the electric field strength at a point in the dielectric is given by

$$\boldsymbol{E} = \frac{\boldsymbol{D}}{\varepsilon_0} + \boldsymbol{E}_{\text{pol}} = \frac{\boldsymbol{D}}{\varepsilon_0} - \frac{\boldsymbol{P}}{\varepsilon_0} \tag{17.3}$$

and

$$\boldsymbol{D} = \varepsilon_0\boldsymbol{E} + \boldsymbol{P}$$

As we have seen, polarization in a dielectric material will not result in the production of a net additional electric flux in the material. That is, the electric flux density at any point within the material is unaffected by the presence or nature of the material and is the same as the electric flux density from the same sources in the absence of the dielectric (that is, if the material was displaced by vacuum). From the point of view of the macroscopic model considered here, polarization charges produced in a dielectric in response to the applied electric field do *not* act as sources or sinks of electric flux unlike, for example, real ('free') charges embedded in or located outside the dielectric

[1]From the macroscopic viewpoint, real charges, which are sources of electric flux are often referred to as 'free charges'. On the other hand, 'fictitious' polarization charges (such as those on the end faces of the cylindrical element in Figure 17.5) are usually called 'bound charges'. The latter terminology arises because, as we will see in Section 17.7, from a microscopic perspective such charges are indeed real charges in polarised molecules or atoms.

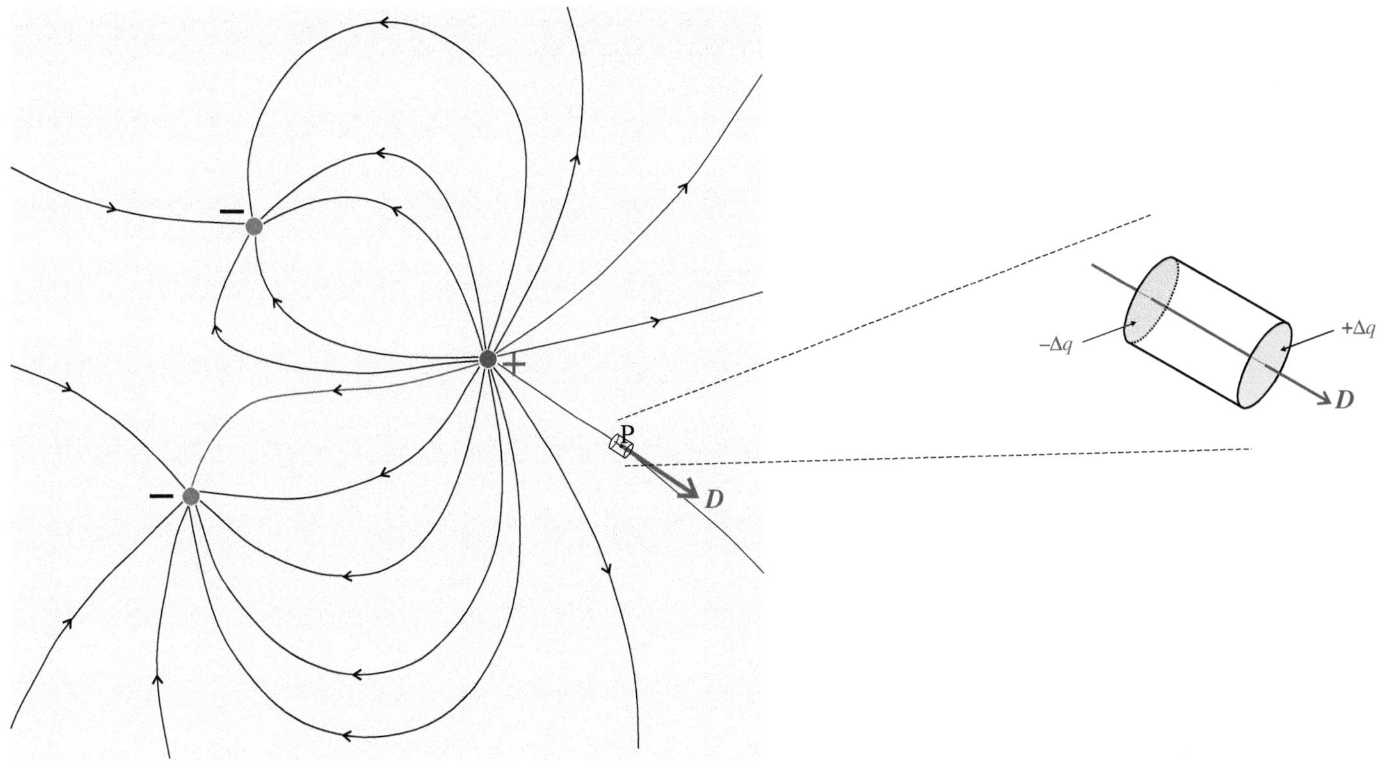

Figure 17.6. Polarization surface charges on the end faces of a cylindrical volume element in the body of a dielectric material due to an electric field.

which do act as sources/sinks. Thus whenever Gauss's law, $\iint_S \boldsymbol{D}\cdot d\boldsymbol{A} = \sum_{\text{inside S}} q_i$, is applied in the context of dielectrics, care must be taken to ensure that the q_is do not include surface charges due to polarization, which arise directly from the response of the medium to the applied field.

17.3 Electric susceptibility

In dielectric materials, the degree of polarization at a point in the medium depends on the strength of the field $\boldsymbol{E}$ at that point. Experimentally, for most dielectric materials, it is found that $|\boldsymbol{E}_{\text{pol}}|$ is directly proportional to $|\boldsymbol{E}|$ over a reasonably wide range of field strength values, that is

$$\boldsymbol{E}_{\text{pol}} \propto -\boldsymbol{E} \rightarrow \boldsymbol{E}_{\text{pol}} = -(constant)\boldsymbol{E} = -\chi_e \boldsymbol{E}$$

The constant of proportionality χ_e is called the **electric susceptibility** and is characteristic of the material involved. Note that, since $\boldsymbol{P} = -\varepsilon_0 \boldsymbol{E}_{\text{pol}}$, the electric polarization is

$$\boldsymbol{P} = \varepsilon_0 \chi_e \boldsymbol{E} \quad \text{and} \quad \chi_e > 0.$$

Thus, using Equation (17.3), the electric field strength at a point within a dielectric material is given by

$$\boldsymbol{E} = \frac{\boldsymbol{D}}{\varepsilon_0} - \chi_e \boldsymbol{E}$$

Hence

$$\boldsymbol{D} = \varepsilon_0(1 + \chi_e)\boldsymbol{E} = \varepsilon \boldsymbol{E}$$

and we can now interpret the permittivity of the medium to be

$$\varepsilon = (1 + \chi_e)\varepsilon_0$$

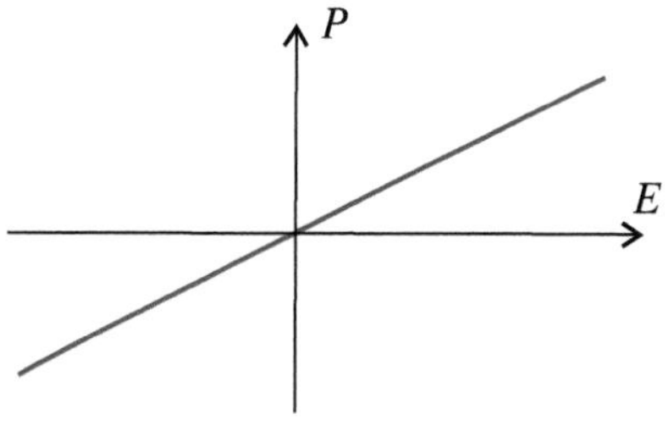

Figure 17.7. Plot of polarization versus electric field strength for a linear material. The slope of the straight line is $\varepsilon_0 \chi_e$.

Values of permittivity are often given in terms of **relative permittivity**, the value of the permittivity relative to that of vacuum, that is

$$\varepsilon_r := \frac{\varepsilon}{\varepsilon_0} = 1 + \chi_e$$

For most dielectric materials in typical electric fields, therefore, a plot of $\boldsymbol{D}$ versus $\boldsymbol{E}$ (or $\boldsymbol{P}$ versus $\boldsymbol{E}$) yields a straight line (Figure 17.7). Such substances are described as *linear* materials.

In the discussion above it has been assumed that substances under discussion were linear, isotropic and homogeneous (that is, *l.i.h.* materials). Examples of materials with non-linear electrical properties will be discussed in Section 17.5.

17.4 Boundaries between dielectric media

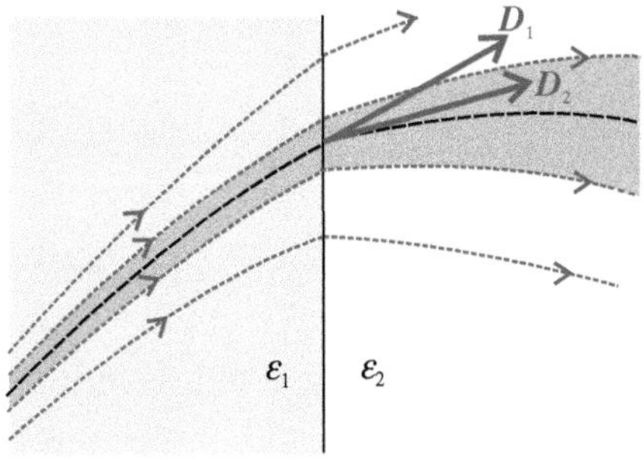

Figure 17.8. The change in the direction of the electric field at the interface between two different dielectric media.

In the discussion presented in earlier sections of this chapter it has been assumed that all sources of electric field have either been in vacuum or embedded in an effectively infinite non-conducting medium. In the case of situations involving two or more media of different electric permittivities, however, care must be exercised in applying the results obtained. Issues do not generally arise when the electric field is applied at right angles to the surfaces of the specimen. If, however, the electric field makes an angle with the interface between the media (as in Figure 17.8), the direction of the electric flux density (and hence the electric field strength) will be different on either side of the interface.

The definition of electric flux requires that the net flux within any 'tube' (shaded in Figure 17.8) must remain unchanged across the interface – effectively a statement of Gauss's law. Consider a narrow tube of flux of cross-sectional area ΔA_1 incident on an interface between medium 1 and medium 2 as shown (shaded) in Figure 17.9(a). Since ΔA_1 is small, the interface can be considered to be planar over the extent of the tube of flux. On entering medium 2 the tube is 'bent' as indicated and the cross-sectional area of the narrow tube is now ΔA_2.

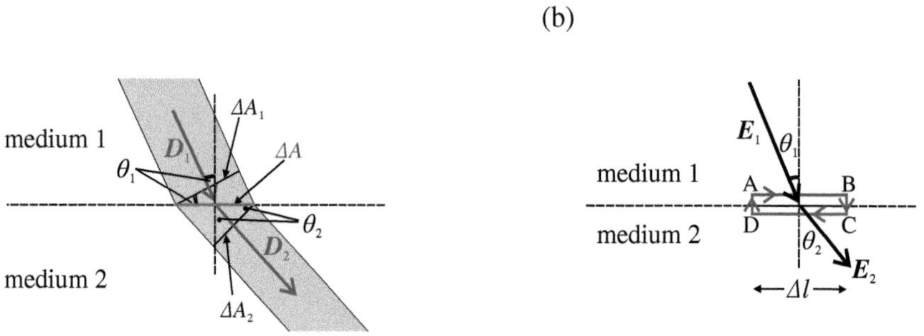

Figure 17.9. A narrow tube of electric flux passing through the interface between two different dielectric media.

(a) Invoking the continuity of electric flux across interface

$$\Delta\Phi_E = |\boldsymbol{D}_1|(\Delta A_1) = |\boldsymbol{D}_2|(\Delta A_2)$$

But $\Delta A_1 = (\Delta A)\cos\theta_1$ and $\Delta A_2 = (\Delta A)\cos\theta_2$, where the angles are as indicated in the figure, and thus

$$|\boldsymbol{D}_1|\cos\theta_1 = |\boldsymbol{D}_2|\cos\theta_2$$

$$(\boldsymbol{D}_1)_\perp = (\boldsymbol{D}_2)_\perp$$

(b) Since the electric field is conservative, we can apply Equation (16.24) to the path ABCDA in Figure 17.9(b)

$$\oint_{ABCDA} \boldsymbol{E}\cdot d\boldsymbol{l} = 0 \rightarrow \int_{AB} \boldsymbol{E}\cdot d\boldsymbol{l} + \int_{BC} \boldsymbol{E}\cdot d\boldsymbol{l} + \int_{CD} \boldsymbol{E}\cdot d\boldsymbol{l} + \int_{DA} \boldsymbol{E}\cdot d\boldsymbol{l} = 0$$

But

$$\int_{BC} \boldsymbol{E}\cdot d\boldsymbol{l} = -\int_{DA} \boldsymbol{E}\cdot d\boldsymbol{l} \rightarrow \int_{AB} \boldsymbol{E}\cdot d\boldsymbol{l} + \int_{CD} \boldsymbol{E}\cdot d\boldsymbol{l} = 0$$

$$\rightarrow \quad |\boldsymbol{E}_1|\sin\theta_1(\Delta l) = |\boldsymbol{E}_2|\sin\theta_2(\Delta l)$$

$$\rightarrow \quad (\boldsymbol{E}_1)_\parallel = (\boldsymbol{E}_2)_\parallel$$

Combining results (a) and (b) together, we obtain

$$|\varepsilon_1\boldsymbol{E}_1|\cos\theta_1 = |\varepsilon_2\boldsymbol{E}_2|\cos\theta_2 \quad \text{and} \quad |\boldsymbol{E}_1|\sin\theta_1 = |\boldsymbol{E}_2|\sin\theta_2$$

and hence,

$$\varepsilon_1\cot\theta_1 = \varepsilon_2\cot\theta_2$$

Thus, the fact that the electric flux is continuous across an interface, together with the fact that the electric field is conservative, leads to the following boundary conditions at the interface

$$\boxed{(\boldsymbol{D}_1)_\perp = (\boldsymbol{D}_2)_\perp} \quad \text{and} \quad \boxed{(\boldsymbol{E}_1)_{\parallel} = (\boldsymbol{E}_2)_{\parallel}}$$

where the subscripts ⊥ and ∥ refer, respectively, to the components of the fields perpendicular and parallel to the interface.

For problems based on the material presented in this section visit up.ucc.ie/17/ and follow the link to the problems.

17.5 Ferroelectricity and paraelectricity; permanently polarised materials

Ferroelectricity

If an electric field is applied to certain materials, it is observed that these materials remain permanently polarised after the field has been removed (for example, barium titanate $BaTiO_3$, lead titanate $PbTiO_3$). Such materials are described as **ferroelectric**. If the temperature of a specimen is raised sufficiently, however, the material loses its ferroelectric properties. The temperature at which a particular material makes the transition to or from ferroelectric behaviour is called the **Curie temperature** for that material.

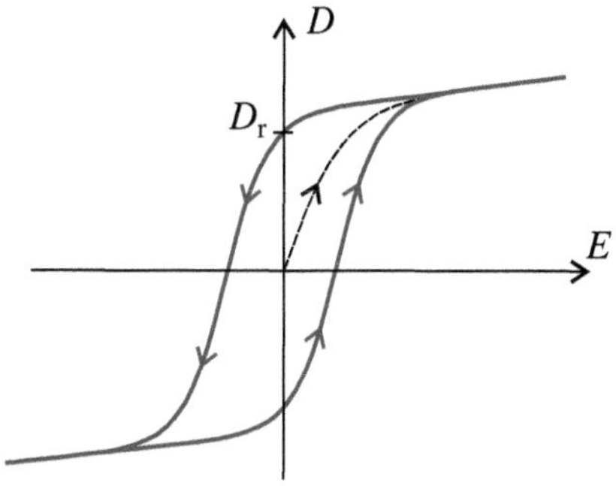

Figure 17.10. A *D—E* plot for a ferroelectric material.

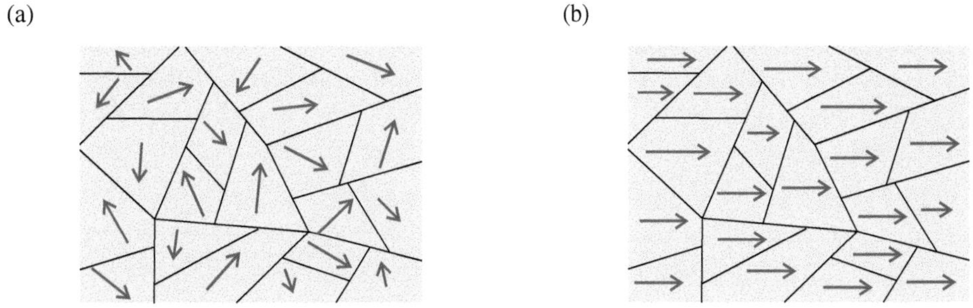

Figure 17.11. Illustration of ferroelectric domains in an unpolarised specimen (left) and a fully polarised specimen (right).

There is a highly *non-linear* relationship between $\boldsymbol{P}$ and $\boldsymbol{E}$ (and, equivalently, $\boldsymbol{D}$ and $\boldsymbol{E}$) in ferroelectric materials, such as that illustrated in Figure 17.10. If the external field applied to a sample is increased continuously the resulting polarisation gradually becomes fully saturated. When the external field is removed, a residual electric flux density (D_r) inside the specimen remains. Ferroelectric materials, therefore, retain a 'memory' of their previous polarization and are said to exhibit *hysteresis*; the *D—E* plots are examples of 'hysteresis loops'.

When ferroelectric substances are studied using a high resolution microscope, such as an electron microscope, it is observed that such materials comprise a large number of small permanently polarised areas called **domains.** Normally the orientation of polarization of individual domain is random (as illustrated in Figure 17.11(a)) but under the influence of an external electric field, the polarization of individual domains tends to reorient along the direction of the field. As the external field is increased more domains line up until the specimen becomes saturated, as indicated in Figure 17.11(b).

Paraelectricity

Certain materials (mostly ceramics) are ferroelectric at low temperatures, but exhibit *paraelectric* behaviour above their Curie temperatures. In this case hysteresis is not observed — a specimen loses its polarization when the applied field is removed — although the *P versus E* plot is usually non-linear (Figure 17.12).

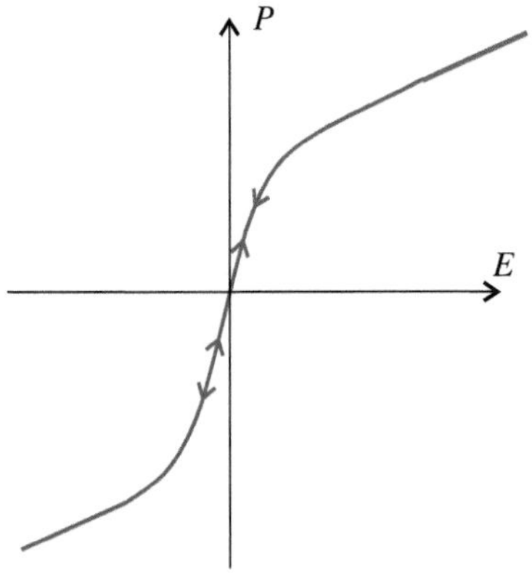

Figure 17.12. A *P versus E* plot of a paraelectric material.

17.6 Uniformly polarised rod; the 'bar electret'

While the study of ferroelectric materials is important in its own right, for example in its application to piezoelectric and other devices (considered at the end of this section), it is important to note that the discussion presented in the next two sections also provides a valuable foundation for the treatment and analysis of corresponding magnetic systems in Chapters 18 and 19.

Permanently polarised specimens ('**electrets**') may be fabricated from ferroelectric substances. Consider the example of a cylindrical rod of uniform cross-section made from a ferroelectric material. Suppose that the rod has been fully polarised uniformly throughout, that is, $\boldsymbol{P}$ = constant everywhere — a very idealised requirement in practice. Using a similar argument to that of Section 17.2, we see that the effect of the polarization leads to the appearance of a net charge on the surface at each end of the rod as indicated in Figure 17.13.

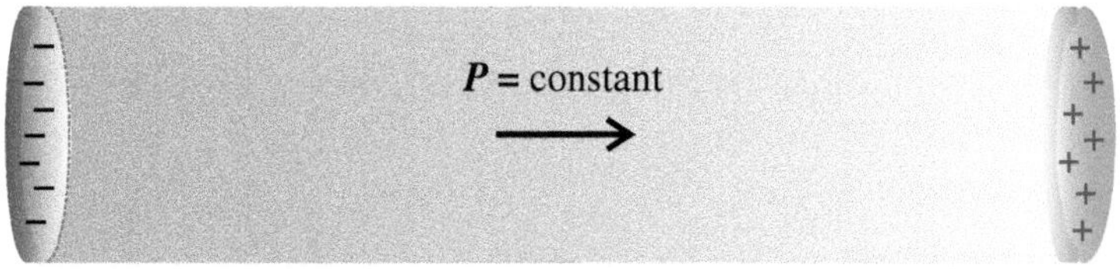

Figure 17.13. A uniformly polarised cylindrical rod.

Thus, for example, the electric field of a polarised uniform rod with circular cross-section can be modelled reasonably well by two uniformly charged circular discs of opposite signs, one at each end. The electric field strength can be considered to be due to the two charged discs in vacuum; the electric field lines originate on the positive polarization charges and terminate on the negative ones.

Flux lines and field lines computed numerically for a uniformly polarised bar electret are shown in Figure 17.14. Note that the electric flux is continuous across the end faces of the cylinder, since flux lines can only terminate on real ('free') charges which do not exist in this case. From the point of view of an observer outside the specimen, however, the end faces may be considered to be sources/sinks of electric flux.

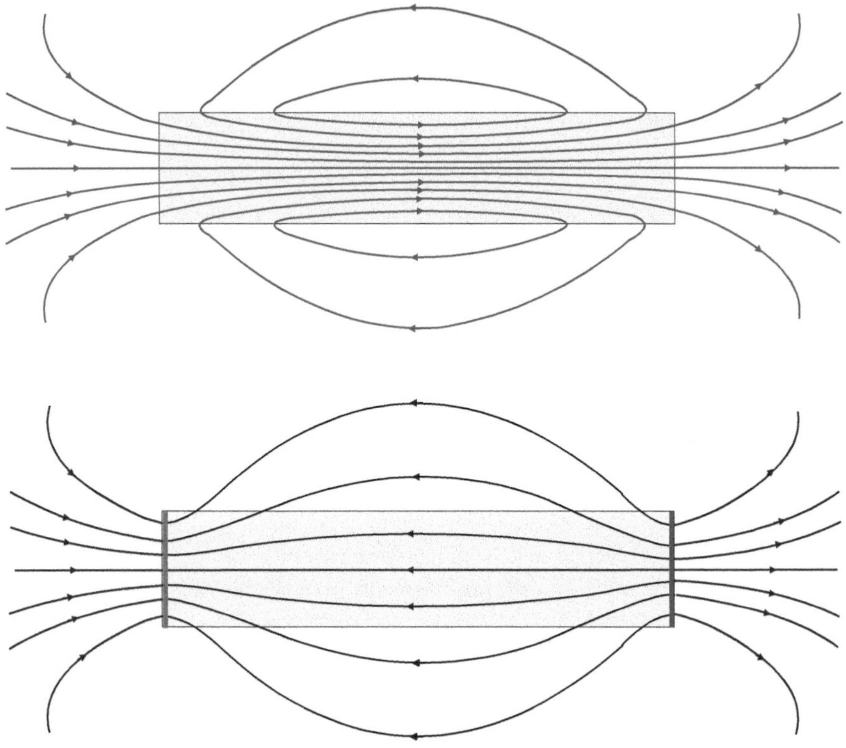

Figure 17.14. Electric flux lines (above - blue) and electric field lines (below - black) for a uniformly polarised rod.

Inside the specimen the electric field strength is in the opposite direction to that of the flux density and the electric field lines are discontinuous at the end faces. They originate from the positive polarization charges at one face and terminate on the negative charges at the other.

Electric field on the axis of a uniformly polarised cylinder

Figure 17.15 shows the two uniformly charged discs (surface charge density $-\sigma$ and σ) which, as discussed above, can be considered as replacing the uniformly polarised cylinder in Figure 17.13. The electric field strength a point P on the axis of the electret is the sum of the field strengths arising from the positively and negatively charged discs.

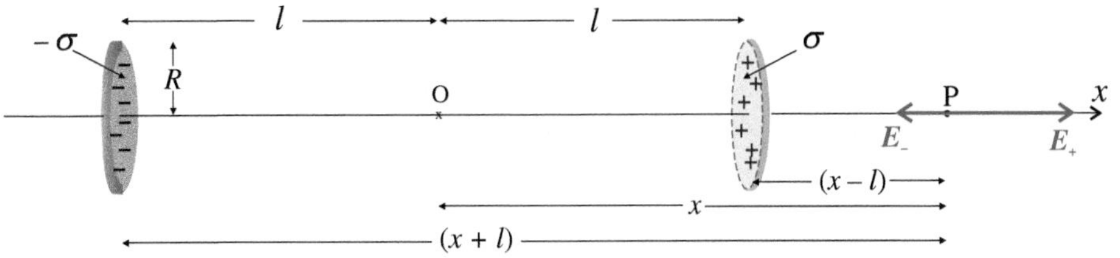

Figure 17.15. The electret in Figure 17.14 is replaced by two oppositely charged discs. The electric field strength at the external point P on the axis is the sum of the contributions from each of the two uniformly charged discs.

If the point P is outside the cylinder, as indicated in Figure 17.15, the field strength vector from the positive disc (E_+ in the figure) is in the $+x$ direction while that from the negative disc ($E\_$) is in the $-x$ direction. The electric field strength at a point P on the axis at a distance x from the centre of a uniformly charged disc is given by Equation (16.19), namely

$$\boldsymbol{E}(x) = \frac{\sigma}{2\varepsilon_0}\left(1 - \frac{x}{\sqrt{x^2 + R^2}}\right)\boldsymbol{i}$$

where $\boldsymbol{i}$ is a unit vector in the $+x$-direction.

Applying this to the situation in shown Figure 17.15,

$$\boldsymbol{E}(x) = \frac{\sigma}{2\varepsilon_0}\left(1 - \frac{x-l}{\sqrt{(x-l)^2 + R^2}}\right)\boldsymbol{i} - \frac{\sigma}{2\varepsilon_0}\left(1 - \frac{x+1}{\sqrt{(x+1)^2 + R^2}}\right)\boldsymbol{i}$$

$$= \frac{\sigma}{2\varepsilon_0}\left(\frac{x+l}{\sqrt{(x+l)^2 + R^2}} - \frac{x-l}{\sqrt{(x-l)^2 + R^2}}\right)\boldsymbol{i}$$

We now apply this result to two special cases.

Case 1: Short flat cylinder (disc electret)
In the case of a short flat specimen, $R \gg l$, and hence

$$\boldsymbol{E}(x) \to \frac{\sigma}{2\varepsilon_0}\left(\frac{x+l}{\sqrt{x^2 + R^2}} - \frac{x-l}{\sqrt{x^2 + R^2}}\right) \to \frac{\sigma}{\varepsilon_0}\frac{l}{\sqrt{x^2 + R^2}} \to 0$$

Case 2: Long thin polarised rod
In the case of a long narrow specimen, the 'electret' looks from the outside as if it has a point charge near each end (Figure 17.16). Thus, the electric field is akin to that of an electric dipole (Section 16.6), the electric dipole moment being $2ql$ directed along the axis, and the charges can be treated as point sources/sinks of electric flux.

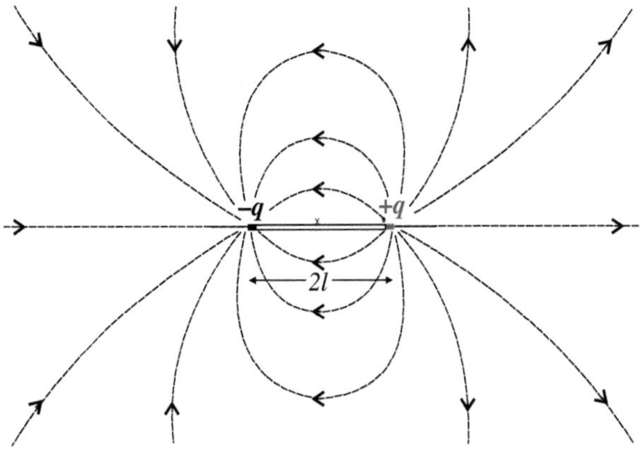

Figure 17.16. Electric field of a long thin electret.

It is important to note that, in this case, these charges of equal and opposite signs cannot be decoupled from one another. The magnitude of such charge may be described as the '*electric pole strength*' of the dipole. A long bar electret is the electric equivalent of the familiar bar magnet that we will encounter in Chapter 18. Unlike in the corresponding case of bar magnets, however, it is difficult to maintain the polarization of bar electrets for long periods. Furthermore, in a normal environment each pole of the electret will attract dust, ions, *etc* of the opposite charge which tend to screen the electret so that no external electric field is observed.

Piezoelectric effect in ferroelectric materials

Electrets, in common with certain other materials, exhibit the phenomenon of *piezoelectricity* in which the external application of mechanical stress to a sample of ferroelectric material gives rise to a change in the electric polarization inside an electret which, in turn, results in a measurable potential difference across the specimen (Figure 17.17). Conversely, when the specimen is subjected to an external electric field this leads to mechanical stresses or strains in the material.

Electrets can be used as the basis of low-cost microphones. When a soundwave is incident on a sheet of ferroelectric material placed between two thin conducting films it induces stress in the electret which results in the development of a potential difference between

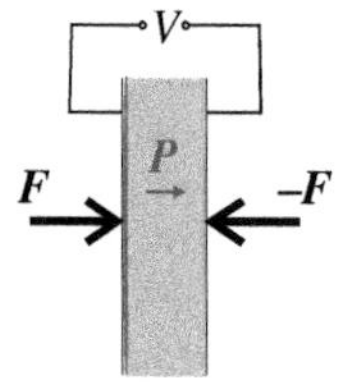

Figure 17.17. A potential difference is developed between the end faces of an electret in response to an applied force.

the conductors. This electrical signal will have an amplitude and frequency proportional to that of the incident sound; the output voltage may then be amplified and fed to a loudspeaker or recorded as required.

An electret microphone is an example of a **transducer**, a device that transforms energy in the form of variations in a physical system (pressure in this case) into a proportional electric signal. Ferroelectric materials are used in a great variety of devices such as pressure sensors, force sensors, ultrasound sensors and hydrophones.

We have treated the topics of ferroelectricity and electrets here in some detail because, as will be encountered in Chapters 18 and 19, the ideas involved also provide a valuable introduction to the treatment of equivalent magnetic properties.

For problems based on the material presented in this section visit up.ucc.ie/17/ and follow the link to the problems.

17.7 Microscopic models of electric polarization

As we have seen, when a piece of non-conducting (dielectric) material is placed in an electric field there is no migration of free charges as occurs in the case of conducting media. Nevertheless, the application of the external electric field has an effect at the microscopic level which is responsible for the phenomenon of polarization discussed above. Even if the molecular charge density distributions are spherically symmetric with no separation of charge (that is *non-polar molecules* such as that illustrated in Figure 17.18), the electric field produces a small displacement of the negative electronic charge clouds relative to the positive charges which are fixed in the solid.

Figure 17.18. Polarization of a non-polar molecule.

The effect of this is to induce electric dipoles in the medium which are aligned along the field direction, as illustrated in Figure 17.19; such dipoles at the atomic/molecular level are responsible for polarization of the medium.

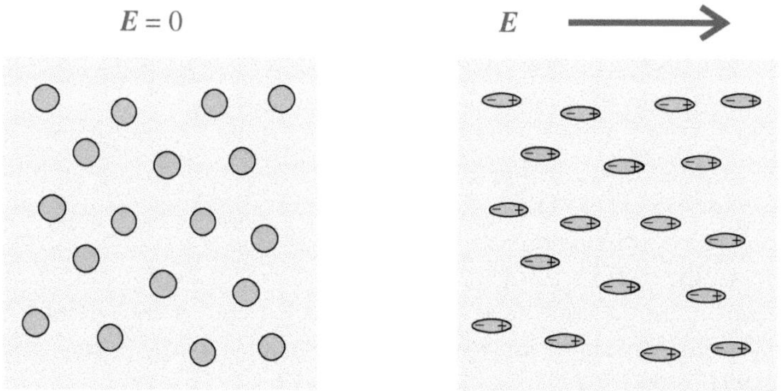

Figure 17.19. The application of an external electric field causes a dielectric medium of non-polar molecules (left) to become polarised (right).

In some materials the molecular structure is such that charge is separated so that the molecules have permanent electric dipole moments (*polar molecules,* to be discussed further in Section 25.1). The water molecule, illustrated in Figure 17.20, is one such example.

In the absence of an external electric field, these molecules are oriented randomly at normal temperatures due to thermal agitation so that they have no net effect. The application of an external field to such a polar material, however, tends to align the dipoles along the general direction of the field as indicated in Figure 17.21. If there is also a displacement of the electronic charge clouds, as in non-polar materials, this will add to the molecular dipole moments.

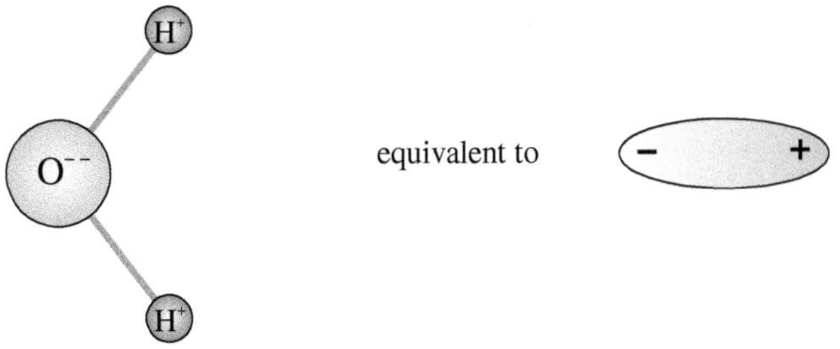

Figure 17.20. The permanent electric dipole moment of the H_2O molecule.

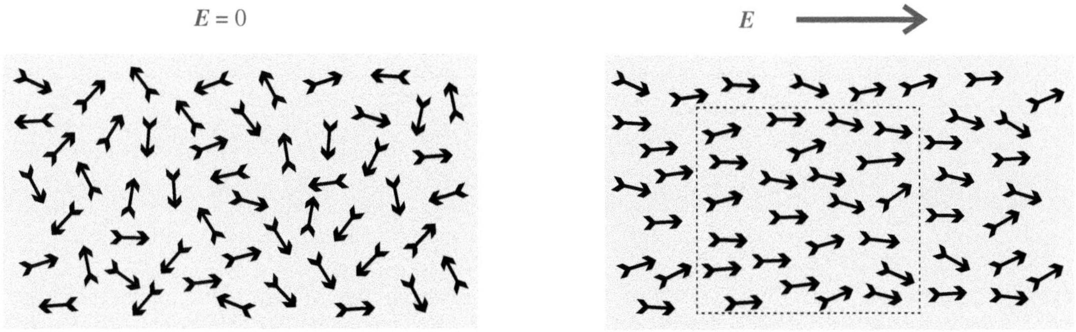

Figure 17.21. Unlike the material illustrated in Figure 17.19, some dielectric media comprise polar molecules. In the absence of an external electric field these molecules are randomly oriented by thermal agitation (left) but when an external electric field is applied the medium becomes polarised (right) with the dipoles aligned preferentially parallel to the field.

In the case of both polar and non-polar molecules, the degree of alignment is proportional to the strength of the applied field over a wide range of field strengths, giving rise to the linear nature of dielectrics.

It is important to note that the notion of a volume element in the context of any macroscopic model of matter (for example, the volume element discussed in Section 17.2, illustrated in Figure 17.5 and subsequent figures), presumes that such an element still contains a very large number of microscopic (molecular) dipoles. Polarization charges which were treated as 'fictitious' in the macroscopic model in Section 17.2 are seen, from a microscopic perspective, as real charges arising from the unpaired charges of aligned molecular dipoles near the edges of any arbitrary volume, such as those at the left- and right-hand edges of the dashed box in Figure 17.21.

17.8 Capacitors

A capacitor in its simplest form comprises two conductors, usually called 'plates'; when the capacitor is charged, the plates have equal but opposite electric charge (as illustrated in Figure 17.22(a). This is most easily arranged by connecting one of the conductors to ground (zero potential) and by placing a charge on the other; this has the effect of *inducing* a charge on the grounded plate which is equal and opposite to that on the charged plate (Figure 17.22(b)). Charging can also be accomplished by connecting a power supply across the plates; if the power supply is then disconnected, the plates retain their (equal and opposite) charge.

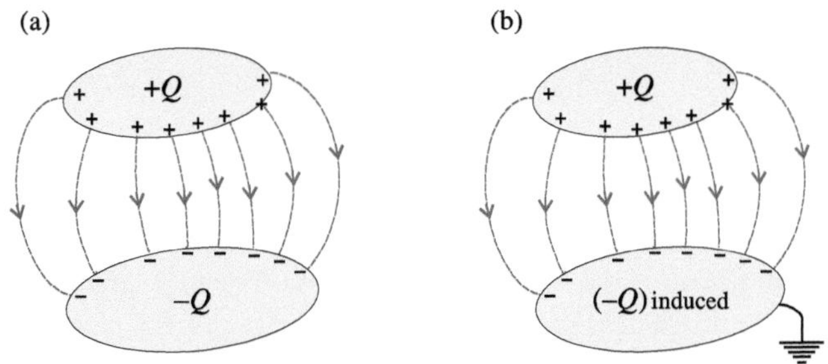

Figure 17.22. A capacitor usually comprises two conductors. (a) In most applications the conductors have equal and opposite charges. (b) If one conductor is grounded, an equal but opposite charge is induced on it by the charge on the other conductor.

Because of the mutual attraction between the unlike charges on the plates, the charges distribute themselves on the surface of the conductors so that they are confined to those regions of the conductors nearest to each other. Hence the electric flux from the positive plate to the negative plate is confined mainly to the space between the plates. The electric field between the plates must be such that the conductors are equipotential surfaces and the field lines close to the surface of a conductor will be perpendicular to that surface.

Suppose that the charges on the 'plates' of a capacitor are $+Q$ and $-Q$ and that the corresponding potential difference between the plates is V. Suppose now that the magnitude of the charges on the plates is increased by a factor k, that is to $+kQ$ and $-kQ$, and hence that each surface element on the conductor has its charge increased from δq to $k(\delta q)$. Since the electric potential at any point due to an element of charge δq is directly proportional to δq, as can be seen from Equation (16.25), the electric potential at any point must also increase by a factor k and hence the potential difference between the plates is correspondingly increased by the factor k. This shows that the potential difference between the plates of a charged capacitor is directly proportional to the magnitude of the charge applied to each plate, that is $V \propto Q$ or

$$Q = (constant)V$$

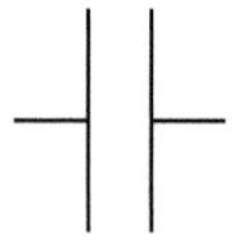

Figure 17.23. The conventional symbol for a capacitor.

The constant of proportionality, which is characteristic of the particular capacitor involved, is called the **capacitance** of the capacitor and is defined by

$$\boxed{C := \frac{Q}{V}}$$

The SI unit of capacitance is the coulomb per volt; 1 C V^{-1} is called 1 farad = 1 F. The electrical circuit symbol for a capacitor is shown in Figure 17.23.

Energy stored in a charged capacitor

If the plates of a charged capacitor are joined by a piece of conducting wire, a current flows in the wire and the capacitor is 'discharged'. Obviously energy is stored in a charged capacitor; the stored energy is released and appears as heat in the wire when the capacitor is discharged in the manner described. The energy stored is equal to the work done in charging the capacitor.

Let ΔW be the work done in charging a capacitor from a charge q to a charge $q + \Delta q$, that is $\Delta W = \mathcal{V}(\Delta q)$, where $\mathcal{V}$ is the potential difference between the plates when the charge on the plates is q. Thus the energy stored in the capacitor, that is the work done in charging a capacitor from zero to a charge Q, is given by

$$W_{\text{str}} = \int_0^Q \mathcal{V}dq = \int_0^Q \frac{q}{C}dq = \frac{1}{C}\int_0^Q qdq = \frac{Q^2}{2C}$$

or, in terms of the potential difference between the plates, $V\left(=\dfrac{Q}{C}\right)$,

$$\boxed{W_{\text{str}} = \frac{1}{2}QV = \frac{1}{2}CV^2} \qquad (17.4)$$

17.9 Examples of capacitors with simple geometry

The value of C for a particular capacitor depends on the geometrical shape and arrangement of the plates and on the electrical properties of the insulating medium (the 'dielectric') in which the plates are immersed.

When the geometry of the plates exhibits a sufficient degree of symmetry it is relatively easy to derive an expression for the capacitance of the system. The general procedure should be clear from the two examples which follow.

(a) *Parallel plate capacitor*

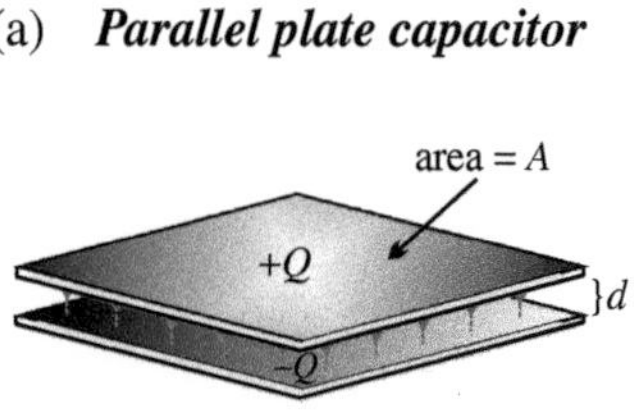

Figure 17.24. A parallel plate capacitor.

The conductors of a parallel plate capacitor, as the name suggests, are flat plates which are evenly separated as indicated in Figure 17.24. Let the area of each plate be A and the separation between the plates be d. If the area of the plates is sufficiently large (dimensions of a plate $\gg d$), the charge Q will be distributed uniformly over the surface of the plates and, hence, the electric field between the plates will be uniform; that is, any variation from uniformity near the edges of the plates ('edge effects') can be neglected. If the (constant) electric field strength in the space between the plates is E, the magnitude of the potential difference between the plates is given by

$$V = \int Edl = E\int dl = Ed$$

and hence

$$E = \frac{V}{d} \qquad (17.5)$$

If the magnitude of the charge on each plate is Q, the total electric flux emanating from the positively charged plate is $\Phi_E = Q$ and hence the electric flux density in the space between the plates is given by

$$D = \frac{\Phi_E}{A} = \frac{Q}{A}$$

independent of the material filling the space between the plates. Now since $D = \varepsilon E$, where ε is the permittivity of that material, we get

$$\frac{Q}{A} = \varepsilon \frac{V}{d}$$

and hence the capacitance of the parallel plate capacitor is given by

$$C = \frac{Q}{V} = \varepsilon \frac{A}{d} \quad (17.6)$$

From Equation (17.6), we see that ε has the dimensions of capacitance per unit length and, indeed, values of permittivity are usually given in such units; for example, the permittivity of vacuum $\varepsilon_0 = 8.85 \times 10^{-12}$ F m^{-1}.

To get a feeling for the magnitude of the capacitance of a simple parallel plate capacitor, consider the case of such a capacitor for which $A = 10$ cm^2, $d = 1$ mm and the space between the plates is filled with air, $\varepsilon_{air} \approx \varepsilon_0$. In this case, therefore,

$$C = (8.85 \times 10^{-12}\ \text{F m}^{-1}) \frac{10 \times 10^{-4}\ \text{m}^2}{10^{-3}\ \text{m}} \approx 10^{-11}\ \text{F} = 10\ \text{pF}$$

If such a capacitor were charged so that the potential difference between the plates was (say) 12 V, the magnitude of the charge on a plate would be

$$Q = CV = (10^{-11}\text{F})(12\ \text{V}) = 1.2 \times 10^{-10}\ \text{C} = 0.12\ \text{nC}$$

and, from Equation (17.4), the energy stored would be

$$W_{str} = \frac{1}{2}CV^2 = \frac{1}{2}(10^{-11}\ \text{F})(12\ \text{V})^2 = 7.2 \times 10^{-10}\ \text{J} = 720\ \text{pJ}$$

(b) ***Cylindrical capacitor***

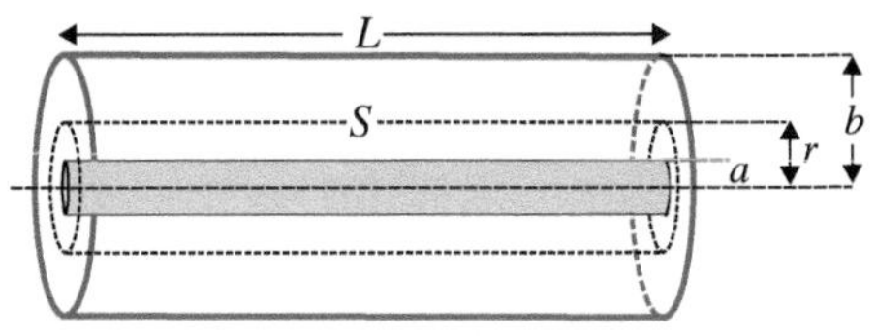

Figure 17.25. A cylindrical capacitor.

Capacitors used in everyday electrical circuits are often cylindrical in shape. Consider the capacitor shown in Figure 17.25 which comprises an inner cylindrical conductor of radius a and an outer coaxial cylindrical conductor of radius b (where $b > a$). The space between these conductors is filled with a dielectric material of permittivity ε. Both cylinders have the same length L and we assume that $L \gg a, b$ so that, again, edge effects may be neglected. Suppose that the inner conductor ('plate') has a charge Q and the outer conductor has a charge $-Q$ and thus the total electric flux from the inner to the outer cylinder is confined to the region between the plates and is given by

$$\Phi_E = Q$$

We can use Gauss's law to determine an expression for the electric flux density, and hence the electric field strength, at points at a distance r from the axis of the system ($a < r < b$). Because of the cylindrical symmetry, the field is radial in the plane perpendicular to the axis of the cylinder. Hence the total flux passing outwards through the surface S (indicated by the dashed lines in Figure 17.25 — a so-called 'Gaussian surface' — which in this case is a cylindrical surface of radius r and length L) is given by

$$\Phi_E = \int\!\!\int_S \mathbf{D} \cdot d\mathbf{A} = |D|(2\pi r)L = \sum_i q_i = Q$$

from which we can write $D(r) = \dfrac{Q}{2\pi L}\dfrac{1}{r}$ and

$$E(r) = \frac{Q}{2\pi\varepsilon L}\frac{1}{r}$$

Thus, the potential difference between the plates is given by

$$V = -\int \mathbf{E} \cdot d\mathbf{l} = -\int_b^a \frac{Q}{2\pi\varepsilon L}\frac{1}{r}dr = \frac{Q}{2\pi\varepsilon L}\int_a^b \frac{dr}{r} = \frac{Q}{2\pi\varepsilon L}\ \ln\frac{b}{a}$$

where the path of integration has been taken along a radius. From this it follows that the capacitance of the system is given by

$$C = \frac{Q}{V} = 2\pi\varepsilon\frac{L}{\ln\frac{b}{a}}$$

As an illustration, let us determine the capacitance per unit length of a piece of coaxial TV cable for which $a = 1.2$ mm, $b = 3.5$ mm and the insulating material between the conductors has a permittivity of 2.5×10^{-11}F m^{-1}. In this case

$$\frac{C}{L} = 2\pi(2.5 \times 10^{-11}\,\mathrm{F\,m^{-1}})/\ln\left(\frac{3.5}{1.2}\right) \approx 1.5 \times 10^{-10}\,\mathrm{F\,m^{-1}} = 150\,\mathrm{pF\,m^{-1}}.$$

A convenient way to obtain a specific low capacitance is to use an appropriate length of coaxial cable.

Edge effects and stray capacitance

In practice in many situations, edge effects cannot be completely neglected. Figure 17.26(a) represents the case of a parallel plate capacitor where the separation between the plates is not very much less the dimensions of the plates. In this case, the charge will not remain uniform over the full extent of the plates causing a distortion of the flux near the edges as shown. Thus, the electric field between the plates will be reduced towards the sides of the plates.

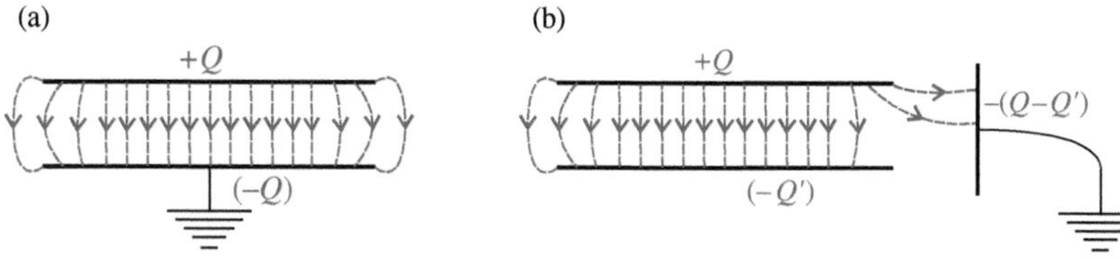

Figure 17.26. Illustration of (a) edge effects and (b) stray capacitance in the case of a parallel plate capalcitor.

A further complication may arise if there is another earthed object in the vicinity of the capacitor. In this situation some of the flux may end up on this object instead of the earthed plate of the capacitor (Figure 17.26(b)), with the result that capacitance is reduced from its parallel plate value.

Study Worked Examples 17.2 and 17.3

For problems based on the material presented in this section visit up.ucc.ie/17/ *and follow the link to the problems.*

17.10 Energy stored in an electric field

We recall that the energy stored in a capacitor, when there is a potential difference V between the plates, is given by $W^{\mathrm{E}}_{\mathrm{str}} = \frac{1}{2}CV^2$ (Equation (17.4)). In the simple case of a parallel plate capacitor where the capacitance is given by Equation (17.6), namely

$$C = \varepsilon\frac{A}{d}$$

this becomes

$$W^{\mathrm{E}}_{\mathrm{str}} = \frac{1}{2}\varepsilon\frac{A}{d}V^2 = \frac{1}{2}\varepsilon\frac{A}{d}(Ed)^2$$

where we have used Equation (17.5) to replace the potential difference by Ed, where E is the electric field strength between the plates. Thus

$$W^{\mathrm{E}}_{\mathrm{str}} = \frac{1}{2}\varepsilon(Ad)E^2 = \frac{1}{2}\varepsilon(\mathcal{V})E^2 = \frac{1}{2}\varepsilon E^2\mathcal{V}$$

where $\mathcal{V}$ is the volume of the space between the plates. The energy per unit volume, usually called the **energy density**, is given by

$$w^{\mathrm{E}}_{\mathrm{str}} = \frac{dW^{\mathrm{E}}_{\mathrm{str}}}{d\mathcal{V}} = \frac{1}{2}\varepsilon E^2 = \frac{1}{2}ED$$

The latter result is independent of the source of the field since any uniform field may be considered as having been produced by an 'equivalent' parallel plate capacitor. We can also assume that it may be extended to the more general case of the energy density at a point in a non-uniform electric field since, in real situations, the field may be considered to be sensibly uniform over a small region surrounding the point.

17.11 Capacitors in series and in parallel

Capacitors in series

Consider two capacitors which are connected together as shown in Figure 17.27; capacitors joined in this way are said to be connected *in series*. We wish to determine the capacitance of the single capacitor which is equivalent to the combination of the two capacitors in series; we can then use this single value to represent the combined effect of the two capacitors in an electrical circuit. If a charge Q is deposited on the left hand plate of the first capacitor (capacitance = C_1), then a charge $-Q$ will be induced on the opposite plate which, in turn, leaves a charge $+Q$ on the left hand plate of the second capacitor (capacitance = C_2) which finally induces a charge $-Q$ on the right hand plate of this capacitor. Now the potential difference across the combination is equal to the sum of the potential differences across each component, that is

$$V = V_1 + V_2$$

and hence, from the definition of capacitance

$$V = \frac{Q}{C} = \frac{Q}{C_1} + \frac{Q}{C_2} \quad \Rightarrow \quad \frac{1}{C} = \frac{1}{C_1} + \frac{1}{C_2}$$

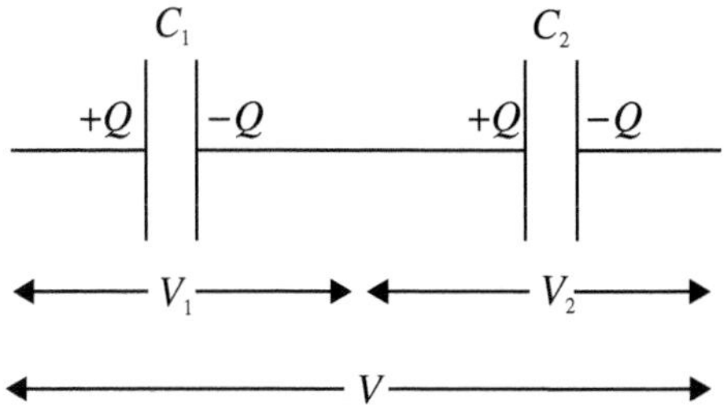

Figure 17.27. Capacitors connected in series. The single capacitor C is equivalent to the series combination C_1 and C_2.

More generally, for any number of capacitors connected in series, we see that the reciprocal of the capacitance of the combination is equal to the sum of the reciprocals of the individual capacitances.

$$\boxed{\frac{1}{C} = \frac{1}{C_1} + \frac{1}{C_2} + \frac{1}{C_3} + \ldots}$$

Capacitors in parallel

Figure 17.28 shows two capacitors connected *in parallel*; the left hand plates of both capacitors are joined together, as are the two right hand plates. In this case the potential difference between the plates is the same for both capacitors. When a charge Q is applied to the plates, it distributes itself between the two plates with Q_1 on the first capacitor and Q_2 on the second capacitor so that $Q = Q_1 + Q_2$ and hence

$$CV = C_1V + C_2V$$

where V is the common potential difference across the plates of each capacitor. Thus $C = C_1 + C_2$ and extending the analysis to any number of capacitors connected in parallel

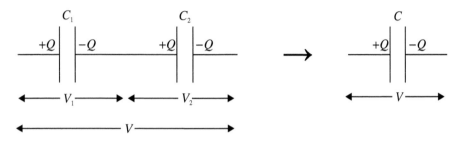

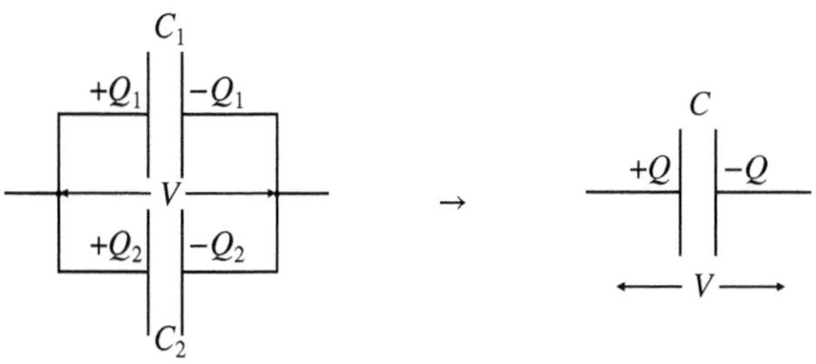

Figure 17.28. Capacitors connected in parallel.

$$\boxed{C = C_1 + C_2 + C_3 + \ldots}$$

Note the contrast between the relationships for capacitors in series and parallel and the corresponding relationships for resistors, Equations (15.6) and (15.12).

For problems based on the material presented in this section visit up.ucc.ie/17/ and follow the link to the problems.

17.12 Charge and discharge of a capacitor through a resistor

Discharge of a capacitor through a resistor

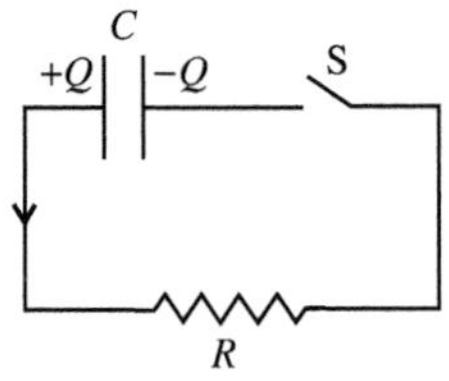

Figure 17.29. Discharge of a capacitor through a resistor.

Consider a capacitor with charges $+Q$ and $-Q$ on its plates as shown in Figure 17.29. Initially, when the switch S is open, the potential difference across the plates is $V = \frac{Q}{C}$ and the potential difference across the resistor is zero. At some instant $t = 0$ the switch is closed and the capacitor discharges as a current flows through the resistor R. At a time t later, the charge on the capacitor will have dropped to some value $q < Q$ so that the potential difference across the plates is now $\frac{q}{C}$ and a potential difference of RI has developed across the resistor due to the current I flowing through it at time t. Since there is no source of emf in the circuit the net potential difference around the complete circuit is zero (this can be thought of as an extension of Kirchhoff's loop rule (Section 15.11) to take into consideration potential differences developed across capacitors), that is

$$-\frac{q}{C} + RI = 0$$

Since $I = -\frac{dq}{dt}$ we can write

$$\frac{q}{C} + R\frac{dq}{dt} = 0 \quad \text{or} \quad \frac{dq}{dt} = -\frac{1}{RC}q$$

This differential equation can be integrated as follows

$$\int \frac{dq}{q} = -\frac{1}{RC}\int dt' \quad \rightarrow \quad \ln q = -\frac{1}{RC}t + constant.$$

Invoking the initial condition that $q = Q$ at $t = 0$, we obtain $\ln Q = 0 + constant$ and hence

$$\ln q = -\frac{1}{RC}t + \ln Q \quad \rightarrow \quad \ln\frac{q}{Q} = -\frac{1}{RC}t$$

$$\rightarrow \quad q(t) = Qe^{-\frac{t}{RC}}$$

Thus, the charge on the plates of the capacitor falls off exponentially with time as shown in Figure 17.30. The time taken for the charge to fall by a factor $\frac{1}{e}$ is RC; this characteristic discharge time is called the **time constant** of the circuit. This type of circuit (called an 'RC circuit') is often used in electronic networks to produce waveforms with particular time characteristics; a specific value of the time constant can be selected by choosing particular values of R and C. If, for example, $C = 1$ μF and $R = 1$ kΩ, the time constant is $(10^3\ \Omega)(10^{-6}\text{F}) = 1$ ms.

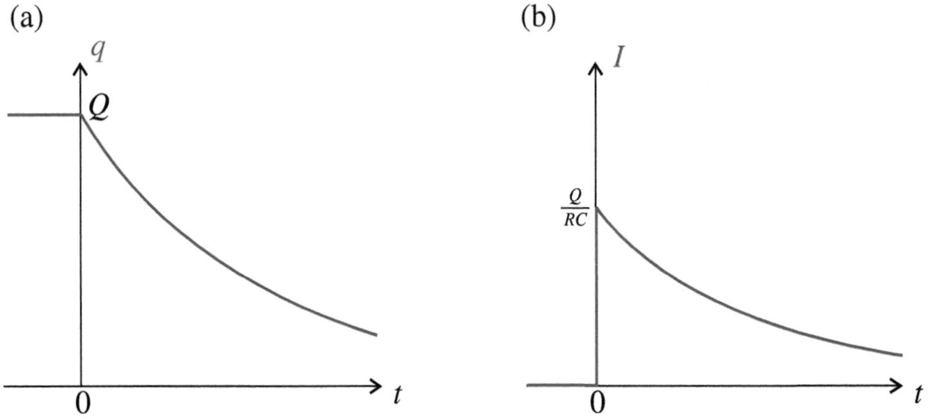

Figure 17.30. The charge on the capacitor plates (blue curve) and the current in the resistor (black curve) as functions of time after the switch in Figure 17.29 is closed.

The current through the resistor appears discontinuously as soon as the switch is closed, as indicated in Figure 17.30, and falls off exponentially thereafter, since

$$I(t) = -\frac{dq}{dt} = \frac{Q}{RC}e^{-t/RC} = I(0)e^{-t/RC}$$

Charging a capacitor through a resistor

Consider the circuit in Figure 17.31 where a capacitor is being charged through a resistor R using a power supply of emf E. Suppose that the capacitor is initially uncharged and that the switch S is closed at $t = 0$. Let the charge on the capacitor plates at some later time t

be q. Thus for $t \leq 0$, $q = 0$ and for $t > 0$ the emf is equal to the sum of the potential differences around the circuit (again, this is the generalisation of Kirchhoff's loop rule to take electrostatic potential differences into account). Thus for $t > 0$

$$E = RI + \frac{q}{C} \quad \text{or} \quad R\frac{dq}{dt} + \frac{q}{C} - E = 0.$$

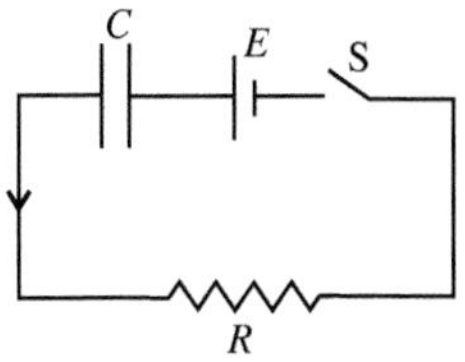

Figure 17.31. Charging a capacitor through a resistor.

If we write this as $RC\frac{dq}{dt} = EC - q$ and change the variable to θ where $\theta = EC - q$, then $\frac{d\theta}{dt} = -\frac{dq}{dt}$ and the differential equation can be written as $-RC\frac{d\theta}{dt} = \theta$. Integration of this equation yields

$$\int \frac{d\theta}{\theta} = -\frac{1}{RC}\int dt'$$

so that

$$\ln \theta = -\frac{1}{RC}t + constant$$

Applying the initial condition that $q = 0$ at $t = 0$ or $\theta = EC$ at $t = 0$, we obtain

$$\ln(EC - q) = -\frac{1}{RC}t + \ln(EC) \quad \text{or} \quad \ln\left(1 - \frac{q}{EC}\right) = -\frac{1}{RC}t$$

and thus

$$1 - \frac{q}{EC} = e^{-t/RC}$$

$$\rightarrow \quad q(t) = EC\left(1 - e^{-\frac{t}{RC}}\right)$$

where EC is the charge on the capacitor as $t \rightarrow \infty$. The charging curve for a capacitor, as determined by this equation, is shown in Figure 17.32(a); as before, the time constant RC determines the characteristic charging time.

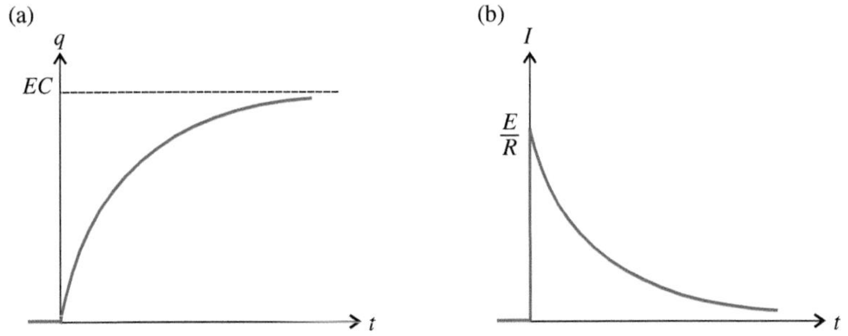

Figure 17.32. (a) The charge on the capacitor plates as a function of time after the switch in Figure 17.31 is closed and (b) the corresponding current in the circuit.

The current in the circuit (Figure 17.32(b)) depends on time as follows

$$I(t) = -\frac{dq}{dt} = EC\left(\frac{1}{RC}e^{-\frac{t}{RC}}\right) = \frac{E}{R}e^{-\frac{t}{RC}} = I(0)e^{-\frac{t}{RC}}$$

Study Worked Example 17.4

For problems based on the material presented in this section visit up.ucc.ie/17/ *and follow the link to the problems.*

17.13 Measurement of permittivity

Measurement of capacitance

The discharging curve (Figure 17.30) or the charging curve (Figure 17.32(a)) for a particular capacitor and resistor can be determined experimentally by connecting a high-resistance voltmeter, which can record potential differences which vary with time across the capacitor as it discharges or charges. Standard curve fitting software can be applied to this data to determine the 'best fit' value of the constant RC. If the value of R is known, a value for the capacitance C can be determined.

Measurement of permittivity

The value of the permittivity of any material which can be used as the medium between the plates of a capacitor of known geometry, such as the parallel plate or cylindrical capacitors discussed in Section 17.9, can be determined by measuring the capacitance of a capacitor in the manner described above.

Relative permittivities of some common insulators are given in Table 17.1.

Table 17.1 Relative permittivities of some common materials

material	relative permittivity (ε_r)
vacuum*	1
air	1.0005
teflon	2.1
polystyrene	2.17
lucite	2.8
Plexiglass (perspex)	3.4
glass	4 – 10
porcelain	5 – 7
mica	2.5 – 7
silicon	12
germanium	117
water	80

*Permittivity of vacuum (ε_0) = $8.854\,187\,816 \times 10^{-12}$ F m^{-1}

Further reading

Further material on some of the topics covered in this chapter may be found in *Electromagnetism* by Grant and Phillips (See BIBLIOGRAPHY for book details)

WORKED EXAMPLES

Worked Example 17.1: A charge q is located at a distance a from the centre of grounded conducting sphere of radius R. Determine the magnitude and location of a point image charge q' which, when substituted for the sphere, replicates the electric potential at all points previously occupied by the surface of the sphere.

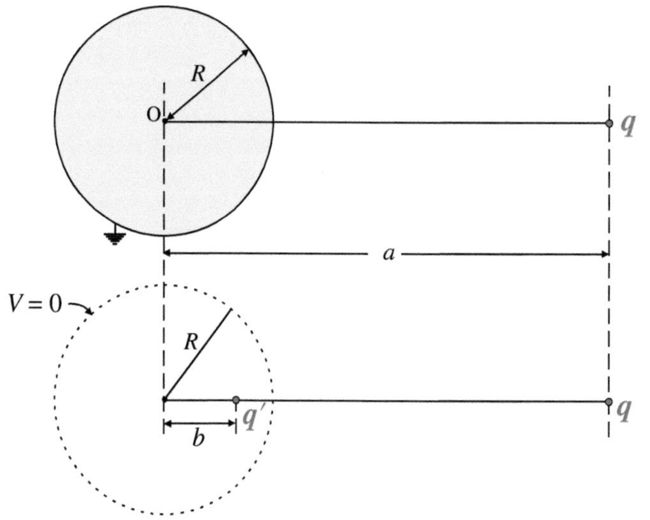

Figure 17.33. Worked Example 17.1: Upper figure: A point charge q is placed at a distance a from the centre of an earthed conducting sphere. The sphere is replaced (lower figure) by a point charge at a distance b ($< a$) as shown.

The upper part of Figure 17.33 shows a point charge placed at a distance a from the centre of an earthed conducting sphere of radius R.

The lower part of the figure shows the earthed sphere replaced by a point charge q' as indicated; this is one possible arrangement of point charges that results in zero potential at all points previously occupied by the surface of the sphere. Our objective is to determine values of q' and b such that the electrical potential at every point on this surface is zero. Let the point P in Figure 17.34 be one such point.

The electrical potential at P due to the two charges is given by

$$\varphi_P = \frac{q}{4\pi\varepsilon r} + \frac{q'}{4\pi\varepsilon r'} = 0$$

whence

$$\frac{q'}{q} = -\frac{r'}{r}$$

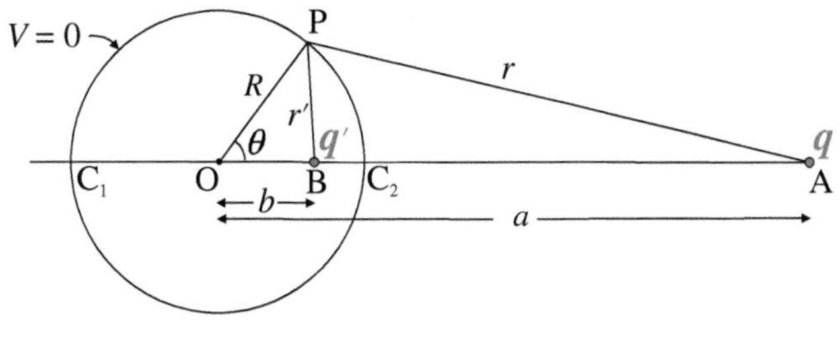

Figure 17.34. Worked Example 17.1.

The electrical potential at the points C_1 and C_2 is also zero so, similarly,

$$\frac{q'}{q} = -\frac{R+b}{R+a} \quad \text{and} \quad \frac{q'}{q} = -\frac{R-b}{a-R}$$

which yields $R^2 = ab$ and

$$\frac{r'}{r} = -\frac{q'}{q} = \frac{R+b}{R+a} = \frac{R+\frac{R^2}{a}}{R+a} = \frac{R}{a}$$

$$\rightarrow \quad q' = -\frac{R}{a}q \quad \text{and} \quad b = \frac{R^2}{a}$$

Thus, the magnitude (q') and location (b) of the 'image charge' are determined and the electric field at any point outside the conducting sphere is determined by the combined contribution from this and the charge q.

Worked Example 17.2:

(a) Derive an equation for the capacitance of a spherical capacitor, comprising concentric spheres of radii a and b where $b > a$ and the medium between the shells has permittivity ε.

(b) Show that, when $(b-a) \ll a$, the equation reduces to that for a parallel plate capacitor.

(c) The inner and outer shells of a spherical capacitor have radii of 5.0 and 7.0 cm and are at potentials of 10.0 V and 6.0 V relative to earth, respectively. Determine the electric field strength at a point halfway between the shells.

(a) The potential drop between $r = a$ and $r = b$ is given by Equation (16.22),

$$\mathcal{V} = -\int_a^b \boldsymbol{E}\cdot d\boldsymbol{r} \text{ where } E = -\frac{Q}{4\pi\varepsilon r^2}\hat{\boldsymbol{r}}$$

Thus

$$V = \frac{Q}{4\pi\varepsilon}\int_a^b \frac{dr}{r^2} = \frac{Q}{4\pi\varepsilon}\left[-\frac{1}{r}\right]_a^b = \frac{Q}{4\pi\varepsilon}\left[\frac{1}{a}-\frac{1}{b}\right]$$

and hence, from the definition of capacitance (Section 17.8)

$$C = \frac{Q}{V} = 4\pi\varepsilon\left(\frac{ab}{b-a}\right) \tag{17.7}$$

(b) When $(b-a) \ll a$, $a \sim b$ and $4\pi ab \sim 4\pi a^2 \sim 4\pi b^2$, which is approximately the surface area A of either sphere. The distance between the plates $d = b - a$, so that in this case Equation (17.7) may be written $C = \dfrac{\varepsilon(4\pi a^2)}{b-a} = \dfrac{\varepsilon A}{d}$, which is Equation (17.6) for the capacitance of a parallel plate capacitor.

(c) $C = 4\pi\varepsilon\left(\dfrac{0.05 \times 0.07}{0.07 - 0.05}\right) \rightarrow \quad Q = CV = 4\pi\varepsilon\,(0.0035/0.02)\,(10.0 - 6.0) = 0.7(4\pi\varepsilon)$

At a point halfway between the shells ($r = 0.06$ m),

$$E = -\frac{Q}{4\pi\varepsilon r^2} = \frac{0.7\ 4\pi\varepsilon}{r^2\ 4\pi\varepsilon} = \frac{0.7}{(0.06)^2} = 194\ \text{V m}^{-1}.$$

Worked Example 17.3: A parallel plate capacitor is constructed by placing square metal sheets of side length 30 cm on either side of a sheet of polythene which is 20 mm thick (Figure 17.35).

(a) If the permittivity of polythene is 2.2×10^{-11} F m^{-1}, what is the capacitance of the system?

(b) If a 50 V power supply is connected across the plates calculate (i) the charge on the plates and (ii) the energy stored in the capacitor.

(c) If the polythene is removed but the separation of the plates is maintained at 2.0 mm, what is the charge on the plates?

(d) If the power supply had been disconnected from the plates prior to the removal of the polythene, what would have been the potential difference between the plates?

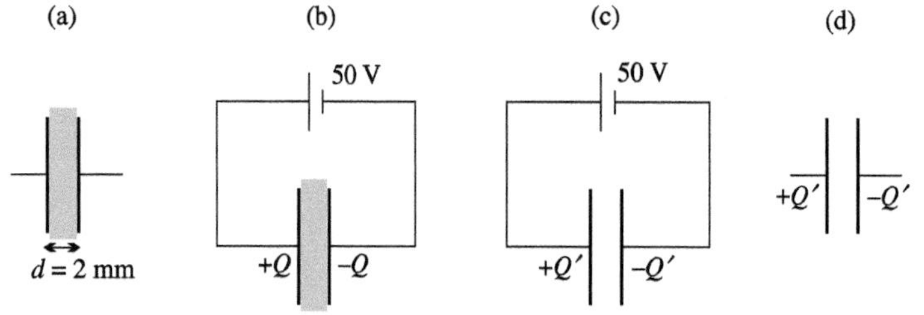

Figure 17.35. Worked Example 17.3.

(a) $C = \varepsilon\dfrac{A}{d} = (2.2\times10^{-11})\dfrac{(0.3)^2}{2\times10^{-2}} = 9.9\times10^{-10}\,\text{F} \approx 1\,\text{nF}$

(b) (i) $Q = CV = (9.9\times10^{-10})(50) = 4.95\times10^{-8}\,\text{C} \approx 50\,\text{nC}$

(ii) $E_{str} = \dfrac{1}{2}CV^2 = \dfrac{1}{2}(9.9\times10^{-10})(50)^2 = 1.24\times10^{-6}\,\text{J} = 1.24\,\mu\text{J}$

(c) $Q' = \dfrac{C'}{C}Q = \dfrac{\varepsilon_0}{\varepsilon}Q = \dfrac{8.85}{22}Q = 1.99\times10^{-8}\,\text{C} = 20\,\text{nC}$

(d) $V' = \dfrac{C}{C'}V = \dfrac{\varepsilon}{\varepsilon_0}V = \dfrac{22}{8.85}(50\,\text{V}) = 124\,\text{V}$

Worked Example 17.4: A capacitor, resistance and switch are connected in series. When the switch is closed ($t = 0$ s) the potential difference across the capacitor is 100 V. At $t = 10$ s the potential difference across the capacitor is measured to be 1.0 V. Determine

(a) the time constant of the circuit and
(b) the potential difference across the capacitor at $t = 17$ s.
By what factor should the resistance be changed to triple the potential difference at $t = 17$ s?

(a) The circuit is illustrated in Figure 17.29 and the equation describing the variation of the charge on the plates of the capacitor as a function of time is derived in the first part of Section 17.12, namely

$q(t) = Qe^{-\frac{t}{RC}}$, where RC is the time constant of the circuit.

At $t = 0$ s, the potential difference across the capacitor, $V_0 = \dfrac{Q}{C}$

At time t, $$V(t) = \frac{q(t)}{C} \rightarrow V(t) = V_0 e^{-\frac{t}{RC}} \qquad (17.8)$$

$V_0 = 100$ V and $V(t) = 1.0$ V when $t = 10$ s, $\rightarrow 1.0 = 100e^{-\frac{10}{RC}} \rightarrow 100 = e^{\frac{10}{RC}} \rightarrow \ln 100 = \dfrac{10}{RC} \rightarrow RC = \dfrac{10}{\ln 100} = \dfrac{10}{4.605} = 2.17$ s.

(b) At $t = 17$ s, Equation (17.8) gives $V(17) = 100e^{-\frac{17}{2.17}} = 0.04$ V

For $V(17) = (3\times0.04) = 0.12$ V, Equation (17.8) becomes $V(17) = 100e^{-\frac{17}{R'C}} = 0.12$ V

Thus $\ln\dfrac{0.12}{100} = -\dfrac{17}{R'C} \rightarrow R'C = 2.53$ s

To produce this increase in the time constant, R must increase by a factor $\dfrac{2.53}{2.17} = 1.17$

PROBLEMS

For problems based on the material covered this chapter visit up.ucc.ie/17/ and follow the link to the problems.

18

Magnetic fields

AIMS

- to investigate and ascertain the laws – *Ampere's law* and the *Biot-Savart law* – which allow us to predict the strength of the magnetic field generated by any configuration of electric currents
- to determine relationships which can be used to predict the strength of the magnetic field generated by a system of magnets
- to develop a general formalism which enables a description of the relationship between magnetic fields and their sources to be made

18.1 Magnetism

For at least two thousand years, magnetic phenomena have been well known and put to practical use. Pieces of naturally occurring ore (called lodestone or loadstone) were known to attract or repel one another and to be able to pick up small pieces of other materials, notably objects containing iron, nickel or cobalt. In time, methods were devised to make artificial permanent magnets which behaved in a similar fashion, but which exhibited much stronger effects. The simplest such man-made magnet is the familiar **bar magnet**, which is essentially the magnetic equivalent of the bar electret (Section 17.6).

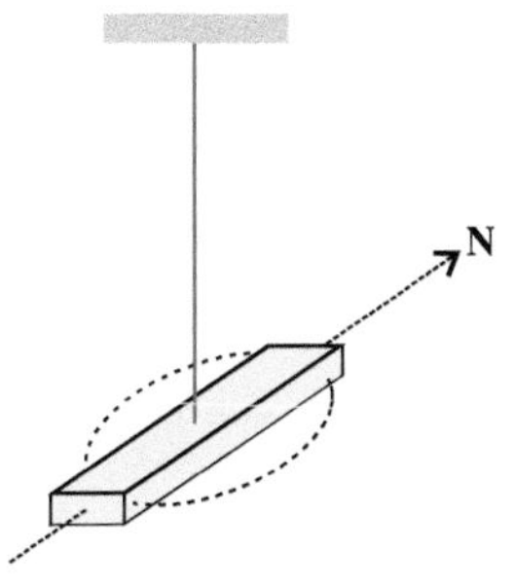

Figure 18.1. A bar magnet suspended so that it is free to rotate in a horizontal plane aligns along a north-south direction.

If a bar magnet is set up so that it is free to rotate in a horizontal plane, for example by hanging it from a thread (illustrated in Figure 18.1) or by placing it on a piece of wood which is floating in water, it will be noticed that the magnet rotates until its long axis is aligned generally in a north–south direction. This observation gave rise to the magnetic compass as used in navigation, the 'needle' of a compass being a small light bar magnet.

Further understanding of the behaviour of permanent magnets may be gleaned from an experiment along the following lines. A sheet of glass is placed on top of a bar magnet lying on a horizontal surface. Iron filings (small thin needles of iron) are scattered as uniformly as possible over the glass. The filings tend to congregate near each end of the magnet and tend to orient themselves towards two points inside the magnet, implying that the magnetic effect is strongest near these two points (Figure 18.2(a)).

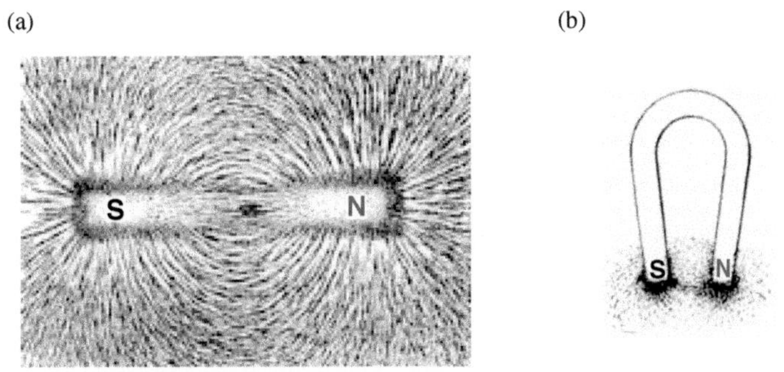

Figure 18.2. (a) Iron filings scattered uniformly around a bar magnet distribute themselves as shown. The points, near either end of the magnet, towards which the nearby iron filings point are called magnetic poles. (b) Iron filings around a horseshoe magnet.

Understanding Physics, Third Edition. Michael Mansfield and Colm O'Sullivan.

© 2020 John Wiley & Sons Ltd. Published 2020 by John Wiley & Sons Ltd.

The magnetic pole model

The behaviour described above suggests a *model* of a bar magnet which is indicated in Figure 18.2 The magnet can be visualised as comprising two *magnetic poles*, one near each end. The pole which tends to point northwards (that is, the 'north seeking' pole) is called the N pole and the other is called the S pole.

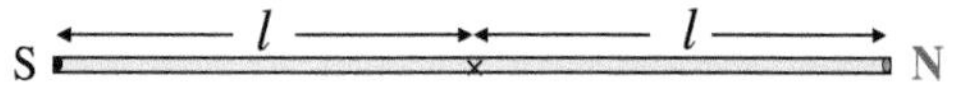

Figure 18.3. A magnetic needle has opposite poles at each end.

Permanent magnets can also be made in a great variety of shapes and forms; a particularly common and useful device is the 'horseshoe' magnet (Figure 18.2(b)), which may be thought of as a bar magnet which has been bent so that the two poles are brought close together.

The torque acting on the magnet which causes it to align along the north–south direction can be thought of as arising from forces on each pole, the N pole being pulled northward and the S pole southward. If a bar magnet is placed on a piece of wood which is then floated on a dish of water, it will rotate until the axis is aligned approximately in a north-south direction, but there is never a tendency for the magnet as a whole to move either northward or southward. This experimental observation, that there is no net translational force, implies that the forces on each pole must be equal and opposite and that the poles must be considered to have, in some sense, equal strength. This is exactly analogous to the bar electret discussed in Section 17.6 but with, in this case, an equal and opposite 'magnetic charge' (magnetic pole) instead of the equal and opposite electric charge of the electret.

Similar to the case of long thin electrets (Case 2 of Section 17.6), long thin permanent magnets ('*magnetic needles*') can be fabricated, in which case there will be effectively a **point** magnetic pole at each end (Figure 18.3) forming a **magnetic dipole**.

Experiments may be performed with these needles to study the forces between such magnetic poles (indeed, Coulomb performed experiments along these lines using the same torsion balance that he used when studying the force between electric charges – recall Section 16.3). If the N pole of one such magnet is brought close to N pole of the other, they tend to push each other apart (Figure 18.4). This repulsive force is also observed between two S poles, but the force between an S pole and an N pole is found to be attractive. In other words, *like poles repel and unlike poles attract each other*. There is a clear parallel here with the electrostatic force between charges (Section 16.1); the formalism that will be developed in this chapter to describe magnetic systems will have strong analogies with that used in electrostatics.

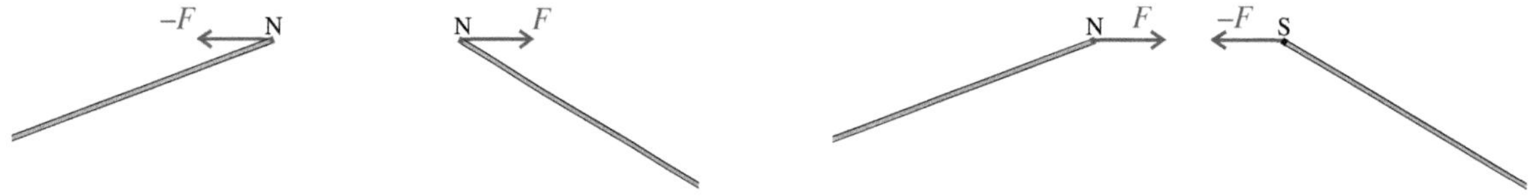

Figure 18.4. Forces between magnetic poles.

Note that the **magnetic pole model** used here does not require that individual magnetic poles can exist as separate entities; no free single pole (magnetic monopole) has ever been observed. This does not mean that the idea of a magnetic pole should be abandoned immediately; indeed, as we shall see, the model proves to be a powerful tool in the study of forces between magnets, the behaviour of compass needles, etc. It also proves a very convenient way of defining some of the physical quantities that we will use to formulate the theory of magnetism. It does not, however, provide a complete picture and will be reinterpreted in Section 19.9 after which the pole model will be discarded in favour of a more sophisticated picture.

Magnetic field

Further consideration of Figure 18.2, for example, shows that those iron filings which are not very near one or other of the poles, take up orientations under the combined influence of both poles. The influence is seen clearly to reach out to points which are quite distant from the magnet. This is yet another example of 'action at a distance'. Following an approach similar to that adopted for gravitation and electrostatics, we can define *the region of influence of a magnet* as the magnetic force field or the *magnetic field* of the magnet. This field is clearly strongest near the poles and decreases with increasing distance from the magnet. Again, we can introduce the idea of field lines to describe the 'shape' of the field. By convention, a *magnetic field line* is defined as a line along which a free N pole would move if released from rest. The fact that free magnetic poles have never been observed and most probably do not exist, in no way invalidates this definition.

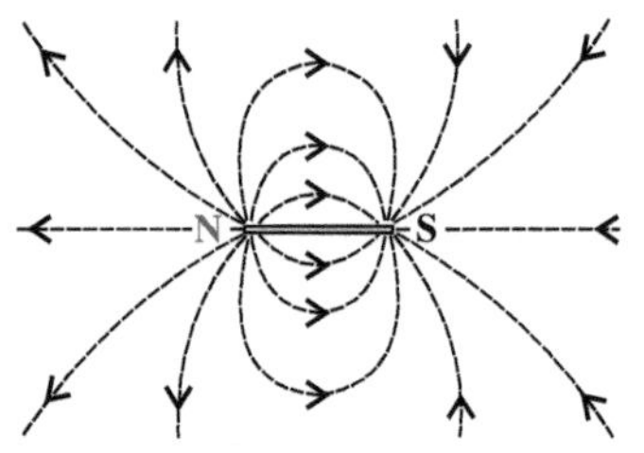

Figure 18.5. Magnetic fields lines around a magnetic needle.

The orientations of the iron filings in Figure 18.2 give a good indication of the direction of the field lines. A practical way of plotting magnetic field lines is to use a short magnetic compass; such a compass is essentially a magnetic needle suspended at its centre of gravity and free to rotate in a horizontal plane. The compass needle will line up along (tangentially to) the local field line. By marking the position of each end of the needle and then moving the compass so that the tail of the needle is placed where the head had been previously and repeating this process, individual field lines can be identified and plotted. Figure 18.5 shows some of the field lines in the neighbourhood of a long thin bar magnet. That the shape of this field looks similar to that of an electric dipole (Section 16.6) is not a coincidence and indicates a parallel between electrostatic and magnetostatic phenomena which will discussed in more detail later.

Terrestrial magnetism

The magnetic field of the Earth is believed to be due to the motion of molten material in the core and to the existence of magnetic materials in the crust. The resulting magnetic field is complicated and non-static. It turns out, however, that a reasonable description can be given by considering the field to be due to a simple magnetic dipole (Figure 18.6(a) and Figure 18.6(b)), which has a S pole in the northern hemisphere and a N pole in the southern hemisphere. The field lines at any location on the surface of the Earth will make an angle with the horizontal (called the 'angle of dip' or the 'inclination' — δ in Figure 18.6(a)) which is zero at the magnetic equator and increases with increasing latitude until it reaches 90° at the north magnetic pole.

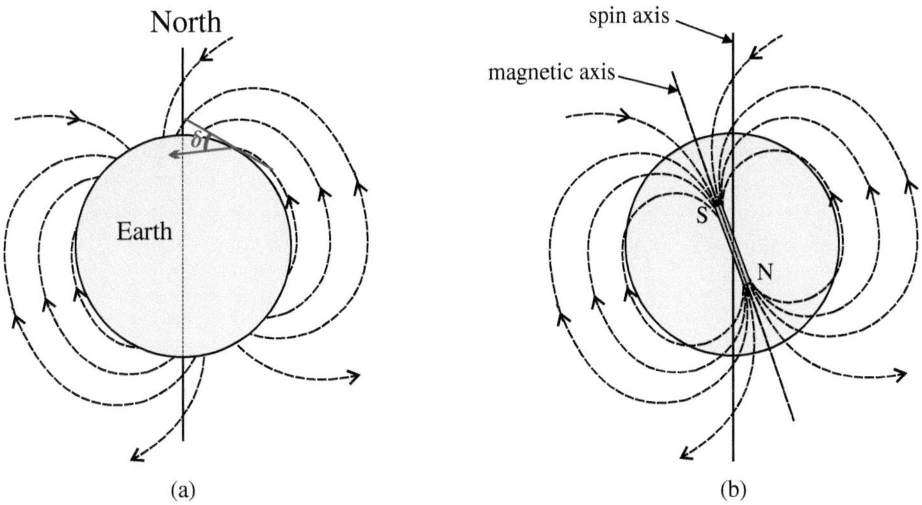

Figure 18.6. (a) Magnetic field lines around the Earth. (b) Dipole model of the Earth's magnetic field. The view shown is from outside the Earth at approximately 0° longitude in the equatorial plane.

Since the axis of the dipole does not coincide with the spin axis of the Earth, magnetic north is different from true (geographic) north. Furthermore, both the strength of the Earth's magnetic field and the orientation of its dipole axis vary on timescales of thousands of years.

A bar magnet or compass needle which is allowed to rotate in a horizontal plane will be affected only by the horizontal component of the Earth's magnetic field, and so it will tend to rotate until its magnetic axis lies in a vertical plane called the *magnetic meridian*. At most locations on the surface of the Earth there is a small but significant angle (called the *magnetic variation* or *declination*) between the magnetic and geographic meridians. The value of the magnetic variation must be known if a compass is to be used for navigation; it is usually given on maps.

The magnetic pole model will prove to be very effective in developing a description of the behaviour of permanent magnets; in particular, it will enable us to describe the forces between magnets in terms of the forces between their poles. At this point, however, the model remains purely qualitative and, before we can achieve a proper explanation of magnetic phenomena, it will be necessary to define appropriate physical quantities and to develop a theoretical formalism. It turns out, of course, that permanent magnets are not the only sources of magnetic effects; we have already come across an example of a magnetic field that is not produced by a permanent magnet, namely that surrounding a wire carrying an electric current encountered in Chapter 15. In that case in particular, we saw that the magnetic field lines surrounding a very long straight wire are closed loops in the form of concentric circles.

18.2 The work of Ampère, Biot, and Savart

In a remarkable five year period, between 1820 and 1825, the basic laws which describe the magnetic effects produced by electric currents were developed following initial discoveries by Hans Oersted (1777–1851). Most of this development was carried out in France where Jean Baptiste Biot (1774–1862), Felix Savart (1791–1841), and André-Marie Ampère (1775–1836) performed a range of careful experiments on which they built most of the theory which forms the basis of this chapter.

Figure 18.7(a) illustrates the sort of experiment attempted by Biot, Savart, and Ampère. A magnetic needle is fixed to a horizontal platform and a long straight wire carrying an electric current passes vertically through a hole in the platform. The whole platform is free to rotate about the wire. As we have seen (Section 15.1), the magnetic field lines are concentric circles centred on the wire (Figure 18.7(b) gives the view from above). Let us consider the forces on the magnetic needle. The force $\boldsymbol{F}_N$ on the N pole due to the influence of the magnetic field, tangential to the local field line, will tend to rotate the platform anticlockwise while the force $\boldsymbol{F}_S$ on the S pole, also tangential to its local field line, will tend to rotate the platform clockwise.

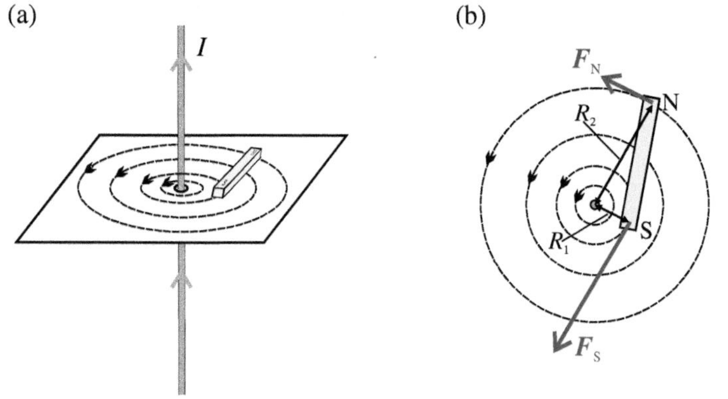

Figure 18.7. (a) A magnetic needle is attached to a horizontal platform and a long straight current carrying wire runs vertically through a hole at the centre of the platform. The platform is free to rotate about the wire in a horizontal plane. The field lines are concentric circles around the wire. (b) The magnetic needle in (a), viewed from above, showing the direction of the forces acting on each pole.

The net moment (Section 5.11) about the wire, therefore, will be

$$|\boldsymbol{F}_N|R_2 - |\boldsymbol{F}_S|R_1$$

where R_1 and R_2 are the perpendicular distances of the S and N poles, respectively, from the wire, as indicated in Figure 18.7(b).

Depending on how the force on a magnetic pole depends on distance from the wire, this net moment could tend to rotate the platform either anticlockwise (if $F_N R_2 > F_S R_1$) or clockwise (if $F_N R_2 < F_S R_1$). The investigators discovered that no experiment, however sensitive, exhibits such a rotation, implying that the net moment is always zero in this situation, that is

$$F_N R_2 = F_S R_1 \quad \text{or} \quad \frac{F_N}{F_S} = \frac{R_1}{R_2}$$

The conclusion which must be drawn from the experimental observations, therefore, is that the force on a magnetic pole at a distance R from a very long straight current carrying wire is inversely proportional to R, that is

$$F \propto \frac{1}{R}$$

The force on a magnetic pole will also depend on the strength of the electric current flowing in the wire. Indeed, it is clear from the following argument that the force must be directly proportional to the current. If we double the current, the effect is the same as placing another wire carrying the same current very close and parallel to the original wire. Each current will produce exactly the same force on the pole so that the net force is twice that produced by either wire alone. Thus, the force on a magnetic pole at a distance R from a very long straight wire carrying a current I is given by

$$F = k\frac{I}{R} \tag{18.1}$$

where k is a constant. The moment experienced by that pole about the wire is given by

$$M = FR = kI \tag{18.2}$$

and hence is independent of the location of the pole on the platform.

In the sections that follow we adopt the following strategy, designed to develop a formalism to describe the effects of currents and magnets. First we shall define a quantity (magnetic pole strength) which can be used as a measure of the strength of a magnet. This, in turn, will allow us to represent the magnetic effects of magnetic poles and electric currents in terms of the force fields produced by them. This will then allow us to analyse magnetic effects quantitatively.

18.3 Magnetic pole strength

Ampère, in particular, realised that, had the net moment on the magnet about the current carrying wire not been zero (that is, had the moments on the N and S poles not cancelled exactly), the experimental arrangement described above would have provided a convenient way of using the magnetic field to generate mechanical energy.

The mechanical work done by the magnetic field in moving a pole on a circular path, of radius R centred on the wire, once around the wire (ending up in its original position) is given by

$$W = \oint_C \boldsymbol{F} \cdot \boldsymbol{dl} = |F|(2\pi R) = M(2\pi) = (kI)(2\pi) = 2\pi kI$$

where we have used Equation (18.2) above. Note that $\oint_C \boldsymbol{F} \cdot \boldsymbol{dl}$ represents integration over a complete closed path. Since the field is conservative, the work done is independent of R and hence on the path taken. Of course, the work done against the field by the opposite pole is $-2\pi kI$ so that no net work is done if the magnet ends up in the position from which it started.

It is clear that, for a fixed current I, the value of the force F, and hence the work done depends only on the strength of the magnet, so that the constant k in Equation (18.1) is a measure of this strength. The quantity $2\pi k$ is defined as the **magnetic pole strength** of the magnet. Thus the magnetic pole strength (denoted by the symbol p) can be defined as the work done per unit current in bringing a pole once around the current, that is

$$\boxed{p := \frac{W}{I}} \tag{18.3}$$

and the constant k in equations (18.1) and (18.2) can be replaced by $\frac{p}{2\pi}$. The force on a magnetic pole of strength p at a distance R from a very long wire carrying a current I can then be written, from Equation (18.1), as

$$F = \frac{pI}{2\pi R} \tag{18.4}$$

and is directed along the tangent to the field line passing through the pole (Figure 18.7(b)).

Magnetic pole strength can be defined as an algebraic quantity if the following *convention of signs* is also adopted:

$$p > 0 \text{ for N poles} \quad \text{and} \quad p < 0 \text{ for S poles}$$

From the definition (18.3), the SI unit of magnetic pole strength, called the weber, is J A^{-1} (that is, 1 J A^{-1} = 1 weber = 1 Wb).

The magnetic pole strength of a bar magnet is the analogue in magnetostatics of the magnitude of the bound electric charge at the ends of a bar electret which we called 'electric pole strength' in Section 17.6. As in the case of the charges in the electret, the N and S magnetic poles are of equal and opposites strengths and cannot be separated.

18.4 Magnetic field strength

We can now define a physical quantity which is a measure of the strength of the field at a point in a magnetic field, analogous to the concept of electric field strength in Section 16.2. The *magnetic field strength* (sometimes called the 'magnetic field intensity') at a point in a magnetic field is defined as being the force per unit N pole at that point. Thus, magnetic field strength is defined in such a way that, if the end of a magnetic needle with pole of strength p were to be placed at a point in a magnetic field, the force experienced by the pole of the needle (Figure 18.8) is given by

$$F = pH$$

so that the **magnetic field strength** at a point $\boldsymbol{r}$ is defined by

$$\boxed{\boldsymbol{H}(\boldsymbol{r}) := \frac{\boldsymbol{F}(\boldsymbol{r})}{p}} \tag{18.5}$$

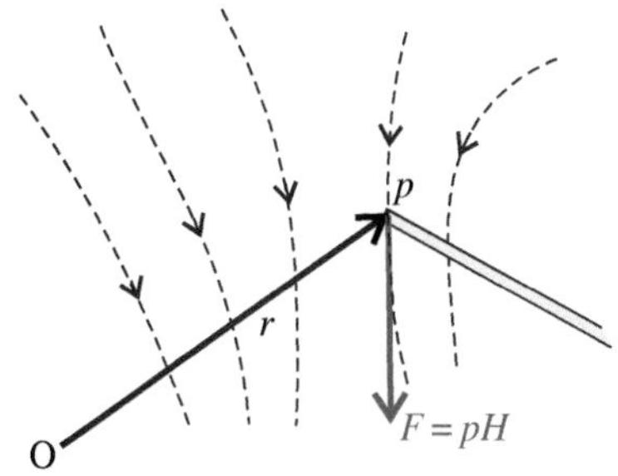

Figure 18.8. The force on the pole of a magnetic needle in a magnetic field is $p\boldsymbol{H}$, where $\boldsymbol{H}$ is the magnetic field strength at the position of the pole (at the end of the needle). The direction of the force is tangential to the magnetic field line passing through the point where the pole is located.

Since $\boldsymbol{F}$ is a vector, $\boldsymbol{H}$ is also a vector and a complete description of the field is provided if the value of $\boldsymbol{H}$ is known at all points in the field.

From the definition (18.5) of magnetic field strength, the SI unit of this quantity is N Wb^{-1}. Note, however, that we can write N Wb^{-1} = N/(J A^{-1}) = (N A)/N m = A m^{-1}. The latter form is more commonly used as the unit of magnetic field strength.

From Equation (18.4), we can see that the magnetic field strength at a point which is at a distance R from a very long straight wire carrying a current I is given by

$$H = \frac{I}{2\pi R} \tag{18.6}$$

The direction of $\boldsymbol{H}$ at any point is tangential to the local field line through the point in the sense given by the right-hand rule for the direction of current flow (Figure 15.4).

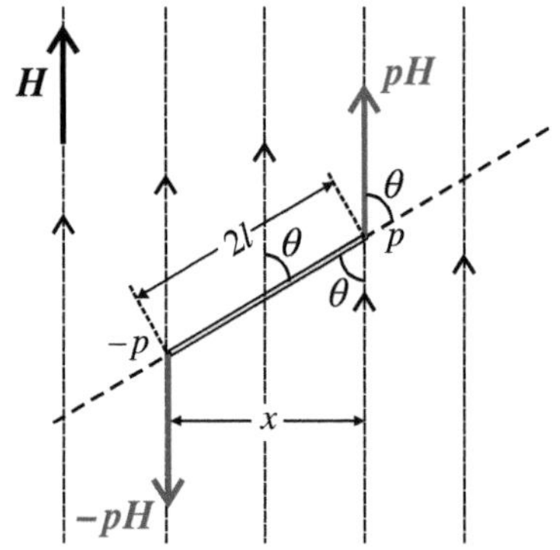

Figure 18.9. Torque on a bar magnet in a uniform magnetic field.

Torque on a magnet in a magnetic field

Consider a magnetic needle in a uniform magnetic field of strength $\boldsymbol{H}$, as shown in Figure 18.9. The magnet experiences the effect of equal and opposite forces of magnitude pH acting on each pole; in other words, the magnet experiences a torque. This torque, the moment of the couple, on the magnet is $(pH)x = pH(2l \sin\theta)$, where θ is the angle between the magnetic axis and the direction of the field. This can be written (Section 7.3) in vector form as

$$\boldsymbol{T} = 2pl\boldsymbol{n} \times \boldsymbol{H} \tag{18.7}$$

where $\boldsymbol{n}$ is a unit vector directed along the magnetic axis from the S to the N pole.

Study Worked Example 18.1

For problems based on the material presented in this section visit up.ucc.ie/18/ *and follow the link to the problems*

18.5 Ampère's law

Ampère went one important step beyond the ideas discussed above. Further experiments convinced him that no net mechanical work could be done in moving a magnet around any current (not just in the case of the long straight wire) provided that the magnet always completes a closed path; that is, it ends up in the exact position from which it started.

This led Ampère to postulate that the work done in moving a magnetic pole once around a current along any closed path is the same, independent of the geometry of the current carrying electric circuit. Thus the work done in moving a magnetic pole of strength p once around the closed path C in Figure 18.10 is the same as for the long straight wire and is given by

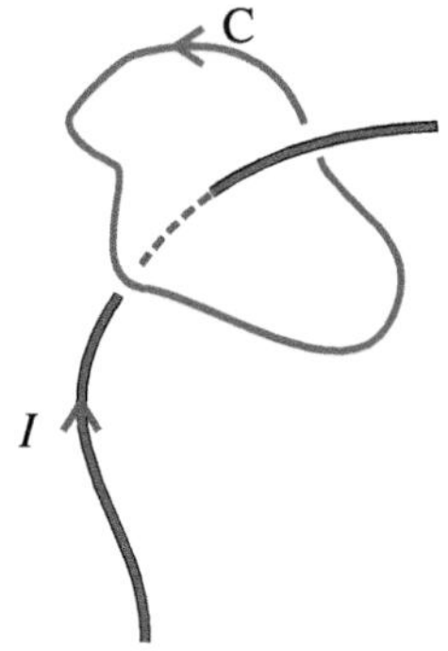

Figure 18.10. Ampère's law, Equation (18.8), holds for any closed path C around a current.

$$W = \oint_{\mathrm{C}} \boldsymbol{F} \cdot d\boldsymbol{l} = pI$$

Since, at any point, $\boldsymbol{F} = p\boldsymbol{H}$, this can be written as

$$\oint_{\mathrm{C}} \boldsymbol{H} \cdot d\boldsymbol{l} = I \tag{18.8}$$

Equation (18.8) is known as **Ampère's law**. Note that, for any circuit carrying a current I, the law holds for any and every possible closed path C. This means that, in principle, Ampère's law determines the magnetic field strength $\boldsymbol{H}$ at every point in space due to the current in the circuit. The complex geometry of circuits makes such an approach impractical for determining field strengths in most cases. However, as will be seen in the two examples below (Applications (a) and (b)), Ampère's law can be used to determine the magnetic field strength in situations in which the geometry of the circuit has a high degree of symmetry.

The current I in Equation (18.8) is the total current flowing through the closed loop C. In a case where the current is distributed throughout a conducting medium, Ampère's law can be generalised in terms of the current density (Section 15.2) as follows

$$\oint_{\mathrm{C}} \boldsymbol{H} \cdot d\boldsymbol{l} = \iint_{\mathrm{S}} \boldsymbol{J} \cdot d\boldsymbol{A} \tag{18.8a}$$

where C is any closed path and S is any surface bounded by C (Figure 18.11). As before, the convention for the positive direction of $\Delta\boldsymbol{A}$ must be consistent with the choice of positive sense of traversal around the closed path C.

Applications of Ampère's law

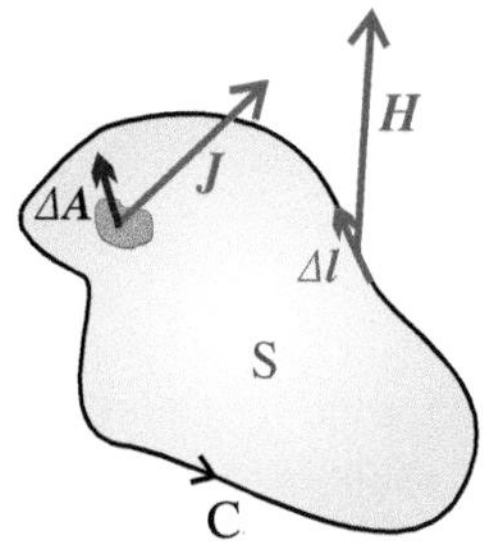

Figure 18.11. Illustration of the meaning of the terms used in Equation (18.8a). $\boldsymbol{H}$ is the magnetic field strength at the line element $\boldsymbol{\Delta l}$ on the curve C and $\boldsymbol{J}$ is the current density at the area element $\boldsymbol{\Delta A}$ on the surface S.

(a) ***Magnetic field due to uniform current in a cylindrical wire***
Consider a long straight wire of radius a, illustrated in Figure 18.12, carrying a current I which is uniformly distributed over the cross-sectional area of the wire, that is the current density $J = \dfrac{I}{\pi a^2} = constant$. If we apply Ampère's law to a circular path centred on the axis of the wire, then, from symmetry, the magnitude of the magnetic field strength is the same at any point on such a path.

Case 1: $R < a$: Applying Ampère's law to the circular path C inside the wire in Figure 18.12, we obtain

$$\oint_C \boldsymbol{H} \cdot d\boldsymbol{l} = I = \text{current inside radius } R = J(\pi R^2) = \frac{\pi R^2 I}{\pi a^2} = \frac{R^2 I}{a^2}$$

$$\rightarrow \quad |\boldsymbol{H}|(2\pi R) = \frac{R^2 I}{a^2} \quad \rightarrow \quad |\boldsymbol{H}| = \frac{I}{2\pi a^2} R$$

Case 2: $R > a$: Applying Ampère's law to the circular path C' outside the wire in Figure 18.12 we obtain

$$\oint_C \boldsymbol{H} \cdot d\boldsymbol{l} = |\boldsymbol{H}|(2\pi R) = I$$

and thus

$$H = \begin{cases} \dfrac{I}{2\pi a^2} R & \text{for } R < a \\[2ex] \dfrac{I}{2\pi R} & \text{for } R \geq a \end{cases}$$

The direction of the magnetic field strength at any point is tangential to the circle of radius R through that point, consistent with the convention adopted for the direction of current flow (Section 15.2).

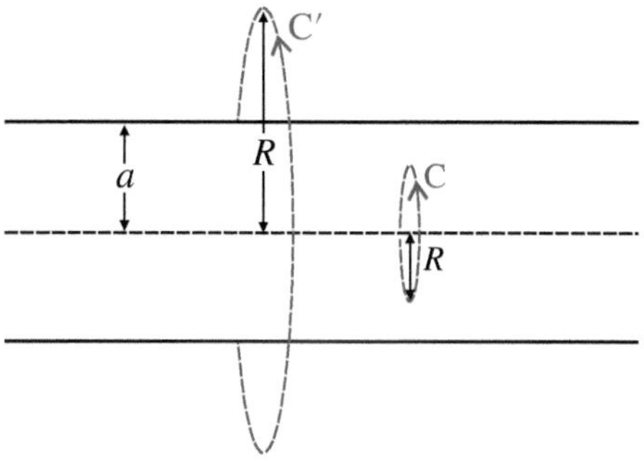

Figure 18.12. Determination of the expressions for the magnetic field strength at points inside and outside a wire of circular cross-section which carries a uniform current.

(b) ***Magnetic field in a long cylindrical solenoid***
Figure 18.13 shows a long solenoid comprising tightly wound turns of insulated wire in a cylindrical shape such that there are n_l turns per unit length. Let the current flowing in the wire be I. Consider now the rectangular path ABCDA shown, where the points A and D are at a very large distance from the solenoid. From Ampère's law

$$\oint_C \boldsymbol{H} \cdot d\boldsymbol{l} = n_l x I$$

since the total current enclosed by the path is $n_l x I$ where x is the distance BC. Thus

$$\int_A^B \boldsymbol{H} \cdot d\boldsymbol{l} + \int_B^C \boldsymbol{H} \cdot d\boldsymbol{l} + \int_C^D \boldsymbol{H} \cdot d\boldsymbol{l} + \int_D^A \boldsymbol{H} \cdot d\boldsymbol{l} = n_l x I$$

If A and D are so far from the solenoid that the magnetic field due to the solenoid can be neglected, then $\int_D^A \boldsymbol{H} \cdot d\boldsymbol{l} = 0$. From the symmetry of the situation, $\int_A^B \boldsymbol{H} \cdot d\boldsymbol{l} = -\int_C^D \boldsymbol{H} \cdot d\boldsymbol{l}$ and hence $\int_B^C \boldsymbol{H} \cdot d\boldsymbol{l} = n_l x I$.

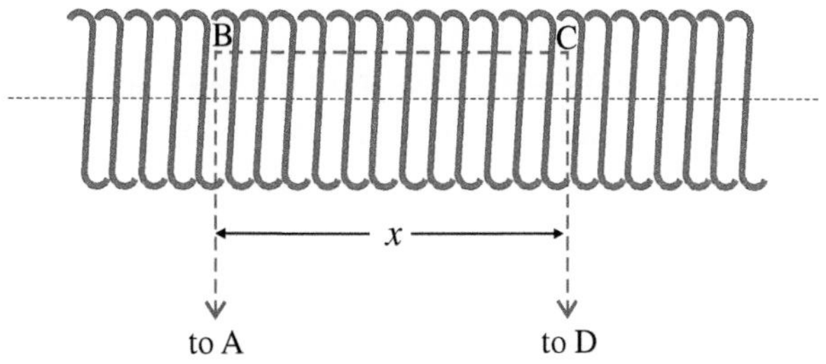

Figure 18.13. Determination of an expression for the magnetic field strength inside a long cylindrical solenoid.

If the solenoid is infinitely long, then (again from symmetry) H must be constant over the path BC and so

$$\int_{\mathrm{B}}^{\mathrm{C}} \boldsymbol{H}\cdot d\boldsymbol{l} = n_l x I \quad \rightarrow \quad H\int_{\mathrm{B}}^{\mathrm{C}} dl = Hx = n_l x I$$

whence

$$H = n_l I \tag{18.9}$$

To get a feel for the strength of the field produced in this way let us estimate the value of H inside a solenoid which is 2 m long and has 6000 turns carrying a current of 1 A. Provided that the diameter of the solenoid is very much less than its length (not more than a few centimetres in this case), the magnetic field strength is $\dfrac{6000}{2\mathrm{m}} \times 1\ \mathrm{A} = 3000\ \mathrm{A\,m^{-1}}$, which is over 200 times stronger than the geomagnetic field at the Earth's surface.

Note that, since this result is independent of the distance of BC from the axis of the solenoid, we must conclude that the magnetic field strength is the same everywhere inside an infinitely long solenoid; in other words, the field inside the solenoid is uniform. Note also that, by taking the points B and C outside the solenoid, we can deduce that the magnetic field strength outside the solenoid is zero, since no current is then enclosed by the path ABCDA.

Study Worked Example 18.2

For problems based on the material presented in this section visit up.ucc.ie/18/ *and follow the link to the problems*

18.6 The Biot-Savart law

As we have seen, Ampère's law can be applied satisfactorily to determine the magnetic field strength due to an electric current when the geometry of the circuit exhibits a high degree of symmetry. To make further progress, however, we need a prescription which would enable the magnetic field strength due to a current in a circuit of arbitrary geometrical shape to be determined at any point in space, at least in principle. This problem was also addressed by Ampère and by Biot and Savart. To formulate a rule which can be applied to a circuit of any shape, we need to know how to calculate the contribution to the magnetic field strength from any small element of the circuit. The field strength due to the whole circuit can then be determined by adding up the contributions from all such small elements (in other words, by integration over the whole circuit).

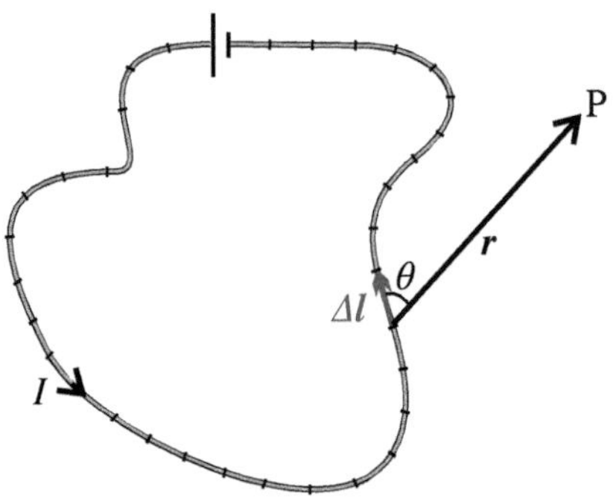

Figure 18.14. A circuit carrying a current I is broken up into a large number of small elements. The Biot-Savart law gives an expression for the contribution to the magnetic field strength at P due to the current I in the element $\boldsymbol{\Delta l}$.

Consider the situation illustrated in Figure 18.14. Let $\boldsymbol{\Delta H}$ be the contribution of the magnetic field strength at the point P due to the current I flowing in the element of wire $\boldsymbol{\Delta l}$. Applying the argument which we used in Section 18.2, we can conclude that $|\boldsymbol{\Delta H}| \propto I$. Furthermore in the limit $|\boldsymbol{\Delta l}| \rightarrow 0$, the contribution to the field from two identical adjacent small elements must be twice that from either one alone and thus we conclude that $|\boldsymbol{\Delta H}| \propto |\boldsymbol{\Delta l}|$. The question of how $\boldsymbol{\Delta H}$ depends on $\boldsymbol{r}$, the displacement of P relative to the element, is a more complicated issue, as we shall see.

Let us begin by assuming that the dependence of $\boldsymbol{\Delta H}$ on the distance r has the form of a power law (that is, we assume $\boldsymbol{\Delta H} \propto r^n$) and leave the question of its dependence on θ (the angle between $\boldsymbol{\Delta l}$ and $\boldsymbol{r}$) open for the time being. Thus, we will assume that the magnitude of the contribution to $\boldsymbol{H}(\boldsymbol{r})$ due to the current flowing in $\boldsymbol{\Delta l}$ is given by

$$|\boldsymbol{\Delta H}| = IF(\theta)r^n|\boldsymbol{\Delta l}| \tag{18.10}$$

where n is an (as yet) undetermined constant and $F(\theta)$ is an (as yet) unknown function of θ. Now, whatever the dependence on r and θ, there is one touchstone on which we can rely: when the contributions from all the elements are added up in the case of an infinitely long straight wire, the result must be that given by Equation (18.6), that is

$$\int_{-\infty}^{\infty} IF(\theta)r^n dx = \frac{I}{2\pi R} \tag{18.11}$$

The first step in evaluating the integral is to write each of the three variables r, θ and x in terms of one single variable. For this purpose, the most convenient variable to use is the angle φ as illustrated in Figure 18.15. The change of variables can be accomplished as follows:

(a) $\cos\varphi = \dfrac{R}{r} \rightarrow r = \dfrac{R}{\cos\varphi}$

(b) $\tan\varphi = \dfrac{x}{R} \rightarrow x = R\tan\varphi \rightarrow \dfrac{dx}{d\varphi} = R\sec^2\varphi \rightarrow dx = \dfrac{R}{\cos^2\varphi}d\varphi$

(c) $F(\theta) = f(\varphi)$

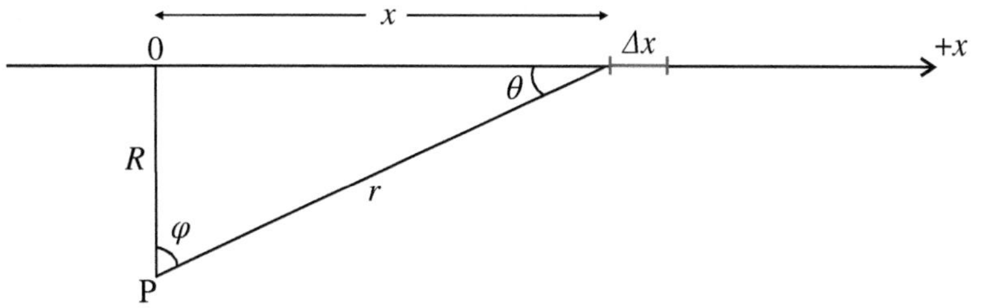

Figure 18.15. Variables used in the evaluation of the integral in Equation (18.11).

The integral on the left-hand side of Equation (18.11), therefore, becomes

$$I\int_{-\infty}^{\infty} F(\theta)r^n dx = I\int_{-\frac{\pi}{2}}^{\frac{\pi}{2}} f(\varphi)\frac{R^n}{\cos^n\varphi}\frac{R}{\cos^2\varphi}d\varphi = IR^{n+1}\int_{-\frac{\pi}{2}}^{\frac{\pi}{2}}\frac{f(\varphi)}{(\cos\varphi)^{n+2}}d\varphi$$

By equating this expression with the right hand side of Equation (18.11), we see that the only possible value of n is such that $R^{n+1} = R^{-1}$ and hence $n = -2$ in which case the denominator in the integral becomes $(\cos\varphi)^{n+2} = (\cos\varphi)^0 = 1$

We are left, finally, with the question of what function $F(\theta)$, or $f(\phi)$, should be adopted. Putting the value $n = -2$ in the above expression, we obtain the requirement that

$$I\int_{-\infty}^{\infty} F(\theta)r^{-2}dx = \frac{I}{R}\int_{-\frac{\pi}{2}}^{\frac{\pi}{2}} f(\varphi)d\varphi = \frac{I}{2\pi R} \quad \Rightarrow \quad \int_{-\frac{\pi}{2}}^{\frac{\pi}{2}} f(\varphi)d\varphi = \frac{1}{2\pi} \tag{18.12}$$

There is, in fact, a whole family of different functions which satisfy this condition equally well – a consequence of the fact that many different functions can yield the same result when integrated over a closed path. Historically Ampère, on the one hand, and Biot and Savart, on the other, proposed different solutions to this problem. The version due to Biot and Savart turns out to be more convenient in practical situations and is now adopted almost universally.

Since $\int_{-\frac{\pi}{2}}^{\frac{\pi}{2}} \cos\varphi d\varphi = \sin\left(\frac{\pi}{2}\right) - \sin\left(-\frac{\pi}{2}\right) = 2$, then $f(\varphi) = \frac{1}{4\pi}\cos\varphi$ will satisfy Equation (18.12). In terms of the angle $\theta = \frac{\pi}{2} - \varphi$ rather than φ, $F(\theta) = \frac{1}{4\pi}\sin\theta$ and therefore Equation (18.10) can be written

$$|\Delta \boldsymbol{H}| = \frac{I\sin\theta(\Delta l)}{4\pi r^2} \tag{18.13}$$

This relationship is usually known as the **law of Biot and Savart**, but it must be remembered that ***ΔH*** represents the contribution to ***H*** from the element ***Δl*** and hence the right-hand side of Equation (18.13) is meaningful only in the context of an integrand in an integral over some complete circuit.

The direction of ***ΔH***, given by the 'right hand rule' (Figure 15.4), is perpendicular to the plane defined by ***Δl*** and ***r*** (Figure 18.14), that is tangential to the field line through P produced by the current in ***Δl***. Accordingly, both the magnitude and direction of ***ΔH*** can be expressed by the single vector relationship

$$\Delta \boldsymbol{H} = \frac{I(\Delta \boldsymbol{l})\times\hat{\boldsymbol{r}}}{4\pi r^2} \tag{18.14}$$

where $\hat{\boldsymbol{r}}$ is a unit vector in the direction of ***r***. Note that the use of the cross product ensures that Equation (18.14) incorporates the right hand rule for the direction of current flow.

The magnetic field strength at the point P due to a complete circuit is given by the following vector form of the Biot–Savart law

$$\boldsymbol{H} = \frac{I}{4\pi}\oint_{\text{circuit}}\frac{d\boldsymbol{l}\times\hat{\boldsymbol{r}}}{r^2} \tag{18.14a}$$

In many cases the circuit is confined to a plane, in which case $\Delta\boldsymbol{l}\times\hat{\boldsymbol{r}}$ is perpendicular to the plane for all the elements. Thus, ***H*** at any point in the plane of the circuit is also perpendicular to the plane and its direction can be determined easily from the right hand rule for the direction of current flow.

It should be noted that the Biot-Savart law carries no more information than Ampère's law (Equation (18.8)) – they are just two different formulations of the same law of nature. The Biot-Savart law, however, is more useful for determining the magnetic field strength arising from circuits of known geometry but have insufficient symmetry to enable Ampère's law to be used.

18.7 Applications of the Biot-Savart law

(*a*) ***Magnetic field strength at the centre of a circular loop***

A very simple but important practical application of the Biot-Savart law is the case of a plane circular loop of wire of radius a carrying a current I as in Figure 18.16. In this case both $\theta \left(= \frac{\pi}{2}\right)$ and $r\,(= a)$ are constants for all elements $\boldsymbol{\Delta l}$. Thus we can write Equation (18.13) as

$$|\boldsymbol{\Delta H}_{\mathrm{C}}| = \frac{I \sin\frac{\pi}{2}(\Delta l)}{4\pi a^2} = \frac{I}{4\pi a^2}(\Delta l)$$

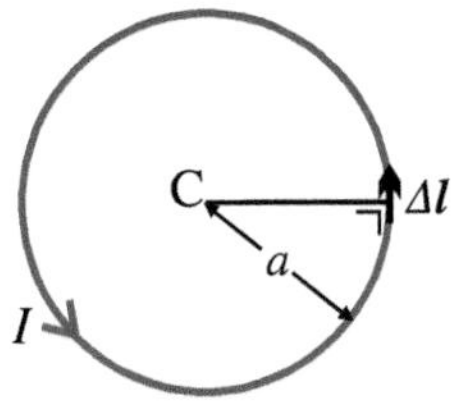

Figure 18.16. Determination of an expression for the magnetic field strength at the centre of a circular current carrying loop.

Adding up the contributions from all elements around the circuit (the circumference of the circle) yields

$$|\boldsymbol{H}_{\mathrm{C}}| = \frac{I}{4\pi a^2}\oint_{\text{circle}} dl = \frac{I}{4\pi a^2}(2\pi a) = \frac{I}{2a}$$

From the vector form of the Biot-Savart law or from the right hand rule, we can deduce that the direction of $\boldsymbol{H}_{\mathrm{C}}$ is perpendicular to the plane of the loop and, in the case of the current flow shown in Figure 18.16, is directed out of the plane of the page.

(*b*) ***Magnetic field strength at the centre of a rectangular coil***

It can be seen readily that the magnetic field strength at the point M, the centre of the rectangle in Figure 18.17, is the sum of the contributions from each of the four straight sections (AB, BC, CD and DA). Let us first determine the contribution from the top side AB, that is

$$|\boldsymbol{H}_{\mathrm{AB}}| = \frac{I}{4\pi}\int_{\mathrm{AB}} \frac{\sin\theta}{r^2}dx$$

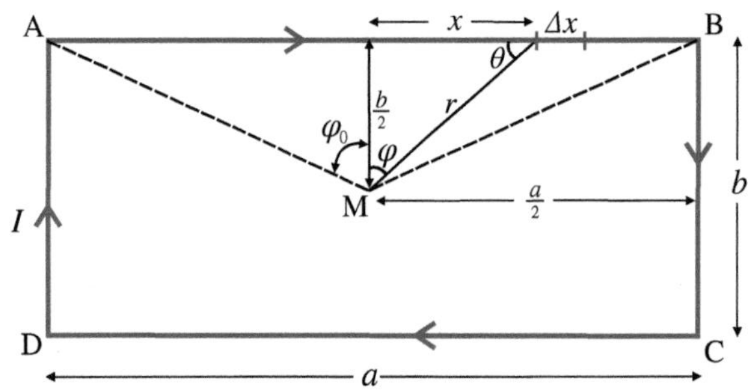

Figure 18.17. Determination of an expression for the magnetic field strength at the centre of a rectangular current carrying coil.

Making the same change in variable as in Section 18.6, that is $\sin\theta \to \cos\varphi$; $\quad r \to \dfrac{\frac{b}{2}}{\cos\varphi}$; $\quad dx \to \dfrac{\frac{b}{2}}{\cos^2\varphi}d\varphi$

$$|\boldsymbol{H}_{\mathrm{AB}}| = \frac{I}{4\pi}\int_{\mathrm{AB}} \cos\varphi \frac{\cos^2\varphi}{\left(\frac{b}{2}\right)^2}\frac{\frac{b}{2}}{\cos^2\varphi}d\varphi = \frac{I}{2\pi b}\int_{-\varphi_0}^{\varphi_0} \cos\varphi d\varphi = \frac{I}{2\pi b}[\sin\varphi_0 - \sin(-\varphi_0)]$$

$$= \frac{I}{\pi b}\sin\varphi_0 = \frac{I}{\pi b}\frac{a}{\sqrt{a^2+b^2}}$$

Now, by symmetry, $|\boldsymbol{H}_{\mathrm{CD}}| = |\boldsymbol{H}_{\mathrm{AB}}| = \dfrac{I}{\pi b}\dfrac{a}{\sqrt{a^2+b^2}}$

By interchanging a and b, this result can be applied to the sections BC and DA, that is

$$|\boldsymbol{H}_{\mathrm{BC}}| = |\boldsymbol{H}_{\mathrm{DA}}| = \frac{I}{\pi a}\frac{b}{\sqrt{a^2+b^2}}$$

Hence $$|\boldsymbol{H}_{\mathrm{M}}| = |\boldsymbol{H}_{\mathrm{AB}}| + |\boldsymbol{H}_{\mathrm{BC}}| + |\boldsymbol{H}_{\mathrm{CD}}| + |\boldsymbol{H}_{\mathrm{DA}}| = \frac{2I}{\pi\sqrt{a^2+b^2}}\left(\frac{a}{b}+\frac{b}{a}\right) = \frac{2I\sqrt{a^2+b^2}}{\pi ab}$$

In Figure 18.17, $\boldsymbol{H}_{\mathrm{M}}$ is directed perpendicular to the plane of the page and inwards.

Study Worked Example 18.3

For problems based on the material presented in this section visit up.ucc.ie/18/ *and follow the link to the problems*

18.8 Magnetic flux and magnetic flux density

The Biot_Savart law provides a procedure which allows us to determine the magnetic field strength at a point in the field produced by an electric current flowing in a circuit. Our next objective is to discover a procedure which will enable us to determine the magnetic field strength at a point in the field produced by a magnet or system of magnets. As in the case of gravitational and electrostatic fields, the concept of **magnetic flux** will prove very helpful in reaching this objective.

An examination of the field lines observed around long thin magnets (for example Figure 18.5) suggests a model in which the N poles act as sources of magnetic influence and S poles act as sinks. While nothing is actually emitted from N poles, the concept of a 'flux' emanating from N poles and 'flowing' to S poles proves useful in this context, just as electric flux proved useful in the treatment of the fields of electric charges in Chapter 16.

As before, the total flux from a source is also used as a measure of the strength of the source. Thus the flux concept is carried over to the case of magnetic fields simply by defining the total magnetic flux (Φ_{M}) from a pole as being equal to the magnetic pole strength, that is

$$\Phi_{\mathrm{M}} := p$$

Even when the magnet is not long and thin (such as in Figure 18.2(a)), the pole strength can be defined as the total flux emanating from the N pole end (or sinking at the S pole).

The intensity at some point in a magnetic field can be measured in terms of the magnetic flux per unit area at that point. Thus a vector quantity, called the **magnetic flux density** is defined by

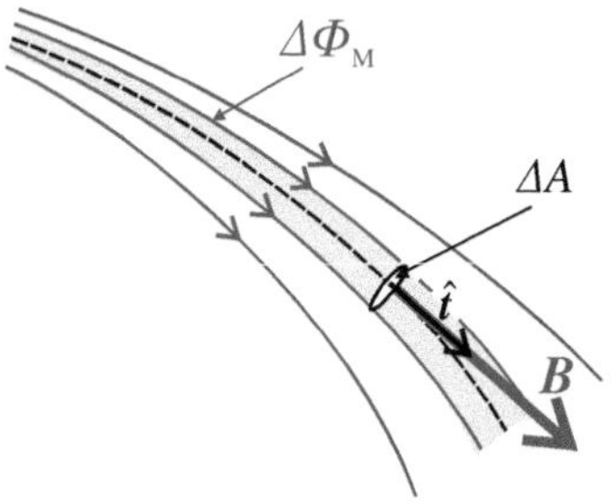

Figure 18.18. $\Delta\Phi_{\mathrm{M}}$ is the magnetic flux through the area ΔA. The magnetic flux density $\boldsymbol{B}$ is directed perpendicularly to the area ΔA, that is, tangentially to the local magnetic field line.

$$\boldsymbol{B} := \lim_{\Delta A \to 0} \frac{\Delta\Phi_{\mathrm{M}}}{\Delta A}\,\hat{\boldsymbol{t}}$$

where $\Delta\Phi_{\mathrm{M}}$ is the magnetic flux through an area ΔA which is perpendicular to the field lines through the point and $\hat{\boldsymbol{t}}$ is a unit vector tangential to the field lines, that is perpendicular to ΔA, as illustrated in Figure 18.18. The SI unit of magnetic flux density is Wb m^{-2}, called the tesla (after Nikola Tesla 1856–1943), that is, 1 tesla = 1 Wb m^{-2} = 1 T.[1]

From the definitions presented above, both the magnetic flux density $\boldsymbol{B}$ and the magnetic field strength $\boldsymbol{H}$ at a point in a non-conducting medium are directed along the tangent to the local magnetic field line through that point ($\hat{\boldsymbol{t}}$ in Figure 18.18). Thus $\boldsymbol{B}$ is parallel to $\boldsymbol{H}$, that is

$$\boldsymbol{B} = \mu\boldsymbol{H}$$

The value of μ depends on the nature of the medium in which the currents and/or magnetic poles which give rise to the magnetic field and the field itself are immersed. The quantity μ, therefore, is a property of the medium and is called the **permeability** of the medium. The distinction between $\boldsymbol{B}$ and $\boldsymbol{H}$ will become clearer in Section 19.9.

Strictly speaking permeability is constant only for linear, isotropic, homogeneous media. In a *non-isotropic* medium, such as inside a crystalline material, $\boldsymbol{B}$ may not be parallel to $\boldsymbol{H}$, in which case the permeability is a tensor quantity and $\boldsymbol{B}$ is related to $\boldsymbol{H}$ as follows

$$B_i = \sum_{i=1}^{i=3} \mu_{ij} H_j$$

where i = 1, 2, 3 stand for the x-, y- and z-directions. As in the case of dielectrics, we shall not consider anisotropic media in this book.

[1]Many textbooks call the quantity $\boldsymbol{B}$ the 'magnetic field' or the 'B-field' but these terms are not recommended by the International Union of Pure and Applied Physics (see Cohen and Giacomo (1987), Section 4.5, pp. 31–3).

18.9 Magnetic fields of permanent magnets; magnetic dipoles

Single magnetic pole

Consider the case of a single isolated magnetic pole of strength p, as illustrated in Figure 18.19 for the case of an N pole. In the case of a single pole, the magnetic field lines emerge radially outward from a N pole ($p > 0$) or converge inward on a S pole ($p < 0$). By symmetry the magnitude of the magnetic flux density is the same at all points at a distance r from the pole and is equal to the total flux from the pole divided by the surface area of a sphere of radius r. Thus

$$|\boldsymbol{B}| = \frac{\Phi_{\mathrm{M}}}{\text{surface area of sphere}} = \frac{p}{4\pi r^2}$$

or, in vector form,

$$\boldsymbol{B} = \frac{p}{4\pi r^2}\hat{\boldsymbol{r}} \tag{18.15}$$

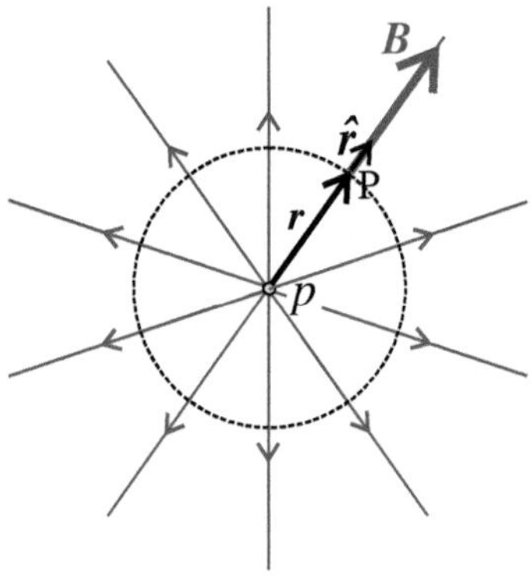

Figure 18.19. Magnetic field at a displacement $\boldsymbol{r}$ from a single pole of strength p — the field lines are drawn for an N pole ($p > 0$). By symmetry, the magnitude of the magnetic flux density is the same at all points on a sphere of radius r.

Assuming that the pole is immersed in a linear, isotropic, homogeneous medium of permeability μ, the magnetic field strength at a distance r from a single pole is given by

$$\boldsymbol{H} = \frac{\boldsymbol{B}}{\mu} = \frac{p}{4\pi\mu r^2}\hat{\boldsymbol{r}} \tag{18.16}$$

Magnetic dipole

Single poles do not exist in isolation but real situations, such as bar magnets or more complicated combinations of poles, can be dealt with by adding up the contributions to the flux density from each pole (the addition process being vectorial, of course). Hence the magnetic flux density at some arbitrary point P in the field of a magnetic needle (Figure 18.20) can be determined by applying Equation (18.15) to each pole as follows

$$\boldsymbol{B} = \frac{p}{4\pi r_+^2}\hat{\boldsymbol{r}}_+ + \frac{(-p)}{4\pi r_-^2}\hat{\boldsymbol{r}}_- \tag{18.17}$$

where, as shown in the figure, $\hat{\boldsymbol{r}}_+$ and $\hat{\boldsymbol{r}}_-$ are the unit vectors from the N pole and S pole, respectively, to P.

Equation (18.17) is formally identical to Equation (16.9) for the electric flux density from an electric dipole but with electric charge (q) replaced by magnetic pole strength (p). Accordingly, we can invoke the results derived in Section 16.6 and make the substitution $q \to p$, in which case we find that the magnetic flux density at the point P is given by

$$\boldsymbol{B}(\boldsymbol{r}) = \frac{2pl}{4\pi r^3}[(3\cos^2\theta - 1)\boldsymbol{i} + 3\sin\theta\cos\theta\boldsymbol{j})]$$

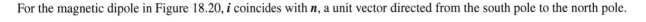

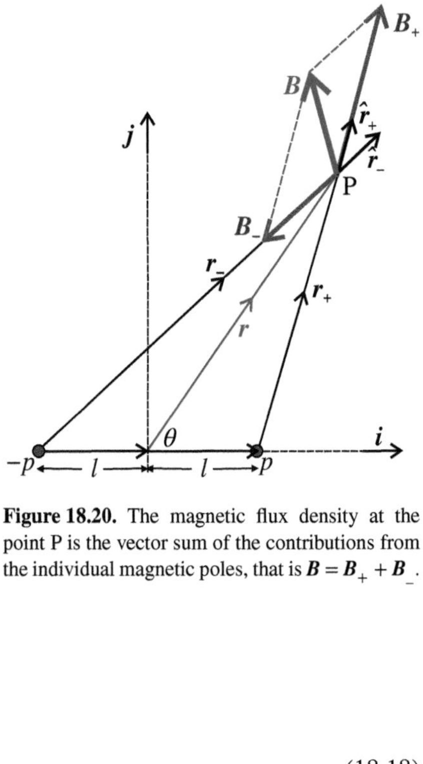

Figure 18.20. The magnetic flux density at the point P is the vector sum of the contributions from the individual magnetic poles, that is $\boldsymbol{B} = \boldsymbol{B}_+ + \boldsymbol{B}_-$.

The magnetic field strength at the point can then be determined by dividing $\boldsymbol{B}$ by μ, that is

$$\boldsymbol{H}(\boldsymbol{r}) = \frac{2pl}{4\pi\mu r^3}[(3\cos^2\theta - 1)\boldsymbol{i} + 3\sin\theta\cos\theta\boldsymbol{j})]$$

or

$$\boldsymbol{H}(\boldsymbol{r}) = \frac{|\boldsymbol{m}|}{4\pi r^3}[(3\cos^2\theta - 1)\boldsymbol{i} + 3\sin\theta\cos\theta\boldsymbol{j})] \tag{18.18}$$

where the quantity $\boldsymbol{m}$ is the **magnetic dipole moment** of the magnet[2] which is defined as

$$\boxed{\boldsymbol{m} := \frac{2pl}{\mu}\boldsymbol{n}} \tag{18.19}$$

For the magnetic dipole in Figure 18.20, $\boldsymbol{i}$ coincides with $\boldsymbol{n}$, a unit vector directed from the south pole to the north pole.

[2]The symbol $\boldsymbol{m}$ for magnetic dipole moment will always be indicated as a vector (that is, in bold type) thus avoiding any danger of confusion with the symbol for mass.

Using definition (18.19), the torque on a magnetic dipole in a uniform magnetic field given by Equation (18.7) can be rewritten as

$$\boldsymbol{T} = \boldsymbol{m} \times \boldsymbol{B} \tag{18.20}$$

For problems based on the material presented in this section visit up.ucc.ie/18/ and follow the link to the problems

18.10 Forces between magnets; Gauss's law for magnetism

Forces between magnets can be understood as resulting from the forces between the poles of the different magnets. An expression for the force of interaction between two poles of strength p_1 and p_2 and separated by a displacement $\boldsymbol{r}$ (Figure 18.21) in an (infinite) medium of permeability μ, can be found readily. From Equation (18.16) the magnetic field strength due to the pole p_1 at the point occupied by p_2 is given by

$$\boldsymbol{H} = \frac{p_1}{4\pi\mu r^2}\hat{\boldsymbol{r}}$$

and hence the force on p_2 is given by

$$\boxed{\boldsymbol{F} = p_2\boldsymbol{H} = \frac{p_1 p_2}{4\pi\mu r^2}\hat{\boldsymbol{r}}} \tag{18.21}$$

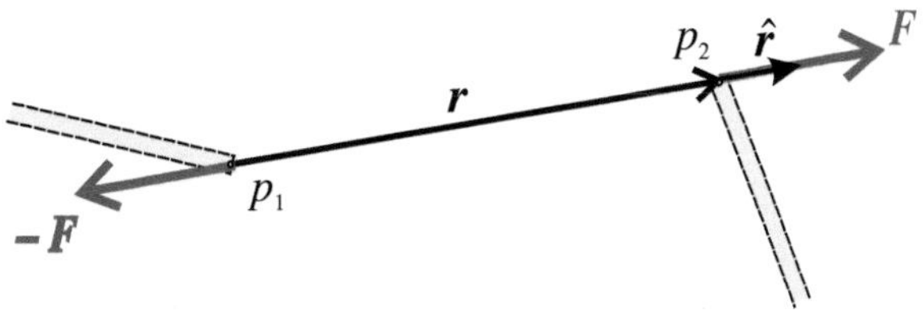

Figure 18.21. The force between magnetic poles at a distance r apart is given by Equation (18.21). The directions of the forces shown in the figure are for the case in which the poles are either both positive or both negative (repulsive interaction). For unlike poles the interaction is attractive and the direction of the forces would be reversed.

Note that $\boldsymbol{F}$ is in the $+\hat{\boldsymbol{r}}$ direction (repulsive) if p_1 and p_2 have the same sign but is in the $-\hat{\boldsymbol{r}}$ direction (attractive) if the poles strengths are of opposite sign. This vector equation, therefore, incorporates the experimental observation (Section 18.1) 'like poles repel, unlike poles attract'.

The expression for the interaction between magnetic poles (Equation (18.21)) was derived directly from the concept of magnetic flux. It must be remembered, however, that when the magnetic flux model was introduced we had no guarantee that it would prove to be a fruitful concept. All flux models (gravitational flux, light flux, etc.) automatically give rise to an inverse square law of force from simple geometric considerations. Conversely, the concept of flux is only useful in the context of inverse square law fields. Thus, as in the case of electric flux, an entirely equivalent statement to Equation (18.21) is the following:

> the magnetic flux through any closed surface in a magnetic field is equal to the algebraic *sum of the magnetic poles within the surface*

that is

$$\underset{\substack{\text{closed}\\\text{surface}}}{\oint\!\!\oint} \boldsymbol{B}\cdot d\boldsymbol{A} = \sum_{\substack{\text{inside}\\\text{surface}}} p_i \tag{18.22}$$

Equation (18.22) is called **Gauss's law for magnetism**. Since no free magnetic poles exist and assuming that the surfaces do not intersect magnets, the right hand side of Equation (18.22) is zero for all practical purposes, that is,

$$\underset{\substack{\text{closed}\\\text{surface}}}{\oint\!\!\oint} \boldsymbol{B}\cdot d\boldsymbol{A} = 0 \tag{18.23}$$

For problems based on the material presented in this section visit up.ucc.ie/18/ and follow the link to the problems

18.11 The laws of magnetostatics in differential form

Following the same procedure used in the case of electrostatics in Section 16.14, it is shown in Section 18.11 on the *Understanding Physics* website (up.ucc.ie/18/11/) that the fundamental laws of magnetostatics can be expressed in differential form as follows:

Ampère's law, Equation (18.8a),

$$\nabla \times \boldsymbol{H} = \boldsymbol{J}$$

Gauss's law for magnetostatics, Equation (18.23),

$$\nabla \cdot \boldsymbol{B} = 0$$

As before, these expressions relate the strength of a magnetic field at every point to the sources of the field.

Further reading

Further material on some of the topics covered in this chapter may be found in *Electromagnetism* by Grant and Phillips (See BIBLIOGRAPHY for book details)

WORKED EXAMPLES

Worked Example 18.1: A bar magnet of pole strength 35 μWb has a square cross section of side length 3 cm and a magnetic length 30 cm. It is suspended from a thread so that it is free to rotate in a horizontal plane about the suspension as axis (Figure 18.22(a) and (b)).

(i) If the horizontal component of the Earth's magnetic field strength is 14.3 A m^{-1}, what is the torque experienced by the magnet when it makes an angle of 10° with the magnetic meridian?
(ii) Show that if the magnet is displaced through a small angle from its equilibrium position and released, it will oscillate with simple harmonic motion.
(iii) Calculate the period of these oscillations if the physical length of the magnet is 32 cm and it has a mass of 50 g.

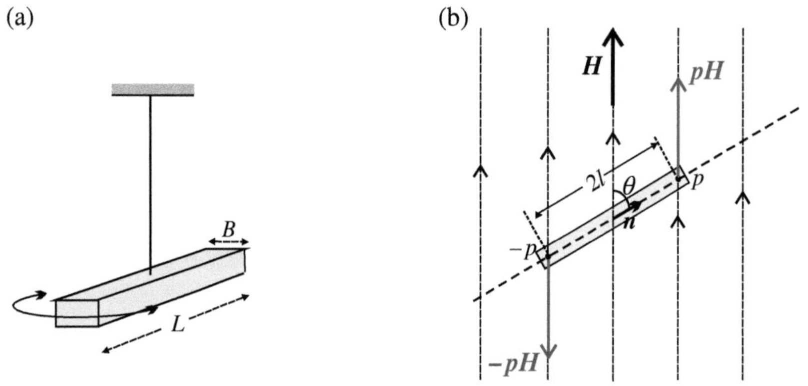

Figure 18.22. Worked Example 18.1: (a) A bar magnet suspended from a thread so that it is free to rotate in a horizontal plane. (b) The forces on a bar magnet in a uniform magnetic field.

(i) Torque on magnet $= \boldsymbol{T} = 2pl\boldsymbol{n} \times \boldsymbol{H}$
$\rightarrow |\boldsymbol{T}| = 2plH\sin\theta = 2(35 \times 10^{-6})(0.15)(14.3)\sin 10^{\circ} = 2.6 \times 10^{-5}$ m N
(ii) Again $T(\theta) = -2plH\sin\theta \approx -2plH\theta \propto -\theta \Rightarrow$ simple harmonic motion

$\rightarrow$ Restoring torque per unit angular displacement $= \dfrac{T}{\theta} = 2plH \;\Rightarrow\;$ Period $= 2\pi\sqrt{\dfrac{I}{2plH}}$

where I is the moment of inertia about a vertical axis through the centre of mass (recall Section 7.10).

(iii) Now $I = \dfrac{1}{12}M(L^2 + B^2)$ (recall, Worked Example 7.3) and hence the period of oscillations is

$$2\pi\sqrt{\frac{M(L^2 + B^2)}{24plH}} = 2\pi\sqrt{\frac{(0.05)(0.32^2 + 0.03^2)}{24(35 \times 10^{-6})(0.15)(14.3)}} = 10.6\text{ s}$$

Worked Example 18.2: Derive expressions for the magnetic field strength as a function of the distance r from the axis of a coaxial cable, as illustrated in Figure 18.23(a) and (b), when a current I flows in the central conductor and returns in the opposite direction in the outer conductor.

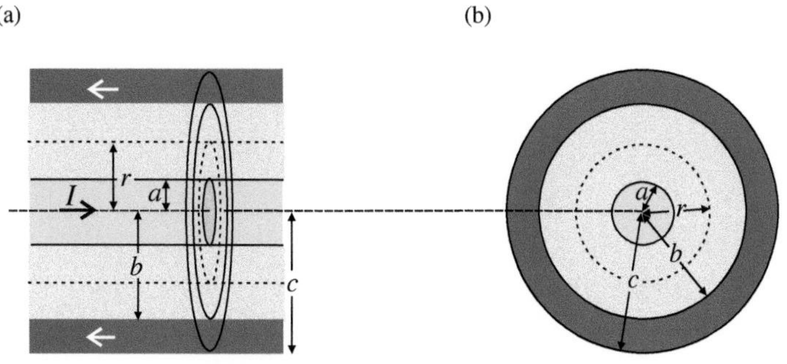

Figure 18.23. Worked Example 18.2.

Applying Ampere's law to a circular path of radius r in each region as follows.

(i) $r < a$: $|\boldsymbol{H}|(2\pi r) = J(\pi r^2) = \frac{I}{\pi a^2}\pi r^2 \quad \rightarrow \quad H = \frac{I}{2\pi a^2} r$

(ii) $a < r < b$: $|\boldsymbol{H}|(2\pi r) = I \quad \rightarrow \quad H = \frac{I}{2\pi r}$

(iii) $b < r < c$ $|\boldsymbol{H}|(2\pi r) = I - J(\pi r^2 - \pi b^2) = I - \frac{I}{\pi(c^2 - b^2)}(\pi r^2 - \pi b^2) \rightarrow H = \frac{I}{2\pi(c^2 - b^2)}\frac{c^2 - r^2}{r}$

(iv) $r > c$: $|\boldsymbol{H}|(2\pi r) = I - I = 0 \quad \rightarrow \quad H = 0$

Worked Example 18.3: Derive an expression for the magnetic field strength at a point on the axis of a circular loop of wire of radius a and carrying a current I, as a function of the distance x measured along the axis from the centre of the loop.

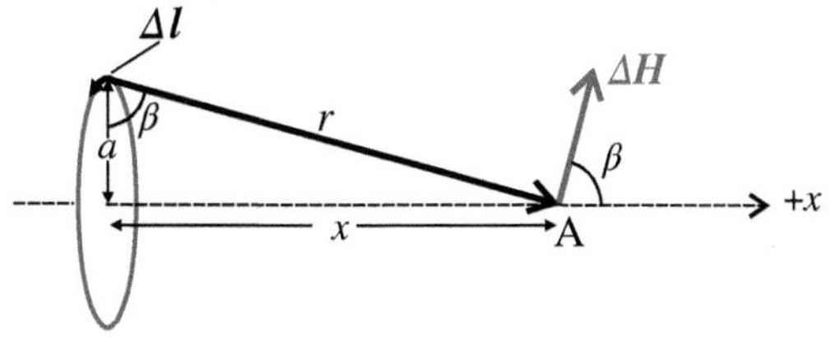

Figure 18.24. Worked Example 18.3.

Take the plane of the loop to be perpendicular to the page, as in Figure 18.24, so that $\boldsymbol{\Delta l}$ is perpendicular to the page and outwards.

The contribution to $\boldsymbol{H}_A$ from the element $\boldsymbol{\Delta l}$ is $\boldsymbol{\Delta H} = \frac{I\boldsymbol{\Delta l} \times \hat{\boldsymbol{r}}}{4\pi r^2} \rightarrow |\boldsymbol{\Delta H}| = \frac{I \sin\theta(\Delta l)}{4\pi r^2}$, with $\boldsymbol{\Delta H}$ in the plane of the page and perpendicular to $\boldsymbol{r}$, as indicated in Figure 18.24.

Since $\theta = \frac{\pi}{2}$, $|\boldsymbol{\Delta H}| = \frac{I}{4\pi r^2}\Delta l$ in the direction indicated in the figure.

When the contributions from all the $\boldsymbol{\Delta l}$s are added up (vectorially) there will be no net contribution from components perpendicular to the x-axis. Thus we need only consider x-components of $\boldsymbol{\Delta H}$, namely

$$|\boldsymbol{\Delta H}| \cos\beta = \frac{I}{4\pi r^2}\cos\beta \Delta l = \frac{I}{4\pi r^2}\frac{a}{r}\Delta l = \frac{Ia}{4\pi r^3}\Delta l$$

Hence

$$|\boldsymbol{H}_A| = \oint_{\text{loop}} |\boldsymbol{\Delta H}| \cos\beta = \frac{Ia}{4\pi r^3}\oint dl = \frac{Ia}{4\pi r^3}(2\pi a) = \frac{Ia^2}{2r^3} \quad \rightarrow \quad H_A(x) = \frac{Ia^2}{2(x^2 + a^2)^{3/2}}$$

PROBLEMS

For problems based on the material covered this chapter visit up.ucc.ie/18/ and follow the link to the problems.

19

Interactions between magnetic fields and electric currents; magnetic materials

AIMS

- to understand the forces experienced by electric currents in magnetic fields
- to show how such forces may be utilised in instrument design
- to study the forces on charges which are moving in magnetic fields
- to investigate the nature of magnetic materials

19.1 Forces between currents and magnets

Consider the situation shown in Figure 19.1. The current I in the wire produces a magnetic field at the location occupied by a pole of strength p. The contribution to the force on p due to the current I flowing in the element $\boldsymbol{\Delta l}$ is directed perpendicularly to the page inwards and, from Equation (18.14), is given by

$$\boldsymbol{\Delta F} = p\boldsymbol{\Delta H} = p\frac{I(\boldsymbol{\Delta l}) \times \hat{\boldsymbol{r}}}{4\pi r^2}$$

From Newton's third law, there will be an equal and opposite force *on* the element $\boldsymbol{\Delta l}$ due to the influence of the pole p, which is given by

$$\boldsymbol{\Delta F} = -p\frac{I(\boldsymbol{\Delta l}) \times \hat{\boldsymbol{r}}}{4\pi r^2} \tag{19.1}$$

and is directed perpendicularly to the page outwards. Let us now change the origin relative to which $\boldsymbol{r}$ is defined so that it is located at the pole p rather than at the element $\boldsymbol{\Delta l}$, that is $\boldsymbol{r} \to -\boldsymbol{r}$ in Equation (19.1). The force on the element $\boldsymbol{\Delta l}$ due to the pole p is then given by

$$\boldsymbol{\Delta F} = \frac{pI(\boldsymbol{\Delta l}) \times \hat{\boldsymbol{r}}}{4\pi r^2} = I(\boldsymbol{\Delta l}) \times \left(\frac{p}{4\pi r^2}\hat{\boldsymbol{r}}\right)$$

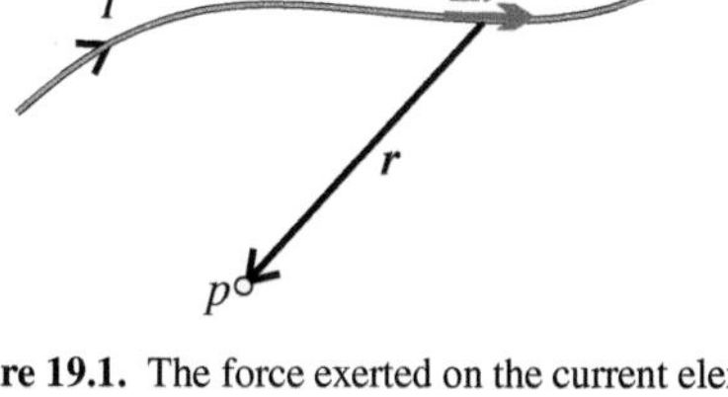

Figure 19.1. The force exerted on the current element $\boldsymbol{\Delta l}$ by the magnetic pole p is directed perpendicularly to the page outwards if p is an N pole.

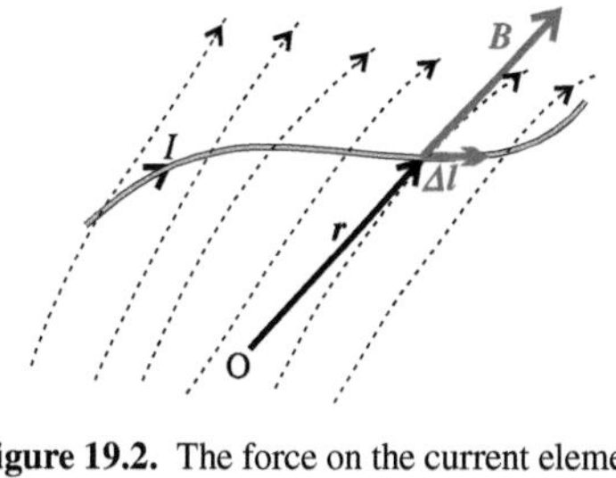

Figure 19.2. The force on the current element $\boldsymbol{\Delta l}$ due to a magnetic field of flux density B at the element is given by Equation (19.2).

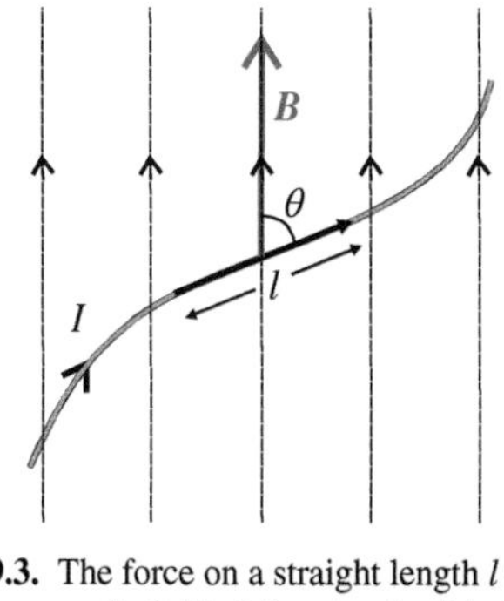

Figure 19.3. The force on a straight length l of wire in a uniform magnetic field of flux density B is $IlB\sin\theta$, where I is the current in the wire and θ is the angle the wire makes with the direction of the magnetic field.

Understanding Physics, Third Edition. Michael Mansfield and Colm O'Sullivan.
© 2020 John Wiley & Sons Ltd. Published 2020 by John Wiley & Sons Ltd.

Now the term $\frac{p}{4\pi r^2}$ in this equation is the magnetic flux density $\boldsymbol{B}$ at the element $\Delta \boldsymbol{l}$ due to the pole p (recall Equation (18.15)) and hence we may write

$$\boxed{\Delta \boldsymbol{F} = I(\Delta \boldsymbol{l}) \times \boldsymbol{B}} \tag{19.2}$$

This result describes equally well the force on a current element due to **any** source of magnetic field since, in the limit $\Delta \boldsymbol{l} \to 0$, the flux density at the element will be the same as that due to an equivalent single pole (see Figure 19.2).

Equation (19.2) is an important result with many practical applications. In particular, in the case of a straight piece of current carrying wire in a uniform magnetic field (Figure 19.3) the force on a length l of the wire is given by $\boldsymbol{F} = I\boldsymbol{l} \times \boldsymbol{B}$, and thus

$$|\boldsymbol{F}| = IlB \sin \theta$$

directed along $\boldsymbol{l} \times \boldsymbol{B}$ which is perpendicular to the page and outward in the figure.

For problems based on the material presented in this section visit up.ucc.ie/19/ *and follow the link to the problems.*

19.2 The force between two long parallel wires

Consider two long parallel wires embedded in a linear, isotropic, homogeneous medium, as shown in Fig, 19.4. The magnetic field strength at the location of the element $\Delta \boldsymbol{l}$ due to the current I_1 flowing in the upper wire, is given by Equation (18.6), that is

$$|\boldsymbol{H}_1| = \frac{I_1}{2\pi r}$$

directed perpendicularly into the page. The corresponding magnetic flux density is

$$|\boldsymbol{B}_1| = \mu \frac{I_1}{2\pi r}$$

where μ is the permeability of the medium in which the wires are embedded.

Thus the force on Δl is given by $\Delta \boldsymbol{F} = I_2 \Delta l \times \boldsymbol{B}_1$ (Equation (19.2) which is in the direction indicated in Figure 19.4 and hence, as noted in Section 15.1, appears as a repulsion between the wires when the currents I_1 and I_2 flow in the directions shown. Thus

$$|\Delta \boldsymbol{F}| = I_2 B_1 (\Delta l) = I_2 \left(\mu \frac{I_1}{2\pi r} \right) (\Delta l) \quad \to \quad \frac{dF}{dl} = \frac{\mu}{2\pi} \frac{I_1 I_2}{r}$$

or, if the wires are in a vacuum,

$$\frac{dF}{dl} = \frac{\mu_0}{2\pi} \frac{I_1 I_2}{r}$$

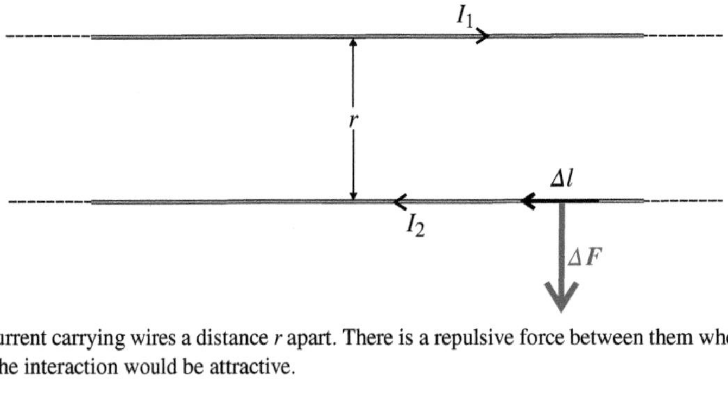

Figure 19.4. Two infinitely long parallel current carrying wires a distance r apart. There is a repulsive force between them when the currents flow in the directions shown. If one of the currents were to be reversed, the interaction would be attractive.

From 1948 [9th CGPM] until 2019, the ampere was defined so that, if a constant current of 1 A were to be maintained in two straight parallel wires of infinite length placed 1 metre apart in a vacuum, the force between these conductors would be exactly 2×10^{-7} newton per metre of length. Thus, prior to 2019, in SI units $\frac{dF}{dl} = 2 \times 10^{-7} \frac{I_1 I_2}{r}$ and hence the permeability of vacuum had the value of $\mu_0 = 4\pi \times 10^{-7}$ N A^{-2}. With the 2019 definition of the ampere (Section 15.3), the value of μ_0 is no longer exact and must be determined experimentally; nevertheless the value remains close to $4\pi \times 10^{-7}$ N A^{-2}.

The **relative permeability** of a medium is defined as $\mu_r := \frac{\mu}{\mu_0}$, that is $\mu = \mu_r \mu_0$.

19.3 Current loop in a magnetic field

Figure 19.5 shows a plane loop of wire of arbitrary shape in a constant uniform magnetic field of flux density $\boldsymbol{B}$. Initially we consider the case in which $\boldsymbol{B}$ lies in the plane of the loop, that is in the x—y plane shown in Figure 19.5(a). If the loop is thought of as being divided up into a large number of narrow strips as shown in the figure, we can consider first the forces on the elements $\Delta \boldsymbol{l}_1$ and $\Delta \boldsymbol{l}_2$, on opposite sides of one of these strips, shaded in Figure 19.5(a). From Equation (19.2), the force on $\Delta \boldsymbol{l}_1$ is $\Delta \boldsymbol{F}_1 = I\Delta \boldsymbol{l}_1 \times \boldsymbol{B}$ and that on $\Delta \boldsymbol{l}_2$ is $\Delta \boldsymbol{F}_2 = I\Delta \boldsymbol{l}_2 \times \boldsymbol{B}$, the two forces being directed in opposite senses along the z-axis (out of the page in the figure).

Now, $$\Delta \boldsymbol{F}_1 = IB\sin\theta_1(\Delta l_1)\boldsymbol{k} = IB(\Delta y)\boldsymbol{k} \quad \text{and} \quad \Delta \boldsymbol{F}_2 = -IB\sin\theta_2(\Delta l_2)\boldsymbol{k} = -IB(\Delta y)\boldsymbol{k}.$$

where $\boldsymbol{k}$ is a unit vector in the z-direction.

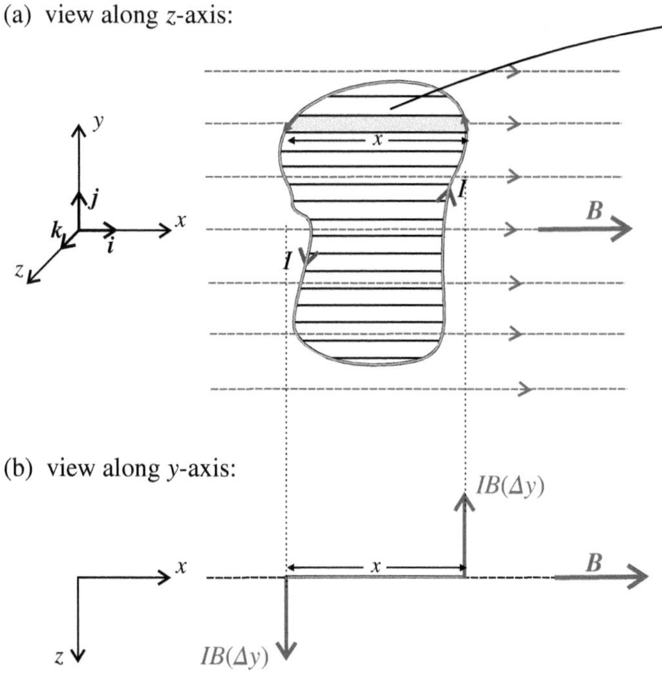

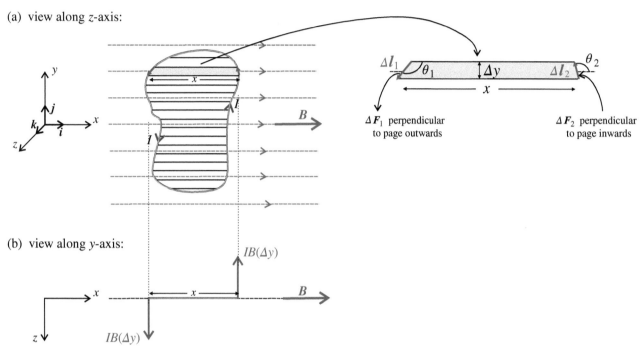

Figure 19.5. (a) A plane current loop in the x–y-plane experiences a uniform magnetic field directed along the x-axis. The z-direction is perpendicular to the page outwards. (b) The coil in figure (a) viewed along the y-axis. The forces on the current elements $\Delta \boldsymbol{l}_1$ and $\Delta \boldsymbol{l}_2$ comprise a couple which, together with similar contributions from other pairs of opposite elements, gives rise to a torque on the loop.

As can be seen from Figure 19.5(b), these two forces constitute a couple (Section 7.3), the torque of which is given by

$$\Delta \boldsymbol{T} = IB(\Delta y)x\boldsymbol{j} = IB(\Delta A)\boldsymbol{j}$$

where ΔA is the area of a strip of length x and width Δy.

Thus, when a current is flowing in the loop it experiences a net torque of

$$\boldsymbol{T} = IBA\boldsymbol{j}$$

where $A = \Sigma(\Delta A)$ is the total area of the loop. This is the maximum value that the torque can have since, if the plane of the loop is rotated through some angle ϕ with respect to the magnetic field, the perpendicular distance between the two forces which produce the torque ΔT is reduced from x to $x\cos\phi$ and the total torque is correspondingly reduced to

$$\boldsymbol{T} = IBA\cos\phi\boldsymbol{j} \tag{19.3}$$

We can rewrite Equation (19.3) in terms of the angle θ between $\boldsymbol{B}$ and the perpendicular to the plane of the loop (that is, $\theta = \frac{\pi}{2} - \phi$), in which case

$$\boldsymbol{T} = IBA\sin\theta\boldsymbol{j} = IA\boldsymbol{k} \times \boldsymbol{B}$$

or, more generally, $$\boldsymbol{T} = IA\boldsymbol{n} \times \boldsymbol{B} \tag{19.4}$$

where $\boldsymbol{n}$ is a unit vector directed perpendicular to the plane of the loop.

In Section 18.4 we derived an expression, Equation (18.7), for the torque on a magnetic needle in a uniform magnetic field, namely $\boldsymbol{T} = 2pl\boldsymbol{n} \times \boldsymbol{H}$, where $\boldsymbol{n}$ is directed along the magnetic axis. We can write this torque in terms of the magnetic flux density $\boldsymbol{B}$ as follows

$$\boldsymbol{T} = \frac{2pl}{\mu}\boldsymbol{n} \times \boldsymbol{B} = \boldsymbol{m} \times \boldsymbol{B}$$

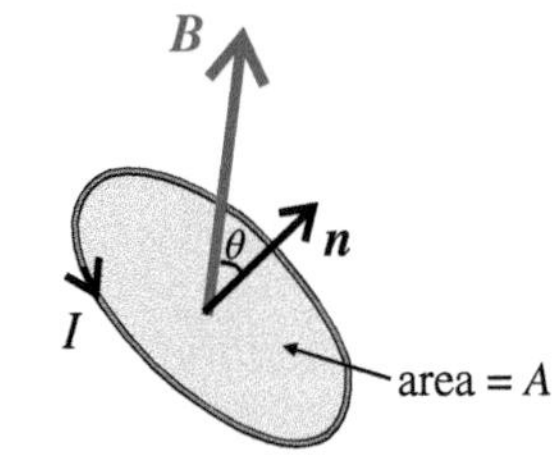

Figure 19.6. The magnetic dipole moment of a plane current loop has magnitude IA and is directed along the unit vector n, which is perpendicular to the plane of the loop in the direction indicated.

and, comparing this with Equation (19.4), we see that $I A\boldsymbol{n}$ plays the same role for a current loop as $\boldsymbol{m} = \frac{2pl}{\mu}\boldsymbol{n}$ does for a magnetic needle. This suggests that, for a plane current loop such as that shown in Figure 19.6, we can define a quantity which is directly equivalent to the magnetic dipole moment of a magnetic needle, defined by Equation (18.19). The **magnetic dipole moment** of a planar current loop is defined as follows

$$\boxed{\boldsymbol{m} := IA\boldsymbol{n}}$$

The direction of $\boldsymbol{n}$, and hence the direction of $\boldsymbol{m}$, is determined by the sense in which the current flows in the loop. In the situation shown in Figure 19.5(a), $\boldsymbol{m}$ is directed in the $+\boldsymbol{k}$-direction; thus, to an observer looking down on the plane of the loop, the magnetic dipole moment is directed towards the observer if the current is flowing in the anticlockwise sense and away from the observer if the current flow is clockwise.

The term 'magnetic dipole' is often used to describe either a magnetic needle or a current loop and the terms 'magnetic moment' or 'electromagnetic moment' are often used interchangeably with 'magnetic dipole moment' in either case.

In the case of either a magnetic needle or of a current loop, therefore, the torque experienced when the dipole is placed in a magnetic field of flux density $\boldsymbol{B}$ (assumed to be uniform over the extent of the magnetic dipole) is given by

$$\boldsymbol{T} = \boldsymbol{m} \times \boldsymbol{B} \tag{19.5}$$

The torque on a current loop is maximum when $\boldsymbol{m}$ is perpendicular to $\boldsymbol{B}$ (as indicated, for example, in Figure 19.5) and the torque is zero when $\boldsymbol{m}$ is parallel to $\boldsymbol{B}$. If it can do so, a magnetic dipole will attempt to align itself along the direction of $\boldsymbol{B}$. Thus the axis of a magnetic needle which is free to rotate aligns along $\boldsymbol{B}$ while, in the case of a current loop, the normal to the plane of the loop aligns along $\boldsymbol{B}$; that is, the plane of the loop aligns perpendicularly to $\boldsymbol{B}$.

While the torque is zero when $\boldsymbol{m}$ is parallel to $\boldsymbol{B}$, there are still forces acting on the elements making up the loop. These forces will tend to distort the shape of the loop — only the rigidity of the loop enables it to retain its shape. We shall see in the next section that atoms and molecules can have magnetic dipole moments and, in these cases, an external magnetic field will distort the shape of such 'elastic' current loops.

Study Worked Example 19.1

Potential energy of a magnetic dipole in a magnetic field

Finally, in this section, we derive an expression for the potential energy of a magnetic dipole of moment $\boldsymbol{m}$ in a magnetic field of flux density $\boldsymbol{B}$ which we assume is uniform over the dimensions of the dipole. We follow the procedure used in Section 16.6 to derive the potential energy of an electric dipole in an electric field.

The potential energy of the dipole when it has been turned through an angle θ from the position when it is aligned along the field direction ($\theta = 0$), against a restoring torque exerted by the field, is given by

$$U(\theta) = \int_0^\theta \boldsymbol{T} \cdot d\boldsymbol{\vartheta} = \int_0^\theta (\boldsymbol{m} \times \boldsymbol{B}) \cdot d\boldsymbol{\vartheta} = \int_0^\theta mB \sin \vartheta d\vartheta = -mB[\cos \vartheta]_0^\theta = mB(1 - \cos \theta)$$

As in Section 16.6, we can choose to fix the zero of potential energy at $\theta = \pi/2$ (when the dipole moment is aligned perpendicularly to the field) by, in this case, subtracting mB from the equation for the potential energy which becomes

$$U(\theta) = mB(1 - \cos \theta) - mB = -mB \cos \theta$$

so that the potential energy may be written

$$U(\theta) = -\boldsymbol{m} \cdot \boldsymbol{B} \tag{19.6}$$

For problems based on the material presented in this section visit up.ucc.ie/19/ *and follow the link to the problems.*

19.4 Magnetic fields due to moving charges

All discussion of magnetic fields so far has centred on electric currents or systems of magnetic poles as the sources of the fields. In Section 15.2, however, electric currents were interpreted as being due to moving charges. Accordingly, we would expect that moving electric charges also should be sources of magnetic field. We now address the question of how the magnetic field produced by a moving charge depends on the position and the velocity of the charge relative to an observer.

The product $I(\Delta l)$ occurs in many of the equations which we derived earlier in this chapter, most notably in the Biot–Savart law (Equation (18.14)),

$$\Delta \boldsymbol{H} = \frac{I(\Delta \boldsymbol{l}) \times \hat{\boldsymbol{r}}}{4\pi r^2}$$

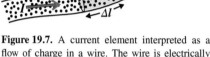

where, $\Delta \boldsymbol{l}$ is an element of a current carrying wire (Figure 19.7). We can write the product $I(\Delta l)$ in terms of the mobile charge density using Equation (15.2a), namely $I = \rho v A$, as follows

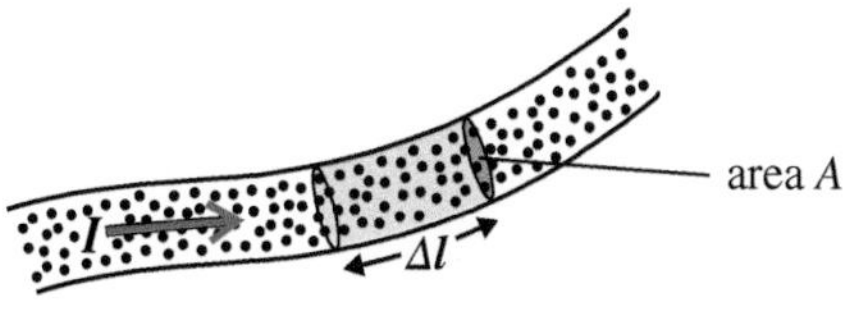

Figure 19.7. A current element interpreted as a flow of charge in a wire. The wire is electrically neutral since the mobile charge flows in a background of equal and opposite fixed charge.

$$I(\Delta l) = \rho v A(\Delta l) = \{\rho A(\Delta l)\}v$$

where v is the velocity of the moving charge.

Now $A(\Delta l)$ is the volume of the element of the wire and hence $\{\rho A(\Delta l)\}$ is the amount of mobile charge within the element which we now write as (ΔQ). Thus $I(\Delta l) = (\Delta Q)v$ and, since $\boldsymbol{v}$ is parallel to Δl, we can write this in vector form

$$I(\Delta \boldsymbol{l}) = (\Delta Q)\boldsymbol{v}$$

Any of the relationships which we derived in terms of the current model can be rewritten in terms of mobile electric charge by replacing $I(\Delta \boldsymbol{l})$ by $(\Delta Q)\boldsymbol{v}$. Thus, for the Biot–Savart law, for example, the contribution to the magnetic field strength at a distance $\boldsymbol{r}$ from a mobile charge element ΔQ is given by

$$\Delta \boldsymbol{H} = \frac{(\Delta Q)\boldsymbol{v} \times \hat{\boldsymbol{r}}}{4\pi r^2}$$

Suppose that, instead of a continuous distribution of charge within the element Δl, there is just a single point charge q, that is $\Delta Q \to q$ moving with velocity $\boldsymbol{v}$ (Figure 19.8), then the instantaneous magnetic field strength at a distance $\boldsymbol{r}$ from this moving charge is given by

$$\boxed{\boldsymbol{H} = \frac{q\boldsymbol{v} \times \hat{\boldsymbol{r}}}{4\pi r^2}} \qquad (19.7)$$

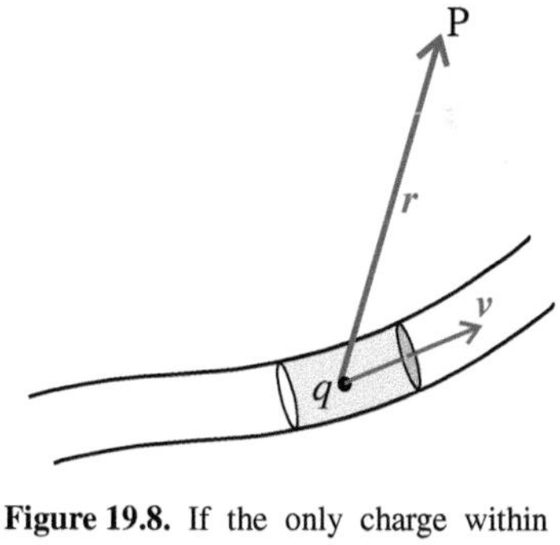

Figure 19.8. If the only charge within a current element is a single moving point charge, the Biot–Savart law can be modified so that it gives the magnetic field strength at the point P due to the moving charge (Equation (19.7)).

which can be considered to be a version of the *Biot–Savart law for a moving charge*. We have thus extended a result, which was originally derived for current elements, to apply to a localised electric charge which is moving relative to an observer who senses the resulting magnetic field.

19.5 Force on a moving electric charge in a magnetic field

Other relationships derived in this chapter for steady electric currents can be extended to apply to moving charges. Equation (19.2) gives the force on an element $\Delta \boldsymbol{l}$ carrying a current I in a magnetic field of flux density $\boldsymbol{B}$, namely

$$\Delta \boldsymbol{F} = I\Delta \boldsymbol{l} \times \boldsymbol{B}$$

Again we can rewrite the result in terms of the mobile charge ΔQ within $\Delta \boldsymbol{l}$ by making the transformation $I\Delta \boldsymbol{l} \to (\Delta Q)\boldsymbol{v}$, as before, and thus

$$\Delta \boldsymbol{F} = (\Delta Q)\boldsymbol{v} \times \boldsymbol{B}$$

In particular, if the element contains just one single point charge q, we can write an expression for the force on a moving charge in a magnetic field as

$$\boxed{\boldsymbol{F} = q\boldsymbol{v} \times \boldsymbol{B}} \tag{19.8}$$

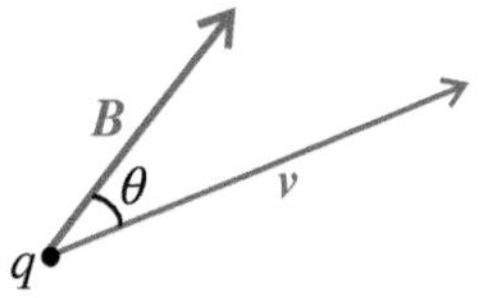

Figure 19.9. The force on a moving charge in a magnetic field is given by Equation (19.8). If the charge in the figure is positive, the force is directed outwards perpendicularly to the page.

Thus $|\boldsymbol{F}| = q|\boldsymbol{v}||\boldsymbol{B}|\sin\theta$ and the force $\boldsymbol{F}$ is directed perpendicularly to the plane of $\boldsymbol{v}$ and $\boldsymbol{B}$ in the direction given by the rule for the vector cross-product (Section 4.11), that is perpendicularly to the page and outwards in Figure 19.9.

Suppose that a particle of charge q and mass m enters a region of constant uniform magnetic field (directed perpendicularly to the page and outwards in Figure 19.10) with its velocity $\boldsymbol{v}$ perpendicular to the direction of the field; that is, $\boldsymbol{v}$ is perpendicular to $\boldsymbol{B}$ so that $\theta = \frac{\pi}{2}$. In this case the particle experiences a force

$$|\boldsymbol{F}| = q|\boldsymbol{v}||\boldsymbol{B}|\sin\frac{\pi}{2} = qvB$$

with $\boldsymbol{F}$ perpendicular to $\boldsymbol{v}$ as indicated in the figure. Since the magnitude of this force is constant and always perpendicular to $\boldsymbol{v}$, it acts as a centripetal force (Section 4.7) which causes the particle to move uniformly in a circular path. Thus,

$$|\boldsymbol{F}| = qvB = ma_c = m\frac{v^2}{R}$$

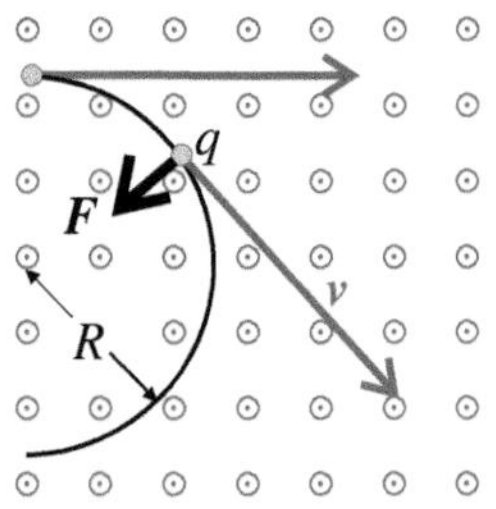

Figure 19.10. If a charged particle is moving in a uniform magnetic field so that its velocity is perpendicular to the field lines, it will execute circular motion as shown. The coloured dots with circles (⊙) represent the magnetic field lines of a uniform field directed perpendicularly to the page and outwards.

with the radius of the circle given by

$$\boxed{R = \frac{mv}{qB} = \frac{p}{qB}} \tag{19.9}$$

If the particle is not confined to the plane perpendicular to the magnetic field, that is if it has a component of the velocity parallel to the field, then this constant velocity component is superimposed on the circular motion and the particle follows a helical path centred on a field line. In these circumstances, Equation (19.9) can still be applied to the projection of the motion onto the plane perpendicular to the magnetic field. The above result has many important applications, some of which will be encountered in the next section and others later in the book (for example in Figure 27.22).

Study Worked Examples 19.2 and 19.3

For problems based on the material presented in this section visit up.ucc.ie/19/ *and follow the link to the problems.*

19.6 Applications of moving charges in uniform magnetic fields; the classical Hall effect

Mass spectrometry

Equation (19.9) is the basis, for example, of the important experimental technique of mass spectrometry. If different isotopes of the same element having the same degree of ionisation are accelerated from a source S across a potential difference V before entering a uniform magnetic field of flux density $\boldsymbol{B}$ (Figure 19.11), they will enter the field with speeds which are given (Section 16.13) by

$$v_i = \sqrt{\frac{2qV}{m_i}}$$

where the subscript i refers to a particular isotope. Hence each isotope will travel on a circular path in the magnetic field with radius given by

$$R_i = \frac{m_i v_i}{qB} = \frac{m_i}{qB}\sqrt{\frac{2qV}{m_i}} = \frac{1}{B}\sqrt{\frac{2m_i V}{q}}.$$

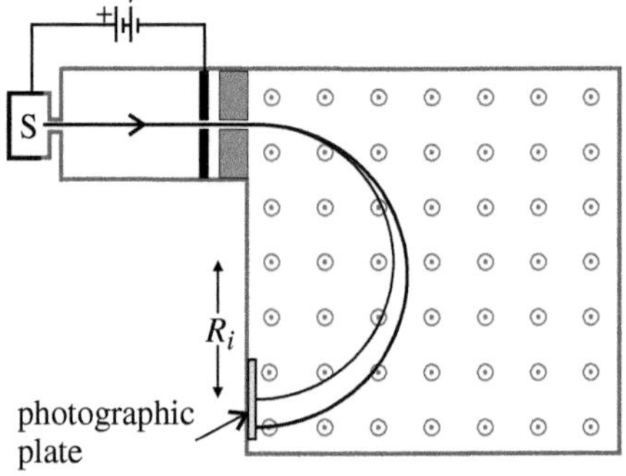

Figure 19.11. A mass spectrometer can be used to separate particles which have the same charge but slightly different masses.

Since the particles all have the same charge q, the radii of their paths will differ only because the different isotopic species have different masses; the radii are proportional to the square root of the particle mass, that is $R \propto \sqrt{m}$. These radii can be measured very accurately, for example by measuring the position of the mark made when the charged particles strike a photographic plate as indicated in Figure 19.11. The same technique can be used for **isotope separation** if the photographic plate is replaced by carefully positioned ion collectors.

As an example, let us estimate the radii of the paths of singly ionised chlorine atoms which enter a magnetic field of flux density 0.5 T, having been first accelerated across a potential difference of 10 kV. For ^{37}Cl the radius is $\frac{1}{0.5\ \mathrm{T}}\sqrt{\frac{2(10^4\,\mathrm{V})(37\times1.66\times10^{-27}\,\mathrm{kg})}{1.6\times10^{-19}\,\mathrm{C}}} = 17.5$ cm, while for ^{35}Cl the radius is $\sqrt{\frac{35}{37}}(17.5\ \mathrm{cm}) = 17.0$ cm.

The Cyclotron

The period T of a charged particle of mass m and charge q which is moving with uniform circular motion in a uniform magnetic field can be determined as follows.

$$T = \frac{2\pi R}{v} = \frac{2\pi}{v}\frac{mv}{qB} = \frac{2\pi m}{qB}$$

The angular frequency of the circular motion is given by

$$\omega_c = 2\pi f = \frac{2\pi}{T} = \frac{qB}{m}$$

Note that the period (and hence the frequency) is *independent of the particle velocity v*. This means that, while higher energy particles execute larger circles, the time taken to complete a circle is the same for all particles of given charge and mass. This result is utilised in a type of charged particle accelerator called a cyclotron and accordingly ω_c is called the **cyclotron** (angular) **frequency**.

The basic elements of a cyclotron are illustrated in Figure 19.12. The particles circulate inside two hollow semicircular metal objects, called 'dees' because of their shape, which have a strong constant uniform magnetic field of flux density B directed perpendicularly to their flat faces, as shown in the figure. An oscillating potential difference is applied across the gap between the dees such that $V = V_0 \sin\omega_c t$, where $\omega_c = \frac{qB}{m}$ is the cyclotron angular frequency for the particles which are to be accelerated.

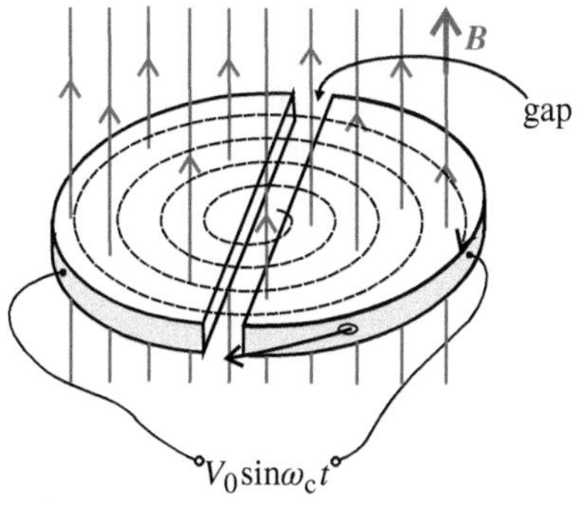

Figure 19.12. A cyclotron is used to accelerate charged particles.

Charged particles injected near the centre of the cyclotron will be accelerated by the electric field in the gap between the dees. On entering a dee, however, a particle will be shielded from the electric field by the metal walls and thus will move in a circular path under the influence of the magnetic field. When the particle emerges again from the dee the electric field will have reversed and the particle will be accelerated once more across the gap and hence will circulate with a larger radius within the second dee.

This process continues with the circulating charge always entering the gap between the dees in step with the applied potential difference. The particles, therefore, acquire larger and larger energy and circulate with correspondingly increasing radii until they reach the edge of the cyclotron, at which point they may be extracted, through a small aperture in the curved face of one of the dees, in the form of a beam which can be projected at a target for experimental studies in high energy physics and other areas.

The Synchrotron

Equation (19.9) was derived on the assumption that the particles move at non-relativistic speeds. In fact, the equation for the radius given in the form $R = \frac{p}{qB}$ is also valid relativistically provided that the relativistic expression for the momentum (Equation (9.34)) $p = mu\Big/\sqrt{1-\frac{u^2}{c^2}}$ is used. In this case, the period of a circulating charge is given by $T = \frac{2\pi m}{qB}\Big/\sqrt{1-\frac{u^2}{c^2}}$ and is *not* independent of the velocity u. Thus, a cyclotron, in which the frequency of the oscillating potential difference is fixed, cannot continue to accelerate charged particles once they reach relativistic speeds.

This difficulty is overcome in particle accelerators called synchrotrons. Both the magnetic flux density B and the oscillator frequency are varied during the accelerating process in such a way that the circulating charges continue to remain in step with the oscillator frequency and also in such a way that the particles follow circular paths. For further details the reader may consult Alonso and Finn (1992) pp. 581–583.

The classical Hall effect

When moving charge carriers in a conductor or semiconductor are subjected to a magnetic field they experience a force given by Equation (19.8), namely

$$\boldsymbol{F} = q\boldsymbol{v} \times \boldsymbol{B}$$

The behaviour of charge carriers in response to a uniform magnetic field which is perpendicular to their direction of motion produces the **Hall effect**. As described below, the effect can be used to determine the drift velocity, density and polarity of charge carriers.

Consider a metal strip of width w and of thickness t which is connected to a source of emf, as shown in Figure 19.13. An electric field in the metal strip produces a current I to the right.

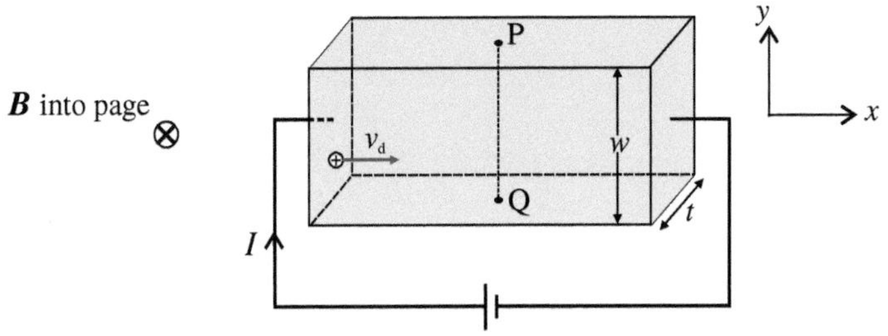

Figure 19.13. Experimental arrangement to measure the Hall effect voltage. An emf and a uniform magnetic field are applied to a metal strip.

When a uniform magnetic field of flux density $\boldsymbol{B}$ is applied perpendicularly to a surface of the strip (into and perpendicular to the page in Figure 19.13) the force on a positive charge q is

$$\boldsymbol{F}_B = q\boldsymbol{v}_\mathrm{d} \times \boldsymbol{B} = qv_\mathrm{d}B \qquad \text{in the direction} \qquad \mathrm{Q} \to \mathrm{P} \tag{19.10}$$

where v_d is the drift velocity and P and Q are two points at the edge of the strip such that PQ is perpendicular to v_d.

$\boldsymbol{F}_B$ causes the positive charges to move towards the top face (towards P) until the resulting electric field E_y prevents further movement of charge in this direction. The potential difference which, in consequence, develops between P and Q is the known as the **Hall voltage** and is given by

$$V_\mathrm{H} = V_\mathrm{P} - V_\mathrm{Q} = E_y w \tag{19.11}$$

In equilibrium, the force due to the magnetic field $\boldsymbol{F}_B$ balances $\boldsymbol{F}_E$, the force produced by the electric field E_y. Thus $qE_y = qv_\mathrm{d}B$ and hence $E_y = v_\mathrm{d}B$. Substituting for E_y in Equation (19.11) yields

$$V_\mathrm{H} = v_\mathrm{d}Bw \tag{19.12}$$

Thus, by measuring V_H, B and w, the drift velocity of the charge carriers can be determined. Furthermore, using Equation (15.3a), which for positive charge may be written

$$J = \frac{I}{A} = nqv_\mathrm{d}$$

where the area $A = wt$ and n is the charge carrier density. By substituting for v_d from Equation (19.12), we obtain

$$I = \frac{nqAV_\mathrm{H}}{Bw}$$

Hence n can be determined by measuring I, A, V_H, B and w.

We have assumed in our analysis that the charge carriers are positive and, therefore, that v_d is to the right in Figure 19.13. If the charge carriers are negative, v_d is to the left and both q and v_d change sign in Equation (19.10). Hence, for negative carriers, $\boldsymbol{F}_B$ acts in the Q → P direction, so that negative charges accumulate at P which is at a lower electric potential than Q. The sign of the Hall voltage, therefore, gives the polarity of the charge carriers.

In a *Hall probe*, the Hall effect is used to measure B. From Equation (19.12), V_H is proportional to B so that a calibrated voltmeter can give direct measurements of B.

For problems based on the material presented in this section visit up.ucc.ie/19/ *and follow the link to the problems.*

19.7 Charge in a combined electric and magnetic field; the Lorentz force

If a particle of charge q moving with a velocity $\boldsymbol{u}$ experiences, simultaneously, both an electric and a magnetic field, two apparently distinct forces are exerted on it, namely (i) the force $q\boldsymbol{E}$ due to the electric field of strength $\boldsymbol{E}$ and (ii) the force due to its motion in the magnetic field given by Equation (19.8), namely $\boldsymbol{F} = q\boldsymbol{u} \times \boldsymbol{B}$. Thus the total force on the charge is given by

$$\boxed{\boldsymbol{F} = q\boldsymbol{E} + q\boldsymbol{u} \times \boldsymbol{B}} \tag{19.13}$$

where $\boldsymbol{u}$ is the velocity of the charge relative to a frame in which the magnetic flux density is $\boldsymbol{B}$. The combined force given by Equation (19.13) is called **the Lorentz force**.

Since $\boldsymbol{u}$ depends on the frame of reference in which the fields are observed, Equation (19.13) shows that the magnitude and the direction of the Lorentz force depend on the choice of reference frame. A frame in which the field is purely electric can always be found (that is a coordinate frame in which the charge q is at rest). In our discussion so far we have assumed that the electric field is frame independent but, as we shall see, this assumption is naive to say the least. Consider two inertial frames of reference as indicated in Figure 19.14 where the K′-frame is moving with constant velocity $\boldsymbol{V}$ relative to the K-frame along their common x-axis, that is $\boldsymbol{V} = V\boldsymbol{i}$. The Lorentz force in the K-frame is given by Equation (19.13) and the Lorentz force observed in the K′-frame is

$$\boldsymbol{F}' = q\boldsymbol{E}' + q\boldsymbol{u}' \times \boldsymbol{B}'$$

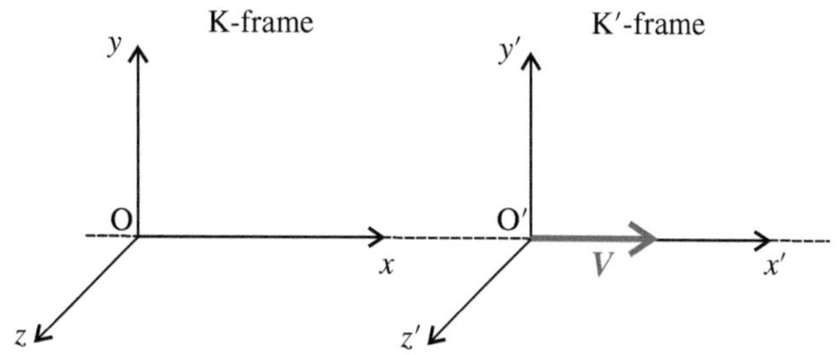

Figure 19.14. Two different reference frames: the K′-frame is moving with velocity V with respect to the K-frame along their common x-axis.

If the velocities involved are non-relativistic, we can invoke the Galilean velocity transformation Equation (8.14), namely $\boldsymbol{u}' = \boldsymbol{u} - \boldsymbol{V}$, from which we obtain

$$\boldsymbol{F}' = q\boldsymbol{E}' + q(\boldsymbol{u} - \boldsymbol{V}) \times \boldsymbol{B}'$$

If observers in each frame see the charge as experiencing the same force, then

$$q\boldsymbol{E}' + q(\boldsymbol{u} - \boldsymbol{V}) \times \boldsymbol{B}' = q\boldsymbol{E} + q\boldsymbol{u} \times \boldsymbol{B}$$

or

$$\boldsymbol{E}' + (\boldsymbol{u} - \boldsymbol{V}) \times \boldsymbol{B}' = \boldsymbol{E} + \boldsymbol{u} \times \boldsymbol{B}$$

This implies that either or both the $\boldsymbol{E}$- and $\boldsymbol{B}$- fields must depend on both $\boldsymbol{E}'$ and $\boldsymbol{B}'$, in other words $\boldsymbol{E} = \boldsymbol{E}(\boldsymbol{E}', \boldsymbol{B}', \boldsymbol{V})$ and/or $\boldsymbol{B} = \boldsymbol{B}(\boldsymbol{E}', \boldsymbol{B}', \boldsymbol{V})$, *even at non-relativistic velocities.*

To proceed along this line of thinking, in an attempt to discover the appropriate rules for transforming $\boldsymbol{E}$ and $\boldsymbol{B}$ fields between reference frames, would not be productive since the Galilean transformation is not applicable to electromagnetism. Instead the specific transformations for velocities and forces required by the special theory of relativity must be used (see Section 19.14 below).

19.8 Magnetic dipole moments of charged particles in closed orbits

We saw in Section 19.5 that a charge moving with constant speed in a uniform magnetic field executes circular motion if its velocity is directed perpendicularly to the field lines, the radius of the circle being given by Equation (19.9). A charge which is moving in a closed planar path such as this is equivalent to a current loop, at least as far as its time-average behaviour is concerned. Hence such a system will have a magnetic dipole moment, defined for current loops in Section 19.3 by $\boldsymbol{m} := IA\boldsymbol{n}$, where $\boldsymbol{n}$ is a unit vector perpendicular to the plane of the loop of area A. In the case of an orbiting charged particle, the average current is given by $I = \frac{q}{T}$ where $T = \frac{2\pi R}{v}$ is the period of the orbit and thus

$$|\boldsymbol{m}| = \frac{q}{T}A = \frac{qv}{2\pi R}(\pi R^2) = \frac{1}{2}qvR = \frac{q}{2m}mvR = \frac{q}{2m}|\boldsymbol{L}|$$

where we have identified the quantity mvR as the angular momentum of the particle of mass m moving with speed v in a circular orbit of radius R (Section 5.12).

A similar situation arises in the case of a charged particle which is in orbit in the field of a point charge of opposite sign (Figure 19.15), for example an electron moving in the electric field of a proton in a hydrogen atom. In this case the centripetal force (Section 4.7) is provided by the Coulomb attraction between the charges, Equation (16.6), that is

$$\boldsymbol{F} = \frac{q_1 q_2}{4\pi\varepsilon_0 r^2}\hat{\boldsymbol{r}} = m\boldsymbol{a}_c = -m\frac{v^2}{R}\hat{\boldsymbol{r}}$$

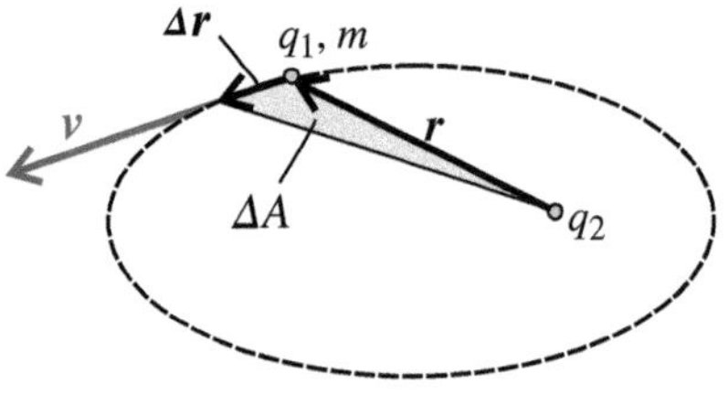

Figure 19.15. A charged particle in orbit under the influence of its Coulomb interaction with a particle of opposite charge.

The problem is identical to the Kepler problem for gravitational forces which we studied in Sections 5.12 and 5.13 except that $-GMm$ is replaced by $\frac{q_1 q_2}{4\pi\varepsilon_0}$ in this case. As we saw in Section 5.13, since the angular momentum of the system is conserved ($\boldsymbol{L}$ = constant), the area of the orbit is proportional to the orbital period. Invoking Equation (5.25) we see that

$$\frac{A}{T} = \frac{|\boldsymbol{L}|}{2m} = constant$$

Thus, the magnitude of the magnetic dipole moment of the current loop produced by the orbiting charge is given by

$$|\boldsymbol{m}| = IA = \frac{q}{T}A = q\left(\frac{|\boldsymbol{L}|}{2m}\right) = \frac{q}{2m}|\boldsymbol{L}|$$

which is the same result as derived earlier in this section for a charge circulating in a uniform magnetic field.

In either case we can write

$$\boxed{\boldsymbol{m} = \frac{q}{2m}\boldsymbol{L}} \qquad (19.14)$$

Equation (19.14) is quite general and has important applications in atomic (Section 24.9), nuclear and particle physics.

19.9 Polarisation of magnetic materials; magnetisation, magnetic susceptibility

Just as the application of an electric field caused polarisation of microscopic electric dipoles (Section 17.7) in dielectric materials, the application of a magnetic field to a magnetic material causes polarisation of microscopic magnetic dipoles in the material. In fact, as we will see, most of the relationships derived in the case of electrostatics can be carried over to magnetostatics simply by replacing $\boldsymbol{E}$ by $\boldsymbol{H}$ and $\boldsymbol{D}$ by $\boldsymbol{B}$.

There is, however, one significant difference between the electric and magnetic behaviour: the Biot–Savart law tells us that the magnetic field strength ($\boldsymbol{H}$) at a point in a material due to an electric current (such as the current loop in Figure 19.16) does not depend on the permeability of the material in which it is embedded. When a specimen is polarised, therefore, it is the magnetic flux density at a point within the medium that is modified, being the sum of the flux density due to the external field alone ($\mu_0\boldsymbol{H}$) and an additional magnetic flux density arising from the average effect of the aligned dipoles ($\boldsymbol{B}_{pol}$). Thus, the magnetic flux density in any small macroscopic region within a magnetic specimen, such as the cylindrical element in Figure 19.16, is given by

$$\boldsymbol{B} = \mu_0\boldsymbol{H} + \boldsymbol{B}_{pol} \qquad (19.15)$$

The magnitude and direction of $\boldsymbol{B}_{pol}$ depend on the microscopic nature of the material involved, as a result of which $\boldsymbol{B}_{pol}$ can be either parallel or antiparallel to $\boldsymbol{H}$.

Magnetisation

Each of the end faces of the cylindrical element shown in Figure 19.16 has a net magnetic pole strength of magnitude $\Delta p = \Delta\Phi = |\boldsymbol{B}_{pol}|(\Delta A)$. The magnetic dipole moment of the element may be obtained from Equation (18.19), $|\Delta\boldsymbol{m}| = \frac{2(\Delta p)l}{\mu}$, where $2l = \Delta z$.

Thus, in this case,

$$|\Delta\boldsymbol{m}| = \frac{|\boldsymbol{B}_{pol}|(\Delta A)\,(\Delta z)}{\mu_0} = \frac{|\boldsymbol{B}_{pol}|(\Delta V)}{\mu_0}.$$

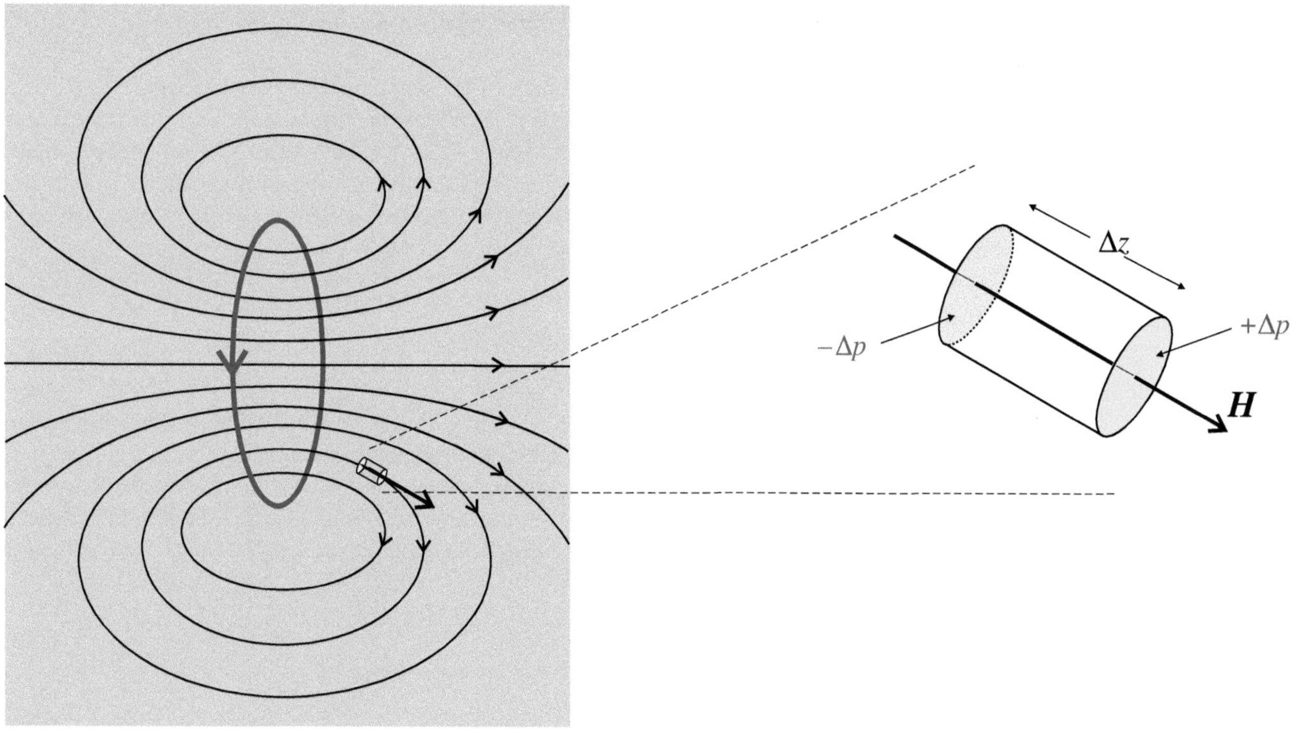

Figure 19.16. Polarisation of a magnetic material by the field due to an electric current embedded in the material. The source of the field, however, could have been any system of currents and/or permanent magnets so embedded or external to a sample of material of finite size.

The **magnetisation** at a point in a polarised specimen is defined as

$$\boxed{\boldsymbol{M} := \frac{d\boldsymbol{m}}{dV}}$$

namely, the magnetic dipole moment per unit volume (that is, the magnetic equivalent of polarisation $\boldsymbol{P}$ in electric materials).

Thus $\boldsymbol{M} := \frac{d\boldsymbol{m}}{dV} = \frac{\boldsymbol{B}_{\text{pol}}}{\mu_0}$ or $\boldsymbol{B}_{\text{pol}} = \mu_0 \boldsymbol{M}$ and Equation (19.15) becomes

$$\boldsymbol{B} = \mu_0 \boldsymbol{H} + \mu_0 \boldsymbol{M}$$

Magnetic susceptibility

The value of $\boldsymbol{B}_{\text{pol}}$ depends on the strength of the applied field but, over a wide range of field strengths, the magnitude of $\boldsymbol{B}_{\text{pol}}$ is proportional to the magnitude of $\boldsymbol{H}$, that is

$$\boldsymbol{B}_{\text{pol}} = (constant)\boldsymbol{H} = \chi_{\text{m}} \mu_0 \boldsymbol{H},$$

where the (dimensionless) constant χ_{m} is called the **magnetic susceptibility** of the material and may be positive or negative. Thus

$$\boldsymbol{B} = \mu_0 \boldsymbol{H} + \chi_{\text{m}} \mu_0 \boldsymbol{H} = (\mathbf{1} + \chi_{\text{m}}) \mu_0 \boldsymbol{H} = \mu \boldsymbol{H}$$

where we have now interpreted the permeability of a medium to be $\mu = (1 + \chi_{\text{m}})\mu_0$ so that the relative permittivity is

$$\mu_{\text{r}} = (1 + \chi_{\text{m}})$$

Magnetic materials mostly fall into three general classes: (i) paramagnetic materials (small positive susceptibilities), (ii) diamagnetic materials (small negative susceptibilities) and (iii) ferromagnetic materials (large positive susceptibilities and for which, as we will see in Section 19.12, $\boldsymbol{B}$ is not linearly proportional to $\boldsymbol{H}$).

19.10 Paramagnetism and diamagnetism

Paramagnetism

When a paramagnetic specimen becomes polarised due to an external magnetic field, the magnetic flux density inside the material is observed to be larger than that which would be observed in the absence of the specimen ($\chi_{\text{m}} > 0$). That is, $\mu > \mu_0$ for paramagnetic materials although, because χ_m is very small (typically 10^{-4} or 10^{-5}), the difference between μ and μ_0 may be insignificant in many cases.

Since the magnetic susceptibility χ_m is a measure of the degree of alignment of the permanent dipoles in the medium, it is to be expected that χ_m for paramagnetic materials will depend on temperature. Since the randomising effect of thermal motions reduces the degree of alignment, χ_m should decrease with increasing temperature and this is indeed observed. The general behaviour of paramagnetic materials with varying temperature was studied by Pierre Curie (1859–1906) who found that χ_m is inversely proportional to temperature, as long as the applied magnetic field is not too large, or the temperature is not too low. This result, known as **Curie's law**, can be stated as follows:

$$\boxed{\chi_m = \frac{C}{T}}$$

where C, the Curie constant, is characteristic of the particular paramagnetic material.

From a microscopic perspective, paramagnetic materials can be considered as comprising large numbers of permanent microscopic magnetic dipoles (such dipoles arise from a combination of the orbital motion and the intrinsic dipole moment of atomic electrons which will be discussed further in Section 24.13). In most materials, at room temperature, thermal motion tends to keep the dipoles randomly oriented unless an external magnetic field is applied (Figure 19.17(a)). When a magnetic field is applied to a paramagnetic specimen the dipoles will develop, on average, a preferential alignment along the direction of the field (Figure 19.17(b)). As soon as the field is removed thermal agitation re-establishes the random orientation.

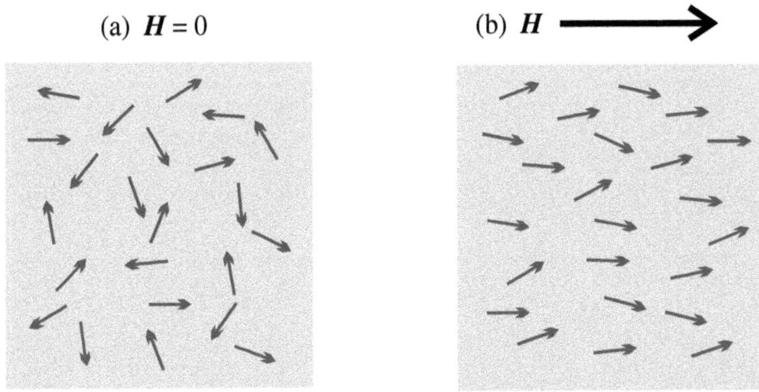

Figure 19.17. (a) Paramagnetic materials comprise permanent microscopic magnetic dipoles which, in the absence of an external magnetic field, are randomly oriented by thermal agitation. (b) When an external magnetic field is applied to a paramagnetic specimen, the material becomes magnetised with the dipoles preferentially aligned.

Diamagnetism

When an external magnetic field is applied to a diamagnetic material the result is a small *decrease* in the magnetic flux density inside the specimen compared to that which would be observed in the absence of the medium. In other words, in such materials, χ_m is negative and $\mu < \mu_0$ although the difference between μ and μ_0 is very small ($\sim$ one part in 10^5), as it is in the case of paramagnetism. Since thermal agitation does not play a significant role in this case, χ_m is essentially independent of temperature in diamagnetic materials. While all materials can show diamagnetic behaviour in principle, the effect is small and is dominated by paramagnetism or ferromagnetism (see below) in media having permanent dipoles.

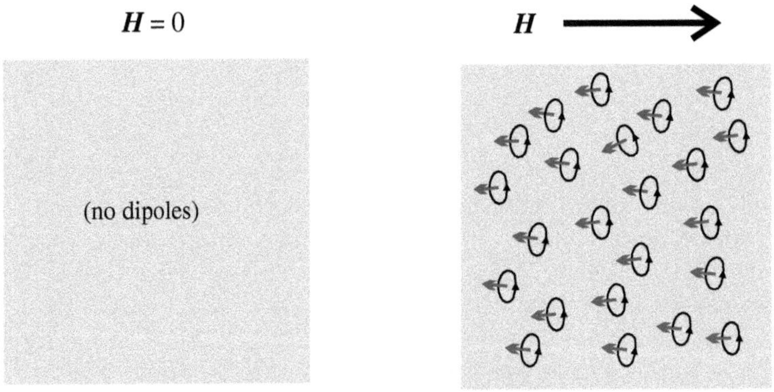

Figure 19.18. In the absence of a magnetic field there are no magnetic dipoles in diamagnetic materials. Current loops are induced in when an external magnetic field is applied. The dipole moments of these current loops preferentially align antiparallel to the applied field.

The electron configuration (described in Section 24.13) in atoms of **diamagnetic materials** is such that there are no permanent dipoles, which is why these materials cannot show paramagnetic behaviour. Nevertheless, an external magnetic field applied to these materials can *induce* current loops in the atoms but, in this case, the atomic current loops will always line up in the opposite direction to the applied field (Figure 19.18). Hence, there is a small decrease in the magnetic flux density inside the specimen compared to that which would be observed in the absence of the medium.

19.11 Boundaries between magnetic media

As was the case for electric fields in dielectric materials (Section 17.4), if a magnetic field near the interface between two linear, homogeneous, and isotropic magnetic materials makes an angle with the interface, the direction of the magnetic flux density $\boldsymbol{B}$ (and hence the magnetic field strength $\boldsymbol{H}$) will be different on either side of the interface.

As in other flux models discussed previously, a magnetic flux is defined such that the flux within any 'tube' delineated by magnetic field lines must remain unchanged across the interface. Thus, the treatment of boundaries between dielectric materials in Section 17.4 may also be applied to boundaries between magnetic materials, leading to the following boundary conditions at the interface.

$$\boxed{(\boldsymbol{B}_1)_\perp = (\boldsymbol{B}_2)_\perp} \quad \text{and} \quad \boxed{(\boldsymbol{H}_1)_\parallel = (\boldsymbol{H}_2)_\parallel}$$

where the subscripts $\perp$ and $\parallel$ refer to the components of the fields perpendicular and parallel to the interface, respectively.

19.12 Ferromagnetism; permanent magnets revisited

If a magnetic field is applied to certain materials, it is observed that these materials remain permanently magnetised after the field has been removed, a similar behaviour to that observed in the case of the ferroelectric materials discussed in Sections 17.5 and 17.6. Such materials are described as **ferromagnetic**; iron, nickel, cobalt, and alloys containing these elements behave in this way, even at room temperature. If the temperature of a specimen is raised sufficiently, however, the material loses its ferromagnetic properties and becomes paramagnetic. Furthermore, materials which are paramagnetic at room temperature may become ferromagnetic if cooled. The temperature at which a particular material makes the transition from paramagnetic to ferromagnetic behaviour (or vice versa) is called the **Curie temperature** for that material.

As is the case for $\boldsymbol{D}$ and $\boldsymbol{E}$ in ferroelectric materials (Section 17.5), there is a highly non-linear relationship between $\boldsymbol{B}$ and $\boldsymbol{H}$ in ferromagnetic materials, illustrated in Figure 19.19(a) and (b). As the external magnetic field is increased all the elementary dipoles gradually become fully aligned, at which point the magnetisation is said to be saturated and B can be of the order $10^3 \mu_0 H$. When the external field is removed, the residual magnetic flux density inside the specimen remains at a high value. Ferromagnetic materials, therefore, retain a 'memory' of their previous magnetisation and exhibit *hysteresis*; the B—H plots are 'hysteresis loops'. The intercept on the B-axis (B_r — called the *residual flux density*) determines how much magnetism remains when the H-field is removed while the intercept on the H-axis (H_c — the *coercive field strength*) determines how large a reverse field is required to fully demagnetise the specimen.

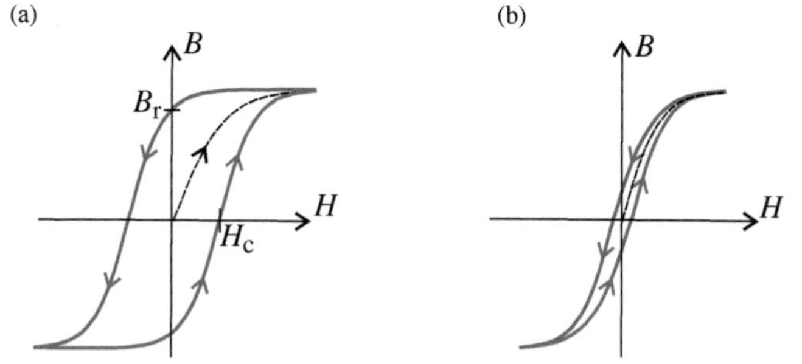

Figure 19.19. (a) Hysteresis loop of a magnetically hard material discussed in text. (b) Hysteresis loop of a magnetically soft material. In both plots the dashed line applies to the initial application of the field.

When ferromagnetic substances are studied a under high resolution microscope it is observed that such materials comprise a large number of small permanently polarised **domains.** The formation of domains is attributable to quantum mechanical effects which will be described at the end of Section 24.13. Normally the orientations of individual domains are random (Figure 19.20(a)) but, under the influence of an external magnetic field, the orientations of individual domains tend to align along the direction of the field. As the external field is increased more domains line up until the specimen becomes saturated, as indicated in Figure 19.20(b).

Permanent magnets are constructed from ferromagnetic materials with broad hysteresis loops (Figure 19.19(a)). It takes a significant amount of energy to magnetise a magnetically 'hard' specimen, but it is equally difficult to demagnetise the specimen.

The discussion in Section 17.6 of a uniformly polarised dielectric rod may be applied similarly to a uniformly magnetised cylinder (an example of a **bar magnet**) with $\boldsymbol{H}$ replacing $\boldsymbol{E}$ and $\boldsymbol{B}$ replacing $\boldsymbol{D}$. The diagram in Figure 17.14 can represent the magnetic case on the understanding that the field lines are magnetic flux lines (above) and magnetic field lines (below), respectively.

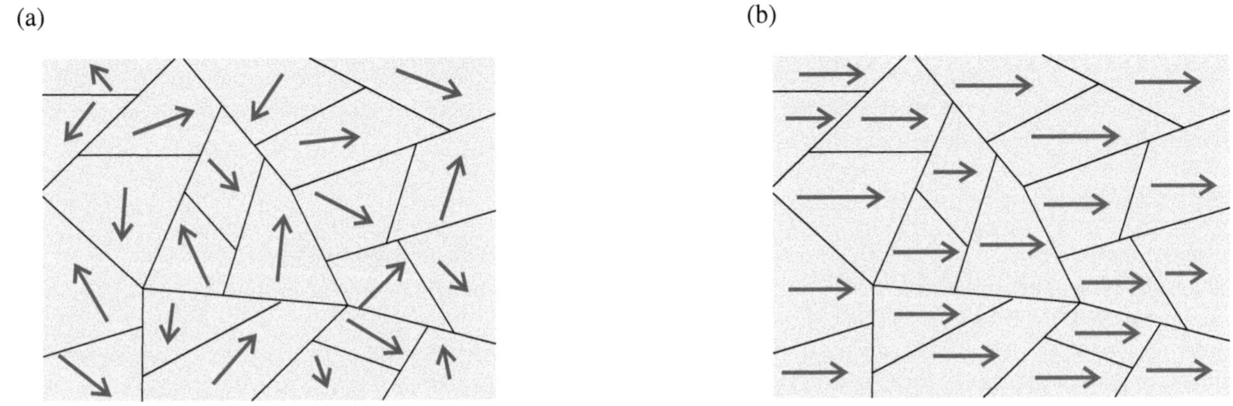

Figure 19.20. Illustration of ferromagnetic domains in (a) an unmagnetised specimen and (b) a fully magnetised specimen.

Magnetic needles

In the case of a long narrow specimen, where the dipoles are aligned along its length, the specimen looks from the outside as if it has a point magnetic pole near each end – in other words, it is what we have called a '**magnetic needle**'. As in the case of the electret, these poles cannot be decoupled from the specimen.

Soft magnetic materials

Some materials, while being fully ferromagnetic, do not remain permanently magnetised after the magnetic field has been removed; in other words, they have narrow hysteresis loops (Figure 19.19(b)). The magnetically 'soft' iron used in transformer cores and the ferrite rods used as aerials in radios are examples of such materials. In an electromagnet (Figure 19.21) the space inside a solenoid is filled with a soft ferromagnetic material. The magnetic field strength inside the solenoid is given by Equation (18.9) and hence the magnetic flux density is approximately $10^3 \mu_0 n_l I$. The electromagnet, therefore, can act as a strong magnetic needle, the 'pole strength' of which is one thousand times that of the solenoid without the soft iron core. The electromagnet has the advantage of being capable of being switched on or off; that is, it provides a controllable magnetic field.

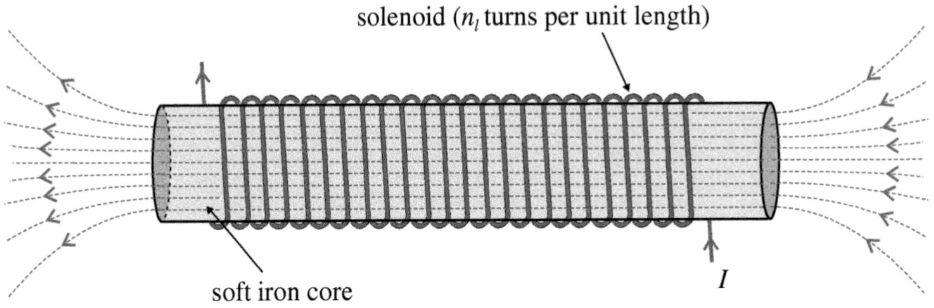

Figure 19.21. The magnetic flux density inside a solenoid is increased approximately one thousandfold by the introduction of a ferromagnetic core. In an electromagnet, a soft magnetic material is used for the core, so that the magnetic field disappears as soon as the current is switched off.

19.13 Moving coil meters and electric motors

Moving coil meters

The torque experienced by a current loop in a magnetic field, given by Equation (19.5), is also the basic principle which governs the design of moving coil instruments. In its simplest form, indicated in Figure 19.22(a), a moving coil meter or galvanometer consists of a planar coil of wire in which a current can flow. The coil is held by pivots which allow it to rotate about an axis in its plane and, when no current flows, the coil is held with its plane parallel to a uniform magnetic field provided by the poles of a permanent magnet.

When a current is passed through the coil, it experiences a torque which tends to turn it about the axis of rotation. This torque is opposed by either a hairspring (like that used on the balance wheel of an old style watch) or an elastic suspension (similar to that used in a torsion balance, described in Section 7.5). In either case the coil will rotate initially but will then settle down in equilibrium between two opposing torques, namely the torque $\boldsymbol{T} = IA\boldsymbol{n} \times \boldsymbol{B}$ due to the current and the restoring torque due to the hairspring/suspension, $c\phi$, where ϕ is the angle through which the coil has rotated (Figure 19.22(b)) and c is the restoring torque per unit angle of rotation, which

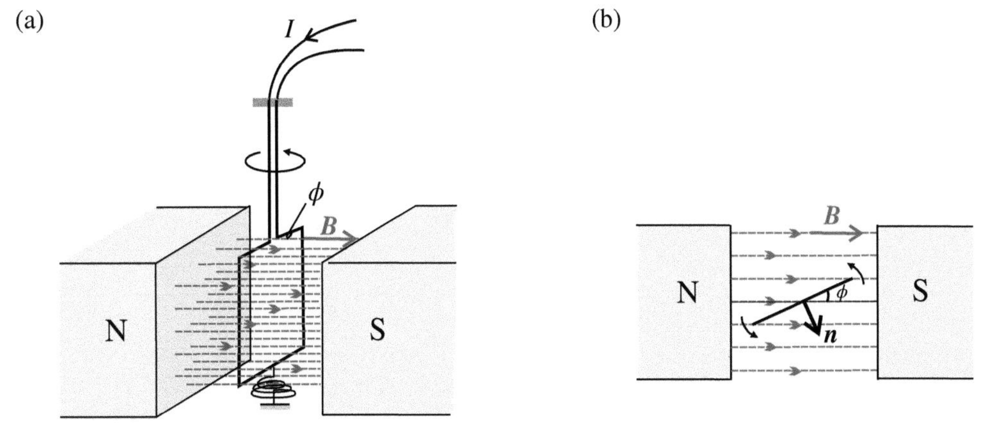

Figure 19.22. (a) A simple moving coil meter comprising a plane coil made from a single turn of wire placed in a uniform magnetic field. When a current flows, the coil experiences a torque which tends to rotate it about a vertical axis. This torque is counteracted by an opposing torque exerted by the suspension so that the coil settles down in equilibrium having rotated through an angle which depends on the strength of the current. (b) The moving coil instrument in (a), viewed from above; the torque is directed perpendicularly to the page and outwards.

depends on the sensitivity of the hairspring/suspension. Thus the angle through which the coil turns as a result of a steady current I flowing in it is determined from Equation (19.5), that is

$$|\boldsymbol{T}| = c\phi = |\boldsymbol{m} \times \boldsymbol{B}|$$

$$c\phi = IBA \sin\phi = IBA \cos\phi$$

Since B, A, and c can all be measured, the current can be determined from

$$I = \frac{c}{BA} \frac{\phi}{\cos\phi}$$

A serious disadvantage of such a design is that the response of the instrument is very non-linear (that is, I is not directly proportional to ϕ) except for small angles for which $\cos\phi \to 1$; it would be more useful as a measuring instrument if I was directly proportional to ϕ. A more practical moving coil meter, which is designed to have a linear relationship between I and ϕ, is shown in Figure 19.23(a). The magnetic field is produced by a permanent magnet with pole pieces shaped as shown. The effect of the soft iron cylinder on which the coil is wound is to 'suck in' the magnetic flux as indicated by the field lines in Figure 19.23(b) and this, in turn, produces a radial field in the gap between the soft iron cylinder and the curved pole pieces. Thus, as the coil rotates, the angle between $\boldsymbol{m}$ and $\boldsymbol{B}$ remains $\frac{\pi}{2}$, that is the torque due to a current I on a single turn of wire is IBA at all values of ϕ. To increase the sensitivity of the instrument the coil may comprise a number of turns N. For this type of instrument, known as a *d'Arsonval galvanometer*, the condition for equilibrium becomes

$$NIBA = c\phi \quad \text{or} \quad I = \frac{c}{NBA}\phi$$

so that the meter has a linear response. For maximum sensitivity, N and B are made as large as possible. Many practical laboratory instruments, especially older models, are based on the moving coil galvanometer, as illustrated in Worked Example 19.4.

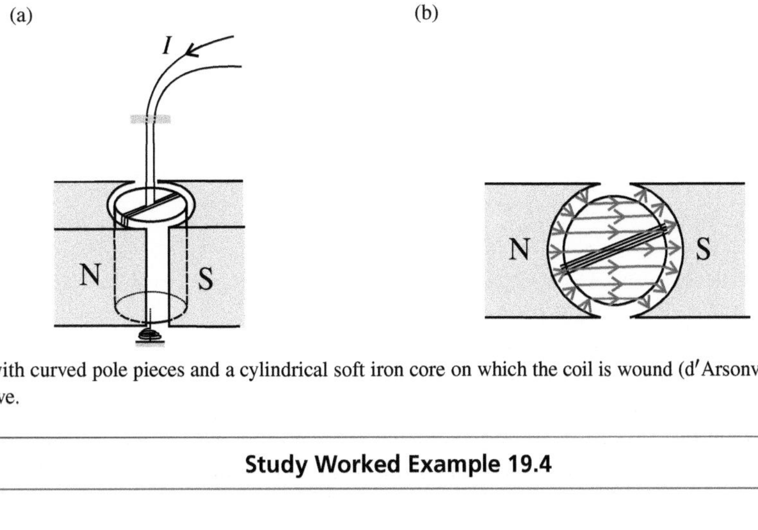

Figure 19.23. A moving coil meter with curved pole pieces and a cylindrical soft iron core on which the coil is wound (d'Arsonval galvanometer). (b) The moving coil galvanometer in (a), viewed from above.

Study Worked Example 19.4

Electric motors

If the plane of the current carrying coil in Figure 19.22(a) is initially parallel to the magnetic field, the coil experiences a torque which decreases in magnitude until it becomes zero at $\phi = \frac{\pi}{2}$. The inertia of the coil causes it to rotate beyond $\phi = \frac{\pi}{2}$ but thereafter it experiences a torque which is in the opposite direction to the original torque and this causes the coil to slow down. If, however, the direction of the current could be reversed every time the coil passes through the $\phi = \frac{\pi}{2}, \frac{3\pi}{2}$, etc. positions, the torque exerted on the coil, although varying periodically in magnitude, would always be in the same direction and the coil would continue to rotate for as long as the electrical energy is applied. This is the basis of the electric (d.c.) motor, the principle of which is indicated in Figure 19.24. The reversal of the current in the coil at the appropriate moment is achieved by the use of a split-ring commutator, as shown in the figure. Contact between the split-ring and the source of current is maintained by two carbon 'brushes', a different side of the split ring making contact with the positive terminal of the power supply during each half cycle.

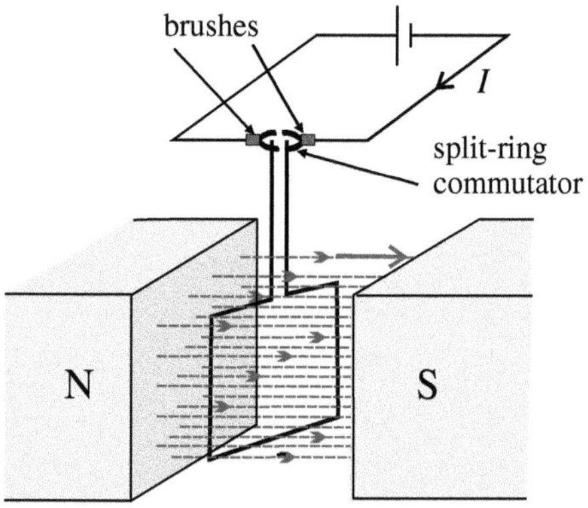

Figure 19.24. A simple electric motor.

As it stands, the motor shown in Figure 19.24 will not start if $\phi = \frac{\pi}{2}$ when the current is switched on. This can be overcome by having one or more extra coils whose plane(s) are offset from the original coil by an acute angle; in this case the commutator has more than two sections.

Naively, one might expect the rotating coil to accelerate indefinitely; however, not only are there inevitably mechanical frictional forces involved but, as we shall see in the next chapter, damping forces of electromagnetic origin also arise.

For problems based on the material presented in this section visit up.ucc.ie/19/ *and follow the link to the problems.*

19.14 Electric and magnetic fields in moving reference frames

As was mentioned in Section 19.7, the Galilean rules for transforming between reference frames are not applicable in electromagnetism. Instead the appropriate equations are the relativistic velocity transformations (9.29) to (9.31)

$$u_x' = \frac{u_x - V}{1 - \frac{u_x V}{c^2}}; \quad u_y' = \frac{u_y}{\gamma\left(1 - \frac{u_x V}{c^2}\right)}; \quad u_z' = \frac{u_z}{\gamma\left(1 - \frac{u_x V}{c^2}\right)}, \quad \text{where} \quad \gamma = \frac{1}{\sqrt{1 - \frac{V^2}{c^2}}}$$

These, and the corresponding force transformations (web section up.ucc.ie/9/11/1/), are used in Section 19.14 on the *Understanding Physics* website at up.ucc.ie/19/14/ to show how electric and magnetic fields must be transformed when observed in reference frames moving with relative velocities.

Further reading

Further material on some of the topics covered in this chapter may be found in *Electromagnetism* by Grant and Phillips (See BIBLIOGRAPHY for book details)

WORKED EXAMPLES

Worked Example 19.1: A wire carrying a current of 3.5 A divides into two paths which form a rectangular framework of dimensions 3 cm x 4 cm as shown in Figure 19.25. One of the 4 cm lengths of wire has a resistance of 2.0 Ω and the other has a resistance of 5.0 Ω. The resistances of the 3 cm lengths are negligible. If the plane of the rectangle is parallel to a uniform magnetic field of flux density 2.4 T, what is the net moment on the framework about the axis A′A″ shown in the figure?

Referring to Figure 19.25: $l = 4$ cm and $d = 1.5$ cm

$$\text{Resistance of left hand vertical side} = R_1 = 2.0\ \Omega; \quad \text{current in left hand side} = I_1$$

$$\text{Resistance of right hand vertical side} = R_2 = 5.0\ \Omega; \quad \text{current in right hand side} = I_2$$

Now $R_1 I_1 = R_2 I_2 \rightarrow 2I_1 = 5I_2$ and $I_1 + I_2 = 3.5$ A $\rightarrow I_1 = 2.5$ A; $I_2 = 1.0$ A

Using Equation (19.2), the force on a length l of straight wire carrying a current I in a flux density $\boldsymbol{B}$ is given by $F = BIl \sin\theta$. Since, in both cases, $\theta = \frac{\pi}{2}$

$$\text{force on left hand side} = F_1 = BI_1 l \text{ (out of page)}$$

$$\text{force on right hand side} = F_2 = BI_2 l \text{ (out of page)}$$

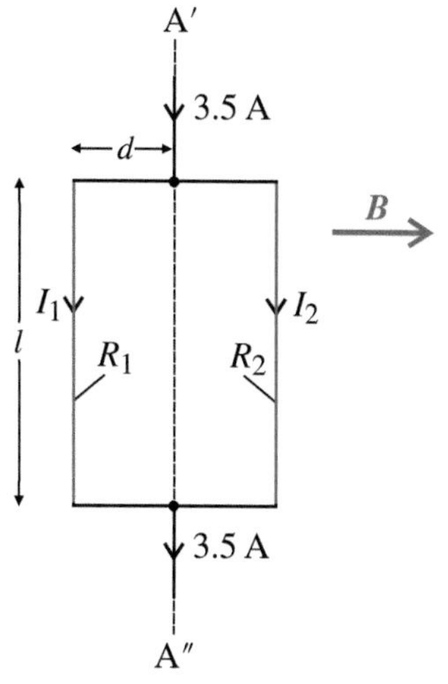

Figure 19.25. Worked Example 19.1.

Net moment on framework about axis AA′ $= F_1d - F_2d = BI_1ld - BI_2ld = Bld(I_1 - I_2) = (2.4)(0.04)(0.015)(2.5 - 1.0) = 2.2$ mN m
The moment is anticlockwise when viewed along the direction of the current.

Worked Example 19.2: Two protons are moving in a vacuum along parallel paths with equal and opposite velocities of magnitude v. If F_E and F_M are, respectively, the magnitudes of the electric and the magnetic forces between the protons when they pass each other, show that $\frac{F_M}{F_E} = \mu_0\varepsilon_0 v^2$.

If d is distance between the protons when they pass each other, the magnitude of the electric force is given by Equation (16.6) (Coulomb's Law) $F_E = \frac{e^2}{4\pi\varepsilon_0 d^2}$
The magnitude of the magnetic force is given by Equation (19.8), $F_M = evB = ev\mu_0 H$,

where, from (Equation (19.7)), $|H| = \frac{ev}{4\pi d^2}$, so that $F_M = ev\mu_0\frac{ev}{4\pi d^2}$

Thus

$$\frac{F_M}{F_E} = \left(ev\mu_0\frac{ev}{4\pi d^2} \Big/ \frac{e^2}{4\pi\varepsilon_0 d^2}\right) = \mu_0\varepsilon_0 v^2$$

When we study electromagnetic waves in Section 21.2 we will find that $c^2 = \frac{1}{\mu_0\varepsilon_0}$, where c is the velocity of light in a vacuum. The force ratio can therefore be written $\frac{F_M}{F_E} = \frac{v^2}{c^2}$, from which we can conclude that, unless v is comparable in magnitude to c, the magnetic force is much smaller than the electric force.

Worked Example 19.3: What is the radius of curvature of the path of an electron travelling near the surface of a neutron star at a velocity of 10^7 m s^{-1} perpendicular to the magnetic flux density which is approximately 10^8 T? Estimate the energy and the speed of the electron if the radius of curvature is to be greater than the radius of the neutron star ($\approx$ 10 km).

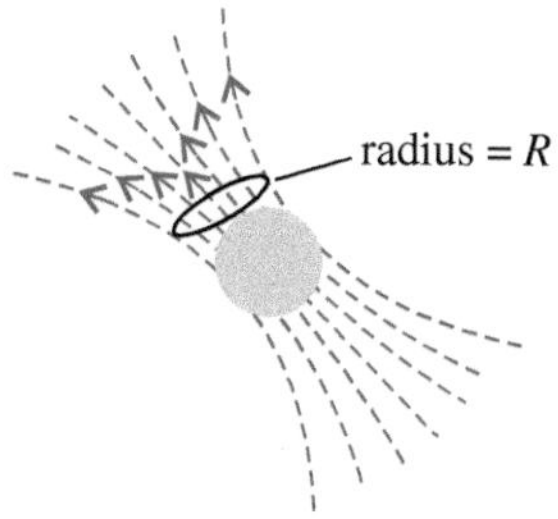

Figure 19.26. Worked Example 19.3.

Referring to Figure 19.26, we see from Equation (19.9) that

$$R = \frac{mu}{qB} = \frac{(9.1\times10^{-31})(10^7)}{(1.6\times10^{-19})(10^8)} = 5.7\times10^{-13}\ \text{m}$$

For $R = 10$ km $\quad p = qBR = (1.6\times10^{-19})(10^8)(10^4) = 1.6\times10^{-7}$ kg m s^{-1}

A classical calculation of the speed $\left(\frac{p}{m} = 10^{23}\text{m s}^{-1}!\right)$ indicates that a relativistic treatment is required. Using the relativistic energy-momentum relationship (Equation 9.40), $E = \sqrt{c^2p^2 + m^2c^4}$, where $cp = (3 \times 10^8)(1.6 \times 10^{-7}) = 48\text{ J} = \frac{48}{1.6 \times 10^{-19}}\text{ eV} = 3 \times 10^{20}\text{ eV}$ is very much greater than the rest energy of the electron, $m_ec^2 = (511\text{ eV}/c^2)c^2 = 511\text{ eV}$. Thus $E \approx cp$ and we get, using Equation (9.41),

$$u = \frac{c^2p}{E} = \frac{cp}{E}c \approx c$$

Worked Example 19.4: A moving coil meter of internal resistance 10.0 Ω gives a full-scale deflection when a 100 μA current passes through it. Determine the value of the resistances which, when connected in series or parallel with the meter will enable it to measure

(i) currents up to 1.0 A or
(ii) potential differences up to 10.0 V.

(i) If a resistor R_p is connected in parallel with the moving coil meter R_g, as illustrated in Figure 19.27, only a fixed proportion of the 1.0 A current to be measured will pass through the meter. We therefore seek a value of R_p which will produce a current of 1.0×10^{-4} A through R_g, with the remaining $[1.0 - (1.0 \times 10^{-4})]$ A passing through R_p.

The potential difference V is the same across the two resistors.

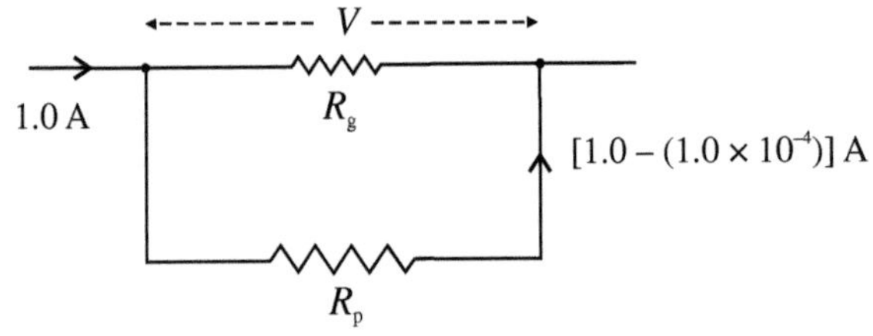

Figure 19.27. Worked Example 19.4.

Thus $V = (1.0 \times 10^{-4})10.0 = [(1.0 - (1.0 \times 10^{-4})]R_p \rightarrow R_p = \frac{10^{-3}}{0.999} \approx 10^{-3}\ \Omega$, in parallel with the meter.

(ii) If a resistor R_s is connected in series with the moving coil meter R_g, as illustrated in Figure 19.28, only a fixed proportion of the potential difference to be measured will be applied across the meter. We therefore seek a value of R_s which will produce a current of 1.0×10^{-4} A through R_g when 10.0 V is applied across the two resistances.

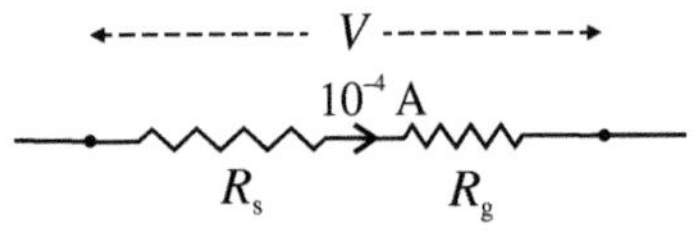

Figure 19.28. Worked Example 19.4.

The total potential drop across the two resistances, $10.0\text{ V} = (R_s + 10.0)10^{-4}\text{ V} \rightarrow R_s = \frac{10.0}{10^{-4}} - 10 \approx 10^5\ \Omega$, in series with the meter.
Thus, by connecting suitably chosen resistors in parallel or in series, a moving coil meter may be used to measure a wide range of currents or potential differences.

PROBLEMS

For problems based on the material covered this chapter visit up.ucc.ie/19/ and follow the link to the problems.

20

Electromagnetic induction: time-varying emfs

AIMS

- to show how, whenever there is a changing magnetic flux through a circuit, an emf is induced in the circuit
- to demonstrate how the response of a circuit to a time varying emf depends on a property of the circuit called *inductance*
- to show how a coil which is rotating at a constant rate in a uniform magnetic field generates a sinusoidally varying emf and how such sinusoidal emfs give rise to *alternating currents*
- to show that the combined effect of resistive, inductive and capacitive components on currents and potential differences in alternating current circuits are very different from those produced by steady currents, as discussed in Chapter 15

20.1 The principle of electromagnetic induction

As we saw in Section 19.5, an electric charge moving in a magnetic field experiences a force. Consequently, when an electrical conductor is moved in a magnetic field, the charge carriers in the conductor may experience a force as a result of its motion relative to the field. First let us consider what happens in the simple case, illustrated in Figure 20.1, in which a straight piece of conducting wire is moved through a constant uniform magnetic field. Let the flux density $\boldsymbol{B}$ be directed outwards perpendicularly to the page and let the velocity $\boldsymbol{v}$ of the wire be directed in such a way that the wire remains confined to the plane of the page. Every charge carrier in the wire has the same velocity as the wire itself and hence each such charge q will experience a force given by Equation (19.8), namely

$$\boldsymbol{F} = q\boldsymbol{v} \times \boldsymbol{B}$$

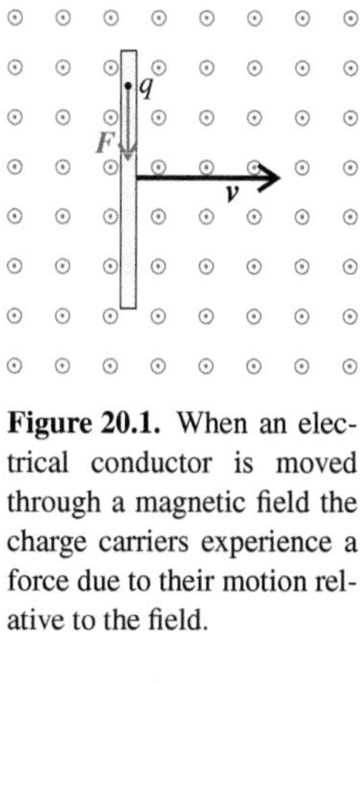

Figure 20.1. When an electrical conductor is moved through a magnetic field the charge carriers experience a force due to their motion relative to the field.

Since, in the case illustrated in Figure 20.1, $\boldsymbol{v}$ is perpendicular to $\boldsymbol{B}$, $|\boldsymbol{F}| = qvB$ and is directed along the wire as shown in the figure.

If the wire is part of a complete circuit this force causes a current to flow. This could be arranged, in principle at least, as indicated in Figure 20.2, where a wire AB of length l lies on two smooth conducting parallel rails along which it can be moved so that the ends of the wire, A and B, remain in electrical contact with the rails. The circuit is completed by a resistor R which joins the ends of the tracks as shown. A charge q can move from A to B under the influence of the force qvB and the mechanical work done in this process is given by

$$W_{AB} = \int \boldsymbol{F} \cdot d\boldsymbol{l} = |\boldsymbol{F}|l\cos\theta = |\boldsymbol{F}|y = qvBy$$

where y is the perpendicular distance between the rails, as shown in Figure 20.2. Thus, the work done per unit charge is $\dfrac{W_{AB}}{q} = Byv$. This result can be shown to hold also if the wire AB is not straight.

The motion of the conductor relative to the magnetic field, therefore, has induced an emf whose magnitude is equal to the potential difference between the points A and B which, from Equation (16.23), is given by

$$V_{AB} = \frac{W_{AB}}{q} = Byv \tag{20.1}$$

Understanding Physics, Third Edition. Michael Mansfield and Colm O'Sullivan.
© 2020 John Wiley & Sons Ltd. Published 2020 by John Wiley & Sons Ltd.

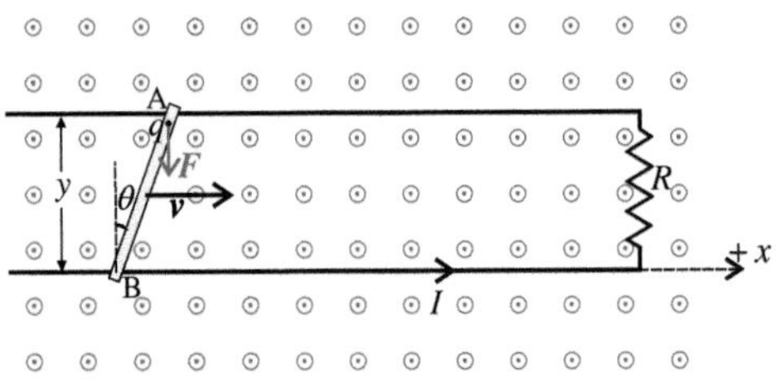

Figure 20.2. When the piece of wire AB is caused to slide along fixed parallel rails, an electric current flows in the circuit.

Assuming that the wire and rails have negligible resistance, the current induced in the circuit by the motion of the wire is

$$I = \frac{V_{AB}}{R} = \frac{Byv}{R}$$

If the direction of motion of the wire is taken as the $+x$-direction as indicated in Figure 20.2, Equation (20.1) can be written as

$$V_{AB} = Byv = By\frac{dx}{dt} = B\frac{dA}{dt}$$

where we have recognised that $y\frac{dx}{dy} = \frac{d(xy)}{dt} = \frac{dA}{dt}$ is the rate at which area is being swept out by the moving wire. Since B is the magnetic flux per unit area, $\Phi_M = \iint BdA = BA$, is the magnetic flux through the closed circuit of area A and

$$V_{AB} = B\frac{dA}{dt} = \frac{d(BA)}{dt} = \frac{d\Phi_M}{dt} \tag{20.2}$$

Thus, the induced emf is equal to the rate of change of the magnitude of the magnetic flux through the circuit.

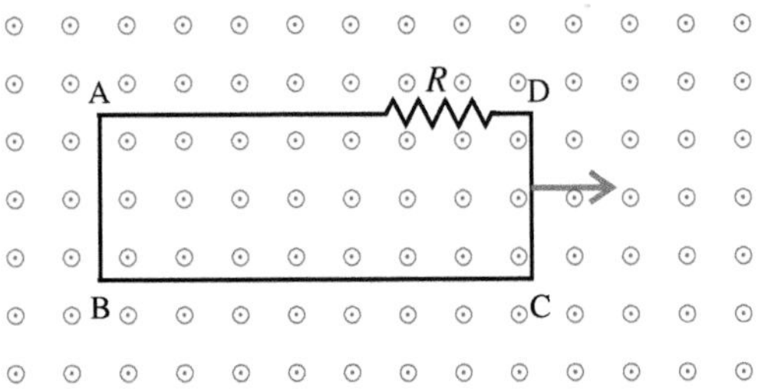

Figure 20.3. No current is induced when a complete loop is moved as indicated.

In the above discussion the magnetic field was both constant and uniform and was perpendicular to the plane of the circuit in which the induced current flows. In this case the change in the magnetic flux through the circuit arises from the changing area of the circuit. Note that if a loop of wire of *fixed area* is moved in a uniform field in such a way that the plane of the loop remains perpendicular to the field, no emf is induced by the motion since $\frac{d\Phi_M}{dt} = 0$; that is, the total flux through the loop is the same throughout its motion. For the rigid rectangular loop illustrated in Figure 20.3, for example, the emf induced in the side AB is equal and opposite to the emf induced in the side DC when the magnetic field is uniform.

On the other hand, if the field is *non-uniform*, $\frac{d\Phi_M}{dt} \neq 0$ so that, in general, a net emf is induced in the loop in the course of its motion. Furthermore, since $\frac{d\Phi_M}{dt} = \frac{d(BA)}{dt} = B\frac{dA}{dt} + A\frac{dB}{dt}$, an emf can be also be induced by changing $\boldsymbol{B}$. Thus, for example, if a loop is held fixed relative to a magnetic field which is uniform but *time-varying*, then the emf induced is $\frac{d\Phi_M}{dt} = A\frac{dB}{dt}$.

Thus in all cases, wherever there is a changing magnetic flux through a complete circuit, there will be an emf induced in the circuit given by

$$E_{ind} = \left|\frac{d\Phi_M}{dt}\right| \tag{20.3}$$

Relationship (20.3), which states that *the induced emf in a circuit due to a changing magnetic flux is equal to the rate of change of the magnetic flux through the circuit,* is a statement of the **principle of electromagnetic induction**. It is also known as **Faraday's law** after its discoverer Michael Faraday (1791–1867).

Equation (20.3) is stated in terms of the magnitude of the rate of change of flux because, at this stage, we do not yet know in which sense around the circuit will the emf be positive.

If the resistance of the circuit is R_{circ}, the corresponding current induced is

$$I_{ind} = \frac{1}{R_{circ}}\left|\frac{d\Phi_M}{dt}\right|$$

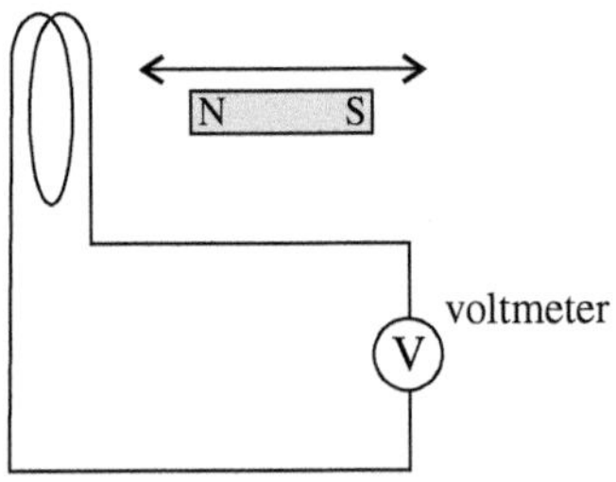

Figure 20.4. The induced emf due to the motion of the magnet relative to the coil can be measured by the voltmeter.

and the power generated is given by

$$P = E_{ind}I_{ind} = \left|\frac{d\Phi_M}{dt}\right| I_{ind} = \frac{1}{R_{circ}}\left|\frac{d\Phi_M}{dt}\right|^2 \tag{20.4}$$

The principle of electromagnetic induction can be demonstrated by the simple experiment which is illustrated in Figure 20.4. The 'voltmeter' used in this case is any instrument which can record and display a voltage—time plot, such as a computerised data acquisition system. If the magnet is moved back and forth relative to the coil, or if the coil is moved relative to the magnet, the corresponding induced emf is observed on the 'voltmeter' screen (as shown in Figure 20.5, for example).

The same effect is observed if the source of magnetic field is a current in a second coil (coil 2 in Figure 20.6), the induced emf appearing when there is relative motion between the coils. An induced emf is also observed if both coils in Figure 20.6 are held fixed and the magnetic flux through coil 1 is changed by varying the current in coil 2, for example by adjusting the variable resistor in a continuous manner. This illustrates that the principle of electromagnetic induction remains valid whatever the cause of the changing magnetic flux through the circuit.

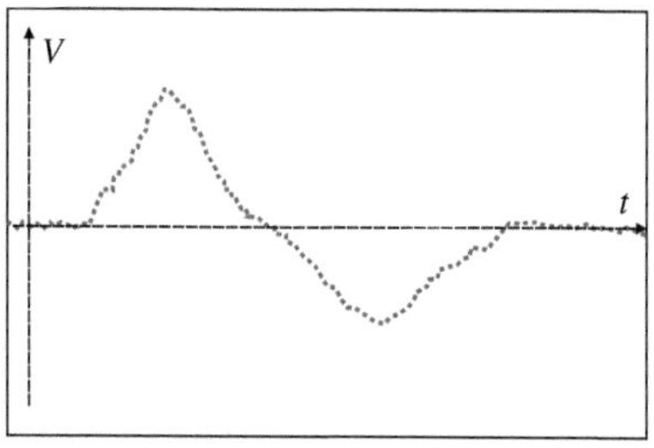

Figure 20.5. A plot of induced emf as a function of time for the experiment illustrated in Figure 20.4.

Energy considerations

We have seen (Equation (20.4)) that the thermal power generated in a circuit by an induced emf depends on that emf in the same way as that generated by a conventional power supply. One may well ask if we are getting power for nothing in this case, particularly if we ensure that there are no friction-like mechanical forces involved. Looking again at Figure 20.2, however, we see that when an induced current flows there must be a force on the conductor AB because it carries a current which will experience a force due to the magnetic field, as discussed in Section 19.1. For a segment $\boldsymbol{\Delta l}$, this force is given by Equation (19.2), namely $\boldsymbol{\Delta F} = I(\boldsymbol{\Delta l})\times \boldsymbol{B}$, and hence the force on the conductor AB is given by $|\boldsymbol{F}| = IlB$ in the direction shown in Figure 20.7, that is perpendicular to AB and in the plane of the loop.

This force must be overcome to keep AB moving with velocity $\boldsymbol{v}$ and the mechanical power generated in these circumstances (recall Section 5.8) is given by Equation (5.19), that is

$$P = \boldsymbol{F}\cdot\boldsymbol{v} = \{(IlB)\cos\theta\}v = (IyB)v = IBy\frac{dx}{dt} = IB\frac{dA}{dt} = \left|\frac{d\Phi_{\mathrm{M}}}{dt}\right| I = E_{\mathrm{ind}}I$$

which is exactly the same value as the thermal power given by Equation (20.4). Thus we see that the power generated, when a current is induced in a circuit by a changing magnetic field, is at the expense of the power that is generated by the forces which must be applied to overcome the resistance to motion exerted when the induced current flows in the magnetic field. In this case, the direction of the induced emf is such that the force exerted on the induced current by the magnetic field opposes the motion of the wire AB. Similarly, when the magnet is moved relative to the coil in Figure 20.4, the resulting current induced in the coil generates a magnetic field which exerts a force on the magnet which opposes its motion. This is a special case of a useful general rule, known as **Lenz's 'law'**, which can be used to identify the sense in which an emf is induced in a circuit by electromagnetic induction, namely

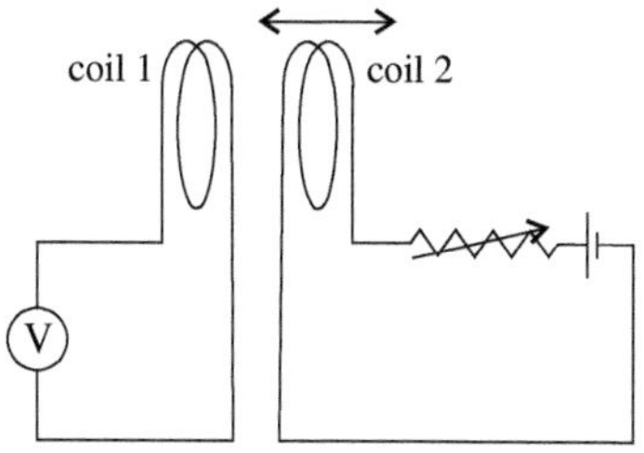

Figure 20.6. An emf is induced in coil 1 due to the relative motion of the current carrying coil 2 or if the current in coil 2 is varied.

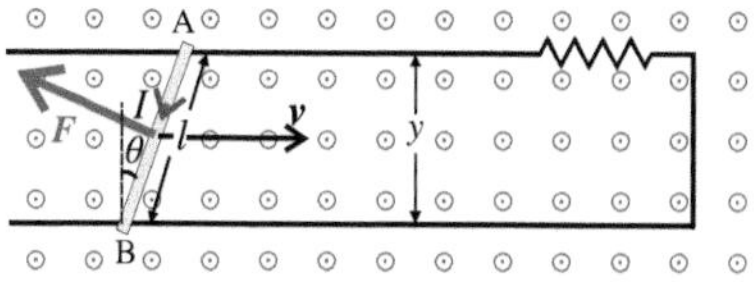

Figure 20.7. Because of the effect of the magnetic field on the induced current flowing from A to B, the moving wire will experience a force $\boldsymbol{F}$ which must be overcome to keep the wire moving at constant speed.

> *the emf induced in a circuit is such that the magnetic field produced by the induced current opposes the change in the magnetic field inducing the current*

In other words, the direction of the induced current is such that it tends to conserve the net magnetic flux through the circuit.

Incorporating Lenz's law in Faraday's law

Faraday's law as stated by Equation (20.3) gives the magnitude of the induced emf due to a changing magnetic flux in a circuit. The direction around the circuit in which this emf acts can be identified from Lenz's law (assuming prior knowledge of the Biot-Savart law or, equivalently, the rule for direction of current flow). These two aspects of the underlying fundamental law of nature can be incorporated into a single equation provided we are clear on the use of the appropriate convention of signs.

As we saw in Section 16.9, the emf in a circuit can be expressed as $\oint_{\mathrm{C}} \boldsymbol{E}\cdot d\boldsymbol{l}$, where C is the curve delineating the circuit and $\boldsymbol{E}$ represents the electric field strength at each point in the circuit. Accordingly, the left hand side of Equation (20.3) can be replaced by $\oint_{\mathrm{C}} \boldsymbol{E}\cdot d\boldsymbol{l}$; the algebraic sign of this induced emf being determined by the choice of which direction around C is chosen to be the positive sense. The total flux through the circuit, given by $\Phi_M = \iint_{\mathrm{S}} \boldsymbol{B}\cdot d\boldsymbol{A} = \iint_{\mathrm{S}} \boldsymbol{B}\cdot\boldsymbol{n}dA$, is also an algebraic quantity, the sign of which is determined by the choice of positive $\boldsymbol{n}$. Any statement of Faraday's law which incorporates Lenz's law (effectively conservation of energy)

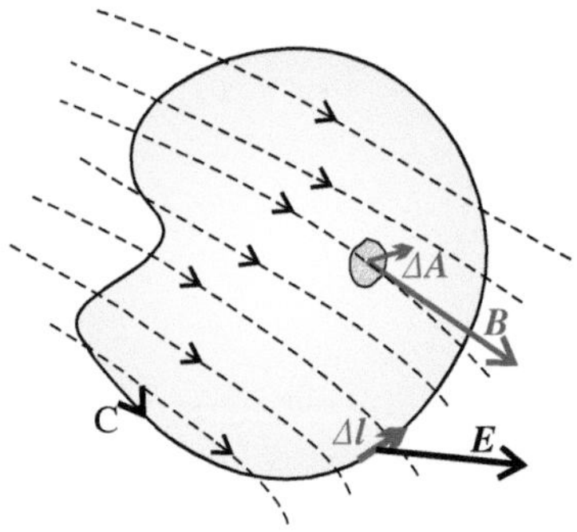

Figure 20.8. A changing flux through the circuit C causes an emf to be induced in the circuit. The choice of positive direction around C must be the same as that around the area element ΔA which, in turn, gives the positive direction for the vector $\boldsymbol{\Delta A}$.

requires that both of the 'conventions of signs' referred to here be consistent[1] and this, in turn, necessitates the inclusion of a minus sign. In such circumstances Faraday's law can be stated more generally as follows:

$$\oint_C \boldsymbol{E}\cdot d\boldsymbol{l} = -\frac{d}{dt}\iint_S \boldsymbol{B}\cdot d\boldsymbol{A}$$

where S is the surface delineated by the circuit C, as indicated in Figure 20.8. It is important to understand, in this context, that the choice of positive sense around C also determines the positive sense around the area element and hence the positive direction for the vector $\boldsymbol{\Delta A}$. Thus Faraday's law can also be expressed in terms of an algebraic emf $\mathcal{E}_{ind}$ as

$$\mathcal{E}_{ind} = -\frac{d\Phi_M}{dt} \qquad (20.3A)$$

provided the choice of direction of positive emf and current is consistent with the above. In these circumstances $E_{ind} = |\mathcal{E}_{ind}|$.

Study Worked Example 20.1

For problems based on the material presented in this section visit up.ucc.ie/20/ *and follow the link to the problems.*

20.2 Simple applications of electromagnetic induction

Figure 20.9 shows a typical plot, similar to that shown in Figure 20.5, of the emf induced in a coil as a function of time for the case in which a magnet is moved to and fro by hand relative to the coil. At any instant the emf in the coil is given by Equation (20.3), that is

$$E(t) = \left|\frac{d\Phi_M}{dt}\right|$$

and hence the change of flux through the coil between any two times t_1 and t_2 is

$$\Delta\Phi_M = \int_{t_1}^{t_2} E(t)dt$$

which is the area under the emf—time plot, shown shaded in Figure 20.9.

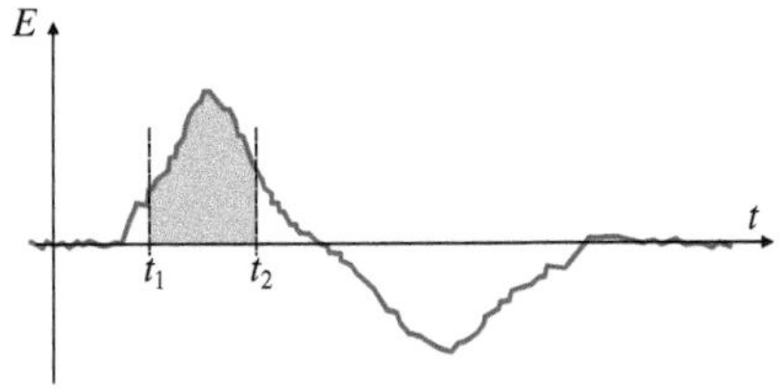

Figure 20.9. The change of flux through the coil during any time interval in the experiment shown in Figure 20.4 is equal to the area under emf—time curve in that interval.

Note that if the single-turn coil were to be replaced by a coil with N turns, the induced emf would be N times the rate of change of flux through each single turn, that is

$$E(t) = N\left|\frac{d\Phi_M}{dt}\right|$$

and the observed area under the curve would be increased correspondingly.

Computerised 'voltmeters', mentioned above, are ideally suited to measurements of this type because, as well as being able to display voltage—time plots, like that of Figure 20.9, they can also determine directly the value of an area such as that shaded in the figure. Two practical applications of such measurements will now be described.

Measurement of magnetic field strength

Suppose that we wish to measure the strength of the uniform magnetic field between the poles of a magnet as indicated in Figure 20.10. A small flat circular coil of wire which has N turns each of radius a can be placed in the field with its plane perpendicular to the direction of the field, as indicated in the figure. The magnetic flux through each turn in the coil in this position is $\Phi_M = BA = B(\pi a^2)$, where B is the magnetic flux density of the field. The ends of the coil are connected to a computerised voltmeter as shown.

[1] In strict mathematical language this means that the 'induced orientation' of the surface S must be consistent with the choice of 'orientation' of C.

If the coil is removed slowly from the field the variation of the induced emf with time while the coil is being removed can be recorded and plotted; Figure 20.11 shows an example of such a plot. The area under the plotted curve, shown shaded in the figure, is given by

$$\textit{area under curve} = \int_{t_1}^{t_2} E(t)dt = \int_{t_1}^{t_2} N\frac{d\Phi_{\mathrm{M}}}{dt}dt = \int_{t_1}^{t_2} Nd\Phi_{\mathrm{M}} = N\int_{t_1}^{t_2} d\Phi_{\mathrm{M}} = NB(\pi a^2) - 0$$

$$\rightarrow \quad B = \frac{\textit{area under voltage—time plot}}{\pi N a^2}$$

assuming that the flux through the coil is negligible when it is removed from the field. Thus the magnetic field strength between the poles of the magnet can be calculated (by the computer) from

$$H = \frac{B}{\mu_0} = \frac{\textit{area under voltage—time plot}}{\pi \mu_o N a^2}$$

Note that, to give the field strength in A m^{-1}, the area under the voltage—time plot must be measured in volt second.

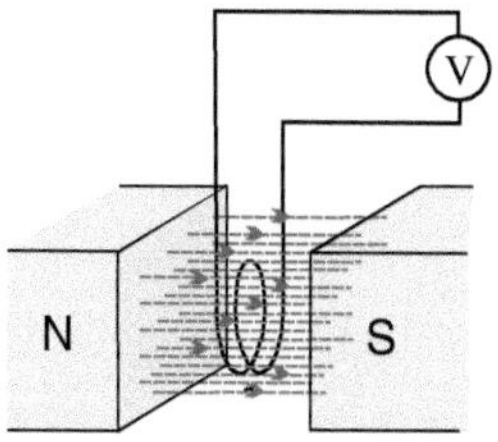

Figure 20.10. Experimental measurement of the magnetic field strength between the pole pieces of a magnet.

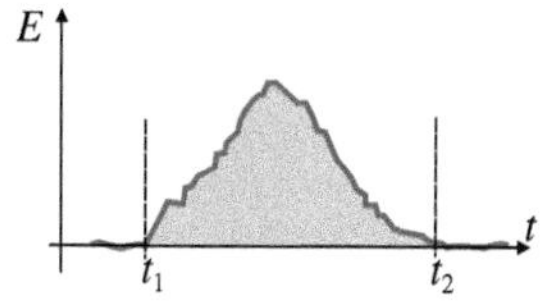

Figure 20.11. Plot of measured emf as a function of time as the coil in Figure 20.10 is removed from the field.

Measurement of magnetic pole strength

The magnetic pole strength of a long thin bar magnet can be estimated using a similar experimental technique. In this case a small coil of wire of N turns is wound around the centre of the bar magnet as shown in Figure 20.12. Assuming that the length of the magnet is much larger than the diameter of the coil, the change in magnetic flux through the coil as the magnet is completely removed from the coil in the direction indicated in Figure 20.12, is

$$\int_{t_1}^{t_2} E(t)dt = \int_{t_1}^{t_2} N\frac{d\Phi_{\mathrm{M}}}{dt}dt = \int_{t_1}^{t_2} Nd\Phi_{\mathrm{M}} = N\int_{t_1}^{t_2} d\Phi_{\mathrm{M}} = N(\Delta\Phi_M) = Np$$

where p is the pole strength of the magnet (equal to the total flux emanating from a pole) and thus

$$p = \frac{\textit{area under voltage—time plot}}{N}$$

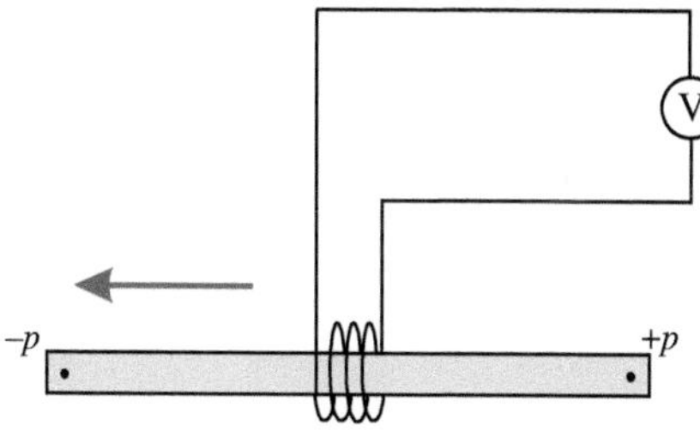

Figure 20.12. Experimental measurement of the pole strength of a bar magnet.

For problems based on the material presented in this section visit up.ucc.ie/20/ and follow the link to the problems.

20.3 Self-inductance

Figure 20.13 illustrates a simple circuit which comprises a single power supply whose terminals can be joined through a loop of resistance wire by closing a switch. When the switch is open, no current flows and there is no magnetic flux through the circuit. After the switch has been closed, however, a current begins to flow in the wire which produces a magnetic field, as indicated. Let Φ_{M} be the total magnetic flux through the circuit at an instant when the current is I, that is

$$\Phi_{\mathrm{M}} = \oint_{\text{circuit}} \mathbf{B}\cdot d\mathbf{A} = \mu_0 \oint_{\text{circuit}} \mathbf{H}\cdot d\mathbf{A}$$

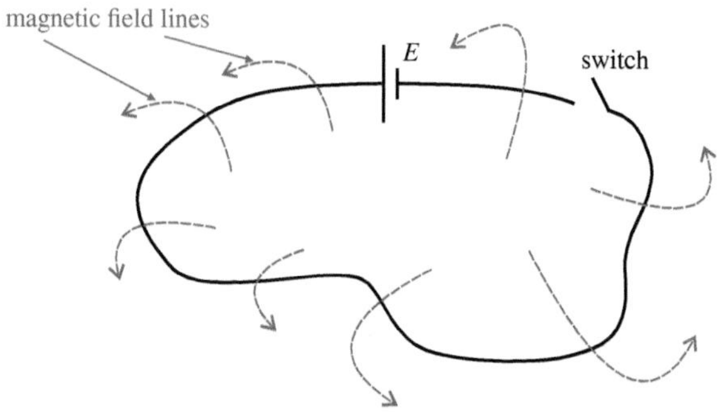

Figure 20.13. When the switch is closed, allowing a current to flow in the circuit, a magnetic field appears. During the 'switching on' process, the magnetic flux through the circuit is changing and hence an emf is induced.

Now, from the Biot-Savart law (Section 18.6), $|\boldsymbol{H}| \propto I$, and hence $\Phi_M \propto I$ or

$$\Phi_M = (constant)I$$

After the switch is closed, the magnetic flux through the circuit increases from zero to some maximum value; a steady current $I_S = \frac{E}{R_{circ}}$ then flows in the circuit. During this short 'switching on' time there is a *changing* flux through the circuit, $\Phi_M = \Phi_M(t)$. The principle of electromagnetic induction (Section 20.1) tells us that, while the flux is changing, there is an induced emf which is directed so as to oppose the current flowing. The net emf is

$$E - E_{ind} = E - \left|\frac{d\Phi_M}{dt}\right|$$

where, in this case, E_{ind} is called the '**back emf**'. Thus, the current flowing in the circuit at any instant is given by

$$I = \frac{E - E_{ind}}{R_{circ}}$$

The back emf only appears while the flux is changing; once the current settles down to its steady value

$$\Phi_M = constant \quad \Rightarrow \quad \frac{d\Phi_M}{dt} = 0 \quad \Rightarrow \quad E_{ind} = 0$$

and

$$I \quad \rightarrow \quad I_S = \frac{E}{R_{circ}}$$

Since, as we have seen, $\Phi_M \propto I$, the induced emf is proportional to the time derivative of I, that is

$$E_{ind} \propto \frac{dI}{dt}$$

or

$$E_{ind} = (constant)\frac{dI}{dt}$$

The value of the constant of proportionality depends on the geometry of the circuit and is called the **self-inductance** of the circuit and is denoted by the symbol L, that is

$$\boxed{E_{ind} = L\frac{dI}{dt}}$$

The SI unit of self-inductance is V s A^{-1} and 1 V s A^{-1} = 1 henry = 1 H.

Energy stored in an inductance

The current flowing in the circuit at any instant during the 'switching on' interval is determined by the equation

$$E - E_{ind} = R_{circ}I$$

which we can write as a differential equation

$$E = R_{circ}I + L\frac{dI}{dt} \tag{20.5}$$

Thus the power generated, when the current flowing is I, is given by

$$P = EI = \left(R_{circ}I + L\frac{dI}{dt}\right)I = R_{circ}I^2 + LI\frac{dI}{dt}$$

and hence the additional power required to overcome the back emf is given by

$$\frac{dW}{dt} = LI\frac{dI}{dt}$$

From this we see that the additional energy needed to increase the current from I to $I + \Delta I$ is $\Delta W = LI(\Delta I)$ and hence the work done in overcoming the induced emf in order to achieve a steady current I_S in the circuit is

$$W_{stored} = \int_0^{I_S} LIdI = L\int_0^{I_S} IdI = \frac{1}{2}LI_S^2 \tag{20.6}$$

This result is directly analogous to the expression $W_{stored} = \frac{1}{2}CV^2$ for the energy stored in a charged capacitor, Equation (17.4). In the case of the inductance, the energy will be released if the current is switched off, that is if the switch is opened. In circuits which have high self-inductance, the release of stored energy in this way can give rise to a spark at the switch which may cause damage to the switch. On the other hand, this effect may be used in a constructive way; for example, to provide the ignition spark needed to run an internal combustion engine.

Lumped inductance

The self-inductance discussed above is that of a complete single loop circuit and the inductance is distributed throughout the circuit. If a component in a circuit is designed in such a way that it is responsible for most of the self-inductance of the circuit, it is referred to as a *lumped inductance* and we can attribute the self-inductance to the component rather than to the circuit as a whole. Such a component, that is one with fixed self-inductance, is called an **inductor**. An example of a component with relatively high self-inductance is a tightly wound solenoid which, as we saw in Section 18.5, comprises many turns of wire wound on a cylindrical shape.

If a time-varying current flows in a solenoid with N turns, the back emf is N times the emf induced in each single turn.

The standard electrical symbol for an inductor, shown in Figure 20.14, reflects the fact that the many common inductors have solenoidal shape.

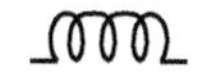

Figure 20.14. The standard symbol for an inductor.

Study Worked Example 20.2

The self-inductance of a long solenoid

We now derive an expression for the self-inductance of a solenoid of length l with N closely wound turns and cross-sectional area A, in the case where the solenoid is very long (that is, $l \gg \sqrt{A}$) and the inside of the solenoid is filled with a material of permeability μ. A typical solenoid is made by winding a wire on a cylinder of soft iron, as illustrated in Figure 20.15.

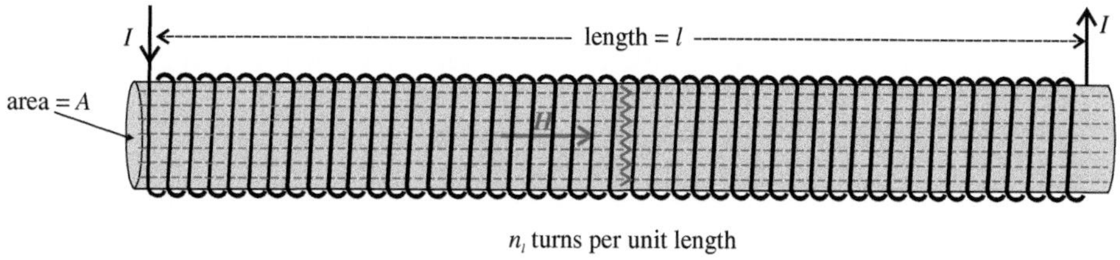

Figure 20.15. Magnetic field inside a long solenoid.

If the current flowing in the solenoid is I, we know from Equation (18.9) that the magnetic field strength inside the solenoid is given by

$$H = n_l I = \frac{N}{l} I$$

and hence the corresponding flux density is $B = \mu \frac{N}{l} I$, where μ is the permeability of the material which fills the inside of the solenoid. Thus the magnetic flux through a single turn is

$$\phi_M = BA = \left(\mu \frac{N}{l} I\right) A$$

and hence, as I is varied, the induced emf in each turn is $\left|\frac{d\phi_M}{dt}\right| = \frac{\mu NA}{l}\frac{dI}{dt}$. The total induced emf in the solenoid is N times the emf induced in a single turn and is given by

$$E_{\text{ind}} = N\left|\frac{d\phi_M}{dt}\right| = \frac{\mu N^2 A}{l}\frac{dI}{dt}$$

Recalling the definition of self-inductance, $E_{\text{ind}} = L\frac{dI}{dt}$, we can write the self-inductance of the solenoid as

$$L = \frac{\mu N^2 A}{l} = \mu\left(\frac{N}{l}\right)^2 lA = \mu n_l^2 lA \qquad (20.7)$$

where $n_l = \frac{N}{l}$ is the number of turns per unit length. Thus, filling the interior of a solenoid with soft iron produces a high inductance (since $\mu \approx 10^3 \mu_0$) as does winding the turns tightly (since L depends on the square of the number of turns per unit length).

It is assumed in the above that all of the magnetic flux stays within the solenoid. If the wire is not tightly wound, however, some of the flux may emerge between the windings, an example of what is called ‘flux leakage’.

Study Worked Example 20.3

Energy stored in a magnetic field

We may use the result which we derived above for the self-inductance of a very long solenoid to determine an expression for the energy density in a magnetic field. The energy stored when a current I flows in an inductor of self-inductance L is given by Equation (20.6), $W^{M}_{str} = \frac{1}{2}LI^2$. From Equation (20.7) we know that the self-inductance of a very long solenoid of length l with n_l turns per unit length and of cross-sectional area A is $\mu n_l^2 lA$ and hence $W^{M}_{str} = \frac{1}{2}(\mu n_l^2 lA)I^2$ or, in terms of the magnetic field strength inside the solenoid, $H = n_l I$

$$W^{M}_{str} = \frac{1}{2}\mu(Al)H^2 = \frac{1}{2}\mu H^2 \mathcal{V}$$

where $\mathcal{V}$ is the volume inside the solenoid of length l.

Hence the magnetic energy density (energy per unit volume) at a point in a magnetic field is given by

$$w^{M}_{str} = \frac{dW^{M}_{str}}{d\mathcal{V}} = \frac{1}{2}\mu H^2 = \frac{1}{2}HB$$

For problems based on the material presented in this section visit up.ucc.ie/20/ and follow the link to the problems.

20.4 The series *L-R* circuit

The circuit of Figure 20.13 is represented again in Figure 20.16, this time using standard electrical symbols. The current flowing in the circuit during the 'switching on' time is determined by the differential Equation (20.5) which, in this case takes the form

$$E = RI + L\frac{dI}{dt}$$

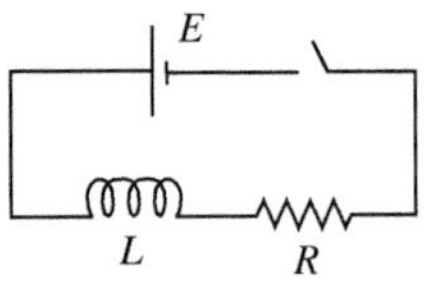

Figure 20.16. A series *L-R* circuit.

Rewriting this equation as follows

$$\frac{L}{R}\frac{dI}{dt} = \frac{E}{R} - I$$

we can change the variable by temporarily replacing $\left(\frac{E}{R} - I\right)$ by x. If $x = \frac{E}{R} - I$, then $\frac{dx}{dt} = -\frac{dI}{dt}$ so that we can write

$$-\frac{L}{R}\frac{dx}{dt} = x \quad \Rightarrow \quad \frac{dx}{dt} = -\frac{R}{L}x.$$

Separating the variables and integrating we obtain

$$\int \frac{dx}{x} = -\frac{R}{L}\int dt$$

which, performing the integration, becomes

$$\ln x = -\frac{R}{L}t + constant$$

Before the switch was closed the current was zero; that is, $I = 0$ for $t \le 0$. Thus, at $t = 0$, $x = \frac{E}{R} - 0$ and hence $\ln\left(\frac{E}{R}\right) = 0 + constant$, which determines the value of the constant of integration. Thus,

$$\ln x = -\frac{R}{L}t + \ln\left(\frac{E}{R}\right)$$

or, re-substituting $\frac{E}{R} - I$ for x,

$$\ln\left(\frac{E}{R} - I\right) = -\frac{R}{L}t + \ln\left(\frac{E}{R}\right)$$

$$\rightarrow \quad \ln\left(1 - \frac{RI}{E}\right) = -\frac{R}{L}t$$

$$\rightarrow \quad I = \frac{E}{R}\left(1 - e^{-\frac{R}{L}t}\right)$$

This function is plotted in Figure 20.17 which shows how the current builds up from zero to its steady value $I_s = \frac{E}{R}$. The quantity $\frac{L}{R}$ is called the time constant of the circuit (recall the definition of a time constant in Section 17.12).

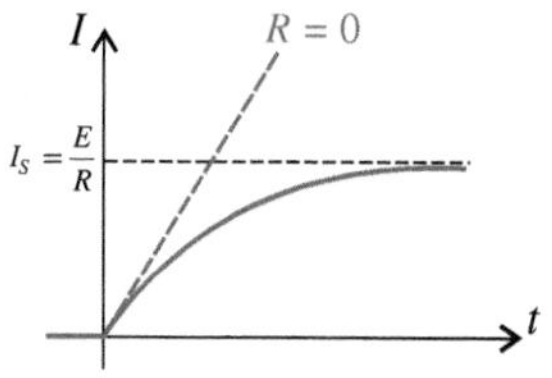

Figure 20.17. Build up of current as a function of time after the switch in Figure 20.16 is closed.

It is left as an exercise for the reader to show that, when the steady current $I_s = \frac{E}{R}$ is switched off (say by changing the switch in Figure 20.18 from position (i) to position (ii)), the current decays according to

$$I = I_s e^{-\frac{R}{L}t}$$

that is, it decays with a time constant of $\frac{L}{R}$.

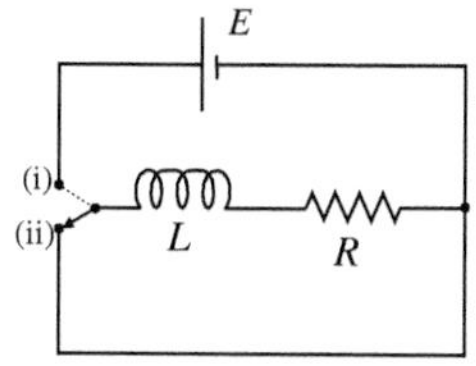

Figure 20.18. Experimental arrangement to observe the decay of current in a series *L-R* circuit.

Figure 20.19 shows how the energy which is stored while a steady current is flowing is released when the current is switched off; the current 'lost' due to the back emf during the switching on interval A reappears at B after the circuit is broken.

It is interesting to investigate what would happen to the current in a case in which the resistance is very small ($R \to 0$ or, more strictly, $Rt \ll L$). The exponential function can be expanded in a power series (Appendix A.9.1(i)) as follows

$$e^{-\frac{R}{L}t} = 1 - \frac{R}{L}t + \frac{R^2}{2L^2}t^2 + \ldots$$

which gives
$$I = \frac{E}{R}\left(1 - 1 + \frac{R}{L}t - \frac{R^2}{2L^2}t^2 + \ldots\right) = \frac{E}{R}\left(\frac{R}{L}t - \frac{R^2}{2L^2}t^2 + \ldots\right)$$
$$= \frac{E}{L}t - \frac{ER}{2L^2}t^2 + \text{terms in } R^2t^3, \; etc$$

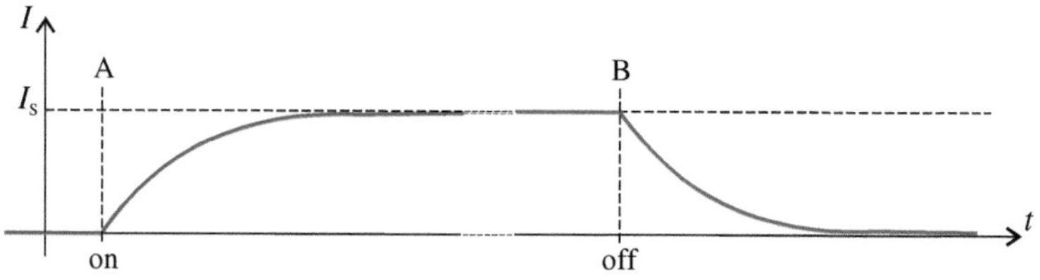

Figure 20.19. A current—time plot for the circuit in Figure 20.16 when the switch is first closed (A) and opened at a later time (B).

Thus for $R \to 0$ the current is given by $I = \frac{E}{L}t$ which indicates that the current will increase linearly without limit (blue dashed line in Figure 20.17) unless there is some resistance in the circuit.

For problems based on the material presented in this section visit up.ucc.ie/20/ *and follow the link to the problems.*

20.5 Discharge of a capacitor through an inductor and a resistor

Consider the circuit shown in Figure 20.20 in which, when the switch is open, a capacitor has been charged so that it has a charge $+Q$ on one plate and an equal and opposite charge on the other. If the switch is closed at an instant $t = 0$, the capacitor will proceed to discharge through the inductor and resistor. At some later time t, the charge on the plates will have dropped to a value q and the current in the circuit at this time will be $I = -\frac{dq}{dt}$. Since there is no source of emf in the circuit, the net potential difference around the complete loop is zero; that is, taking the positive sense around the circuit as that of the direction of the current shown in the figure,

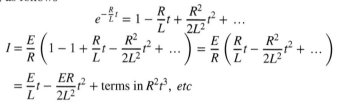

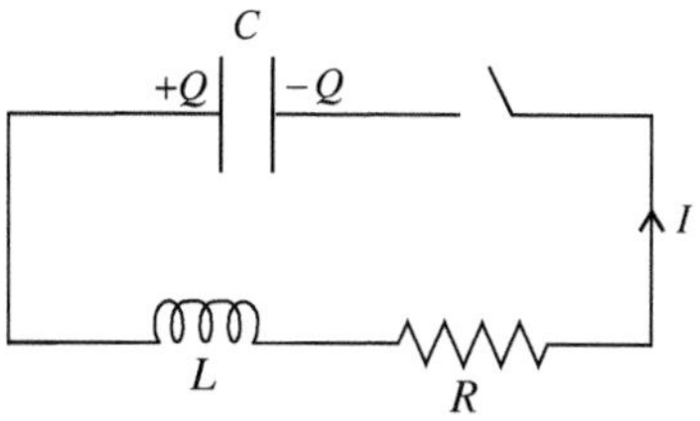

Figure 20.20. When the switch is closed the charged capacitor discharges through the inductor and resistor.

$$-\frac{q}{C} + L\frac{dI}{dt} + RI = 0$$

Replacing I by $-\frac{dq}{dt}$, we obtain the following differential equation, which determines the charge on the plates at any instant t

$$L\frac{d^2q}{dt^2} + R\frac{dq}{dt} + \frac{q}{C} = 0 \qquad (20.8)$$

We have encountered a very similar differential equation previously, namely Equation (3.19) in Section 3.11 which described damped simple harmonic motion. Indeed Equation (20.8) is formally equivalent to Equation (3.19) with $m \to L$, $b \to R$ and $k \to \frac{1}{C}$ so we need not repeat the analysis; the solution is the same as that given in Section 3.11 with these transformations ($m \to L$, etc.). Thus, the charge on the plates at time t is given by

$$q = Ae^{-\frac{R}{2L}t}\sin(\omega t + \phi) \quad \text{where } \omega = \sqrt{\frac{1}{LC} - \frac{R^2}{4L^2}}$$

where A and ϕ are constants which must be determined from the initial conditions.

To simplify the algebra, we temporarily let $\gamma = \frac{R}{2L}$ and $\omega_0 = \sqrt{\frac{1}{LC}}$ in which case we can write

$$q = Ae^{-\gamma t}\sin(\omega t + \phi) \quad \text{where } \omega = \sqrt{\omega_0^2 - \gamma^2}$$

The initial conditions in this case are (i) $q = Q$ at $t = 0$ and (ii) $I = \frac{dq}{dt} = 0$ at $t = 0$ which we apply, in turn, as follows:

$$\text{(i)} \to \quad Q = A\sin\phi \quad \to \quad A = \frac{Q}{\sin\phi}$$

Before we can apply the second initial condition, we must differentiate q with respect to time to determine an expression for the current in the circuit, that is

$$I = \frac{dq}{dt} = -\gamma Ae^{-\gamma t}\sin(\omega t + \phi) + \omega Ae^{-\gamma t}\cos(\omega t + \phi)$$

$$\text{(ii)} \to \quad 0 = -\gamma A\sin\phi + \omega A\cos\phi \quad \to \quad \gamma\sin\phi = \omega\cos\phi \quad \to \quad \tan\phi = \frac{\omega}{\gamma}.$$

We may also write $\gamma^2\sin^2\phi = \omega^2\cos^2\phi = \omega^2(1 - \sin^2\phi) \quad \to \quad \sin^2\phi(\gamma^2 + \omega^2) = \omega^2$ and hence

$$\sin\phi = \frac{\omega}{\sqrt{\omega^2 + \gamma^2}} = \frac{\omega}{\omega_0} \quad \text{and} \quad \cos\phi = \frac{\gamma}{\sqrt{\omega^2 + \gamma^2}} = \frac{\gamma}{\omega_0}$$

Thus the two constants in the expression $q = Ae^{-\gamma t}\sin(\omega t + \phi)$ are given by

$$A = \frac{Q}{\sin\phi} = \frac{\omega_0}{\omega}Q \quad \text{and} \quad \phi = \tan^{-1}\frac{\omega}{\gamma}$$

Substituting these values we obtain the following expression for the charge on the capacitor plates as a function of time

$$q = \frac{\omega_0}{\omega}Qe^{-\gamma t}\sin(\omega t + \phi)$$

or, in terms of R, C and L,

$$q = \frac{Q}{\sqrt{1 - \frac{R^2C}{4L}}}e^{-\frac{R}{2L}t}\sin(\omega t + \phi) \tag{20.9}$$

where $\phi = \tan^{-1}\sqrt{\frac{4L}{R^2C} - 1}$ and $\omega = \sqrt{\frac{1}{LC} - \frac{R^2}{4L^2}}$, as before.

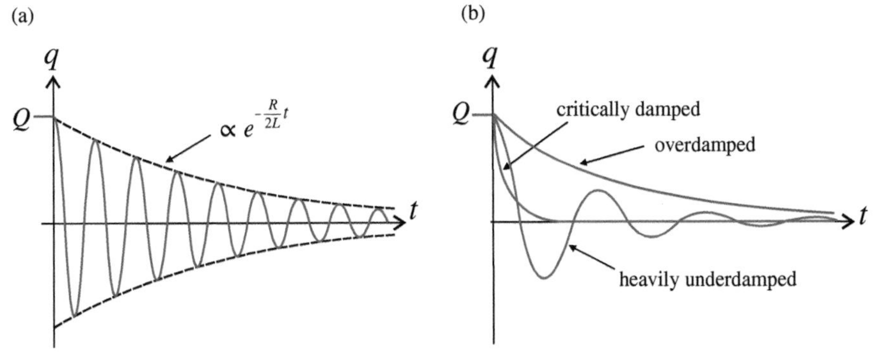

Figure 20.21. Plot of the charge on the capacitor in Figure 20.20 as a function of time after the switch is closed (a) showing the underdamped case and (b) the cases of critical damping, overdamping and heavy underdamping.

The variation of q with time described by Equation (20.9) is sketched in Figure 20.21(a) for a case where ω is real ($4L > R^2C$). The oscillations are modulated by an envelope which decays exponentially; the decay time (or time constant) of the envelope is $\frac{2L}{R}$. The cases of critical damping ($\omega = 0$) and overdamping (ω imaginary) which we encountered in Section 3.11 may also arise here and are illustrated in Figure 20.21(b).

Note that when $R \to 0$, Equation (20.8) is formally equivalent to simple harmonic motion and, as one would expect in such circumstance, Equation (20.9) predicts that the oscillations which have been set up will continue indefinitely with characteristic frequency $\omega = \omega_0 = \frac{1}{\sqrt{LC}}$. In this case the charge on the plates is given by $q = Q \sin \omega_0 t$ and the potential difference across the plates of the capacitor varies according to

$$V = \frac{q}{C} = \frac{Q}{C}\sin \omega_0 t = \frac{Q}{C}\sin\left(\frac{1}{\sqrt{LC}}t\right)$$

Energy swaps back and forth between magnetic energy in the inductor and electrical energy in the capacitor, in direct analogy with the periodic exchange of potential and kinetic energy in a harmonic oscillator (Section 3.10).

Such oscillations could be (and are) used as a source of sinusoidally varying emf (an '*L-C oscillator*'). For reasons which will be explained in Chapter 21, such a system continuously loses energy despite the fact that there is negligible resistance in which heat can be generated. Thus, the use of such an oscillator as a continuous source of oscillating voltage requires a continuous input of energy.

For problems based on the material presented in this section visit up.ucc.ie/20/ and follow the link to the problems.

20.6 Time-varying emfs: mutual inductance: transformers

If the magnetic flux through a circuit varies continuously, a time varying emf will be induced. Figure 20.22 shows a simple *L-R* circuit in which an inductor and a resistor are connected in series across a time-varying power supply of emf $E = E(t)$. While the emf is changing there will always be a time-varying 'back emf' in the circuit and the current at any time is determined by

$$E(t) - E_{\text{ind}}(t) = RI \tag{20.10}$$

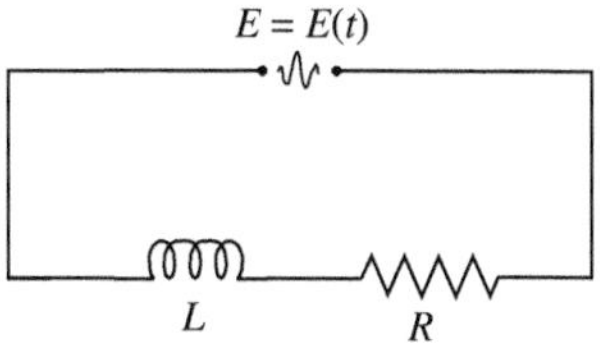

Figure 20.22. A time-varying emf connected across a series combination of an inductor and a resistor.

The dependence of I on t in this case will depend, of course, on the specific time dependence of E.

Consider next the situation shown in Figure 20.23. Here the inductor is a solenoid with N turns which are wound on a soft iron core. As we saw in Section 20.3, the use of soft iron produces a high self-inductance. If the resistance of the circuit is very small, $E_{\text{ind}} \gg RI$ in Equation (20.10) and

$$E(t) - E_{\text{ind}} \to 0$$

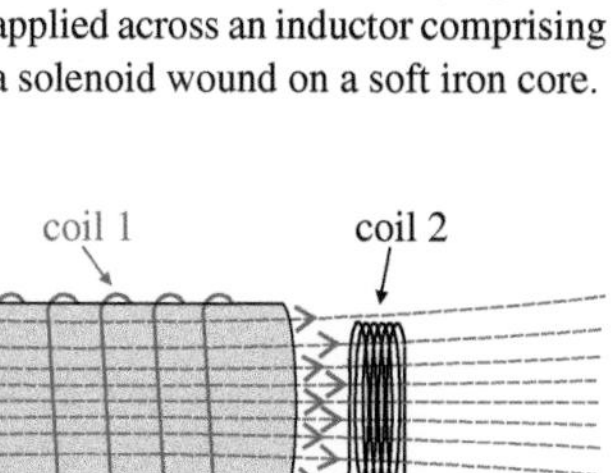

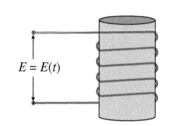

Figure 20.23. A time-varying emf applied across an inductor comprising a solenoid wound on a soft iron core.

Thus the induced emf at any instant is given by $E_{\text{ind}} = E(t)$, that is in the limit $R \to 0$ the 'back emf' exactly balances the applied emf. Now, from Faraday's law, $E_{\text{ind}} = N\left|\frac{d\Phi_M}{dt}\right|$ where $\Phi_M = \Phi_M(t)$ is the magnetic flux in the core at the instant t.

Hence
$$\left|\frac{d\Phi_M}{dt}\right| = \frac{E(t)}{N} \quad \to \quad \Phi_M(t) = \frac{1}{N}\int E(t)dt.$$

Mutual inductance

Next consider the situation illustrated in Figure 20.24 in which two coils are placed near each other. A continuously varying current I_1, driven by a time-varying emf $E_1 = E_1(t)$, flows in coil 1, which has N_1 turns. Some of the resulting (continuously changing) magnetic flux passes through the second coil, coil 2 with N_2 turns, thereby inducing an emf in it which is given by

$$E_2(t) = N_2\left|\frac{d\Phi_2}{dt}\right|$$

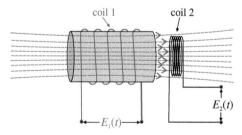

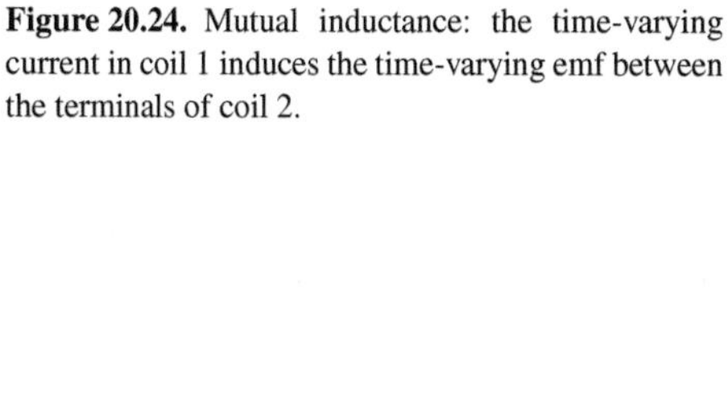

Figure 20.24. Mutual inductance: the time-varying current in coil 1 induces the time-varying emf between the terminals of coil 2.

where $\Phi_2 = \Phi_2(t)$ is the magnetic flux through each turn of coil 2 at time t. Now, since (from the Biot-Savart law, Section 18.6) the magnetic field strength at any point is proportional to I_1, Φ_2 is proportional to I_1, that is

$$N_2\Phi_2 = (constant)I_1 = MI_1 \tag{20.11}$$

and hence
$$\boxed{E_2(t) = M\frac{dI_1}{dt}}$$

If the varying emf had been applied to coil 2, instead of to coil 1 in Figure 20.24, the emf induced in coil 1 would be given by

$$E_1(t) = M'\frac{dI_2}{dt}$$

It can be shown that $M' = M$ — the proof of this requires the use of vector relationships which are beyond the scope of this book.[2] The constant M depends only on geometrical factors such as the areas of the coils, the number of turns in each, the distance apart, *etc.* and is defined as the **mutual inductance** of the pair of coils.

From the definition of M, Equation (20.11), we see that the SI unit of mutual inductance is the same as the SI unit of self-inductance, namely the henry.

Transformers

Consider two coils of insulated wire which are wound on a common soft iron core as illustrated in Figure 20.25. The coil across which the continuously changing emf ($E_p = E_p(t)$) is applied, called the primary coil, has N_p turns while the other coil, the secondary coil, has N_s turns. For the primary circuit, Equation (20.10) becomes

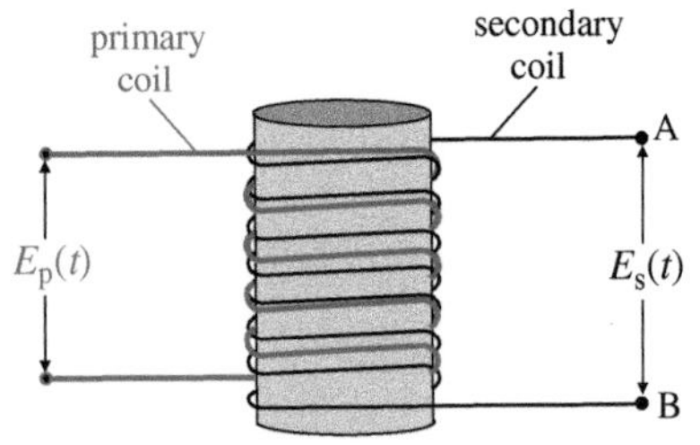

Figure 20.25. A transformer in which two coils are wound on the same soft iron core. A time-varying emf applied across the primary coil will induce a corresponding time-varying emf in the secondary coil whose magnitude depends on the ratio of the number of turns in the coils.

$$E_p(t) - N_p\frac{d\Phi}{dt} = RI_p$$

where Φ is the magnetic flux in the core (and hence the flux through any single turn) at time t.

In the limit $R \to 0$ we see that the rate of change of flux in the core is given by

$$\frac{d\Phi}{dt} = \frac{E_p(t)}{N_p}$$

Since, in this case, all of this changing flux threads the secondary coil an emf will be induced between the points A and B in the figure which is given by

$$E_s(t) = N_s\frac{d\Phi}{dt} = N_s\frac{E_p(t)}{N_p}$$

or

$$\boxed{E_s(t) = \frac{N_s}{N_p}E_p(t)} \tag{20.12}$$

Thus, an emf is induced in the secondary which has the same time dependence as the emf applied to the primary but is larger, or smaller, at any instant by a factor $\frac{N_s}{N_p}$. We distinguish two cases as follows:

(i) *Step-up transformer:* If $N_s > N_p$ then $E_s > E_p$ and hence the emf induced in the secondary coil is greater than the emf applied across the primary coil.

(ii) *Step-down transformer:* If $N_s < N_p$ then $E_s < E_p$ and the secondary emf is less than the primary emf.

In practice, one does not need to wind both coils on the same part of the core as illustrated in Figure 20.25. More usually the coils are wound on opposite sides of a closed ferromagnetic core, as shown in Figure 20.26. Since, as discussed in Section 19.12, the magnetic field is much stronger inside the ferromagnetic material than outside, most of the magnetic flux is 'trapped' inside the core, as shown, and hence the flux through any turn in the primary coil is effectively the same as that through any turn of the secondary.

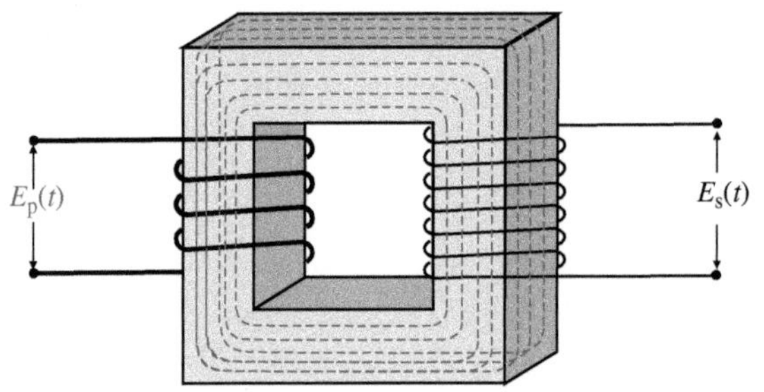

Figure 20.26. A transformer in which the coils are wound on opposite sides of a continuous soft iron core. The dashed blue lines represent the (time-varying) magnetic flux in the core.

[2]Interested readers may wish to consult a more advanced text on electromagnetism, for example Cheng (1989) pp 273–4.

Note that Equation (20.12) assumes that the resistance in the primary circuit is negligible. This requirement can be achieved in heavy duty transformers where the coil can be made from thick copper wire, but for small compact transformers the windings are much thinner so that the coil resistance can be of the order of ohms.

We shall return to a discussion of transformers and describe some important applications in Section 20.8

For problems based on the material presented in this section visit up.ucc.ie/20/ and follow the link to the problems.

20.7 Alternating current (a.c.)

We now consider what happens when a plane loop of wire is rotated at constant angular speed in a constant uniform magnetic field. In the situation depicted in Figure 20.27(a), the magnetic field is directed outwards perpendicularly to the plane of the page and the loop lies in this plane. The same situation is shown in Figure 20.27(b), but, in this case, it is viewed looking down from above. When the plane of the coil is perpendicular to the magnetic field, as in these figures, the magnetic flux through the loop is at its maximum value and is given by

$$\Phi_{max} = BA$$

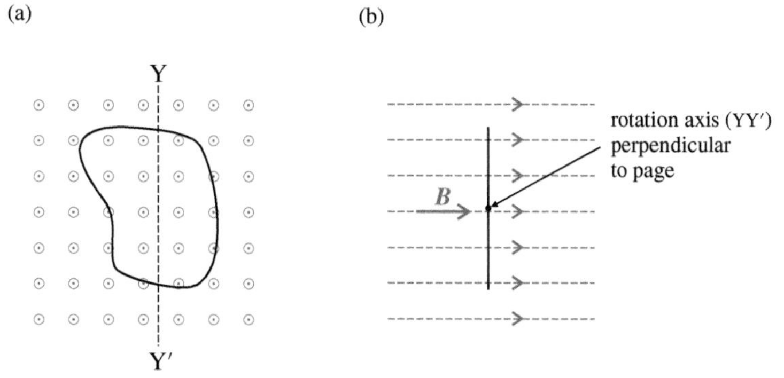

Figure 20.27. (a) A plane loop of wire placed in a constant uniform magnetic field such that the field lines are perpendicular to the plane of the loop. (b) The same arrangement as in (a) viewed from above.

where A is the area of the loop and $\boldsymbol{B}$ is the magnetic flux density.

Suppose now that the loop is rotated about the axis YY′, as is indicated in Figure 20.28(a). When, as shown in Figure 20.28(b), the plane of the loop makes an angle θ with respect to the position where the flux through the loop is maximum, the area presented to the magnetic flux becomes $A\cos\theta$ and the flux through the loop is reduced to

$$\Phi = \Phi_{max}\cos\theta = BA\cos\theta.$$

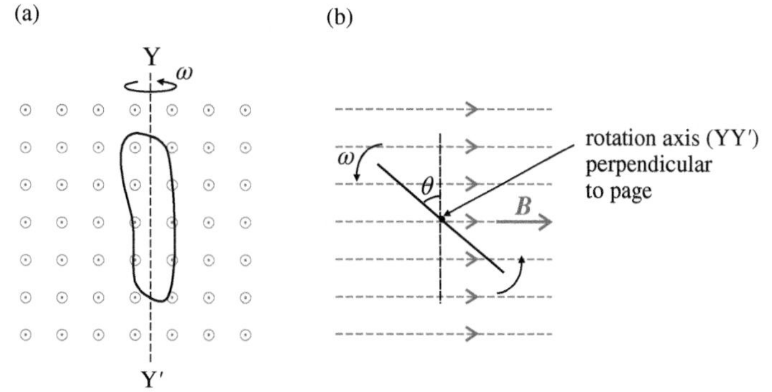

Figure 20.28. (a) If the coil in Figure 20.27 is rotated about an axis, an emf will be induced in the loop because of the changing magnetic flux through it. (b) The rotating coil in (a) viewed from above.

When the plane of the loop is parallel to $\boldsymbol{B}$, $\theta = \frac{\pi}{2}$ and $\Phi = 0$.

If the loop rotates with constant angular velocity ω, then $\theta = \omega t$. If, also, we take the time $t = 0$ to be an instant when $\Phi = \Phi_{max}$ (that is $\theta = 0$), the flux through the loop varies with time according to

$$\Phi = \Phi_{max}\cos\omega t = BA\cos\omega t.$$

The principle of electromagnetic induction in the form of Equation (20.3A) tells us that, as the loop rotates, the changing flux will induce an emf in the loop which is given by

$$E_{\text{ind}} = -\frac{d\Phi}{dt} = \omega BA \sin \omega t = E_0 \sin \omega t$$

where E_0 is the maximum (or 'peak') value of the induced emf.

The magnitude of the induced emf depends on the strength of the field, the area of the loop and the rate of rotation. For a single turn of wire of area 0.1 m^2 rotating at one revolution per second in a magnetic field of flux density 1.5 T, the magnitude of E_0 is $(2\pi \text{ rad s}^{-1})(1.5\text{ T})(0.1\text{ m}^2) \approx 1$ V.

To use a coil which is rotating in a magnetic field as a source of time varying current we need an arrangement which ensures that, as the coil rotates, contact is maintained with the circuit through which the current is to flow, without the wires wrapping around each other. One such arrangement is indicated in Figure 20.29 where metal rings are connected to the wires leading to the rotating coil. Sliding contacts, such as carbon 'brushes', are used to connect the coil to the stationary external circuitry. The points A and B in Figure 20.29 can be considered as the terminals of a power supply which provides a sinusoidally varying emf, that is

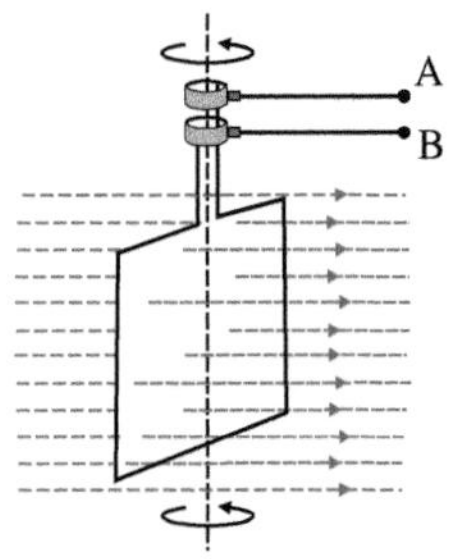

Figure 20.29. While the coil rotates in the magnetic field, a sinusoidal emf is generated between the terminals A and B. Electrical contact with a coil rotating in a magnetic field is achieved by carbon 'brushes' which maintain contact with metal rings connected to the wires from each end of the coil, as indicated.

$$E_{AB} = E(t) = \frac{d\Phi}{dt} = E_0 \sin \omega t$$

If the coil has N turns, the maximum value (or the 'peak' value) of the emf is given by

$$E_0 = N\omega BA$$

If one or more resistors are connected to the points A and B as shown in Figure 20.30, the current which flows in the circuit at any instant is given by

$$I = I(t) = \frac{E(t)}{R_{\text{circ}}} = \frac{E_0 \sin \omega t}{R_{\text{circ}}} = \frac{E_0}{R_{\text{circ}}} \sin \omega t = I_0 \sin \omega t$$

where R_{circ} is the effective resistance of the circuit and I_0 is the maximum (peak) value of the current.

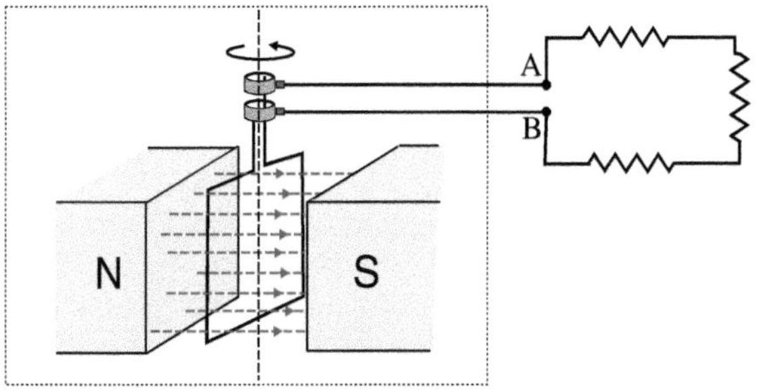

Figure 20.30. A simple alternating current (a.c.) power supply. The points A and B may be considered to be the terminals of source of sinusoidally varying emf. If a resistive load is connected across the terminals, as shown, a sinusoidally varying current will flow in the resistors (see Figure 20.31).

Thus, a sinusoidally varying current is generated: Figure 20.31 shows how the emf and the current vary with time. Note that the current has the same frequency as the applied emf and is in phase with it. Such a sinusoidal current is called an **alternating current** (a.c.) and an arrangement such as that described in Figure 20.30 is called an *a.c generator.*

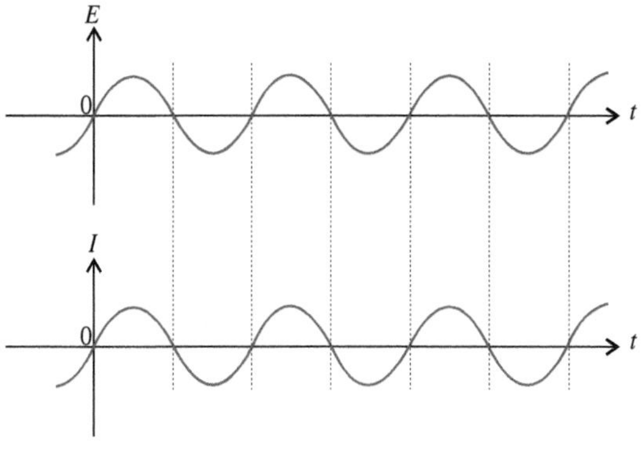

Figure 20.31. The variation of emf (above) and current (below) with time in an a.c. circuit.

The potential difference developed across a resistor through which an alternating current flows also depends sinusoidally on time since, at any instant, $V = V(t) = RI(t) = RI_0 \sin \omega t$.

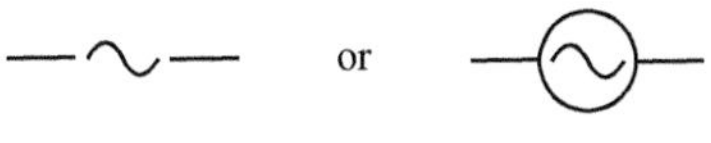

Figure 20.32. Standard symbols for an a.c. power supply.

The frequency of the alternating emf, current and potential difference depends on the angular frequency of the rotating coil. The maximum frequency that can be obtained in the manner described is limited by mechanical considerations. To generate higher a.c. frequencies we have to resort to non-mechanical, often much simpler, ways of constructing alternating current sources; for example, a.c. can be generated using simple *L-C* circuits as described at the end of Section 20.5. Standard symbols for an a.c. power supply are given in Figure 20.32.

The current in an a.c. circuit changes direction every half cycle so there is no net transfer of charge carriers in this case. Also the average value of emf, current and potential difference are all zero, that is

$$\langle E \rangle = \langle E_0 \sin \omega t \rangle = E_0 \langle \sin \omega t \rangle = 0, \quad \langle I \rangle = I_0 \langle \sin \omega t \rangle = 0 \quad \text{and} \quad \langle V \rangle = V_0 \langle \sin \omega t \rangle = 0.$$

On the other hand, thermal power *is* generated since

$$\langle P_{\text{circ}} \rangle = \langle E(t)I(t) \rangle = \langle E_0 \sin \omega t \times I_0 \sin \omega t \rangle = E_0 I_0 \langle \sin^2 \omega t \rangle = \frac{1}{2} E_0 I_0$$

Similarly, as illustrated in Figure 20.33, while the average alternating current flowing in a resistance R is zero, the average power generated in it is non-zero and is given by

$$\langle P_{\text{R}} \rangle = \langle RI^2 \rangle = R \langle I^2 \rangle = R \langle I_0^2 \sin^2 \omega t \rangle = RI_0^2 \langle \sin^2 \omega t \rangle = \frac{1}{2} RI_0^2$$

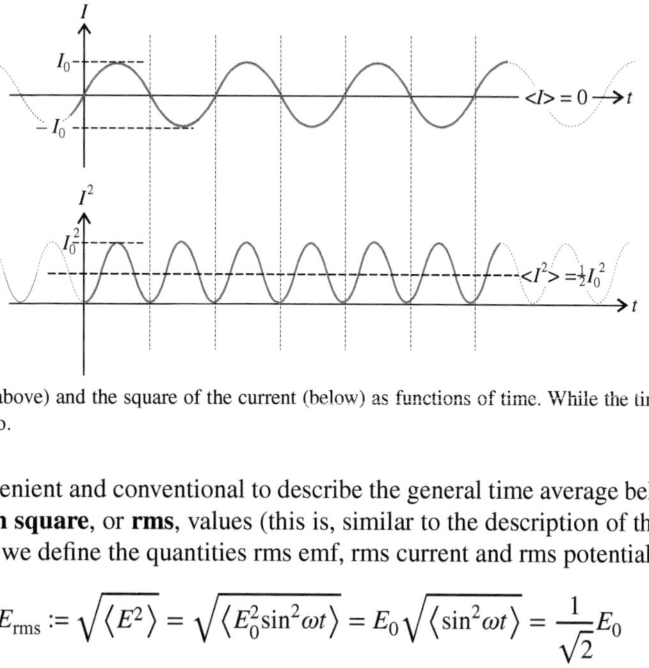

Figure 20.33. Plot of an alternating current (above) and the square of the current (below) as functions of time. While the time average value of the current is zero, the time average value of the (current)^2 is non-zero.

In dealing with a.c. circuits, it is convenient and conventional to describe the general time average behaviour of the sinusoidally varying quantities in terms of their **root mean square**, or **rms**, values (this is, similar to the description of the average speeds of gas molecules which we used in Chapter 12). Thus, we define the quantities rms emf, rms current and rms potential difference as follows:

$$E_{\text{rms}} := \sqrt{\langle E^2 \rangle} = \sqrt{\langle E_0^2 \sin^2 \omega t \rangle} = E_0 \sqrt{\langle \sin^2 \omega t \rangle} = \frac{1}{\sqrt{2}} E_0$$

$$I_{\text{rms}} := \sqrt{\langle I^2 \rangle} = \sqrt{\langle I_0^2 \sin^2 \omega t \rangle} = I_0 \sqrt{\langle \sin^2 \omega t \rangle} = \frac{1}{\sqrt{2}} I_0$$

$$V_{\text{rms}} := \sqrt{\langle V^2 \rangle} = \sqrt{\langle V_0^2 \sin^2 \omega t \rangle} = V_0 \sqrt{\langle \sin^2 \omega t \rangle} = \frac{1}{\sqrt{2}} V_0$$

We see directly from these definitions that the average thermal power generated in an a.c. circuit is given by $\langle P_{\text{circ}} \rangle = \frac{1}{2} E_0 I_0 = E_{\text{rms}} I_{\text{rms}}$ and that the average thermal power generated in a resistor is given by $\langle P_{\text{R}} \rangle = \frac{1}{2} RI_0^2 = RI_{\text{rms}}^2$. Indeed, it is easy to show that when rms values are used, all of the results derived in Chapter 15 for direct current (d.c.) circuits (recall summary in Section 15.7) carry over directly to a.c. When dealing exclusively with a.c. circuits the subscript rms is usually dropped.

For problems based on the material presented in this section visit up.ucc.ie/20/ and follow the link to the problems.

20.8 Alternating current transformers

Step-up and step-down transformers, which we described in Section 20.6, can be used to increase or decrease a.c. voltages. Consider the transformer shown in Figure 20.34 where the number of turns in the primary and secondary coils are N_p and N_s, respectively. An alternating voltage $E_p = E_0 \sin \omega t$ is applied across the primary coil as indicated. If there is no flux leakage and the resistances of the voltage source and the primary coil are negligible, the emf induced in the secondary coil is given, from Equation (20.12), by

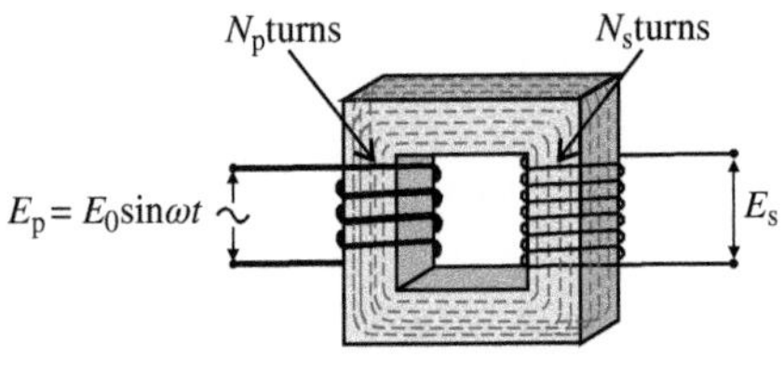

Figure 20.34. An a.c. transformer.

$$E_s(t) = \frac{N_s}{N_p}E_p = \frac{N_s}{N_p}E_0 \sin \omega t$$

or, in terms of rms values,

$$(E_s)_{rms} = \frac{N_s}{N_p}(E_p)_{rms}$$

For brevity, we now drop the subscript rms but the reader must bear in mind that for the rest of this section E_p, E_s, I_p and I_s refer to the rms values of these quantities.

The sinusoidally changing current in the primary coil gives rise to a sinusoidally changing magnetic flux in the core which reverses its direction every half cycle. Thus, the magnetisation of the material of the core is continuously reversing which causes a certain amount of heating in the core. While, with a careful choice of the material used for the transformer core, this heating can be reduced, it can never be completely eliminated. In other words, the total transfer of electrical power from primary to secondary is impossible so that the power generated in the secondary is always less than that in the primary coil, that is $E_s I_s < E_p I_p$ or

$$E_s I_s = \varepsilon E_p I_p$$

where $\varepsilon(<1)$ is called the **efficiency** of the transformer and I_s is the current drawn by whatever load is connected across the secondary coil (for a simple resistive load, $I_s = \frac{E_s}{R}$, where R is the total resistance of the secondary circuit.). In general, the value of ε is not constant but depends on the load.

The current which flows in a primary circuit depends on what load is connected across the secondary coil. When an rms current I_s flows in the secondary circuit, the rms current in the primary coil is given by

$$I_p = \frac{1}{\varepsilon}\frac{E_s}{E_p}I_s = \frac{1}{\varepsilon}\frac{N_s}{N_p}I_s = \frac{1}{\varepsilon}\frac{1}{R}\left(\frac{N_s}{N_p}\right)^2 E_p$$

As before, we distinguish the two cases

(i) *Step-up transformer*: $N_s > N_p$, $E_s > E_p$ and $I_p > I_s$ (higher voltage, lower current in secondary than in primary)
(ii) *Step-down transformer*: $N_s < N_p$, $E_s < E_p$ and $I_p < I_s$ (lower voltage, higher current in secondary than in primary)

These results have important implications for the transmission of electrical energy over long distances such as via a national electricity grid. The ohmic (RI^2) power losses per unit length in overhead or underground cables depend on the current flowing in the cable so that, in order to keep such power losses to a minimum, the current flowing in the cable should be as small as possible. Thus, step-up transformers are used at the generating station (Figure 20.35) so that the distribution over the national grid takes place at high voltage and low current. For domestic uses, however, low voltages are required for safety and high currents may be needed to operate domestic appliances such as electric kettles, heaters, etc. Accordingly, step-down transformers are used (on the right in Figure 20.35) to provide the 'mains' voltage for domestic use; further step-down transformers are often used to provide even lower and safer voltages (for example, in mobile phone chargers, etc.). For industrial applications, step-down transformers are also used but in this case the voltages do not have to be as low as in domestic situations.

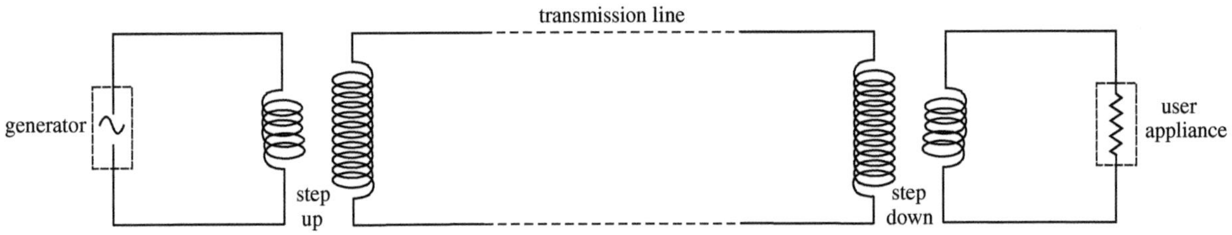

Figure 20.35. Transmission of power over an electricity grid is carried out at high voltage to minimise power losses. A step-up transformer at the generating station allows for high voltage and low current transmission. A step-down transformer is required to provide safer voltages and the higher currents needed for domestic appliances.

For problems based on the material presented in this section visit up.ucc.ie/20/ and follow the link to the problems.

20.9 Resistance, capacitance, and inductance in a.c. circuits

We have already considered, in Section 20.7, what happens when resistors are connected to an a.c. power supply. In general, there is very little difference from the behaviour of these components in dc circuits except that the current and potential differences vary with time and follow the variation in the applied emf. If the circuit contains a capacitor or an inductor, however, the situation is quite different. To understand the effect that each type of component has in an a.c. circuit, we first consider in turn the three simple cases of (i) a single resistor, (ii) a single capacitor and (iii) a single inductor, connected across an a.c. power supply.

(i) ***Resistance in an a.c. circuit***

As we saw in Section 20.7, when a resistor is connected to an a.c. supply as shown in Figure 20.36, the current in the circuit at any time t is given by

$$I = \frac{E(t)}{R} = \frac{E_0}{R}\sin\omega t$$

The current is in phase with the applied emf, as illustrated in Figure 20.31.

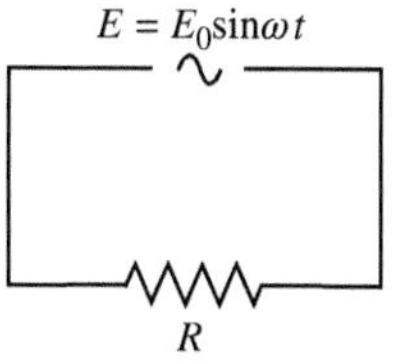

Figure 20.36. A resistor connected directly across an a.c. power supply.

(ii) ***Capacitance in an a.c. circuit***

We know, from our study of capacitors in Chapter 17, that if there is a capacitor in a single loop dc circuit, the capacitor acts as a break in the circuit ('open circuit') and no current can flow once the capacitor is charged. In an a.c. circuit (Figure 20.37), however, the continuously changing charge on one plate induces a corresponding change of charge on the other plate. The charge continuously builds up on, and discharges from, the capacitor in a sinusoidal manner and, hence, a sinusoidally changing current flows in the circuit. In this sense, the alternating current can be thought of as 'passing through' the capacitor. Thus a capacitor can be used as a 'd.c. filter'; that is, it allows the time varying part of a voltage signal to pass through but not any steady voltage component.

From the definition of capacitance (Section 17.8), the charge on a plate of the capacitor at any instant t is given by

$$q = q(t) = CV(t) = CE_0\sin\omega t$$

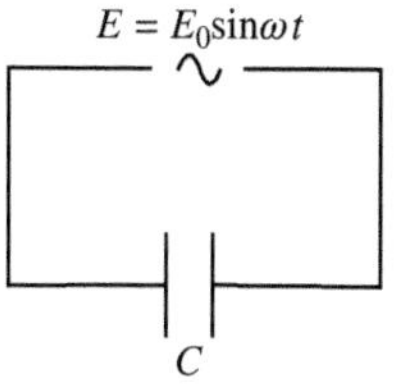

Figure 20.37. A capacitor connected directly across an a.c. power supply.

and hence the current flowing at that instant is given by the time derivative of q, namely

$$I = \frac{dq}{dt} = \omega CE_0\cos\omega t$$

Since $\sin\left(\omega t + \frac{\pi}{2}\right) = \cos\omega t$, the current can be rewritten as

$$I = \frac{E_0}{{}^1\!/_{\omega C}}\sin\left(\omega t + \frac{\pi}{2}\right)$$

Clearly, in this case, the current is not in phase with the applied emf but rather *leads* the emf by $\frac{\pi}{2}$, that is the current reaches maximum (for example) at a time $\frac{T}{4}$ *before* the emf is maximum, such that $\Delta(\omega t) = \frac{\pi}{2}$, as can be seen from the plots in Figure 20.38.

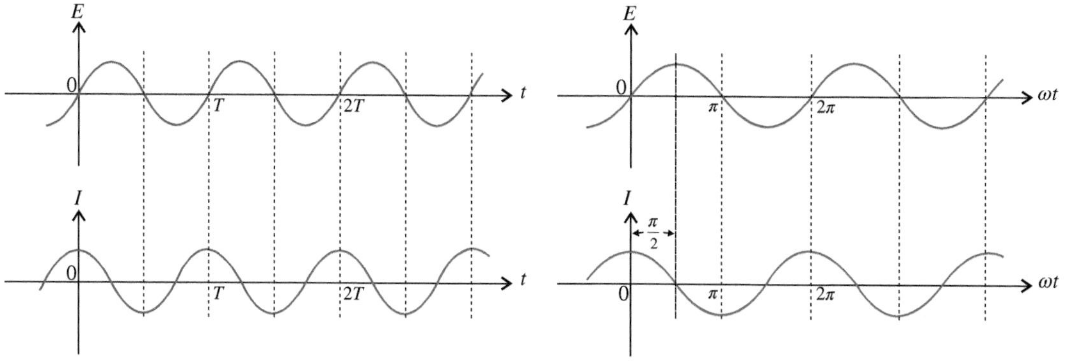

Figure 20.38. A plot of the current in the circuit in Figure 20.37 as a function of time (left) and of ωt (right) compared with the applied emf (above) plotted on the same timescale. Note that the current is not in phase with the emf but leads it by $\frac{\pi}{2}$.

The rms value of the current is given by $I_{rms} = \frac{E_{rms}}{1/\omega C}$. The factor $\frac{1}{\omega C}$ in this case plays a role similar to that played by resistance in case (i) above and is called the **reactance** of the capacitor, defined as

$$\boxed{X_C := \frac{1}{\omega C}}$$

(iii) ***Inductance in an a.c. circuit***

Figure 20.39 shows a single inductor connected across the terminals of an a.c. power supply. As we saw in Section 20.3, the net emf in this case is $E - L\frac{dI}{dt}$ and, in the limit $R \to 0$, we have $E - L\frac{dI}{dt} = 0$ or $\frac{dI}{dt} = \frac{E}{L} = \frac{E_0}{L}\sin\omega t$.

Thus the current at an instant t is the integral of this last term and is given by

$$I = I(t) = \int dI = \frac{E_0}{L}\int \sin\omega t' dt' = -\frac{E_0}{\omega L}\cos\omega t$$

which can be rewritten as

$$I = \frac{E_0}{\omega L}\sin\left(\omega t - \frac{\pi}{2}\right)$$

Again the current is not in phase with the applied emf but, in this case, it *lags* behind the emf by $\frac{\pi}{2}$, that is the current reaches maximum (for example) at a time $\frac{T}{4}$ *after* the emf was maximum, as indicated in Figure 20.40. The rms current is given by $I_{rms} = \frac{E_{rms}}{\omega L}$ and the factor ωL is seen to play a role similar to that played by R in case (i) or $\frac{1}{\omega C}$ in case (ii) and is called the **reactance** of the inductor, defined as

$$\boxed{X_L := \omega L}$$

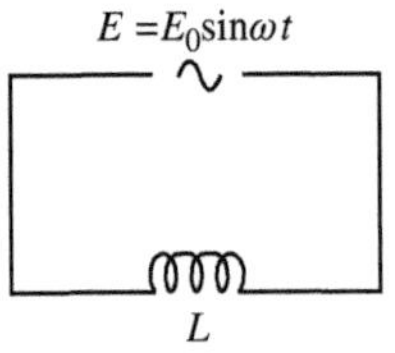

Figure 20.39. An inductor connected directly across an a.c. power supply.

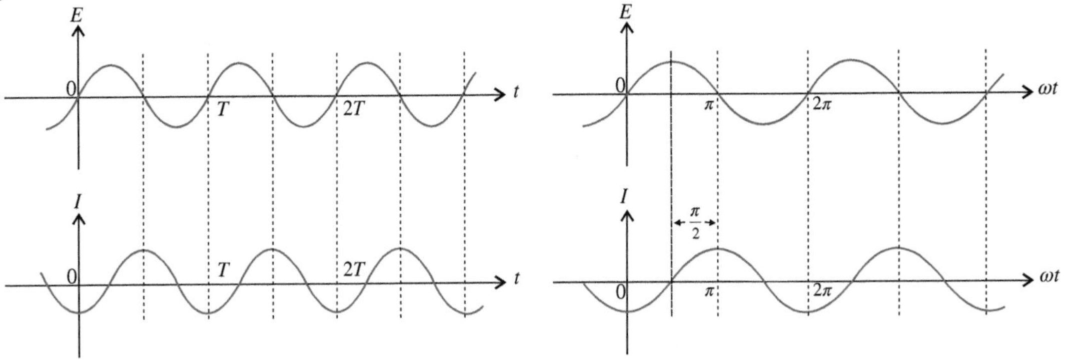

Figure 20.40. A plot of the current in the circuit in Figure 20.39 as a function of time (left) and of ωt (right) compared with the applied emf (above) plotted on the same timescale. Note that, in this case, the current lags behind the emf by $\frac{\pi}{2}$.

Impedance

The three simple cases outlined above can be covered by a single expression as follows:

$$I = \frac{E_0}{|Z|}\sin(\omega t + \phi)$$

where

$$|Z| = \begin{cases} R \\ \frac{1}{\omega C} \\ \omega L \end{cases} \quad \text{and} \quad \phi = \begin{cases} 0 \\ +\frac{\pi}{2} \\ -\frac{\pi}{2} \end{cases} \quad \text{for a} \begin{cases} \text{resistor} \\ \text{capacitor} \\ \text{inductor} \end{cases}$$

The quantity $|Z|$ here is a generalisation of the concept of resistance to the case of a.c. circuits and is the magnitude of a *complex* quantity called **impedance** which is defined as

$$\boxed{Z := R + iX}$$

where $i = \sqrt{-1}$ and $X = X_C - X_L = \left(\frac{1}{\omega C} - \omega L\right)$.

Note that, while the impedance of a resistance is constant, the impedance of a capacitor or of an inductor depends on the frequency of the a.c. source; that is, $Z = Z(\omega)$ in general. The rms current is determined by the magnitude of the impedance in each case since $I_{\rm rms} = \dfrac{E_{\rm rms}}{|Z|}$, where $|Z| = \sqrt{R^2 + X^2}$ (Appendix A.8).

Some examples of the magnitude of the impedance of various resistors, capacitors and inductors at particular frequencies are given in Table 20.1.

Table 20.1 Magnitude of the impedances of different components at various frequencies.

	resistor			capacitor			inductor		
$f = \dfrac{\omega}{2\pi}$	1 Ω	1 kΩ	1 MΩ	1 pF	1 nF	1 μF	1 μH	1 mH	1 H
d.c.	1 Ω	1 kΩ	1 MΩ	∞	∞	∞	0	0	0
50 Hz	1 Ω	1 kΩ	1 MΩ	3.2 GΩ	3.2 MΩ	3.2 kΩ	0.31 mΩ	0.31 Ω	314 Ω
1 kHz	1 Ω	1 kΩ	1 MΩ	160 MΩ	160 kΩ	160 Ω	6.3 mΩ	6.3 Ω	6.3 kΩ
1 MHz	1 Ω	1 kΩ	1 MΩ	160 kΩ	160 Ω	0.16 Ω	6.3 Ω	6.3 kΩ	6.3 MΩ
1 GHz	1 Ω	1 kΩ	1 MΩ	160 Ω	0.16 Ω	0.16 mΩ	6.3 kΩ	6.3 MΩ	6.3 GΩ

Note: $\omega = 0$ for d.c.

Study Worked Example 20.5

For problems based on the material presented in this section visit up.ucc.ie/20/ *and follow the link to the problems.*

20.10 The series *L-C-R* circuit: phasor diagrams

As an example of how capacitive and inductive components behave in an a.c. circuit, we consider the case, illustrated in Figure 20.41, in which a resistor, a capacitor and an inductor are connected in series across an a.c. power supply. In this case the sum of the potential differences across the components at any instant is equal to the emf of the power supply at that instant, that is

$$V_R + V_C + V_L = E$$

or

$$RI + \frac{q}{C} + L\frac{dI}{dt} = E_0 \sin\omega t \qquad (20.13)$$

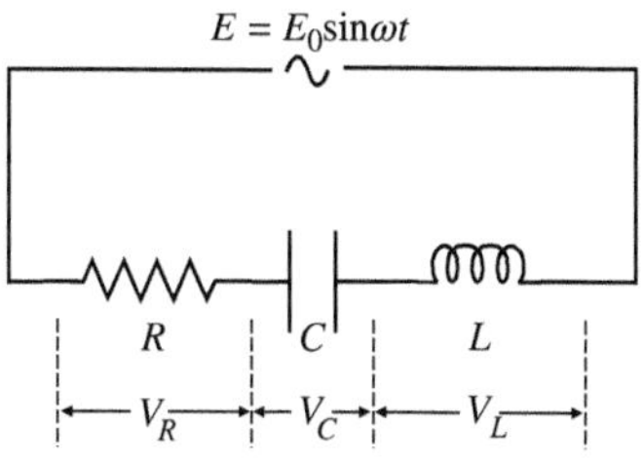

Figure 20.41. A series *L-C-R* a.c. circuit.

From the discussion in the previous section we would expect in general that the current in the circuit would not be in phase with the applied voltage and would depend on time according to the following relationship

$$I(t) = I_0 \sin(\omega t + \phi) \qquad (20.14)$$

where I_0 (the maximum current) and ϕ (the phase difference between the current and the applied emf) are, as yet, undetermined constants.

Now

$$q = \int I(t)dt = I_0 \int \sin(\omega t + \phi)dt = -\frac{I_0}{\omega}\cos(\omega t + \phi)$$

$$\rightarrow \quad V_C = \frac{q}{C} = -\frac{I_0}{\omega C}\cos(\omega t + \phi) = \frac{I_0}{\omega C}\sin\left(\omega t + \phi - \frac{\pi}{2}\right)$$

$$\frac{dI}{dt} = \omega I_0 \cos(\omega t + \phi) \quad \rightarrow \quad V_L = \omega L I_0 \cos(\omega t + \phi) = \omega L I_0 \sin\left(\omega t + \phi + \frac{\pi}{2}\right)$$

and $V_R = RI \rightarrow V_R = RI_0 \sin(\omega t + \phi)$. Thus, while the current I is the same in all components, the voltage across the capacitor lags I by $\frac{\pi}{2}$ and the voltage across the inductor leads I by $\frac{\pi}{2}$.

Substituting these expressions for V_R, V_C and V_L into Equation (20.13) we obtain

$$RI_0\sin(\omega t + \phi) - \frac{I_0}{\omega C}\cos(\omega t + \phi) + \omega L I_0\cos(\omega t + \phi) = E_0\sin\omega t \qquad (20.15)$$

or

$$RI_0\sin(\omega t + \phi) + \frac{I_0}{\omega C}\sin\left(\omega t + \phi - \frac{\pi}{2}\right) + \omega L I_0\sin\left(\omega t + \phi + \frac{\pi}{2}\right) = E_0\sin\omega t \qquad (20.16)$$

Equation (20.15) holds for all t and, in particular, must hold for the following two special cases.

Case 1: $\omega t = 0$:

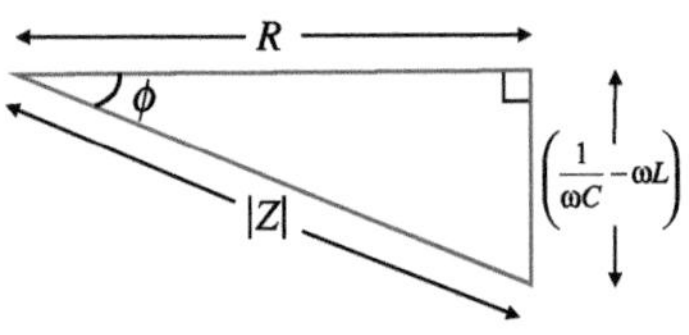

Figure 20.42. Geometrical representation of how the phase angle ϕ depends on the values of the components and the frequency of the source for a series *L-C-R* circuit (Figure 20.41).

Equation (20.15) $\rightarrow$ $I_0\left[R\sin\phi + \left(\omega L - \dfrac{1}{\omega C}\right)\cos\phi\right] = 0$ and, since $I_0 \neq 0$, $\tan\phi = \dfrac{\dfrac{1}{\omega C} - \omega L}{R}$

This shows that the current in the circuit is out of phase with the applied emf by a phase angle ϕ which depends on the values of the components involved as well as on the frequency of the voltage source. The right-angled triangle in Figure 20.42 shows, diagrammatically, how ϕ depends on the values of R, C, L and ω.

In particular, we see from inspection that

$$\sin\phi = \frac{\dfrac{1}{\omega C} - \omega L}{|Z|} \quad \text{and} \quad \cos\phi = \frac{R}{|Z|}, \quad \text{where} \quad |Z| = \sqrt{R^2 + \left(\frac{1}{\omega C} - \omega L\right)^2}$$

where $|Z|$ is the hypotenuse of the triangle, which we identify as the magnitude of the impedance of the circuit.

Case 2: $\omega t = \dfrac{\pi}{2}$:

Equation (20.15) $\rightarrow$ $I_0\left[R\sin\left(\phi + \frac{\pi}{2}\right) + \left(\omega L - \frac{1}{\omega C}\right)\cos\left(\phi + \frac{\pi}{2}\right)\right] = E_0$

Using the trigonometric relationships $\sin\left(\phi + \frac{\pi}{2}\right) = \cos\phi$ and $\cos\left(\phi + \frac{\pi}{2}\right) = -\sin\phi$ (Appendix A.3.2), we obtain

$$I_0\left[R\cos\phi - \left(\omega L - \frac{1}{\omega C}\right)\sin\phi\right] = E_0$$

and, using the expressions for $\sin\phi$ and $\cos\phi$ above, we get

$$I_0 = \frac{E_0}{R\cos\phi + \left(\dfrac{1}{\omega C} - \omega L\right)\sin\phi} = \frac{E_0|Z|}{R^2 + \left(\dfrac{1}{\omega C} - \omega L\right)^2} = \frac{E_0}{|Z|}$$

Returning to the general case and substituting this expression for I_0 back into Equation (20.14) we see that the current in the circuit at any instant t is given by

$$I = \frac{E_0}{\sqrt{R^2 + \left(\dfrac{1}{\omega C} - \omega L\right)^2}}\sin(\omega t + \phi)$$

that is,

$$I(t) = \frac{E_0}{|Z|}\sin(\omega t + \phi) \tag{20.17}$$

where

$$|Z| := \sqrt{R^2 + \left(\frac{1}{\omega C} - \omega L\right)^2} \quad \text{and} \quad \phi = \tan^{-1}\left(\frac{\dfrac{1}{\omega C} - \omega L}{R}\right)$$

Note that $I_{\text{rms}} = \dfrac{E_{\text{rms}}}{|Z|}$, showing again that $|Z|$ plays a role in an a.c. circuit which is the counterpart of resistance in a dc circuit. The analysis here is formally identical to that in the study of forced oscillations covered in Section 3.12 at up.ucc,ie/3/12/ on the *Understanding Physics* website. Indeed, the results obtained above could equally well have been derived by direct analogy with that analysis.

Phasor diagrams

Equation (20.16) is an example of the addition of a number of sinusoidal functions having the same frequency but different relative phases. We encountered a similar situation in Chapter 13 in the context of the superposition of sinusoidal wave disturbances (Section 13.6). Such situations can be analysed conveniently by means of a technique known as *addition of phasors*.

Consider the case of the addition of two sinusoidal quantities which differ by a phase of ϕ, that is

$$y_1(x, t) = A_1\sin\omega t \quad \text{and} \quad y_2(x, t) = A_2\sin(\omega t + \phi).$$

The sum of y_1 and y_2 is

$$y_1(x,t) + y_2(x,t) = A_1 \sin \omega t + A_2 \sin(\omega t + \phi)$$

Each of the terms on the right of this equation can be seen to be the vertical component of a vector (y_1 is the vertical component of $\boldsymbol{A}_1$ and y_2 is the vertical component of $\boldsymbol{A}_2$, as illustrated in Figure 20.43). We have already encountered an example of a sinusoidal motion being treated in this way; recall the geometrical interpretation of simple harmonic motion in Figure 3.11. The vertical component of the vector sum $\boldsymbol{A}_1 + \boldsymbol{A}_2$ is the sum of the vertical components of the individual vectors. In this context, the vectors are called **phasors**.

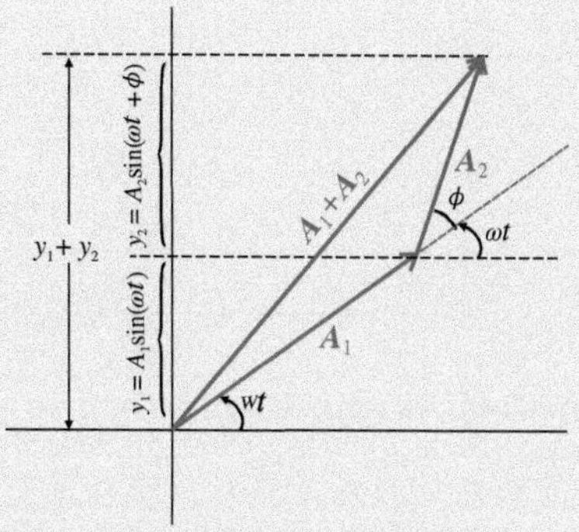

Figure 20.43. Addition of phasors.

Figure 20.43 shows the addition of two phasors at some instant t. As time evolves, the angle ωt increases at a steady rate and each phasor (and consequently the whole addition diagram) rotates counterclockwise about O with the same constant angular velocity ω and so the phasors maintain their relative orientation and the vertical components oscillate.

Any number of phasors may be added using the above procedure. We will make considerable use of addition of phasors in Chapter 22.

Returning to the *L-C-R* a.c. circuit, we can treat the voltages across each component and the applied emf as phasors. Equation (20.16) is interpreted in terms of phasor addition in Figure 20.44 where three *voltage phasors* have been drawn to represent, respectively, the three quantities RI_0, $\dfrac{1}{\omega C}I_0$ and $\omega L I_0$. The relative orientations of the phasors are determined by the phase relationship between them as established in the previous section; the directions of the $\omega L I_0$ and $\dfrac{1}{\omega C}I_0$ phasors relative to the RI_0 phasor represent, respectively, the $\dfrac{\pi}{2}$ lead and $\dfrac{\pi}{2}$ lag which we have discussed earlier. As shown in the figure, the magnitudes of the three phasors are RI_0, $\dfrac{1}{\omega C}I_0$ and $\omega L I_0$ and these can be added vectorially to give a resultant phasor the magnitude of which can be seen, by reference to Figure 20.44, to be $E_0 = |Z| I_0$ where $|Z| = \sqrt{R^2 + \left(\dfrac{1}{\omega C} - \omega L\right)^2}$ as before.

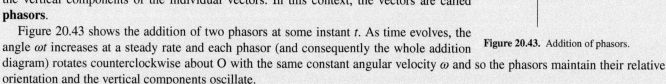

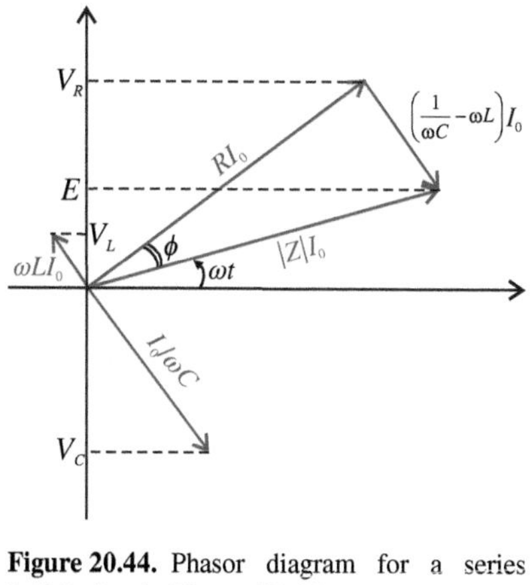

Figure 20.44. Phasor diagram for a series *L-C-R* circuit (Figure 20.41). V_R, V_C, V_L and E vary with time as the blue vectors rotate at constant angular velocity. The phase difference between the current in the circuit and the applied emf is ϕ (Note that in many texts the phases of the voltages across the components are given relative to the current rather than, as in this case, relative to the emf).

The potential difference across the resistor is in phase with the current which, in turn, is out of phase with the applied emf (represented by the phasor $|Z|I_0$) by an angle

$$\phi = \tan^{-1} \frac{\dfrac{1}{\omega C} - \omega L}{R}$$

The rms potential differences are proportional to the lengths of the corresponding phasors; rms values carry no phase information. Worked Example 20.6 demonstrates the use of the phasor method just described. Phasor diagrams prove particularly valuable in the analysis of a.c. circuits which are more complicated than those discussed in this chapter.

Study Worked Example 20.6

Resonance in series L-C-R circuits

An examination of Equation (20.17) shows that the current in a series *L-C-R* circuit increases with decreasing Z. Since Z is a function of frequency, the current varies with the frequency of the voltage source for fixed R, C and L, the current being at a maximum when Z is at a minimum. Thus, as in the case of forced oscillations of a mechanical harmonic oscillator (Section 3.12 on the *Understanding Physics* website at up.ucc.ie/3/12/), there is a particular frequency (the **resonant frequency**), at which the current has a maximum value. This occurs at a frequency $\omega = \omega_0$ which is such that

$$\frac{1}{\omega_0 C} - \omega_0 L = 0 \quad \text{or} \quad \omega_0^2 = \frac{1}{LC}$$

$$\Rightarrow \quad \boxed{\omega_0 = \frac{1}{\sqrt{LC}}}$$

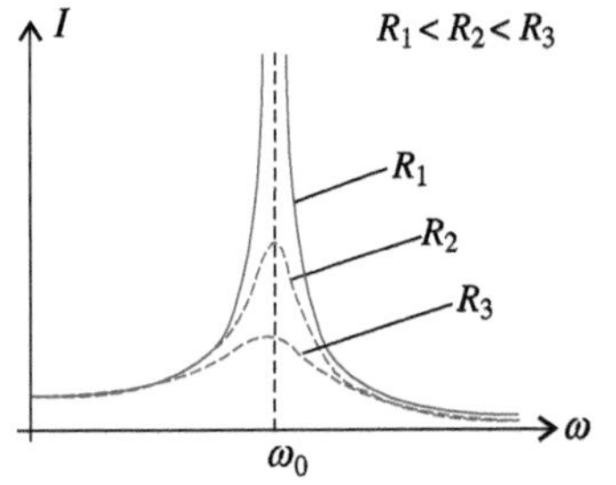

Figure 20.45. Resonance in a series *L-C-R* circuit. Plots of current in the circuit as a function of frequency for a number of different values of resistance. Note that the resonance is sharper for the lower values of resistance.

Note that the phase angle ϕ is zero at resonance and that the resonant frequency is independent of R.

The main features of a resonance can be studied by plotting the rms current as a function of frequency in the region of the resonant frequency. A number of such 'resonance curves', corresponding to different values of the resistance R, are shown in Figure 20.45. The sharpness of the resonance is represented by the width of the resonance curve; note that the resonance is sharpest when the resistance in the circuit is lowest.

20.11 Power in an a.c. circuit

The power in an a.c. circuit is dissipated in the resistive components only (recall the definition of resistance in Section 15.4). At any instant, of course, energy may be stored in capacitive and inductive components but this energy remains constant on average and is released (or remains stored on charged capacitors) when the power supply is disconnected or switched off.

If the current drawn from a power supply of emf $E = E_0 \sin \omega t$ is $I = I_0 \sin(\omega t + \phi)$, the instantaneous power generated in the circuit will be given by

$$\begin{aligned} P = EI &= [E_0 \sin \omega t][I_0 \sin(\omega t + \phi)] = E_0 I_0 \sin \omega t \sin(\omega t + \phi)) \\ &= E_0 I_0 \sin \omega t[\sin \omega t \cos \phi + \cos \omega t \sin \phi] \\ &= E_0 I_0 \sin^2 \omega t \cos \phi + E_0 I_0 \sin \omega t \cos \omega t \sin \phi \\ &= E_0 I_0 \cos \phi \sin^2 \omega t + \frac{1}{2} E_0 I_0 \sin \phi \sin 2\omega t \end{aligned}$$

Thus, the **average power** generated in the circuit is

$$\langle P \rangle = E_0 I_0 \cos \phi \left\langle \sin^2 \omega t \right\rangle + \frac{1}{2} E_0 I_0 \sin \phi \left\langle \sin 2\omega t \right\rangle$$

and hence, since $\left\langle \sin^2 \omega t \right\rangle = \frac{1}{2}$ and $\langle \sin 2\omega t \rangle = 0$,

$$\langle P \rangle = \frac{1}{2} E_0 I_0 \cos \phi = E_{rms} I_{rms} \cos \phi \tag{20.18}$$

where $\cos\phi$ is called the **power factor** of the circuit.

To deliver power in the most efficient (least energy consuming) way to the resistive components in an a.c. circuit, ϕ must be as close to zero ($\cos\phi \to 1$) as possible. If a circuit has components of large inductance, the addition of an appropriate capacitance can reduce the phase and bring the power factor close to unity, a technique known as *phase compensation*.

Note that the expression for the average power, Equation (20.18), is an example of a case in which the d.c. result ($P = EI$) cannot be carried over to a.c. In terms of the resistance of the circuit, however, the average power is given by

$$\langle P \rangle = \left\langle R_{circ} I^2 \right\rangle = R_{circ} \left\langle I_0^2 \sin^2(\omega t + \phi) \right\rangle = \frac{1}{2} R_{circ} I_0^2 = R_{circ} I_{rms}^2$$

which does have the same form as the d.c. case. Combining this result with Equation (20.18) we obtain

$$E_{rms} \cos \phi = R_{circ} I_{rms}$$

In the case of the simple series *L-C-R* circuit (Figure 20.41) discussed in the previous section, $R_{circ} = R$ and the power factor is given by $\cos \phi = \frac{R}{Z}$. Thus, in that case, the average power in the circuit, as expected, is given by

$$\langle P \rangle = E_{rms} I_{rms} \cos \phi = E_{rms} I_{rms} \frac{R}{|Z|} = \frac{E_{rms}}{|Z|} R I_{rms} = R I_{rms}^2$$

Further reading

Further material on some of the topics covered in this chapter may be found in *Electromagnetism* by Grant and Phillips. (See BIBLIOGRAPHY for book details)

WORKED EXAMPLES

Worked Example 20.1: A magnetic field, directed perpendicularly to the x—y plane, is uniform in the y-direction but increases with increasing x according to $\boldsymbol{B}(x) = \beta x\boldsymbol{i}$, where $\boldsymbol{i}$ is a unit vector in the x-direction and $\beta = 30$ mT m^{-1}. A plane rectangular loop of wire of dimensions 15 cm x 20 cm is pulled through the field in the $+x$-direction with constant velocity of 5 m s^{-1}. The plane of the loop is kept in the x—y plane with the longer side in the x-direction (Figure 20.46). (a) What is the emf of the loop? (b) What is the current in the loop if its resistance is 3 Ω? (c) What force has to be applied to the loop to keep the speed constant, assuming that there is no friction or other mechanical resistance to the motion?

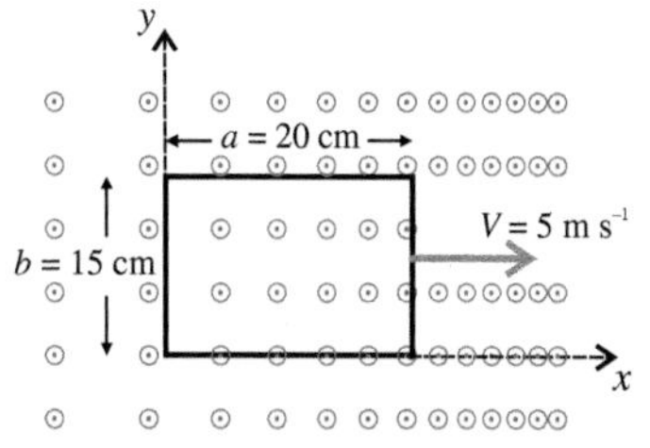

Figure 20.46. Worked Example 20.1.

$$\boldsymbol{B} = \boldsymbol{B}(x) = \beta x\boldsymbol{i}, \quad \text{where} \quad \beta = 30 \text{ mT m}^{-1}; \quad a = 0.2 \text{ m}; b = 0.15 \text{ m}$$

(a) $E_{\text{ind}} = \dfrac{d(BA)}{dt} = A\dfrac{dB}{dt} = A\dfrac{d(\beta x)}{dt} = A\beta\dfrac{dx}{dt} = A\beta v = (0.2)(0.15)(30 \times 10^{-3})(5) = 4.5 \times 10^{-3} = 4.5 \text{ mV}$

(b) $I = \dfrac{E_{\text{ind}}}{R} = (4.5 \times 10^{-3})/3 = 1.5 \text{ mA}$

(c) Power generated $= P = \boldsymbol{F} \cdot \boldsymbol{v} = Fv = E_{\text{ind}} I \quad \rightarrow \quad F = \dfrac{E_{\text{ind}} I}{v} = \dfrac{(4.5 \times 10^{-3})(1.5 \times 10^{-3})}{5} = 1.35 \text{ μN}$

Alternatively, using Equation (19.2), that is, $\Delta\boldsymbol{F} = I\Delta\boldsymbol{l} \times \boldsymbol{B}$, $\rightarrow \quad F = Ib\beta(x + a - x) = I\beta ba = (1.5 \times 10^{-3})(30 \times 10^{-3})(0.2)(0.15) = 1.35$ μN

Worked Example 20.2: The current through an inductor (Figure 20.47) of self-inductance L varies with time as $I = I_0 e^{-\gamma t}$, where I_0 and γ are constants. Derive an expression for the energy released as a function of time, assuming that the resistance in the circuit is negligible. Show explicitly, for this case, that the total energy released over a very long time is independent of γ.

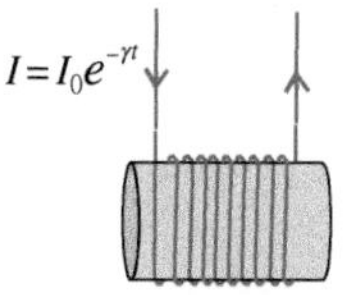

Figure 20.47. Worked Example 20.2.

Using Equation (20.3A) $\quad E_{\text{ind}} = -L\dfrac{dI}{dt} \quad \rightarrow \quad$ Power generated $= P(t) = E_{\text{ind}} I = -LI\dfrac{dI}{dt} = -LI_0^2 e^{-\gamma t}(-\gamma)e^{-\gamma t} = \gamma L I_0^2 e^{-2\gamma t}$

Energy released from $t = 0$ to time t is $\displaystyle\int_0^t P(t')dt' = \gamma L I_0^2 \int_0^t e^{-2\gamma t'} dt' = \frac{1}{2} L I_0^2 \int_0^t e^{-2\gamma t'} d(2\gamma t')$

Letting $2\gamma t' = \tau$. the limits of the integral become $\tau = 0$ and $\tau = 2\gamma t$

$$\rightarrow \quad \text{Energy released} = \frac{1}{2} L I_0^2 \int_0^{2\gamma t} e^{-\tau} d\tau = \frac{1}{2} L I_0^2 (1 - e^{-2\gamma t})$$

As $t \rightarrow \infty$ energy released $\rightarrow \quad \dfrac{1}{2} L I_0^2$

Worked Example 20.3: Derive an expression for the self-inductance per unit length of coaxial cable, having inner radius a and outer radius b, when the material in the space between the conductors has permeability μ.

Current I flows through the inner conductor and returns through the outer conducting screen. Consider the magnetic flux through the grey shaded rectangular area of length l indicated in Figure 20.48. The magnetic field strength at a distance r from the axis is given by $H = \dfrac{I}{2\pi r}$ and hence $B = \dfrac{\mu I}{2\pi r}$. Thus the magnetic flux through the strip of area $l\Delta r$ is $\dfrac{\mu I}{2\pi r}(l\Delta r)$ and the magnetic flux through the whole shaded area is given by

$$\Phi_{\text{M}} = \frac{\mu I l}{2\pi} \int_a^b \frac{dr}{r} = \frac{\mu I l}{2\pi} \ln\left(\frac{b}{a}\right) \quad \rightarrow \quad L = \frac{\Phi_{\text{M}}}{I} = \frac{\mu l}{2\pi} \ln\left(\frac{b}{a}\right) \quad \rightarrow \quad \frac{L}{l} = \frac{\mu}{2\pi} \ln\left(\frac{b}{a}\right)$$

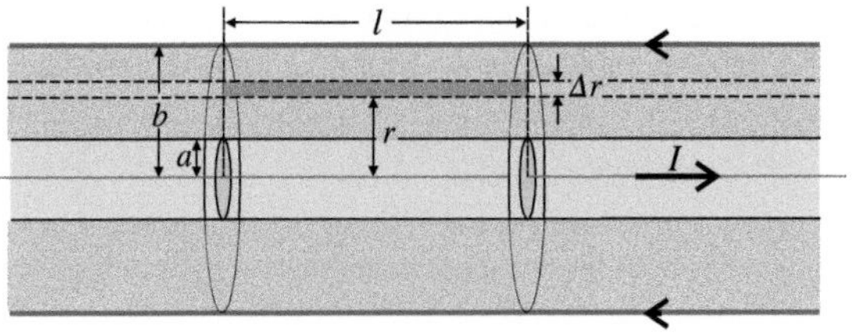

Figure 20.48. Worked Example 20.3.

Worked Example 20.4: A long solenoid in air has a radius of 3 cm and 12,000 turns per metre. A coil having 400 turns is tightly wound over the outside of the solenoid, that is it has essentially the same cross-sectional area (Figure 20.49). If the current in the long solenoid changes at a constant rate of 15 A s^{-1}, what is the emf induced in the other coil?

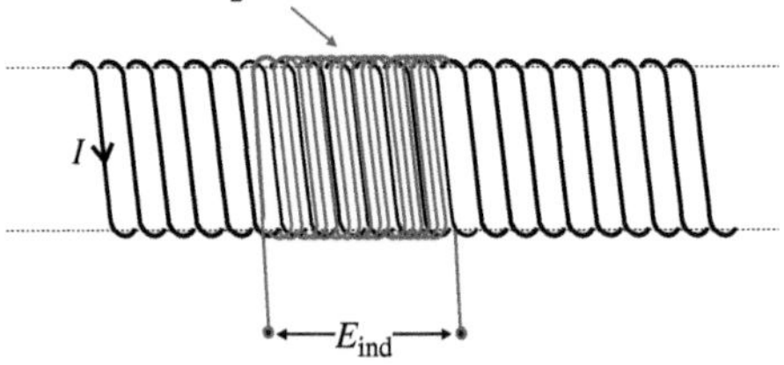

Figure 20.49. Worked Example 20.4.

The emf induced in the second coil is given by $E_{ind} = N_2 \frac{d\phi}{dt}$, where ϕ is the flux inside the solenoid.

The rate of change of flux through the long solenoid is $N_1 \frac{d\phi}{dt} = L_1 \frac{dI_1}{dt}$ and thus $E_{ind} = N_2 \frac{L_1}{N_1} \frac{dI_1}{dt}$.

From Equation (20.7) $L_1 = \frac{\mu N_1^2 A}{l}$

and hence $E_{ind} = \mu N_2 \frac{N_1}{l} A \frac{dI_1}{dt} = 4\pi \times 10^{-7}(400)(12000)(\pi)(0.03)^2(15) = 0.26\text{ V}$

Worked Example 20.5: A 50 Ω resistor, a 10 mH inductor and a 0.5 μF capacitor are connected in series with a 20 V, 1.6 kHz a.c. power supply (Figure 20.50). (a) Determine (i) the impedance of the circuit, (ii) the rms current in the circuit, (iii) the phase difference between the current and the applied emf. (b) Calculate the resonant frequency of the circuit.

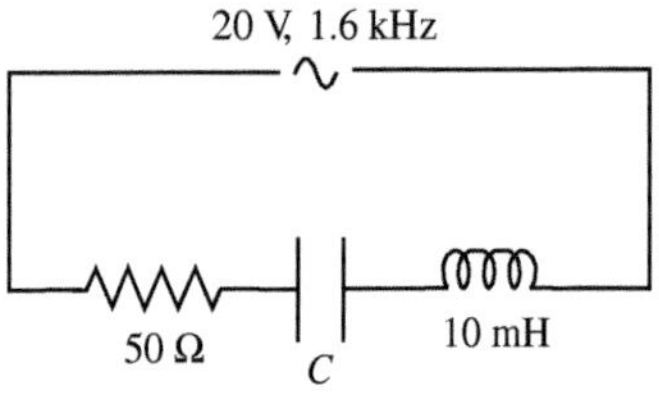

Figure 20.50. Worked Example 20.5.

$$\frac{1}{\omega C} = 199\ \Omega;\quad \omega L = 100.5\ \Omega;$$

(a) (i) $Z = \sqrt{R^2 + \left(\frac{1}{\omega C} - \omega L\right)^2} = \sqrt{(50)^2 + (199 - 100.5)^2} = \sqrt{(50)^2 + (98.5)^2} = 110\ \Omega$

(ii) $I = \frac{E}{Z} = \frac{20}{110} = 0.18\text{ A}$

(iii) $\tan\phi = \frac{98.5}{50} = 1.97 \rightarrow \phi = 63°$

(b) $\omega_0 = \frac{1}{\sqrt{LC}} = \frac{1}{\sqrt{(10^{-2})(0.5 \times 10^{-6})}} = 1.4 \times 10^4\text{ rad s}^{-1} \rightarrow f = \frac{\omega_0}{2\pi} = 2.25\text{ kHz}$

Worked Example 20.6: A 350 Ω resistor, a 10 μF capacitor and a coil are connected in series across a 240 V, 50 Hz a.c. power supply (Figure 20.51(a)). The rms potential difference measured across the resistor is 180 V and that measured across the series combination of capacitor and coil is 120 V. What is the resistance of the coil if its self-inductance is 0.15 H?

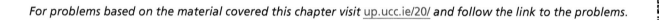

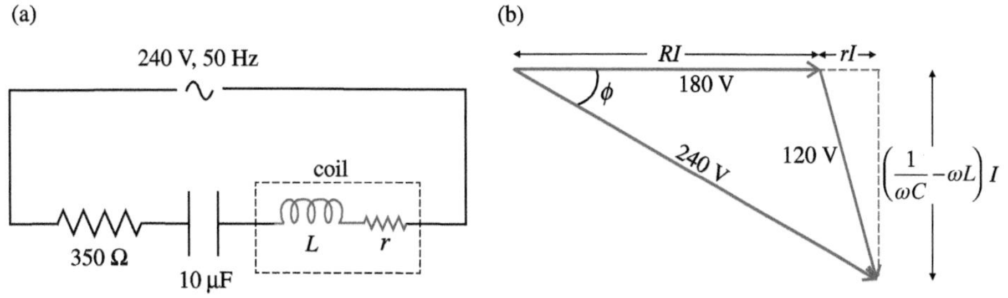

Figure 20.51. (a) Worked Example 20.6. (b) Phasor diagram for Worked Example 20.6.

$$\omega = 2\pi f = 100\pi \text{ rad/s}; \quad R = 350\ \Omega; \quad 1/\omega C = 318\ \Omega; \quad \omega L = 47\ \Omega; \quad 1/\omega C - \omega L = 271\ \Omega.$$

Let r be the resistance of the coil. Note that $1/\omega C > \omega L$.

Applying the cosine rule (Appendix A.3.3) to the phasor diagram Figure 20.51(b)

$$\cos\phi = \frac{240^2 + 180^2 - 120^2}{2 \times 240 \times 180} = 0.875 \quad \rightarrow \quad \phi = 29°$$

Also, from Figure 20.51, $\tan\phi = \dfrac{\left(\frac{1}{\omega C} - \omega L\right)}{R + r} = \dfrac{271}{350 + r} \quad \rightarrow \quad r + 350 = 490 \quad \rightarrow \quad r = 140\ \Omega$

PROBLEMS

For problems based on the material covered this chapter visit up.ucc.ie/20/ and follow the link to the problems.

21

Maxwell's equations: electromagnetic radiation

AIMS

- to generalise the laws of electromagnetism so that certain difficulties in the laws, as they were formulated in previous chapters, can be overcome
- to show that the generalised forms of the laws (*Maxwell's equations*) enable the emission and propagation of electromagnetic waves to be understood
- to consider energy, momentum and angular momentum in electromagnetic waves
- to re-examine the photon model in the light of wave–particle duality

21.1 Reconsideration of the laws of electromagnetism: Maxwell's equations

Faraday's law revisited

The principle of electromagnetic induction (Faraday's law) was stated in Section 20.1 as follows $\oint_{C} \boldsymbol{E} \cdot d\boldsymbol{l} = -\frac{d}{dt}\iint_{S} \boldsymbol{B} \cdot d\boldsymbol{A}$, where S is the surface delineated by the circuit C. Applied to the circuit shown in Figure 21.1, this equation states that the changing magnetic flux through the circuit produces an electric field strength $\boldsymbol{E}$ everywhere in the wire which is responsible for the induced emf and will cause an electric current to flow in the closed loop of wire.

One might expect, however, that the electric field produced by the changing magnetic flux exists *whether or not a conducting loop is present*. This suggests that we can generalise the principle of electromagnetic induction to apply to any closed path in space and hence Faraday's law could be restated as follows

$$\oint_{C} \boldsymbol{E} \cdot d\boldsymbol{l} = -\frac{d}{dt}\iint_{S} \boldsymbol{B} \cdot d\boldsymbol{A} \qquad (21.1)$$

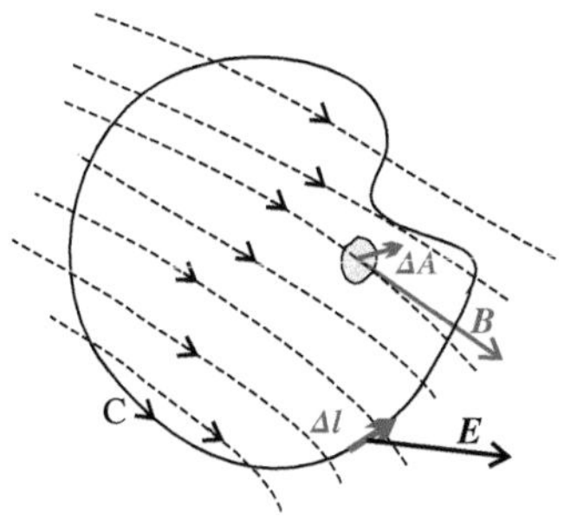

Figure 21.1. Faraday's law implies that a changing magnetic flux through a circuit produces an electric field strength at each point in the circuit which is responsible for the induced emf. The principle can be generalised to apply to any closed curve C where the rate of change of magnetic flux involved is that through any surface bounded by C.

where now C is ***any*** *closed curve* and S *is* ***any*** *surface bounded by* C. Thus a changing magnetic flux is seen to give rise to an electric field which is determined by Equation (21.1) where $\boldsymbol{E}$ and $\boldsymbol{B}$ are the field strengths as observed in the rest frame of the curve C.

Let us now summarise the fundamental laws of electromagnetism which we have encountered up to this point.

(a) ***Ampère's law*** (Equation (18.8)): $\oint_{C} \boldsymbol{H} \cdot d\boldsymbol{l} = I$ (equivalent to the Biot-Savart law)

(b) ***Gauss's law for magnetism*** (Equation (18.22)): $\oiint_{S} \boldsymbol{B} \cdot d\boldsymbol{A} = \sum_{i} p_i$

Understanding Physics, Third Edition. Michael Mansfield and Colm O'Sullivan.
© 2020 John Wiley & Sons Ltd. Published 2020 by John Wiley & Sons Ltd.

(c) ***Gauss's law for electrostatics*** (Equation (16.14)): $\oiint_S \boldsymbol{D} \cdot d\boldsymbol{A} = \sum_i q_i$ (equivalent to Coulomb's law)

(d) ***Faraday's law*** (Equation (21.1)): $\oint_C \boldsymbol{E} \cdot d\boldsymbol{l} = -\frac{d\Phi_M}{dt} = -\frac{d}{dt}\iint_S \boldsymbol{B} \cdot d\boldsymbol{A}$ (equivalent to the Lorentz force law)

There is a clear symmetry between Gauss's law for magnetism and Gauss's law for electrostatics, although, at a microscopic level, the right hand side of the former equation will be zero since no free magnetic monopoles have ever been observed ($\sum_i p_i = 0$).

Nevertheless, if magnetic monopoles did exist, one would expect that the motion of free magnetic poles would generate 'magnetic currents' which would in turn be expected to give rise to electric fields, just as electric currents give rise to magnetic fields. To take such magnetic currents into consideration, Equation (21.1) would have to contain an additional term (I_M), analogous to Ampère's law, as follows

$$\oint_C \boldsymbol{E} \cdot d\boldsymbol{l} = I_M - \frac{d\Phi_M}{dt} = I_M - \frac{d}{dt}\iint_S \boldsymbol{B} \cdot d\boldsymbol{A}$$

With this modification, let us look again at the four fundamental laws

$$\oint_C \boldsymbol{H} \cdot d\boldsymbol{l} = I \tag{21.2a}$$

$$\oiint_S \boldsymbol{B} \cdot d\boldsymbol{A} = \sum_i p_i \tag{21.2b}$$

$$\oiint_S \boldsymbol{D} \cdot d\boldsymbol{A} = \sum_i q_i \tag{21.2c}$$

$$\oint_C \boldsymbol{E} \cdot d\boldsymbol{l} = I_M - \frac{d}{dt}\iint_S \boldsymbol{B} \cdot d\boldsymbol{A} \tag{21.2d}$$

Ampère's law revisited

A closer look at Ampère's law (21.2a) shows up a difficulty, at least in the case of time-varying currents. Consider the case of a simple *R-C* circuit shown in Figure 21.2(a), where a time varying current $I = I(t)$ flows in the circuit. Writing Ampère's law in terms of the current density, Equation (18.8a), namely $\oint_C \boldsymbol{H} \cdot d\boldsymbol{l} = \iint_S \boldsymbol{J} \cdot d\boldsymbol{A}$, let us apply the law to the curve C, a circular path around the wire as shown in the figure. The current through the surface S (the plane circular disc enclosed by C) is clearly $I(t)$ and hence, as expected,

$$\oint_C \boldsymbol{H} \cdot d\boldsymbol{l} = \iint_S \boldsymbol{J} \cdot d\boldsymbol{A} = I(t)$$

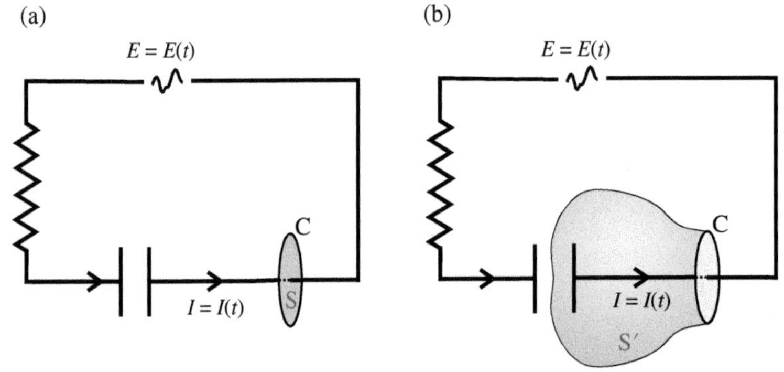

Figure 21.2. Time-varying current in a simple *R-C* circuit. (a) Ampère's law applies to the surface S bounded by the closed curve C. (b) Same circuit as in (a) but Ampère's law fails to hold for the surface S' bounded by the same curve C as in (a).

The surface S in Ampère's law, however, is *any* surface bounded by the closed curve C and hence should apply equally to the surface S' in Figure 21.2(b). There is, of course, no movement of charge between the plates of the capacitor and, hence, there is *no* movement of charge in or out through this surface, that is

$$\oint_C \boldsymbol{H} \cdot d\boldsymbol{l} = \iint_{S'} \boldsymbol{J} \cdot d\boldsymbol{A} = 0$$

Clearly, since $I \neq 0$, Ampère's law as formulated fails in this particular context and some generalisation is required.

A clue to the solution may be found by considering the remaining lack of symmetry in Equations (21.2a–d). The equations would be completely symmetric in the electric and magnetic fields if Ampère's law had included an additional term of the form $\frac{d\Phi_E}{dt} = \frac{d}{dt}\iint_S \boldsymbol{D}\cdot d\boldsymbol{A}$. The existence of such a term would imply that *a changing electric flux would give rise to a magnetic field* just as, according to Faraday's law, a changing magnetic flux gives rise to an electric field. The proposal by James Clerk Maxwell (1831–1879), that just such a term be included in Ampère's law, has profound theoretical and practical implications. In particular, as we shall see, it turns out that the emission and propagation of electromagnetic radiation can be explained on this basis. Modern telecommunications technology rests on this discovery.

Maxwell's suggestion, therefore, was that Ampère's law should be extended as follows

$$\oint_C \boldsymbol{H}\cdot d\boldsymbol{l} = I + I_D$$

where I is the usual conduction current and the additional term, which Maxwell called **the displacement current**, is given by

$$I_D = \frac{d\Phi_E}{dt} = \frac{d}{dt}\iint_S \boldsymbol{D}\cdot d\boldsymbol{A}$$

Thus, the generalisation of Ampère's law (which we will call the **Ampère-Maxwell law**) can be written as

$$\oint_C \boldsymbol{H}\cdot d\boldsymbol{l} = I + \frac{d}{dt}\iint_S \boldsymbol{D}\cdot d\boldsymbol{A} \qquad (21.3)$$

We can get some understanding of the nature of the displacement current by considering again the simple R-C circuit in Figure 21.2(a). At any instant t the total electric flux between the capacitor plates is given by $\Phi_E = \iint_S \boldsymbol{D}\cdot d\boldsymbol{A} = q(t)$ = the charge on the plates at time t. Hence the displacement current at that instant is $I_D = \frac{d\Phi_E}{dt} = \frac{dq}{dt} = I(t)$ which is the conduction current in the circuit at time t. Thus we conclude that the displacement current in the space between the plates of the capacitor is identical to the conduction current in the rest of the circuit. The Ampère–Maxwell law (21.3) implies that a magnetic field is produced by a displacement current just as a magnetic field is produced by a conduction current. In Worked Example 21.1 it is shown that the magnetic field generated by the displacement current in a parallel plate capacitor with circular plates depends on the magnitude of the current and the geometry of the system in exactly the same way as that generated by a conduction current in a cylindrical wire which has the same diameter as that of the capacitor plates.

Maxwell's equations

The basic laws of classical electromagnetism can now be summarised by the four equations

$$\oiint_S \boldsymbol{B}\cdot d\boldsymbol{A} = \sum_i p_i \qquad \oint_C \boldsymbol{H}\cdot d\mathbf{l} = I + \frac{d\Phi_E}{dt} = I + \frac{d}{dt}\iint_S \boldsymbol{D}\cdot d\boldsymbol{A}$$

$$\oiint_S \boldsymbol{D}\cdot d\boldsymbol{A} = \sum_i q_i \qquad \oint_C \boldsymbol{E}\cdot d\mathbf{l} = I_M - \frac{d\Phi_M}{dt} = I_M - \frac{d}{dt}\iint_S \boldsymbol{B}\cdot d\boldsymbol{A}$$

Note the symmetry between $\boldsymbol{B}$ and $\boldsymbol{D}$ and between $\boldsymbol{H}$ and $\boldsymbol{E}$ when the equations are written in this form.

Taking into consideration that no free magnetic monopoles have been observed in nature (that is, $p_i = 0$ for all i and $I_M = 0$) the four equations are usually simplified as follows:

Ampere-Maxwell law:

$$\oint_C \boldsymbol{H}\cdot d\boldsymbol{l} = I + \frac{d}{dt}\iint_S \boldsymbol{D}\cdot d\boldsymbol{A} \qquad (21.4a)$$

Gauss's law for magnetism:

$$\oiint_S \boldsymbol{B}\cdot d\boldsymbol{A} = 0 \qquad (21.4b)$$

Gauss's law for electrostatics:

$$\oiint_S \boldsymbol{D}\cdot d\boldsymbol{A} = \sum_i q_i \qquad (21.4c)$$

Faraday's law:

$$\oint_C \boldsymbol{E}\cdot d\boldsymbol{l} = -\frac{d}{dt}\iint_S \boldsymbol{B}\cdot d\boldsymbol{A} \qquad (21.4d)$$

The four relationships (21.4a–d) between the electric and magnetic fields and their sources are known as **Maxwell's equations in integral form**.

Study Worked Example 21.1

For problems based on the material presented in this section visit up.ucc.ie/21/ and follow the link to the problems.

21.2 Plane electromagnetic waves

One of the most important consequences of Maxwell's formulation of the laws of electromagnetism is that it allows for the existence of electromagnetic radiation produced by an accelerating charge. The 'disturbance' in the case of an electromagnetic wave is both an electric field and a magnetic field. The basic laws of electromagnetism, as given by equations (21.2a–d), prove inadequate on their own to explain the existence of electromagnetic waves. While moving charges do indeed produce both electric and magnetic fields – as we know, electric fields are caused by both stationary and moving charges while moving charges also give rise to magnetic fields – the net transfer of energy through space by such fields cannot be explained in the absence of the displacement current term in Equation (21.4a).

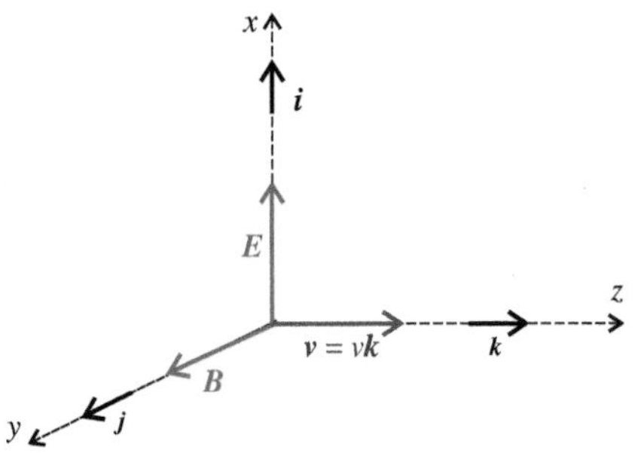

Figure 21.3. A plane electromagnetic wave propagating in the $+z$-direction with wave speed $\boldsymbol{v}$. The $\boldsymbol{E}$ and $\boldsymbol{B}$ fields are in the x- and y-directions, respectively.

A full explanation of how an accelerating charge produces electromagnetic waves is beyond the scope of this book.[1] We confine our analysis here to considering the special case where $\boldsymbol{E} \perp \boldsymbol{B}$ and both are perpendicular to the direction of propagation of the wave in an infinite, non-conducting, isotropic medium far from the source of the wave. We will show that Maxwell's equations allow the existence of electromagnetic waves in this case. At large distances from the source, as we recall from Section 13.4, the waves will be **plane waves**; in this section, and for most of this chapter, the discussion will be confined to plane waves but many of the results derived can be shown to hold more generally.

If we take the direction of propagation of the wave to be the z-direction, we can write $\boldsymbol{E} = E(z, t)\boldsymbol{i}$ and $\boldsymbol{B} = B(z, t)\boldsymbol{j}$ and the wave velocity can be written as $\boldsymbol{v} = v\boldsymbol{k}$, where the unit vectors $\boldsymbol{i}, \boldsymbol{j}$ and $\boldsymbol{k}$ are as indicated in Figure 21.3.

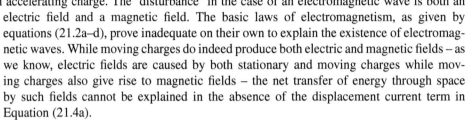

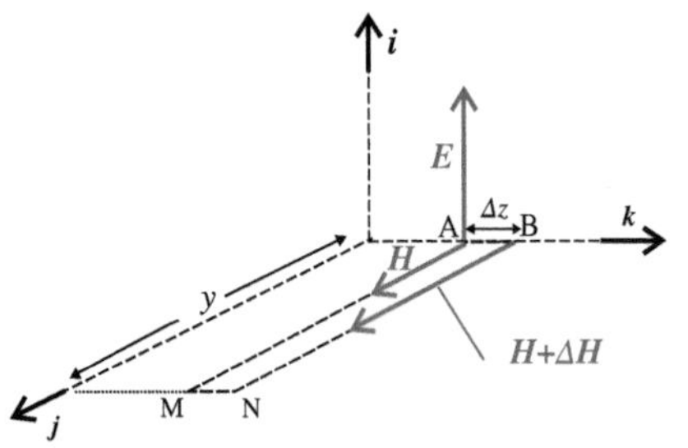

Figure 21.4. Application of Equation (21.4a) to the closed path AMNBA in the y-z plane.

Since, for now, we are interested in the behaviour of the wave only in regions of non-conducting media where there are no sources (that is $I = 0$ and $q_i = 0$ for all i), the Maxwell equations (21.4a) and (21.4d) can be written, respectively, as

$$\oint_C \boldsymbol{H} \cdot d\boldsymbol{l} = \frac{d}{dt} \iint_S \boldsymbol{D} \cdot d\boldsymbol{A} \quad \text{and} \quad \oint_C \boldsymbol{E} \cdot d\boldsymbol{l} = -\frac{d}{dt} \iint_S \boldsymbol{B} \cdot d\boldsymbol{A}$$

We apply the first of these equations to the particular closed path C (AMNBA) indicated in Figure 21.4. Let the magnitudes of the magnetic field strength at z and $z + \Delta z$ be H and $H + \Delta H$, respectively. Taking the integral over the path in the direction indicated we obtain

$$\oint_C \boldsymbol{H} \cdot d\boldsymbol{l} = \int_{AM} \boldsymbol{H} \cdot d\boldsymbol{l} + \int_{MN} \boldsymbol{H} \cdot d\boldsymbol{l} + \int_{NB} \boldsymbol{H} \cdot d\boldsymbol{l} + \int_{BA} \boldsymbol{H} \cdot d\boldsymbol{l} = Hy + 0 - (H + \Delta H)y + 0 = -(\Delta H)y$$

Now $\dfrac{d}{dt} \iint_S \boldsymbol{D} \cdot d\boldsymbol{A} = \dfrac{d(DA)}{dt}$ and, since the path C is fixed and A is constant, we can write

$$\frac{d}{dt} \iint_S \boldsymbol{D} \cdot d\boldsymbol{A} = A\frac{\partial D}{\partial t} = (\Delta z)y\frac{\partial D}{\partial t}$$

where, in this case, we have used the partial derivative notation to express differentiation with respect to time while keeping x, y and z constant (we do this because, in what follows, we are also interested in spatial variation of the fields). Thus Equation (21.4a) gives

$$-(\Delta H) = (\Delta z)\frac{\partial D}{\partial t}$$

$$\Rightarrow \boxed{\frac{\partial H}{\partial z} = -\frac{\partial D}{\partial t}} \qquad (21.5)$$

[1]The interested reader might consult Alonso and Finn (1992) pp 790–7 or Grant and Phillips (1990) Chapter 13

Let us now apply the equation $\oint_C \boldsymbol{E} \cdot d\boldsymbol{l} = -\frac{d}{dt}\iint_S \boldsymbol{B} \cdot d\boldsymbol{A}$ to a different path C (ABPQA) shown in Figure 21.5. Let the magnitudes of the electric field strengths at z and $z + \Delta z$ be E and $E + \Delta E$, respectively. Taking the integral over the path in the direction indicated we obtain

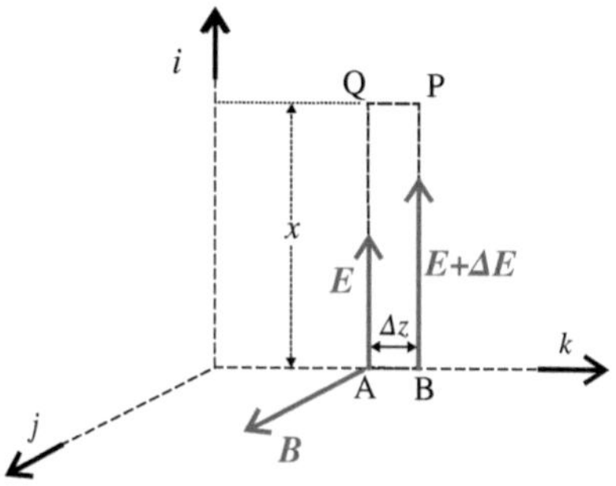

Figure 21.5. Application of Equation (21.4b) to the closed path ABPQA in the x-z plane.

$$\oint_C \boldsymbol{E} \cdot d\boldsymbol{l} = \int_{AB} \boldsymbol{E} \cdot d\boldsymbol{l} + \int_{BP} \boldsymbol{E} \cdot d\boldsymbol{l} + \int_{PQ} \boldsymbol{E} \cdot d\boldsymbol{l} + \int_{QA} \boldsymbol{E} \cdot d\boldsymbol{l} = 0 + (E + \Delta E)x + 0 + (-E)x = (\Delta E)x$$

and since $\frac{d}{dt}\iint_S \boldsymbol{B} \cdot d\boldsymbol{A} = \frac{\partial(BA)}{\partial t} = A\frac{\partial B}{\partial t} = (\Delta z)x\frac{\partial B}{\partial t}$, Equation (21.4d) gives

$$(\Delta E) = -(\Delta z)\frac{\partial B}{\partial t}$$

$$\Rightarrow \boxed{\frac{\partial E}{\partial z} = -\frac{\partial B}{\partial t}} \tag{21.6}$$

Equations (21.5) and (21.6), which link the time and spatial derivatives of the fields at any point, are **Maxwell's equations in differential form** applied to this special *one-dimensional case*. The three-dimensional differential form of Maxwell's equations is derived on the *Understanding Physics* website at up.ucc.ie/21/11/.

If we now differentiate Equation (21.5) with respect to t and Equation (21.6) with respect to z we get

$$\frac{\partial}{\partial t}\frac{\partial H}{\partial z} = -\frac{\partial^2 D}{\partial t^2} = -\varepsilon\frac{\partial^2 E}{\partial t^2} \quad \text{and} \quad \frac{\partial^2 E}{\partial z^2} = \frac{\partial}{\partial z}\frac{\partial E}{\partial z} = \frac{\partial}{\partial z}\left(-\frac{\partial B}{\partial t}\right) = -\mu\frac{\partial}{\partial z}\frac{\partial H}{\partial t}$$

Since it does not matter in which order the differentiation is performed (Appendix A.6.3),

$$\frac{\partial}{\partial t}\frac{\partial H}{\partial z} = \frac{\partial}{\partial z}\frac{\partial H}{\partial t}$$

and thus

$$\boxed{\frac{\partial^2 E}{\partial z^2} = \mu\varepsilon\frac{\partial^2 E}{\partial t^2}} \tag{21.7}$$

which we immediately recognise as the one-dimensional *wave equation* (13.27). It is easy to show that the magnetic field strength H, and hence B, also obeys the wave equation, that is

$$\boxed{\frac{\partial^2 H}{\partial z^2} = \mu\varepsilon\frac{\partial^2 H}{\partial t^2}} \tag{21.8}$$

Thus, we see that Maxwell's equation allow a wave behaviour of the electromagnetic fields. Note that, if the displacement current term in Equation (21.4a) is set to zero, the term on the right-hand side of Equation (21.5) becomes zero. In this case Equation (21.7) takes the form $\frac{\partial^2 E}{\partial z^2} = 0$ which does not admit wave-like solutions.

Comparing Equations (21.7) or (21.8) with Equation (13.27) shows that the wave velocity is given by $\boldsymbol{v} = \frac{1}{\sqrt{\mu\varepsilon}}\boldsymbol{k}$, where μ and ε are, respectively, the permeability and permittivity of the medium through which the wave is travelling.

We know (Section 13.10) that a solution to Equation (21.7) is $E(z, t) = E_0 \sin(kz - \omega t)$, where E_0 is the maximum value (also called the 'peak value' or the 'amplitude') of the electric field strength. Similarly, a solution to Equation (21.8) is $H(z, t) = H_0 \sin(kz - \omega t + \phi)$, where H_0 is the maximum value of the magnetic field strength and ϕ allows for a possible phase difference between the electric and magnetic fields. Thus, the electric and magnetic field strengths at position z and time t are given, respectively, by

$$\boldsymbol{E}(z, t) = E(z, t)\boldsymbol{i} = E_0 \sin(kz - \omega t)\boldsymbol{i} \tag{21.9a}$$

$$\boldsymbol{B}(z, t) = B(z, t)\boldsymbol{j} = B_0 \sin(kz - \omega t + \phi)\boldsymbol{j} \tag{21.9b}$$

The relationship between E and B in an electromagnetic wave can be obtained by substituting the expressions (21.9a) and (21.9b) into the differential Equation (21.6). Thus, we get

$$kE_0 \cos(kz - \omega t) = -(-\omega)B_0 \cos(kz - \omega t + \phi)$$

or

$$E_0 \cos(kz - \omega t) = \frac{\omega}{k}B_0 \cos(kz - \omega t + \phi) = vB_0 \cos(kz - \omega t + \phi).$$

where we have used Equation (13.5) to replace $\frac{\omega}{k}$ by v. This result holds for all z and t and hence, in particular, will hold for $kz - \omega t = \frac{\pi}{2}$ in which case

$$E_0 \cos\frac{\pi}{2} = vB_0 \cos\left(\frac{\pi}{2} + \phi\right) \quad \Rightarrow \quad 0 = -vB_0 \sin\phi \quad \Rightarrow \quad \phi = 0 \ (\text{or } 2\pi, 4\pi, 6\pi \ \ldots \text{ etc}).$$

This shows that the electric and magnetic fields of an electromagnetic wave are **in phase** and thus that $E_0\cos(kz - \omega t) = vB_0\cos(kz - \omega t)$ and hence

$$\boxed{E_0 = vB_0}$$

Hence equations (21.9a) and (21.9b) can also be written as

$$\boldsymbol{E}(z,t) = E(z,t)\boldsymbol{i} = vB_0 \sin(kz - \omega t)\boldsymbol{i} = \boldsymbol{B} \times \boldsymbol{v} \tag{21.10a}$$

$$\boldsymbol{B}(z,t) = B(z,t)\boldsymbol{j} = B_0 \sin(kz - \omega t)\boldsymbol{j} = \frac{1}{v}\boldsymbol{k} \times \boldsymbol{E} \tag{21.10b}$$

Note also that $\dfrac{\langle E^2 \rangle}{\langle B^2 \rangle} = v^2$.

If the source of waves is a simple single frequency generator, then the electromagnetic radiation will be monochromatic (ω fixed) as in the preceding discussion. More complicated sources will produce a spectrum of wavelengths and, as we know from Section 13.15, this can be considered as a superposition of monochromatic sinusoidal signals. In general the permittivity of a medium depends on frequency, that is $\varepsilon = \varepsilon(\omega)$ and hence the wave speed is frequency dependent, $v = v(\omega)$, and the wavelength λ also depends on the electromagnetic properties of the medium. Thus the wave exhibits the phenomenon of **dispersion** – different frequency components travel at different speeds as discussed in Section 13.15. In vacuum, however, all electromagnetic waves have the same speed

$$v \rightarrow c = \frac{1}{\sqrt{\mu_0 \varepsilon_0}} = 299\,792\,458 \text{ m s}^{-1}$$

and, as we know from Chapter 9, this value is the same in all reference frames.

For problems based on the material presented in this section visit up.ucc.ie/21/ and follow the link to the problems.

21.3 Experimental observation of electromagnetic radiation

A plane electromagnetic wave, whose source is an accelerating point electric charge, has the ***E***-field at any point P in space confined to the plane which is defined by the direction of the acceleration ***a*** of the charge and the line of sight from P to the charge; the wave propagates along the latter line (Figure 21.6). The ***B***-field is perpendicular to ***E***, both vectors lying in a plane perpendicular to the direction of propagation of the wave, that is electromagnetic waves are transverse. This last observation is not confined to plane waves – electromagnetic waves in infinite non-conducting isotropic media are always transverse with ***E*** perpendicular to ***B***.

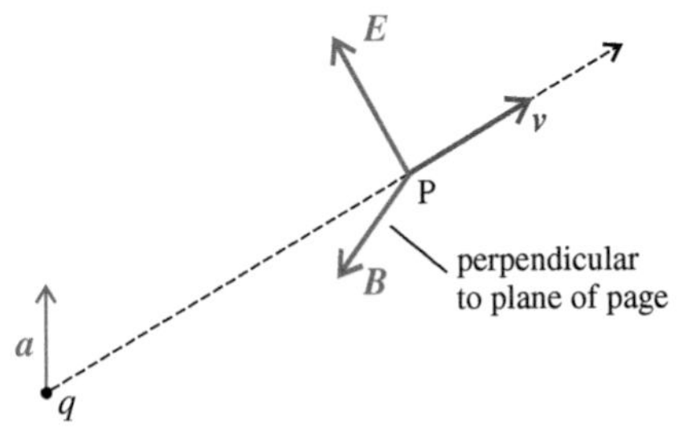

Figure 21.6. An electromagnetic wave emitted by an accelerating electric charge has its ***E***-field in the plane defined by the acceleration ***a*** and the wave velocity ***v***. The ***B***-field is perpendicular to that plane, as indicated.

One way of providing accelerating charges is to drive a time-varying electric current through a straight conducting wire. Consider the experimental arrangement illustrated in Figure 21.7. The a.c. generator could be a rotating coil in a magnetic field, like that described in Section 20.7, provided that the coil rotates very rapidly but, more usually, it is an *L-C* oscillator of the kind mentioned at the end of Section 20.5 or an oscillating 'piezoelectric' crystal of the type used in watches and clocks. The oscillator is connected to adjacent ends of two straight sections of wire, as indicated in Figure 21.7 to form what is called a '*centre-fed linear antenna*'. For efficient wave propagation, the length of the antenna must be chosen to match the wavelength of the electromagnetic radiation produced by the oscillator so that lower frequency sources require longer antennae.

An identical antenna (or 'aerial') is placed some distance away and is aligned so that it is parallel to the first antenna, as shown. If this second antenna is connected to a sensitive voltmeter which can detect rapidly varying signals, the voltmeter will detect a sinusoidally varying voltage of the same frequency as the oscillator which is connected to the first antenna. This experimental observation confirms that energy has been transmitted from the first antenna to the second.

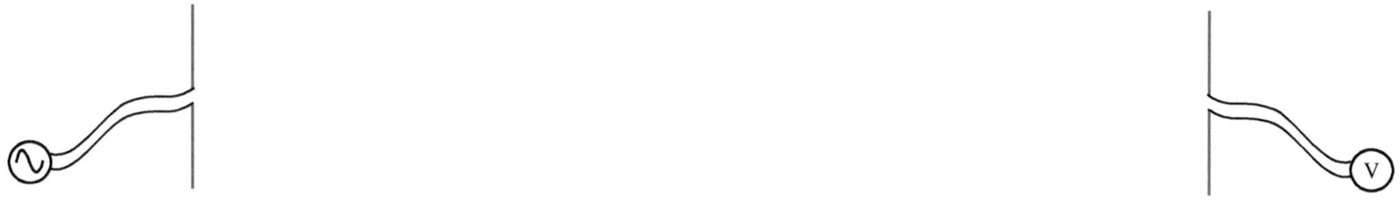

Figure 21.7. A sinusoidal voltage source connected to a centre-fed linear antenna. Electromagnetic waves emitted from the antenna are detected by a similar antenna, aligned parallel to the transmitting antenna, which is connected to a sensitive voltage detector.

The first direct experimental test of Maxwell's theory was carried out by Heinrich Hertz (1857–1894) in 1887 (see, for example, Fishbane, Gasiorowicz, and Thornton (1993) for details). It is also interesting to note that the first indications of the photoelectric effect (Section 14.5) showed up in these experiments.

The transmission of energy in this way is not confined to sinusoidal sources; no such effect is observed when d.c. is fed to the antenna, but any time-varying signal will be transmitted, although it is not practicable to transmit or detect electromagnetic waves at very low frequencies (<1 kHz).

What is happening in the experiment which we have described is that the oscillator is driving charge carriers back and forth along the transmitting antenna in a sinusoidal manner. These execute simple harmonic motion and so are accelerating in a sinusoidal fashion, the effect of which is to generate a sinusoidally varying electric field around this antenna. The electric field is not isotropic – the electric field strength is maximum in the plane perpendicular to the antenna and through its centre; at all points in this plane the $\boldsymbol{E}$-vector is directed parallel to the line of the antenna. The electric field strength falls off with increasing angle with respect to this plane and is observed to be zero at all points in line with the antenna; Figure 21.8 shows a typical radiation pattern from a linear antenna (this is an example of what is called a 'polar plot' — transmitted energy is plotted as a function of the angle θ that the line of sight makes with the antenna).

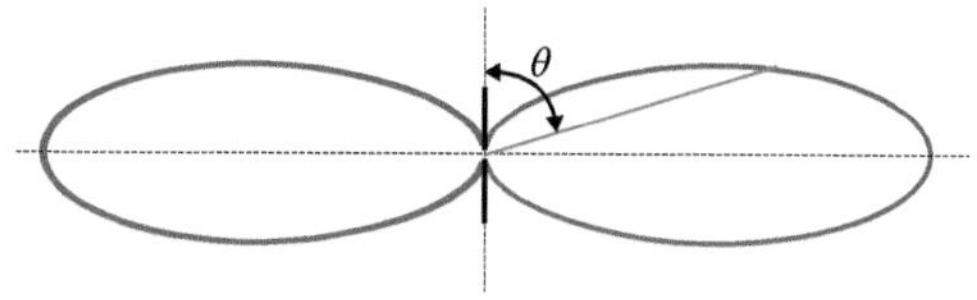

Figure 21.8. The radiation pattern from a linear dipole antenna. The radiation is isotropic in the plane perpendicular to the page. Note that no radiation is transmitted along the line of the antenna.

The charge carriers in the receiving antenna, on the right in Figure 21.7, experience a force due to the sinusoidally varying electric field of the radiation and thus are driven up and down the antenna with simple harmonic motion. This motion of the charges generates a sinusoidal potential difference across the input of the voltmeter. This is the basic principle underlying the transmission and reception of radio and TV signals. The production of electromagnetic waves at higher frequencies requires, as we shall see in the next section, very different types of 'oscillators' and 'antennas', but one important feature is common – accelerating charges are always involved.

The use of a different type of receiving antenna, namely a ferrite rod like that used as an aerial in modern radios, indicates that electromagnetic radiation exhibits a magnetic field as well as the electric field. In this way it can be confirmed that the magnetic flux density vector $\boldsymbol{B}$ is perpendicular to $\boldsymbol{E}$ and, furthermore, that both $\boldsymbol{E}$ and $\boldsymbol{B}$ are perpendicular to the direction of propagation of the wave as indicated in Figure 21.9, confirming that electromagnetic waves are *transverse*.

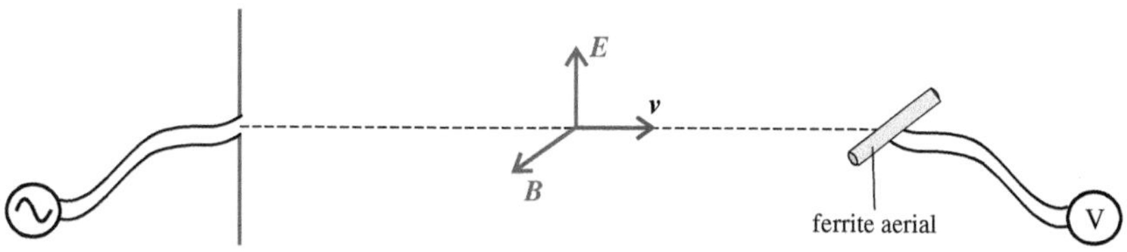

Figure 21.9. The receiving antenna in Figure 21.7 may be replaced by a ferrite rod which detects the $\boldsymbol{B}$-field of the electromagnetic wave.

21.4 The electromagnetic spectrum

As was suggested above, radio and TV signals are electromagnetic waves which are generated by driving charges back and forth in antennas of various shapes and sizes. It is found that electromagnetic waves are generated most efficiently when the length of the transmitting antenna is of the same order of magnitude as the wavelength of the radiation produced. At higher frequencies, the wavelength will be smaller and the lengths of the antennas needed for efficient transmission and detection will be correspondingly smaller. By exploiting a wide variety of sources and sensors, electromagnetic radiation can be produced and detected across a spectrum which ranges over at least thirty decades in frequency. Figure 21.10 shows the electromagnetic spectrum divided into frequency and wavelength ranges on a logarithmic scale. Despite the continuous nature of the spectrum, it is convenient to consider it as divided into the following frequency ranges, distinguished by the different effects the radiation in each range produces when it interacts with matter.

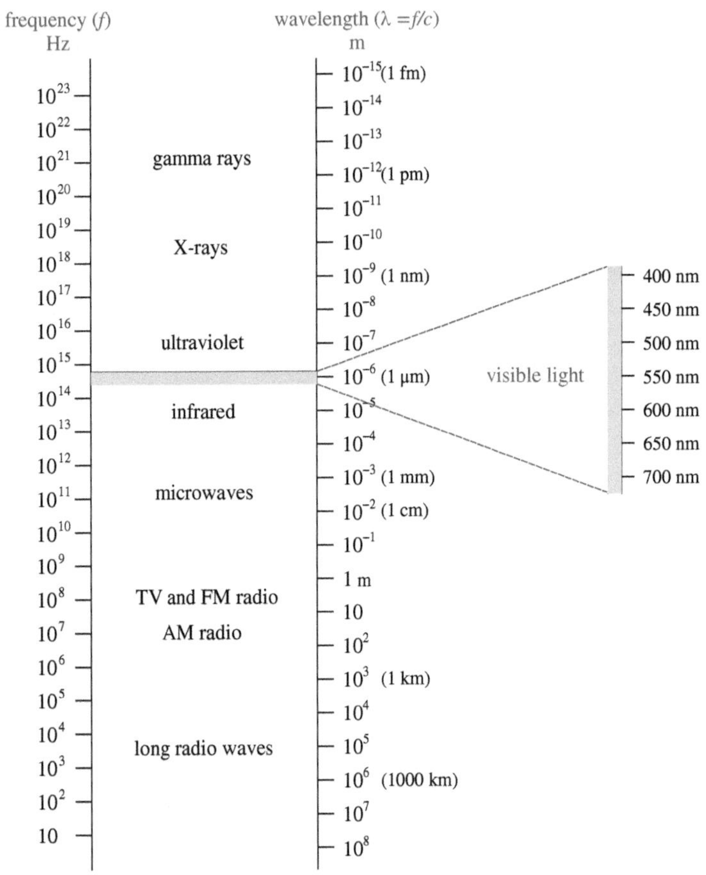

Figure 21.10. The electromagnetic spectrum.

(*a*) ***Radio-frequency waves***: At frequencies from a few kHz up to about 1 GHz (wavelengths in vacuum from many kilometres to less than a metre), antennas of various types can be used. This range covers VLF (very low frequency) navigation signals, long-, medium- and short-wave radio, VHF and UHF radio and TV, mobile radio, police and air traffic control communications. VLF radio signals require antennas of kilometre dimensions which are as long as is practicable; UHF signals, on the other hand, can be received by aerials which are fractions of a metre in length.

(*b*) ***Microwaves***: Above 1 GHz and up to over 100 GHz (wavelengths from less than 1 m to 1 mm), electromagnetic waves may be generated by electronic devices based on vacuum tube technology (such as magnetrons or klystrons) or their semiconductor based equivalents. The transmitting and receiving antennas usually include metal horns or concave spherical surfaces ('dishes'); these are used to focus the incoming signal on to the receiver or to transmit the outgoing signal as a parallel beam. Electromagnetic waves in this region of the spectrum are called microwaves and, first used in radar, are now also used for satellite TV and telecommunications, terrestrial TV and telephone distribution. Microwaves also have important applications in experimental studies of atomic and molecular systems, as well as their familiar use in domestic cooking appliances.

(*c*) ***Radiant heat/Infra-red radiation***: Electromagnetic waves are detected as radiant heat or infra-red radiation at frequencies between about 10^{12} Hz and 4×10^{14} Hz (1 mm to nearly 800 nm wavelength). This radiation is produced by 'antennae' of molecular dimensions, the radiation in this case coming from accelerating charges such as oscillating electrons, molecular vibrations and/or rotations, (such as arise in hot solids, for example), etc. As we know from Chapter 14, quantum physics must be applied to such systems; the radiation is emitted when the molecules make transitions from higher to lower energy states.

(*d*) ***Visible radiation (light)***: The very narrow range of frequencies to which the human eye is sensitive extends from about 4×10^{14} Hz to about 8×10^{-14} Hz (from just under 800 nm to around 400 nm wavelength). This radiation is produced by bodies which are sufficiently hot that the maxima of their emission spectra appear at these frequencies (for example, filament lamps), from transitions from excited atomic or ionic states (for example, ionised gases) or by charged particles which are moving at great speeds on curved paths (the so-called 'synchrotron radiation').

(*e*) ***Ultra-violet radiation***: The de-excitation of very highly ionised atoms produces ultra-violet radiation which corresponds to frequencies in the range 8×10^{14} Hz to around 10^{17} Hz (400 nm to around 1 nm wavelength). The energies involved here are very close to that required for atomic ionisation or molecular dissociation (for $f = 8 \times 10^{14}$ Hz, $E = hf \approx 5.3 \times 10^{-19}$ J ≈ 3 eV) which accounts for the chemical effects caused by ultra-violet radiation.

(*f*) ***X-rays***: X-rays are produced by the excitation of the innermost electrons in atoms or by the linear deceleration of electrons as they pass through matter ('bremsstrahlung' — Section 14.6). The frequencies involved range from 10^{17} Hz to 10^{19} Hz (10^{-19} m to 10^{-11} m).

(g) ***γ-rays***: This very short wavelength radiation arises from excited nuclei which are produced, for example, in radioactive decay processes (Section 27.5) and in collisions between particles at high energies. Gamma-ray frequencies range between 10^{18} Hz, overlapping the X-ray range, to over 10^{33} Hz (wavelengths of 10^{-10} m to 10^{-25} m); the very highest frequencies have been observed only in cosmic radiation.

Despite the different effects produced when electromagnetic waves of different frequencies interact with matter, the evidence clearly shows that electromagnetic radiation constitutes a single phenomenon.

Spectroscopy

The study of the range of frequencies emitted by a source of electromagnetic radiation is called **spectroscopy**. The particular experimental technique used to split the radiation emitted by a source into its different component frequencies (wavelengths) depends on the frequency range involved. In the case of visible light, a diffraction grating (to be described in Section 22.5) can separate (resolve) frequencies which are very close together and enable the spectrum to be analysed accurately and in great detail. Spectroscopy can be used to determine the characteristic wavelengths emitted by atoms and molecules which may then be used to identify the constituents of a material or to study the structure of the atoms and molecules involved, as will be described in Chapter 24.

Most sources of electromagnetic radiation emit a range of frequencies but the appearance of the spectrum can be quite different. A *continuous spectrum* of electromagnetic frequencies (Figure 21.11(a)) is emitted by sources such as hot solids (for example, a filament lamp), decelerating charges (for example, an X-ray tube) and rotating charges (for example, radiogalaxies).

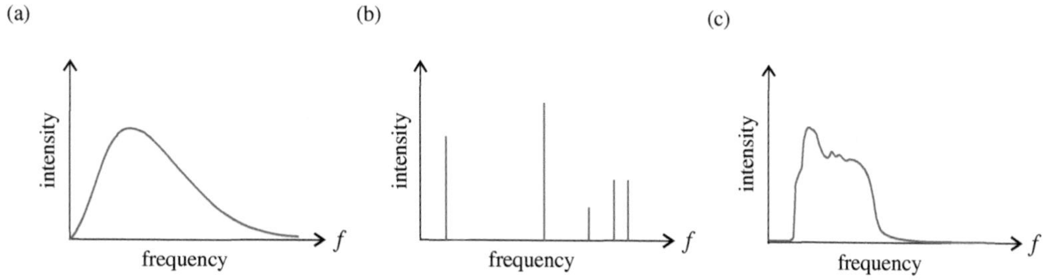

Figure 21.11. Examples of (a) a continuous spectrum, (b) a line spectrum and (c) a band structure spectrum.

In a gas discharge tube, on the other hand, atoms and ions are excited through inelastic collisions in which their kinetic energy is used to excite electrons in the atoms to higher energy states which de-excite and release electromagnetic radiation. This process is considered in more detail in Section 24.7. In contrast to Figure 21.11(a), the spectrum of the electromagnetic radiation emitted in this case is *discrete* — it occurs only at certain frequencies which are characteristic of the particular atoms involved. Thus a *line spectrum* is observed, such as that illustrated in Figure 21.11(b).

Very many sources of electromagnetic radiation are more complicated. Not only may they emit a combination of continuous and discrete frequencies but may also, particularly in the case of certain molecules and solids, show band structure (Figure 21.11(c)), that is continuous over a limited range of frequencies.

21.5 Polarisation of electromagnetic waves

As we saw in Section 21.3, when electromagnetic waves are emitted from a simple linear antenna, the electric field strength vector $\boldsymbol{E}$ at any point in space is confined to a plane which is defined by the line of the antenna and the line of sight from the antenna to the point in question The wave propagates along the latter line with the $\boldsymbol{B}$-field perpendicular to $\boldsymbol{E}$, each vector lying in a plane perpendicular to the direction of propagation of the wave. Such a wave is said to be **linearly polarised**; by convention the direction of the $\boldsymbol{E}$-field is taken as the direction of polarisation.

Suppose that, instead of a single antenna, there are two centre-fed linear antennas at right angles to each other, as indicated in Figure 21.12, each driven by an ac source of identical frequency but not necessarily in phase with each other. Let us consider the nature of the electric field strength at a point P on the line perpendicular to the plane of the two antennas. In this case we choose the direction of propagation of the wave as the $\boldsymbol{k}$-direction and we let $\boldsymbol{i}$ be a unit vector parallel to one antenna and $\boldsymbol{j}$ be a unit vector parallel to the other. The net electric field strength at P is the vector sum of the contributions from the individual antennas, namely

$$E = E_1 \sin(kz - \omega t)\boldsymbol{i} + E_2 \sin(kz - \omega t + \phi)\boldsymbol{j} \tag{21.11}$$

where ϕ is the phase difference between the two sources. The behaviour of the electromagnetic field at the point P depends on the value of this phase difference.

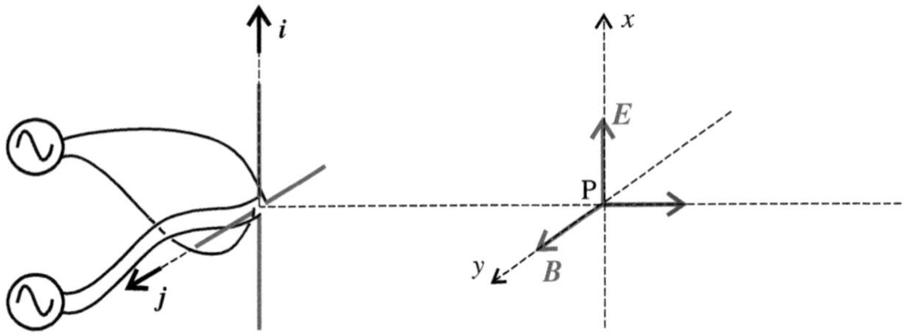

Figure 21.12. Two centre-fed linear antennas at right angles to each other. Each antenna is driven by a sinusoidal source of the same frequency but not necessarily in phase.

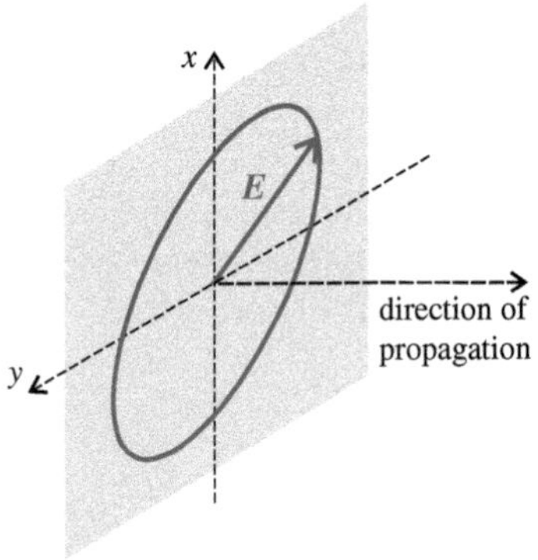

Figure 21.13. Elliptically polarised electromagnetic wave.

Let (x,y) be the coordinates of the tip of the $\boldsymbol{E}$-vector in the x—y plane as indicated in Figure 21.13, that is $x = E_1 \sin(kz - \omega t)$ and $y = E_2 \sin(kz - \omega t + \phi)$. For simplicity we will let $E_1 = a$, $E_2 = b$ and $kz - \omega t = \theta$ and the right hand side of (21.11) becomes $x\boldsymbol{i} + y\boldsymbol{j}$, where

(i) $x = a \sin\theta$ and
(ii) $y = b\sin(\theta + \phi) = b \sin\theta \cos\phi + b\cos\theta \sin\phi$ (Appendix A.3.5(viii)).

We can eliminate θ from (i) and (ii) as follows.

$$\text{(i)} \rightarrow \quad \sin\theta = \frac{x}{a} \quad \text{and} \quad \cos\theta = \frac{\sqrt{a^2 - x^2}}{a} = \sqrt{1 - \frac{x^2}{a^2}}$$

$$\text{(ii)} \rightarrow \quad y - b\sin\theta\cos\phi = b\cos\theta\sin\phi \quad \rightarrow \quad y - b\frac{x}{a}\cos\phi = b\sin\phi\sqrt{1 - \frac{x^2}{a^2}}$$

Squaring both sides we can write

$$y^2 - 2yb\frac{x}{a}\cos\phi + b^2\frac{x^2}{a^2}\cos^2\phi = b^2\sin^2\phi - b^2\frac{x^2}{a^2}\sin^2\phi$$

Dividing across by b^2 and noting that $\sin^2\phi + \cos^2\phi = 1$, we obtain

$$\frac{y^2}{b^2} - 2\frac{y}{b}\frac{x}{a}\cos\phi + \frac{x^2}{a^2} = \sin^2\phi \tag{21.12}$$

Since the phase difference between the two sources must be constant (that is, the sources are coherent $\rightarrow \phi =$ constant), Equation (21.12) represents an ellipse in the x—y plane such that its major axis makes an angle $\tan^{-1}\frac{b}{a} = \tan^{-1}\frac{E_2}{E_1}$ with the x-axis (Appendix A.10.2(ii)). The net electric field is never zero, in general; it varies between a maximum value along the major axis and a minimum value along the minor axis of the ellipse, the orientation of the major and minor axes relative to the x- and y-axes being determined by the values of $a\ (= E_1)$, $b\ (= E_2)$ and ϕ.

The state of polarisation of an electromagnetic wave is described by the locus of the tip of the $\boldsymbol{E}$-vector. In the situation which we have just considered, the $\boldsymbol{E}$-vector traces out an ellipse in the plane perpendicular to the direction of propagation of the wave (Figure 21.13), in which case the wave is said to be **elliptically polarised**. We now consider a number of special cases.

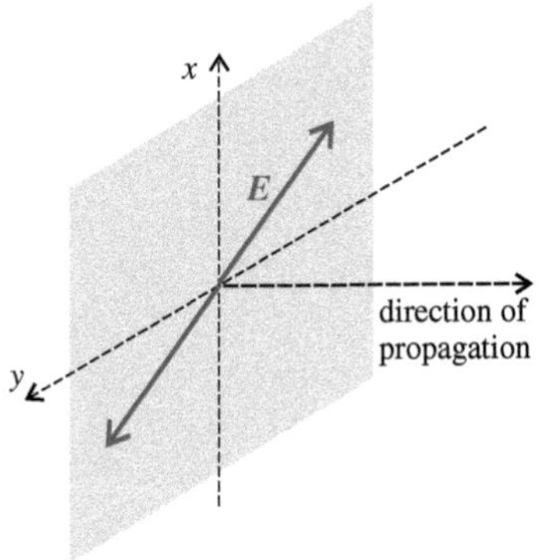

Figure 21.14. Linearly polarised electromagnetic wave.

Case 1: $\phi = 0$: Linearly polarised wave
Putting $\cos\phi = 1$ and $\sin\phi = 0$ in Equation (21.12) we obtain

$$\frac{y^2}{b^2} - 2\frac{y}{b}\frac{x}{a} + \frac{x^2}{a^2} = 0 \quad \rightarrow \quad \left(\frac{y}{b} - \frac{x}{a}\right)^2 = 0 \quad \rightarrow \quad y = \frac{E_2}{E_1}x$$

which shows that the tip of the $\boldsymbol{E}$-vector is confined to a straight line in the x—y plane. Thus if there is no phase difference between the two sources, the net electric field strength will be **linearly polarised** such that the direction of polarisation makes an constant angle of $\tan^{-1}\frac{E_2}{E_1}$ with the x-axis, as indicated in Figure 21.14.

Case 2: $\phi = \pm\frac{\pi}{2}$ and $E_1 = E_2$*: Circular polarisation*

If the phase difference between the sources is $\pm\frac{\pi}{2}$, then $\cos\phi = 0$ and $\sin\phi = \pm 1$ and Equation (21.12) becomes $\frac{y^2}{b^2} + \frac{x^2}{a^2} = 1$. If, in addition, $E_1 = E_2 = E$ (say), that is $a = b = E$, the equation takes the form $y^2 + x^2 = E^2$ which we recognise as the equation of a circle (Appendix A.10.2(i)). In this case Equation (21.11) takes the form

$$\boldsymbol{E} = E\sin(kz - \omega t)\boldsymbol{i} + E\sin\left(kz - \omega t \pm \frac{\pi}{2}\right)\boldsymbol{j} = E\sin(kz - \omega t)\boldsymbol{i} \pm E\cos(kz - \omega t)\boldsymbol{j}$$

so that the tip of the $\boldsymbol{E}$-vector traces out a circle in the x-y plane, rotating clockwise or counter-clockwise depending on whether the sign of the phase difference is plus or minus. If the tip of the $\boldsymbol{E}$-vector rotates clockwise when viewed looking towards the source the wave is said to be **right-circularly polarised** (Figure 21.15(a)) while if the rotation is counter-clockwise one has **left-circular** polarisation (Figure 21.15(b)).[2]

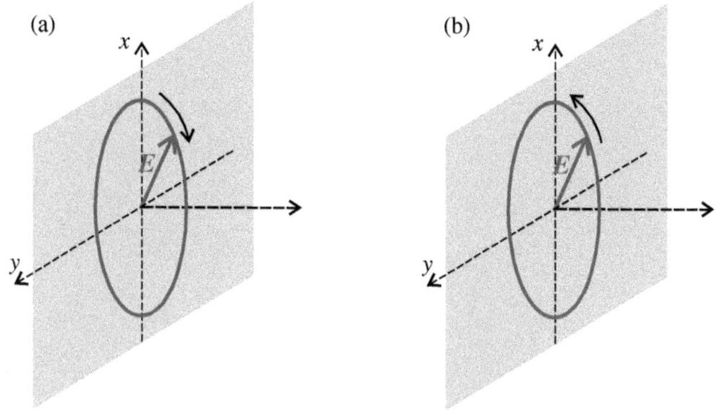

Figure 21.15. (a) Right circularly polarised electromagnetic wave. (b) Left circularly polarised electromagnetic wave.

Case 3: Unpolarised and partially polarised radiation
In many situations, the phase difference between the x- and y-components of the electric field varies in a random manner. This is the case, for example, where the source of radiation comprises a large number of independent 'antennas', for example in molecular, atomic or nuclear sources. In such cases the radiation is said to be **unpolarised**; a combination of polarised and unpolarised radiation is said to be **partially polarised**.

An unpolarised wave can be turned into a linearly polarised wave by passing it through a region in which the component of the electric field vector in a particular direction is absorbed. Unpolarised microwaves, for example, can be passed through a wire grid (Figure 21.16) which has the effect of absorbing that part of the radiation which has the $\boldsymbol{E}$-field parallel to the wires. This is because the free electrons in a wire can be driven along the wire by the force exerted by the component of the $\boldsymbol{E}$-field parallel to the wires, thereby absorbing energy from the beam by Joule heating. On the other hand, the diameter of the wires ($\sim$ mm $\ll \lambda$) is too small to allow efficient absorption of the energy associated with the component of the $\boldsymbol{E}$-field in the direction perpendicular to the wires of the grid. The transmitted beam, therefore, will be linearly polarised with the direction of polarisation at right angles to the wires of the grid, that is parallel to the line XX' in the figure. The wire grid, in this case, functions as a *polariser* and the line XX' is called the *axis* of the polariser.

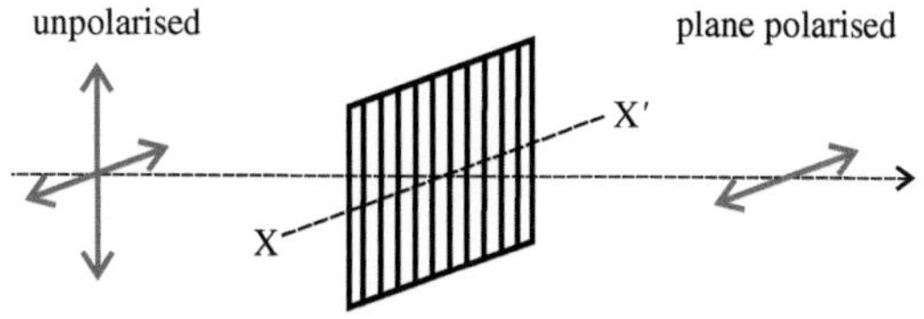

Figure 21.16. When an unpolarised microwave beam is incident on a wire grid, the component of the electric field strength parallel to the wires is absorbed.

Consider now a *linearly polarised* electromagnetic wave which is incident normally on a linear polariser in such a way that the $\boldsymbol{E}$-field makes an angle θ with the axis of the polariser, as indicated in Figure 21.17. If we choose the x-direction to be along the axis of the polariser, we can write the incident electric field strength in terms of its components, namely $|\boldsymbol{E}|\cos\theta\boldsymbol{i} + |\boldsymbol{E}|\sin\theta\boldsymbol{j}$. Since only the

[2]The convention given here for right/left circular polarisation is that adopted in optics. It should be noted, however, that the opposite convention is used in nuclear physics for the polarisation states of gamma rays.

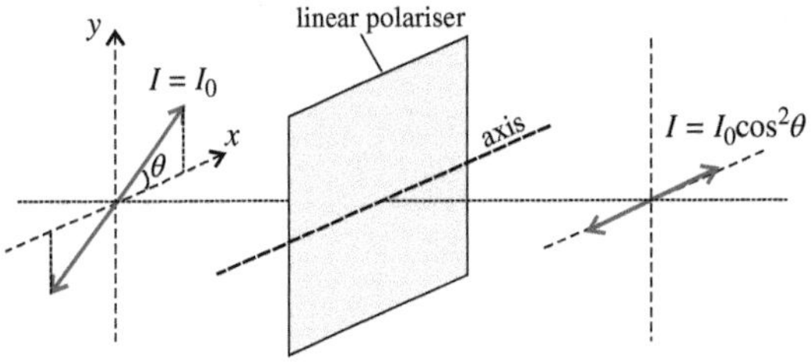

Figure 21.17. Only the component of the electromagnetic wave whose E-field is parallel to the axis is transmitted through a linear polariser.

x-component will pass through the polariser, the transmitted electric field strength will be $|\boldsymbol{E}|\cos\theta\boldsymbol{i}$. Like any wave, the intensity of an electromagnetic wave is proportional to the square of the amplitude (Section 13.3). We will see in the next section that, in the case of electromagnetic waves, the intensity is proportional to $|\boldsymbol{E}|^2$ (and $|\boldsymbol{B}|^2$) and hence the intensity of the transmitted beam is related to the incident intensity I_0 by

$$I = I_0\cos^2\theta$$

a result known as **Malus' law** (Etienne Louis Malus, 1775–1812). In particular, note that if $\theta = \dfrac{\pi}{2}$, the transmitted intensity is zero.

When an unpolarised wave passes through a linear polariser, it emerges linearly polarised.

In Chapter 22, we will encounter further examples of polarisation of electromagnetic radiation in the context of light waves.

Study Worked Example 21.2

For problems based on the material presented in this section visit up.ucc.ie/21/ *and follow the link to the problems.*

21.6 Energy, momentum and angular momentum in electromagnetic waves

Energy in an electromagnetic wave

The energy in an electromagnetic wave resides in the electric and magnetic fields. The idea of energy being stored in such fields is not new to us if we recall that the energy in a charged capacitor and in a current carrying inductor is stored in an electric field in the first case and in a magnetic field in the second. Indeed we have already investigated how stored energy depends on the strength of the field in each case; in the case of an electric field the *energy density* is given by $w^{\mathrm{E}} = \frac{1}{2}\varepsilon E^2$ (Section 17.10) while that for a magnetic field is $w^{\mathrm{M}} = \frac{1}{2}\mu H^2$ (Section 20.3). When electric and magnetic fields are present simultaneously, such as in an electromagnetic wave, the total instantaneous energy density is

$$w = w^{\mathrm{E}} + w^{\mathrm{M}} = \frac{1}{2}\varepsilon E^2 + \frac{1}{2}\mu H^2$$

or, using the vector form of the field strengths,

$$\boxed{w = \frac{1}{2}\boldsymbol{E}\cdot\boldsymbol{D} + \frac{1}{2}\boldsymbol{H}\cdot\boldsymbol{B} = \frac{1}{2}(\boldsymbol{E}\cdot\boldsymbol{D} + \boldsymbol{H}\cdot\boldsymbol{B})}$$

Note, also, that the energy of an electromagnetic wave in a non-conducting medium is equally divided between electric energy and magnetic energy since

$$\frac{w^{\mathrm{E}}}{w^{\mathrm{M}}} = \frac{\frac{1}{2}\varepsilon E^2}{\frac{1}{2}\mu H^2} = \frac{\mu\varepsilon E^2}{B^2} = \mu\varepsilon v^2 = 1$$

where we have used the result $E = vB$ derived in Section 21.2.

Consider now a plane electromagnetic wave which is travelling in the z-direction, as indicated in Figure 21.18. In a small time interval Δt, the energy crossing an area A perpendicular to the direction of propagation of the wave is equal to the energy within the volume element shown which has cross-sectional area A and length $v\Delta t$, where v is the speed of the wave. Thus, the energy crossing A in time Δt is given by

$$\Delta W = w[A(v\Delta t)] = wv(A\Delta t) \qquad (21.13)$$

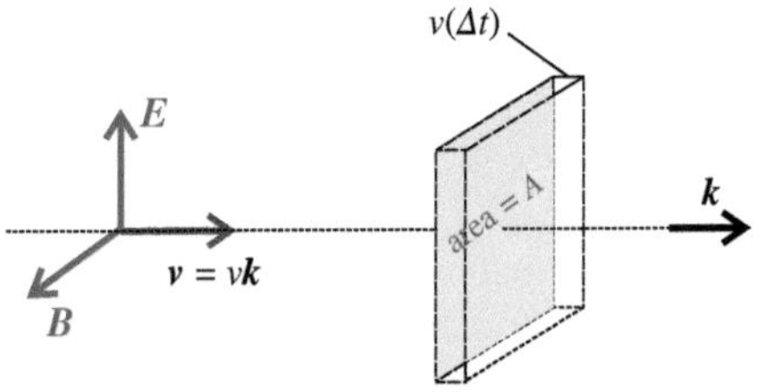

Figure 21.18. The electromagnetic energy crossing area A in time Δt is equal to the electromagnetic energy in a volume of cross-sectional area A and thickness $v(\Delta t)$, where v is the wave velocity.

and hence the energy flowing across unit area in unit time is given by

$$S = \frac{\Delta W}{A(\Delta t)} = wv = \left(\frac{1}{2}\varepsilon E^2 + \frac{1}{2}\mu H^2\right) v$$

The energy density is

$$w = \frac{S}{v}$$

and its average value is

$$\langle w \rangle = \frac{\langle S \rangle}{v}$$

For a sinusoidal wave, the average flow of energy per unit area per unit time is

$$\langle S \rangle = \left(\frac{1}{2}\varepsilon\langle E^2\rangle + \frac{1}{2}\mu\langle H_0^2\rangle\right) v = \frac{1}{2}\varepsilon v E_0^2 \langle \sin^2(kz - \omega t)\rangle + \frac{1}{2}\mu v H_0^2 \langle \sin^2(kz - \omega t)\rangle$$

Using $E_0 = vB_0 = v\mu H_0$ and $\mu\varepsilon v^2 = 1$ and, recalling $\langle \sin^2 \rangle = \frac{1}{2}$, we can write

$$\langle S \rangle = \frac{1}{4}\varepsilon v^2 \mu H_0 E_0 + \frac{1}{4}E_0 H_0 = \frac{1}{2}E_0 H_0 = E_{\text{rms}} H_{\text{rms}}$$

If we write the field strengths in vector form, this last result can be written as follows

$$\langle \boldsymbol{S} \rangle = \langle \boldsymbol{E} \times \boldsymbol{H} \rangle = E_{\text{rms}} H_{\text{rms}} \boldsymbol{k}$$

since $\boldsymbol{E}$ is perpendicular to $\boldsymbol{H}$. The vector $\boldsymbol{S} = \boldsymbol{E} \times \boldsymbol{H}$ which represents the instantaneous *energy flow per unit area per unit time* is known as the **Poynting vector**. Note that the average value of $\boldsymbol{S}$ can also be written solely in terms of the amplitude of the electric field strength, or of the magnetic field strength, as follows

$$\langle \boldsymbol{S} \rangle = \frac{1}{2}E_0 H_0 \boldsymbol{k} = \frac{1}{2}E_0 \left(\frac{E_0}{v\mu}\right) \boldsymbol{k} = \frac{E_0^2}{2\mu v}\boldsymbol{k} = \frac{1}{2}\mu v H_0^2 \boldsymbol{k} = \frac{vB_0^2}{2\mu}\boldsymbol{k} \tag{21.14}$$

That is, $\langle \boldsymbol{S} \rangle \propto vB_0^2$; note that this is the electromagnetic radiation equivalent of Equation (13.8).

In all of the above discussion it is assumed that the electromagnetic wave is propagating in a linear homogeneous isotropic medium. For an electromagnetic wave in vacuum

$$v \to c \quad \text{and} \quad \mu\varepsilon \to \mu_0\varepsilon_0$$

Study Worked Example 21.3

Momentum in an electromagnetic wave

Consider a plane electromagnetic wave which is incident normally on the surface of a conducting medium. We first discuss the effect of the electric and magnetic fields on the mobile charges within a thin slice of conductor, of cross-sectional area ab and thickness Δz, just inside the surface of the conductor, as indicated in Figure 21.19. We will assume that the frequency ω of the wave is sufficiently small that the electric field may be considered to behave as a quasistatic field, that is that the period of the wave oscillations $T = \frac{2\pi}{\omega} >> \tau$, where τ is the time between collisions of the charge carriers with the lattice ions.

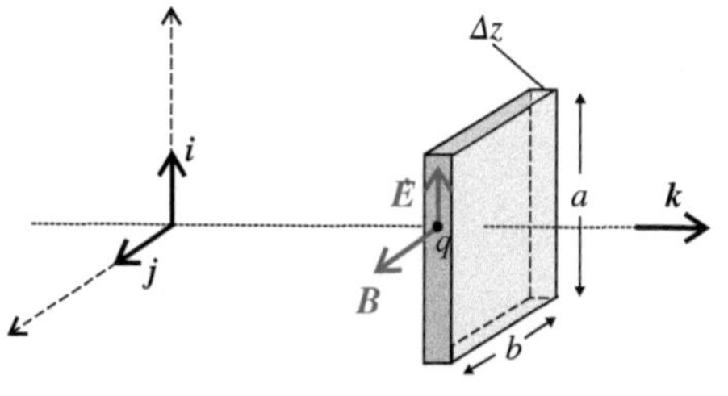

Figure 21.19. Electromagnetic wave incident normally on a conducting medium.

Over some small-time interval which is much smaller than T, the electric field can be considered constant, in which case the charge carriers will move with fixed drift velocity in the $\boldsymbol{i}$-direction. Thus, Ohm's law can be applied to the currents in the conductor, namely the current per unit area is given by Equation (16.30), $\boldsymbol{J} = \sigma\boldsymbol{E}$ where σ is the conductivity of the conducting material. The effect of the electric field, therefore, is to cause a current which is given by

$$I = b(\Delta z)|\boldsymbol{J}| = b\sigma|\boldsymbol{E}|(\Delta z)$$

to flow in the slice. This current will flow in the direction of the $\boldsymbol{E}$-vector and hence will reverse every half cycle of the wave.

This current produced by the electric field experiences a force due to the magnetic flux density $\boldsymbol{B}$ of the wave. This force can be calculated from Equation (19.2), namely

$$\Delta \boldsymbol{F} = I \Delta \boldsymbol{l} \times \boldsymbol{B}$$

and is given by

$$\boldsymbol{F} = I(a\boldsymbol{i}) \times (|\boldsymbol{B}|\boldsymbol{j}) = Ia|\boldsymbol{B}|\boldsymbol{k} = b\sigma|\boldsymbol{E}|(\Delta z)a|\boldsymbol{B}|\boldsymbol{k} = (ab)\sigma|\boldsymbol{E}||\boldsymbol{B}|(\Delta z)\boldsymbol{k}$$

Note that this force is in the $+\boldsymbol{k}$ direction, irrespective of whether $\boldsymbol{E}$ is in the $+\boldsymbol{i}$ or $-\boldsymbol{i}$ direction.

Thus the strip of thickness Δz experiences a force per unit area – or a pressure – due to the incident radiation which is given by

$$P_{\text{rad}} = \left\langle \frac{|\boldsymbol{F}|}{ab} \right\rangle = \sigma\langle|\boldsymbol{E}||\boldsymbol{B}|\rangle(\Delta z) = \sigma E_0 B_0 <\sin^2(kz - \omega t)> (\Delta z) = \frac{1}{2}\sigma E_0 B_0(\Delta z)$$

The detailed behaviour of the electric and magnetic fields as the wave penetrates deeper into the conductor is beyond the scope of this book but it is clear, at least qualitatively, that when the electromagnetic wave is absorbed it exerts a pressure on the conductor. Such a pressure is called **radiation pressure** and indicates that the electromagnetic wave carries *momentum* as well as energy.

Suppose that the energy in a plane electromagnetic wave which is incident normally on a solid surface is completely absorbed. The total amount of energy so deposited in a time Δt in a section of area A and thickness $v(\Delta t)$ is given by Equation (21.13), namely

$$\Delta W = wvA(\Delta t) = w\frac{dz}{dt}A(\Delta t) = wA(\Delta z)$$

where w is the energy density and v is the wave speed. The work done by the force ($P_{\text{rad}}A$) to move this surface through a distance Δz in the $\boldsymbol{k}$-direction is

$$\Delta W = (P_{\text{rad}}A)(\Delta z) = wA(\Delta z)$$

from which we can see that the radiation pressure is equal to the energy density

$$P_{\text{rad}} = w \quad \text{(normal incidence)}$$

Radiation pressure is usually very small for ordinary electromagnetic wave intensities. For example, for sunlight incident on some parts of the Earth's surface, the magnitude of the Poynting vector is approximately one kW m^{-2} and hence the corresponding radiation pressure is of the order of $\dfrac{1 \times 10^3}{3 \times 10^8} < 10^{-5}$ Pa. For a laser beam with a power density of 10^{12} W m^{-2}, on the other hand, the radiation pressure is about 10^4 Pa ≈ 0.1 atm.

The force $P_{\text{rad}}A$ is the rate of change of momentum over an area A of surface and hence the *rate of flow of momentum per unit area* – the momentum flux – in an electromagnetic wave may be written as

$$\boldsymbol{G} = P_{\text{rad}}\boldsymbol{k} = w\boldsymbol{k} = \frac{\boldsymbol{S}}{v} \tag{21.15}$$

where we have invoked the result $w = \dfrac{S}{v}$ derived in the previous section.

The average momentum flux is given by

$$\langle|\boldsymbol{G}|\rangle = \langle w \rangle = \frac{1}{2}\langle \boldsymbol{E}\cdot\boldsymbol{D} + \boldsymbol{H}\cdot\boldsymbol{B}\rangle$$

Note that, while the results above were derived by considering the special case of the absorption of an electromagnetic wave in an ohmic conductor, we have associated the momentum flux $\boldsymbol{G}$ with the incident plane wave in a general way.

Angular momentum in an electromagnetic wave

We saw above that, when a plane electromagnetic wave is absorbed by a conductor, it exerts a force on the conductor; indeed, since all electromagnetic waves carry momentum, such a force must be exerted when electromagnetic radiation is absorbed, irrespective of the state of polarisation of the wave. In addition to this force, elliptically polarised radiation incident normally on an absorbing surface of a body also exerts a *torque* on the body. This effect is most pronounced in circularly polarised radiation where right-circularly polarised waves exert a clockwise torque (Figure 21.20(a)) and left-circularly polarised waves an anticlockwise torque (Figure 21.20(b)), when viewed from the source of the radiation.

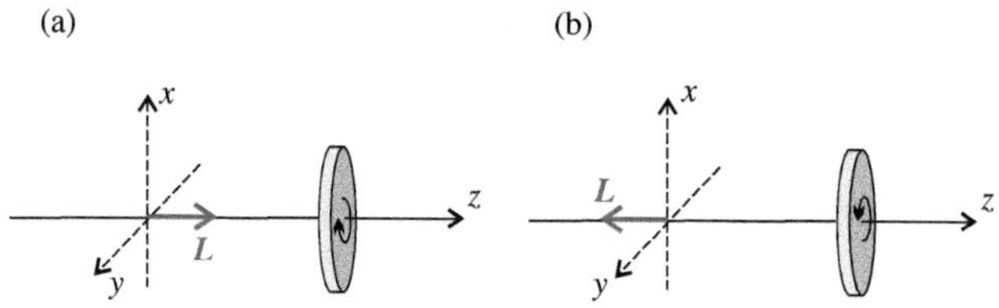

Figure 21.20. (a) A right circularly polarised electromagnetic wave exerts a torque on a conducting disc in the direction indicated. This shows that a right circularly polarised wave carries angular momentum (**L**). (b) A left circularly polarised electromagnetic wave exerts a torque on a conducting disc in the opposite direction to that exerted by a right circularly polarised wave.

Thus, electromagnetic waves are seen to carry angular momentum as well as energy and (linear) momentum. A full treatment of this effect is not possible here; we simply note in passing that electromagnetic theory predicts that the average *angular momentum flux* of right/left circularly polarised radiation is found to be $\pm\frac{\langle \boldsymbol{S}\rangle}{\omega}$, where $\langle \boldsymbol{S}\rangle$ is the average energy flux.

Study Worked Example 21.4

For problems based on the material presented in this section visit up.ucc.ie/21/ *and follow the link to the problems.*

21.7 The photon model revisited

In Section 14.7 the photon concept was introduced to explain the particle-like behaviour of electromagnetic radiation. The energy of a photon of frequency f (angular frequency $\omega = 2\pi f$) is given by $E = hf = \hbar\omega$. Using Equation (14.11), $E = cp$ for the photon, the corresponding momentum of the photon can be written as $\boldsymbol{p} = \frac{E}{c}\boldsymbol{k} = \frac{hf}{c}\boldsymbol{k} = \frac{\hbar\omega}{c}\boldsymbol{k}$, where $\boldsymbol{k}$ is a unit vector in the direction of propagation of the photon. For a beam of photons, such that the particle flux (the number of photons crossing unit area per unit time) is Φ, the corresponding energy flux is given by

$$S = E\Phi = hf\Phi \qquad (21.16)$$

and, from Equation (21.15), the momentum flux is given by

$$\boldsymbol{G} = \frac{\boldsymbol{S}}{c} = \frac{\Phi hf}{c}\boldsymbol{k}$$

Comparing these relationships for the photon model with the results derived in Section 21.6 for the energy and momentum of electromagnetic waves, the correspondence between the particle and wave pictures becomes clear. Comparing Equation (21.14), which relates the energy flux and the amplitude of the electric field strength, namely $\langle \boldsymbol{S}\rangle = \frac{E_0^2}{2\mu_0 c}\boldsymbol{k}$ in vacuum, with Equation (21.16) above we obtain $hf\Phi = \frac{E_0^2}{2\mu_0 c}$ and hence the amplitude of the electric field strength is related to the photon flux Φ by

$$E_0 = \sqrt{2\mu_0 chf\Phi}$$

As an example, let us use this result to estimate the amplitude of the electric field strength in a He-Ne laser beam ($\lambda = 633$ nm) in which the flux of photons is $10^{22}\ \mathrm{m^{-2}\ s^{-1}}$. In this case

$$E_0 = \sqrt{2\mu_0 chf\Phi} = \sqrt{\frac{2\mu_0 c^2 h\Phi}{\lambda}} = \sqrt{\frac{2h\Phi}{\varepsilon_0\lambda}} = \sqrt{\frac{2(6.63\times10^{-34})(10^{22})}{(8.85\times10^{-12})(6.33\times10^{-7})}} = 1.54\ \mathrm{kV\ m^{-1}}.$$

The corresponding peak magnetic flux density is given by $B_0 = E_0/c \approx 5\ \mu\mathrm{T}$.

Angular momentum of photons

As indicated in Section 21.6, classical electromagnetism requires that the angular momentum flux in a beam of circularly polarised electromagnetic waves is $\pm\frac{\langle S\rangle}{\omega}$, that is the energy flux divided by the angular frequency. In terms of the photon model this suggests that the photon must have angular momentum directed along the direction of propagation given by the energy divided by ω, that is

$$L_z = \pm\frac{E}{\omega} = \pm\frac{hf}{\omega} = \pm\frac{\hbar\omega}{\omega} = \pm\hbar$$

In quantum mechanical language we can represent the angular momentum state of a left circularly polarised photon by ψ_+ and that of a right circularly polarised photon by ψ_-, such that

$$L_z\psi_+ = +\hbar\psi_+ \quad \text{and} \quad L_z\psi_- = -\hbar\psi_-$$

The state of a general elliptically polarised beam can be represented by a linear combination of the two states ψ_+ and ψ_-, that is $\psi = a\psi_+ + b\psi_-$. For a linearly polarised beam $a = \pm b = \pm\frac{1}{\sqrt{2}}$.

Study Worked Example 21.5

For problems based on the material presented in this section visit up.ucc.ie/21/ *and follow the link to the problems.*

21.8 Reflection of electromagnetic waves at an interface between non-conducting media

When an electromagnetic wave is incident on a plane interface between two different non-conducting media, the boundary conditions derived in Sections 17.4 and 19.11 must be satisfied. As a consequence part of the incident energy is reflected at the interface; the incident wave is split into a reflected wave and a transmitted wave in a manner that depends on the permittivities of the two media. This phenomenon is discussed in Section 21.8 on the *Understanding Physics* website at up.ucc.ie/21/8/ for the special case of normal incidence.

21.9 Electromagnetic waves in a conducting medium

In all of our discussion on electromagnetic wave up to this point, we have assumed that the waves propagate in non-conducting media. In conducting media, however, the electric field will cause movement of free charges and hence the current density can be non-zero. This causes a phase difference between the electric and magnetic field strengths of a wave in a conducting medium with significant consequences, some of which are discussed in Section 21.9 on the *Understanding Physics* website (up.ucc.ie/21/9/).

21.10 Invariance of electromagnetism under the Lorentz transformation

Maxwell's equations in their general form are invariant under a change of reference frame (the reader will recall from Chapter 9 that this effect prompted Einstein to formulate the Special Theory of Relativity). A full proof of this result is beyond the scope of this book but it is relatively easy to reach this conclusion in the special one-dimensional case treated in this chapter. In Section 21.10 on the *Understanding Physics* website (up.ucc.ie/21/10/) we show that the one-dimensional differential forms of Maxwell's equations, equations (21.5) and (21.6), are invariant under a Lorentz transformation.

21.11 Differential form of Maxwell's equations

In Sections 16.14 and 18.11 the three-dimensional differential forms of the laws of electrostatics and magnetostatics, respectively, were presented. In Section 21.11 on the *Understanding Physics* website (up.ucc.ie/21/11/) these are combined with the conclusions in Section 21.1 to derive differential forms of Maxwell's equations.

Further reading

Further material on some of the topics covered in this chapter may be found in *Electromagnetism* by Grant and Phillips (See BIBLIOGRAPHY for book details).

WORKED EXAMPLES

Worked Example 21.1: A parallel plate capacitor has coaxial circular plates of radius R separated by a distance $d(\ll R)$ and a time-varying emf $V = V(t)$ is applied across the plates (Figure 21.21). Derive expressions for

(a) the total displacement current in the capacitor at time t and
(b) the magnetic field strength at a radial distance r from the centre of the capacitor at time t for the two cases (i) $r \le R$ and (ii) $r \ge R$.

It may be assumed that there are no edge effects, that is, that the electric field between the plates is uniform for $r < R$ and zero for $r > R$.

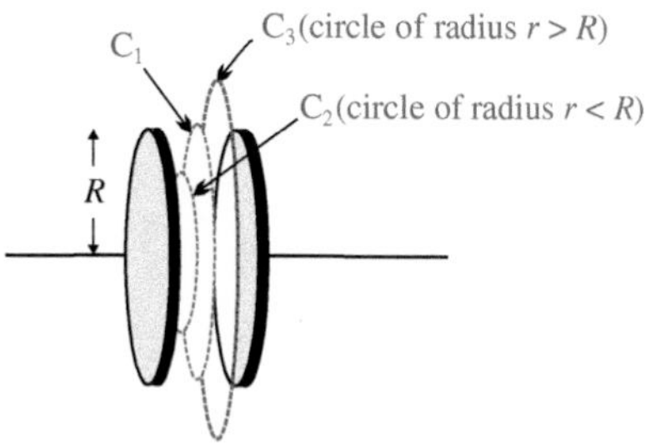

Figure 21.21. Worked Example 21.1.

(a) Displacement current $= \dfrac{d\Phi_E}{dt} = \dfrac{d}{dt}\iint_{S_1} \boldsymbol{D}\cdot d\boldsymbol{A}$, where S_1 is the surface bounded by the circular path C_1 of radius R in Figure 21.21. Since $\boldsymbol{D} = \varepsilon\boldsymbol{E}$ is uniform and $E = \dfrac{V}{d}$ we obtain

$$I_D = \frac{d}{dt}\iint_{S_1} \varepsilon\boldsymbol{E}\cdot d\boldsymbol{A} = \frac{d}{dt}\left[\left(\varepsilon\frac{V}{d}\right)\int dA\right] = \frac{d}{dt}\left[\left(\varepsilon\frac{V}{d}\right)(\pi R^2)\right] = \frac{\pi\varepsilon R^2}{d}\frac{dV}{dt}$$

(b) *Case (i)* $r < R$: Using the Ampere-Maxwell law, Equation (21.2a), $\oint_{C_2} \boldsymbol{H}\cdot d\boldsymbol{l} = \dfrac{d}{dt}\iint_{S_2} \boldsymbol{D}\cdot d\boldsymbol{A}$, where C_2 is a circular path of radius r ($< R$) and S_2 is the surface enclosed by C_2, namely the circle of radius r shown in the figure. By symmetry, the magnitude of $\boldsymbol{H}$ is the same everywhere on C_2 and thus

$$|\boldsymbol{H}|(2\pi r) = \frac{d}{dt}\left(\varepsilon\frac{V}{d}\right)(\pi r^2) \quad\Rightarrow\quad H(r,t) = \frac{\varepsilon}{2d}\frac{dV}{dt}r$$

Case (ii) $r > R$: Again $\oint_{C_3} \boldsymbol{H}\cdot d\boldsymbol{l} = \dfrac{d}{dt}\iint_{S_3} \boldsymbol{D}\cdot d\boldsymbol{A}$, where C_3 and S_3 relate to the circular path with $r > R$ shown in the figure. Since $E = 0$ for $r > R$ we obtain

$$|\boldsymbol{H}|(2\pi R) = \frac{d}{dt}\left[\left(\varepsilon\frac{V}{d}\right)\int_0^R dA\right] = \frac{d}{dt}\left[\left(\varepsilon\frac{V}{d}\right)(\pi R^2)\right]$$

$$\Rightarrow\quad H(r,t) = \frac{\varepsilon R^2}{2d}\frac{dV}{dt}\frac{1}{r}$$

Combining cases (i) and (ii) and using the result of part (a) we obtain

$$H(r) = \begin{cases} \dfrac{I_D}{2\pi R^2}r & \text{for } r < R \\[2ex] \dfrac{I_D}{2\pi r} & \text{for } r \ge R \end{cases}$$

Note that this is formally identical to the result derived in application (a) in Section 18.5 for the magnetic field due to a uniform conduction current in a cylindrical wire of radius R.

Worked Example 21.2: A linearly polarised plane light wave is incident normally on two Polaroid sheets. The axis of the first sheet makes an angle of 50° with respect to the direction of polarisation of the wave and the axis of the second sheet makes an angle of 60° with respect to the axis of the first sheet. By what factor is the intensity of the beam reduced by passing through the Polaroid sheets?

Let I_0 be the intensity of the incident beam. Using Malus' law (end of Section 21.5), the intensity of the beam after passing through first sheet is given by $I_1 = I_0\cos^2 50°$ and thus the intensity after passing through the second sheet is given by

$$I_2 = I_1\cos^2 60° = I_0\cos^2 50°\cos^2 60° = (0.25)(0.41)I_0 = (0.1)I_0$$

$$\rightarrow \text{intensity reduced by a factor of 10}$$

Worked Example 21.3: Estimate the peak electric and magnetic field strengths at a distance of 1 m from a 100 W light bulb. For the purpose of your estimate you may assume that the bulb is a point source which radiates isotropically.

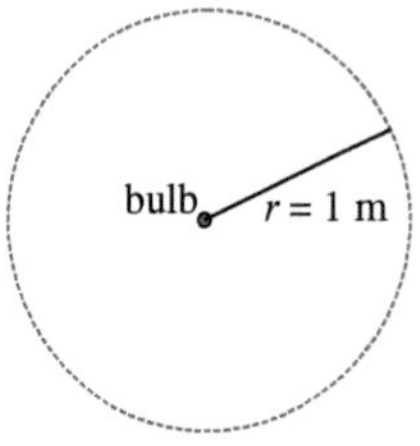

Figure 21.22. Worked Example 21.3.

The energy flux at 1 m from the source (Figure 21.22) is given by

$$\langle S\rangle = \frac{P_{\text{bulb}}}{4\pi r^2} = \frac{100}{4\pi(1)^2} = \frac{25}{\pi}\ \text{W m}^{-2} = \frac{E_0^2}{2\mu_0 c} = \frac{cB_0^2}{2\mu_0}$$

$$\rightarrow \quad E_0 = \sqrt{2\mu_0 c\langle S\rangle} = \sqrt{2\times 4\pi\times 10^{-7}\times 3\times 10^8\times\left(\frac{25}{\pi}\right)} = 77.5\ \text{kV m}^{-1}$$

$$\rightarrow \quad B_0 = \frac{E_0}{c} = \frac{77.5}{3\times 10^8} = 0.26\ \mu\text{T}$$

Worked Example 21.4: A dust grain in interplanetary space experiences a force due to the radiation pressure exerted by sunlight (Figure 21.23). If the total power radiated by the Sun is 4×10^{26} W, estimate the force due to radiation pressure on a spherical dust grain of 1 μm diameter at a distance from the Sun equal to the mean radius of the Earth's orbit (1.5×10^{11} m), assuming that all the incident radiation is absorbed. Compare your result with the force on the dust grain due to the gravitational attraction of the Sun if the density of the dust is 1.5×10^3 kg m^{-3} and the mass of the Sun is 2×10^{30} kg

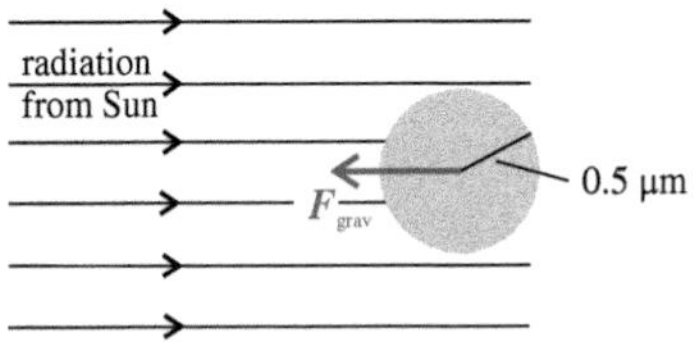

Figure 21.23. Worked Example 21.4.

$$\text{Energy flux at } 1.5\times 10^{11}\text{ m from the sun} = \frac{P_{\text{sun}}}{4\pi r^2} = \frac{4\times 10^{26}}{4\pi(1.5\times 10^{11})^2} = 1.4\ \text{kW m}^{-2}$$

$$\text{Radiation pressure at this distance} = \frac{\langle S\rangle}{c} = \frac{1400}{3\times 10^8}\ \text{Pa}$$

Since the area presented by the dust grain is $A = \pi\times(0.5\times 10^{-6})^2$, the force on the dust grain due to solar radiation is

$$F_{\text{rad}} = \frac{\langle S\rangle A}{c} = \frac{1400\times\pi\times(0.5\times 10^{-6})^2}{3\times 10^8} = 3.7\times 10^{-18}\ \text{N}$$

$$F_{\text{grav}} = G\frac{M_{\text{sun}}m}{r^2} = \frac{GM_{\text{sun}}\rho V}{r^2} = \frac{4\pi(6.67\times 10^{-11})(2\times 10^{30})(0.5\times 10^{-6})^3(1.5\times 10^3)}{3(1.5\times 10^{11})^2} = 4.7\times 10^{-18}\ \text{N}$$

Note: The radiation pressure exerted by sunlight has important consequences on the dynamics of small particles in the solar system (see Problem 21.6.12 on the website up.ucc.ie/21/), including the formation of the dust tails of comets.

Worked Example 21.5: If the total power radiated by the Sun is 4×10^{26} W, estimate (to within an order of magnitude) the number of photons emitted per second, assuming the average wavelength of the photons is 500 nm. Hence estimate the photon flux at the top of the Earth's atmosphere, assuming that the Earth-Sun distance is 1.5×10^{11} m.

$$\text{Energy of a 500 nm photon} = hf = \frac{hc}{\lambda} = \frac{6.6 \times 10^{-34} \times 3 \times 10^{8}}{5 \times 10^{-7}} \approx 4 \times 10^{-19}\ \text{J}$$

$$\Rightarrow \text{Number of photons/second emitted by Sun} = \frac{P_{\text{sun}}}{hf} = \frac{4 \times 10^{26}}{4 \times 10^{-19}} \approx 10^{45}\ \text{photons per second}$$

$$\Rightarrow \text{Photon flux at } 1.5 \times 10^{11}\ \text{m} \approx \frac{10^{45}}{4\pi(1.5 \times 10^{11})^2} \approx 4 \times 10^{21}\ \text{photons m}^{-2}\ \text{s}^{-1}$$

PROBLEMS

For problems based on the material covered this chapter visit up.ucc.ie/21/ and follow the link to the problems.

22

Wave optics

AIMS

- to discuss light as a form of electromagnetic radiation
- to describe the phenomenon of coherence and its application to lasers
- to discuss diffraction and interference of electromagnetic waves, particularly of light and X-rays
- to discuss the diffraction grating and to describe its use in spectroscopy

22.1 Electromagnetic nature of light

We saw in Section 21.4 that what humans observe as visible light comprises a part, albeit small, of the electromagnetic spectrum. Sources of accelerating charges with the appropriate frequency to generate light waves include incandescent lamps, fluorescent discharge tubes, hot solids, ionised gases, etc. Light from a star, such as the Sun, arises from the ionised gases in the hot outer atmosphere of the star. Being a form of electromagnetic radiation, it is expected that light will exhibit all the general characteristics of waves including polarisation and interference.

The collisions which produce electromagnetic radiation from most sources of light occur at random. The different charges are not accelerated in phase and the resulting electromagnetic radiation is random in the time at which it is emitted. The orientations of the oscillating charges, which can be considered to behave like independent microscopic antennas, are also aligned at random as they emit radiation, and hence the directions and polarisation of the emitted electromagnetic waves vary in a random manner, that is, they are **unpolarised**. Such radiation is described as **incoherent** and interference between independent sources cannot normally be observed.

On the other hand, more exotic sources such as so-called 'synchrotron radiation' and lasers (discussed below) will generate light in a more organised form.

Polarised light

Linear polarisers

Unpolarised light waves can be linearly polarised (recall Section 21.5) by passing a light beam through a sheet of Polaroid®. Polaroid sheets are made from long polymer molecules which are aligned parallel to one another. The response of the sheet to the electric field of a light wave is similar to that of the wire grid in the case of microwaves; that is, radiation which has its $\boldsymbol{E}$-field parallel to the direction of alignment of the polymer molecules is absorbed while the component at right angles to this is unaffected. The axis of a Polaroid sheet, therefore, is perpendicular to the direction of alignment of the molecules and only the component of the incident electric field strength along the direction of the axis will be transmitted (Figure 22.1). Thus Malus' law (Section 21.5) applies to polarised light passing through a Polaroid. No light is transmitted through two Polaroid sheets if their axes are at right angles to each other.

Polarisation by scattering

A beam from a flashlight or searchlight is not easily visible from the side when it passes through clear air but is seen more clearly if the air contains small dust particles or fog droplets. In the latter case the radiation is said to be *scattered* by the particles and the scattering mechanism can be understood qualitatively as follows. The sinusoidally varying electric field of the electromagnetic radiation in the beam (Figure 22.2) sets charges in the scattering particles oscillating with simple harmonic motion along the direction of the $\boldsymbol{E}$-field at the frequency of the incident radiation (at least in cases where this frequency is much less than the natural frequency of the bound charges). These oscillating charges, in turn, act as harmonic sources of radiation. Since the electric field is transverse, such oscillations will take place in the plane perpendicular to the direction of propagation of the incident beam.

Understanding Physics, Third Edition. Michael Mansfield and Colm O'Sullivan.
© 2020 John Wiley & Sons Ltd. Published 2020 by John Wiley & Sons Ltd.

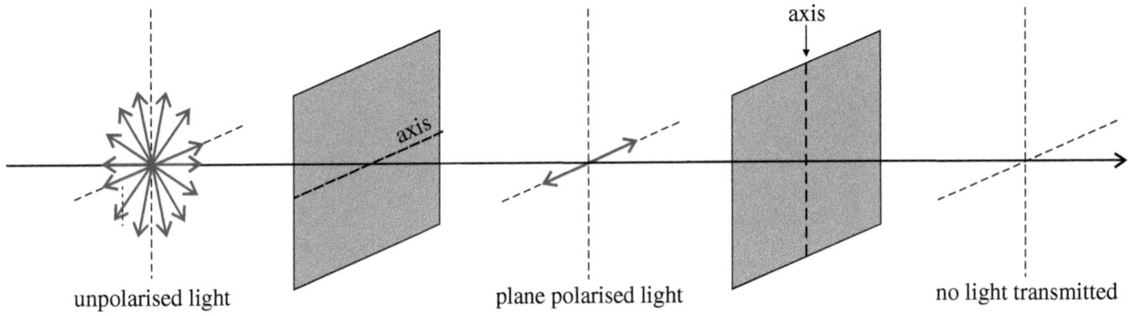

Figure 22.1. Passage of light through Polaroids.

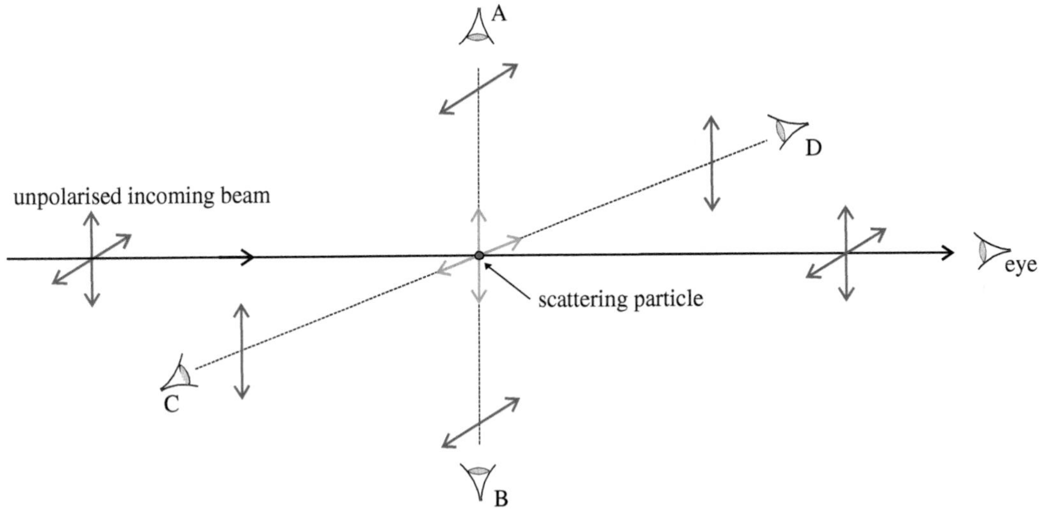

Figure 22.2. Polarisation of light by scattering from small particles. The arrows indicate the directions of the *E*-field oscillations.

If the incoming radiation is unpolarised, these oscillations can be along any line in the perpendicular plane. Each oscillating charge acts like a linear antenna and an observer viewing along a line perpendicular to the beam (that is from any of the positions A, B, C, or D in the figure) will see the radiation from those oscillations which have a component perpendicular to the line of sight but will receive no radiation from oscillations vibrating towards or away from the observer. Thus, the radiation which is scattered exactly at right angles to the beam is linearly polarised. Radiation scattered at other angles is partially polarised. To an observer looking directly into the beam (labelled 'eye' in the figure), the radiation remains unpolarised – neither perpendicular component is affected.

The scattering of light from bound electrons was investigated by Lord Rayleigh who sought to explain why the sky appears blue. A quantitative treatment of scattering of electromagnetic radiation requires a more advanced knowledge of electromagnetic theory.[1] Such treatment shows that the power scattered is proportional to the fourth power of the frequency, that is

$$P_{\text{scat}} \propto \omega^4$$

Thus high frequency (short wavelength) components of the radiation are scattered more efficiently than low frequency (longer wavelength) components. This means that blue light from the Sun is scattered more than red light, thus providing an explanation of the blueness of the daytime sky. Similar scattering from dust particles in the air explains redness of the sky observed at sunset/sunrise.

The Doppler effect in light

Like any wave phenomenon, electromagnetic waves show the Doppler effect (Section 13.9), a full treatment of which must be relativistic in this case (see below). For light waves from a moving source where the speed of the source $v_s \ll c$, the Doppler effect produces a frequency shift Δf which can be calculated using Equation (13.22),

$$\Delta f = f' - f = \frac{c}{c - v_s} f - f$$

where v_s is positive if the source is approaching the observer.

[1] A reader interested in the scattering of sunlight by the atmosphere and other optical scattering phenomena might consult Heavens and Ditchburn (1991).

Thus,
$$\frac{\Delta f}{f} = \frac{c - c + v_s}{c - v_s} \approx \frac{v_s}{c} \quad \text{if} \quad v_s \ll c$$

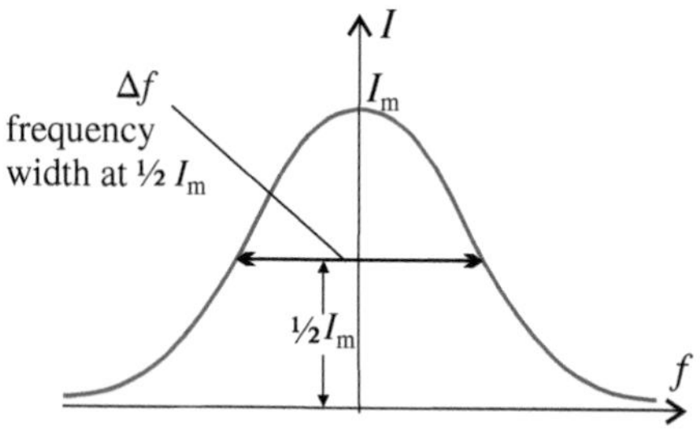

Figure 22.3. A plot of the intensity against frequency of a spectral line, showing the spread of frequencies due to the Doppler effect.

The frequency shift is to the blue (higher frequency) when the light source is approaching the observer and to the red (lower frequency) when the source is moving away from the observer. This effect poses a problem when high resolution spectroscopy is used to study the characteristic spectra of atoms and molecules. Wavelengths are shifted at random towards both the red and blue due to the thermal motion of the atoms and molecules, leading to a broadening of spectral lines, and hence difficulties in resolving closely spaced lines. The Boltzmann distribution function for molecular velocities (Section 12.5) can be used (see Worked Example 22.1) to show that the frequency width at half the maximum intensity of a spectral line, as illustrated in Figure 22.3, is given by

$$\Delta f = \frac{2f}{c}\sqrt{\frac{2kT \ln 2}{m}} \tag{22.1}$$

where m is the atomic or molecular mass. For a neon atom at room temperature (300 K) this equation gives $\Delta f = 1.7$ GHz for the characteristic neon spectral line of wavelength 632.8 nm. This may be compared with the natural line width of 12 MHz which was calculated for this line in Worked Example 14.4. The Doppler line broadening of this line is, therefore, roughly 100 times larger than its natural width. Equation (22.1) can be used to determine the temperature of a gas from the *Doppler breadth* of a spectral line.

The Doppler effect for light waves is important in astrophysics, where measurements of the wavelength shift of known spectral lines may be used to determine the velocities of distant stars and galaxies. In the case of distant galaxies, however, their velocities can approach the velocity of light so that a relativistic treatment is required.

The relativistic Doppler effect

In Section 9.11, we derived the energy transformation equation (last of Equations (9.43))

$$E' = \gamma(E - vp_x)$$

We can apply this result to a light wave emitted by a galaxy as it recedes from the Earth with velocity v. As was noted, the energy of a photon is related to its momentum through the Equation (14.11), that is $E = cp$. For a photon travelling towards the Earth along the x-axis we can write

$$E' = \gamma\left(E - \frac{Ev}{c}\right) = \gamma E\left(1 - \frac{v}{c}\right)$$

where E and E' are, respectively, the energies of the photon in the Earth and galaxy frames. Invoking the Einstein postulate, (Equation (14.8), we see that, if $f = \frac{E}{h}$ and $f' = \frac{E'}{h}$ are the frequencies of the electromagnetic wave in the frames of the receding galaxy and of the Earth, respectively, we can write $f' = \gamma f\left(1 - \frac{v}{c}\right)$

Thus
$$\left(\frac{f'}{f}\right)^2 = \frac{\left(1 - \frac{v}{c}\right)^2}{1 - \frac{v^2}{c^2}} = \frac{(c - v)^2}{(c + v)(c - v)}$$

and hence
$$f' = \sqrt{\frac{c - v}{c + v}}f$$

Study Worked Example 22.1

For problems based on the material presented in this section visit up.ucc.ie/22/ and follow the link to the problems.

22.2 Coherence: the laser

Two light sources are said to be **coherent** if the waves emitted by the two sources are of the same frequency and are linked in phase; that is, the phase difference between waves from the sources is constant. In this case interference effects may be observed, as discussed in Section 13.7. However, as noted above the light emitted from most sources is **incoherent** so that interference between independent sources cannot normally be observed.

It is possible, however, to produce coherent light using a **laser** (Light Amplification by Stimulated Emission of Radiation). As an example, we consider a low pressure gas laser.

In a gas laser, an electrical discharge is used to excite an atomic gas in a vacuum tube which has been filled to a low pressure. Mirrors of high reflectivity are placed at each end of the tube, as illustrated in Figure 22.4, so that, if l is the distance between the mirrors, the arrangement forms a *tuned cavity* for electromagnetic waves whose wavelengths satisfy the standing wave condition $\lambda = \frac{2l}{n}$ (Equation (13.16)). In this case, λ is a characteristic wavelength of the light emitted by excited gas atoms when they make transitions to lower energy states (a process which we will discuss in Chapter 24). The corresponding (resonant) frequencies of the standing waves are given by Equation (13.17), $f_n = \frac{nc}{2l}$ where $n = 1, 2, 3\ldots$ and c is the velocity of light.

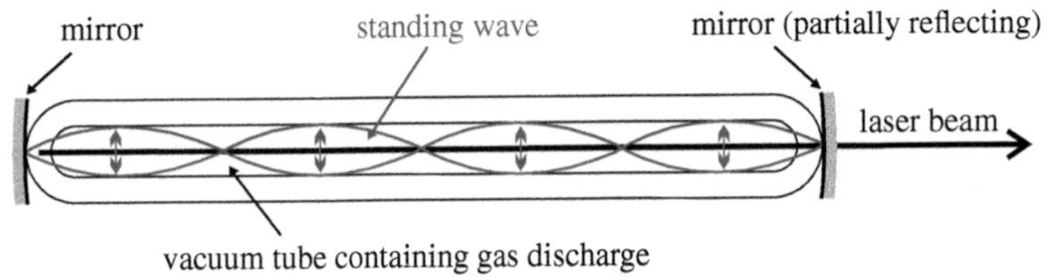

Figure 22.4. Standing electromagnetic waves produced in the tuned cavity of a laser.

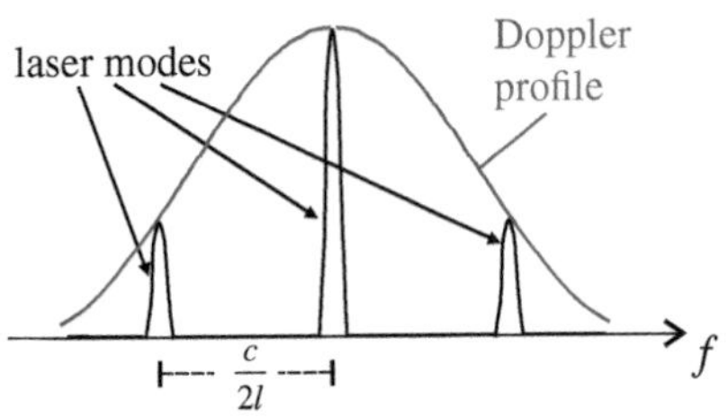

Figure 22.5. The laser modes (standing waves in a laser cavity) shown within the Doppler broadened profile of the characteristic laser frequency.

Without the tuned cavity, the emission of light by atoms excited by collisions in such a discharge occurs through *spontaneous emission*. This process occurs randomly in time and direction so that, as noted in the previous section, the light produced is incoherent. However, by placing the discharge in a cavity and by devising a means of arranging *population inversion*, whereby there are more atoms in the required excited state than in the lower energy state, a situation can be created in which the light produced by radiation from the excited states *stimulates* other excited atoms or ions in the cavity to radiate at the same frequency. This radiation, known as *stimulated emission,* has the characteristic frequency of the atomic transition and, because excitation of each gas atom is driven by the same standing wave, its emission is **coherent**, that is, co-ordinated in time and direction. The intensity of the standing light waves builds up in the cavity — light is amplified at the resonant frequency. The light is emitted through a partially reflecting mirror on one side of the cavity, as illustrated in Figure 22.4.

The emission of electromagnetic waves by a laser may be compared with the emission of radio waves by electrons in a transmitting antenna, emission that is co-ordinated by the a.c. voltage fed into the antenna.

The frequencies of standing waves in the laser cavity, f_n, are known as *laser modes*. Several laser modes can often be seen within the Doppler broadened profile of a characteristic laser frequency, as illustrated in Figure 22.5.

Laser light, therefore, is 'organised light' and has a number of remarkable properties when compared with the light produced by a conventional (incoherent) source of electromagnetic radiation.

- Laser light is very intense due to the amplification process in the cavity. To produce an energy density equal to that produced by some laser beams, a hot object would have to be at a temperature of 10^{30} K.
- Light energy can be delivered in pulses of very short duration so that remarkably high powers can be achieved.
- Laser light is highly coherent. It is monochromatic, to a very high degree, at the characteristic frequency of the emitting atoms to which the cavity has been tuned.
- The electromagnetic waves produced by a laser are plane waves with very little divergence. They can travel over large distances with little loss of intensity. As noted in Section 3.4, laser light can be detected after travelling from the Earth to the Moon and back.
- The waves are usually plane polarised.

Laser action can be produced in other devices, in addition to the gas laser. One such device, the semiconductor laser, will be described in Section 26.8.

Lasers have a wide and diverse range of applications. Some notable uses are in communications, in the reading of DVD's, CD's and bar codes, in welding tools, in remote sensing instruments, in carrying out surgery less invasively, in the production of holograms and in a wide range of scientific tools.

For problems based on the material presented in this section visit up.ucc.ie/22/ *and follow the link to the problems.*

22.3 Diffraction at a single slit

We consider first the diffraction of light from a distant source which is incident on a long slit of width a in an otherwise opaque barrier; the size of a should not be more than one order of magnitude greater than λ (recall Section 13.7). For simplicity, we assume that the light strikes the slit perpendicularly, as illustrated in Figures 22.6 and 22.7. To analyse diffraction at this aperture we divide the wavefront into N strips, each of width $\frac{a}{N}$. According to Huygen's principle (Section 13.5), each strip acts as an independent source. Let us consider the superposition of waves from two adjacent strips, S_1 and S_2, when viewed at some angle θ to the direction of the incident wave, as illustrated in Figure 22.6.

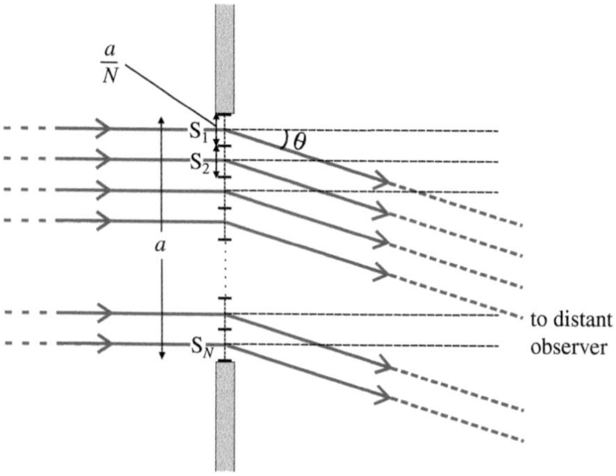

Figure 22.6. The diffraction of a plane wave which is incident perpendicularly on a single slit (aperture): the long dimension of the slit is perpendicular to the page.

The path difference between wave crests from S_1 and S_2 is $\frac{a}{N}\sin\theta$, as indicated in Figure 22.7. We have seen in Section 13.6 that a path difference of one wavelength, λ, between two wave crests, corresponds to a phase difference of 2π. A path difference of $\frac{a}{N}\sin\theta$ therefore corresponds to a phase difference which is given by

$$\phi = 2\pi\left[\frac{\frac{a}{N}\sin\theta}{\lambda}\right] \tag{22.2}$$

We must now determine the net effect of superposing N waves of equal amplitude but with phases which increase by equal angles between adjacent waves. For this case we can use the *phasors addition* technique, discussed in Section 20.10, which enables us to determine the effect of superposing two waves of amplitudes A_1 and A_2 which differ in phase by ϕ. This is achieved by representing the wave amplitudes by two vectors, $\boldsymbol{A}_1$ and $\boldsymbol{A}_2$, with an angle ϕ between their directions (Figure 22.8). The amplitude resulting from superposing these two amplitudes is then the magnitude of the vector $\boldsymbol{A}_1 + \boldsymbol{A}_2$.

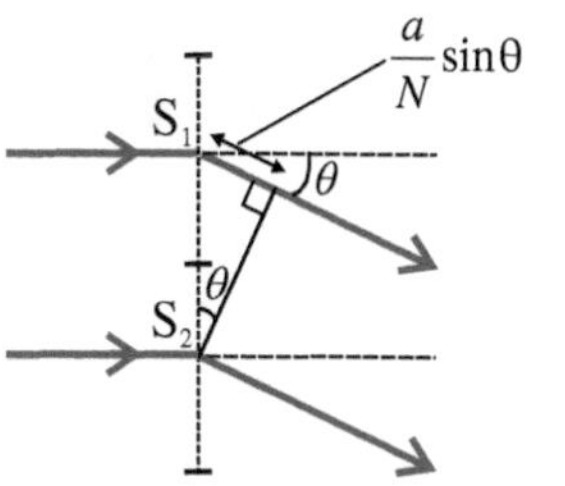

Figure 22.7. The superposition of waves from adjacent strips S_1 and S_2 of a single slit (aperture).

Thus the result of adding N waves of equal amplitude A' but with phases which increase by ϕ between adjacent waves, is given by the resultant vector $\boldsymbol{A}$ in Figure 22.9.

If we now let the number of strips become very large ($N \to \infty$, $a \to 0$) so that A' and ϕ become infinitesimally small, the line representing the addition of the amplitudes becomes an arc of a circle, as illustrated in Figure 22.10. The resultant vector $\boldsymbol{A}$ is then the chord RQ and thus

$$\frac{A_\theta}{A_0} = \frac{\text{chord length RQ}}{\text{arc length RQ}}$$

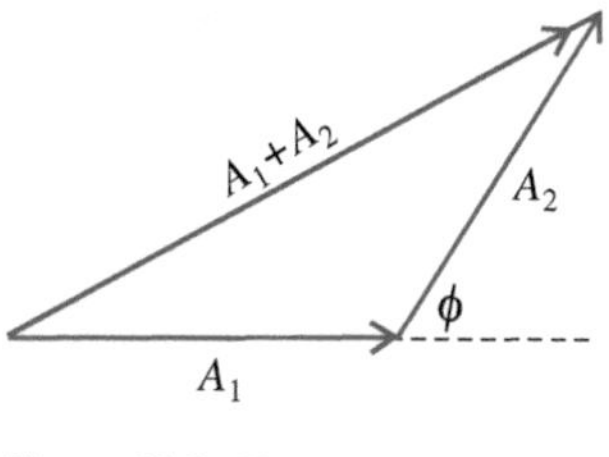

Figure 22.8. Phasor addition for two waves of amplitudes A_1 and A_2.

where A_θ is the resultant amplitude in the direction θ and A_0 is the amplitude at $\theta = 0$.

The phase difference 2α between the first and Nth strip is represented by the angle between the tangents at R and Q (the factor of 2 is chosen for convenience). The total phase change between R and Q is $N\phi$ so that

$$2\alpha = N\phi \tag{22.3}$$

The length of the chord RQ gives A_θ, the resultant wave amplitude in the direction θ. Note that when $\theta = 0$ (the straight through direction) and hence, from Equation (22.2), $\phi = 0$, the resultant amplitude A_0 is a straight line of length NA', as indicated in Figure 22.11. Note also that the length of this line is equal to the length of the arc RQ in Figure 22.10.

If, as illustrated in Figure 22.12, C is the centre of the circle and CD is the perpendicular bisector of RQ, we can write $\sin\alpha = \frac{\text{RD}}{\text{RC}}$.

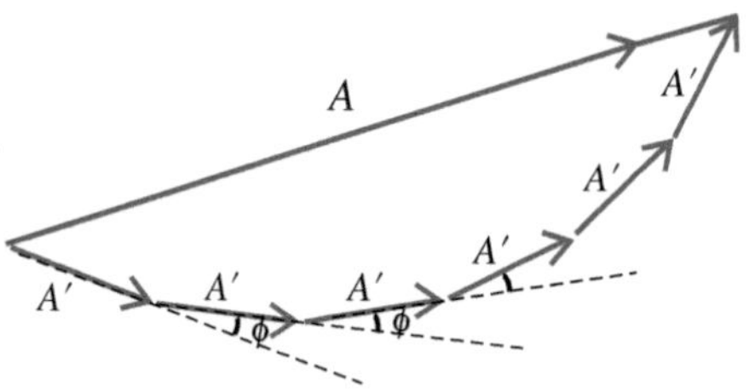

Figure 22.9. Single slit diffraction: the addition of six waves each of equal amplitude A', with phases which increase by ϕ between adjacent waves, viewed in a direction which makes an angle θ with the incident wave, as illustrated in Figures 22.6 and 22.7.

Thus RD = (RC)sin α and RQ = 2(RC) sin α.

From the definition of the radian (Appendix A.3.1), the arc length RQ = (RC)(2α)

Thus

$$\frac{A_\theta}{A_0} = \frac{\text{chord length RQ}}{\text{arc length RQ}} = \frac{2(\text{RC})\sin\alpha}{2(\text{RC})\alpha} = \frac{\sin\alpha}{\alpha} \tag{22.4}$$

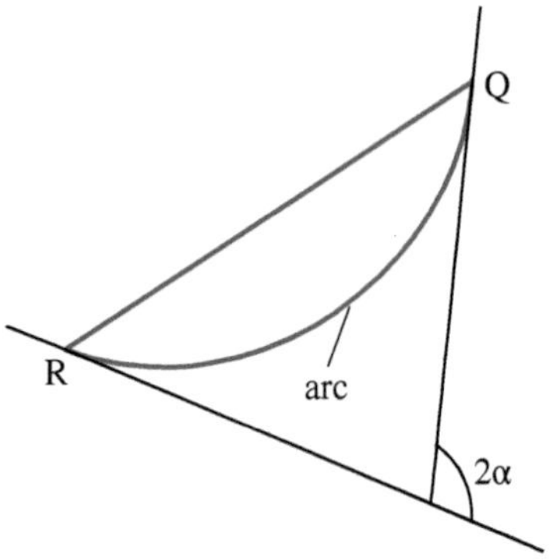

Figure 22.10. Single slit diffraction: the addition of N waves each of equal amplitude A', with phases which increase by ϕ between adjacent waves, in the limit $N \to \infty$, $A' \to 0$ and $\phi \to 0$, viewed in a direction which makes an angle θ with the incident wave, as illustrated in Figures 22.6 and 22.7.

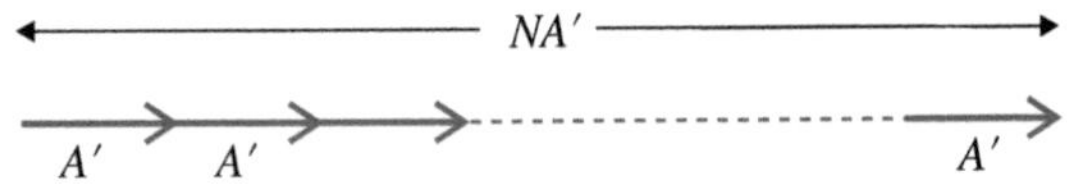

Figure 22.11. Single slit diffraction: the addition of N waves each of amplitude A' when $\theta = 0$ (the straight through direction).

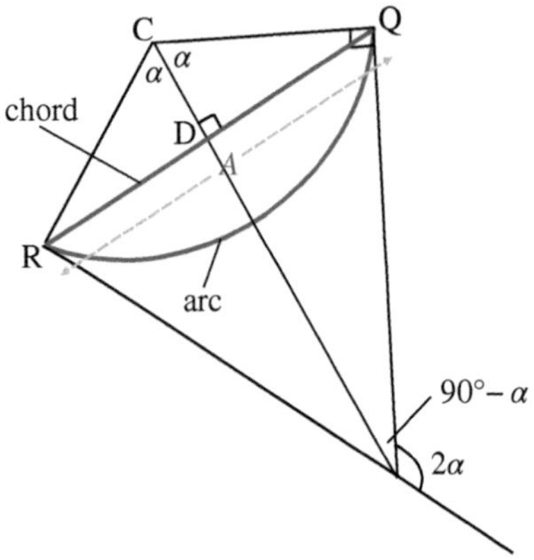

Figure 22.12. Single slit diffraction: the resultant amplitudes, in the direction θ (blue chord) and in the straight through direction (blue arc).

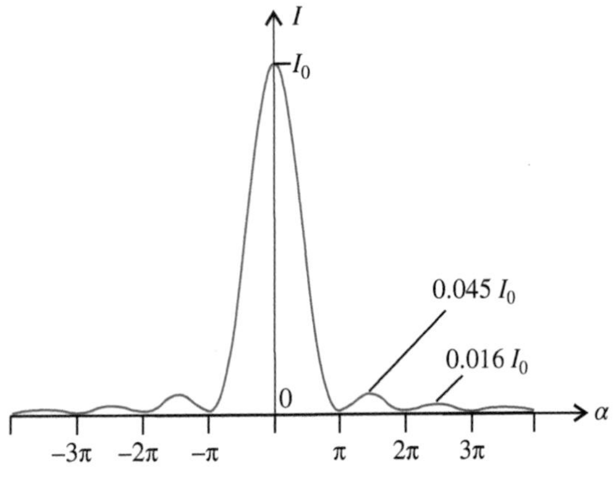

Figure 22.13. The single slit diffraction pattern plotted as a function of $\alpha = \frac{\pi a}{\lambda}\sin\theta$.

From Equation (13.8) the intensity of a wave is proportional to the square of its amplitude, that is $\frac{I(\theta)}{I_0} = \frac{A_\theta^2}{A_0^2}$, where I_0 is the intensity in the 'straight through' direction ($\theta = 0$). We can therefore write the intensity observed in a direction at an angle θ to that of the incident beam as follows:

$$I(\theta) = I_0 \frac{A_\theta^2}{A_0^2} = I_0 \frac{\sin^2\alpha}{\alpha^2} \tag{22.5}$$

where, from Equations (22.2) and (22.3),

$$\alpha = \frac{N\phi}{2} = \left(\frac{N}{2}\right) 2\pi \left[\frac{(a/N)\sin\theta}{\lambda}\right] = \frac{\pi a}{\lambda}\sin\theta$$

and thus

$$I(\theta) = I_0 \frac{\sin^2\alpha}{\alpha^2} = I_0 \frac{\sin^2\left(\frac{\pi a}{\lambda}\sin\theta\right)}{\left(\frac{\pi a}{\lambda}\sin\theta\right)^2}$$

Since $\lim_{\alpha\to 0}\left(\frac{\sin\alpha}{\alpha}\right) = 1$, $I(\theta)$ has maximum value at $\alpha = 0$; that is, $I(\theta) = I_0$ when $\theta = 0$. As α, and hence θ, increases from zero, $I(\theta)$ varies, so that a plot of $I(\theta)$ versus θ shows further maxima and minima as illustrated in Figure 22.13.

Minima $\left(I(\theta) \propto \frac{\sin\alpha}{\alpha} \to 0\right)$ occur when $\alpha = n\pi$ ($n = 1, 2, 3, \ldots$). The positions of further maxima on either side of the central maximum — known as *subsidiary maxima* — can be determined graphically from a plot such as that in Figure 22.13. They occur at $\alpha = 0, 1.43\pi, 2.46\pi, 3.47\pi, \ldots$, which can be represented approximately by $\alpha = \left(n + \frac{1}{2}\right)\pi$, where n is an integer. The equivalent appropriate values of $\sin\theta$ are then given by

$$\sin\theta = \frac{\lambda\alpha}{\pi a} = \left(n + \frac{1}{2}\right)\frac{\lambda}{a} \tag{22.6}$$

Intensities of subsidiary maxima (for which $\sin\alpha = 1$ and $\alpha = \left(\frac{2n+1}{2}\right)\pi$) are given by

$$I(\theta) = I_0 \frac{\sin^2\alpha}{\alpha^2} = \frac{4}{(2n+1)^2\pi^2} I_0$$

Thus the intensities of the first and second subsidiary maxima ($n = 1, 2$) are given by $I_1 = 0.045 I_0$ and $I_2 = 0.016 I_0$, respectively, as indicated in Figure 22.13.

The intensities of the subsidiary maxima are small compared with that of the central maximum but, nonetheless, as illustrated in Figure 22.14, they can produce appreciable intensity in the geometric shadow region of the slit.

According to Equation (22.6), the first subsidiary maximum ($n = 1$) occurs when $\sin\theta = \frac{3}{2}\frac{\lambda}{a}$, so that, when $a < \frac{3}{2}\lambda$, no value of θ less than or equal to $\frac{\pi}{2}$ satisfies this equation and diffraction is not observed. When $a \gg \lambda$ the angles at which maxima occur outside the central maximum, that is $\theta = \sin^{-1}\left[\left(n + \frac{1}{2}\right)\frac{\lambda}{a}\right]$, are all very small so that diffraction effects at the edge of such an aperture are difficult to observe. The wavelength of visible light ranges from 400 to 750 nm so that apertures in optical instruments (for example, lens and mirror diameters) are much larger than the wavelength of light. Consequently, in our everyday experience, diffraction effects are generally negligible for light waves and we would be justified in treating light, to a first approximation, as if it travels in straight lines — as will be assumed in the next chapter — except where apertures involved are very small ($\sim\lambda$). This contrasts strongly with the behaviour of sound waves whose

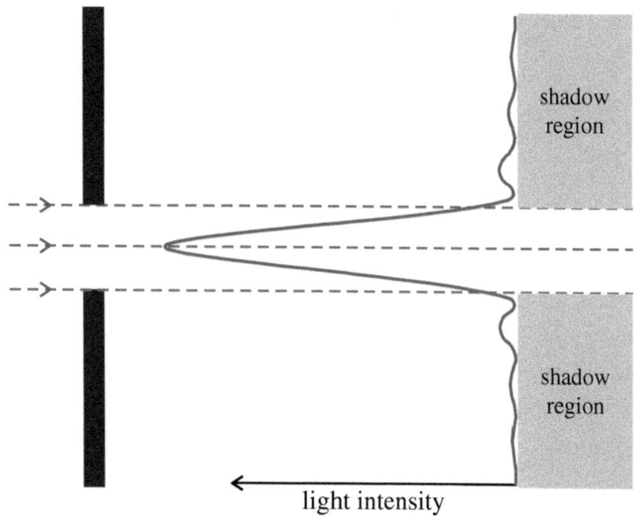

Figure 22.14. Diffraction of a plane wave from a single slit, as viewed on a screen placed behind the slit.

wavelengths range from centimetres to metres, and are generally comparable with the dimensions of most apertures they encounter. Diffraction effects, therefore, can be substantial for sound waves (for example, when sound waves enter a room through a doorway, the sound is heard throughout the room, not just in the geometric shadow of the doorway).

In optical instruments, such as telescopes and microscopes, diffraction effects tend to blur images. This means that the accuracy with which an image can be located depends on the wavelength of the light used. Consider the light which is emitted from two adjacent point sources (S_1 and S_2 in Figure 22.15) when it is observed through a slit of width a. The sources could be, for example, two adjacent stars viewed through a telescope whose smallest lens diameter is a. Each star on its own would give rise to a distinct diffraction pattern on a screen placed behind the aperture; the resulting intensity pattern is the resultant of these two slightly displaced patterns.

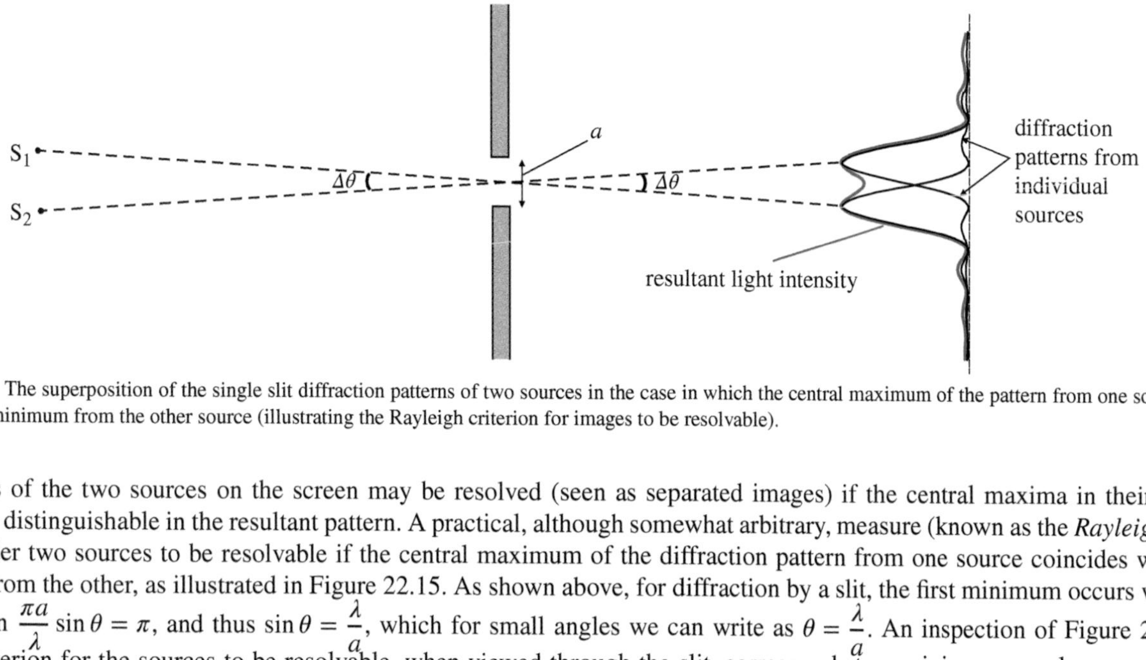

Figure 22.15. The superposition of the single slit diffraction patterns of two sources in the case in which the central maximum of the pattern from one source coincides with the first minimum from the other source (illustrating the Rayleigh criterion for images to be resolvable).

The images of the two sources on the screen may be resolved (seen as separated images) if the central maxima in their diffraction patterns are distinguishable in the resultant pattern. A practical, although somewhat arbitrary, measure (known as the *Rayleigh criterion*) is to consider two sources to be resolvable if the central maximum of the diffraction pattern from one source coincides with the first minimum from the other, as illustrated in Figure 22.15. As shown above, for diffraction by a slit, the first minimum occurs when $\alpha = \pi$, that is when $\frac{\pi a}{\lambda}\sin\theta = \pi$, and thus $\sin\theta = \frac{\lambda}{a}$, which for small angles we can write as $\theta = \frac{\lambda}{a}$. An inspection of Figure 22.15 shows that the criterion for the sources to be resolvable, when viewed through the slit, corresponds to a minimum angular separation which is given by

$$\Delta\theta \approx \frac{\lambda}{a}$$

For a circular aperture of diameter d the Rayleigh condition for minimum angular resolution can be shown[2] to be

$$\Delta\theta = 1.22\frac{\lambda}{d}$$

[2]see, for example Heavens and Ditchburn (1991) Sections 6.9 and 19.11.

Thus, to obtain high resolution (small $\Delta\theta$), short wavelength light (small λ) and a large aperture should be used.

For problems based on the material presented in this section visit up.ucc.ie/22/ and follow the link to the problems.

22.4 Two slit interference and diffraction: Young's double slit experiment

It might be expected that, if a screen is illuminated simultaneously by two sources of light waves (for example two filament lamps), an interference pattern similar to that described in Section 13.6 (Figure 13.18) would be observed. For reasons discussed in Section 22.2, however, it is not possible to control the phase of a light source, except in the special case of a laser. Consequently, it is not possible to construct two independent conventional light sources which operate in phase or with a constant phase difference between them.

We can, however, produce two sources which oscillate with the same phase starting with a single source. By using a two slit arrangement, as illustrated in Figure 22.16, it is possible to create two secondary light sources (the slits) which, because they are equidistant from a single (primary) source, oscillate in phase and thus produce coherent waves at the same frequency as the source. The **Young's double slit** experiment, named after Thomas Young (1773–1829), uses such an arrangement. Light waves of a fixed frequency and wavelength (*monochromatic* light) from a primary source are directed onto two narrow parallel slits. Even if the primary source is not equidistant from the slits, they still act as coherent sources since there is a constant phase difference between them.

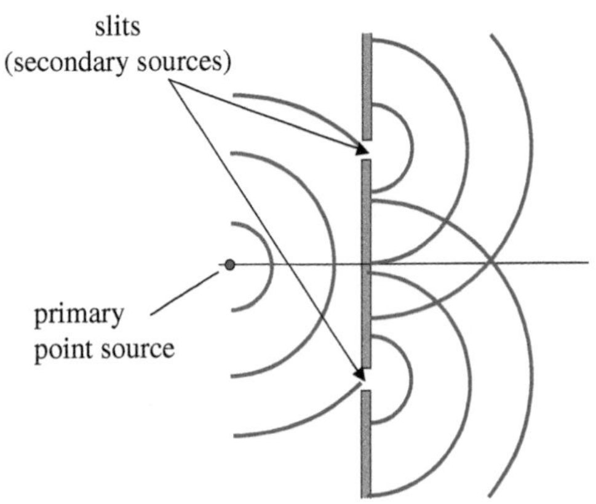

Figure 22.16. A double-slit arrangement: the slits act as two sources oscillating in phase.

We now analyse the diffraction pattern produced in the case of a plane wave which is incident normally on two parallel slits (narrow openings in an opaque screen), each of width a, which are separated by a distance b, as illustrated in Figure 22.17.

Again, we divide each slit into identical infinitesimal strips and add the contribution from each strip using a vector diagram. Consider first the slit AB. Figure 22.18 represents the addition of amplitudes from the infinitesimal strips which make up AB and is essentially the same as the single slit diagram (Figure 22.12) and thus, as before, $\alpha = \frac{\pi a}{\lambda}\sin\theta$, where 2α is the phase difference between the first and the last strip in the slit.

We now add the contribution from each of the two slits AB and CD, each of amplitude A_1, to produce the diagram shown in Figure 22.19. Note that there is an opaque section BC between the two single slit diagrams. As indicated in Figure 22.17 the path difference between waves from point A on slit AB and point C on slit CD is $(a+b)\sin\theta$, so that the phase difference between waves from these two points can be written

$$\frac{2\pi}{\lambda}(a+b)\sin\theta = 2\gamma$$

Thus,

$$\gamma = \frac{\pi(a+b)}{\lambda}\sin\theta.$$

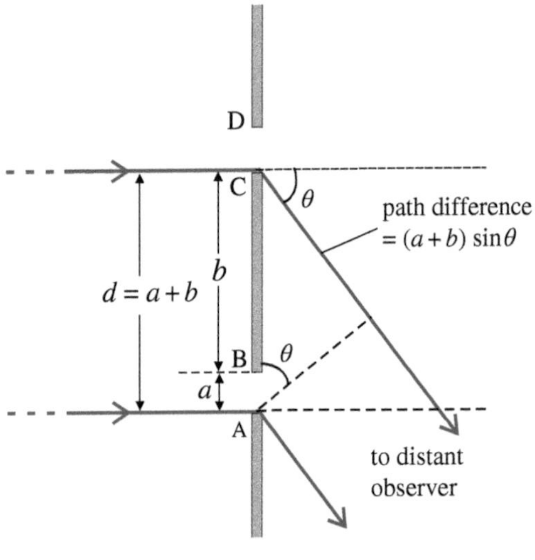

Figure 22.17. Two slit diffraction: the diffraction of a plane wave which is incident perpendicularly on two narrow parallel slits.

The angle between the tangents at A and C in Figure 22.19 is therefore 2γ. Usually $b \gg a$ so that $\gamma \gg \alpha$.

To determine the resultant intensity of waves from the two slits when viewed at an angle θ, we add the amplitudes represented by the chords AB and CD, both of magnitude A_1, using the vector diagram Figure 22.20. The total resultant amplitude A is given by the cosine rule (Appendix A.3.3), $A^2 = A_1^2 + A_1^2 + 2A_1^2\cos 2\gamma = 2A_1^2(1+\cos 2\gamma)$

Using the trigonometric identity $\cos 2\theta = 2\cos^2\theta - 1$ (Appendix A.3.5(v)) we can write A^2 as

$$A^2 = 4A_1^2\cos^2\gamma$$

and thus

$$A = 2A_1\cos\gamma$$

In the previous section we showed (Equation (22.4) and Figure 22.12) that, for a single slit,

$$\frac{A_\theta}{A_0} = \frac{\text{chord length RQ}}{\text{arc length RQ}} = \frac{\sin\alpha}{\alpha}$$

We can apply this result to either slit to give

$$\frac{A_1}{A_0} = \frac{\sin\alpha}{\alpha} \tag{22.7}$$

The total straight through ($\theta = 0$, $\alpha = 0$) amplitude for the two slits is $2A_0$ so that we can write the total resultant amplitude at angle θ as

$$A = 2A_1\cos\gamma = 2A_0\frac{\sin\alpha}{\alpha}\cos\gamma$$

The relative intensity at this angle can be determined from

$$\frac{I(\theta)}{I_0} = \frac{A^2}{(2A_0)^2} = \frac{\sin^2\alpha}{\alpha^2}\cos^2\gamma$$

where I_0 is the straight through intensity from the two slits together.

The term $\frac{\sin^2\alpha}{\alpha^2}$ can be recognised as corresponding to the single slit diffraction pattern (Equation (22.5)) while the term $\cos^2\gamma$ is the two source interference pattern discussed in Section 13.6.

Thus the intensity of the light observed at an angle θ to the incident beam is given by

$$I(\theta) = I_0\frac{\sin^2\alpha}{\alpha^2}\cos^2\gamma = I_0\left[\frac{\sin\left(\frac{\pi a}{\lambda}\sin\theta\right)}{\frac{\pi a}{\lambda}\sin\theta}\right]^2\cos^2\left(\frac{\pi(a+b)}{\lambda}\sin\theta\right)$$

The second factor in the term on the right will have maximum values when when $\cos^2\left(\frac{\pi(a+b)}{\lambda}\sin\theta\right) = 1$, that is, when $\gamma = \left(\frac{\pi(a+b)}{\lambda}\sin\theta\right) = n\pi$, where n is an integer. This condition can be written $(a+b)\sin\theta = n\lambda$ or

$$\boxed{d\sin\theta = n\lambda} \tag{22.8}$$

where $d = a + b$ is the separation between equivalent points, such as A and C or B and D, on the slits (Figure 22.17). Equation (22.8) is the Young's double slit interference condition. As one would expect, this is the same as the condition for maximum intensity from two synchronous point sources separated by a distance d (Equation (13.14)).

When $b \gg a$, which is usually the case in practice, the second (interference) factor in the equation for $I(\theta)$ varies more rapidly with θ than the first. The resulting pattern, illustrated in Figure 22.21, is a rapidly varying $\cos^2\gamma$ two-slit interference pattern modulated by a single slit diffraction envelope defined by the $\frac{\sin^2\alpha}{\alpha^2}$ term.

Hence, in a two slit experiment with an incident plane wave, the light intensity can vanish at certain values of θ, such as θ_m in Figure 22.21, and can fall off quite quickly at larger angles unless the slit width is very small. The disappearance of interference maxima at certain values of θ is further illustrated in Worked Example 22.2.

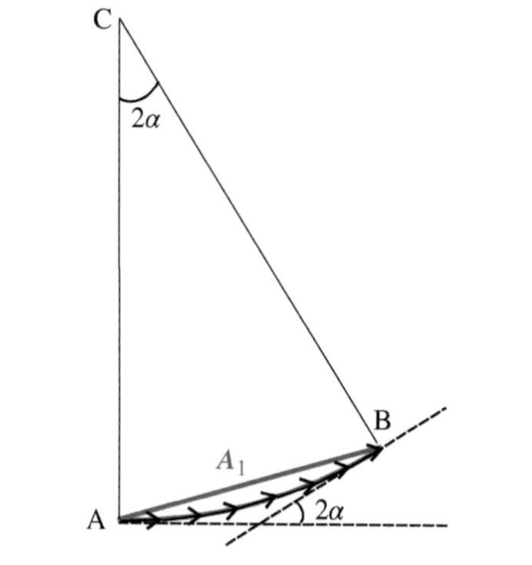

Figure 22.18. The addition of waves from the slit AB, viewed in the direction θ, to produce a resultant amplitude A_1.

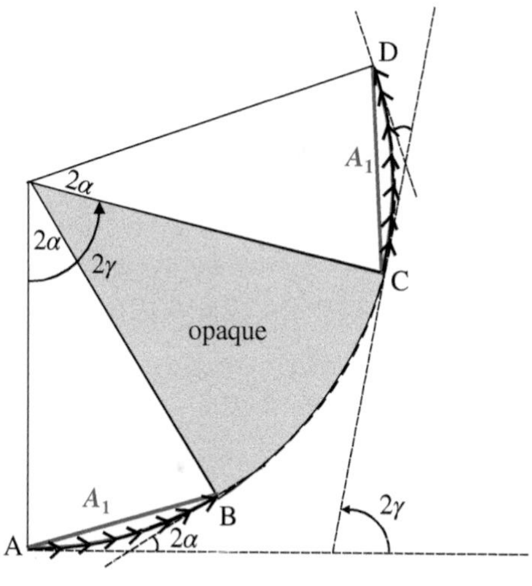

Figure 22.19. Two slit diffraction: the resultant amplitude A_1 from each slit, viewed in the direction θ.

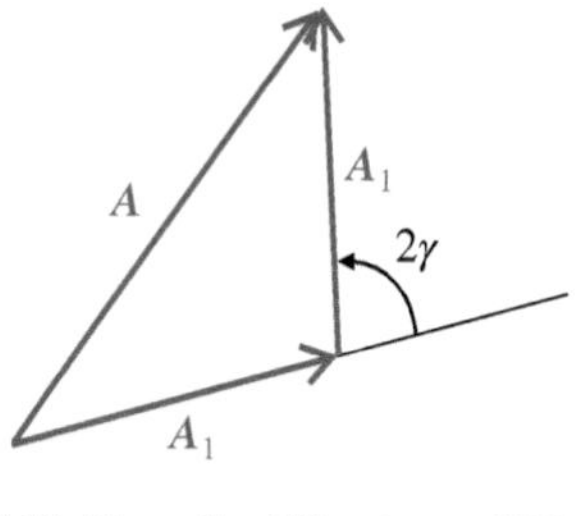

Figure 22.20. Two slit diffraction: addition of the resultant amplitudes from the two slits, viewed in the direction θ, to produce a total resultant amplitude A.

The Young's double slit interference pattern may be observed on a screen which is placed to the right of the slits in Figure 22.17. If d is known, the wavelength of the light may be determined from Equation (22.8) if measurements are made of the angular separation of successive maxima or minima. However, because the maxima and minima are not sharply defined in Figure 22.21, it is difficult to make accurate measurements of the angular separation, and therefore of λ.

Note that interference and diffraction are in fact aspects of the same phenomenon. The term *diffraction* is usually used to describe the spreading of a wave pattern due to interference between adjacent points of a wavefront and *interference* is used to describe interference between waves from independent coherent sources or from portions of a wavefront from a single source which have been separated and then recombined.

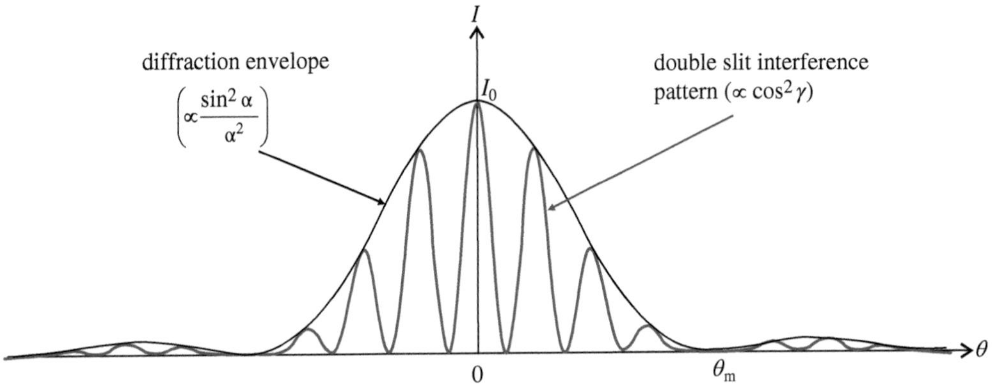

Figure 22.21. The two slit diffraction pattern is plotted as a function of θ. The two slit interference pattern (blue) is modulated by a single slit diffraction envelope (black).

Study Worked Example 22.2

For problems based on the material presented in this section visit up.ucc.ie/22/ *and follow the link to the problems.*

22.5 Multiple slit interference: the diffraction grating

As noted in the previous section, a Young double slit arrangement is not ideal for making accurate measurements of wavelength. A more useful arrangement for this purpose is the *diffraction grating*, which is illustrated in Figure 22.22.

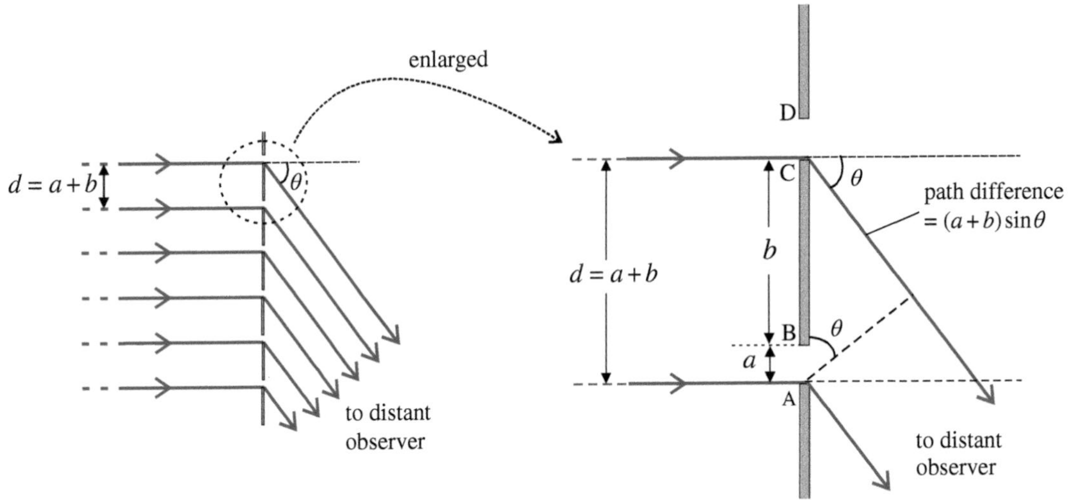

Figure 22.22. Multiple slit interference (diffraction grating): plane waves are incident perpendicularly on a large number of identical narrow parallel slits.

The diffraction grating comprises a large number N (usually well in excess of one thousand and often orders of magnitude larger) of narrow slits, which are regularly spaced so that the distance between any two adjacent slits, d, is the same. The slits illustrated in Figure 22.22 are each of width a and are separated by a distance b so that $d = a + b$. Plane waves are incident normally on the plane of the grating.

Waves emerging from any slit in the grating interfere with the waves emerging from all the other slits, a phenomenon known as *multiple slit interference* and is similar to the interference from a linear array of synchronous sources discussed in Section 13.6. As before, we determine the resultant amplitude as a function of the angle θ by using a phasor diagram (Figure 22.23).

Figure 22.23 is an extension of the two-slit diagram Figure 22.19, to N slits. As before, for any slit, $A_1 = A_0 \frac{\sin\alpha}{\alpha}$ where $\alpha = \frac{\pi a}{\lambda}\sin\theta$ and A_0 is the amplitude at $\theta = 0$. The phase difference between any two adjacent slits is given by $\gamma = \frac{\pi(a+b)}{\lambda}\sin\theta$, as in the two-slit case.

The addition of the amplitudes of all slits, each of amplitude A_1, is illustrated in Figure 22.24, extending the analysis in the previous section to the N-slit case.

For a large number of slits, we can approximate the line which represents the addition of amplitudes from all slits to an arc PQ of a circle (just as we approximated Figure 22.9 by Figure 22.10) and can write A_θ, the total resultant amplitude for the N slits, in terms of R, the radius of the circle. From the geometry of Figure 22.24, we see that

$$\sin N\gamma = \frac{\frac{1}{2}A_\theta}{R} \quad \text{and} \quad \sin\gamma = \frac{\frac{1}{2}A_1}{R}$$

and hence

$$A_\theta = 2R\sin N\gamma \quad \text{and} \quad A_1 = 2R\sin\gamma$$

so that

$$\frac{A_\theta}{A_1} = \frac{\sin N\gamma}{\sin\gamma}$$

For a single slit we can write (Equation (22.4))

$$A_1 = \frac{\sin\alpha}{\alpha}A_0$$

Thus

$$A_\theta = \left(\frac{\sin\alpha}{\alpha}\right)\left(\frac{\sin N\gamma}{\sin\gamma}\right)A_0$$

The total straight through amplitude $A_{0T} = NA_0$ so that we can also write

$$A_\theta = \frac{A_{0T}}{N}\left(\frac{\sin\alpha}{\alpha}\right)\left(\frac{\sin N\gamma}{\sin\gamma}\right)$$

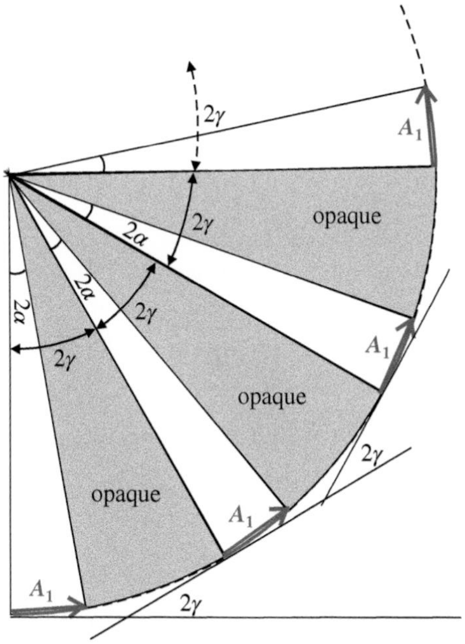

Figure 22.23. Multiple slit interference: the resultant amplitudes from each slit, viewed in the direction θ, with opaque sections between the slits.

and, therefore, the intensity observed at an angle θ with respect to the direction of the incident plane wave is given by

$$I(\theta) = \frac{I_0}{N^2}\left(\frac{\sin\alpha}{\alpha}\right)^2\left(\frac{\sin N\gamma}{\sin\gamma}\right)^2 \tag{22.9}$$

where I_0 is the total straight through intensity. Note that as $\theta \to 0$, both $\alpha \to 0$ and $\gamma \to 0$ and $\frac{\sin\alpha}{\alpha} \to 1$ and $\frac{\sin N\gamma}{\sin\gamma} \to 1$.

As in the two slit case, the term $\left(\frac{\sin\alpha}{\alpha}\right)^2$ is the single slit diffraction pattern, Equation (22.5), while the term $\left(\frac{\sin N\gamma}{\sin\gamma}\right)^2$, where $\gamma = \frac{\pi(a+b)}{\lambda}\sin\theta$, gives the multiple slit interference pattern. Usually $b \gg a$ so that, $\gamma \gg \alpha$ and, as in the two slit case, the latter pattern

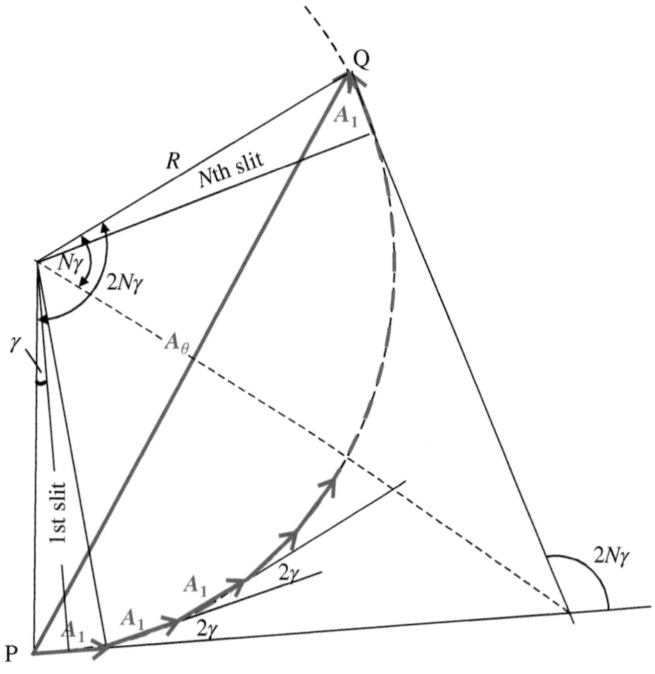

Figure 22.24. Multiple slit interference: the addition of the amplitudes from N slits, viewed in the direction θ, to produce the total resultant amplitude A_θ.

is modulated by a more slowly varying diffraction envelope. This multiple slit interference term is a fairly complicated function of θ with many maxima and minima. Note, however, that $\dfrac{\sin N\gamma}{\sin\gamma} \to 1$ for any value of γ such that $\gamma = n\pi$ $(n = 0, \pm 1, \pm 2\ldots)$ and these correspond to the strongest maxima of intensity. These occur when $\gamma = \dfrac{\pi(a+b)}{\lambda}\sin\theta = n\pi$ which can be written as $(a+b)\sin\theta = n\lambda$ or

$$\boxed{d\sin\theta = n\lambda} \tag{22.10}$$

This equation for the strongest interference maxima is the **diffraction grating equation**. It turns out to be the same as that obtained for the Young's double slit (Equation (22.8). In the multiple slit case weaker, subsidiary, maxima occur at other values of γ, such as approximately at $\gamma = \dfrac{3\pi}{2N}$, as illustrated in Figure 22.25.

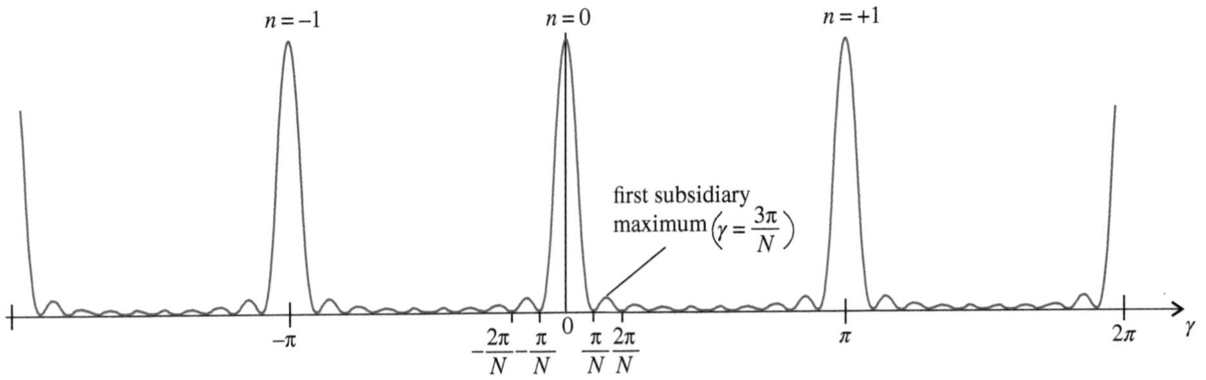

Figure 22.25. The multiple slit diffraction pattern plotted as a function of $\gamma = \sin\left(\dfrac{\pi(a+b)}{\lambda}\right)\sin\theta$, showing principal and subsidiary maxima.

For large N, the intensities of subsidiary maxima are very small compared to those of the principle maxima which in turn can be very narrow.

Writing (22.9) explicitly in terms of the angle θ, we get

$$I(\theta) = \frac{I_0}{N^2}\left[\frac{\sin\left(\frac{\pi a}{\lambda}\sin\theta\right)}{\frac{\pi a}{\lambda}\sin\theta}\right]^2\left[\frac{\sin\left(\frac{N\pi(a+b)}{\lambda}\right)\sin\theta}{\sin\left(\frac{\pi(a+b)}{\lambda}\right)\sin\theta}\right]^2 \tag{22.11}$$

which gives the observed intensity pattern produced by light passing through a diffraction grating where the principle maxima, located as given by Equation (22.10), are modulated by a single slit diffraction envelope (Figure 22.26). When $a \sim \lambda$, however, the fall-off in intensity due to single slit diffraction effects is negligible.

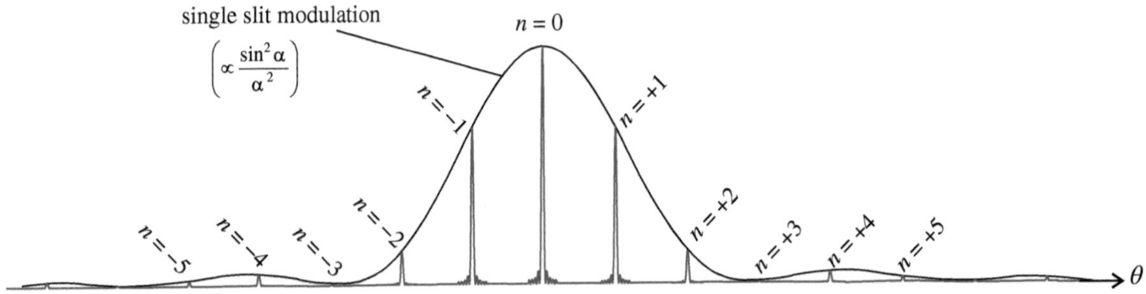

Figure 22.26. The multiple slit diffraction pattern (blue) plotted as a function of θ, showing how the intensities of the principal maxima are modulated by a single slit diffraction envelope. Note the possible occurrence of 'missing orders' (at $n = \pm 3$ in the case illustrated).

For practical purposes the important feature of the diffraction grating intensity pattern is the sharpness of the maxima. The maxima, which correspond to different values of n in Equation (22.10) (known as different **orders**), are sharp and, as indicated in Figure 22.27, are separated by regions of very low light intensity so that the corresponding values of θ in Equation (22.10), and consequently the wavelength, may be determined very accurately.

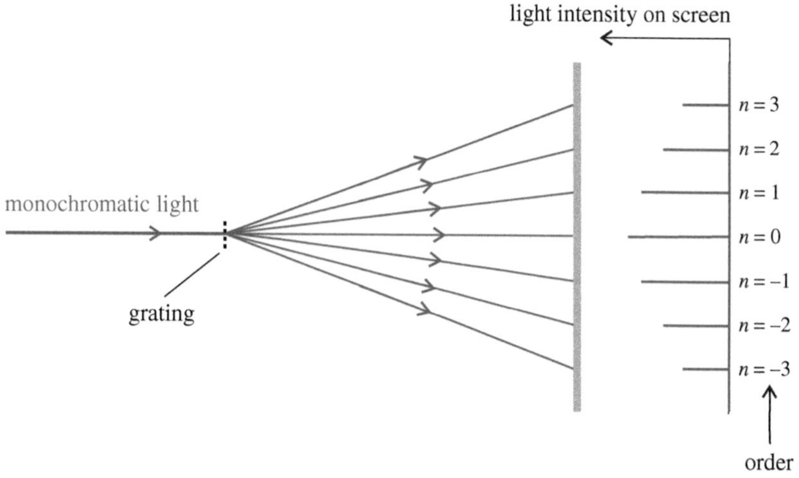

Figure 22.27. The multiple slit diffraction pattern, as viewed on a screen placed behind the diffraction grating.

Note that accurate measurements of wavelengths are achieved more easily when large values of the angle θ are used. Because $\sin\theta = n\frac{\lambda}{d}$ (Equation (22.10)), the separation of successive slits, d, should be of the same order of magnitude as the wavelength of light, which is about 500 nm for visible light.

Study Worked Example 22.3

For problems based on the material presented in this section visit up.ucc.ie/22/ and follow the link to the problems.

22.6 Diffraction of X-rays: Bragg scattering

When we discussed the electromagnetic spectrum in Section 21.4 we noted that X-rays, like light rays, are electromagnetic waves but that they have wavelengths of the order of 10^{-10} m, about one thousandth of the wavelength of visible light. When X-rays are incident on a diffraction grating with very small slit spacing, say $d = 10^{-6}$ m, the grating Equation (22.10) tells us that the angle of diffraction of the first order ($n = 1$) maximum is given by,

$$\theta = \sin^{-1}(\lambda/d) = \sin^{-1}(10^{-4}) \sim 10^{-4}\ \text{rad} = 0.006°$$

It is very difficult to measure an angle as small as this accurately. For accurate measurements, therefore, we must seek a grating with a slit separation of the order of 10^{-10} m if we wish to measure wavelengths in the X-ray region. While it is not possible to manufacture such a grating, as we saw in the context of electron and neutron diffraction (Section 14.8), nature has provided us with just such an arrangement — the periodic arrangement of the atoms or molecules in a crystal.

When a beam of monochromatic X-rays is incident on a crystal, as illustrated in Figure 22.28, X-rays penetrate deep into the crystal structure. The electrons located at each lattice site experience the sinusoidal electric field strength of the X-ray beam (Sections 21.2 to 21.4) and hence oscillate with simple harmonic motion driven by the field force $e\boldsymbol{E}$. Thus each lattice site acts as a source (an 'antenna') of electromagnetic waves radiating in all directions with the same frequency as the incident waves. Because these 'scattering centres' are aligned at regular intervals along the axes of the crystal, it is expected that the interference effects observed will resemble that from an array of coherent point sources of waves discussed in Section 13.6, but with the array being three dimensional in this case.

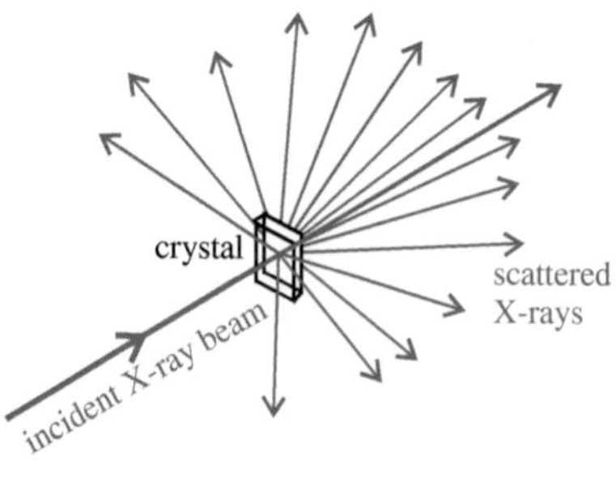

Figure 22.28. Diffraction of a beam of monochromatic X-rays by a crystal.

To avoid the complexity of a full treatment we will confine ourselves to the case of orthorhombic crystals (that is, crystals in which the three crystal axes are all perpendicular to each other). Figure 22.29 shows one plane of such a crystal where the inter-atomic spacing along the x-axis

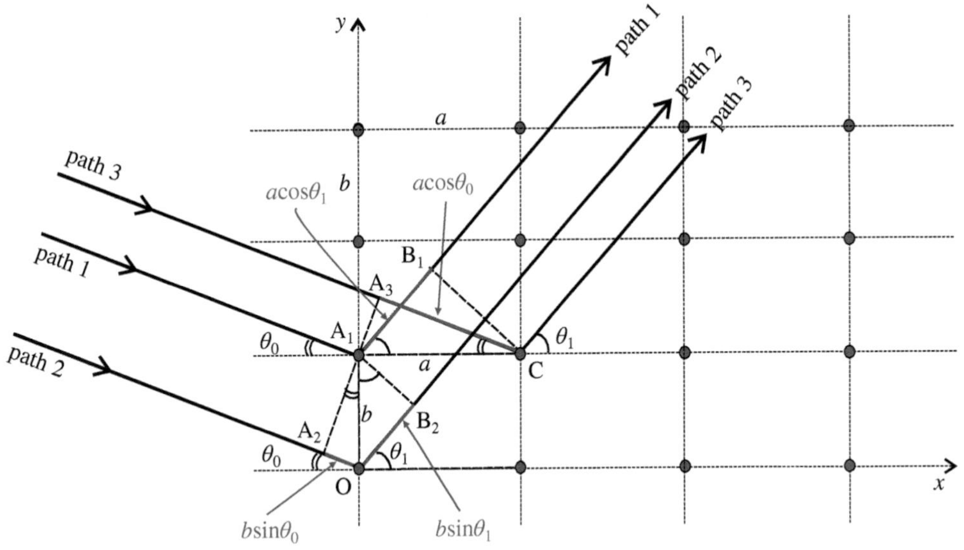

Figure 22.29. The scattering of X-rays by atoms atomic or molecular scattering centres in a crystal lattice.

is a and that along the y-axis is b. We assume that the atoms are fixed, that is we ignore lattice vibrations. We will also assume that the plane wavefronts of the incident X-ray beam are perpendicular to the x–y plane.

In Figure 22.29, the beam strikes the crystal so that the direction of the incident beam makes an angle θ_0 with the x-axis. We now consider the intensity of the scattered radiation observed at some arbitrary angle θ with respect to the x-axis at a point which is at a large distance from the crystal compared to the crystal dimensions, confining our attention to the x–y plane.

For constructive interference to be observed at this angle, two conditions must be satisfied simultaneously, namely, (a) the radiation scattered from all of the sources in a y-array must be in phase and (b) the radiation scattered from all of the sources in an x-array must also be in phase. In other words (a) the path difference between the paths followed by the radiation scattered by any two adjacent atoms along the y-axis (such as paths 1 and path 2 in Figure 22.29) must be equal to an integral number of wavelengths and (b) the path difference between the paths followed by the radiation scattered by any two adjacent atoms along the x-axis (such as paths 1 and path 3 in the figure) must be equal to an integral number of wavelengths. Accordingly, the criteria for constructive interference at the angle θ_1 are

(a) $\text{path2} - \text{path1} = A_2O + OB_2 = b\sin\theta_0 + b\sin\theta_1 = h'\lambda \ (h' = 0, \pm1, \pm2, \pm3, \ldots)$

(b) $\text{path3} - \text{path1} = A_3C - A_1B_1 = a\cos\theta_0 - a\cos\theta_1 = k'\lambda \ (k' = 0, \pm1, \pm2, \pm3, \ldots)$

Thus, for fixed θ_0, all possible values of θ_1 which give rise to intensity maxima are given by

$$b(\sin\theta_0 + \sin\theta_1) = h'\lambda \tag{22.12}$$

$$a(\cos\theta_0 - \cos\theta_1) = k'\lambda \tag{22.13}$$

In Figure 22.30, a plane PQ (solid blue line) perpendicular to the plane of the page is set up so that the angle θ that the incident wave makes with this plane is equal to the angle the re-radiated wave makes with the same plane — simulating specular reflection from the plane, that is

$$\theta = \theta_0 + \alpha = \theta_1 - \alpha$$

where α is the angle between PQ and the x-axis.

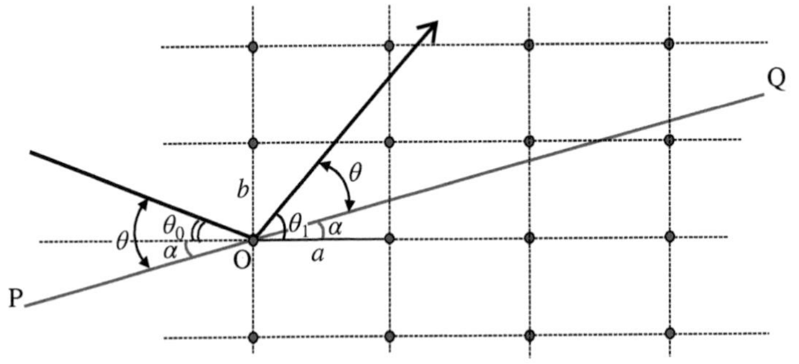

Figure 22.30. The 'reflection' of X-rays from a plane.

Now, invoking identities A.3.5(xi) and A.3.5(xii) in Appendix A, one gets

$$\sin\theta_0 + \sin\theta_1 = 2\sin\theta\cos\alpha \quad \text{and} \quad \cos\theta_0 - \cos\theta_1 = 2\sin\theta\sin\alpha$$

and thus (22.12) $\rightarrow$ $2b\sin\theta\cos\alpha = h'\lambda$ (22.12a)

and thus (22.13) $\rightarrow$ $2a\sin\theta\sin\alpha = k'\lambda$ (22.13a)

Let n be the highest common factor of h' and k', that is $h' = nh$ and $k' = nk$, where n is any integer and h and k are integers with no common factor. Thus,

(22.12a) $\rightarrow$ $2b\sin\theta\cos\alpha = nh\lambda$

and (22.13a) $\rightarrow$ $2a\sin\theta\sin\alpha = nk\lambda$

It follows directly from these equations that $2\left\{\frac{b}{h}\cos\alpha\right\}\sin\theta = n\lambda$ and $2\left\{\frac{a}{k}\sin\alpha\right\}\sin\theta = n\lambda$ and, furthermore, that $\frac{b}{h}\cos\alpha = \frac{n\lambda}{2\sin\theta} = \frac{a}{k}\sin\alpha$. Hence $\left\{\frac{a}{k}\sin\alpha\right\} = \left\{\frac{b}{h}\cos\alpha\right\}$ which we now give the common value d.

We conclude, therefore, that the two criteria (22.12) and (22.13) for intensity maxima are equivalent to the single requirement

$$2d\sin\theta = n\lambda \tag{22.14}$$

where

$$d = \frac{a}{k}\sin\alpha = \frac{b}{h}\cos\alpha \tag{22.15}$$

Since $\sin^2\alpha + \cos^2\alpha = \frac{d^2k^2}{a^2} + \frac{d^2h^2}{b} = 1$, then $d = \left(\frac{k^2}{a^2} + \frac{h^2}{b^2}\right)^{-\frac{1}{2}} = \text{constant.}$

We can interpret d (and h and k) geometrically by inspecting Figure 22.31. The solid blue line represents the plane passing through O and the scattering centre at $x = 3a$ and $y = 2b$ (in general, any plane can be drawn through O and a lattice point $x = ha$ and $y = kb$). The dashed blue lines represent all the adjacent parallel planes through other lattice points. It can be seen that these planes divide the x- and y- sides of every basic cell (Figure 22.31(b)) into k parts and h parts, respectively. From Figure 22.31(b), we see that $\sin\alpha = \frac{hd}{b}$ and $\cos\alpha = \frac{kd}{a}$ where d is the perpendicular distance between adjacent planes (the '*inter-planar spacing*'). It is clear that the constant d in Equation (22.14) above is also the inter-planar spacing of the set of $\{h, k\}$ planes (the $\{3, 2\}$ planes in the case of Figure 22.31).

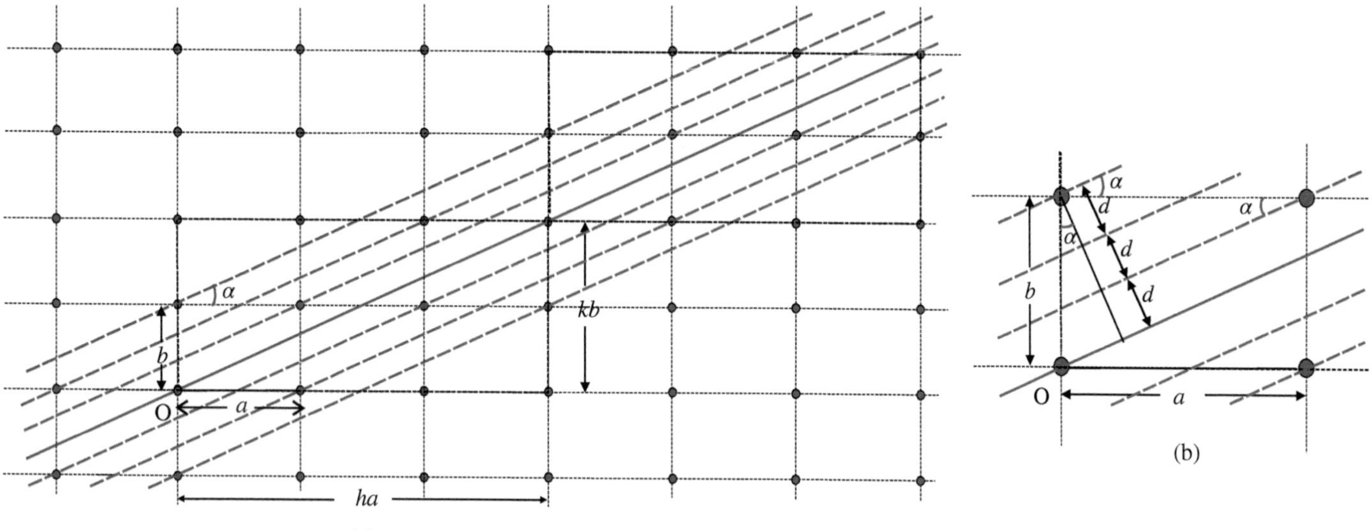

Figure 22.31. (a) Some of the set of 'reflecting' planes for the special case of $h = 3$ and $k = 2$. (b) Intersection of the axes of a basic cell by the planes. The approach is easily extended to arbitrary values of h and k.

The condition for intensity maxima (22.14), namely

$$\boxed{2d\sin\theta = n\lambda} \tag{22.16}$$

applies to all reflecting planes corresponding to all possible values of h and k where d is the relevant inter-planar spacing. Equation (22.16) is known as the **Bragg condition** after William Henry Bragg (1862–1942) and William Lawrence Bragg (1890–1971) who showed that this condition enabled a much simpler approach to be made to the analysis of X-ray diffraction patterns than in the original theory of Max von Laue (1879–1960).

In any crystal (not necessarily orthorhombic) there will be a large number of such planes, with diverse orientations and spacings, each characterised by specific values of h, k and l (the latter index associated with the third direction in the crystal). The Bragg condition applied to all such planes, therefore, specifies the directions at which all possible intensity maxima will be observed.

Note that the Bragg condition may be interpreted as resulting from the constructive interference that would be observed due to the 'reflection' of X-rays from a family of parallel planes. From this perspective, the path difference between the X-rays reflected at one plane and that at the next adjacent plane should be an integral number of wavelengths. From inspection of Figure 22.32, it is evident that the path difference is $2d \sin \theta$.

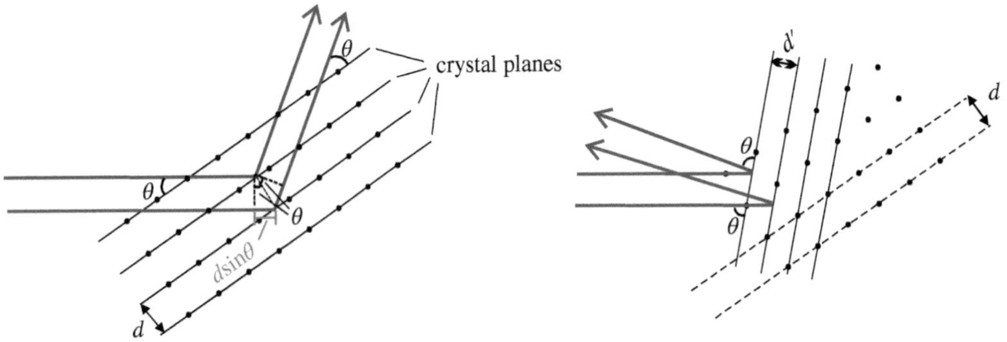

Figure 22.32. The scattering of X-rays by two different sets of crystal planes: the angle θ between the direction of the scattered X-rays and the crystal planes is equal to the angle between the incident X-ray beam and the plane.

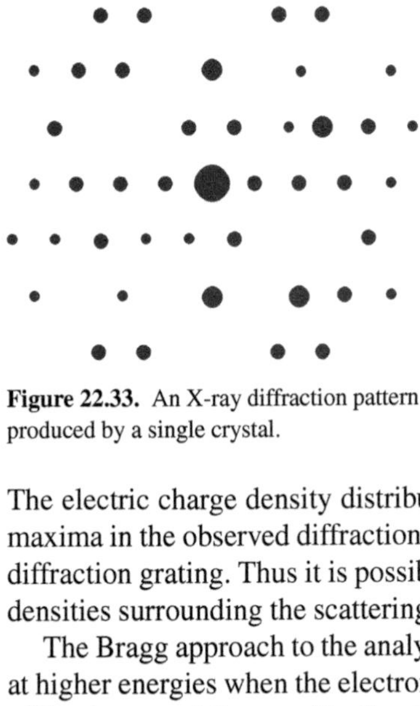

Figure 22.33. An X-ray diffraction pattern produced by a single crystal.

An X-ray diffraction pattern from a single crystal, therefore, is more complicated than that obtained with a one-dimensional grating. If the crystal is rotated about one of its axes while a narrow beam of monochromatic X-rays is directed at the crystal perpendicular to the rotation axis, a pattern of the type shown in Figure 22.33 may be observed on photographic film or other X-ray detection system. Each dot is produced by a different set of crystal planes which satisfy the Bragg condition.

X-ray crystallography is usually used in either of two ways: (a) if d, θ and n are known, λ may be determined; that is, X-ray wavelengths may be determined using a *crystal spectrometer* or (b) if we know θ, n and λ, we can determine d, the inter-planar spacing in the crystal and the corresponding lattice parameters.

X-ray diffraction techniques have proved to be a very valuable tool in establishing the structure of crystals and molecules in materials science, chemical and biological applications. In the discussion above it was assumed that the scattering centres were essentially point sources. In practice, particularly in the case of complex atoms or molecules located at the lattice points, this is not the case. The electric charge density distribution around each atom or molecule can have a significant effect on the intensity of the interference maxima in the observed diffraction pattern, in much the same way as the slit width affects the relative intensity of the orders in an optical diffraction grating. Thus it is possible to use the measured intensities of maxima in a diffraction pattern to reconstruct the electric charge densities surrounding the scattering sites. The helical structure of the DNA molecule, for example, was established using this technique.

The Bragg approach to the analysis of X-ray diffraction patterns must also be applied to neutron diffraction and to electron diffraction at higher energies when the electron beam can penetrate a number of atomic planes. Worked Example 22.4 shows how the original result of Davisson and Germer (Section 14.8) may be analysed in terms of Bragg scattering.

Study Worked Example 22.4

For problems based on the material presented in this section visit up.ucc.ie/22/ and follow the link to the problems.

22.7 The SI unit of luminous intensity, the candela

We saw in the study of waves in Chapter 13 (Section 13.4) that the wavefronts generated by a point source were concentric spheres centred on the source and that the intensity of (spherical) waves emitted from a point source at a distance r from the source is given by

$$I(r) = \frac{P}{4\pi r^2}$$

where P is the strength of the source. Thus, the wave intensity of spherical waves falls off with the inverse square of distance from the source.

While in general light intensity can be measured in units of W m^{-2}, a special SI base unit the **candela** is defined to measure the **luminous intensity** of a source over a specific frequency distribution of emitted light waves. The candela is defined for specific technical purposes which we will not need to consider in this book but we give its definition below[a] to complete our account of SI base units.

The candela, symbol cd, is the SI unit of luminous intensity in a given direction. It is defined by taking the fixed numerical value of the luminous efficacy of monochromatic radiation of frequency 540 × 10^{12} Hz, K_{cd}, to be 683 when expressed in the unit lm W^{-1}, which is equal to cd sr W^{-1}, or cd sr kg^{-1} m^{-2} s^3, where the kilogram, metre and second are defined in terms of h, c and $\Delta\nu_{Cs}$

Further reading

Further material on some of the topics covered in this chapter may be found in *Insight into Optics* by Heavens and Ditchburn and in *Optics* by Smith and Thomson (See BIBLIOGRAPHY for book details).

WORKED EXAMPLES

Worked Example 22.1: Derive Equation (22.1), the equation that gives the frequency width of a (Doppler broadened) spectral line at half the maximum line intensity.
[Assume that atomic velocities, observed along the line of sight to an observer, follow a Maxwell-Boltzmann distribution (Section 12.5). You may also assume that the atomic velocities are non-relativistic].

Applying the Doppler effect equation (13.22) to light emitted from an atom which is moving towards a stationary observer at a velocity v, we obtain

$$f' = \frac{c}{c-v}f \rightarrow f' - f = \left(\frac{c}{c-v} - 1\right)f = \left(\frac{v}{c-v}\right)f.$$

when $v \ll c$, this may be written

$$f' - f = \left(\frac{v}{c}\right)f \rightarrow v = \frac{(f'-f)}{f}c \qquad (22.17)$$

The equation for $P_x(v_x)$ in Section 12.5 gives the velocity distribution of atoms moving in one dimension, defined in this case by the line to sight to the observer. Thus

$$P(v)\Delta v = \sqrt{\frac{m}{2\pi kT}}e^{-\frac{\frac{1}{2}mv^2}{kT}}\Delta v \qquad (22.18)$$

The ratio of I_v, the light intensity when the velocity is v, to I_0 the intensity at the centre of the spectral line (where $v = 0$), equals the ratio of the numbers of emitting atoms with these velocities, that is

$$\frac{I_v}{I_0} = \frac{P(v)}{P(v=0)}$$

Applying Equation (22.18),

$$\frac{I_v}{I_0} = \frac{\sqrt{\frac{m}{2\pi kT}}e^{-\frac{\frac{1}{2}mv^2}{kT}}}{\sqrt{\frac{m}{2\pi kT}}e^{-0}} = e^{-\frac{\frac{1}{2}mv^2}{kT}}.$$

Substituting for v from Equation (22.17),

$$\frac{I_v}{I_0} = e^{-\frac{\frac{1}{2}m(f'-f)^2c^2}{kTf^2}}.$$

This distribution of intensity (identified as a *gaussian* function in Section 12.5) is illustrated in Figure 22.34.

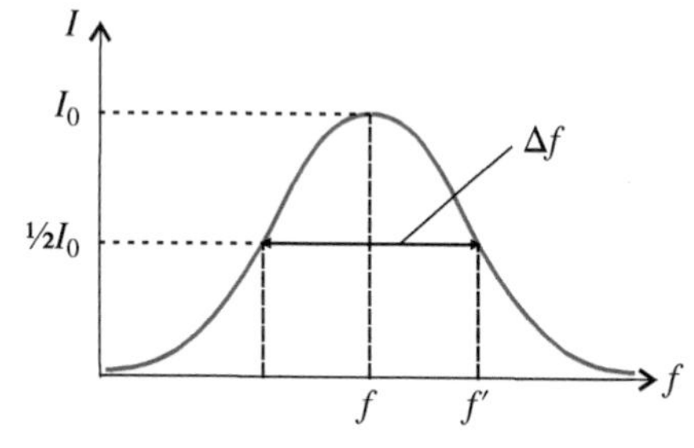

Figure 22.34. Worked Example 22.1.

At half maximum intensity,

$$\frac{I_v}{I_0} = \frac{1}{2} = e^{-\frac{\frac{1}{2}m(f'-f)^2c^2}{kTf^2}} \rightarrow \frac{\frac{1}{2}m(f'-f)^2c^2}{kTf^2} = \ln 2 \rightarrow (f'-f)^2 = \frac{2kTf^2}{mc^2}\ln 2 \rightarrow (f'-f) = \frac{f}{c}\sqrt{\frac{2kT}{m}\ln 2}$$

As indicated in Figure 22.34, the full width of the line at half maximum intensity, $\Delta f = 2(f' - f)$, so we can write $\Delta f = \frac{2f}{c}\sqrt{\frac{2kT}{m}\ln 2}$, which is Equation (22.1), as required.

[a]The steradian (sr) is a unit of solid angle.

Worked Example 22.2: For what value of $\frac{a}{b}$ (with a and b as indicated in Figure 22.17) does the third maximum of the two slit interference pattern disappear. Which other maxima disappear?

The equation for light intensity as a function of θ, for the two slit pattern, is derived in Section 22.4,

$$I(\theta) = I_0\left[\frac{\sin\left(\frac{\pi a}{\lambda}\sin\theta\right)}{\frac{\pi a}{\lambda}\sin\theta}\right]^2 \cos^2\left(\frac{\pi(a+b)}{\lambda}\right)\sin\theta.$$

We seek the value of $\sin\theta$ for which the third maximum of the interference term, $\cos^2\left(\frac{\pi(a+b)}{\lambda}\right)\sin\theta$, coincides with the first minimum of the diffraction term, $\left[\frac{\sin\left(\frac{\pi a}{\lambda}\sin\theta\right)}{\frac{\pi a}{\lambda}\sin\theta}\right]^2$. For the interference term the condition is $\frac{\pi(a+b)}{\lambda}\sin\theta = 3\pi$ and for the diffraction term it is $\frac{\pi a}{\lambda}\sin\theta = \pi$. If both conditions are satisfied simultaneously, $3a = a + b \quad \rightarrow \quad 2a = b \quad \rightarrow \quad \frac{a}{b} = 2.$

The condition for further maxima to disappear is that the interference maximum coincides with a diffraction minimum. This occurs when

$$\sin\theta = \frac{n_1\lambda}{a+b} = \frac{n_2\lambda}{a} \quad \rightarrow \quad \frac{n_1}{n_2} = \frac{a+b}{b} = 3, \text{ where } n_2 = 1, 2, 3, \text{ etc.}$$

Thus every third interference maximum is missing.

Worked Example 22.3: If the wavelength of the light used to produce the multiple slit diffraction pattern shown in Figure 22.26 is 590 nm and the first order line is observed at an angle of 7.8°, (a) determine the periodicity of the grating (the distance between corresponding points on adjacent slits), (b) determine the angle at which the second order line is observed and (c) estimate the width of an individual slit (hint: note that the third order line is missing in the case illustrated in Figure 22.26).

(a) $n\lambda = d\sin\theta \quad \rightarrow \quad (1)(590\times10^{-9}) = d\sin\theta_1 = d\sin(7.8°) \quad \rightarrow \quad d = \frac{590\times10^{-9}}{0.136} = 43\ \mu\text{m}$

(b) $2\lambda = d\sin\theta_2 \quad \rightarrow \quad \theta_2 = \sin^{-1}\left(\frac{2\lambda}{d}\right) = \sin^{-1}(2\sin\theta_1) = \sin^{-1}(0.2714) = 15.7°$

(c) For single slit interference, Intensity $\propto \left(\frac{\sin\alpha}{\alpha}\right)^2$ has minima when $\alpha = \pm\pi,\ \pm 2\pi,\ \pm 3\pi, \ldots$

Thus, for the first missing order ($n = 3$), $\alpha_3 = \frac{\pi a}{\lambda}\sin\theta_3 = \left(\frac{\pi a}{\lambda}\right)\left(\frac{3\lambda}{d}\right) = \pi \quad \rightarrow \quad a = \frac{d}{3} = 1.43\ \mu\text{m}.$

Worked Example 22.4: Show how Bragg's treatment of diffraction by a single crystal accounts for the result of the Davisson and Germer experiment described in Section 14.8. Note that the relevant interplanar spacing for nickel is 9.2×10^{-10} m.

The experimental arrangement used by Davisson and Germer is illustrated in Figure 14.13. A beam of electrons strikes a nickel crystal and the intensity of the scattered electrons is measured as a function of the angle ϕ. Figure 14.14 shows the observed variation of intensity with ϕ for a beam of incident electrons of energy 8.8×10^{-18} J. The figure shows clearly that a maximum of intensity is detected at $\phi = 50°$. The occurrence of a maximum is qualitative evidence of interference — a wave effect. In Section 14.8, we showed that the de Broglie wavelength of an electron of energy 8.8×10^{-18} J is 1.65×10^{-10} m. We now investigate whether the observed maximum corresponds quantitatively to diffraction at this wavelength.

Applying the Bragg condition $\lambda = 2d\sin\theta$ (Equation (22.16) to the observed angle $\theta = (180° - \phi)/2 = 65°$ (from Figure 14.14), and using the relevant interplanar spacing for nickel, the wavelength of the diffracted waves is calculated to be

$$\lambda = 2\times9.2\times10^{-11}\sin 65° = 1.65\times10^{-10}\ \text{m}$$

The 8.8×10^{-18} J electrons therefore exhibit the diffraction pattern expected of electron waves of this de Broglie wavelength.

PROBLEMS

For problems based on the material covered this chapter visit up.ucc.ie/22/ and follow the link to the problems.

23

Geometrical optics

AIMS

- to introduce the ray model and to ascertain how it explains the phenomena of reflection and refraction
- to describe how the ray model explains image formation in mirrors and lenses
- to explain the operation of certain optical instruments
- to discuss dispersion of light in transparent media

23.1 The ray model: geometrical optics

Consider a geometrical area which is a long way from a source of waves as shown shaded in Figure 23.1. The spherical wavefronts that pass through such an area emerge almost parallel. These waves, therefore, can be approximated to plane waves and, in the limit of the shaded area in Figure 23.1 becoming very small, may be considered to travel in an infinitesimally narrow beam.

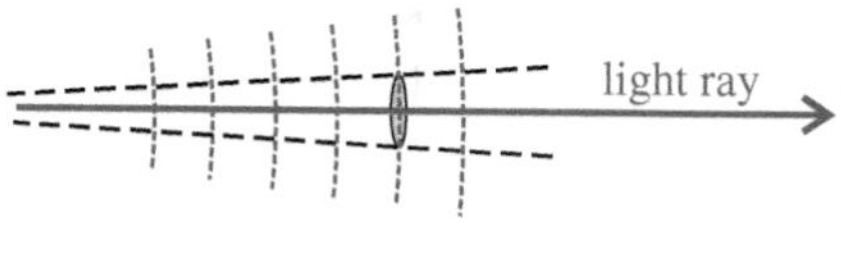

Figure 23.1. A ray of light as the limit of a very narrow beam of plane waves.

In such a situation the energy of the wave may be considered to propagate in straight lines, as it does in the case of particles. This model of wave propagation leads to the concept of a light ray in geometrical optics, the ray being always perpendicular to the wavefront. Thus, a point source of light may be considered to be a source of **rays** emanating radially from the source in all directions (Figure 23.2). Note that the point source need not be intrinsically self-luminous; a point on any illuminated object may be considered as a point source of light in the context of geometrical optics.

In this **ray model**, light energy is considered to travel in straight lines unless it encounters an interface between two media, in which case its progress may be deduced from the laws of *reflection* and *refraction* (Equations (13.9) and (13.10)), which were derived from Huygens' principle in Section 13.5.

The laws of refraction and reflection, together with simple geometry, may be used to describe how light rays behave in a wide variety of situations involving reflection from mirrors or refraction through interfaces between transparent media through which the rays are travelling. It must be remembered, however, that all apertures or obstructions encountered must have dimensions very much larger than the wavelength of the light; otherwise diffraction effects (Section 13.7 and Chapter 22) become important and the ray model cannot be employed.

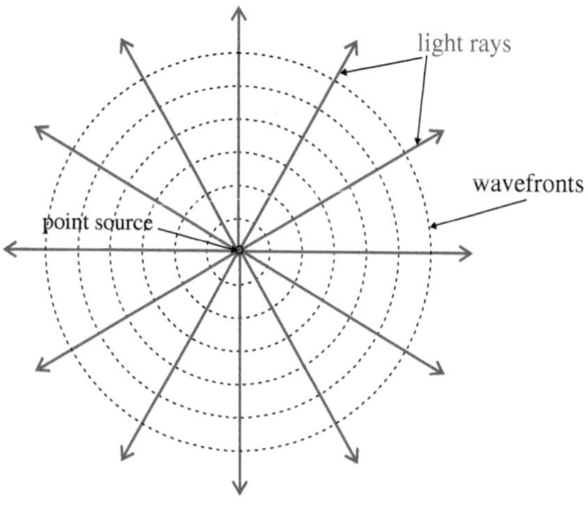

Figure 23.2. Rays from a point source of light; the figure shows just a few of the infinite number of such rays envisaged in the ray model.

23.2 Reflection of light

The law of reflection

The law of reflection (Section 13.5) may be restated in terms of the ray model as follows. A ray of light will be reflected from a surface such that (a) the angle which a reflected ray makes with the normal to the surface at the point of incidence is equal to the angle between the incident ray and the same normal and (b) the reflected ray will be confined to the plane defined by the incident ray and the normal to the surface at the point of incidence ($\theta_i = \theta_r$ in Figure 23.3).

Understanding Physics, Third Edition. Michael Mansfield and Colm O'Sullivan.

© 2020 John Wiley & Sons Ltd. Published 2020 by John Wiley & Sons Ltd.

Reflection of light from a polished or silvered surface (a mirror) may be used to manipulate the path of light for practical purposes. In this section we will see some examples of how certain types of mirror can be used to form images of illuminated or self-luminous objects.

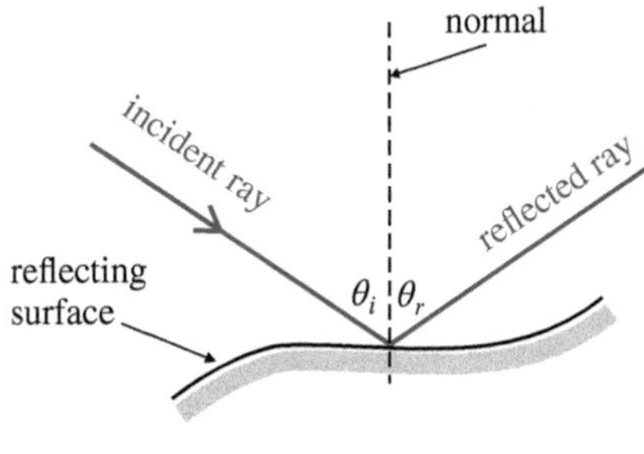

Figure 23.3. The reflection of a ray of light from a reflective surface; the reflected ray is in the plane defined by the incident ray and the normal.

Image formation by reflection at mirrors

Consider the plane mirror with its reflective surface facing left, as illustrated in Figure 23.4. Light is radiated in all directions from the point source S but only those rays striking the mirror surface contribute to image formation. Initially we consider only two such rays. The ray which strikes the mirror perpendicularly (point M, angle of incidence = 0) will be reflected back along its own path. A ray striking the mirror at some arbitrary point P will be reflected according to the law of reflection as shown (angle of incidence = angle of reflection = θ). The two rays diverge after reflection but if projected backwards they appear to intersect at some point I.

Looking at the two triangles Δ SMP and Δ IMP, simple geometric considerations tell us that $\angle\text{SMP} = \frac{\pi}{2} = \angle\text{IMP}$ and $\angle\text{MSP} = \theta = \angle\text{MIP}$. The side MP is common to both triangles and hence SM = MI and

$$y = x$$

Since this is independent of θ, the above discussion will hold true for any choice of the point P so that every ray striking the mirror will be reflected as if it had emanated from the point I (Figure 23.5).

For anyone looking into the mirror, his/her eye will detect the diverging pencil of reflected rays (or, equivalently, spherical wavefronts) and will be unable to distinguish the situation from that of a point source located at I. This is our first example of **image formation**. In this case the image of a point source is a point located at the same perpendicular distance behind the mirror as the source is in front of it.

If, instead of a single point, one had an extended object (SS′ in Figure 23.6), then each point on the source will give rise to a point image. As a result, a sharp extended image (II′ in Figure 23.6), which is the same size as the object, is formed behind the mirror. Note again that the object need not be self-illuminated; in most cases of image formation, the object is illuminated by some external source (sunlight, lamp, etc.). In these circumstances, each point on the object scatters light in all possible directions and accordingly is effectively a point source of light. Such a point source is sometimes described as a 'point object'.

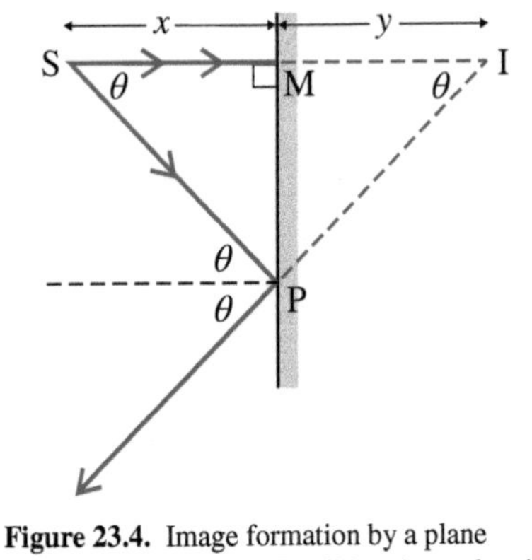

Figure 23.4. Image formation by a plane mirror: every ray from S striking the surface is reflected so that it appears to emanate from I.

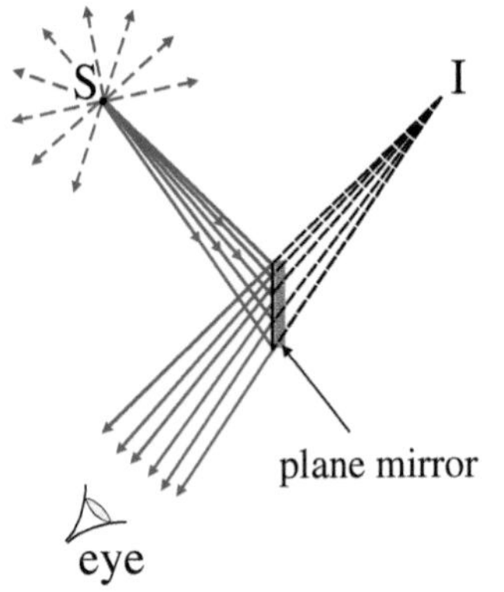

Figure 23.5. Image formation by a plane mirror: the light entering the eye is in the form of a narrow diverging pencil of rays emanating from I.

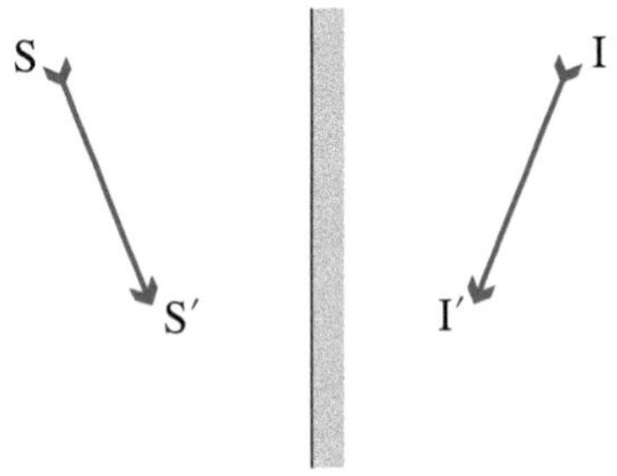

Figure 23.6. Image of an extended object in a plane mirror.

23.3 Image formation by spherical mirrors

Spherical mirrors are common everyday objects and come in two forms: *convex* (reflective surface curving outwards – Figure 23.7(a)) and *concave* (reflecting surface curving inwards — Figure 23.7(b)).

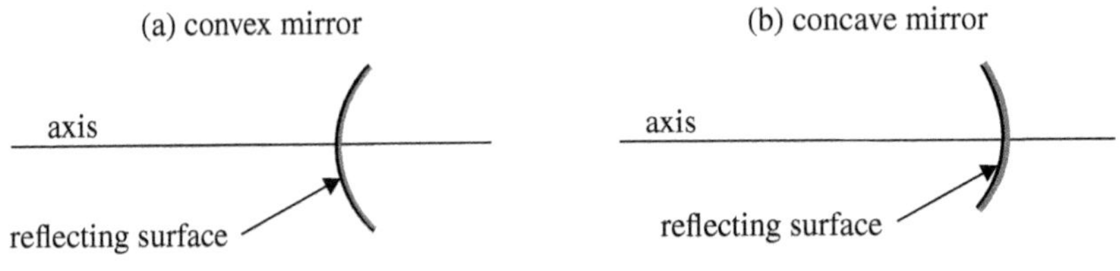

Figure 23.7. The two types of spherical mirror.

Convex mirror

We consider first the possibility of image formation by a convex mirror. The mirror illustrated in Figure 23.8 has its reflective surface facing left and has a radius of curvature ρ. A point source S placed on the axis of the mirror will radiate light rays in all directions but, again, only those rays striking the mirror surface contribute to image formation. Again we consider just two rays. In the first place, it is clear that the ray striking the centre M of the mirror, where the angle of incidence is zero, will be reflected back along its own path. Secondly, a ray striking the mirror at some arbitrary point P at a height h above the axis will be reflected away from the axis as shown in the figure (angle of incidence = angle of reflection = θ). When the direction of this ray is projected backwards it intersects the axis at a point which we denote as I.

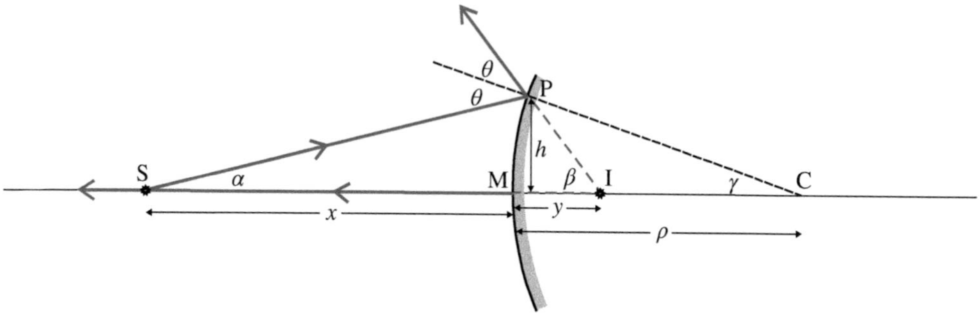

Figure 23.8. Image formation in a convex mirror.

The normal to the surface at the point P intersects the axis at the point C which is the centre of curvature of the mirror; that is, PC = MC = ρ. Thus, the angle between this normal and the mirror axis is given by γ, where $\tan\gamma = \frac{h}{\rho}$. If $h \ll \rho$ (that is, γ small) then

$$\gamma = \frac{h}{\rho}$$

For the triangle Δ IPC, the exterior angle $\beta = \theta + \gamma$. Similarly for the triangle Δ IPS, $2\theta = \beta + \alpha$. Thus, eliminating θ, we find $\beta = \alpha + 2\gamma$.

Now, if the diameter of the mirror $\ll \rho$ ('small aperture mirror') and x is not too much less than ρ, the angles α and β will be small, that is, we can approximate these angles as follows

$$\alpha \approx \frac{h}{x} \text{ and } \beta \approx \frac{h}{y}$$

in which case

$$\frac{h}{y} = \frac{h}{x} + 2\frac{h}{\rho}$$

and hence

$$\frac{1}{x} - \frac{1}{y} = -\frac{2}{\rho} \tag{23.1}$$

or $y = \frac{x\rho}{2x + \rho}$, which shows that the position of I is independent of h (the choice of the point P on the surface of the mirror) in this situation.

Thus, within the approximation made, all rays from S that strike the mirror will be reflected so that they diverge from a single point I (Figure 23.9). In other words, the approximation demands that all rays remain in a pencil forming a small angle with the axis of the mirror (so called '*paraxial rays*'). If the angles are larger (that is, if the mirror has larger aperture), a point source will not give rise to a point image and the image of any extended object will be distorted ('*spherical aberration*').

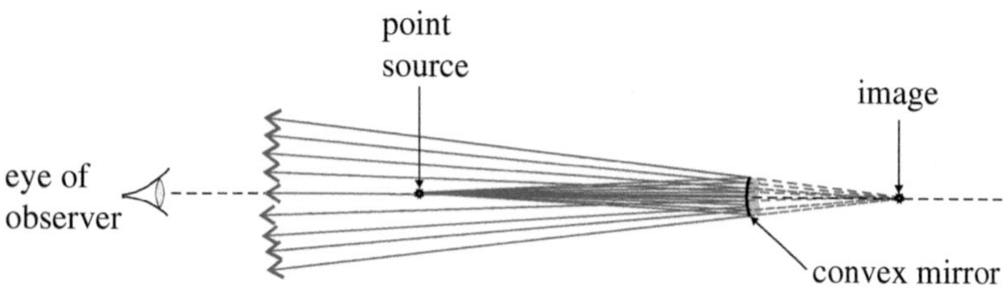

Figure 23.9. Image formation in a convex mirror showing the diverging pencil of rays which gives rise to an image behind the mirror.

Concave mirror

We now turn our attention to image formation by a concave mirror. In this case two distinct possibilities arise depending on how far the source is from the mirror. When the source is near the mirror (Figure 23.10), the reflected rays diverge as before and an image is formed behind the mirror. Note, however, that following the same analysis that led to Equation (23.1), we find that the expression that determines the location of the image in terms of the position of the source and the radius of curvature is different from that obtained above in the case of a convex mirror.

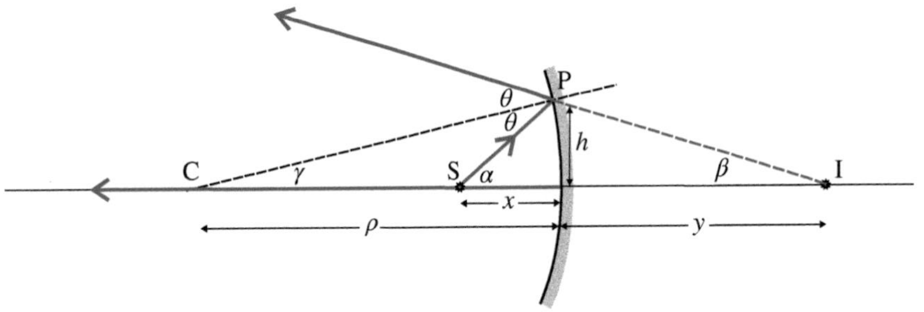

ΔIPS: $2\theta = \beta + \alpha$

ΔIPC: $\theta = \beta + \gamma$

$\rightarrow \alpha - \beta = 2\gamma$

$\rightarrow \frac{1}{x} - \frac{1}{y} = \frac{2}{\rho}$

Figure 23.10. Image formation in a concave mirror in the case where the reflected rays do not cross the axis.

When the source is located further away from the source (Figure 23.11) a wholly different situation arises. When the laws of reflection are applied in this case, one sees that the light ray is reflected towards the axis and intersects the axis at some point I. Furthermore, the expression that determines the location of the image is different yet again from the corresponding expressions obtained in the two previous cases discussed above.

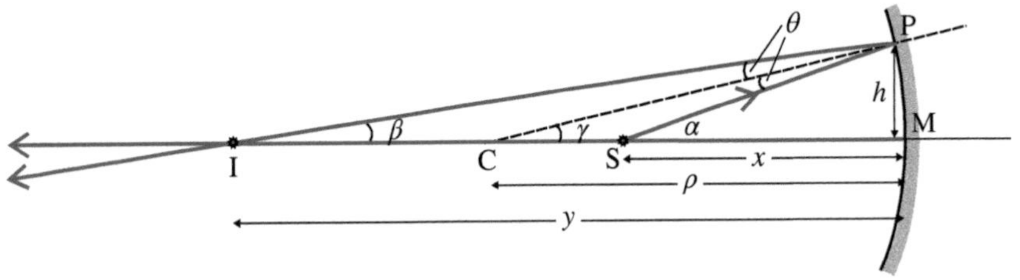

ΔIPS: $\alpha = \beta + 2\theta$

ΔIPC: $\gamma = \beta + \theta$

$\rightarrow \alpha + \beta = 2\gamma$

$\rightarrow \frac{1}{x} + \frac{1}{y} = \frac{2}{\rho}$

Figure 23.11. Image formation in a concave mirror in the case where the source is sufficiently far from the mirror so that reflected rays cross the axis.

The way that the image is formed in this case is rather different from all of the earlier examples of image formation. The eye of an observer detects a diverging pencil of rays emanating from the point I but, in this situation, the rays first converge and actually pass physically through the point I before they start to diverge (Figure 23.12). The type of image formed in this case is said to be a **real image**; it is possible to cast an image of the source on a screen placed perpendicular to the axis at the point I. On the other hand, in all three of the cases studied earlier, namely the plane mirror, the convex mirror and the concave mirror with the source near the mirror, the light appears to come from a point through which no physical light energy has passed. No image will be cast on a screen placed at I and, in such cases, the image is known as a **virtual image**.

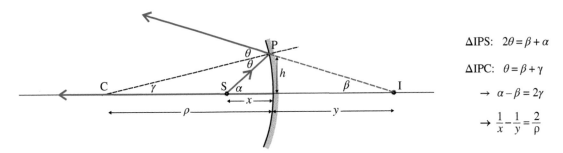

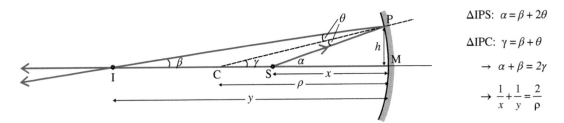

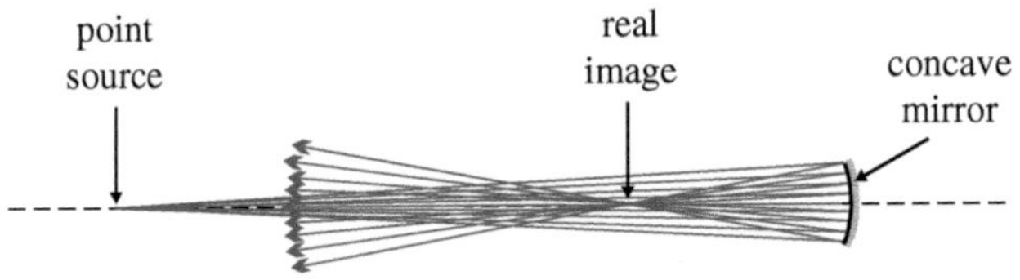

Figure 23.12. Formation in a real image in a concave mirror.

It is unsatisfactory to have to use a different expression to determine the location of the image for each possible way an image can be formed by a spherical mirror. An alternative approach is to adopt an appropriate **Convention of Signs** which, as we shall see, will enable us to cover all cases by a single relationship. What this involves is choosing to treat variables representing distances of images and sources from a mirror as well as the radius of curvature of the mirror as **algebraic quantities**. Symbols used to represent these quantities, therefore, represent a sign, magnitude and unit (recall Section 2.2) unlike the variables x, y, and ρ used above which are always positive quantities. An inspection of the results derived in this section implies that all possible cases can be covered by the single relationship

$$\boxed{\frac{1}{u} + \frac{1}{v} = \frac{2}{R}} \tag{23.2}$$

where

u is the distance from the source to the centre of the mirror
v is the distance of the image to the centre of the mirror and
R is the radius of curvature of the mirror

provided the following **Convention of Signs** for spherical mirrors is adopted

- real images are at positive distances from the mirror (that is, $v = +y, y > 0$)
- virtual images are at negative distances from the mirror (that is, $v = -y, y > 0$)
- concave mirrors have positive radii of curvature (that is, $R = +\rho, \rho > 0$)
- convex mirrors have negative radii of curvature (that is, $R = -\rho, \rho > 0$)

When using Equation (23.2), therefore, one must be careful to attach the appropriate sign to numerical values being substituted. It follows that a negative result of a calculation for an image distance means that the image is virtual while a negative result for a radius of curvature means that the mirror is convex.

Study Worked Example 23.1

For problems based on the material presented in this section visit up.ucc.ie/23/ *and follow the link to the problems.*

23.4 Refraction of light

The law of refraction

The law of refraction discussed in Section 13.5, when applied to geometrical optics, tells us that a ray of light will be refracted at a plane interface in such a way that (a) the refracted ray lies in the plane defined by the incident ray and the normal to the interface at the point of incidence and (b) the angle in the second medium (the 'angle of refraction' θ_2) is related to the angle in the first medium (the 'angle of incidence' θ_1) as follows

$$\frac{\sin\theta_1}{v_1} = \frac{\sin\theta_2}{v_2} \tag{23.3}$$

where v_1 and v_2 are the velocities of light in the respective media (Figure 23.13).

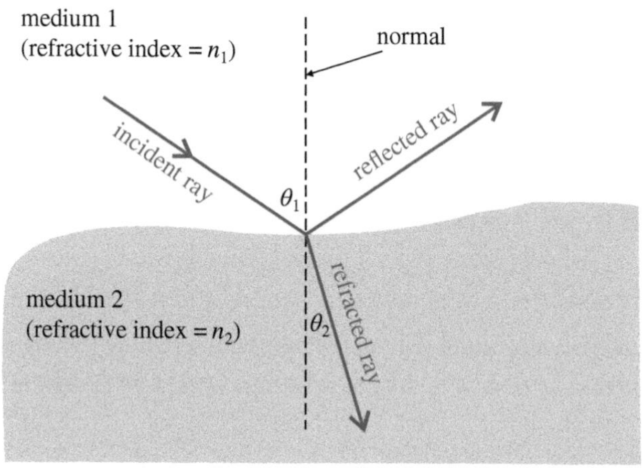

Figure 23.13. The refraction of a ray of light as it passes through the interface between two transparent media; note that some light is also reflected. The refracted ray is in the plane defined by the incident ray and the normal.

The ratio of the speed of an electromagnetic wave in vacuum to its speed in a medium is called the **refractive index** of the medium that is

$$n := \frac{c}{v}$$

and, using the value for wave speed, ($v = \frac{1}{\sqrt{\mu\varepsilon}}$) derived in Section 21.2, we get

$$n = \frac{\sqrt{\mu\varepsilon}}{\sqrt{\mu_0\varepsilon_0}} = \sqrt{\mu_r\varepsilon_r}$$

where μ and ε are, respectively, the permeability and permittivity of the medium. Note that in non-ferromagnetic media $\mu \approx \mu_0$ and hence $n \approx \sqrt{\varepsilon_r}$.

Now $n_1 = \frac{c}{v_1}$ and $n_2 = \frac{c}{v_2}$ and so we can write (23.3) as $\frac{c}{n_1}\sin\theta_2 = \frac{c}{n_2}\sin\theta_1$

or

$$n_1 \sin\theta_1 = n_2 \sin\theta_2 \tag{23.4}$$

Stated in the latter form, the law of refraction is usually known as **Snell's law**. This law is also credited to Abu Said Ibn Sahl (940–1000) of Baghdad.

Note that, by definition, the refractive index of vacuum is 1 exactly. Refractive index usually increases with the density of the medium and also varies with the frequency of the radiation, that is $n = n(f)$ or $n = n(\omega)$. The refractive index of air is approximately 1.0003 and can be safely approximated by unity in most situations.

Study Worked Example 23.2

Refraction at a plane interface

Consider the passage of a light ray from a medium with higher index of refraction, n_2, to a medium with lower index of refraction, n_1 — usually this means from a dense medium (such as glass) to a rare medium (such as air). The diagram on the left of Figure 23.14 shows two rays emitted from a point source S. One ray strikes the interface perpendicularly and, accordingly, passes through undeviated. The second ray strikes the interface at some arbitrary point P and is refracted as shown so that, if projected backwards, it would intersect the first ray at the point I. Applying Snell's law (23.4)

$$n_1\left(\frac{\mathrm{PC}}{\mathrm{PI}}\right) = n_2\left(\frac{\mathrm{PC}}{\mathrm{PS}}\right) \quad \text{or} \quad n_1\left(\frac{h}{\sqrt{y^2+h^2}}\right) = n_2\left(\frac{h}{\sqrt{x^2+h^2}}\right)$$

Anyone looking into the medium from above will only observe a narrow pencil of rays (right hand diagram in Figure 23.14) due to the limited aperture of the human eyes so that, effectively, only rays for which P is close to C will be involved. In other words, $h \ll x, y$ so that $\frac{n_1}{y} = \frac{n_2}{x}$. Thus the distance below the surface at which the image is formed is given by

$$y = \frac{n_1}{n_2}x$$

where x is the distance of the source below the surface. Since $n_1 < n_2$ then $y < x$, which explains why, for example, if one looks into a river or pool it appears to be shallower than it really is and why a straight rod partially immersed in the water appears to have a bend in it.

Total internal reflection

Consider again the passage of a light ray from a medium with an index of refraction n_1 to a medium with a lower index of refraction n_2. Maxwell's equations require that there is always some reflection at interfaces between non-conducting media (see Section 21.8 on the *Understanding Physics* website at up.ucc.ie/21/8/) even at normal incidence (~ 4% for ordinary glass in that case). As illustrated in Figure 23.15, because $n_1 < n_2$ in Equation (23.4), the angle of refraction is greater than the angle of incidence; that is, $\theta_1 > \theta_2$. As θ_2 is increased, a point is reached at which θ_1 reaches its maximum possible value, namely $\frac{\pi}{2}$. The corresponding value of θ_2 is known as the **critical angle,** θ_c in Figure 23.15(b)

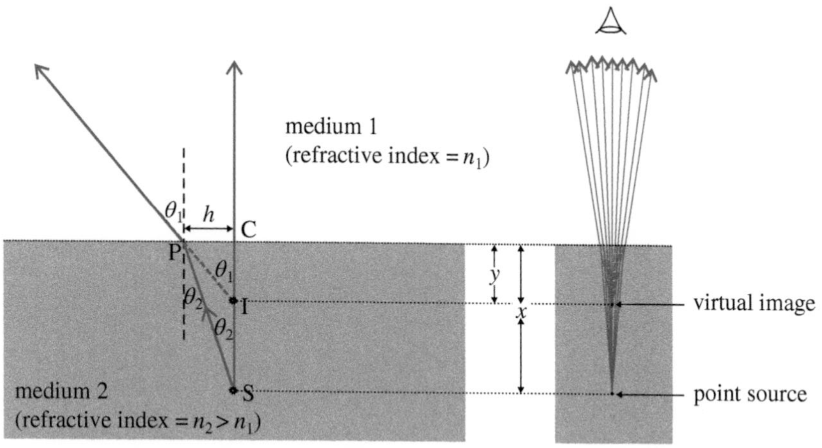

Figure 23.14. Image formation due to the refraction of light travelling from one transparent medium into another less dense one.

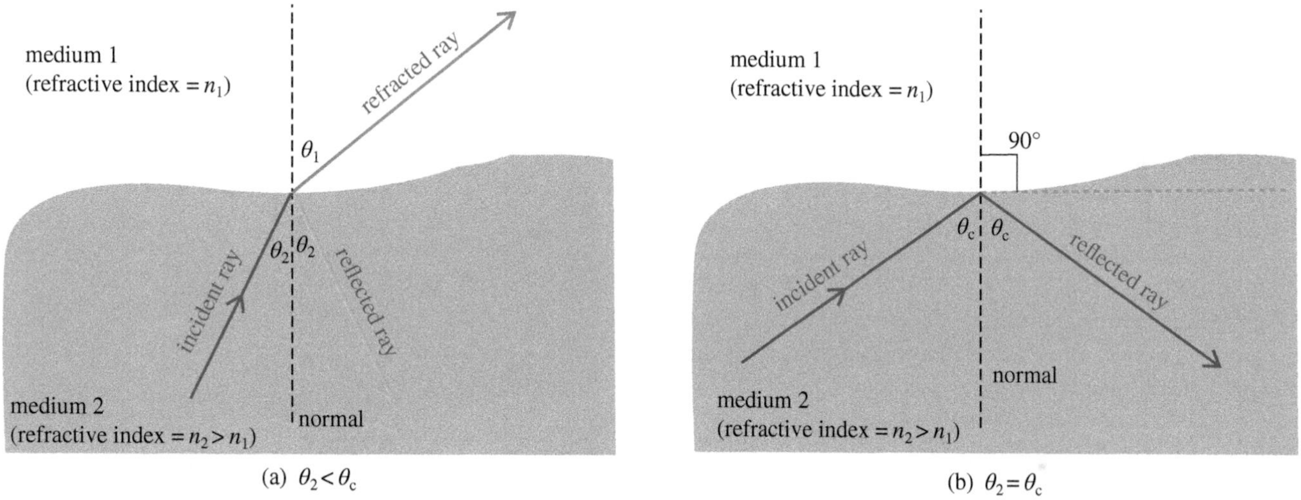

Figure 23.15. (a) Partial reflection of ray in the more dense medium. (b) Total internal reflection occurs when the angle of incidence in the more dense medium exceeds the critical angle.

From Equation (23.4) we have $n_2 \sin \theta_c = n_1 \sin \left(\frac{\pi}{2}\right)$

and thus

$$\theta_c = \sin^{-1} \left(\frac{n_1}{n_2}\right)$$

where n_1 ($<n_2$) is the refractive index of the less dense medium. When $\theta_2 > \theta_c$ the light ray cannot emerge. The refracted ray vanishes and the reflected ray carries the full intensity of the incident ray back into the dense medium — this phenomenon is known as **total internal reflection**. Experimental measurement of the critical angle can be used to determine the ratio of the refractive indices or, if one of the media is air or vacuum, to determine the refractive index of the other medium.

Total internal refection has an important practical application in *fibre optics*. As illustrated in Figure 23.16, an optical fibre comprises a cylindrical tube of glass which is clad with a material of lower refractive index. The rays which enter the fibre and strike the fibre walls at angles which are greater than θ_c cannot escape from the fibre. The light, therefore, is channelled along the fibre with only very small losses through the cladding. Optical fibres are of major importance in the communication of optical signals in the telecommunications industry and in enabling physicians to view sites within the human body (endoscopy) with minimal disturbance of the patient.

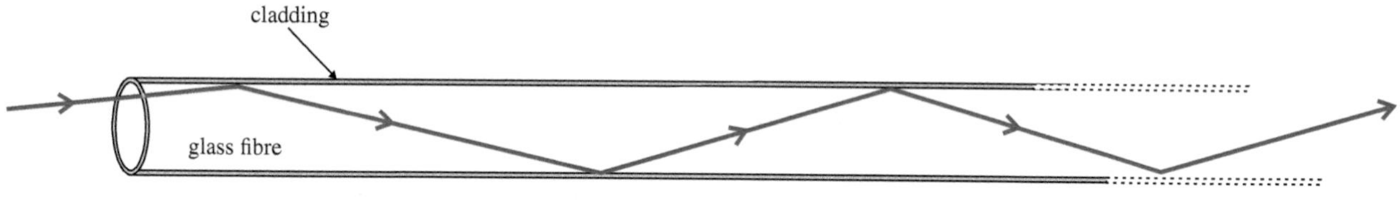

Figure 23.16. The channelling of a light ray through total internal reflection in an optical fibre.

In Section 14.17, *frustrated internal reflection* was mentioned as an example of tunnelling of waves. This phenomenon can now be illustrated. Figure 23.17(a) shows a ray of light in glass striking a plane glass-air interface at an angle greater than the critical angle; under normal circumstance the light will undergo total internal reflection. If, however, a second glass block is brought very close to the glass/air interface (Figure 23.17(b)), sufficiently close that the gap between the two blocks is comparable to the wavelength of the light, some of the light waves can pass across the gap into the second block. Total internal reflection is 'frustrated' by tunnelling of the light wave.

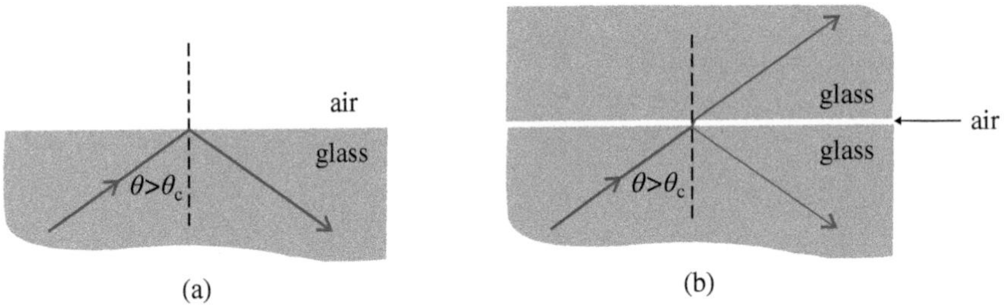

Figure 23.17. (a) Total internal reflection of a light ray. (b) Frustrated total internal reflection; some of the intensity of the light ray is transmitted through the air gap.

Polarisation of light by reflection

When unpolarised radiation is reflected from a plane surface, the reflected waves are observed to be partially polarised. We can obtain a qualitative understanding of this effect by considering the reflection of an unpolarised plane wave at a plane interface as illustrated in Figure 23.18. The surface at which the incident radiation is reflected is in the plane perpendicular to the page. The unpolarised incident radiation may be considered as having components of the electric field strength (a) parallel to the reflecting surface (that is perpendicular to the page, indicated by circles in the figure) and (b) in the plane of the page (the direction of the vibration being indicated by the two-headed arrows).

As we have seen above, on striking the interface, some of the radiation is reflected and some is transmitted across the interface and undergoes refraction. We can consider the reflected wave as being caused by the effect of the vibration of charges within the medium which oscillate under the influence of the refracted wave in a manner not unlike the scattering process discussed in Section 22.1. Thus, an observer viewing the reflected wave sees the full effect of the radiation from the oscillations vibrating perpendicularly to the page but observes a reduced field strength resulting from the oscillations in the plane of the page. Thus, the reflected radiation is partially polarised.

If the incident angle (θ_B in Figure 23.19) is such that the reflected wave is perpendicular to the refracted wave, the ***E***-field of the reflected wave will have no component along the line of sight and thus the reflected wave will be linearly polarised, the direction of polarisation being parallel to the interface. From the figure we see that

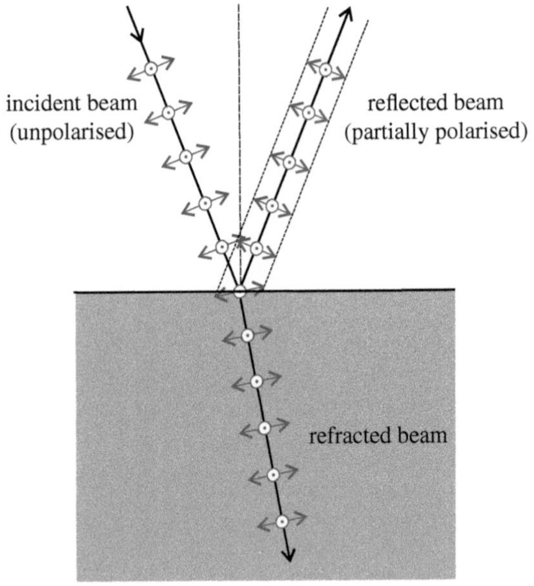

Figure 23.18. The light reflected from a plane interface between two media is partially polarised.

$$\theta_B + \frac{\pi}{2} + \phi_B = \pi \quad \text{or} \quad \phi_B = \frac{\pi}{2} - \theta_B$$

Invoking Snell's law (Equation (23.4)) for the case where the incident and reflected rays are in air, we obtain

$$1 \times \sin\theta_B = n \sin\phi_B = n \sin\left(\frac{\pi}{2} - \theta_B\right) = n\cos\theta_B$$

$$\Rightarrow \quad \tan\theta_B = n$$

The incident angle for which the reflected wave is linearly polarised is called *Brewster's angle* (David Brewster 1781–1868).

Let us consider, as an example, at what angle light should be incident on a plane glass surface so that the reflected light be 100% polarised. Typically, the refractive index of glass is about 1.5 and hence

$$\theta_B = \tan^{-1}(1.5) = 56.3°$$

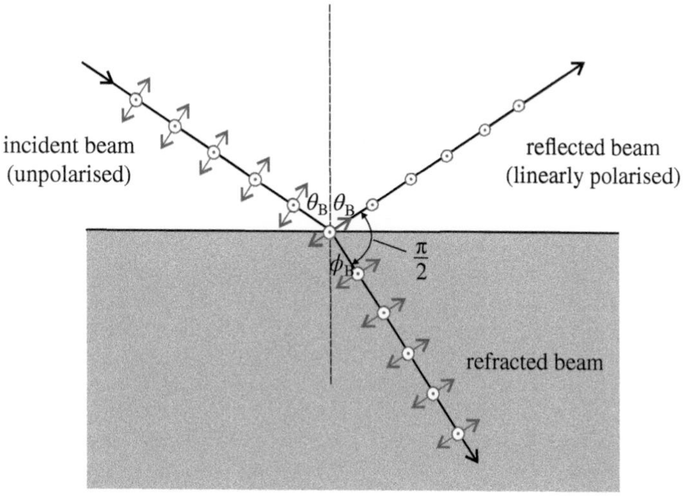

Figure 23.19. When a beam of light is incident on a plane interface at Brewster's angle, the reflected light is linearly polarised.

The measurement of Brewster's angle for light reflected from a polished surface of an *opaque* medium enables the refractive index of such a medium to be determined.

We can now see why Polaroid sunglasses are so useful in reducing glare from sunlight reflected from water and other horizontal shiny surfaces; this is because the light reflected from such surfaces is partially polarised. Since the Polaroid sheets are inserted in the sunglasses with the polarisation axis vertical, a significant fraction of the reflected light is filtered out.

For problems based on the material presented in this section visit up.ucc.ie/23/ *and follow the link to the problems.*

23.5 Refraction at successive plane interfaces

Parallel interfaces

Figure 23.20 shows a material of higher refractive index wedged between material of lower refractive index in such a way that that the interface planes are parallel to one another; for example, a sheet of glass surrounded by air. A ray of light striking an interface perpendicularly will pass straight through without its path being affected. If the ray strikes the interface at some angle θ ($\neq 0$) with the normal, it will be refracted as indicated in the figure. Since the interface planes are parallel, the angle of incidence at the second interface is equal to the angle of refraction φ at the first. Application of Snell's law shows that the ray will emerge parallel to its original direction, albeit with some lateral translation (to the right in the figure).

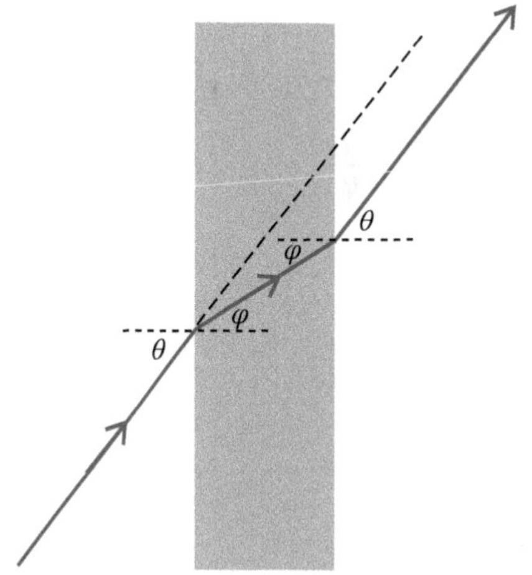

Figure 23.20. Refraction of a light ray on passing through a medium with parallel interfaces, such as a block of glass in air.

Inclined interfaces; prisms

If the interfaces between the media are not parallel as in Figure 23.20 but rather inclined at some angle A as in Figure 23.21, a different situation arises. In this case, as can be seen from the figure, an incident ray, having been refracted at each interface in turn, emerges in a direction different from that of its original path. The net effect is to bend the ray through some angle (δ in the figure). As we shall see, this deviation of light rays can be used in a controlled way to facilitate image formation.

A ray of light incident on the first (that is, left-hand) interface in Figure 23.21 strikes that interface making an angle of θ_1 with the normal to the interface. Let the refractive index of the outside medium be n_o and that of the inside medium be n ($n > n_o$). From Snell's law (23.4), we know that $n_o \sin\theta_1 = n \sin\varphi_1$ where φ_1 is the angle between the ray in the medium of refractive index n and the normal to the interface. Similarly, for refraction at the second interface, $n \sin\varphi_2 = n_o \sin\theta_2$. Since the angle between the two normals to the interface must be the same as the angle between the two interfaces, we see that the exterior angle of the triangle containing φ_1 and φ_2 must also be A and hence

$$A = \varphi_1 + \varphi_2$$

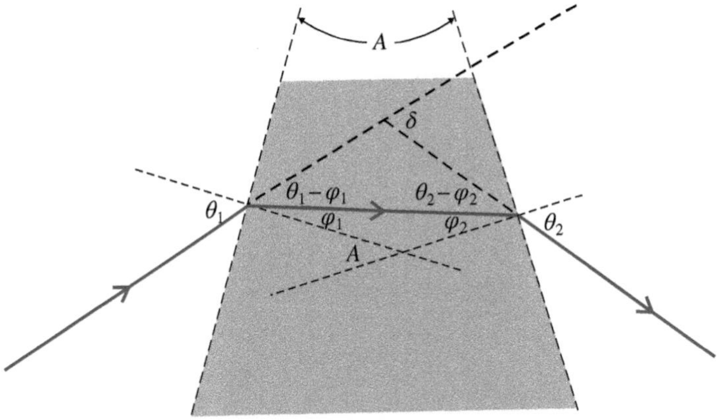

Figure 23.21. The deviation of a light ray on passing through a medium with inclined interfaces.

The deviation of the ray resulting from its passage through both interfaces, therefore, is given by

$$\delta = (\theta_1 - \varphi_1) + (\theta_2 - \varphi_2) = (\theta_1 + \theta_2) - A$$

An important special case arises when all angles are small (Figure 23.22); that is, A and θ_1 are small and, hence, φ_1, φ_2 and θ_2 are also small. Using the approximation for small angles ($\sin\theta \approx \theta$) and applying Snell's law, we get $n_o\theta_1 = n\varphi_1$ and $n\varphi_2 = n_o\theta_2$ and hence

$$\delta = \left(\frac{n}{n_o}\varphi_1 + \frac{n}{n_o}\varphi_2\right) - A = \left(\frac{n}{n_o} - 1\right)A$$

If, as is usually the case in what follows below, the outside medium is air or vacuum ($n_o \to 1$), the angle of deviation is given by

$$\delta = (n - 1)A \qquad (23.5)$$

When constructed from glass, the element shown shaded in Figure 23.21 is a section of what is usually called a glass 'prism' and, in the case of Figure 23.22, is called a 'thin prism'.

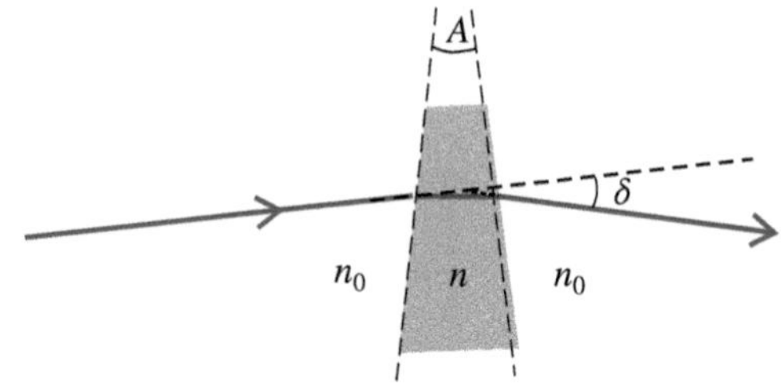

Figure 23.22. Small angle deviation of a light ray in passing through a medium where the interfaces are inclined at a small angle (a thin prism).

Image formation by lenses

As can be seen from Equation (23.5), the deviation of a ray is directly proportional to the angle between the interfaces. This result can be used to design a simple imaging device, the basic structure of which is indicated in Figure 23.23. The device comprises a large number of small elements (thin prisms) made from a transparent material; the interfaces of the element on the axis are parallel and the angle between the interfaces of the other elements increases with distance from the centre as shown. Thus, a ray of light will be bent more by outside elements compared with those nearer the centre.

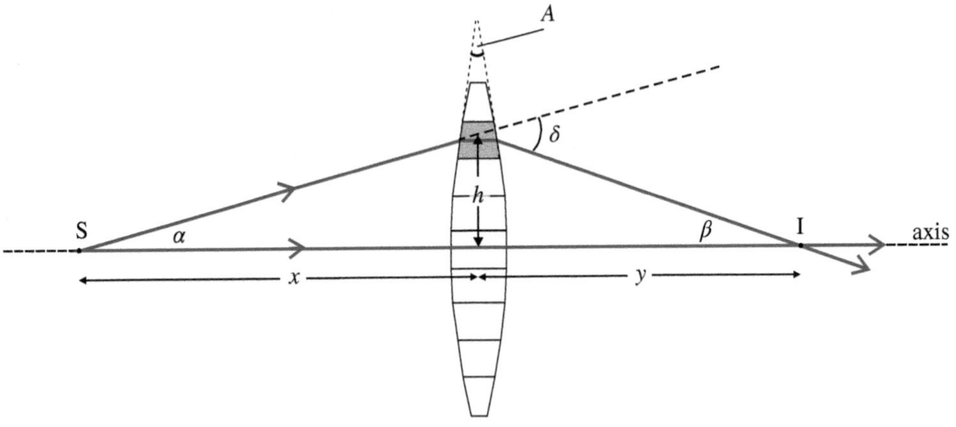

Figure 23.23. Image formation of light from a point source by an optical system comprising stacked thin prisms elements of varying angle.

Consider light emitted from a point source S located on the axis of the system as shown in Figure 23.23. A ray of light from S travelling along the axis will strike the central element and pass through the system undeviated. A ray of light striking an element at some arbitrary distance h from the centre will be deviated by an angle δ. Depending on the distance of S from the centre of the system, this ray may (or may not) cross the axis. We will consider here only the case where the ray intersects the axis at a point I. For the formation of a clear image, one requires that all rays from S that strike the system are also bent so that they pass through I, irrespective of the value of h. We now investigate how this may be arranged in the case of small aperture system (that is, $h \ll x, y$ — *paraxial* rays).

From Figure 23.23, $\delta = \alpha + \beta \approx \frac{h}{x} + \frac{h}{y}$, that is, $\delta \propto h$ and since, from Equation (23.5), $\delta \propto A$ we see that the criterion for a point image (all rays from S passing through I) is that A be directly proportional to h. The image formation device so constructed is an example of a **lens**, similar to that known as a *Fresnel lens*, originally developed for lighthouses and now commonly used in automobile headlights, overhead projectors, projection televisions, low-cost magnifying glasses, etc.

The **power of a lens** is defined as the angular deviation per unit distance from the axis, that is

$$\boxed{P := \frac{\delta}{h}}$$

Thus,

$$P = \frac{\delta}{h} = \frac{1}{x} + \frac{1}{y} = (n-1)\frac{A}{h} = \text{constant}$$

and the position of the image of a point source at a distance x from the centre of the lens is given by $y = \frac{x}{xP - 1}$.

From the definition, the SI unit of lens power is rad m^{-1}, known as the **dioptre** and denoted by the symbol D.

23.6 Image formation by spherical lenses

Spherical lenses form part of many everyday objects such as reading glasses, magnifying glasses, binoculars, etc. or scientific instruments such as microscopes and telescopes. As we shall see, image formation in spherical lenses can occur in a similar way to that discussed in the previous section. Consider, as an example, the double convex (bi-convex) lens shown in Figure 23.24. The left-hand face is part of a sphere of radius ρ_1 with the centre of curvature at C_1 while the right-hand face has radius of curvature ρ_2 with centre at C_2. In the figure, the radii of curvature are shown as measured to the centre of the lens rather than to the corresponding faces; since all lenses to be discussed are thin, this approximation has negligible consequences.

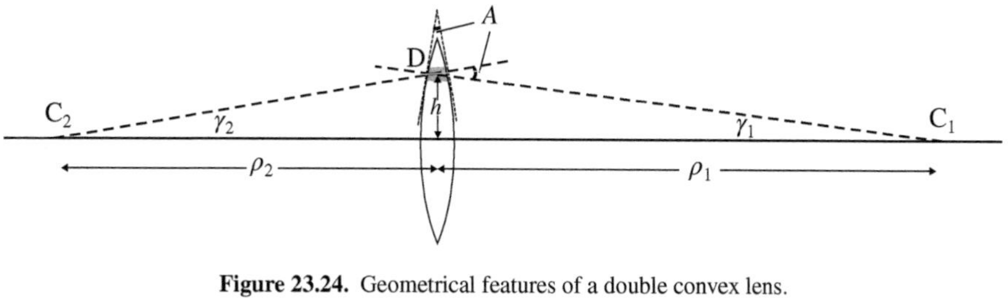

Figure 23.24. Geometrical features of a double convex lens.

The dashed lines which have been drawn perpendicular to each face in Figure 23.24 have to pass through the corresponding centres of curvature. The angle A between these lines (each normal to the surface of a face) is the same as the angle between the corresponding tangent lines drawn at the same points. A sufficiently small horizontal slice (shaded element in the figure) will behave like an element (thin prism) of the lens discussed in the previous section. Inspecting the $\Delta\, C_1DC_2$, we see that

$$A = \gamma_1 + \gamma_2 = \frac{h}{\rho_1} + \frac{h}{\rho_2} \tag{23.6}$$

which shows that A is directly proportional to h again in this case.

Let us now consider what happens when a ray of light from a point source on the axis passes through this lens (Figure 23.25). The analysis is identical to that applied to Figure 23.23 above. A ray of light from S travelling along the axis will pass through the centre of the lens undeviated. A ray of light striking the lens at some arbitrary distance h from the axis will pass through the shaded element and be deviated by an angle $\delta = (n-1)A$. Clearly, there are two possibilities depending on the distance of the source from the centre of the lens: either (i) the ray will be deviated sufficiently so that it crosses the axis as shown in Figures 23.25 and 23.26, forming a real image, or (ii) it is not deviated sufficiently so that it emerges from the lens directed away from the horizontal, in which case a virtual image will be formed (Figure 23.27 — Case 1b).

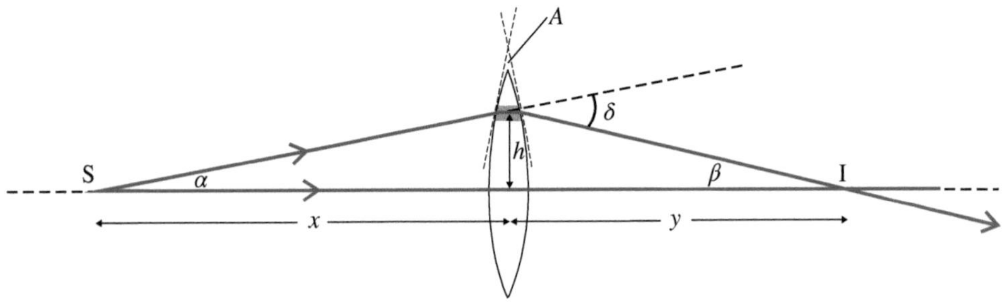

Figure 23.25. Image formation by a converging lens where the point source S is sufficiently far from the lens so that light rays are deviated so as to cross the axis.

In the case shown in Figure 23.25, applying the same geometrical considerations as for Figure 23.23 above we have $\delta = \alpha + \beta \approx \frac{h}{x} + \frac{h}{y}$ (if α and β are small). Thus the power of the lens is given by

$$P = \frac{\delta}{h} = \frac{1}{x} + \frac{1}{y} = (n-1)\frac{A}{h} = (n-1)\left(\frac{1}{\rho_1} + \frac{1}{\rho_2}\right) \tag{23.7}$$

where Equation (23.6) has been used to eliminate A. Since, in the small angle approximation used, the value of y does not depend on h, all rays from S striking the lens will cross the axis at the same point thus forming a real point image at I (Figure 23.26).

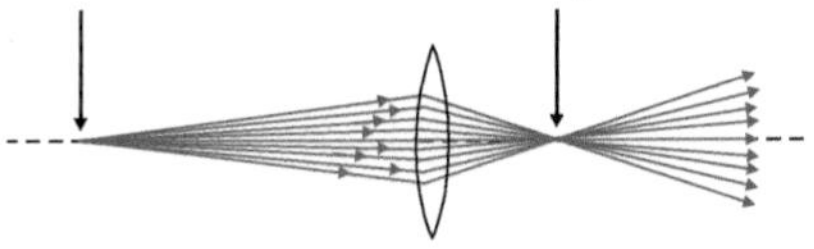

Figure 23.26. Formation of a real image in a converging lens.

The double-convex lens discussed above tends to make any incident light converge, even though a real image may not be formed in all situations; such a lens is called a *converging lens*. Other kinds of lens also exist, however, where one or both of the faces may be concave (or even planar – infinite radius of curvature). A lens that tends to make incident light diverge is classified as a *diverging lens*. Each case must be analysed to determine the appropriate relationship corresponding to Equation (23.7) above. Figure 23.27 summarises the relevant cases, each of which gives rise to a different relationship. Note that the small angle approximation ($\sin\theta \approx \theta$) has been made in all cases which means that the relationships derived apply only to *thin* lenses of *small aperture* (diameter of lens $\ll \rho_1$, ρ_2 and x not too close to the lens — paraxial rays).

Distortion of images will occur if the 'paraxial rays' condition is not satisfied giving rise to spherical aberration (as in Section 23.3 for spherical mirrors).

As in the case of spherical mirrors, it is possible to unify all possible cases of image formation by lenses in a single relationship between algebraic variables by adopting an appropriate convention of signs. This relationship (often called the '*lens maker's formula*') is as follows.

$$\boxed{\frac{1}{u} + \frac{1}{v} = (n-1)\left(\frac{1}{R_1} + \frac{1}{R_2}\right) = P} \tag{23.8}$$

where

u is the distance from the source to the centre of the lens
v is the distance of the image to the centre of the lens and
R_1 and R_2 are the radii of curvature of the two faces of the lens

provided the following **Convention of Signs** is adopted

- real images are at positive distances from the centre of the lens (that is, $v = +y$, $y > 0$)
- virtual images are at negative distances from the centre of the lens (that is, $v = -y$, $y > 0$)
- convex faces have positive radii of curvature (that is, $R = +\rho$, $\rho > 0$)
- concave faces have negative radii of curvature (that is, $R = -\rho$, $\rho > 0$)

The convention implies that the power of a lens can also be treated as an algebraic quantity with

$P > 0$ for converging lenses and $P < 0$ for diverging lenses.

In the foregoing we have considered only situations in which a diverging pencil of rays is incident on the lens. A somewhat different situation is illustrated in Figure 23.28. Here a converging pencil of rays is incident on the right-hand lens (say, from some other optical system to the left). In the absence of the lens, this converging pencil would have formed a real image at A but instead is intercepted by the lens so that a real image is formed at B. Such circumstances can also be handled by the lens maker's formula but some care is needed in the application. If the rays in Figure 23.28 were reversed, we see that this is the same as the formation of a virtual image by a converging

CONVERGING LENSES

1. Double-convex lens

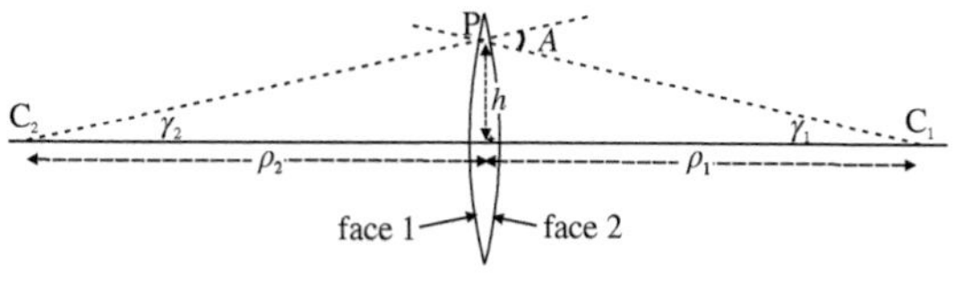

ΔC_1PC_2: $A = \gamma_1 + \gamma_2 \approx \dfrac{h}{\rho_1} + \dfrac{h}{\rho_2}$

Case 1a: Real image formation

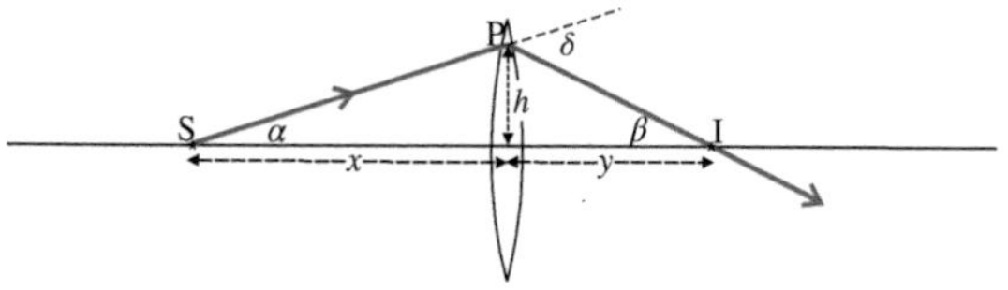

ΔSPI: $\delta = \alpha + \beta \approx \dfrac{h}{x} + \dfrac{h}{y}$

$\delta = (n-1)A \rightarrow \dfrac{1}{x} + \dfrac{1}{y} = (n-1)\left(\dfrac{1}{\rho_1} + \dfrac{1}{\rho_2}\right)$

Case 1b: Virtual image formation

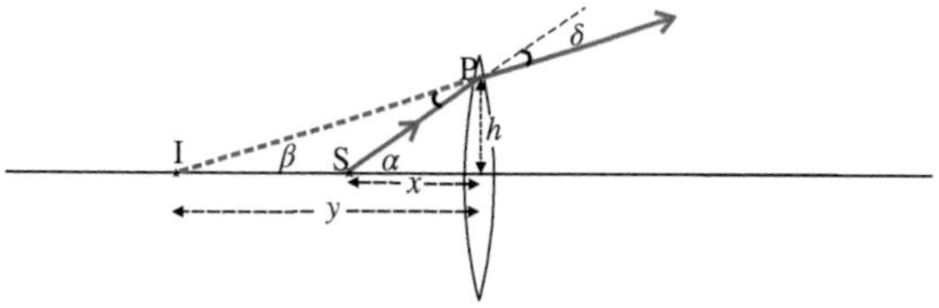

ΔSPI: $\alpha = \beta + \delta \rightarrow \delta = \alpha - \beta \approx \dfrac{h}{x} - \dfrac{h}{y}$

$\delta = (n-1)A \rightarrow \dfrac{1}{x} - \dfrac{1}{y} = (n-1)\left(\dfrac{1}{\rho_1} - \dfrac{1}{\rho_2}\right)$

2. Concavo-convex lens

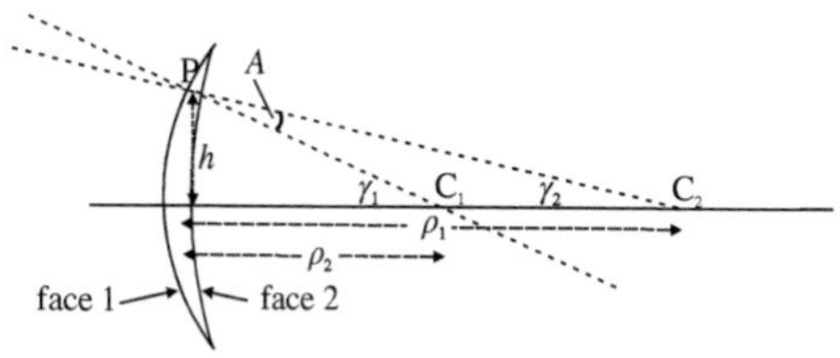

ΔC_1PC_2: $A = \gamma_1 - \gamma_2 \approx \dfrac{h}{\rho_1} - \dfrac{h}{\rho_2}$

Case 2a: Real image formation

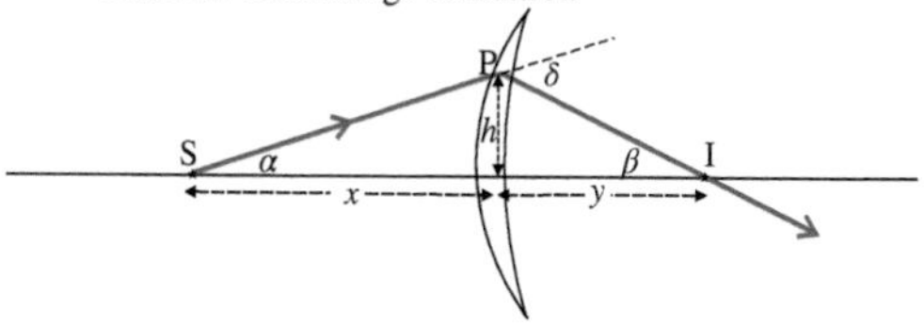

ΔSPI: $\delta = \alpha + \beta \approx \dfrac{h}{x} + \dfrac{h}{y}$

$\delta = (n-1)A \rightarrow \dfrac{1}{x} + \dfrac{1}{y} = (n-1)\left(\dfrac{1}{\rho_1} - \dfrac{1}{\rho_2}\right)$

Case 2b: Virtual image formation

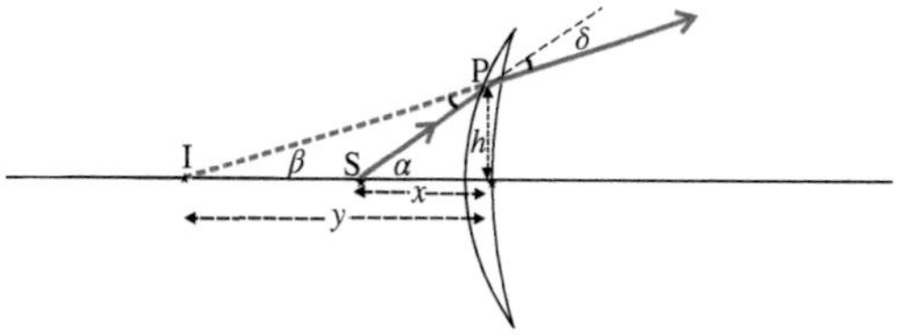

ΔSPI: $\alpha = \beta + \delta \rightarrow \delta = \alpha - \beta \approx \dfrac{h}{x} - \dfrac{h}{y}$

$\delta = (n-1)A \rightarrow \dfrac{1}{x} - \dfrac{1}{y} = (n-1)\left(\dfrac{1}{\rho_1} + \dfrac{1}{\rho_2}\right)$

DIVERGING LENSES

1. Double-concave lens

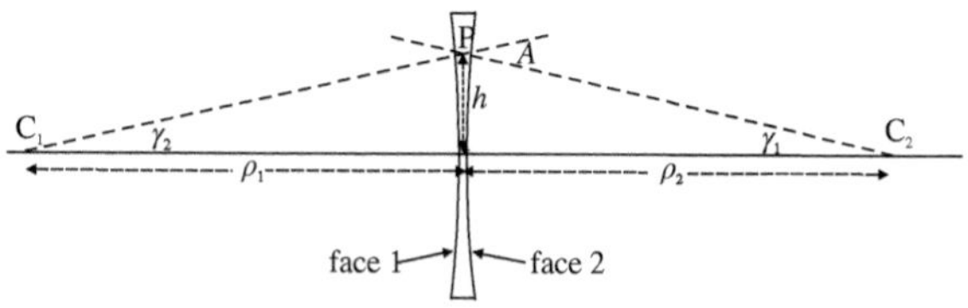

ΔC_1PC_2: $A = \gamma_1 + \gamma_2 \approx \dfrac{h}{\rho_1} + \dfrac{h}{\rho_2}$

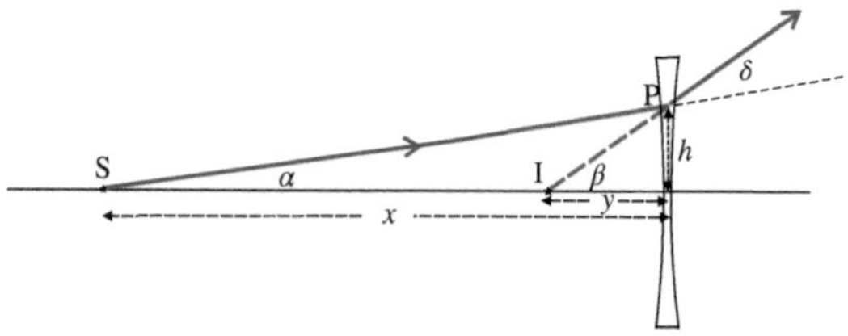

ΔSPI: $\beta = \alpha + \delta \rightarrow \delta = \beta - \alpha \approx \dfrac{h}{y} - \dfrac{h}{x}$

$\delta = (n-1)A \rightarrow \dfrac{1}{x} - \dfrac{1}{y} = (n-1)\left(-\dfrac{1}{\rho_1} + \dfrac{1}{\rho_2}\right)$

2. Convexo-concave lens

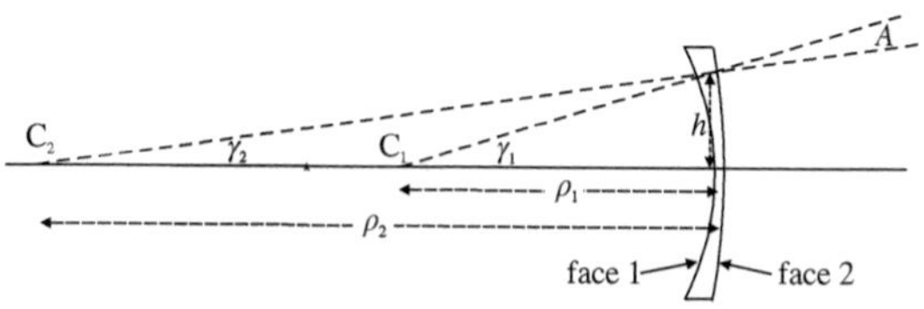

ΔC_1PC_2: $\gamma_1 = \gamma_2 + A \rightarrow A = \gamma_1 - \gamma_2 \approx \dfrac{h}{\rho_1} - \dfrac{h}{\rho_2}$

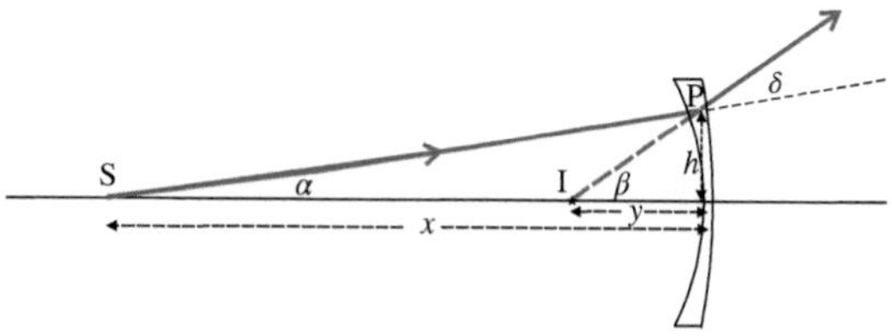

ΔSPI: $\beta = \alpha + \delta \rightarrow \delta = \beta - \alpha \approx \dfrac{h}{y} - \dfrac{h}{x}$

$\delta = (n-1)A \rightarrow \dfrac{1}{x} - \dfrac{1}{y} = (n-1)\left(-\dfrac{1}{\rho_1} - \dfrac{1}{\rho_2}\right)$

Figure 23.27. Different cases of image formation by spherical lenses.

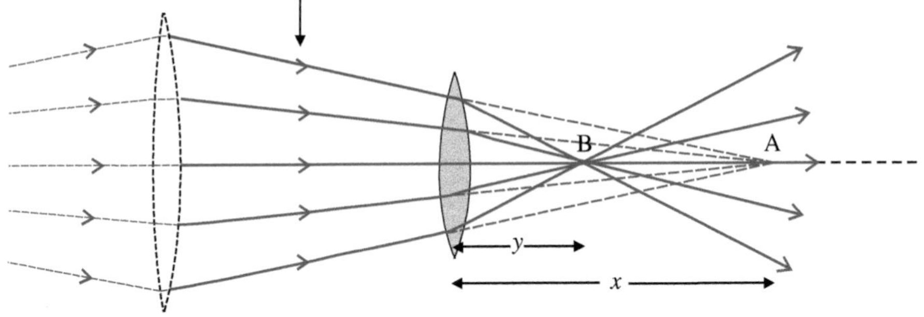

Figure 23.28. The point A must be treated as a virtual source in the context of Equation (23.24).

lens (Figure 23.27 – Case 1b). Applying Equation (23.8) we get $\frac{1}{(-x)} + \frac{1}{y} = P$, in other words $u = -x$ and $v = +y$. Thus, the situation described can be analysed using the lens-makers formula by treating the point A as a '*virtual source*'; that is, its distance from the centre of the lens is negative.

Another situation where care is required in applying Equation (23.8) is illustrated in Figure 23.29. Here a virtual image I_1 of the point source S is produced by the converging lens on the left. The diverging beam emerging from the lens strikes a second converging lens to the right which converges the beam to form a real image I_2. In applying Equation (23.8) to the right-hand lens, the (virtual) image I_1 must be treated as a *real source* ($u = +x$).

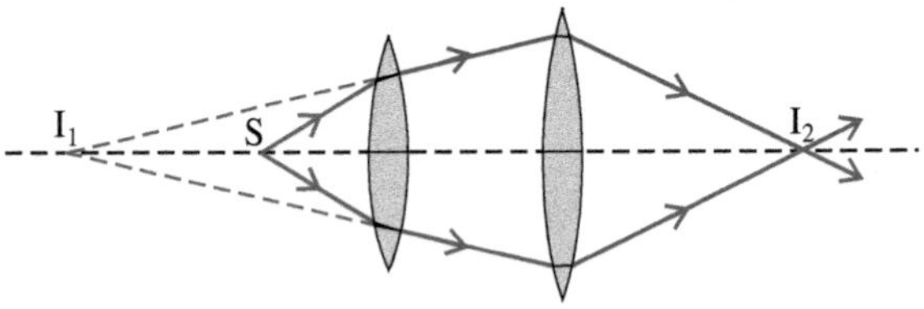

Figure 23.29. The image I_1 formed by the first lens must be treated as a real source when applying Equation (23.8) to the right hand lens.

Focal length

In Figure 23.30 images are illustrated for cases in which a parallel beam (light rays from a source at infinity) are incident on a lens. In this case the image lies at the **focal point** or **focus**, F, of the lens and the distance from the focal point to the lens centre is known as the **focal length**, f. The plane perpendicular to the axis at the focus is called the **focal plane**. If δ is the angle through which the ray is deviated by a lens and h is the distance from the ray to the axis of the lens then, for the cases shown in Figure 23.30, we can write, for small angle deviations, $\delta \approx \sin\delta \approx \frac{h}{f}$ and the power of the lens is given by

$$P = \frac{\delta}{h} = \frac{1}{f}$$

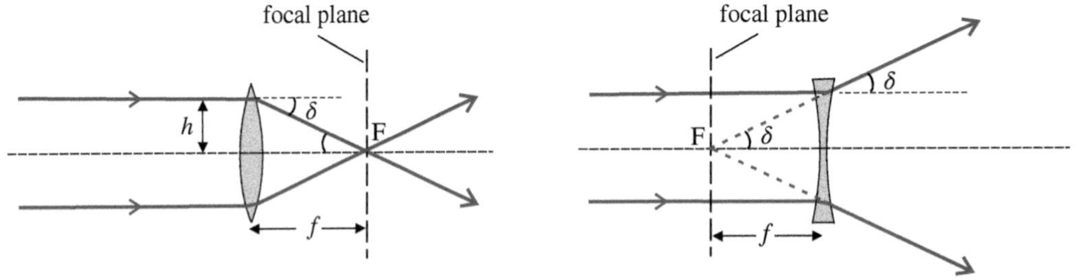

Figure 23.30. Foci, focal lengths and focal planes of spherical lenses; note that, since light can also be incident from the right, there is also a focus and a focal plane on the opposite side in each case.

To avoid distortion of the image (spherical aberration) δ must be a small angle, so the aperture of the lens, or mirror, must be much less than its focal length.

Similarly a focal length can be defined for a spherical mirror. By putting $u = \infty$ and $v = f$ in Equation (23.2), we see that $f = \frac{1}{2}R$ for spherical mirrors and the power of the mirror is given by

$$P = \frac{1}{f} = \frac{2}{R}$$

Note that, in terms of power (or focal length), the same general equation applies to both spherical mirrors and lenses, that is

$$\frac{1}{u}+\frac{1}{v}=P=\frac{1}{f}=\begin{cases}\dfrac{2}{R} & \text{for spherical mirrors}\\[2ex] (n-1)\left(\dfrac{1}{R_1}+\dfrac{1}{R_2}\right) & \text{for spherical lenses}\end{cases}$$

provided the appropriate Convention of Signs is adopted in each case.

Two thin lenses in contact

If two thin lenses are placed touching each other so that their axes coincide (Figure 23.31), the combination can be considered to be equivalent to a single lens. A ray of light from S, striking the first lens at a point h above the axis, is deviated through an angle δ_1 by the first lens and, on striking the second lens, is deviated by a further angle δ_2. Since both lenses are thin, it can be assumed that h is the same for both lenses. The net deviation by the combination of the two lenses is $\delta = \delta_1 + \delta_2$ and, hence, the power of the equivalent single lens is given by

$$P=\frac{\delta}{h}=\frac{\delta_1}{h}+\frac{\delta_2}{h}=P_1+P_2 \tag{23.9}$$

where P_1 and P_2 are the powers of the individual lenses, respectively. Note that if one of the lenses is diverging, the corresponding angle must be subtracted. Thus Equation (23.9) is an algebraic expression; that is, P_1 and P_2 can stand for positive or negative values depending on whether a component lens is converging or diverging.

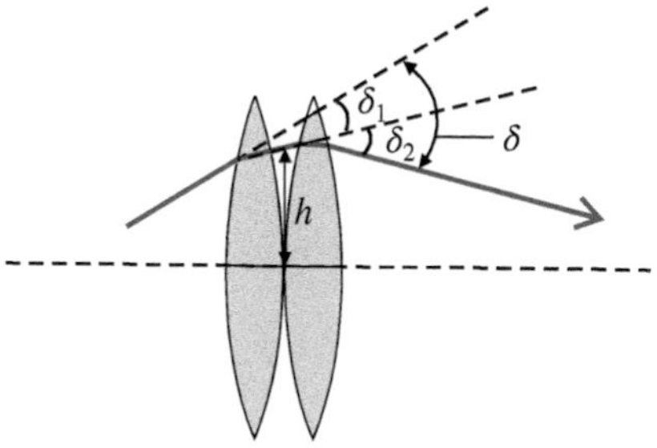

Figure 23.31. Deviation of a light ray by two thin lenses in contact.

Study Worked Example 23.3

For problems based on the material presented in this section visit up.ucc.ie/23/ *and follow the link to the problems.*

23.7 Image formation of extended objects: magnification; telescopes and microscopes

In our discussion so far on image formation by mirrors and lenses, we have confined the analysis to point sources on the axis of the optical system. To broaden the analysis to *extended objects*, however, we need to consider how an image is formed of an object slightly displaced from the axis (retaining the paraxial rays approximation).

In Figure 23.32(a) a small object SS′ is placed so that it is perpendicular to the axis of a converging lens (the size of the object is exaggerated for clarity of the diagram). The location of I, the image of S, can be found from Equation (23.8), as before. The location

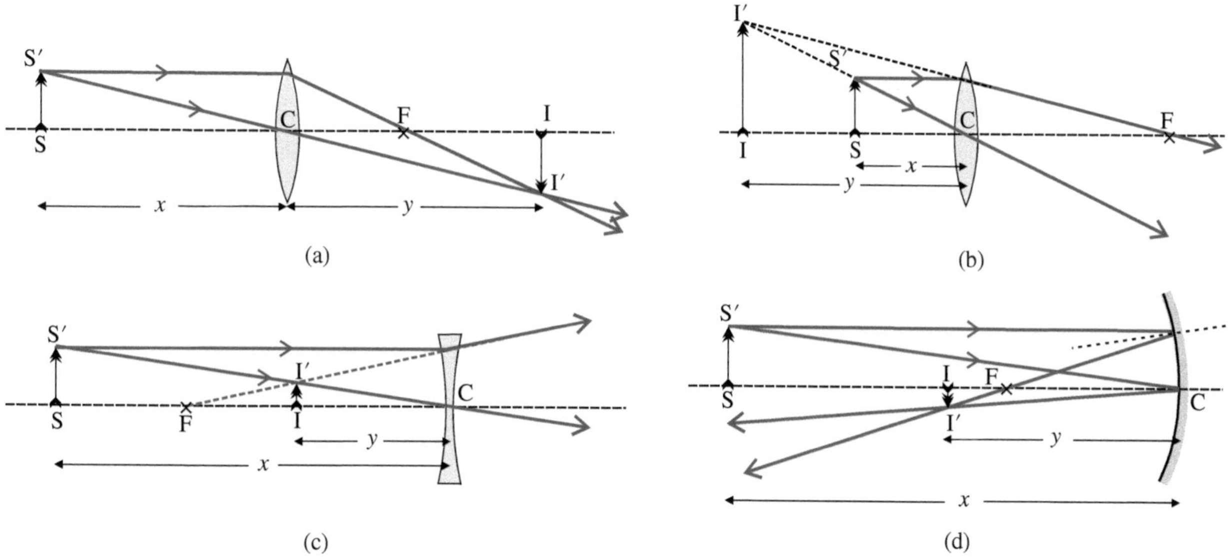

Figure 23.32. Examples of image formation of an extended object by spherical lenses and mirrors.

of I′, the image of S′, can be determined from the following geometrical construction. Consider two particular rays emanating from the point S′, (i) the ray which passes through the centre of the lens and emerges undeviated from its original direction and (ii) the ray which travels parallel to the axis and, on striking the lens, will be deviated to pass through the focal point. The point where these two rays intersect is the image I′; all other rays from S′ which strike the lens will also pass through this point.

For the case illustrated in Figure 23.32(a), the image of the extended object II′ is inverted and magnified. Note that Δ SS′C and Δ II′C are similar triangles and, hence (Appendix A.2.2), the magnification (the ratio of the length of the image to the length of the object) is

$$\frac{\text{II}'}{\text{SS}'} = \frac{\text{IC}}{\text{SC}} = \frac{y}{x} \tag{23.10}$$

As one would expect at this stage, many different situations can arise where the image may be upright or inverted, diminished or magnified with respect to the object (some further examples are shown in Figure 23.32) but it is clear from the geometry in each case that Equation (23.10) still applies. Using the sign conventions for mirrors and lenses adopted in previous sections, we define the (algebraic) **magnification** as follows.

$$\boxed{m := -\frac{v}{u}}$$

With this definition, a positive magnification denotes an upright image while negative magnification denotes an inverted image in all cases.[1]

In many optical instruments, combinations of lenses and/or mirrors, in which the image of one optical component serves as the object of the next, are used to produce final images of the desired magnification and type. Some examples are outlined below.

Refracting telescope

A simple version of a refracting telescope is shown in Figure 23.33. It comprises two lenses, one (the 'objective') has long focal length while the focal length of the other (the 'eyepiece') is shorter. The lenses are placed coaxially so that their focal planes coincide. Light from a distant object striking the objective lens will form an image II′ in this plane. This image is observed through the eyepiece and, since it lies in the focal plane of the eyepiece, will form a final (inverted) image at infinity.

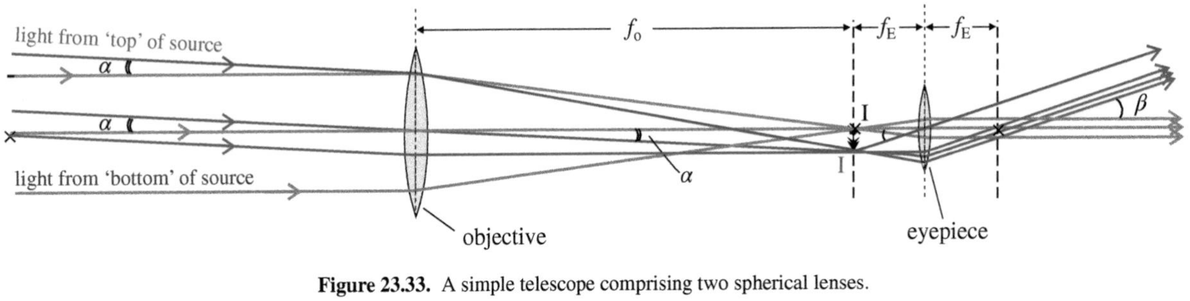

Figure 23.33. A simple telescope comprising two spherical lenses.

In Figure 23.33, the telescope is pointed so that one side (the 'bottom' in the figure) of the distant object lies on the axis of the system. Light from that side enters the objective as a beam parallel to the axis of the system (grey rays in the figure) while light from the other side of the object (blue rays) enters in a parallel beam making a small angle α with the axis. The blue beam emerges from the eyepiece at an angle β with the axis. The ratio of these two angles is called the *angular magnification* (sometimes called *magnifying power*) and is given by

$$\frac{\beta}{\alpha} = \frac{\text{II}'/f_o}{\text{II}'/f_e} = \frac{f_o}{f_e}$$

Thus for large magnifying power, a telescope needs a long focal length (low power) objective and a short focal length (high power) eyepiece. Usually, the eyepiece can be moved along the axis (focused) to position the final image where required. In practice, in order to avoid image distortion due to dispersion both objectives and eyepieces are usually more complex than indicated here.

[1]Note that other texts may use a different sign convention for magnification.

Reflecting telescope

For use in situations where the intensity of light is low, such as in astronomical observations, it is necessary to make the aperture and thus the light collecting capacity of the objective lens as large as possible. There are practical limits to how large a lens may be made and supported. It is easier, however, to construct very large mirrors so that reflecting telescopes are more commonly used for astronomy (also avoiding dispersion effects in the objective described in the next section). Since the image is normally very small, the placement of the eyepiece in the incident beam does not obstruct much of the light; alternatively a mirror may be used to deflect the light sideways before it comes to a focus so the image may be viewed from the side through an eyepiece. Very often the mirror is parabolic rather than spherical to reduce spherical aberration of light from a distant object.

Microscope

A simple microscope (Figure 23.34) comprises two lenses but in this case the objective lens has very short focal length. The object to be examined OO′ is placed just outside the focal plane of the objective thus forming an enlarged real image JJ′ as shown in the figure. This intermediate image is observed through an eyepiece in such a way as to form a virtual image II′ (inverted with respect to the object). The overall magnification, therefore, is given by

$$m = \frac{II'}{OO'} = \frac{II'}{JJ'}\frac{JJ'}{OO'} = (m_{\text{eyepiece}})(m_{\text{objective}})$$

where m_{eyepiece} and $m_{\text{objective}}$ are, respectively, the magnifications of the eyepiece and objective lenses.

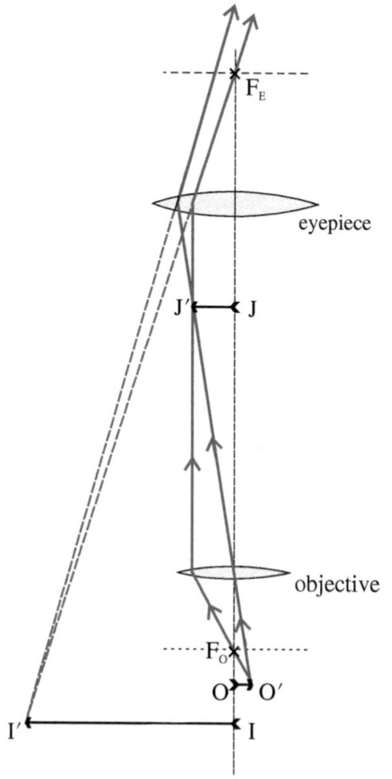

Figure 23.34. A simple microscope comprising two spherical lenses.

Other telescopes and microscopes

Image formation by mirrors and lenses is not confined to the visible part of the electromagnetic spectrum. Radio waves, radiant heat, infra-red, ultra-violet, X-, and gamma-rays can all be reflected, refracted or otherwise focused to form images, although the reflecting surfaces or 'lenses' involved can be quite different from those used for light.

As a consequence of wave-particle duality (Section 14.8), particles can also be used for image formation. As we saw in Section 22.3, the minimum angular resolution obtainable is given by $\Delta\theta \propto \frac{\lambda}{a}$ so, for example, in the **electron microscope** the very short de Broglie wavelengths involved enable very high resolution to be achieved.

Study Worked Example 23.4

For problems based on the material presented in this section visit up.ucc.ie/23/ and follow the link to the problems.

23.8 Dispersion of light

The speed of light in a medium, and hence the refractive index of the medium, varies with the frequency of the light, that is $n = n(f)$. For many transparent materials, the dependence of refractive index on wavelength can be described reasonably well by the following empirical relationship known as Cauchy's Equation (Augustin Louis Cauchy 1789–1857).

$$n(\lambda) = A + \frac{B}{\lambda^2}$$

where A and B are constants characteristic of the particular medium involved.

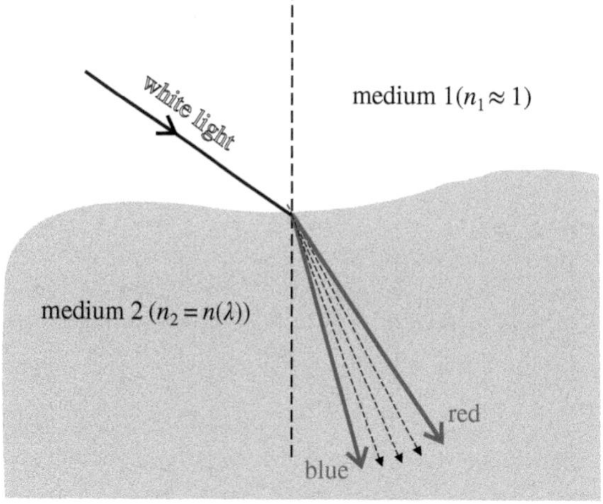

Figure 23.35. The dispersion of white light into its constituent colours by refraction at an interface.

Thus, a light ray which contains different frequencies (colours) may be split by refraction into its component frequencies (Figure 23.35), the phenomenon of **dispersion** already discussed in Chapter 13. The degree of dispersion can differ from one transparent material to another and is measured by a quantity called the **dispersive power** which is characteristic of each medium. Dispersive power is defined in terms of the refractive indices of the material for three specific colours, namely the F, C and D Fraunhofer spectral lines — certain blue, red and yellow characteristic lines, respectively, first designated as such by Joseph von Fraunhofer (1787–1826) when he observed them in the spectrum of the Sun.

The dispersive power of a medium is defined as follows

$$\boxed{\omega := \frac{n_F - n_C}{n_D - 1}}$$

where n_F, n_C and n_D are, respectively, the refractive indices of the medium for the F, C and D Fraunhofer lines. In most practical situations the quantity n_D can be replaced by the average refractive index (that is, values measured with white light).

Dispersion of light is most easily observed by passing light through a triangular prism (Figure 23.36). This technique is used in a '*prism spectrometer*', as an alternative to a diffraction grating, to measure the wavelengths of the different component colours emitted from a light source. Note, however, that blue light is deviated through a larger angle than red light, in contrast to the case of a diffraction grating (as predicted by Equation (22.10)).

The phenomenon of dispersion affects the formation of images by lenses (but not mirrors) if the light from the sources is not monochromatic. If the source in Figure 23.37, for example, emits a mixture of light between red and blue, the different colours will be deviated by different amounts on passing through the lens and, as a result, the blue light converges to form an image slightly nearer the lens than the corresponding red image (**chromatic aberration**).

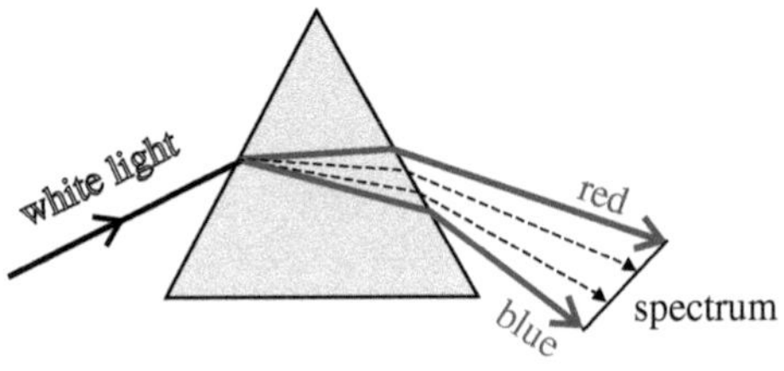

Figure 23.36. Dispersion of light rays with a mixture of frequencies (that is, different colours) through a triangular prism.

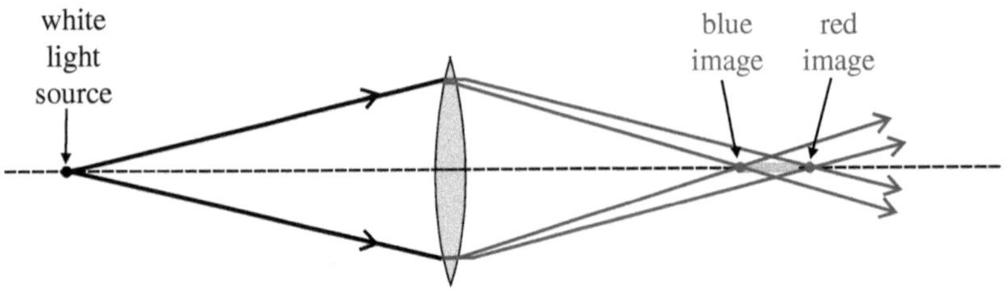

Figure 23.37. Chromatic aberration in a spherical lens: the image is distorted because of the different refractive indices for different colours. The effect is greatly exaggerated in the figure.

Worked Example 23.5 shows how chromatic aberration may be eliminated by constructing a combination of two thin lenses in contact each made from transparent materials with different dispersive properties (a so-called *achromatic lens*).

Study Worked Example 23.5

For problems based on the material presented in this section visit up.ucc.ie/23/ and follow the link to the problems.

Further reading

Further material on some of the topics covered in this chapter may be found in *Insight into Optics* by Heavens and Ditchburn and in *Optics* by Smith and Thomson (See BIBLIOGRAPHY for book details).

WORKED EXAMPLES

Worked Example 23.1: When a point source of light is placed on the axis of a concave mirror 1.5 m from the mirror, it is found that the image is located at exactly the same point as the source. (a) If the source is now moved along the axis to a point 0.5 m nearer to the mirror, where will the image be located? (b) Where will the image be located if the source is moved a further 0.5 m closer to the mirror? In each case, determine whether the image is real or virtual.

Note if $u = v = x$, $\frac{1}{u} + \frac{1}{v} = \frac{2}{R} \rightarrow \frac{1}{x} + \frac{1}{x} = \frac{2}{R} \rightarrow R = x = 1.5\text{m} \Rightarrow$ the source is at the centre of curvature of the mirror.

(a) $u = 1.0\text{ m}: \quad \frac{1}{1.0} + \frac{1}{v} = \frac{2}{1.5} \quad \rightarrow \quad \frac{1}{v} = \frac{2}{1.5} - \frac{1}{1.0} = \frac{4}{3} - 1 = \frac{1}{3} \quad \rightarrow \quad v = 3.0\text{ m (real)}$

(b) $u = 0.5\text{ m}: \quad \frac{1}{0.5} + \frac{1}{v} = \frac{2}{1.5} \quad \rightarrow \quad \frac{1}{v} = \frac{2}{1.5} - \frac{1}{0.5} = \frac{4}{3} - 2 = -\frac{2}{3} \quad \rightarrow \quad v = -1.5\text{ m (virtual)}$

Worked Example 23.2: If the wavelength of a spectral line is λ_{vac} when measured in a vacuum derive an equation which gives λ_{air}, the wavelength of the same line when measured in air.

From the definition of refractive index $n = \dfrac{c}{v_{air}}$ where n is the refractive index of air, c is the velocity of light in a vacuum and v_{air} is the velocity of light in air.

Thus

$$c = nv_{air}$$

If f is the frequency of the spectral line we can write $\dfrac{c}{f} = \dfrac{(nv_{air})}{f}$

so that using $c = f\lambda_{vac}$ and $v_{air} = f\lambda_{air}$ we obtain $\lambda_{vac} = n\lambda_{air}$
(Thus, to convert vacuum wavelengths to wavelengths in air, we must multiply by n)

Worked Example 23.3: A plane mirror is placed on the floor and a thin equi-convex lens (both faces have the same curvature) is placed on top of it. It is found that the image of a small object on the axis of the lens, located 60 cm above the lens, coincides exactly with the position of the object. If the space between the lens and the mirror is now filled with water it is found that the object and image coincide if the object is moved to a point on the axis 90 cm from the lens. If the refractive index of the glass of the lens is 1.5, determine (a) the radius of curvature of the faces of the lens and (b) the refractive index of water.

(a) Lens on plane mirror before the introduction of water is equivalent to two identical lenses in contact (equivalent two-lens system is shown on the right in Figure 23.38).
Power of combination $= P + P_1 + P_1 = 2P_1$

$$\frac{1}{0.6} + \frac{1}{0.6} = \frac{2}{0.6} = P = 2P_1 \quad \rightarrow \quad P_1 = \frac{1}{0.6} = 1.67\,\text{D}$$

The lens maker's formula (23.8) $\rightarrow (n_{glass} - 1)\left(\frac{2}{R}\right) = (0.5)\left(\frac{2}{R}\right) = 1.67\,\text{D} \rightarrow R = \frac{1}{1.67} = 0.6\,\text{m}$

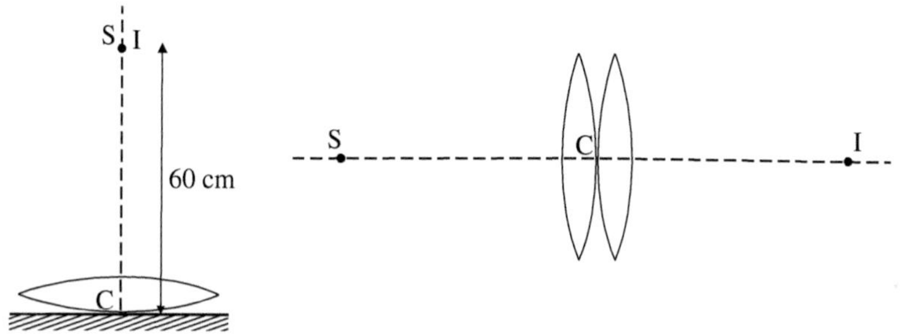

Figure 23.38. Worked Example 23.3

(b) After water is introduced, the system is equivalent to three thin lenses in contact, as illustrated on the right of Figure 23.39

$$\frac{1}{0.90} + \frac{1}{0.90} = P = P_1 + P_2 + P_1 = 2P_1 + P_2 = 2 \times 1.67 + P_2 \quad \rightarrow \quad P_2 = \frac{2}{0.90} - 2 \times 1.67 = -1.11\,\text{D}$$

$$P_2 = (n_{water} - 1)\left(\frac{1}{R_1} + \frac{1}{R_2}\right) = (n_{water} - 1)\left(-\frac{2}{R}\right) = (n_{water} - 1)\left(-\frac{2}{0.6}\right) = -1.11\,\text{D}$$

$$\rightarrow n_{water} - 1 = \frac{1.11 \times 0.6}{2} = 0.33$$

$$\rightarrow n_{water} = 1.33$$

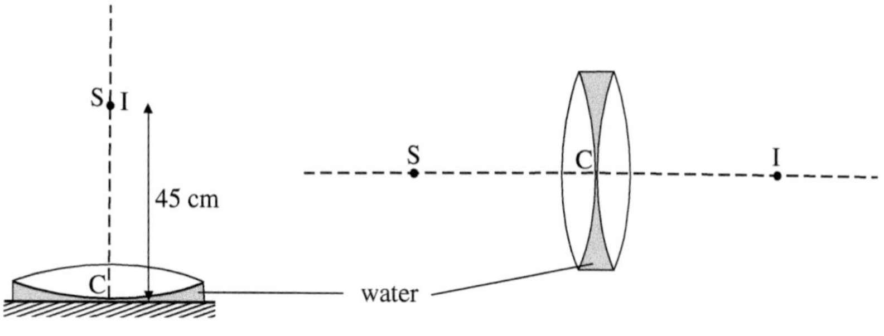

Figure 23.39. Worked Example 23.3 – water between lens and mirror

Worked Example 23.4: A converging lens of power +5 D and a diverging lens of power −5 D are placed 0.5 m apart so that their axes coincide. A small rod is placed at right angles to the axis at a distance of 0.4 m from the converging lens and on the opposite side from the second lens (Figure 23.40). (a) Where will the image of this object be located when viewed along the axis through both lenses? (b) What will be the magnification and will the image be real or virtual?

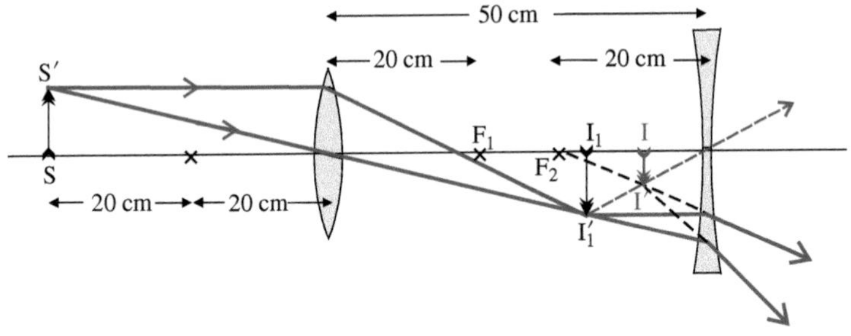

Figure 23.40. Worked Example 23.4

(a) Converging lens ($P = +5\,D$) : $u = 0.4$ m

$$\frac{1}{u}+\frac{1}{v}=P \rightarrow \frac{1}{0.4}+\frac{1}{v}=5 \rightarrow \frac{1}{v}=5-\frac{1}{0.4}=2.5 \rightarrow v=0.4\text{ m}$$

(Note: $m=-\frac{v}{u}=-\frac{0.4}{0.4}=-1 \rightarrow \text{II}'=\text{SS}'$)

Diverging lens ($P = -5\,D$) : $u = 0.5 - 0.4 = 0.1$ m

$$\frac{1}{u}+\frac{1}{v}=P \rightarrow \frac{1}{0.1}+\frac{1}{v}=-5 \rightarrow \frac{1}{v}=-5-\frac{1}{0.1}=-15 \rightarrow v=-0.067\text{ m}$$

(b) No magnification by first lens. For second lens, v is negative so the image $I_1I'_1$ is virtual. Thus

$$m=-\frac{v}{u}=-\frac{-0.067}{0.1}=0.67$$

Worked Example 23.5: An achromatic lens is to be constructed from two thin lenses made from two different types of glass (Figure 23.41(a)). One of the glasses has refractive index of 1.57 and dispersive power 0.017 while the other has refractive index 1.63 and dispersive power 0.027. The power of the achromatic combination is to be 1 D (that is, focal length =1.0 m). (a) Determine the required power of each component lens. (b) If the faces in contact were to be planar, as in Figure 23.41(b), determine the radius of curvature of the other face of each lens.

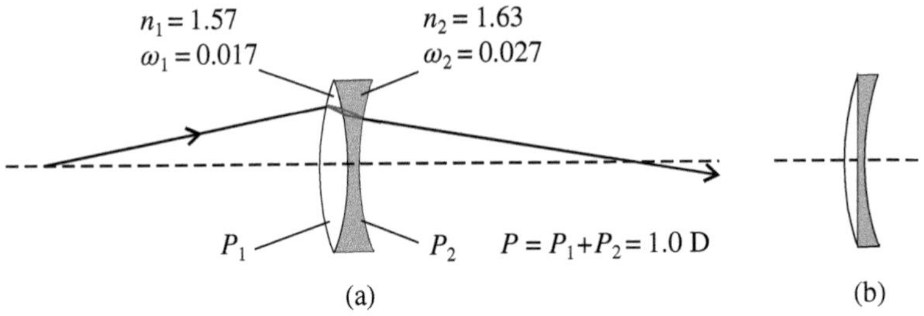

Figure 23.41. Worked Example 23.5

(a) For any lens:

$$P_F=(n_F-1)\left(\frac{1}{R_1}+\frac{1}{R_2}\right)\text{ (blue light)} \quad\text{and}\quad P_C=(n_C-1)\left(\frac{1}{R_1}+\frac{1}{R_2}\right)\text{ (red light)}$$

$$\rightarrow\quad P_F-P_C=(n_F-n_C)\left(\frac{1}{R_1}+\frac{1}{R_2}\right)=\omega(n-1)\left(\frac{1}{R_1}+\frac{1}{R_2}\right)=\omega P\ \left(\text{since, by definition, }\omega=\frac{(n_F-n_C)}{n-1}\right)$$

The condition that there be no chromatic aberration is that $(P_1)_F+(P_2)_F=(P_1)_C+(P_2)_C$

or

$$(P_1)_F-(P_1)_C=-\{(P_2)_F-(P_2)_C\} \rightarrow \boxed{\omega_1P_1=-\omega_2P_2}$$

This, together with the design requirement $P_1+P_2=P$, yields

$$P_1=\frac{\omega_2}{\omega_2-\omega_1}P \quad\text{and}\quad P_2=-\frac{\omega_1}{\omega_2-\omega_1}P$$

Thus

$$P_1=\frac{0.027}{0.027-0.017}(1\text{ D})=2.7\text{ D} \quad\text{and}\quad P_2=-\frac{0.017}{0.027-0.017}(1\text{ D})=-1.7\text{ D (Note that }P_1+P_2=1.0\text{ D, as required)}$$

(b) If the faces in contact were planar (Figure 23.41(b)):

$$P_1 = (n_1 - 1)\left(\frac{1}{R_1} + \frac{1}{R_2}\right) = (n_1 - 1)\left(\frac{1}{R_1}\right) \left(\text{since } \frac{1}{R_2} = \frac{1}{\infty} = 0\right) \rightarrow 2.7 = (0.57)\frac{1}{R_1} \rightarrow R_1 = 0.21\text{ m}$$

$$P_2 = (n_2 - 1)\left(\frac{1}{R_1'} + \frac{1}{R_2'}\right) = (n_1 - 1)\left(0 + \frac{1}{R_2'}\right) \rightarrow -1.7 = (0.63)\frac{1}{R_2'} \rightarrow R_2' = -0.37\text{ m}$$

PROBLEMS

For problems based on the material covered this chapter visit up.ucc.ie/23/ and follow the link to the problems.

24

Atomic physics

AIMS

- to develop a model of atomic structure
- to show that, in order to explain the observed spectrum of the hydrogen atom, arbitrary quantum postulates – *the Bohr postulates* – must be added to the classical model of atomic structure
- to show how a fully quantum mechanical treatment of the hydrogen atom removes the arbitrary aspects of the Bohr postulates and accounts for further properties of one-electron atoms
- to interpret the magnetic properties of atoms in terms of the properties and behaviour of atomic electrons, including an intrinsic electron angular momentum called *electron spin*
- to show that the behaviour of electrons in multielectron systems must conform to quantum mechanical requirements of indistinguishability of identical particles, as stated in the *Pauli exclusion principle*
- to describe how the structure and properties of multielectron atoms, and hence the form of the periodic table, may be accounted for by the Pauli exclusion principle and the quantum mechanical solutions of the one-electron atom

24.1 Atomic models

When an electric current is passed through an atomic gas, the atoms are ionised – the electrically neutral atoms divide into (negatively charged) electrons and positively charged ions. It is reasonable, therefore, to suppose that atoms contain electrons. We also know that the mass of an electron is considerably smaller than that of an atom so that it is also likely that the positive charge in an atom is associated with most of the mass.

This much was known about the atom at the beginning of the twentieth century. The question which then arose was, "how are mass and charge distributed in an atom? – that is, what is the structure of an atom?" The answers to this question – and thus the basis of a model for the atom – were provided by experiments performed in 1910 by Ernest Rutherford (1871–1937), and by Hans Geiger (1882–1945) and Ernest Marsden (1889–1970), in which thin foils of gold and silver were probed through bombardment by α-particles ($^4_2$He nuclei). The *scattering angles*, the angles through which the positively charged α-particles were deflected by Coulomb interactions with charges in the atoms, were measured and used to deduce the atomic charge distribution. In some respects the experiment can be compared with an investigation of the size and shape of an object by subjecting it to bombardment by a succession of balls. By studying the angles through which different balls are deflected (scattered) after hitting the object it is possible to deduce its size and shape.

If the positive charge, and thus the associated mass, in an atom is distributed uniformly throughout the space occupied by the atom we would expect the α-particles to experience only small net deflections due to their Coulomb interaction with positive charge as they pass through the atom. The Rutherford experiment, however, revealed a very different picture. While most α-particles were hardly scattered at all, some (about one in 10^4) were scattered through more than 90°. A significant number of α-particles were even deflected through angles close to 180°. If the charge is distributed uniformly throughout the atom, the chance of successive scatterings adding up to a total scattering angle of more than 90° is calculated to be only one in 10^{3500}. In Rutherford's own words: "It was quite the most incredible event that has ever happened to me in my life. It was almost as if you fired a 15-inch shell at a piece of tissue paper and it came back and hit you".

The clear implication of Rutherford's experiment, as illustrated in Figure 24.1, is that all the positive charge and almost all the mass of an atom are concentrated in a very small volume at the centre of the atom with most of the volume of the atom comprising empty space. The radius of the small volume at the centre can be deduced from the observed distribution of scattering angles. For all atoms, it is found to be of the order of 10^{-14} m which is four orders of magnitude smaller than a typical atomic radius (10^{-10} m). The small volume in which the atomic mass and positive charge is concentrated is known as the **atomic nucleus**.

Understanding Physics, Third Edition. Michael Mansfield and Colm O'Sullivan.
© 2020 John Wiley & Sons Ltd. Published 2020 by John Wiley & Sons Ltd.

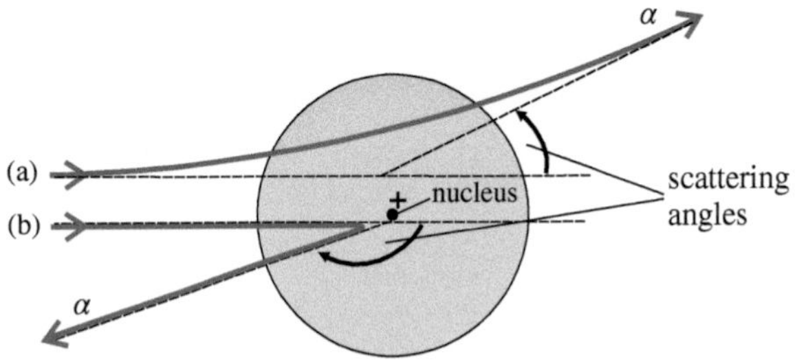

Figure 24.1. The scattering of an α-particle as it passes through an atom in which the positive charge (and the mass) is concentrated in a very small volume at the centre of the atom (the atomic *nucleus*). α-particle trajectories are shown for cases in which the α-particle passes (a) close to the nucleus and (b) very close to the nucleus.

The picture of the atom which emerges from the Rutherford experiment, therefore, is of a massive positively charged nucleus which is surrounded by a much larger volume in which the electrons are somehow situated. Such a model immediately prompts the question: "why does the positively charged nucleus not draw the electrons into it, causing the atom to collapse?" An answer to this question is suggested when the situation is compared to that of the planets in the solar system. As described in Sections 5.12 and 5.13, the planets avoid being drawn into the very large mass of the Sun by gravitational attraction because they are performing orbital motion around the Sun. The comparison suggests a **planetary** model of the atom, a microscopic solar system in which the centripetal force needed to sustain orbital motion is provided by Coulomb attraction rather than gravitational attraction.

The planetary model

The planetary model of the simplest atom, hydrogen, comprises a single electron of charge $-e$ which is orbiting around a nucleus of charge $+e$ (a proton) at a distance r, as illustrated in Figure 24.2.

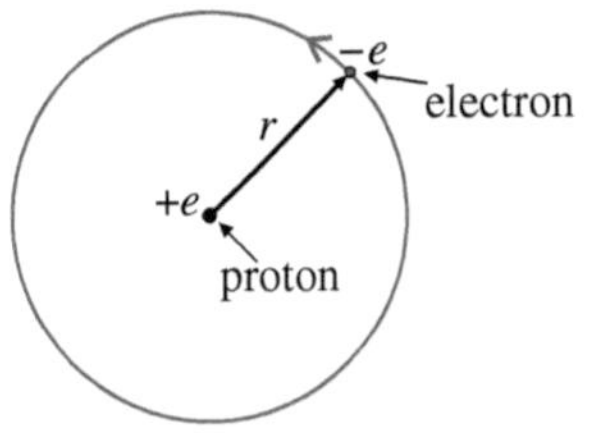

Figure 24.2. The planetary model of the hydrogen atom.

We can equate the centripetal force to the attractive Coulomb force to obtain

$$\frac{e^2}{4\pi\varepsilon_0 r^2} = \frac{mv^2}{r},$$

which we can write as

$$mv^2 = \frac{e^2}{4\pi\varepsilon_0 r} \tag{24.1}$$

The total energy of the electron is given by

$$E = \text{kinetic energy} + \text{(electrostatic) potential energy (obtained from Equation 16.25)}$$
$$= \frac{1}{2}mv^2 - \frac{e^2}{4\pi\varepsilon_0 r}$$

Substituting for mv^2 from Equation (24.1), this becomes

$$E = \frac{e^2}{8\pi\varepsilon_0 r} - \frac{e^2}{4\pi\varepsilon_0 r} = -\frac{e^2}{8\pi\varepsilon_0 r} \tag{24.2}$$

Alternatively, again using Equation (24.1), we can write E in terms of v

$$E = -\frac{1}{2}mv^2 \tag{24.3}$$

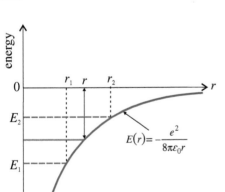

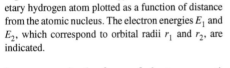

Figure 24.3. The total energy of an electron in a planetary hydrogen atom plotted as a function of distance from the atomic nucleus. The electron energies E_1 and E_2, which correspond to orbital radii r_1 and r_2, are indicated.

As expected in a bound system, the total energy of the electron is negative.

Attractive though the planetary model of the atom is, a major problem arises when the theory of electromagnetism is taken into consideration. As noted in Section 21.2, an accelerating charge produces electromagnetic radiation. In the planetary atom the electrons constantly experience centripetal acceleration and would be expected therefore to lose energy in the form of electromagnetic radiation. As indicated by Equation (24.2) and, as illustrated by a plot of the total energy function (Figure 24.3), it follows that as the electron loses energy, from E_2 to E_1 for example, the radius of its orbit must decrease (from r_2 to r_1). The electron would quickly spiral inwards towards the nucleus – the atom would collapse. The planetary atom would not be viable.

Measurements of an important atomic property, the ionisation energy, also seem to be at odds with a planetary model. The ionisation energy is the energy needed to remove an electron completely from the atom. It is a positive energy which cancels the negative binding energy, thus freeing the electron. It is found to have the same value for any atom of a given element, so that different atoms of the same element cannot be distinguished from each other by their ionisation energies.

This result is not expected classically because the energy needed to remove an electron from an atom – its binding energy as given by Equation (24.2) and plotted in Figure 24.3 – can take any value if r can take any value. The implication is that electrons in atoms can have only certain orbital radii. This situation, in which electron orbital radii, and thus binding energies, are fixed, is reminiscent of the quantisation of energy in a bound system – a very general result in quantum mechanics (Section 14.15). A full analysis of the structure of the atom will therefore require quantum mechanics, as we might expect for such an obviously microscopic system.

Before applying quantum mechanics fully, however, we begin by showing how the planetary model (a classical model) can be modified to overcome the difficulties discussed above. The resulting model – the **Bohr model** – is described as a *semi-classical* model because quantum concepts are grafted on to the classical model somewhat arbitrarily to ensure that the theory produces the correct answers. This approach follows the historical development of atomic theory and is similar to that used to introduce quantum mechanics in Chapter 14. It has the advantage of introducing quantum mechanical concepts in a manner which is more readily understandable in terms of our classical view of the world.

Although in due course we shall describe a fully quantum mechanical treatment of the hydrogen atom, we shall often find it helpful to analyse and discuss atomic properties in semi-classical terms. Any semi-classical result which we use can be justified by a fully quantum mechanical treatment. The full quantum mechanical treatment will indicate clearly that electrons in atoms cannot be regarded as point particles which are executing precise orbits. Nonetheless, many atomic properties can be interpreted in terms of such a model. In many respects, atoms behave *as if* their component electrons behaved in such a way.

The application of quantum mechanical methods to atoms has been one of its most effective and important applications and, historically, has gone hand in hand with the development of quantum mechanics which we have described in Chapter 14. As we shall see in the next section, atomic properties can be determined with considerable precision using spectroscopic methods so that atomic physics has been, and continues to be, an important testing ground of the theory of quantum mechanics.

24.2 The spectrum of hydrogen: the Rydberg formula

When an electric current is passed through an atomic gas or vapour, the atoms absorb energy through collisions with electrons and ions and may then release this energy as light – electromagnetic radiation. The emitted light may be broken into its component wavelengths, using a diffraction grating spectrometer (Section 22.5). The **spectrum** which is obtained in this way comprises **discrete** lines – light of well defined wavelengths which are characteristic of a particular atom. The spectrum may be used to identify which atoms are present in a system under study – an important analytical technique known as **spectroscopy,** as noted in Section 21.4. The technique is particularly valuable in situations in which it is not possible to take samples from the system under investigation, as is clearly the case for astrophysical studies.

In the case of the hydrogen atom, a particularly simple spectrum is observed, as illustrated in Figure 24.4. Note that only the three (possibly four), longer wavelength lines shown in Figure 24.4 can be detected by the human eye so that another form of detection, such as a photographic emulsion or photoelectric detection (see Section 26.7), must be used to observe the full spectrum shown in this figure.

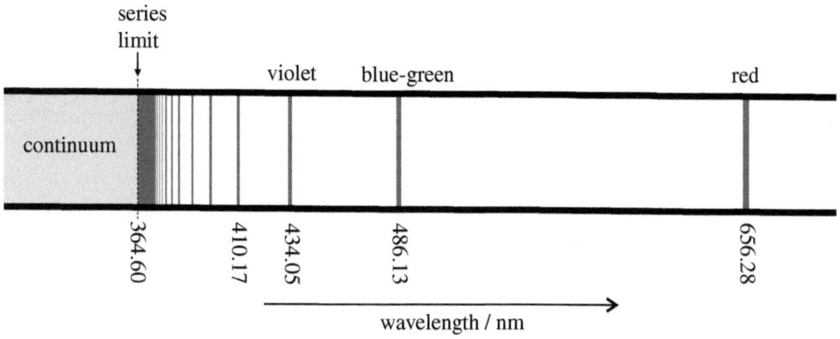

Figure 24.4. The observed spectrum of atomic hydrogen in the visible and ultraviolet wavelength regions (the Balmer series of spectral lines).

The hydrogen spectrum shows remarkable regularity in both the wavelengths of the lines and their intensities. The lines form a **series** in which the separation between successive lines decreases with decreasing wavelength (increasing frequency) until a point is reached, the **series limit**, at which emission becomes continuous with wavelength. Given the regularity of the hydrogen spectrum, it is not surprising that a general empirical mathematical formula was soon devised to describe the wavelengths of the visible series of hydrogen – known as the **Balmer** series, after Johann Balmer (1825–1898). The formula, known as a **Rydberg formula** (after Johannes Rydberg, 1854–1919), was derived more than 20 years before the planetary model was proposed. The Rydberg formula, which gives the reciprocals of the wavelengths, for the Balmer series of hydrogen is as follows

$$\frac{1}{\lambda} = \frac{f}{c} = R_{\mathrm{H}}\left[\frac{1}{4} - \frac{1}{n^2}\right] \tag{24.4}$$

where R_H is a constant known as the **Rydberg constant** for hydrogen. From spectroscopic measurements of wavelengths, which can be very accurate, its value is determined to be 10 967 757.6 ± 1.2 m^{-1}. The integer n in Equation (24.4) can take the values 3, 4, 5, ... etc. for the Balmer series, each value of n representing a member of the series, that is a spectral line. Thus $n = 3$ gives $\lambda = 656.47$ nm, the wavelength of the Balmer red line, the first member of the series. Substitution of $n = 4$ gives $\lambda = 486.27$ nm, the Balmer blue-green line, and so on for further members of the series. The series limit corresponds to $n = \infty$, for which Equation (24.4) gives $\lambda = 364.71$ nm. The limit is in the ultraviolet region of the electromagnetic spectrum.

Note that Equation (24.4) gives wavelengths as measured in vacuum so that the values calculated above are slightly larger than the wavelengths indicated in Figure 24.4, which are the wavelengths of these lines when measured in air. To convert a vacuum wavelength, λ_{vac}, to a wavelength in air, λ_{air}, we use $\lambda_{vac} = n\lambda_{air}$ (derived in Worked Example 23.2) where n is the refractive index of air (1.000292). Application of this equation brings the calculated wavelengths into agreement with the values given in Figure 24.4.

24.3 The Bohr postulates

Niels Bohr (1885–1962) sought to explain the Rydberg formula by modifying the planetary model. He showed that this can be achieved by adopting the following postulates, known as the **Bohr postulates**, for circular orbits of the electron in a hydrogen atom.

Bohr postulate (i). Instead of the infinite number of orbits, with different radii, which are possible classically, the electron in a hydrogen atom can only be in orbits in which the magnitude of its angular momentum, L, satisfies the equation

$$L = mvr = \frac{nh}{2\pi} = n\hbar \qquad \text{where } n = 1, 2, 3\ldots \text{ and } h \text{ is the Planck constant}$$

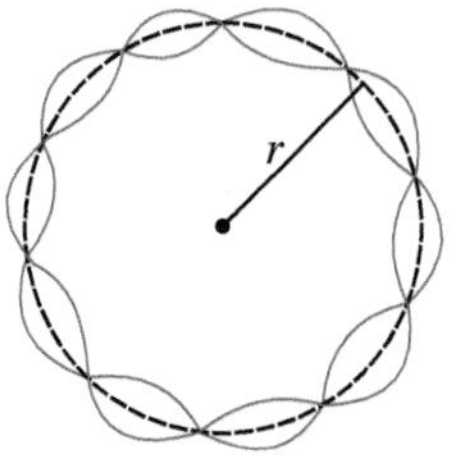

Figure 24.5. A standing de Broglie electron wave around a Bohr orbit.

Note that by writing postulate (i) in terms of p, the magnitude of the linear momentum of the electron,

$$L = pr = \frac{nh}{2\pi}$$

and by substituting $p = \frac{h}{\lambda}$ from the de Broglie hypothesis (Section 14.8) we obtain

$$L = pr = \frac{h}{\lambda}r = \frac{nh}{2\pi} = n\hbar$$

which simplifies to

$$n\lambda = 2\pi r$$

Postulate (i) can be interpreted, therefore, as a statement that the circumference of an allowed Bohr orbit must correspond to an integral number of de Broglie electron wavelengths. As illustrated in Figure 24.5, this is equivalent to the statement that an electron orbit is allowed only if a de Broglie standing wave can form around its circumference. While this view of an 'orbiting electron wave' is not consistent with the three dimensional quantum mechanical analysis which we shall present in Section 24.5, it does show that Bohr postulate (i) can be regarded as evidence of standing electron waves in the atom.

We can also use the orbiting electron wave concept to suggest why electrons are not found in orbits which do not conform to postulate (i). If the circumference of an orbit does not correspond to an integral number of de Broglie electron wavelengths, at any point on the orbit the wave will have a different phase on each traversal, which will give rise to destructive interference as illustrated in Figure 24.6. We saw in Section 14.12 that the average intensity of a particle wave at a point is a measure of the probability of the particle being at that point. Zero average intensity therefore means that the electron cannot be in such an orbit, as stated in postulate (i).

Figure 24.6. De Broglie electron waves around an orbit in a case in which the electron wavelength does not satisfy the standing wave condition. The waves interfere to produce a net cancellation.

Bohr postulate (ii). Contrary to classical electromagnetic theory, electrons in allowed orbits do not produce electromagnetic radiation.

This postulate arbitrarily suspends classical electromagnetic theory to ensure that electrons in allowed orbits do not lose energy and therefore do not spiral into the nucleus.

> ***Bohr postulate (iii).*** *Electrons can transfer from one allowed orbit to another allowed orbit of lower energy. When this occurs, electromagnetic radiation is produced with a precise frequency, f, which is given by*
>
> $$hf = E_{\mathrm{i}} - E_{\mathrm{f}}$$
>
> *where E_{i} and E_{f} are, respectively, the energies of the electron in its initial and final orbits.*

We show in the next section that the Bohr postulates predict the Rydberg formula for hydrogen (Equation (24.4)).

24.4 The Bohr theory of the hydrogen atom

From Equations (24.2) and (24.3), the total energy of the orbiting electron in the hydrogen atom is given by

$$E = -\frac{1}{2}mv^2 = -\frac{e^2}{8\pi\varepsilon_0 r} \tag{24.5}$$

Since, for circular orbits $L = mvr$ (Section 5.12), the Bohr postulate (i) can be written $v = \frac{n\hbar}{mr}$. Substituting for v in Equation (24.5), we obtain

$$E = -\frac{1}{2}m\left(\frac{n\hbar}{mr}\right)^2 = -\frac{e^2}{8\pi\varepsilon_0 r}$$

which can be simplified to give

$$r = \frac{4\pi\hbar^2\varepsilon_0}{e^2 m}n^2$$

which we write as

$$r = a_0 n^2 \tag{24.6}$$

where

$$\boxed{a_0 = \frac{4\pi\hbar^2\varepsilon_0}{e^2 m}} \tag{24.7}$$

The quantity a_0, a combination of fundamental constants with the value 5.3×10^{-11} m, is known as the **Bohr radius** and is often used as a convenient measure of length on the atomic scale. The diameter of the $(n = 1)$ electron orbit in hydrogen, $2a_0$, is thus about 10^{-10} m – the typical atomic scale given in Table 1.1. Bohr theory therefore produces a value for the atomic 'size' which agrees with that deduced from measurements on crystals and gases. The theory also predicts the radii of other allowed orbits, $4a_0$, $9a_0$, etc. but radii which do not conform to Equation (24.6) are not allowed.

The Bohr postulate (i), namely the quantisation of angular momentum, has therefore led directly to quantisation of the radius of the electron orbit.

Substitution of r, from Equation (24.6), into Equation (24.5) gives

$$E = -\frac{e^2}{8\pi\varepsilon_0 r} = -\frac{e^2}{8\pi\varepsilon_0 a_0 n^2} = -\frac{e^2}{8\pi\varepsilon_0}\frac{e^2 m}{4\pi\hbar^2\varepsilon_0}\frac{1}{n^2} = -\frac{e^4 m}{32\pi^2\hbar^2\varepsilon_0^2}\frac{1}{n^2}$$

which we write as

$$\boxed{E_n = -\frac{E_0}{n^2}} \tag{24.8}$$

where

$$E_0 = \frac{e^4 m}{32\pi^2\hbar^2\varepsilon_0^2} = 2.17 \times 10^{-18}\ \mathrm{J} = 13.60\ \mathrm{eV}. \tag{24.9}$$

Thus the Bohr postulate (i) has also led to quantisation of energy. The lowest energy state, E_1, corresponds to $n = 1$ in Equation (24.8), so that $E_1 = -E_0$. This state is known as the **ground state** of hydrogen, and has the most negative energy; it is thus the most tightly bound state. Atomic hydrogen is usually found in this state. The next state, corresponding to $n = 2$, has energy $E_2 = -\frac{1}{4}E_0$ and is known as the *first excited state*.

In Figure 24.7 the values of the quantised energies given by Equation (24.8) are compared with a plot of the total energy function $E(r) = -\frac{e^2}{8\pi\varepsilon_0 r}$ (Equation (24.2)).

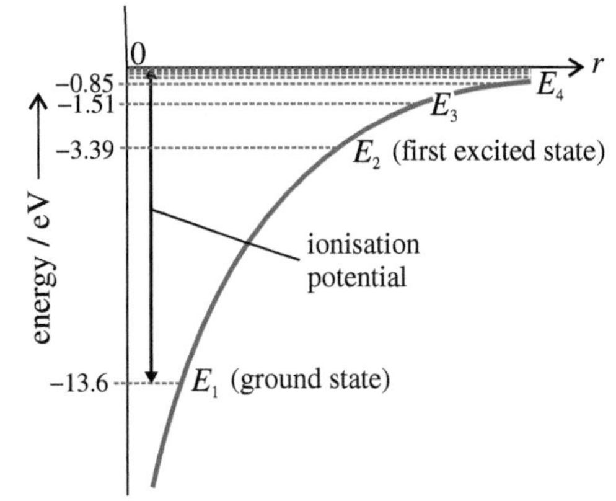

Figure 24.7. The total energy function of an electron in a hydrogen atom, with the quantised Bohr electron energies and the ionisation energy indicated.

Note that the ionisation energy of hydrogen, the energy needed to completely remove an electron from the ground state ($n = 1$) to $n = \infty$, is given by $E_\infty - E_1 = 0 - (-E_0) = E_0 = 13.6$ eV, as indicated in Figure 24.7. According to the Bohr postulates, the ionisation energy can only have this value, which is determined by fundamental constants. The Bohr postulates therefore explain why the ionisation potential is the same for all hydrogen atoms, an experimental finding we noted in the first section of this chapter. The measured ionisation energy of atomic hydrogen in its ground state, 13.60 eV, matches the theoretical value, E_0, very well. Note that ionisation energy is often expressed in volts as the *ionisation potential*; this is the electrical potential difference needed to give an electron the ionisation energy. Thus the ionisation potential of hydrogen is 13.60 V.

The Bohr postulate (ii) allows the electron to occupy one of these allowed energy states without loss of energy through radiation. The Bohr postulate (iii) applies when the electron makes a transition between two allowed states. It states that electromagnetic radiation of frequency f is produced, where f is given by

$$f = \frac{E_\mathrm{i} - E_\mathrm{f}}{h} \tag{24.10}$$

If n_i and n_f are values of n – the *quantum numbers* – associated with E_i and E_f, respectively, so that $E_\mathrm{i} = -\frac{E_0}{n_\mathrm{i}^2}$ and $E_\mathrm{f} = -\frac{E_0}{n_\mathrm{f}^2}$, we can substitute for E_i and E_f in Equation (24.10) to obtain

$$f = \frac{1}{h}\left[-\frac{E_0}{n_\mathrm{i}^2} - \left(-\frac{E_0}{n_\mathrm{f}^2}\right)\right] = \frac{E_0}{h}\left(\frac{1}{n_\mathrm{f}^2} - \frac{1}{n_\mathrm{i}^2}\right)$$

and therefore

$$\frac{1}{\lambda} = \frac{f}{c} = \frac{E_0}{hc}\left(\frac{1}{n_\mathrm{f}^2} - \frac{1}{n_\mathrm{i}^2}\right) = R_\infty\left(\frac{1}{n_\mathrm{f}^2} - \frac{1}{n_\mathrm{i}^2}\right) \tag{24.11}$$

where

$$R_\infty = \frac{E_0}{hc} = \frac{e^4 m}{64\pi^3\hbar^3\varepsilon_0^2 c} = 10973731\ \mathrm{m^{-1}} \tag{24.12}$$

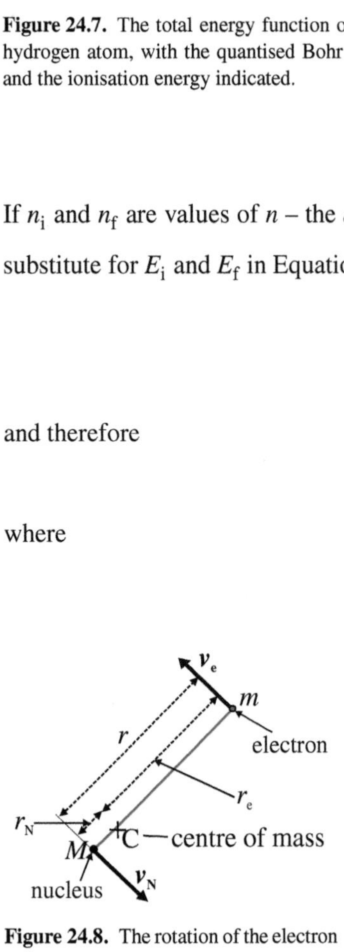

Figure 24.8. The rotation of the electron and nucleus about the centre of mass of the hydrogen atom.

Comparing Equation (24.11) with the Rydberg formula (24.4), we can see that the two equations are identical if $n_\mathrm{f} = 2$ and $n_\mathrm{i} = n$, although there is a small but significant difference between the measured constant $R_\mathrm{H} = 10\ 967\ 758\ \mathrm{m^{-1}}$ (Section 24.2) and the theoretical constant $R_\infty = 10\ 973\ 731\ \mathrm{m^{-1}}$. The difference arises because, in deriving R_∞, we have assumed that the electron performs uniform circular motion with the nucleus at the centre of the motion. This assumes that the centre of mass of the system is located at the nucleus which is true only if the mass of the nucleus is infinitely greater than that of the electron. This is why the symbol R_∞ is used to describe the constant in Equation (24.11). Although the mass of the nucleus (the proton mass) is greater than the electron mass by a factor of 1836, it is not infinitely greater, so that, as illustrated in Figure 24.8, the centre of mass of the system, C, about which both the nucleus and electron orbit, is close to, but not at, the centre of the nucleus. The system can be compared to the Earth–Moon system in which both the Earth and Moon rotate about a point which is 4640 km from the centre of the Earth.

Such a system – a two-body system – has already been considered in Section 6.7 where it is shown that the system can be treated as a single particle system if the mass of the electron is replaced by the *reduced mass* of the system, given by Equation (6.19),

$$m_\mathrm{r} = \frac{mM}{m + M} \tag{24.13}$$

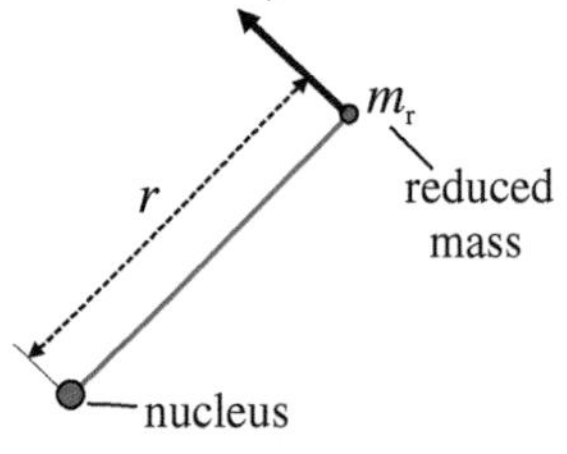

Figure 24.9. The rotation of the reduced mass of the electron about a stationary nucleus of infinite mass – a one-body system which is equivalent to the two-body system shown in Figure 24.8.

We can correct for the finite mass of the nucleus, therefore, simply by replacing m by m_r in the Bohr Equations (24.6) to (24.11). The Bohr postulate (i) becomes

$$L = m_\mathrm{r} v r = n\hbar$$

where v is the velocity and r the displacement of the electron relative to the nucleus. This equation is now a statement that the total angular momentum of the system, rather than just that of the electron, is quantised. Thus the two-body problem is equivalent to the one body problem illustrated in Figure 24.9.

For hydrogen, $m_r = \dfrac{m_e m_p}{m_e + m_p} = \dfrac{m_p/m_e}{1+\dfrac{m_p}{m_e}} m_e = \dfrac{1836}{1837} m_e$ so that, by substituting m_r for m in Equation (24.9), E_0 is reduced by this fraction and the theoretical **Rydberg formula for hydrogen** (24.11) becomes

$$\frac{1}{\lambda} = \frac{f}{c} = \frac{E_0}{hc}\left(\frac{1}{n_f^2} - \frac{1}{n_i^2}\right) = R_H\left(\frac{1}{n_f^2} - \frac{1}{n_i^2}\right) \tag{24.14}$$

where $R_H = \dfrac{1836}{1837} R_\infty$. The theoretical value of R_H now agrees with the observed value for hydrogen within experimental uncertainties, which are in fact very small.

Note that the reduced mass is different for deuterium and tritium, the isotopes of hydrogen, because in these cases the nuclear mass, M, is different in the reduced mass Equation (24.13). For deuterium M is $3672m_e$. This means that the Rydberg constant for deuterium, R_D, is slightly smaller than R_H. Consequently all lines of the deuterium Balmer series are shifted slightly to shorter wavelengths compared with the equivalent hydrogen Balmer lines, an effect known as *isotope shift*. For the Balmer red line, the shift is 0.18 nm (as calculated in Worked Example 24.1).

The hydrogen spectrum

The complete hydrogen spectrum can now be understood from the Bohr energy level diagram of hydrogen which is shown in Figure 24.10. As indicated in this figure, the **Balmer** series corresponds to transitions from excited levels ($n_i = 3, 4, 5, \ldots$) to the $n_f = 2$ level of hydrogen. The Balmer series is observed in the visible/ultraviolet region of the electromagnetic spectrum, with a series limit at 364.7 nm.

Further Rydberg series in hydrogen may be identified from Figure 24.10:

- Transitions from excited levels, $n_i = 2, 3, 4, \ldots.$, to the $n_f = 1$ (ground) level. These are known as the **Lyman** series (after Theodore Lyman, 1874–1954) and are observed in the far ultraviolet region of the electromagnetic spectrum (Section 21.4), with a series limit at 91.1 nm.
- Transitions from excited levels, $n_i = 4, 5, 6 \ldots.$, to the $n_f = 3$ level are known as the **Paschen** series and are observed in the near infra-red region with a series limit at 820 nm.
- Series of transitions from the $n_i = 5, 6, 7 \ldots.$ levels and from the $n_i = 6, 7, 8, \ldots.$ levels to the $n_f = 4$ and $n_f = 5$ levels, respectively, also occur in the infra-red region and are known as the **Brackett** and **Pfund** series, respectively, with series limits at 14.59 and 22.79 μm.

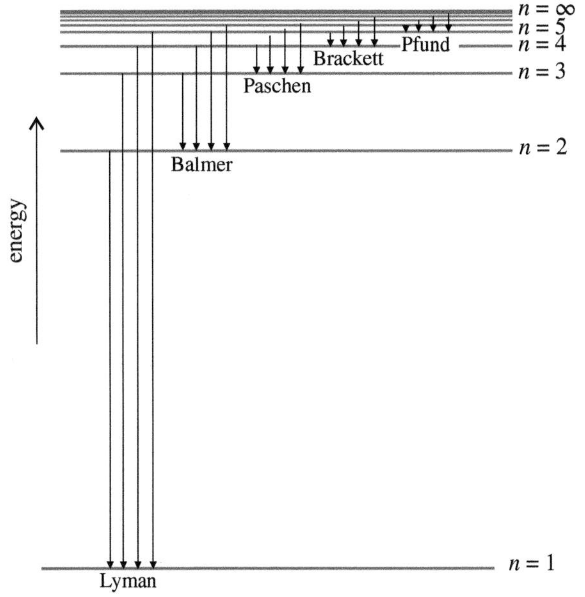

Figure 24.10. An energy level diagram for the hydrogen atom showing the transitions which make up the Lyman, Balmer, Paschen, Brackett and Pfund series of spectral lines.

The wavelengths of all members of these series can be predicted precisely by the general Rydberg formula (24.14). In each case the series limit is obtained by inserting $n_i = \infty$ into the formula.

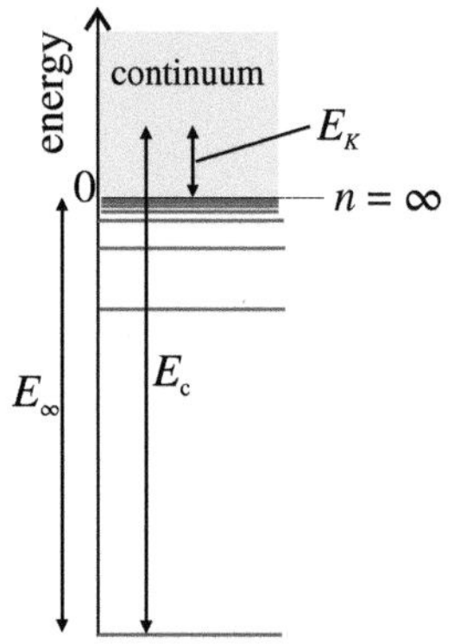

Figure 24.11. Ionisation of a hydrogen atom. The electron is excited from the ground state into the continuum of energy levels, corresponding to unbound states, which are available above the ionisation energy.

As indicated in Figure 24.11, when an electron in the ground ($n = 1$) state of hydrogen is given an energy E_c, which is greater than the ionisation energy of hydrogen, E_∞, the electron is removed entirely from the atom and the excess energy, $E_k = E_c - E_\infty$, is used to give kinetic energy to the now free electron. As noted in the discussion of free particles in Section 14.13, the energy of such an electron is not quantised – a continuum of energy values is available to free electrons.

The velocity of an electron in the first ($n = 1$) Bohr orbit of hydrogen may be estimated from the Bohr model

From Equations (24.3) and (24.8) we can write

$$E_1 = -E_0 = -\frac{1}{2}mv^2 \quad \text{where} \quad E_0 = 2.17 \times 10^{-18}\ \text{J}$$

It follows that $$v^2 = \frac{2E_0}{m} \quad \text{and thus} \quad v = 2.19 \times 10^6\ \text{m s}^{-1}$$

Note that this velocity is almost 1% of the velocity of light, indicating that a complete treatment of the hydrogen atom should include relativistic effects, a point to which we shall return in Section 24.10.

In Section 14.10 we showed that, as a consequence of the Heisenberg uncertainty principle, the uncertainty in the position of an electron which is travelling with a linear velocity of 2.2×10^6 m s^{-1} is of the order of 10^{-10} m. This distance is of the same order as the radius of the first Bohr orbit, 5.3×10^{-11} m, a typical atomic size, so that, while in many respects electrons behave as if they are performing orbits around the nucleus, the Bohr model should not be carried to the extent of treating electrons as point particles which are located at definite positions within the hydrogen atom. Thus, while it is reasonable to assign orbital properties, such as energy and angular momentum, to a Bohr orbit, because these are constants of the motion, that is conserved quantities, it is not reasonable to assign a position and velocity to the electron which is executing this orbit.

Other one-electron atoms

The Bohr model and its results may be extended easily to other one-electron atoms. We have already indicated above how it can be applied to isotopes of hydrogen, with suitable adjustments to account for the modified reduced masses. It can also be applied to ions such as He^+, Li^{2+}, Be^{3+} etc. in which a single electron orbits a nuclear charge $+Ze$. These one-electron ions – known as *hydrogen-like ions* – can be treated in Bohr theory by replacing $+e$ in Equation (24.1) by $+Ze$ (where $Z = 2$ for He etc.) and by modifying the reduced mass to take account of the increased nuclear mass. As a result the energy level Equation (24.9) becomes $E_0 = \dfrac{Z^2e^4m_r}{32\pi^2\hbar^2\varepsilon_0^2}$ with a consequent change in the **Rydberg formula** (24.11), which now becomes

$$\frac{1}{\lambda} = \frac{f}{c} = \frac{E_0}{hc}\left(\frac{1}{n_f^2} - \frac{1}{n_i^2}\right) = RZ^2\left(\frac{1}{n_f^2} - \frac{1}{n_i^2}\right) \tag{24.15}$$

where $R = \dfrac{e^4m_r}{64\pi^3\hbar^3\varepsilon_0^2c}$. Note that as the nuclear mass M increases, the value of $m_r = \dfrac{m_eM}{m_e + M}$ approaches that of m_e and the system more closely resembles the infinite nuclear mass model which we used initially. As Z increases and we move to more massive one-electron ions, the value of R therefore approaches R_∞. In Worked Example 24.2, Bohr theory is applied to the hydrogen-like ion Li^{2+}.

Study Worked Examples 24.1 and 24.2

For problems based on the material presented in this section visit up.ucc.ie/24/ *and follow the link to the problems.*

24.5 The quantum mechanical (Schrödinger) solution of the one-electron atom

While the semi-classical Bohr theory is very successful in predicting the energy levels of one-electron atoms, and hence the wavelengths of the spectral lines which are observed when an electron makes a transition between energy levels, it has a number of unsatisfactory features, namely:

(a) It is successful only for one-electron atoms. It fails when applied to helium and other multielectron atoms.
(b) The theory gives no insights into important atomic properties such as line intensities, for example it predicts the energies of transitions but not the rates at which transitions take place – determined by the *transition probabilities*.

(c) The postulates are somewhat arbitrary, and can be contrary to classical physics. For example, the second postulate suspends the classical theory of electromagnetism. The postulates are formulated to agree with experimentally observed results but with little deeper justification.

To extend our understanding of atoms and to place atomic theory on a firmer basis, we now apply a fully quantum mechanical treatment to the one-electron atom, using the method described in Section 14.11. As usual, the first step is to write down the Schrödinger equation for the system by substituting the appropriate potential energy function into the general Schrödinger equation. The potential energy function for the one-electron atom is the Coulomb potential

$$U(r) = -\frac{1}{4\pi\varepsilon_0}\frac{Ze^2}{r} \tag{24.16}$$

which, as illustrated in Figure 24.12, can be written in Cartesian (x, y, z) coordinates as

$$U(x, y, z) = -\frac{1}{4\pi\varepsilon_0}\frac{Ze^2}{\sqrt{x^2+y^2+z^2}} \tag{24.17}$$

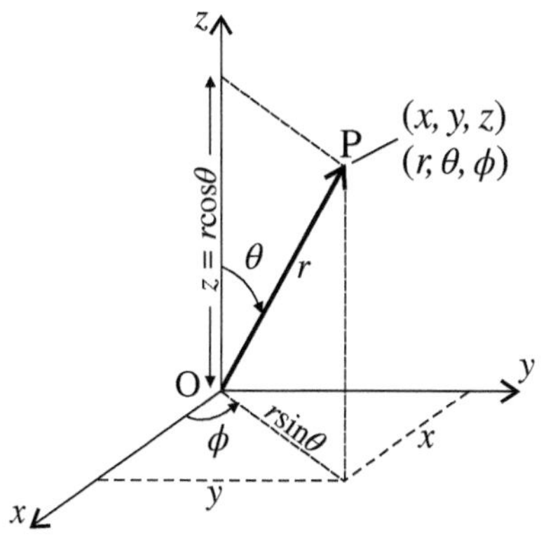

Figure 24.12. The coordinates of the electron in a one-electron atom in the Cartesian (x, y, z) and spherical polar (r, θ, ϕ) coordinate systems.

Equations (24.16) and (24.17) are written for the general one-electron atom case in which the nuclear charge is $+Ze$. Hydrogen results may be obtained at any stage of the following analysis simply by substituting $Z = 1$.

The potential energy function is time independent so that, as described in Section 14.14, the time and spatial variables can be separated. Our task therefore reduces to the solution of the time-independent Schrödinger equation. For this (three-dimensional case) the one-dimensional Equation (14.43),

$$-\frac{\hbar^2}{2m}\frac{d^2\psi(x)}{dx^2} + U(x)\psi(x) = E\psi(x)$$

becomes

$$-\frac{\hbar^2}{2m}\left(\frac{\partial^2\psi}{\partial x^2}+\frac{\partial^2\psi}{\partial y^2}+\frac{\partial^2\psi}{\partial z^2}\right) + U(x, y, z)\psi = E\psi \tag{24.18}$$

which, using Laplacian notation (Appendix A.13.1), $\nabla^2 = \left(\frac{\partial^2\psi}{\partial x^2}+\frac{\partial^2\psi}{\partial y^2}+\frac{\partial^2\psi}{\partial z^2}\right)$, can be written as

$$-\frac{\hbar^2}{2m}\nabla^2\psi + U(x, y, z)\psi = E\psi$$

The electron is free to move in three dimensional space – so that the wavefunction ψ is a function of the three variables x, y and z; hence the need for partial derivatives in Equation (24.18).

There are two principal differences between the one-electron atom and the simple systems which we analyzed quantum mechanically in Chapter 14. These are:

(a) We are no longer dealing with a single particle system; there are two bodies, the electron and the nucleus. However we can reduce the problem to an equivalent one-body problem by using the concept of reduced mass, as described in the previous section. To achieve this, we simply substitute m_r for m in Equation (24.18).
(b) The problem is not one-dimensional. The time-independent Schrödinger equation is now a partial differential equation in three variables, a feature which complicates its solution.

Equation (24.18) is solved through separation of the variables, the technique which we used to separate the space and time variables in the time-dependent Schrödinger equation (Section 14.14). The technique is somewhat complicated to apply when the potential function is expressed as a function of three variables, as it is in Equation (24.18). Since the potential function can be written in terms of the single variable r (Equation (24.16)), it is convenient to use spherical polar coordinates (Appendix A.13.3). To convert Equation (24.18) to these coordinates, the term $\nabla^2\psi$ is written in spherical polar co-ordinates, as indicated in Appendix A.13.3. Equation (24.18) then becomes

$$-\frac{\hbar^2}{2m_r}\left[\frac{1}{r^2}\frac{\partial}{\partial r}\left(\frac{r^2\partial\psi}{\partial r}\right)+\frac{1}{r^2\sin\theta}\frac{\partial}{\partial\theta}\left(\sin\theta\frac{\partial\psi}{\partial\theta}\right)+\frac{1}{r^2\sin^2\theta}\frac{\partial^2\psi}{\partial\phi^2}\right] + U(r)\psi = E\psi \tag{24.19}$$

where ψ is now a function of r, θ and ϕ.

The change of coordinates will allow us to find a solution of the form

$$\psi(r, \theta, \phi) = R(r)\Theta(\theta)\Phi(\phi) \tag{24.20}$$

that is, solutions in which the eigenfunction $\psi(r, \theta, \phi)$ is written as the product of three independent functions, each of one variable.

For brevity, we write Equation (24.20) as $\psi = R\Theta\Phi$

Substituting Equation (24.20) into Equation (24.19). we obtain

$$-\frac{\hbar^2}{2m_r}\left[\frac{1}{r^2}\frac{\partial}{\partial r}\left(r^2\frac{\partial(R\Theta\Phi)}{\partial r}\right)+\frac{1}{r^2\sin\theta}\frac{\partial}{\partial\theta}\left(\sin\theta\frac{\partial(R\Theta\Phi)}{\partial\theta}\right)+\frac{1}{r^2\sin^2\theta}\frac{\partial^2(R\Theta\Phi)}{\partial\phi^2}\right]+U(r)R\Theta\Phi = ER\Theta\Phi$$

Carrying out the partial differentiations $\frac{\partial(R\Theta\Phi)}{\partial r} = \Theta\Phi\frac{dR}{dr}$ etc., and multiplying throughout by $-\frac{2m_r r^2\sin^2\theta}{\hbar^2 R\Theta\Phi}$ this equation becomes

$$\frac{\sin^2\theta}{R}\frac{d}{dr}\left(r^2\frac{dR}{dr}\right)+\frac{1}{\Phi}\frac{d^2\Phi}{d\phi^2}+\frac{\sin\theta}{\Theta}\frac{d}{d\theta}\left(\sin\theta\frac{d\Theta}{d\theta}\right) = -\frac{2m_r r^2\sin^2\theta}{\hbar^2}(E-U)$$

which we can write as

$$\frac{1}{\Phi}\frac{d^2\Phi}{d\phi^2} = -\frac{\sin^2\theta}{R}\frac{d}{dr}\left(r^2\frac{dR}{dr}\right)-\frac{\sin\theta}{\Theta}\frac{d}{d\theta}\left(\sin\theta\frac{d\Theta}{d\theta}\right)-\frac{2m_r r^2\sin^2\theta}{\hbar^2}(E-U) \tag{24.21}$$

We have separated the term in ϕ on the left hand side of the equation; the left hand side does not depend on r and θ whereas the right hand side does not depend on ϕ. Such an equation can only be satisfied if the common value of the two sides is a constant. In this case it is convenient to write this constant, known as the *separation constant* as, $-m_l^2$. Thus we obtain

$$\frac{1}{\Phi}\frac{d^2\Phi}{d\phi^2} = -m_l^2 \text{ which can be written } \frac{d^2\Phi}{d\phi^2} = -m_l^2\Phi \tag{24.22}$$

We can also equate the right hand side of Equation (24.21) to $-m_l^2$

$$-\frac{\sin^2\theta}{R}\frac{d}{dr}\left(r^2\frac{dR}{dr}\right)-\frac{\sin\theta}{\Theta}\frac{d}{d\theta}\left(\sin\theta\frac{d\Theta}{d\theta}\right)-\frac{2m_r r^2\sin^2\theta}{\hbar^2}(E-U) = -m_l^2$$

Dividing this equation throughout by $-\sin^2\theta$ and rearranging the terms we obtain

$$\frac{1}{R}\frac{d}{dr}\left(r^2\frac{dR}{dr}\right)+\frac{2m_r r^2}{\hbar^2}(E-U) = \frac{m_l^2}{\sin^2\theta}-\frac{1}{\Theta\sin\theta}\frac{d}{d\theta}\left(\sin\theta\frac{d\Theta}{d\theta}\right)$$

We have now separated the variables r and θ and can equate both sides of the equation to a separation constant which, in this case, it is convenient to call $l(l+1)$. Thus we obtain two ordinary differential equations, in θ and r respectively

$$-\frac{1}{\sin\theta}\frac{d}{d\theta}\left(\sin\theta\frac{d\Theta}{d\theta}\right)+\frac{m_l^2\Theta}{\sin^2\theta} = l(l+1)\Theta \tag{24.23}$$

and

$$\frac{1}{r^2}\frac{d}{dr}\left(r^2\frac{dR}{dr}\right)+\frac{2m_r}{\hbar^2}(E-U)R = l(l+1)\frac{R}{r^2} \tag{24.24}$$

Equations (24.22), (24.23) and (24.24) are three independent ordinary differential equations in the variables ϕ, θ and r, respectively, each of which can be solved analytically, that is each solution can be expressed as a mathematical function.

If a solution to the Schrödinger equation is to be physically meaningful, it must be both single valued and finite everywhere. The solution to Equation (24.22) is readily shown to be $\Phi = e^{im_l\phi}$. The condition that Φ be single valued means that the solution must be the same at $\phi = 0, 2\pi, \ldots$. etc. because, from Figure 24.12, these angles correspond to the same position in space. Thus

$$e^{im_l 0} = e^{im_l 2\pi}$$

which can be written (using Appendix A.8) as $1 = \cos m_l 2\pi + i\sin m_l 2\pi$

This can only be satisfied if $m_l = 0, \pm 1, \pm 2, \ldots$

The acceptable (physically meaningful) solutions of Equation (24.22) are therefore

$$\Phi_{m_l}(\phi) = e^{im_l\phi} \tag{24.25}$$

where $m_l = 0, \pm 1, \pm 2, \ldots$

The general solutions of both Equations (24.23) and (24.24) involve power series and tabulated polynomials. The procedures used to obtain these solutions are described in the websection up.ucc.ie/24/5/1/. The web section up.ucc.ie/24/5/2/ shows how Equation (24.24) — the *radial Schrödinger equation* for a one-electron atom — can be solved for the lowest energy state of hydrogen and, in Section 24.6 we state and discuss the general solutions.

The Θ solutions of Equation (24.23) are found to be acceptable (finite) only if l is an integer with values, $|m_l|$, $|m_l| + 1$, $|m_l| + 2, \ldots$ that is, if $l \geq |m_l|$. The form of the solution depends on the values of l and m_l so that Θ solutions are labelled $\Theta_{lm_l}(\theta)$.

The form of the R solutions of the radial Equation (24.24) are found to depend on the values of l and of n, where n is an integer which can take the values 1, 2, 3... as long as $n > l$. R solutions are therefore labelled $R_{nl}(r)$.

Energies

The corresponding **eigenvalues (energies) of the one-electron atom** also follow from the solution of Equation (24.19) for the case $U(r) = -\frac{Ze^2}{4\pi\varepsilon_0 r}$. They are found to be given by

$$E_n = -\frac{Z^2 e^4 m_r}{32\pi^2\hbar^2\varepsilon_0}\frac{1}{n^2} = -\frac{Z^2 E_0}{n^2} \tag{24.26}$$

The $n = 1$ solution is derived in the web section up.ucc.ie/24/5/2/.

Note that, when $Z = 1$, Equation (24.26) is identical to Equation (24.8), the equation for total energy which we obtained for hydrogen from the Bohr Theory. Thus Figure 24.7, in which the quantised energy values are shown on a plot of the energy function, applies equally well to the full quantum mechanical treatment. It also follows that results which are derived from the Bohr energies, such as the Rydberg formula, are identical in the Schrödinger treatment. The theoretical basis for these results is, however, no longer arbitrary. It is the direct result of the postulates of quantum mechanics and the requirement that solutions are physically meaningful.

Quantum numbers

Three integers (quantum numbers), n, l and m_l, emerge from the analysis, one for each degree of freedom, a general result in quantum mechanics, The relationships between the quantum numbers, which we have just described, can be restated in the following ways:

- $n = 1, 2, 3\ldots$. n is known as the **principal** quantum number, because, as we have noted above (Equation (24.26)), it specifies the total energy of the atom.
- l can take integral values such that $l < n$. Thus $l = 0, 1, \ldots (n - 1)$. n values of l are therefore possible for each n. l is known as the **orbital** quantum number because, as we shall see in Section 24.8, it specifies the orbital angular momentum of the atom.
- For a given value of l, m_l can take integral values such that $|m_l| \leq 1$. For a given l, therefore, m_l can take the values $-l, \ldots -1, 0, +1, \ldots + l$; thus $(2l + 1)$ values of m_l are possible for each value of l. m_l is known as the **magnetic** quantum number because, as we shall see in Section 24.9, it specifies the orientation of the orbital angular momentum of the electron in an external magnetic field.

The full eigenfunction solutions to Equation (24.19), the time-independent Schrödinger equation for the one-electron atom, can therefore be written

$$\psi_{nlm_l}(r, \theta, \phi) = R_{nl}(r)\Theta_{lm_l}(\theta)\Phi_{m_l}(\phi) \tag{24.27}$$

In the next section we shall examine the form of these complicated functions for the lowest energy states of the one-electron atom but, before we consider them in detail, we should examine the corresponding eigenvalues and the labels which are used to specify the different states of the atom.

Note that although, as indicated above, many different values of l and m_l and thus many different eigenfunctions are possible for a given value of n, the energies given by Equation (24.26) depend only on n. The different values of l and m_l all correspond to the same energy E_n. This phenomenon, whereby different eigenfunctions, which are specified by different sets of quantum numbers, correspond to the same energy is known as **degeneracy**.

The following system is used to label one-electron eigenfunctions.

For the lowest energy state ($n = 1$), the only possible value of both l and m_l is zero. There is only one set of quantum numbers $(n, l, m_l) = (1, 0, 0)$ and, thus, only one eigenfunction, which we label ψ_{100}. This state is not degenerate.

For the next energy state ($n = 2$), l can take either of two values, 0 or 1. When $l = 0$, m_l can only be 0 but, when $l = 1$, m_l can take any of three values, $+1$, 0 or -1. In total there are four different sets of the three quantum numbers – four eigenfunctions – which we label ψ_{200}, ψ_{211}, ψ_{210}, and ψ_{21-1}, respectively, and which contain different functions of θ and ϕ, that is different angular dependencies. The $n = 2$ state is therefore four times degenerate.

The l-values of a state of a one-electron atom are often given in the following notation – known as *spectroscopic notation*

$l =$ 0,	1,	2,	3,	4,	5,	6, …	states are labelled
s,	p,	d,	f,	g,	h,	i, …	etc., respectively,

The first four letters follow labels originally used by spectroscopists to describe the appearances of certain groups of spectral lines, namely sharp, principal, diffuse, fundamental.

Thus, in this notation, the $(n,l) = (1,0)$, $(2,0)$, $(2,1)$, $(3,0)$, $(3,1)$ and $(3,2)$ states are labelled 1s, 2s, 2p, 3s, 3p and 3d, respectively.

24.6 Interpretation of the one-electron atom eigenfunctions

We have described how the eigenfunctions of the one-electron atom $\psi_{nlm_l}(r, \theta, \phi)$ may be obtained and how they are labelled. We now interpret their form in more detail, examining the information which the eigenfunctions give us concerning the structure and properties of one-electron atoms.

The general form of the $\Phi_{m_l}(\phi)$ solution to Equation (24.22) has already been stated, namely Equation (24.25)

$$\Phi_{m_l}(\phi) = e^{im_l\phi}$$

The general forms of the $\Theta_{lm_l}(\theta)$ and $R_{nl}(r)$ solutions to Equations (24.23) and (24.24) are, respectively,

$$\Theta_{lm_l}(\theta) = (\sin^{|m_l|}\theta)F_{l|m_l|}(\cos\theta)$$

and

$$R_{nl}(r) = e^{-\frac{Zr}{na_0}}\left(\frac{Zr}{a_0}\right)^l G_{nl}\left(\frac{Zr}{a_0}\right)$$

$\Theta_{lm_l}(\theta)$ is known as an associated Legendre function where $F_{l|m_l|}(\cos\theta)$ is a polynomial in $\cos\theta$ (as described in the web section up.ucc.ie/24/5/1). The exact forms of the full $n = 1, 2, 3$ eigenfunctions are listed in Table 24.1. The constant preceding each eigenfunction is a normalisation factor, chosen to ensure that the probability of finding the electron somewhere in all space is 1, that is that

$$\int_{\text{all space}} \psi^*_{nlm_l}\psi_{nlm_l}dV = 1$$

where dV is a volume element. Note that, as indicated in Table 24.1, the $l = 0$ eigenfunctions $\psi_{n\,00}$ are independent of the angles θ and ϕ; these solutions are therefore spherically symmetric.

In going down Table 24.1, θ dependence first appears in the ψ_{210} eigenfunction. In this case, in which $m_l = 0$, the $\sin^{|m_l|}\theta$ term is 1 and the polynomial in $(\cos\theta)$, $F_{l|m_l|}(\cos\theta)$, has the simple form $\cos\theta$. For the $\psi_{21\pm1}$ eigenfunctions, $m_l = \pm1$ so that $\sin^{|m_l|}\theta = \sin\theta$ and the polynomial $F_{l|m_l|}(\cos\theta)$ has the value 1. ϕ dependence – the $e^{im_l\phi}$ term – appears only when m_l is non-zero; that is, it first appears in the $\psi_{21\pm1}$ eigenfunctions.

A wavefunction or an eigenfunction (the time-independent part of the wavefunction) is not directly observable but the square of the wavefunction represents a quantity which, in principle, may be determined through observation, that is the probability of finding a particle per unit volume. We therefore study the one-electron eigenfunctions in terms of their associated probability densities. In the one-dimensional case, probability density is given by

$$P(x)dx = \psi^*(x)\psi(x)dx$$

In the present three-dimensional case this becomes, from Equation (24.20), the probability of finding the electron within a volume dV located at (r, θ, ϕ),

$$P_{nlm_l}(r,\theta,\phi)dV = [R^*_{nl}(r)R_{nl}(r)][\Theta^*_{lm_l}(\theta)\Theta_{lm_l}(\theta)][\Phi^*_{m_l}(\phi)\Phi_{m_l}(\phi)]dV$$

The probability density function, therefore, is (like the eigenfunction) composed of three distinct parts, one radial and two angular. Let us first investigate the radial dependence of probability density.

Table 24.1 Normalised eigenfunctions of the one-electron atom for $n = 1$, 2 and 3. (note: for hydrogen $Z = 1$)

n	l	m_l	$R_{nl}(r)$ $\Theta_{lm_l}(\theta)$ $\Phi_{m_l}(\phi)$			
1	0	0	$\frac{1}{\sqrt{\pi}}\left(\frac{Z}{a_0}\right)^{\frac{3}{2}}$	$e^{-\frac{Zr}{a_0}}$	1	1
2	0	0	$\frac{1}{4\sqrt{2\pi}}\left(\frac{Z}{a_0}\right)^{\frac{3}{2}}$	$\left(2-\frac{Zr}{a_0}\right)e^{-\frac{Zr}{2a_0}}$	1	1
2	1	0	$\frac{1}{4\sqrt{2\pi}}\left(\frac{Z}{a_0}\right)^{\frac{3}{2}}$	$\frac{Zr}{a_0}e^{-\frac{Zr}{2a_0}}$	$\cos\theta$	1
2	1	± 1	$\frac{1}{8\sqrt{2\pi}}\left(\frac{Z}{a_0}\right)^{\frac{3}{2}}$	$\frac{Zr}{a_0}e^{-\frac{Zr}{2a_0}}$	$\sin\theta$	$e^{\pm i\phi}$
3	0	0	$\frac{1}{81\sqrt{3\pi}}\left(\frac{Z}{a_0}\right)^{\frac{3}{2}}$	$\left(27-18\frac{Zr}{a_0}+2\frac{Z^2r^2}{a_0^2}\right)^2 e^{-\frac{Zr}{3a_0}}$	1	1
3	1	0	$\frac{\sqrt{2}}{81\sqrt{\pi}}\left(\frac{Z}{a_0}\right)^{\frac{3}{2}}$	$\left(6-\frac{Zr}{a_0}\right)\frac{Zr}{a_0}e^{-\frac{Zr}{3a_0}}$	$\cos\theta$	1
3	1	± 1	$\frac{1}{81\sqrt{\pi}}\left(\frac{Z}{a_0}\right)^{\frac{3}{2}}$	$\left(6-\frac{Zr}{a_0}\right)\frac{Zr}{a_0}e^{-\frac{Zr}{3a_0}}$	$\sin\theta$	$e^{\pm i\phi}$
3	2	0	$\frac{1}{81\sqrt{6\pi}}\left(\frac{Z}{a_0}\right)^{\frac{3}{2}}$	$\frac{Z^2r^2}{a_0^2}e^{-\frac{Zr}{3a_0}}$	$(3\cos^2\theta - 1)$	1
3	2	± 1	$\frac{1}{81\sqrt{\pi}}\left(\frac{Z}{a_0}\right)^{\frac{3}{2}}$	$\frac{Z^2r^2}{a_0^2}e^{-\frac{Zr}{3a_0}}$	$\sin\theta\cos\theta$	$e^{\pm i\phi}$
3	2	± 2	$\frac{1}{162\sqrt{\pi}}\left(\frac{Z}{a_0}\right)^{\frac{3}{2}}$	$\frac{Z^2r^2}{a_0^2}e^{-\frac{Zr}{3a_0}}$	$\sin^2\theta$	$e^{\pm 2i\phi}$

The radial dependence of probability density

By integrating the probability density over the volume enclosed between two spherical shells of radii r and $r + dr$, as illustrated in Figure 24.13, we obtain the probability of finding the electron at a distance between r and $r + dr$ from the centre of the atom.

$$P_{nl}(r)dr = R_{nl}^*(r)R_{nl}(r)4\pi r^2 dr \tag{24.28}$$

Where, as illustrated in Figure 24.13, the volume enclosed between the shells is $4\pi r^2 dr$. The $P_{nl}(r)$ functions for the $n = 1$, 2 and 3 eigenfunctions are plotted in Figure 24.14.

The $P_{10}(r)$ eigenfunction (1s in spectroscopic notation) has a single maximum. This can be understood when the ψ_{100} eigenfunction of Table 24.1 is substituted into the probability density Equation (24.28)

$$P_{10}(r) = \frac{1}{\pi}\left(\frac{Z}{a_0}\right)^3 e^{-\frac{2Zr}{a_0}} 4\pi r^2 = 4\left(\frac{Z}{a_0}\right)^3 r^2 e^{-\frac{2Zr}{a_0}}$$

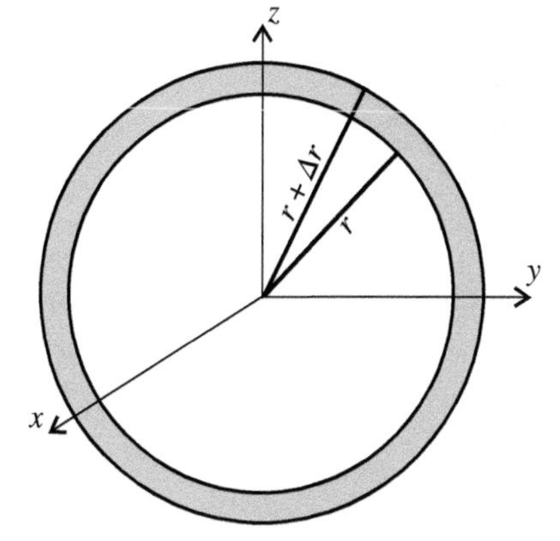

Figure 24.13. The volume enclosed between two spherical shells of radii r and $r + dr$.

When $r \ll \frac{a_0}{2Z}$, $e^{-\frac{2Zr}{a_0}} \approx 1$ so that, initially, $P_{10}(r)$ rises in proportion to r^2. However, as r increases further, $2Zr$ approaches a_0 and the exponential term $e^{-\frac{2Zr}{a_0}}$ becomes smaller, bringing $P_{10}(r)$ to zero at large r. $P_{10}(r)$ shows a maximum at $r = a_0$, as shown in Worked Example 24.4. All one-electron eigenfunctions contain an exponential term, $e^{-\frac{Zr}{na_0}}$, which means that the probability of finding the electron at $Zr \gg na_0$ is small. The exponential term ensures that the radial probability function never has a significant value far beyond the Bohr electron orbit.

For the ψ_{200} eigenfunction (2s in spectroscopic notation), the polynomial component of the $R(r)$ function, $G_{nl}\left(\frac{Zr}{a_0}\right) = \left(2 - \frac{Zr}{a_0}\right)$ so that $P_{20}(r)$ is proportional to $r^2\left(2 - \frac{Zr}{a_0}\right)^2$. In consequence, as illustrated in Figure 24.14, this function has two maxima so that the electron has a significant probability of being close to the nucleus and also a significant probability of being at larger values of r.

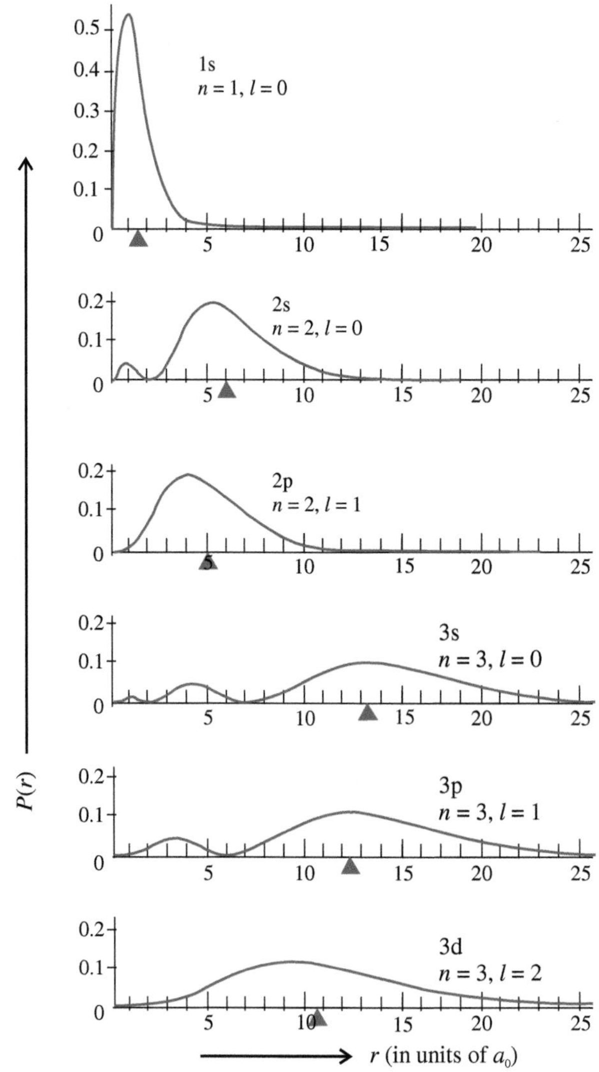

Figure 24.14. Radial probability functions $P_{nl}(r)$ for the n = 1, 2 and 3 states of the one-electron atom. In each case the expectation value of r, $\langle r_{nl} \rangle$ is indicated by a triangle ▲.

$P_{nl}(r)$ plots for the $n = 3$ eigenfunctions are also shown in Figure 24.14. Note that, for a given value of n, the probability density functions for the lower l states have additional maxima close to the nucleus. In these states, electrons are more likely to be found close to the nucleus than in higher l states. At the same time the expectation values of r,

$$< r_{nl} >= \int R^*_{nl}(r) r R_{nl}(r) 4\pi r^2 dr,$$

which are indicated by triangles (▲) in Figure 24.14, decrease with increasing l for a given n. The expectation value of r in the ground state, $\langle r_{10} \rangle$, is shown, in Worked Example 24.3, to be $\frac{3a_0}{2}$.

The angular dependence of probability density

We now investigate the angular dependence of the $n = 1, 2$ eigenfunctions of Table 24.1. The general ϕ solution is $\Phi_{m_l}(\phi) = e^{im_l\phi}$ so that the probability function $\Phi_{m_l}(\phi)\Phi^*_{m_l}(\phi) = e^{im_l\phi}e^{-im_l\phi} = 1$ for all one-electron eigenfunctions. This means that none of the one-electron probability density functions shows any ϕ dependence. They do not change as ϕ varies from 0 to 2π which, referring to Figure 24.12, means that they all are symmetric to rotation about the z-axis.

θ dependence may be represented by polar plots in which, as illustrated in Figure 24.15, the distance of a point on the plot from the origin is proportional to $\Theta^*_{lm_l}(\theta)\Theta_{lm_l}(\theta)$. Polar plots of θ dependence of the $n = 1, 2$ probability densities are shown in Figure 24.16.

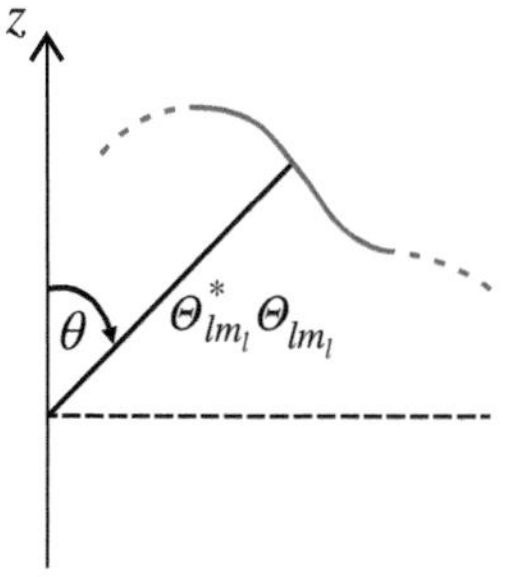

Figure 24.15. A polar plot. The distance of a point on the plot from the origin is proportional to $\Theta^*_{lm_l}(\theta)\Theta_{lm_l}(\theta)$.

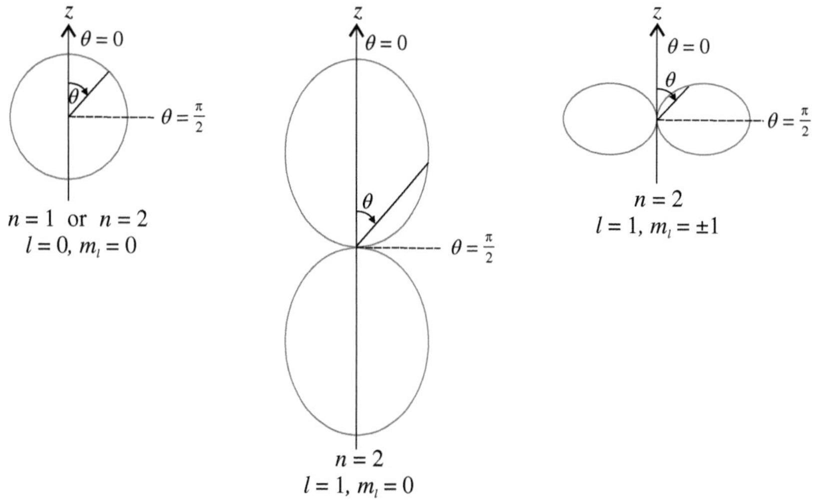

Figure 24.16. Polar plots showing the θ dependence of the $n = 1, 2$ probability densities.

The ψ_{100} (1s) and ψ_{200} (2s) functions are independent of θ so that $\Theta^*_{00}(\theta)\Theta_{00}(\theta) = 1$ and consequently the polar plots are simply circles. For the ψ_{210} (2p) eigenfunction $\Theta^*_{10}(\theta)\Theta_{10}(\theta)$ is proportional to $\cos^2\theta$ so that, as shown in Figure 24.16, the polar plots show maxima close to the z-axis when $\theta \to 0$. For the $\psi_{21\pm1}$ (also 2p) eigenfunctions $\Theta^*_{1\pm1}(\theta)\Theta_{1\pm1}(\theta)$ is proportional to $\sin^2\theta$ so these plots show probability maxima in the x–y plane when $\theta \to \frac{\pi}{2}$, again as illustrated in Figure 24.16.

In Figure 24.17 a polar plot is given for a $n = 3$ eigenfunction, the $\psi_{32\pm1}$ eigenfunction. This plot shows how further maxima in the angular probability functions (preferred directions) appear for higher l values.

In each case because, as noted above, all distributions are symmetric under rotation about the z-axis, the full three-dimensional angular dependence may be obtained by rotating the polar plots about the z -axis. The circle for $l = 0,\ m_l = 0$ in Figure 24.16 becomes a sphere, the maxima along the z-axis for $l = 1,\ m_l = 0$ become two 'egg' shapes, centred along the z-axis, and the maxima in the x–y plane for $l = 1,\ m_l = \pm 1$ becomes a 'ring doughnut' shape.

Atomic charge densities $\rho_{nlm_l}(r, \theta, \phi)$ are directly related to electron probability densities through

$$\rho_{nlm_l}(r, \theta, \phi) = -eP_{nlm_l}(r, \theta, \phi) = -e\psi^*_{nlml}(r, \theta, \phi)\psi_{nlm_l}(r, \theta, \phi)$$

where – e is the electron charge. Electron probability densities, as represented by the full three-dimensional probability functions (known as *atomic orbitals*), can be viewed, therefore, as three dimensional charge clouds, a picture which is more consistent with the uncertainty principle than the Bohr model of electron orbits.

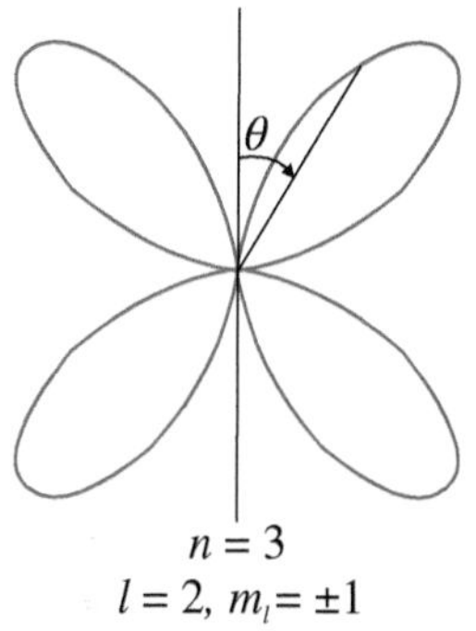

Figure 24.17. A polar plot showing the θ dependence of the $l = 2,\ m_l = \pm 1$ probability density.

Study Worked Examples 24.3 and 24.4

For problems based on the material presented in this section visit up.ucc.ie/24/ and follow the link to the problems.

24.7 Intensities of spectral lines: selection rules

At the beginning of Section 24.5 we noted that a weakness of the Bohr theory is its inability to predict the intensities of spectral lines. The quantum mechanical theory, which we have now introduced, can overcome this shortcoming. As outlined below, the probability of a transition taking place may be calculated with a knowledge of the initial and final state wavefunctions. The transition probability is related directly to the line intensity.

An atom which is undergoing a transition may be viewed as possessing a probability density distribution (charge cloud) which is oscillating between the initial and final state distributions. The oscillating charge cloud is not spherically symmetric; there is a net separation of positive and negative charge and this separation varies as the cloud oscillates, as illustrated in Figure 24.18. Oscillation of charge involves acceleration of charge and, as described in Section 21.3, an accelerating electric charge produces electromagnetic radiation.

The mechanism which produces the largest transition probability, and hence the greatest rate of emission of electromagnetic energy, in an atom is the oscillation of its electric dipole moment (Section 16.6). The atomic electric dipole moment is given by $\boldsymbol{p} = -e\boldsymbol{r}$ where $\boldsymbol{r}$ is the displacement of the negative charge relative to the positive charge, as illustrated in Figure 24.18. The rate of emission of electromagnetic energy through this process is proportional to p^2 (see web section up.ucc.ie/24/7/1/). In the web section up.ucc.ie/24/7/2/ it is shown that

$$\langle p\rangle \propto \int_{\text{all space}} \Psi_{\text{f}}^{*}(r,\theta,\phi,t)(-e\boldsymbol{r})\Psi_{\text{i}}(r,\theta,\phi,t)dV$$

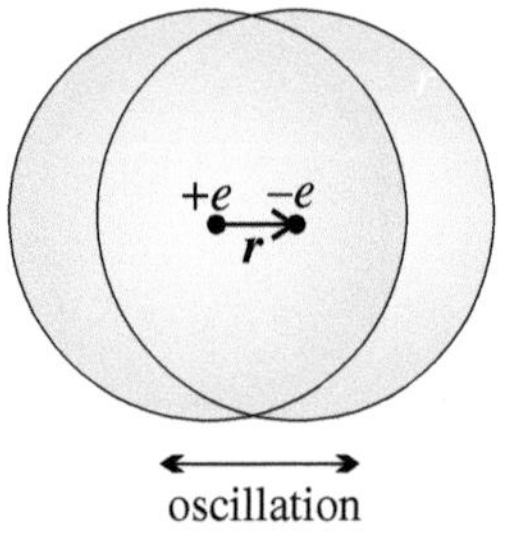

Figure 24.18. A schematic illustration of the oscillation of the electron probability density in an atom which is undergoing a transition.

where $\Psi_{\text{i}}(r,\theta,\phi,t) = \psi_{\text{i}}(r,\theta,\phi)e^{-\frac{iE_{\text{i}}t}{\hbar}}$ and $\Psi_{\text{f}}(r,\theta,\phi,t) = \psi_{\text{f}}(r,\theta,\phi)e^{-\frac{iE_{\text{f}}t}{\hbar}}$ are, respectively, the initial and final state wavefunctions and E_{i} and E_{f} are, respectively, the final and initial state energies. $\Psi_{\text{f}}^{*}(r,\theta,\phi,t) = \psi_{\text{f}}^{*}(r,\theta,\phi)e^{\frac{iE_{\text{f}}t}{\hbar}}$ so that we can write the integral as

$$\int_{\text{all space}} \psi_{\text{f}}^{*}(r,\theta,\phi)e^{\frac{iE_{\text{f}}t}{\hbar}}(-e\boldsymbol{r})\psi_{\text{i}}(r,\theta,\phi)e^{-\frac{iE_{\text{i}}t}{\hbar}}dV = e^{\frac{i(E_{\text{f}}-E_{\text{i}})t}{\hbar}}\int_{\text{all space}} \psi_{\text{f}}^{*}(r,\theta,\phi)(-e\boldsymbol{r})\psi_{\text{i}}(r,\theta,\phi)dV$$

The $e^{\frac{i(E_{\text{f}}-E_{\text{i}})t}{\hbar}}$ term is an oscillatory function with a frequency which is given by

$$\omega = 2\pi f = \frac{E_{\text{f}} - E_{\text{i}}}{\hbar}$$

The frequency of the electromagnetic radiation emitted is therefore given by

$$hf = E_{\text{f}} - E_{\text{i}}$$

This is the Bohr postulate (iii) (Section 24.3), now placed on a firmer theoretical basis.

The rate at which electromagnetic radiation is produced is determined by the *electric dipole integral,*

$$\int_{\text{all sapce}} \psi_{\text{f}}^{*}(r,\theta,\phi)(-e\boldsymbol{r})\psi_{\text{i}}(r,\theta,\phi)dV$$

It is found that the value of this integral depends critically on the symmetry properties of the ψ_{f} and ψ_{i} eigenfunctions. The symmetry of an eigenfunction can be shown to be determined by the value of the l quantum number. If l_{f} and l_{i} are, respectively, the values of l for the final and initial states, it is found that contributions to the electric dipole integral from different parts of the atom cancel unless $\Delta l = l_{\text{f}} - l_{\text{i}} = \pm 1$.

Thus electric dipole radiation is produced only when $\Delta l = \pm 1$, a condition which is known as a **selection rule** for the transition. Let us consider the implications of this rule for the Lyman series of hydrogen.

As noted in Section 24.4, the Lyman series corresponds to transitions from excited levels with $n_{\text{i}} = 2, 3, 4, \ldots.$ to the ground level ($n_{\text{f}} = 1$) of hydrogen. The ground level of hydrogen is the 1s ($l = 0$) state, so that, according to the $\Delta l = \pm 1$ selection rule, transitions to the ground state are possible only from excited levels with $l = 1$, namely the 2p, 3p 4p, ... levels. Transitions from $l = 0$ levels (that is 2s, 3s, 4s, ...) or $l = 2$ levels (3d, 4d. 5d...), although possible from energy considerations, are forbidden by the electric dipole radiation selection rule.

Application of the $\Delta l = \pm 1$ selection rule to the Balmer series shows that each line actually comprises three transitions, that is the $n_i = 3 \rightarrow n_{\text{f}} = 2$ red line is made up of 3p → 2s, 3s → 2p and 3d → 2p transitions.

For problems based on the material presented in this section visit up.ucc.ie/24/ *and follow the link to the problems.*

24.8 Quantisation of angular momentum

As shown in Figures 24.16 and 24.17, the quantum mechanical solution of the one-electron atom leads not only to energy quantisation but also to quantisation of the orientations of the probability densities relative to the z-axis. When $l \neq 0$, the probability of finding an electron in certain directions relative to the z-axis is zero. This effect is reminiscent of the nodes which occur in the one-dimensional standing waves which are the eigenfunctions of particles in bound systems (Section 14.15). In the present three-dimensional case we find that there are nodal directions rather than nodal points. The eigenfunctions of the one-electron atom are therefore three-dimensional standing waves, known as *spherical harmonics*, with nodal directions which are imposed by the boundary conditions of the one-electron bound system.

As noted in Section 24.6, the quantum numbers which specify the orientations of the probability distributions are l and m_l. A quantum mechanical analysis (as given in web section up.ucc.ie/24/8/1/) shows that l is directly associated with $|\boldsymbol{L}|$, the magnitude of the orbital angular momentum of the atom through the equation

$$L^2 = l(l+1)\hbar^2 \qquad (24.29)$$

This result supports the Bohr postulate (i), namely $L = n\hbar$, in as much as it states that the angular momentum is quantised. Note, however, that the form of the quantisation is different from that given by the Bohr postulate. Only at large values of l, that is when $\sqrt{l(l+1)}\hbar \to l\hbar \to |\boldsymbol{L}|$, does Equation (24.29) produce the same values as the Bohr postulate. Note also that Equation (24.29) allows the possibility of zero angular momentum (for $l = 0$), a result which is not allowed by the semi-classical Bohr theory.

When $l \neq 0$, quantisation of angular momentum limits the $\boldsymbol{L}$ vector to certain directions (orientations) with respect to the z-axis, an effect known as *space quantisation*. The allowed directions are those for which L_z, the component of $\boldsymbol{L}$ along the z- axis, satisfies the equation

$$L_z = m_l\hbar \qquad (24.30)$$

The allowed angles between $\boldsymbol{L}$ and the z-axis are thus given by

$$\cos\theta = \frac{L_z}{|\boldsymbol{L}|} = \frac{m_l\hbar}{\sqrt{l(l+1)\hbar^2}} = \frac{m_l}{\sqrt{l(l+1)}}$$

For any given value of l, therefore, the direction of $\boldsymbol{L}$ is specified by the value of the quantum number m_l.

The effect of this quantisation condition is illustrated in Figure 24.19 for a case in which $l = 2$. As noted in Section 24.5, in this case m_l can take only the values $-2, -1, 0, +1, +2$ so that L_z can take only the values $-2\hbar, -\hbar, 0, +\hbar, +2\hbar$. Note that the magnitude of $\boldsymbol{L}$ is the same for each L_z, namely (from Equation (24.29)) $|\boldsymbol{L}| = \sqrt{2(2+1)}\hbar = \sqrt{6}\hbar$.

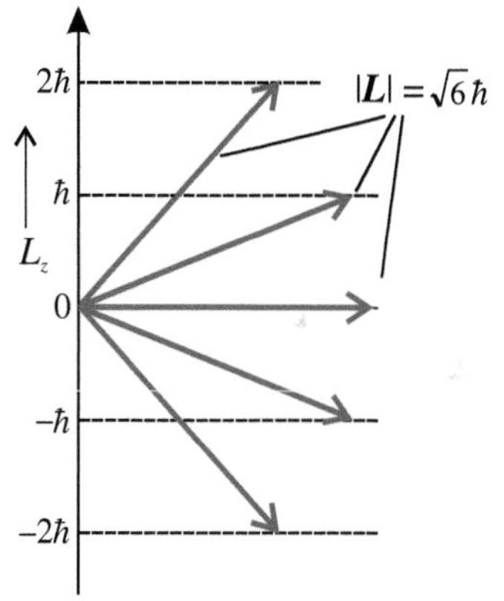

Figure 24.19. The allowed orientations, and corresponding L_z components, of the angular momentum of an atom for which $l = 2$.

We can observe energy quantisation experimentally by studying the energy levels of hydrogen for example, but we cannot observe space quantisation in a free atom experimentally. To do this we would need a reference direction such as the z-axis. The problem is that the z-axis is not a unique direction in a free atom – it is a mathematical convenience which was introduced as an aid to analysis. There is no preferred direction in a (spherically symmetric) free atom. If we perform an experiment to study the spatial distributions of electron probability density in a collection of free hydrogen atoms in, for example, the E_2 ($n = 2$) energy state, we are dealing with a collection of randomised z-axes and can detect only the average electron distribution for the four $n = 2$ states, $\psi_{200}, \psi_{21-1}, \psi_{210}$ and ψ_{211}.

This is given by $\frac{1}{4}[\psi_{200}\psi^*_{200} + \psi_{21-1}\psi^*{}_{21-1} + \psi_{210}\psi^*_{210} + \psi_{211}\psi^*_{211}]$.

As shown in Worked Example 24.5, when we substitute the $n = 2$ eigenfunctions of Table 24.1 into this equation we find that the θ dependence of the average probability density reduces to a term in $\left[\frac{1}{2}\sin^2\theta + \cos^2\theta + \frac{1}{2}\sin^2\theta\right]$ which always equals one.

The average probability density distribution in the $n = 2$ state is therefore independent of angle – it is spherically symmetric. We can go further and deduce that because, as already noted in Section 24.5, the ψ_{200} eigenfunction is spherically symmetric, the average $P_{20}(r)$ and $P_{21}(r)$ distributions must be spherically symmetric independently. This result can be shown to be quite general. The average probability density function of a collection of one-electron atoms with the same values of n, l is spherically symmetric. This means that space quantisation is not detectable in free one-electron atoms.

We can, however, observe space quantisation if a unique direction is defined in an atom through, for example, the application of a uniform external magnetic field. The direction of the field can define the z-axis, allowing space quantisation to be detected. In the next section we consider the behaviour of an atomic electron in an external magnetic field.

Study Worked Example 24.5

For problems based on the material presented in this section visit up.ucc.ie/24/ and follow the link to the problems

24.9 Magnetic effects in one-electron atoms: the Zeeman effect.

We shall examine magnetic effects in atoms using the Bohr model, noting, however, that the results which we obtain can be verified by a full quantum mechanical analysis.

In the Bohr model, the electron behaves as though it is executing an orbit around the nucleus. It therefore forms a current loop with a magnetic dipole moment $\boldsymbol{m}$. Such atomic current loops give rise to the microscopic magnetic dipoles which were introduced in the discussion of magnetic materials in Section 19.10.

In Section 19.8 we analysed the case of an orbiting electric charge showing that $\boldsymbol{m}$ is directly related to the orbital angular momentum $\boldsymbol{L}$ of the charge (Equation (19.14)). In the case of an orbiting electron, Equation (19.14) becomes

$$\boldsymbol{m} = -\frac{e}{2m_e}\boldsymbol{L} \tag{24.31}$$

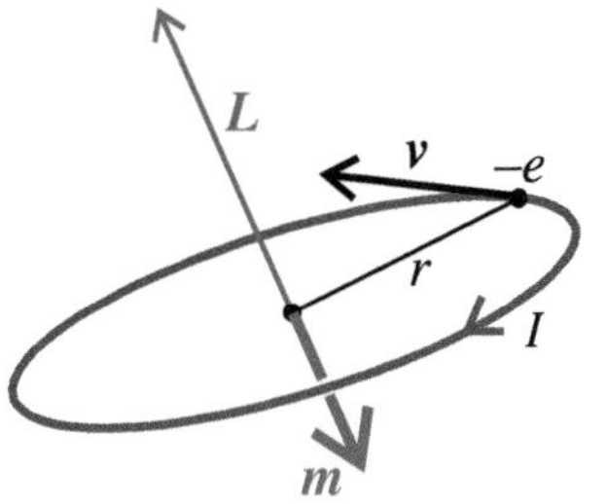

Figure 24.20. The orbital angular momentum $\boldsymbol{L}$ and the orbital magnetic dipole moment $\boldsymbol{m}$ of a Bohr atom.

The minus sign arises because the electron is negatively charged; $\boldsymbol{m}$ is therefore antiparallel to $\boldsymbol{L}$, as illustrated in Figure 24.20. The ratio $\frac{e}{2m_e}$ which relates orbital magnetic moment to orbital angular momentum is known as the **orbital gyromagnetic ratio.**

Let us consider now how an atomic current loop behaves in the presence of an external magnetic field. We know from Equation (19.5) that when a magnetic dipole, such as a current loop, is placed in an external magnetic field it experiences a torque which is given by

$$\boldsymbol{T} = \boldsymbol{m} \times \boldsymbol{B}$$

which acts to align $\boldsymbol{m}$ with $\boldsymbol{B}$, as illustrated in Figure 24.21. The potential energy associated with the orientation of $\boldsymbol{m}$ with respect to $\boldsymbol{B}$ is given by Equation (19.6).

$$U = -\boldsymbol{m} \cdot \boldsymbol{B} = -|\boldsymbol{m}||\boldsymbol{B}|\cos\theta \tag{24.32}$$

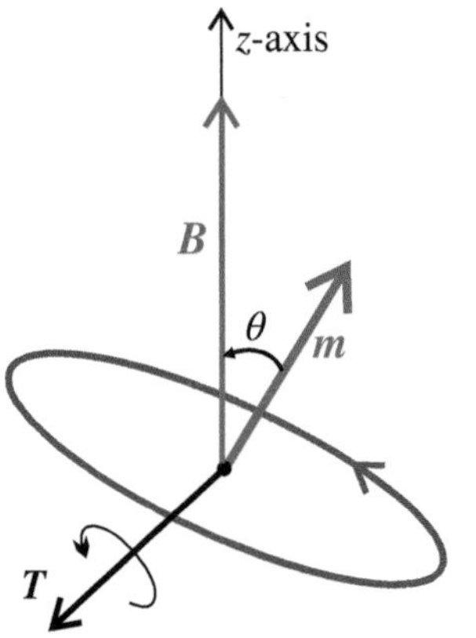

Figure 24.21. The orientation of the atomic orbital magnetic diploe moment $\boldsymbol{m}$ of an atom in an external magnetic field $\boldsymbol{B}$.

This energy is at a minimum (that is most negative) when $\boldsymbol{m}$ is parallel to $\boldsymbol{B}$, that is when $\theta = 0$.

Let us now apply these results to the case of an atomic magnetic dipole in a uniform external magnetic field whose direction defines the z-axis. In the atom the orientation of $\boldsymbol{L}$, and thus (through Equation (24.31)) $\boldsymbol{m}$, is quantised relative to the z-axis (the $\boldsymbol{B}$ direction) through Equation (24.30), namely

$$L_z = m_l\hbar$$

Thus, using Equation (24.31), the potential energy of the dipole in the field is given by

$$U = -|\boldsymbol{m}||\boldsymbol{B}|\cos\theta = \frac{e}{2m_e}|\boldsymbol{L}||\boldsymbol{B}|\cos\theta = \frac{e}{2m_e}|\boldsymbol{B}|L_z = \frac{e}{2m_e}|\boldsymbol{B}|m_l\hbar \tag{24.33}$$

The quantity $\frac{e\hbar}{2m_e}$, with the value 0.927×10^{-23} A m^2, is often used as a unit for atomic magnetic dipole moments, and is known as the **Bohr magneton,** μ_B. Thus Equation (24.33) can be written,

$$U = |\boldsymbol{B}|m_l\mu_B$$

The (orientational) potential energy of the atomic magnetic dipole in the external magnetic field is quantised, therefore, with a value which is determined by the quantum number m_l. m_l can be positive or negative so that this potential energy adds to, or subtracts from, the quantised energies E_n given by Equation (24.26). In an external magnetic field, each atomic energy level splits into a number of components corresponding to different values of m_l. Hence m_l degeneracy is removed by the external magnetic field; this is why m_l is called the *magnetic* quantum number. Because there are $(2l+1)$ values of m_l for each l, the level splits into $(2l+1)$ components.

The Zeeman effect

Because the frequency of a spectral line corresponds to the energy difference between the initial and final energy levels (Bohr postulate (iii)), in an external field the line splits into a number of components, an effect known as the **Zeeman Effect** after Pieter Zeeman (1865–1943). In Figure 24.22, the effect is illustrated for a transition between a $l = 1$ state and a $l = 0$ state.

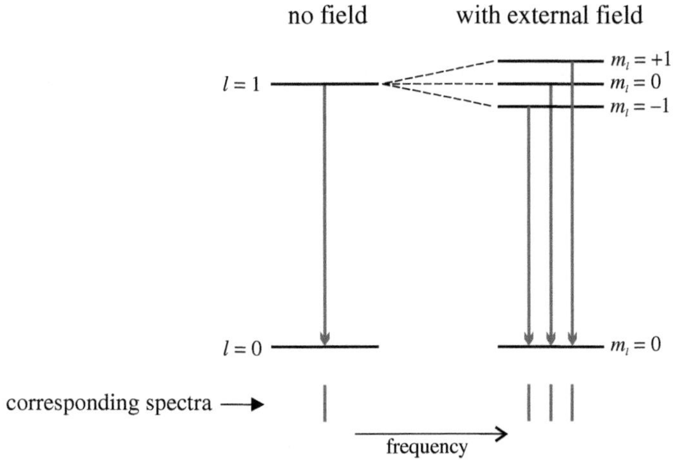

Figure 24.22. A schematic representation of the Zeeman effect splitting of a transition between $l = 1$ and $l = 0$ states. The Zeeman splittings of the $m_l = 1, 0, -1$ states are greatly exaggerated in comparison to the energy difference between the $l = 0, 1$ levels.

While the spectral lines of some atoms split in an external magnetic field in the manner shown in Figure 24.22, providing clear evidence of space quantisation, the number of components observed in the case of hydrogen is greater than predicted. The reason for this departure from theory is that there is a further magnetic dipole moment which we have not yet taken into account in our discussion of the one-electron atom. Further evidence of this phenomenon is provided by the experiment which we will describe in the next section.

For problems based on the material presented in this section visit up.ucc.ie/24/ *and follow the link to the problems*

24.10 The Stern-Gerlach experiment: electron spin

Consider a current loop in a magnetic field. If the field is uniform and perpendicular to the plane of the loop, as illustrated in Figure 24.23, the forces on each element of the current loop, which are given by Equation (19.2),

$$\Sigma\,\boldsymbol{\Delta F} = \Sigma\, I\boldsymbol{\Delta l} \times \boldsymbol{B}$$

cancel in one circuit around the loop so that there is no net force on it.

If, however, $\boldsymbol{B}$ is not uniform and, as illustrated in Figure 24.24, increases in the direction perpendicular to the plane of the loop (the z – direction), while the components of $\boldsymbol{\Delta F}$ in the plane of the loop still cancel, $\boldsymbol{\Delta F}$ is no longer in the plane of the loop and consequently the components perpendicular to the loop contribute to a net (translational) force $\boldsymbol{F}$ in the direction in which $\boldsymbol{B}$ is increasing, that is in the z-direction.

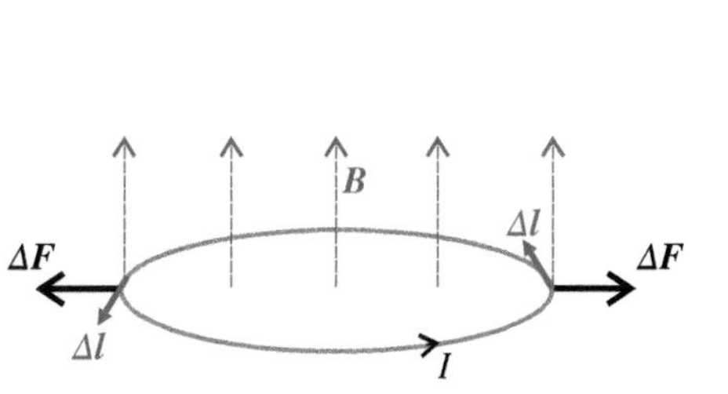

Figure 24.23. The forces on a current loop in a uniform magnetic field. Note that there is no net force on the loop.

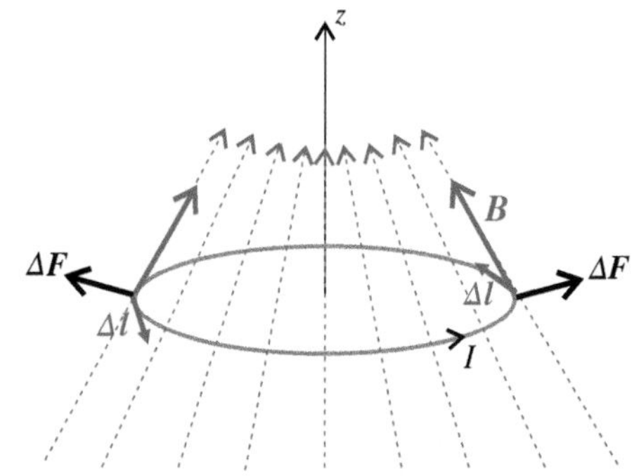

Figure 24.24. The forces on a current loop in a non-uniform magnetic field. There is a net (translational) force on the loop in the z-direction.

The Stern-Gerlach experiment

In the Stern-Gerlach (Otto Stern, 1888–1969 and Walter Gerlach, 1899–1981) experiment, the net force $\boldsymbol{F}$ is used to separate the various m_l components in a beam of atoms. While Stern and Gerlach used silver atoms in their original experiment, the experiment can be performed with hydrogen atoms, which in their ground (1s) state have the quantum numbers ($n = 1$, $l = 0$, $m_l = 0$). For simplicity, we consider the hydrogen case here.

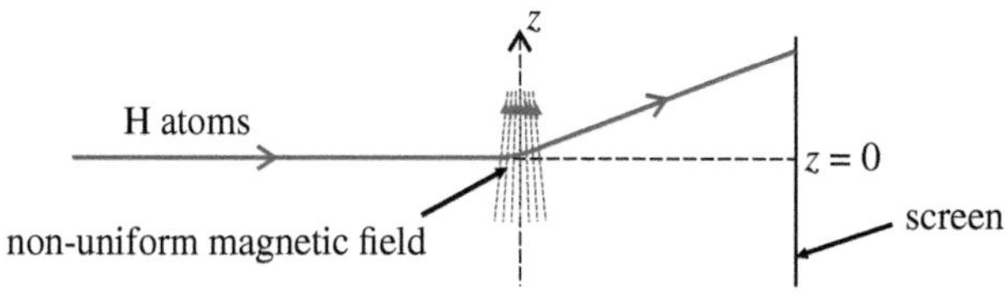

Figure 24.25. Schematic diagram of the Stern-Gerlach experiment.

As shown in Figure 24.25 the hydrogen atoms pass through a non-uniform field where they experience a force in the z-direction. As described in Section 5.7, this force may be determined by differentiating the potential energy function. Thus,

$$\boldsymbol{F} = -\left(\frac{\partial U}{\partial x}\boldsymbol{i} + \frac{\partial U}{\partial y}\boldsymbol{j} + \frac{\partial U}{\partial z}\boldsymbol{k}\right) = -\nabla U \qquad \text{(Equation (5.16))}$$

The potential energy is given by Equation (24.33) $U = \frac{e}{2m_e}|\boldsymbol{B}|L_z$ so that we obtain

$$\boldsymbol{F} = -\left(\frac{\partial B}{\partial x}\boldsymbol{i} + \frac{\partial B}{\partial y}\boldsymbol{j} + \frac{\partial B}{\partial z}\boldsymbol{k}\right)\frac{e}{2m_e}L_z$$

In the case considered here $\boldsymbol{B}$ changes as a function of z only so that this equation simplifies to

$$\boldsymbol{F} = -\frac{e}{2m_e}\frac{\partial B}{\partial z}\boldsymbol{k}L_z$$

Thus, $\boldsymbol{F}$ is proportional $-L_z$. Atoms are deflected by amounts which are determined by their L_z values.

Classically, without space quantisation, L_z can have any value between $-|\boldsymbol{L}|$ and $+|\boldsymbol{L}|$ so that in the classical case a continuous band of deflections would be expected between these two extremes, as illustrated in Figure 24.26(a).

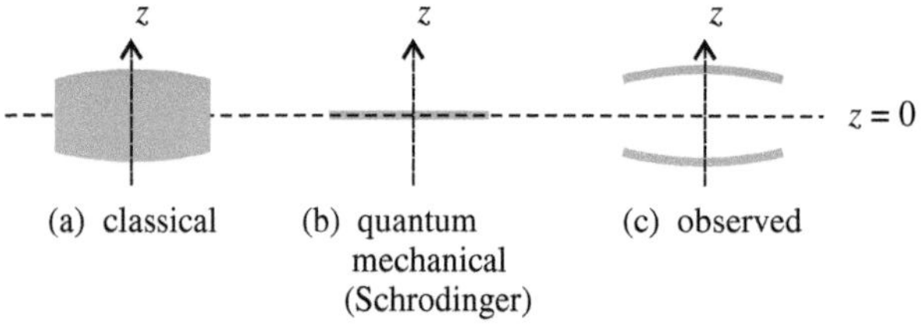

Figure 24.26. The Stern-Gerlach experiment in atomic hydrogen. **(a)** the continuous band of deflections expected classically, **(b)** the single undeflected beam expected quantum mechanically and **(c)** the two deflections actually observed.

Quantum mechanically however, as a result of space quantisation, we expect L_z to be quantised according to Equation (24.30), namely $L_z = m_l\hbar$.

Thus only certain deflections, determined by the $(2l + 1)$ allowed values of m_l, are expected. In the case of hydrogen atoms in their ground state, for which $l = m_l = 0$, this means that only one, undeflected, beam is expected, as illustrated in Figure 24.26(b).

Experimentally, however, as shown in Figure 24.26(c), the Stern-Gerlach experiment produces two quantised deflections, confirming qualitatively that space quantisation is occurring but showing also that it produces double the expected number of components. It is particularly surprising that an even number of components is observed because the number of m_l values, $(2l + 1)$, is always odd.

Electron spin

The clear implication of the Stern-Gerlach experiment is that, in addition to the orbital magnetic dipole moment, the hydrogen atom must possess a further magnetic dipole moment which has two possible quantised orientations.

This dipole moment is the *intrinsic* magnetic dipole moment of the electron $\boldsymbol{m}_S$ which is associated with an *intrinsic angular momentum* $\boldsymbol{S}$ in much the same way as orbital magnetic dipole moment is associated with orbital angular momentum (Equation (24.31)). Note, however, that the *spin gyromagnetic ratio*, which relates the intrinsic magnetic dipole moment of an electron to its intrinsic angular momentum is found to be almost exactly double the orbital gyromagnetic ratio. Thus

$$\boldsymbol{m}_S = -2 \times \frac{e}{2m_e}\boldsymbol{S} = -\frac{e}{m_e}\boldsymbol{S} \qquad (24.34)$$

To account for the two observed deflections in the Stern-Gerlach experiment, the quantum number associated with this angular momentum, s, is assigned the value $\frac{1}{2}$ so that the number of components along the z-axis, which is $(2s+1)$ by analogy with the $(2l+1)$ components of l, has the value 2. The two components along the z-axis are identified by the quantum number m_s, which can take either of the values $+\frac{1}{2}$ or $-\frac{1}{2}$. By analogy with Equations (24.29) and (24.30) the quantum numbers s and m_s are related to the intrinsic angular momentum of the electron, S, and its component along the z-axis, S_z, through the equations

$$S^2 = s(s+1)\hbar^2$$

and

$$S_z = m_s\hbar$$

For historical reasons, the intrinsic angular momentum of the electron is called **electron spin** and the corresponding quantum number s is called the **spin quantum number**. It is tempting to visualise electron spin classically by comparing it to the spin of a planet about an axis through its centre as it orbits the Sun. Such a model has the attraction that it introduces a fourth degree of freedom, rotation about an axis through the electron, so that the introduction of a fourth quantum number is in line with the comment in Section 24.5 to the effect that each degree of freedom has an associated quantum number. The model does not, however, stand up to close scrutiny as a classical model. If we try to justify the measured spin magnetic dipole moment of the electron by calculating the magnetic dipole moment of a rotating sphere whose charge – the electron charge – is distributed uniformly over its surface, we find that the electron would either have to be as large as the atom or would have to rotate so fast that its surface would be travelling faster than the speed of light. Electron spin in fact emerges as a consequence of the relativistic quantum theory of Dirac (Paul Dirac, 1902–1984), along with the rules which govern its behaviour. We can, however, graft it on to the non-relativistic (Schrödinger) treatment of the one-electron atom, which we have described in Section 24.5, as an additional *ad hoc* concept.

With the added concept of electron spin, four quantum numbers, n, l, m_l and m_s, are needed to specify a state of a one-electron atom. Two different sets of these quantum numbers are now possible for the ground (1s) state of hydrogen, namely, $(n=1,\ l=0,\ m_l=0,\ m_s=+\frac{1}{2})$ and $(n=1,\ l=0,\ m_l=0,\ m_s=-\frac{1}{2})$.

A full treatment of the Zeeman effect should include the spin magnetic dipole moment which also aligns with $\boldsymbol{B}$ in a quantised manner. The total magnetic dipole moment of an atom arises from a combination of the magnetic moment due to the orbital motion of the electrons and their intrinsic magnetic dipole moments. The treatment given in the previous section applies only to special cases in which the total electron spin of an atom is zero, a situation which can occur in certain states of multielectron atoms in which electron spin components cancel. The treatment can also be applied to the *pionic hydrogen atom* – an important system in nuclear physics – in which a pion, a particle of zero intrinsic spin (discussed further in Section 27.11), substitutes for the electron. In these special cases, in which the spin is zero, the effect is known as the *normal* Zeeman effect.

In the more general case – known, perhaps inappropriately, as the *anomalous* Zeeman effect – the total orbital and spin angular momenta of an atom are both non zero and the splitting of atomic energy levels, and thus transition energies, in weak magnetic fields is more complex. This is why, as noted in the previous section, the Zeeman splittings in hydrogen do not show a normal Zeeman effect pattern.

The anomalous effect may be analysed by introducing a gyromagnetic ratio with a value which is determined by the values of both the orbital and spin angular momenta (as described in the web section up.ucc.ie.24/10/1/).

24.11 The spin-orbit interaction

Figure 24.27 (a) illustrates the motion of an electron in a Bohr orbit around the nucleus of a hydrogen atom, as viewed from the nucleus. In Figure 24.27 (b), we show the same motion as viewed from the electron. From the viewpoint of the electron, the nucleus is in orbit around the electron just as, from the Earth, the Sun appears to be moving around the Earth.

As illustrated in Figure 24.27 (b), the nucleus, a charge $+e$ moving with velocity $-\boldsymbol{v}$ in the frame of the electron, constitutes a current loop with the electron at its centre.

The magnetic flux density at an electron in a hydrogen atom due to the motion of the nucleus, may be calculated from the Biot-Savart law for a moving charge (Equation (19.7))

$$\boldsymbol{B} = \mu_0\boldsymbol{H} = \frac{\mu_0 e[(-\boldsymbol{v})\times\boldsymbol{r}]}{4\pi r^3} = -\frac{\mu_0 e[\boldsymbol{v}\times\boldsymbol{r}]}{4\pi r^3} \qquad (24.35)$$

The orbital angular momentum of the electron about the nucleus is given in Section 5.11, $\boldsymbol{L} = m_e[\boldsymbol{r}\times\boldsymbol{v}] = -m_e[\boldsymbol{v}\times\boldsymbol{r}]$ so we can write Equation (24.35) as

$$\boldsymbol{B} = \frac{\mu_0 e}{4\pi m_e r^3}\boldsymbol{L} \qquad (24.36)$$

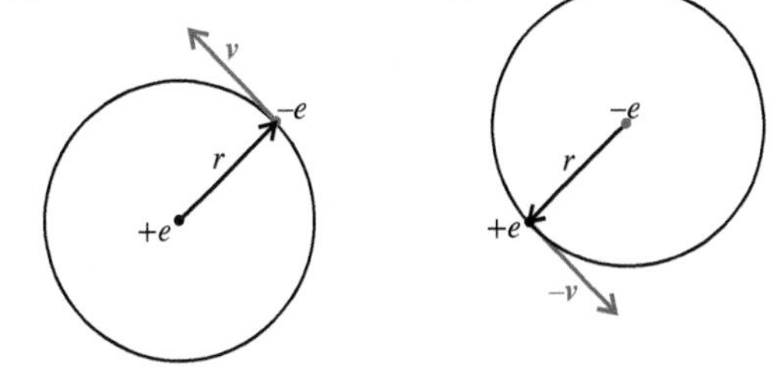

Figure 24.27. (a) The motion of an electron around the nucleus of a hydrogen atom. (b) The same motion, as viewed from the electron.

Note that this is an *internal* magnetic field of the atom, always present and independent of any external magnetic field which may be applied.

In Worked Example 24.6, the magnetic flux density produced at the nucleus by an electron in the 2p ($n = 2$, $l = 1$) state of hydrogen is calculated to be nearly 0.74 T, a flux density which is comparable to that produced by a fairly strong laboratory magnet. The magnetic flux density produced at the electron by the motion of the nucleus has the same magnitude.

The intrinsic spin magnetic dipole moment of the electron $\boldsymbol{m}_s$ experiences the internal magnetic field $\boldsymbol{B}$ given by Equation (24.36) and is thus subject to a torque $\boldsymbol{T} = \boldsymbol{m}_s \text{ x } \boldsymbol{B}$ (Equation 19.5). The (orientational) potential energy of the electron magnetic dipole due to this torque is given by Equation (19.6),

$$U = -\boldsymbol{m}_s \cdot \boldsymbol{B}$$

Note that we have evaluated this energy in a (non-inertial) reference frame in which the electron is at rest, with the nucleus orbiting around it. However laboratory observations are made in the frame of the nucleus, which can be considered to be of infinite mass if the reduced mass of the electron is used, as discussed in Section 24.4. The relativistic transformation of the electron's velocity in a succession of rest frames, as it orbits around the nucleus, into the nuclear rest frame (the *Thomas precession*, described in web section up.ucc.ie/24/11/1/) results in a reduction of the potential energy by a factor of two. This becomes

$$U = -\frac{1}{2}\boldsymbol{m}_s \cdot \boldsymbol{B}$$

Substituting for $\boldsymbol{B}$ from Equation (24.36) and for $\boldsymbol{m}_s$ from Equation (24.34), we obtain the following equation for the potential energy of the electron magnetic dipole in the hydrogen atom

$$U = \frac{\mu_0 e^2}{8\pi m_e^2 r^3} \; \boldsymbol{S} \cdot \boldsymbol{L} \tag{24.37}$$

This energy is proportional to $\boldsymbol{S} \cdot \boldsymbol{L}$, that is to the scalar product of the electron spin and electron angular momentum vectors, and is therefore called the **spin-orbit** energy.

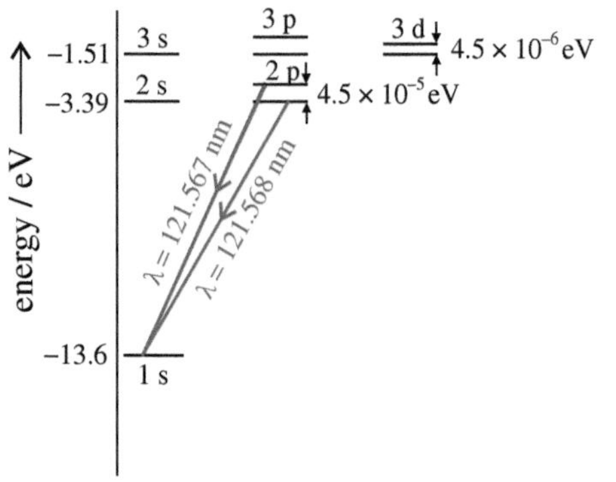

Figure 24.28. The energy levels of the hydrogen atom, including spin-orbit splitting. Note that, for the purposes of illustration, the spin-orbit splittings are greatly exaggerated.

As in the case of the orbital magnetic dipole in an external field (Section 24.7), the allowed orientations of $\boldsymbol{m}_s$ with respect to $\boldsymbol{B}$ are quantised, There are two possible spin directions, which we have labelled $m_s = \pm\frac{1}{2}$ and therefore two possible orientations of $\boldsymbol{m}_s$. The spin-orbit energy, given by Equation (24.37) adds to the quantised energies E_n given by Equation (24.26). m_s degeneracy is removed, leading to a doubling of those hydrogen energy levels for which l, and hence $\boldsymbol{L}$, is non-zero. Thus all levels, except the 1s, 2s etc. levels, are split into two. The energy levels of hydrogen, with the spin orbit interaction included, are shown schematically in Figure 24.28.

Note that the spin-orbit splitting is very small in hydrogen, typically only 10^{-4} of the energy difference between successive levels of different n. The spin-orbit splitting of levels is therefore often referred to as *fine structure*. As indicated in Figure 24.28, the splitting of the 2p level leads to a very narrow splitting in the energy of 2p →1s transition, the first member of the Lyman series of hydrogen (Section 24.4). A diagram, such as Figure 24.28, on which atomic energy levels and possible transition energies or wavelengths are indicated, is called a *Grotrian* diagram (after Walter Grotrian 1890–1954).

While spin-orbit energies are very small in light atoms, they become larger with increasing Z. On close examination, the characteristic yellow line of sodium at 589 nm is found to comprise two lines at 589.0 and 589.6 nm, the splitting between the lines being due to the spin-orbit interaction in the lowest excited state of the outermost electron of sodium.

We now have a comprehensive description of the one-electron atom and will proceed to extend our results to multielectron atoms in Sections 24.13 and 24.14. First, however, we must give some thought to some special considerations which apply when quantum mechanics is applied to a system, such as a multielectron atom, which contains more than one identical particle.

Study Worked Example 24.6

For problems based on the material presented in this section visit up.ucc.ie/24/ and follow the link to the problems

24.12 Identical particles in quantum mechanics: the Pauli exclusion principle

Classically, identical particles, such as two electrons in a box, can be labelled and in principle may be distinguished from one another at any time by following their trajectories – by tracking them. In the quantum mechanical description of such a situation, however, if the particles are close enough to one another – if they are part of the same system and their probability distributions overlap – it is not possible to distinguish between them.

The *indistinguishability* of identical particles is a fundamental feature of any quantum mechanical analysis and must be built into the solution. Observable properties of a system must not depend on the labels which are assigned to the particles. The solutions for the system – the wavefunctions – must be such that observable results must not change if the labels of identical particles are exchanged. Let us consider the implications of this statement for a system of two identical particles, such as the two electrons in the helium atom.

Let us label the two particle states a and b. In the case of the one-electron atom each label refers to a particular set of values of the quantum numbers n, l, m_l and m_s. Let us also label the coordinates of the two particles (1) and (2) respectively, for example (1) referring to a particular set of values of x-, y- and z-coordinates. Thus $\psi_a(1)$ specifies the eigenfunction of a particle in state a and at a position which is specified by the set of coordinates (1).

The total eigenfunction ψ_T of a system which comprises two particles is given by the product of the eigenfunctions of the individual particles. The two possible total eigrnfunctions are therefore

$$\psi_T = \psi_a(1)\psi_b(2) \tag{24.38}$$

or

$$\psi_T = \psi_a(2)\psi_b(1) \tag{24.39}$$

Indistinguishability requires that the values of observable quantities such as the probability density shall not change when we exchange the particles, that is when they exchange their position coordinates so that $1 \to 2$ and $2 \to 1$. This requirement can be stated as

$$P(1, 2) = P(2, 1)$$

Let us consider whether Equation (24.38) satisfies this requirement. The probability density is given by

$$P(1, 2) = \psi_a^*(1)\psi_b^*(2)\psi_a(1)\psi_b(2)$$

Exchange of particles $1 \to 2$ and $2 \to 1$ produces

$$P(2, 1) = \psi_a^*(2)\psi_b^*(1)\psi_a(2)\psi_b(1)$$

However, in general, $P(1, 2) \neq P(2, 1)$ because, for example, $\psi_a^*(2)$ is evaluated at the coordinates specified by (2) whereas $\psi_a^*(1)$ is evaluated at the coordinates specified by (1). Equation (24.38) does not satisfy the requirement of indistinguishability. Similarly, neither does Equation (24.39).

It is possible, however, to satisfy the requirement of indistinguishability, $P(1, 2) = P(2, 1)$, by constructing linear combinations of Equations (24.38) and (24.39) with specific symmetries. The Schrödinger equation is a *linear* differential equation in $\psi(x, t)$, as described in Section 3.13, so that any linear combination of its solutions is also a solution. There are two possibilities, the first being to construct a *symmetric* eigenfunction

$$\psi_B = \frac{1}{\sqrt{2}}[\psi_a(1)\psi_b(2) + \psi_a(2)\psi_b(1)] \tag{24.40}$$

This function is unchanged when the particles are exchanged $1 \to 2$ and $2 \to 1$. Consequently $P(1, 2) = P(2, 1)$. The factor $\frac{1}{\sqrt{2}}$ ensures that ψ_B is normalised if $\psi_a(1)\psi_b(2)$ and $\psi_a(2)\psi_b(1)$ are normalised. The second possibility is to construct an *antisymmetric* eigenfunction

$$\psi_F = \frac{1}{\sqrt{2}}[\psi_a(1)\psi_b(2) - \psi_a(2)\psi_b(1)] \tag{24.41}$$

This function changes sign when the particles are exchanged $1 \to 2$ and $2 \to 1$. Note, however, that, because $P(1, 2) = \psi_F^*\psi_F$ and $P(2, 1) = (-\psi_F^*)(-\psi_F) = \psi_F^*\psi_F$ and thus $P(1, 2) = P(2, 1)$, the requirement of indistinguishability is satisfied; again, $\frac{1}{\sqrt{2}}$ is a normalisation factor.

There are two distinct types of particle behaviour in quantum mechanical systems of identical particles. Each type of behaviour occurs and is associated with a distinct species of particle. The two species of particles are **bosons,** for which the total wavefunction is always symmetric (Equation (24.40)), and **fermions,** for which the total wavefunction is always antisymmetric (Equation (24.41)). Experimental observations have established that all particles with integral spin, such as photons or α-particles, are bosons and that all particles with half integral spin, such as electrons, protons and neutrons, are fermions.

Consider a case in which both the particles are in the same state, that is a = b. The symmetric (boson) wavefunction (24.40) becomes

$$\psi_B = \frac{1}{\sqrt{2}}[\psi_a(1)\psi_a(2) + \psi_a(2)\psi_a(1)] = \frac{2}{\sqrt{2}}[\psi_a(1)\psi_a(2)]$$

The probability that both bosons be in the same state, $\psi_B^*\psi_B = \frac{4}{2}[\psi_a^*(1)\psi_a(1)\psi_a^*(2)\psi_a(2)]$, is therefore twice its value for distinguishable particles.

On the other hand, when a = b, the antisymmetric (fermion) wavefunction (24.41) becomes

$$\psi_F = \frac{1}{\sqrt{2}}[\psi_a(1)\psi_a(2) - \psi_a(2)\psi_a(1)] = 0$$

It is impossible, therefore, for both fermions to be in the same quantum state. This result can be generalised to systems of many identical fermions, such as electrons in a multielectron atom. To satisfy the requirements of indistinguishability the total eigenfunction of the system must be antisymmetric. When the eigenfunction of the system is constructed from the eigenfunctions of the component fermions of the system it is found that it can be antisymmetric only if no two particles are in the same quantum state.

This finding is usually stated as the **Pauli exclusion principle** (after Wolfgang Pauli,1900–1958), namely:

no two electrons in a system can be in the same quantum state

We shall see in the next section and in the next chapter that this principle is fundamental to the behaviour of electrons in multielectron atoms and in solids.

Bosons, on the other hand, are not restricted by the Pauli exclusion principle. Any number of bosons can be in the same quantum state and, in fact, this situation is favoured energetically. This means that bosons tend to congregate in the lowest energy state of a system. As we shall see in Sections 25.7 and 25.8, this tendency can lead to remarkable behaviour by systems of bosons.

24.13 The periodic table: multielectron atoms

In the periodic table of the atomic elements, which is reproduced in Appendix E, elements are ordered according to observed periodicities in their physical and chemical properties. Elements in a given column (known as a *group*) have similar properties. As described below, the structure of the periodic table can be explained by using the results which we have derived for one-electron atoms in conjunction with the Pauli exclusion principle.

Electrons in multielectron atoms are assigned (n, l) values in the same way as in one-electron atoms. Electrons with the same n values are described as belonging to the same *shell* and those with the same (n, l) values as belonging to the same *subshell*. Spectroscopic notation (1s, 2s, 2p etc.) is used to label subshells. In Section 24.5 we noted that $(2l+1)$ values of m_l are possible for each l and in Section 24.7 that two values of m_s are possible for each of these m_l states. It follows from the Pauli exclusion principle that each subshell can contain up to $2(2l+1)$ electrons. The number of electrons in a subshell is known as the *occupancy* of the subshell so that the occupancies of filled s, p, d and f subshells, given by $2(2l+1)$, are 2, 6,10 and 14 respectively.

The periodic table of multielectron atoms in their ground states (Appendix E) has been constructed by sequentially adding one unit of nuclear charge and one electron to atomic systems in accordance with the Pauli exclusion principle and so as to minimise the total energy of the system.

The order of filling of the subshells, illustrated in Figure 24.29, is

$$1s, 2s, 2p, 3s, 3p, 4s, 3d, 4p, 5s, 5p, 6s, 4f, 5d, 6p, 7s, 5f, 6d \ldots$$

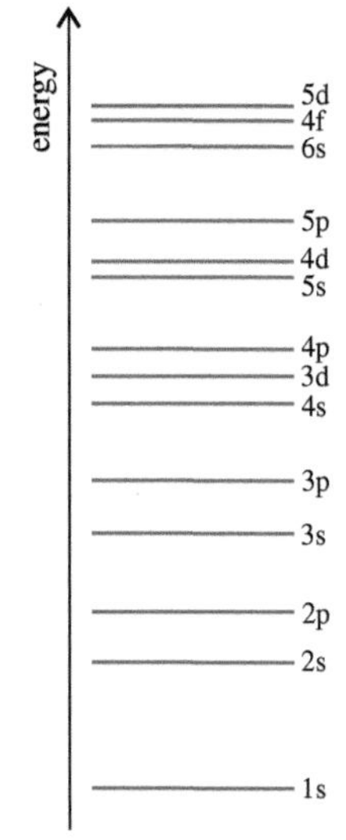

Figure 24.29. The order of filling of the energy levels of the subshells for the outermost electron of a multielectron atom.

Figure 24.29 shows that, in general, the energy of an electron in a multielectron atom increases with n, as it does in a one-electron atom, but that, unlike in the one-electron atom, the energy also increases substantially with increasing l. The l-dependence in multielectron atoms is due to the average Coulomb field produced by other electrons. We shall return to this effect shortly when we examine the alkali (group Ia) atoms in detail.

Let us consider, as an example, the sodium atom which contains 11 electrons. The first two electrons can occupy the 1s subshell, the second two the 2s subshell, the next six the 2p subshell leaving a single electron in the 3s subshell. The electron structure (known as the electron **configuration**) of the sodium atom in its ground state is therefore written, with occupancies given as superscripts,

$$1s^2 2s^2 2p^6 3s$$

The electron configurations of the elements in their ground states are indicated in Appendix E. The physical and chemical properties of the elements follow directly from their electronic structures, as illustrated by the following examples.

The inert gases (group O of the periodic table)

The inert gases are atoms in which all subshells are completely filled, in which case they are known as *closed* subshells. An example is neon ($Z = 10$) with the electron configuration $1s^2 2s^2 2p^6$. When we discussed the one-electron probability densities in Section 24.6, we noted that the average probability density of the electrons in a collection of one-electron atoms in the same (n, l) state is spherically symmetric. It follows that the probability density of the electrons in a closed subshell of a multielectron atom, $\Sigma P_{nl}(r)$, and thus the charge density, is also spherically symmetric. As shown in Section 16.8, when viewed from outside, a uniform spherical charged distribution behaves as if all the charge is located at the centre of the sphere.

When viewed from a point outside the atom, such as P in Figure 24.30, an inert gas atom appears to be a neutral point particle without any magnetic dipole moment. This is because, in a closed (filled) subshell, for each electron with a non-zero value of m_l or m_s, there is another with opposite values. The net orbital angular momentum, and thus the magnetic dipole moment, of a closed subshell is therefore zero, as noted in the discussion of diamagnetism in Section 19.10. Inert gas atoms are very stable because, as indicated in Figure 24.29, a relatively large energy is needed to excite or remove an electron from the outermost p-subshell to the next subshell, which is a s-subshell. As we shall see in Section 25.1, the usual mechanisms which allow atoms to bond to one another to form molecules and solids involve electron exchange or sharing, or interaction between electric or magnetic dipole moments, and are therefore not available to inert gases.

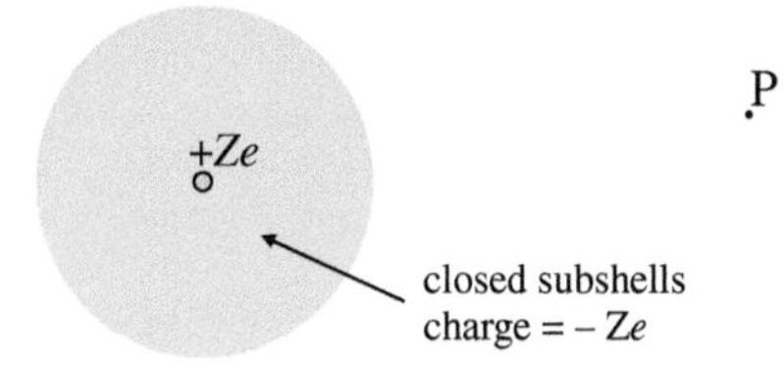

Figure 24.30. The spherical distribution of charge in an inert gas atom.

Alkali atoms (group Ia)

An alkali atom comprises an inert gas core of $(Z-1)$ electrons in closed subshells and a single outer s-electron, as illustrated in Figure 24.31. The outer electron sees the point nuclear charge, $+Ze$, surrounded by – *screened by* – a spherically symmetric electron distribution of charge $-(Z-1)e$ which behaves like a point charge of value $-(Z-1)e$ at the nucleus. The outer electron, therefore, sees a net charge of $+Ze-(Z-1)e = +e$, at the nucleus.

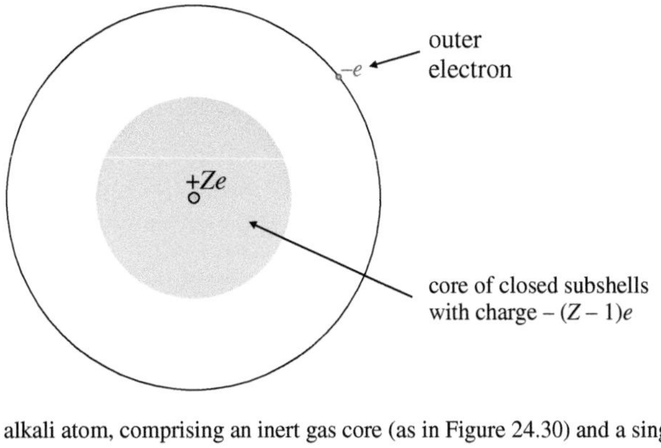

Figure 24.31. An alkali atom, comprising an inert gas core (as in Figure 24.30) and a single outer electron.

The charge distribution seen by the outer electron is very much like that seen by the electron in a hydrogen atom and the system can be analysed in a very similar way. The electron experiences a spherically symmetric central field, although, due to presence of the closed subshells, the form of the potential energy function differs from that of a point charge (Equation (24.16)), particularly at small values of r. Because the field is symmetric, however, the radial and angular eigenfunctions can be separated – as in the one-electron atom (Section 24.5) – so that the one-electron angular solutions can be carried directly into the multielectron atom. Thus, one-electron quantum numbers n, l, m_l, m_s are used to label the states of an alkali atom.

The energy levels of the sodium atom, namely the excited energy levels of the outer 3s electron, are compared with those of hydrogen in Figure 24.32. The most striking difference between the hydrogen and sodium cases is that the sodium energies are strongly dependent on l. This dependence can be explained through an inspection of the radial probability densities of Figure 24.14. The probability density functions of low l electrons have maxima in the region close to the nucleus and such electrons are less effectively shielded by the core of closed subshells. Consequently, low l electrons see more nuclear charge and are more strongly bound than those with high l. This effect also explains the l-dependence in the order of filling subshells in the periodic table (Figure 24.29).

As indicated in Figure 24.29, the 3s electron in a sodium atom is much less strongly bound than any of the electrons in inner filled subshells. Its binding energy is comparable to that of the electron in the hydrogen atom so that the spectra produced by transitions between the energy levels of Figure 24.32 lie in the visible region of the electromagnetic spectrum. Transitions involving the tightly bound inner

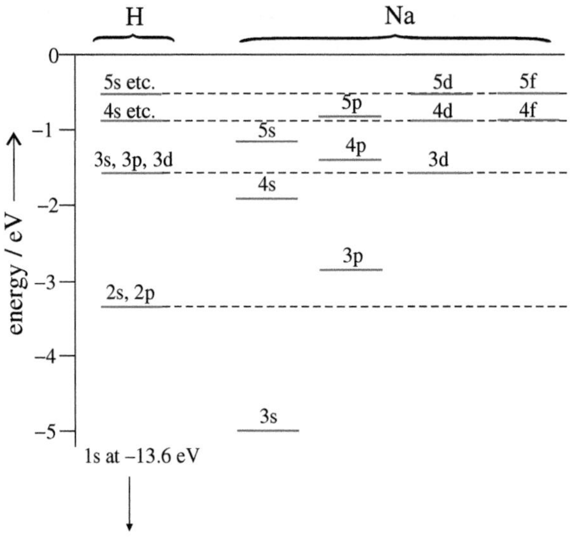

Figure 24.32. A comparison of the energy levels of sodium and hydrogen atoms.

subshells of atoms lie at much higher energies and are generally observed in the X-ray region. Such transitions are responsible for the discrete emission lines which are superposed on the X-ray continuum, as shown in Figure 14.8.

Because the outer electron of an alkali atom is weakly bound, it interacts readily with other atoms in its immediate environment. Alkali atoms are therefore highly reactive chemically.

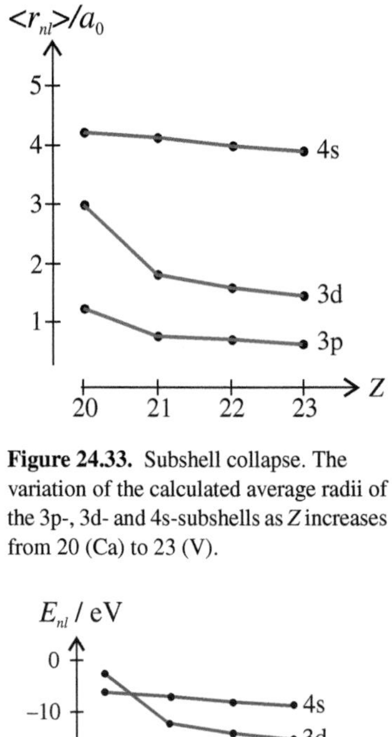

Figure 24.33. Subshell collapse. The variation of the calculated average radii of the 3p-, 3d- and 4s-subshells as *Z* increases from 20 (Ca) to 23 (V).

Atoms with partly filled p-subshells (groups II to VII)

The chemical and physical properties of atoms with partly filled outer p-subshells are acutely dependent on the number of electrons in the outer subshell – known as *valence* electrons. As noted above, the angular solutions of the one-electron atom can be carried directly into multielectron atoms. As shown in Figure 24.16, the probability density distributions of p ($l = 1$) electrons are highly non-spherical, with preferred directions in space. This leads to preferred directions for the bonds which are formed through sharing of such electrons (to be described in Section 25.1). Consequently, molecules formed in this way have characteristic structures in which the orientations of the component atoms follow the orientations of the $l = 1$ probability density functions. The rich structural chemistry of the carbon atom (with ground configuration $1s^2\ 2s^2\ 2p^2$) provides many examples of this effect.

Partly filled subshells have a net angular momentum and hence a net magnetic dipole moment. As noted in Section 19.10, the alignment of such atomic magnetic dipole moments in an applied field is responsible for paramagnetism.

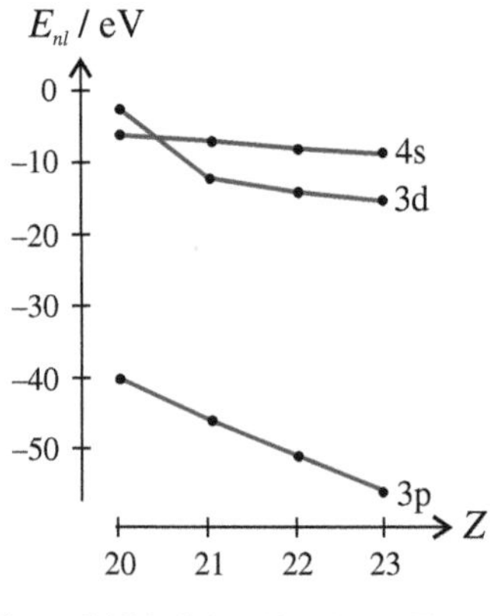

Figure 24.34. Subshell collapse. The variation of the calculated binding energies of the 3p-, 3d- and 4s-subshells as *Z* increases from 20 (Ca) to 23 (V).

Transition row elements (atoms with partly filled d- or f-subshells); ferromagnetism

The transition elements correspond to the filling of the d- and f-subshells in the periodic table (Appendix E), for example the filling of the 3d-subshell between Sc ($Z = 21$) and Ni ($Z = 28$). The binding energies of the high l (high angular momentum) electrons in these subshells are very sensitive to small changes in the potential energy function. Consequently, the binding energy of d- or f-subshell electrons can increase dramatically with a small increase of Z, leading to a sudden reduction of the average radius of such subshells when compared with the average radii of other (low l) subshells. This effect, which is often described as a *collapse* of the subshell into the atomic core, is illustrated in Figures 24.33 and 24.34 where the variation of calculated average radii and binding energies, respectively, of the 3p-, 3d- and 4s-subshells are compared for Ca ($Z = 20$) to V ($Z = 23$).

Unlike the s- or p-subshells, which are filled when they are the outer subshells of the atom, collapsed d- and f-subshells are filled as inner subshells. As such, these electrons are shielded from the environment of the atom by electrons in the outer subshells and have little effect on the chemical properties, and most of the physical properties, of the atom. These properties are determined by the outer electrons in the valence s-subshell and do not vary much along a transition row.

A partly filled d- or f-subshell has a net angular momentum and hence a net magnetic dipole moment. Magnetic dipole moments of inner subshells, primarily (intrinsic) electron spin magnetic dipole moments, are responsible for **ferromagnetism** (Section 19.11). The occurrence of domains, which were introduced in that section to account for ferromagnetic behaviour, may be attributed to a quantum mechanical effect involving pairs of electron, known as the *exchange interaction.*

The **exchange interaction** is a consequence of the Pauli exclusion principle which requires that electrons in nearby atoms in a ferromagnetic crystal which are in the same spin state (identified by m_s values) cannot have the same spatial quantum numbers. This means that electrons with parallel spins tend to have spatial distributions which are further apart – in effect the electrons avoid each other – so that the (positive) energy due to their mutual coulomb repulsion is reduced and energies are lower when the electron spins are parallel. The reduction in energy due to the exchange interaction can be many orders of magnitude greater than the reduction of energy due to the magnetic dipole-dipole interaction (which tends to align spin dipoles so that they are antiparallel).

As a result, *magnetic domains*, in which electron spins are aligned parallel, are formed in ferromagnetic materials. The range of the exchange interaction is shorter, however than that of the dipole-dipole interaction so that, over large distances, the dipole-dipole interaction dominates, limiting the size of the domains.

For problems based on the material presented in this section visit up.ucc.ie/24/ and follow the link to the problems

24.14 The theory of multielectron atoms

In the previous section we have shown how the structure of multielectron atoms can be described qualitatively using results which were originally obtained for the one-electron atom. The implication is that, to a first approximation, each electron in a multielectron atom behaves as though it does not see the other electrons in the atom – it behaves as though it is an *independent particle* in a central field. The independent particle model works because most electrons in multielectron atoms are in closed subshells with spherically symmetric charge distributions. In this section we review the basic principles underlying the theory of multielectron atoms. A more detailed account of how these principles are applied may be found in atomic physics texts such as in Chapters 9 and 10 of Eisberg and Resnick (1985).

The model of the multielectron atom used in theoretical calculations relies heavily on the spherical symmetry of closed subshells. Except in the lightest atoms it is impracticable to treat the Coulomb repulsion between each pair of electrons individually. Each electron interacts with every other electron in the atom and the number of variables grows quickly to unmanageable proportions. Because of the spherical symmetry of closed subshells, however, we can deal with the gross (averaged) effect of Coulomb repulsion simply by modifying the r dependence of the central field. Such a model is known as a *central field approximation.* Interactions which are not included in the averaged field, such as Coulomb and magnetic interactions between electrons which are not in closed subshells, are then introduced as second-order effects. Because of the spherical symmetry of the situation, the radial and angular solutions can be separated, and the angular solutions of the one-electron case are carried directly into the multielectron case.

Multielectron wavefunctions are not calculated analytically. In general, they are obtained through *self-consistent field* techniques. The first step in the self consistent field technique is to guess a plausible radial potential energy function – that is, a one-electron atomic function with screening by the inner subshells – and to use this to calculate the wavefunctions. The calculated wavefunctions can then be used to calculate the charge distribution and hence a potential energy function which can then be compared with the original guess. The potential energy function is changed progressively by small amounts until the potential energy function which is fed in matches the function which is calculated from the wavefunctions – that is until *self-consistency* is achieved. This technique, known as the *Hartree-Fock* technique, involves repeated sequences of calculations requiring numerical solutions of differential equations and integrations and is therefore ideally suited to computers. Accurate theoretical calculations of the structure and properties of multielectron atoms, such as the order of filling of the subshells (Figure 24.29), can be achieved with such methods.

The states of a multielectron atom are labelled in terms of quantum numbers which are associated with its total angular momentum. This is not surprising, given the spherical symmetry of the system. Classically, angular momentum is conserved in a system with no external torques. Correspondingly, angular momentum is quantised in quantum mechanical solutions. We have already encountered the quantisation of the orbital angular momentum of the electron in a one-electron atom (Equation (24.29). Observations of atomic properties tend to produce labels which are associated with conserved quantities, such as n (energy) and l (angular momentum) in the one-electron case, which match the theoretically calculated angular momentum quantum numbers.

24.15 Further uses of the solutions of the one-electron atom

In this chapter, we have devoted considerable space to the analysis of the one-electron atom and to the interpretation of its solutions. We have done so because the one-electron atom is of considerable importance in physics, both as a test of theory and as a basis for the analysis of more complex systems. We conclude this chapter by outlining briefly some further uses of the theory of the one-electron atom.

The quantum mechanical solution of the one-electron atom is an exact analytical solution of a real physical situation. As such, it has a central place in physics as a testing ground of the theory of quantum mechanics. Every detail of the hydrogen spectrum – which can be measured with considerable accuracy using optical methods – should be calculable. We have already noted in Section 24.4 that, strictly, the hydrogen atom should be treated relativistically. The relativistic quantum theory of Dirac, from which electron spin emerges naturally, also accounts for detailed differences (that is, the fine structure splittings) between the observed energy levels and the values obtained from Schrödinger theory. The effects of quantum electrodynamics – the interaction of the electron with its own quantised electromagnetic field (see Section 27.12) – also produces measurable modifications to the hydrogen spectrum so that the hydrogen spectrum has been used, and continues to be used, to test this theory quantitatively.

As analytical solutions, the one-electron wavefunctions are also used as a basis for the analysis of more complex situations. To a first approximation, an electron in a complex system is treated as though it is an independent particle in a one-electron atom, the effects of the more complex situations being added as second order effects. In the previous section we have described how such an approach is used in the theory of multi-electron atoms. The gross effects of interactions between electrons in the closed subshells of atoms are included by modifying the central radial field. Other effects, due to electrostatic and magnetic interactions between electrons which are not in closed subshells, are then introduced as corrections.

Atomic probability density distributions (or *orbitals* as they are often called in molecular physics), as derived from one-electron wavefunctions, may be used as a basis for describing the electronic structure of molecules. Molecular orbitals are derived from combinations of the atomic orbitals of the atoms which make up the molecule. Molecules are not spherically symmetric and the angular momentum quantum numbers used to label the states of atoms cannot be carried directly into molecules. Molecules often exhibit symmetry about an axis or plane, however, so that the components of the angular momenta of atomic orbitals, with reference to such axes or planes, can be used to label molecular states.

As noted in Section 6.9, conservation principles in physics are intimately connected with underlying symmetries. In quantum mechanics the value of a conserved quantity which is associated with a symmetry is determined by a knowledge of the corresponding quantum numbers. It is to be expected, therefore, that the labels which are used to describe and categorise the observed states of systems will correspond to the quantum numbers which emerge from quantum mechanical analyses. Thus, the labelling systems which have been developed by experimenters to identify the states of atoms, molecules etc., tend to reflect underlying symmetries in the systems under study.

In solids, the more tightly bound electrons in the inner subshells of atoms behave, to a first approximation, as though they are in isolated atoms. The energy states of solids, too, can be described using the quantum numbers which originate from the quantum mechanical solution of one-electron atoms.

Further reading

Further material on the topics covered in this chapter may be found in *Quantum Physics* by Eisberg and Resnick and in *Introduction to Quantum Mechanics* by Phillips (See BIBLIOGRAPHY for book details).

WORKED EXAMPLES

Worked Example 24.1: Calculate the isotope shift of the first member of the deuterium Balmer series ($n = 3$ to $n = 2$ transition), given that the wavelength of this line in hydrogen is 656.28 nm.

The wavelengths of the $n = 3$ to $n = 2$ transitions in hydrogen and deuterium are given by the Rydberg formula (Equation (24.14))

$$\frac{1}{\lambda_H} = R_H\left(\frac{1}{2^2} - \frac{1}{3^2}\right) \quad \text{and} \quad \frac{1}{\lambda_D} = R_D\left(\frac{1}{2^2} - \frac{1}{3^2}\right)$$

Thus

$$\frac{\lambda_H}{\lambda_D} = \frac{R_D}{R_H}$$

R_D and R_H are given by Equation (24.12), suitably modified to take account of the reduced masses of hydrogen and deuterium, m_H and m_D, respectively. Thus

$$R_H = \frac{e^4 m_H}{64\pi^3\hbar^3\varepsilon_0^2 c} \quad \text{and} \quad R_D = \frac{e^4 m_D}{64\pi^3\hbar^3\varepsilon_0^2 c} \quad \text{so that} \quad \frac{R_D}{R_H} = \frac{m_D}{m_H}$$

From Equation (24.13),

$$m_H = \frac{Mm_e}{m_e + M} \quad \text{and} \quad m_D = \frac{2Mm_e}{m_e + 2M}$$

where M is the mass of a nucleon (proton or neutron) and m_e is the electron mass.

Thus

$$\frac{m_D}{m_H} = \frac{2(m_e + M)}{m_e + 2M}$$

Using $M = 1837m_e$ we obtain

$$\frac{\lambda_H}{\lambda_D} = \frac{R_D}{R_H} = \frac{m_D}{m_H} = \frac{3676}{3675}$$

$\lambda_H = 656.28$ nm so that $\lambda_D = 656.10$ nm and the isotope shift $(\lambda_H - \lambda_D) = 0.18$ nm

Worked Example 24.2: Calculate the Rydberg constant and the ionisation energy (in eV) of doubly ionised $^6_3$Li.

From Equation (24.12) the Rydberg constant for $^6_3$Li,
$$R_{Li} = \frac{e^4 m_{Li}}{64\pi^3\hbar^3\varepsilon_0^2 c} = \frac{m_{Li}}{m_e}R_\infty$$

where m_{Li} is the reduced mass of the electron in doubly ionised Li, given by Equation (24.13),

$$m_{Li} = \frac{6Mm_e}{m_e + 6M} = \left(\frac{11022}{11023}\right)m_e$$

where M is the mass of the nucleus.

Thus
$$R_{Li} = \frac{m_{Li}}{m_e}R_\infty = \left(\frac{11022}{11023}\right)10973731 = 10972735\ \text{m}^{-1}$$

The wavelength corresponding to the ionisation energy of $^6_3$Li is given by the Rydberg formula (Equation (24.15)) for Li with $n_f = 1$ and $n_i = \infty$. Thus

$$\frac{1}{\lambda} = \frac{f}{c} = Z^2 R_{Li} = 3^2 \times 10972735 \rightarrow f = 2.963 \times 10^{16}\ \text{Hz},$$

and the ionisation energy, $E = hf = 1.964 \times 10^{-17}$ J = 122.76 eV

Worked Example 24.3: Calculate the expectation value of the radial coordinate of an electron in the ground ($n = 1$ to $l = 0$) state of hydrogen.

From Table 24.1 the radial ground state eigenfunction of hydrogen ($Z = 1$), $R_{10}(r) = \frac{1}{\sqrt{\pi a_0^3}} e^{-\frac{r}{a_0}}$

Thus
$$\langle r_{10}\rangle = \int_0^\infty R*_{10}(r) r R_{10}(r) 4\pi r^2 dr = \int_0^\infty \frac{4}{a_0^3} r^3 e^{-\frac{2r}{a_0}} dr$$

Let $x = \frac{2}{a_0}r$ so that $dx = \frac{2}{a_0}dr$ and we can write $\langle r_{10}\rangle = \frac{a_0}{4}\int_0^\infty x^3 e^{-x} dx$

Using $\int_0^\infty x^n e^{-x} dx = n!$ (Appendix A, table A.3), we obtain $\langle r_{10}\rangle = \frac{a_0}{4}3! = \frac{3a_0}{2}$, as indicated by the blue triangle for the 1s state in Figure 24.14.

Worked Example 24.4: Calculate the value of r for which the probability density is at a maximum for hydrogen in its ground state. Compare your answer with the expectation value of the radius as calculated in Worked Example 24.3 and comment on any difference.

The radial probability density $P_{nl}(r) = R^*_{nl}(r)\, R_{nl}(r) 4\pi r^2 dr$

For the hydrogen ground state, from Table 24.1, $P_{10}(r)dr = \frac{1}{\pi a_0{}^3} e^{-2r/a_0} 4\pi r^2 dr$

The maximum occurs when $\frac{dP_{10}(r)}{dr} = \frac{1}{4a_0{}^3}\left(2re^{-2r/a_0} - \frac{2}{a_0}r^2 e^{-2r/a_0}\right) = 0$

that is, when $r = \frac{r^2}{a_0} \rightarrow r = a_0$.

$\langle r\rangle$ is greater than r_{max} because $P_{10}(r)$ is not symmetric about its maximum (Figure 24.14).

Worked Example 24.5: Evaluate the average of the probability density functions for the $n = 2$ states of the one-electron atom.

The average $n = 2$ probability density is given by $\frac{1}{4}(\psi^*_{200}\psi_{200} + \psi*_{21-1}\psi_{21-1} + \psi^*_{210}\psi_{210} + \psi^*_{211}\psi_{211})$

Substituting for ψ_{200} etc., from Table 24.1, this becomes

$$\frac{1}{128\pi}\left(\frac{Z}{a_0}\right)^3 e^{-\frac{Zr}{a_0}}\left[\left(2 - \frac{Zr}{a_0}\right)^2 + \left(\frac{Zr}{a_0}\right)^2\left(\frac{1}{2}\sin^2\theta + \cos^2\theta + \frac{1}{2}\sin^2\theta\right)\right]$$

which simplifies to
$$\frac{1}{128\pi}\left(\frac{Z}{a_0}\right)^3 e^{-\frac{Zr}{a_0}}\left[\left(2 - \frac{Zr}{a_0}\right)^2 + \left(\frac{Zr}{a_0}\right)^2\right]$$

This function is independent of θ or ϕ. Thus the average probability density is spherically symmetric. The 2s (ψ_{200}) function is always spherically symmetric so that it follows that the average of the three 2p probability density functions is also always spherically symmetric.

Worked Example 24.6: Calculate the magnitudes of (a) the magnetic dipole moment and (b) the magnetic flux density at the centre of a hydrogen atom in a 2p ($n = 2$, $l = 1$) state.
[note: For the $n = 2$, $l = 1$ state $\left\langle \frac{1}{r^3} \right\rangle = \frac{1}{24a_0^3}$]

(a) From Equation (24.31) $\boldsymbol{m} = -\frac{e}{2m}\boldsymbol{L}$, where, from Equation (24.29), for $l = 1$, $|\boldsymbol{L}| = \sqrt{l(l+1)}\hbar = \sqrt{2}\hbar$

Thus
$$|\boldsymbol{m}| = -\frac{e}{2m}\sqrt{2}\hbar = 1.31 \times 10^{-23}\ \mathrm{A\ m^2}$$

(b) The magnetic flux density is given by Equation (24.36)
$$\boldsymbol{B} = -\frac{\mu_0 e}{4\pi m_e r^3}\boldsymbol{L}$$

Again $|\boldsymbol{L}| = \sqrt{2}\hbar$ and using $\left\langle \frac{1}{r^3} \right\rangle = \frac{1}{24a_0^3}$, we obtain $B = 0.74$ T

PROBLEMS

For problems based on the material covered this chapter visit up.ucc.ie/24/ and follow the links to the problems.

25

Electrons in solids: quantum statistics

AIMS

- to examine bonding mechanisms in molecules and solids
- to interpret macroscopic properties of solids, such as electrical conductivity and specific heat capacity, in terms of the microscopic behaviour of electrons in the solid, showing that such behaviour is governed by quantum mechanical considerations
- to show how the quantum behaviour of a system of bosons (such as a gas of photons) is very different from that of a system of fermions (such as electrons in a solid) and how the phenomenon of *superconductivity* may be accounted for in terms of the behaviour of bosons

25.1 Bonding in molecules and solids

As the temperature of a system of atoms or molecules decreases, they arrange themselves in a manner which minimises the total energy of the system. We have seen, (Section 5.10) and as illustrated in Figure 25.1, that when two particles which are bound to each other by an attractive force are brought closer together the potential energy of the system $U(r)$ becomes more negative. In order to minimise energy, therefore, the atoms or molecules arrange themselves so as to strengthen the attractive forces (*bonds*) between them. The resulting arrangement, in which each atom or molecule is held rigidly by bonds to its neighbours, is called a *solid*. If the system is cooled quickly, the ordering of the molecules is short range – it extends over only a few atoms to form an *amorphous* solid such as glass. If the system is cooled slowly, the ordering can be long-range, extending over thousands of atoms to form a *crystal*.

We describe below five types of bonding between atoms and molecules, each of which originates in Coulomb interactions between electric charges in the atoms and molecules. In real systems, bonding usually involves a mixture of these mechanisms.

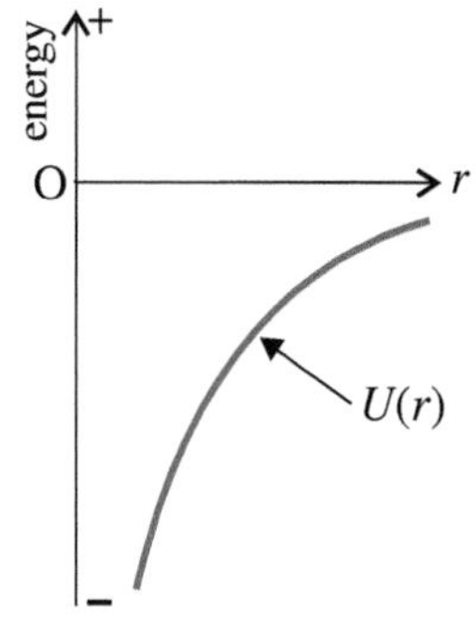

Figure 25.1. An attractive (binding) potential energy function.

Ionic bonding

We noted in Section 24.13 that the outer electron of an alkali (group IA) atom is bound only weakly to the atom. For example, an energy of 5.1 eV is sufficient to remove the 3s electron from a sodium atom. In contrast, group VII (halogen) atoms have a high electron affinities – that is energy is released when an electron is captured so that a halogen atom is more stable when it has acquired an additional electron to fill its outer subshell. Chlorine, for example, which has a $3p^5$ outer subshell configuration, is more stable when it has captured a sixth electron to fill this subshell; an energy of 3.6 eV is released in the process.

For a net expenditure of $5.1 - 3.6 = 1.5$ eV, therefore, an electron can be transferred from a sodium atom to a chlorine atom to form free Na^+ and Cl^- ions. In Figure 25.2, the potential energy of a bound system, comprising a Na^+ ion and a Cl^- ion, is plotted as a function of r, the separation of the two nuclei. As the two ions are brought together from infinity, a potential minimum occurs at $r = 0.25$ nm. This is the equilibrium position of the system; at this value of r the potential energy of the system is 4.9 eV lower than its energy when the Na^+ and Cl^- ions are completely separate. As demonstrated in Worked Example 25.1, this energy can be estimated quite well through a simple Coulomb energy calculation. 4.9 eV is substantially more than the net energy, 1.5 eV, needed to transfer an electron from Na to Cl so that, if the separation of the two nuclei becomes close to 0.25 nm, it is favourable in energy terms for an electron to transfer from the Na atom to the Cl atom. In this situation the ions are bound together – an *ionic bond* is formed. The total energy released when a pair of neutral Na and

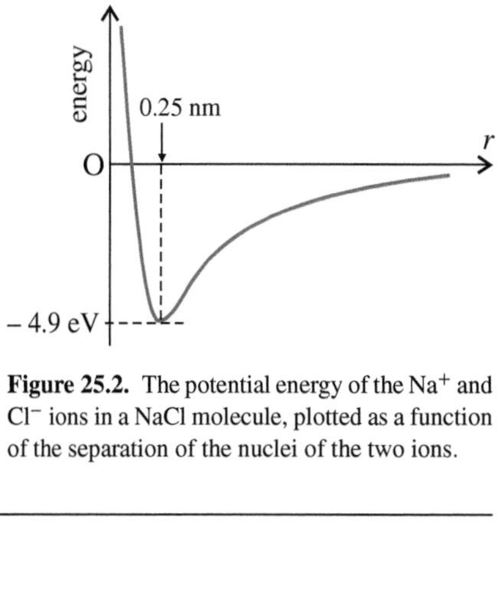

Figure 25.2. The potential energy of the Na^+ and Cl^- ions in a NaCl molecule, plotted as a function of the separation of the nuclei of the two ions.

Understanding Physics, Third Edition. Michael Mansfield and Colm O'Sullivan.
© 2020 John Wiley & Sons Ltd. Published 2020 by John Wiley & Sons Ltd.

Cl atoms exchange an electron and form a bound ionic Na^+Cl^- molecule is therefore 4.9 – 1.5 = 3.4 eV. This energy is known as the *dissociation energy* of NaCl, because it is the energy needed to return a Na^+Cl^- molecule to free Na and Cl atoms.

The occurrence of the potential minimum at $r = 0.25$ nm in Figure 25.2 can be explained as follows. At large values of r the electron distributions of the two ions do not overlap. The two ions can be treated as two point charges, $+e$ and $-e$, so that the potential energy of the system can be determined from Equation (16.25), namely

$$U(r) = Q'V(r) = Q'\frac{Q}{4\pi\varepsilon_0 r} \quad \text{where} \quad Q = +e \quad \text{and} \quad Q' = -e$$

Thus

$$U(r) = \frac{e^2}{4\pi\varepsilon_0 r} \tag{25.1}$$

As r becomes smaller than 0.25 nm, however, the charge distributions of the two ions begin to overlap and two effects occur which increase the energy of the system. First the shielding of the nuclei by their surrounding electron distributions is reduced, leading to increased internuclear repulsion and, secondly, the electrons from the two ions begin to form a single system to which the Pauli exclusion principle must be applied. In the extreme case in which the two nuclei coincide ($r \to 0$), if the exclusion principle did not apply, we would have a single system containing two sets of filled 1s-, 2s-, and 2p-subshells. To satisfy the exclusion principle, half of these electrons would have to move into higher energy states, thereby increasing the energy of the system.

In an ionic molecule, such as NaCl, the electrons in the component ions are all in closed subshells and have spherically symmetric distributions. The ionic bonding mechanism, therefore, can operate for any mutual orientation of the ions – it is not directional.

Ionic crystals

Ionic bonding of a pair of Na^+ and Cl^- ions can be extended to cover a very large number of ions. In this case a cubic ionic crystal is formed, as illustrated in Figure 25.3. In the crystal, the distance between successive Na^+ and Cl^- ions, the *lattice spacing*, R, is 0.28 nm, a little larger than the equilibrium spacing between the ions in a single molecule.

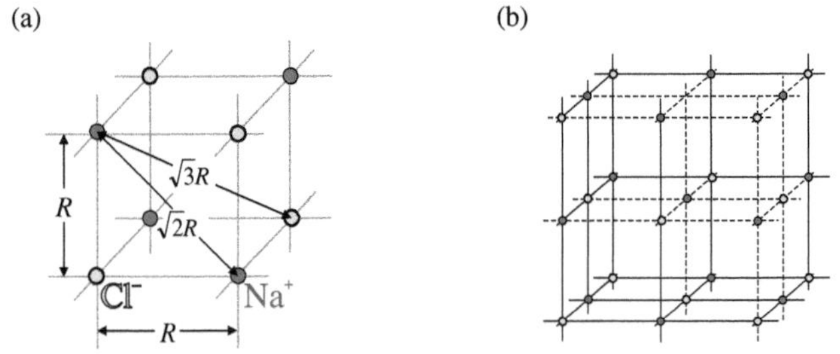

Figure 25.3. The cubic structure of the Na^+Cl^- crystal.

In the ionic crystal, each Na^+ ion experiences a Coulomb attraction from the six nearest Cl^- ions, which are at a distance R from it. It also experiences a Coulomb repulsion from the twelve nearest Na^+ ions, which, as indicated in Figure 25.3, are at a distance $\sqrt{R^2 + R^2} = \sqrt{2}R$ and then an attraction from eight Cl^- ions at a distance $\sqrt{R^2 + R^2 + R^2} = \sqrt{3}R$, as also indicated in Figure 25.3, etc.

The result is that the Na^+ ion experiences a net binding energy which is given by

$$E_{\text{total}} = -6\frac{1}{4\pi\varepsilon_0}\frac{e^2}{R} + 12\frac{1}{4\pi\varepsilon_0}\frac{e^2}{\sqrt{2}R} - 8\frac{1}{4\pi\varepsilon_0}\frac{e^2}{\sqrt{3}R} + \cdots - \cdots$$

This series can be summed to give

$$E_{\text{total}} = -1.7476\frac{1}{4\pi\varepsilon_0}\frac{e^2}{R} \tag{25.2}$$

where the constant, with the value 1.7476 for NaCl, is known as the *Madelung constant* (after Erwin Madelung 1881–1972). Substitution of $R = 0.28$ nm into Equation (25.2) yields a value of E_{total} of about –9 eV. Note, however, that, in deriving Equation (25.2), we have considered the interaction of a Na^+ ion with all other ions and have assigned all the interaction energy to the Na^+ ion. When we calculate the electrostatic energy of two charges $+e$ and $-e$, using the equation $U(r) = \frac{1}{4\pi\varepsilon_0}\frac{e^2}{R}$, we calculate the total energy of the two charges. The electrostatic potential energy associated with one charge is half this amount. The binding energy of a Na^+ ion in a NaCl crystal is therefore $\frac{E_{\text{total}}}{2}$, that is about 4.5 eV.

The bond energy in a NaCl crystal is considerably greater than the typical thermal energy of the ions at room temperature ($kT \approx 0.025$ eV). Thermal vibration at room temperature is therefore very unlikely to disturb the bonds in such a crystal. The bonds are strong and rigid and consequently ionic crystals are very hard and possess high melting points (for example 528 K for NaCl). Because electrons are bound to particular atoms as part of the ionic bonding mechanism, there are no mobile charge carriers to carry electric current in the crystal. Ionic crystals are therefore good electrical insulators, a property which is discussed further in Section 26.2.

Covalent bonding

Covalent bonding occurs when atoms share electrons in order to produce closed subshells. Let us consider a simple case, the H_2 molecule, in which the 1s electrons of the hydrogen atoms in their ground states are shared to produce a closed $1s^2$ subshell.

When the two hydrogen atoms are brought close together, the 1s eigenfunctions overlap. Because the 1s electrons are indistinguishable, we cannot associate a particular electron with a particular atom – they are shared and are members of a single system to which the Pauli exclusion principle applies.

If two hydrogen nuclei are brought together, we find that the total energy of the H_2 molecule depends strongly on whether the electron spins are parallel or antiparallel, as illustrated in Figure 25.4.

The dependence of the total energy on the mutual spin orientation of the electrons can be understood in terms of the Pauli exclusion principle. The requirement that no two electrons in a system can be in the same quantum state can be interpreted as a requirement that two electrons which are in the same spin state cannot occupy the same region of space. If the spins of the two 1s electrons are parallel as the atoms are brought together, the electrons avoid each other and the probability of finding an electron in the region between the nuclei is reduced. There is less negative charge between the two nuclei so that the shielding of the nuclei from each other, by their surrounding electron distributions, is reduced. In this case, therefore, Coulomb repulsion between the nuclei increases rapidly with decreasing r, raising the total energy of the molecule, as shown in Figure 25.4. When the spins are parallel, the energy plot shows no minimum so that bonding cannot occur in this case.

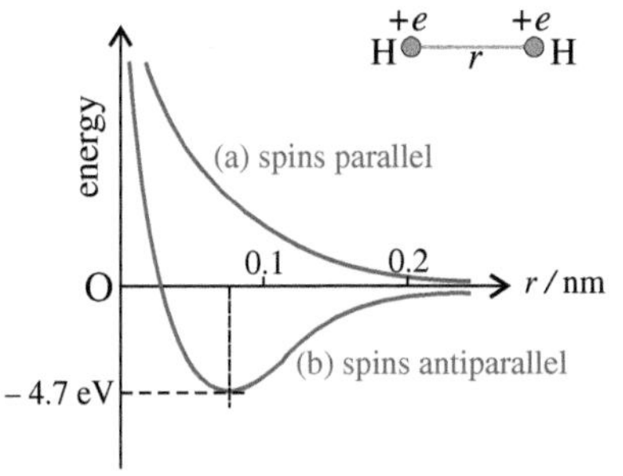

Figure 25.4. The total energy of the H_2 molecule, plotted as a function of the separation of the two nuclei, (a) when the electron spins are parallel and (b) when they are antiparallel.

If, however, the electron spins are opposed as the atoms are brought together, the electrons are in different quantum states and can occupy the same region of space, as they do in the ground state of the helium atom, for example. In this case, both electrons can occupy the region between the nuclei. Their wavefunctions superpose and the probability of finding an electron in this region increases. Consequently, the shielding of the nuclear charges from each other increases and the positive energy due to this repulsive interaction decreases. The total energy of the molecule, therefore, reduces with decreasing r, producing the minimum shown in Figure 25.4. This is the equilibrium separation of the nuclei, at which distance covalent bonding can occur. As r is reduced below 0.07 nm, the potential energy increases due to increasing internuclear repulsion.

Covalent bonding involves *pairing* of electrons with opposing spins from partly filled subshells of atoms, with each atom contributing one electron. Covalent bonding clearly requires that the electron distributions overlap. We saw in Section 24.6 that p-electron distributions show pronounced angular dependence so that covalent bonding through p-electrons is highly directional. The orientation of the atoms is well-defined in a covalent bond which involves p-electrons, unlike the non-directional bonds of the ionic case.

The group IV elements (carbon, silicon, germanium) have four vacancies in their outer p-subshells. As indicated in Figure 24.29, the binding energies of outer s- and p-subshells are similar, so that all four electrons in the s- and p-subshells can take part in the covalent bonding process. The carbon atom, which has a $1s^2 2s^2 2p^2$ ground state configuration (Appendix E), tends to form four covalent bonds which are directed towards the corners of a tetrahedron as illustrated in Figure 25.5. The bonding of carbon atoms with other carbon atoms in a tetrahedral geometry produces the diamond crystal lattice. The covalent bonding energy of group IV elements is very large, 7.4 eV in the case of diamond. As in ionic crystals, electrons are bound to particular atoms as part of the bonding mechanism. There are no mobile charge carriers so that group IV elements are insulators.

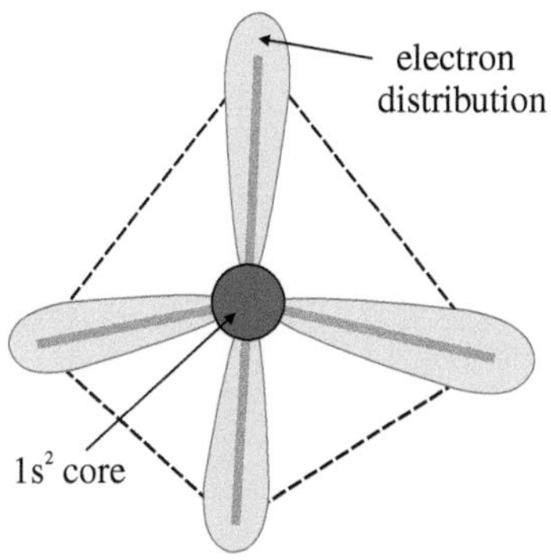

Figure 25.5. The tetrahedral structure of the four covalent bonds of the carbon atom.

The metallic bond

When a crystal is formed from atoms which possess weakly bound valence electrons, such as sodium atoms, these electrons are shared in a form of the covalent bonding mechanism in which the sharing extends from nearest neighbours to all positive ions in the crystal. The metallic bonding mechanism is therefore between a 'sea' of free electrons and an array of fixed positive ions. The attraction between them provides the bonding energy which holds the metal together.

Like the ionic bond, the metallic bond is not directional but, unlike the ionic bond, the metallic bond does not act between specific neighbours. The attractive Coulomb force comes from all directions. Thus, when a metal is deformed, it does not fracture because the metallic bonds – the cohesive forces – continue to act from all directions. Pure metals are therefore ductile and malleable.

Electrons in metals are not bound to particular atoms and are free to move anywhere within the bounds of the material. Metals contain very large numbers of mobile charges and are therefore excellent conductors of both heat and electricity.

Polar (molecular) bonds

In some molecules, known as *polar* molecules (previously encountered in Section 17.7), the sharing of electrons leads to a separation of charge and hence to a permanent electric dipole moment in the molecule. The water molecule, illustrated in Figure 25.6, is an example of such a molecule.

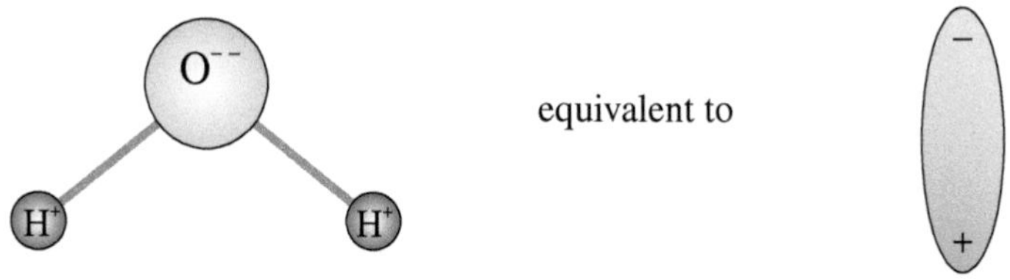

Figure 25.6. The permanent electric dipole moment of the H_2O molecule (a polar molecule).

In solids, polar molecules align in such a way that the positive components of the molecular dipoles are close to the negative components of other dipoles, as illustrated in Figure 25.7. The net result is increased Coulomb binding.

Although, as noted in the discussion of the inert gases in the Inert Gases subsection of Section 24.13, atoms with closed subshells do not have electric or magnetic dipole moments, the electric field produced by a polar molecule can induce charge separation and thus an electric dipole moment in an inert gas molecule. A polar molecule can thus attract and form a bond with an inert gas atom. The process is analogous to the attraction of an unmagnetised piece of iron by a bar magnet.

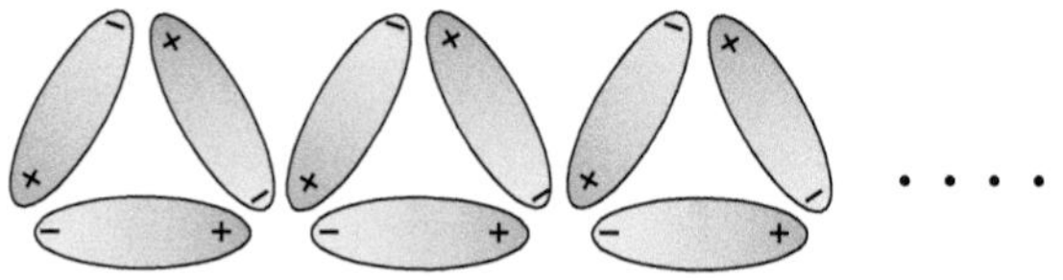

Figure 25.7. Polar bonding between polar molecules.

The Van der Waals bond

Inert gas atoms can also bond with one another. Although the electron distribution in an inert gas atom is spherically symmetric on average, electron distributions can be correlated between neighbouring atoms. The situation may be represented roughly through the Bohr model. Consider two helium atoms, each with two electrons in a circular orbit, as illustrated in Figure 25.8.

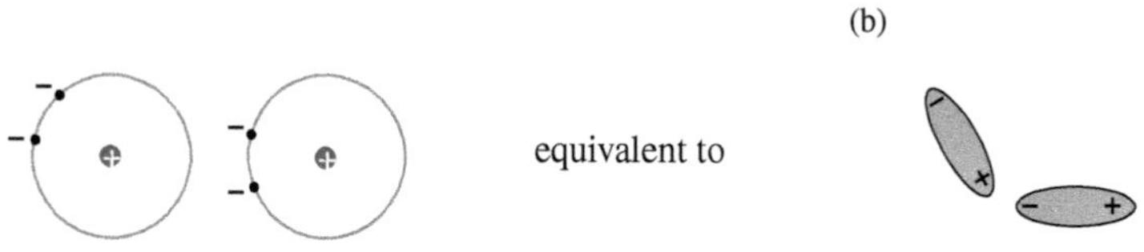

Figure 25.8. The van der Waals bond. The fluctuating electric dipole moments of two helium atoms, as observed at a fixed time.

On average, the electric dipole moment of each of the atoms is zero because each of the two electrons is as likely to be on one side of the atom as it is to be on the other. However, at any instant of time, as illustrated in Figure 25.8, the dipole moment is non-zero. This fluctuating dipole moment produces an electric field which tends to polarise the charge distribution in neighbouring atoms thus inducing correlated, aligned dipole moments as illustrated in Figure 25.8(b). The force between aligned electric dipoles is attractive and hence the atoms bond. This mechanism is known as the **Van der Waals** bond.

In Table 25.1 typical bonding energies are given for the bonding mechanisms which have been described in this section. Note that the Van der Waals bond is very weak and is therefore noticeable only in atoms in which the ionic or covalent mechanisms are absent, such as the inert gases. Inert gases crystallise only at low temperatures where thermal energies are too small to break the weak Van der Waals bonds.

Study Worked Example 25.1

For problems based on the material presented in this section visit up.ucc.ie/25/ *and follow the link to the problems*

Table 25.1 **Summary of bonding mechanisms.**

Name	Mechanism	Typical bond energy (eV) per atom or molecule
1. ionic	transfer of electrons	5
2. covalent	shared electron pairs	10
3. metallic	free-electron gas	3
4. polar (molecular)	attraction of molecular dipoles	0.4
5. van der Waals	correlation of electron distributions	0.1

25.2 The classical free electron model of solids

In this section we seek to account for the macroscopic behaviour of electrons in ohmic solids (to which Ohm's law applies) in terms of a microscopic model called the *free electron model* – also known as the *Drude model* after Paul Drude (1863–1906). The model assumes that the solid conductor comprises an array of fixed positive ions in the space between which weakly bound electrons are free to move anywhere within the confines of a solid. The latter are the 'free electrons', also known as the *conduction* electrons, of the model. At the boundaries of the solid the electrons encounter a large potential barrier which confines them within the conductor. Note that the description of electron behaviour in the Drude model is very similar to that suggested by the metallic bond mechanism in the previous section.

Electrical conductivity

First we attempt to use this model to account for the electrical conductivity of metals σ, defined through Equation (16.30),

$$J = \sigma E, \tag{25.3}$$

a statement of Ohm's Law.

The model assumes that the laws of classical mechanics apply. When an electron collides with a fixed ion it will be scattered randomly and will move in a straight line until it makes a collision with another ion as illustrated in Figure 25.9.

The electrons do not interact with one another and achieve thermal equilibrium only by interaction with the ionic core of the crystal lattice in contrast to the behaviour of molecules in the kinetic theory of gases. Nevertheless, it is assumed that Maxwell–Boltzmann statistics can be applied with the root-mean-square velocity of an electron at temperature T is given by Equation (12.3)

$$v_{\text{rms}} = \sqrt{\frac{3kT}{m_{\text{e}}}} \tag{25.4}$$

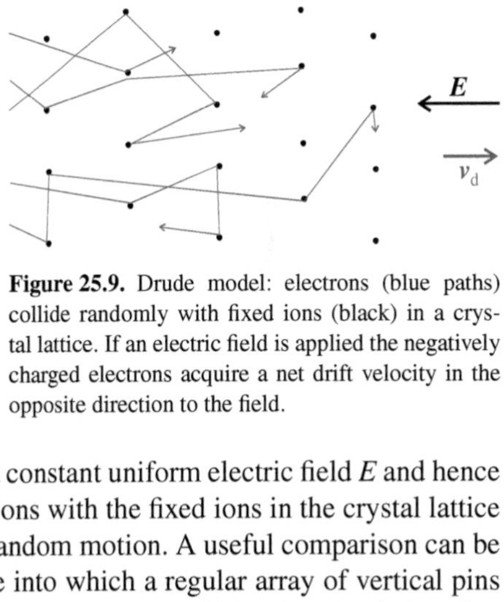

Figure 25.9. Drude model: electrons (blue paths) collide randomly with fixed ions (black) in a crystal lattice. If an electric field is applied the negatively charged electrons acquire a net drift velocity in the opposite direction to the field.

For an electron in a metal at room temperature (300 K), $v_{\text{rms}} = 1.2 \times 10^5$ m s^{-1}.

When a fixed potential difference is applied across a conductor, the electrons experience a constant uniform electric field E and hence a force $F = qE = -eE$ which gives rise to an acceleration. In the Drude model, the interactions with the fixed ions in the crystal lattice are assumed to give rise to a net drift velocity of the electrons which is superposed on their random motion. A useful comparison can be made with an experiment in which a large number of ball-bearings are dropped onto a table into which a regular array of vertical pins has been set (somewhat similar to a pin-ball machine). When the table is horizontal a random motion of the ball-bearings is observed as they collide with the pins. If the table is slightly tilted, the random motion is still observed but there is, additionally, a slow net drift of ball-bearings down the slope.

In Section 15.3 the drift velocity v_{d} for the typical case of a 1 A current which is flowing in a copper wire of cross-sectional area 1 mm^2 was calculated to be about 10^{-4} m s^{-1}. Note that this velocity is about nine orders of magnitude less than the root-mean-square velocity (calculated above) of the electrons as they execute random motion at room temperature. Because $v_{\text{d}} \ll v_{\text{rms}}$, the drift velocity has very little influence on the rate at which electrons collide; at room temperature the collision rate is determined almost entirely by v_{rms}.

The drift motion of the electrons determines the macroscopic current carried by the wire. In Section 15.2 it was shown that current density is related to mobile charge density ρ though the Equation (15.3), that is

$$J = \rho v$$

where v is the speed of the moving charge.

In the present case, in which the charge carriers are electrons of charge $-e$ travelling with a drift velocity v_d, this becomes

$$J = -n_e e v_d \tag{25.5}$$

where n_e is the free electron density — the number of free electrons per unit volume.

The force on an electron is $F = -eE = ma$, so that the acceleration is given by

$$a = -\frac{eE}{m}$$

which is constant if E is constant.

To simplify analysis we assume that the electron is brought to rest by each interaction with a lattice ion. If τ is the average time between collisions (known as the *scattering time*) and assuming that an electron is scattered randomly on collision, the drift velocity is given by Equation (2.18), that is $v = v_0 + At$. Thus

$$v_d = 0 + \frac{-eE}{m}\tau = -\frac{eE}{m}\tau \tag{25.6}$$

Substituting this value of v_d into relationship for J above we obtain

$$J = \frac{n_e e^2 \tau}{m} E \tag{25.7}$$

Comparing this equation with Ohm's law in the form stated in Equation (25.3), it is clear that Equation (25.7) has the form of Ohm's law with the conductivity given by

$$\sigma = \frac{n_e e^2 \tau}{m} \tag{25.8}$$

While classical theory predicts the form of Ohm's law correctly, attempts to account for measured values of σ, using the theory, fail badly, as demonstrated below for the case of copper.

Recalling that the collision rate is determined by the root-mean-square velocity, as given by kinetic theory, we can estimate the scattering time τ from the equation

$$\tau = \frac{l}{v_{rms}} \tag{25.9}$$

where l is the *mean free path* between collisions.

The measured conductivity of copper is approximately $6 \times 10^7\ \Omega^{-1}\text{m}^{-1}$. In Worked Example 25.2, the electron density in copper, n_e, is calculated to be $8.4 \times 10^{28}\ \text{m}^{-3}$. Substituting these values into Equation (25.8) we obtain

$$6 \times 10^7 = \frac{(8.4 \times 10^{28})(1.6 \times 10^{-19})^2\ \tau}{9.1 \times 10^{-31}}$$

Thus $\tau = 2.5 \times 10^{-14}$ s. Using our earlier estimate of v_{rms} at room temperature, namely $1.2 \times 10^5\ \text{m s}^{-1}$, we can substitute for τ and v_{rms} in Equation (25.9) to obtain $l = 3$ nm. This value is more than an order of magnitude greater than the interatomic spacing in copper (0.209 nm) so that the classical model of conductivity, in which the drag force on electrons is provided by collisions with each lattice ion, is not supported by the measured conductivity.

The classical theory also fails to explain the observed temperature dependence of conductivity. From Equation (25.4), $v_{rms} = \sqrt{\frac{3kT}{m}}$, we would expect v_{rms} to be proportional to $T^{\frac{1}{2}}$. From Equations (25.9) and (25.8) we would then expect τ, and thus σ, to be proportional to $T^{-\frac{1}{2}}$. Experimentally, however, σ is found to be proportional to T^{-1}.

Molar specific heat capacity of a conductor

An even more striking difference between the predictions of the classical free electron model and experimental observation is encountered when the theory is used to calculate the molar specific heat capacity of a conductor due to its conduction electrons.

We showed, in Section 12.1, that the molar specific heat capacity at constant volume of an ideal gas which is made up of particles which are moving with three degrees of freedom, is given by

$$c_{Vm} = \frac{3}{2}R \tag{25.10}$$

where R is the molar gas constant. According to the classical free electron model, this equation should also apply to conduction electrons in a solid. The heat capacity of the conduction electrons should be added to the heat capacity of the crystal lattice (Section 12.8). Experiment, however, shows that the heat capacity of metals is not noticeably different from that of other crystals, the implication being that the contribution from conduction electrons is considerably smaller than predicted by classical theory. The measured heat capacity of a metal due to conduction electrons is in fact four orders of magnitude smaller than predicted by Equation (25.10) and, furthermore, shows temperature dependence, again contrary to Equation (25.10).

The lower value of the experimental specific heat capacity indicates that some mechanism, which has not yet been taken into consideration in the free electron model, is preventing most conduction electrons from absorbing energy. We have seen, in Einstein's theory of specific heat capacities (Section 14.4) for example, how quantum effects control the way in which energy is absorbed, so that it is likely that the missing ingredient is quantum mechanical in its origin. In the next section we apply quantum considerations to the behaviour of conduction electrons in a solid and thus are able, in Sections 25.5 and 25.6, to account for the differences between theory and experiment which we have noted above.

Study Worked Example 25.2

25.3 The quantum mechanical free electron model: the Fermi energy

Quantum mechanics treats the conduction electrons in a solid as a single bound system of fermions (Section 24.12), not unlike the electrons in a multielectron atom except that the bounds of the system are now set by the macroscopic dimensions of the solid. The system therefore contains enormous numbers of fermions ($\sim 10^{28}$ m^{-3}). The additional features which are brought to the free electron model by a quantum mechanical treatment are therefore familiar from our study of multielectron atoms (Section 24.13). They are:

(a) Because the electrons are in a bound system their energy levels are quantised.
(b) Electrons must obey the Pauli exclusion principle. No two electrons can have the same set of quantum numbers. Energy levels are filled in order of increasing energy.

As we shall see these new features change the statistical distribution of electron energies in a conductor profoundly. A classical free electron gas obeys Maxwell–Boltzmann statistics (Section 12.7) in which all electrons can have the same energy (zero energy) at 0 K, in contradiction to the exclusion principle. We need therefore a new statistical distribution which conforms to the quantum mechanical rules which govern the behaviour of a system of identical fermions (Section 24.12). This affects the way in which an electron gas can acquire energy (and thus its specific heat). The wave nature of the electrons also affects the manner in which they travel through the crystal lattice – and thus its electrical conductivity.

Otherwise the quantum mechanical free electron model makes the same assumptions as the classical model. At the edges of the solid the electrons encounter a large potential barrier which may be represented approximately by a three-dimensional infinite potential well – a three-dimensional box. We now proceed to analyse this situation quantum mechanically.

In Section 14.15 we solved the one-dimensional infinite potential square well – the one-dimensional box as illustrated in Figure 25.10 – quantum mechanically.

The solution (Equation (14.44)) to the time-independent one-dimensional Schrödinger equation is

$$\psi(x) = B\sin kx \qquad \text{where} \qquad k = \sqrt{\frac{2mE}{\hbar^2}}$$

The boundary conditions impose quantisation (standing wave solutions) such that k can take only the values $k_n = \frac{n\pi}{a}$, where $n = 1, 2, 3 \ldots$

Consequently, energies are quantised according to Equation (14.46)

$$E_n = n^2 E_1 \qquad \text{where} \qquad E_1 = \frac{\hbar^2\pi^2}{2ma^2} \tag{25.11}$$

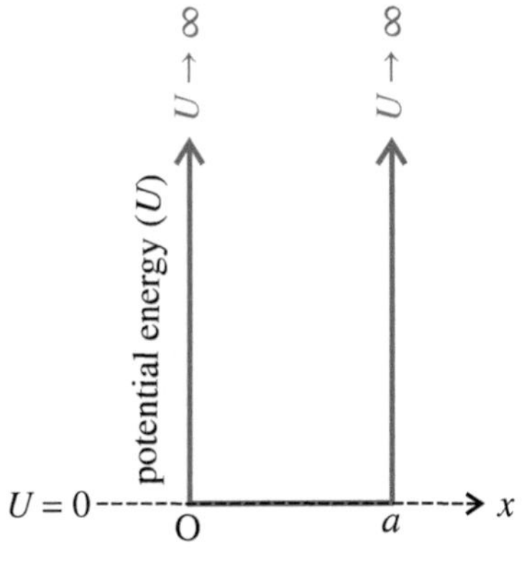

Figure 25.10. The one-dimensional infinite square well potential.

We now generalise this result to three dimensions. The three-dimensional time-independent Schrödinger equation for a particle in a region in which $U = 0$ is

$$-\frac{\hbar^2}{2m}\left(\frac{\partial^2\psi}{\partial x^2} + \frac{\partial^2\psi}{\partial y^2} + \frac{\partial^2\psi}{\partial z^2}\right) = E\psi \tag{25.12}$$

The solution of this equation requires separation of the variables, the technique used to solve the other three-dimensional Schrödinger equation which we have encountered – the one-electron atom (Section 24.5). As shown below, the three variables x, y and z can be separated to produce three ordinary differential equations. The eigenfunction can be written, therefore, as the product of three independent functions of x, y and z.

$$\psi(x,y,z) = \psi_1(x)\psi_2(y)\psi_3(z) \tag{25.13}$$

Substitution of Equation (25.13) into Equation (25.12) yields

$$-\frac{\hbar^2}{2m}\left(\psi_2\psi_3\frac{\partial^2\psi_1}{\partial x^2} + \psi_1\psi_3\frac{\partial^2\psi_2}{\partial y^2} + \psi_1\psi_2\frac{\partial^2\psi_3}{\partial z^2}\right) = E\psi_1\psi_2\psi_3$$

Thus

$$-\frac{\hbar^2}{2m}\left(\frac{1}{\psi_1}\frac{\partial^2\psi_1}{\partial x^2} + \frac{1}{\psi_2}\frac{\partial^2\psi_2}{\partial y^2} + \frac{1}{\psi_3}\frac{\partial^2\psi_3}{\partial z^2}\right) = E$$

and

$$-\frac{\hbar^2}{2m}\left(\frac{1}{\psi_1}\frac{\partial^2\psi_1}{\partial x^2}\right) = E + \frac{\hbar^2}{2m}\left(\frac{1}{\psi_2}\frac{\partial^2\psi_2}{\partial y^2} + \frac{1}{\psi_3}\frac{\partial^2\psi_3}{\partial z^2}\right)$$

The term in x has been separated on the left-hand side of the equation. The equation can be satisfied only if both sides are equal to the same constant, C, the separation constant. The equation in x is therefore

$$-\frac{\hbar^2}{2m}\frac{\partial^2\psi_1}{\partial x^2} = C\psi_1$$

This is the differential equation which we solved in the case of the one-dimensional infinite square potential well (Section 14.15). The solution is $\psi_1(x) = A(\sin k_1 x)$. The terms in y and z can be separated in exactly the same way, as expected from the symmetry of Equation (25.12). The solutions are $\psi_2(y) = B(\sin k_2 y)$ and $\psi_3(z) = D(\sin k_3 z)$

Equation (25.13) can be written, therefore, as

$$\psi(x,y,z) = A'(\sin k_1 x)(\sin k_2 y)(\sin k_3 z), \quad \text{where} \quad A' = ABD$$

If the box is a cube of side length a, the boundary conditions give $k_1 = \dfrac{n_1\pi}{a}, k_2 \dfrac{n_2\pi}{a}$ and $k_3 = \dfrac{n_3\pi}{a}$, where n_1, n_2 and n_3 can each take the values 1, 2, 3, … and the eigenvalues – the quantised energies – are given by

$$E_{n_1n_2n_3} = E_1(n_1^2 + n_2^2 + n_3^2) \quad \text{where} \quad E_1 = \frac{\hbar^2\pi^2}{2ma^2} \tag{25.14}$$

Note that, as before, there is one quantum number for each degree of freedom.

Let us now consider the occupancy of these energy levels at 0 K. From Equation (25.14) it is clear that the lowest energy state occurs when $n_1 = n_2 = n_3 = 1$. We will label states of the three-dimensional well (n_1, n_2, n_3) so that, in this notation, this lowest state is the (111) state. The energy of the (111) state is therefore $3E_1$ (from Equation (25.14)). Taking into consideration the two possible components of the electron spin in this state, $m_s = \frac{1}{2}$ or $-\frac{1}{2}$, the occupancy of this state is two, from the Pauli exclusion principle.

The next state in order of increasing energy, (211), has an energy $E_1(2^2 + 1^2 + 1)^2 = 6E_1$. The (121) and (112) states have the same energy. For each of these three states, two spin options are possible so that in total the occupancy of this energy level is six – there is six-fold degeneracy at this energy.

The next energy state (221) has energy $9E_1$. It has the same energy as the (212) and (122) states so that again six electrons can have this energy.

Let us now consider a specific example – an electron gas comprising 40 electrons. In Table 25.2 we examine the filling of the states of this gas, applying the Pauli exclusion principle in the manner indicated above.

Table 25.2 shows that, for a gas of 40 electrons at 0 K, all energy levels up to and including $17E_1$ are filled. Note that this result is very different from the classical result at $T = 0$. In the classical case the average energy of an electron is $\frac{3}{2}kT = 0$ so that all 40 electrons would have zero energy.

The Fermi Energy

The energy of the highest occupied level, $17E_1$, in the 40 electron gas, is known as the **Fermi energy** (after Enrico Fermi, 1901–1954), E_F, of the gas at 0 K.

Table 25.2 Filling of states in a gas of 40 electrons.

State remaining (n_1, n_2, n_3)	Energy $E_1\,(n_1^2+n_2^2+n_3^2)$	Occupancy	Number of electrons remaining from original 40
(111)	$3E_1$	2	38
(211)	$6E_1$	6	32
(221)	$9E_1$	6	26
(311)	$11E_1$	6	20
(222)	$12E_1$	2	18
(321)	$14E_1$	12*	6
(322)	$17E_1$	6	0

[a] The (321) energy level has an occupancy of 12 because six different combinations, each with two spin options, produce this energy, namely (123), (132), (213), (231), (312) and (321).

In Figure 25.11 the energy distribution function, $F(E)$, the probability that an energy level is filled, is plotted against energy for this case. This shows simply that $F(E) = 1$ when $E < E_F$ and $F(E) = 0$ when $E > E_F$

The function $F(E)$ is known as the **Fermi–Dirac** distribution function at 0 K. The Fermi–Dirac function applies to all systems of fermions and, as we have already noted, differs considerably from the Maxwell–Boltzmann distribution function.

While the 40 electron gas, which we have just considered, provides a useful illustration of the manner in which the exclusion principle governs the energy distribution of a system of fermions in a three-dimensional box, the technique which we have used to deduce the Fermi energy is not practical for a real conductor which contains 10^{28} electrons m^{-3}. In the next section we show how the Fermi energy can be calculated in such a case.

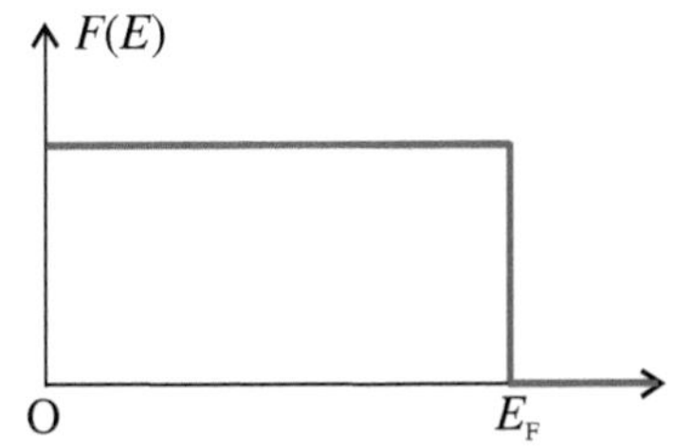

Figure 25.11. The electron energy distribution function $F(E)$ in an electron gas at 0 K. The value of the Fermi energy E_F is indicated.

For problems based on the material presented in this section visit up.ucc.ie/25/ and follow the link to the problems

25.4 The electron energy distribution at 0 K

First we need to devise a means of counting the number of energy states of a system of fermions as a function of energy. It is easier in the first instance to count the states according to their n values. This is achieved using the three dimensional space which is illustrated in Figure 25.12. In this space the three dimensions, plotted along the three coordinate axes, are the values of the quantum numbers n_1, n_2 and n_3. Plots of the values of these numbers are therefore plots in an imaginary space – *n-space*, a concept which we have already used to count standing waves in a cavity (Section 13.8). Each point in n-space corresponds to a unique set of the three quantum numbers. As an example, the point in n-space which corresponds to the (111) state of the three-dimensional box is shown in Figure 25.12.

A plot of the energy states of the three-dimensional box therefore produces a three dimensional lattice with one state per unit volume. The energy of a state is given by Equation (25.14).

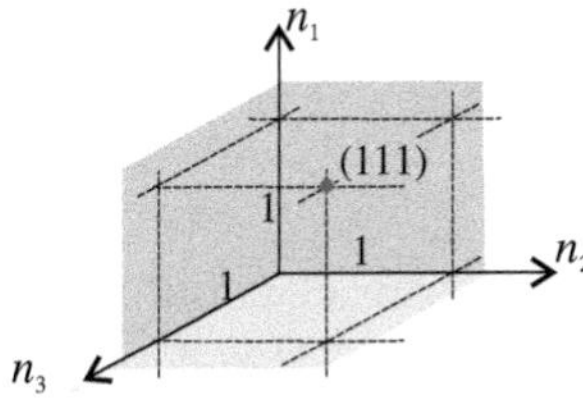

Figure 25.12. The (111) state of the three-dimensional box, plotted in n-space.

$$E_{n_1n_2n_3} = E_1(n_1^2 + n_2^2 + n_3^2)$$

which we can write as

$$E = E_1n^2, \quad \text{where} \quad n^2 = ({n_1}^2 + n_2^2 + n_3^2) \tag{25.15}$$

n is therefore the 'distance' from the origin in n-space. For the (111) state, for example, $n = \sqrt{3}$. The square of the distance from the origin, n^2, may be used to measure of the energy of a state. Degenerate states such as (211), (121) and (112) are therefore equidistant from the origin.

Energy states are filled until the Fermi level $E_F(0)$, the highest occupied level at $T = 0$ K, is reached. In n-space this will correspond to a particular value of n, namely n_F, which is given by the equation

$$E_F(0) = E_1{n_F}^2 \tag{25.16}$$

n_F is therefore the distance from the origin of the furthest occupied points in n-space. These points lie on the surface of a sphere of radius n_F. n_1, n_2 and n_3, and thus n, can take only positive values so that, in calculating the volume of n-space which is occupied, we should consider only the positive octant, the portion of the sphere which is illustrated in Figure 25.13. The volume of the positive octant is $\frac{1}{8}$ of the total volume of the sphere, namely $\frac{1}{8} \times \frac{4}{3}\pi n_F^{\,3} = \frac{1}{6}\pi n_F^{\,3}$

The total number of electrons in the system, N, is given by (the volume of the octant) × (number of states per unit volume of n-space) × (number of spin states for each point).

There are two spin states for each point and, as noted above, there is one state per unit volume of n-space. We can therefore write

$$N = 2 \times \frac{1}{6}\pi n_F^{\,3} = \frac{1}{3}\pi n_F^{\,3}$$

In a three-dimensional box of volume a^3 the total number of electrons

$$N = n_e a^3 \tag{25.17}$$

where n_e is the electron density, the number of electrons per unit volume of real space.

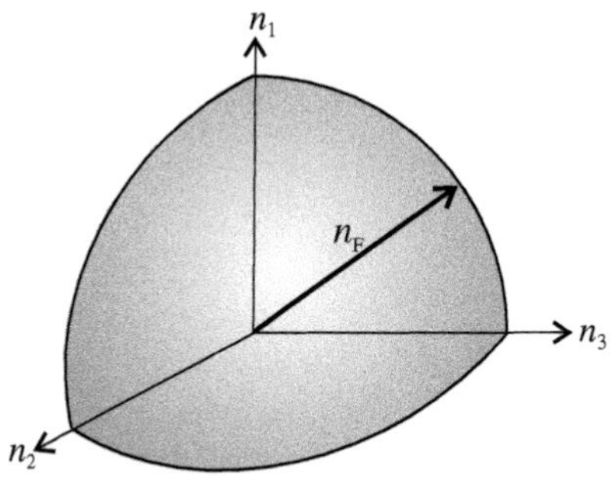

Figure 25.13. Occupied states in n-space. States are occupied within the positive octant of a sphere of radius n_F.

We can therefore write $N = n_e a^3 = \dfrac{1}{3}\pi n_F^{\,3}$ to obtain $n_F = \left(\dfrac{3n_e a^3}{\pi}\right)^{\frac{1}{3}}$

We can now substitute this expression for n_F in Equation (25.16) to obtain an equation which gives the Fermi energy in terms of n_e

$$E_F(0) = E_1 n_F^{\,2} = E_1\left(\frac{3n_e a^3}{\pi}\right)^{\frac{2}{3}}$$

Recalling that $E_1 = \dfrac{\hbar^2\pi^2}{2ma^2}$ (Equation (25.11)), we can write the **Fermi energy at 0 K** as

$$\boxed{E_F(0) = \frac{\hbar^2}{2m}(3n_e\pi^2)^{\frac{2}{3}}} \tag{25.18}$$

Although we have been considering the special case of a system of electrons in a box, the Equation (25.18) which we have derived is very general in its application. It may be applied to any system of fermions to give the Fermi energy in terms of the fermion density. In Worked Example 25.3 the equation is applied to neutrons (which are also fermions) in a neutron star.

As calculated in Worked Example 25.2, $n_e = 8.4 \times 10^{28}\ \text{m}^{-3}$ for solid copper. Substitution into Equation (25.18) yields $E_F(0) \approx 7$ eV. The second column of Table 25.2 indicates that the average separation between adjacent energy levels lies between $2E_1$ and $3E_1$. Taking the typical separation of adjacent energy levels to be approximately $2.5E_1$, we can estimate the separation between successive energy levels in copper,

$$\Delta E \approx 2.5E_1 = 2.5\frac{\hbar^2\pi^2}{2ma^2}$$

For a cube of copper metal of side length 1 cm this equation gives $\Delta E \approx 4 \times 10^{-14}$ eV. The separation between adjacent levels is therefore fourteen orders of magnitude less than the Fermi energy. The picture which is emerging is of electron energy levels which are so close together that they are indistinguishable from a continuum. They are said to form a *quasi-continuum*.

The density of states function

In describing the electron energy levels of a conductor it is therefore impractical to list the energies of individual levels. Instead we define a *density of states function*, $g(E)$, a function which gives the number of states per unit energy range as a function of energy. Thus

$$g(E)\Delta E := \text{the number of states with energy between } E \text{ and } E + \Delta E$$

The form of the function $g(E)$ may be derived as follows – E and $(E + \Delta E)$ can be written in terms of n using Equation (25.15),

$$E = E_1 n^2 \quad \text{and} \quad (E + \Delta E) = E_1(n + \Delta n)^2$$

The number of states between n and $(n + \Delta n)$ can be calculated from our earlier analysis. As indicated in Figure 25.14, the number of states between n and $(n + \Delta n)$ corresponds to the volume between the shells of the octant of radii n and $(n + \Delta n)$. This is

$$\tfrac{1}{8}\,4\pi n^2 \Delta n = \tfrac{1}{2}\,\pi n^2 \Delta n$$

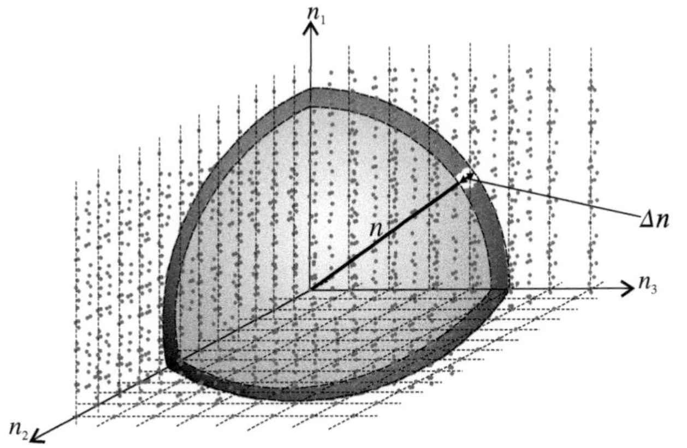

Figure 25.14. The states between shells of radii n and $n + \Delta n$ in n-space.

Thus, allowing for the two spin options,

$$g(E)\Delta E = 2\tfrac{1}{2}\pi n^2 \Delta n = \pi n^2 \Delta n \qquad (25.19)$$

The next step is to express the right-hand side of Equation (25.19) as a function of E. From Equation (25.15)

$$n = \sqrt{\frac{E}{E_1}} \quad \text{and thus, by differentiation,} \quad \Delta n = \frac{1}{2}\sqrt{\frac{1}{E_1}}\frac{\Delta E}{\sqrt{E}} = \frac{1}{2}\frac{\Delta E}{\sqrt{EE_1}}$$

Equation (25.19) can now be written

$$g(E)\Delta E = \pi\frac{E}{E_1}\frac{1}{2}\frac{\Delta E}{\sqrt{EE_1}} = C\sqrt{E}\Delta E \qquad (25.20)$$

where $C = \dfrac{\pi}{2E_1^{3/2}}$ which, using Equation (25.11), can be written as $C = \dfrac{a^2(2m)^{3/2}}{2\hbar^3\pi^2}$. Thus, C is a constant for a given conductor. $g(E)$ is plotted as a function of E in Figure 25.15.

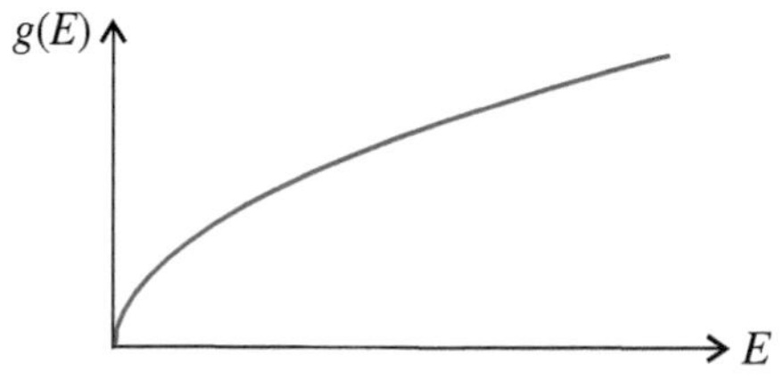

Figure 25.15. The density of states function $g(E)$ plotted as a function of E.

To convert $g(E)$ to a function of the number of electrons per unit energy at each energy, $N(E)$, we need to take into account the probability of a state being occupied. This is given by the Fermi–Dirac distribution function, $F(E)$ at 0 K which we have already plotted in Figure 25.11. In mathematical terms, the number of electrons with energy between E and $E + \Delta E$ can be written as

$$N(E)\Delta E = g(E)F(E)\Delta E \qquad (25.21)$$

The effect of multiplying $g(E)$ by $F(E)$ is simply to introduce a cut-off in $N(E)$ at the Fermi level, as illustrated in Figure 25.16.

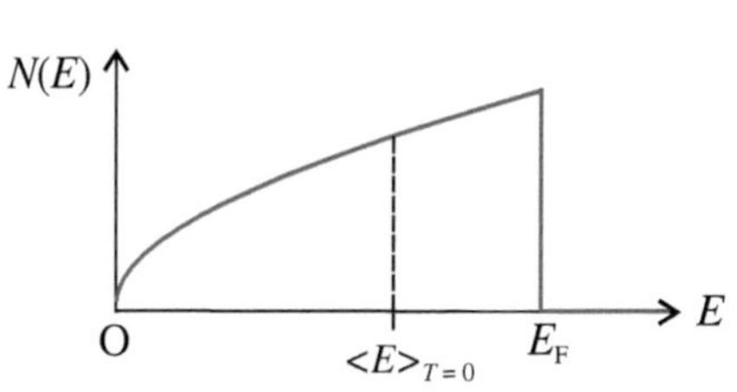

Figure 25.16. $N(E)$, the number of electrons with energy between E and $E + \Delta E$, plotted as a function of E at 0 K. The dashed line marks the average energy at $T = 0$ K.

Knowing $N(E)$ we can calculate the average value of a quantity at $T = 0$ K by integrating the product of the quantity with its distribution function (as explained in Section 12.4). In this case the distribution function, the probability that an electron has an energy between E and $E + \Delta E$, is $\dfrac{N(E)}{N}$, where N is the total number of electrons. This method is used in Worked Example 25.4 to calculate the average energy of an electron at $T = 0$ K, which is found to be $\frac{3}{5}E_F$. The average electron energy is therefore three fifths of the Fermi energy, as indicated on Figure 25.16.

So far, we have restricted our examination of electron energy distributions to the special case in which $T = 0$. We now describe how the analysis can be extended to higher temperatures.

Study Worked Examples 25.3 and 25.4

For problems based on the material presented in this section visit up.ucc.ie/25/ and follow the link to the problems

25.5 Electron energy distributions at $T > 0$ K

When $T > 0$ K, the thermal energy available enables electrons to make transitions to states above those they occupy at 0 K. Note that, as indicated in Figure 25.17, the thermal energy at room temperature, $kT \approx 0.025$ eV, is considerably less than the typical Fermi energy in a metal, which we estimated in the previous section to be a few eV (≈ 7 eV for copper).

In consequence, only those electrons with energies within kT of the Fermi energy E_F have empty states above them into which they can make a transition. Electrons with smaller energies cannot make transitions because all states which are kT above them are already filled. Thus, only a small fraction, approximately $\frac{kT}{E_F}$, of electrons in a conductor can be excited thermally at room temperature.

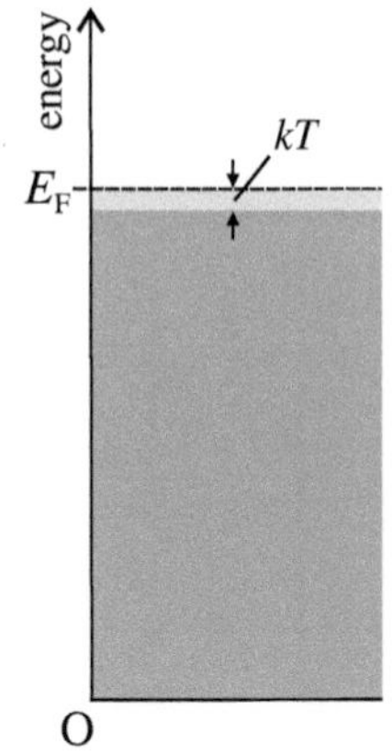

Figure 25.17. An energy level diagram for electrons in a metal at $T > 0$ K, comparing the thermal energy kT with the Fermi energy.

The distribution function of occupied states at $T > 0$ K, the **Fermi–Dirac distribution function at $T > 0$ K**, is not very different from the Fermi–Dirac function at 0 K, which we plotted in Figure 25.11. The probability that an electron will have energy E in a system at a temperature T may be shown (web section up.ucc.ie/25/5/1) to be given by

$$F(E) = \frac{1}{e^{(E-E_F)/kT} + 1} \tag{25.22}$$

Equation (25.22) is plotted in Figure 25.18, where it is compared with the distribution function at $T = 0$, as previously shown in Figure 25.11.

Note that when $T = 0$ K, Equation (25.22) reduces to the Fermi–Dirac function at 0 K, namely,

$$F(E) = 1 \quad \text{for} \quad E < E_F \quad \text{and} \quad F(E) = 0 \quad \text{for} \quad E > E_F$$

If $E \leq E_F - kT$, which is true for most electron energy levels, then $F(E) \approx 1$. The Fermi–Dirac distribution at $T > 0$ K differs significantly from the distribution at 0 K only at energies which are approximately within kT of E_F. This point may be illustrated by substituting $E = E_F - kT$ and $E = E_F + kT$, respectively, into Equation (25.22). We find that $E = E_F - kT$ yields $F(E) = 0.73$ and $E = E_F + kT$ yields $F(E) = 0.27$, as indicated in Figure 25.18. Some further illustrations of the use of the Fermi–Dirac function are given in Worked Example 25.5.

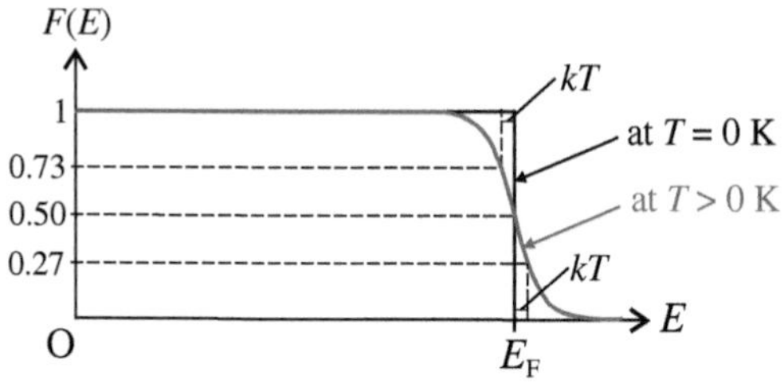

Figure 25.18. A comparison of Fermi–Dirac distribution functions (Equation (25.22)) at $T > 0$ K (blue line) and at $T = 0$ K (black line).

In the previous section we defined the Fermi energy E_F to be the energy of the highest occupied level at 0 K. We must now broaden the definition of E_F to include cases in which $T > 0$ K. E_F is defined, more generally, as the energy at which the probability of a state being occupied is 50 %. Note that substitution of $E = E_F$ into Equation (25.22) yields $F(E) = \frac{1}{2}$, consistent with this definition.

We are now in a position to show how the quantum mechanical free electron model can overcome the failures of the classical model which we noted in Section 25.2, namely the failure to account for the measured values of electron specific heat capacity and conductivity.

Study Worked Example 25.5

For problems based on the material presented in this section visit up.ucc.ie/25/ and follow the link to the problems.

25.6 Specific heat capacity and conductivity in the quantum free electron model

Specific heat capacity

In the previous section we noted that a major consequence of the application of quantum theory to the free electron model is that only a small fraction of electrons, $\approx \frac{kT}{E_F}$, are able to gain thermal energy. This contrasts sharply with the classical model in which all electrons can gain thermal energy. The ability of the electrons to absorb heat energy is therefore greatly reduced in the quantum case. We can estimate the quantum mechanical molar specific heat capacity of the free electrons simply by multiplying the classical value, $\frac{3}{2}R$, by the fraction $\frac{kT}{E_F}$. This gives $c_{Vm} \approx \frac{3}{2}\frac{kTR}{E_F}$.

A more exact treatment changes the constant in this equation from $\frac{3}{2}$ to $\frac{\pi^2}{2}$. Substituting a typical Fermi energy of a few eV into this equation, we obtain $c_{Vm} \approx (10^{-4})RT$. The quantum mechanical value of c_{Vm} is four orders of magnitude less than the classical value given by Equation (25.10) and is also temperature dependent, in agreement with the experimental results noted at the end of Section 25.2.

Conductivity

Although the exclusion principle prevents most free electrons in a conductor from absorbing thermal energy it does not prevent all free electrons from responding to an applied electric field. Let us consider how the electron velocity distribution changes when an electric field is applied. We can write the density of states equation $g(E)\Delta E = C\sqrt{E}\Delta E$ (Equation (25.20)), in terms of velocity using $E = \frac{1}{2}mv^2$. Thus we substitute $\sqrt{E} = \pm\sqrt{\frac{m}{2}}v$ and $dE = mvdv$, to obtain a distribution function $G(v)$ for electron velocities.

$$G(v)\Delta v \propto v^2\Delta v$$

The distribution function for the x-component of v, that is, the number of electrons with x-component of velocity v_x per unit velocity range, is plotted (solid line) in Figure 25.19. Note that velocity states are occupied to $\pm v_F$, the Fermi velocity, which is given by

$$v_F = \pm\sqrt{\frac{2E_F}{m}} \tag{25.23}$$

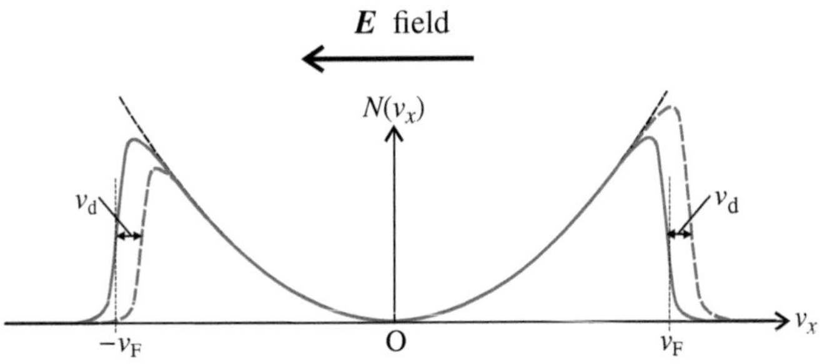

Figure 25.19. The distribution function of the x-component of the electron velocity in a conductor with no applied electric field (solid line) and with an applied electric field E (dashed line).

When an electric field is applied, all electrons acquire a drift velocity, in the $+v_x$ direction, which is given by Equation (25.6), $v_d = \frac{-eE}{m}\tau$. If E is in the $-v_x$ direction, v_d is in the $+v_x$ direction and the entire velocity distribution function is shifted in the $+v_x$ direction, as indicated by the dashed line in Figure 25.19.

In raising electrons from higher energy occupied states near the Fermi velocity to unoccupied states, the electric field creates vacancies at just the rate needed to accommodate electrons which are excited from lower occupied states. There is, therefore, always a vacant state ready to receive an electron which is changing its state in response to the applied field. All free electrons increase their velocities together and hence all are able to gain momentum and energy at the same rate.

The excitation of electrons with energies close to E_F therefore regulates the process for all electrons. Hence the scattering time τ is determined by v_F, the velocity of an electron at the Fermi level, rather than v_{rms}, as assumed in the classical treatment (Equation (25.9)).

We can use Equation (25.23) to estimate v_F in copper, which has a Fermi energy of about 7 eV. We obtain $v_F = 1.56 \times 10^6$ m s^{-1}, which is roughly a factor of 12 greater than v_{rms} at room temperature (estimated to be 1.2×10^5 m s^{-1} in Section 25.2).

The quantum mechanical treatment of conductivity produces the same equations as the classical treatment (Section 25.2), except that v_F replaces v_{rms} in Equation (25.9) for the mean free path. This therefore becomes

$$l = v_F\tau$$

Using the value of τ which we derived from the measured conductivity of copper in Section 25.2, namely 2.5×10^{-14} s, the mean free path becomes $(1.56 \times 10^3) \times (2.5 \times 10^{-14}) = 39$ nm.

The quantum mechanical treatment, therefore, yields a mean free path which is an order of magnitude greater than the classical value (3 nm) and which is, therefore, even further removed from the interatomic spacing in copper (0.209 nm). The clear implication of this result is that electrons are not scattered by every ion in the crystal lattice, a conclusion which we can understand if we calculate the de Broglie wavelength, λ_F, of an electron at the Fermi level. Using Equation (14.19) we obtain

$$\lambda_F = \frac{h}{p_F} = \frac{h}{mv_F} = 0.465 \text{ nm}.$$

The electron wavelength $\lambda_F > 2d$ so the Bragg condition for intensity maxima, $2d \sin\theta = n\lambda$ (Equation (22.16)) cannot be satisfied for any angle of incidence other than $\theta = 0$. Electrons cannot be scattered by a perfect crystal lattice. They are in fact scattered due to lattice defects and vibrations of ions in the lattice. In an ideal lattice in which the lattice ions are visualised as point particles, the vibrating ions offer an effective cross-sectional area for scattering of πA^2, where A is the amplitude of vibration as illustrated in Figure 25.20. The larger the cross-sectional area, the greater the probability of scattering and the shorter the mean free path. Thus,

$$l \propto \frac{1}{\pi A^2}$$

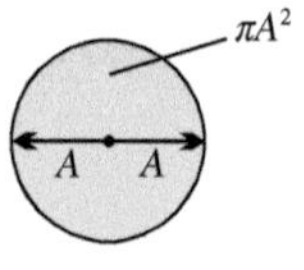

Figure 25.20. The scattering cross-section of a vibrating lattice ion.

In Section 3.10 we showed that the energy of vibration of a classical oscillator is proportional to the square of the amplitude of the vibration, A^2. From the kinetic theory of solids (Section 12.8), the energy of vibration of an atom or molecule in a solid is proportional to T, the temperature, so that $A^2 \propto T$.

We can therefore write,

$$\tau \propto l \propto \frac{1}{\pi A^2} \propto \frac{1}{T}$$

From Equation (25.8) it follows that the electrical conductivity $\sigma \propto \frac{1}{T}$, which explains the observed temperature dependence of σ.

For problems based on the material presented in this section visit up.ucc.ie/25/ and follow the link to the problems

25.7 Quantum statistics: systems of bosons

In Section 12.7 we showed that, for a system of identical particles at temperature T, the fraction of particles per unit energy range at energy E is given by the **Maxwell–Boltzmann distribution function** (Equation (12.10)),

$$F_{\text{M-B}}(E) = Ae^{-\frac{E}{kT}}$$

It is implicit in the (classical) Maxwell–Boltzmann treatment that all particles are distinguishable and behave independently. This requires that particle wave functions do not overlap appreciably, a reasonable assumption for a system of gas molecules in a container but not a reasonable assumption for a system of electrons in an atom or in a metal. In the Maxwell–Boltzmann approach, the probability distribution for each particle is treated independently of all the other particles and consequently the overall shape of the distribution function is the same for one particle as it is for a large number of particles.

When we discussed systems of identical particles in quantum mechanics (Section 24.12), we noted the quantum mechanical requirement of indistinguishability of identical particles leads to a requirement that there should be no observable change in the system when particle labels are exchanged. As a consequence, the behaviour of any particle in a quantum mechanical system is related to that of all other particles. Particles cannot be treated as independent. The system must be treated as a whole, leading to distribution functions which depend on the total number of particles in the system. This point was illustrated in Section 25.3 when we considered how the Pauli exclusion principle regulates the occupation of energy states in an electron gas. An energy state is only available to an electron when it is not occupied by another electron. The behaviour of any electron is clearly influenced by that of the other electrons in the system. The shape of the distribution function for a quantum mechanical system therefore depends on the number of particles in the system.

In Section 24.12 we noted that the requirement that no observable property should change when particle labels are exchanged imposes certain symmetries on the total eigenfunction of a system. Two distinct types of particle were identified, namely *bosons*, for which the total eigenfunction has to be symmetric and which, in consequence, tend to congregate in the same state, and *fermions*, for which the total eigenfunction has to be antisymmetric and which cannot occupy the same state (the Pauli exclusion principle). Different distribution functions are needed therefore for bosons and fermions.

The distribution function for fermions, the **Fermi–Dirac distribution function** is

$$\boxed{F_{\text{F-D}}(E) = \frac{1}{\alpha e^{E/kT} + 1}} \tag{25.24}$$

The value of α depends on the number of particles in the system. We have already stated and used this function as Equation (25.22) when considering the behaviour of electrons in a metal at $T > 0$ K,

$$F(E) = \frac{1}{e^{(E-E_F)/kT} + 1}$$

In this form of the equation α has been set to be equal to $e^{-\frac{E_F}{kT}}$ to ensure that $F(E) = \frac{1}{2}$ when $E = E_F$.

The distribution function for bosons, the **Bose–Einstein distribution function** (after Satyendra Nath Bose, 1894–1974), is

$$F_{\text{B-E}}(E) = \frac{1}{\alpha e^{E/kT} - 1} \tag{25.25}$$

Again, the value of α depends on the number of particles in the system.

Electromagnetic waves in a cavity can be treated as a system of bosons (a photon gas) and hence the Bose–Einstein distribution function can be used to derive the Planck blackbody radiation Equation (Equation (14.4)). Derivations of the Fermi–Dirac and Bose–Einstein distribution functions may be found in the web section up.ucc.ie/25/5/1/), The Maxwell–Boltzmann, Fermi–Dirac, and Bose–Einstein distribution functions for $\alpha = \frac{1}{e}$ are plotted and compared as functions of energy in Figure 25.21.

Note that when $E \gg kT$ the Fermi–Dirac and Bose–Einstein functions reduce to a Maxwell–Boltzmann function, $\frac{1}{\alpha}e^{-\frac{E}{kT}}$. Thus, all three functions coincide at high energies but differ substantially at lower energies. Note that the Fermi–Dirac function never exceeds one (in accordance with the exclusion principle) and is substantially lower than the Maxwell–Boltzmann function. In contrast, the Bose–Einstein function is considerably greater than the Maxwell–Boltzmann function at low energy. Bose–Einstein statistics favour multiple occupancy of the lowest energy states by bosons. Consequently bosons tend to congregate in the lowest energy state of a system. Thus, all bosons in a system tend to have the same properties.

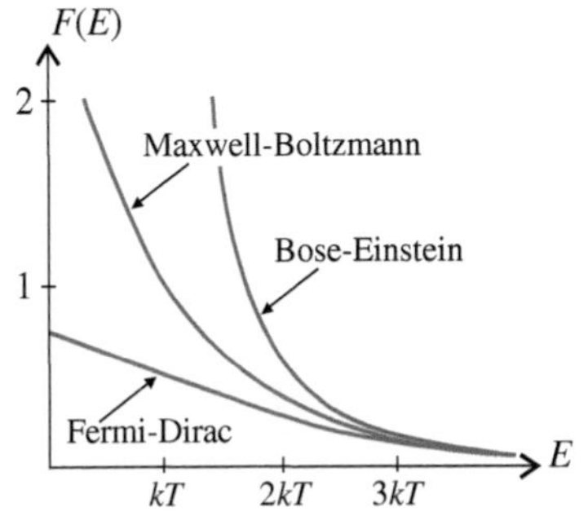

Figure 25.21. A comparison of the Maxwell–Boltzmann, Fermi–Dirac and Bose Einstein distribution functions when $\alpha = \frac{1}{e}$.

We can illustrate the characteristic properties of a system of bosons by considering the properties of the photons which are produced by a laser (Section 22.2). As bosons, photons tend to congregate in the same momentum, energy and phase state.

Because the photons tend to be in the same momentum state, they all travel in the same direction leading to a beam of photons with very little divergence – known as a *collimated* beam. Secondly, because laser photons are all in the same energy state and thus, from $E = hf$, the same frequency state, laser light is highly monochromatic. Thirdly, the photons all have the same phase and are therefore highly coherent. The observed properties of laser photons, which we have noted previously (Section 22.2), are therefore characteristic of a system of bosons.

In the next section we consider a further example of boson behaviour, the phenomenon of superconductivity.

25.8 Superconductivity

Superconductors are materials which show remarkable conducting properties when cooled below a certain temperature known as the *critical temperature* T_c. The value of the critical temperature is a characteristic of the material.

Consider, for example, the resistivity of frozen mercury as a function of temperature, as plotted in Figure 25.22. The resistivity of mercury drops sharply to zero below 4.15 K, the critical temperature for mercury. Mercury becomes a **superconductor** at this temperature and conducts electricity without any heat generation. Many other metals exhibit superconductivity at low temperatures. Because there is no power loss when the resistance is zero, a current, once established, can flow indefinitely in a superconducting wire as long as it is kept below its critical temperature.

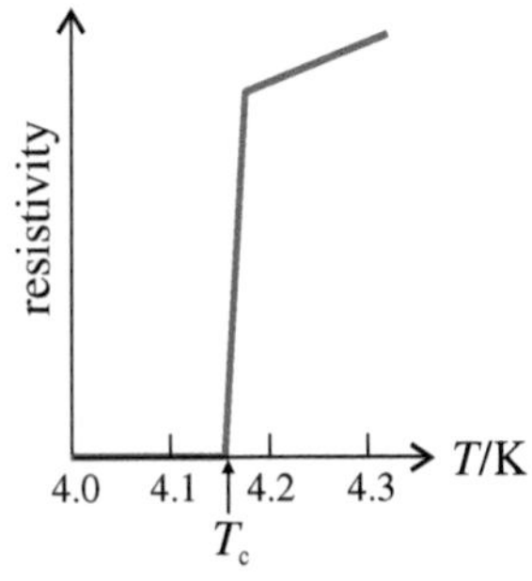

Figure 25.22. Superconductivity. A plot of the resistivity of mercury as a function of temperature, showing the critical temperature below which the resistivity is zero.

As a superconductor is cooled below its critical temperature, magnetic fields are expelled from its interior, an effect known as the *Meissner Effect* (after Walther Meissner 1882–1974). Superconductivity arises from interactions between electrons and vibrating lattice ions which are described in BCS theory (named after John Bardeen 1908–1991, Leon Cooper 1930– and John Schrieffer 1931–2019). An electron which passes close to a lattice ions can, through the Coulomb interaction, impart some momentum to the ions which causes them to move slightly together, producing a region of enhanced positive charge density. Because of the elastic properties of the lattice, this region can propagate as a wave of enhanced charge density which can attract a second electron. The net result is that the two electrons have exchanged momentum. This leads to a weak attraction between electrons which are close to the Fermi energy.

As a consequence of this effect, the interacting electrons tend to group in pairs, known as *Cooper pairs*, with their electron spins opposed so as to produce an entity with zero total spin. Cooper pairs therefore have integral spin and behave as bosons. Cooper pairs formed in this way may be a long way apart in the metal (~100 nm) so that their bonds tend to entangle and they form part of a single system of bosons. The pairs have lower energies than the free electrons because they are in bound states – energy is required to separate a pair. This leads to an energy gap in the conduction band close to the Fermi energy as shown in Figure 25.23. There are no allowed levels in the gap.

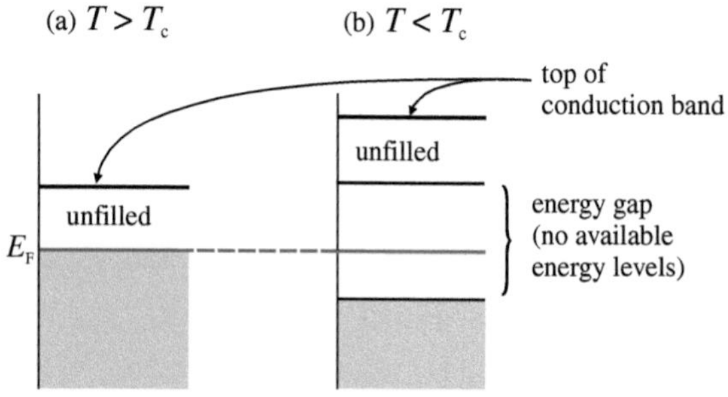

Figure 25.23. The formation of the energy gap as a metal makes a transition from a normal state (a) to a superconducting state (b).

Bardeen, Cooper and Schrieffer pointed out that because the Cooper pairs are bosons, they congregate in the same quantum state. The pairs condense into this state when the temperature of the metal falls below the critical temperature, T_c. This boson state is highly coherent so that the pairs move together in a coordinated way when subjected to an electric field, thus producing the phenomenon of superconductivity. The entire coherent state would have to be broken up for it to lose energy. The usual mechanism of electrical resistance – interaction with vibrating lattice ions, as discussed in Section 25.6 – cannot provide the energy needed to slow down a macroscopic number of electron pairs at the same instant. There is therefore no mechanism available through which the electron pairs can lose energy; they continue to flow indefinitely in a superconducting circuit.

The Meissner effect is a further consequence of the coherence of the superconducting state. Because all the Cooper pairs act together, even a very small change in the magnetic flux can induce the current needed to cancel the inducing field.

Until 1986 the material with the highest known critical temperature was the alloy Nb_3Ge, with a critical temperature of 23.2 K. In 1986 a cuprate compound (an oxide of copper, barium and lanthanum) was found to have a critical temperature of 30 K. Further materials with critical temperatures above 100 K have since been found. The mechanism of superconductivity in such materials is again believed to be the formation of Cooper pairs. However the usual lattice interaction is generally thought to be too weak to bind pairs at such high temperatures and the source of the pairing interaction in such materials has yet to be established clearly. So far there is no widely accepted theory to explain the properties of high temperature superconductors.

WORKED EXAMPLES

Worked Example 25.1: The equilibrium separation between the atomic nuclei in a NaCl molecule is 0.251 nm. Assuming that the Na^+ and Cl^- ions have spherically symmetric charge distributions which do not overlap, estimate the potential energy of the molecule at equilibrium separation, with reference to its value when the two ions are an infinite distance apart.

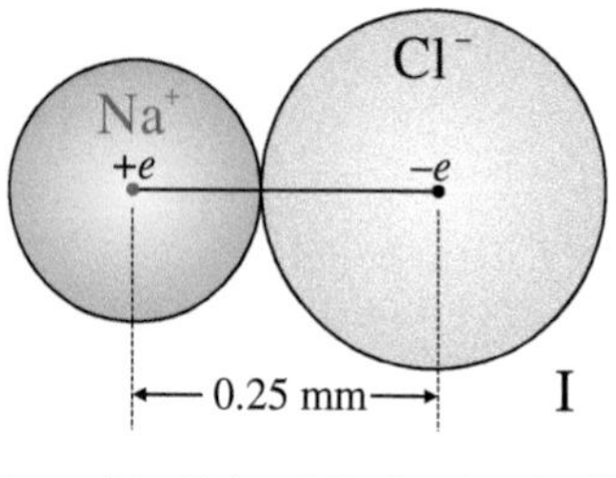

Figure 25.24. The charge distributions of the Na^+ and Cl^- ions in a NaCl molecule (Worked Example 25.1).

Because the Na^+ and Cl^- ions have spherically symmetric charge distributions, they can each be treated as point charges (Section 16.8), as illustrated in Figure 25.24. The potential energy at equilibrium is then the energy needed to increase the separation of two charges, $+e$ and $-e$, from 0.251 nm to infinity. This energy is given by Equation (25.1)

$$\frac{e^2}{4\pi\varepsilon_0 r} = \frac{(1.6\times10^{-19})^2}{4\pi(8.85\times10^{-12})\times(2.51\times10^{-10})} = 9.17\times10^{-19}\ \text{J} = 5.7\ \text{eV}$$

This value compares quite well with the measured value of 4.9 eV. The discrepancy arises because the assumption that the charge distributions do not overlap is only approximately true.

Worked Example 25.2: Given that the density of the monovalent metal copper is 8960 kg m^{-3} and that the molar mass of copper (the mass of one mole of copper atoms) is 0.064 kg mol^{-1}, calculate the density of conduction electrons in copper.

Because copper is monovalent, each atom contributes one free electron to the metal. The number of free electrons per unit volume, n_e, is equal, therefore to the number of atoms per unit volume, N which is given by (the number of moles per m^3) × (number of atoms per mole). The number of atoms per mole is the Avogadro constant, $N_A = 6.02\times10^{23}$ atoms mol^{-1}

Thus

$$n_e = N = \frac{8960}{0.064}\times6.02\times10^{23} = 8.4\times10^{28}\ \text{electrons m}^{-3}.$$

Worked Example 25.3: A neutron star can be considered to be a giant nucleus which is composed only of neutrons. Neutrons, like electrons, are fermions and obey the Pauli exclusion principle. They follow the Fermi–Dirac distribution of energies. If a neutron star has a mass of 4×10^{30} kg and a radius of 10 km, calculate the average energy of neutrons in the star.

In Worked Example 25.4 (below) we show that the average energy of a fermion is $\frac{3}{5}E_F$. E_F may be estimated from Equation (25.18) so that we need to know the neutron density.

The number of neutrons in the star is $\frac{\text{total mass}}{\text{neutron mass}} = \frac{4 \times 10^{30}}{1.67 \times 10^{-27}} = 2.40 \times 10^{57}$

The volume of the star $= \frac{4}{3}\pi r^3 = \frac{4}{3}\pi(10^4)^3 = (4.19 \times 10^{12})\ \text{m}^3$

and therefore the neutron density $n = \frac{2.4 \times 10^{57}}{4.19 \times 10^{12}} = 5.72 \times 10^{44}\ \text{m}^{-3}$

Substituting into Equation (25.18), using the value for the neutron mass given in Appendix B, yields

$$E_F = \left(\frac{6.63 \times 10^{-34}}{2\pi}\right)^2 \frac{1}{2 \times 1.67 \times 10^{-27}}(3 \times 5.72 \times 10^{44}\pi^2)^{\frac{2}{3}} = 2.20 \times 10^{-11}\ \text{J} = 1.37 \times 10^8\ \text{eV}.$$

Thus, the average energy of a neutron $= \frac{3}{5}E_F = 8.22 \times 10^7$ eV.

Worked Example 25.4: Calculate the average energy of an electron in an electron gas at $T = 0$ K in terms of the Fermi energy of the gas.

As explained in Section 12.4, the average value of a quantity may be calculated by integrating the product of the quantity with its distribution function. The distribution function, the probability of an electron having an energy E, is $\frac{N(E)}{N}dE$ where $N(E)dE$ is the number of electrons with energy between E and $E + dE$ and N is the total number of electrons. Thus,

$$\langle E \rangle_{T=0} = \int_0^\infty \frac{N(E)EdE}{N}$$

Substituting for $N(E)$, $g(E)$ and N using equations (25.21), (25.20) and (25.17), this becomes

$$\langle E \rangle_{T=0} = \frac{1}{n_e a^3}\int_0^{E_F} C\sqrt{E}EdE = \frac{2}{5}\frac{C}{n_e a^3}E_F^{\frac{5}{2}} \text{ since } F(E) = 1 \text{ in the range 0 to } E_F$$

Substituting $E_F(0) = \frac{\hbar^2}{2m}(3n_e\pi^2)^{\frac{2}{3}}$ (Equation (25.18)) and $C = \frac{a^3(2m)^{\frac{3}{2}}}{2\hbar^3\pi^2}$ (Section 25.4)

we obtain $$\langle E \rangle_{T=0} = \frac{3}{5}E_F$$

Worked Example 25.5: The Fermi energy of copper is 7.0 eV. If the temperature is 300 K, calculate the probability that a state be occupied at (a) 6.0 eV (b) 6.9 eV (c) 7.1 eV and (d) 8.0 eV.

Using the Fermi–Dirac function (Equation (25.22)) for the case when

$$kT = (1.38 \times 10^{-23}) \times 300 = (4.41 \times 10^{-21})\ \text{J}$$

(a) $E - E_F = -1.0\ \text{eV} = -1.6 \times 10^{-19}$ J. Thus $F(E) = \frac{1}{e^{-\frac{1.6\times10^{-19}}{4.4\times10^{-21}}} + 1} \approx 1$

(b) $E - E_F = -0.1\ \text{eV} = -1.6 \times 10^{-20}$ J. Thus $F(E) = \frac{1}{e^{-\frac{1.6\times10^{-20}}{4.4\times10^{-21}}} + 1} = 0.979$

(c) $E - E_F = +0.1\ \text{eV} = +1.6 \times 10^{-20}$ J. Thus $F(E) = \frac{1}{e^{\frac{1.6\times10^{-20}}{4.4\times10^{-21}}} + 1} = 0.021$

(d) $E - E_F = +1.0\ \text{eV} = +1.6 \times 10^{-19}$ J. Thus $F(E) = \frac{1}{e^{\frac{1.6\times10^{-19}}{4.4\times10^{-21}}} + 1} = 1.6 \times 10^{-17}$

PROBLEMS

For problems based on the material presented in this chapter visit up.ucc.ie/25/ and follow the link to the problems

26

Semiconductors

AIMS

- to show how the *band theory* of solids may be used to extend the quantum treatment of electrons in solids from conductors to insulators and semiconductors
- to describe how the conductivity of a semiconductor can change substantially when energy is absorbed or impurities are added
- to describe how such effects are put to practical use when semiconductors of different types are placed in contact to form *diodes* and *transistors*

26.1 The band theory of solids

When, in Section 25.1, we described the covalent bonding mechanism in hydrogen, we showed (Figure 25.4) that, as two hydrogen atoms are brought together, the curve which represents the total potential energy of the system splits in two, depending on the relative orientation of the electron spins. Thus when two atoms form a single system the ground state energy splits in two. This argument can be extended to many-atom systems as follows.

When N atoms are brought together each energy level splits into N energy levels. This effect is illustrated in Figure 26.1 for a system of five hydrogen atoms arranged in a straight line.

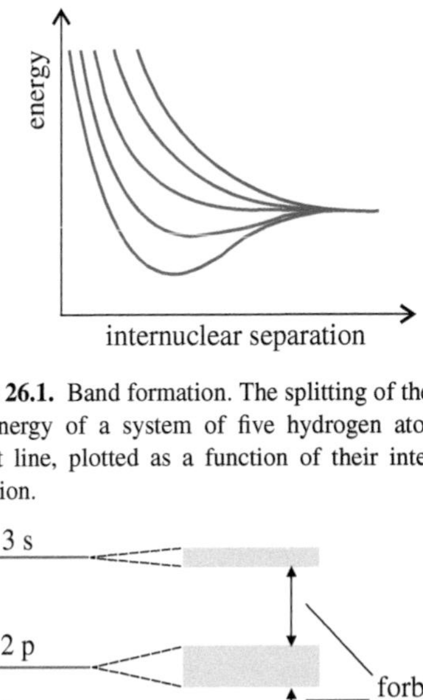

Figure 26.1. Band formation. The splitting of the ground state energy of a system of five hydrogen atoms in a straight line, plotted as a function of their internuclear separation.

One cubic centimetre of a typical solid contains of the order of 10^{22} atoms so that, in the solid, the atomic energy levels split into this number of component energy levels. The net result is an energy **band**, a set of energy levels which are grouped so closely together that they appear to form a continuum of energies. In Figure 26.2 the energy levels of atomic and metallic sodium are compared schematically. The gaps between the bands are known as *forbidden energy bands*. There are no allowed energy levels in the forbidden bands.

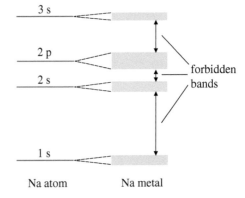

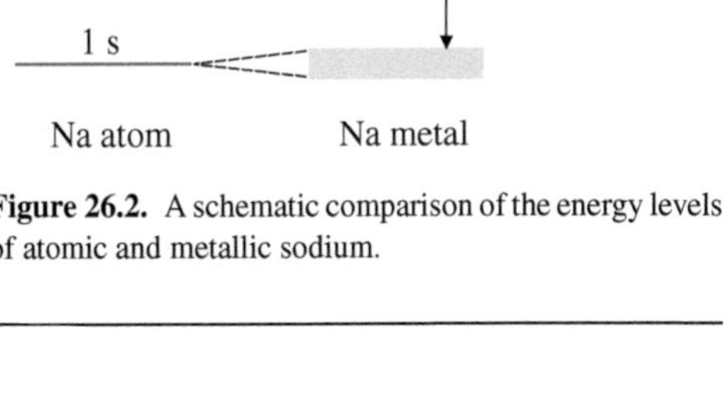

Figure 26.2. A schematic comparison of the energy levels of atomic and metallic sodium.

In a solid at 0 K, the electrons occupy the lowest energy states available. The lower energy bands are therefore completely filled, and the uppermost band is either completely filled or partially filled, depending on the number of electrons in the system and the number of energy states available. The highest energy band which contains electrons is known as the **valence band**. If this band is only partially filled it is also known as the **conduction band** because electrons in this band are responsible for conduction processes. As we shall see in the next section, whether a solid behaves as a conductor, a semiconductor or an insulator depends on whether the uppermost band is completely filled or partially filled and, when the valence band is filled, on the energy gap between the valence and conduction bands since, in general, unoccupied bands can serve as conduction bands if electrons are promoted to them from the valence band.

Although an electron in a metal is not bound to any particular atom, it experiences an attractive force as it passes near each ion. The net effect of all the ions on an electron which is moving through the metal along a straight line is a periodically repeating potential energy function as illustrated in Figure 26.3.

Classically, an electron which encounters a potential such as that illustrated in Figure 26.3 speeds up as it approaches each ion and slows down after passing it.

Understanding Physics, Third Edition. Michael Mansfield and Colm O'Sullivan.
© 2020 John Wiley & Sons Ltd. Published 2020 by John Wiley & Sons Ltd.

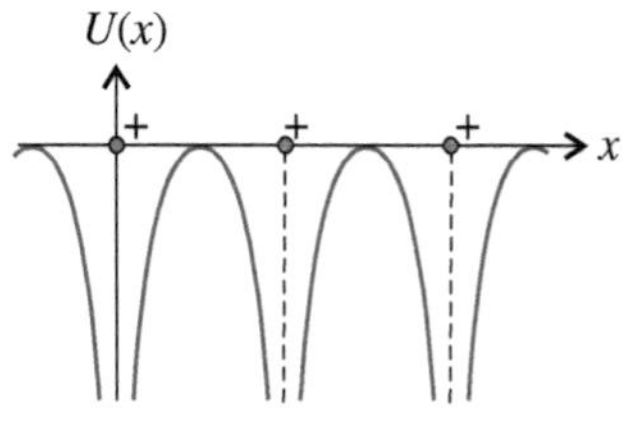

Figure 26.3. The periodic potential energy function experienced by an electron as it moves in a straight line through a metal.

Considering the electron as a wave, however, we find that there are certain ranges of de Broglie wavelengths for which this motion is not possible. These correspond to the forbidden energy bands of Figure 26.2. However, for an electron of low momentum, and therefore long wavelength, the electron motion is that of a free particle. This is an example of a very general wave property, noted in Section 13.7, that waves are not disturbed when they encounter obstacles of dimensions which are small compared with their wavelength. For example, the propagation of a wave motion across a water surface changes very little when the wavefronts encounter an obstacle whose size is much smaller than the wavelength of the waves.

26.2 Conductors, insulators and semiconductors

Conductors

In a **conductor**, the uppermost band is only partly filled, as illustrated in Figure 26.4. In sodium metal, for example, this band is half filled. The Fermi energy is therefore in the middle of this band. When the conductor is subjected to an electric field, the electrons are able to respond – that is acquire kinetic energy – by transferring to higher energy unoccupied states in the same band in the manner described in Section 25.6. Note that, although in the band model the energy levels are not continuously distributed from zero to E_F, as assumed in the free electron model (Figure 25.17), the exclusion principle ensures that only those electrons with energies which are close to E_F are important in determining the conductivity, as explained in the Section 25.6. As far as the conduction electrons are concerned, the energy bands shown in Figure 26.4 are equivalent to the single quasi-continuum of the free electron model.

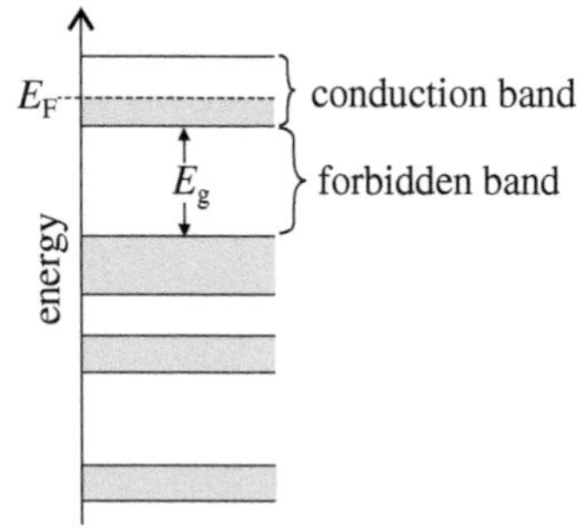

Figure 26.4. The filling of the energy bands of a conductor.

Insulators

In an **insulator** the valence band is completely filled, as illustrated in Figure 26.5. The gap between the valence band and the next unoccupied band is very large, for example $E_g \approx 6$ eV in the case of diamond. This energy is considerably larger than the additional kinetic energy which may be given to an electron through the application of a typical macroscopic electric field. Electrons, therefore, are locked into the filled energy bands and cannot respond to an applied electric field. This is why electrons in an insulator cannot carry an electric current.

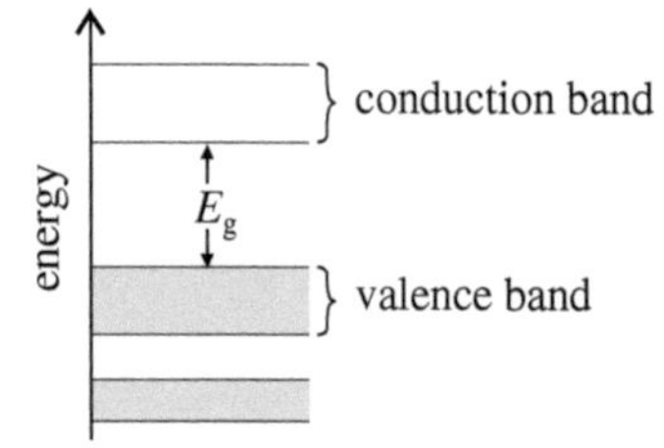

Figure 26.5. The filling of the energy bands of a insulator, indicating the energy gap E_g between the valence and conduction bands.

Semiconductors

In a pure **semiconductor** such as silicon, the valence band is completely filled, as in an insulator. However, as illustrated in Figure 26.6, the gap between the valence band and the next higher band, the conduction band, is typically about 1 eV. When $T > 0$ K some valence band electrons can acquire enough thermal energy to make the transition to the conduction band fairly easily and can then carry an electric current as in a conductor. These electrons are known as **n-carriers** (negatively charged carriers). Note that a semiconductor can also carry a current in its valence band as well as its conduction band because, as illustrated in Figure 26.6, electrons which transfer to the conduction band leave behind unoccupied states, known as **holes**, in the valence band.

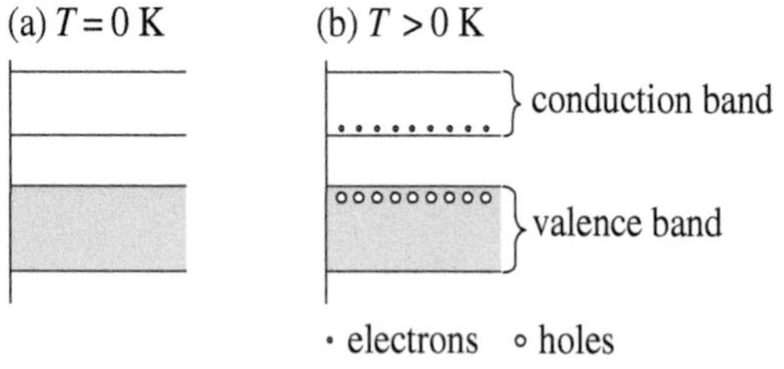

Figure 26.6. The filling of the energy bands of a semiconductor (a) at $T = 0$ K and (b) at $T > 0$ K. When $T > 0$ K some electrons can make a transition to the conduction band, leaving holes in the valence band.

A hole represents a missing electron, a missing negative charge, which can be regarded as equivalent to an additional positive charge. A hole may move through the conductor in the following manner, illustrated in Figure 26.7 which shows successive positions of the hole as a function of time.

When a hole is filled by a nearby electron, this electron leaves a hole in its former position which, in turn, is filled by another electron, etc. etc. As valence band electrons move by jumping into available holes (moving to the left in Figure 26.7) the hole moves in the opposite direction, equivalent to a positive current in that direction (to the right in Figure 26.7). The holes are therefore known as **p-carriers** (positively charged carriers). The process may be compared to the upward motion of air bubbles in a liquid. An air bubble may be regarded as a hole in the liquid so that as portions of the liquid move downwards (under the influence of the Earth's uniform gravitational field) to fill holes, the holes move in the opposite direction – upwards.

The conductivity of a semiconductor increases with temperature because, as illustrated in Figure 26.6(b) ($T > 0$ K), thermal disturbances excite some electrons from the valence to the conduction band, leaving corresponding holes in the valence band, and thus enabling the semiconductor to carry more electric current in both the conduction and valence bands as the temperature increases. The conductivity of a semiconductor therefore increases with increasing temperature, in contrast to the behaviour of a conductor whose conductivity decreases with increasing temperature due to the increased vibration of lattice ions, as described in Section 25.6. In a semiconductor, the increase of conductivity with temperature due to excitation of electrons from the valence band overwhelms any tendency towards reduced conductivity due to lattice ion vibrations.

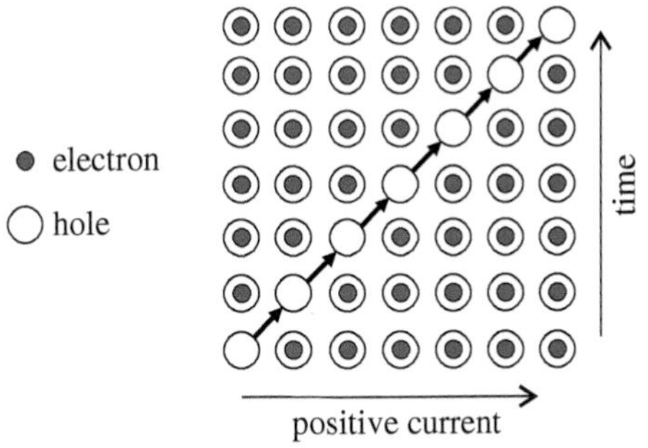

Figure 26.7. The motion of a hole through a (one dimensional) semiconductor. Successive positions of the hole are shown as a function of time.

The conductivity of a semiconductor can change greatly when additional energy is introduced, for example through heating or by subjecting the semiconductor to a beam of photons or energetic particles. Worked Example 26.1 illustrates the sensitivity of the probability of occupation of a state at the bottom of the conduction band to a small increase in temperature. The behaviour of a semiconductor can also be modified substantially by the introduction of impurities. Because the conductivity of a semiconductor can be manipulated in such a sensitive manner, semiconductor devices have a wide range of applications in electronics. In the remainder of this chapter we discuss the techniques used to manipulate the properties of semiconductors and indicate some of their applications.

Study Worked Example 26.1

For problems based on the material presented in this section visit up.ucc.ie/26/ *and follow the link to the problems*

26.3 Intrinsic and extrinsic (doped) semiconductors

If, as illustrated in Figure 26.8, E_c is the lowest energy of the conduction band and E_v is the highest energy in the valence band, the **gap energy** in a semiconductor is given by

$$E_g = E_c - E_v$$

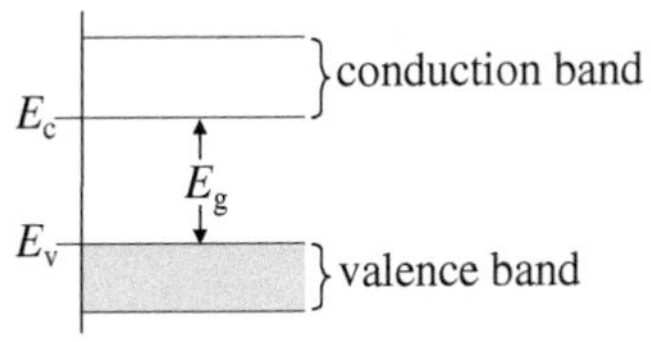

Figure 26.8. The gap energy of a semiconductor.

At $T = 0$, the highest occupied energy level is at the top of the valence band so that the Fermi energy $E_F = E_v$. However, when $T > 0$ the exact definition of E_F is not obvious. The energy of the highest occupied energy level would seem to lie somewhere between E_c and E_v, but there are no allowed energy levels in the gap. Let us consider how the definition of E_F may be extended to cover this situation.

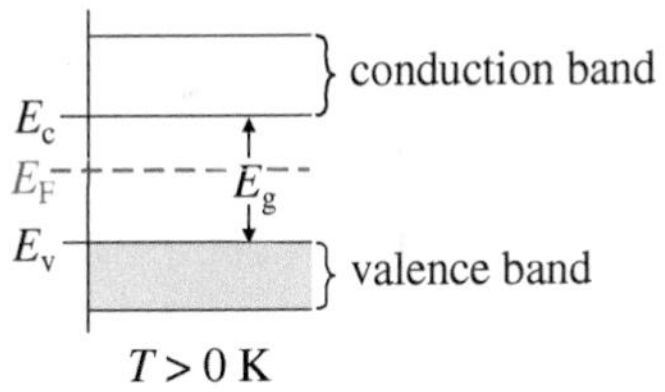

Figure 26.9. The position of the Fermi energy in an intrinsic semiconductor when $T > 0$ K.

Intrinsic semiconductors

In an **intrinsic semiconductor**, a solid which is a semiconductor in its pure state, each valence band electron which transfers to the conduction band leaves a hole in the valence band. The number of electrons in the conduction band, therefore equals the number of holes in the valence band, that is the number of electrons with energies near E_c equals the number of holes with energies near E_v. We can express this in terms of the Fermi–Dirac function (Equation (25.22)) by equating $F(E_c)$, the probability of finding an electron in the conduction band with energy near E_c, to $[1 - F(E_v)]$, the probability of *not* finding an electron in the valence band with energy near E_v, Thus

$$\frac{1}{e^{(E_c - E_F)/kT} + 1} = 1 - \frac{1}{e^{(E_V - E_F)/kT} + 1} \qquad (26.1)$$

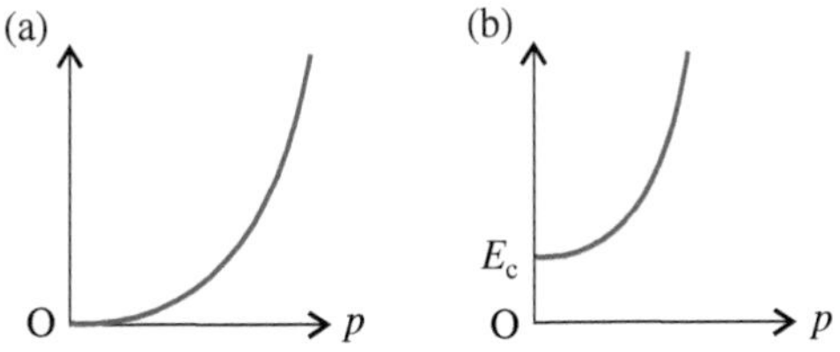

Figure 26.10. The energy of particle, plotted as a function of its momentum (a) for a free particle (b) for an electron in the conduction band of a semiconductor.

The solution to this equation is $E_F = \dfrac{E_c + E_v}{2}$, as may be verified by substitution. Thus, in an intrinsic semiconductor the Fermi energy, as used in the Fermi–Dirac distribution function, lies in the middle of the energy gap as illustrated in Figure 26.9.

As for any non-relativistic free particle, the relationship between the energy E and the momentum p of a free electron is given by

$$E = \frac{p^2}{2m}$$

A plot of energy against momentum therefore has the shape of a parabola, as illustrated in Figure 26.10(a).

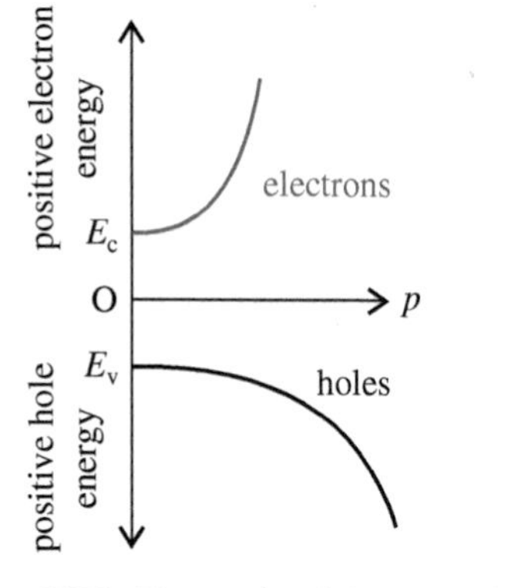

Figure 26.11. The energies of electrons and holes in a semiconductor, plotted as functions of momentum. Note that, for holes, increasing positive hole energy is plotted downwards on the energy axis.

In a semiconductor, because of interactions between the electrons and the lattice ions, the relationship between the energy and the momentum of an electron differs from that for the free electron in that both the minimum energy and the steepness of the parabola are different, as illustrated in Figure 26.10(b). In this case the new relationship between E and p can be represented by the equation

$$E = E_0 + \frac{p^2}{2m^*}$$

where m^* is called the *effective mass* of the electron, and E_0 is its minimum energy, namely the minimum energy of an electron in the conduction band, E_c, as indicated in Figure 26.10(b).

An electron moves under the influence of the external applied field and the internal field due to the ions. The introduction of m^* takes care of the internal field and the problem reduces to the motion of a particle m^* in the external field. The value of m^* is determined by the lattice and for small values of p, $m^* = m$. As m^* decreases, p increases.

Usually $m^* < m$, e.g. for the semiconductor gallium arsenide (GaAs) $\frac{m^*}{m} \sim 0.067$. The interaction between holes in the valence band and the lattice ions is different from that between electrons in the conduction band and lattice ions so that m^* is different for holes, as indicated in Figure 26.11 in which energy – momentum parabolae are compared for electrons and holes. Note that, for holes, E_0 becomes E_v, the minimum energy of a hole, as indicated in Figure 26.11.

Extrinsic semiconductors: doping

Additional n- or p-carriers may be introduced into a semiconductor through the controlled addition of a small number of impurity ions, a process known as **doping**. Such doped semi-conductors are known as **extrinsic** or impurity semiconductors.

Consider, for example, the addition of arsenic atoms to a germanium crystal. As indicated in Appendix E, the germanium atom has four valence electrons ($4s^24p^2$), whereas the arsenic atom has five ($4s^24p^3$). Each arsenic atom, therefore, brings an additional electron to the germanium crystal – it is known as a *donor* impurity. This kind of doped material is described as an **n-type semiconductor**.

The additional electron is only weakly bound to the arsenic ion. We can understand why this is so by comparing its binding energy to a typical atomic binding energy, for example as given by the Bohr model in Equations (24.8) and (24.9),

$$E = -\frac{me^4}{32\pi^2\hbar^2\varepsilon_0^2 n^2} \tag{26.2}$$

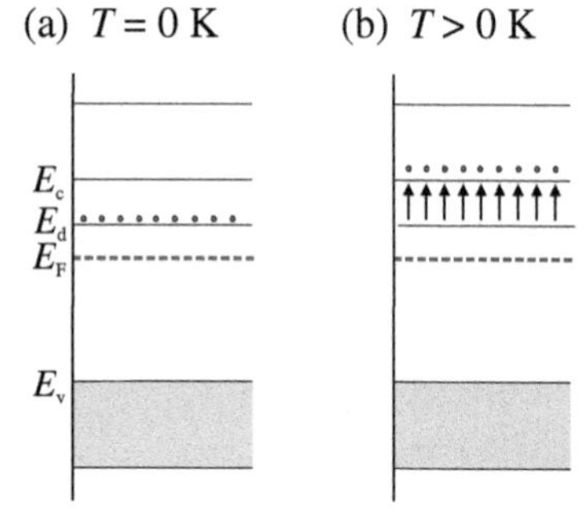

Figure 26.12. The energy bands and donor energy level of an n-type semiconductor (a) at $T = 0$ K and (b) at $T > 0$ K.

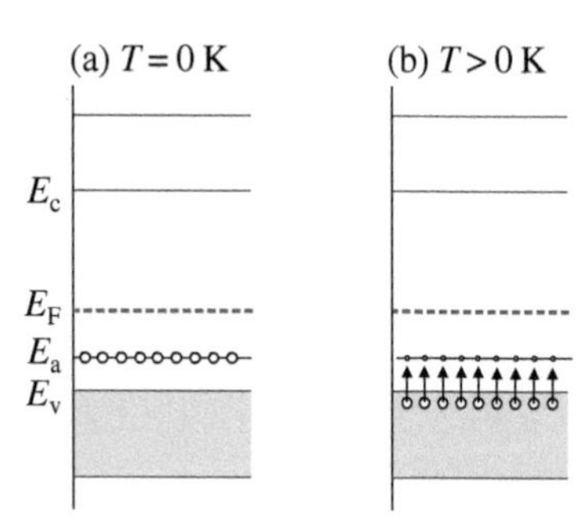

Figure 26.13. The energy bands and acceptor energy level of a p-type semiconductor (a) at $T = 0$ K and (b) at $T > 0$ K.

In the case of the additional electron in the germanium lattice, the energy is reduced because m is reduced to m^*. Moreover, because the radius of the electron orbit, $r = \frac{4\pi\hbar^2\varepsilon_0}{e^2m}n^2$ (Equations (24.6) and (24.7)) increases when $m^* < m$, the electron passes through a significant amount of the surrounding germanium material so that ε, the electric permittivity of germanium replaces ε_0 in the equation for the binding energy E. Typically ε is an order of magnitude larger than ε_0, reducing the binding energy E (Equation (26.2)) further. The additional weakly bound electrons therefore appear to occupy additional energy levels, donor levels with energy E_d, which, for $T = 0$ K, are just below the bottom of the conduction band E_c, as illustrated in Figure 26.12(a). The additional electrons are within kT of the conduction band so that they are very easily excited to that band by thermal agitation at room temperature, as indicated in Figure 26.12(b) ($T > 0$ K).

It is also possible to produce holes in the valence band through the addition of impurities with less electrons in their outer shell than the lattice atoms. These are known as *acceptor* impurities. Boron, for example, has three valence electrons $2s^22p$ (Appendix E) and brings additional holes to a germanium crystal, producing what is known as a **p-type semiconductor**. The net result is additional unoccupied energy levels, acceptor levels of energy E_a, which are just above E_v, as illustrated in Figure 26.13. Electrons from the valence band are easily excited to these levels, known as acceptor levels, at room temperature, leaving holes in the valence band which can act as p-carriers of electric current.

The position of the Fermi energy in an extrinsic semiconductor depends on the relative densities of the n- and p-type carriers. The densities are no longer equal as they are in an intrinsic semiconductor. As indicated in Figure 26.12, E_F is higher in a n-type semiconductor, than in a pure semiconductor, lying just below E_c. In a p-type semiconductor (Figure 26.13) however E_F is lower, lying just above E_v.

A semiconductor can absorb a photon only if the photon has enough energy to excite an electron from the valence band (or from an acceptor or donor level in the case of an extrinsic semiconductor) to an unoccupied level of the conduction band. Thus the photon frequency, f, must satisfy the condition

$$hf > E_g$$

where E_g is the gap energy from the valence band (or from an acceptor or donor level in the case of an extrinsic semiconductor) to the conduction band.

If $hf < E_g$, the photon cannot be absorbed by this mechanism and passes through the semiconductor. The semiconductor is transparent to such frequencies.

Photodetectors

When a semiconductor is exposed to photons with energy $hf > E_g$, increased numbers of n- and p-carriers are produced leading to increased conductivity. Semiconductors can therefore be used as *photodetectors*, devices which, for example, can measure light intensities in exposure meters in cameras or may be used to switch lamps on at dusk and off at dawn.

For problems based on the material presented in this section visit up.ucc.ie/26/ and follow the link to the problems

26.4 Junctions in conductors

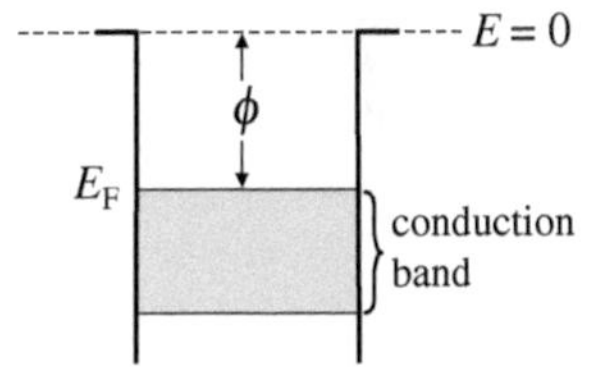

Figure 26.14. An energy diagram for electrons in the conduction band of a metal showing the work function ϕ and the Fermi energy E_F.

When we discussed the photoelectric effect in Section 14.5, we defined the *work function* of a metal ϕ as the binding energy of an electron in the metal – the minimum energy needed to remove an electron. In the context of the band theory of conduction electrons we can now interpret the work function as the energy difference between the Fermi energy and the energy of the electron when it is just able to escape from the metal. The situation is illustrated in Figure 26.14 where the potential energy outside the metal – outside the box – is taken as $E = 0$.

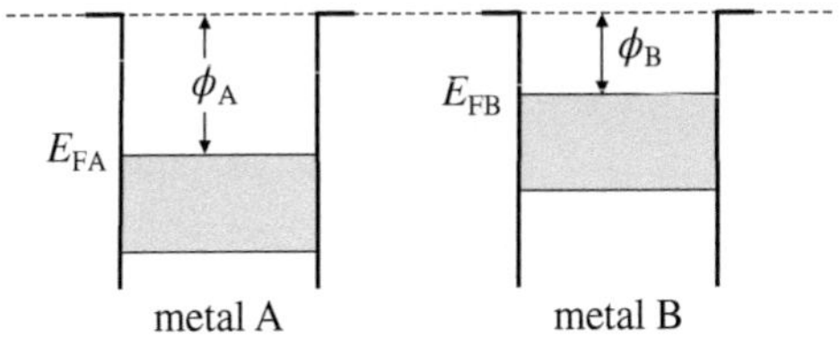

Figure 26.15. Energy diagrams for two dissimilar metals, with different work functions and Fermi energies, when not in contact.

Thermionic emission

We can understand the phenomenon of **thermionic emission**, the emission of electrons into a vacuum from metal surfaces, by considering Figure 26.14. At high temperatures (large values of kT) the distribution of electrons among the available energy states in a metal extends well above E_F. If T is sufficiently high, some electrons may acquire an energy greater than $E_F + \phi$ and can escape from the metal. Thermionic emission is of practical importance because it is responsible for the emission of electrons from heated filaments in vacuum tubes and certain nanomaterials.

By calculating the fraction of electrons that have enough energy to escape from a metal surface, it can be shown (Worked Example 26.2) that the photoelectric effect is effectively temperature independent.

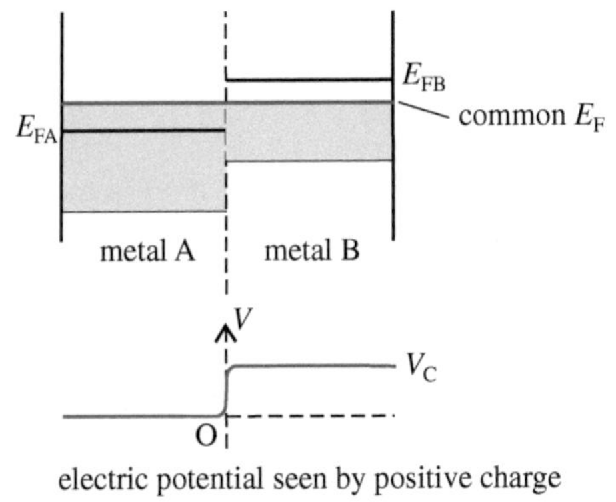

Figure 26.16. The Fermi energy of the two metals shown in Figure 26.15 when they are placed in contact. The electric potential seen by a positive charge as it moves across the junction is plotted at the bottom of the figure.

Conductors in contact: the contact potential

We now consider the behaviour of the electrons in two dissimilar solids when the solids are placed in contact. Let us consider first the case of two conductors in contact.

We consider two metals A and B with work functions ϕ_A and ϕ_B such that $\phi_A > \phi_B$. When not in contact their conduction bands are as shown in Figure 26.15.

When the metals are placed in contact, electrons seek to occupy the lowest energy states so that there is a net diffusion of electrons from metal B to metal A across the *junction* – the boundary between the two metals. The process continues until the energy of the highest occupied level is the same in both metals, that is until both metals have a common Fermi level E_F. This situation is represented in Figure 26.16. In Worked Example 26.3 it is shown that the Fermi energy in equilibrium of two materials in contact must be the same in the two materials.

When equilibrium has been reached metal A will have a net negative charge and metal B a net positive charge. Energy is needed to take a positive charge from metal A to metal B so that there is an electrical potential difference, V_C, across the junction between the two metals, as indicated by the potential plot at the foot of Figure 26.16. V_C is known as the **contact potential**, a property of the two metals which are in contact.

Note that the contact potential is not measurable with a voltmeter connected across the metals A and B, because, in addition to the contact potential across the junction of A and B, contact potential differences develop where the leads to the voltmeter make contact with the metals. These potential differences cancel the contact potential across the junction between the metals; otherwise a net current would be produced without energy input, in violation of basic thermodynamical principles.

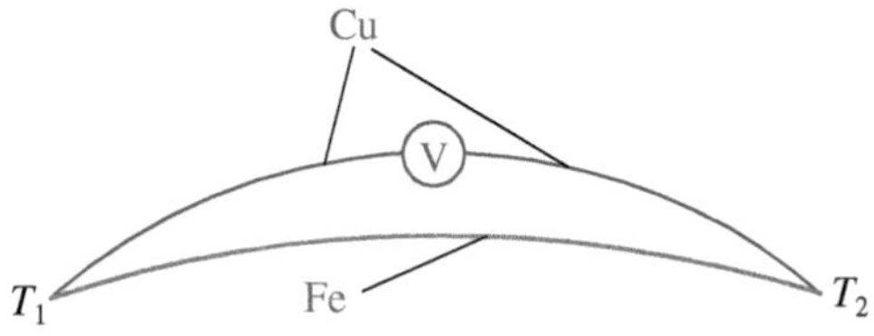

Figure 26.17. A thermocouple – comprising two junctions between dissimilar metals (Fe and Cu).

Note that the potential step seen by an electron in moving from B to A in Figure 26.16, which is $-e(-V_C) = +eV_C$, is the same as that seen by a positive charge $+e$ in moving from A to B. Thus the positive potential steps experienced by positive charges at junctions, such as V_C in Figure 26.16, also present positive potential steps to electrons which are moving in the opposite direction. This is a general result which we shall use frequently in this chapter. The potential step, shown in Figure 26.16, is a step up in potential for both positive charges (such as holes) which are moving to the right and for negative charges (such as electrons) which are moving to the left. Similarly, it is a potential drop for both positive charges which are moving to the left and for negative charges which are moving to the right.

The thermocouple

Contact potentials change with temperature so that, when two junctions between dissimilar metals, such as copper and iron, are at different temperatures, as illustrated in Figure 26.17, the contact potentials no longer cancel and a net emf can be detected by a voltmeter. The energy needed to produce the current is provided by heat energy. The emf produced depends on the temperature difference between the junctions and thus the device – known as a *thermocouple* – can be used to measure temperature.

Study Worked Examples 26.2 and 26.3

For problems based on the material presented in this section visit up.ucc.ie/26/ and follow the link to the problems

26.5 Junctions in semiconductors; the p–n junction

The equalisation of the Fermi levels of two conductors in contact is a general phenomenon which can be shown to apply equally to semiconductors in contact. It is an example of the very general thermodynamical principle that temperatures (internal energies) equalise in a system in equilibrium. The principle may be applied to two different intrinsic semiconductors in contact or to a single semiconductor, one part of which has been doped with donor impurities and the other with acceptor impurities.

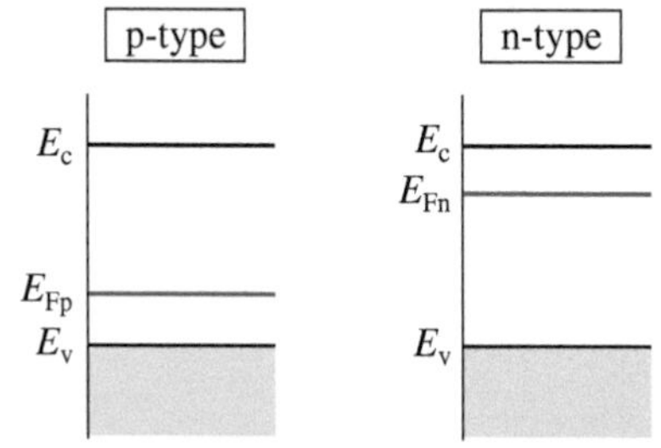

Figure 26.18. The band energies and Fermi energies of a p- and n-type semiconductor when they are not in contact.

When an n-type semiconductor is in contact with a p-type semiconductor, a **p-n junction** is formed. Let us consider a p-n junction in which the n- and p-type semiconductors are manufactured by suitable doping of a piece of intrinsic semiconductor.

Band energies in the two semiconductors, if they existed separately, are shown in Figure 26.18. The Fermi energies, E_{Fn} and E_{Fp}, are different in the n- and p-type semiconductors, as has already been indicated in Figures 26.12 and 26.13. The gap energies are the same, however, because the semiconductors are manufactured by doping the same intrinsic semiconductor.

When the junction is formed, both the conduction and valence bands move so that the internal potential energy eV_C brings the Fermi energies in the two regions together. This is achieved by holes on the p-side diffusing to the n-side and by electrons on the n-side diffusing to the p-side until the electric field set up by the separation of charge stops further diffusion.

Thus, if E_{Fn} and E_{Fp} are the Fermi energies in the two regions when they are separate (Figure 26.18), eV_C, the energy needed to bring E_{Fp} up to E_{Fn}, is given by

$$E_{Fn} - E_{Fp} = eV_C$$

After the junction is formed, as illustrated in Figure 26.19, if E_{cp} and E_{cn} are the lowest energies in the conduction bands of the p- and n-type regions, respectively

$$E_{cp} - E_{cn} = eV_C$$

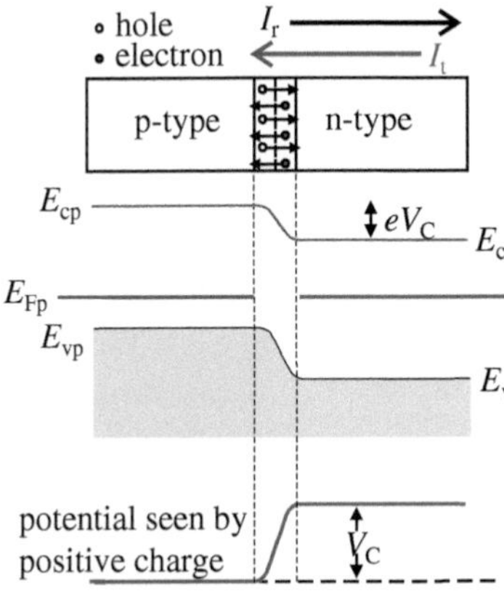

Figure 26.19. The p-n junction. The band energies and Fermi energy of the p- and n-type semiconductors, shown in Figure 26.18, when they are placed in contact. The directions of the recombination and thermal currents are shown at the top of the figure and the electric potential seen by a positive charge as it moves to the right across the junction is plotted at the bottom of the figure.

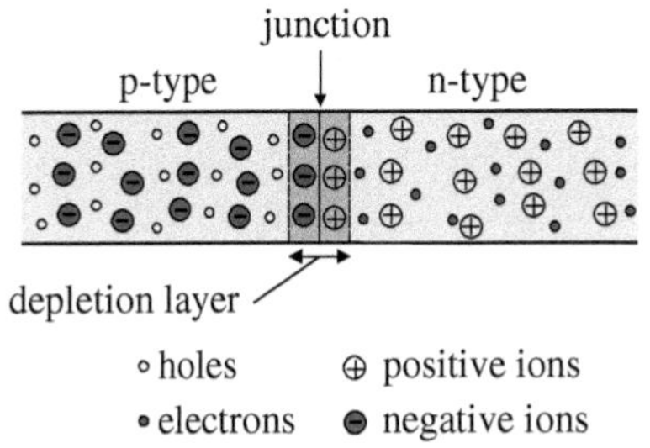

Figure 26.20. The depletion layer of a p-n junction.

The internally generated potential V_C may be considered to act across a region of finite width, known as the *depletion layer*. As illustrated in Figure 26.20, this is the region into which the holes and electrons have moved to recombine, the process which sets up the internal potential difference and equalises the Fermi energies. Although V_C is small – of the order of a volt – because the depletion layer is so narrow – about 10^{-6} m – the electric field in the depletion layer, $E = \frac{V}{d} \approx 10^6\ \mathrm{V\,m^{-1}}$, can be very large. Note that in real materials the edges of the depletion layer are not sharp as indicated in Figure 26.20.

In addition to the n- and p-carriers which are produced by doping, a small number of electron-hole pairs form spontaneously in both the p- and n-type regions due to thermal agitation in the semiconductor. The distribution of energies of the charge carriers allows some holes generated in the p-region to have sufficient energy to overcome the potential barrier V_C and to enter the n-side where they quickly recombine with the plentiful supply of free electrons. Similarly some electrons generated on the n-side cross the barrier and recombine with holes on the p-side, This flow of charge constitutes a current to the right in Figure 26.19 known as the *recombination current* I_r, as indicated at the top of the figure. In equilibrium, the recombination current is balanced by a *thermal current* I_t to the left, as also indicated in Figure 26.19. This is due to holes which are generated near the depletion layer on the n-side and then 'fall down' the potential drop V_C to the p-side, and to electrons which are generated on the p-side and then move up V_C (a potential drop for electrons) to the n-side. The thermal current increases with temperature but is independent of V_C.

Note that, as in metals, V_C, the internal potential difference or contact potential, is not imposed externally. It is a property of the junction. Let us now consider what happens when an external potential V_{ext} is applied to a p-n junction. The junction is then said to be **biased**.

26.6 Biased p-n junctions; the semiconductor diode

Forward biasing

Let us first consider a case in which, as shown in Figure 26.21, V_{ext} is applied so as to reduce the potential difference between the n- and p-sides from V_C to $V_C - V_{ext}$, a form of biasing known as *forward biasing*. The resulting band energies are shown in Figure 26.21. Note that when the junction is biased there is a net flow of current across it. We no longer have equilibrium and the Fermi energy is no longer the same throughout the material. Without equilibrium the Fermi energies in the p- and n-regions, E_{Fp} and E_{Fn}, respectively, are different, as shown in Figure 26.21.

As shown in the potential energy plot at the foot of Figure 26.21, forward biasing reduces the potential step seen by holes moving from left to right. Similarly electrons diffuse more easily from the n-side to the p-side. There is therefore a net positive current from the p-side to the n-side which increases rapidly as V_{ext} increases and, therefore, $V_C - V_{ext}$ decreases. There is, additionally, a very small opposing current due to thermally generated holes on the n-side and electrons on the p-side which fall down the potential barrier, but this current is negligible compared to the current produced by the forward bias. Effectively, the forward bias increases the recombination current but leaves the thermal current unchanged, so that there is a net current from the p- to the n-region.

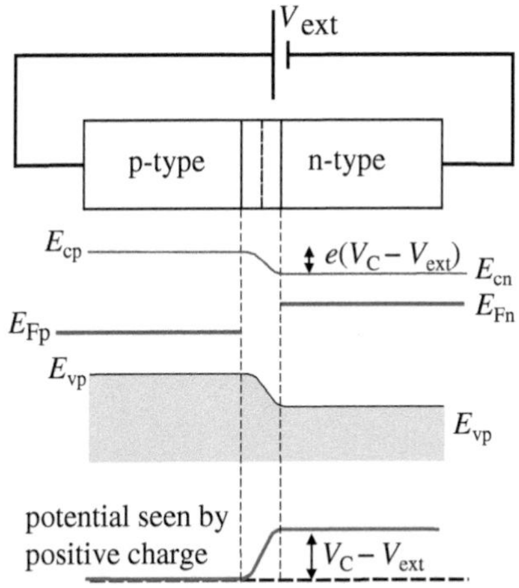

Figure 26.21. A forward-biased p-n junction showing the band energies and the Fermi energies of the p- and n-type semiconductors. The electric potential seen by a positive charge as it moves across the junction is plotted at the bottom of the figure.

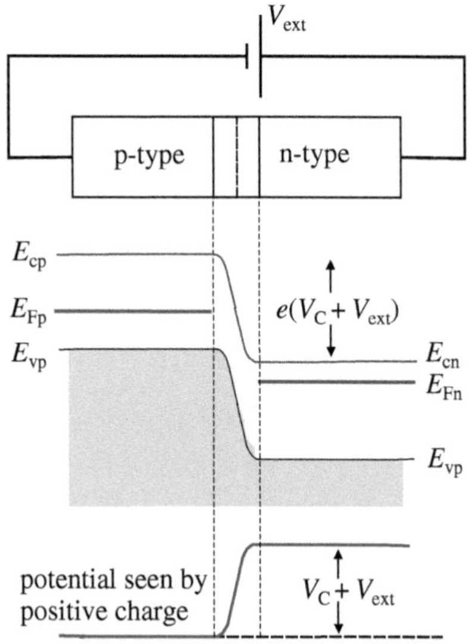

Figure 26.22. A reverse-biased p-n junction showing the band energies and the Fermi energies of the p- and n-type semiconductors. The electric potential seen by a positive charge as it moves across the junction is plotted at the bottom of the figure.

Reverse biasing

When, as illustrated in Figure 26.22, V_{ext} is applied so as to increase the potential difference between the n- and p-sides from V_C to $V_C + V_{ext}$ the junction is said to be *reverse biased*. As illustrated in the potential energy plot at the foot of Figure 26.22, both holes moving from left to right and electrons moving from right to left have a larger potential step to overcome and there is very little current due to these carriers. Note, however, that there is a very small thermal current I_t (which is independent of bias) due to holes produced on the n-side and electrons produced on the p-side which fall down the potential step. Note also that the width of the depletion layer increases when a reverse bias is applied, that is as the potential difference across the junction increases.

The currents (I) produced by forward and reverse biasing of a p-n junction are plotted against applied potential difference (V) in Figure 26.23. This known as the *$I - V$ characteristic* of the p-n junction. The resistance of the p-n junction, which is the $\frac{V}{I}$ ratio at any point on the characteristic, is small in the forward bias case but very large for reverse bias. A p-n junction is clearly non-ohmic, that is, its resistance changes as V changes. Note that, in order to show both the forward and reverse currents on the same figure, a more sensitive scale is used for the much smaller reverse bias current.

The semiconductor diode

The $I - V$ characteristic shown in Figure 26.23 shows that, effectively, current can pass through a p-n junction in only one direction. A device which restricts current flow to one direction is known as a **diode**. The characteristic of an *ideal diode* is shown in Figure 26.24.

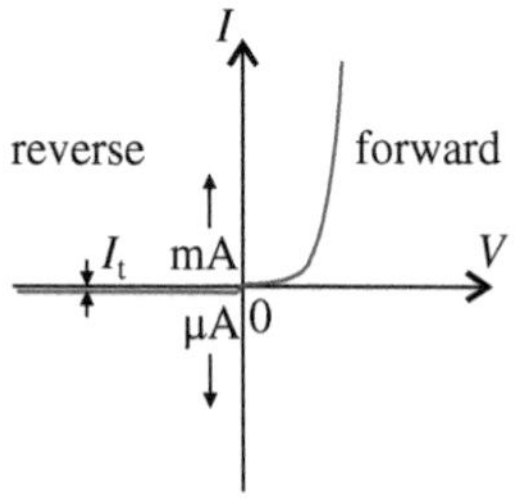

Figure 26.23. The $I - V$ characteristic of a p-n junction. Note that a more sensitive scale is used for the reverse current.

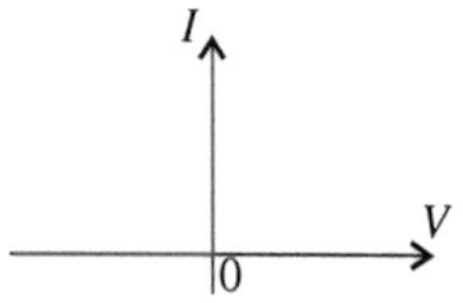

Figure 26.24. The $I - V$ characteristic of an ideal diode.

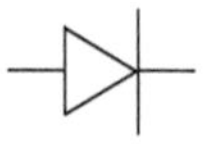

Figure 26.25. The circuit symbol for a diode. Positive current can flow in the direction indicated by the triangle.

Real diodes, such as p-n junctions, only approximate to ideal diodes but in many cases their behaviour in circuits can be described satisfactorily by considering them to be ideal diodes. The circuit symbol for a diode is shown in Figure 26.25. The triangle points in the direction in which positive current can flow.

Semiconductor diodes have a very wide range of applications, as illustrated in the next three sections.

26.7 Photodiodes, particle detectors and solar cells

The photodiode

The production of electron-hole pairs by light which falls near the depletion layer of p-n junction is used, in a **photodiode**, to detect light. As indicated in Figure 26.26, when the p-n junction is reverse biased the dependence of the *photocurrent*, I_p, on light intensity is linear so that I_p may be used to measure light intensity.

The particle detector

Electron-hole pairs can also be produced by charged particles when they pass through the depletion layer. This effect is used in solid state **particle detectors** to detect charged particles, for example those produced in radioactive decays (Section 27.6).

The solar cell

In a **solar cell** the photocurrent produced by light falling on the depletion layer is used to generate electric power from sunlight. Solar cells operate without any external power supply, relying on optical power to generate current and voltage.

When a photon of frequency $f > \frac{E_g}{h}$, where E_g is the gap energy (Figure 26.8), falls near the depletion layer of a p-n junction an electron from the valence band can be excited to the conduction band, producing a hole and an electron. Holes produced in this way in the n-region near the junction and, similarly, electrons produced in the p-region, fall down the potential barrier and constitute a current I_p — the photocurrent — which adds to already existing thermal current I_t (described in Section 26.5). This process produces a net positive charge on the p-side and a net negative charge on the n-side, so that the potential barrier is reduced to $V_C - V$, where V is a potential difference which develops across the diode (the terminal voltage) and is measurable with a voltmeter. Because the potential barrier at the junction is lower, the recombination current I_r (described in Section 26.5) increases and, as indicated in Figure 26.27, equilibrium is reached when $I_r = I_t + I_p$,

If the intensity of the incident light and the temperature is kept constant, both I_p and I_t are constant, independent of the barrier height. I_r, on the other hand, increases as the barrier height is reduced. When current is drawn from the cell the terminal voltage V is reduced, the barrier height, $V_C - V$, is increased so that I_r is decreased but I_p and I_t are unaffected. Thus $I_p + I_t > I_r$ and this current can deliver power to an external load resistance. Note that when the load resistance is zero (short circuit) $V \rightarrow 0$ and the barrier height rises to its maximum value, that is V_C. In this case I_r is at a minimum value and the cell is not able to deliver power.

Solar cells are constructed in such a way that solar photons are absorbed near the depletion layer, as illustrated in Figure 26.28. Semiconductors with very small band gaps are used in solar cells to ensure that even the lowest frequency (longest wavelength) sunlight is absorbed.

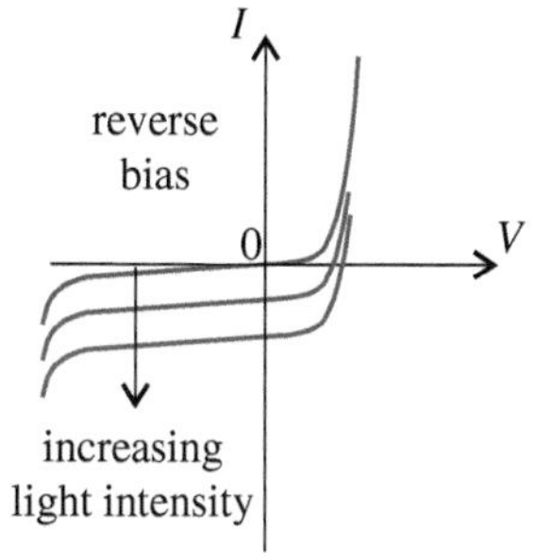

Figure 26.26. $I - V$ characteristics of a photodiode showing how the (negative) photocurrent increases with increasing light intensity.

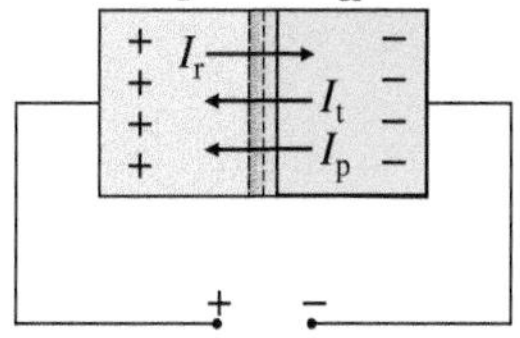

Figure 26.27. A schematic diagram showing the currents in a solar cell.

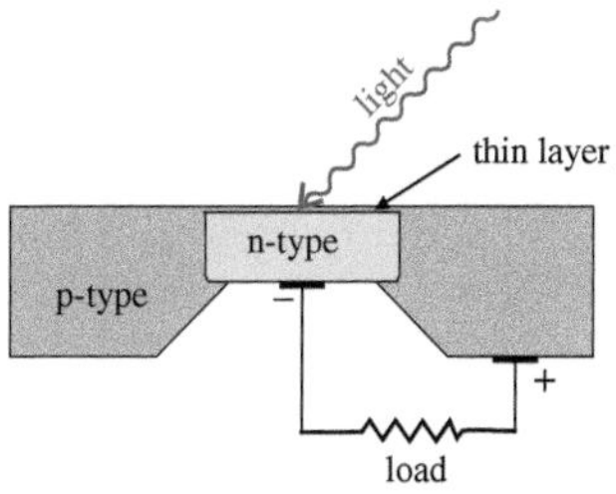

Figure 26.28. Schematic diagram of the construction of a solar cell.

26.8 Light emitting diodes; semiconductor lasers

The light emitting diode

A **light emitting diode** (LED) is essentially a solar cell which operates in reverse. When a forward bias is applied across a p-n junction, electrons flow from the n-side to the p-side and holes from the p-side to the n-side. When electrons arrive at the p-side they recombine with the available holes just outside the depletion layer, releasing their energy as photons (light). Similarly, holes arriving at the n-side combine with electrons and also produce light. LEDs can be used in a very wide range of applications in lighting and in electronic displays. With suitable choices of doping and transition levels LEDs can emit light in any region of the visible and near visible spectrum. They are compact, durable, use little energy and may be switched on and off rapidly.

The semiconductor diode laser

The semiconductor laser, like the LED, is forward biased p-n junction in which photons are produced when electrons from the conduction band of the n-side recombine with holes in the valence band of the p-side. The plentiful supply of electrons in the n-side and holes in the p-side provides the population inversion needed to produce laser operation, as described in Section 22.2. If the current through the junction is sufficiently large, the photons produced stimulate emission of further photons of the same frequency and a chain reaction occurs in which stimulated emission dominates. As illustrated in Figure 26.29, opposite faces of the p-n junction are made parallel and polished so that the laser junction layer, typically 1 μm thick, forms an optical cavity. Thus laser light emerges parallel to the junction.

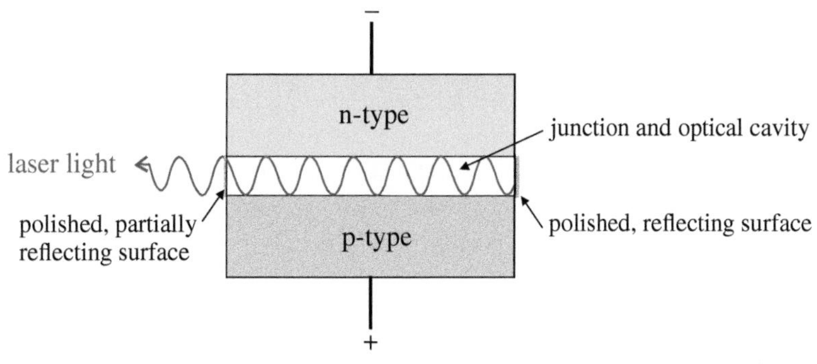

Figure 26.29. Schematic construction of a semiconductor laser. Note that the width of the junction layer is greatly exaggerated.

Stacking of the laser diodes can increase the power produced considerably and can also enable multiple wavelengths to be generated. Laser diodes are small, compact, inexpensive and have a very wide range of applications. They are used, for example, in CD/DVD/Blue ray readers, in barcode scanners and in fibre optic communications.

26.9 The tunnel diode

One type of semiconductor diode, the **tunnel diode**, provides a good example of the quantum mechanical phenomenon of tunnelling through a potential barrier, as discussed in Section 14.17. In the tunnel diode, the n- and p-type regions are heavily doped producing the energy band structure shown in Figure 26.30(a).

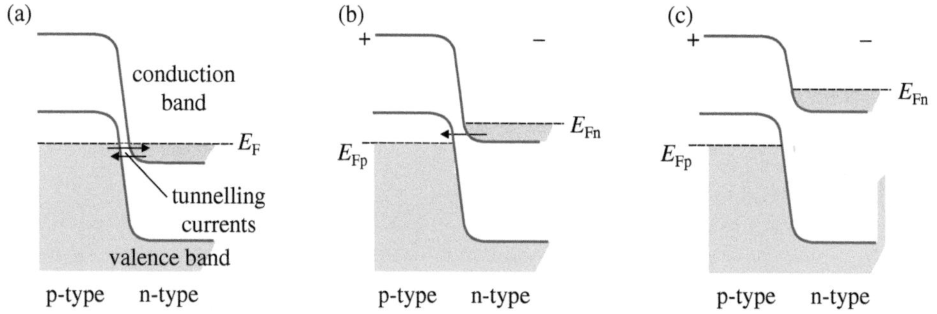

Figure 26.30. The tunnel diode. (a) When there is no bias voltage the bottom of the n-conduction band overlaps the top of the p-valence band. Electrons tunnel in both directions. (b) The band structure and Fermi energies when a small forward bias voltage is applied. Electrons can tunnel only from the n-type to the p-type region. (c) The band structure and Fermi energies when a large forward bias voltage is applied. No tunnelling can occur.

The depletion layer is so narrow in a tunnel diode ($\sim 10^{-9}$ m) that the bottom of the n-conduction band overlaps the top of the p-valence band. As illustrated in Figure 26.30(a), due to the high concentration of doping impurities, the donor levels merge with the bottom of the n-conduction band, moving the Fermi energy up into the conduction band. Similarly acceptor levels merge with the top of the p-valence band, lowering the Fermi energy of the n-type semiconductor below the top of the band.

Because the width of the depletion layer is comparable to the de Broglie wavelength of an electron in the semiconductor, electrons can tunnel through the forbidden band, as indicated in Figure 26.30(a). Without biasing, electrons tunnel in both directions. An equilibrium is set up, with the Fermi energy the same throughout the diode.

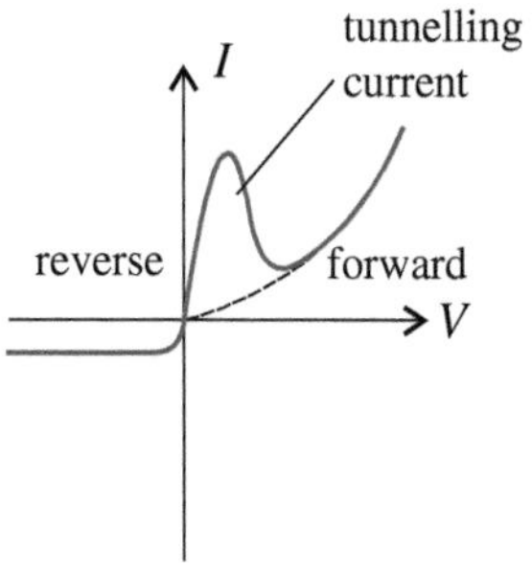

Figure 26.31. The $I - V$ characteristic of a tunnel diode showing the enhanced current due to tunnelling at low forward bias voltages.

When a small forward bias voltage is applied, the band structure changes to that shown in Figure 26.30(b). The filled part of the n conduction band is now opposite the empty part of the p valence band, so that tunnelling by electrons is from the n to p regions only, equivalent to a positive current to the right in the figure.

When the forward bias is increased the bands no longer overlap, as shown in Figure 26.30(c). The tunnel current ceases and the diode behaves like a conventional p-n junction (as described in Figure 26.23).

The $I - V$ characteristic of the tunnel diode is shown in Figure 26.31. At low forward bias voltages there is an enhanced current due to tunnelling.

The most important practical feature of the tunnelling current is the speed with which electrons can tunnel. The process is considerably faster than diffusion of electrons and holes through the depletion layer, the process which is used in conventional p-n diodes. Tunnel diodes therefore respond very quickly to voltage changes and can be used as fast switches in computing circuits.

26.10 Transistors

Transistors are semiconductor structures with three terminals. A current flowing between one pair of terminals can be controlled by a potential placed across another pair. We describe below the two main types of transistor, the **bipolar junction transistor** and the **field effect transistor (FET)**.

The bipolar junction transistor

The bipolar junction transistor is available in two types, the n-p-n type in which a *thin* layer of p-type semiconductor is sandwiched between two n-type semiconductors and the p-n-p type in which a thin layer of n-type semiconductor is sandwiched between two p-type semiconductors. The two types are illustrated in Figure 26.30. The bipolar transistor is so called because both electrons and holes act as current carriers.

A bipolar transistor, therefore, comprises two p-n junctions. The three terminals, to which electrical connections are made, are known as the emitter, base and collector, as indicated in Figure 26.32. The circuit symbols for bipolar transistors are also shown in this figure.

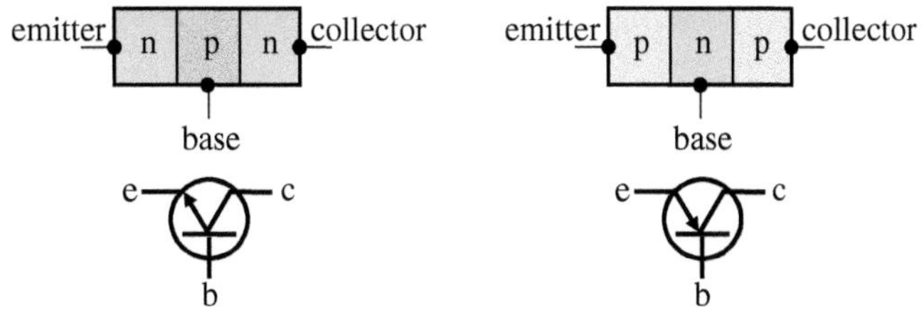

Figure 26.32. Schematic diagrams of n-p-n and p-n-p bipolar junction transistors. The circuit symbols for each device are shown beneath each diagram.

The band structure of an n-p-n bipolar transistor without external biasing is shown in Figure 26.33. The band energies adjust to maintain a constant Fermi level throughout the device in the manner we have already described for p-n junctions (Figure 26.19).

Figure 26.34 shows how the band structure of a n-p-n transistor changes when a forward bias V_{eb} is applied across the emitter-base junction and a reverse bias V_{bc} is applied across the base-collector junction, an arrangement which is known as *common base* biasing. The band energies adjust in the manner we have described previously for forward and reverse biased p-n junctions (Figures 26.21 and 26.22). The emitter region is more heavily doped than the base so that most of the current is made up of electrons moving from left to right (that is from emitter to base). For this reason the potential seen by an electron, as well as that seen by a positive charge, is shown at the foot of Figure 26.34.

Because the base region is so thin with a low concentration of holes (light doping), the operation of the bipolar n-p-n transistor is not simply that of two independent p-n diodes which are connected back to back. The emitter-base junction is forward biased so that a substantial positive current I_e flows from the base to the emitter; that is, there is a substantial flow of electrons into the base region. Because the base region is so thin and the concentration of holes is low there, most electrons do not recombine in the base region but diffuse through it to the base-collector junction where they 'fall down' the potential step into the collector. The small number of electrons which recombine in the base are represented by a small base current I_b, as indicated in Figure 26.35. Most current into the emitter I_e therefore appears as the collector current I_c and we can write

$$I_e = I_b + I_c \tag{26.3}$$

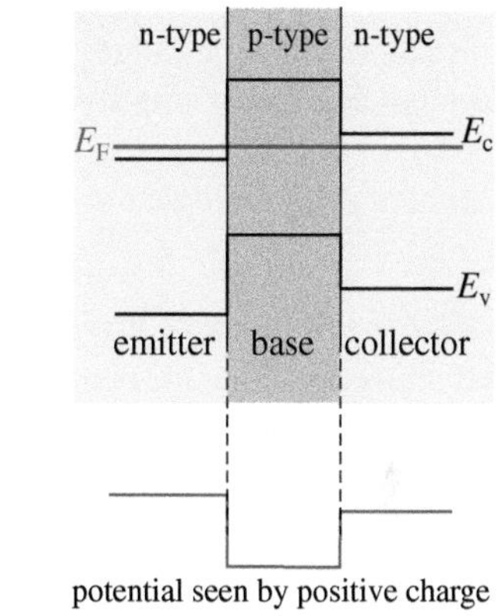

Figure 26.33. The band structure and Fermi energy in a bipolar n-p-n transistor without external biasing. The electric potential seen by a positive charge, as it moves to the right across the junction, is plotted at the bottom of the figure.

The current flows due to the movement of holes and electrons in a n-p-n bipolar transistor when it is biased in common base mode, are shown in Figure 26.35. The *current gain* in common base mode is defined as

$$\alpha := \frac{I_c}{I_e},$$

As noted above, I_c is only a little less than I_e so that α is close to unity, typically 0.97 to 0.98.

Another important method of biasing is the *common emitter* mode, illustrated in Figure 26.36 for a n-p-n transistor. In this case the biasing potentials are placed across the base-emitter and emitter-collector junctions. Again $I_e = I_b + I_c$

The current gain in common emitter mode is defined as $\beta = \frac{I_c}{I_b}$

We can use Equation (26.3) to write

$$\frac{I_e}{I_c} = \frac{I_c + I_b}{I_c} = 1 + \frac{I_b}{I_c} \quad \text{and hence} \quad \frac{1}{\alpha} = 1 + \frac{1}{\beta}$$

which it follows that

$$\beta = \frac{\alpha}{1 - \alpha}$$

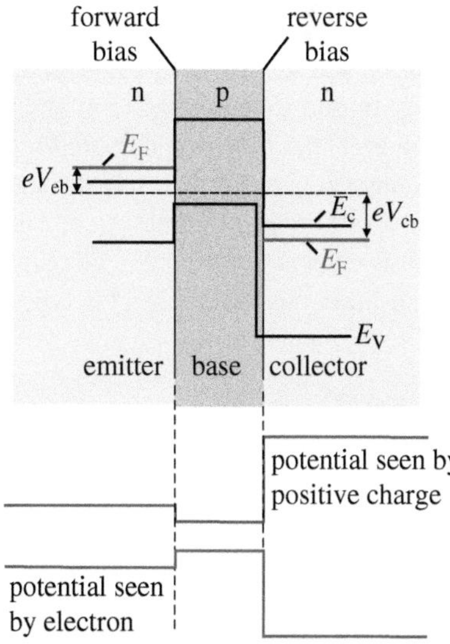

Figure 26.34. The band structure and Fermi energies in a bipolar n-p-n transistor in which the emitter-base junction is forward biased and the base-collector junction is reverse biased (known as *common base* biasing). The electric potentials seen by positive charges and electrons, respectively, as they move across the junction, are plotted at the bottom of the figure.

As noted above, α is typically 0.97 to 0.98 so that β, the current gain in common emitter mode, can be very large, typically 30 to 100. In this mode a small current (I_b) controls a large current (I_e) so that the transistor can be used as a current amplifier – small changes in the input (base) current can produce large changes in the output (collector) current.

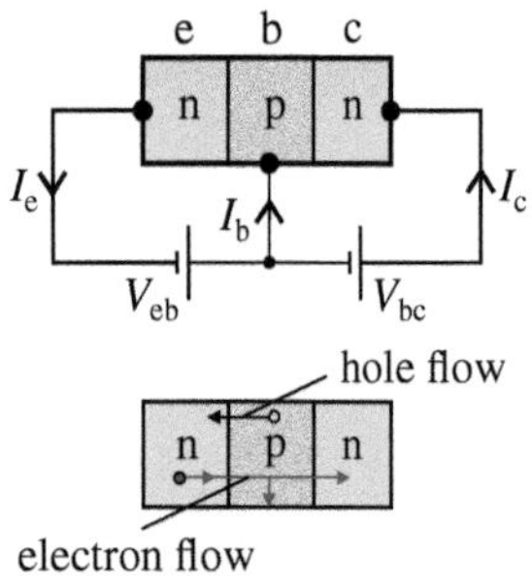

Figure 26.35. The biasing and current directions in a common base biased n-p-n transistor.

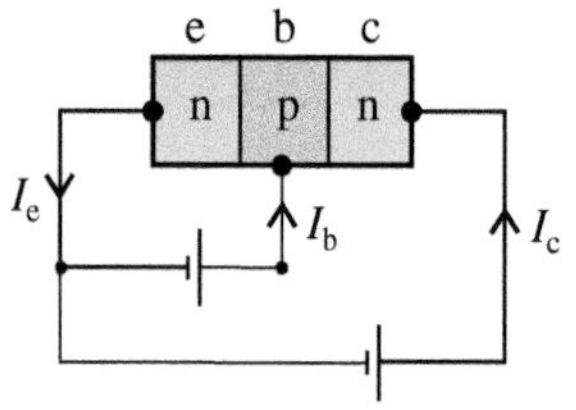

Figure 26.36. The biasing and current directions in a common emitter biased n-p-n transistor.

The transistor has a large range of applications as a switch or an amplifier. With a suitable choice of resistors in the biasing circuit, the transistor can be used to amplify a.c. voltages or as a switch in which the voltage drop across an external resistance which is connected to the collector can be switched on or off by small changes in the forward bias voltage across the emitter-base junction. Such switches are used extensively in logic circuits.

The field effect transistor (FET)

As noted in Section 26.6, the resistance of a forward biased p-n junction is low. The input impedance of the junction transistor in common base mode is therefore low. The input impedance is higher in common emitter mode but is not high enough for many applications so that an alternative transistor structure which has a much higher input impedance, the field effect transistor (FET), is often used. As shown in Figure 26.37, a n-channel FET can be constructed from a block of n-type material with terminals, called the *source* and *drain* at either end, and with a strip of p-type material, called the *gate* fixed along one side. The circuit symbol for the FET is also shown in this figure.

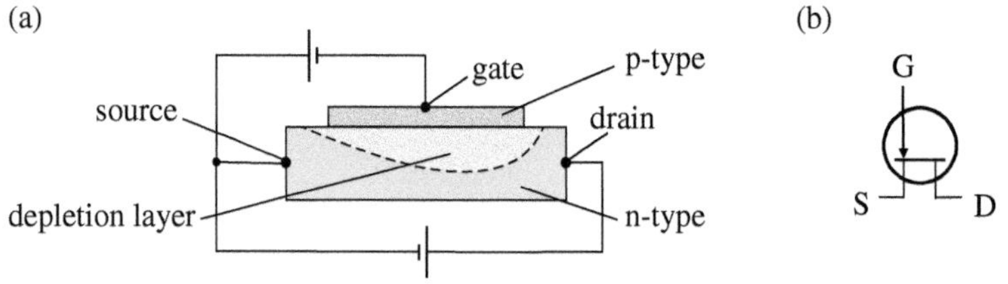

Figure 26.37. (a) A schematic diagram of a n-channel field effect transistor (FET). (b) The circuit symbol for this device.

When biased as shown in Figure 26.37, electrons move from the source to the drain through the n-type channel. The p-n junction is reverse biased so that both the n- and p-type semiconductors near the junction are depleted of charge carriers. The higher the reverse bias on the gate, the more the depleted region extends into the n-type channel, increasing its resistance, and the lower the current becomes. Thus, the gate voltage controls the source-drain current. Very little current flows through the gate due to its reverse bias so that the device has a very high input impedance. The field effect transistor is therefore a voltage-controlled device in contrast to the bipolar transistor which is current controlled. The current is transported by carriers of one polarity only (the majority carriers) which are electrons in the case of the n-channel FET. This type of transistor is therefore also called a unipolar transistor.

Integrated circuits

In practice, transistors are not manufactured by joining separate pieces of doped semiconductors but by diffusing acceptor or donor atoms from a gas into a single thin crystal substrate of semiconductor – known as a *wafer*. The layout of the doped regions can be controlled by masking selected areas of the crystal from the gas. In this way *integrated circuits*, which contain millions of transistors and other electronic components are manufactured using a single crystal – a *chip* – which is only a few millimetres across.

Electronic circuits are now usually constructed from integrated circuits rather than from individual transistors. An important example is the OP-AMP (Operational Amplifier) a highly versatile, high voltage gain, amplifier with a wide range of applications in consumer, industrial and scientific electronics. Integrated circuits provide complete electronic circuits for counters, registers, memories etc. and, as such, are the main components of computer manufacture. Detailed and comprehensive descriptions of the practical uses of semiconductor devices may be found *The Art of Electronics* by Horowitz and Hill and further material on the topics covered in this chapter may be found in *Solid State Physics* by Hook and Hall (See BIBLIOGRAPHY for book details).

WORKED EXAMPLES

Worked Example 26.1: Determine the probability that a state at the bottom of the conduction band is occupied at 300 K for each of the following values of the band gap energy (i) 0.10 eV, (ii) 1.0 eV (typical of a semiconductor) (iii) 6.0 eV (typical of an insulator).
Assume, in each case, that the Fermi energy lies in the middle of the band gap.
By what factors do these probabilities increase when the temperature increases by 10 K?

Applying Equation (25.22),
$$F(E) = \frac{1}{e^{\frac{(E-E_F)}{kT}}+1}.$$

(i) $E = 0.10\,\text{eV}, E - E_F = 0.05\,\text{eV}, T = 300\,\text{K} \rightarrow F(E) = 0.126$

(ii) $E = 1.0\,\text{eV}, E - E_F = 0.5\,\text{eV}, T = 300\,\text{K} \rightarrow F(E) = 4.05 \times 10^{-9}$

(iii) $E = 6.0\,\text{eV}, E - E_F = 3.0\,\text{eV}, T = 300\,\text{K} \rightarrow F(E) = 4.43 \times 10^{-51}$

Note the sensitivity of the probability of occupation to the width of the band gap.
Repeating the calculations for a temperature of 310 K

(i) $E = 0.10\,\text{eV}, E - E_F = 0.05\,\text{eV}, T = 310\,\text{K} \rightarrow F(E) = 0.13,$ increased by a factor $\frac{0.134}{0.126} = 1.06$

(ii) $E = 1.0\,\text{eV}, E - E_F = 0.5\,\text{eV}, T = 310\,\text{K} \rightarrow F(E) = 7.56 \times 10^{-9},$ increased by a factor $\frac{7.56 \times 10^{-9}}{4.05 \times 10^{-9}} = 1.87$

(iii) $E = 6.0\,\text{eV}, E - E_F = 3.0\,\text{eV}, T = 310\,\text{K} \rightarrow F(E) = 1.87 \times 10^{-49},$ increased by a factor $\frac{1.87 \times 10^{-49}}{4.43 \times 10^{-51}} = 42.2$

Note the sensitivity of the probability of occupation to a small increase in temperature, particularly in cases (ii) (a semiconductor) and (iii) (an insulator).

Worked Example 26.2: Given that the work function of copper is 4.1 eV, estimate the fraction of conduction electrons which have enough energy to escape from a copper surface at 1000 K.
The photoelectric effect is found to be effectively temperature independent. Comment on this finding.
Note: The melting point of copper is 1358 K.

The fraction of electrons with enough energy to escape,
$$F(E) = \frac{1}{e^{(E-E_F)/kT}+1}$$

where the work function $= E - E_F = 4.1\text{ eV} = 6.56 \times 10^{-19}$ J and $kT = 1.38 \times 10^{-20}$ J.

Thus
$$F(E) = \frac{1}{e^{47.5}+1} = 2.3 \times 10^{-21}$$

The number of electrons with energies significantly greater than $(E - E_F)$ is very small at temperatures up to the melting point of copper. Hence the photoelectric effect changes little when the temperature changes.

Worked Example 26.3: Show that, if two materials are in contact so that electrons can pass between them, the Fermi energy in thermal equilibrium must be the same in the two materials.
Note:

1) In thermal equilibrium there is no net energy flow and no net charge flow at any energy.
2) The density of occupied states at energy E is given by $n(E)F(E)$, where $n(E)$ is the number density of available energy states and $F(E)$ is the probability of occupation, that is the Fermi–Dirac distribution function (Equation (25.22)).

The probability of a state being vacant is one minus the probability of that state being occupied.
Thus the density of unoccupied states = (density of available states) × (probability of vacancy) = $n(E)[1 - F(E)]$
Denoting the two materials by A and B, the flow of electrons from A to B at energy E is proportional to the number of electrons present in A multiplied by the number of empty states present in B
Thus, the flow of electrons from A to B at energy E

$$\propto [n_A(E)F_A(E)]\{n_B(E)[1 - F_B(E)]\} \tag{26.4}$$

Similarly, the flow of electrons from B to A at energy E

$$\propto [n_B(E)F_B(E)]\{n_A(E)[1 - F_A(E)]\} \tag{26.5}$$

In thermal equilibrium Equations (26.4) and (26.5) may be equated

$$[n_A(E)F_A(E)]\{n_B(E)[1 - F_B(E)]\} = [n_B(E)F_B(E)]\{n_A(E)[1 - F_A(E)]\}$$
$$\rightarrow F_A(E) - F_A(E)\,F_B(E) = F_B(E) - F_A(E)\,F_B(E)$$
$$\rightarrow F_A(E) = F_B(E) \quad (26.6)$$

Examining Equation (25.22), namely, $F(E) = \frac{1}{e^{(E-E_F)/kT} + 1}$, we note that Equation (26.6) can only be satisfied when E_F is the same in the two materials.

PROBLEMS

For problems based on the material presented in this chapter visit up.ucc.ie/26/ *and follow the link to the problems*

27

Nuclear and particle physics

AIMS

- to develop models of nuclear structure (in particular the *liquid drop* and *shell* models) which can account for observed regularities in the properties of atomic nuclei
- to describe *radioactivity* – the decay of unstable nuclei – through a radioactive decay law and to interpret the three naturally occurring decay processes, α-, β- and γ-decay, in terms of nuclear models
- to describe how further nuclear properties may be deduced through analyses of artificially produced nuclear reactions and how two nuclear processes, fission and fusion, can be used to generate large quantities of energy
- to classify, according to their properties, the *subnuclear* particles which are produced in large numbers in very high energy nuclear reactions, with the aim of obtaining evidence of an underlying particle structure
- to describe a model of particle structure in which a small number of basic building blocks (*quarks* and *leptons*) interact through fundamental forces

27.1 Properties of atomic nuclei

In Section 24.1 we described how, by scattering α-particles from gold and silver atoms, Rutherford and Geiger and Marsden were able to show that the mass of an atom is concentrated in a very small volume of diameter $\sim 10^{-14}$ m, which is known as the *nucleus*. We now investigate the properties, and hence the structure, of the atomic nucleus.

The size and the structure of atomic nuclei may be determined through bombardment by electrons whose de Broglie wavelength is less than the typical nuclear size. In Worked Example 9.4, we showed that the momentum of a 250 MeV electron is 250 MeV/c, so that such an electron, with a Broglie wavelength of about 5×10^{-15} m, is suitable as a nuclear probe. Electrons interact with nuclear charge (protons) through the Coulomb interaction and thus reveal the proton distribution within the nucleus. The characteristic mean radius R of any nucleus is found to be given by the equation

$$R = R_0 A^{\frac{1}{3}} \tag{27.1}$$

where A is the mass number of the atom, defined in Section 1.2 as the total number of nucleons (protons and neutrons) in the nucleus, and R_0 is a constant with a value of about 1.1 fm. The femtometre (10^{-15} m) is a convenient measure of length on the nuclear scale and is sometimes called the *fermi* in the nuclear context. Assuming that nuclei are spherical, with volumes given by $\frac{4\pi r^3}{3}$, Equation (27.1) tells us that the volume of any nucleus is proportional to A, the total number of nucleons which it contains. This result suggests that nucleon density is approximately the same in all nuclei and that the density of the nuclear matter in any nucleus, ρ_N, can be estimated using Equation (27.1),

$$\rho_N = \frac{\text{nuclear mass}}{\text{nuclear volume}} = \frac{AM}{\frac{4}{3}\pi R^3} = \frac{3AM}{4\pi R_0^3 A} = \frac{3M}{4\pi R_0^3}$$

where M is the mass of a nucleon, which is 1.67×10^{-27} kg. Thus $\rho_N \approx 3 \times 10^{17}$ kg m^{-3}, a density which is more than fourteen orders of magnitude greater than that of typical solid matter.

Understanding Physics, Third Edition. Michael Mansfield and Colm O'Sullivan.
© 2020 John Wiley & Sons Ltd. Published 2020 by John Wiley & Sons Ltd.

The occurrence of stable nuclei

In Section 1.2, we noted that *nuclides* (atomic nuclei) are labelled using the notation ${}^{A}_{Z}X$ where X is the chemical symbol for the element, Z is the atomic number – the number of protons in the nucleus – and A is the mass number, that is the total number of nucleons in the nuclide. The number of neutrons in a nucleus is therefore given by $N = A - Z$.

We also mentioned in Section 1.2 that nuclides with the same Z but with different numbers of neutrons, and thus with different A, are known as *isotopes*. Isotopes of an element have the same number of electrons and therefore have the same chemical properties. However the nuclear properties of isotopes of the same element are very different. In general, only a few isotopes of each element – often only one in the case of light nuclides – are stable. The Z and N values of 284 known stable nuclides are plotted in Figure 27.1.

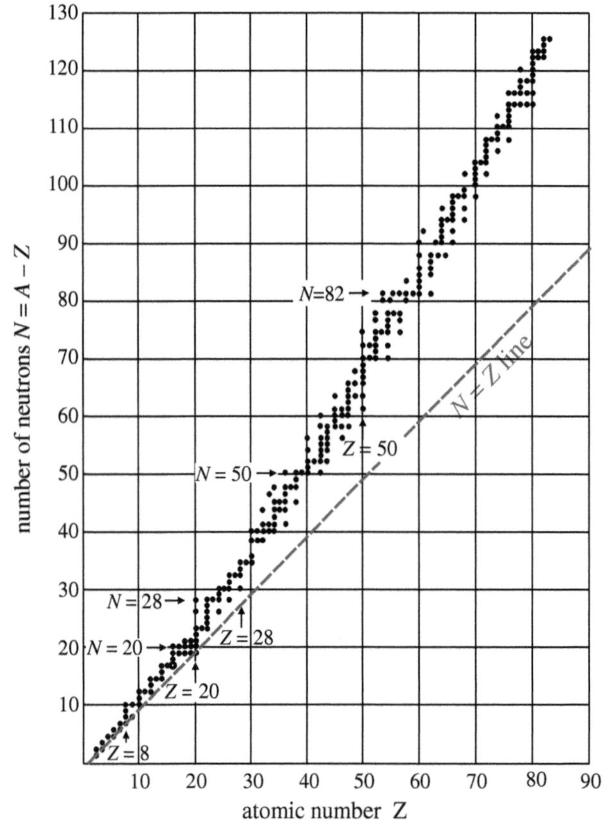

Figure 27.1. The curve of stability. A plot of neutron number N against proton number Z for stable nuclides.

Note that stable nuclides show a preference for certain values of Z and N, as indicated in Figure 27.1, for example $Z = 8$, 20, 28, 50 and 82. Furthermore, there is a clear tendency for stable nuclides to occur when N and/or Z are even. Of about 284 known stable nuclides, 166 have even values of both N and Z, 57 have even N and odd Z, 53 have odd N and even Z, but only 8 have odd values of both N and Z.

In addition to the stable nuclides which are plotted in Figure 27.1, a large number of unstable nuclides – known as *radionuclides* – can be produced in nuclear decays (discussed in Sections 27.4 and 27.5) and in nuclear reactions (discussed in Section 27.7). Radionuclides have average lifetimes which range from nanoseconds to billions of years. They decay through a variety of processes which will be described in more detail in Section 27.5.

Note that the stable nuclei of light (low A) elements tend to contain equal numbers of protons and neutrons and therefore lie close to the $N = Z$ line which is plotted in Figure 27.1. For higher values of A, however, the plot of stable nuclides – known as *the curve of stability* – moves away from the $N = Z$ line. As Z increases these nuclei contain progressively more neutrons than protons. We shall explain why this is so when the liquid drop model is considered in Section 27.3.

Nuclear masses

The masses of nuclei can be determined with accuracies as high as one part in 10^8 using the technique of mass spectrometry, outlined in Section 19.6. These measurements show clearly that the mass of any nucleus is smaller – typically 1% smaller – than the sum of the masses of its constituent nucleons when they are free. We have already explained this effect in Section 9.14 in terms of the very strong binding energies which are associated with the strong nuclear force (Section 1.4), the force which holds nucleons together in bound systems – that is in nuclei. These large negative energies reduce the total energy – and thus the mass – of a system of nucleons significantly when they are bound together in a nucleus. Measurements of the amounts by which the total mass of the system is reduced in the nucleus provides important information on nuclear structure. This effect is discussed in more detail in Section 27.2.

Nuclear spin

Experiments show that many nuclei have an intrinsic angular momentum, which is known as *nuclear spin,* and a magnetic dipole moment. The nuclear magnetic dipole moment of a nucleus can be measured by studying the interaction between the nuclear magnetic dipole moment and the atomic magnetic field. This interaction is similar to the spin-orbit interaction, the interaction between the electron spin magnetic dipole moment and the atomic magnetic field which produces fine structure in atomic spectral lines, as discussed in Section 24.11. The interaction with the nuclear magnetic dipole moment produces much smaller, but measurable, splittings in atomic spectral lines – known as *hyperfine structure*. Hyperfine splittings are smaller than fine structure splittings because the magnetic dipole moments of nuclei are considerably smaller than those of electrons, a result which is expected from examination of Equation (24.31), the equation which relates the magnetic dipole moment of an atomic electron to its orbital angular momentum, $\boldsymbol{m} = -\dfrac{e}{2m_e}\boldsymbol{L}$.

In the nuclear case, the nucleon mass replaces the electron mass, m_e, in this equation. In consequence nuclear magnetic dipole moments are generally smaller than atomic magnetic dipole moments by a factor of about 2000.

Nuclear spectra

Nuclei, like atoms, emit photons with characteristic spectra, although, in the nuclear case, the wavelengths of the photons lie in the γ-ray, rather than the visible or ultraviolet, region of the electromagnetic spectrum. This provides clear evidence that nuclei, like atoms, possess a large range of quantised energy states. A photon is emitted when a nucleus makes a transition between excited energy states or from an excited state to the ground state, a process which we discuss further in Section 27.5.

27.2 Nuclear binding energies

We noted in the previous section that the mass of a nucleus is significantly smaller than the sum of the masses of its constituent nucleons, when they are free, and attributed the reduction in mass to the negative contribution of nuclear binding energy to the mass of the bound system. The amount by which a nuclear mass is less than the sum of the masses of its constituent nucleons is therefore a measure of the binding energy of the nucleus and provides important information on nuclear structure.

Measurements of the mass loss effect are summarised in Figure 27.2 in which nuclear mass per constituent nucleon, measured in MeV/c^2 units, is plotted against the number of nucleons A. If nuclear binding had been insignificant, this plot would simply have been that of the average free nucleon mass $m_{av} = \dfrac{Zm_p + Nm_n}{A}$

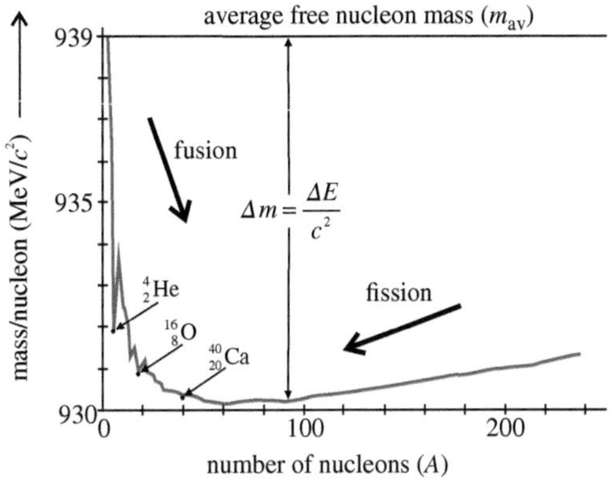

Figure 27.2. A plot of nuclear mass per constituent nucleon against number of nucleons (given by the mass number A).

where m_p is the proton mass (938.3 MeV/c^2), m_n is the neutron mass (939.6 MeV/c^2) and N is the number of neutrons. A plot of m_{av} against A is nearly a straight line parallel to the A axis, as indicated at the top of the figure. Slight deviations from this straight line occur because the relative proportions of protons and neutrons in nuclei vary with A. For our present purposes, however, the plot can be considered to be a straight line.

Using Equation (9.56), $\Delta E = (\Delta m)c^2$, the nuclear binding energy per nucleon, ΔE (a negative energy) for any nucleus may be obtained from Δm, the mass per nucleon for that nucleus minus the average free nucleon mass, as indicated in Figure 27.2.

Figure 27.2 shows that the mass per nucleon is roughly the same for most nuclei. The binding energy per nucleon is therefore also fairly constant (between −7 and −9 MeV) for most nuclei. Note, however, that certain nuclei, for example $^{4}_{2}$He, $^{16}_{8}$O and $^{40}_{20}$Ca, are more stable (that is they have higher binding energies) than their neighbours in the plot[1].

[1]Note that in many books binding energy per nucleon is plotted against A as a positive quantity. Such a plot is therefore an inversion of the plot of Figure 27.2 about the average free nucleon mass line.

For light (low A) nuclei, the mass per nucleon plot rises sharply as A decreases. There is also an increase in mass per nucleon, and hence a decrease in binding energy per nucleon, for heavy (high A) nuclei as A increases. The nuclei of elements of medium A are the most strongly bound and therefore the most stable. Maximum binding energy (the minimum of the mass per nucleon plot of Figure 27.2) occurs in the region $A \approx 55 - 60$, corresponding to nuclides of iron. We can thus identify two general types of nuclear process which are 'downhill' in energy terms in Figure 27.2 and can therefore release energy. These are **nuclear fusion,** in which the nucleons of light nuclides rearrange (that is fuse together) to form nuclei of medium A, and **nuclear fission,** in which the heavy nuclei break up, with their nucleons rearranging to form nuclei of medium A. Both processes release very large amounts of energy and may be used to generate energy, which is usually referred to as nuclear energy. Fission and fusion are discussed in more detail in Sections 27.8 to 27.10.

In the next section we attempt to interpret, and hopefully explain, the nuclear properties outlined above. This will be achieved through the construction of models, essentially through comparisons of the behaviour and properties of nuclei with those of more familiar systems which we can visualise and analyse more easily.

For problems based on the material presented in this section visit up.ucc.ie/27/ and follow the link to the problems

---

27.3 Nuclear models

The liquid drop model

The liquid drop model attempts to account for the energy per nucleon plot of Figure 27.2 by adding a series of corrective terms to the average free nucleon mass (shown at the top of the figure). The use of the liquid drop model is suggested by two of the nuclear properties which we noted above, namely the tendency for the nucleon density to be the same in all nuclei and the tendency for the mass per nucleon, and hence the binding energy per nucleon, to be approximately the same for most nuclei.

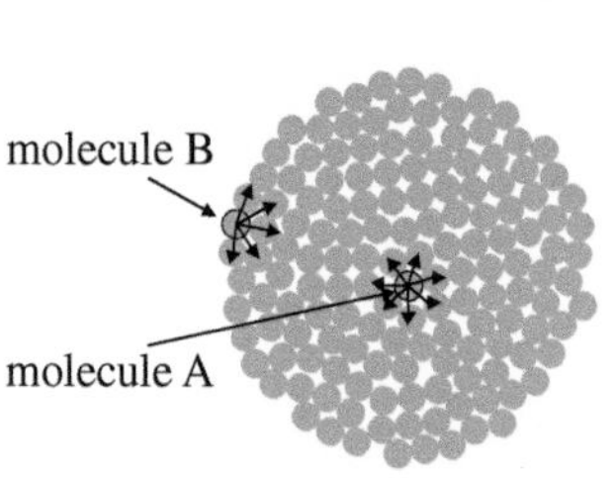

Figure 27.3. A liquid drop, showing the forces acting on a molecule (A) in the bulk of the liquid and on a molecule (B) at the surface of the drop.

These properties are also characteristic of molecules in a classical liquid drop. Molecular density in a drop, like nucleon density in a nucleus, is uniform throughout the drop. Each molecule in the bulk of the liquid is held in the liquid by attractive interactions with its nearest neighbours. As illustrated for molecule A in Figure 27.3, such forces act equally in all directions and are short range, as is the strong nuclear force. They are not affected by the addition of further molecules to the outside of the drop. Each molecule in the bulk of the drop therefore experiences the same binding energy, as do nucleons in a nucleus.

Applying this model to the nucleus leads to a (negative) binding energy which is proportional to the volume of the nucleus, that is proportional to R^3 and thus, from Equation (27.1) to the number of nucleons, A. The binding energy per nucleon produced by this effect is therefore proportional to $\frac{A}{A}$, that is, it is independent of A, and, as shown in Figure 27.4, the addition of such a term – the *volume term* – to the average free nucleon line reduces the total energy by a constant amount.

In a classical liquid drop, we need to take account of the fact that molecules at the surface of the drop, such as molecule B in Figure 27.3, do not have neighbours on all sides and are therefore less strongly bound. This effect reduces the total binding energy, making it less negative, and thus makes a positive contribution (which is proportional to the surface area of the drop, $4\pi R^2$) to the total energy. In the nuclear case it follows, from Equation (27.1), that this positive energy contribution is proportional to $\left(A^{\frac{1}{3}}\right)^2 = A^{\frac{2}{3}}$. When this term – the *surface term* – is added as a binding energy per nucleon term, proportional to $+\frac{A^{\frac{2}{3}}}{A} = +\frac{1}{A^{\frac{1}{3}}}$, to the average free nucleon plot of Figure 27.4, it produces a substantial rise in energy for low values of A.

Unlike a classical liquid drop, the nucleus has an electric charge, namely $+ Ze$. We must therefore add a further classical term – the *Coulomb term* – to represent the positive energy due to Coulomb repulsion between all the protons in the drop. The Coulomb energy is the energy released when the entire nuclear charge is dispersed to infinity. It is a 'self-energy' like gravitational 'self-energy' (Section 5.7), the energy needed to disperse the mass in a uniform sphere to infinity, although, in the nuclear case, the self-energy is repulsive. In Worked Example 5.2, we showed that the gravitational self-energy of a uniform sphere, of mass M and radius R, is proportional to $-\frac{M^2}{R}$. By direct analogy, the nuclear Coulomb self-energy is proportional to $+\frac{(Ze)^2}{R}$, that is to $+\frac{Z^2}{R}$ or, using Equation (27.1), to $+\frac{Z^2}{A^{\frac{1}{3}}}$. The Coulomb term becomes important for high Z, and therefore high A, nuclei so that addition of this term, a mass per nucleon term which is proportional to $+\frac{Z^2}{AA^{\frac{1}{3}}} = +\frac{Z^2}{A^{\frac{4}{3}}}$ produces a rise in the plot at high values of A, as indicated in Figure 27.4.

The Coulomb term explains the tendency of high Z nuclei to contain an excess of neutrons and thus their departure from the $N = Z$ line, as shown in Figure 27.1. For very small (low Z) nuclei, each proton which is added to a nucleus interacts attractively with all other nucleons, through the strong nuclear force (Section 1.4), and repulsively with all protons, through the much weaker Coulomb force. For

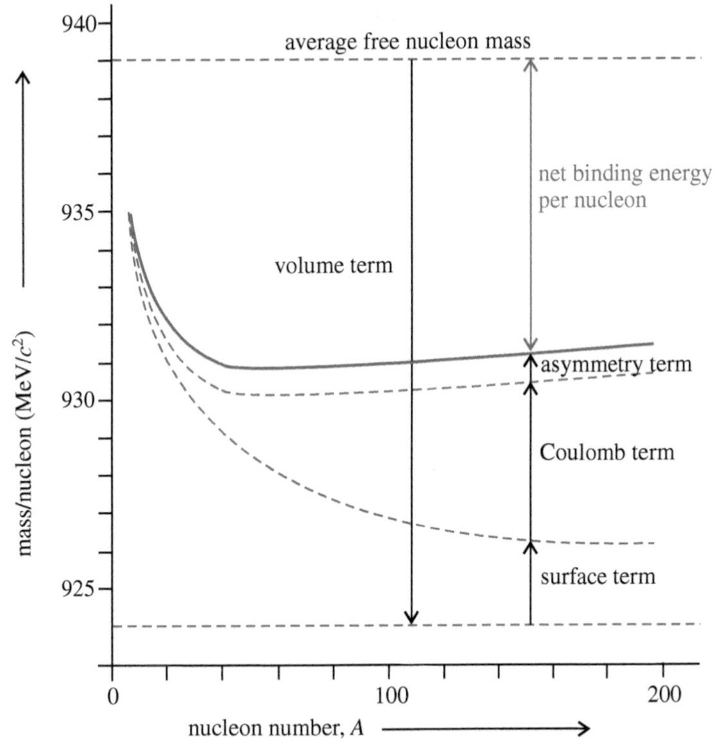

Figure 27.4. The liquid drop model. A plot of nuclear mass per constituent nucleon against the nucleon number, showing how the experimental plot (Figure 27.2) can be reproduced by adding successive corrective terms, summarised in formula (27.2), to the average free nucleon mass.

higher Z nuclei however, because the nuclear force is short range, additional protons interact attractively with only a fixed number of other nucleons but interact repulsively with all other protons through the long range Coulomb force, producing a repulsive energy which, as noted above, increases in proportion to Z^2. As Z increases therefore, it is more favourable to add neutrons rather than protons and thus, typically, $N \approx 0.6A$ for high Z nuclei. Eventually (when $Z > 92$) the positive contribution of the Coulomb term exceeds the negative contribution of the strong nuclear force and stable nuclei can no longer form.

Two further empirical terms are needed in the liquid drop model to obtain good agreement between the plot of Figure 27.4 and the observed mass per nucleon plot of Figure 27.2. These terms, called the *asymmetry* and *pairing* terms, have no analogy in the classical liquid drop because they originate in the quantum mechanical behaviour of the nucleons in the nucleus.

The *asymmetry* term is proportional to $+\dfrac{\left[Z-\frac{A}{2}\right]^2}{A}$ and is introduced empirically to account for the observed tendency of stable nuclides to contain equal numbers of protons and neutrons. When $N = Z$, and thus $Z = \dfrac{A}{2}$, this term is zero but as Z increases and $N - Z$, and hence $\left(\dfrac{A}{2} - Z\right)$, increases, this term makes an increasing positive contribution, as shown in Figure 27.4.

The *pairing* term is proportional to $+\ \alpha A^{-\frac{1}{2}}$, where α can take the values 0, ±1, accounts empirically for the observed tendency, which we have noted in Section 27.1, for Z and N to be even. Thus $\alpha = -\ 1$ (increased binding) when both Z and N are even, $\alpha = 0$ (no change) when one of N or Z is even and the other odd and $\alpha = +\ 1$ (decreased binding) when both N and Z are odd.

The liquid drop model is summed up in the **semi-empirical mass formula** (also known as the Bethe-Weizsacker formula after Hans Bethe (1906–2005) and Carl von Weizsacker (1912–2007)), which gives the mass of any nucleus in terms of Z and A

$$M_{Z,A} = Zm_p + (A-Z)m_n - a_1A + a_2A^{\frac{2}{3}} + a_3\frac{Z^2}{A^{\frac{1}{3}}} + a_4\frac{\left[Z-\frac{A}{2}\right]^2}{A} + \begin{Bmatrix} -1 \text{ (for both } Z \text{ and } N \text{ even)} \\ 0 \text{ (for one only of } N \text{ or } Z \text{ even)} \\ +1 \text{ (for both } Z \text{ and } N \text{ even)} \end{Bmatrix} a_5A^{-\frac{1}{2}} \tag{27.2}$$

where m_p and m_n are the proton and neutron masses, respectively, and the parameters a_1, a_2, a_3, a_4 and a_5 are chosen empirically to give the best fit to the observed mass per nucleon curve. The values of these parameters, as listed below, give $M_{Z,A}$ in atomic mass units, u (Section 10.12) (where 1 u = 1.661 × 10^{-27} kg = 931.5 MeV/c^2).

$$a_1 = 0.01691 \text{ u},\ a_2 = 0.01911 \text{ u},\ a_3 = 0.000763 \text{ u},\ a_4 = 0.10175 \text{ u and } a_5 = 0.012 \text{ u}.$$

The formula is very useful in predicting the values of nuclear masses. As we shall see in Worked Example 27.3, the masses of the particles taking part in a nuclear decay or reaction are very important in determining whether the process is viable in energy terms.

The shell (or independent particle) model: the Fermi gas model

When we discussed the plot of stable nuclei (Figure 27.1) in Section 27.1, we noted that a large number of stable nuclides occur for particular values of N and/or Z. Moreover these same values of N and Z tend to produce nuclei with unusually large binding energies, examples of which, as shown in Figure 27.2, are ${}^{4}_{2}\mathrm{He}$ $(N = Z = 2)$, ${}^{16}_{8}\mathrm{O}$ $(N = Z = 8)$ and ${}^{40}_{20}\mathrm{Ca}$ $(N = Z = 20)$. These values of N and Z, namely 2, 8, 20, 28, 50, 82 and 126, are known as *magic numbers* and correspond to particularly stable groupings of nucleons. Nuclides for which both N and Z are magic, such as ${}^{4}_{2}\mathrm{He}$, ${}^{16}_{8}\mathrm{O}$, ${}^{40}_{20}\mathrm{Ca}$ and ${}^{208}_{126}\mathrm{Pb}$ $(N = 126,\ Z = 82)$ are known as *doubly magic* and are exceptionally stable.

This situation is reminiscent of closed shells of electrons in atoms where, for certain values of Z, namely 2, 10, 18, 36, 54 and 86, corresponding to the inert gases (group 0 atoms of Section 24.13), electronic structures are exceptionally stable. The comparison suggests a shell model for nuclei. The similarity with the atomic model is confirmed by experiments in which the energy needed to remove the least strongly bound nucleon from a nucleus is determined. It is found that the last nucleon is most easily removed when the N or Z value of the nucleus is one more than a magic number, for example ${}^{17}_{8}\mathrm{O}$ $(N = 8 + 1)$, ${}^{87}_{36}\mathrm{Kr}$ $(N = 50 + 1)$. These nuclei are the nuclear equivalents of the alkali atoms (group Ia atoms of Section 24.13).

The atomic model, which we outlined for multielectron atoms in Section 24.14, is an independent particle model. Electrons move independently in the central field which is produced by the nucleus and the average effect of other electrons. Interactions with individual electrons are treated as secondary effects. At first sight this picture, if adopted for a nucleus, would seem to be in direct contradiction to the picture presented by the liquid drop model, in which nucleons interact strongly with their nearest neighbours. The two models can be reconciled, however, if we remember that the protons and the neutrons in a nucleus are systems of fermions which are subject to the Pauli exclusion principle. The systems of protons and neutrons comprise two independent Fermi gases, as discussed in Sections 25.3 and 25.4. When two identical fermions interact in a Fermi gas they cannot change their energy states because there are no empty states available to receive them unless they are close to the Fermi energy. The only available mechanism for energy change is exchange of states. However, exchange of identical particles produces no observable effect. Consequently, although we would not be justified in assuming that nucleons in a nucleus do not interact with one another, we can reasonably assume that, because such interactions have no observable effects, the nucleons behave *as though* they are independent particles.

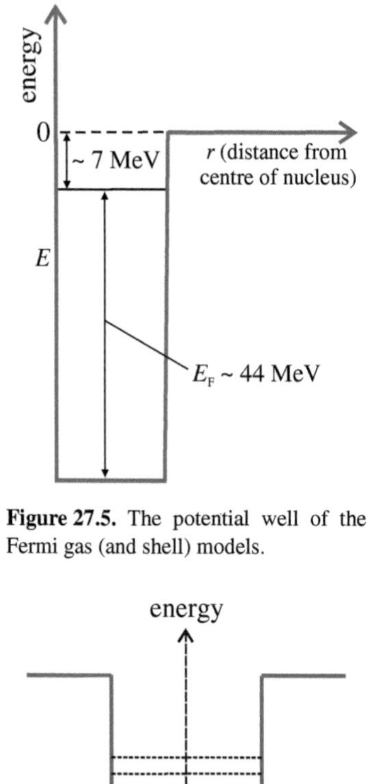

Figure 27.5. The potential well of the Fermi gas (and shell) models.

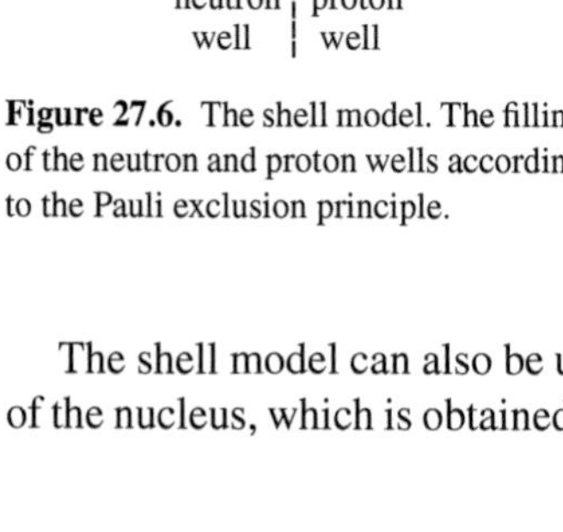

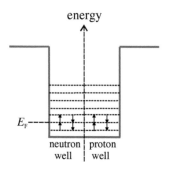

Figure 27.6. The shell model. The filling of the neutron and proton wells according to the Pauli exclusion principle.

As shown in Worked Example 27.1, by estimating the neutron or proton density in a nucleus we can use the **Fermi gas model** (Equation (25.18)) to estimate the Fermi energy of the neutrons or protons in a nucleus. A value of about 44 MeV is obtained. As illustrated in Figure 27.5, given that the binding energy of most nucleons is about 7 MeV (as can be deduced from the mass per nucleon plot, Figure 27.2), this estimate indicates the depth of the nuclear potential to be about 50 MeV. This estimate of the well depth proves very useful in quantitative applications of the shell model.

The Fermi gas model also explains why nuclei tend to contain equal numbers of protons and neutrons when Coulomb effects are negligible, as they are in light nuclei. The strong nuclear force acts in exactly the same way for protons and neutrons. The potential wells for protons and neutrons are therefore identical but independent as far as the Pauli exclusion principle is concerned. In each well, levels are filled until the Fermi energy is reached, as illustrated in Figure 27.6, thus leading to equal numbers of protons and neutrons. Note that each level of the proton or neutron well can accommodate two particles, because the intrinsic spin $\left(s = \frac{1}{2}\right)$ of a proton or neutron, like that of an electron in a multi-electron atom (Section 24.13) or in a conductor (Section 25.3), can be in either of two S_z-states, as indicated by the 'up' and 'down' arrows in Figure 27.6.

The shell model of the nucleus therefore assumes that the protons and neutrons form two separate systems of fermions which behave as independent particles in a spherically symmetric central field which is due to the average effect of strong nuclear interactions with all other nucleons. The net nuclear potential, $U(r)$, for this short range force is assumed to approximate to a square well, as illustrated in Figures 27.5 and 27.6. Apart from the shape of the potential energy function, the model is similar to that used for the multi-electron atom and is analysed by solving the Schrödinger equation for this potential. The analysis is, however, necessarily more approximate than that used in atomic physics because the nuclear potential, unlike the atomic potential (Section 24.14), cannot be deduced from self-consistent field theory.

By trying many different forms of $U(r)$ and by introducing further properties of the strong nuclear force which are suggested by comparisons between calculated and observed energy levels, in particular interaction between the orbital and spin angular momenta of nucleons, an energy level diagram can be calculated for nuclei in their ground states. As in atoms, the levels are filled in order of increasing energy according to the Pauli exclusion principle. The observed values of magic numbers can be predicted from such a diagram. They correspond to the total occupancy of the nucleus in cases in which there is an exceptionally large gap between the filled energy levels and the next higher energy level, as in inert gas atoms (Section 24.13).

The shell model can also be used to predict the spins of nuclei in their ground states. The *nuclear spin* is the total angular momentum of the nucleus, which is obtained, as in the multi-electron atom, by adding the angular momenta of both protons and neutrons as predicted by the shell model, according to rules for adding angular momenta which are specified by quantum mechanics. The use of the name nuclear spin to describe the total angular momentum of a nucleus is, perhaps, misleading because nuclear spin includes both the spin and orbital angular momenta of the nucleons.

We would also expect the shell model to predict nuclear magnetic dipole moments accurately because they should be directly related to nuclear spins through Equation (24.31). Predictions of nuclear magnetic dipole moments using the shell model are, however, very poor, indicating that we are encountering a limitation of the shell model. In the shell model it is assumed that, as in the atom, the nuclear spins and magnetic dipole moments of nucleons with opposite values of these quantities cancel one another out – they *pair off* to produce zero net effect, not unlike the pairing of electrons in covalent bonding (Section 25.1) or in Cooper pairs (Section 25.8). Departures from this assumption do not matter in the case of nuclear spin because this quantity is quantised. When paired nucleons have non-zero total spin, any unpaired nucleon in the nucleus must assume an angular momentum which produces a quantised value of the nuclear spin, as predicted by the shell model. However the nuclear magnetic dipole moment is not quantised. To predict nuclear magnetic dipole moments, we need a development of the shell model known as the collective model.

A more detailed description of the shell model and of its results and applications may be found in Section 2.3 of Lilley (2001).

The collective model

In the collective model, as in the shell model, it is assumed that nucleons in unfilled shells move independently in the net nuclear potential which is produced by the core of filled subshells. However, this potential is not spherically symmetric; it can undergo deformations which represent the collective (correlated) motion of the core nucleons, a phenomenon which is associated more with the liquid drop model. Consider, for example, a nucleon which is moving in a circular orbit of large radius close to the surface of a nucleus. As it moves, it distorts the core through strong interactions with the core nucleons. Bulges follow it, leading to a distortion of the potential which greatly complicates the solution of the Schrödinger equation but enables many more phenomena, such as the observed magnetic dipole moments, to be described accurately.

Nuclear physics provides a good example of the use of models in physics. Some simple models, which are proposed following comparisons with more familiar situations encountered in other areas of physics, can be useful in interpreting certain properties of the system and can be used to make predictions. However, they must not be considered to represent the system fully and are limited in their range of application. More sophisticated models give a more comprehensive account of the properties of the system but they too must be expected to have their limitations

---

Study Worked Example 27.1

For problems based on the material presented in this section visit up.ucc.ie/27/ *and follow the link to the problems*

27.4 Radioactivity

As noted in Section 27.1, most nuclei are unstable. They decay by emitting a particle spontaneously, thus producing in the process another nucleus, known as a *residual* or *daughter nucleus*. The initial nucleus is often called the *parent* nucleus. The radioactive decay process is random, depending, as we shall see in the next section, on quantum mechanical processes such as tunnelling. It is not possible to predict exactly when a particular nucleus will decay. We can only describe the process statistically.

The statistical nature of the decay process is described by the **radioactive decay law** which states that, if a sample contains N radioactive nuclei, the rate at which the nuclei decay, the *activity*, $-\frac{dN}{dt}$, is proportional to N. The minus sign indicates that there is a decrease in the number of nuclei with time. Thus

$$-\frac{dN}{dt} = \lambda N \tag{27.3}$$

where λ, the *disintegration constant*, is characteristic of the nucleus concerned. It is determined by nuclear processes which we discuss in the next section.

We can write Equation (27.3) as $\frac{dN}{N} = -\lambda dt$

If there are N_0 nuclei at $t = 0$, integration of this equation gives

$$\int_{N_0}^{N} \frac{dN}{N} = -\int_0^t \lambda dt$$

$$\ln N - \ln N_0 = -\lambda t$$

Thus

$$\boxed{N = N_0 e^{-\lambda t}} \tag{27.4}$$

Equation (27.4), known as the **radioactive decay equation**, gives N, the number of nuclei remaining undecayed at any time t. Radioactive decay therefore follows an exponential curve, as illustrated in Figure 27.7.

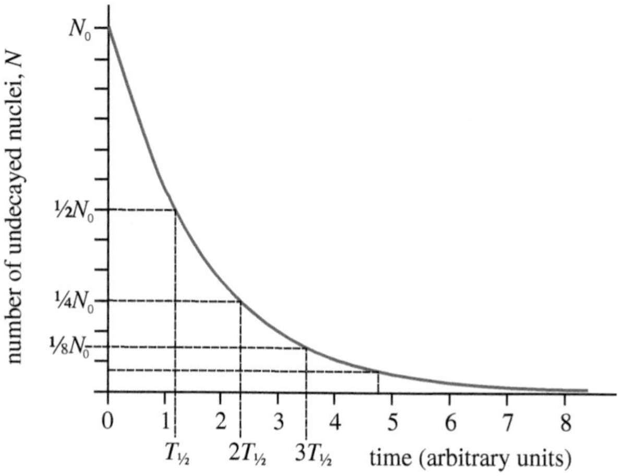

Figure 27.7. The exponential decay of radioactive nuclei.

The decay of a radionuclide can be characterised by its **half-life**, $T_{\frac{1}{2}}$. The half-life is the time taken for half the nuclei in any sample to decay, that is $N = \frac{N_0}{2}$ when $t = T_{\frac{1}{2}}$. Substitution in Equation (27.4) therefore yields

$$\frac{1}{2} = e^{-\lambda T_{\frac{1}{2}}}$$

and thus

$$\boxed{T_{\frac{1}{2}} = \frac{\ln 2}{\lambda} = \frac{0.693}{\lambda}} \tag{27.5}$$

The decay of a radionuclide may also be characterised by its **mean lifetime**, T_m, the average time a radionuclide survives before decay. Using Equation (27.4) we can show how T_m is related to $T_{\frac{1}{2}}$,

$$T_m = \frac{\text{sum of all lifetimes}}{\text{number of radionuclides}} = \frac{\int_0^\infty N(t)dt}{N_0} = \int_0^\infty e^{-\lambda t} dt = -\frac{1}{\lambda}[e^{-\lambda t}]_0^\infty = -\frac{1}{\lambda}[0-1] = \frac{1}{\lambda} = \frac{T_{\frac{1}{2}}}{0.693}$$

Carbon dating

The age of an object can often be estimated if it contains a radionuclide with a known half-life. The radionuclide ${}^{14}_{6}C$ with a half-life of 5730 years is frequently used for this purpose, a procedure known as **carbon dating**.

${}^{14}_{6}C$ is produced in the upper atmosphere of the Earth as a result of the bombardment of atmospheric nitrogen by cosmic rays. In the atmosphere the ${}^{14}_{6}C$ produced mixes with the stable isotope of carbon ${}^{12}_{6}C$ in the form of CO_2 molecules. Generally, one CO_2 molecule in 10^{13} contains a ${}^{14}_{6}C$ nucleus. Living organisms, such as plants and animals, absorb carbon from the atmosphere and therefore contain this proportion of ${}^{14}_{6}C$. When a living organism dies, however, it no longer absorbs carbon and the proportion of ${}^{14}_{6}C$ decays according to the radioactive decay law. The proportion of ${}^{14}_{6}C$ to ${}^{12}_{6}C$ in a sample of wood, for example, can be used to estimate when the tree, from which the wood was obtained, died and, thus, can be used to date the wood. Care must be taken in using this technique. For example, it is not reasonable to assume that the rate at which ${}^{14}_{6}C$ is produced has always been the same, but corrections can be made for this effect. Carbon dating has proved a very valuable dating technique for archaeologists.

There are three principal types of natural radioactive decay processes, α-, β- and γ- decay, which are named according to the particles emitted in the decays. We describe each process in more detail in the next section.

For problems based on the material presented in this section visit up.ucc.ie/27/ *and follow the link to the problems*

27.5 α-, β- and γ-decay

α-decay

α-decay occurs when a nucleus emits a ${}^{4}_{2}He$ nucleus (an α- particle). An example of such a decay is

$${}^{238}_{92}U \rightarrow {}^{234}_{90}Th + {}^{4}_{2}He$$

If the total mass of the product particles (the thorium nucleus and the α-particle) is less than the mass of the parent nucleus, the process is viable in energy terms, that is there is a net release of kinetic energy. The mass difference Δm, which may be calculated using the semi-empirical mass formula (Equation (27.2)), can be used to calculate the *decay energy*, $E = (\Delta m)c^2$, of the α-decay process. As shown in Section 6.4, most of the decay energy is released as kinetic energy of the lighter particle, the α-particle in the case of α-decay. The decay energy of the α-particle is always the same when a given radionuclide decays to a particular state of the daughter nucleus. In the case of the decay of $^{238}_{92}U$ it is 4.25 MeV. As shown in Section 6.4, this result is characteristic of a process in which the initial nucleus splits into only two fragments (a two-body decay). In such a case the laws of conservation of momentum and energy require that the decay energy be divided between the product particles in a unique way.

The potential barrier seen by an α-particle within or near a nucleus is illustrated in Figure 27.8. Within the nucleus the attractive potential due to the strong nuclear force, as discussed in the description of the shell model in Section 27.3, dominates. Outside the nucleus, this short range force disappears rapidly and the positive potential due to Coulomb repulsion between the positively charged α-particle and the residual nucleus takes over. The net potential due to these two effects is thus a barrier. Classically the α-particle cannot escape but, as we have already noted in Section 14.17, quantum mechanically the α–particle can tunnel through the barrier.

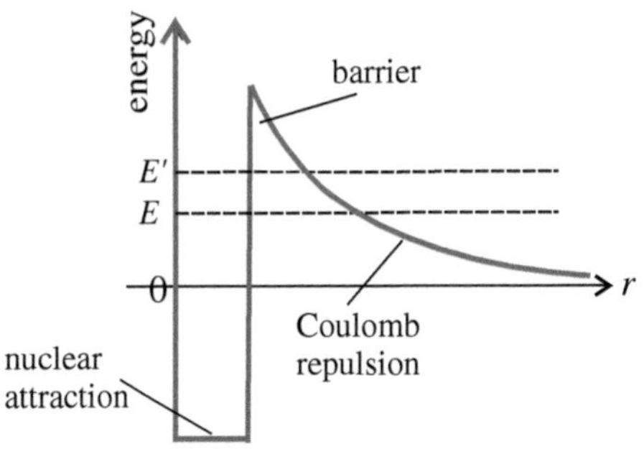

Figure 27.8. α-decay. The tunnelling of an α-particle through the potential barrier at the edge of a nucleus.

The probability that an α-particle will escape from a nucleus is governed therefore by the probability of tunnelling through the barrier, a quantum process which can only be treated statistically. The quantum mechanical treatment of α-particle decay in 1928 by Gamow (George Gamow, 1904–1968), Condon and Turley was one of the most convincing quantitative demonstrations of the need for quantum mechanics.

As illustrated in Figure 27.8, the higher the decay energy, E, the smaller the barrier seen by the α-particle, the greater the probability of decay and hence the shorter the half-life. The spread of α-decay energies of different nuclei, and hence their half-lives, is immense; measured half-lives range from 3×10^{-7} s to 1.4×10^{17} s (4.4×10^{9} years).

The radionuclides $^{232}_{90}Th$, $^{238}_{92}U$ and $^{235}_{92}U$ have very low decay energies and hence very long half-lives. Their half-lives, which are listed in Table 27.1, are comparable to the age of the Earth ($\sim 10^{10}$ years) so that significant quantities of these radionuclides, which were present when the Earth was formed, are still present on the Earth today.

Table 27.1 Radioactive series.

Series	Parent	Half-life of parent/years	End product
4n	$^{232}_{90}Th$	1.41×10^{10}	$^{208}_{82}Pb$
4n + 1	$^{237}_{93}Np$	2.14×10^{6}	$^{209}_{83}Bi$
4n + 2	$^{238}_{92}U$	4.51×10^{9}	$^{206}_{82}Pb$
4n + 3	$^{235}_{92}U$	7.13×10^{8}	$^{207}_{82}Pb$

$^{232}_{90}Th$, $^{238}_{92}U$ and $^{235}_{92}U$ decay to other radionuclides which, in turn, decay, forming a sequence of decays known as a *radioactive series*. These series are the origin of much of the radioactivity which occurs naturally on the Earth today. As indicated in Worked Example 27.2, the relative abundances of $^{238}_{92}U$ and $^{235}_{92}U$ can be used to estimate the age of the Earth. The radioactive series finally terminates when a stable nuclide is reached, often an isotope of lead (for which $Z = 82$, a magic number). The nuclide $^{208}_{82}Pb$, with $N = 126$, is doubly magic and thus exceptionally stable, as noted in Section 27.3. The $^{238}_{92}U$ series is shown in Figure 27.9.

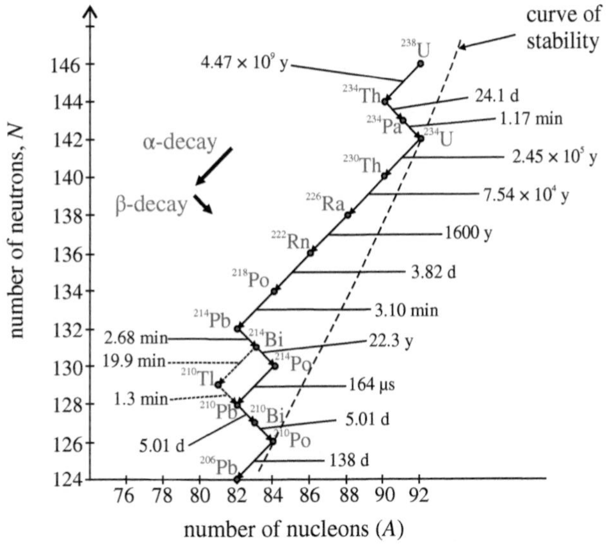

Figure 27.9. The $^{238}_{92}U$ radioactive series ($A = 4n + 2$).

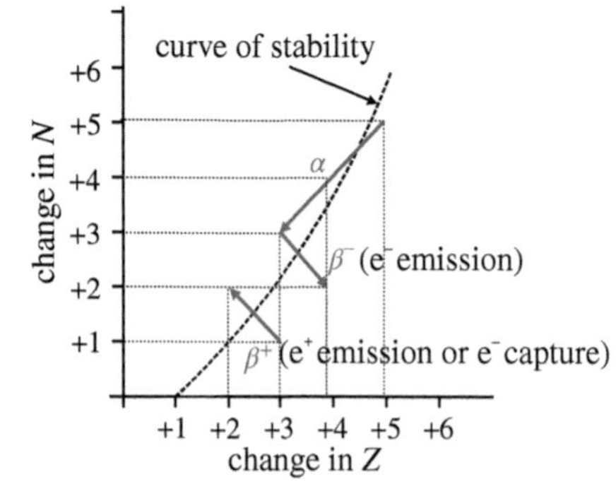

Figure 27.10. The changes in N and Z produced by α- or β-decay processes. β-decay can return a radioactive series to the curve of stability after it has been taken away from the curve by α-decay.

Each α-decay in a series reduces the mass number A by 4. Thus, as indicated in the first column of Table 27.1, the A values of the members of a particular series can be described by a simple formula, for example $(4n + 2)$ for the $^{238}_{92}$U series, where $n = 59$ for the first member of the series, $n = 58$ for the second member, $^{234}_{90}$Th, etc. Note that a fourth series, the $^{237}_{93}$Np series, is also included in Table 27.1. The half-life of $^{237}_{93}$Np is substantially less than the age of the Earth so that radionuclides of this series do not occur naturally on the Earth today.

The daughter nucleus which is formed following α-decay has N and Z values which are two less than that of the parent nucleus. As indicated in Figure 27.10 this trend, $(Z, A) \to (Z - 2, A - 4)$, if maintained, takes the series away from the curve of stability of Figure 27.1.

β-decay

In β-decay an electron is emitted, thus increasing the positive charge of the nucleus by $+e$, while keeping the total number of nucleons A the same. In effect a neutron is converted into a proton so that the Z and A values change as follows in a β-decay, $(Z, A) \to (Z + 1, A)$.

β-decay, therefore, takes the sequence back towards the curve of stability as illustrated in Figures 27.9 and 27.10. Note that, in some cases, such as $^{214}_{83}$Bi in Figure 27.9, alternative α- or β-decay mechanisms are available; some nuclei decay by α-emission, others by β-emission and the series branches. The total decay rate of the nuclide $\frac{dN}{dt}$ is then given by the sum of the decay rates of the two mechanisms

$$\frac{dN}{dt} = \frac{dN_1}{dt} + \frac{dN_2}{dt}$$

which, using Equation (27.3), can be written in terms of the total disintegration constant λ and the disintegration constants for the two mechanisms, λ_1 and λ_2,

$$-\lambda N = -\lambda_1 N - \lambda_2 N$$

Thus $\lambda = \lambda_1 + \lambda_2$ and the half-life of the nuclide is given by a generalisation of Equation (27.5), namely

$$T_{\frac{1}{2}} = \frac{\ln 2}{\lambda_1 + \lambda_2}$$

This equation can be extended easily to cover three or more alternative decay mechanisms.

In addition to the β-decay process, in which an electron (e^-) is emitted, two further types of β-decay can occur, *positron emission* and *electron capture*. In positron emission a positron (e^+), a particle which is identical to the electron except that it carries a positive electric charge, $+e$, is emitted. In this case the positive charge of the nucleus decreases by $+e$, while keeping the total number of nucleons A the same. In electron capture an atomic electron is absorbed into the nucleus, reducing its charge by one. The process is most likely to occur when the eigenfunction, and thus the probability density, of the atomic electron has a significant value in the region of the nucleus. 1s shell electrons are therefore most likely to be captured. For both positron emission and electron capture, Z and A values change as follows, $(Z, A) \to (Z - 1, A)$. The discussion of β-decay which follows may be applied to all three types of β-decay.

The condition for β-decay to occur is simply that, as for α-decay, the total mass of the product particles must be less than that of the initial particles. Again, the semi-empirical mass formula (Equation (27.2)) may be used to predict whether or not a certain β-decay can occur.

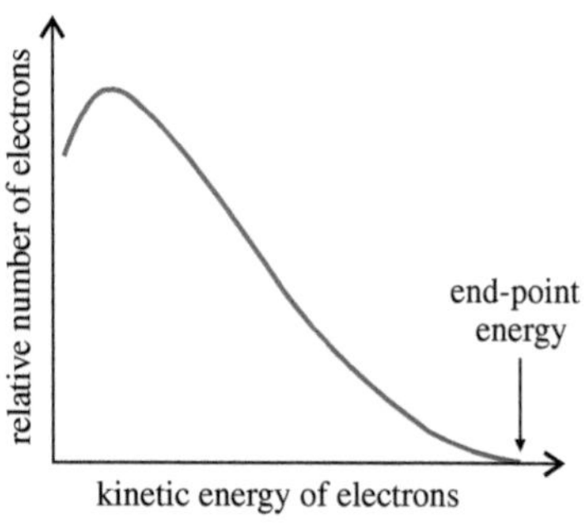

Figure 27.11. A β-decay energy spectrum, showing the end-point energy.

One of the most notable features of β-decay is that the β-particles produced, e^- or e^+, exhibit a continuous spectrum of energies, as illustrated in Figure 27.11. The maximum energy, the *end-point energy*, is found to be equal to the reaction energy, as given by the mass difference.

The continuous spectrum of the β-particles cannot be produced by decay into two fragments, since, as noted in Section 6.4, two body decays can produce only a single discrete energy (as in α-decay). The clear implication is that more than two particles are produced in the β-decay process. The third particle which is produced in β-decay is a hitherto unknown particle, postulated by Pauli in 1930. This is the *neutrino,* ν_e, in the case of positron emission and electron capture, and the *antineutrino*, $\overline{\nu_e}$, in the cases of electron emission. The antineutrino is an example of an antiparticle, to be discussed in Section 27.11.

The full β-decay processes should therefore be written

Electron emission $\quad {}^A_Z\mathrm{X} \to {}^{\;\;A}_{Z+1}\mathrm{X}' + e^- + \overline{\nu_e} \quad$ for example $\quad {}^1_0\mathrm{n} \to {}^1_1\mathrm{H} + e^- + \overline{\nu_e}$, or the decay of ${}^{14}_{6}C$, used in carbon dating (Section 27.4), for example ${}^{14}_{6}C \rightarrow {}^{14}_{7}N + e^{-} + \overline{\nu_e}$

$$\text{Positron emission} \quad {}^{A}_{Z}X \rightarrow {}_{Z-1}{}^{A}X' + e^{+} + \nu_e \quad \text{for example} \quad {}^{11}_{6}C \rightarrow {}^{11}_{5}B + e^{+} + \nu_e$$

$$\text{Electron capture} \quad {}^{A}_{Z}X + e^{-} \rightarrow {}_{Z-1}{}^{A}X' + \nu_e \quad \text{for example} \quad {}^{64}_{29}Cu + e^{-} \rightarrow {}^{64}_{28}Ni + \nu_e$$

The inclusion of the neutrino in the process also solves a difficulty which emerged in explaining the observed nuclear spins of the initial and residual nuclei. In neutron decay, the example of electron emission which is given above, the total angular momentum before decay is determined by the spin of the neutron $\left(s = \frac{1}{2}\right)$. According to the quantum mechanical rules for adding angular momenta, the addition of the half integral spins of the proton (${}^{1}_{1}H$) and the electron can only give integral spin so that, without a $\overline{\nu_e}$, with half-integral spin, angular momentum could not be conserved in the decay. The neutrino has half-integral spin and is thus a fermion. It is also clear that the neutrino must have zero charge (to ensure that charge is conserved in the decay) and zero mass, or at least negligible mass, to explain cases in which the β-particle is observed to take all the decay energy, that is at the end-point energy of the β-decay spectrum of Figure 27.11. At the end-point there is no energy available to produce even a stationary neutrino of significant mass.

A free neutron is unstable and decays by β-decay ${}^{1}_{0}n \rightarrow {}^{1}_{1}H + e^{-} + \overline{\nu_e}$, with a mean lifetime of 881 s. Neutrons within stable nuclei do not decay because, in this case, the conversion of a neutron to a proton would increase the total rest energy of the nucleus by more than the energy provided by the β-decay. The process is not viable therefore in energy terms in a stable nucleus.

β-decay does not take place through the strong nuclear force but through a second type of nuclear force the called the *weak nuclear force* (Section 1.4). Observation of β-decay enables the properties of the weak force, such as its strength and range, to be estimated. The weak force is weaker than the strong force by a factor of 10^{13} and is therefore negligible in situations in which either the strong force or the electromagnetic force is also present. Neutrinos can interact with matter only through the weak interaction and are consequently very difficult to detect. They were eventually detected by Clyde Cowan (1919–1974) and Frederick Reines (1918–1998) in 1956, more than 20 years after they had been postulated and more than 60 years after β-decay had been first observed.

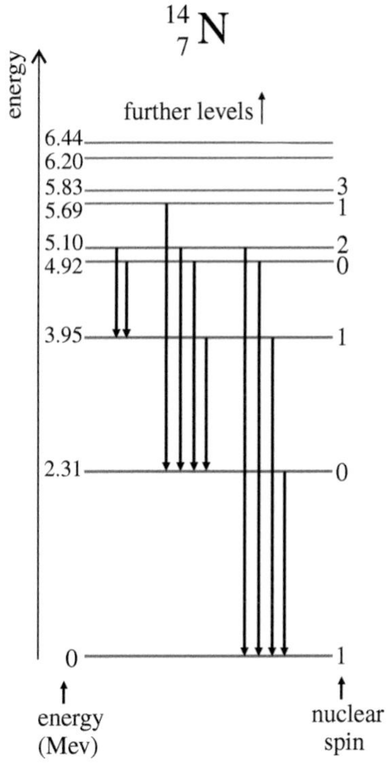

Figure 27.12. Some energy levels of the ${}^{14}_{7}N$ nucleus, indicating γ-decay transitions.

The probability of a β-decay interaction taking place, and thus the half-life of nuclei which decay through the β-decay interaction, is determined by the nuclear eigenfunctions of the initial and final states in a manner which is similar to that in which atomic initial and final state eigenfunctions determine the probability of photon emission from an atom (Section 24.7). In the case of β-decay, the mechanism involves the weak interaction rather than the electromagnetic interaction of the atomic case.

γ-decay

The residual nucleus which is formed following α- or β-decay is often in an excited state, which we denote by an asterisk, for example ${}^{12}_{6}C^{*}$. Often the energy of the state is not sufficiently great to enable further α- or β-decays to take place – the process is not viable from energy considerations – so that the excited state decays to a lower energy state of the same nucleus through the emission of a photon. The process is very similar to the emission of photons from atoms (Section 24.7). Both take place through the electromagnetic interaction and, as in the atomic case, the probability that the process will occur depends on the initial and final state eigenfunctions.

Nuclei, like atoms, can have an extensive set of excited energy states. Some energy levels of the ${}^{14}_{7}N$ nucleus are shown in Figure 27.12.

For some nuclei, particularly those whose *N* and *Z* values are near magic numbers, useful predictions of the energies and nuclear spins of low energy excited states can be obtained from the shell model (Section 27.3). Because nuclear energies, which involve the strong nuclear force, are so much larger than atomic energies, which involve the Coulomb force, nuclear energy levels are of the order of MeV rather than eV. This also means that the energies of photons produced in transitions between nuclear levels are very large. The photons which are emitted therefore have energies which place them in the γ-ray region of the electromagnetic spectrum.

Study Worked Example 27.2

For problems based on the material presented in this section visit up.ucc.ie/27/ *and follow the link to the problems*

27.6 Detection of radiation: units of radioactivity

It is instructive, at this point, to summarise and compare the ionising and penetrating powers of the various types of natural radiation.

The *α-particle* is a ${}^{4}_{2}He$ nucleus and therefore has substantial mass and an electric charge of $+2e$. Consequently, it interacts strongly with any matter it encounters, producing to a very high degree of ionisation in the matter. This readiness to interact also means that, while

α-particles can cause substantial damage, they have very little penetrating power. An α-particle is easily stopped by a sheet of paper and will not travel more than 3 to 5 cm in air.

The *β-particle* is an electron or positron and therefore has a mass which is smaller than that of an α-particle by a factor of $4 \times 1840 \sim 7000$ and a charge of $\pm e$ (depending on whether it is a positron or electron). It therefore produces considerably less ionisation than an α-particle. β-particles can travel through air but are stopped by thin sheets of aluminium.

A *γ-ray* produces no direct ionisation because it is electrically neutral. It interacts with matter through the photoelectric effect (Section 14.5), the Compton effect (Section 14.7) or through pair production (which will be described in Section 27.11). γ-rays travel easily through air and through most light (low Z) matter. Substantial thicknesses of lead are needed to absorb γ-rays.

Detectors

Radioactivity can be detected using a wide variety of techniques. We outline below three devices that have been used traditionally. Some can also be used to measure the energies of radioactive particles.

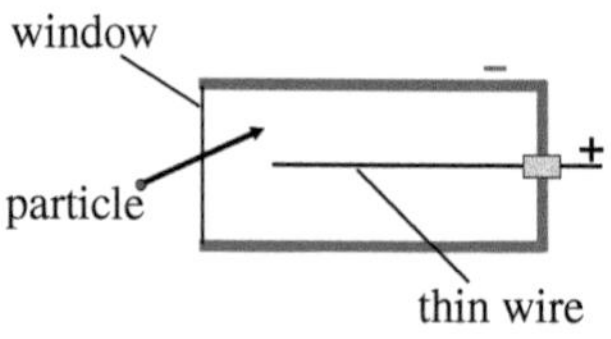

Figure 27.13. A schematic diagram of a Geiger-Muller tube.

The *Geiger-Muller* tube, illustrated in Figure 27.13, comprises a thin wire which is held along the axis of a metal cylinder. The cylinder is filled with a low-pressure gas mixture and the central wire (the anode) is placed at a positive electrical potential of several hundred volts relative to the cylinder.

When a particle, released for example in a radioactive decay, passes through the window at the end of the tube, it ionises the gas mixture. The negative ions which are produced move towards the anode where they encounter a strong electric field. They are accelerated in this region producing a cascade of negative ions which are collected at the anode. After each cascade, a pause, typically of about 0.5 ms, is needed for the gas mixture to recover, that is to de-ionise. Each particle therefore produces a current pulse which can be counted. The Geiger-Muller tube is well suited to the counting of β-particles and can also be used to count α-particles if a thin window is used at the end of the tube.

Solid state detectors use reverse biased p-n junctions, as described in Section 26.7. Energetic particles can produce electron-hole pairs in the depletion region which increases the reverse current for a short period. These current bursts can be used to count particles. Solid state detectors can be used to detect α-, β- or γ-particles/rays.

In a *scintillation counter,* the window at the end of a photomultiplier tube (a photon detector) is coated with a scintillating material, for example sodium iodide with Tl impurity. Particles which hit the scintillator produce photons with ultraviolet frequencies which are converted into electrons by the photomultiplier. Each electron produces a cascade of electrons in the photomultiplier which is measured as a current pulse. Scintillation counters may be used to detect α-, β- or γ-particles/rays.

Radiation detectors can detect very small amounts of radioactive emissions quantitatively and non-intrusively. This is the key to the use of radionuclides in a wide range of medical, industrial and environmental applications. The progress of very small quantities of radionuclides – *trace elements* – can be followed closely with minimal disturbance of the process under investigation.

Units of radioactivity

Four different units are used to measure the effects of radioactive sources. As described below, some of these units take account of the effects produced by the radiation on materials they encounter as well as the basic activity of the source. The different units therefore have different applications.

The **becquerel** (Bq) named after Antoine Henri Becquerel (1852–1908) measures the *activity* of a radioactive source, the number of disintegrations per second $\frac{dN}{dt}$, which is given by λN (Equation (27.3)). One becquerel = one disintegration per second. The nature of

Table 27.2 Some typical dose equivalents (average annual doses).

Source	mSv
Cosmic rays	0.330
Terrestial	0.350
In the body	0.250
Radon (inhaled)	1.30
Occupational	0.006
Nuclear weapons testing	0.006
Nuclear power	0.001
Transatlanic flight	0.08
Medical X-rays	0.380
Nuclear medicine	0.14
Total	2.7

Notes: Maximum annual exposure for a member of the public: 1.0 mSv
Maximum annual exposure for a radiation worker: 20.0 mSv (averaged over 5 years)
$LD_{60/50}$ (dose lethal to 50% of those exposed to it within 60 days of exposure) is somewhere between 3000 and 5000 mSv
Probable fatal dose (whole body short-term exposure): 5000 mSv

the source and the types of radiation produced have no bearing on this measure of radioactivity. A non-SI unit for activity, the **curie** (Ci), named after Marie Curie (1867–1934), is sometimes used. One curie = 3.7×10^{10} disintegrations per second (1 Ci = 3.7×10^{10} Bq).

The **roentgen** (R), named after Wilhelm Röntgen (1845–1923), measures *exposure*, the ability of a beam of radiation to deliver energy to material it passes through. One roentgen is defined as the ability to deliver 8.78 mJ of energy to 1 kg of dry air in specified conditions. It does not, however, tell us whether or not the energy will actually be absorbed.

The **gray** (Gy) measures the energy actually absorbed from any type of radiation by a specific object. The *absorbed dose* is one gray when 1 J/kg has been delivered to the object by ionizing radiation. A non-SI unit, the *rad* (radiation absorbed dose), which equals 0.01 Gy, is sometimes used to measure absorbed dose.

The **sievert** (Sv) is a measure of *dose equivalent*. It takes account of the fact that although different types of radiation may deliver the same energy per unit mass to a human body they do not all have the same biological effect. The dose in Sv is found by multiplying the absorbed dose (in grays) by the appropriate RBE (relative biological effectiveness) factor. For X-rays, γ-rays and electrons RBE $\approx$ 1. For α-particles RBE $\approx$ 10. Devices which monitor the dosages received by personnel exposed to radiation are therefore designed to measure dose equivalents in Sv. A non-SI unit the *rem* (roentgen equivalent in man) which equals 0.01 Sv is sometimes used to measure dose equivalent. Table 27.2 lists some typical dose equivalents encountered in everyday life.

For problems based on the material presented in this section visit up.ucc.ie/27/ and follow the link to the problems

27.7 Nuclear reactions

We now extend our examination of nuclear processes from decays to nuclear reactions in general. As noted in Section 27.5, naturally occurring nuclear decays are usually part of natural radioactive series of decays which originate with very long-lived radionuclides (present when the Earth formed) and end on stable nuclides. It is also possible to produce a wealth of radionuclides artificially through bombardment of nuclei by a wide range of particles with a wide range of energies. Studies of these reactions – which are collision processes, as considered in Sections 6.4, 6.5, 8.3 and 9.12 – provide information about excited states of nuclei which supplements the information provided by the study of decay processes.

Any long-lived nucleus can be used as the bombarding particle. Typically the residual nucleus is stable if its $\frac{Z}{A}$ ratio is comparable to that of the target nucleus. Reactions are usually characterised in terms of the bombarding and product particles which are stated in brackets and placed between the target and residual nuclides. Three examples of this notation are given below.

1. $^{14}_{7}$N (n, p) $^{14}_{6}$C. The bombarding particle in this case is a neutron and the reaction produces a proton as well as the residual nucleus. Large fluxes of energetic neutrons may be obtained from nuclear reactors (to be discussed in Section 27.9).
2. $^{9}_{4}$Be (p, n) $^{9}_{5}$B. The bombarding particle is a proton which can be accelerated to the required energy in a cyclotron or synchrotron (as described in Section 19.6). The reaction produces a neutron as well as the residual nucleus.
3. $^{27}_{13}$Al (d, α) $^{25}_{12}$Mg. The bombarding particle is a deuteron (d) – a deuterium ($^{2}_{1}$H) nucleus – which can be accelerated to the desired energy in a cyclotron (Section 19.6). The reaction produces an α-particle as well as the residual nucleus.

Nuclear reactions can be analysed by applying conservation laws for (a) total relativistic energy (b) linear momentum (c) angular momentum (d) electric charge and (e) the number of nucleons, together with further conservation laws which are suggested by systematic studies of nuclear reactions.

Consider for example the general nuclear interaction, A (a,b) B, that is, a + A → B + b, as illustrated in Figure 27.14. A and B are, respectively, the target and residual nuclei and a and b are the bombarding and product particles.

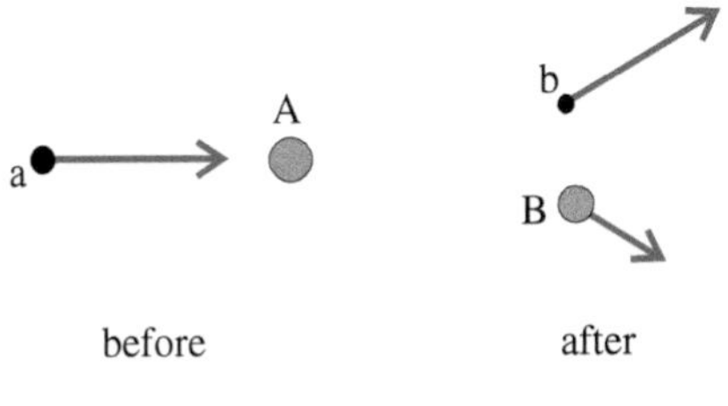

Figure 27.14. A general nuclear interaction A (a,b) B. The bombarding particle a collides with the target nucleus A to produce a residual nucleus B and a product particle b.

By applying conservation of relativistic energy (Section 9.12) in the laboratory frame we obtain

$$(T_a + m_a c^2) + m_A c^2 = (T_B + m_B c^2) + (T_b + m_b c^2)$$

where T_a etc. denote the relativistic kinetic energies, and m_a etc. the masses, of the four particles and it is assumed that A is initially at rest.

We can therefore write
$$T_B + T_b - T_a = (m_a + m_A - m_B - m_b)c^2 = Q$$

where Q is known as the *Q-value* of the reaction, as discussed in Section 6.5. Q can be positive (an *exothermic* reaction) or negative (an *endothermic* reaction). If the Q-value can be measured, the mass of any one of the particles can be determined if the other three rest masses are known. While, in principle, the Q-value can be determined by measuring T_a,T_b and T_B, the last quantity is usually difficult to measure, so that the principle of conservation of momentum is used to generate additional equations from which T_B can be eliminated. Knowing Q, the unknown mass can then be determined from $(m_a + m_A - m_B - m_b)c^2 = Q$

While conservation of angular momentum does not affect the energy balance of a reaction, it does affect its probability, that is the rate at which the reaction takes place. Studies of reaction rates can therefore be used to deduce further nuclear properties such as angular momentum.

Nuclear cross-sections

Reaction rates in collision processes are often represented using the concept of *cross-section*. The cross-section of a process is an imaginary area around the target particle which is perpendicular to the direction of the bombarding particles, as illustrated in Figure 27.15.

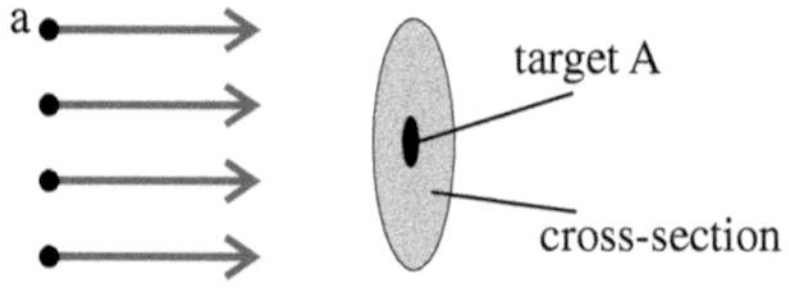

Figure 27.15. The cross-section of a target particle.

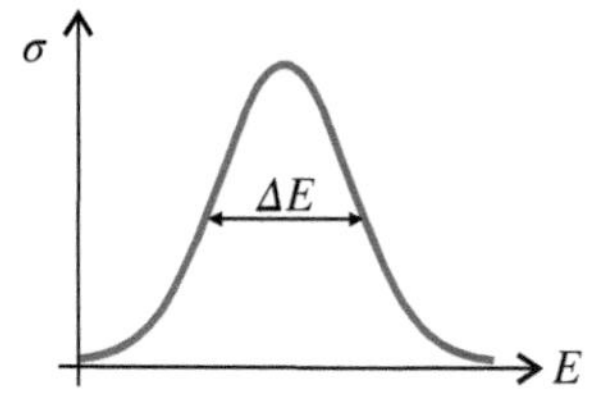

Figure 27.16. A resonance in the cross-section of a reaction plotted against collision energy.

The larger the cross-section the greater the probability of collision. Nuclear collision cross-sections are often stated in terms of a unit called the *barn*, written b, where $1\text{ b} = 10^{-28}\text{ m}^2$.

The cross-section of a reaction can vary rapidly as a function of the collision energy. As illustrated in Figure 27.16, a plot of cross-section, σ, against collision energy, E, can show a pronounced peak – a *resonance*. A resonance corresponds to an excited state of the product particles. The reaction takes place more rapidly when the collision energy is tuned to the energy of this excited state. The energy width of a resonance, ΔE, can be interpreted using the Heisenberg uncertainty principle, as stated in Equation (14.26),

$$(\Delta E)(\Delta t) \geq \frac{\hbar}{2}$$

The mean lifetime of the excited state, τ, is the time in which the energy of the state can be measured and is hence the uncertainty in time, Δt. Thus $\Delta t \sim \tau$ and $\Delta E \sim \dfrac{\hbar}{2\tau}$. In this way the energy width of a resonance can be used to determine the lifetime of a state, provided that the width is greater than the experimental resolution.

For problems based on the material presented in this section visit up.ucc.ie/27/ *and follow the link to the problems*

27.8 Nuclear fission and nuclear fusion

When we discussed the plot of nuclear mass per nucleon against number of nucleons (Figure 27.2), we noted that two general types of nuclear reaction are expected to produce a net release of energy, that is two general types of reaction have positive Q-values. These are (a) nuclear fusion, in which light nuclei fuse together to form nuclei of medium A, and (b) nuclear fission, in which heavy nuclei break up to form nuclei of medium A. Both processes are potential sources of useful energy. Many of the advantages and disadvantages of the two processes may be inferred from an inspection of Figure 27.2, which is reproduced as Figure 27.17, with some specific nuclei of interest marked on the plot.

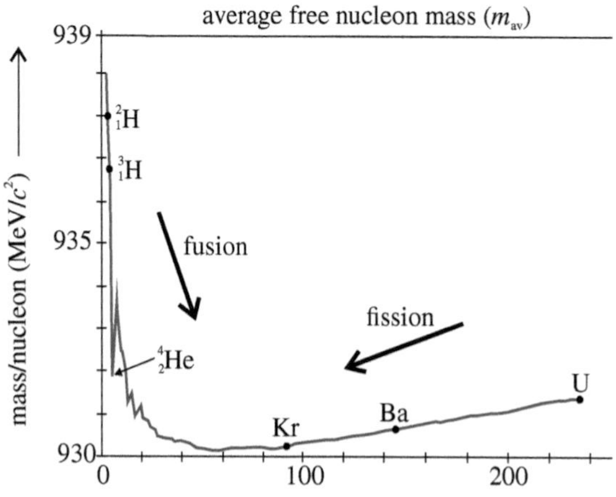

Figure 27.17. A plot of nuclear mass per constituent nucleon against nucleon number (as in Figure 27.2), indicating some nuclei of interest in the nuclear fission and fusion processes.

Nuclear fusion uses as its fuel the nuclei of very light elements, the elements which are the most abundant in the Universe and are also abundant in the Earth's surface. The most promising interaction, in terms of energy released, uses the two isotopes of hydrogen, deuterium $^2_1$H and tritium $^3_1$H,

$$^2_1\text{H} + {}^3_1\text{H} \rightarrow {}^4_2\text{He} + {}^1_0\text{n} + 17.6\text{ MeV}$$

One part in 6000 of the hydrogen found in sea water is in the form of the stable isotope, deuterium. Tritium, a radioactive isotope with a half life of 12.3 years, does not occur naturally but, as we shall see, when we discuss fusion in more detail in Section 27.10, it can be produced readily from lithium, a very abundant element in the Earth's surface. It is estimated that there is sufficient deuterium and lithium present in the Earth's surface to satisfy the current energy needs of the world for the next 10^8 years, provided the energy can be harnessed efficiently.

The end product – the waste – of the deuterium-tritium reaction is ${}_2^4\mathrm{He}$, an element which is chemically inert and which has a highly stable nucleus. The waste product of the deuterium-tritium reaction is therefore very safe.

Nuclear fission, on the other hand, uses isotopes of the heavy element uranium as its fuel. Uranium is a rare element in the Earth's surface. Current nuclear reactors, the thermal reactors which we discuss in the next section, use the rarer isotope ${}_{92}^{235}\mathrm{U}$. The world's supplies of ${}_{92}^{235}\mathrm{U}$ would probably be exhausted in a couple of hundred years if they were to be used to satisfy current energy needs. The reason that heavy elements are rare may be inferred from Figure 27.17. The formation of heavy nuclei from the fusion of intermediate A nuclei is an 'uphill' process in energy terms. The fusing nuclei must have very high energies to make the processes viable. Such conditions only occur in cataclysmic events in the Universe, for example in supernovae.

A typical fission reaction of ${}_{92}^{235}\mathrm{U}$ is ${}_0^1\mathrm{n} + {}_{92}^{235}\mathrm{U} \rightarrow {}_{56}^{141}\mathrm{Ba} + {}_{36}^{92}\mathrm{Kr} + 3{}_0^1\mathrm{n}$, ${}_{56}^{141}\mathrm{Ba}$ and ${}_{36}^{92}\mathrm{Kr}$ being just two examples of the possible end products of this fission reaction. As can be seen in Figure 27.1, the $\frac{N}{Z}$ ratio for Ba or Kr is smaller than for U so that, as well as two large fragments, some neutrons are generally produced in fission reactions to bring the products closer to the curve of stability. Note, however, that even after some neutrons have been shed, the barium and krypton nuclides produced are not stable isotopes of these elements. They are highly unstable radionuclides with excess neutrons and with half-lives of 18 minutes and 2.3 s, respectively. Each therefore gives rise to a series of radioactive decays. The end products of fission reactions are therefore highly active radionuclides which can be very hazardous biologically and, in some cases, very long-lived. The safe storage of this waste over very long periods of time poses a major problem.

On the basis of this brief examination, fusion, which uses an abundant fuel and produces a safe waste product, seems to have all the advantages. Yet all current commercial exploitation of nuclear energy for electricity generation is based on fission. The reason is that self-sustaining fission reactions are very much easier to produce technologically than self-sustaining fusion reactions, indeed self-sustaining fusion reactions have yet to be achieved. In the next two sections we outline the main technologies which are used to generate (or to attempt to generate in the case of fusion) electricity from nuclear energy.

27.9 Fission reactors

Nuclear fission can occur spontaneously in nuclides with $Z \geq 92$ where it competes with α-decay as a decay mechanism. (α-decay, in which a nucleus splits into two atomic nuclei, may be regarded as a form of fission). Fission occurs because heavy nuclei are unable to retain spherical shapes. As illustrated in Figure 27.18, the shapes of heavy nuclei can oscillate between squashed and elongated ellipsoids, a behaviour which would be expected in a large liquid drop. Occasionally – since it is a random process – the shape of the nucleus can distort (Figure 27.18(c)) to the point at which, due to the increased surface area and the Coulomb repulsion between the new centres of charge which are forming, it is energetically favourable for the nucleus to split in two (Figure 27.18(d)).

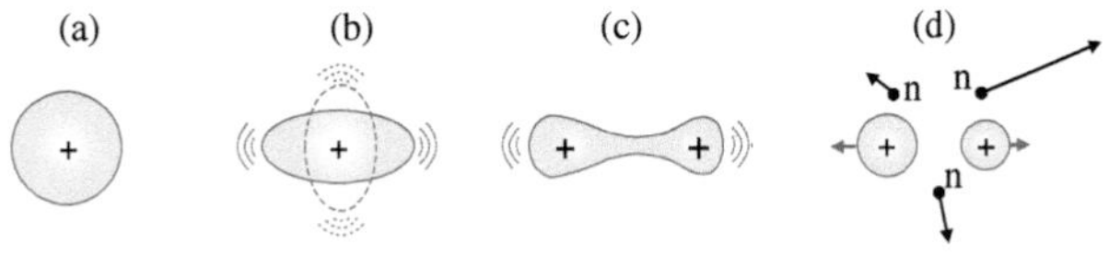

Figure 27.18. Nuclear fission. The shape of the nucleus (a) vibrates (b) until the distortion reaches a critical state (c) in which it is energetically favourable for the nucleons to split into two particles (d).

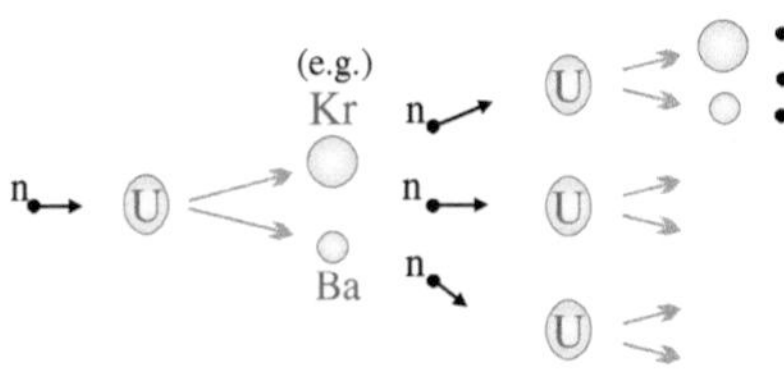

Figure 27.19. A cascade of fission reactions.

The fission process can occur very rapidly following absorption of a neutron by the nucleus – *induced* fission. The nucleus breaks quickly into two major fragments and a number of neutrons, the latter being the key to the use of fission as a self-sustaining process. Each fission produces 2 to 3 neutrons so that, as illustrated in Figure 27.19, these neutrons can induce further fissions in nearby uranium nuclei, leading to a cascade of fission reactions.

The 'core' of a fission reactor comprises a certain mass of uranium. Neutron losses from the core are proportional to its surface area (that is $\propto r^2$), whereas the number of fission reactions induced by fission neutrons is proportional to its volume (that is $\propto r^3$). Thus, by increasing the uranium mass, a critical mass is reached for which the rate at which fissions are induced by fission neutrons exceeds the rate of neutron loss and a self-sustaining fission process is produced. An important parameter in the process is k, the neutron multiplication factor, the average number of fission neutrons produced which induce further fissions. If this is less than one, the process is described as *sub-critical*. When k exceeds one it becomes *critical*.

Two kinds of neutrons – known as *prompt* and *delayed* neutrons – are produced in the fission process. The prompt neutrons are emitted directly in the fission process (within 10^{-13} s). The delayed neutrons are emitted in subsequent decays in which the products, such as ${}_{56}^{141}\mathrm{Ba}$ and ${}_{36}^{92}\mathrm{Kr}$, which have an excess of neutrons, attempt to reach the curve of stability. Delayed neutrons can be emitted many seconds after a fission. The fission process goes out of control very quickly when it is governed by the prompt neutrons – as it is in an atomic bomb. As we shall see below, the delayed neutrons play a very important role in the control of fission in thermal reactors.

Natural uranium comprises two isotopes, 99.3% $^{238}_{92}U$ and 0.7% $^{235}_{92}U$. $^{235}_{92}U$ is fissile to (can be broken up by) slow (thermal) neutrons (of energy ~ 0.025 eV) as well as to fast neutrons (with energies greater than 1 MeV). $^{238}_{92}U$, on the other hand, is fissile only to fast neutrons with energies greater than 1.1 MeV. In Worked Example 27.3, we show how these properties can be deduced by applying the semi-empirical mass formula (Equation (27.2)) to the fission of these isotopes. $^{238}_{92}U$, with even values of both N and Z, is particularly stable and the energy brought in by slow neutrons is not enough to make fission viable in energy terms. As indicated by the following decay sequence, $^{238}_{92}U$ can capture slow neutrons but decays without fission,

$$^{1}_{0}n + ^{238}_{92}U \rightarrow ^{239}_{92}U^{*} \rightarrow ^{239}_{93}Np + e^{-} + \overline{\nu_e},$$

followed by the decay

$$^{239}_{93}Np \rightarrow ^{239}_{94}Pu + e^{-} + \overline{\nu_e} \tag{27.6}$$

The thermal reactor

A *thermal reactor*, the type used in almost all commercial production of nuclear energy today, uses fission of $^{235}_{92}U$ by slow neutrons. The structure of a typical thermal reactor is illustrated in Figure 27.20.

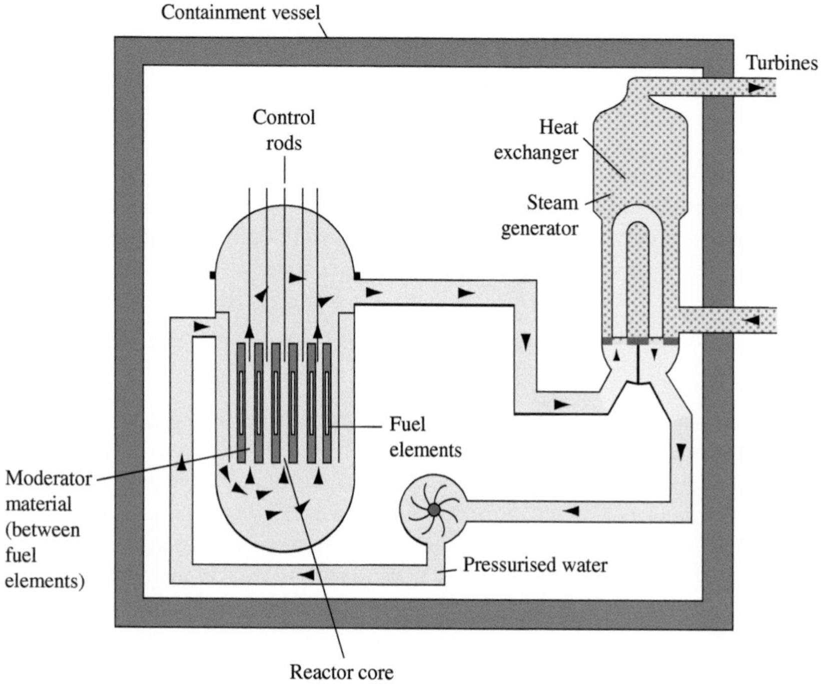

Figure 27.20. Schematic structure of a typical thermal fission reactor. Illustrated for a pressurised water reactor.

In a thermal reactor, almost all neutrons are slowed down to thermal energies by the introduction of a *moderator*, usually D_2O or graphite, between the fuel elements. With the moderator present, the average time between the fissions produced by thermalised prompt neutrons is about 1 ms. When delayed neutrons are included the average neutron life-time between fissions is about 0.1 s. By inserting or withdrawing *control rods*, which are made of a material which strongly absorbs neutrons, such as cadmium or boron, the average time between fissions is increased or decreased so that the balance between critical and subcritical conditions (that is the value of the neutron multiplication factor k) can be carefully controlled. The increase in the average time between fissions due to delayed neutrons is an essential ingredient in the design of the mechanical control system.

The energy produced by fission reactions heats the reactor core and is extracted through a heat exchanger (using a coolant) which boils water to drive steam turbines which then produce electricity, as in a conventional fossil fuel power station. Many different types of thermal reactor, using different fuel mixtures, different moderators and different coolants, are used. Although there are differences in costs and efficiency between the different reactor types, the physical principles on which they are based are essentially the same.

The fast (breeder) reactor

A weakness of all thermal reactors is their inefficient use of uranium. Only the rarer isotope $^{235}_{92}U$ is used. As noted in the previous Section, $^{235}_{92}U$ is a limited resource. Another type of fission reactor, the *fast reactor*, attempts to overcome this problem by using fission of $^{238}_{92}U$ by fast neutrons. The uranium is enriched with 20% $^{239}_{94}Pu$, a by-product of thermal reactors, as indicated in the reaction described above by Equation (27.6), which is more fissile to neutrons than $^{238}_{92}U$ and also produces on average more neutrons per fission.

The fast reactor has no moderator. It is controlled by moving the fuel elements or boron nitride control rods. The core is much more compact and operates at higher temperatures and hence at greater thermal efficiencies. As a consequence, the coolant system has to be very effective. Liquid sodium has been used but this is a highly reactive material chemically and therefore very difficult to handle.

An important feature of the fast reactor is that it produces a high neutron flux at the edge of the core so that, by surrounding the core with a blanket of $^{238}_{92}U$, it can breed its own fuel through the reaction (27.6). The fast reactor is therefore often known as the *fast breeder* reactor. This reactor makes much more efficient use of its fuel, using 50% of natural uranium as against about 1% in thermal reactors. The fast reactor is also useful as a means of consuming plutonium, a particularly hazardous by-product of thermal reactors.

Although fast reactors have been used for commercial electricity generation, their technology has not yet been shown to be economically competitive with that of thermal reactors. Safe handling of the highly toxic $^{239}_{94}Pu$ fuel and of the liquid sodium coolant is particularly challenging.

Study Worked Example 27.3

27.10 Thermonuclear fusion

Self-sustaining nuclear fission is achieved fairly easily because the bombarding particle, the neutron, is electrically neutral and experiences no Coulomb repulsion as it approaches a positively charged heavy nucleus. In the case of the fusion reaction, however, the fusing nuclei are both positively charged and experience a very strong Coulomb repulsion before they are close enough to each other for the attractive strong nuclear force to take over. We have already considered the potential seen by fusing nuclei – a potential barrier – in Section 14.17. The barrier is reproduced in Figure 27.21.

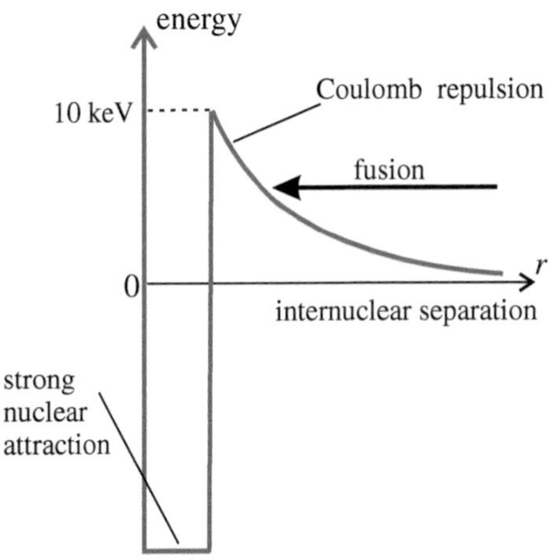

Figure 27.21. Nuclear fusion. The repulsive Coulomb barrier between the two nuclei must be overcome for fusion to occur.

The kinetic energy which an ion needs to overcome the barrier is of the order of 10^4 eV. Such energies are easily produced in accelerators but this is not a viable method of generating energy through fusion because the energy needed to run an accelerator is greater than the energy produced by fusion reactions. We need instead a medium which is made up of ions with average energies which exceed 10^4 eV. A medium in which all atoms are fully ionised is known as a *plasma*, sometimes called 'the fourth state of matter' (in addition to the solid, liquid and gaseous states). If the ion energy in a plasma is sufficiently large, fusion reactions can generate enough energy to sustain the plasma at the energies needed for fusion. Such a self-sustaining plasma is known as a *thermonuclear plasma.*

In a viable thermonuclear plasma, ions must collide with sufficient energy and sufficiently often for a net gain of energy to be achieved. The plasma parameter which describes the average ion energy is the *ion temperature* T_i. The plasma parameters which determine the collision rate and the number of collisions which take place before the energy of an ion is lost are the *ion density*, n_i, and the *energy confinement time*, τ_E, the average time for the energy of an ion to be held in the plasma before escaping as radiation or through particle loss.

The conditions for self-sustaining thermonuclear fusion are known as the *Lawson criteria*. For the deuterium–tritium reaction,

$$^2_1H + ^3_1H \rightarrow ^4_2He + ^1_0n + 17.6\,MeV,$$

the Lawson criteria are $T_i > 5\,keV = 8 \times 10^{-16}$ J and $n_i\tau_E > 10^{20}\,m^{-3}\,s$

The average energy of plasma ions is often expressed as a temperature in kelvin using $\left\langle \frac{1}{2}mv^2 \right\rangle = kT$. Thus 8×10^{-16} J is stated as 6×10^7 K.

Plasmas cannot be constrained by solid containers because, as soon as a plasma touches a solid wall, it cools and is extinguished. Other means must be devised to contain a thermonuclear plasma long enough for the Lawson criteria to be satisfied.

Thermonuclear plasmas are easily sustained in astrophysical environments such as stars. Stars are plasmas with sufficient mass for gravitational attraction to contain the ions and nuclear fusion is the energy process by which stellar energy is produced. In this sense fusion energy, which we receive on Earth as radiation from the Sun, is the primary source of most terrestrial energy. Wind, and hence wave, energy is largely attributable to the heating effect of the Sun. Energy from plants and fossil fuels can be regarded as stored energy which originates in fusion processes in the Sun.

A mass similar to that of the planet Jupiter is needed to hold a thermonuclear plasma together through gravitational confinement, so clearly this is not a viable process on the terrestrial scale. Two other methods of confinement – *magnetic* and *inertial* confinement – are used therefore to attempt to contain thermonuclear plasmas on the terrestrial scale.

Magnetic confinement

Magnetic confinement seeks to satisfy the second Lawson criterion, $n_i\tau_E > 10^{20}\,m^{-3}$ s, by using magnetic fields to contain a plasma of density $10^{20}\,m^{-3}$ for more than one second. We noted, in Section 19.5, that a charged particle which has a velocity component perpendicular to a magnetic field line performs uniform circular motion around the field line. When this motion is combined with any component of velocity which the particle has along the field line the net result is helical (corkscrew) motion about the field line, as illustrated in Figure 27.22.

Figure 27.22. Confinement of a charged particle along a magnetic field line. (a) The charged particle performs helical motion about the field line. (b) The radius of the helix reduces as the magnetic field becomes stronger.

The radius of the helix is given by Equation (19.9), $R = \frac{mv}{qB}$, so that, as the magnetic field becomes stronger, the helix becomes tighter, as illustrated in Figure 27.22(b). Hence any motion of a charged particle with a component along the field direction ends up as net motion along the field line and the particles are constrained to move along the field lines unless they are deflected by collisions with other particles. They are described as 'frozen' onto the field lines. This effect can be observed in solar flares – streams of charged particles which move along disruptions in the magnetic field lines of the Sun.

Plasma ions, therefore, can be confined to move along magnetic field lines. We must consider next how ion losses at the end of the field lines can be prevented. The most usual method of solving this problem is to bend the field lines back on themselves – thus forming hoops. The plasma is then confined to a toroidal or 'ring doughnut' shape. The most successful magnetic confinement device has been the *tokamak* (from the Russian acronym for a toroidal magnetic chamber). The construction of the tokamak is illustrated in Figure 27.23.

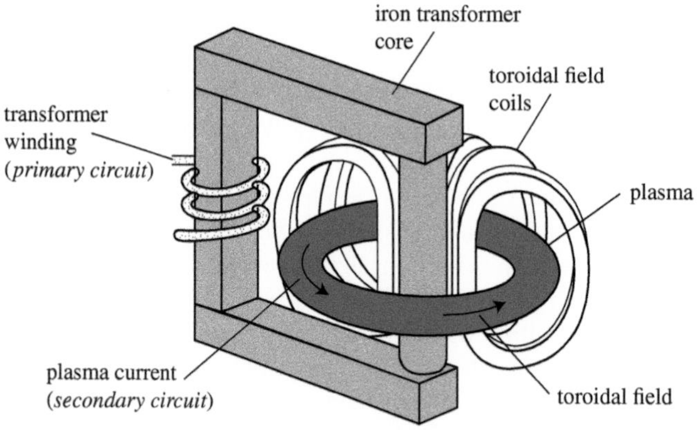

Figure 27.23. The structure of the tokamak.

In a tokamak, strong toroidal magnetic fields confine the plasma to the torus. As shown in Figure 27.23, the toroidal plasma is the secondary circuit of a transformer. Plasmas, which comprise highly mobile charged particles, are excellent electrical conductors so that the plasma can be heated rapidly by a strong toroidal current in the secondary circuit.

Magnetic confinement of thermonuclear plasmas has been the subject of a major international research programme for more than 60 years. A plasma behaves like a fluid and can easily develop instabilities. The technical problems of confining a plasma for a significant time are considerable but the present generation of machines have attained the Lawson criteria separately (that is $T_\mathrm{i} > 4 \times 10^7$ K, $n_\mathrm{i} > 10^{20}$ m^{-3} and $\tau_E > 1$ s) although not simultaneously. The indications are, however, that controlled thermonuclear fusion can be achieved with a modest scaling up of the size of magnetic confinement machines. It is too early to say whether fusion will be a commercially viable energy source. This will depend, amongst other things, on how efficiently the energy can be extracted, a topic which we consider below, following a discussion of inertial confinement.

Inertial confinement

Inertial confinement seeks to satisfy the second Lawson criterion, $n_\mathrm{i}\tau_E > 10^{20}$ m^{-3} s, by creating very large ion densities ($\sim 10^{32}$ m^{-3}), many orders of magnitude greater than the density of solid matter, with containment times of 10^{-12} s. The plasma is compressed by implosion of matter so that the confinement time is determined by the delay in dispersion of the plasma due to its inertia.

To produce a highly compressed plasma, solid pellets of deuterium-tritium are bombarded from all sides simultaneously and on a very short time scale. The technique is illustrated in Figure 27.24, where the energy needed to compress the pellet is delivered by synchronised laser beams. Electron and ion beams may also be used.

When the energy is delivered to the pellet, material from the surface is ablated (blasted off) outwards, as illustrated in Figure 27.24. In reaction (that is from Newton's third law), the inside of the pellet is compressed inwards. Inertial confinement, like magnetic confinement, of thermonuclear plasmas has been the subject of a large research programme but results to date have fallen short of the Lawson criteria.

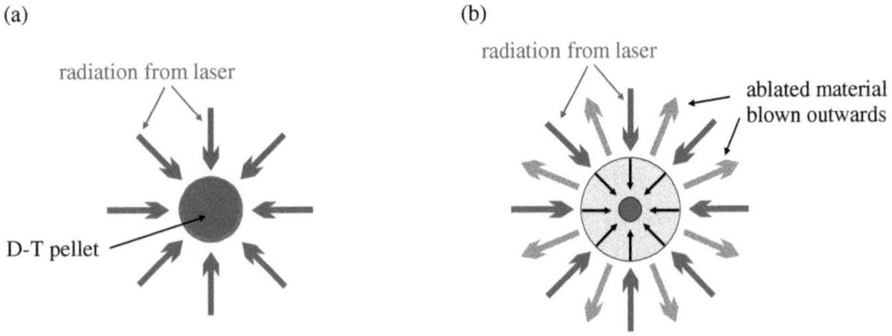

Figure 27.24. Nuclear fusion through inertial confinement.

Although controlled inertial confinement has yet to be achieved, an uncontrolled fusion explosion is produced in a hydrogen bomb through the compression of deuterium by a fission explosion.

The fusion reactor

In a fusion reactor, whether using magnetic or inertial confinement, the vacuum chamber in which the plasma is formed will be surrounded with a blanket of lithium. The blanket has two purposes. First neutrons from the fusion plasma are absorbed, thus extracting thermonuclear heat energy. Secondly the blanket produces tritium fuel through the reactions with either of the stable isotopes of Li,

$$^1_0\text{n} + {}^6_3\text{Li} \rightarrow {}^3_1\text{H} + {}^4_2\text{He} + 4.8\,\text{MeV} \quad \text{or} \quad {}^1_0\text{n} + {}^7_3\text{Li} \rightarrow {}^3_1\text{H} + {}^4_2\text{He} + {}^1_0\text{n} - 2.5\,\text{MeV}$$

Although, as stated in Section 27.8, the waste product of the deuterium-tritium reaction is the very safe nuclide ^4_2He, there are aspects of the process which produce hazardous waste.

First, although ideally the tritium fuel is entirely consumed in the process, if the vacuum vessel were to be punctured while the machine is in operation, this hazardous radioactive gas could be released. Note, however, that only grams of tritium will be used in the machine when it is in operation.

The second potentially hazardous aspect of a fusion reactor is activation of the walls and surroundings of the reactor as a result of intense neutron bombardment. Activation is the process in which stable nuclides absorb neutrons and thus form radioactive isotopes. Activation in a fusion reactor, however, will be considerably less than in a fission reactor. When a fission reactor is switched off typically 200 MW of 'afterheat' remains due to the radioactivity of the fission products and activation of the reactor structure. In the case of a fusion reactor, it is estimated that only 5 kW of afterheat will be produced when it is switched off.

Much further research is needed before nuclear fusion can be properly assessed as a commercially viable energy source. If progress is satisfactory the first fusion reactors are expected to become available from the middle of this century.

Cold fusion

Finally, we consider briefly the possibility of achieving *cold fusion,* a process which was mentioned in the discussion of tunnelling in Section 14.17. In cold fusion, instead of attempting to give the fusing nuclei enough energy to surmount the barrier, a means of enabling room temperature ions to circumvent the barrier is sought. One possibility, discussed in Section 14.17, is to produce enhanced tunnelling through the barrier of Figure 27.21 by arranging situations in which the fusing nuclei are packed more closely together. Another process, known as *muon catalysed fusion*, uses intermediate reactions in which negatively charged muons substitute for electrons in the fusing atoms.

The muon is a subnuclear particle with properties which are similar to those of the electron except that it is unstable (with a mean lifetime of 2.2×10^{-6} s) and is 207 times more massive than the electron. We discuss the muon further in the next two sections.

The radius of the first Bohr orbit of the electron in the hydrogen atom is given by Equation (24.7), $r = \dfrac{4\pi\hbar^2\varepsilon_0}{e^2 m}$, so that when m increases by a factor of 207, the radius of the muon orbit is reduced by this factor. The 'muonic hydrogen' atom is therefore 207 times smaller than the electronic hydrogen atom and, consequently, muonic atoms can approach the nuclei of other atoms very closely as neutral entities before they experience Coulomb repulsion. Molecules are often formed. The potential barrier of Figure 27.21 becomes much narrower, considerably enhancing the probability of tunnelling. While muon catalysed fusion has been observed, it is unlikely that the energy produced in the process could ever exceed the energy required to accelerate the muons in the first place. Many attempts to produce fusion reactions at room temperature have been, and continue to be, made but the viability of cold fusion as a process for energy production on a commercial scale has yet to be demonstrated.

A more detailed and more advanced description of the topics covered in the first ten sections of this chapter may be found in Lilley (2001).

For problems based on the material presented in this section visit up.ucc.ie/27/ and follow the link to the problems

27.11 Sub-nuclear particles

One of the most attractive features of the atomic and nuclear models which we have described in Chapter 24 and in Section 27.3 is their basic simplicity. The structure of all atoms and of all nuclei – stable and unstable – can be described in terms of only three basic building blocks, the electron, the neutron and the proton. When they were discovered, these particles seemed to be the smallest elements of matter and were often called elementary or fundamental particles. As we shall see in the next section, other particles now have a better claim to this name so that we shall refer to particles such as the electron, the neutron and the proton as *sub-nuclear* particles or often simply as particles.

In the course of this chapter a number of further sub-nuclear particles, the positron, the neutrino and the muon have already emerged to upset the simple description of matter outlined above. Studies of nuclear reactions, using very high energy particles as probes, reveal an abundance of particles.

The techniques used to analyse particle reactions are broadly those described in Section 27.7. High energy accelerators, such as the cyclotron and synchrotron, as described in 19.6, and their modern equivalents, are needed to produce these particles and sophisticated detection techniques are needed to determine their properties.

If the particles are placed in a uniform magnetic field B, the radii of curvature of their tracks can be used to determine their momenta using Equation (19.9), $R = \frac{mv}{qB}$. By applying conservation laws for quantities with which we are familiar, such as total relativistic energy, linear momentum, angular momentum and electric charge, together with further conservation laws which are suggested by systematic studies of a large number of high energy interactions, well defined properties can be established for the newly observed particles. A review of the experimental methods used in particle physics may be found in Chapter 4 of Martin and Shaw (2019).

Classification of subnuclear particles: isospin

The task, therefore, is to establish order among the hundreds of sub-nuclear particles which have been observed and, hopefully, to uncover a new underlying simplicity. A sensible first step in this process is to group the particles according to their properties, in the hope of observing regularities and hence evidence of underlying structure. This approach may be compared with the construction of the periodic table – the arrangement of the elements according to their chemical properties by Dmitri Mendeléev (1832–1907) in the nineteenth century. His approach must have seemed haphazard and arbitrary to many chemists of the time but, from it, patterns and order emerged. As we now know (Section 24.13) the patterns which emerged reflect the electronic structure of atoms.

As a first step, particles may be grouped according to their masses. The following three divisions were apparent from the early days.

- **Leptons**, (from the Greek word for thin or small) particles with masses in the approximate range 0 – 130 MeV/c^2. Examples are the neutrino, ν (~ 0), electron, e (0.5) and muon, μ (106).
- **Mesons**, (from the Greek word for middle) particles with masses in the approximate range 130 – 900 MeV/c^2. Examples are the pion, π (140) and the kaon, K (500).
- **Baryons**, (from the Greek word for heavy) particles with masses in the range from 900 MeV/c^2 upwards. Examples are the proton, p (938), neutron, n (940) and the lambda particle, Λ (1116).

Observations of other properties of sub-nuclear particles, such as intrinsic spin, reinforce this classification. Leptons are all found to have half-integral spin $\left(\frac{1}{2}\right)$, mesons integral spin (0 or 1) and baryons half integral spin $\left(\frac{1}{2} \text{ or } \frac{3}{2}\right)$. The mass classification described above is not followed strictly; there is some overlap in mass ranges. For example, the tauon, τ, with mass 1784 MeV/c^2, has spin $\frac{1}{2}$ and other properties which identify it as a lepton.

Particles are also classified according to the types of interaction (electromagnetic, strong nuclear or weak nuclear) through which particle reactions and decays take place. The type of interaction involved in a decay can be deduced from the time scale over which it takes place. For the strong interaction, the timescale is 10^{-23} to 10^{-20} s, for the electromagnetic interaction 10^{-18} to 10^{-15} s and for the weak interaction 10^{-10} s to 15 minutes. Only baryons and mesons are found to take part in strong interactions so that they are grouped together as **hadrons** (from the Greek word for strong) – strongly interacting particles. All particles interact through the weak interaction although this interaction is negligible when the strong interaction is also present.

Some of the more important particles are listed in Table 27.3, together with some of their properties. Many of these properties, such as rest mass, lifetime, charge and intrinsic spin are familiar. Other properties, such as isospin, strangeness and hypercharge emerge from systematic analyses of nuclear reactions through conservation laws. These properties are observed only on the particle scale and have no macroscopic equivalents. It is not the first time that we have come across such a particle property. In Section 24.10 we noted that electron spin cannot be interpreted as a classical spin angular momentum, its origins being in relativistic quantum mechanics.

Isospin is particularly important in revealing patterns in the properties of sub-nuclear particles. The concept of isospin is rooted in the manner in which the strong nuclear interaction treats particles. As noted in Section 27.3, the strong interaction treats protons and neutrons in exactly the same way – as nucleons. To the strong interaction the proton and neutron are two states of the nucleon. The nucleon is labelled with a new quantum number – isospin T – which is treated in the same way as ordinary spin mathematically, but is not an angular momentum. We saw in Section 24.10 that the spin quantum number of the electron has two components $m_s = \pm\frac{1}{2}$, which

Table 27.3 Properties of subnuclear particles.

Name	Symbol	Mass	Mean Lifetime (s)	Charge (+e)	Intrinsic Spin (*s*)	Lepton Number (L_e, L_μ, L_τ)	Baryon number (*B*)	Isospin (*T*)	Isospin component (T_z)	Strangeness (*S*)	Hypercharge ($Y = B + S$)	Principal decay
leptons												
electron neutrino	ν_e	0	stable	0	½	+1	0					
muon neutrino	ν_μ	0	stable	0	½	+1	0					
tauon neutrino	ν_τ	0	stable	0	½	+1	0					
electron	e^-	0.511	stable	−1	½	+1	0					
Muon	μ^-	105.7	2.2×10^{-6}	−1	½	+1	0					$e^- + \overline{\nu_e} + \nu_\mu$
tauon	τ^-	1777	2.9×10^{-13}	−1	½	+1	0					$e^- + \overline{\nu_e} + \nu_\tau$
mesons												
pion	π^+	139.6	2.6×10^{-8}	+1	0		0	1	+1	0	0	$\mu^+ + \nu_\mu$
	π^0	135.0	8×10^{-17}	0	0		0	1	0	0	0	$\gamma + \gamma$
	π^-	139.6	2.6×10^{-8}	−1	0		0	1	−1	0	0	$\mu^- + \overline{\nu_\mu}$
kaon	K^+	493.6	1.2×10^{-8}	+1	0		0	½	+½	+1	+1	$\mu^+ + \nu_\mu$
	K^0	497.6	8.9×10^{-11}	0	0		0	½	−½	+1	+1	$\pi^+ + \pi^-$
$\overline{K^0}$	$\overline{K^0}$	497.6	5.1×10^{-8}	0	0		0	½	+½	−1	−1	$\pi^+ + \pi^-$
	K^-	493.6	1.2×10^{-8}	−1	0		0	½	−½	−1	−1	$\mu^- + \overline{\nu_\mu}$
eta	η	548	5×10^{-19}	0	0		0	0	0	0	0	$\gamma + \gamma$
baryons												
proton	p	938.3	stable	+1	½		1	½	+½	0	1	
neutron	n	939.6	881	0	½		1	½	−½	0	1	$p + e^- + \overline{\nu_e}$
lambda	Λ^0	1116	2.6×10^{-10}	0	½		1	0	0	−1	0	$p + \pi^-$
sigma	Σ^+	1189	8.0×10^{-11}	+1	½		1	1	+1	−1	0	$p + \pi^0$
	Σ^0	1192	7×10^{-20}	0	½		1	1	0	−1	0	$\Lambda^0 + \gamma$
	Σ^-	1197	1.5×10^{-10}	−1	½		1	1	−1	−1	0	$n + \pi^-$
xi	Ξ^0	1315	2.9×10^{-10}	0	½		1	½	+½	−2	−1	$\Lambda^0 + \pi^0$
	Ξ^-	1322	1.6×10^{-10}	−1	½		1	½	−½	−2	−1	$\Lambda^0 + \pi^-$
omega	Ω^-	1672	8.2×10^{-11}	−1	3/2		1	0	0	−3	−2	$\Lambda^0 + K^-$

correspond to the two possible directions of the spin vector in real space. Similarly the isospin quantum number T of the nucleon can have two components, $T_z = \pm\frac{1}{2}$, which correspond to the two possible directions of the isospin vector in isospin space, which is an imaginary space (compare with n-space, which we considered in Section 25.4). As indicated in Table 27.3, the proton corresponds to isospin up $\left(T_z = +\frac{1}{2}\right)$ and the neutron to isospin down $\left(T_z = -\frac{1}{2}\right)$. The proton and neutron are described as members of an *isospin doublet*. All reactions which take place through the strong interaction conserve isospin.

Antiparticles: particle annihilation and pair creation

For each lepton and baryon there is a corresponding anti-particle with the same rest mass and lifetime but with opposite values of lepton number, baryon number, isospin component, strangeness, charge and hypercharge. We have already come across two antiparticles, the positron, which is the antiparticle of the electron, and the anti-neutrino, which is the antiparticle of the neutrino. An antiparticle is usually denoted by the particle symbol with a bar above it, for example $\overline{p}$ (the anti-proton) and $\overline{\nu_e}$ (the anti-neutrino).

A particle can combine with its antiparticle to produce energy, a process called *annihilation*, for example

$$e^- + e^+ \rightarrow \gamma + \gamma$$

In this type of reaction all rest mass is converted into energy so that very large photon energies can be produced. In the case of electron–positron annihilation, the minimum energy released is $2m_ec^2 = 1.02\,\text{MeV} = 1.6 \times 10^{-13}\,\text{J}$, which produces two photons (γ-rays) of energy 0.51 MeV and wavelength 2.4×10^{-12} m.

Conversely, the energy of a photon can create particle–antiparticle pairs, for example $\gamma \rightarrow e^+ + e^-$, a process known as *pair production*. Note that the process can take place only above a threshold energy which is given by the total rest mass of the pair which is created, that is 1.02 MeV in the case of electron–positron production. Note also that, as shown in Worked Example 27.4, the process cannot take place spontaneously in empty space. Interaction of the photon with another particle is needed to satisfy momentum conservation.

Let us now consider further some properties of the leptons and hadrons (mesons and baryons) listed in Table 27.3.

Leptons

There are three kinds of lepton number, L_e, L_μ and L_τ, each of which must be conserved separately in any interaction, for example

$$\mu^- \rightarrow e^- + \overline{\nu_e} + \nu_\mu$$

$$L_e \quad 0 = 1 + (-1) + 0 \quad \text{(conservation of } L_e\text{)}$$

$$L_\mu \quad 1 = 0 + 0 + 1 \quad \text{(conservation of } L_\mu\text{)}$$

Lepton conservation therefore rules out decays such as

$$p \rightarrow e^+ + \gamma$$

(which is possible in terms of energy, spin and charge conservation)

because $$L_e \quad 0 \neq -1 + 0$$

Leptons are non-strongly interacting particles with no evidence of internal structure.

Hadrons (mesons and baryons)

Any number of mesons can be generated in a reaction if enough energy is available. They are not conserved and, in this respect, mesons are like photons. The property *strangeness*, denoted by the quantum number S, is suggested by interactions involving K mesons. Conservation of strangeness explains why K mesons and other particles with strangeness are always produced in pairs, one with a strangeness of +1 the other with strangeness −1.

For example,

$$p + p \rightarrow p + p + K^0 + \overline{K^0}$$

$$S \quad 0 \quad 0 \quad = \quad 0 \quad 0 \quad +1 - 1$$

Baryon number B is always conserved. Baryons can have strangeness so that the decay

$$\Sigma^0 \rightarrow \Lambda^0 + \gamma$$

$$S \quad -1 \qquad -1 + 0$$

can occur (through the electromagnetic interaction) but not

$$\begin{array}{lccc} & \Lambda^0 & \rightarrow & n+\gamma \\ S & -1 & \neq & 0+0 \end{array}$$

because

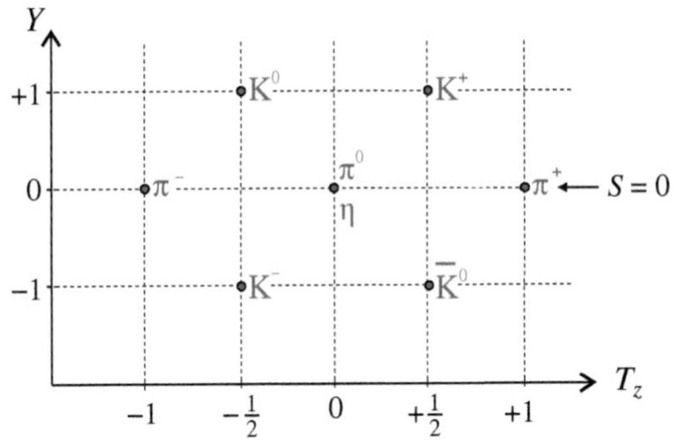

Figure 27.25. The spin 0 meson octet. A plot of hypercharge against isospin component.

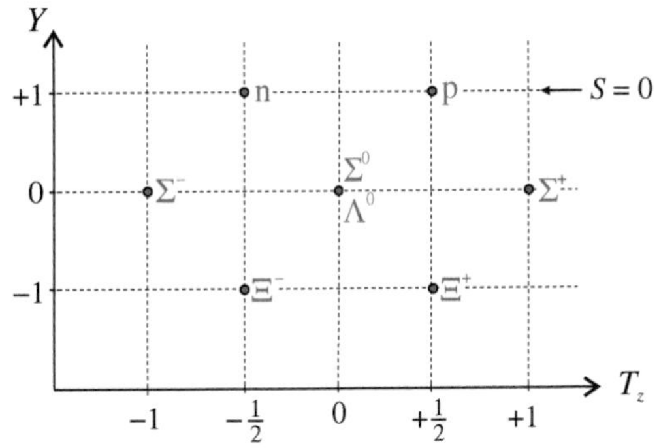

Figure 27.26. The spin $\frac{1}{2}$ baryon octet. A plot of hypercharge against isospin component.

There is clear evidence that baryons possess structure on the 10^{-16} m scale. The internal structures of the proton and neutron can be investigated through electron scattering experiments in which the electrons interact with charges in the nucleons. Such experiments reveal how charge is distributed within the proton and neutron.

Regularities in the properties of sub-nuclear particles may be investigated using the information summarised in Table 27.3. We seek to construct a 'periodic table' for sub-nuclear particles. Properties which reveal these regularities in a particularly striking manner are *hypercharge* Y, the sum of the baryon number B and strangeness S, and isospin component T_z. When Y is plotted against T_z for the spin 0 mesons (Figure 27.25) and for the spin $\frac{1}{2}$ baryons (Figure 27.26), octet patterns emerge. Note that the proton–neutron isospin doublet, which we have described above, is a component of the octet symmetry of the spin $\frac{1}{2}$ baryons.

In Figure 27.27 a similar plot is shown for spin $\frac{3}{2}$ baryons which, except for the Ω^-, are not listed in Table 27.3. In this case a decuplet pattern emerges.

In the next section we will see how the observed octet and decuplet patterns of particle properties suggest an underlying particle structure.

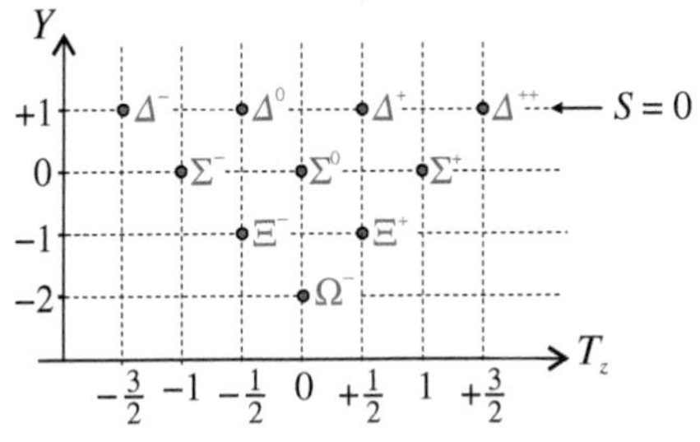

Figure 27.27. The spin $\frac{3}{2}$ baryon decuplet. A plot of hypercharge against isospin component.

Study Worked Example 27.4

For problems based on the material presented in this section visit up.ucc.ie/27/ *and follow the link to the problems*

27.12 The quark model

A major clue to the underlying structure of sub-nuclear particles is obtained when the observed multiplet symmetries of hadrons are compared with results from group theory, a mathematical framework used to describe the symmetry properties of systems. A certain group, which is generated by rotations in isospin space, matches exactly the observed octet and decuplet symmetries of the $Y - T_z$ plots of Figures 27.25 to 27.27.

The generation of this group involves 'shuffling' three entities, suggesting that the $Y - T_z$ plots are evidence of an underlying structure based on three fundamental particles called **quarks**. This model also accounts for the observed charge structure of the proton and neutron.

Up, down and strange quarks

The original quark model of Murray Gell-man (1929–2019) and Yuval Ne'eman (1925–2006) proposed three quarks, known as the *up* (u), *down* (d) and *strange* (s) quarks, to account for the strangely interacting particles known at that time (the 1960's). We will encounter further quarks later in this section. The properties of the up, down and strange quarks are summarised in Table 27.4. For each quark there is a corresponding antiparticle ($\overline{u}$, $\overline{d}$ and $\overline{s}$) with opposite electric charge, baryon number, strangeness, isospin component and hypercharge.

Table 27.4 Up, down and strange quarks.

Name	Symbol	Charge Q	Spin s	Baryon number B	Strangeness S	Isospin component T_z	Hypercharge $Y = B + S$
up	u	$+\frac{2}{3}$	$\frac{1}{2}$	$+\frac{1}{3}$	0	$\frac{1}{2}$	$+\frac{1}{3}$
down	d	$-\frac{1}{3}$	$\frac{1}{2}$	$+\frac{1}{3}$	0	$-\frac{1}{2}$	$+\frac{1}{3}$
strange	s	$-\frac{1}{3}$	$\frac{1}{2}$	$+\frac{1}{3}$	-1	0	$-\frac{2}{3}$

The quark model proposes the following structure for hadrons. Mesons are composed of quark-antiquark combinations (with $B = \frac{1}{3} - \frac{1}{3} = 0$ from Table 27.4) and baryons are three quark combinations (with $B = \frac{1}{3} + \frac{1}{3} + \frac{1}{3} = 1$). Antibaryons are combinations of three antiquarks. The quark structure of the spin 0 mesons and of the spin $\frac{1}{2}$ and spin $\frac{3}{2}$ baryons of Figures 27.25 to 27.27 are therefore as shown in Figures 27.28 to 27.30. The quark model does not propose any quark structure for leptons. They are considered to be fundamental particles without internal structure.

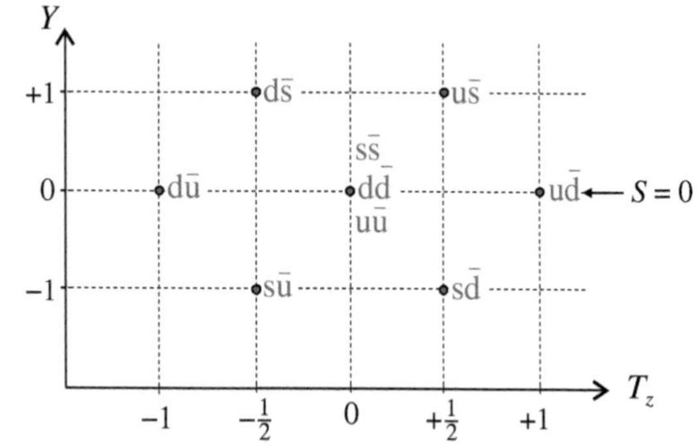

Figure 27.28. The quark-antiquark compositions of the particles which comprise the spin 0 octet (Figure 27.25).

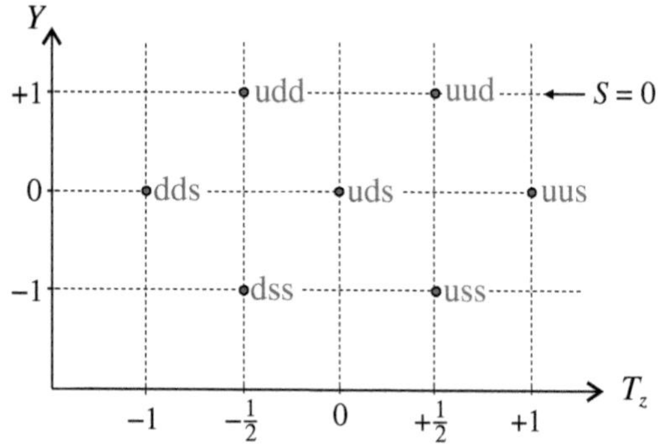

Figure 27.29. The three quark compositions of the particles which comprise the spin $\frac{1}{2}$ octet (Figure 27.26).

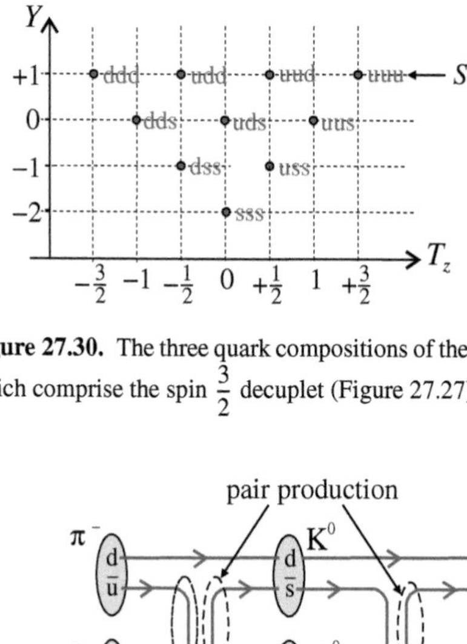

Figure 27.30. The three quark compositions of the particles which comprise the spin $\frac{3}{2}$ decuplet (Figure 27.27).

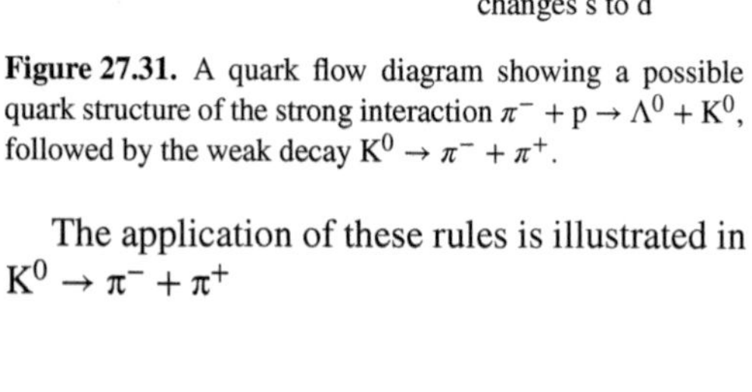

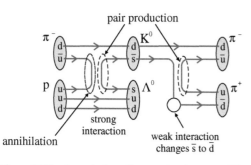

Figure 27.31. A quark flow diagram showing a possible quark structure of the strong interaction $\pi^- + p \rightarrow \Lambda^0 + K^0$, followed by the weak decay $K^0 \rightarrow \pi^- + \pi^+$.

Note that three different quark-antiquark combinations, $s\bar{s}$, $d\bar{d}$ and $u\bar{u}$, correspond to $Y = 0$, $T_z = 0$ in Figure 27.28. The π^0 and η particles are therefore considered to be a mixture of these combinations. Similarly, for the spin $\frac{1}{2}$ baryons, mixtures – linear combinations of the eigenfunctions, as allowed by quantum mechanics – of the uds, dsu and sud combinations at $Y = 0$, $T_z = 0$ in Figure 27.29 produce the Σ^0 and Λ^0 particles.

The non-strange particles, the electron, neutrino, proton, neutron, photon and the three pions, which comprise ordinary matter or appear in its low energy interactions, can be described in terms of the up and down quarks, the electron and the neutrino. Strangeness is only needed to describe the unstable particles and resonances produced in high energy experiments.

The quantum theory of quarks does much more than explain the multiplet patterns – it predicts many particle properties, for example rest masses, decay modes, lifetimes and magnetic moments. The properties of the Ω^- particle, a member of the spin $\frac{3}{2}$ decuplet (Figure 27.27), were predicted correctly before the Ω^- was observed, providing strong support for the quark model.

We are now in a position to analyse particle reactions and decays of particles in terms of their quark structure. Such analyses, known as *quark flow diagrams*, follow two rules:

(a) Quark-antiquark pairs can be created if sufficient energy is available. They can also annihilate to produce energy, for example $d + \bar{d} \rightarrow$ energy (released as γ's or other particles)

(b) The weak interaction can change quarks from one type to another but the strong and electromagnetic interactions cannot.

The application of these rules is illustrated in Figure 27.31 for the strong interaction $\pi^- + p \rightarrow \Lambda^0 + K^0$, followed by the weak decay $K^0 \rightarrow \pi^- + \pi^+$

Quark colour

As noted in Table 27.4, quarks have spin $\frac{1}{2}$ and are therefore fermions which are subject to the Pauli exclusion principle (Section 24.12). Baryons, however, are composed of three quarks in which two or even three can be of the same kind (for example the Ω^- with three strange quarks), in apparent violation of the exclusion principle. A further quark property – **'colour'** – is proposed to overcome this problem.

There are six possible colours, red, green and blue for quarks and anti-red, anti-green and anti-blue for antiquarks. In the colour hypothesis, each baryon is composed of three quarks of different colours which can only combine to produce white, which is regarded as colour neutral. Each meson is composed of a coloured quark and an antiquark of the corresponding anticolour so that again the colours cancel to produce a particle which is colour neutral.

Both baryons and mesons are therefore always colour neutral so that, while colour is important within hadrons, it is not observable from the outside.

The quark colour theory is much more than a means of circumventing the exclusion principle. On a more fundamental level we can consider the strong interaction to be based on colour just as the electromagnetic interaction is based on electric charge. The quantitative theory of the strong interaction is called *quantum chromodynamics*.

Charmed, top and bottom quarks: particle generations

The original quark model, which proposed three quarks, accounted for the subnuclear particles known at that time. Additional quarks have since been added to account for the properties of particles which have been discovered as the energies of particle collision experiments have increased and from general symmetry arguments. The additional quarks, the charmed (c), top (t) and bottom (b) quarks, are listed in Table 27.5, an extension of Table 27.3. The six different species of quarks are referred to as quark *flavours*, as indicated in the table.

Table 27.5 Quark properties.

Flavour	Symbol	Charge Q	Isospin T	Strangeness S	Charm C	Bottomness B	Topness T
up	u	$+\frac{2}{3}$	$\frac{1}{2}$	0	0	0	0
down	d	$-\frac{1}{3}$	$\frac{1}{2}$	0	0	0	0
strange	s	$-\frac{1}{3}$	0	−1	0	0	0
charmed	c	$+\frac{2}{3}$	0	0	1	0	0
bottom	b	$-\frac{1}{3}$	0	0	0	−1	0
top	t	$+\frac{2}{3}$	0	0	0	0	1

The charmed quark was originally proposed from symmetry considerations, to form a pair with the strange quark which matched the up-down quark pair. Charm, like strangeness, was shown to influence certain hadron decays. As indicated in Table 27.6, the top and bottom quarks have large masses so that hadrons containing these quarks can be produced only at the highest energies available. The existence of the bottom quark was demonstrated in 1977 and that of the top quark in 1994.

Table 27.6 Quark and lepton generations.

Quarks		Generation	Leptons	
u $\sim 4 \times 10^{-30}$	d $\sim 8 \times 10^{-30}$	first	e 9.1×10^{-31}	ν_e $<2 \times 10^{-35}$
c $\sim 2 \times 10^{-27}$	s $\sim 2 \times 10^{-28}$	second	μ 1.9×10^{-28}	ν_μ $<4 \times 10^{-31}$
t $\sim 3 \times 10^{-25}$	b $\sim 8 \times 10^{-27}$	third	τ 3.2×10^{-27}	ν_τ $<6 \times 10^{-29}$

Notes

[a] Quark and lepton masses are indicated, under their symbols, in kg. Quark masses cannot be determined directly so that the values given here are theoretical estimates known as *current masses*.

[b] The neutrino masses may in fact be zero.

In total, three generations of particles, each containing two quarks and two leptons, have been proposed, as indicated in Table 27.6. As noted earlier in this section, the first generation is sufficient to describe everyday matter. The second generation describes most unstable particles and resonances while the third generation particles are revealed only at the extremely high energies now accessible in particle collision experiments. Observations of the energy width of the resonance corresponding to a particular particle, the Z^0, can be used to estimate the number of particle generations. In Section 27.7 (Figure 27.16) we noted that the energy width of a resonant state is directly related to the lifetime of the state. The lifetime of the Z^0 depends on the number of channels available into which it can decay which, in turn, depends on the number of particle generations available. The observed breadth of the Z^0 resonance indicates strongly that there are only three generations of quarks and leptons.

On the basis of the quark properties presented in Table 27.5, it should be possible to observe free quarks. Quarks should be stable in their free state because there is nothing for them to decay to - they cannot decay to ordinary particles because these have integral charge. The observation of a free quark would provide conclusive evidence for the quark model. However free quarks have not been found, despite exhaustive searches. As argued below, there is good reason to suppose that quarks cannot escape from hadrons.

Gluons: virtual particles

According to quantum chromodynamics, the fundamental force between quarks, the interaction which holds them together in hadrons, is due to the exchange of **gluons**, particles which are emitted and absorbed by colour charges. This picture is similar to the description of the electromagnetic interaction which is given in *quantum electrodynamics*, namely that the Coulomb interaction is due to the exchange of a photon between electric charges. As noted in Section 14.7, the photon, although massless, can carry momentum so that the exchange of a photon leads to the exchange of momentum and thus a force between the charges – the Coulomb force (Section 16.3).

In the quantum electrodynamical description, a charged particle can spontaneously emit and absorb a *virtual* photon of energy ΔE, as long as this takes place in a time interval Δt, which conforms to the uncertainty principle (Equation (14.26),

$$(\Delta E)(\Delta t) \geq \frac{\hbar}{2}$$

The photon has no mass so that ΔE can be almost zero and Δt almost infinite. Thus, the virtual photon exchanged by two charges in the electromagnetic interaction can travel almost an infinite distance in Δt; hence the infinite range of the electromagnetic interaction.

In a similar description of the strong interaction, Hideki Yukawa (1907–1981) interpreted the strong interaction between nucleons in terms of the exchange of a virtual particle – the pion. The pion has mass m_π (given in Table 27.3) so that its minimum (rest) energy is $m_\pi c^2$. This sets a minimum value of ΔE and hence, from the uncertainty principle, a maximum value of $\Delta t = \dfrac{\hbar}{m_\pi c^2}$. If we assume that pions travel at velocities close to c (an assumption which is borne out by experiment), this, in turn sets a maximum value of the range of the strong nuclear force, $r' = c\Delta t = \dfrac{\hbar}{m_\pi c}$; hence the short range of the strong nuclear force. Yukawa was able to use the known range of the strong nuclear force (~ 2 fm) to estimate the mass of the pion, more than ten years before the particle was observed.

Gluons, like photons, are massless but, unlike photons which do not carry electric charge, gluons carry colour charge and anticolour charge so that emission or absorption of a gluon always changes the quark colour. As discussed below, this property of the gluon means that the attractive force between two quarks rises rapidly as the quarks are separated. In consequence free quarks are unlikely to be observed, a phenomenon known as 'quark confinement'.

Colour field lines are pulled together by interactions between gluons which, being colour charged, attract one another. The colour field lines draw closer together as separation of the colour charge increases. The gluons line up along the field lines with colours attracted to anticolours. As a result a coloured quark sees more anti-colour, an effect that increases with distance between quarks because more gluons appear so that the effective colour charge seen by each quark increases with distance. Consequently, the force between the quarks increases rapidly with distance and quarks are confined to hadrons.

The gluon force is a basic manifestation of the strong interaction. In this sense the strong interaction between baryons can be compared with the forces which bind atoms in molecules (as discussed in Section 25.1). These forces result from interactions between electrons in adjacent atoms and are thus manifestations of the basic interaction which holds electrons in atoms – the Coulomb force. Gluons, however carry colour charge, which cannot be observed outside a hadron, so that they cannot be exchanged directly between quarks in adjacent baryons. Instead they produce colour neutral quark–antiquark combinations – that is mesons, such as the pion – which can be exchanged between baryons. This is the interaction which we call the strong nuclear force, as described in the Yukawa theory.

The sum of the current masses (noted in Table 27.6) of the three quarks which make up the proton is substantially less than the measured mass of the proton. In fact, quark masses account for only about 1% of the proton mass. The remaining mass is accounted for by the gluons which, although massless, can spontaneously produce quark-antiquark pairs. This sea of quark-antiquark pairs accounts for the remaining hadron mass.

The situation can also be viewed in terms of binding energies. In Section 9.14, we noted that the contribution of the atomic binding energy to the atomic rest mass is negligible. In the nuclear case (discussed in Section 27.2), the binding energy becomes significant – it is about 1% of the rest mass of the nucleus. In the case of the hadron, the binding energy – the energy associated with the gluons – accounts for 99% of the total rest mass.

As noted in Section 1.4, theories to account for the four fundamental interactions have been undergoing a process of unification, with the strong nuclear interaction corresponding to the colour (gluon) interaction at a fundamental level. A model known as the *standard model* has been developed to provide a theory of three of these interactions – electromagnetic, strong nuclear and weak nuclear – and of the elementary particles described in this section.

A more advanced and more comprehensive description of the topics covered in Sections 27.11 and 27.12 may be found in Martin and Shaw (2019).

For problems based on the material presented in this section visit up.ucc.ie/27/ and follow the link to the problems

Astrophysics

We saw in Chapter 5 how astronomy has motivated the development of some of the most fundamental ideas in physics; in particular the contributions of Galileo and Newton were critically influenced by astronomical observations. The concepts, models and processes of nuclear and particle physics are also at the centre of theories of the origin of the Universe and of the birth and development of stars. The *Physics of Stars* by Phillips (Wiley 1999) describes how stellar structure and evolution can be understood in terms of fundamental physics.

Further reading

Nuclear Physics, Principles and Applications, by Lilley.
Nuclear and Particle Physics, an Introduction, 3rd edition, by Martin and Shaw. (See BIBLIOGRAPHY for book details).

WORKED EXAMPLES

Worked Example 27.1: Estimate the Fermi energy of a typical nucleus.

The Fermi energy of any group of fermions is given by Equation (25.18),

$$E_F(0) = \frac{\hbar^2}{2m}(3n_F\pi^2)^{\frac{2}{3}}$$

where n_F is the fermion density (the number of fermions m^{-3}) and m is the fermion mass.
Consider a Fermi gas of neutrons in a spherical nucleus of radius R where (Equation (27.1))

$$R = R_0A^{\frac{1}{3}} \quad \text{and} \quad R_0 \approx 1.1 \text{ fm}$$

In a typical nucleus the number of neutrons, $N \approx 0.60$A. Thus

$$n_F = \frac{0.60A}{\frac{4}{3}\pi R_0^3 A} = \frac{0.60}{1.33\pi(1.1\times10^{-15})^3} = 1.08\times10^{44}\text{ m}^{-3}$$

From Equation (25.18)

$$E_F = \frac{\hbar^2}{2\times1.67\times10^{-27}}(3\times1.08\times10^{44}\times\pi^2)^{\frac{2}{3}} = 45\text{ MeV}$$

Worked Example 27.2: At the present time 99.3% of the uranium found on the Earth comprises the isotope $^{238}_{92}$U with the remaining 0.7% comprising $^{235}_{92}$U. Using the half-life data given in Table 27.1 and, assuming that both isotopes were present in equal abundance when the Earth was formed, estimate the age of the Earth.

If N is the number of nuclei of either isotope present when the Earth was formed and t is the age of the Earth, the numbers of $^{235}_{92}$U and $^{238}_{92}$U nuclei today are (from Equation (27.4))

$$N_{235} = Ne^{-\lambda_{235}t} \quad \text{and} \quad N_{238} = Ne^{-\lambda_{238}t}$$

where λ_{235} and λ_{238} are, respectively, the decay constants of $^{235}_{92}$U and $^{238}_{92}$U

From Equation (27.5), $\lambda_{235} = \dfrac{0.693}{7.13\times10^8} = 9.72\times10^{-10}\text{ yr}^{-1}$ and $\lambda_{238} = \dfrac{0.693}{4.51\times10^9} = 1.53\times10^{-10}\text{ yr}^{-1}$

Thus the present abundance ratio, $$\frac{N_{235}}{N_{238}} = e^{-\lambda_{235}t+\lambda_{238}t} = 0.007$$

which we can write as $$\ln\ 0.007 = t[-(9.72\times10^{-10}) + (1.53\times10^{-10})]$$

which gives $$t \approx 6\times10^9 \text{ years.}$$

This value is in reasonable agreement with that obtained from more sophisticated geological and cosmological arguments.

Worked Example 27.3: An excitation energy of about 6 MeV is needed to provoke a vibration which will lead to fission in a uranium nuclide. Use the semi-empirical mass formula to show that $^{235}_{92}$U is fissile to thermal neutrons whereas $^{238}_{92}$U is not.

The excitation energies of the product nuclei after neutron capture – the energy available to produce fission – are given by

$$[(M_{235} + m_n) - M_{236}]c^2 \text{ in the } ^{235}_{92}\text{U case}$$

and

$$[(M_{238} + m_n) - M_{239}]c^2 \text{ in the } ^{238}_{92}\text{U case}$$

Substituting for M_{235} and M_{236} from the semi-empirical mass formula (Equation (27.2)) and using the values of the fitted parameters given in Section 27.3, we obtain, for the $^{235}_{92}\text{U}$ case,

$$\left\{92m_p + 143m_n + m_n - a_1(235) + a_2(235)^{\frac{2}{3}} + a_3\frac{92^2}{(235)^{\frac{1}{3}}} + a_4\frac{\left(92 - \frac{235}{2}\right)^2}{235}\right\}c^2$$

$$-\left\{92m_p + 144m_n - a_1(236) + a_2(236)^{\frac{2}{3}} + a_3\frac{92^2}{(236)^{\frac{1}{3}}} + a_4\frac{\left(92 - \frac{236}{2}\right)^2}{236} - \frac{a_5}{(236)^{\frac{1}{2}}}\right\}c^2$$

The proton and neutron masses cancel. Substituting for the coefficients a_1 etc., we obtain $(0.00720\ u)c^2 = 6.71$ MeV which is sufficient energy to lead to fission without any contribution from the kinetic energy of the neutron. Capture of a thermal neutron will produce fission.

A similar calculation for $^{238}_{92}\text{U}$ yields 4.88 MeV. To induce fission a contribution of a least 1.15 MeV is needed from the kinetic energy of the neutron. A fast neutron is therefore required. The difference in excitation energies in the two cases is mostly due to contributions from the pairing term. This is zero for $^{235}_{92}\text{U}$ and $^{239}_{92}\text{U}$ but is negative for $^{236}_{92}\text{U}$ and $^{238}_{92}\text{U}$.

Worked Example 27.4: Show that pair production cannot take place spontaneously in empty space, that is that interaction with another particle is needed to satisfy momentum conservation.

Application of energy conservation to spontaneous pair production in vacuum gives

$$hf = 2mc^2 \tag{27.7}$$

where f is the frequency of the photon and m is the mass of the particle or antiparticle

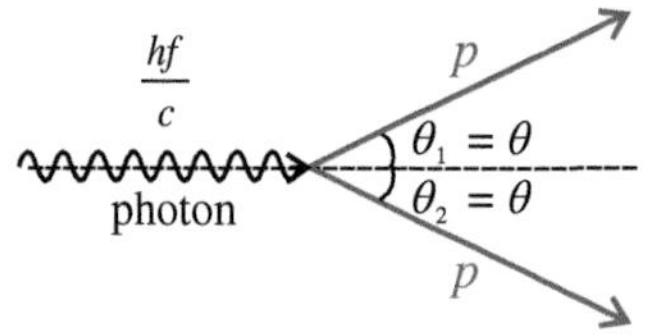

Figure 27.32. Particle-antiparticle pair production (Worked Example 27.4).

If, as shown in Figure 27.32, θ_1 and θ_2 are the angles between the photon direction and directions of the particle and antiparticle which are produced, momentum conservation perpendicular to the photon direction requires that θ_1 and θ_2 have the same value, which we call θ.

If p is the momentum of either the particle or antiparticle, momentum conservation in the photon direction gives

$$\frac{h}{\lambda} = \frac{hf}{c} = 2p\cos\theta \quad \rightarrow \quad hf = 2pc\cos\theta$$

Just above threshold, the kinetic energies of the particle and antiparticle are small so that we can write the momentum (non-relativistically) as $p = mv$

Thus
$$hf = 2mvc\cos\theta = 2mc^2\frac{v}{c}\cos\theta$$

Now $\frac{v}{c} < 1$ and $\cos\theta \le 1$ so that $hf < 2mc^2$

which violates the energy conservation principle, as outlined in Equation (27.7) above. It is impossible, therefore, for pair production to satisfy both energy and momentum conservation unless another particle is available to take some of the photon momentum.

PROBLEMS

For problems based on the material presented in this chapter visit up.ucc.ie/27/ and follow the link to the problems

Appendix A

Mathematical rules and formulas*

A.1 Perimeters, areas and volumes

(i) **Triangle** (base b, perpendicular height h): area $= \frac{1}{2}bh$

(ii) **Circle** (radius r): perimeter $= 2\pi r$
area $= \pi r^2$

(iii) **Ellipse** (semiaxes a and b, see Fig. A.14): area $= \pi ab$

(iv) **Right circular cylinder** (radius r, height h): volume $= \pi r^2 h$
surface area $= 2\pi rh + 2(\pi r^2)$

(v) **Right circular cone** (radius of base r, height h): volume $= \frac{1}{3}\pi r^2 h$
surface area $= \pi r\sqrt{r^2 + h^2} + \pi r^2$

(vi) **Sphere** (radius r): volume $= \frac{4}{3}\pi r^3$
surface area $= 4\pi r^2$

A.2 Plane geometry

Some useful theorems:

Theorem A.2.1 *(Pythagorean Theorem)*

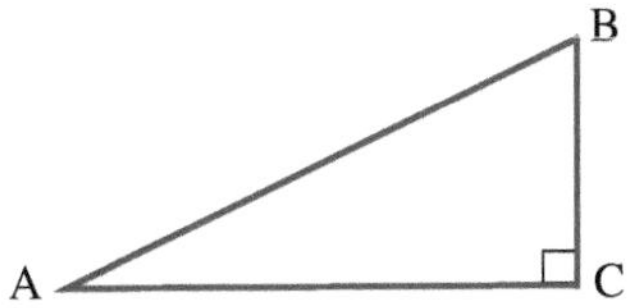

For right angled triangles:
$AC^2 + BC^2 = AB^2$

Figure A.1.

Theorem A.2.2

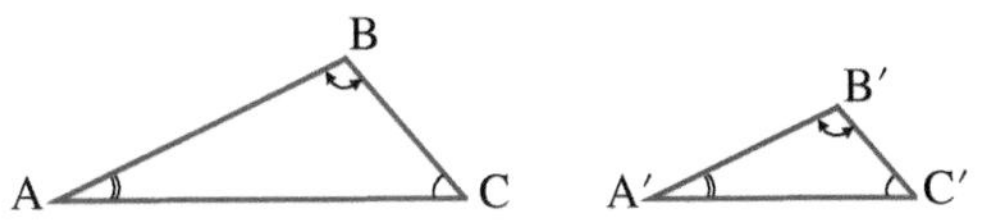

For similar triangles:
$$\frac{AB}{A'B'} = \frac{BC}{B'C'} = \frac{CA}{C'A'}$$

Figure A.2.

*For more details on any of the topics covered in this Appendix, the reader may wish to consult Lambourne and Tinker *Basic Mathematics for the Physical Sciences* (John Wiley & Sons, 2000) or, for more advanced treatments, its companion volume Tinker and Lambourne *Further Mathematics for the Physical* Sciences (John Wiley & Sons, 2000).

Understanding Physics, Third Edition. Michael Mansfield and Colm O'Sullivan.
© 2020 John Wiley & Sons Ltd. Published 2020 by John Wiley & Sons Ltd.

Theorem A.2.3

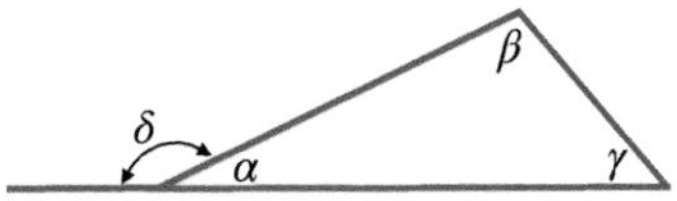

Figure A.3.

For any triangle:
$\alpha + \beta + \gamma = \pi$
$\delta = \beta + \gamma$

Theorem A.2.4

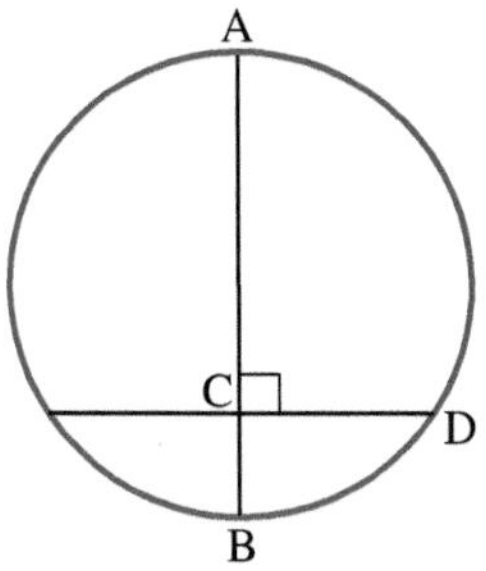

Figure A.4.

For diameter and chord of circle as shown:
AC. CB = CD^2

Theorem A.2.5

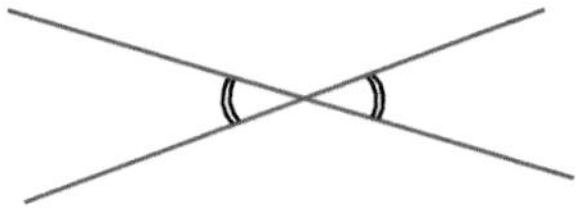

Figure A.5.

For intersecting lines:
vertically opposite angles are equal

Theorem A.2.6

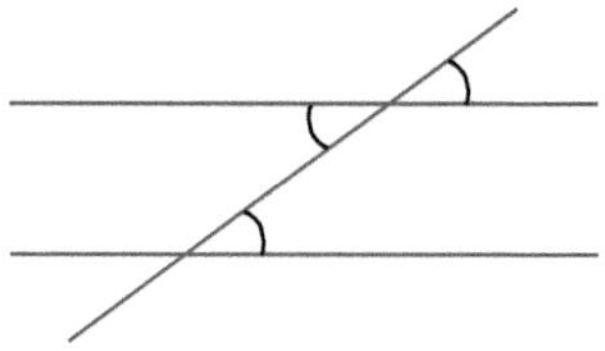

Figure A.6.

For parallel lines:
alternate and corresponding angles are equal

A.3 Trigonometry

A.3.1 A.3.1 Definition of the radian:

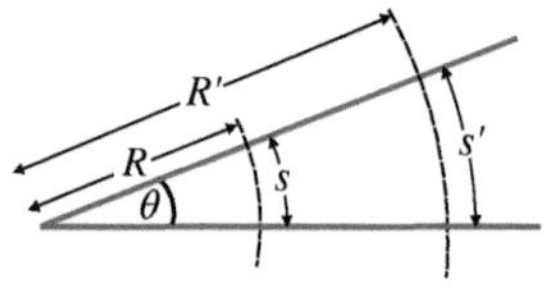

Figure A.7.

$$\theta := \frac{s}{R} = \frac{s'}{R'}$$
When $s = 2\pi R$, $\theta = 2\pi$ rad $= 360°$

[See **Section 2.7** for a discussion of angular measure]

A.3.2 A.3.2 Trigonometric functions

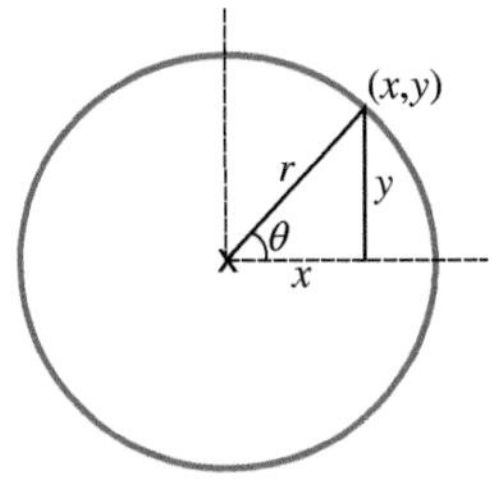

Figure A.8.

Definitions: (see Figure A.8)

$$\sin\theta := \frac{y}{r} \qquad \cos\theta := \frac{x}{r} \qquad \tan\theta := \frac{y}{x} \qquad \text{cosec}\,\theta := \frac{r}{y} \qquad \sec\theta := \frac{r}{x} \qquad \cot\theta := \frac{x}{y}$$

(where x and y are the coordinates of points on a circle of radius r)

From the definitions:

$$\tan\theta = \frac{\sin\theta}{\cos\theta} \qquad \text{cosec}\theta = \frac{1}{\sin\theta} \qquad \sec\theta = \frac{1}{\cos\theta} \qquad \cot\theta = \frac{1}{\tan\theta}$$

Note that

(i) $\sin\left(\frac{\pi}{2} \pm \theta\right) = \cos\theta$ (ii) $\cos\left(\frac{\pi}{2} \pm \theta\right) = \mp\sin\theta$

(iii) $\sin(\pi \pm \theta) = \mp\sin\theta$ (iv) $\cos(\pi \pm \theta) = -\cos\theta$

(v) $\sin(-\theta) = -\sin\theta$ (vi) $\cos(-\theta) = \cos\theta$

Inverse trigonometric functions:

If $\sin\theta = u, \cos\theta = v$ and $\tan\theta = w$, then $\theta = \sin^{-1}u = \cos^{-1}v = \tan^{-1}w$ (or $\theta = \arcsin u = \arccos v = \arctan w$)

A.3.3 The cosine rule

$$c^2 = a^2 + b^2 - 2ab\cos\gamma$$

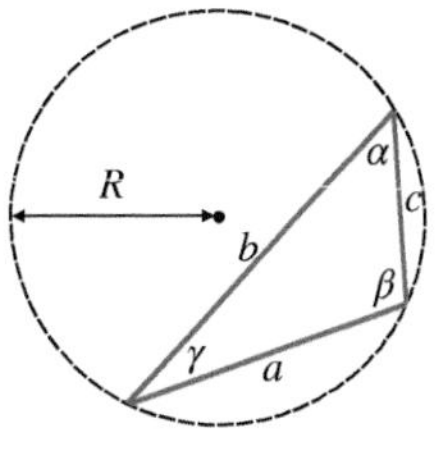

Figure A.9.

A.3.4 The sine rule

$$\frac{a}{\sin\alpha} = \frac{b}{\sin\beta} = \frac{c}{\sin\gamma} = 2R \text{ where } R \text{ is the radius of the excribed circle}$$

A.3.5 Some trigonometric identities:

(i) $\sin^2\theta + \cos^2\theta = 1$ (ii) $\sec^2\theta = 1 + \tan^2\theta$ (iii) $\text{cosec}^2\theta = 1 + \cot^2\theta$

(iv) $\sin 2\theta = 2\sin\theta\cos\theta$ (v) $\cos 2\theta = 2\cos^2\theta - 1 = 1 - 2\sin^2\theta = \cos^2\theta - \sin^2\theta$

(vi) $\sin\frac{1}{2}\theta = \sqrt{\frac{1}{2}(1 - \cos\theta)}$ (vii) $\cos\frac{1}{2}\theta = \sqrt{\frac{1}{2}(1 + \cos\theta)}$

(viii) $\sin(\alpha \pm \beta) = \sin\alpha\cos\beta \pm \cos\alpha\sin\beta$

(ix) $\cos(\alpha \pm \beta) = \cos\alpha\cos\beta \mp \sin\alpha\sin\beta$

(x) $\tan(\alpha \pm \beta) = \dfrac{\tan\alpha \pm tan\beta}{1 \mp \tan\alpha\tan\beta}$

(xi) $\sin\alpha + \sin\beta = 2\sin\left[\frac{1}{2}(\alpha - \beta)\right]\cos\left[\frac{1}{2}(\alpha - \beta)\right]$; $\sin\alpha - \sin\beta = 2\cos\left[\frac{1}{2}(\alpha + \beta)\right]\sin\left[\frac{1}{2}(\alpha - \beta)\right]$

(xii) $\cos\alpha + \cos\beta = 2\cos\left[\frac{1}{2}(\alpha + \beta)\right]\cos\left[\frac{1}{2}(\alpha - \beta)\right]$; $\cos\alpha - \cos\beta = -2\sin\left[\frac{1}{2}(\alpha + \beta)\right]\sin\left[\frac{1}{2}(\alpha - \beta)\right]$

(xiii) $\sin\theta = \dfrac{2t}{1 + t^2}$ (xiv) $\cos\theta = \dfrac{1 - t^2}{1 + t^2}$ (xv) $\tan\theta = \dfrac{2t}{1 - t^2}$ $\left(\text{where } t = \tan\frac{1}{2}\theta\right)$

A.4 Algebraic identities

A.4.1 Some useful identities

(i) $a(b+c) = ab + ac$ (ii) $(a \pm b)^2 = a^2 \pm 2ab + b^2$
(iii) $(ab)^n = a^n b^n$ (iv) $a^n a^m = a^{n+m}$ (v) $(a^n)^m = a^{nm}$
(vi) $a^2 - b^2 = (a+b)(a-b)$ (vii) $a^3 - b^3 = (a-b)(a^2 + ab + b^2)$

A.4.2 Solution of quadratic equations

$$\text{If } ax^2 + bx + c = 0 \quad \text{then} \quad x = \frac{-b \pm \sqrt{b^2 - 4ac}}{2a}$$

A.5 The logarithmic and exponential functions

A.5.1 The logarithmic function

The **logarithm** of a number x to the base b, denoted by $\log_b x$, is defined such that $b^{\log_b x} = x$ (for $b > 0, b \neq 1$)

(i) $\log_b(xy) = \log_b x + \log_b y$ (ii) $\log_b\left(\frac{x}{y}\right) = \log_b x - \log_b y$
(iii) $\log_b(x^n) = n\log_b x$ (n real) (iv) $\log_b x = \frac{\log_c x}{\log_c b}$
(v) $\log_b 1 = 0$ (vi) $\log_b b = 1$

A.5.2 The exponential function

The **exponential function** is defined by the infinite series

$$e^x := \exp(x) := 1 + x + \frac{x^2}{2!} + \frac{x^3}{3!} + \frac{x^4}{4!} + \cdots$$

where $n!$ is called factorial n and is defined as

$$n! := n(n-1)(n-2)(n-3)\ldots\ldots\ldots 4.3.2.1$$

The **number** $e = e^1 = 1 + \frac{1}{1!} + \frac{1}{2!} + \frac{1}{3!} + \frac{1}{4!} + \cdots = 2.718282\ \ldots$
The **natural logarithm** of x is denoted by $\ln x$ and is the logarithm of x to base e; that is, $\ln x = \log_e x \rightarrow x = e^{\ln x}$

A.6 Differentiation

Definition: If $f = f(x)$, the derivative of f with respect to x is defined as

$$\frac{df}{dx} := f'(x) := \operatorname*{limit}_{\Delta x \to 0} \frac{\Delta f}{\Delta x}$$

The derivatives of a number of common functions are given in Table A.1.

A.6.1 Rules for differentiation

Rule (i) If $f = f(x)$ and c is constant then $\frac{d}{dx}(cf) = c\frac{df}{dx}$

For example: $\frac{d}{dx}(5x^3) = 5\frac{d(x^3)}{dx} = 15x^2$

Rule (ii) If $f = f(x)$ and $g = g(x)$ then $\frac{d}{dx}(f+g) = \frac{df}{dx} + \frac{dg}{dx}$

For example: $\frac{d}{dx}(5x^3 + 4x^2 + 6) = \frac{d(5x^3)}{dx} + \frac{d(4x^2)}{dx} + \frac{d6}{dx} = 15x^2 + 8x + 0$

Rule (iii) If $f = f(x)$ and $g = g(x)$ then $\frac{d}{dx}(fg) = f\frac{dg}{dx} + g\frac{df}{dx}$

For example: $\frac{d}{dx}(x^3 \ln x) = x^3\frac{d(\ln x)}{dx} + \ln x\frac{d(x^3)}{dx} = x^3\frac{1}{x} + (\ln x)(3x^2) = x^2(1 + 3\ln x)$

Rule (iv) If $f = f(u)$ and $u = u(x)$ then $\frac{df}{dx} = \frac{df}{du}\frac{du}{dx}$ (the '**chain rule**')

For example: $\frac{d}{d\theta}(\sin 3\theta) = \frac{d(\sin 3\theta)}{d(3\theta)}\frac{d(3\theta)}{d\theta} = 3\cos(3\theta)$

[See **Section 2.4** for an introduction to the use of differentiation in the analysis of physics]

Table A.1 Some common derivatives

$f(x)$	$\frac{df}{dx}$
x^n [n integer]	nx^{n-1}
$\ln x$	$\frac{1}{x}$
$x \ln x$	$1 + \ln x$
$\sin x$	$\cos x$
$\cos x$	$-\sin x$
$e^{\pm x}$	$\pm e^{\pm x}$
$\tan x$	$\sec^2 x$
$\cot x$	$-\mathrm{cosec}^2 x$
$\sec x$	$\tan x \sec x$
$\mathrm{cosec}\, x$	$-\cot x\, \mathrm{cosec}\, x$
$\sin^{-1}x\ [-\frac{\pi}{2} < \sin^{-1}x < \frac{\pi}{2}]$	$\frac{1}{\sqrt{1-x^2}}$
$\cos^{-1}x\ [-\pi < \cos^{-1}x < \pi]$	$-\frac{1}{1-x^2}$
$\tan^{-1}x$	$\frac{1}{1+x^2}$

A.6.2 Maxima and minima

If $\frac{df}{dx} = 0$ at $x = a$ then the point $(x = a, f(x) = f(a))$ is either a maximum or a minimum of $f(x)$

If $\frac{d^2f}{dx^2} < 0$ at $x = a$ then the point is a maximum

If $\frac{d^2f}{dx^2} > 0$ at $x = a$ then the point is a minimum

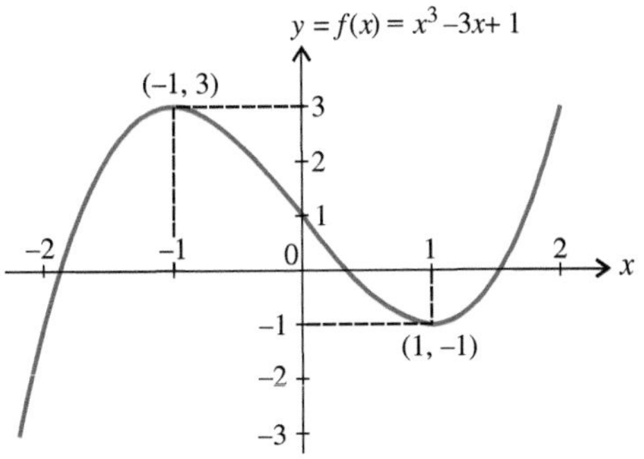

Figure A.10.

For example, if $f(x) = x^3 - 3x + 1$ (see Figure A.10)

$$\frac{df}{dx} = 3x^2 - 3 \quad \text{and thus} \quad \frac{df}{dx} = 0 \Rightarrow 3x^2 = 3 \rightarrow x = \pm 1$$

$$f(-1) = 3 \quad \text{and} \quad f(+1) = -1$$

Now $\quad \frac{d^2f}{dx^2} = 6x \;\Rightarrow\; \text{at } x = -1 \quad \frac{d^2f}{dx^2} = -6 \;\Rightarrow\; (-1, 3)$ is a maximum value of the function

and $\quad \text{at } x = +1 \quad \frac{d^2f}{dx^2} = +6 \;\Rightarrow\; (1, -1)$ is a minimum value

[See **Section 2.6** for an introduction to the use of maximum/minimum techniques in physics]

A.6.3 Partial derivatives

Definition: If $f = f(x, y, z)$, the partial derivative of f with respect to x is defined as

$$\frac{\partial f}{dx} := \underset{\Delta x \to 0}{\text{limit}} \left[\frac{f(x + \Delta x, y, z) - f(x, y, z)}{\Delta x} \right]$$

with similar definitions for $\frac{\partial f}{\partial y}$ and $\frac{\partial f}{\partial z}$.

Thus the partial derivative of a function of a number of independent variables with respect to one of those variables can be obtained using the rules for ordinary differentiation and treating all the other variables as constants.

For example, if $f(x, y, z) = ax^2z + bxy - cyz^3$

$$\rightarrow \quad \frac{\partial f}{\partial x} = 2axz + by, \quad \frac{\partial f}{\partial y} = bx - cz^3 \quad \text{and} \quad \frac{\partial f}{\partial z} = ax^2 - 3cyz^2$$

If $f = f(x, y, z)$, then a small change in f due to simultaneous small changes Δx, Δy and Δz in x, y and z is given by

$$\Delta f = \frac{\partial f}{\partial x}\Delta x + \frac{\partial f}{\partial y}\Delta y + \frac{\partial f}{\partial z}\Delta z$$

If $F = GH$, where $G = G(x, y, z)$ and $H = H(x, y, z)$, then

$$\Delta F = G\Delta H + H\Delta G$$

In general, physical quantities are 'well behaved functions', for example they are finite and continuous everywhere (see Fig. 14.19 for examples of functions which are not well behaved in this sense). For such well behaved functions

$$\frac{\partial}{\partial y}\left(\frac{\partial f}{\partial x}\right) = \frac{\partial^2 f}{\partial y \partial x} = \frac{\partial^2 f}{\partial x \partial y} = \frac{\partial}{\partial x}\left(\frac{\partial f}{\partial y}\right)$$

[See **Section 5.7** for an example of the use of partial derivatives in physics]

A.7 Integration

Definition:

The **integral function**, or the **antiderivative**, of a function $f(x)$ is defined as a function $F(x)$ such that $\frac{dF}{dx} = f(x)$ and the **integral** of $f(x)$ is given by $\int f(x)dx = F(x) + c$ where c is an arbitrary constant. An integral in this form of is called an **indefinite integral**. If the value of the integral is known at a particular value of x, the value of c can be determined.

Integral functions of a number of commonly used functions are given in Table A.2.

The **integral** of a function $f(x)$ between two points $x = x_A$ and $x = x_B$ can be defined as

$$\int_{x=x_A}^{x=x_B} f(x)dx := \underset{\Delta x_i \to 0}{\text{limit}}[f(x_1)\Delta x_1 + f(x_2)\Delta x_2 + \ldots \ldots \ldots \ldots] = \sum_i f(x_i)\Delta x_i$$

Such an integral (that is, an integral between limits) is called a **definite integral** and can be evaluated in terms of the values of the function $F(x)$ at the two limits such that

$$\int_{x=x_A}^{x=x_B} f(x)dx := F(x_B) - F(x_A)$$

The values of some useful definite integrals are given in Table A.3.

A.7.1 Rules for integration

Rule (i) If $f = f(x)$ and c is constant then $\int_a^b cf(x)dx = c\int_a^b f(x)dx$

For example, $$\int_0^2 3x^3\,dx = 3\int_0^2 x^3\,dx = 3\left[\frac{x^4}{4}\right]_0^2 = 12$$

Rule (ii) If $f = f(x)$ and $g = g(x)$ then

$$\int_a^b \{f(x) + g(x)\}dx = \int_a^b f(x)dx + \int_a^b g(x)dx$$

For example, $$\int_{-1}^{+1} (x^4 + 3x^2)dx = \int_{-1}^{+1} x^4\,dx + 3\int_{-1}^{+1} x^2\,dx = \left[\frac{x^5}{5}\right]_{-1}^{+1} + 3\left[\frac{x^3}{3}\right]_{-1}^{+1} = \frac{12}{5}$$

Rule (iii) If $f = f(x)$ then $\int_a^b f(x)dx = \int_a^c f(x)dx + \int_c^b f(x)dx$

and $\int_a^b f(x)dx = -\int_b^a f(x)dx$

Table A.2 Some common indefinite integrals

$f(x)$	$\int f(x)dx$
$x^n (n \text{ constant} \neq -1)$	$\frac{x^{n+1}}{n+1}$
$\frac{1}{x}$	$\ln\lvert x\rvert$
$\sin x$	$-\cos x$
$\cos x$	$\sin x$
$\tan x$	$-\ln \cos x$
$\sec x$	$\ln \lvert\sec x + \tan x\rvert$
$\mathrm{cosec}\, x$	$\ln \lvert\mathrm{cosec}\, x + \cot x\rvert$
$\cot x$	$\ln \lvert\sin x\rvert$
$\sin^2 x$	$\frac{1}{2}(x - \sin x \cos x) = \frac{1}{2}\left(x - \frac{1}{2}\sin 2x\right)$
$\cos^2 x$	$\frac{1}{2}(x + \sin x \cos x) = \frac{1}{2}\left(x + \frac{1}{2}\sin 2x\right)$
$\sin^3 x$	$\frac{1}{3}(\cos^3 x - 3\cos x)$
$\cos^3 x$	$\frac{1}{3}(3\sin x - \sin^3 x)$
$\tan^2 x$	$\tan x - x$
$x \sin x$	$\sin x - x\cos x$
$\sin x \cos x$	$\frac{1}{2}\sin^2 x$
$x\sin^2 x$	$\frac{1}{4}\left(x^2 - x\sin 2x - \frac{1}{2}\cos 2x\right)$
$e^{\pm x}$	$\pm e^{\pm x}$
$\ln x$	$x\ln x - x$
$\frac{1}{1+x}$	$\ln(1+x)$
$\frac{1}{1-x^2}$	$\frac{1}{2}\ln\left\lvert\frac{x+1}{x-1}\right\rvert$
$\frac{1}{\sqrt{x^2 \pm 1}} (x > 1)$	$\ln(x + \sqrt{x^2 \pm 1})$
$\frac{1}{\sqrt{1-x^2}} \quad (x < 1)$	$\sin^{-1} x$
$\sqrt{1-x^2}$	$\frac{1}{2}(x\sqrt{1-x^2} + \sin^{-1} x)$
$\frac{1}{1+x^2}$	$\tan^{-1} x$
$\frac{1}{x\sqrt{1 \pm x^2}}$	$-\ln\left(\frac{1+\sqrt{1 \pm x^2}}{x}\right)$

Table A.3 Some useful definite integrals

$\int_0^\infty x^n e^{-x} dx = n! \quad (n \text{ integer} > 0)$	
$\int_0^\infty \frac{x}{e^x - 1} dx = \frac{\pi^2}{6}$	$\int_0^\infty \frac{x^3}{e^x - 1} dx = \frac{\pi^4}{15}$
$\int_0^\infty \frac{dx}{1+e^x} = \ln 2$	$\int_0^\infty \frac{e^{-x}}{\sqrt{x}} dx = \sqrt{\pi}$
$\int_0^\infty e^{-x^2} dx = \frac{1}{2}\sqrt{\pi}$	$\int_0^\infty x e^{-x^2} dx = \frac{1}{2}$
$\int_0^\infty x^2 e^{-x^2} dx = \frac{1}{4}\sqrt{\pi}$	$\int_0^\infty x^3 e^{-x^2} dx = \frac{1}{2}$
$\int_0^\infty x^4 e^{-x^2} dx = \frac{3}{8}\sqrt{\pi}$	$\int_0^\infty x^5 e^{-x^2} dx = 1$
$\int_0^\infty x^6 e^{-x^2} dx = \frac{15}{16}\sqrt{\pi}$	$\int_0^\infty x^7 e^{-x^2} dx = 3$
$\int_0^a x\sin^2\left(\frac{\pi x}{a}\right) dx = \left(\frac{a}{\pi}\right)^2 \int_0^\pi \xi \sin^2\xi\, d\xi = \frac{a^2}{4}$	

Rule (iv) If $f = f(u)$ and $u = u(x)$ then $\int_a^b f(u)du = \int_{x(a)}^{x(b)} f(u)\frac{du}{dx}dx$

For example, $\int \sin x \cos x dx = \int \sin x \, d(\sin x) = \frac{1}{2}\sin^2 x + \text{constant}$

or, for example, $\int_0^\pi \cos^2 u \sin u \, du = ?$

Let $\cos u = x \quad \rightarrow \quad \frac{dx}{du} = -\sin u \quad \rightarrow \quad \frac{du}{dx} = -\frac{1}{\sin u}$

$$\Rightarrow \int_0^\pi \cos^2 u \sin u \, du = -\int_{+1}^{-1} x^2 dx = \left[\frac{x^3}{3}\right]_{-1}^{+1} = \frac{2}{3}$$

Rule (v) (**Integration by parts**) If $u = u(x)$ and $v = v(x)$ then

$$\int u\frac{dv}{dx}\,dx = uv - \int v\frac{du}{dx}\,dx \rightarrow \int_a^b u\,dv = [uv]_a^b - \int_a^b v\,du$$

For example,

$$\int xe^{ax}dx = \frac{1}{a}\int x\frac{de^{ax}}{dx}\,dx = \frac{1}{a}xe^{ax} - \frac{1}{a}\int e^{ax}\,dx = \frac{1}{a}xe^{ax} - \frac{1}{a^2}e^{ax}\text{ (+ an arbitrary constant)}$$

or, for example,

$$\int_0^2 xe^{3x}dx = \frac{1}{3}[xe^{3x}]_0^2 - \frac{1}{3}\int_0^2 e^{3x}\,dx = \frac{1}{3}[xe^{3x}]_0^2 - \frac{1}{9}[e^{3x}]_0^2 = \left[\frac{1}{3}e^{3x}\left(x - \frac{1}{3}\right)\right]_0^2 = \frac{5e^6+1}{9}$$

[See **Section 2.5** for an introduction to the use of integration in the analysis of problems in physics]

A.8 Complex numbers

Definitions:

$z := x + iy$ (where x and y are real numbers and $i := \sqrt{-1}$)
real part of $z := \mathrm{Re}(z) := x$
imaginary part of $z := \mathrm{Im}(z) := y$
complex conjugate of $z := z^* := x - iy$
square modulus of $z := |z|^2 := z^*z = x^2 + y^2$
modulus (or magnitude) of $z := |z| = \sqrt{x^2 + y^2}$

Some useful identities

(i) $e^{\pm ix} = \cos x \pm i\sin x$

(ii) $\sin x = \frac{1}{2i}(e^{ix} - e^{-ix})$

(iii) $\cos x = \frac{1}{2}(e^{ix} + e^{-ix})$

(iv) $(\cos x + i\sin x)^n = \cos nx + i\sin nx$

(v) $e^x = \sinh x + \cosh x$

(vi) $\sinh x = \frac{1}{2}(e^x - e^{-x})$

(vii) $\cosh x = \frac{1}{2}(e^x + e^{-x})$

(viii) $\sin(ix) = i\sinh x$
$\sinh(ix) = i\sin x$

A.9 Series expansions

A.9.1 Series expansions of some common functions

(i) $e^x := \exp(x) := 1 + x + \frac{x^2}{2!} + \frac{x^3}{3!} + \frac{x^4}{4!} + \frac{x^5}{5!} + \ldots$ (recall definition in section A.5.2)

(ii) $\sin x = x - \frac{x^3}{3!} + \frac{x^5}{5!} + \ldots = \frac{1}{2i}(e^{ix} - e^{-ix})$

(iii) $\cos x = 1 - \frac{x^2}{2!} + \frac{x^4}{4!} + \ldots = \frac{1}{2}(e^{ix} + e^{-ix})$

(iv) $\tan^{-1}x = x - \frac{x^3}{3} + \frac{x^5}{5} - \cdots$

(v) $\ln(1+x) = x - \frac{x^2}{2} + \frac{x^3}{3} - \cdots$

A.9.2 The binomial theorem

$$(1+x)^n = 1 + \frac{n}{1!}x + \frac{n(n-1)}{2!}x^2 + \frac{n(n-1)(n-2)}{3!}x^3 + \cdots$$

A.9.3 The Taylor series

$$f(x) = f(a) + \left[\frac{df}{dx}\right]_{x=a}(x-a) + \frac{1}{2!}\left[\frac{d^2f}{dx^2}\right]_{x=a}(x-a)^2 + \frac{1}{3!}\left[\frac{d^3f}{dx^3}\right]_{x=a}(x-a)^3 + \cdots$$

where $\left[\frac{df}{dx}\right]_{x=a} := f'(a)$ is the value of the derivative of $f(x)$ at $x = a$,

where $\left[\frac{df^2}{dx^2}\right]_{x=a} := f''(a)$ is the value of the second derivative of $f(x)$ at $x = a$, etc.

Note: $f(x + \Delta x) = f(x) + \frac{df}{dx}(\Delta x) + \frac{1}{2!}\frac{d^2f}{dx^2}(\Delta x)^2 + \ldots.$

A.9.4 Arithmetic progression

$$a + (a+h) + (a+2h) + \cdots + [a + (n-1)h] = \sum_{k=0}^{k=n-1}(a+kh) = \frac{n}{2}[2a + (n-1)h]$$

A.9.5 Geometric progression

$$a + ar + ar^2 + \cdots + ar^{n-1} = \sum_{k=0}^{k=n-1} ar^k = \frac{a(r^n - 1)}{r-1}$$

Sum to infinity For $-1 < r < 1$ $\quad \sum_{k=0}^{k=\infty} ar^k = \frac{a}{1-r}$

A.10 Analytical geometry

A.10.1 The line

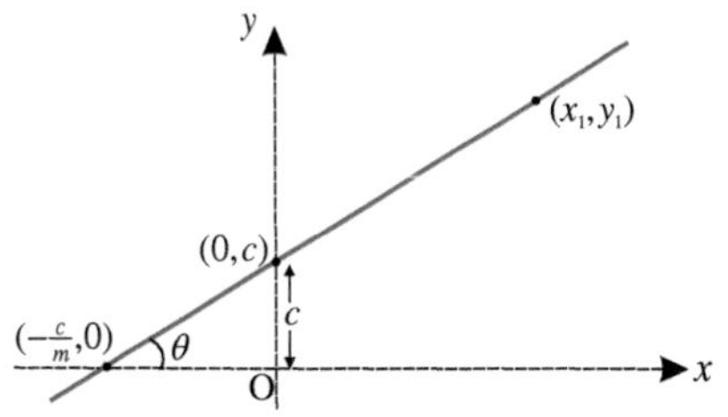

Figure A.11.

The equation of a (straight) **line** (Figure A.11) in plane Cartesian coordinates is

$$y = mx + c$$

where $m = \tan\theta = constant$ is the **slope** of the line and $c = constant$ is the **intercept** the line makes with the y-axis. Alternatively, the equation of a line can be written as

$$y - y_1 = m(x - x_1)$$

where (x_1, y_1) is any point on the line.

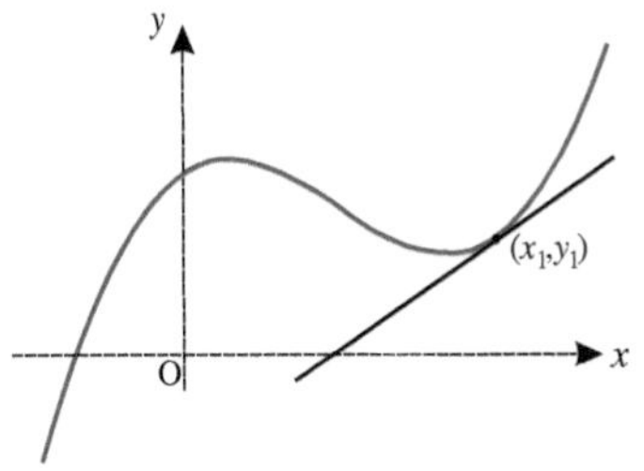

Figure A.12.

The slope of the *tangent* line to a curve $y = y(x)$ at a point (x_1, y_1) (see Figure A.12) is given by

$$m = \left[\frac{dy}{dx}\right]_{x=x_1}$$

A.10.2 Conic sections in plane Cartesian coordinates

A.10.2(i) The circle

The equation of a **circle** with its centre at the point (h,k) (Figure A.13) is

$$(x-h)^2 + (y-k)^2 = a^2$$

where a (= constant) is the radius of the circle.

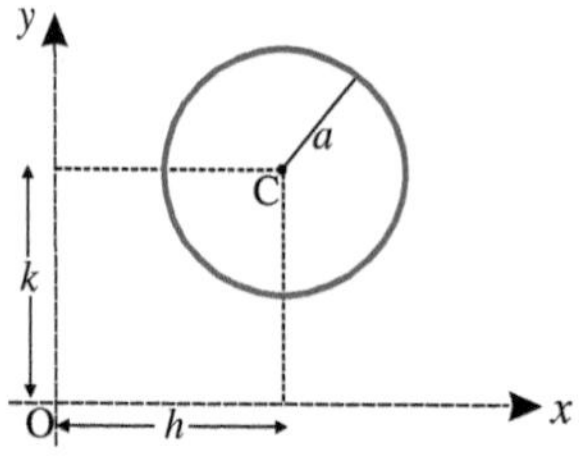

Figure A.13.

If the origin of coordinates is moved to the centre of the circle the equation becomes

$$x^2 + y^2 = a^2$$

which can also be written in parametric form as follows $x = a\cos t$; $y = a\sin t$

A.10.2(ii) The ellipse

The equation $$\frac{(x-h)^2}{a^2} + \frac{(y-k)^2}{b^2} = 1$$

describes an **ellipse** with its centre at the point (h,k), as indicated in Figure A.14
The constants a and b are the semimajor and semiminor axes, respectively.
If the origin of coordinates is at the centre of the ellipse the equation becomes

$$\frac{x^2}{a^2} + \frac{y^2}{b^2} = 1$$

and the corresponding parametric form of this equation is

$$x = a\cos t; \quad y = b\sin t$$

Note that a circle is a special case of an ellipse with $b = a$.

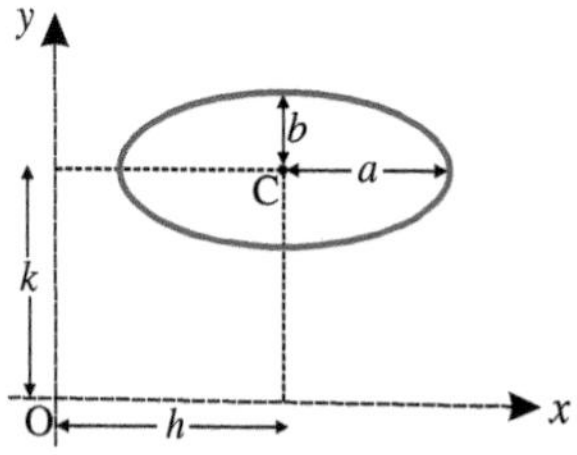

Figure A.14.

A.10.2(iii) The parabola

The equation $$y - y_1 = 4a(x - x_1)^2$$

where a is a contant, describes a **parabola** with its vertex at the point $V(x_1, y_1)$ as indicated in Figure A.15. The corresponding parametric form of this equation is

$$x = at^2; \quad y = 2at$$

If the origin of coordinates is at the vertex of the parabola, the equation becomes

$$y = 4ax^2$$

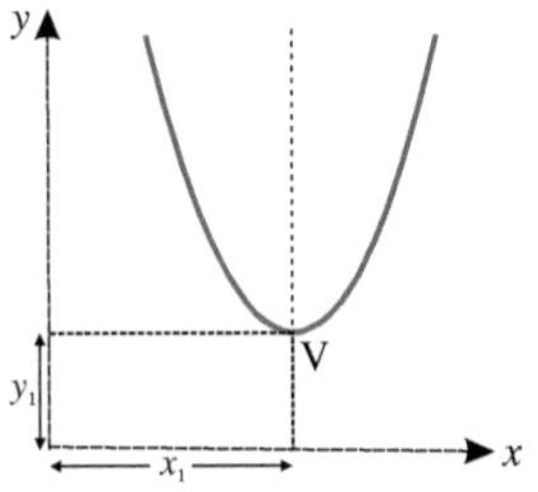

Figure A.15.

A.10.2(iv) The hyperbola

The equation
$$\frac{x^2}{a^2} - \frac{y^2}{b^2} = 1$$
where a and b are constants, describes a **hyperbola** as in Figure A.16. The parametric form of this equation is

$$x = a \cosh t; \quad y = b \sinh t$$

The dashed lines, which coincide with the curve where $x, y \to \infty$, are called the **asymptotes** and have equations $y = \pm\frac{b}{a}x$.

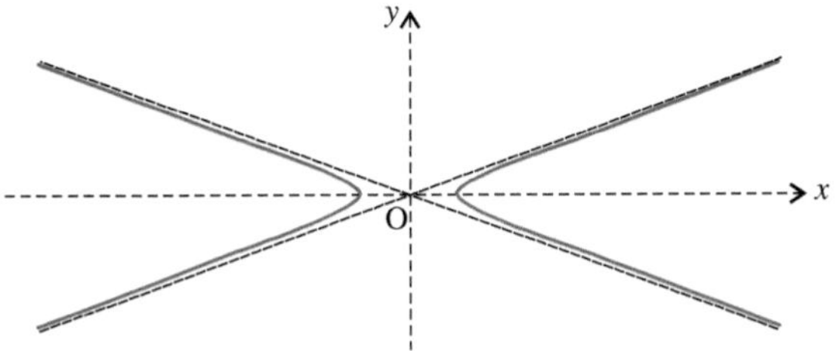

Figure A.16.

A.10.2(v) The rectangular hyperbola

A hyperbola with $b = a$ has asymptotes given by $y = x$ and $y = -x$, respectively, which are lines making angles at 45° with the coordinate axes (Figure A.17). This special case, known as a **rectangular hyperbola**, is described by the equation

$$x^2 - y^2 = a^2$$

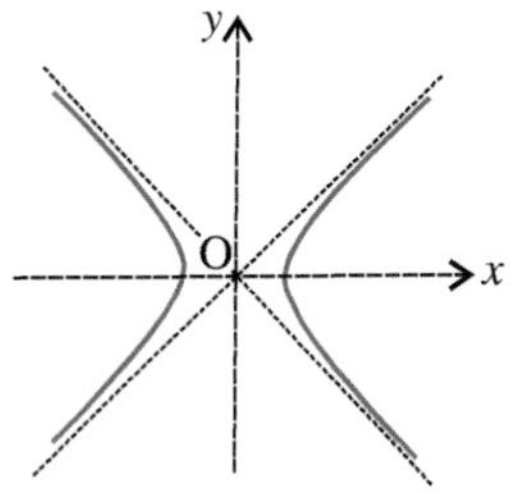

Figure A.17.

If the coordinate system is rotated through 45° so that the asymptotes are the x- and y-axes (Figure A.18(e)), the equation of a rectangular hyperbola becomes $xy = c^2$ or

$$y = \frac{c^2}{x}$$

where c is a constant. The corresponding parametric equations are $x = ct;\ y = \frac{c}{t}$.

Note: The class of curves discussed in this section are known as **conic sections** (or **conics**) since they can all be generated by the intersection of a plane and a double right circular cone.

A.10.3 Conic sections in plane polar coordinates

The curves discussed in section A.10.2 above can all be described by the following single equation when polar (r, θ) coordinates are used (plane polar coordinates are related to Cartesian coordinates by $x = r\cos\theta$ and $y = r\sin\theta$, see Figure A.18)

$$r = \frac{\alpha}{1 + \varepsilon \cos\theta}$$

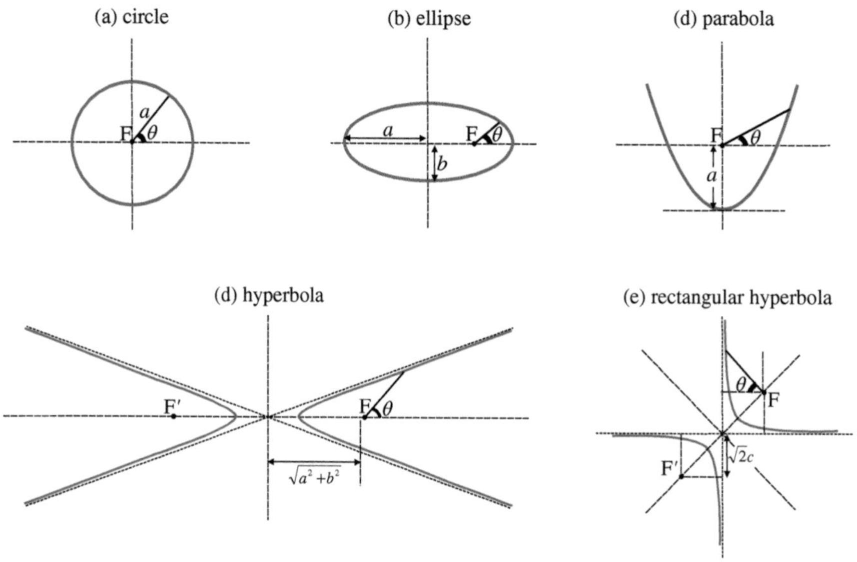

Figure A.18. Note that the rectangular hyperbola (e) has been rotated by 45°

where α and ε are constants. The particular curve obtained depends on the value of ε (which is called the **eccentricity** of the curve) as follows.

(a) $\varepsilon = 0$ for a circle

(b) $\varepsilon = \dfrac{\sqrt{a^2 - b^2}}{a} < 1$ for an ellipse

(c) $\varepsilon = 1$ for a parabola

(d) $\varepsilon = \dfrac{\sqrt{a^2 + b^2}}{a} > 1$ for a hyperbola

(e) $\varepsilon = \sqrt{2}$ for a rectangular hyperbola

The origin of the coordinate system, relative to which r and θ in the polar equation above are given, is called a **focus** of the conic section. The location of the foci of the different curves are indicated in Figure A.18. Note that ellipses and hyperbolae have two such foci.

A.11 Vector algebra

[See **Sections 4.2 and 4.11** for an introduction to the use of vector techniques]

A.11.1 Addition of vectors

Two vectors, $\boldsymbol{A}$ and $\boldsymbol{B}$ may be added to form a *resultant* vector $\boldsymbol{C} = \boldsymbol{A} + \boldsymbol{B}$ by either the *triangle* method (Figure A.19a) or the *parallelogram* method (Figure A.19b).

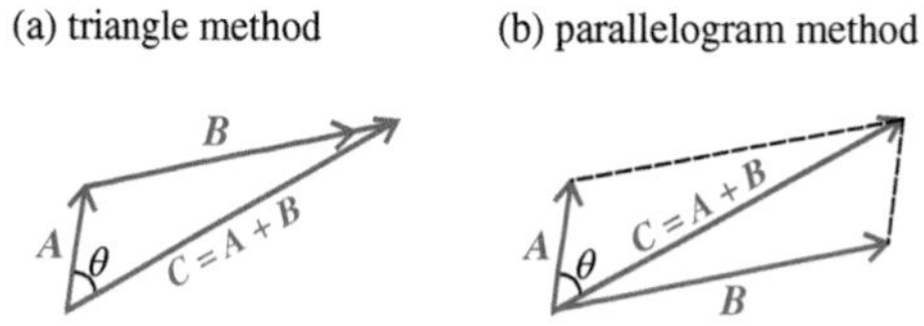

Figure A.19.

The cosine rule (A.3.3) and/or the sine rule (A.3.4) may be used to determine the magnitude and the direction of $\boldsymbol{C}$.

A.11.2 Negative vectors and subtraction of vectors

The **negative vector** $-\boldsymbol{A}$ is a vector which has the same magnitude as the vector $\boldsymbol{A}$ but which is in the opposite direction to $\boldsymbol{A}$. Thus

$$\boldsymbol{A} + (-\boldsymbol{A}) = 0$$

The **difference** of vector $\boldsymbol{A}$ and vector $\boldsymbol{B}$ is defined as

$$C := \boldsymbol{A} - \boldsymbol{B} := \boldsymbol{A} + (-\boldsymbol{B})$$

A.11.3 The magnitude of a vector

The **magnitude** of the vector $\boldsymbol{A}$ is denoted by $|\boldsymbol{A}|$ and is a positive quantity equal to the 'length' of the vector.

A.11.4 Multiplication of vectors by scalars

The multiplication of the vector $\boldsymbol{A}$ by the scalar c produces a vector in the same direction as $\boldsymbol{A}$ but with a magnitude which is c times the magnitude of $\boldsymbol{A}$. That is

$$\boldsymbol{B} = c\boldsymbol{A} \quad \text{where} \quad |\boldsymbol{B}| = c|\boldsymbol{A}| \quad \text{and} \quad \boldsymbol{B} \text{ is parallel to } \boldsymbol{A}$$

A.11.5 Unit vectors and components of vectors

A **unit vector** is a vector of magnitude unity. For example the unit vector in the direction of a vector $\boldsymbol{A}$ is defined as

$$\boldsymbol{a} := \frac{\boldsymbol{A}}{|\boldsymbol{A}|}$$

If we define $\boldsymbol{i}, \boldsymbol{j}$ and $\boldsymbol{k}$ to be unit vectors in the x-, y-and z-directions, respectively, we can write

$$C = C_x\boldsymbol{i} + C_y\boldsymbol{j} + C_z\boldsymbol{k} = |C| \cos\theta_x\boldsymbol{i} + |C| \sin\theta_y\boldsymbol{j} + |C| \sin\theta_z\boldsymbol{k}$$

where θ_x, θ_y and θ_z are the angles the vector $\boldsymbol{C}$ makes with the positive x-, y- and z-axes, respectively.
The quantities C_x, C_y and C_z are *algebraic* quantities called, respectively, the x-, y- and z-**components** of $\boldsymbol{C}$.
In general, the component of $\boldsymbol{C}$ in the x- direction (say) is given by

$$C_x = |C| \cos\theta_x$$

and

$$|C|^2 = C_x^2 + C_y^2 + C_z^2$$

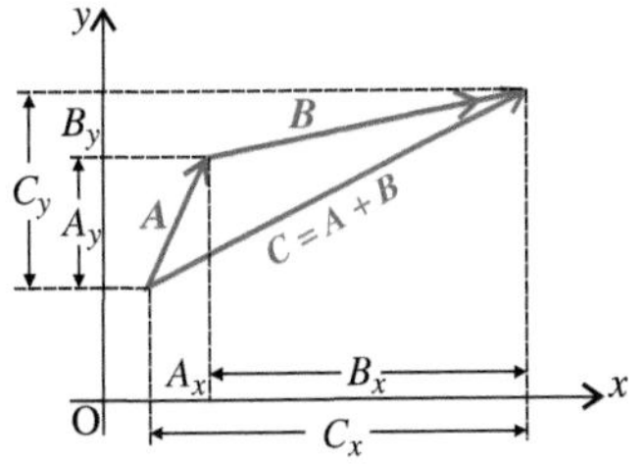

Figure A.20.

While vectors add according to the triangle rule, the *components* of vectors are added *algebraically*, that is if $\boldsymbol{C} = \boldsymbol{A} + \boldsymbol{B}$ then, from Figure A.20,

$$C_x = A_x + B_x \quad \text{and} \quad C_y = A_y + B_y$$

A.11.6 The scalar product

The **scalar product** (or 'dot product') of two vectors $\boldsymbol{A}$ and $\boldsymbol{B}$ is a *scalar* denoted by $\boldsymbol{A} \cdot \boldsymbol{B}$ (pronounced "A dot B") and defined by

$$\boldsymbol{A} \cdot \boldsymbol{B} := |\boldsymbol{A}||\boldsymbol{B}| \cos\theta$$

where θ is the angle between the directions of $\boldsymbol{A}$ and $\boldsymbol{B}$ (Figure A.21).

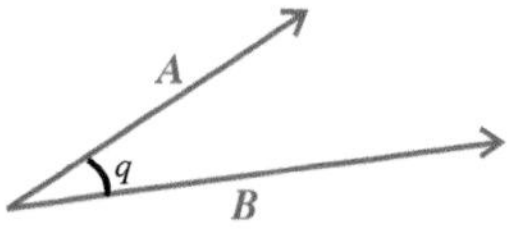

Figure A.21.

[The scalar product is introduced in **Section 4.2**]

Some properties of scalar products:

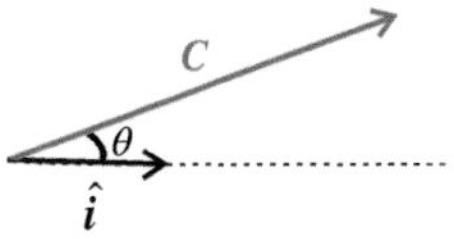

Figure A.22.

(i) The scalar product of a vector $\boldsymbol{C}$ and a unit vector is the component of $\boldsymbol{C}$ in the direction defined by the unit vector (Figure A.22), that is $C_x = \boldsymbol{C} \cdot \boldsymbol{i} = |\boldsymbol{C}||\boldsymbol{i}| \cos\theta = |\boldsymbol{C}| \cos\theta$.
(ii) The scalar product of a vector with itself is the square of the magnitude of the vector, since $\boldsymbol{V} \cdot \boldsymbol{V} := |\boldsymbol{V}||\boldsymbol{V}| \cos 0 = |\boldsymbol{V}||\boldsymbol{V}| = |\boldsymbol{V}|^2 = V^2$. In the case of unit vectors $\boldsymbol{i} \cdot \boldsymbol{i} = \boldsymbol{j} \cdot \boldsymbol{j} = \boldsymbol{k} \cdot \boldsymbol{k} = 1$
(iii) The scalar product of two vectors at right angles to each other is zero since, for example, $\boldsymbol{i} \cdot \boldsymbol{j} = |\boldsymbol{i}||\boldsymbol{j}| \cos\frac{\pi}{2} = 0$ and, similarly, $\boldsymbol{j} \cdot \boldsymbol{k} = \boldsymbol{k} \cdot \boldsymbol{i} = 0$.
(iv) $\boldsymbol{A} \cdot \boldsymbol{B} = \boldsymbol{B} \cdot \boldsymbol{A}$

A.11.7 The vector product

The **vector product** (or 'cross product') of two vectors $\boldsymbol{A}$ and $\boldsymbol{B}$ is a *vector* $\boldsymbol{C}$ denoted by $\boldsymbol{C} = \boldsymbol{A} \times \boldsymbol{B}$ (pronounced "A cross B") such that (*i*) the magnitude of $\boldsymbol{C}$ is given by

$$|\boldsymbol{C}| = |\boldsymbol{A}||\boldsymbol{B}| \sin\theta$$

where θ is the smaller angle between $\boldsymbol{A}$ and $\boldsymbol{B}$ (Figure A.23), and (*ii*) the direction of $\boldsymbol{C}$ is defined as being perpendicular to the plane containing $\boldsymbol{A}$ and $\boldsymbol{B}$ and in the direction of advance of a right-hand screw rotated from $\boldsymbol{A}$ to $\boldsymbol{B}$ (that is, perpendicular to the page and inward in the case illustrated in Figure A.23).

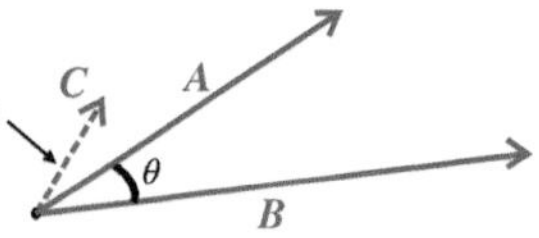

Figure A.23.

[The vector product is introduced in **Section 4.11**]

Some properties of vector products:

(i) $\boldsymbol{A} \times \boldsymbol{B} = -\boldsymbol{B} \times \boldsymbol{A}$
(ii) The vector product of two parallel vectors is zero, since if $\boldsymbol{A}$ is parallel to $\boldsymbol{B}$, then $|\boldsymbol{A} \times \boldsymbol{B}| = |\boldsymbol{A}||\boldsymbol{B}| \sin 0 = 0$
(iii) The magnitude of the vector product of two vectors at right angles to each other is equal to the product of their magnitudes, since if $\boldsymbol{A}$ is perpendicular to $\boldsymbol{B}$, then

$$|\boldsymbol{A} \times \boldsymbol{B}| = |\boldsymbol{A}||\boldsymbol{B}| \sin\frac{\pi}{2} = |\boldsymbol{A}||\boldsymbol{B}|$$

(iv) $\boldsymbol{i} \times \boldsymbol{j} = \boldsymbol{k};\quad \boldsymbol{j} \times \boldsymbol{k} = \boldsymbol{i};\quad \boldsymbol{k} \times \boldsymbol{i} = \boldsymbol{j}$ and $\boldsymbol{i} \times \boldsymbol{i} = \boldsymbol{j} \times \boldsymbol{j} = \boldsymbol{k} \times \boldsymbol{k} = 0$ where $\boldsymbol{i}, \boldsymbol{j}$ and $\boldsymbol{k}$ are orthogonal unit vectors.

A.11.8 Some useful vector identities

(i) $\boldsymbol{A} \times (\boldsymbol{B} \times \boldsymbol{C}) = (\boldsymbol{A} \cdot \boldsymbol{C})\boldsymbol{B} - (\boldsymbol{A} \cdot \boldsymbol{B})\boldsymbol{C}$
(ii) $\boldsymbol{A} \cdot (\boldsymbol{B} \times \boldsymbol{C}) = \boldsymbol{B} \cdot (\boldsymbol{C} \times \boldsymbol{A}) = \boldsymbol{C} \cdot (\boldsymbol{A} \times \boldsymbol{B})$
(iii) $(\boldsymbol{A} \times \boldsymbol{B}) \cdot (\boldsymbol{C} \times \boldsymbol{D}) = (\boldsymbol{A} \cdot \boldsymbol{C})(\boldsymbol{B} \cdot \boldsymbol{D}) - (\boldsymbol{A} \cdot \boldsymbol{D})(\boldsymbol{B} \cdot \boldsymbol{C})$
(iv) $(\boldsymbol{A} \times \boldsymbol{B}) \times (\boldsymbol{C} \times \boldsymbol{D}) = [\boldsymbol{A} \cdot (\boldsymbol{B} \times \boldsymbol{D})]\boldsymbol{C} - [\boldsymbol{A} \cdot (\boldsymbol{B} \times \boldsymbol{C})]\boldsymbol{D}$

A.12 Vector calculus

A.12.1 Derivatives of vector quantities

The derivative of a vector function $\boldsymbol{F} = \boldsymbol{F}(x)$ with respect to the (scalar) variable x is defined as

$$\frac{d\boldsymbol{F}}{dx} := \operatorname*{limit}_{\Delta x \to 0} \frac{\Delta \boldsymbol{F}}{\Delta x}$$

If $\boldsymbol{A}$ and $\boldsymbol{B}$ are vector functions of x

$$\frac{d(\boldsymbol{A} \cdot \boldsymbol{B})}{dx} = \frac{d\boldsymbol{A}}{dx} \cdot \boldsymbol{B} + \boldsymbol{A} \cdot \frac{d\boldsymbol{B}}{dx}$$

$$\frac{d(\boldsymbol{A} \times \boldsymbol{B})}{dx} = \frac{d\boldsymbol{A}}{dx} \times \boldsymbol{B} + \boldsymbol{A} \times \frac{d\boldsymbol{B}}{dx}$$

A.12.1(i) The gradient

If $\Phi = \Phi(\boldsymbol{r}) = \Phi(x, y, z)$ is a scalar function of $\boldsymbol{r}$, the **gradient** of $\Phi(\boldsymbol{r})$ is defined as

$$\nabla\Phi := \frac{\partial\Phi}{\partial x}\boldsymbol{i} + \frac{\partial\Phi}{\partial y}\boldsymbol{j} + \frac{\partial\Phi}{\partial z}\boldsymbol{k}$$

A.12.1(ii) The divergence

If $\boldsymbol{D} = \boldsymbol{D}(\boldsymbol{r}) = \boldsymbol{D}(x, y, z)$ is a vector function of $\boldsymbol{r}$, the **divergence** of $\boldsymbol{D}(\boldsymbol{r})$ is defined as

$$\nabla\cdot\boldsymbol{D} := \frac{\partial D_x}{\partial x} + \frac{\partial D_y}{\partial y} + \frac{\partial D_z}{\partial z}$$

where $\boldsymbol{D}_x$, $\boldsymbol{D}_y$ and $\boldsymbol{D}_z$ are the x-, y- and z-components of $\boldsymbol{D}$.

A.12.1(iii) The curl

If $\boldsymbol{D} = \boldsymbol{D}(\boldsymbol{r}) = \boldsymbol{D}(x, y, z)$ is a vector function of $\boldsymbol{r}$, the **curl** of $\boldsymbol{D}(\boldsymbol{r})$ is defined as

$$\nabla\times\boldsymbol{D} := \left(\frac{\partial D_z}{\partial y} - \frac{\partial D_y}{\partial z}\right)\boldsymbol{i} + \left(\frac{\partial D_x}{\partial z} - \frac{\partial D_z}{\partial x}\right)\boldsymbol{j} + \left(\frac{\partial D_y}{\partial x} - \frac{\partial D_x}{\partial y}\right)\boldsymbol{k}$$

where $\boldsymbol{D}_x$, $\boldsymbol{D}_y$ and $\boldsymbol{D}_z$ are the x-, y- and z-components of $\boldsymbol{D}$.

A.12.2 Line integrals

When the value of a vector fuction $\boldsymbol{F}(\boldsymbol{r})$ is defined at points on a curve C, the line integral of the function $\boldsymbol{F}(\boldsymbol{r})$ between two points A and B on C (Figure A.24) is defined as

$$\int_A^B \boldsymbol{F}(\boldsymbol{r})\cdot d\boldsymbol{s} := \underset{\Delta s_i\to 0}{\text{limit}}\,[\boldsymbol{F}(\boldsymbol{r}_1)\cdot\Delta\boldsymbol{s}_1 + \boldsymbol{F}(\boldsymbol{r}_2)\cdot\Delta\boldsymbol{s}_2 + \ldots]$$

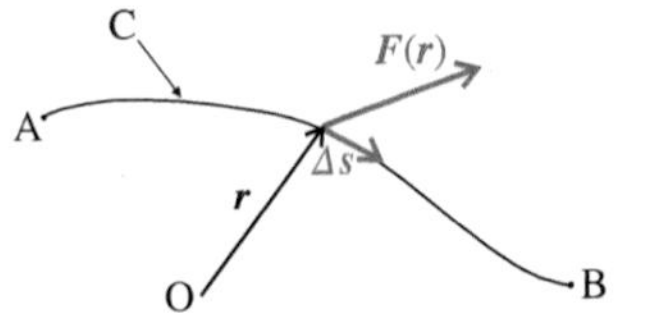

Figure A.24.

In contrast to the simple integrals in section A.7 above, the value of a line integral between two points A and B cannot be determined, in general, in terms of the evaluation of an integral function at the points A and B but depends on the path taken from A to B.
[See **Section 5.7** for a discussion of the use of line integrals in the treatment of force fields]

A.12.3 Surface integrals

When the value of a vector fuction $\boldsymbol{D}(\boldsymbol{r})$ is defined at points on a surface S, the surface integral of the function $\boldsymbol{D}(\boldsymbol{r})$ over the surface (Figure A.25) is defined as

$$\int_S \boldsymbol{D}(\boldsymbol{r})\cdot d\boldsymbol{A} := \underset{\Delta A_i\to 0}{\text{limit}}\,[\boldsymbol{D}(\boldsymbol{r}_1)\cdot\Delta\boldsymbol{A}_1 + \boldsymbol{D}(\boldsymbol{r}_2)\cdot\Delta\boldsymbol{A}_2 + \ldots]$$

where $\boldsymbol{\Delta A}_i := (\Delta A_i)\boldsymbol{n}_i$ $(i = 1, 2, 3, \ldots)$ and ΔA_i is the area of the ith element of the surface and $\boldsymbol{n}_i$ is a unit vector normal to the ith area element.

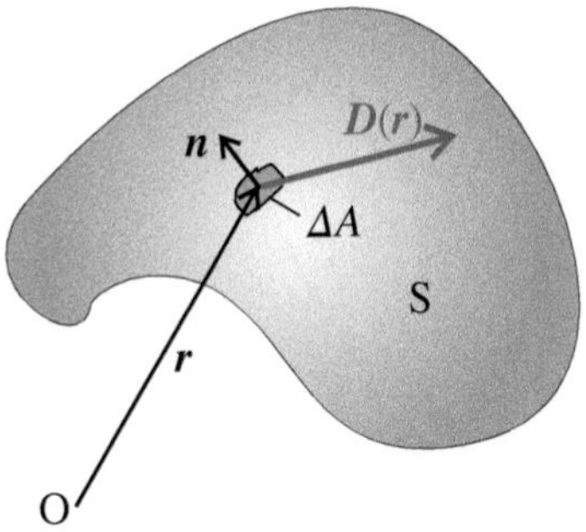

Figure A.25.

[See **Section 5.4** for an example of the use of surface integrals in physics]

A.12.4 Volume integrals

When the value of a scalar function $\Phi(\boldsymbol{r})$ is defined at points in a volume V, the volume integral of the function $\Phi(\boldsymbol{r})$ over the volume (Figure A.26) is defined as

$$\int_V \Phi(\boldsymbol{r})dV := \lim_{\Delta V_i \to 0}[\Phi(r_1)\Delta V_1 + \Phi(r_2)\Delta V_2 + \dots \dots \dots \dots]$$

where $\Phi(r_i)$ is the value of the function at the ith volume element ΔV_i.

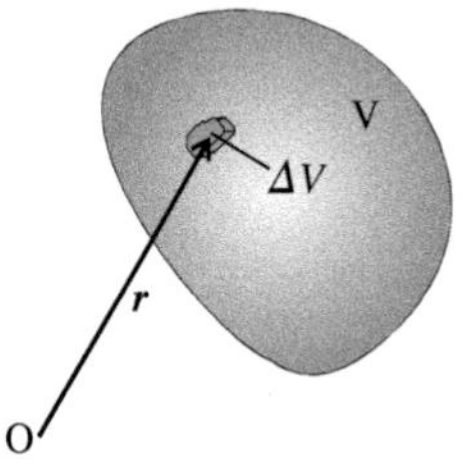

Figure A.26.

[See **Section 7.7** for an example of the use of volume integrals]

A.12.5 The divergence theorem

If S is a smooth surface enclosing a volume V, for any vector field $\boldsymbol{D}(\boldsymbol{r})$

$$\oiint_S \boldsymbol{D}\cdot d\boldsymbol{A} = \oiiint_V (\nabla\cdot\boldsymbol{D})dV$$

where $\nabla.\boldsymbol{D}$ is the divergence of the field as defined in Section A.12.1(ii) above.

A.12.6 Stokes' theorem

If C is a closed smooth boundary bounding a smooth surface S, for any vector function $\boldsymbol{D}(\boldsymbol{r})$

$$\oint_C \boldsymbol{D}\cdot d\boldsymbol{s} = \iint_S (\nabla\times\boldsymbol{D})\cdot d\boldsymbol{A}$$

where $\nabla\times\boldsymbol{D}$ is the curl of the field as defined in Section A.12.1(iii) above.

A.13 Orthogonal coordinate systems in three dimensions

This section shows how some common quantities are represented in different coordinate systems.

A.13.1 Rectangular (Cartesian) coordinates

A general point P(x, y, z) has displacement relative to the origin of $\boldsymbol{r} = x\boldsymbol{i} + y\boldsymbol{j} + z\boldsymbol{k}$, where $\boldsymbol{i}, \boldsymbol{j}$ and $\boldsymbol{k}$ are unit vectors along the x-, y- and z-axes, respectively (see Figure A.27).

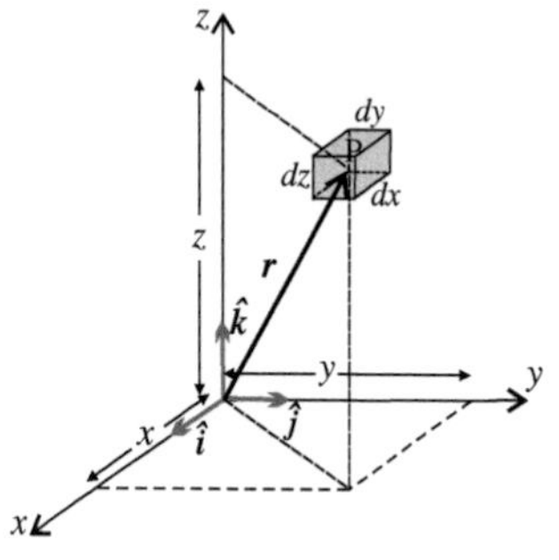

Figure A.27.

Scalar function: $\Phi = \Phi(x, y, z)$

Vector function: $\boldsymbol{F} = \boldsymbol{F}(x, y, z) = F_x\boldsymbol{i} + F_y\boldsymbol{j} + F_z\boldsymbol{k}$

Volume element: $\Delta V = \Delta x \Delta y \Delta z$

Gradient: $\nabla\Phi := \dfrac{\partial\Phi}{\partial x}\boldsymbol{i} + \dfrac{\partial\Phi}{\partial y}\boldsymbol{j} + \dfrac{\partial\Phi}{\partial z}\boldsymbol{k}$

Laplacian: $\nabla^2\Phi := \dfrac{\partial^2\Phi}{\partial x^2} + \dfrac{\partial^2\Phi}{\partial y^2} + \dfrac{\partial^2\Phi}{\partial z^2}$

A.13.2 Cylindrical polar coordinates

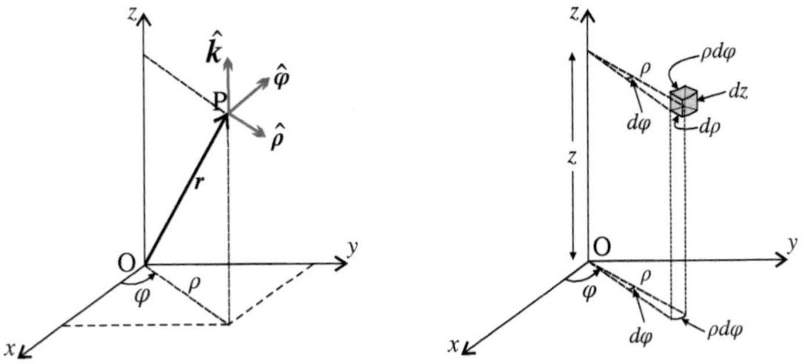

Figure A.28.

General point P(ρ, φ, z) is related to rectangular coordinates (see Figure A.28) as follows.

$$x = \rho\cos\varphi; \quad y = \rho\sin\varphi; \quad z = z \quad (0 \le \varphi \le 2\pi)$$

Scalar function: $\Phi = \Phi(\rho, \varphi, z)$

Vector function: $\boldsymbol{F} = \boldsymbol{F}(\rho, \varphi, z) = F_\rho\hat{\boldsymbol{\rho}} + F_\varphi\hat{\boldsymbol{\varphi}} + F_z\hat{\boldsymbol{k}}$

Volume element: $\Delta V = \rho\Delta\rho\Delta\varphi\Delta z$

Gradient: $\nabla\Phi := \dfrac{\partial\Phi}{\partial\rho}\hat{\boldsymbol{\rho}} + \dfrac{1}{\rho}\dfrac{\partial\Phi}{\partial\rho}\hat{\boldsymbol{\varphi}} + \dfrac{\partial\Phi}{\partial z}\hat{\boldsymbol{k}}$

Laplacian: $\nabla^2\Phi := \dfrac{\partial^2\Phi}{\partial\rho^2} + \dfrac{1}{\rho}\dfrac{\partial\Phi}{\partial\rho} + \dfrac{1}{\rho^2}\dfrac{\partial^2\Phi}{\partial\varphi^2} + \dfrac{\partial^2\Phi}{\partial z^2}$

A.13.3 Spherical polar coordinates

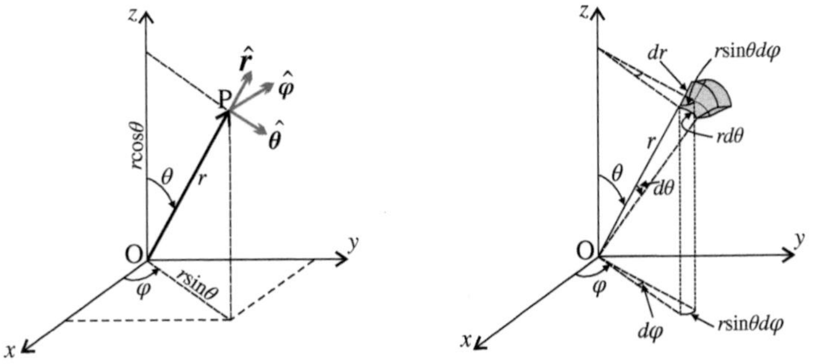

Figure A.29.

General point P(r, θ, φ) is related to rectangular coordinates (see Figure A.29) as follows.

$$x = r\sin\theta\cos\varphi; \quad y = r\sin\theta\sin\varphi; \quad z = r\cos\theta \quad (0 \le \theta \le \pi, \quad 0 \le \varphi \le 2\pi)$$

Scalar function: $\boldsymbol{\Phi} = \boldsymbol{\Phi}(r, \theta, \varphi)$

Vector function: $\boldsymbol{F} = \boldsymbol{F}(r, \theta, \varphi) = F_r \hat{\boldsymbol{r}} + F_\theta \hat{\boldsymbol{\theta}} + F_\varphi \hat{\boldsymbol{\varphi}}$

Volume element: $\Delta V = r^2 \sin\theta \, \Delta r \Delta\theta \Delta\varphi$

Gradient: $\nabla\boldsymbol{\Phi} := \frac{\partial \boldsymbol{\Phi}}{\partial \rho}\hat{\boldsymbol{r}} + \frac{1}{r}\frac{\partial \boldsymbol{\Phi}}{\partial \theta}\hat{\boldsymbol{\theta}} + \frac{1}{r\sin\theta}\frac{\partial \boldsymbol{\Phi}}{\partial \varphi}\hat{\boldsymbol{\varphi}}$

Laplacian:
$$\nabla^2\boldsymbol{\Phi} := \frac{1}{r^2}\frac{\partial}{\partial r}\left(r^2\frac{\partial \boldsymbol{\Phi}}{\partial r}\right) + \frac{1}{r^2 \sin\theta}\frac{\partial}{\partial \theta}\left(\sin\theta\frac{\partial \boldsymbol{\Phi}}{\partial \theta}\right) + \frac{1}{r^2\sin^2\theta}\frac{\partial^2 \Phi}{\partial \varphi^2}$$
$$= \frac{\partial^2 \boldsymbol{\Phi}}{\partial r^2} + \frac{2}{r}\frac{\partial \boldsymbol{\Phi}}{\partial r} + \frac{1}{r^2 \sin\theta}\frac{\partial}{\partial \theta}\left(\sin\theta\frac{\partial \boldsymbol{\Phi}}{\partial \theta}\right) + \frac{1}{r^2\sin^2\theta}\frac{\partial^2 \boldsymbol{\Phi}}{\partial \varphi^2}$$

Appendix B

Some fundamental physical constants[1]

quantity	symbol	value	relative uncertainty

B.1 Universal constants

quantity	symbol	value	relative uncertainty
speed of light in vacuum	c	$299\,792\,458$ m s^{-1}	(exact)
permeability of vacuum	μ_0	$1.256\,637\,062\,12 \times 10^{-6}$ N A^{-2}	1.5×10^{-10}
permittivity of vacuum	ε_0	$8.854\,187\,813 \times 10^{-12}$ F m^{-1}	1.5×10^{-10}
gravitational constant	G	$6.674\,30 \times 10^{-11}$ m^3 kg^{-1}s^{-2}	2.2×10^{-5}
Planck constant	h	$6.626\,070\,15 \times 10^{-34}$ J s	(exact)
		$4.135\,667\,696 \ldots\ldots \times 10^{-15}$ eV s	(exact)
$\hbar = \dfrac{h}{2\pi}$	$\hbar$	$1.054\,571\,817 \ldots\ldots \times 10^{-34}$ J s	(exact)
		$6.582\,119\,569 \ldots\ldots \times 10^{-16}$ eV s	(exact)

B.2 Electromagnetic constants

quantity	symbol	value	relative uncertainty
elementary charge	e	$1.602\,176\,634 \times 10^{-19}$ C	(exact)
Bohr magneton	μ_B	$9.274\,010\,0783 \times 10^{-24}$ J T^{-1}	3×10^{-10}
$\left[\dfrac{e\hbar}{2m_e}\right]$		$5.788\,381\,8060 \times 10^{-5}$ eV T^{-1}	3×10^{-10}
nuclear magneton	μ_N	$5.050\,783\,7461 \times 10^{-27}$ J T^{-1}	3.1×10^{-10}
$\left[\dfrac{e\hbar}{2m_p}\right]$		$3.152\,451\,258\,44 \times 10^{-8}$ eV T^{-1}	3.1×10^{-10}

[1]Data taken from the *2018 CODATA recommended values* (https://physics.nist.gov/cuu/Constants/index.html)

Understanding Physics, Third Edition. Michael Mansfield and Colm O'Sullivan.
© 2020 John Wiley & Sons Ltd. Published 2020 by John Wiley & Sons Ltd.

quantity	symbol	value	relative uncertainty

B.3 Atomic constants

quantity	symbol	value	relative uncertainty
fine-structure constant $\left[\frac{\mu_0 ce^2}{2h}\right]$	α	$7.297\ 352\ 5693 \times 10^{-3}$	1.5×10^{-10}
inverse fine-structure constant	α^{-1}	137.035 999 084	1.5×10^{-10}
Rydberg constant $\left[\frac{m_e c\alpha^2}{2h}\right]$	R_∞	$10\ 973\ 731.568\ 160\ \text{m}^{-1}$	1.9×10^{-12}
Bohr radius	a_0	$5.291\ 772\ 109\ 03 \times 10^{-11}$ m	1.5×10^{-10}
electron mass	m_e	$9.109\ 383\ 7015 \times 10^{-31}$ kg	3×10^{-10}
		$5.485\ 799\ 090\ 65 \times 10^{-4}$ u	2.9×10^{-11}
		$0.510\ 998\ 950\ 00\ \text{MeV}/c^2$	3×10^{-10}
electron magnetic moment	μ_e	$-9.284\ 764\ 7043 \times 10^{-24}\ \text{J T}^{-1}$	3×10^{-10}
muon mass	m_μ	$1.883\ 531\ 627 \times 10^{-28}$ kg	2.2×10^{-8}
		0.113 428 9259 u	2.2×10^{-8}
		$105.658\ 3755\ \text{MeV}/c^2$	2.2×10^{-8}
muon magnetic moment	μ_μ	$-4.490\ 448\ 30 \times 10^{-26}\ \text{J T}^{-1}$	2.2×10^{-8}
proton mass	m_p	$1.672\ 621\ 923\ 69 \times 10^{-27}$ kg	3.1×10^{-10}
		1.007 276 466 621 u	5.3×10^{-11}
		$938.272\ 088\ 16\ \text{MeV}/c^2$	3.1×10^{-10}
proton magnetic moment	μ_p	$1.410\ 606\ 797\ 36 \times 10^{-26}\ \text{J T}^{-1}$	4.2×10^{-10}
neutron mass	m_n	$1.674\ 927\ 498\ 04 \times 10^{-27}$ kg	5.7×10^{-10}
		1.008 664 915 95 u	4.8×10^{-10}
		$939.565\ 420\ 52\ \text{MeV}/c^2$	5.7×10^{-10}
neutron magnetic moment	μ_n	$-9.662\ 3651 \times 10^{-27}\ \text{J T}^{-1}$	2.4×10^{-7}
deuteron mass	m_d	$3.343\ 583\ 7724 \times 10^{-27}$ kg	3.0×10^{-10}
		2.013 553 212 745 u	2×10^{-11}
		$1875.612\ 942\ 57\ \text{MeV}/c^2$	3.0×10^{-10}
deutron magnetic moment	μ_d	$4.330\ 735\ 094 \times 10^{-27}\ \text{J T}^{-1}$	2.6×10^{-9}
α-particle mass	m_α	$6.644\ 657\ 3357 \times 10^{-27}$ kg	3.0×10^{-10}
		4.001 506 179 127 u	1.6×10^{-11}
		$3727.379\ 4066\ \text{MeV}/c^2$	3.0×10^{-10}

B.4 Physico-chemical constants

quantity	symbol	value	relative uncertainty
Avogadro constant	N_A	$6.022\ 140\ 76 \times 10^{23}\ \text{mol}^{-1}$	(exact)
atomic mass constant $\left[\frac{1}{12}m(^{12}C)\right]$	m_u	$1.660\ 539\ 066\ 60 \times 10^{-27}$ kg	3.0×10^{-10}
		$931.494\ 102\ 42\ \text{MeV}/c^2$	3.0×10^{-10}
molar gas constant $[N_A k]$	R	$8.314\ 462 \ldots\ldots\ \text{J mol}^{-1}\text{K}^{-1}$	(exact)
Boltzmann constant	k	$1.380\ 649 \times 10^{-23}\ \text{J K}^{-1}$	(exact)
		$8.617\ 333 \ldots\ldots \times 10^{-5}\ \text{eV K}^{-1}$	(exact)
Stefan-Boltzmann constant $\left[\frac{\pi^2 k^4}{60\hbar^3 c^2}\right]$	σ	$5.670\ 374 \times 10^{-8}\ \text{W m}^{-2}\text{K}^{-4}$	(exact)

Appendix C

Some astrophysical and geophysical data

C.1 The Sun

mass	1.99×10^{30} kg
mean radius	6.96×10^{8} m
mean density	1.41×10^{3} kg m^{-3}
rotation period	37 d (poles); 26 d (equator)
distance from centre of galaxy	2.2×10^{20} m
luminosity	3.90×10^{26} W
surface temperature	5.8×10^{3} K
central temperature	1.6×10^{7} K

C.2 The Earth

mass	5.98×10^{24} kg
mean radius	6.37×10^{6} m
mean density	5.52×10^{3} kg m^{-3}
surface gravity	9.81 m s^{-2}
rotation period	23 h 56 m
mean distance from Sun	1.50×10^{11} m

Geomagnetic data

Approximate values of the total magnetic flux density at various latitudes and longitudes.

Longitude	**Latitude**						
	60° *N*	**40° *N***	**20° *N***	**0°**	**20° *S***	**420° *S***	**60° *S***
0 °	50 μT	44 μT	36 μT	31 μT	29 μT	27 μT	33 μT
90 ° *E*	61 μT	55 μT	44 μT	42 μT	49 μT	56 μT	59 μT
180 °	53 μT	43 μT	35 μT	35 μT	44 μT	55 μT	64 μT
90 ° *W*	62 μT	57 μT	44 μT	32 μT	28 μT	33 μT	44 μT

C.3 The Moon

mass	7.36×10^{22} kg
mean radius	1.74×10^{6} m
mean density	3.34×10^{3} kg m^{-3}
surface gravity	1.67 m s^{-2}
rotation period	27.3 d
mean distance from Earth	3.84×10^{8} m

Understanding Physics, Third Edition. Michael Mansfield and Colm O'Sullivan.
© 2020 John Wiley & Sons Ltd. Published 2020 by John Wiley & Sons Ltd.

C.4 The planets

	mass	mean density	equatorial radius	rotation period	semimajor axis	eccentricity of orbit	orbital period
	$\times 10^{24}$ kg	$\times 10^3$ kg m^{-3}	$\times 10^6$ m	day	$\times 10^{12}$ m		year
Mercury	0.329	5.43	2.44	58.7	0.0579	0.206	0.241
Venus	4.87	5.25	6.05	243	0.1082	0.007	0.615
Earth	5.98	5.52	6.38	0.997	0.1496	0.017	1.00
Mars	0.640	3.95	3.39	1.026	0.2279	0.093	1.88
Jupiter	1901	1.33	71.5	0.410	0.7783	0.048	11.9
Saturn	569.2	0.69	60.3	0.426	1.427	0.056	29.5
Uranus	86.9	1.29	25.6	0.746	2.871	0.046	84.0
Neptune	102.5	1.63	24.8	0.800	4.497	0.010	165

C.5 The Milky Way galaxy

mass	$>1 \times 10^{42}$ kg
mean radius	4×10^{20} m
thickness of disk	2×10^{19} m
distance of Sun from centre	2.4×10^{20} m
rotation period (about galactic centre)	2×10^8 y
orbital speed of Sun	2.3×10^5 m s^{-1}

C.6 Cosmological data

Hubble constant (H_0) (NASA WMAP mission)	73.8 ± 2.4 km s^{-1} Mpc^{-1} (1 pc = 3.09×10^{16} m)
age of the universe	$1.377 \pm 0.006 \times 10^{10}$ y
temperature of the cosmic background radiation	2.725 K

Appendix D

The international system of units[1] — SI

Système International d'Unités had its genesis in the signing of the Convention of the Metre in 1875 and a proposal by Giovanni Georgi in 1901 for a system comprising four base units (the kilogram, metre, second, and ampere). In 1935, the International Electrotechnical Commission (IEC) had adopted Georgi's 'rationalised MKSA' system which was then widely embraced in engineering.

In 1960, the 11th Conférence Générale des Poids et Mesures (CGPM) agreed the name 'Système International d'Unités', with the international abbreviation ***SI***, which is the practical system of units adopted by international agreement throughout most of the world today. The Conference also established rules for prefixes, derived units and other matters thus providing a comprehensive specification for all units of measurement.

The SI base units

SI defined seven base units, those for the physical quantities **time, length, mass, electric current, thermodynamic temperature, amount of substance,** and **luminous intensity**. All other units are then obtained as products of powers of these units, which involve no numerical factors; these are called 'coherent derived units'. The seven base units in SI are as follows:

Table D.1 The seven traditional SI base units.

quantity	time	length	mass	electric current	thermodynamic temperature	amount of substance	luminous intensity
unit	second	metre	kilogram	ampere	kelvin	mole	candela
symbol	s	m	kg	A	K	mol	cd

Over the years, SI has employed three different methods of defining base units: (a) using international prototype artifacts (for example, kilogram prior to 2019), (b) using specific physical phenomena (for example, caesium atomic clock to define the second from 1967) and (c) fixing the value of fundamental physical constants (for example, fixing the value of the speed of light in vacuum c to define the metre in 1983). In most cases the definition required a specific experimental approach to make an absolute measurement based on the definition (such an approach is called a 'practical realisation of the unit').

The CGPM resolved that, subsequent to May 2019, SI would be entirely defined using the third of the above approaches. Seven fundamental physical constants, whose values were known with sufficient accuracy in 2019, are assigned *exact* values, similar to (and including) assigning a fixed value to c as had been done in the case of the 1983 definition of the metre. The effect of this is to provide definitions for seven unit quantities equivalent to the seven traditional base units based on this set of defining constants.

The 2019 redefinitions of SI base units in terms of fixed fundamental constants

On 16 November 2018, the 26th CGPM voted unanimously in favour of revised definitions of the SI base units which the International Committee for Weights and Measures (CIPM) had proposed earlier that year. The new definitions came into force on 20 May 2019.

[1]For more details consult the latest edition of the *SI Brochure* on the BIPM website https://www.bipm.org/en/measurement-units/rev-si/#communication.

Understanding Physics, Third Edition. Michael Mansfield and Colm O'Sullivan.
© 2020 John Wiley & Sons Ltd. Published 2020 by John Wiley & Sons Ltd.

Table D.2 The seven defining constants of SI and their corresponding units.

Defining constant	Symbol	Numerical value	Unit
hyperfine splitting of Cs	$\Delta\nu_{Cs}$	9 192 631 770	s^{-1} (= Hz)
speed of light in vacuum	c	299 792 458	m s^{-1}
Planck constant	h	$6.626\ 070\ 15 \times 10^{-34}$	kg m^2 s^{-1} (= J s)
elementary charge	e	$1.602\ 176\ 634 \times 10^{-19}$	A s (= C)
Boltzman constant	k	$1.380\ 649 \times 10^{-23}$	kg m^2 s^{-2} K^{-1} (= J K^{-1})
Avogadro constant	N_A	$6.022\ 140\ 76 \times 10^{23}$	mol^{-1}
luminous efficacy	K_{cd}	683	kg^{-1} m^{-2} s^3 cd sr (see Note)

Note: The steradian, symbol sr, is the SI unit of solid angle. The unit kg^{-1} m^{-2} s^3 cd sr = lm W^{-1}

The **second**, symbol s, is the SI unit of time. It is defined by taking the fixed numerical value of the caesium frequency $\Delta\nu_{Cs}$, the unperturbed ground-state hyperfine transition frequency of the caesium–133 atom, to be 9 192 631 770 when expressed in the unit Hz, which is equal to s^{-1}.

The **metre**, symbol m, is the SI unit of length. It is defined by taking the fixed numerical value of the speed of light in vacuum c to be 299 792 458 when expressed in the unit m s^{-1}, where the second is defined in terms of the caesium frequency $\Delta\nu_{Cs}$.

The **kilogram**, symbol kg, is the SI unit of mass. It is defined by taking the fixed numerical value of the Planck constant h to be $6.626\ 070\ 15 \times 10^{-34}$ when expressed in the unit J s, which is equal to kg m^2 s^{-1}, where the metre and the second are defined in terms of c and $\Delta\nu_{Cs}$.

The **ampere**, symbol A, is the SI unit of electric current. It is defined by taking the fixed numerical value of the elementary charge e to be $1.602\ 176\ 634 \times 10^{-19}$ when expressed in the unit C, which is equal to A s, where the second is defined in terms of $\Delta\nu_{Cs}$.

The **kelvin**, symbol K, is the SI unit of thermodynamic temperature. It is defined by taking the fixed numerical value of the Boltzmann constant k to be $1.380\ 649 \times 10^{-23}$ when expressed in the unit J K^{-1}, which is equal to kg m^2 s^{-2} K^{-1}, where the kilogram, metre and second are defined in terms of h, c and $\Delta\nu_{Cs}$.

The **mole**, symbol mol, is the SI unit of amount of substance. One mole contains exactly $6.022\ 140\ 76 \times 10^{23}$ elementary entities. This number is the fixed numerical value of the Avogadro constant, N_A, when expressed in the unit mol^{-1} and is called the Avogadro number. The amount of substance, symbol n, of a system is a measure of the number of specified elementary entities. An elementary entity may be an atom, a molecule, an ion, an electron, any other particle or specified group of particles.

The **candela**, symbol cd, is the SI unit of luminous intensity in a given direction. It is defined by taking the fixed numerical value of the luminous efficacy of monochromatic radiation of frequency 540×10^{12} Hz, K_{cd}, to be 683 when expressed in the unit lm W^{-1}, which is equal to cd sr W^{-1}, or cd sr kg^{-1} m^{-2} s^3, where the kilogram, metre and second are defined in terms of h, c and $\Delta\nu_{Cs}$.

This approach represents the simplest and most fundamental way to define SI. Strictly speaking, no distinction is made between base units and derived units. It also effectively decouples the definition and the practical realisation[2] of the units. While definitions may well remain unchanged over a long period of time, practical realisations can be established by many different experiments as technology evolves, including the possibility of totally new experiments not yet devised.

Defining base units by statements

Alternatively, but entirely equivalently, SI can be defined by statements that explicitly characterise the seven individual base units. Such statements are as follows.

The **second** is the duration of 9 192 631 770 periods of the radiation corresponding to the transition between the two hyperfine levels of the unperturbed ground state of the caesium 133 atom.

The **metre** is the length of the path travelled by light in vacuum during a time interval of 1/299 792 458 of a second.

[2]The practical realisation of a unit (*mise en pratique*) is a set of instructions that allows secondary standards to be calibrated at the highest level of accuracy based on the legal definition of the unit. This enables such standards to be disseminated in a practical way, for example, to provide national standards in individual countries.

The definition of the **kilogram** given above effectively defines a unit (kg m^2 s^{-1} — the unit of angular momentum) which together with the definitions of the second and the metre leads to a definition of the kilogram expressed in terms of the value of the Planck constant.

The **ampere** is the electric current corresponding to the flow of $1/(1.602\ 176\ 634\ 10^{-19})$ elementary charges per second.

The **kelvin** is equal to the change of thermodynamic temperature that results in a change of thermal energy kT by $1.380\ 649\ 10^{-23}$ J.

The **mole** is the amount of substance of a system that contains $6.022\ 140\ 76\ 10^{23}$ specified elementary entities.

The **candela** is the luminous intensity, in a given direction, of a source that emits monochromatic radiation of frequency 540×10^{12} Hz and that has a radiant intensity in that direction of (1/683) W sr^{-1}.

Base units expressed in terms of the defining constants[3]

Since $\nu_{=} = 9\ 192.631\ 770\ s^{-1}$, we can write the unit second as $s = \dfrac{9\ 192\ 631\ 770}{\nu_{=}}$

and, similarly, since $c = 299\ 792\ 458\ m\ s^{-1}$, the unit metre can be written as $m = \dfrac{c}{299\ 792\ 458}\ s$

or

$$m = \frac{c}{299\ 792\ 458} \times \frac{9\ 192\ 631\ 770}{\nu_{=}} = 30.66 \ldots \frac{c}{\nu_{=}}.$$

In this way all the base units can be stated in terms of combinations of the defining constants as follows.

unit **second**:	$s = \dfrac{9\ 192\ 631\ 770}{\nu_{=}}$
unit **metre**:	$m = 30.66 \ldots\ldots \dfrac{c}{\nu_{=}}$
unit **kilogram**:	$kg = 1.475 \ldots\ldots \dfrac{h\nu_{=}}{c^2}$
unit **kelvin**:	$K = 2.226 \ldots\ldots \dfrac{h\nu_{=}}{k}$
unit **ampere**:	$A = 6.789 \ldots\ldots \nu_{=} e$
unit **mole**:	$mol = \dfrac{6.022\ 140\ 76 \times 10^{23}}{N_A}$
unit **candela**:	$cd = 2.614 \ldots\ldots h\nu_{=} K_{cd}$

[3]The official SI symbol for Cs-133 hyperfine transition frequency is $\Delta\nu_{Cs}$ but a more compact (unofficial) symbol $\nu_{=}$ is used instead below.

Bibliography

Alonso, A. and Finn, E.J. (1992) *Physics*, Addison-Wesley.

Cohen, E.R. and Giacomo, P. (1987) *Symbols, Units, Nomenclature and Fundamental Constants: Document I.U.P.A.P.-25 (SUMAMCO 87-1),* IUPAP (reprinted from *Physica* 146A (1987) 1-68).

Eisberg, R. and Resnick, R. (1985) *Quantum Physics, second edition*, John Wiley & Sons, New York.

Flowers, B.H. and Mendoza, E. (1970) *Properties of Matter*, John Wiley & Sons, Chichester.

Grant, I.S. and Phillips, W.R. (1990) *Electromagnetism, second edition*, John Wiley & Sons, Chichester.

Halliday, D., Resnick, R. and Walker, J. (2013) *Fundamentals of Physics, tenth edition*, John Wiley & Sons, New York.

Heavens, O.S. and Ditchburn, R.W. (1991) *Insight into Optics*, John Wiley & Sons, Chichester.

Hook, J.R. and Hall, H.E. (1992) *Solid State Physics, second edition*, John Wiley & Sons, Chichester.

Horowitz, P. and Hill, W. (2005) *The Art of Electronics, third edition*, Cambridge University Press.

King, G.C. (2009) *Vibrations and Waves*, John Wiley & Sons, Chichester.

Lambourne, R. and Tinker, M. (2000), *Basic mathematics for the physical sciences* John Wiley & Sons, Chichester.

Lilley, J. (2001) *Nuclear Physics, Principles and Applications*, John Wiley & Sons, Chichester.

Mandl, F. (1988) *Statistical Physics, second edition*, John Wiley & Sons, Chichester.

Mandl, F. (1992) *Quantum Mechanics*, John Wiley & Sons, Chichester.

Martin, B.R. and Shaw, G. (2019) *Nuclear and Particle Physics, an Introduction, third edition*, John Wiley & Sons, Chichester.

Pain, H.J. (2005) *The Physics of Vibrations and Waves, sixth edition*, John Wiley & Sons, Chichester.

Phillips, A.C. (1999) *The Physics of Stars, second edition*, John Wiley & Sons, Chichester.

Phillips, A.C. (2003) *Introduction to Quantum Mechanics*, John Wiley & Sons, Chichester.

Smith, F.G. and Thomson, J.H. (1988) *Optics*, John Wiley & Sons, Chichester.

Tinker, M. and Lambourne, R. (2000) *Further mathematics for the physical sciences*, John Wiley & Sons, Chichester.

Understanding Physics, Third Edition. Michael Mansfield and Colm O'Sullivan.

© 2020 John Wiley & Sons Ltd. Published 2020 by John Wiley & Sons Ltd.

Index

Note: Page numbers in *italics* refer to Figures; those in **bold** to Tables.

Understanding Physics, Third Edition. Michael Mansfield and Colm O'Sullivan.
© 2020 John Wiley & Sons Ltd. Published 2020 by John Wiley & Sons Ltd.